AF323393

Comprehensive Organometallic Chemistry II
A Review of the Literature 1982–1994

Comprehensive Organometallic Chemistry II

A Review of the Literature 1982–1994

Editors-in-Chief

Edward W. Abel
University of Exeter, UK

F. Gordon A. Stone
Baylor University, Waco, TX, USA

Geoffrey Wilkinson
Imperial College of Science, Technology and Medicine, London, UK

Volume 4

SCANDIUM, YTTRIUM, LANTHANIDES AND
ACTINIDES, AND TITANIUM GROUP

Volume Editor

Michael F. Lappert
University of Sussex, Brighton, UK

PERGAMON

UK   Elsevier Science Ltd., The Boulevard, Langford Lane, Kidlington, Oxford
            OX5 1GB, UK

USA   Elsevier Science Inc., 660 White Plains Road, Tarrytown, New York
            10591-5153, USA

JAPAN   Elsevier Science Japan, Tsunashima Building Annex, 3-20-12 Yushima,
            Bunkyo-ku, Tokyo 113, Japan

First edition 1995

Library of Congress Cataloging in Publication Data
Comprehensive organometallic chemistry II : a review of the literature
1982–1994 / editors-in-chief, Edward W. Abel, F. Gordon A. Stone,
Geoffrey Wilkinson
     p.   cm.
   Includes indexes.
   1. Organometallic chemistry.   I. Abel, Edward W.   II. Stone, F. Gordon
A.   III.   Wilkinson, Geoffrey.
QD411.C652   1995
547'.05—dc20   95–7030

British Library Cataloguing in Publication Data
A catalogue record for this book is available from the British Library.

ISBN 0–08–040608–4 (set : alk. paper)
ISBN 0–08–042311–6 (Volume 4)

Important note
For safety reasons, readers should always consult the list of abbreviations
on p. xi before making use of the experimental details provided.

Chemical structures drawn by Synopsys Scientific Systems Ltd., Leeds, UK.

Transferred to digital printing 2005

Printed and bound by Antony Rowe Ltd, Eastbourne

Contents

Preface

'Comprehensive Organometallic Chemistry', published in 1982, was well received and remains very highly cited in the primary journal literature. Since its publication, studies on the chemistry of molecules with carbon–metal bonds have continued to expand rapidly. This is due to many factors, ranging from the sheer intellectual challenge and excitement provided by the continuing production of novel results, which demand new ideas, through to the successful application of organometallic species in organic syntheses, the generation of living catalysts for polymerization, and the synthesis of precursors for materials employed in the electronic and ceramic industries. For many reasons, therefore, we judged it timely to update 'Comprehensive Organometallic Chemistry' with a new work.

Due to the scope and depth of this area of chemistry, to have merely revised each of the original nine volumes did not seem the most user-friendly or cost-effective procedure to follow. As a consequence of the sheer bulk of the literature of the subject, a revised edition would necessarily require either the elimination of much chemistry of archival value but which is still important, or the production of a set of volumes significantly larger in number than the original nine. Accordingly, we decided it would be best to use the original work as a basis for new volumes focusing on organometallic chemistry reported since 1982, with reference back to the original work when necessary. For ease of use the new volumes maintain the same general structure as employed previously but reflect the changes in substance and direction the field has undergone in the last ten years. Thus it is not surprising that the largest volume in the new work concerns the role of the transition elements in metal-mediated organic syntheses.

The expansion of organometallic chemistry since the early 1980s also led us to decide that an updating of 'Comprehensive Organometallic Chemistry' would be more effectively accomplished if each volume had one or two editors who would be responsible both for recruiting experts for the Herculean task of writing the many chapters of each volume and for overseeing the content. We are deeply indebted to the volume editors and their authors for the time and effort they have given to the project.

As with the original 'Comprehensive Organometallic Chemistry', published some thirteen years ago, we hope this new version will serve as a pivotal reference point for new work and will function to generate new ideas and perceptions for the continued advance of what will surely continue as a vibrant area of chemistry.

Edward W. Abel
Exeter, UK

F. Gordon A. Stone
Waco, Texas, USA

Geoffrey Wilkinson
London, UK

Preface to 'Comprehensive Organometallic Chemistry'

Although the discovery of the platinum complex that we now know to be the first π-alkene complex, $K[PtCl_3(C_2H_4)]$, by Zeise in 1827 preceded Frankland's discovery (1849) of diethylzinc, it was the latter that initiated the rapidly developing interest during the latter half of the nineteenth century in compounds with organic groups bound to the elements. This era may be considered to have reached its apex in the discovery by Grignard of the magnesium reagents which occupy a special place because of their ease of synthesis and reactivity. With the exception of trimethylplatinum chloride discovered by Pope, Peachy and Gibson in 1907 by use of the Grignard reagent, attempts to make stable transition metal alkyls and aryls corresponding to those of main group elements met with little success, although it is worth recalling that even in 1919 Hein and his co-workers were describing the 'polyphenyl-chromium' compounds now known to be arene complexes.

The other major area of organometallic compounds, namely metal compounds of carbon monoxide, originated in the work starting in 1868 of Schützenberger and later of Mond and his co-workers and was subsequently developed especially by Hieber and his students. During the first half of this century, aided by the use of magnesium and, later, lithium reagents the development of main group organo chemistry was quite rapid, while from about 1920 metal carbonyl chemistry and catalytic reactions of carbon monoxide began to assume importance.

In 1937 Krause and von Grosse published their classic book 'Die Chemie der Metallorganischen Verbindungen'. Almost 1000 pages in length, it listed scores of compounds, mostly involving metals of the main groups of the periodic table. Compounds of the transition elements could be dismissed in 40 pages. Indeed, even in 1956 the stimulating 197-page monograph 'Organometallic Compounds' by Coates adequately reviewed organo transition metal complexes within 27 pages.

Although exceedingly important industrial processes in which transition metals were used for catalysis of organic reactions were developed in the 1930s, mainly in Germany by Reppe, Koch, Roelen, Fischer and Tropsch and others, the most dramatic growth in our knowledge of organometallic chemistry, particularly of transition metals, has stemmed from discoveries made in the middle years of this century. The introduction in the same period of physical methods of structure determination (infrared, nuclear magnetic resonance, and especially single-crystal X-ray diffraction) as routine techniques to be used by preparative chemists allowed increasingly sophisticated exploitation of discoveries. Following the recognition of the structure of ferrocene, other major advances quickly followed, including the isolation of a host of related π-complexes, the synthesis of a plethora of organometallic compounds containing metal–metal bonds, the characterization of low-valent metal species in which hydrocarbons are the only ligands, and the recognition from dynamic NMR spectra that ligand site exchange and tautomerism were common features in organometallic and metal carbonyl chemistry. The discovery of alkene polymerization using aluminium alkyl–titanium chloride systems by Ziegler and Natta and of the Wacker palladium–copper catalysed ethylene oxidation led to enormous developments in these areas.

In the last two decades, organometallic chemistry has grown more rapidly in scope than have the classical divisions of chemistry, leading to publications in journals of all national chemical societies, the appearance of primary journals specifically concerned with the topic, and the growth of annual review volumes designed to assist researchers to keep abreast of accelerating developments.

Organometallic chemistry has become a mature area of science which will obviously continue to grow. We believe that this is an appropriate time to produce a comprehensive review of the subject, treating organo derivatives in the widest sense of both main group and transition elements. Although advances in transition metal chemistry have appeared to dominate progress in recent years, spectacular progress has, nevertheless, also been made in our knowledge of organo compounds of main group elements such as aluminium, boron, lithium and silicon.

In these Volumes we have assembled a compendium of knowledge covering contemporary organometallic and carbon monoxide chemistry. In addition to reviewing the chemistry of the elements individually, two Volumes survey the use of organometallic species in organic synthesis and in catalysis, especially of industrial utility. Within the other Volumes are sections devoted to such diverse topics as the nature of carbon–metal bonds, the dynamic behaviour of organometallic compounds in solution, heteronuclear metal–metal bonded compounds, and the impact of organometallic compounds on the environment. The Volumes provide a unique record, especially of the intensive studies conducted during the past 25 years. The last Volume of indexes of various kinds will assist readers seeking information on the properties and synthesis of compounds and on earlier reviews.

As Editors, we are deeply indebted to all those who have given their time and effort to this project. Our Contributors are among the most active research workers in those areas of the subject that they have reviewed and they have well justified international reputations for their scholarship. We thank them sincerely for their cooperation.

Finally, we believe that 'Comprehensive Organometallic Chemistry', as well as providing a lasting source of information, will provide the stimulus for many new discoveries since we do not believe it possible to read any of the articles without generating ideas for further research.

E. W. ABEL   F. G. A. STONE
Exeter   *Bristol*

G. WILKINSON
London

Contributors to Volume 4

Professor P. Binger
Max-Planck-Institut für Kohlenforschung, Kaiser-Wilhelm-Platz 1, D-45470 Mülheim a.d. Ruhr, Germany

Professor M. Bochmann
School of Chemical Sciences, University of East Anglia, University Plain, Norwich, NR4 7TJ, UK

Professor F. G. N. Cloke
School of Chemistry and Molecular Sciences, University of Sussex, Brighton, BN1 9QJ, UK

Professor F. T. Edelmann
Chemisches Institut der Otto-von-Guericke-Universität Magdeburg, Universitätsplatz 2, Gebäude M, D-39106 Magdeburg, Germany

Professor S. Gambarotta
Chemistry Department, University of Ottawa, Ottawa, Ontario, K1N 6N5, Canada

Dr. A. S. Guram
Department of Chemistry, University of Iowa, Iowa City, IA 52242, USA

Professor E. Hey-Hawkins
Institut für Anorganische Chemie, Universität Leipzig, Talstrasse 35, D-04103 Leipzig, Germany

Professor R. F. Jordan
Department of Chemistry, University of Iowa, Iowa City, IA 52242, USA

Dr. J. Jubb
Chemistry Department, University of Ottawa, Ottawa, Ontario, K1N 6N5, Canada

Dr. S. Podubrin
Max-Planck-Institut für Kohlenforschung, Kaiser-Wilhelm-Platz 1, D-45470 Mülheim a.d. Ruhr, Germany

Dr. D. Richeson
Chemistry Department, University of Ottawa, Ottawa, Ontario, K1N 6N5, Canada

Dr. E. J. Ryan
c/ Cardenal Tenorio 4,1°,C, Alcala de Henares, E-28801 Madrid, Spain

Dr. J. Song
Chemistry Department, University of Ottawa, Ottawa, Ontario, K1N 6N5, Canada

Abbreviations

The abbreviations used throughout 'Comprehensive Organometallic Chemistry II' are consistent with those used in 'Comprehensive Organometallic Chemistry' and with other standard texts in this area. The abbreviations in some instances may differ from those commonly used in other branches of chemistry.

Ac	acetyl
acac	acetylacetonate
AIBN	2,2'-azobisisobutyronitrile
Ar	aryl
arphos	1-(diphenylphosphino)-2-(diphenylarsino)ethane
Azb	azobenzene
9-BBN	9-borabicyclo[3.3.1]nonyl
9-BBN-H	9-borabicyclo[3.3.1]nonane
BHT	2,6-di-t-butyl-4-methylphenol (butylated hydroxytoluene)
bipy	2,2'-bipyridyl
t-BOC	t-butoxycarbonyl
bsa	N,O-bis(trimethylsilyl)acetamide
bstfa	N,O-bis(trimethylsilyl)trifluoroacetamide
btaf	benzyltrimethylammonium fluoride
Bz	benzyl
can	ceric ammonium nitrate
cbd	cyclobutadiene
1,5,9-cdt	cyclododeca-1,5,9-triene
chd	cyclohexadiene
chpt	cycloheptatriene
[Co]	cobalamin
(Co)	cobaloxime [Co(DMG)$_2$] derivative
cod	1,5-cyclooctadiene
cot	cyclooctatetraene
Cp	η^5-cyclopentadienyl
Cp*	pentamethylcyclopentadienyl
18-crown-6	1,4,7,10,13,16-hexaoxacyclooctadecane
CSA	camphorsulfonic acid
csi	chlorosulfonyl isocyanate
Cy	cyclohexyl
dabco	1,4-diazabicyclo[2.2.2]octane
dba	dibenzylideneacetone
dbn	1,5-diazabicyclo[4.3.0]non-5-ene
dbu	1,8-diazabicyclo[5.4.0]undec-7-ene
dcc	dicyclohexylcarbodiimide
dcpe	1,2-bis(dicyclohexylphosphino)ethane
ddq	2,3-dichloro-5,6-dicyano-1,4-benzoquinone
deac	diethylaluminum chloride
dead	diethyl azodicarboxylate
depe	1,2-bis(diethylphosphino)ethane
depm	1,2-bis(diethylphosphino)methane
det	diethyl tartrate (+ or −)

DHP	dihydropyran
diars	1,2-bis(dimethylarsino)benzene
dibal-H	diisobutylaluminum hydride
dien	diethylenetriamine
DIGLYME	bis(2-methoxyethyl)ether
diop	2,3-*O*-isopropylidene-2,3-dihydroxy-1,4-bis(diphenylphosphino)butane
dipt	diisopropyl tartrate (+ or −)
dma	dimethylacetamide
dmac	dimethylaluminum chloride
DMAD	dimethyl acetylenedicarboxylate
dmap	4-dimethylaminopyridine
DME	dimethoxyethane
DMF	*N,N'*-dimethylformamide
DMG	dimethylglyoximate
DMI	*N,N'*-dimethylimidazalone
dmpe	1,2-bis(dimethylphosphino)ethane
dmpm	bis(dimethylphosphino)methane
DMSO	dimethyl sulfoxide
dmtsf	dimethyl(methylthio)sulfonium fluoroborate
dpam	bis(diphenylarsino)methane
dppb	1,4-bis(diphenylphosphino)butane
dppe	1,2-bis(diphenylphosphino)ethane
dppf	1,1'-bis(diphenylphosphino)ferrocene
dpph	1,6-bis(diphenylphosphino)hexane
dppm	bis(diphenylphosphino)methane
dppp	1,3-bis(diphenylphosphino)propane
eadc	ethylaluminum dichloride
edta	ethylenediaminetetraacetate
eedq	*N*-ethoxycarbonyl-2-ethoxy-1,2-dihydroquinoline
en	ethylene-1,2-diamine
Et$_2$O	diethyl ether
F$_6$acac	hexafluoroacetylacetonate
Fc	ferrocenyl
Fp	Fe(CO)$_2$Cp
HFA	hexafluoroacetone
hfacac	hexafluoroacetylacetonate
hfb	hexafluorobut-2-yne
HMPA	hexamethylphosphoramide
hobt	hydroxybenzotriazole
IpcBH$_2$	isopinocampheylborane
Ipc$_2$BH	diisopinocampheylborane
kapa	potassium 3-aminopropylamide
K-selectride	potassium tri-*s*-butylborohydride
LAH	lithium aluminum hydride
LDA	lithium diisopropylamide
LICA	lithium isopropylcyclohexylamide
LITMP	lithium tetramethylpiperidide
L-selectride	lithium tri-*s*-butylborohydride
LTA	lead tetraacetate
mcpba	*m*-chloroperbenzoic acid
MeCN	acetonitrile
MEM	methoxyethoxymethyl
MEM-Cl	β-methoxyethoxymethyl chloride

Mes	mesityl
mma	methyl methacrylate
mmc	methylmagnesium carbonate
MOM	methoxymethyl
Ms	methanesulfonyl
MSA	methanesulfonic acid
MsCl	methanesulfonyl chloride
nap	1-naphthyl
nbd	norbornadiene
NBS	*N*-bromosuccinimide
NCS	*N*-chlorosuccinimide
nmo	*N*-methylmorpholine *N*-oxide
NMP	*N*-methyl-2-pyrrolidone
Nu$^-$	nucleophile
ox	oxalate
pcc	pyridinium chlorochromate
pdc	pyridinium dichromate
phen	1,10-phenanthroline
phth	phthaloyl
ppa	polyphosphoric acid
ppe	polyphosphate ester
[PPN]$^+$	[(Ph$_3$P)$_2$N]$^+$
ppts	pyridinium *p*-toluenesulfonate
py	pyridine
pz	pyrazolyl
Red-Al	sodium bis(2-methoxyethoxy)aluminum dihydride
sal	salicylaldehyde
salen	*N,N'*-bis(salicylaldehydo)ethylenediamine
SEM	β-trimethylsilylethoxymethyl
tas	tris(diethylamino)sulfonium
tasf	tris(diethylamino)sulfonium difluorotrimethylsilicate
tbaf	tetra-*n*-butylammonium fluoride
TBDMS	*t*-butyldimethylsilyl
TBDMS-Cl	*t*-butyldimethylsilyl chloride
TBDPS	*t*-butyldiphenylsilyl
tbhp	*t*-butyl hydroperoxide
TCE	2,2,2-trichloroethanol
TCNE	tetracyanoethene
TCNQ	7,7,8,8-tetracyanoquinodimethane
terpy	2,2':6',2"-terpyridyl
tes	triethylsilyl
Tf	triflyl (trifluoromethanesulfonyl)
TFA	trifluoracetic acid
TFAA	trifluoroacetic anhydride
tfacac	trifluoroacetylacetonate
THF	tetrahydrofuran
THP	tetrahydropyranyl
tipbs-Cl	2,4,6-triisopropylbenzenesulfonyl chloride
tips-Cl	1,3-dichloro-1,1,3,3-tetraisopropyldisiloxane
TMEDA	tetramethylethylenediamine [1,2-bis(dimethylamino)ethane]
TMS	trimethylsilyl
TMS-Cl	trimethylsilyl chloride
TMS-CN	trimethylsilyl cyanide
Tol	tolyl
tpp	*meso*-tetraphenylporphyrin

Tr	trityl (triphenylmethyl)
tren	2,2',2"-triaminotriethylamine
trien	triethylenetetraamine
triphos	1,1,1-tris(diphenylphosphinomethyl)ethane
Ts	tosyl
TsMIC	tosylmethyl isocyanide
ttfa	thallium trifluoroacetate

Contents of All Volumes

1

Zero Oxidation State Complexes of Scandium, Yttrium and the Lanthanide Elements

F. GEOFFREY N. CLOKE

University of Sussex, Brighton, UK

1.1 ALKENE AND ALKYNE COMPLEXES

Europium–ethene species may be synthesized at 12 K by the codeposition of europium atoms with ethene, neat or diluted with argon or xenon.[1] These complexes, which are of the type $Eu(C_2H_4)$ and $Eu(C_2H_4)_n$, decompose above 50 K, and were characterized by UV–visible and IR spectroscopy. Molecular orbital calculations indicate that the ethene ligand in $Eu(C_2H_4)$ is only weakly bonded by back-donation from a europium f orbital into the alkene π^* system, with little forward-donation from the alkene σ or π orbitals.

Initial work on lanthanide metal vapour reactivity examined the reactions of lanthanide atoms towards a variety of small, unsaturated hydrocarbons (Equations (1)–(4)).[2]

$$Ln\ (at.)\ +\ EtC{\equiv}CEt\ \xrightarrow{\ 77\ K\ }\ Ln_2(C_6H_{10})_3\ (Ln = Nd\ or\ Er)\ or\ Ln(C_6H_{10})\ (Ln = Sm\ or\ Yb) \quad (1)$$

$$Ln\ (at.)\ +\ H_2C{=}CHCH{=}CH_2\ \xrightarrow{\ 77\ K\ }\ Ln(C_4H_6)_3 \quad (2)$$
$$Ln = Nd,\ Sm\ or\ Er$$

$$Ln\ (at.)\ +\ H_2C{=}C(Me)C(Me){=}CH_2\ \xrightarrow{\ 77\ K\ }\ Ln(C_6H_{10})_2 \quad (3)$$
$$Ln = La,\ Nd,\ Sm\ or\ Er$$

$$Ln\ (at.) + MeCH{=}CH_2 \xrightarrow{\ 77\ K\ } Ln(C_3H_6)_3 \tag{4}$$

$$Ln = Er$$

Reactions of neodymium or erbium with 3-hexyne gives products of composition Ln_2(3-hexyne)$_3$, whereas with samarium or ytterbium the products are of stoichiometry Ln(3-hexyne). Similarly, with butadiene, neodymium, samarium or erbium afford materials analysing as Ln(butadiene)$_3$, but 2,3-dimethyl substitution of the butadiene ligand yields products of composition Ln(2,3-dimethylbutadiene)$_2$ (Ln = La, Nd, Sm or Er). These products are all intensely coloured and exhibit magnetic moments outside the range usually associated with the respective lanthanide(III) ion values. However, the complex behaviour of these materials in solution indicates that they are oligomeric in nature, and none of the species has been structurally characterized. Hence their precise nature remains unclear, although the products from the lanthanide atom co-condensation reactions with 3-hexyne or 2,3-dimethylbutadiene are found to be catalytically active for the *cis*-hydrogenation of alkynes. In the case of the reactions of lanthanide atoms with ethene, propene or 1,2-propadiene, only erbium is found to yield soluble products. Hydrolytic studies of the complex reaction mixtures indicate that alkene polymerization and insertion are occurring in these reactions, together with C–H bond activation.[2]

1.2 ARENE COMPLEXES

1.2.1 Matrix Isolation Studies

A recent investigation[3] into the reactivity of scandium atoms in hydrocarbon matrices examined the cocondensation reactions of scandium with both inert (adamantane, cyclohexane or deuterocyclohexane) and reactive (benzene, deuterobenzene or cyclohexene) coreactants, using a rotating cryostat at 77 K. ESR spectroscopic measurements indicate that the paramagnetic products formed in the 'inert' matrices are weakly bonded Sc–H species. The codeposition of scandium atoms with benzene leads to two ESR-active species: one possesses a large scandium hyperfine coupling and is believed to be a complex of the type $Sc(\eta\text{-}C_6H_6)_n$, (n = 1 or 2), and the second product is thought to arise from the interaction of a scandium atom with only one double bond of the benzene molecule, since a similar spectrum is obtained with cyclohexene.

1.2.2 Naphthalene and Related Complexes

The lanthanide diiodides LnI_2 (Ln = Sm, Eu or Yb) react with lithium naphthalenide to afford products of variable composition $M_xC_{10}H_8(THF)_y$ (x = 1–1.5, y = 3–4), although in the case of ytterbium the stoichiometry $YbC_{10}H_8(THF)_3$ is established (Equation (5)).[4,5] The nature of these products has been examined by microanalysis and IR spectroscopy, but no structural information is available. Their magnetic moments (μ_{eff}(BM): Sm, 3.5; Eu, 8.2; Yb, 0) are indicative of lanthanide(0) species and they are postulated to contain η^4-naphthalene ligands. $Yb(C_{10}H_8)(THF)_3$ is a catalyst for (i) the room-temperature polymerization of a variety of monomers (styrene, methyl methacrylate, ethyl methacrylate, isoprene or piperylene),[5] (ii) the copolymerization of ethylene oxide with styrene or piperylene,[5] (iii) the conversion of an epoxide and carbon dioxide into an alkylene carbonate,[5] and (iv) the hydrogenation of an alkene (1-hexene, stilbene, isoprene or piperylene).[6] Treatment of $Yb(C_{10}H_8)(THF)_3$ with hydrogen gives a product formulated as $[YbH_2(THF)]_n$, and the same reaction in the presence of diphenylacetylene gives $[YbH_2(THF)_2]_n$ (Equations (6) and (7)).[6]

$$MI_2 + 2\,Li(C_{10}H_8) \xrightarrow{\ 77\ K\ } M(C_{10}H_8)_x(THF)_y \tag{5}$$

$$M = Sm\ or\ Eu,\ x = 1\text{–}2,\ y = 3\text{–}4;\ or\ M = Yb,\ x = 1,\ y = 3$$

$$Yb(C_{10}H_8)(THF)_3 \xrightarrow[\text{THF}]{H_2} [YbH_2(THF)]_n \tag{6}$$

$$Yb(C_{10}H_8)(THF)_3 \xrightarrow[\text{THF}]{H_2,\ Ph_2C_2} [YbH_2(THF)_2]_n \tag{7}$$

$Yb(C_{10}H_8)(THF)_3$ functions as a reducing agent towards metallocenes $[MCp_2]$ (M = V, Cr, Co or Ni), affording $YbCp_2$ and unidentified transition metal fragments, except in the case of vanadium where the structurally characterized intermediate $[\{VCp_2\}_2(\mu,\eta^6,\eta^{6'}\text{-}C_{10}H_8)]$ (**1**) has been isolated (Equation (8)).[7] The reaction of excess $Yb(C_{10}H_8)(THF)_3$ with $[VCp_2]$ gives the novel two-dimensional multidecker complex $[\{VCp\}\{YbCp(THF)\}(\mu,\eta^6,\eta^2\text{-}(C_{10}H_8)]_n$ (**2**), (Equation (9)).[8]

$$Yb(C_{10}H_8)(THF)_3 \ + \ 2\,[VCp_2] \ \xrightarrow{\ THF\ } \qquad\qquad (8)$$

(**1**)

$$\text{excess } Yb(C_{10}H_8)(THF)_3 \ + \ [VCp_2] \ \xrightarrow{\ THF\ } \qquad\qquad (9)$$

(**2**)

1.2.3 Stable Bis(η-arene) Complexes

1.2.3.1 Synthesis and structure

The cocondensation of yttrium or gadolinium vapour with 1,3,5-tri-*t*-butylbenzene affords the stable, crystalline, zero oxidation state sandwich compounds $[M(\eta\text{-}Bu^t_3C_6H_3)_2]$ (M = Y or Gd).[9] Subsequently the reactions of the remaining lanthanide vapours with 1,3,5-tri-*t*-butylbenzene were examined, which yield stable analogues for several of the elements (Equation (10) and Table 1).[10] Physical data for $[Sc(\eta\text{-}Bu^t_3C_6H_3)_2]$ are also included in Table 1, but the unusual products of the reaction between scandium atoms and 1,3,5-tri-*t*-butylbenzene are discussed at the end of this section. The stability trend of the $[Ln(\eta\text{-}Bu^t_3C_6H_3)_2]$ compounds across the lanthanide series is discussed in Section 1.2.3.2.

$$Ln\ (at.) \ + \qquad\qquad \xrightarrow{\ 77\ K\ } \qquad\qquad (10)$$

Ln = Y, (La), Pr, Nd, (Sm), Gd, Tb, Dy, Ho, Er or Lu

The $[Ln(\eta\text{-}Bu^t_3C_6H_3)_2]$ complexes are all extremely air-sensitive, crystalline solids, very soluble in hydrocarbon solvents, and are highly coloured, a feature associated with their characteristic intense band

Table 1 Stability and some physical properties of [Ln(η-Bu$^t_3$C$_6$H$_3$)$_2$] complexes.

Atom	Ground state	Colour	λ_{max} (nm)[a]	Stability[b]
Sc	d^1s^2	orange	495	stable
Y	d^1s^2	purple	529	stable
La	d^1s^2	green	637	dec. >0 °C
Ce	$f^1d^1s^2$			unstable
Pr	f^3s^2	purple	541	dec. >40 °C
Nd	f^4s^2	blue	544	stable
Sm	f^6s^2	green	691	dec. >−30 °C
Eu	f^7s^2			unstable
Gd	$f^7s^2d^1s^2$	purple	542	stable
Tb	f^9s^2	purple	540	stable
Dy	$f^{10}s^2$	purple	556	stable
Ho	$f^{11}s^2$	deep pink	507	stable
Er	$f^{12}s^2$	red	499	stable
Tm	$f^{13}s^2$			unstable
Yb	$f^{14}s^2$			unstable
Lu	$f^{14}d^1s^2$	red-brown	495	stable

[a] Measured in pentane solution. [b] Stable $\Rightarrow$ dec. >100 °C, unstable $\Rightarrow$ unisolable from the metal atom reactor at any convenient temperature.

($\varepsilon > 10^4$) in the visible region of the spectrum. The latter is assigned to a ligand-to-metal charge transfer (LMCT) transition (Section 1.2.4). The f^0 yttrium complex is ESR-silent at room temperature by virtue of its proposed rapidly relaxing ^2E ground state (Section 1.2.3.2), but displays a frozen solution spectrum at 77 K exhibiting a doublet coupling to yttrium, confirming unpaired electron density on the yttrium centre in this genuine yttrium(0) compound. In the case of the f^{14} lutetium analogue, a weak octet (^{175}Lu, 97%, I = 7/2) is observable at room temperature in the ESR spectrum, presumably the result of a somewhat longer relaxation time.

The molecular structure of [Gd(η-Bu$^t_3$C$_6$H$_3$)$_2$] has been determined by x-ray crystallography, and is shown in Figure 1. The structure shows that [Gd(η-Bu$^t_3$C$_6$H$_3$)$_2$] has the classic parallel ring sandwich structure, typical of its d-block analogues. The gadolinium atom lies on a centre of symmetry, with the planar arene rings eclipsed but the substituents staggered in order to minimize t-butyl group interactions. The t-butyl groups are bent out of the plane of the arene rings by 6–10°, as has been observed with ring-substituted bis(η-arene) transition metal complexes; this is attributable to the ring carbons becoming more sp^3 in character on binding to the metal. Two of the three methyl groups of each t-butyl substituent are located between the planes defined by the arene rings, thus surrounding the metal centre by 12 methyl groups, and presumably accounting for the kinetic stability of the compound. The gadolinium–ring carbon distances range from 0.258 5(4) to 0.266 0(4) nm (average 0.263 0(4) nm), and are significantly shorter than the average corresponding distance of 0.289(5) nm found in [Sm(η-C$_6$Me$_6$)(η^2-AlCl$_4$)],[11] despite the fact that samarium(III) is considerably smaller than gadolinium(0). Indeed, the arene–metal bond in [Sm(η-C$_6$Me$_6$)(η^2-AlCl$_4$)] has been described in terms of a weak electrostatic interaction between the π-system of hexamethylbenzene and the samarium(III) centre, whereas the bond in [Gd(η-Bu$^t_3$C$_6$H$_3$)$_2$] is clearly much more akin to that of [Cr(η-C$_6$H$_6$)$_2$] in which the M–ring carbon distance is 0.214 2(2) nm,[12] given the relative sizes of gadolinium and chromium. The bond angles in the arene rings are not perturbed (within standard deviations) from those of the free ligand, but the average ring C–C distances are slightly longer; the latter effect is also observed in [Cr(η-C$_6$H$_6$)$_2$]. The bonding in the [Ln(η-Bu$^t_3$C$_6$H$_3$)$_2$] compounds is discussed in Section 1.2.3.2.

The holmium analogue is isostructural with [Gd(η-Bu$^t_3$C$_6$H$_3$)$_2$]. The one salient difference lies in the average metal–ring carbon distance, that for the holmium compound (0.258 0(3) nm) being 0.005 nm shorter, an effect attributable to the smaller radius of holmium arising from the lanthanide contraction.

The reaction between scandium atoms and 1,3,5-tri-t-butylbenzene gives an unexpected result in yielding not only the scandium(0) sandwich compound [Sc(η-Bu$^t_3$C$_6$H$_3$)$_2$], but also a similarly unique example of scandium(II), [Sc(η-Bu$^t_3$C$_6$H$_3$){η^6,η^1-Bu$^t_2$(CMe$_2$CH$_2$)C$_6$H$_3$}H] (Equation (11)).[13]

[Sc(η-Bu$^t_3$C$_6$H$_3$)$_2$] exhibits ESR spectral behaviour analagous to that of the yttrium analogue: the compound is ESR-silent at room temperature, but displays a frozen solution spectrum showing hyperfine coupling to ^{45}Sc (100%, I = 7/2). While pure [Sc(η-Bu$^t_3$C$_6$H$_3$)$_2$] is obtained by sublimation of the product mixture, [Sc(η-Bu$^t_3$C$_6$H$_3$){η^6,η^1-Bu$^t_2$(CMe$_2$CH$_2$)C$_6$H$_3$}H] decomposes under these conditions. Thus, the assignment of the second, purple, product as the 17-electron scandium(II) complex [Sc(η-

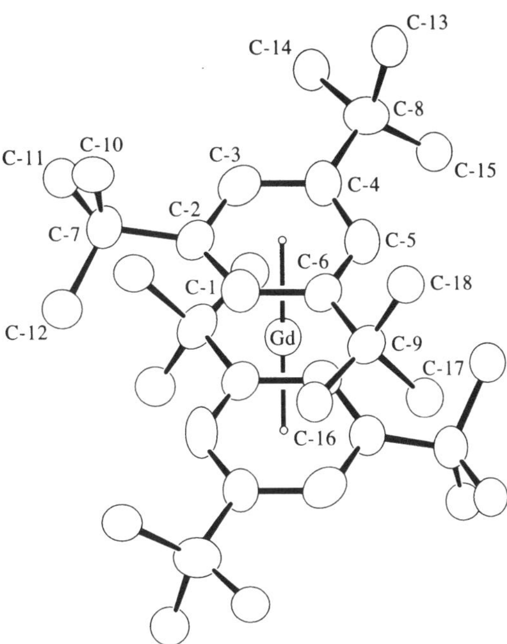

Figure 1   The x-ray structure of $[Gd(\eta\text{-}Bu^t_3C_6H_3)_2]$.

$$Sc\ (at.)\ +\ \cdots \xrightarrow{77\ K} \cdots\ +\ \cdots \qquad (11)$$

$Bu^t_3C_6H_3)\{\eta^6,\eta^1\text{-}Bu^t_2(CMe_2CH_2)C_6H_3\}H]$ was made on the basis of its solution ESR spectrum (octet of doublets of triplets, arising from hyperfine coupling to ^{45}Sc, $-H$ and $-CH_2$, respectively) supported by a simulation, and its visible absorption spectrum ($\lambda_{max} = 557$ nm, $\varepsilon = 10^4$). Surprisingly it was not found possible to interconvert $[Sc(\eta\text{-}Bu^t_3C_6H_3)_2]$ and $[Sc(\eta\text{-}Bu^t_3C_6H_3)\{\eta^6,\eta^1\text{-}Bu^t_2(CMe_2CH_2)C_6H_3\}H]$ either thermally or photochemically, although the latter compound formally arises from insertion of the scandium centre into a methyl group of one of the *t*-butyl ring substituents in $[Sc(\eta\text{-}Bu^t_3C_6H_3)_2]$. The formation of $[Sc(\eta\text{-}Bu^t_3C_6H_3)\{\eta^6,\eta^1\text{-}Bu^t_2(CMe_2CH_2)C_6H_3\}H]$ is ascribed to direct insertion of a naked scandium atom into a methyl group of free 1,3,5-tri-*t*-butylbenzene in the low-temperature matrix.[14]

1.2.3.2 Bonding

The bonding in the $[Ln(\eta\text{-}Bu^t_3C_6H_3)_2]$ compounds has been analysed[14] by analogy with that of the parent bis(η-arene)metal compound, $[Cr(\eta\text{-}C_6H_6)_2]$; the predominantly ring e_{1u} and e_{1g}, and predominantly metal e_{2g} and a_{1g}, molecular orbitals are completely filled in an 18-electron compound. Studies on the 18-electron bis(η-arene)transition metal sandwich complexes have revealed that the HOMOs are the largely metal-based a_{1g} and e_{2g} orbitals. The latter are lower in energy and more delocalized onto the arene rings than the a_{1g} electrons. Application of the same molecular orbital diagram to the 15-electron complexes $[Sc(\eta\text{-}Bu^t_3C_6H_3)_2]$ and $[Y(\eta\text{-}Bu^t_3C_6H_3)_2]$ leads to three electrons in the degenerate e_{2g} levels by successive filling of the energy levels, the compounds having ^2E ground states. This is in accord with their ESR behaviour (Section 1.2.3.1) and effective magnetic moments (Section 1.2.3.3), corresponding to one unpaired electron.

By analogy with $[Sc(\eta\text{-}Bu^t_3C_6H_3)_2]$ and $[Y(\eta\text{-}Bu^t_3C_6H_3)_2]$, it has been proposed that, to a first approximation, a d^1s^2 configuration is required in either the ground or an easily accessible excited state of the lanthanide atom to form a stable bis(η-arene) complex. Figure 2 shows the promotion energies for the $f^n s^2 \to f^{n-1} d^1 s^2$ transition in the lanthanides.

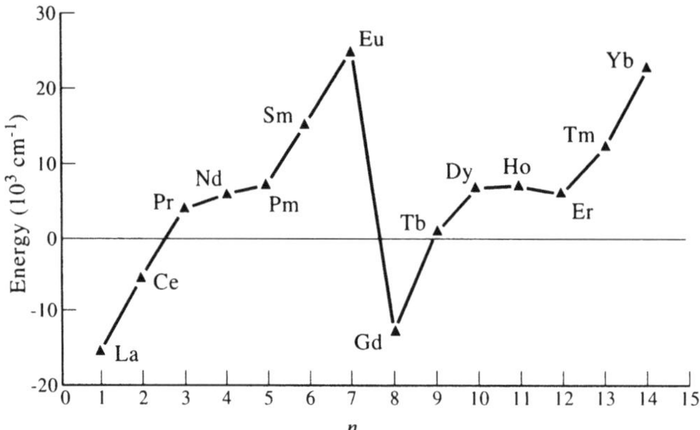

Figure 2   The $f^n_s{}^2 \to f^{n-1} d^1 s^2$ transition in the lanthanide elements.

From Figure 2, it is apparent that samarium, europium, thulium and ytterbium are the least likely lanthanides to form stable bis(η-arene) complexes because their $f^n s^2 \to f^{n-1} d^1 s^2$ promotion energies are relatively high, associated with the stabilities of the f^6, f^7, f^{13} and f^{14} configurations, respectively. The results in Table 1 show that, experimentally, this is the case: the compounds of erbium, thulium and ytterbium are unisolable, and the samarium analogue decomposes to the free metal and ligand above $-30\ ^\circ$C. In these cases the instability may thus be attributed to an insufficient gain in metal–arene bond energy to offset the necessary promotion energy. Table 1 also shows an area of instability at the beginning of the lanthanide series which is not explicable by the former argument, since the ground state of both lanthanum and cerium is $d^1 s^2$ and the promotion energy of praseodymium is small (see Figure 2). The instability of these complexes is attributed to the greater covalent radii of the metals, and consequently the 1,3,5-tri-*t*-butylbenzene ligand is not sufficiently bulky to kinetically stabilize the metal centre.

The mean metal–ligand bond dissociation energies $\overline{D}$ for some of the [Ln(η-Bu$^t_3$C$_6$H$_3$)$_2$] complexes (determined by iodinolytic batch titrational calorimetry[15]) are shown in Table 2, together with values for selected bis(η-arene)transition metal complexes for comparison.

Table 2   Thermochemical data for bis(η-arene)–lanthanide and transition metals.

Complex	$\overline{D}$(Ln-Bu$^t_3$C$_6$H$_3$) (kJ mol^{-1})	Complex	$\overline{D}$(Ln-Bu$^t_3$C$_6$H$_3$) (kJ mol^{-1})
[Y(η-Bu$^t_3$C$_6$H$_3$)$_2$]	301(2)	[Cr(η-C$_6$H$_6$)$_2$]	164.9(1)
[Gd(η-Bu$^t_3$C$_6$H$_3$)$_2$]	285(2)	[Cr(η-C$_6$H$_5$Me)$_2$]	157.0(1)
[Dy(η-Bu$^t_3$C$_6$H$_3$)$_2$]	197(2)	[Cr(η-C$_6$H$_3$Me$_3$)$_2$]	150.7(1)
[Ho(η-Bu$^t_3$C$_6$H$_3$)$_2$]	234(2)	[Mo(η-C$_6$H$_6$)$_2$]	247.0(1)
[Er(η-Bu$^t_3$C$_6$H$_3$)$_2$]	239(2)	[W(η-C$_6$H$_5$Me)$_2$]	303.9(1)

It is evident that the lanthanide–arene bond in these compounds is very strong, indeed stronger than that in bis(η-arene)chromium complexes and approaching that of a third row transition metal (W). The thermochemical data also substantiate the argument presented for the stability trend across the series, in that $\overline{D}$(Ln-Bu$^t_3$C$_6$H$_3$) values parallel the ease of the $f^n s^2 \to f^{n-1} d^1 s^2$ promotion (Gd $\gg$ Er $\geq$ Ho $\approx$ Dy).

1.2.3.3 Magnetism

The experimentally determined solid-state magnetic moments for all the thermally stable [Ln(η-Bu$^t_3$C$_6$H$_3$)$_2$] compounds are shown in Table 3, together with the values for the free atoms, free Ln^{3+} ions, and the calculated values obtained from a theoretical model described previously;[13] the latter also discusses the magnetic data in detail. Overall the divergence of the experimental values from the free-atom values is large, indicating significant perturbation of the atomic f shell on bonding of the arene rings to the lanthanide atom, and the results are taken as corroboration of the bonding model discussed in Section 1.2.3.2.

Table 3 Calculated and experimental magnetic moments for $[Ln(\eta\text{-}Bu^t_3C_6H_3)_2]$ compounds.

		μ_{eff} (BM)		
Metal	*Free atom*	*Ln^{3+}*	*Calculated*	*Experimental*
Sc		0	1.73	1.91
Y		0	1.73	1.74
Nd	2.68	3.62	2.68	1.57[a]
Gd	6.51	7.94	8.94	8.75
Tb	10.64	9.70	10.74	10.69
Dy	10.61	10.64	11.67	11.20
Ho	9.58	10.61	11.63	10.10
Er	7.56	9.58	10.61	11.00[b]
Lu		0	1.73	1.69

[a] The value given is that at <20 K. At temperatures >200 K a value of 5.3 BM is found for μ. [b] The value given is that at <125 K. At temperatures >200 K a value of 9.7 BM is found for μ.

1.2.3.4 Reaction chemistry

(i) Redox reactions

The $[Ln(\eta\text{-}Bu^t_3C_6H_3)_2]$ compounds undergo facile reaction with substrates possessing acidic hydrogens to generate lanthanide(III) species. Thus, reaction of $[Y(\eta\text{-}Bu^t_3C_6H_3)_2]$ with 2,4,6-tri-*t*-butylphenol or hexamethyldisilazane affords the tris(aryloxy)-[16] and tris{bis(trimethylsilyl)amido} yttrium(III)[17] derivatives (Equations (12) and (13)).[13]

$$[Y(\eta\text{-}Bu^t_3C_6H_3)_2] + 3\,Bu^t_3C_6H_2OH \xrightarrow{\text{pentane}} [Y(OC_6H_2Bu^t_3\text{-}2,4,6)_3] + 3/2\,H_2 \qquad (12)$$

$$[Y(\eta\text{-}Bu^t_3C_6H_3)_2] + 3\,(TMS)_2NH \xrightarrow{\text{pentane}} [Y\{N(TMS)_2\}_3] + 3/2\,H_2 \qquad (13)$$

(ii) Ring-displacement reactions

Attempted ring-displacement reactions of the $[Ln(\eta\text{-}Bu^t_3C_6H_3)_2]$ compounds with neutral two-electron donors such as phosphines or carbon monoxide result in decomposition and concomitant deposition of the free metal.[14] However, a photochemically driven reaction between $[Ho(\eta\text{-}Bu^t_3C_6H_3)_2]$ and Bu^t_2-DAB has been found to yield $[Ho(Bu^t_2DAB)_3]$ (Equation (14));[14] the latter is also accessible directly from holmium atoms and Bu^t_2DAB,[18] and its formation from $[Ho(\eta\text{-}Bu^t_3C_6H_3)_2]$ is attributable to the superior π-acceptor abilities of Bu^t_2DAB vs. 1,3,5-tri-*t*-butylbenzene.

(iii) Alkene polymerization

The $[Ln(\eta\text{-}Bu^t_3C_6H_3)_2]$ compounds have been found to be very active homogeneous catalysts for ethene polymerization in hydrocarbon solvents.[14] They produce very high-density, narrow molecular-weight distribution polyethene, with turnover numbers in excess of 10^4. The activity of these systems was found to be dependent on the particular lanthanide, those at the beginning of the lanthanide series giving higher polymerization rates, associated with their greater size. IR and solid-state NMR spectral studies showed there to be minimal branching in the polyethene, which was taken to be indicative of a genuine homogeneous process and to mitigate against a free-radical mechanism.

1.2.4 Reactions of Scandium, Yttrium and Lanthanide Atoms with Other Arenes

The reactions of yttrium atoms with arenes less sterically demanding than tri-*t*-butylbenzene (toluene, mesitylene or 1,3,5-triisopropylbenzene) have been examined.[18] In the case of both toluene and mesitylene, the intense purple matrix formed during the cocondensation experiment does not survive above ~120 K, while bis(η-1,3,5-triisopropylbenzene)yttrium is thermally unstable above 270 K, and can only be characterized in solution by ESR and visible spectroscopy. The stability of the tri-*t*-butylbenzene derivatives is attributed to the steric saturation achieved by the ligand, but also to the increase in metal–arene bond strength arising from the positive inductive effect of the *t*-butyl ring substituents.

The effect of the ring substituents on the stability, charge-transfer band and electronic ground state of bis(η-arene)scandium(0) complexes has been studied in some detail, and the results are shown in Table 4.[19]

Table 4   Selected data for bis(η-arene)scandium(0) complexes.

Arene	*Stability*	λ_{max}[a] (nm)	*ESR*[b]
1,3,5-Me$_3$C$_6$H$_3$ [c]	dec. >–30 °C	523	silent
C$_6$Me$_5$H [c]	dec. >0 °C	534	silent
1,4-Bu$^t_2$C$_6$H$_4$	dec. >100 °C	467	active
C$_6$Me$_6$ [c]	dec. >–40 °C	494	active
1,3,5-Bu$^t_3$C$_6$H$_3$	dec. >120 °C	495	active
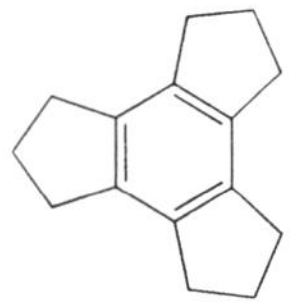 Trindan	dec. >100 °C	500	active
1,3,5-Bu$^t_3$-4-MeC$_6$H$_2$	dec. >120 °C	503	active
2,4,6-Bu$^t_3$C$_5$H$_2$N	dec. >100 °C	470	active

[a] Pentane solution, $\varepsilon > 10^4$. [b] Frozen solution at 77 K. [c] Characterized in solution only.

Overall, there is a good correlation between the bulk of the arene and the thermal stability of the resultant bis(η-arene)scandium(0) complex, and the smaller size of scandium compared with yttrium and the lanthanides results in relatively more stable products even with quite small arenes, for example, mesitylene. There is also a trend for a shift of λ_{max} to longer wavelength (i.e., lower energy) with increasing alkyl substitution (and hence $+I$ effect) of the rings. Thus, this charge-transfer band is assigned as ligand-to-metal in character, and also that in the yttrium and lanthanide analogues (Section 1.2.3.1). Given the electron-rich nature of the metal centres in these compounds, this assignment seems counterintuitive at first sight, but the ^2E ground state of these compounds leaves the d_{z^2} (a_{1g}) orbital empty, and the latter, which is directed towards the centroids of the two rings in a parallel ring sandwich compound, provides a suitable acceptor site for the LMCT transition. However, the trend in λ_{max} is not followed by $[Sc(\eta\text{-}1,3,5\text{-}Me_3C_6H_3)_2]$ or $[Sc(\eta\text{-}C_6Me_5H)_2]$; moreover, the latter are both ESR-silent, even at 4 K. These observations have been rationalized in terms of the adoption of the high-spin ^4A ($e_{2g}^2a_{1g}^1$) ground state by $[Sc(\eta\text{-}1,3,5\text{-}Me_3C_6H_3)_2]$ and $[Sc(\eta\text{-}C_6Me_5H)]$, as opposed to the ^2E state (e_{2g}^3) adopted by the remaining bis(η-arene)scandium(0) complexes.

Table 4 also includes data for $[Sc(\eta\text{-}2,4,6\text{-}Bu^t_3C_5H_2N)_2]$, the first heteroatom-substituted scandium(0) sandwich complex, prepared by cocondensation of scandium vapour with 2,4,6-tri-*t*-butylpyridine (Equation (15)).[19]

$$ \text{(15)} $$

1.3 REFERENCES

1. M. P. Andrews and A. L. Wayda, *Organometallics*, 1988, **7**, 743.
2. W. J. Evans, *Polyhedron*, 1987, **6**, 803 and references therein.
3. J. A. Howard, B. Mile, C. A. Hampson and H. Morris, *J. Chem. Soc., Faraday Trans. 1*, 1989, **85**, 3953.
4. M. N. Bochkarev, A. A. Trifonov, V. K. Cherkasov and G. A. Razuvaev, *Metalloorg. Khim.*, 1988, **1**, 392.
5. M. N. Bochkarev *et al.*, *J. Organomet. Chem.*, 1989, **372**, 217.
6. M. N. Bochkarev, E. A. Fedorova, I. M. Penyagina, O. A. Vasina, A. V. Protchenko and S. Ya Khorshev, *Metalloorg. Khim.*, 1989, **2**, 1317.
7. M. N. Bochkarev, I. L. Fedushkin, H. Schumann and J. Loebel, *J. Organomet. Chem.*, 1991, **410**, 321.
8. M. N. Bochkarev, I. L. Fedushkin, V. K. Cherkasov, V. I. Nevodchikov, H. Schumann and F. H. Goerlitz, *Inorg. Chim. Acta*, 1992, **201**, 69.
9. J. Brennan, F. G. N. Cloke, A. A. Sameh and A. Zalkin, *J. Chem. Soc., Chem. Commun.*, 1987, 1668.
10. D. M. Anderson *et al.*, *J. Chem. Soc., Chem. Commun.*, 1989, 53.
11. F. A. Cotton and W. Schwotzer, *J. Am. Chem. Soc.*, 1986, **108**, 4657.
12. (a) E. Keulin and F. Jellinek, *J. Organomet. Chem.*, 1966, **5**, 490; (b) E. Forster, G. Albrecht, W. Durselen and E. Kurras, *ibid.*, 1969, **19**, 215; (c) A. Haaland, *Acta Chem. Scand.*, 1965, **19**, 41.
13. F. G. N. Cloke, K. Khan and R. N. Perutz, *J. Chem. Soc., Chem. Commun.*, 1991, 1372.
14. F. G. N. Cloke, *Chem. Soc. Rev.*, 1993, 17.
15. W. A. King, T. J. Marks, F. G. N. Cloke, D. M. Anderson and D. J. Duncalf, *J. Am. Chem. Soc.*, 1992, **114**, 9221.
16. P. B. Hitchcock, M. F. Lappert and R. G. Smith, *Inorg. Chim. Acta*, 1987, **139**, 183.
17. D. C. Bradley, J. S. Ghotra and F. A. Hart, *J. Chem. Soc., Dalton Trans.*, 1973, 1021.
18. A. A. Sameh, D. Phil. Thesis, University of Sussex, 1991.
19. K. Khan, D. Phil. Thesis, University of Sussex, 1991.

2
Scandium, Yttrium, and the Lanthanide and Actinide Elements, Excluding their Zero Oxidation State Complexes

FRANK T. EDELMANN

Otto-von-Guericke-Universität Magdeburg, Germany

2.1 INTRODUCTION

The year 1954 marks the beginning of organolanthanide chemistry, when Birmingham and Wilkinson described the tris(cyclopentadienyl)lanthanide complexes Ln(Cp)$_3$.[1] Only two years later, [UCl(Cp)$_3$], the first fully characterized organoactinide complex, was isolated by Reynolds and Wilkinson.[2] However, rapid development of this new area of organometallic chemistry was hampered by the intrinsic instability of organo-f-element compounds towards moisture and oxygen. The highly oxophilic character of the f-elements requires rigorous exclusion of traces of air and moisture during preparation and characterization of organometallic compounds of these metals. With today's more sophisticated experimental and analytical techniques, organolanthanides and -actinides are routinely synthesized and handled and their unique properties are easily studied. Historically, the early discovery of Ln(Cp)$_3$ and [UCl(Cp)$_3$] was followed by a period of relative stagnation that lasted about two decades. The 1970s and 1980s, however, have witnessed an enormous and exciting development of organo-f-element chemistry. This spectacular growth in research activities is mainly due to the fact that many organo-f-element complexes exhibit unique structural and physical properties and many of them are highly active in various catalytic processes. In fact, the catalytic activity of certain organolanthanide hydrocarbyls and hydrides is often dramatically higher than that of comparable d-transition metal catalysts.

This chapter summarizes the present knowledge about the chemistry of organo-f-element complexes. Synthetic methods, structural investigations, and the chemical reactivity of these compounds will be discussed. The prelanthanides scandium, yttrium, and lanthanum will be included because of their similarity in size, charge, and chemical behavior as compared to the lanthanide elements. Generally, this chapter has been restricted to compounds of the lanthanide and actinide elements containing M–C bonds. A brief summary of the general chemical and physicochemical characteristics of the metal ions can be found in *COMC-I*. Thus, in the second part of this chapter we will discuss the chemistry of scandium, yttrium, and lanthanide organometallics, while the third section is devoted to organoactinide chemistry. Each of these sections is subdivided into ligand types and concluded by a discussion of the

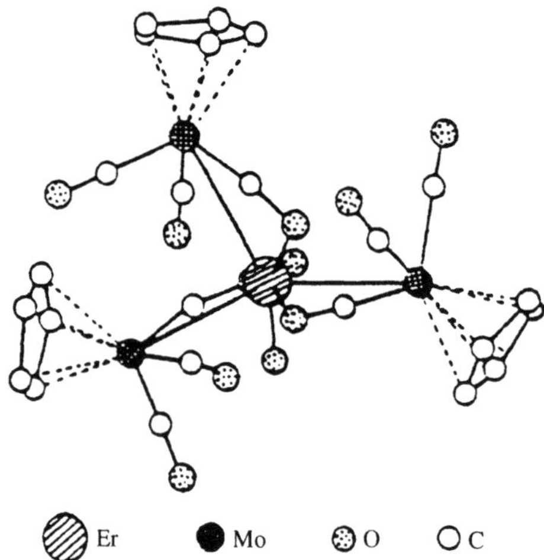

Figure 1   The molecular structure of $[Er(EtOH)(H_2O)_4\{Mo(CO)_3(Cp)\}_3]$.[90]

investigation of the hydrocarbons formed upon hydrolysis indicated in most cases that the products contained direct lanthanide–carbon bonds and the oxidation state of the lanthanide elements was 2+ or 3+. Thus, these compounds will be briefly discussed in this section. Metal vapors of samarium, erbium, and ytterbium have been reacted at −196 °C with various unsaturated hydrocarbons such as ethylene, propylene, and allene. Due to their low solubility the resulting deeply colored products were difficult to analyze. Only $Er(MeCH=CH_2)_3$ was isolated and characterized.[91] Hydrolysis of the products gave mixtures of saturated and unsaturated hydrocarbons like methane, ethane, propane, butane, propylene, butene, allene, and methylacetylene. Similar metal–vapor reactions between samarium, erbium, and ytterbium with 3-hexyne yielded dark brown solids which have been characterized mainly by elemental analysis. The structure of these materials is still unknown, although it can be anticipated that they contain lanthanide–hydrogen bonds as they show some activity as hydrogenation catalysts.[92,93] Similar results were obtained in cocondensation reactions of vaporized lanthanum, neodymium, samarium, and erbium with butadiene or 2,3-dimethylbutadiene at −196 °C, which afforded dark brown paramagnetic solids.[94] Various types of bonding have been discussed for the coordinated diene ligands, but the real nature of the products remains uncertain. The same is true for diene complexes of the type $[(diene)_xLnCl_{3-x}M(L)_n]$ (Ln = La, Pr, Nd, Sm, Gd, Dy; M = $MgCl_2$, LiCl; x, n = 0–3; L = THF, DMSO).[95]

Several lanthanide(II) complexes containing direct lanthanide–carbon bonds with no cyclopentadienyl auxiliary ligands have been mentioned in the literature. In most cases such complexes have not been unambiguously characterized and virtually no structural information is available with the exception of the complexes shown in Scheme 1. Single-crystal x-ray data reported for the 1-aza-allyl complex $[Yb\{N(R)C(Bu^t)CHR\}_2]$ (R = TMS) revealed a monomeric structure in which the *rac*-diastereoisomer predominated with each 1-aza-allyl ligand bound η^3-fashion to the ytterbium atom.[96] Certainly the most thoroughly investigated representative of this class of compounds is bis(pentafluorophenyl)ytterbium(II). This interesting species is easily accessible through a transmetallation route involving treatment of bis(pentafluorophenyl)mercury with elemental ytterbium (Equation (1)). The corresponding europium diaryl has been made analogously.[97–9] Complexes with partially fluorinated phenyl ligands such as $Yb(C_6F_4H-o)_2$ or $Yb(C_6F_4H-p)_2$ are significantly less stable than the pentafluorophenyl derivative. No samarium(II) complex of this type has ever been isolated in a pure state.[98,99] The σ-bonded pentafluorophenyl groups in $Yb(C_6F_5)_2$ can be transferred to transition metals.[99,100] In addition, $Ln(C_6F_5)_2$ (Ln = Sm, Eu) have been demonstrated to be highly versatile precursors for the synthesis of a variety of divalent lanthanide phenoxides and amides.[101]

Substituted carboranes have been used to stabilize divalent organolanthanides containing Ln–C σ-bonds. Several solvated ytterbium(II) derivatives have been prepared by a transmetallation reaction between mercury(II) carboranides and ytterbium powder (Equations (2) and (3)).[102–4]

The homoleptic amide complex $[\{Yb(NR_2)(\mu\text{-}NR_2)\}_2]$[105] has served as the starting material for the preparation of a series of neutral ligand-free homoleptic ytterbium(II) and unique heteroleptic amido ytterbium(II) alkoxides and aryloxides $[\{YbX(\mu\text{-}X)\}_2]$ (X = OAr or $OCBu^t_3$) or $[\{Yb(NR_2)(\mu\text{-}X)\}_2]$ (X = OAr or $OCBu^t_3$) (R = TMS, Ar = $C_6H_2Bu^t_2\text{-}2,6\text{-}Me\text{-}4$), Scheme 2. X-ray structures of $[\{YbX(\mu\text{-}OAr)\}_2]$ and $[\{Yb(NR_2)(\mu\text{-}OCBu^t_3)\}_2]$ were reported.[106]

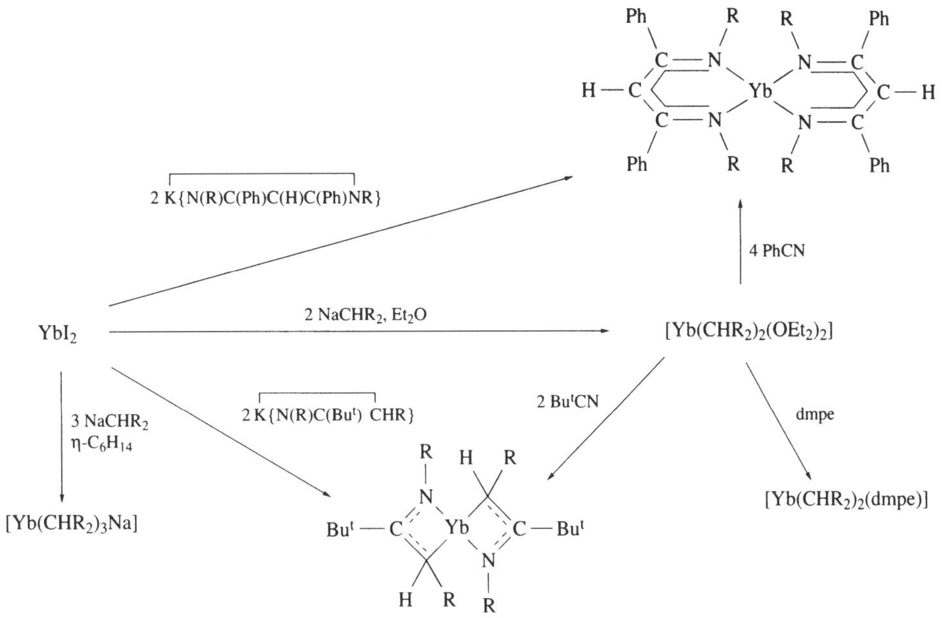

Scheme 1

$$Ln + Hg(C_6F_5)_2 \xrightarrow{THF} Ln(C_6F_5)_2 + Hg \qquad (1)$$

$$Ln = Sm, Eu$$

$$Yb + Hg(o\text{-}C_2H_2B_{10}H_9)_2 \xrightarrow{THF} [Yb(o\text{-}C_2H_2B_{10}H_9)_2(THF)] + Hg \qquad (2)$$

$$Yb + Hg(o\text{-}RC_2B_{10}H_{10})_2 \xrightarrow{THF} [Yb(o\text{-}RC_2B_{10}H_{10})_2(THF)_n] + Hg \qquad (3)$$

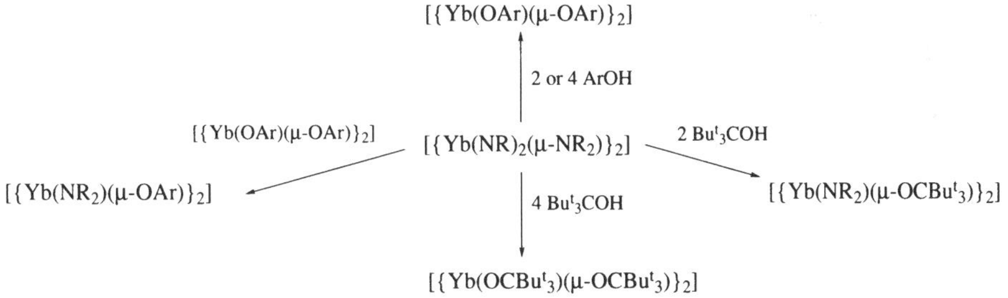

$$R = TMS, \; Ar = C_6H_2Bu^t_2\text{-}2,6\text{-}Me\text{-}4 \; [106]$$

Scheme 2

The preparation of isolable homoleptic lanthanide(III) compounds containing σ-bonded hydrocarbyl ligands was probably one of the most difficult tasks in organolanthanide chemistry. For many years this area of research was dominated by failures, although the first attempts to isolate such simple lanthanide alkyls were made about 50 years ago. The first indication for the possible formation of lanthanide alkyls came from the observation that lanthanum reacts with methyl radicals.[107] However, no simple compound of the type $LnMe_3$ has ever been isolated. Coordinative unsaturation is clearly the main reason for the inherent instability of simple lanthanide trialkyls or triaryls. All early attempts to isolate such compounds failed. For example, an early report on the synthesis of $ScEt_3$ and YEt_3 could never be reproduced.[108,109] A related study showed the inaccessibility of triphenyllanthanum. Treatment of anhydrous $LaCl_3$ with three equivalents of phenyllithium gave biphenyl as the only isolable reaction product.[110] The same result was obtained when lanthanum metal was reacted with diphenylmercury. In the late 1960s the syntheses of triphenylscandium, triphenylyttrium[111,112] and lithium tetraphenyllan-thanate, Li[LaPh$_4$], were published. The products were characterized only on the basis of their IR spectra and elemental analyses as well as quenching reactions with mercury dichloride, CO_2, and benzophenone.

The structures of these materials still await confirmation by x-ray crystallography. In view of the pronounced tendency of the lanthanide elements to adopt high coordination numbers it seems most likely that the homoleptic triphenylmetal derivatives are polymeric materials.[113,114]

By now four different synthetic approaches to well-defined lanthanide hydrocarbyls have been developed. Such compounds can be successfully prepared (a) by formation of stoichiometric adducts with donating solvents such as THF, DME, or TMEDA, (b) by incorporation of donor functions into the organic ligands, (c) by using sterically demanding hydrocarbyl ligands, or (d) by formation of anionic "ate" complexes. Early reports showed that hydrocarbyl ligands such as *t*-butyl, neopentyl, or trimethylsilylmethyl are not sufficiently bulky to permit the isolation of base-free homoleptic organolanthanides. The bis(THF) adducts $[Ln(CH_2TMS)_3(THF)_2]$ are the final products when lanthanide trichlorides are reacted with $LiCH_2TMS$ in THF solution (Equation (4)).[115–18] In the presence of sodium naphthalide the ytterbium derivative has been found to activate N_2.[119]

$$LnCl_3 + 3\,LiCH_2TMS \xrightarrow{\;THF\;} [Ln(CH_2TMS)_3(THF)_2] + 3\,LiCl \qquad (4)$$
$$Ln = Er,\ Tm,\ Yb,\ Lu$$

According to NMR spectral investigations these lanthanide alkyls are trigonal-bipyramidal with the THF ligands occupying the *trans* positions.[117,118] The products are thermally unstable and decompose upon prolonged standing by elimination of THF and tetramethylsilane to give pyrophoric materials of polymeric nature (Equation (5)). These materials have been postulated to be lanthanide carbene complexes, although no structural information is available and the true nature of the products remains unclear.[117]

$$[Ln(CH_2TMS)_3(THF)_2] \xrightarrow{\;\Delta\;} 1/n\,[\{Ln(CH\text{-}TMS)(CH_2TMS)\}_n] + SiMe_4 \qquad (5)$$
$$Ln = Er,\ Lu$$

Similar decomposition reactions and formation of polymeric "carbene" materials and/or hydride species have been observed when $LnCl_3$ (Ln = Y, Nd) was reacted with Bu^nLi, $LiCH_2TMS$, LiBz, or $LiCH_2CMe_2Ph$, respectively. These reactions gave no indications of the formation of isolable homoleptic lanthanide hydrocarbyls.[120–6]

An effective synthetic route leading to monomeric lanthanide(III) hydrocarbyls is the intramolecular saturation of coordination sites by "in-built" donor functions. The earliest reports using this concept described the preparation of homoleptic organolanthanides containing dimethylamino-substituted phenyl or benzyl ligands. The resulting six-coordinated organolanthanides $[Sc\{CH(SiMe_2C_6H_4OMe\text{-}o)_2\}_3]$, $[Ln(CH_2C_6H_4NMe_2\text{-}o)_3]$, and $[Ln(C_6H_4CH_2NMe_2\text{-}o)_3]$ (Ln = Sc, Y, La, Nd, Er, Yb, Lu) are stable without additional THF ligands.[127–30] More recently two other series of lanthanide aryls, $[Ln(C_6H_4Me\text{-}o)_3]$ (Ln = Ce, Pr, Nd, Sm, Gd, Er)[127] and $[Ln(C_6H_4OMe\text{-}o)_3]$ (Ln = Ce, Pr, Nd),[131] as well as the binuclear Yb^{2+}/Yb^{3+} pentaphenyl complex $[Yb(THF)(\mu\text{-}Ph)_3Ph_2Yb(THF)_3]$[132,133] have been described. The latter is a truly remarkable species. It has been prepared by reacting naphthaleneytterbium, $[Yb(THF)_2C_{10}H_8]$ with either Ph_2Hg or Ph_3Bi in THF solution and isolated in the form of red paramagnetic crystals. A crystal structure determination (Figure 2) revealed the presence of an asymmetrical binuclear complex, in which both ytterbium atoms adopt a distorted octahedral coordination geometry. The two ytterbium atoms are bridged by three phenyl groups, which are linked with the first ytterbium atom by an η^1-bond and with the second one by an unsymmetrical η^2-bond. In addition the first ytterbium atom is η^1-bonded to two terminal phenyl groups and one THF ligand, whereas the second ytterbium atom is linked to three THF molecules.[133]

An interesting class of homoleptic lanthanide hydrocarbyls has become available through the use of allyllike diphosphinomethanide ligands. The first member of the series was the lanthanum derivative, $[La\{CH(PPh_2)_2\}_3]$ **(1)**, which was characterized by x-ray crystallography.[134]

(1)

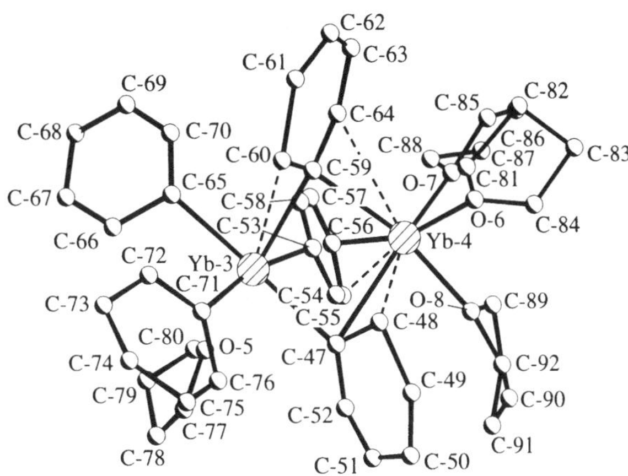

Figure 2 The molecular structure of [Yb(THF)(μ-Ph)$_3$Ph$_2$Yb(THF)$_3$].[133]

Mono- and dinuclear diphosphinomethanides have been reported for samarium(III) (Equations (6) and (7)). Both compounds have been structurally characterized. The disamarium complex is a centrosymmetric dimer in a chair conformation with two diphosphinomethanide ligands bridging via the P-2 and C-1 atoms.[135,136]

$$SmCl_3(THF)_3 + 3\,LiCH(PPh_2)_2 \longrightarrow [Sm\{CH(PPh_2)_2\}_3] + 3\,LiCl + 3\,THF \qquad (6)$$

$$2\,Sm(O_3SCF_3)_3 + 6\,LiCH(PMe_2)_2 \longrightarrow [(Sm\{CH(PMe_2)_2\})_2\{\mu\text{-}\eta^2\text{-}CH(PMe_2)_2\}H] + 6\,LiO_3SCF_3 \qquad (7)$$

Several lanthanide(III) derivatives containing σ-bonded carboranyl ligands have also appeared in the literature. These materials have been prepared by reacting mercury(II) carboranides with elemental lanthanides in THF. This transmetallation reaction directly yields THF adducts of the corresponding lanthanide(III) carboranides (Equation (8)).[107,137]

$$2\,Ln + 3\,Hg(o\text{-}RC_2B_{10}H_{10})_2 \longrightarrow 2\,[Ln(o\text{-}RC_2B_{10}H_{10})_3(THF)_n] + 3\,Hg \qquad (8)$$

$$R = Me;\ Ln = Tm\ (n=3),\ Yb\ (n=2)$$
$$R = Ph;\ Ln = La\ (n=1),\ Tm\ (n=3),\ Yb\ (n=2)$$

Yet another approach to neutral, homoleptic lanthanide hydrocarbyls is the use of bidentate anionic phosphoylide anions of the type $R_2P(CH_2)_2^-$. However, early attempts to obtain monomeric lanthanide complexes with these ligands met with little success. The first such compounds were prepared by a two-step procedure according to Scheme 3.[18]

$$LnCl_3 + 3\,Me_3P{=}CH_2 \longrightarrow [Ln\{CH_2PMe_3Cl\}_3] \xrightarrow[\substack{-3\,C_4H_{10} \\ -3\,LiCl}]{+3\,Bu^nLi} [Ln\{(CH_2)_2PMe_2\}_3]$$

$$Ln = La,\ Pr,\ Nd,\ Sm,\ Gd,\ Ho,\ Er,\ Lu$$

Scheme 3

The lutetium complex [Lu{(CH$_2$)$_2$PBu$^t_2$}$_3$] was made similarly by treating LuCl$_3$ with three equivalents of Li(CH$_2$)$_2$PBu$^t_2$. The ^{1}H NMR spectra of all these complexes indicated a temperature-dependent equilibrium between the monomer and oligomeric and polymeric species in solution. For [Lu{(CH$_2$)$_2$PBu$^t_2$}$_3$] this equilibrium was investigated by ^{1}H, ^{13}C, and ^{31}P NMR spectral studies. The results clearly demonstrated that even the chelating phosphoylide ligand [Bu$^t_2$P(CH$_2$)$_2$]$^-$ is not sufficiently bulky to stabilize low-coordinate homoleptic lanthanide hydrocarbyls.[138]

On summarizing the results discussed above it becomes quite clear that base-free monomeric lanthanide hydrocarbyls can only be synthesized by using very bulky organic ligands. Also, the chances to successfully isolate such complexes should be higher for scandium, yttrium, and the heavier (i.e., smaller) lanthanide ions. In fact, the earliest reports on base-free homoleptic lanthanide hydrocarbyls described the synthesis of scandium and yttrium compounds containing highly sterically demanding disilylated methyl groups.[118] The unsolvated compounds [Y{CH(TMS)$_2$}$_3$][139] and [Sc(CH$_2$SiMe$_2$C$_6$H$_4$OMe-p)$_3$] were synthesized by conventional salt elimination reactions between the metal trichlorides and three equivalents of the corresponding lithium reagents. Obviously in these two cases the ligands were sufficiently bulky to prevent the products from adding solvent molecules or retaining lithium halide.

The first synthesis of base-free lanthanide alkyls by Lappert *et al.* was achieved by combining two useful synthetic strategies. The bis(trimethylsilyl)methyl ligand was employed as a very bulky hydrocarbyl group and the reaction was conducted in the absence of alkali halides by using the 2,4,6-tri-*t*-butylphenoxide as a leaving group (Equation (9)).[140-2]

$$[Ln\{OC_6H_2(Bu^t)_3\text{-}2,4,6\}_3] + 3\,LiCH(TMS)_2 \xrightarrow{\text{pentane}} [Ln\{CH(TMS)_2\}_3] + 3\,LiOC_6H_2(Bu^t)_3\text{-}2,4,6 \quad (9)$$

$$Ln = Y, La, Sm, Lu$$

Under the chosen reaction conditions (pentane solution) the lithium phenoxide by-product precipitates, thus facilitating the isolation of the highly soluble lanthanide hydrocarbyls. Like the homoleptic silylamides $[Ln\{N(TMS)_2\}_3]$ the tris-alkyls adopt a flat pyramidal coordination geometry as has been determined by x-ray crystallography for the lanthanum and samarium derivatives (Figure 3). The average metal–carbon distances are La–C 0.2515(9) nm and Sm–C 0.233(2) nm.

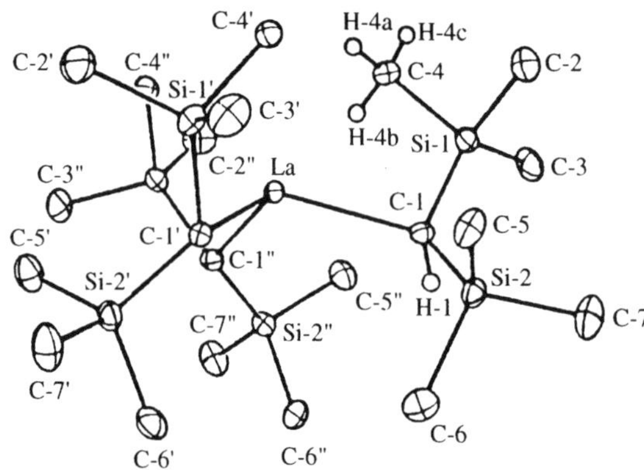

Figure 3 The molecular structure of $[La\{CH(TMS)_2\}_3]$.[140]

Several explanations have been given for the deviation from planarity in $[Ln\{CH(TMS)_2\}_3]$ compounds. Interligand repulsion is thought to be minimized in the pyramidal structure, but more important is the possibility of forming Ln–H–C agostic interactions. In fact, short γ-agostic Ln$\cdots$Me interactions were found in both structurally characterized compounds (La$\cdots$C 0.3121 nm, Sm$\cdots$C 0.285 nm) which are a result of the extreme electronic and coordinative unsaturation of the central lanthanide atoms.

In marked contrast, the more obvious metathetical reaction between lanthanide trichlorides and bis(trimethylsilyl)methyllithium does not afford the homoleptic lanthanide alkyls due to retention of lithium halide (Equation (10)), La–C 0.260(3) nm:[140]

$$LaCl_3 + 3\,LiCH(TMS)_2 \xrightarrow{\text{pmdeta}} [La\{CH(TMS)_2\}_3(\mu\text{-Cl})Li(\text{pmdeta})] + 2\,LiCl \quad (10)$$

$$\text{pmdeta} = N,N,N,'N'',N''\text{-pentamethyldiethylenetriamine}$$

The LiCl adduct contains an almost linear La–Cl–Li unit (La–Cl–Li 165.1°, La–Cl 0.2762 nm, Li–Cl 0.228 nm) and the $[La\{CH(TMS)_2\}_3]$ fragment remains virtually unchanged as compared to the free homoleptic alkyls.[140]

When ytterbium trichloride is used in a similar preparation, the saltlike compound $[Li(THF)_4][YbCl\{CH(TMS)_2\}_3]$ is obtained, which consists of separated ions.[115] Obviously the formation of one or the other type of product depends on a subtle balance between various factors, for example nature of the alkali metal, presence of halide ions, solvent, and ionic radius of the lanthanide ion. A similar situation was recently observed in the chemistry of low-coordinate lanthanide alkoxides and amides. The chloride-bridged neodymium compounds $[Nd\{N(TMS)_2\}_3(\mu\text{-Cl})Li(THF)_3]$ and $[Nd\{(Bu^t)_3CO\}_3(\mu\text{-Cl})Li(THF)_3]$ correspond to the lanthanum alkyl $[La\{CH(TMS)_2\}_3(\mu\text{-Cl})Li(\text{pmdeta})]$, whereas $[Li(THF)_4][NdO\text{-TMS}\{N(TMS)_2\}_3]$[143] can be compared to $[Li(THF)_4][YbCl\{CH(TMS)_2\}_3]$.

The following example may further illustrate how slight changes in the reaction conditions or the choice of reagents can cause interesting differences in the product formation. When the more reactive potassium alkyl $KCH(TMS)_2$ is used instead of $LiCH(TMS)_2$, the reaction with lanthanide trichlorides (in this case $LuCl_3$) can be carried out in diethyl ether instead of THF and affords the solvated KCl adduct $[Lu\{CH(TMS)_2\}_3(\mu\text{-}Cl)K(Et_2O)]$ (Scheme 4). In contrast to the corresponding LiCl adduct the coordinated diethyl ether can easily be removed *in vacuo* to give unsolvated $[Lu\{CH(TMS)_2\}_3(\mu\text{-}Cl)K]$. In this compound the potassium is coordinatively unsaturated and readily adds various solvent molecules such as diethyl ether or even toluene (Scheme 4):[144,145]

$$LuCl_3 + 3\ KCH(TMS)_2 \xrightarrow{\text{ether}} [Lu\{CH(TMS)_2\}_3(\mu\text{-}Cl)K(Et_2O)] \underset{\text{ether}}{\overset{\text{vacuum, }-\text{ether}}{\rightleftharpoons}} [Lu\{CH(TMS)_2\}_3(\mu\text{-}Cl)K]$$

$$[Lu\{CH(TMS)_2\}_3(\mu\text{-}Cl)K(\eta^6\text{-toluene})_2]$$

(vacuum ‖ toluene)

Scheme 4

Especially remarkable is the hydrocarbon-soluble toluene solvate, which has been structurally characterized by x-ray diffraction. As compared to the almost linear Ln–Cl–M (M = Li or K) bridging in $[Lu\{CH(TMS)_2\}_3(\mu\text{-}Cl)K(Et_2O)]$ and $[Nd\{N(TMS)_2\}_3(\mu\text{-}Cl)Li(THF)_3]$ the Lu–Cl–K unit deviates significantly from linearity (145.9°). Two toluene molecules are η^6-coordinated to the potassium ion (Figure 4).

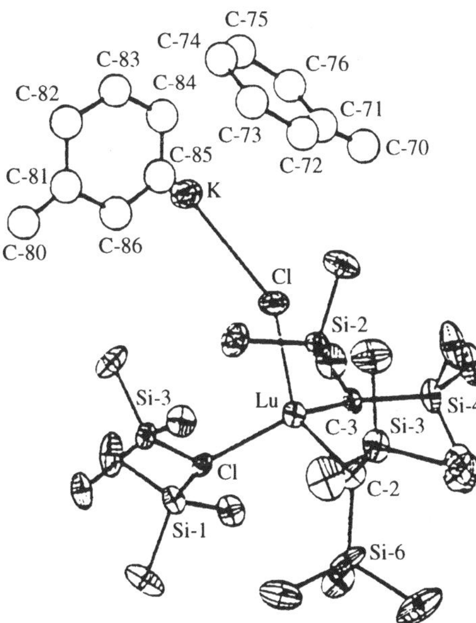

Figure 4 The molecular structure of $[Lu\{CH(TMS)_2\}_3(\mu\text{-}Cl)K(\eta^6\text{-toluene}_2)]$.[144]

Very little has been published on heavier homologues of homoleptic lanthanide hydrocarbyls (i.e., silyls, germyls, stannyls, and plumbyls). Organosilicon and organogermanium complexes of divalent ytterbium have been prepared as shown in Equation (11).[146]

$$2\ Yb + 2\ Ph_3ECl \xrightarrow{\text{THF}} [Yb(EPh_3)_2(THF)_4] + [YbCl_2(THF)_2] \qquad (11)$$

$$E = Si,\ Ge$$

X-ray structural analyses revealed that both compounds have similar centrosymmetrical octahedral structures with the EPh_3 ligands in axial positions.

A crystal structure determination has also been reported for $[Yb\{Sn(CH_2Bu^t)_3\}_2(THF)_2]$. The stannyl derivative was made by reacting YbI_2 with two equivalents of $K[Sn(CH_2Bu^t)_3]$.[147] The compounds $[Ln\{Sn(CH_2TMS)_3\}_3(DME)]$ have been prepared and characterized by spectroscopic methods.[148,149] These compounds can formally be regarded as heavier homologues of lanthanide hydrocarbyls. In this context the homoleptic phosphides and arsenides of the type $Ln\{P(Bu^t)_2\}_3$ and $Ln\{As(Bu^t)_2\}_3$ should also be mentioned, although these are not organometallic compounds in a strict sense. So far no structural information is available on these materials.[150]

2.2.2.2 Anionic homoleptic compounds

Simple lanthanide(III) hydrocarbyls show a strong tendency to add anionic ligands and form "ate" complexes, thereby increasing the formal coordination number around the lanthanide atoms. This phenomenon is frequently encountered when lanthanide hydrocarbyls are prepared from anhydrous lanthanide trichloride and the corresponding lithium alkyl or aryl. Formation of anionic hydrocarbyls is especially favored when the organic ligands are not sufficiently bulky to permit the stabilization of low coordination numbers around the lanthanide atoms. Of special interest are reactions of lanthanide trichlorides with methyllithium, which have been studied in detail by Schumann *et al.* Neutral homoleptic $LnMe_3$ derivatives are coordinatively highly unsaturated and have not been isolated for any member of the lanthanide series. However, treatment of $LnCl_3$ with six equivalents of methyllithium in the presence of a coordinating solvent yields stable, six-coordinated hexamethyllanthanidates, $[LnMe_6]^{3-}$, which can be isolated for all rare-earth elements except promethium and europium.[151] Whereas the promethium reaction has not been investigated, the europium(III) derivative cannot be isolated, because $EuCl_3$ is reduced to an uncharacterized divalent europium species upon treatment with methyllithium.[151] Suitable coligands capable of saturating the coordination sphere of the lithium atoms are N,N,N',N'-tetramethylethylenediamine (TMEDA) and 1,2-dimethoxyethane (DME).[151] The DME derivatives are thermally less stable than the TMEDA adducts (Equations (12) and (13)).

$$LnCl_3 + 6\,MeLi + 3\,TMEDA \xrightarrow{\ Et_2O\ } [Li(TMEDA)]_3[LnMe_6] + 3\,LiCl \qquad (12)$$

$$Ln = Sc,\ Y,\ La–Nd,\ Sm–Lu$$

$$LnCl_3 + 6\,MeLi + 3\,DME \xrightarrow{\ Et_2O\ } [Li(DME)]_3[LnMe_6] + 3\,LiCl \qquad (13)$$

$$Ln = Ce,\ Pr,\ Gd,\ Tb,\ Dy,\ Er,\ Lu$$

The crystalline compounds display the characteristic colors of the lanthanide cations. The molecular structures of $[Li(TMEDA)]_3[LnMe_6]$ (Ln = Ho, Er) and $[Li(DME)]_3[LuMe_6]$ have been determined by x-ray crystallography.[151,152] Figure 5 shows the molecular structure of $[Li(TMEDA)]_3[HoMe_6]$. In these compounds the central lanthanide ion is surrounded by six methyl groups in a slightly distorted octahedral fashion. All Ln–C distances are equal. Each of the three lithium ions is connected to the central lanthanide by two bridging methyl ligands and the tetrahedral coordination sphere around lithium is completed by the chelating TMEDA or DME ligand.[153]

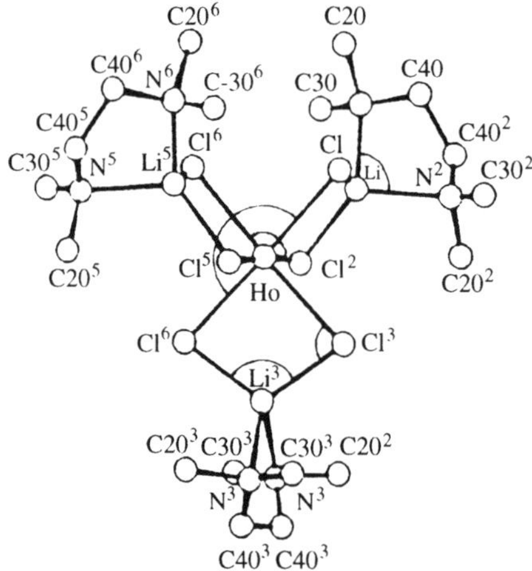

Figure 5   The molecular structure of $[Li(TMEDA)]_3[HoMe_6]$.[151,152]

The derivative chemistry of the anionic hexamethyllanthanidate complexes remains relatively little explored. Besides some reactivity studies toward organic reagents (cf. Section 2.2.14), the samarium derivative $[Li(TMEDA)]_3[SmMe_6]$ has recently been used to prepare the mixed-metal alkoxide $[Li_5Sm(OBu^t)_8]$ via treatment with *t*-butanol in diethyl ether.[154]

With the chelating dimethylaminopropyl ligand a related anionic cerium complex, $Li_3[Ce\{(CH_2)_3NMe_2\}_6]$, has been isolated.[155] An interesting binuclear derivative was obtained by reacting $LuCl_3$ with excess methyllithium in the presence of N,N,N',N'-tetraethylethylenediamine (TEEDA). In this structurally characterized compound there are methyl bridges between the two lutetium atoms as well as between lutetium and lithium.[152]

The low coordination number of 4 around lanthanide elements can be realized in anionic hydrocarbyl complexes containing the bulky t-butyl and trimethylsilylmethyl ligands. Anionic tetra-t-butyl lanthanidates can be obtained by reacting either lanthanide trichlorides or lanthanide tri-t-butoxides with four equivalents of Bu^{t}Li (Equations (14)–(16)).[92,151,156]

$$LnCl_3 + 4\,LiBu^t \xrightarrow{pentane/Et_2O} [Li(Et_2O)_4][LnBu^t_4] + 3\,LiCl \qquad (14)$$
$$Ln = Tb,\ Er,\ Lu$$

$$LnCl_3 + 4\,LiBu^t \xrightarrow{THF} [Li(THF)_n][LnBu^t_4] + 3\,LiCl \qquad (15)$$
$$Ln = Er\ (n = 4),\ Lu\ (n = 3)$$

$$Ln(OBu^t)_3 + 4\,LiBu^t \xrightarrow[TMEDA]{pentane} [Li(TMEDA)_2][LnBu^t_4] + 3\,LiOBu^t \qquad (16)$$
$$Ln = Er,\ Lu$$

The lutetium derivative [Li(TMEDA)$_2$][LuBu$^t_4$] was structurally characterized.[157] The crystal structure consists of separated ions. Other tetrahedrally coordinated lanthanide hydrocarbyl anions have been prepared using the trimethylsilylmethyl ligand. Lithium salts of the [Ln(CH$_2$TMS)$_4$]$^-$ anions have been isolated with diethyl ether or TMEDA as coligands. These salts can be prepared by addition of LiCH$_2$TMS to the preformed homoleptic neutral hydrocarbyls. An alternative preparation involves treatment of lanthanide trichlorides with four equivalents of the lithium reagent in diethyl ether.[115,117,158] The diethyl ether solvates can subsequently be converted into the TMEDA derivatives by exchange of the donor ligands at lithium (Equations (17)–(19)).

$$LnCl_3 + 4\,LiCH_2TMS \xrightarrow{Et_2O} [Li(Et_2O)_4][Ln(CH_2TMS)_4] + 3\,LiCl \qquad (17)$$

$$[Ln(CH_2TMS)_3(THF)_2] + LiCH_2TMS \xrightarrow{THF/TMEDA} [Li(TMEDA)_2][Ln(CH_2TMS)_4] \qquad (18)$$
$$Ln = Y,\ Er,\ Yb$$

$$[Li(Et_2O)_4][Lu(CH_2TMS)_4] + 2\,TMEDA \longrightarrow [Li(TMEDA)_2][Lu(CH_2TMS)_4] + 4\,Et_2O \qquad (19)$$

Similar to the solvated neutral hydrocarbyls, the anionic complexes [Ln(CH$_2$TMS)$_4$]$^-$ are thermally labile even at room temperature and decompose under elimination of tetramethylsilane. The resulting polymeric products have been shown to contain not only the original CH$_2$TMS groups but also CH-TMS and C-TMS units bonded to the lanthanides.[159]

Similar results have been obtained with aryl ligands. A typical example is the reaction of LaCl$_3$ or PrCl$_3$ with phenyllithium which leads to anionic tetraphenylmetallates. Simple homoleptic triphenyl derivatives of these metals are unstable (Equation (20)).[112]

$$LnCl_3 + 4\,PhLi \xrightarrow{THF} Li[LnPh_4] + 3\,LiCl \qquad (20)$$

These compounds are believed to display a tetrahedral coordination geometry. Similar reactions have been carried out with 2,6-dimethylphenyllithium and in the case of [Li(THF)$_4$][Lu(C$_6$H$_3$Me$_2$-2,6)$_4$] the tetrahedral coordination was proved unambiguously by an x-ray structural analysis.[160]

Recently the first anionic lanthanide diene complexes have been prepared and structurally characterized. In the complex [K(THF)$_2$(μ-η^4-Ph$_2$C$_4$H$_4$)$_2$Lu(THF)$_2$]$_n$ (2) lutetium is coordinated by two tetrahapto-diphenylbutadiene ligands and two THF molecules (Lu–C 0.2523 nm, Lu–O 0.2382 nm). Diphenylbutadiene adopts an s-*cis* conformation. In the crystal, a polymeric chain structure results from the fact that potassium is hexahapto-coordinated to two phenyl rings of neighboring diphenylbutadiene molecules.[161]

2.2.2.3 Heteroleptic compounds

Although THF adducts of lanthanide hydrocarbyls are formally heteroleptic compounds, this section will include only lanthanide hydrocarbyls with additional anionic ligands. Reports on unusual compounds like BzLa(H)OCH=CH$_2$·2THF,[162] Me$_2$C=C(Ph)ScCl$_2$(THF)$_3$,[163] LnCl(C$_4$H$_6$) MgCl$_2$(THF)$_n$,[164] Li$_3$[CeMe$_6${Li(acac)}$_3$],[165] Gd(C$_6$H$_4$OMe-o)$_2$Cl,[131] and triphenylmethyl(lanthanide) dichlorides,

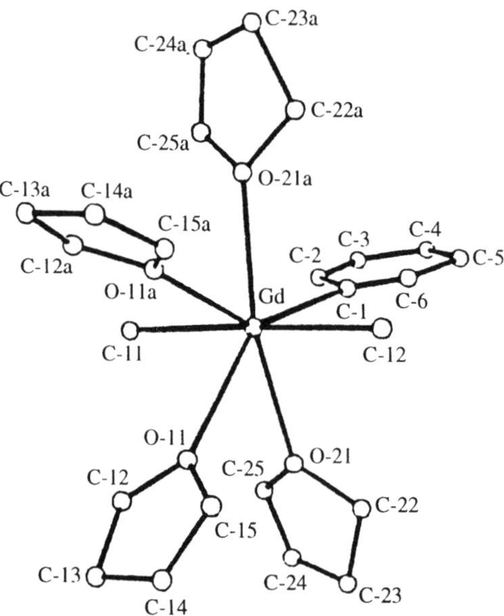

(2)

Ph$_3$CLnCl$_2$(THF)$_n$ (Ln = La, Pr, Nd, Gd, Ho)[163,166,167] still await structural confirmation. Due to the low formal coordination numbers some of these compounds are presumably of oligomeric or polymeric nature. The same is true for some methyl- and phenylcarboranyl lanthanide dichlorides, which have been described for lanthanum, thulium, and ytterbium.[137]

A series of heteroleptic amido-ytterbium(II) and alkyl-ytterbium(II) alkoxides and aryloxides has recently been reported. The amido complexes were discussed in Section 2.2.2.1.[106] Treatment of YbI$_2$ with KCR$_3$ (R = TMS) in Et$_2$O afforded the dimeric alkylytterbium(II) ethoxide [{Yb(CR$_3$)(μ-OEt)(OEt$_2$)}$_2$], which has been structurally characterized.[96] The monomeric aryloxide complex [Yb(CHR$_2$)(OAr)(THF)$_3$] (R = TMS, Ar = C$_6$H$_2$Bu$^t_2$-2,6-Me-4) was obtained upon treatment of [Yb(OAr)$_2$(THF)$_3$] with KCHR$_2$.[96]

A remarkable result in this area was the first structural characterization of a "lanthanide(III) Grignard compound" by Chen *et al.* The σ-phenylgadolinium complex [GdCl$_2$(THF)$_4$Ph] was prepared by a 1:1 reaction of anhydrous gadolinium trichloride and phenyllithium in THF (Figure 6, Gd–C 0.242(2) nm).[168,169]

Figure 6   The molecular structure of [GdCl$_2$(THF)$_4$Ph].[163,169]

Methyl bridging between samarium and lithium was reported for the structurally characterized hydrocarbyl complex [Sm{CH(TMS)$_2$}$_3$(μ-Me)Li(pmdeta)] (Sm–C(Me) 0.233(3) nm, Sm–CH(TMS)$_2$ 0.251(2) nm).[170]

Metallacyclic organolanthanides are exceedingly rare and often insufficiently characterized. Tetraphenylbutadiene lanthanide metallacycles of the composition [Ph$_4$C$_4$Ln(μ-Cl)$_2$Li(THF)$_n$] have been reported for Ln = Nd and Gd.[171] *o*-Dilithiobiphenyl was reported to react with anhydrous lanthanide tribromides to afford metallacyclic species, in which the metal atoms are part of a five-membered ring system. The products were isolated as THF adducts (Equation (21)).[105,172]

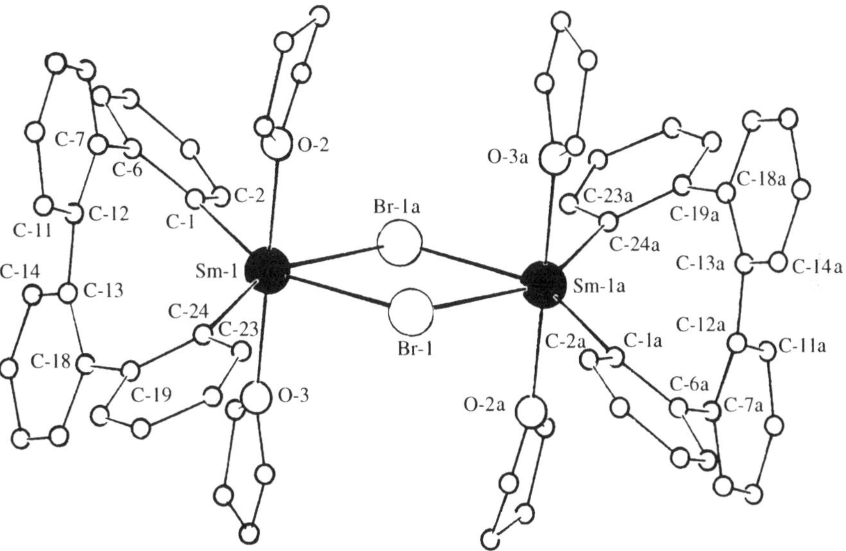

A recent reinvestigation of the reaction of samarium tribromide with *o*-dilithiobiphenyl revealed that at least in this particular case the original formulation of the products with lanthanide-containing five-membered ring systems was incorrect. The structurally characterized dimeric product contained dimetallated *o*-quaterphenyl ligands, which were formed by dimerization of the original biphenyl system (Figure 7).[172] Intercalated in the crystal structure of the samarium metallacycle is the hydrocarbon dibenzo[fg,op]naphthacene (= $C_{24}H_{14}$), which originated from oxidative coupling of two biphenyl dianions. The overall reaction has formulated as shown in Equation (22)) (biph = biphenyl dianion, $C_{24}H_{16}$ = quaterphenyl dianion).

Figure 7   The molecular structure of $[Sm(\mu\text{-}Br)(THF)_2(C_{24}H_{16})]_2 \cdot [C_{24}H_{14}]$ (intercalated $C_{24}H_{14}$ omitted).[172]

$$6\,SmBr_3 + 6\,Li_2(biph) \longrightarrow [Sm(\mu\text{-}Br)(THF)_2(C_{24}H_{16})]_2 \cdot [C_{24}H_{14}] + 2\,Sm + 2\,SmBr_3 + H_2 + 12\,LiBr \quad (22)$$

Alkoxide-stabilized lanthanide hydrocarbyls first became available through the use of the sterically demanding chelating alkoxide ligand 3,3',5,5'-tetra-*t*-butylbiphenyl-2,2'-diolate. The homoleptic lanthanum hydrocarbyl $[La\{CH(TMS)_2\}_3]$ served as an excellent starting material. Scheme 5 outlines the synthesis of solvated and solvent-free alkoxylanthanum hydrocarbyls. The related tris(THF) adduct $[La\{CH(TMS)_2\}\{1,1'\text{-}(2\text{-}OC_6H_2Bu^t_2\text{-}3,5)_2\}(THF)_3]$ was structurally characterized.[173]

Scheme 5

Other alkyl-alkoxylanthanide complexes are of interest as intermediates in the lanthanide-catalyzed polymerization of dienes. An active catalyst system for diene polymerization is $Nd(OPr^i)_3/AlEt_3/AlEt_2Cl$ (cf. Section 2.2.13). From this system the bimetallic alkyl-alkoxide complex $[\{Nd_6Al_3(\mu\text{-}Cl)_6(\mu\text{-}Et)_9Et_5(OPr^i)\}_2]$ was isolated and structurally characterized.[174] A more straightforward route to such complexes involves alkylation of lanthanide alkoxides by aluminum alkyls (Equation (23)).[175]

$$Ln(OBu^t)_3 + 3\,AlMe_3 \longrightarrow [Ln\{(\mu\text{-}OBu^t)(\mu\text{-}Me)AlMe_2\}_3] \qquad (23)$$

$$Ln = Y, Pr, Nd$$

The ^{1}H and ^{13}C NMR spectra of the diamagnetic yttrium derivative showed the presence of both *fac* and *mer* isomers in solution. The neodymium complex was structurally characterized. It exhibits a highly distorted octahedral coordination geometry with the neodymium atom located on a three-fold axis.[175]

Among the nitrogen donor ligands that are able to stabilize lanthanide hydrocarbyls the porphyrin dianions play the most prominent role. Although numerous lanthanide porphyrin complexes are known (e.g., [Ln(OEP)$_2$], [Ln$_2$(OEP)$_3$], [Ln(TPP)(acac)], etc. (OEP = octaethylporphyrinate, TPP = tetraphenyl-porphyrinate)),[176–81] the preparation of σ-alkyl derivatives is a recent development. A suitable scandium precursor, [ScCl(OEP)], has been prepared by reacting [ScCl$_3$(THF)$_3$] with [Li(THF)$_4$][Li(OEP)][182] or [Li$_2$(OEP)(THF)$_2$].[183] The chloride ligand is easily replaced by other ligands such as amides or alkoxides. Alkylation afforded a variety of solvent-free scandium hydrocarbyl derivatives [Sc(OEP)R] (R = Me, CH(TMS)$_2$, CH$_2$Bu^t).[182,183] The molecular structures of [Sc(OEP)Me] and [Sc(OEP)CH(TMS)$_2$] have been determined by x-ray crystallography.[183]

Their relatively small ionic radii also make yttrium and lutetium good candidates for the preparation of porphyrin-stabilized lanthanide hydrocarbyls. The most straightforward preparation involves treatment of the homoleptic hydrocarbyl [Ln$\{$CH(TMS)$_2\}_3$] with octaethylporphyrin in toluene solution. This method avoids any problems associated with salt complexation or solvate formation (Equation (24)).[141]

$$[Ln\{CH(TMS)_2\}_3] + H_2(OEP) \longrightarrow [Ln(OEP)CH(TMS)_2] + 2\,CH_2(TMS)_2 \qquad (24)$$

$$Ln = Y, Lu$$

The lutetium derivative has been structurally characterized (Figure 8). The coordination around lutetium is square-pyramidal.

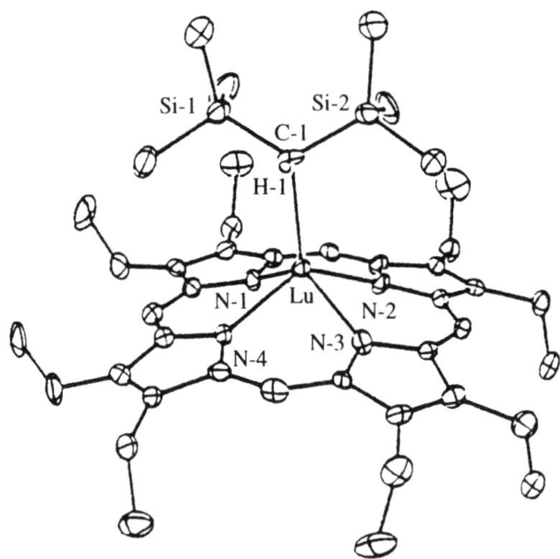

Figure 8   The molecular structure of [Lu(OEP)CH(TMS)$_2$].[141]

A two-step preparation of [Ln(OEP)CH(TMS)$_2$] starts with the homoleptic solvent-free tris-phenoxides [Ln$\{$OC$_6$H$_3$(Bu^t)$_2$-2,6$\}_3$] (Equations (25) and (26)).[141] This route too circumvents complications caused by possible salt incorporation or formation of solvated species.[20,140]

$$[Ln\{OC_6H_3(Bu^t)_2\}_3] + H_2(OEP) \longrightarrow [Ln(OEP)OC_6H_3(Bu^t)_2] + 2\,HOC_6H_3(Bu^t)_2 \qquad (25)$$

$$[Ln(OEP)OC_6H_3(Bu^t)_2\text{-}2,6] + LiCH(TMS)_2 \longrightarrow [Ln(OEP)CH(TMS)_2] + LiOC_6H_3(Bu^t)_2\text{-}2,6 \qquad (26)$$

$$Ln = Y, Lu$$

As other highly reactive lanthanide(III) alkyls, the porphyrin derivatives [Ln(OEP)CH(TMS)$_2$] are interesting precursors for a number of reactions, which are summarized in Scheme 6. Clean hydrolysis upon addition of one equivalent of water affords the dimeric hydroxo complexes [$\{$Ln(μ-OH)(OEP)$\}_2$]

(Ln = Y, Lu). Facile displacement of the hydrocarbyl ligand is achieved with bulky phenols or terminal alkynes to afford the corresponding phenoxide and alkynyl derivatives. Unexpectedly the complexes [Ln(OEP)CH(TMS)$_2$] are resistant to hydrogenolysis. No reaction was observed even under 20 bar of hydrogen. This behavior is in marked contrast to the metallocene lanthanide hydrocarbyls [LnCH(TMS)$_2$(Cp*)$_2$] (Ln = Y, La, Ce, Nd), which easily form the corresponding hydrides upon treatment with hydrogen.[184–6]

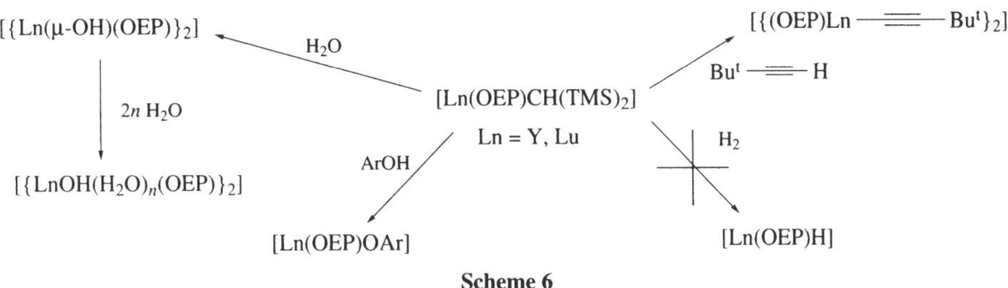

Scheme 6

"Ate" complexes were formed when [Y(OEP)OC$_6$H$_3$(Bu^t)$_2$] was treated with methyllithium and subsequently with two equivalents of Al$_2$Me$_6$ (Scheme 7).[141]

$$[Y(OEP)OC_6H_3(Bu^t)_2\text{-}2,6] + 2\,MeLi \xrightarrow{Et_2O} [Y(OEP)(\mu\text{-}Me)_2Li(Et_2O)] + LiOC_6H_3(Bu^t)_2\text{-}2,6 \xrightarrow{Al_2Me_6}$$

$$[Y(OEP)(\mu\text{-}Me)_2AlMe_2]$$

Scheme 7

The synthetic route to [Y(OEP)(μ-Me)$_2$AlMe$_2$] resembles the multistep preparation of [{Lu(μ-Me)Cp*$_2$}$_2$] (cf. Section 2.2.6.3).[187] However, it was not possible to duplicate the final step in the porphyrin system. Addition of Lewis bases (e.g., pyridine) to [Y(OEP)(μ-Me)$_2$AlMe$_2$] did not yield unsolvated [Y(OEP)Me] but only simple adducts. Oxidation of [Y(OEP)(μ-Me)$_2$AlMe$_2$] with excess O$_2$ at room temperature afforded the bimetallic alkoxide complex [Y(OEP)(μ-OMe)$_2$AlMe$_2$].[141] A somewhat related bimetallic amide, [Yb{N(TMS)$_2$}$_2$(AlMe$_3$)$_2$], was prepared by reacting [Yb$_2${N(TMS)$_2$}$_4$] with two equivalents of Al$_2$Me$_6$. The structurally characterized compound exhibits four Yb–Me–Al interactions and two Yb–Me–Si bridges.[188]

Another nitrogen donor ligand capable of stabilizing heteroleptic lanthanide alkyls is the formally tridentate amido anion [N(SiMe$_2$CH$_2$PMe$_2$)$_2$]$^-$, which was investigated by Fryzuk *et al.* The stepwise preparation of lanthanide hydrocarbyls starting from LnCl$_3$ is illustrated in Scheme 8.[189–91]

$$LnCl_3 + 2\,KN(SiMe_2CH_2PMe_2)_2 \longrightarrow [LnCl\{N(SiMe_2CH_2PMe_2)_2\}_2] \xrightarrow{RLi} [LnR\{N(SiMe_2CH_2PMe_2)_2\}_2]$$

Ln = La, Lu

Scheme 8

The hydrocarbyl derivatives are highly fluxional in solution and undergo hydrocarbon elimination upon heating. A cyclometallated decomposition product, [Y{N(SiMe$_2$CHPMe$_2$)(SiMe$_2$CH$_2$PMe$_2$)}{N(SiMe$_2$CH$_2$PMe$_2$)$_2$}] was structurally characterized.[191] Tridentate polypyrazolylborate anions are equally useful in stabilizing heteroleptic lanthanide alkyls. The sterically demanding HB[3-Bu^t-5-Me-pz]$_3$$^-$ ligand allows the isolation of soluble, monosubstituted ytterbium(II) derivatives. Treatment of the precursor [{HB[3-Bu^t-5-Me-pz]$_3$}YbI(THF)] with alkyllithium reagents provided [{HB[3-Bu^t-5-Me-pz]$_3$}YbCH(TMS)$_2$] and [{HB[3-Bu^t-5-Me-pz]$_3$}YbCH$_2$(TMS)(Et$_2$O)] as the first well-characterized divalent lanthanide hydrocarbyl species.[192]

As in uranium chemistry, bulky heteroallylic ligands of the type [RC$_6$H$_4$C(N-TMS)$_2$]$^-$ have been found to stabilize heteroleptic metal alkyls. The silylated benzamidinate ligands have been demonstrated to be steric equivalents of cyclopentadienyl ligands. Benzamidinate analogues of [YCH(TMS)$_2$(Cp*)$_2$] (cf. Section 2.2.6.3) have been prepared by treatment of either chloride or triflate precursors with LiCH(TMS)$_2$ (Equation (27)).[193–5]

$$[YX(THF)_n\{RC_6H_4C(N\text{-}TMS)_2\}_2] + LiCH(TMS)_2 \longrightarrow [YCH(TMS)_2\{RC_6H_4C(N\text{-}TMS)_2\}_2] + LiX \quad (27)$$

$$X = Cl, n = 1 \qquad\qquad R = H, OMe, CF_3$$
$$X = O_3SCF_3, n = 0$$

A diene-bridged dilanthanum complex, $[\{LaI_2(THF)_3\}_2(\mu\text{-}\eta^4\text{:}\eta^4\text{-PhCH=CHCH=CHPh})]$, was synthesized by reacting finely divided metallic lanthanum with 1,4-diphenylbuta-1,3-diene and iodine in THF (Equation (28)). An x-ray structure determination of the resulting red crystals revealed an inverse sandwich complex with the two lanthanum atoms symmetrically bridged by the diene ligand (Figure 9).[196] The analogous samarium derivative was earlier obtained by a similar synthetic procedure.[197]

$$La + Ph\diagup\diagdown\diagup\diagdown Ph + 2\,I_2 \xrightarrow[\text{THF}]{50\,^{\circ}\text{C}} [\{LaI_2(THF)_3\}_2(\mu\text{-}\eta^4\text{:}\eta^4\text{-PhCH=CHCH=CHPh})] \qquad (28)$$

Figure 9 The molecular structure of $[\{LaI_2(THF)_3\}_2(\mu\text{-}\eta^4\text{:}\eta^4\text{-PhCH=CHCH=CHPh})]$.[196]

Several series of heteroleptic anionic cerium(III) complexes have been prepared by reacting $Ce(acac)_4$ with organolithium reagents. Depending on the stoichiometry of reagents the compounds $Li_3[Ce(acac)_3R]$, $Li_3[Ce(acac)_3R]\cdot L_n$ ($n = 1$, L = DME, TMEDA; $n = 2$, L = THF), and $Li_3[CeMe_6]\cdot 3Li(acac)$ have been isolated.[165] Hexachlorocerate(IV) precursors yielded $[Li_4(RCeCl_6)]$ (R = Me, 1-norbornyl, $(CH_2)_3NMe_2$).[198] The chelating dimethylaminopropyl ligand also allowed the synthesis of anionic complexes of the type $[Li_3[R_3Ln\{(CH_2)_3NMe_2\}_3]]$.[154]

The compounds $[Ln\{N(TMS)_2\}_2\{Sn(CH_2TMS)_3\}(DME)]$ (Ln = Pr, Nd),[148,149] $[Ph_3SnYb(THF)_2(\mu\text{-}Ph)_3Yb(THF)_3]$,[199] and $[\{(PhCO)_2CH\}_2LnSnPh_3]$[200] have been prepared and characterized spectroscopically. These compounds can formally be regarded as heavier homologues of heteroleptic lanthanide hydrocarbyls.

2.2.3 Alkenyl and Alkynyl Compounds

The alkenyl compound bis(β-chlorovinyl)ytterbium(II) was prepared by treatment of metallic ytterbium with the alkenylmercury derivative in THF solution.[201]

Polymeric alkynyllanthanide hydrides have been obtained by co-condensation of samarium, erbium, or ytterbium vapors with 1-hexyne at $-196\,^{\circ}$C. The resulting catalytically active materials are believed to contain $(Bu^nC{\equiv}C)_2SmH$-, $(Bu^nC{\equiv}C)_2ErH$-, or $(Bu^nC{\equiv}C)_3Yb_2H$-units, respectively.[202] Elemental europium and ytterbium have also been found to react with terminal alkynes either in liquid ammonia or THF solution to afford divalent lanthanide alkynides.[203,204] Instead of the rare earth metals the easily accessible bis(pentafluorophenyl)ytterbium(II) can be used as starting material. Treatment of $Yb(C_6F_5)_2$ with terminal alkynes generates the alkynyl derivatives of divalent ytterbium while pentafluorobenzene is eliminated (Equation (29)).[204]

$$Yb(C_6F_5)_2 + 2\,PhC{\equiv}CH \xrightarrow{\text{THF}} Yb(C{\equiv}CPh)_2 + 2\,C_6F_5H \qquad (29)$$

Transmetallation using alkynylmercury reagents is a more effective and generally applicable route to divalent lanthanide alkynyl derivatives. This reaction has been applied to europium and ytterbium derivatives (Equation (30)).[203,204]

$$\text{Ln} + \text{Hg(C}\equiv\text{CR})_2 \xrightarrow{\text{THF}} \text{Ln(C}\equiv\text{CR})_2 + \text{Hg} \tag{30}$$

$$R = Bu^n;\ Ln = Eu,\ Yb$$
$$R = Ph,\ Ln = Yb$$

$Lu(C\equiv CBu^t)_3$ was obtained by reacting the homoletic lutetium hydrocarbyl complex $[Lu(o\text{-}C_6H_4CH_2NMe_2)_3]$ with the terminal alkyne,[129] and $Ln(C\equiv CPh)_3$ have been described for Ln = Pr, Sm, Eu, Gd, Tb, Er, Yb.[205,206] Ytterbium phenylacetylides have been reported to activate N_2 in the presence of sodium naphthalide.[205] Very little structural information is available on the above-mentioned lanthanide alkenyl and alkynyl complexes, although it can be anticipated that all compounds prepared in ethereal solvents are solvated species. In the absence of donor ligands, species like $Ln(C\equiv CR)_2$ and $Ln(C\equiv CR)_3$ are most likely to form polymers.

A well-defined heteroleptic divalent lanthanide alkynyl has become available through the use of a sterically demanding polypyrazolyl ligand. Protonation of $[\{HB[3\text{-}Bu^t\text{-}5\text{-}Me\text{-}pz]_3\}YbN(TMS)_2]$ with phenylacetylene afforded the dark red, monomeric acetylide $[\{HB[3\text{-}Bu^t\text{-}5\text{-}Me\text{-}pz]_3\}YbC\equiv CPh]$ in good yield.[192]

2.2.4 Allyls

The number of well-characterized homoleptic and heteroleptic lanthanide allyls is small, probably because the relatively small allyl ligands alone are not sufficiently bulky to stabilize coordinatively unsaturated lanthanide ions. Reactions of anhydrous lanthanide trichlorides with allyllithium do not yield simple homoleptic tris(allyl)lanthanide compounds. Instead, various anionic complexes have been isolated from such reactions.

Treatment of $LnCl_3$ with excess LiC_3H_5 in THF/dioxane or THF/diethyl ether has been reported to yield "ate" complexes of the general formulae $[Li(dioxane)_n][Ln(C_3H_5)_4]$ (Ln = Ce, Pr, Nd, Sm, Gd, $n = 2$; Ln = Y, $n = 2.5$; Ln = La, $n = 3$)[207,208] and $[Li_2\{Ln(C_3H_5)_5\}(THF)_n]$ (Ln = Y, La, $n = 2.5$; Ln = Ce, Pr, Nd, Sm, $n = 3$),[209] respectively. Work-up of the reaction mixtures with dioxane produced the compounds $[Li_2\{Ln(C_3H_5)_5\}(dioxane)_n]$ (Ln = Ce, Nd, Sm, Gd, Dy, Ho, Er; $n = 1$, 3).[210-12] A crystal structure determination of the cerium derivative showed the compound to have the composition $[\{Li_2(\mu\text{-}C_3H_5)(THF)_3\}\{Ce(\eta^3\text{-}C_3H_5)_4\}]$ with Ce–C bond distances ranging from 0.2712(8) nm to 0.2858(7) nm.[213] Recently the anionic tetraallyl complexes $Li[Ln(C_3H_5)_4]$ have been prepared *in situ* from $LnCl_3$, $[Sn(C_3H_5)_4]$, and Bu^nLi in THF.[214] This preparative method has been utilized to prepare the structurally characterized lanthanum complex $[Li(\mu\text{-}dioxane)_{3/2}][La(\eta^3\text{-}C_3H_5)_4]$ (Equation (31)).[208]

$$LaCl_3 + 2\,Sn(C_3H_5)_4 + 4\,Bu^nLi \xrightarrow[\substack{\text{ii, Et}_2\text{O}\\ \text{dioxane}}]{\substack{\text{i, toluene}}} [Li(\mu\text{-}dioxane)_{3/2}][La(\eta^3\text{-}C_3H_5)_4] + 3\,LiCl + 2\,Sn(C_3H_5)_2(Bu^n)_2 \tag{31}$$

The three η^3-allyl ligands describe a distorted tetrahedron around lanthanum while the lithium atoms are bridged by dioxane molecules to form a two-dimensional layer (Figure 10).

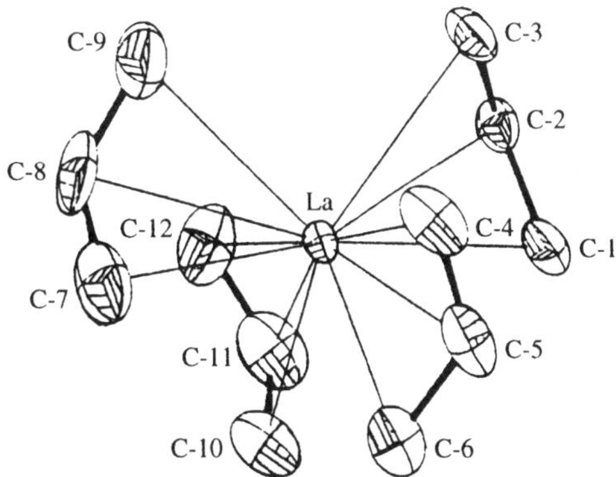

Figure 10 The molecular structure of $[Li(\mu\text{-}dioxane)_{3/2}][La(\eta^3\text{-}C_3H_5)_4]$.[208]

Partial protolysis of the tetraallyllanthanate(III) complex $[Li(dioxane)_{3/2}][La(C_3H_5)_4]$ with one equivalent of a cyclopentadiene derivative and precipitation with diethylether/dioxane provided the mono(cyclopentadienyl)triallyllanthanate complexes $[Li(dioxane)_2][La(C_3H_5)_3Cp']$ ($Cp' = Cp$, Cp^*, C_9H_7, $C_{13}H_9$).[215] More complicated lanthanide magnesium allyls have been obtained by reacting anhydrous $LnCl_3$ with Grignard reagents in the presence of TMEDA to afford complexes of the type $[(\eta^3\text{-}C_3H_5)_2Ln(\mu_3\text{-}Cl)_2(\mu\text{-}Cl)_3Mg_2(TMEDA)_2]$. The compounds $[(\eta^3\text{-}C_3H_5)_2Ce(\mu_3\text{-}Cl)_2(\mu\text{-}Cl)_3\text{-}Mg_2(TMEDA)_2]$ and $[(\eta^3\text{-}C_3H_5)_2Ln(\mu_3\text{-}Br)_2(\mu\text{-}Br)_3Mg_2(Et_2O)_4]$ ($Ln = Ce$, Nd) were characterized by x-ray crystallography.[216]

The tridentate amido ligand $[N(SiMe_2CH_2PMe_2)_2]^-$ has been successfully employed in the stabilization of heteroleptic lanthanide hydrocarbyls. Treatment of $[YCl\{N(SiMe_2CH_2PMe_2)_2\}_2]$ with allylmagnesium chloride, $MgCl(C_3H_5)$, resulted in displacement of one chelating amido ligand to give the structurally characterized chloro-bridged dimer $[\{Y\{N(SiMe_2CH_2PMe_2)_2\}(\mu\text{-}Cl)(\eta^3\text{-}C_3H_5)\}_2]$.[190,191]

2.2.5 Cyclopentadienyl Compounds

2.2.5.1 *M(Cp)X compounds*

Apparently only two compounds of this type have so far been reported. They were prepared either by a simple metathesis reaction using ytterbium dichloride (Equation (32)) or by sodium amalgam reduction of $YbCl_2(Cp)$ (Equation (33)) and characterized spectroscopically.[217]

$$YbCl_2 + NaCp \xrightarrow{\text{L}} [YbCl(L)_nCp] + NaCl \tag{32}$$

$$L = THF, n = 2$$
$$L = DME, n = 1$$

$$YbCl_2Cp + Na/Hg \longrightarrow YbCl(THF)_2Cp + NaCl \tag{33}$$

2.2.5.2 *M(Cp)$_2$ compounds*

Bis(cyclopentadienyl)lanthanide(II) complexes are among the oldest known rare earth organometallics. $Eu(Cp)_2$ was first obtained by reacting the blue solution of elemental europium in liquid ammonia with cyclopentadiene.[218] This reaction afforded an ammonia adduct of bis(cyclopentadienyl)europium as the initial product. The base-free complex was obtained by heating the adduct to 200 °C under high vacuum followed by vacuum sublimation of the thermally highly stable product at 400 °C. The behavior of $Eu(Cp)_2$ nicely illustrates the typical properties of most organometallic lanthanide complexes: the compound is thermally extremely robust while it is rapidly decomposed in the presence of traces of oxygen. $Eu(Cp)_2$ was one of the first compounds studied by ^{151}Eu Mössbauer spectroscopy. $Yb(Cp)_2$ too has been prepared by adopting the same ammonia route (Equation (34)).[18]

$$Ln + 3 C_5H_6 \xrightarrow[\text{ii, Subl.}]{\text{i, NH}_3 \text{ (l.)}} LnCp_2 + C_5H_8 \tag{34}$$

$$Ln = Eu, Yb$$

Several other synthetic routes leading to bis(cyclopentadienyl)lanthanide(II) complexes have been developed. Most of them are carried out in THF solution and initially yield the THF adducts of $Ln(Cp)_2$. An elegant transmetallation reaction starts with the metal powders of samarium, europium, or ytterbium, which are reacted with bis(cyclopentadienyl)mercury. Similarly, ytterbium metal reacts with thallium cyclopentadienide in the presence of catalytic amounts of mercury in DME solution to give $[Yb(DME)(Cp)_2]$ (Equations (35) and (36)).[219] $[Ln(THF)_2(Cp)_2]$ ($Ln = Eu$, Yb),[220] $Yb(MeCN)(Cp)_2$,[220] and $[Yb(THF)(MeC_5H_4)_2]$[221,222] have been prepared similarly.

$$Ln + HgCp_2 \xrightarrow{\text{THF}} [Ln(THF)_nCp_2] + Hg \tag{35}$$

$$Ln = Sm, Eu, Yb$$

$$Yb + 2 TlCp \xrightarrow{\text{DME}} [Yb(DME)Cp_2] + 2 Tl \tag{36}$$

In the solid-state structures of $[Yb(DME)(Cp)_2]$[223] and $[Yb(DME)(MeC_5H_4)_2]$[224] the central ytterbium atom is tetrahedrally surrounded by two cyclopentadienyl ligands and two oxygen atoms of the chelating dimethoxyethane ligand (Yb–C 0.2685 nm). More recently, the THF solvate [Yb(THF)$_2$(Cp)$_2$] was structurally characterized.[225,226] A more complicated polymeric structure was found for the methylcyclopentadienyl derivative [Yb(THF)(MeC$_5$H$_4$)$_2$]. In the solid state this complex forms infinite twisted chains of Yb(THF)(MeC$_5$H$_4$) units which are bridged by the remaining MeC$_5$H$_4$ ligands.[221] Interestingly, bis(cyclopentadienyl)ytterbium has been found to form a stable adduct with triphenylphosphine oxide, [Yb(Ph$_3$PO)$_2$(Cp)$_2$], which was structurally characterized.[227]

The system Sm/Hg(Cp)$_2$ is not as straightforward as originally believed. It was postulated that treatment of activated samarium metal with Hg(Cp)$_2$ in THF/ether would afford [Sm(THF)$_2$(Cp)$_2$], but a careful reinvestigation of this reaction revealed that the product was in fact [Sm(THF)(Cp)$_3$]. However, [Sm(THF)$_2$(Cp)$_2$] might be stable under these conditions in the presence of excess activated samarium metal. Probably the best route to bis(cyclopentadienyl)samarium(II) is the reaction of samarium diiodide with two equivalents of sodium cyclopentadienide as described by Kagan *et al.* (Equation (37)).[228–30]

$$\text{SmI}_2 + 2\,\text{NaCp} \xrightarrow{\text{THF}} \text{SmCp}_2 + 2\,\text{NaI} \tag{37}$$

Lanthanide(II) complexes containing σ-bonded hydrocarbyl ligands are useful starting materials for divalent lanthanide metallocenes. Bis(pentafluorophenyl)derivatives of samarium, europium, and ytterbium as well as bis(phenylethynyl)ytterbium(II) have been used for ligand displacement reactions with cyclopentadiene, methylcyclopentadiene, and indene. This way the organolanthanide(II) complexes [Yb(DME)(Cp)$_2$], [Eu(THF)$_n$(Cp)$_2$] ($n = 0.5$ or 1), [Yb(DME)(MeC$_5$H$_4$)$_2$], [Yb(DME)(C$_5$D$_5$)$_2$], and [Yb(THF)$_2$(C$_9$H$_7$)$_2$] have been prepared.[231,232]

Reduction of various cyclopentadienyllanthanide(III) complexes is another possibility to synthesize bis(cyclopentadienyl)lanthanide(II) derivatives.[18,226,233,234] Either sodium, sodium naphthalide, potassium cyclooctadienide, or metallic ytterbium can be used to reduce YbCl(THF)(Cp)$_2$ to the +2 oxidation state (Equations (38) and (39)).

$$\text{YbCl(THF)Cp}_2 + \text{Na} \xrightarrow{\text{THF}} [\text{Yb(THF)}_n\text{Cp}_2] + 2\,\text{NaCl} \tag{38}$$

$$3\,\text{YbCl(THF)Cp}_2 + \text{Yb} \xrightarrow{\text{THF}} 3\,[\text{Yb(THF)}_n\text{Cp}_2] + \text{YbCl}_3 \tag{39}$$

Even tris(cyclopentadienyl)lanthanides can be used as precursors for lanthanide(II) metallocenes. Purple [Sm(THF)$_n$(Cp)$_2$] was obtained by reductive cleavage of Sm(Cp)$_3$ with potassium naphthalide.[18] Deacon *et al.* have studied the redox chemistry of Yb(Cp)$_3$ and found that this compound can be reduced by metallic ytterbium to give Yb(Cp)$_2$. In turn, Yb(Cp)$_2$ can be oxidized back to Yb(Cp)$_3$ by treatment with TlCp.[228] Reduction of Yb(Cp)$_3$ with sodium naphthalide was originally reported to give Yb(Cp)$_2$ (after sublimation at 400 °C). A recent reinvestigation of this reaction revealed that the sublimed green material was polymeric [{NaYb(Cp)$_3$}$_n$] instead. Thus, the reaction pathway has to be reformulated according to Equation (40).[235]

$$\text{YbCp}_3 + \text{Na} \xrightarrow{\text{THF}} 1/n\,[\{\text{NaYbCp}_3\}_n] \tag{40}$$

In this unusual "ate" complex of divalent ytterbium the different metal atoms are bridged by μ-η^5:η^5-cyclopentadienyl ligands to give a polymeric network. Somewhat surprising is the fact that the compound can be sublimed at very high temperatures without any sign of decomposition. Retention of sodium cyclopentadienide could possibly explain the various color changes associated with Yb(Cp)$_2$ in solution and in the solid state. For solid Yb(Cp)$_2$ a green and a red modification have been described in the literature. Green Yb(Cp)$_2$ has been reported to dissolve in THF to give a red solution from which yellow Yb(THF)(Cp)$_2$ could be crystallized.[18]

Anionic lanthanide(II) complexes of the type [Ln(Cp)$_3$]$^-$ had already been shown to be the products of the reversible electrochemical reduction of Ln(Cp)$_3$ in THF solution. The thermodynamic stability of these anions towards oxidation was shown to decrease in the order Eu > Yb > Sm.[236] An anionic lanthanide(II) hydrocarbyl complex formulated as [Li(THF)$_n$][EuMe(Cp)$_2$] was prepared by the reaction of [Eu(THF)$_n$(Cp)$_2$] with methyllithium.[237]

t-Butyl-substituted cyclopentadienyl ligands are playing an increasing role in organolanthanide chemistry, because the resulting products are often more soluble than the parent cyclopentadienyl derivatives. Ligands such as *t*-butylcyclopentadienyl or 1,3-di-*t*-butylcyclopentadienyl have been successfully employed in the preparation of highly reactive lanthanide(II) organometallics. Simple bis-substituted lanthanide(II) complexes containing these ligands have been prepared by metathetical reactions between the lanthanide diiodide and the corresponding sodium or potassium cyclopentadienide. Monomeric [Yb(THF)$_2$(Bu^{t}C$_5$H$_4$)$_2$] was isolated and characterized by an x-ray analysis (Yb–C (av.) 0.272(2) nm).[238]

Monomeric [Sm(DME)(Bu^{t}C$_5$H$_4$)$_2$] has been made analogously.[239,240] In the case of samarium(II) the use of the sterically more demanding 1,3-di-t-butylcyclopentadienyl ligand leads to the formation of a monosolvate complex. The molecular and crystal structure of [Sm(THF)(Bu$^t_2$C$_5$H$_3$)$_2$] was determined by x-ray diffraction.[241]

In some cases one equivalent of alkali metal cyclopentadienide is retained in the products and "ate" complexes are obtained. This was the case, for example, in the reaction of [SmI$_2$(THF)$_2$] with Na(Bu^{t}C$_5$H$_4$). The reaction yielded the bimetallic complex [{NaSm(THF)(η^5:η^2-Bu^{t}C$_5$H$_4$)$_3$}$_n$], which was structurally characterized by x-ray methods (Figure 11). In the polymeric structure samarium and sodium are bridged by η^5:η^2-Bu^{t}C$_5$H$_4$ units. The average Sm–C distance was found to be 0.289 nm.[241]

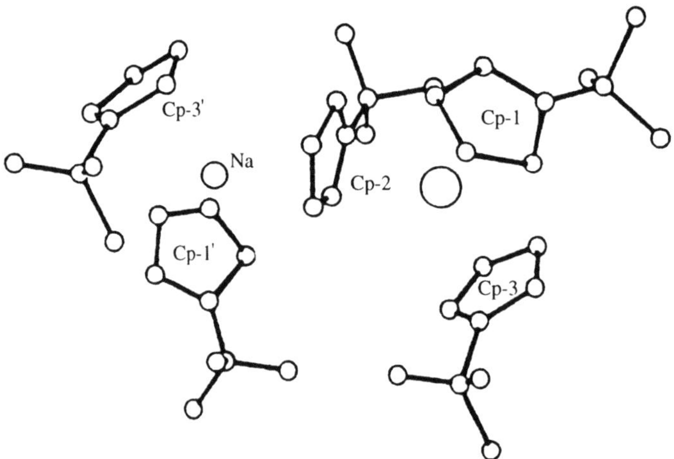

Figure 11   The molecular structure of [{NaSm(THF)(η^5:η^2-Bu^{t}C$_5$H$_4$)$_3$}$_n$].[241]

Trimethylsilyl-substituted cyclopentadienyl ligands are equally useful in the preparation of soluble, highly reactive organolanthanide(II) complexes. Among these ligands the 1,3-bis(trimethylsilyl)cyclopentadienyl anion has turned out to be especially versatile but complexes containing the monosubstituted trimethylsilylcyclopentadienyl ligand are also remarkably stable. For example, dimeric [{Yb(μ-Cl)(TMS-C$_5$H$_4$)$_2$}$_2$] can be reduced by sodium amalgam in THF to give the bis(THF) solvate of Yb(TMS-C$_5$H$_4$)$_2$. The THF ligands can be replaced by TMEDA, and simple heating of the THF adduct yields the green, base-free ytterbium(II) derivative (Scheme 9 and Equation (41)).[242]

$$[\{Yb(\mu\text{-Cl})(TMS\text{-}C_5H_4)_2\}_2] + 2\,Na/Hg \xrightarrow[-Hg]{THF} [Yb(THF)_2(TMS\text{-}C_5H_4)_2] + 2\,NaCl \xrightarrow{\Delta}$$

$$[Yb(TMS\text{-}C_5H_4)_2] + 2\,THF$$

Scheme 9

$$[Yb(THF)_2(TMS\text{-}C_5H_4)_2] + TMEDA \longrightarrow [Yb(TMEDA)(TMS\text{-}C_5H_4)_2] + 2\,THF \qquad (41)$$

[SmI$_2$(THF)$_2$] reacts with two equivalents of K{(TMS)$_2$C$_5$H$_3$} to give the structurally characterized monosolvate [Sm(THF){(TMS)$_2$C$_5$H$_3$}$_2$], which was found to polymerize ethylene. Desolvating the THF adduct gave base-free [Sm{(TMS)$_2$C$_5$H$_3$}$_2$].[243] Similar treatment of [SmI$_2$(THF)$_2$] with K{(TMS)$_3$C$_5$H$_2$} unexpectedly resulted in formation of [Sm(THF){(TMS)$_2$C$_5$H$_3$}{(TMS)$_3$C$_5$H$_2$}], which was crystallographically characterized.[244] The bent metallocenes [Ln{(TMS)$_2$C$_5$H$_3$}$_2$] (Ln = Eu, Yb) have been structurally investigated. Both compounds are polymeric and display unusual intermolecular contacts between the metal centers and γ-methyl groups of neighboring [(TMS)$_2$C$_5$H$_3$] ligands. The europium derivative exhibits an unprecedented structure with a cyclopentadienyl ring bridging η^3:η^5 between two non-equivalent europium atoms. The molecular structures of [{Eu{(TMS)$_2$C$_5$H$_3$}$_2$}$_\infty$] is depicted in Figure 12.[245]

Several donor-substituted cyclopentadienyl ligands have been shown to be very useful in the stabilization of lanthanide(II) metallocenes. A typical example is the methoxyethylcyclopentadienyl ligand, which forms solvated and unsolvated complexes with divalent samarium and ytterbium (Equation (42)):[246]

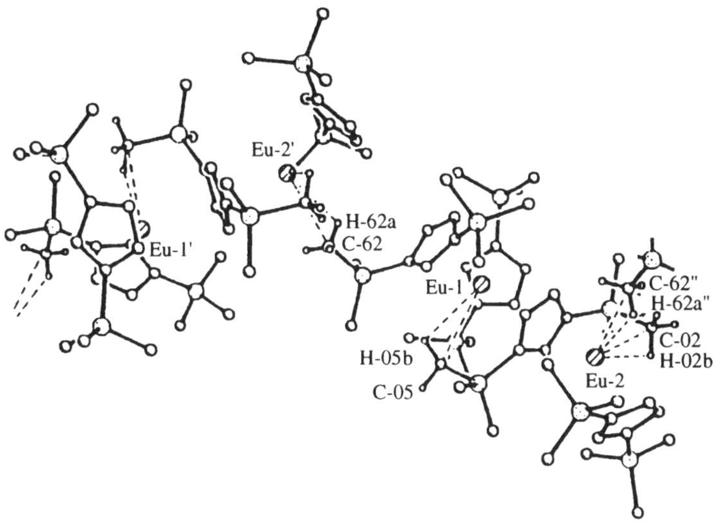

Figure 12 The molecular structure of $[\{Eu\{(TMS)_2C_5H_3\}_2\}_\infty]$.[245]

$$LnI_2 + 2\,K(MeOCH_2CH_2C_5H_4) \xrightarrow{\text{THF}} Ln(MeOCH_2CH_2C_5H_4)_2 + 2\,KI \tag{42}$$

$$Ln = Sm, Yb$$

The THF adduct $[Yb(THF)(MeOCH_2CH_2C_5H_4)_2]$ was crystallographically characterized. The two cyclopentadienyl ring centroids, two ether oxygens, and the oxygen atom of the THF ligand describe a distorted trigonal bipyramid around the central ytterbium atom. Intramolecular chelate stabilization has also been achieved with the amino-functionalized ligand $[C_5Me_4CH_2CH_2NMe_2]^-$. Treatment of samarium diiodide with two equivalents of $Li(C_5Me_4CH_2CH_2NMe_2)$ produced dark green, unsolvated $[Sm(C_5Me_4CH_2CH_2NMe_2)_2]$ (**3**).[247,248]

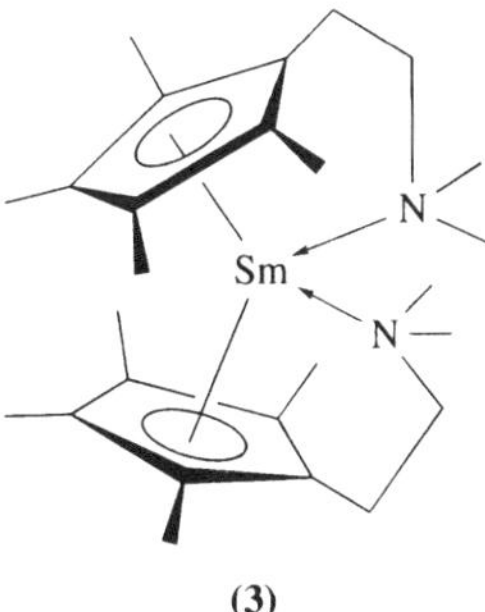

(**3**)

A series of homoleptic lanthanocene(II) complexes was prepared using cyclopentadienyl ligands containing a pendant pyridyl substituent: $[C_5H_4(CMe_2CH_2C_5H_4N\text{-}2)]^-$ ($\equiv {}^-Cp^{py}$), $[C_5H_3(TMS)(CMe_2CH_2C_5H_4N\text{-}2)\text{-}3]^-$ ($\equiv {}^-Cp^{ipy}$), $[C_5H_4(CMe_2C_5H_4N\text{-}2)]^-$ ($\equiv {}^-Cp^{py(s)}$) and $[C_5H_3(TMS)(CMe_2C_5H_4N\text{-}2)\text{-}]^-$ ($\equiv {}^-Cp^{ipy(s)}$). In each case treatment of YbI_2 with two equivalents of the appropriate potassium cyclopentadienyl complex gave the dark green ytterbium(II) complexes $[Yb(Cp^{py})_2]$, $[Yb(Cp^{ipy})_2]$, $[Yb(Cp^{py(s)})_2]$, and $[Yb(Cp^{ipy(s)})_2]$. The crystal structures of $[Yb(Cp^{py})_2]$ and $[Yb(Cp^{ipy(s)})_2]$ showed that in common with (**3**) both of the pyridyl groups in each complex were coordinated to the metal.[249]

A novel aspect of divalent lanthanide metallocene chemistry is the preparation of a "metalloligand." The phosphinoytterbocenes $[Yb(THF)(C_5H_4PPh_2)_2]$ and $[Yb(DME)(C_5H_4PPh_2)_2]$ have been synthesized using different synthetic approaches (Equations (43)–(45)).[250]

$$Yb(C_6F_5)_2 + 2\,CpPPh_2 \xrightarrow{\text{THF}} [Yb(THF)(C_5H_4PPh_2)_2] + 2\,C_6F_5H \tag{43}$$

$$Yb + Hg(C_6F_5)_2 + 2\,CpPPh_2 \xrightarrow{\text{THF}} [Yb(THF)(C_5H_4PPh_2)_2] + Hg + 2\,C_6F_5H \tag{44}$$

applications of the compounds in organic synthesis and homogeneous catalysis. For additional information on topics discussed in this chapter, the reader is referred to a number of recent monographs and detailed review articles that have appeared since the previous edition. Annual reviews covering the lanthanides and actinides have been published regularly in the *Journal of Organometallic Chemistry*[3-7] and other periodicals.[8-15] In addition, a number of review articles have appeared since 1982 covering both organolanthanides and organoactinides.[16] Organolanthanide chemistry has been reviewed by Schumann,[17-20] McCleverty,[21] Deacon,[22] Bochkarev,[23] Shen,[24] Bercaw,[25] Schaverien,[26] and Evans.[27,28] Some review articles have been devoted to organoactinide chemistry,[29,30] more recently those by Marks,[31-3] Takats,[34] Pires de Matos,[35] and Ephritikhine.[36] The *f*-elements have also been included in several other review articles covering more specific topics such as certain ligand types[37,38] and metal–carbon σ-bonds.[39-43] A NATO Advanced Study Institute monograph, *Fundamental and Technological Aspects of Organo-f-Element Chemistry*, was published in 1985.[44] Some representative synthetic procedures in organolanthanide and -actinide chemistry have been collected in a chapter of *Inorganic Syntheses*.[45]

In addition to review articles devoted to all areas of organo-*f*-element chemistry, numerous reviews covering specific synthetic, spectroscopic, and technological aspects of organo-*f*-element compounds have appeared. Among these are articles on low-oxidation-state organolanthanide chemistry,[28,46-8] early lanthanide organometallics,[49,50] *f*-element cyclooctatetraenyl complexes,[51,52] organoactinide chemistry with phosphoylide,[53] and polypyrazolylborate ligands,[54] reactions of cyclopentadienyluranium(IV) complexes with Lewis acids and bases,[55] neptunium and plutonium complexes with monodentate ligands,[56] the organometallic chemistry of neptunium,[57] the magnetochemical properties of organolanthanide complexes,[58,59] thermochemistry of organo-*f*-element complexes,[60-3] bonding properties and electronic structure,[64,65] C–H activation reactions,[66-8] catalytic properties,[69,70] the cryosynthesis of organometallic compounds,[71] the chemical vapor deposition of rare-earth compounds in superconductor synthesis,[72] and structural trends in Ln(Cp)$_3$ and Ln(Cp*)$_2$ complexes.[73-6] Furthermore the crystal structures of over two hundred known cyclopentadienyl lanthanide and actinide complexes have been analyzed in order to find some evidence for covalent bonding in these compounds.[77] Various other articles of both theoretical and experimental nature have been devoted to the question of covalency in organolanthanide[78-87] and -actinide[64,88,89] complexes.

Other reviews dealing with specific synthetic aspects of organo-*f*-element chemistry will be mentioned in the introductions to the particular subsections.

2.2　SCANDIUM, YTTRIUM AND THE LANTHANIDES

2.2.1　Carbonyls

The first and so far only example of a lanthanide(III) complex involving π-coordination of a CO ligand is the structurally characterized erbium complex [Er(EtOH)(H$_2$O)$_4$\{Mo(CO)$_3$(Cp)\}$_3$] (Figure 1). This compound exhibits semi-bridging carbonyl ligands between molybdenum and erbium as well as weak metal–metal bonding.[90]

2.2.2　Hydrocarbyls

2.2.2.1　*Neutral homoleptic compounds*

In this section homoleptic lanthanide hydrocarbyls as well as their THF adducts will be discussed, although the latter are formally heteroleptic complexes. First, low-valent lanthanide hydrocarbyls will be mentioned, followed by a discussion of lanthanide(III) complexes with σ-bonded hydrocarbyl ligands.

Considerable research efforts have been directed in the past towards the synthesis and characterization of lanthanide complexes involving coordination to neutral alkenes, dienes, allenes, or alkynes. However, the tendency of the rare-earth elements to form complexes of this type is extremely low as compared to the corresponding *d*-transition metal chemistry. This can be traced back to the limited radial extension of the 4*f* orbitals and to the highly ionic nature of the bonding in organolanthanide complexes. Metal vapor synthesis has been the main preparative tool to study the reactivity of lanthanide elements towards unsaturated hydrocarbons. The products were generally difficult to characterize and very little is known about their molecular structures. However, an

$$Yb + 2\,TlC_5H_4PPh_2 \xrightarrow{\text{DME}} [Yb(DME)(C_5H_4PPh_2)_2] + 2\,Tl \qquad (45)$$

The "metalloligand" [Yb(THF)(C$_5$H$_4$PPh$_2$)$_2$] has been successfully employed in the preparation of novel heterobimetallic lanthanide complexes (cf. Section 2.2.11.2).[250] Interestingly, a closely related reaction involving the phosphorylcyclopentadienyl ligand C$_5$H$_4$Ph$_2$P(O)$^-$ did not afford the corresponding ytterbium(II) metallocene. As outlined in Equation (46), the resulting product contained the C$_5$H$_4$Ph$_2$P(O)$^-$ coordinated to ytterbium through oxygen instead of the cyclopentadienide ring.[251]

$$[YbCp_2] + [Tl\{C_5H_4Ph_2P(O)\}] + Ph_3PO \xrightarrow{\hspace{2cm}} [Yb\{\eta^1\text{-}C_5H_4Ph_2P(O)\}(Ph_3PO)Cp_2] + Tl \qquad (46)$$

Generally the chemical reactivity of divalent lanthanide metallocenes is quite high. The samarium(II) and ytterbium(II) derivatives especially are powerful one-electron reducing agents. They easily undergo redox reactions and have frequently been used to prepare bis(cyclopentadienyl)lanthanide(III) complexes. For example, [Yb(DME)(Cp)$_2$] can be treated with various mercury, thallium, copper, or silver salts to give the corresponding trivalent YbX(Cp)$_2$ derivatives. The main advantage of such oxidative transformations is that the products can be isolated free of alkali metal halides (Equations (47) and (48)).[252,253]

$$[Yb(DME)Cp_2] + MX \xrightarrow[-DME]{\hspace{2cm}} YbXCp_2 + M \qquad (47)$$

M = 0.5 Hg, Tl, Cu, Ag; X = Cl, Br, C≡CPh, C$_6$F$_5$, O$_2$CMe, O$_2$CC$_6$F$_5$, O$_2$CC$_5$H$_4$N, (MeCO)$_2$CH, (PhCO)$_2$CH

$$2\,[Yb(THF)(C_5H_4PPh_2)_2] + HgCl_2 \xrightarrow{\hspace{2cm}} [\{Yb(\mu\text{-}Cl)(C_5H_4PPh_2)_2\}_2] + Hg + 2\,THF \qquad (48)$$

Despite its low solubility, Sm(Cp)$_2$ has been used in synthetic organic chemistry and its reactivity has been compared to that of samarium diiodide.[254,255] In the presence of sodium naphthalide, Yb(Cp)$_2$ has been found to reduce N$_2$ to ammonia.[119]

2.2.5.3 M(Cp)X$_2$ compounds

The selective preparation of pure mono(cyclopentadienyl)lanthanide derivatives is often a difficult task due to severe coordinative unsaturation which is encountered in these compounds. Unsolvated monomeric complexes cannot be isolated even with the most bulky substituted cyclopentadienyl ligands. Normally, mono(cyclopentadienyl)lanthanide complexes show a strong tendency to increase the coordination number around the metal atom by coordination of solvent molecules and/or incorporation of alkali metal halide. With a few exceptions most lanthanide elements form stable complexes of the type [LnCl$_2$(THF)$_3$(Cp)].[5,18,256] These can be prepared by treatment of anhydrous lanthanide trichlorides with one equivalent of sodium cyclopentadienide or by a ligand distribution reaction between LnCl$_3$ and Ln(Cp)$_3$ in THF or DME solution (Equations (49)–(51)). In some cases (Ln = La, Sm, Eu, Tm, Yb) the tetra-THF solvates [LnCl$_2$(THF)$_4$(Cp)] are also known.[257]

$$LnCl_3 + NaCp \xrightarrow{\text{THF}} LnCl_2(THF)_3Cp + NaCl \qquad (49)$$

$$LnCl_3 + NaCp \xrightarrow{\text{DME}} LnCl_2(DME)_3Cp + NaCl \qquad (50)$$

$$2\,LnCl_3 + LnCp_3 \xrightarrow{\text{THF}} 3\,[LnCl_2(THF)_3Cp] \qquad (51)$$

The crystal and molecular structures of [YCl$_2$(THF)$_3$(Cp)],[258] [NdCl$_2$(THF)$_3$(Cp)],[259] [GdCl$_2$(THF)$_3$(Cp)],[260] [ErCl$_2$(THF)$_3$(Cp)],[261] [YbCl$_2$(THF)$_3$(Cp)],[262] and [YbBr$_2$(THF)$_3$(Cp)][263] have been determined by x-ray crystallography. These isostructural complexes are characterized by a pseudooctahedral coordination geometry around the lanthanide atom with the two halide ligands in *trans* position. The Ln–O distance to the THF ligand *trans* to cyclopentadienyl is significantly longer than those to the other oxygen atoms.

Substituted cyclopentadienyl ligands have also been utilized to prepare monocyclopentadienyllanthanide dihalides. For example, benzylcyclopentadienyl complexes of the type [LnCl$_2$(THF)$_n$(BzC$_5$H$_4$)] (n = 1, 2) have been isolated with neodymium, samarium, and gadolinium.[264] The t-butylcyclopentadienyl complexes [LuCl$_2$(THF)$_2$(Bu^{t}C$_5$H$_4$)] and [SmI$_2$(THF)$_3$(Bu^{t}C$_5$H$_4$)] have been structurally characterized.[239,265] Other examples include [LnCl$_2$(THF)$_n$(TMS-C$_5$H$_4$)] (Ln = Nd, Sm, Gd;

$n = 0$, 1, 2),[266] [LnCl$_2$(THF)$_n$(C$_3$H$_5$C$_5$H$_4$)] (Ln = Nd, Sm, Gd; $n = 0$, 1, 2, 3),[267] and [LnCl$_2$(THF)$_n$(C$_5$H$_9$C$_5$H$_4$)] (Ln = Ce, Pr, Nd, Sm, Gd; $n = 0$, 1, 2, 3).[267,268]

In some cases the reaction of anhydrous lanthanide trichlorides with cyclopentadienyl reagents in a 1:1 stoichiometry yields more complicated products. The formation of these products can be the result of partial hydrolysis and/or retention of alkali metal halide. Typical examples are the compounds [(Cp)Ln(THF)(μ-Cl)$_4$Li$_2$(THF)$_4$] (Ln = La, Nd),[269–71] [{Li(DME)$_2$(THF)}$_2${(Cp)$_4$Nd$_4$(μ$_4$-O)(μ-Cl)$_8$}],[272] and [{Li(THF)$_4$}$_2$(μ$_4$-O){(MeC$_5$H$_4$)NdCl(μ-Cl)NdCl$_2$(MeC$_5$H$_4$)}$_2$],[273] which were structurally characterized by x-ray diffraction. Two structurally characterized cyclopentylcyclopentadienyl complexes of the composition [Er$_4$(μ-Cl)$_6$(μ$_3$-Cl)(μ$_4$-O)(THF)$_3$(C$_5$H$_9$C$_5$H$_4$)$_3$] and [Er(THF)]$_2$(μ-Cl)$_3$(μ$_3$-Cl)$_2$(C$_5$H$_9$C$_5$H$_4$)Na(THF)$_2$·THF] were both obtained by reacting ErCl$_3$ with Na(C$_5$H$_9$C$_5$H$_4$) in a 1:1 molar ratio in THF solution.[274,275] Formation of a tetranuclear compound containing a μ$_4$-oxo bridge was also observed in the reaction of [NdCl(Cp)$_2$·2LiCl·nTHF] with methyllithium in dimethoxyethane solution. The reaction product was shown to be the monocyclopentadienylneodymium complex [Li(-DME)$_3$]$_2$[{Nd(Cp)}$_4$(μ-Me)(μ$_4$-O)(μ-Cl)$_6$] (DME = dimethoxyethane). In this complex the four neodymium atoms form a distorted tetrahedron.[276] Cluster formation is also prevalent in monocyclopentadienyl lanthanide alkoxide chemistry, which was studied by Schumann *et al.*[277] The pentanuclear gadolinium complex [{Gd(Cp)}$_5$(μ-OMe)$_4$(μ$_3$-OMe)$_4$(μ$_5$-O)] was prepared by a reaction of GdCl$_3$ with NaCp and two equivalents of NaOMe in THF solution. Obviously the reaction was aimed to yield the monocyclopentadienyllanthanide bis(alkoxide) derivative [Gd(OMe)$_2$(THF)$_n$(Cp)]. An x-ray crystal structure determination revealed the presence of a cluster alkoxide with five gadolinium atoms forming a tetragonal pyramid with a μ$_5$-oxo ligand inside the pyramid. The four triangular planes of the pyramid are bridged by μ$_3$-methoxide ligands while another four MeO$^-$ ligands act as μ$_2$-bridges between the four gadolinium atoms in the base (Figure 13).

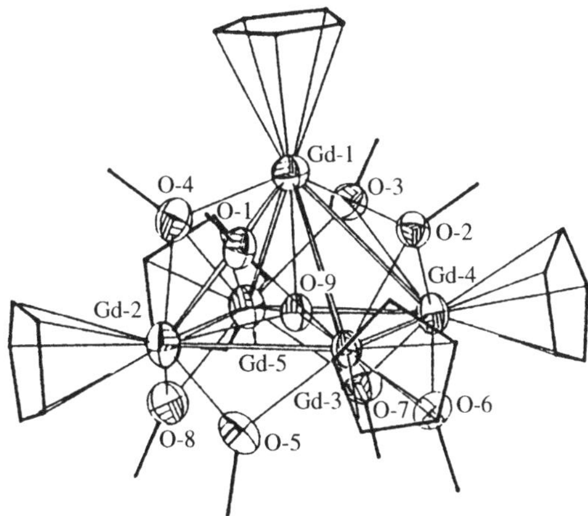

Figure 13　The molecular structure of [{Gd(Cp)}$_5$(μ-OMe)$_4$(μ$_3$-OMe)$_4$(μ$_5$-O)].[277]

Quite interestingly, this type of cluster alkoxide seems to be especially favored in lanthanide alkoxide chemistry as yttrium was reported to form an entirely analogous square-pyramidal Y$_5$O cluster. The compound [{Y(Cp)}$_5$(μ-OMe)$_4$(μ$_3$-OMe)$_4$(μ$_5$-O)] was isolated as a minor by-product when [YCl(THF)(Cp)$_2$], was reacted with potassium methoxide in THF solution. The main product of this reaction was bis(cyclopentadienyl)yttrium methoxide.[278] Degradation of a trimetallic cluster was observed, when [Y$_3$(OBu^t)$_7$Cl$_2$(THF)$_2$] was treated with alkali metal cyclopentadienyl reagents to afford the binuclear alkoxides [{Y(RC$_5$H$_4$)(μ-OBu^t)(OBu^t)}$_2$] (R = H, Me, TMS). The derivatives with R = H and TMS have been structurally characterized.[279]

A novel aspect of mono(cyclopentadienyl)lanthanide chemistry is the use of lanthanide triflates as precursors. By this synthetic variation the problems associated with salt-incorporation can be successfully avoided. The triflate ion has been demonstrated to be an excellent leaving group in organolanthanide chemistry. For example, the mono(cyclopentadienyl)lanthanide(III) bis(triflate) [Lu(O$_3$SCF$_3$)$_2$(THF)$_3$(Cp)] is straightforwardly obtained according to Equation (52).[280]

$$\text{Lu(O}_3\text{SCF}_3)_3 + \text{NaCp} \xrightarrow{\text{THF}} [\text{Lu(O}_3\text{SCF}_3)_2(\text{THF})_3\text{Cp}] + \text{NaO}_3\text{SCF}_3 \qquad (52)$$

A lutetium bis(hydrocarbyl) has been prepared by reacting the triflate precursor with two equivalents of LiCH$_2$TMS (Equation (53)). The product shows no tendency to disproportionate.[280]

$$[Lu(O_3SCF_3)(THF)_3Cp] + 2\ LiCH_2TMS \xrightarrow{\text{THF}} [Ln(CH_2TMS)_2(THF)_3Cp] + 2\ LiO_3SCF_3 \qquad (53)$$

Intramolecular chelation was used in the synthesis of related monohydrocarbyl complexes. The compounds were prepared according to Equations (54) and (55) and [Lu{CH₂CH(Me)CH₂NMe₂}(Cl)(THF)₂Cp] was structurally characterized.[281]

$$LuCl_3 + NaCp + Li[CH_2CH(Me)CH_2NMe_2] \xrightarrow[\substack{-NaCl\\-LiCl}]{\text{THF}} [Lu\{CH_2CH(Me)CH_2NMe_2\}(Cl)(THF)_2Cp] \qquad (54)$$

$$[LuCl_2(THF)_3Cp] + Li(CH_2)_3NMe_2 \xrightarrow[-LiCl]{\text{THF}} [Lu\{(CH_2)_3NMe_2\}(Cl)(THF)_2Cp] \qquad (55)$$

Similarly two mono(cyclopentadienyl)lutetium complexes with intramolecular arsine coordination have been prepared (Equation (56)).[281]

$$[Lu(O_3SCF_3)_2(THF)_3Cp] + 2\ ClMg(CH_2)_3AsR_2 \xrightarrow{\text{THF}} [Lu\{(CH_2)_3AsR_2\}_2Cp] + 2\ [ClMg(O_3SCF_3)] \qquad (56)$$

$$R = Me,\ Bu^t$$

An interesting novel type of mono(cyclopentadienyl)lanthanide hydrocarbyl is represented by the lutetium naphthalene complex [Lu(C₁₀H₈)(DME)(Cp)]. This material was isolated as black, needlelike crystals by reacting [LuCl₂(THF)₂(Cp)] with sodium naphthalide in DME. Formally the complex contains a coordinated $C_{10}H_8^{2-}$ dianion. According to the x-ray structural analysis the bonding between the naphthalene moiety and lutetium can be described as a $2\eta^1{:}\eta^2$ $(2\sigma,\pi)$-interaction (Figure 14).[282] A similar reaction with disodium anthracenide afforded [Lu(C₁₄H₁₀)(THF)₂(Cp)].[283]

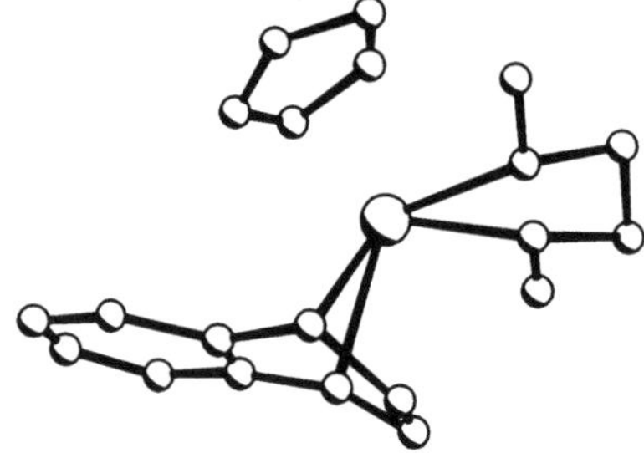

Figure 14   The molecular structure of [Lu(C₁₀H₈)(DME)(Cp)].[282]

Several examples of mono(cyclopentadienyl)lanthanide dicarboxylates have been obtained from Yb(Cp)₃ and free dicarboxylic acids. Protolytic elimination of two cyclopentadienyl ligands has led to the formation of [Yb{O₂CCH=CHCO₂}(Cp)], [Yb{O₂C(CH₂)ₙCO₂}(Cp)] ($n = 0$, 1, 2), and [Yb{o-O₂CC₆H₄CO₂}(Cp)].[284] The molecular structures of these compounds have not been further elucidated. A closely related compound containing a chelating β-aminoketonate ligand was obtained by elimination of cyclopentadiene upon treatment of [{Yb(μ-Cl)(Cp)₂}₂] with the free β-aminoketone (Equation (57)).[18]

$$1/2\ [\{Yb(\mu\text{-}Cl)Cp_2\}_2] + MeC(O)CH=C(NHPh)Me \longrightarrow$$

Related species are the (cyclopentadienyl)lanthanide bis(acetylacetonates) [Ln(acac)₂(Cp)] (Ln = Ce, Nd, Sm, Gd, Dy, Er) and differently substituted diketonates.[285–8] Like other compounds of the type LnL₂(Cp)[289] the bis(acetylacetonates) disproportionate at elevated temperatures to afford Ln(Cp)₃ and [Ln(acac)₃]. The anionic cerium acetylacetonate [Co(Cp)₂][Ce(acac)₃(Cp)] was made by reacting [Ce(THF)(Cp)₃] with [Co(acac)₃].[290] Numerous other LnL₂(Cp) complexes have been prepared in which L is an anionic *O*- or *N*-centered donor ligand. These ligands include the anions of aromatic carboxylic acids,[291] 8-hydroxyquinoline,[292–5] *o*-formylphenols,[295] furfurylalcohol,[295] α-nitroso-β-naphthol,[295] *o*-nitrophenols,[295] benzophenone oxime,[294] and *o*-aminophenols.[292,293] An anionic tris(amido) complex was made by treatment of *in situ* prepared $YbCl_2(Bu^tC_5H_4)$ with three equivalents of lithium diphenylamide and structurally characterized (Equation (58)).[294]

$$YbCl_2(Bu^tC_5H_4) + 3\,LiNPh_2 \longrightarrow [Li(THF)_4][Yb(NPh_2)_3(Bu^tC_5H_4)] + 2\,LiCl \qquad (58)$$

Two scandium sandwich complexes containing the octaethylporphyrin ligand have been reported. $[Sc(OEP)(Cp)]$ and $[Sc(OEP)(MeC_5H_4)]$ were prepared by reacting $[ScCl(OEP)]$ with the appropriate alkali metal cyclopentadienide.[183] In addition, the crystal structure of $[Sc(OEP)(Cp)]$ was determined. Scandium appears to be best suited for these preparations, as due to the small size of the Sc^{3+} ion the reactions are not complicated by salt-incorporation or formation of ill-defined solvates.[296]

The preparation of a mono(cyclopentadienyl)cerium(IV) complex has appeared in the literature, although serious doubts must be cast on this result. The reaction of "$[CeCl_4 \cdot nTHF \cdot xHCl]$" with Na(Cp) at $-40\,°C$ in THF solution was reported to afford a complex formulated as $CeCl_3(Cp) \cdot nTHF \cdot HCl$. This formulation was based only on elemental analysis, IR, and thermogravimetry.[297] The existence of such a compound seems highly unlikely, as all previously described reactions of cerium(IV) compounds with cyclopentadienyl reagents had resulted in the formation of cerium(III) cyclopentadienyl complexes. In addition, the starting material "$[CeCl_4 \cdot nTHF \cdot xHCl]$", also known as the THF adduct of hexachloroceric acid, is a thermally unstable material which has never been fully characterized. Similar results (i.e., formation of cerium(III) complexes) were obtained from reactions of Na(Cp) with ceric ammonium nitrate.[298]

2.2.5.4 $M(Cp)_2X$ compounds

(i) Classical coligands

Bis(cyclopentadienyl)lanthanide fluorides are not generally available due to the insolubility of anhydrous lanthanide trifluorides. Thus, the normal metathetical reactions cannot be applied to these systems, although a few fluorides have been synthesized by other routes. For example, $Yb(MeC_5H_4)_2$ reacts with perfluoroalkenes under defluorination to give $[YbF(THF)(MeC_5H_4)_2]$.[299]

Bis(cyclopentadienyl) derivatives of the type $LnCl(Cp)_2$ play a key role as intermediates in the synthesis of other organolanthanide complexes. Various preparative approaches to these compounds have been investigated, the oldest one being treatment of a lanthanide trichloride with sodium cyclopentadienide in a molar ratio of 1:2.[18] NaCp can be replaced by either thallium cyclopentadienide or $Mg(Cp)_2$.[18] Thallium cyclopentadienide has the advantage of being air-stable and easy to handle. However, the pronounced toxicity of thallium compounds limits the usefulness of this method to some degree. An interesting approach is the so-called "ammonium salt method," which involves treatment of the homoleptic $Ln(Cp)_3$ complex with NH_4Cl in benzene.[18] In some cases replacement of one cyclopentadienyl ligand has also been achieved by reacting the tris(cyclopentadienyl) complex with HCl.[18] Finally, $LnCl(Cp)_2$ derivatives are obtained by comproportionation between $LnCl_3$ and $Ln(Cp)_3$ in the proper stoichiometry or by substitution of other cyclopentadienyl ligands. The synthetic methods leading to bis(cyclopentadienyl)lanthanide chlorides are summarized in Equations (59)–(64).[18]

$$LnCl_3 + 2\,MCp \longrightarrow LnClCp_2 + 2\,MCl \qquad (59)$$
$$M = Na,\ Tl$$

$$ScCl_3 + MgCp_2 \longrightarrow ScClCp_2 + MgCl_2 \qquad (60)$$

$$LnCp_3 + NH_4Cl \longrightarrow LnClCp_2 + C_5H_6 + NH_3 \qquad (61)$$

$$LnCp_3 + HCl \longrightarrow LnClCp_2 + C_5H_6 \qquad (62)$$

$$[ErCl_2(THF)_n(C_5H_9C_5H_4)] \xrightarrow[-NaCl]{NaCp} [ErCl(THF)(C_5H_9C_5H_4)Cp] \qquad (63)$$
$$C_5H_9C_5H_4 = \text{cyclopentylcyclopentadienyl}$$

$$LnCl_3 + 2\,LnCp_3 \longrightarrow 3\,LnClCp_2 \qquad (64)$$

The last reaction is reversible as demonstrated by the thermal rearrangement of $Sm(OAc)(Cp)_2$ and $SmX(Cp)_2$ ($X = Cl$, Br, I). These compounds are in a thermal equilibrium with $Sm(Cp)_3$ and samarium(III) acetate or SmX_3, respectively.[300]

Due to coordinative unsaturation, $LnCl(Cp)_2$ derivatives are often not available for the lighter lanthanide elements lanthanum to neodymium. A corresponding promethium complex has not yet been isolated either. In general, the compounds $LnCl(Cp)_2$ are unstable as monomers. In benzene solution as well as in the solid state, these compounds form chloro-bridged dimers of the type $[\{Ln(\mu\text{-}Cl)(Cp)_2\}_2]$ (**4**).

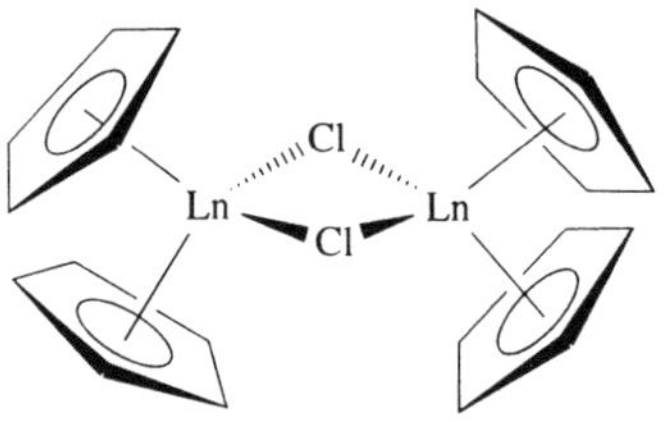

(4)

The structural chemistry of dimeric bis(cyclopentadienyl)lanthanide halides has been studied in detail. Among the crystallographically characterized complexes are $[\{Sc(\mu\text{-}Cl)(Cp)_2\}_2]$,[18] $[\{Y(\mu\text{-}Cl)(Cp)_2\}_2]$,[301] $[\{Ln(\mu\text{-}Cl)(Cp)_2\}_2]$ (Ln = Er, Yb (two modifications), Lu),[302–5] $[\{Ln(\mu\text{-}Br)(Cp)_2\}_2]$ (Ln = Gd, Tb, Dy, Er, Yb),[306–8] $[\{Dy(\mu\text{-}Cl)(Cp)_2\}_n]$,[309] and the tetrametallic $[\{Gd(\mu\text{-}Cl)(Cp)_2\}_4]$.[310] In $[\{Gd(\mu\text{-}Br)(Cp)_2\}_n]$ the bromide ligands act as bridges between infinite double chain polymers. Three bromide bridges form close contacts to each gadolinium atom.[306] The chloride bridges in $[\{Ln(\mu\text{-}Cl)(Cp)_2\}_2]$ are easily cleaved by treatment with THF to give the monomeric 1:1 adducts $[LnCl(THF)(Cp)_2]$.[18] The molecular and crystal structure of $[LuCl(THF)(Cp)_2]$ was determined by x-ray crystallography.[311,312] The molecule exhibits a pseudotetrahedral coordination geometry. In addition, the THF-solvated dimers $[\{Ln(\mu\text{-}Cl)(THF)(Cp)_2\}_2]$ (Ln = Nd, Er, Yb) were structurally characterized.[313–15] THF is not the only donor ligand which forms stable adducts with $LnCl(Cp)_2$ units. For example, $[\{Ln(\mu\text{-}Cl)(Cp)_2\}_2]$ complexes (Ln = Gd, Dy, Ho, Er, Yb, Lu) were found to react with 2,2'-bipyridine to give 1:1 adducts.[316]

In some cases the homoleptic tris(cyclopentadienyl)lanthanides are useful starting materials for the preparation of other $LnX(Cp)_2$ derivatives. A cyclopentadienyl ring is eliminated upon treatment of $Er(Cp)_3$ with elemental iodine yielding $ErI(Cp)_2$. Similarly the cyano complexes $NdCN(Cp)_2$ and $YbCN(Cp)_2$ have been prepared by protolysis of the corresponding tris(cyclopentadienide) with HCN.[18] The molecular structures of these compounds have not been determined, but presumably they are dimeric in the solid state. Monomeric bis(cyclopentadienyl)ytterbium nitrite, $[Yb(NO_2)(THF)(Cp)_2]$, has been structurally characterized and contains a bidentate nitrito ligand.[317] An organocerium nitrate complex, $[Ce(NO_3)_2Na(THF)_2(Cp)_2]$, has been prepared by reacting $[NH_4][Ce(NO_3)_6]$ with sodium cyclopentadienide.[298] Bis(cyclopentadienyl)lanthanide hydrocarbyls can also serve as starting materials for complexes with classical coligands. For example, the heterobimetallic bis(cyclopentadienyl)lanthanide chloride complex $[Yb(\mu\text{-}Cl)_2AlMe_2(Cp)_2]$ was synthesized by reacting dimeric $[\{Yb(\mu\text{-}Me)(Cp)_2\}_2]$ with $MeAlCl_2$. The reaction proceeds via cleavage of the ytterbium–carbon bonds.[18]

Several publications have demonstrated the synthetic advantages of the use of anhydrous lanthanide triflates in the preparation of organolanthanide complexes.[280] Lanthanide triflates are easily obtained by treatment of lanthanide oxides with a stoichiometric amount of triflic acid (Equation (65)).

$$Ln_2O_3 + 3\,CF_3SO_3H \longrightarrow Ln(O_3SCF_3)_3 + 3\,H_2O \tag{65}$$

Acetonitrile extraction of the crude products yields the anhydrous materials. In some cases anhydrous lanthanide triflates have been found to be better precursors than the previously used lanthanide trichlorides. Treatment of $Yb(O_3SCF_3)_3$ with $Na(Cp)$ in a molar ratio of 1:1 produced the bis(cyclopentadienyl)ytterbium triflate $[\{Yb(\mu\text{-}O_3SCF_3)(Cp)_2\}_2]$. An x-ray crystal structure determination of this compound revealed a dimeric structure with bridging triflate ligands.[318]

Mono- and disubstituted cyclopentadienyl ligands have been extensively employed in the preparation of bis(cyclopentadienyl)lanthanide halides. As expected, complexes containing the methylcyclopentadienyl ligand, $MeC_5H_4^-$, are not significantly more stable than the parent $LnCl(Cp)_2$ derivatives.[18] The chloro-bridged dimers $[\{Yb(\mu\text{-}Cl)(MeC_5H_4)_2\}_2]$[319] and $[\{Sm(\mu\text{-}Cl)(THF)(MeC_5H_4)_2\}_2]$ have been structurally characterized.[320] Several disubstituted lanthanide(III) halides containing the trimethylsilyl-cyclopentadienyl ligand have been reported. The known compounds include the base-free dimers $[\{Yb(\mu\text{-}X)(TMS\text{-}C_5H_4)_2\}_2]$ (X = Cl, I)[242] as well as the "ate" complexes $[Y(\mu\text{-}Cl)_2(TMS\text{-}C_5H_4)_2Li(L)]$ (L = DME, 2THF),[112] $[Yb(\mu\text{-}Cl)_2(TMS\text{-}C_5H_4)_2Li(Et_2O)_2]$,[321] and $[Lu(\mu\text{-}Cl)_2(TMS\text{-}C_5H_4)_2Li(TMEDA)]$.[321]

t-Butyl-substituted cyclopentadienyl ligands have been increasingly utilized in organolanthanide chemistry. Ligands such as *t*-butylcyclopentadienyl or 1,3-di-*t*-butylcyclopentadienyl impose an enhanced stability and higher solubility upon the resulting rare-earth derivatives. Several routes to disubstituted organolanthanides containing these ligands have been designed. A lithium chloride adduct of coordinatively unsaturated "$CeCl(Bu^t_2C_5H_3)_2$" has been prepared by treatment of [{Ce(μ-Cl)(Bu$^t_2$C$_5$H$_3$)$_2$}$_2$] with Li[AlH$_4$] in the presence of tetramethylethylenediamine (TMEDA). The product, [Ce(μ-Cl)$_2$(Bu$^t_2$C$_5$H$_3$)$_2$Li(TMEDA)], was characterized by x-ray diffraction.[322] A very similar structure was found in [Ln(μ-Cl)$_2$(Bu^{t}C$_5$H$_4$)$_2$Li(TMEDA)] (Ln = Ce, Sm, Lu)[323-5] and [Lu(μ-Cl)$_2$(Bu^{t}C$_5$H$_4$)$_2$Li(THF)$_2$].[325] Dimeric chloro-bridged molecular structures were established for the structurally characterized *t*-butylcyclopentadienyllanthanide chlorides [{Ln(μ-Cl)(Bu^{t}C$_5$H$_4$)$_2$}$_2$] (Ln = Pr, Gd, Sm, Lu)[305,326] and [{Ln(μ-Cl)(Bu$^t_2$C$_5$H$_3$)$_2$}$_2$] (Ln = Ce, Nd, Lu).[327,328] Other reported preparations and structural investigations include the compounds [SmI(THF)(Bu^{t}C$_5$H$_4$)$_2$],[265] [LnCl(THF)(Bu^{t}C$_5$H$_4$)$_2$] (Ln = Gd, Yb),[329,330] and [LnCl(THF)$_2$(Bu^{t}C$_5$H$_4$)$_2$] (Ln = Pr, Nd).[329]

Especially useful for the synthesis of bis(cyclopentadienyl)lanthanide halides (besides Cp*) is the very bulky 1,3-bis(trimethylsilyl)cyclopentadienyl anion, which was introduced into lanthanide chemistry by Lappert *et al.* With this ligand, two series of disubstituted complexes have been prepared for nearly all elements of the lanthanide group. Treatment of a lanthanide trichloride with two equivalents of lithium 1,3-bis(trimethylsilyl)cyclopentadienide initially affords the solvated lithium chloride adduct [Ln(μ-Cl){(TMS)$_2$C$_5$H$_3$}$_2$Li(THF)$_2$]. Heating above 140 °C yields the unsolvated dimers [{Ln(μ-Cl){(TMS)$_2$C$_5$H$_3$}$_2$}$_2$] (Scheme 10).[18]

$$LnCl_3 + 2\,Li[(TMS)_2C_5H_3] \xrightarrow[-LiCl]{THF} [Ln(\mu\text{-}Cl)\{(TMS)_2C_5H_3\}_2Li(THF)_2] \xrightarrow[-2\,LiCl,\,-4\,THF]{\Delta} [\{Ln(\mu\text{-}Cl)\{(TMS)_2C_5H_3\}_2\}_2]$$

Scheme 10

X-ray structural analyses of [{Ln(μ-Cl){(TMS)$_2$C$_5$H$_3$}$_2$}$_2$] (Ln = Sc, Yb, Lu) revealed the presence of chloro-bridged dimers with pseudotetrahedral coordination around the lanthanide elements.[18,331] These structures very much resemble those of the unsubstituted cyclopentadienyl analogues [{Ln(μ-Cl)(Cp)$_2$}$_2$]. Monomeric anions have been obtained when the dimers were reacted with bulky ammonium or phosphonium salts (Equation (66)).

$$[\{Ln(\mu\text{-}Cl)\{(TMS)_2C_5H_3\}_2\}_2] + 2\,MCl \longrightarrow 2\,M[LnCl_2\{(TMS)_2C_5H_3\}_2] \qquad (66)$$

$$M = PPh_3(CH_2Ph);\ Ln = Nd,\ Dy$$
$$M = PPh_4;\ Ln = Pr,\ Tm$$
$$M = AsPh_4,\ Ln = Nd$$
$$M = N(PPh_3)_2;\ Ln = Y$$

Other ring-substituted bis(cyclopentadienyl)lanthanide halides include a series of cyclopentyl-cyclopentadienyl derivatives [LnCl(THF)$_n$(C$_5$H$_9$C$_5$H$_4$)$_2$] (Ln = Nd, Sm, Er, Yb; n = 0, 1) and [{Ln(μ-Cl)(C$_5$H$_9$C$_5$H$_4$)$_2$}$_2$] (Ln = Er, Yb). The complexes [SmCl(THF)(C$_5$H$_9$C$_5$H$_4$)$_2$] and [{Er(μ-Cl)(C$_5$H$_9$C$_5$H$_4$)$_2$}$_2$] have been structurally characterized.[332] Dimeric [{Nd(μ-Cl){MeC$_5$H$_2$(CH$_2$)$_5$}$_2$}$_2$] contains fused-ring bicyclic cyclopentadienyl ligands (5).[333]

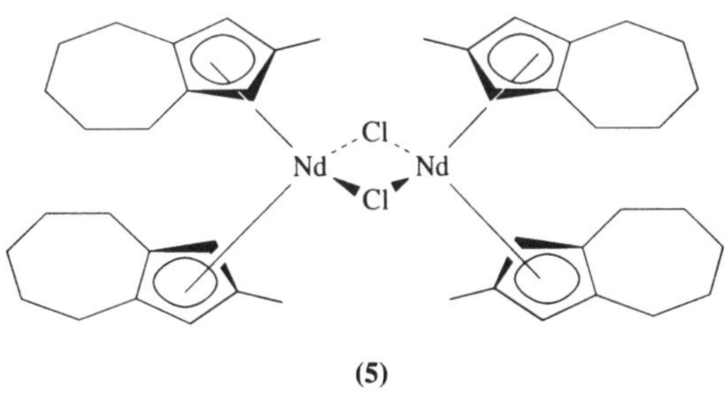

(5)

An interesting route to base-free bis(cyclopentadienyl)lanthanide halides is the use of cyclopentadienyl ligands containing "pendant-arm" donor functions, which are able to coordinatively saturate the lanthanide ion. A typical example of such a ligand is the 3-methoxypropyl-cyclopentadienyl anion, [C$_5$H$_4$(CH$_2$)$_3$OMe]$^-$. The halide-bridged dimers [Ln(μ-Cl){C$_5$H$_4$(CH$_2$)$_3$OMe}$_2$]$_2$ have been prepared by reacting the anhydrous lanthanide trichloride with the sodium salt of the ligand in a 1:2 molar ratio (Equation (67)).[334,335]

$$2\,LnCl_3 \;+\; 4\,Na[C_5H_4(CH_2)_3OMe] \;\xrightarrow{\;THF\;}\; [Ln(\mu\text{-}Cl)\{C_5H_4(CH_2)_3OMe\}_2]_2 \;+\; 4\,NaCl \qquad (67)$$

$$Ln = Y,\ La,\ Pr,\ Nd,\ Gd,\ Ho,\ Er,\ Yb$$

The molecular and crystal structure of the lanthanum derivative has been determined (Figure 15).[334] The $[C_5H_4(CH_2)_3OMe]^-$ anions act as chelating ligands with the dangling methoxy functions being coordinated to the central lanthanum atom. Together with the bridging chloride ligands this coordination mode results in a formal coordination number of nine around the metal atoms. Several members of this series (with $Ln = Y$, La, Nd, Gd, Ho, Er, Yb) have also been studied by x-ray photoelectron spectroscopy.[336]

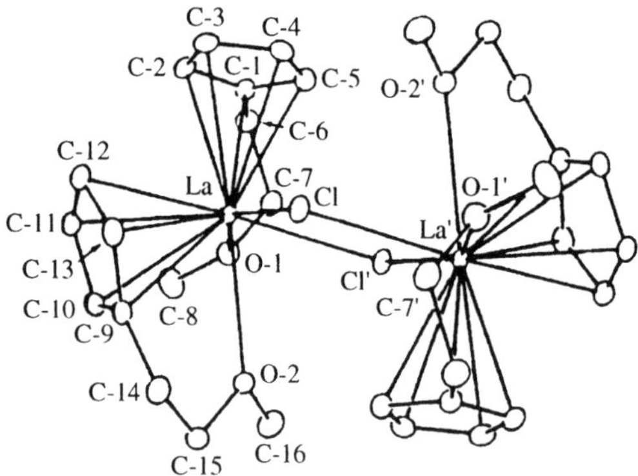

Figure 15   The molecular structure of $[La(\mu\text{-}Cl)\{C_5H_4(CH_2)_3OMe\}_2]_2$.[334]

Various bis(cyclopentadienyl)lanthanide chlorides containing the related 2-methoxyethyl-cyclopentadienyl ligand have been synthesized. In this case the molecular structures of the products depend on the ionic radius of the central metal. The lanthanum derivative $[La(\mu\text{-}Cl)\{C_5H_4(CH_2)_2OMe\}_2]_2$ is a dimer with the formal coordination number 10 around lanthanum. In contrast, the structurally characterized complexes $[LnCl\{C_5H_4(CH_2)_3OMe\}_2]$ (Ln = Dy, Yb) are unsolvated monomers in which the formal coordination number is 9.[337]

The use of the donor-substituted dimethylaminoethyl-cyclopentadienyl ligand has recently allowed the synthesis and structural characterization (Nd–C 0.2731(3)–0.2807(3) nm, Nd–N 0.2788(2) nm, Nd–Cl 0.26989(7) nm) of the first solvent-free, monomeric bis(cyclopentadienyl)lanthanide chloride. $[NdCl(C_5H_4CH_2CH_2NMe_2)_2]$ (**6**) was prepared by reacting $NdCl_3$ with two equivalents of $K(C_5H_4CH_2CH_2NMe_2)$ and was isolated by sublimation.[338]

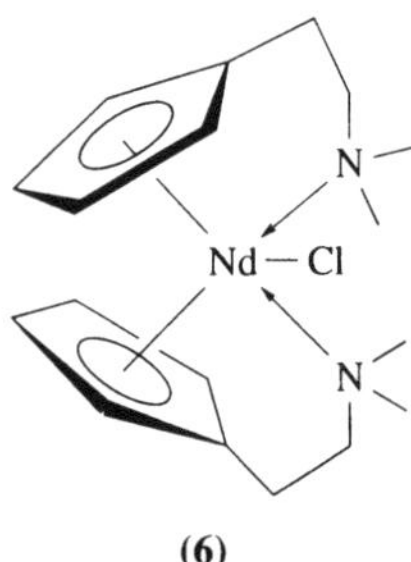

(**6**)

Complexes containing $LnCl(Cp)_2$ units are useful starting materials for the synthesis of various bis(cyclopentadienyl)lanthanide complexes containing bonds between the lanthanides and main group elements. *Ab initio* calculations have been carried out on cations of the type $[Ln(Cp)_2]^+$ (Ln = Sc, Lu) with the result that bent metallocene structures should be favored.[339] Such cations appear to be unknown in the $Ln(Cp)_2$ series, but have been prepared with substituted cyclopentadienyl ligands (cf. Section 2.2.6.3).

In recent years several examples of μ-hydroxo- and μ-oxo-bridged organolanthanide complexes of the type $[\{Ln(\mu\text{-}OH)(Cp)_2\}_2]$ and $[\{LnCp_2\}_2(\mu\text{-}O)]$ have been prepared and structurally characterized. They are generally made by carefully controlled hydrolysis of the corresponding $LnXCp_2$ precursors (X = halide, cyclopentadienyl). One of the first examples was $[\{Y(\mu\text{-}OH)(Cp)_2\}_2]\cdot Ph_2C_2$.[340] The diphenylacetylene in the structure is not coordinated to yttrium but only incorporated in the crystal lattice. Other structurally characterized examples include $[\{Ln(\mu\text{-}OH)(Bu^tC_5H_4)_2\}_2]$ (Ln = Nd, Dy),[341] $[\{Ln(\mu\text{-}OH)(MeOCH_2CH_2C_5H_4)_2\}_2]$ (Ln = Ho, Er),[342] $[\{Yb(\mu\text{-}OH)(TMS\text{-}C_5H_4)_2\}_2]$,[343] and $[\{Ln(\mu\text{-}OH)\text{-}\{(TMS)_2C_5H_3\}_2\}_2]$ (Ln = Sm, Lu).[90,344] The μ-oxo bridged ytterbium complex $[\{Yb(\mu\text{-}OH)(MeC_5H_4)_2\}_2]$ exhibits a linear central Yb–O–Yb unit (Yb–O 0.2915(1) nm, Yb–C 0.2667 nm).[345] The crystallographically characterized lutetium complex $[\{Lu(THF)(Cp)_2\}_2(\mu\text{-}O)]$ was obtained by hydrolysis of $[LuAsPh_2(THF)(Cp)_2]$.[346]

Various dimeric alkoxides of the type $[\{Ln(\mu\text{-}OR)(Cp)_2\}_2]$ (and in some cases monomeric THF solvates $[Ln(OR)(THF)_n(Cp)_2]$ have been prepared by treatment of the chloro complexes with sodium alkoxides[18,347] or by protonation of $Ln(Cp)_3$ with stoichiometric amounts of alcohols.[348–51] The latter method has been utilized to prepare a series of bis(cyclopentadienyl)ytterbium alkoxides containing chiral alkoxide ligands such as (−)-mentholate.[351,352] A typical example of a monomeric species is $[Nd(OC_6H_3Ph_2\text{-}2,6)(THF)_2(Cp)_2]$.[348] Two synthetic methods have been developed to prepare dimeric alkoxides containing μ-OCH=CH$_2$ ligands. The compounds $[\{Ln(\mu\text{-}OCH=CH_2)(RC_5H_4)_2\}_2]$ (Ln = Y, Yb, Lu; R = H, Me) can be made either by treatment of $[\{Ln(\mu\text{-}Cl)(RC_5H_4)_2\}_2]$ with $LiOCH=CH_2$ or via thermolysis of $[\{Yb(\mu\text{-}Me)(Cp)_2\}_2]$ or $[Ln(CH_2TMS)(THF)(RC_5H_4)_2]$ in THF solution in the presence of LiCl. In the latter case the solvent THF acts as an *in situ* precursor for the lithium enolate. The derivatives $[\{Y(\mu\text{-}OCH=CH_2)(MeC_5H_4)_2\}_2]$,[353] $[\{Yb(\mu\text{-}OCH=C=CH_2)(Cp)_2\}_2]$,[350] $[\{Yb(\mu\text{-}OCH_2\text{-}CH=CHMe)(Cp)_2\}_2]$,[352] and $[\{Yb(\mu\text{-}OCH=CH_2)(C_5H_4Me)_2\}_2]$,[350] have been structurally characterized. The latter two complexes were formed by reactions of organolanthanide precursors with THF.[352]

Bis(cyclopentadienyl)lanthanide(III) carboxylates are also dimeric in the solid state. They can be obtained by reacting $[LnCl(THF)(Cp)_2]$ with various sodium carboxylates in THF solution (Equation (68)).[18,252,284,354–9]

$$2\ LnCl(THF)Cp_2 + 2\ NaO_2CR^1 \xrightarrow{\text{THF}} \text{[structure]} + 2\ NaCl \qquad (68)$$

Ln = Sc, Sm, Gd, Tb, Dy, Ho, Er, Tm, Yb, Lu

R^1 = H, Me, CH_2Cl, CCl_3, Et, Bu^n, Bu^t, Ph, $o\text{-}C_6H_4R^2$, C_6F_5, C_5H_4N

An alternative synthetic route involves oxidation of $[Yb(DME)(Cp)_2]$ with Tl^+ or Hg^{2+} carboxylates.[360] Dimeric bis(cyclopentadienyl)ytterbium(III) pentafluorobenzoate was structurally characterized by x-ray crystallography.[252] In this compound each pentafluorobenzoate ligand is bridging two ytterbium ions to give a central eight-membered $Yb_2C_2O_4$ ring.

Similar dimeric structures have been reported for bis(cyclopentadienyl)lanthanide triflates, which have been prepared according to Equation (69). $[\{Yb(\mu\text{-}O_3SCF_3)(Cp)_2\}_2]$ was crystallographically characterized.[318,361]

$$2\ Ln(O_3SCF_3)_3 + 4\ NaCp \longrightarrow [\{Ln(\mu\text{-}O_3SCF_3)Cp_2\}_2] + 4\ NaO_3SCF_3 \qquad (69)$$

Ln = Sc, Yb, Lu

Analogous reactions with dicarboxylic acid ligands gave binuclear complexes of the type $[\{Ln(Cp)_2(\mu\text{-}O_2CRCO_2)\}_2]$ (R = CH_2, CH_2CH_2, $o\text{-}C_6H_4$, $p\text{-}C_6H_4$). These complexes have been made either from $Ln(Cp)_3$ and the free dicarboxylic acids or from $[\{Yb(\mu\text{-}Cl)(Cp)_2\}_2]$ and the corresponding disodium salt.[355] Monomeric bis(cyclopentadienyl)lanthanide carboxylates can be stabilized by addition of bulky Lewis bases, for example $[Ln(O_2CCF_3)(phen)_n(Cp)_2]$ (Ln = La, Ce, Nd, Pr; n = 1, 2; phen = 1,10-phenanthroline).[362]

Related to the carboxylates, although presumably monomeric, are β-diketonate complexes of the type $[Ln\{RC(O)CHC(O)R\}(Cp)_2]$ (R = Me, Ph). These compounds are normally prepared from $[\{Ln(\mu\text{-}Cl)(Cp)_2\}_2]$ and the corresponding sodium β-diketonate (Equation (70), acac = acetylacetonate),[18,285–8,292,295,363] although ytterbium(III) derivatives are also available through oxidation of $[Yb(DME)(Cp)_2]$ with certain metal (Hg, Tl, Ag, Cu) salts of the diketonate ligands.[252] Yet another, more general preparation involves the elimination of cyclopentadiene from a tris(cyclopentadienyl)lanthanide upon treatment with the free β-diketone (Equation (71)).[18,364]

$$LnCl(THF)Cp_2 + Na(acac) \longrightarrow [Ln(acac)Cp_2] + NaCl \tag{70}$$

$$Ln = Sc, Ce, Nd, Sm, Gd, Dy, Er$$

$$YbCp_3 + RC(O)CH_2C(O)R \longrightarrow [Yb\{RC(O)CHC(O)R\}Cp_2] + C_5H_6 \tag{71}$$

$$R = Me, CF_3, Bu^t$$

The molecular structures of $[Yb(acac)(Cp)_2]$[363] and $[Yb\{MeC(O)CHC(O)CF_3\}(Cp)_2]$[364] have been determined. Several acetylacetonates of the type $[Ln(acac)(Cp)_2]$ have been shown to disproportionate at elevated temperatures to give $Ln(Cp)_3$ and $[Ln(acac)_3]$.

Numerous other $LnL(Cp)_2$ complexes have been prepared in which L is a mono- or bidentate anionic *O*- or *N*-centered donor ligand. Monodentate OR ligands usually lead to the formation μ-OR bridged dimers. These ligands include the anions of *o*-nitrophenol,[295] 8-hydroxyquinoline,[293,365] *o*-aminophenols,[293,365] benzophenoneoxime,[366] cyclohexanoneoxime, and α-nitroso-β-naphthol.[295]

An unusual complex containing a tetradentate oxygen ligand was constructed by insertion of carbon monoxide into the reactive lutetium–carbon bond of $[LuBu^t(THF)(Cp)_2]$. The initial product in this reaction is the yellow acyl complex $[Lu\{C(O)Bu^t\}(Cp)_2]$ which can be described by two resonance forms. Further CO insertion yields the final product via postulated ketene and carbene intermediates (Scheme 11).[18]

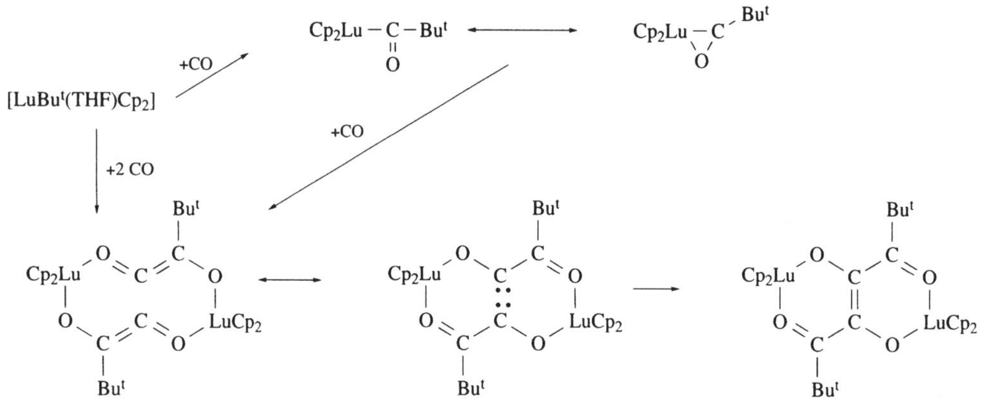

Scheme 11

As in the case of $Ln(Cp)_3$, homoleptic tris(cyclopentadienyl)lanthanide complexes containing substituted cyclopentadienyl ligands can serve as useful starting materials for bis(cyclopentadienyl) derivatives with additional alkoxide or thiolate ligands. In general, one cyclopentadienyl ligand in a lanthanide tris(cyclopentadienyl) is easily cleaved upon treatment with a protic reagent.[297] For example, $[Ce(Bu^tC_5H_4)_3]$ reacts with alcohols and thiols to give the dimers $[\{Ce(\mu\text{-}ER)(Bu^tC_5H_4)_2\}_2]$ (Equation (72)).[367] An alternative synthetic route involves cleavage of E–E bonds in the presence of organolanthanide hydrocarbyls (Equation (73)).[368]

$$2\,[Ce(Bu^tC_5H_4)_3] + 2\,HER \longrightarrow [\{Ce(\mu\text{-}ER)(Bu^tC_5H_4)_2\}_2] + 2\,Bu^tCp \tag{72}$$

$$E = O, S; R = Pr^i, Ph$$

$$[\{Ln(\mu\text{-}Me)(Bu^tC_5H_4)\}_2] + 2\,REER \longrightarrow [\{Ln(\mu\text{-}ER)(Bu^tC_5H_4)\}_2] + 2\,MeER \tag{73}$$

$$Ln = Y, Lu; E = S, Se$$
$$R = Bu^n, Bu^t, Bz, Ph$$

For steric reasons, Bu^tSH does not react with $[Ce(Bu^tC_5H_4)_3]$ but yields the dimer $[\{Ce(\mu\text{-}SBu^t)(MeC_5H_4)_2\}_2]$ upon treatment with $[Ce(MeC_5H_4)_3]$. The molecular structures of $[\{Ce(\mu\text{-}EPr^i)(Bu^tC_5H_4)_2\}_2]$ (E = O, S) have been determined by x-ray methods. Figure 16 shows the structure of the thiolate derivative (Ce–C 0.278(2) nm, Ce–S 0.2870(2) nm).[367] Structurally characterized bis(cyclopentadienyl)lanthanide selenolates are $[Lu(\mu\text{-}SePh)_2(Cp)_2Li(THF)_2]$ (Lu–C 0.260(2) nm, Lu–Se 0.2799(1) nm)[369,370] and $[\{Y(\mu\text{-}SePh)(Bu^tC_5H_4)_2\}_2]\cdot C_6H_6$ (Yb–Se 0.2914(1) nm).[368]

The use of the sterically demanding 1,3-bis(trimethylsilyl)cyclopentadienyl ligand in *f*-element chemistry has been pioneered by Lappert *et al.*[18] This ligand has an even larger cone angle than pentamethylcyclopentadienyl and provides sufficient steric bulk to stabilize even the large lighter lanthanide ions.

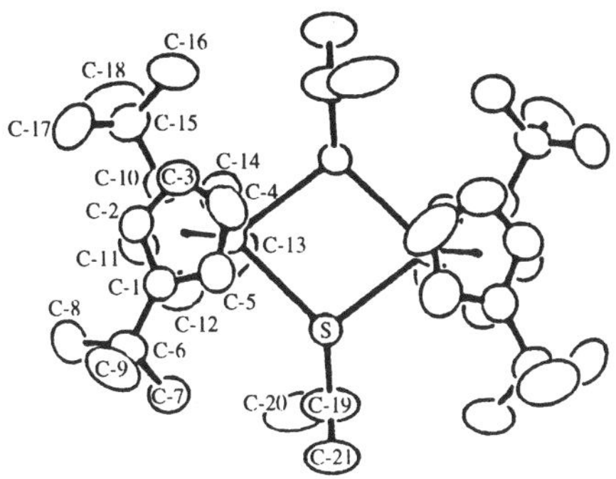

Figure 16 The molecular structure of $[\{Ce(\mu\text{-}SPr^i)(Bu^tC_5H_4)_2\}_2]$.[367]

An interesting modification of the $Ln\{(TMS)_2C_5H_3\}_2$ system is the use of iodide as an effective leaving group. The following two-step procedure has been developed to prepare well-defined solvated iodo complexes of the composition $[Ln(I)(NCMe)_2\{(TMS)_2C_5H_3\}_2]$ (Equations (74) and (75)).[371]

$$[LnI_3(THF)_x] + 2\,K\{(TMS)_2C_5H_3\} \xrightarrow{\text{THF}} [K(THF)_n][LnI_2\{(TMS)_2C_5H_3\}_2] + KI \qquad (74)$$

$$[K(THF)_n][LnI_2\{(TMS)_2C_5H_3\}_2] \xrightarrow{\text{MeCN}} [Ln(I)(NCMe)_2\{(TMS)_2C_5H_3\}_2] + KI \qquad (75)$$

$$Ln = La,\ Ce$$

Several cationic species have been prepared from the iodo complexes by halide abstraction, of which the lanthanum derivative $[La(NCMe)(DME)\{(TMS)_2C_5H_3\}_2][BPh_4]\cdot0.5DME$ was structurally characterized (Scheme 12).[371]

$$[Ln(I)(NCMe)_2\{(TMS)_2C_5H_3\}_2] + Ag[BPh_4] \xrightarrow[-AgI]{\text{THF}} [Ln(THF)_2\{(TMS)_2C_5H_3\}_2][BPh_4] \xrightarrow{\text{MeCN, DME}}$$

$$[Ln(NCMe)(DME)\{(TMS)_2C_5H_3\}_2][BPh_4]\cdot0.5DME$$

$$Ln = La,\ Ce$$

Scheme 12

Bis(cyclopentadienyl)lanthanide complexes with Ln–N bonds can be prepared by two different routes. One is the metathesis of bis(cyclopentadienyl)lanthanide chlorides with metal amide compounds and the other one involves the protolysis of lanthanide–carbon bonds by N–H functions. A typical example for the metathetical preparation is the synthesis of the lanthanide pyrrolyl complexes $[Ln(NC_4H_4)(Cp)_2]$ (Ln = Sm, Dy, Yb, Lu) and $[Lu(NC_4H_2Me_2)(THF)(Cp)_2]$ from $LnCl(THF)(Cp)_2$ and the corresponding sodium pyrrolyl. The dimethylpyrrolyl derivative has been established by x-ray crystallography (Lu–N 0.2289(4) nm).[289,372]

The anionic bis(cyclopentadienyl)lanthanide hydrocarbyl derivatives $[LnMe_2(Cp)_2]^-$ served as starting materials for lanthanide diorganoamides. Treatment of the lithium salts with two equivalents of diphenylamine afforded the terminal amido complexes $[Li(TMEDA)][Sm(NPh_2)_2(Cp)_2]$ and $[Li(THF)_4][Lu(NPh_2)_2(Cp)_2]$. The lutetium derivative was crystallographically characterized (Lu–N 0.2290(7) nm and 0.2293(7) nm).[373] A second structurally established complex of this series is $[Li(DME)_3][Lu(NPh_2)_2(Cp)_2]$.[374]

Organolanthanide complexes containing unsubstituted amido ligands have only rarely been described,[375] although the first compound of this type was reported as early as 1963. $[\{Er(\mu\text{-}NH_2)(Cp)_2\}_2]$ was prepared from the corresponding chloro complex and $NaNH_2$.[18] $[YbNH_2(Cp)_2]$ was made by thermal decomposition of $[Yb(NH_3)(Cp)_3]$ at 250 °C. This amide too is presumably a dimer with bridging $\bar{N}H_2$ ligands.[18] The compound $[\{Yb(\mu\text{-}NH_2)(MeC_5H_4)_2\}_2]$ has been structurally characterized (Yb–C 0.259(1)–0.266(1) nm, Yb–N 0.229(1) nm and 0.232(1) nm).

An anionic bis(cyclopentadienyl)yttrium amide, $Li[Y(NMeCH_2CH_2NMe_2)_2(Cp)_2]$, has been prepared by reacting $[YCl(THF)(Cp)_2]$ with two equivalents of lithium N,N,N'-trimethylethylenediamide.[376] Both oxygen and nitrogen coordination has been observed in the recently synthesized complex bis(μ-acetoneoximato)bis(cyclopentadienyl)digadolinium, $[\{Gd(\mu\text{-}\eta^2\text{-}ONCMe_2)(Cp)_2\}_2]$. The compound was prepared by reacting $Gd(Cp)_3$ with acetoneoxime in THF solution (Equation (76)). An x-ray crystal structure analysis revealed an unusual structure with the N–O fragment of the acetoneoximato ligand acting as both a bridging and a side-on donating group. This bonding situation represents a novel coordination mode of an oximato ligand to a metal.[377]

$$2\,GdCp_3 + 2\,HON{=}CMe_2 \xrightarrow{\ \ THF\ \ } [\{Gd(\mu\text{-}\eta^2\text{-}ONCMe_2)Cp_2\}_2] + 2\,C_5H_6 \qquad (76)$$

Among the group 15 elements nitrogen donor ligands are best suited for coordination to lanthanides. In agreement with the hard Lewis acid character of the lanthanide(III) cations, the number of stable organolanthanide complexes containing Ln–P, Ln–As, Ln–Sb, and Ln–Bi bonds is quite small. The first example of an "organolanthanide–phosphine" was reported by Schumann *et al.* The phosphido-bridged lutetium complex $[Lu(\mu\text{-}PPh_2)_2(Cp)_2Li(TMEDA)]\cdot1/2$toluene was characterized by an x-ray analysis (Figure 17, Lu–P 0.2782(1) nm and 0.2813(2) nm).[369] $[Lu(\mu\text{-}AsPh_2)_2(Cp)_2Li(TMEDA)]$ was prepared analogously (Lu–C 0.259(1) nm, Lu–As 0.288(1) nm).[369,378,379]

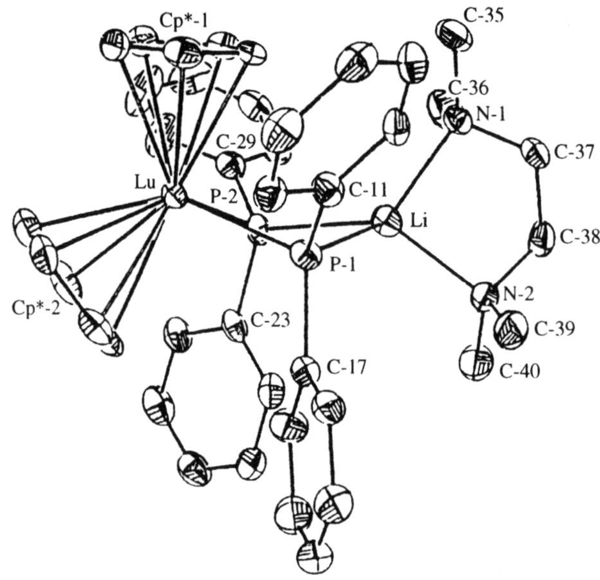

Figure 17　The molecular structure of $[Lu(\mu\text{-}PPh_2)_2(Cp)_2Li(TMEDA)]$.[369]

Other bis(cyclopentadienyl)lanthanide phosphides have been prepared according to Equation (77) by treatment of halide precursors with lithium diorganophosphides.[18,380]

$$LnCl(THF)Cp_2 + LiPR^1R^2 \xrightarrow{\ \ THF\ \ } [LnPR^1R^2(THF)Cp_2] + LiCl \qquad (77)$$

$$Ln = Tb,\ Ho,\ Er,\ Tm,\ Yb,\ Lu$$
$$R^1 = R^2 = Bu^t,\ c\text{-}C_6H_{11},\ Ph$$
$$R^1 = Bu^t,\ R^2 = Ph$$

Other synthetic routes to such complexes have been developed and some of these compounds are sufficiently stable to be isolated and characterized. The lutetium hydrocarbyl $[Lu(CH_2TMS)(THF)(Cp)_2]$ has been used to prepare compounds containing Lu–P and Lu–As bonds. Treatment of the precursor with diphenylphosphine or diphenylarsine resulted in the formation of $[LuPPh_2(THF)(Cp)_2]$ and $[LuAsPh_2(THF)(Cp)_2]$ respectively (Equation (78)). The preparation is very clean because volatile tetramethylsilane is the only by-product.[346]

$$[Ln(CH_2TMS)(THF)Cp_2] + HEPh_2 \longrightarrow [LnEPh_2(THF)Cp_2] + SiMe_4 \qquad (78)$$

$$E = P,\ As$$

On attempted recrystallization of $[LuPPh_2(THF)(Cp)_2]$ from THF solution, the unusual alkoxide-bridged dimer $[\{Lu\{\mu\text{-}O(CH_2)_4PPh_2\}(Cp)_2\}_2]$ was formed via cleavage of the THF. Under similar conditions the arsenic derivative $[LuAsPh_2(THF)(Cp)_2]$ gave the hydrolysis product $[(\mu\text{-}O)\{Lu(THF)(Cp)_2\}_2]$ (see above).[346]

(ii) Hydrocarbyls

In marked contrast to the pronounced instability of simple homoleptic lanthanide(III) hydrocarbyls, complexes containing a LnR(Cp)$_2$ unit (R = alkyl, aryl) are quite stable and easily accessible. Bis(cyclopentadienyl)lanthanide hydrocarbyls are normally prepared from the corresponding LnCl(Cp)$_2$ precursors. A general synthetic route involves treatment of bis(cyclopentadienyl)lanthanide chlorides with equimolar amounts of lithium alkyls or aryls in THF solution (Equation (79)). By this method a large number of THF adducts of the composition [LnR(THF)(Cp)$_2$] have been prepared.[18,381,382] When R is a small alkyl group, the products have only been characterized in solution by NMR spectroscopy. The stability of the lanthanide hydrocarbyls increases significantly if sterically demanding alkyl or aryl ligands are used. Alternative but not so widely used preparations involve treatment of lanthanide tris(cyclopentadienyls) with LiR or oxidation of lanthanide(II) metallocenes with HgR$_2$ (Equations (80) and (81)).[383–6]

$$[\text{LnCl(THF)Cp}_2] + \text{LiR} \xrightarrow[-78\,^\circ\text{C}]{\text{THF}} [\text{LnR(THF)Cp}_2] + \text{LiCl} \tag{79}$$

Ln = Y, Nd, Sm, Dy, Er, Tm, Yb, Lu

R = Me, Et, Pr^i, Bu^n, Bu^t, CH$_2$Bu^t, CH$_2$TMS, Bz,

Ph, C$_6$H$_4$Me-*p*, C$_6$H$_4$Cl-*p*, CH$_2$PMe$_2$

$$[\text{Ln(THF)Cp}_3] + \text{LiR} \longrightarrow [\text{LnR(THF)Cp}_2] + \text{LiCp} \tag{80}$$

Ln = Nd, Lu

R = Bu^s, Bu^t

$$2\,[\text{Yb(DME)Cp}_2] + \text{HgR}_2 \longrightarrow 2\,[\text{YbRCp}_2] + \text{Hg} \tag{81}$$

R = C$_6$F$_5$, C$_6$Cl$_5$, C≡CPh

The molecular structure of [LuCH$_2$TMS(THF)(Cp)$_2$] has been determined by x-ray diffraction.[347] The coordination geometry in this complex can be described as pseudotetrahedral. One THF ligand, the centers of the cyclopentadienyl rings, and the σ-bonded carbon atom form a distorted tetrahedron around the central lutetium atom. A similar structure was found for [YbMe(THF)(Cp)$_2$] (Yb–C(σ) 0.236(1) nm).[387]

The stabilization of monomeric [LnR(Cp)$_2$] species without coordinated solvent molecules normally requires the use of a chelating ligand that allows the intramolecular saturation of reactive coordination sites at the metal atom. A typical example is the scandium complex [Sc(C$_6$H$_4$CH$_2$NMe$_2$-*o*)(Cp)$_2$], which contains the chelating *o*-dimethylaminomethylphenyl ligand.[127,388] The formal coordination number in this complex is increased to 8 through the intramolecular coordination of the dimethylamino group. Other complexes with intramolecular chelation are the dimethylaminopropyl derivatives [Lu(CH$_2$CH$_2$CH$_2$NMe$_2$)(Cp)$_2$] (Lu–C(σ) 0.2221 nm, Lu–N 0.2371 nm) and [Lu{CH$_2$CH(Me)CH$_2$NMe$_2$}(Cp)$_2$].[281] Unsolvated [LnCH(TMS)$_2$(Cp)$_2$] (Ln = Sm, Lu) complexes are stable due to the steric bulk of the hydrocarbyl ligand.[384]

Monomeric lanthanide hydrocarbyls containing bridging methyl groups between the lanthanide atoms and lithium are obtained when LnCl(Cp)$_2$ precursors are treated with two equivalents of methyllithium. Coligands such as THF or TMEDA have been used in these reactions to saturate the coordination sphere around lithium. The resulting products can be regarded as hydrocarbyl analogues of the well known lithium chloride adducts [Ln(μ-Cl)$_2$Li(THF)$_2$(Cp)$_2$] and [Ln(μ-Cl)$_2$Li(TMEDA)(Cp)$_2$] (Equations (82) and (83)).[389,390]

$$\text{LnCl(THF)Cp}_2 + 2\,\text{MeLi} + \text{TMEDA} \xrightarrow{\text{THF}} [\text{Ln(μ-Me)}_2\text{Li(TMEDA)Cp}_2] + \text{LiCl} \tag{82}$$

Ln = Er, Lu

$$\text{LnCl(THF)Cp}_2 + 2\,\text{MeLi} \xrightarrow{\text{THF}} [\text{Ln(μ-Me)}_2\text{Li(THF)}_2\text{Cp}_2] + \text{LiCl} \tag{83}$$

Ln = Er, Lu

The crystal and molecular structure of [Er(μ-Me)$_2$Li(TMEDA)(Cp)$_2$][389] and its lutetium analogue[390] have been determined. Erbium is pseudotetrahedrally coordinated and connected to lithium via two bridging methyl groups. The lithium atom is coordinatively saturated by coordination of a chelating TMEDA ligand.

A two-step procedure has been designed for the preparation of unsolvated $LnR(Cp)_2$ species. In the first step, dimeric bis(cyclopentadienyl)lanthanide chlorides are reacted with lithium tetraalkylaluminates, $Li[AlR_4]$, to give the bimetallic complexes $[Ln(\mu-R)_2AlR_2(Cp)_2]$ (Equation (84)).[390] From these intermediates the trialkylaluminum can be abstracted by addition of pyridine, which forms an insoluble 1:1 adduct of the type $R_3AlNC_5H_5$. This reaction provides the unsolvated $LnR(Cp)_2$ derivatives in the form of alkyl-bridged dimers (Equation (85)). In some cases simple metathetical reactions also lead to the μ-Me bridged dimers (Equation (86)).[391]

$$[\{Ln(\mu\text{-}Cl)Cp_2\}_2] + 2\,Li[AlR_4] \xrightarrow{\text{toluene}} 2\,[Ln(\mu\text{-}R)_2AlR_2Cp_2] + 2\,LiCl \qquad (84)$$

$$R = Me;\ Ln = Sc,\ Y,\ Gd,\ Dy,\ Ho,\ Er,\ Tm,\ Yb$$
$$R = Et;\ Ln = Sc,\ Y,\ Ho$$

$$2\,[Ln(\mu\text{-}Me)_2AlMe_2Cp_2] + 2\,C_5H_5N \xrightarrow{\text{toluene}} [\{Ln(\mu\text{-}Me)Cp_2\}_2] + 2\,Me_3AlNC_5H_5 \qquad (85)$$

$$[\{Nd(\mu\text{-}Cl)(Bu^tC_5H_4)_2\}_2] + 2\,MeLi \longrightarrow [\{Nd(\mu\text{-}Me)(Bu^tC_5H_4)_2\}_2] + 2\,LiCl \qquad (86)$$

The x-ray crystal structures of $[\{Yb(\mu\text{-}Me)(Cp)_2\}_2]$,[390] $[\{Yb(\mu\text{-}Me)\}_2(MeC_5H_4)_2]$[392] and $[\{Nd(\mu\text{-}Me)(Bu^tC_5H_4)_2\}_2]$[391] have been reported. In contrast to the pentamethylcyclopentadienyl analogues $[ScMe(Cp^*)_2]$ and $[\{LnMe(Cp^*)_2\}_2]$ (Ln = Y, Lu) (see below) the dimers $[\{Ln(\mu\text{-}Me)(Cp)_2\}_2]$ (Ln = Y, Lu) do not react with methane.

Related anionic complexes of the type $[LnR_2(Cp)_2]^-$ can be prepared by reacting $[\{Yb(\mu\text{-}Cl)(Cp)_2\}_2]$ with excess alkyllithium. A structurally characterized example is $[Li_2(DME)_2(dioxane)][Y(CH_2TMS)_2(Cp)_2]_2$, which contains isolated $[Y(CH_2TMS)_2(Cp)_2]^-$ ions.[393]

As a result of its high Lewis-acidity, the bimetallic scandium derivative reacts with donor ligands such as THF or pyridine to give the monomeric solvated complexes $[ScMe(S)(Cp)_2]$ and trimethylaluminum (Equation (87)).[390]

$$2\,[Sc(\mu\text{-}Me)_2AlMe_2Cp_2] + 2\,S \xrightarrow{\text{toluene}} 2\,[ScMe(S)Cp_2] + Al_2Me_6 \qquad (87)$$

$$S = THF,\ C_5H_5N$$

A number of bis(cyclopentadienyl)lanthanide alkynyl complexes have been reported. Compounds of the type $[\{Ln(\mu\text{-}C\equiv CPh)(Cp)_2\}_2]$ are normally prepared by metathetical reactions of $[LnCl(THF)(Cp)_2]$ with alkali metal derivatives of phenylacetylene (Equation (88)). An alternative preparation involves treatment of tris(cyclopentadienyl)lanthanides with free terminal alkynes (Equation (89)). In this case volatile cyclopentadiene is formed as the only by-product. Differently substituted organolanthanide alkynyl complexes have been prepared according to this route.[18,26]

$$2\,LnCl(THF)Cp_2 + 2\,MC\equiv CPh \longrightarrow [\{Ln(\mu\text{-}C\equiv CPh)Cp_2\}_2] + 2\,MCl \qquad (88)$$

$$Ln = Sc,\ Gd,\ Ho,\ Er,\ Yb$$
$$M = Li,\ Na,\ K$$

$$2\,LnCp_3 + 2\,HC\equiv CR \longrightarrow [\{Ln(\mu\text{-}C\equiv CR)Cp_2\}_2] + 2\,C_5H_6 \qquad (89)$$

$$Ln = Nd,\ Yb$$
$$R = Bu^n, C_6H_{13}^n,\ c\text{-}C_6H_{11}, Ph,\ CpFeC_5H_4$$

An alternative preparation involves C–H activation of terminal alkynes with lanthanide(II) metallocenes.[240] X-ray structural investigations of $[\{Er(\mu\text{-}C\equiv CBu^t)(Cp)_2\}_2]$,[18] $[\{Sm(\mu\text{-}C\equiv CBu^t)(MeC_5H_4)_2\}_2]$,[280] and $[\{Sm(\mu\text{-}C\equiv CPh)(Bu^tC_5H_4)_2\}_2]$[240] revealed the dimeric nature of these complexes with bridging alkynyl ligands. Monomeric alkynyls are $[LnC\equiv CPh(phen)(Cp)_2]$ (Ln = La, Nd; phen = 1,10-phenanthroline).[394]

Neutral and anionic phosphoylides have been occasionally shown to be versatile ligands in organo-f-element chemistry. Bis(cyclopentadienyl)lanthanide derivatives containing neutral phosphoylides as ligands have been prepared by displacement of THF by the ylide. Both halides and hydrocarbyls can be used as precursors in these ligand exchange reactions (Equations (90) and (91)).[18,395,396]

$$LuCl(THF)Cp_2 + Ph_3P=CH_2 \longrightarrow [LuCl(CH_2PPh_3)Cp_2] + THF \qquad (90)$$

$$[LuR^1(THF)Cp_2] + R^2_3P=CHR^3 \longrightarrow [LuR^1(CHR^3PR^2_3)Cp_2] + THF \qquad (91)$$

$$R^1 = Bu^t,\ R^2 = Ph,\ R^3 = H$$
$$R^1 = CH_2TMS,\ R^2 = Ph,\ R^3 = H$$
$$R^1 = Bu^t,\ R^2 = Me,\ R^3 = TMS$$

An interesting intramolecular metallation of a phenyl ring was observed when the phosphoylide adduct $[LuCl(CH_2PPh_3)(Cp)_2]$ was treated with strong bases such as MeLi or sodium hydride (Scheme 13).[18,394]

$$[LuCl(CH_2PPh_3)Cp_2] \xrightarrow[-\ LiCl]{+\ MeLi} [LuMe(CH_2PPh_3)Cp_2] \xrightarrow{-CH_4}$$

Scheme 13

The cyclometallated product can also be obtained directly by reacting $[LuCl(THF)(Cp)_2]$ with the phosphoylide $Ph_3P=CH_2$ in the presence of t-butyllithium.[18,396]

Especially useful for the stabilization of lanthanide–carbon bonds are chelating anionic phosphoylide complexes of the type $R_2P(CH_2)_2^-$. Several bis(cyclopentadienyl)lanthanide derivatives of these ligands have been reported. In contrast to the homoleptic lanthanide(III) phosphoylides $[Ln\{(CH_2)_2PR_2\}_3]$ (see above) the cyclopentadienyl complexes $[Ln\{(CH_2)_2PR_2\}(Cp)_2]$ are monomeric in benzene solution. Their preparation is easily achieved by reacting bis(cyclopentadienyl)lanthanide chlorides with lithiated phosphoylides (Equations (92) and (93)).[18,138,397]

$$ScClCp_2 + Li(CH_2)_2PPh_2 \xrightarrow{THF} [Sc\{(CH_2)_2PPh_2\}Cp_2] + LiCl \qquad (92)$$

$$[LnCl(THF)(C_5H_5)_2] + [Li(CH_2)_2PBu^t] \xrightarrow{THF} [Lu\{(CH_2)_2PBu^t_2\}(C_5H_5)_2] + LiCl \qquad (93)$$

Allyl complexes of the type $[Ln(allyl)(Cp)_2]$ have been described for scandium, neodymium, samarium, holmium, and erbium. These compounds are accessible by reacting the chloride precursors with $MgBrC_3H_5$.[18,398] NMR spectral studies showed the allyl ligand to be fluxional. The reaction of $[SmCl(THF)(Cp)_2]$ with sodium allyloxypropynylide ($NaC\equiv CCH_2OCH_2CH=CH_2$) in THF solution was reported to afford $[Sm(\eta^5\text{-}C_6H_5)(THF)(Cp)_2]$ by an unknown mechanism. According to the x-ray structural analysis the complex contains the novel highly strained anionic cyclohexen-4-yl ligand.[399]

Recently two examples of novel anionic bis(cyclopentadienyl)lanthanide hydrocarbyls have been reported. First, neodymium trichloride reacts with two equivalents of the lithium salt of 9-(cyclopentadienyl-1-methylethyl)-9,10-dihydroanthracene to give the "ate" complex $[Li(THF)_4][Nd\{(C_5H_4)CMe_2(C_{14}H_{10})\}_2]$ which has been characterized by an x-ray crystal structure determination. The central neodymium atom is in a pseudotetrahedral coordination environment. It is η^5-bonded to the two cyclopentadienyl rings and σ-bonded to the C-10 atoms of the two dihydroanthracene moieties (Figure 18).[400]

The second unusual anionic $Ln(Cp)_2$ hydrocarbyl complex is $[Na(diglyme)_2][Lu(C_{14}H_{10})(Cp)_2]$, which contains a coordinated dihydroanthracene dianion.[401,402] This complex was prepared by reacting $[LuCl(THF)(Cp)_2]$ with either radical-ionic or dianionic sodium anthracenides and isolated after recrystallization from diglyme. The molecular structure consists of isolated ions (Figure 19).

Heavier homologues of the bis(cyclopentadienyl)lanthanide hydrocarbyls (i.e., silyls, germyls, etc.) have been studied only in a very small number. The anionic silyl complexes $[Li(DME)_3]$-$[Ln(TMS)_2(Cp)_2]$ were prepared by Schumann *et al.* according to (Equation (94)) and the lutetium derivative was structurally characterized (Lu–Si 0.2888(2) nm).[369,402-5]

$$LnClCp_2 + 2\,Li\text{-}TMS \xrightarrow{DME} [Li(DME)_3][Ln(TMS)_2Cp_2] + LiCl \qquad (94)$$

$$Ln = Sm,\ Dy,\ Ho,\ Er,\ Tm,\ Lu$$

Even the higher group 14 homologues germanium and tin have been successfully incorporated in isolable organolanthanide complexes. These were produced by reactions of bis(cyclopentadienyl)-lanthanide chloride precursors with MPh_3^- (M = Ge, Sn) (Equation (95)).[18]

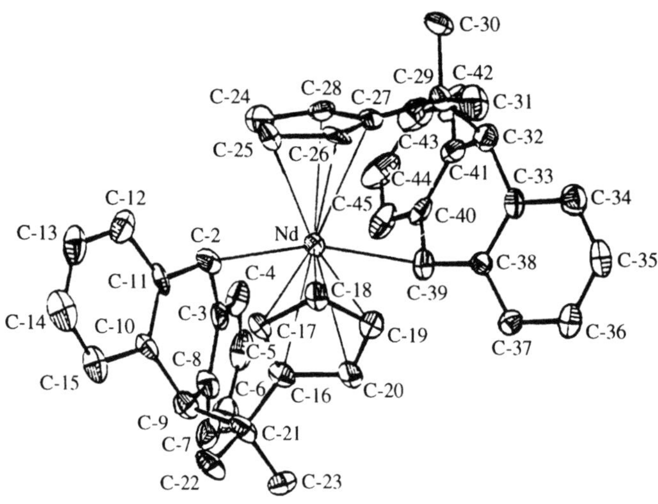

Figure 18   The molecular structure of the anion in $[Li(THF)_4][Nd\{(C_5H_4)CMe_2(C_{14}H_{10})\}_2]$.[400]

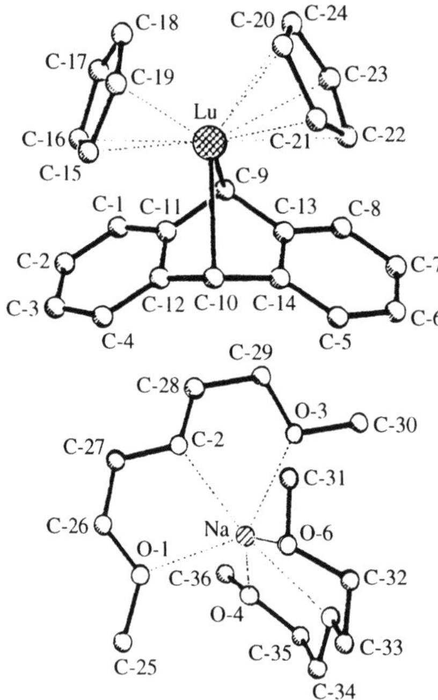

Figure 19   The molecular structure of $[Na(diglyme)_2][Lu(C_{14}H_{10})(Cp)_2]$.[402]

$$LnCl(THF)Cp_2 + LiMPh_3 \xrightarrow{\text{THF}} LnMPh_3Cp_2 + LiCl \qquad (95)$$

A similar synthetic route was adapted to prepare a series of monomeric bis(cyclopentadienyl)-scandium silyl and germyl complexes (Equation (96)).[406]

$$ScClCp_2 + LiER_3(THF)_3 \longrightarrow [Sc(ER_3)(THF)Cp_2] + LiCl + 2\,THF \qquad (96)$$

$$ER_3 = Si(TMS)_3,\ Si(TMS)_2Ph,\ Si(Bu^t)Ph_2,\ SiPh_3,\ Ge(TMS)_3$$

The molecular structure of $[Sc\{Si(TMS)_3\}(THF)(Cp)_2]$ was established by x-ray crystallography (Sc–Si 0.2863(2) nm). Like organolanthanide hydrocarbyls, the scandium silyls and germyls polymerize ethylene and undergo facile migratory insertion reactions with CO and xylyl isocyanide.[406]

Generally the chemical reactivity of bis(cyclopentadienyl)lanthanide hydrocarbyls is very high. They easily undergo hydrogenolysis to form the corresponding organolanthanide hydrides. With carbon monoxide and isonitriles, facile insertion reactions are observed. A typical example is the CO insertion into the Lu–C bond of $LuBu^t(Cp)_2$ (cf. Section 2.2.5.4). The initial product can be described by a resonance hybrid of an acyl complex and the mesomeric alkoxycarbene (Scheme 14).[18]

$$[LuBu^t(THF)Cp_2] \xrightarrow{\ +CO\ } Cp_2Lu-\underset{\underset{O}{\|}}{C}-Bu^t \longleftrightarrow Cp_2Lu-\underset{\underset{O}{\diagup}}{\overset{}{C}}-Bu^t$$

Scheme 14

(iii) Hydrides

Bis(cyclopentadienyl)lanthanide hydrides play an important role in organolanthanide chemistry as these complexes are highly reactive and offer a great potential in homogeneous catalysis. Theoretical studies on the bonding in hydride bridged $Ln(Cp)_2$ complexes have been carried out by Hoffmann *et al.*[407] Subsequently several of these hydrides have been synthesized and structurally characterized. Structural types include dimeric and trimeric complexes of the types $[\{Ln(\mu\text{-}H)(Cp)_2\}_2]$ and $[\{Ln(\mu\text{-}H)(Cp)_2\}_3]$, the trimer anions $[(\mu_3\text{-}H)\{Ln(\mu\text{-}H)(Cp)_2\}_3]^-$ as well as heterobimetallic hydrides, including tetrahydroborates and tetrahydroaluminates. These compounds will also be discussed in this section.

Two main synthetic routes leading to bis(cyclopentadienyl)lanthanide hydrides have been developed. One of them is hydrogenolysis of suitable $Ln(Cp)_2$ hydrocarbyl precursors[18,382] and the other involves metathetical reactions of the corresponding halide precursors with metal hydrides, including $Na[BH_4]$ and $Li[AlH_4]$. Hydride species have also been discussed as intermediates; for example, exchange reactions between $Nd(Cp)_3$ and Bu^nLi are thought to proceed via intermediate organoneodymium hydrides.[408]

The reaction of $Lu(Cp)_2$ alkyl and aryl complexes with H_2 produced the bis(cyclopentadienyl)-lutetium hydride $[\{Lu(\mu\text{-}H)(Cp)_2\}_2]$, while $[\{Lu(\mu\text{-}D)(Cp)_2\}_2]$ was formed when the hydrocarbyls were treated with D_2 (Equation (97)).

$$[LuRCp_2] + H_2/D_2 \longrightarrow [\{Lu(\mu\text{-}H/D)Cp_2\}_2] + RH/D \tag{97}$$

$$R = \text{alkyl, aryl}$$

THF solvates of the dimeric hydrides are obtained when the hydrogenolysis of bis(cyclopentadienyl)-lanthanide hydrocarbyls is carried out in the presence of THF (Equations (98) and (99)).[18,26,382,409,410]

$$2\,[LnR(THF)Cp_2] + 2\,H_2 \xrightarrow{\ \text{toluene}\ } [\{Ln(\mu\text{-}H)(THF)Cp_2\}_2] + 2\,RH \tag{98}$$

$$R = \text{Me, Bu}^t, \text{Ph; Ln} = \text{Y, Er, Lu}$$
$$R = \text{Bu}^t, \text{CH}_2\text{Bu}^t, \text{CH}_2\text{TMS; Ln} = \text{Lu}$$

$$2\,[LnR(THF)(MeC_5H_4)_2] + 2\,H_2 \xrightarrow{\ \text{toluene}\ } [\{Ln(\mu\text{-}H)(THF)(MeC_5H_4)_2\}_2] + 2\,RH \tag{99}$$

$$R = \text{Me, Bu}^t; \text{Ln} = \text{Y, Er, Lu}$$

The formation of bis(cyclopentadienyl)lanthanide hydrides is also observed when certain alkyls of the type $[LnR(THF)(Cp)_2]$ (Ln = Nd, Lu; R = Bu^t, CHMeEt) thermally decompose via β-hydride elimination.[384] The formation of cyclopentadienylneodymium hydride intermediates was postulated for the reaction of $Nd(Cp)_3$ with *n*-butyllithium in benzene, although no well-defined reaction products were isolated.[408]

The molecular and crystal structures of $[\{Y(\mu\text{-}H)(MeOCH_2CH_2C_5H_4)_2\}_2]$,[411] $[\{Ln(\mu\text{-}H)(THF)(MeC_5H_4)_2\}_2]$ (Ln = Y, Er),[382,392] and $[\{Lu(\mu\text{-}H)(THF)(Cp)_2\}_2]$,[412,413] have been determined. The complexes are dimerized via two hydride ligands and the coordination sphere around each lanthanide atom is completed by an additional THF ligand or the pendant methoxy function. In the absence of a coordinating solvent, the cyclotrimeric hydride $[\{Y(\mu\text{-}H)(MeC_5H_4)_2\}_3]$ has been isolated and structurally characterized. The compound was made by hydrogenolysis of $[\{Y(\mu\text{-}Me)(Cp)_2\}_2]$.[392]

Bulky cyclopentadienyl ligands such as *t*-butylcyclopentadienyl or 1,3-di-*t*-butylcyclopentadienyl allow the preparation of unsolvated dimeric organolanthanide hydrides. Structurally characterized examples are $[\{Lu(\mu\text{-}H)(Bu^tC_5H_4)_2\}_2]$[414] and $[\{Ln(\mu\text{-}H)(Bu^t_2C_5H_3)_2\}_2]$ (Ln = Ce, Sm).[414,415]

The hydride-bridged dimers have been found to be highly reactive. For example, $[\{Er(\mu\text{-}H)(THF)(MeC_5H_4)_2\}_2]$ reacts with Bu^tNC to give the insertion product $[\{Er(\mu\text{-}HC=NBu^t)(MeC_5H_4)_2\}_2]$.[416] Similar treatment of $[\{Ln(\mu\text{-}H)(Cp)_2\}_2]$ with Bu^tNC afforded the dimeric insertion products $[\{Ln(\mu\text{-}HC=NBu^t)(Cp)_2\}_2]$ (Ln = Y, Er) with bridging formimidoyl ligands. The structures of both complexes were determined by x-ray diffraction.[416,417]

[{Y(μ-H)(THF)(Cp)$_2$}$_2$] and [{Y(μ-H)(THF)(MeC$_5$H$_4$)$_2$}$_2$] have also been found to react with alkenes, alkynes, allene, nitriles, and pyridine to afford the new compounds [Y(Et)(THF)(RC$_5$H$_4$)$_2$], [Y(Pr^n)(THF)(RC$_5$H$_4$)$_2$], [Y(η^3-allyl)(THF)(RC$_5$H$_4$)$_2$], [{Y(μ-C≡CBu^{t}RC$_5$H$_4$)$_2$)}$_2$], [Y{C(R^2)=CHR^2}-(THF)(R^{1}C$_5$H$_4$)$_2$] (R^2 = Et, Ph), [{Y(μ-NCHBu^t)(RC$_5$H$_4$)$_2$}$_2$], [{Y(μ-H)(NC$_5$H$_5$)(RC$_5$H$_4$)$_2$}$_2$], and [Y(NC$_5$H$_6$)(NC$_5$H$_5$)(RC$_5$H$_4$)$_2$] (R = H, Me).[409] The dimeric molecular structure of [{Y(μ-NCHBu^t)(RC$_5$H$_4$)$_2$}$_2$] was determined by x-ray methods. Scheme 15 summarizes the reactivity of the bis(cyclopentadienyl)yttrium hydrides [{Y(μ-H)(THF)(RC$_5$H$_4$)$_2$}$_2$] (R = H, Me) towards various substrates.

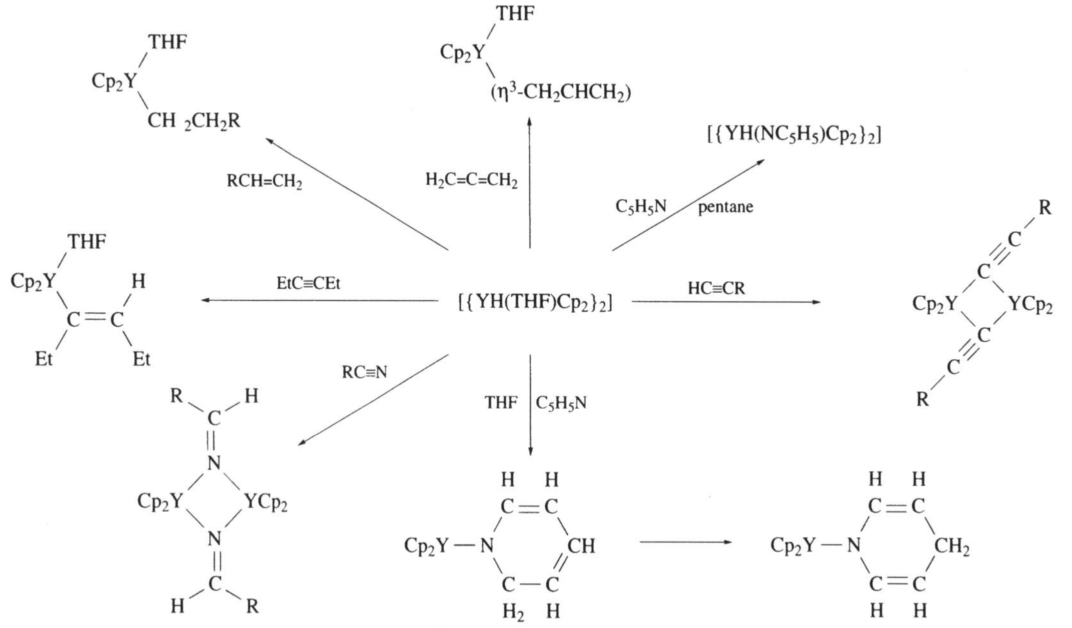

Scheme 15

[{Lu(μ-H)(THF)(Cp)$_2$}$_2$] has been reported to react with benzaldehyde to yield the alkoxide derivative [LuOBz(THF)(Cp)$_2$].[410]

A more complex though common type of bis(cyclopentadienyl)lanthanide hydrides contains a μ$_3$-hydride ligand in a trinuclear anion. Several reactions leading to these anionic hydride complexes have been reported. Their formation is frequently observed during decomposition of certain bis(cyclopentadienyl)lanthanide hydrocarbyls via β-elimination. This way the compound [Li(THF)$_3$][(μ$_3$-H){Lu(μ-H)(Cp)$_2$}$_3$] was synthesized.[418] The lutetium derivative [Na(THF)$_6$][(μ$_3$-H){Lu(μ-H(Cp)$_2$}$_3$] was made by reacting [LuCl(THF)(Cp)$_2$] with metallic sodium in THF.[412] An improved synthesis of [Li(THF)$_4$][(μ$_3$-H){Y(μ-H)(Cp)$_2$}$_3$] involves treatment of [{Y(μ-H)(THF)(Cp)$_2$}$_2$] with either Bu^{t}Li, MeLi, or LiH. In the case of Bu^{t}Li the product was obtained in 76% yield along with small amounts of [Y(Bu^t)(THF)(Cp)$_2$] and [Y(THF)(Cp)$_3$]. The perdeuterated analogue was made by using [{Y(μ-D)(THF)(Cp)$_2$}$_2$] as starting material.[419] One of the bridging hydrides in these trinuclear complexes can be selectively replaced by a chloride ligand. Figure 20 shows the molecular structure of the erbium derivative [Li(THF)$_4$][(μ$_3$-H)(μ-H)$_2$(μ-Cl){Er(Cp)$_2$}$_3$].[418] Various alkoxy hydride anions of the type [(μ$_3$-H){Y(μ-H)(Cp)$_2$}$_x${Y(μ-OMe)(Cp)$_2$}$_{3-x}$]$^-$ (x = 0, 1, 2) have been prepared by the reaction of [Li(THF)$_4$][(μ$_3$-H){Y(μ-H)(Cp)$_2$}$_3$] with methanol.[420] A neutral trinuclear hydride is the structurally characterized lutetium complex [{Lu(Cp)$_2$}$_3$(μ-H)(μ$_3$-H)].[421]

Borohydride complexes of the type [Ln(BH$_4$)(Cp)$_2$] have been prepared and studied by IR spectroscopy. These compounds are accessible by several synthetic routes (Equations (100)–(103)).[18,335,347,422-4]

$$[LnCl(THF)Cp_2] + Na[BH_4] \xrightarrow{\text{THF}} [Ln(BH_4)(THF)Cp_2] + NaCl \tag{100}$$

$$Ln = Sm, Er, Yb, Lu$$

The reaction of [{Y(μ-Cl)(Cp)$_2$}$_2$] with Na[BH$_4$] was solvent-dependent. In THF solution the THF solvate [Y(BH$_4$)(THF)(Cp)$_2$] was obtained, while the analogous reaction in DME afforded

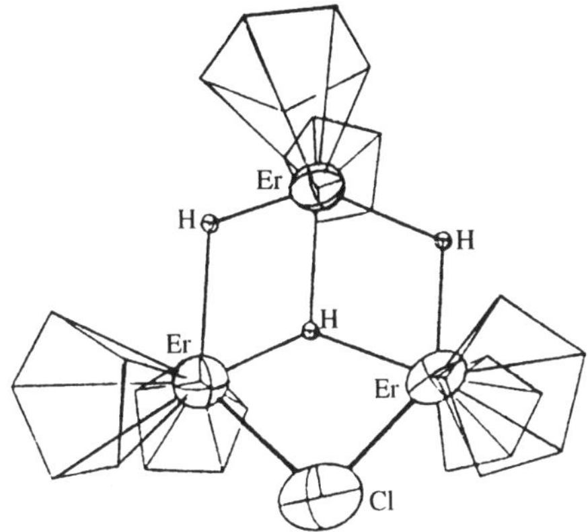

Figure 20　The molecular structure of $[Li(THF)_4][\{Er(Cp)_2\}_3(\mu_3\text{-}H)(\mu\text{-}H)_2(\mu\text{-}Cl)]$.[419]

$$[LnCl(THF)\{(TMS)_2C_5H_3\}_2] + Na[BH_4] \xrightarrow{\text{THF}} [Ln(BH_4)(THF)\{(TMS)_2C_5H_3\}_2] + NaCl \qquad (101)$$

$$Ln = Sc,\ Y,\ La,\ Pr,\ Nd,\ Sm,\ Yb$$

$$[LnCl(MeOCH_2CH_2C_5H_4)_2] + Na[BH_4] \xrightarrow{\text{THF}} [Ln(BH_4)(MeOCH_2CH_2C_5H_4)_2] + NaCl \qquad (102)$$

$$Ln = Y,\ Yb$$

$$[Sc(BH_4)_3(THF)_2] + 2\,Na(RC_5H_4) \xrightarrow{\text{THF}} [Sc(BH_4)(RC_5H_4)_2] + 2\,Na[BH_4] \qquad (103)$$

$$R = H,\ Me$$

$Na[Y(BH_4)_2(Cp)_2]$.[3] For $[Ln(BH_4)(THF)\{(TMS)_2C_5H_3\}_2]$ (Ln = Y, Yb) the IR data indicated a bidentate chelating coordination of the borohydride ligand. This was confirmed by an x-ray structural analysis of unsolvated $[Sc(BH_4)\{(TMS)_2C_5H_3\}_2]$.[422] According to the IR spectra, the coordination of the $[BH_4]^-$ ligand is tridentate in all other complexes of this type. Crystal structure determinations have also been reported for $[Ln(BH_4)(MeOCH_2CH_2C_5H_4)_2]$ (Ln = Y, Yb).[335]

Various reactions of bis(cyclopentadienyl)lanthanide chlorides with neutral and anionic aluminum hydrides have been reported.[414,425-7] Besides mononuclear compounds such as $[Y(AlH_4)(THF)(Cp)_2]$ and $[Y(AlH_4)(Et_2O)(Cp)_2]$,[428] several more complex heterobimetallic hydrides have been isolated and structurally characterized. For example, $[\{Y(\mu\text{-}Cl)(Cp)_2\}_2]$ reacts with AlH_3 in diethyl ether or triethylamine to give the bimetallic hydrido species $[YCl(Cp)_2]_2\cdot AlH_3\cdot Et_2O$ and $[YCl(Cp)_2\cdot AlH_3\cdot NEt_3]_2$, respectively.[429,430] The molecular structures of $[\{Y(\mu_3\text{-}H)(\mu\text{-}H)AlH_2(THF)(Cp)\}_2]$ and $[\{Y(Cp)_2\}_2(\mu_3\text{-}H)(\mu\text{-}H)_2AlH_2(\mu\text{-}H)_2AlH(OEt_2)]$[428] are depicted in Figures 21 and 22. Similar treatment of $LuCl(Cp)_2$ with $Li[AlH_4]$ and NEt_3 afforded the binuclear complex $[\{Lu(\mu\text{-}H)_2(Cp)_2AlH_2(NEt_3)\}_2]$, which upon dissociation forms the monomeric $[Lu(\mu\text{-}H)(Cp)_2AlH_3(NEt_3)]$. In the latter an $[AlH_4]^-$ ligand is coordinated to lutetium in an unusual monodentate fashion.[426]

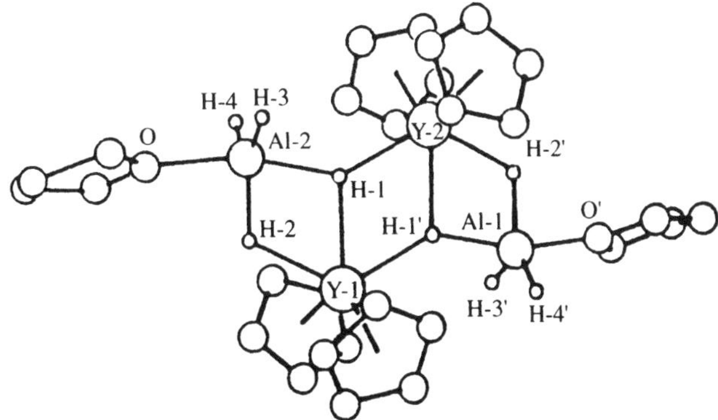

Figure 21　The molecular structure of $[\{Y(\mu_3\text{-}H)(\mu\text{-}H)(Cp)_2AlH_2(THF)\}_2]$.[428]

Several interesting bimetallic hydrides were made starting with a samarium(II) derivative of *t*-butylcyclopentadiene. The compound $[\{Sm(Bu^tC_5H_4)_2\}_2\{(\mu_3\text{-}H)(\mu\text{-}H)_2AlH(THF)\}_2]$ was obtained by

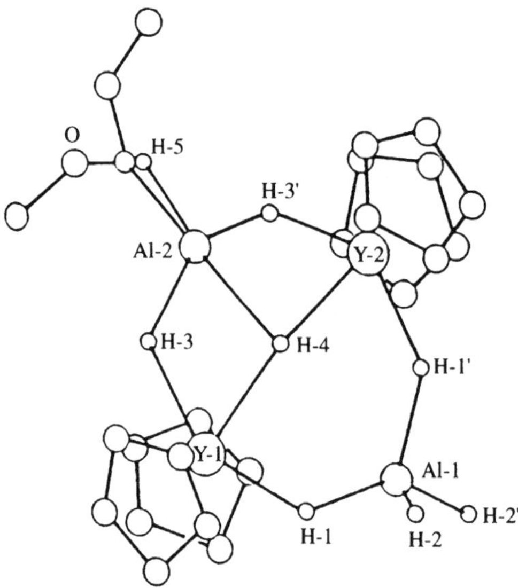

Figure 22   The molecular structure of $[\{Y(Cp)_2\}_2(\mu_3\text{-}H)(\mu\text{-}H)_2AlH_2(\mu\text{-}H)_2AlH(OEt_2)]$.[428]

reacting $\{NaSm(\eta^5:\eta^2\text{-}Bu^tC_5H_4)_3(THF)\}_n$ with $AlH_3\cdot THF$ and characterized by an x-ray crystal structure determination.[322] Lutetium forms the polynuclear aluminohydride $[LuH(AlH_4)(Bu^tC_5H_4)_2]_4(Et_2O)_2$.

Polymetallic bis(cyclopentadienyl)lanthanide hydrides form a large and particularly fascinating class of structurally very diverse compounds. Table 1 summarizes the structurally characterized examples.

Table 1   Structurally characterized polymetallic lanthanide hydrides.

Formula	Ref.
Binuclear complexes	
$[Ce\{\mu:\eta^4\text{-}(\eta^3\text{-}H)_2B(\mu\text{-}H)_2\}(Bu^t_2C_5H_3)_2]_2$ (Figure 23)	431
$[Y(\mu_3\text{-}H)(\mu\text{-}H)AlH_2(THF)(Cp)_2]_2$	432
$[Y(\mu_3\text{-}H)(\mu\text{-}H)AlH_2(NEt_3)(Cp)_2]_2$	428
$[Y(Cp)_2]_2(\mu_3\text{-}H)(\mu\text{-}H)_2AlH_2(\mu\text{-}H)_2AlH(OEt_2)$	428
$[Sm(\mu\text{-}BH_4)\{(TMS)_2C_5H_3\}_2]_2$	415
$[Sm(Bu^tC_5H_4)_2]_2\}(\mu_3\text{-}H)(\mu\text{-}H)_2AlH(THF)\}_2$	322
$[Sm(Bu^tC_5H_4)_2]_2(\mu\text{-}H)\mu\text{-}[(\mu_3\text{-}H)_2Al(\mu\text{-}H)_2(TMEDA)]$ (Figure 25)	421
$[Sm(Cp)_2(\mu_3\text{-}H)]_2[(\mu\text{-}H)AlH_2(NEt_3)]_2$	421
$[Yb(Cp)_2(\mu_3\text{-}H)]_2[(\mu\text{-}H)AlH_2(NEt_3)]_2\cdot C_6H_6$	422
$[Lu(\mu\text{-}H)(Cp)_2]_2[(\mu\text{-}H)AlH(NEt_3)]_2\cdot C_6H_6$ (Figure 24)	422
Tetranuclear complexes	
$Sm(Bu^t_2C_5H_3)[(\mu\text{-}H)_2(\mu_3\text{-}H)_2Al(TMEDA)]_2[Sm(Bu^t_2C_5H_3)H]_2[(\mu\text{-}H)_3Al(\mu\text{-}H)Al(\mu\text{-}H)_3]$- $[Sm(\mu_3\text{-}H)_2(Bu^t_2C_5H_3)_2]$	433
$[Lu(Bu^tC_5H_4)_2H(AlH_4)]_4(Et_2O)_2\cdot Et_2O$	434
$[Lu(Bu^t_2C_5H_3)_2H]_4[AlH_4(OEt_2)]_2[AlH_4]_2$	435

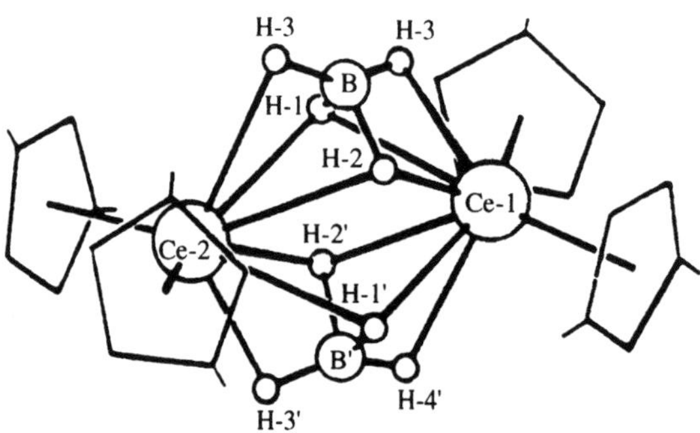

Figure 23   The molecular structure of $[Ce\{\mu:\eta^4\text{-}(\eta^3\text{-}H)_2B(\mu\text{-}H)_2\}(Bu^t_2C_5H_3)_2]_2$.[329]

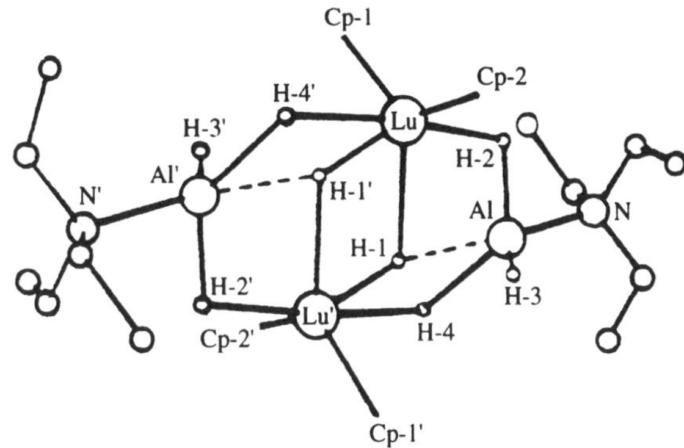

Figure 24　The molecular structure of [Lu(μ-H)(Cp)$_2$]$_2$[(μ-H)AlH(NEt$_3$)]$_2$·C$_6$H$_6$.[433]

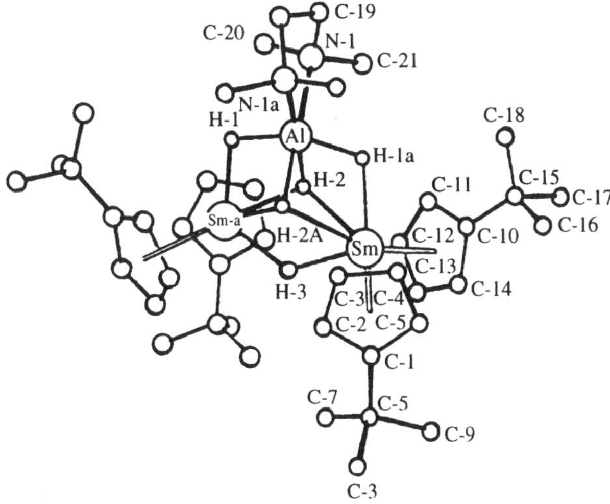

Figure 25　The molecular structure of [Sm(Bu^{t}C$_5$H$_4$)$_2$]$_2$(μ-H)μ-[(μ_3-H)$_2$Al(μ-H)$_2$(TMEDA)].[422]

2.2.5.5 *M(Cp)$_3$ compounds*

Homoleptic complexes of the type Ln(Cp)$_3$ were the first organolanthanide compounds ever made.[1] They have been synthesized for all lanthanide elements including promethium. Until now they represent the most thoroughly studied class of organolanthanide complexes. Classical synthetic routes to these compounds involve treatment of anhydrous lanthanide trichlorides with the stoichiometric amounts of cyclopentadienyl reagents such as NaCp or KCp.[18] Radioactive tris(cyclopentadienyl)promethium was first obtained in 1967 by neutron bombardment of Nd(Cp)$_3$ and subsequently characterized in the matrix of unreacted neodymium starting material. On a preparative scale the promethium derivative was made by treatment of PmCl$_3$ with Be(Cp)$_2$ or Mg(Cp)$_2$ in the melt (Equations (104) and (105)).[18]

$$\text{LnCl}_3 + 3\,\text{MCp} \xrightarrow{\text{THF}} \text{LnCp}_3 + 3\,\text{MCl} \tag{104}$$

$$\text{M} = \text{Na, K}$$
$$\text{Ln} = \text{Sc, Y, La, Ce, Pr, Nd, Sm, Gd, Tb, Dy, Ho, Er, Tm, Yb, Lu}$$

$$2\,\text{PmCl}_3 + 3\,\text{MCp}_2 \longrightarrow 2\,\text{PmCp}_3 + 3\,\text{MCl}_2 \tag{105}$$

$$\text{M} = \text{Be, Mg}$$

Deacon *et al.* have shown that thallium cyclopentadienides can also serve as useful cyclopentadienyl transfer reagents in organolanthanide chemistry. The compounds Ln(Cp)$_3$ (Ln = Er, Yb), [Ln(THF)(Cp)$_3$] (Ln = Ce, Nd, Sm, Gd, Er), [Ln(py)(Cp)$_3$] (py = pyridine) and Ln(MeC$_5$H$_4$)$_3$ have been prepared in a straightforward manner by reacting the lanthanide metals with thallium cyclopentadienides, Tl(RC$_5$H$_4$) (R = H, Me), in THF. This appears to be a general synthetic method for the preparation of a wide range of lanthanide tris(cyclopentadienyls).[220,222]

As mentioned above, the tris(cyclopentadienyl)lanthanides are probably the most intensively investigated group of organolanthanide complexes. Their physical properties, molecular structures, and chemical behavior are well documented. With the exception of the thermolabile europium derivative[18] all tris(cyclopentadienyl)lanthanides are thermally extremely stable though highly air-sensitive. As in other series of organolanthanide complexes the cerium derivative is the most air-sensitive species. The melting points of the $Ln(Cp)_3$ complexes fall in the range of 240–435 °C. The purest samples are normally isolated by vacuum sublimation. For several tris(cyclopentadienyl)lanthanides the vapor pressure, density, viscosity, and surface tension as well as their thermal decomposition have been studied.[436–8] Some indication of fixation of molecular nitrogen by $Ln(Cp)_3$ complexes was reported, although to a much lesser extent than that found for lanthanide(II) metallocenes, and no isolable dinitrogen adducts have been obtained.[439] However, in the presence of sodium naphthalide the complexes $Ln(Cp)_3$ (Ln = La, Ce, Pr, Nd, Sm, Eu) reduce N_2 to NH_3.[440] Other investigations include photoluminescence and chemiluminescence measurements[441–3] as well as mass spectrometry[444] and detailed ^{1}H-NMR spectral studies.[445] From spectroscopic studies a mostly ionic bonding between the lanthanide(3+) ions and the negatively charged cyclopentadienyl ligands was deduced.[159] This is in agreement with theoretical studies on the bonding in $Ln(Cp)_3$ and related complexes which suggest that the $4f$ and $6s$ orbitals do not play an important role in the bonding to the cyclopentadienyl ligands.[79–82,86,87,446] The electronic structures of tris(cyclopentadienyl)lanthanides and their Lewis base adducts have been thoroughly investigated by Amberger, Edelstein, and Kanellakopulos. MCD spectra and calculated crystal field splitting patterns have been reported for a large number of complexes.[447–58] Due to the strong tendency of the lanthanide ions to adopt high coordination numbers, the solid state structures of the tris(cyclopentadienyl)lanthanides do not consist of isolated $Ln(Cp)_3$ molecules (7).

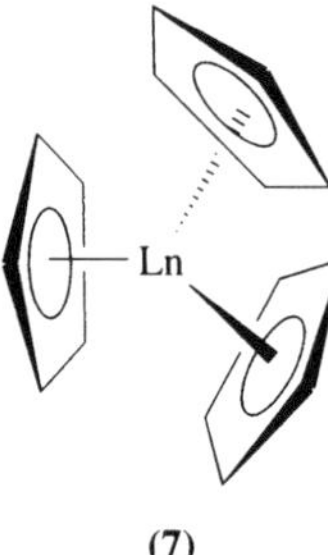

(7)

Despite their relatively high formal coordination number, $Ln(Cp)_3$ units are coordinatively unsaturated and associate to give oligomers or (more often) polymeric chains. Thus far, three different structural types have been found. Cyclopentadienyl ligands can bridge two lanthanide atoms through either one (μ-η^1) or two (μ-η^2) ring carbon atoms. μ-η^5:η^2-Bridging cyclopentadienyl ligands are found in the complexes of the largest lanthanide ions lanthanum(III) and praseodymium(III) (i.e., those which tend to adopt the highest coordination numbers). The molecular structures of $La(Cp)_3$ and $Pr(Cp)_3$ consist of polymeric zig-zag chains in which each lanthanide atom is surrounded by three η^5-coordinated cyclopentadienyl rings. One of these rings is involved in an additional η^2-coordination to a neighboring lanthanide atom. Figure 26 shows the molecular structure of the lanthanum complex as derived from an x-ray analysis.[459] The La–C distances in this complex vary between 0.2560 and 0.3034 nm. In this structural environment the formal coordination number around the metal atoms is 11.

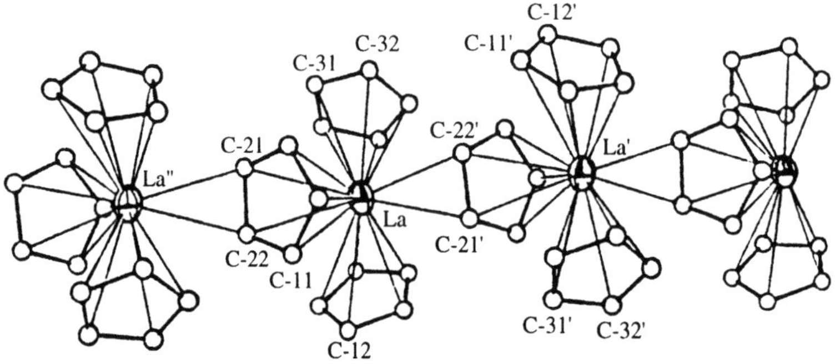

Figure 26   The molecular structure of unsolvated $La(Cp)_3$.[459]

η^1-Bridging was found in the tetrameric structures of La(MeC$_5$H$_4$)$_3$[460] and Nd(MeC$_5$H$_4$)$_3$[18] and in the polymeric chain structures of Y(Cp)$_3$,[461] Sm(Cp)$_3$ (two modifications),[18,462] Er(Cp)$_3$,[463] Tm(Cp)$_3$,[463] and Yb(Cp)$_3$.[464] The latter three complexes are isostructural. Here the formal coordination number around the central atoms is 10, reflecting the smaller ionic radii of these lanthanide ions. With M$\cdots$C distances to the bridging cyclopentadienyl rings of more than 0.314 nm, these interactions must be considered fairly weak. The molecular structure of Tm(Cp)$_3$ is shown in Figure 27.

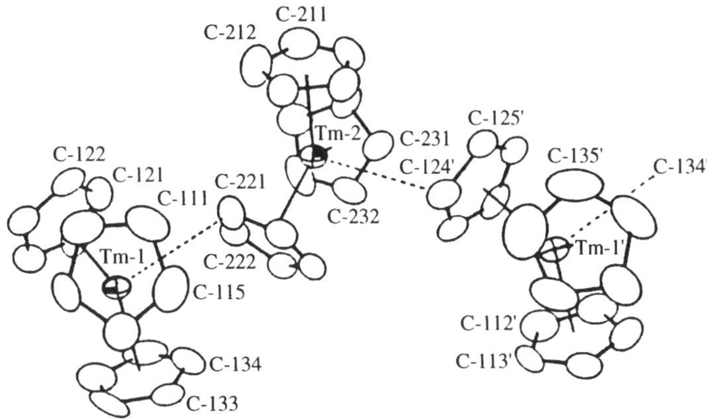

Figure 27 The molecular structure of unsolvated Tm(Cp)$_3$.[463]

In Sc(Cp)$_3$[18] and Lu(Cp)$_3$[465] the formal coordination number is reduced to 8. As a result of the lanthanide contraction these two complexes are isostructural. In the polymeric chain structures only two cyclopentadienyl rings are pentahapto coordinated, while the third one bridges two lanthanide atoms via two different carbon atoms of the ring. Thus, these compounds can be described as [{Ln(μ-η^5:η^2-Cp)(Cp)$_2$}$_n$]. The molecular structure of the lutetium derivative is depicted in Figure 28.

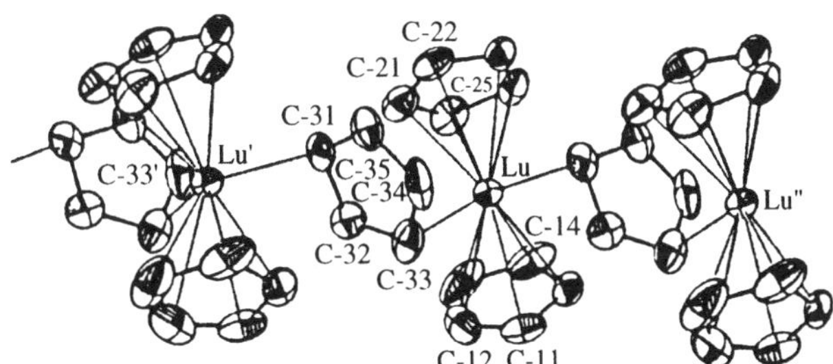

Figure 28 The molecular structure of unsolvated Lu(Cp)$_3$.[465]

Tris(methylcyclopentadienyl) complexes have been described for almost all elements of the lanthanide series. They are accessible through the usual synthetic route by treatment of anhydrous lanthanide trichlorides with three equivalents of sodium methylcyclopentadienide.[18] [Sm(MeC$_5$H$_4$)$_3$] has also been prepared by reacting [Sm(C$_6$F$_5$)$_2$] with cyclopentadiene. Similar reactions with europium and ytterbium produced the metallocene(II) derivatives.[231] [Sm(MeC$_5$H$_4$)$_3$] can be reduced electrochemically to give the anion [Sm(MeC$_5$H$_4$)$_3$]$^-$.[236] Several Ln(MeC$_5$H$_4$)$_3$ complexes have been investigated by x-ray crystallography. Tetrameric structures have been found for [{La(MeC$_5$H$_4$)$_3$}$_4$][468] and [{Ce(MeC$_5$H$_4$)$_3$}$_4$],[466] while [Yb(MeC$_5$H$_4$)$_3$] is monomeric in the solid state.[467,468] Homoleptic complexes of the type [Ln(Pr^{i}C$_5$H$_4$)$_3$] have been described for Ln = Ce, La, Pr, and Nd.[18,469] Complexes containing the *i*-propyl and *i*-butylcyclopentadienyl ligands have recently gained renewed interest as molecular precursors for lanthanide based materials.[238,338,470] Complexes of the type [Ln(Pr^{i}C$_5$H$_4$)$_3$] are among the most volatile in the tris(cyclopentadienyl)lanthanide series and thus can be used as precursors in MOCVD processes (cf. Section 2.2.15). They can be prepared conventionally be reacting the anhydrous metal trichlorides with the sodium salt of *i*-propylcyclopentadiene. Other examples of readily sublimable substituted tris(cyclopentadienyl)lanthanides are [Nd(Bu^{i}C$_5$H$_4$)$_3$],[338] structurally characterized [Nd(Bu^{i}CH$_2$C$_5$H$_4$)$_3$],[338] and [Ln(Bu^{i}C$_5$H$_4$)$_3$] (Ln = Nd, Dy, Tm).[341] [Nd(Bu^{i}C$_5$H$_4$)$_3$] is the most volatile complex in this series.[338]

Sterically crowded tris(cyclopentadienyl)lanthanide complexes containing bulky cyclopentadienyl ligands such as [Bu^{i}C$_5$H$_4$]$^-$, [(Bu^i)$_2$C$_5$H$_3$]$^-$, [TMS-C$_5$H$_4$]$^-$, or [1,3-(TMS)$_2$C$_5$H$_3$]$^-$ were not accessible via classical metathetical routes. This reflects the increased steric hindrance in the products. Only recently some sterically very crowded complexes of this type have been prepared. A number of cerium complexes containing these ligands have been investigated by Andersen et al.[367,466] The compounds [Ce(Bu^{t}C$_5$H$_4$)$_3$] (as well as [La(Bu^{t}C$_5$H$_4$)$_3$][239]) and [Ce(TMS-C$_5$H$_4$)$_3$] were prepared by reacting CeCl$_3$ with K(RC$_5$H$_4$) (R = Bu^t, TMS) and structurally characterized. Unsolvated [Ce(Bu^{t}C$_5$H$_4$)$_3$] is a tetramer in the solid state. Each cerium is coordinated to two terminal *t*-butylcyclopentadienyl ligands. In addition, two cerium atoms are bridged by a [Bu^{t}C$_5$H$_4$]$^-$ ligand in an η^5:η^1 fashion (Figure 29). The tetrameric unit is disrupted upon adduct formation, for example with Bu^{t}CN. The complex [Ce(Bu^{t}CN)(Bu^{t}C$_5$H$_4$)$_3$] was found to be a monomer in the solid state.

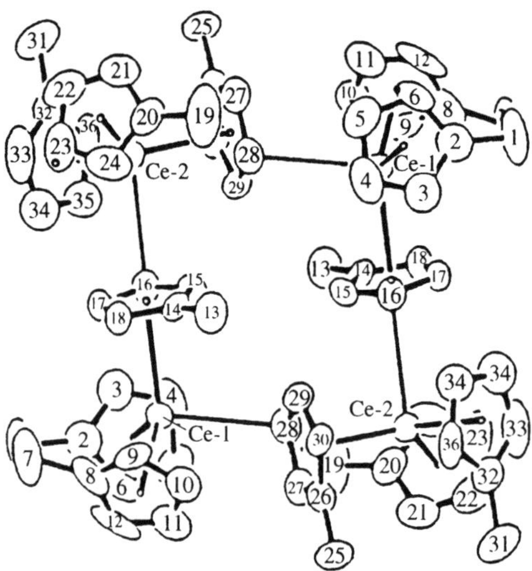

Figure 29   The molecular structure of [{Ce(Bu^{t}C$_5$H$_4$)$_3$}$_4$].[466]

The use of the even bulkier 1,3-bis(trimethylsilyl)cyclopentadienyl ligand allows the synthesis of unsolvated homoleptic lanthanide complexes which are monomeric in the solid state. The average cerium–carbon distance in [Ce{1,3-(TMS)$_2$C$_5$H$_3$}$_3$] is 0.283(4) nm. This compound is not available via a classical metathesis route by reacting cerium trichloride with the cyclopentadienyl–alkali metal reagent. Instead, the complex was prepared by protolysis of [Ce{N(TMS)$_2$}$_3$] with 1,3-bis(trimethylsilyl)cyclopentadiene (Equation (106)).[367,466]

$$[Ce\{N(TMS)_2\}_3] + 3\ 1{,}3\text{-}(TMS)_2C_5H_4 \longrightarrow [Ce\{1{,}3\text{-}(TMS)_2C_5H_3\}_3] + 3\ HN(TMS)_2 \qquad (106)$$

Evans et al. obtained the homoleptic samarium derivative [Sm{1,3-(TMS)$_2$C$_5$H$_3$}$_3$] by two different synthetic routes.[243] The complex was prepared either by the reaction of samarium trichloride with three equivalents of K{(TMS)$_2$C$_5$H$_3$} or via treatment of [Sm{(TMS)$_2$C$_5$H$_3$}$_2$] with carbon monoxide under pressure (620 kPa) and the molecular structure was determined by x-ray diffraction. Large differences in the Sm–C distances (av. 0.276(4) nm) are due to the fact that the carbon atoms bonded to silicons are tilted away from the central samarium atom.

Pentakis(methoxycarbonyl)cyclopentadiene is an easily accessible and highly acidic cyclopentadiene derivative. There is only one report dealing with lanthanide complexes containing this ligand. Treatment of lanthanide carbonates with C$_5$(CO$_2$Me)$_5$H afforded the trisubstituted complexes [Ln{C$_5$(CO$_2$Me)$_5$}$_3$], which surprisingly turned out to be air-stable and soluble in water. Although no structural information is available on these interesting materials, IR and ^{1}H NMR studies as well as conductivity measurements indicated some Ln–O coordination. This is of course not surprising in view of the highly oxophilic nature of the rare-earth elements.[471]

Recently a rare example of a homoleptic lanthanide(III) complex containing fused-ring bicyclic cyclopentadienyl ligands has been reported. The neodymium compound [Nd{MeC$_5$H$_2$(CH$_2$)$_5$}$_3$] (**8**) was prepared and structurally characterized.[333]

Another novel aspect is the preparation of "metalloligands" by using phosphino-substituted cyclopentadienyls. The first example in the tris(cyclopentadienyl)lanthanide series was [La(C$_5$H$_4$PPh$_2$)$_2$(THF)(Cp)], which was obtained via two different routes and structurally characterized (Scheme 16). The compound may find use as a potentially bidentate ligand for transition metals.[472]

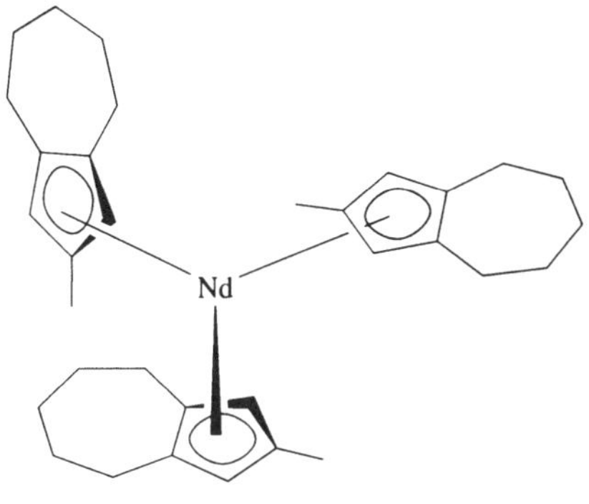

(8)

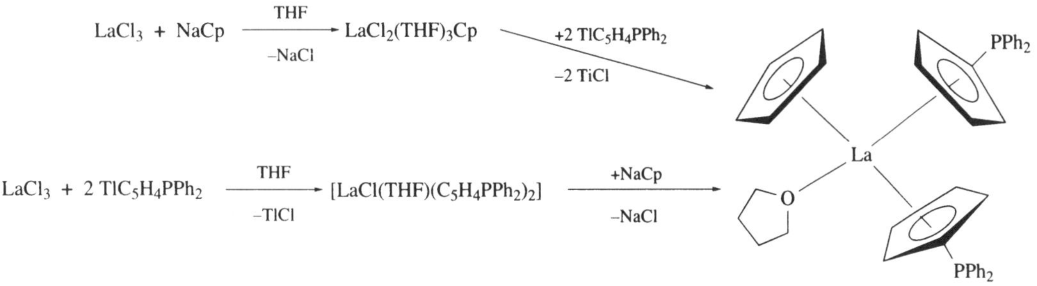

Scheme 16

Finally, a number of tris(cyclopentadienyl)lanthanide complexes containing "pendant-arm" cyclopentadienyl ligands have been prepared and characterized. The complex $[Pr(C_5H_4CH_2CH_2OMe)_3]$ has been structurally characterized by x-ray diffraction.[473] The complexes $[Nd(C_5H_4CH_2CH_2OMe)_3]$ and $[Pr(C_5H_4CH_2CH_2NMe_2)_3]$ have been found to be quite volatile and can be readily isolated by sublimation.[338] A one-pot reaction of $LnCl_3$ with two equivalents of $Na(Cp)$ and one equivalent of $[Na(C_5H_4CH_2CH_2OMe)]$ afforded the mixed tris(cyclopentadienyl)lanthanide complexes $[Ln(C_5H_4CH_2CH_2OMe)(Cp)_2]$ (Ln = Y, Lu, Sm, Gd, Er, Yb). In the structurally characterized yttrium derivative the three ring centroids and the ether oxygen atom from a distorted tetrahedron around yttrium.[474]

2.2.5.6 *[M(Cp)₃L] compounds*

Monomeric unsolvated tris(cyclopentadienyl)lanthanide complexes are coordinatively unsaturated. In the absence of solvents this leads to the formation of polymeric structures in the solid state. In these polymers bridging cyclopentadienyl ligands increase the coordination number around the metal atoms. The $\mu\text{-}\eta^1$ or $\mu\text{-}\eta^2$-cyclopentadienyl bridges are easily cleaved upon addition of coordinating solvents. 1:1-Adducts of the tris(cyclopentadienyl)lanthanide complexes with various donor ligands form a large and well-investigated class of organolanthanide compounds. The most prominent examples are the THF adducts.[18,475] They are generally obtained as the initial products by reacting $LnCl_3$ with three equivalents of $Na(Cp)$ in THF solution. Another generally applicable method is the reaction of lanthanide metals with $Tl(Cp)$ in THF (cf. Section 2.2.5.5).[476] An alternative preparation of $[Ce(THF)(Cp)_3]$ involves treatment of $[NH_4][Ce(NO_3)_6]$ with sodium cyclopentadienide.[290,298] Other suitable ligands can be coordinated to the metal center through oxygen, nitrogen, carbon, or phosphorus. Typical examples include lithium halides,[477] triethyl phosphate, ammonia,[18] THF-d₈,[478] furan,[479] pyridine,[18,220] acetonitrile, propionitrile,[220,480–2] isonitriles,[18] tetramethylurea,[483] phosphoylides,[484] carboxylic acid esters,[485–7] and phosphines.[18] The adducts $[Ln(MeCN)(Cp)_3]$ (Ln = Nd, Sm, Yb) and $[Ln(C_5H_5N)(Cp)_3]$ can be prepared directly by carrying out the transmetallation with $Tl(Cp)$ in acetonitrile or pyridine.[220] The Lewis acidity of unsolvated tris(cyclopentadienyl)lanthanide complexes is so strong that they may even react with transition metal carbonyl or nitrosyl derivatives to form isocarbonyl-linked species. Dipole moments and dielectric constants have been measured for various complexes of this type.[482,488] Numerous studies have appeared on the electronic structure[447–58,478,485,489,490] and magnetochemical properties[489] of tris(cyclopentadienyl)lanthanide adducts. The crystallizing properties of 150 adducts derived from $Ln(Cp)_3$ and $Ln(MeC_5H_4)_3$ have been examined.[485]

Especially the molecular structures of the THF adducts [Ln(THF)(Cp)$_3$] have been extensively studied. In all adducts [LnL(Cp)$_3$] the three cyclopentadienyl ligands are η^5-coordinated and the coordination geometry of the adducts is pseudotetrahedral. This structure was shown to exist for the whole series of lanthanide elements including yttrium and lanthanum. All compounds of the type [Ln(THF)(Cp)$_3$] are isostructural (Ln = Y,[491] La,[491] Ce,[492] Pr,[493] Nd,[493] Sm,[494,495] Gd,[496] Dy,[488,495,497] Er,[492] Lu[498]). Other structurally characterized adducts are [Lu(Et$_2$O)(Cp)$_3$],[499] [Ln(MeCN)(Cp)$_3$],[485,500] [Sm(CD$_3$CN)(Cp)$_3$],[501] [Ln(EtCN)(Cp)$_3$] (Ln = La, Pr, Yb),[481,502] [Ln(C$_5$H$_5$N)(Cp)$_3$] (Ln = Sm, Nd),[499] [Ce(Bu^{i}CN)(MeC$_5$H$_4$)$_3$],[466] [Ce(L)(Cp)$_3$] (L = quinuclidine or P(OCH$_2$)$_3$CEt),[503] [Ln(TMU)(Cp)$_3$] (Ln = Ce, Nd; TMU = tetramethylurea),[483] [Ln(butyl acetate)(Cp)$_3$] (Ln = La, Pr),[485,486,485] and [Ce(PMe$_3$)(MeC$_5$H$_4$)$_3$] (Ce–C 0.282(4) nm, Ce–P 0.3072(4) nm).[504] Homoleptic tris(cyclopentadienyl)-lanthanide complexes containing bulky, substituted cyclopentadienyl ligands form adducts with donor ligands as well, as in [La(THF)(Bu^{t}C$_5$H$_4$)$_3$][239] and [Nd(μ-Br)Li(THF)$_3$(Bu^{t}C$_5$H$_4$)$_3$].[477] The tetrameric structure of [{CeBu^{t}C$_5$H$_4$)$_3$}$_4$] is disrupted upon addition of pivalonitrile to give the structurally characterized monomeric adduct [Ce(Bu^{t}CN)(Bu^{t}C$_5$H$_4$)$_3$].[367,466] In this compound the average Ce–C distance has been found to be 0.279(3) nm.

The complex [Ce{1,3-(TMS)$_2$C$_5$H$_3$}$_3$], which is already a monomer in the solid state, also forms a stable 1:1 adduct with Bu^{t}NC (Figure 30).[367,466] Due to the more severe steric hindrance the average Ce–C distance in this compound (0.283(4) nm) is slightly longer than that in the *t*-butylcyclopentadienyl derivative [Ce(Bu^{t}NC)(Bu^{t}C$_5$H$_4$)$_3$].

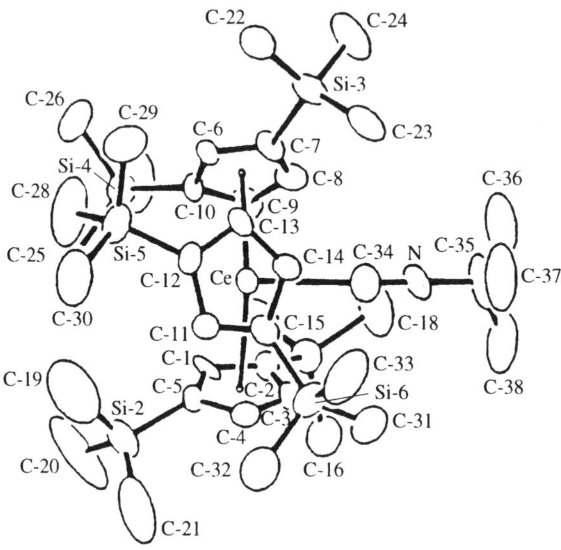

Figure 30   The molecular structure of [Ce(Bu^{t}NC){1,3-(TMS)$_2$C$_5$H$_3$}$_3$].[367,466]

Perhaps the most unusual adducts of tris(cyclopentadienyl)lanthanides are the aquo complexes [Y(H$_2$O)(Cp)$_3$] and [Ho(H$_2$O)(MeC$_5$H$_4$)$_3$], which are the first stable water adducts of organo-*f*-element compounds. These materials were obtained in small quantities by partial hydrolysis of [Y{O(CH$_2$CH$_2$C$_5$H$_4$)$_2$}(Cp)] or [Ho{O(CH$_2$CH$_2$C$_5$H$_4$)$_2$}(MeC$_5$H$_4$)], respectively. The holmium derivative was crystallographically characterized (Ho–O 0.231(2) nm). The existence of these two complexes is quite remarkable in view of the extreme hydrolytic instability of organolanthanides which is generally encountered.[505]

Several adducts of tris(cyclopentadienyl)lanthanum have been studied by ^{139}La NMR spectroscopy.[506] Among the compounds investigated are [La(DMF)(Cp)$_3$], [La(DMSO)(Cp)$_3$], [La{OP(OMe)$_3$}(Cp)$_3$] and [La(OCMe$_2$)(Cp)$_3$].[507]

A small number of anionic adducts of the tris(cyclopentadienyl)lanthanides has also been reported. The reaction of Lu(Cp)$_3$ with NaH or NaD in THF gave the anionic hydride species [Na(THF)$_6$][(Cp)$_3$Lu(μ-H)Lu(Cp)$_3$] or its deuterium analogue, respectively (Equation (107)).[418]

$$2\ [\text{LuCp}_3] + \text{NaH/NaD} \xrightarrow{\text{THF}} [\text{Na(THF)}_6][\text{Cp}_3\text{Lu}(\mu\text{-H/D})\text{LuCp}_3] \tag{107}$$

The x-ray structure determination of [Na(THF)$_6$][(Cp)$_3$Lu(μ-H)Lu(Cp)$_3$] showed a single hydride ligand bridging the two lutetium atoms. Related anionic complexes are the structurally characterized salts [{Li(DME)$_3$}{(Cp)$_3$Sm(μ-Cl)}{Sm(Cp)$_3$}][369] and [{Li(DME)$_3$}{(Cp)$_3$Sm(μ-N$_3$)Sm(Cp)$_3$}] (Figure 31),[508] which were prepared by reacting [Sm(Cp)$_3$] with Cl$^-$ or N$_3^-$ in DME solution.

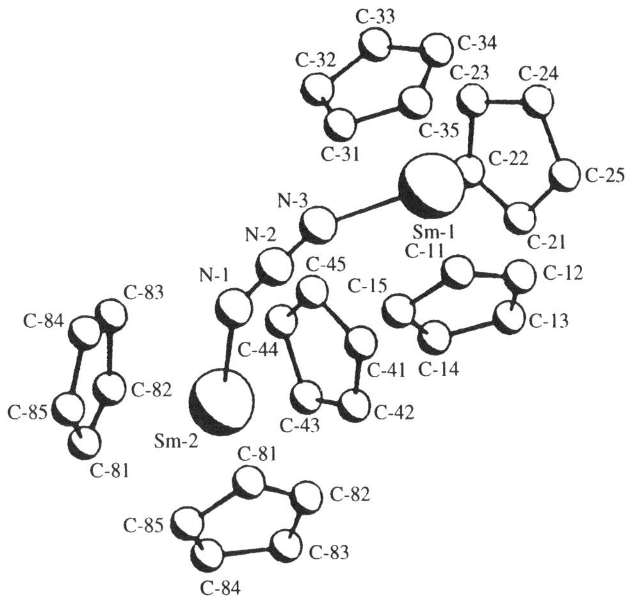

Figure 31 The molecular structure of the anion of $[\text{Li(DME)}_3][\text{Cp}_3\text{Sm}(\mu\text{-N}_3)\text{Sm(Cp)}_3]$.[508]

A rare example of an anionic hydrocarbyl adduct is $[\text{Li(DME)}_3][\text{NdPh(Cp)}_3]$, which was made from Nd(Cp)_3 and phenyllithium in DME solution.[509] The crystallographically determined structure consists of separated ion pairs with the phenyl group σ-bonded to neodymium.[510]

2.2.5.7 [M(Cp)₃L₂] compounds

The first trigonal bipyramidal adducts of tris(cyclopentadienyl)lanthanides with two neutral nitrogen donor ligands were described by Fischer and co-workers.[511] The complexes $[\text{Ln(MeCN)}_2(\text{Cp})_3]$ (Ln = La, Ce, Pr) and $[\text{La(EtCN)}_2(\text{Cp})_3]$[480] were prepared according to Equations (108)–(110). Obviously only the largest lanthanide ions leave enough room to accommodate two additional ligands in their coordination sphere.

$$\text{LnCp}_3 \ + \ \text{excess RCN} \ \xrightarrow{\text{THF}} \ [\text{Ln(RCN)}_2\text{Cp}_3] \qquad (108)$$

$$[\text{Ln(THF)Cp}_3] \ + \ \text{excess RCN} \ \xrightarrow{\text{THF}} \ [\text{Ln(RCN)}_2\text{Cp}_3] \qquad (109)$$

$$\text{Ln(RCN)Cp}_3 \ + \ \text{excess RCN} \ \xrightarrow{\text{pentane}} \ [\text{Ln(RCN)}_2\text{Cp}_3] \qquad (110)$$

In some cases the formation of mixtures of 1:1 and 1:2 adducts was observed. The isolation of pure 1:2 adducts usually requires that the neutral ligand is used in large excess. This can be best achieved, for example, by recrystallization of the starting Ln(Cp)_3 complexes from neat acetonitrile or propionitrile. X-ray analyses showed these compounds to be isostructural. The structurally characterized compounds include $[\text{Ln(MeCN)}_2(\text{Cp})_3]$ (Ln = La, Ce, Pr)[511] and $[\text{La(EtCN)}_2(\text{Cp})_3]$.[26,480,481] In the formally eleven-coordinate complexes the nitriles are in the axial position while the three cyclopentadienyl rings occupy the equatorial sites. Isocyanides are also capable of forming trigonal bipyramidal 1:2 adducts with tris(cyclopentadienyl)lanthanides. The complex $[\text{La(CN-}c\text{-C}_6\text{H}_{11})(\text{Cp})_3]$ was prepared from La(Cp)_3 and cyclohexyl isocyanide in toluene and characterized by ^{139}La NMR spectroscopy.[512] In addition, the electronic structure of $[\text{LnL}_2(\text{Cp})_3]$ complexes has been the subject of thorough investigations.[513]

2.2.5.8 [M(Cp)₃X] compounds

In the lanthanide series the oxidation state +4 is remarkably stable in the case of cerium, because cerium(IV) has the electronic configuration of xenon. Thus, it can be anticipated that cerium is better suited to form stable organolanthanide(IV) compounds than all other rare-earth elements. However, cerium(IV) is a strong oxidizing agent, and numerous early attempts to generate stable cerium(IV) organometallics have failed. In particular, reactions of $[pyH]_2[CeCl_6]$ (pyH = pyridinium cation) with sodium cyclopentadienide and related reagents invariably led to formation of cerium(III) cyclopentadienyl complexes. Early reports about the syntheses of tetrakis(cyclopentadienyl)-cerium(IV)[514] and tetrakis(indenyl)cerium(IV)[206] have subsequently been refuted.[515] Careful repetition of the original preparations revealed that $[Ce(THF)(Cp)_3]$ and $[Ce(THF)(C_9H_7)_3]$ were the only isolable products when $[pyH]_2[CeCl_6]$ was treated in THF solution with four equivalents of sodium cyclopentadienide or sodium indenide, respectively. There is little or no doubt now that virtually all previously reported complexes of the type $[CeX(Cp)_3]$ and $[CeX_2(C_9H_7)_2]$ (X = H, alkyl, aryl, BH_4, Cl, NH_2, N_3, CN, NCO, NCS, NO_2, NO_3, OR, SR, O_2CR) simply do not exist. All these materials are in fact cerium(III) species.[18]

Thus far only two examples of authentic cyclopentadienyl cerium(IV) complexes have been reported (Equations (111) and (112)).[516,517]

$$Ce(OPr^i)_4 + 3\,SnMe_3Cp \longrightarrow [Ce(OPr^i)Cp_3] + 3\,SnOPr^iMe_3 \tag{111}$$

$$[Ce(OBu^t)(NO_3)_3(THF)] + 3\,NaCp \longrightarrow [Ce(OBu^t)Cp_3] + 3\,NaNO_3 \tag{112}$$

Both organocerium(IV) alkoxides are isolate as black, crystalline solids. Theoretical calculations and PE spectra provided solid evidence for the presence of true organocerium(IV) compounds. $[Ce(OPr^i)(Cp)_3]$ is a relatively strong oxidizing agent as shown by cyclic voltammetry measurements. The compound shows a reversible one-electron reduction step.[516] The molecular structure of $[Ce(OBu^t)(Cp)_3]$ is shown in Figure 32 (Ce–C 0.276(2) nm).

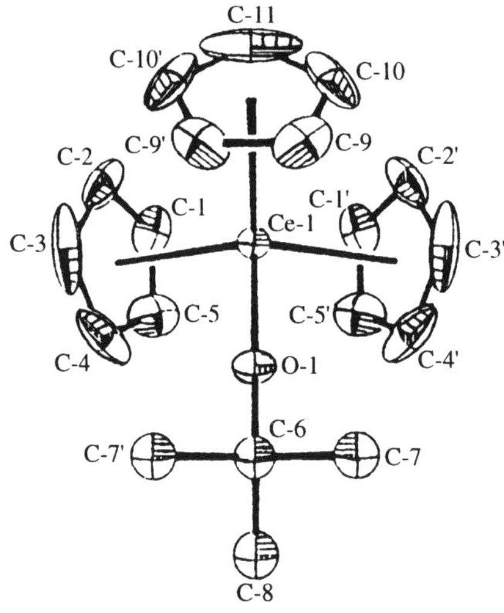

Figure 32   The molecular structure of $[Ce(OBu^t)(Cp)_3]$.[517]

2.2.5.9 *[M(Cp)$_4$]$^-$ compounds*

Anionic tetrakis(cyclopentadienyl)lanthanate complexes of the type $[Na(THF)_n][Ln(Cp)_4]$ (Ln = La, Ce, Nd, Pr; n = 0, 1) have been prepared by Thiele *et al.* $[Na(THF)_n][Ln(Cp)_4]$ and $Na[Pr(Cp)_4]$ were made by reacting the anhydrous lanthanide trichlorides with four equivalents of Na(Cp). $[Na(THF)][Ln(Cp)_4]$ (Ln = Ce, Nd) have been prepared analogously using either $[NH_4]_2[Ce(NO_3)_6]$ or $[(NH_4)_2][Nd(NO_3)_5]$ as precursors. According to their spectroscopic data these compounds contain three pentahapto cyclopentadienyl and one $[\eta^1\text{-}Cp]^-$ ligand.[290,518]

The cerium compound $[Na(THF)][Ce(Cp)_4]$ has been investigated as a cyclopentadienyl transfer reagent (Equation (113)). Treatment of this salt with the tribenzylmetal chlorides of tin and titanium results in the formation of $[M(Bz)_3Cp]$ (M = Sn, Ti).[519]

$$[Na(THF)][CeCp_4] + RCl \xrightarrow{THF} CeCp_3 + RCp + NaCl \qquad (113)$$

$$R = MeC(O), Bz, CPh_3, N[C(O)CH_2]_2$$

2.2.6 Modified Cyclopentadienyl Ligands

2.2.6.1 Ring-bridged cyclopentadienyls

Ring-bridged bifunctional cyclopentadienyl ligands play an important role in the preparation of disubstituted lanthanide derivatives. It was found that complexes containing a $LnCl(Cp)_2$ unit are not stable for the early lanthanide elements lanthanum to neodymium. This problem can be circumvented with the use of bridged cyclopentadienyl ligands. Several ligand systems have been developed, in which the cyclopentadienyl rings are connected through simple hydrocarbon chains. More sophisticated ligands have donor atoms such as oxygen or nitrogen incorporated in the bridge. These allow the isolation of unsolvated bis(cyclopentadienyl)lanthanide halide derivatives. The first ligand of this type to be employed in organolanthanide chemistry was the dianion of 1,1'-trimethylene-bis(cyclopentadiene), $^-C_5H_4(CH_2)_3C_5H_4{}^-$. The disodium salt of this ligand was reacted with anhydrous lanthanide trichlorides to give exclusively disubstitution products of the type $[LnCl(THF)\{C_5H_4(CH_2)_3C_5H_4\}]$ (Equation (114)). The formation of stable products with the lighter lanthanide elements was in marked contrast to the corresponding reaction with NaCp. The coordinated THF molecules can be replaced by other ligands such as 2,2'-bipyridine.[520-2]

$$LnCl_3 + NaC_5H_4(CH_2)_3C_5H_4Na \xrightarrow{THF} [LnCl(THF)\{C_5H_4(CH_2)_3C_5H_4\}] + 2\,NaCl \qquad (114)$$

$$Ln = La, Ce, Pr, Nd, Gd, Dy, Ho, Er, Yb, Lu$$

Further treatment of the halide precursors with alkyllithium or aryllithium afforded a series of new lanthanide hydrocarbyls containing the ring-bridged cyclopentadienyl ligands. Hydrogenolysis afforded the dimeric hydrides (Scheme 17).[521-3]

$$[LnCl(THF)\{C_5H_4(CH_2)_3C_5H_4\}] + RLi \longrightarrow [Ln(R)(THF)\{C_5H_4(CH_2)_3C_5H_4\}] + LiCl \xrightarrow{H_2}$$

$$Ln = Y, La, Pr, Dy, Er, Lu$$

$$R = Bu^t, CH_2Bu^t, Ph, p\text{-}C_6H_4Me$$

$$[\{Ln(\mu\text{-}H)(THF)\{C_5H_4(CH_2)_3C_5H_4\}\}_2] + RH$$

$$Ln = Y, Dy, Er, Lu$$

Scheme 17

Similar reaction sequences have been reported for the corresponding 1,1'-pentamethylene-bis(cyclopentadienyl) complexes, for example $[LnCl(THF)\{C_5H_4(CH_2)_5C_5H_4\}]$ (Ln = Y, Sm, Gd, Dy, Er, Lu).[524] A related series of complexes, $[LnCl(THF)\{C_5H_4CH_2C_6H_4CH_2C_5H_4\}]$ (Ln = La, Pr, Nd, Dy, Er, Yb) has been prepared by using the xylylene-bridged bis(cyclopentadienyl) ligand.[525]

The trimethylene-bridged bis(cyclopentadienyl) ligand has also been successfully employed in the preparation of lanthanide(II) *ansa*-metallocenes. Treatment of $LnCl_2$ with the disodium salt of the ligand in THF afforded the solvated species $[Ln(THF)_2\{C_5H_4(CH_2)_3C_5H_4\}]$ (Ln = Sm, Yb). The ytterbium(II) derivative was structurally characterized (Yb–C 0.267(2)–0.272(2) nm, ring centroid–Yb–ring centroid 127(1)°).[526]

Several complexes containing the tetramethylethylene-bridged dicyclopentadienyl ligand have been reported. Direct reaction of activated ($HgCl_2$) lanthanide metal powders with 6,6-dimethylfulvene in THF solution afforded the tetramethylethylene bridged lanthanide(II) metallocenes $[Ln\{Me_4C_2(C_5H_4)_2\}]$ (Ln = Sm, Yb; Equation (115)).[527]

The *ansa*-samarocene derivative $[SmMe_4C_2(C_5H_3Bu^t)_2]$ was prepared analogously from 6,6-dimethyl-3-t-butylfulvene.[528] The lanthanide(II) complexes $[Ln\{Me_4C_2(C_5H_4)_2\}]$ (Ln = Sm, Yb) served as precursors for ring-bridged lanthanide(III) derivatives with various sulfur and selenium ligands, for example $[Ln(SeR)\{Me_4C_2(C_5H_4)_2\}]$, $[Ln(S_2CNMe_2)\{Me_4C_2(C_5H_4)_2\}]$, and $[Ln\{S_2P(OMe)_2\}\{Me_4C_2(C_5H_4)_2\}]$ (Ln = Sm, Yb).[528] The synthesis and structural characterization of $[\{Sm(\mu\text{-}Cl)(THF)\{Me_4C_2(C_5H_4)_2\}\}_2]$ $[Mg_2Cl_3(THF)_6][YbCl_2\{Me_4C_2(C_5H_4)_2\}]$ and $[Mg_2Cl_3(THF)_6][SmCl_2\text{-}$

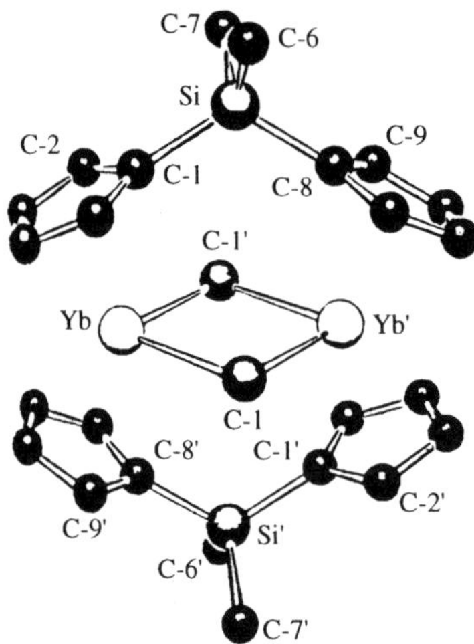

$\{Me_4C_2(C_5H_4)_2\}]$ have also been described.[529-32] Treatment of $[Mg_2Cl_3(THF)_6][SmCl_2\text{-}\{Me_4C_2(C_5H_4)_2\}]$ with one equivalent of Na(Cp) in THF afforded $[Sm(Cp)(THF)\text{-}\{Me_4C_2(C_5H_4)_2\}]$, which was structurally characterized.[533]

A novel coordination mode for ring-bridged cyclopentadienyl ligands was reported by Fischer and co-workers according to Equation (116).[304,534]

$$2\,LnX_3 + 2\,Na_2\{Me_2Si(C_5H_4)_2\} \xrightarrow{\text{THF}} [\{Ln(\mu\text{-}X)\{Me_2Si(C_5H_4)_2\}\}_2] + 4\,NaCl \qquad (116)$$

$$Ln = Y;\ X = Cl$$
$$Ln = Yb;\ X = Cl,\ Br$$

In the resulting dinuclear complexes the ligands are not chelating but bridge the two metal centers instead (Figure 33). Mass spectrometry has been shown to be a useful tool to determine between the chelating or metal-bridging nature of bidentate cyclopentadienyl ligands.[535,536]

Figure 33   The molecular structure of $[\{Yb(\mu\text{-}Cl)\{Me_2Si(C_5H_4)_2\}\}_2]$.[534]

Treatment of $[\{Y(\mu\text{-}Cl)\{Me_2Si(C_5H_4)_2\}\}_2]$ with sodium hydride afforded the dimeric μ-hydrides with the $[Me_2Si(C_5H_4)_2]^{2-}$ ligands in a chelating coordination mode, whereas for M = Yb the metal-bridged dimers $[(Yb(THF)\{Me_2Si(C_5H_4)_2\})_2(\mu\text{-}H)(\mu\text{-}Cl)]$ were isolated. In this case metal-bridging was clearly established by an x-ray crystal structure determination of $[(Yb(THF)\{Me_2Si(C_5H_4)_2\})_2(\mu\text{-}H)(\mu\text{-}Cl)]$.[536]

A similar coordination mode with a bifunctional cyclopentadienyl ligand spanning two lanthanide centers was found for the dimeric hydrides $[\{Ln(\mu\text{-}H)\{Me_2Si(C_5H_4)(C_5Me_4)\}\}_2]$ (Ln = Y, Lu).[537] In this particular ligand one cyclopentadienyl and one tetramethylcyclopentadienyl group are connected by a Me_2Si unit. The hydrocarbyl precursors $[LnCH(TMS)_2\{Me_2Si(C_5H_4)(C_5Me_4)\}]$ (Ln = Y, Lu) are monomeric in the solid state.[175] They undergo slow hydrogenation to afford the dimeric hydrides $[\{Ln(\mu\text{-}H)\{Me_2Si(C_5H_4)(C_5Me_4)\}\}_2]$ (Ln = Y, Lu), which have the same bridging structures as shown in Figure 61. The hydrides have been found to react with α-alkenes $RCH=CH_2$ (R = H, Me, Bu^n) to afford the μ-H, μ-alkyl complexes $[(\mu\text{-}H)(\mu\text{-}CH_2CH_2R)\{LnMe_2Si(C_5H_4)(C_5Me_4)\}_2]$. An x-ray structure determination was reported for the μ-ethyl derivative.[537]

Important results have been obtained with bifunctional ligands in which tetramethylcyclopentadienyl units are connected by Me_2Si or Me_2Ge[538,539] units. These compounds, especially the hydrocarbyl and hydride derivatives, show interesting catalytic behavior. They can be regarded as "tied-back" analogues of the $(Cp^*)_2Ln$ derivatives. Halides, hydrocarbyls, and hydrides containing these ligands can be synthesized by the methods established in pentamethylcyclopentadienyl chemistry (Scheme 18).

$$\text{LnCl}_3 + \text{Li}_2\{\text{Me}_2\text{Si}(\text{C}_5\text{Me}_4)_2\} \xrightarrow[-\text{LiCl}]{\text{Et}_2\text{O}} [\text{Ln}(\mu\text{-Cl})_2\text{Li}(\text{Et}_2\text{O})_2\{\text{Me}_2\text{Si}(\text{C}_5\text{Me}_4)_2\}] \xrightarrow[-2\,\text{LiCl}]{\text{LiCH(TMS)}_2}$$

$$[\text{LnCH(TMS)}_2\{\text{Me}_2\text{Si}(\text{C}_5\text{Me}_4)_2\}]$$

Ln = Nd, Sm, Lu

Scheme 18

The "tied-back" organolanthanide hydrocarbyls [NdCH(TMS)$_2$\{Me$_2$Si(C$_5$Me$_4$)$_2$\}],[540] [LuCH(TMS)$_2$\{Me$_2$Si(C$_5$Me$_4$)(C$_5$H$_4$)\}],[537] [HoCH(TMS)$_2$\{Me$_2$Ge(C$_5$Me$_4$)$_2$\}] (Figure 34)[538] and [Me$_2$Si(C$_5$Me$_4$)\{C$_5$H$_3$CH$_2$CH$_2$P(Bu^t)$_2$\}ScCH(TMS)$_2$][541] as well as the halide precursors [Ln(μ-Cl)$_2$Li(THF)$_2$\{Me$_2$Ge(C$_5$Me$_4$)$_2$\}] (Ln = Sm, Lu)[538] and [Y(μ-Cl)$_2$Li(THF)$_2$[Me$_2$Si\{C$_5$H$_2$(Bu^t-4)(TMS-2)\}$_2$][542] have been structurally characterized, with all hydrocarbyls exhibiting a characteristic γ-C–H–Ln agostic interaction.[543]

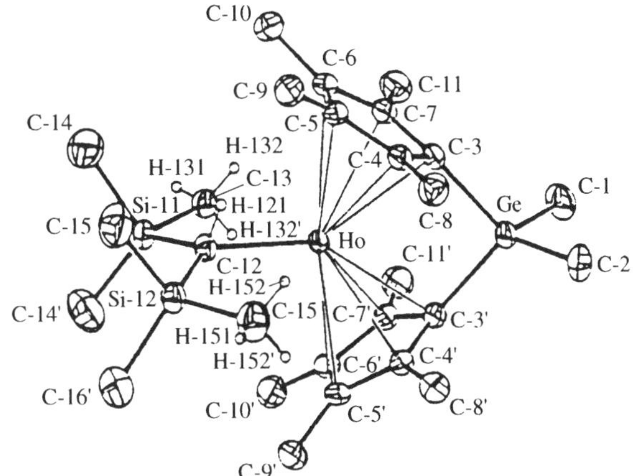

Figure 34 The molecular structure of [HoCH(TMS)$_2$\{Me$_2$Ge(C$_5$Me$_4$)$_2$\}].[538]

Subsequent hydrogenolysis of the hydrocarbyls with H$_2$ affords the corresponding hydride dimers, which have been shown to polymerize ethylene and oligomerize propylene and 1-hexene.[540]

The monomeric "tied-back" bis(cyclopentadienyl)scandium hydride [ScH\{Me$_2$Si(C$_5$H$_3$Bu^t)$_2$\}] was synthesized in order to study its reactions with unsaturated hydrocarbons. It was found that the hydride reacts with butadiene and several nonconjugated dienes to give η^3-allyl complexes of the type [Sc(η^3-allyl)\{Me$_2$Si(C$_5$H$_3$Bu^t)$_2$\}]. The same complexes were obtained from reactions of the hydride with methylenecyclopropane or substituted conjugated dienes. The reactivity of [ScH\{Me$_2$Si(C$_5$H$_3$Bu^t)$_2$\}] is summarized in Scheme 19.[544]

Various related Me$_2$Si-bridged difunctional cyclopentadienyl ligands and their scandium derivatives have been reported by Bercaw *et al.*[25]

Especially useful for the stabilization of unsolvated and alkali halide-free bis(cyclopentadienyl)-lanthanide halides are ring-bridged cyclopentadienyl ligands containing an oxygen- or nitrogen-donor function in the bridge. The chemistry of these ligands has been developed by Qian *et al.* A typical example is the 1,1'-(3-oxapentamethylene)dicyclopentadienyl ligand, which forms stable unsolvated complexes with most of the lanthanide anions (Schemes 20 and 21).[545–9]

All these compounds are stabilized by intramolecular coordination of the oxo function of the bridge. Subsequent treatment of the halide derivatives with Na(RC$_5$H$_4$) afforded the bridged tris(cyclopentadienyl)lanthanides [Ln\{O(CH$_2$CH$_2$C$_5$H$_4$)$_2$\}(RC$_5$H$_4$)] (R = H, Ln = Y, Nd, Gd, Er, Yb, Lu; R = Me, Ln = Y, Yb).[505,550] Partial hydrolysis of [Y\{O(CH$_2$CH$_2$C$_5$H$_4$)$_2$\}(Cp)] afforded the μ-OH bridged dimer [\{Y\{O(CH$_2$CH$_2$C$_5$H$_4$)$_2$\}(μ-OH)\}$_2$], which was crystallographically characterized.[505] Internal oxygen-coordination was also found in lanthanide chlorides of the type [\{Ln\{C$_4$H$_2$O(CH$_2$C$_5$H$_4$)$_2$\}(μ-Cl)\}$_2$] (Ln = Y, Nd, Sm, Yb) containing a furan-bridged bis(cyclopentadienyl) ligand.[551] A related tetramethyldisiloxane-bridged ligand provided the lanthanocene chlorides [PrCl(THF)$_n$-\{O(Me$_2$SiC$_5$H$_4$)$_2$\}] (n = 1, 2) and [\{Yb(μ-Cl)\{O(Me$_2$SiC$_5$H$_4$)$_2$\}\}$_2$]. The ytterbium dimer was structurally characterized. In this case the oxygen atoms in the bridge did not participate in the coordination to ytterbium.[552] Nitrogen-chelation was found in the chloro-bridged dimers of the type [\{Ln(μ-Cl)\{MeN(CH$_2$CH$_2$C$_5$H$_4$)$_2$\}\}$_2$] (**9**, Ln = Y, Nd, Sm, Yb) and [\{Ln(μ-Cl)\{C$_5$H$_3$N(CH$_2$C$_5$H$_4$)$_2$-2,6\}\}$_2$] (Ln = Y, Pr, Nd, Sm, Dy, Er, Yb, Lu) which have been made by using amine-linked bis(cyclopentadienyl) ligands.[553,554]

Scheme 19

$$LnCl_3 + Na_2\{O(CH_2CH_2C_5H_4)_2\} \xrightarrow[-2\ NaCl]{THF} [LnCl\{O(CH_2CH_2C_5H_4)_2\}] \xrightarrow[-LiCl]{Bu^tLi}$$

$$[Ln(Bu^t)\{O(CH_2CH_2C_5H_4)_2\}] \xrightarrow[-Bu^tH]{H_2} [\{Ln(\mu\text{-}H)\{O(CH_2CH_2C_5H_4)_2\}\}_2]$$

Ln = Y, Gd, Er, Yb, Lu

Scheme 20

$$[LnCl\{O(CH_2CH_2C_5H_4)_2\}] + NaN_2C_3HMe_2 \xrightarrow{THF} [Ln(N_2C_3HMe_2)\{O(CH_2CH_2C_5H_4)_2\}] + NaCl \xrightarrow{H_2O}$$

$$1/2\ [[Ln\{O(CH_2CH_2C_5H_4)_2\}]_2(\mu\text{-}OH)(\mu\text{-}N_2C_3HMe_2)]$$

Ln = Y, Lu
$N_2C_3HMe_2$ = dimethylpyrazolyl

Scheme 21

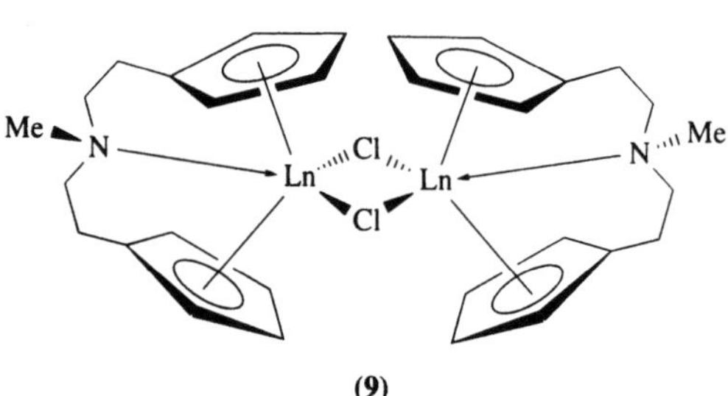

(9)

2.2.6.2 Indenyl and related compounds

A samarium(II) indenyl complex, $[Sm(THF)_3(C_9H_7)_2]$, was prepared by reacting $[SmI_2(THF)_2]$ with two equivalents of potassium indenide, and structurally characterized. The formation of a tris(THF) solvate indicates a more open coordination environment than in $[Sm(THF)_2(Cp^*)_2]$.[555]

Mono- and bis(indenyl)lanthanide(III) chlorides can be prepared by reacting $LnCl_3$ (Ln = Pr, Nd, Sm, Gd) with the stoichiometric amounts of $Na(C_9H_7)$ in THF.[556–8] Such complexes are generally isolated as THF adducts. The crystal structure determination of $[GdCl_2(THF)_3(C_9H_7)]$ has been reported.[558] A structurally characterized mono(indenyl) yttrium bis(alkoxide), $[\{Y(\mu\text{-}OBu^t)OBu^t(C_9H_7)\}_2]$, was prepared by reacting $[YCl_3(THF)_x]$ with two equivalents of $NaOBu^t$ and one equivalent of sodium indenide.[279] Various complexes of the type $[Ln(L)_{3-n}(C_9H_7)_n]$ (n = 1, 2) with chelating O- or N-donor ligands have been prepared by reacting $[Ln(THF)(C_9H_7)_3]$ with the protonated form of the chelating ligand. The ligands include the anions of 8-hydroxyquinoline and o-aminophenols.[293,559] An interesting new indenyl sandwich complex is $[Sc(C_9H_7)(OEP)]$ (OEP = octaethylporphyrin dianion). The compound was made by reacting $[ScCl(OEP)]$ with sodium indenide in THF and was structurally characterized. The indenyl ligand is pentahapto-coordinated to scandium.[183]

Homoleptic tris(indenyl)lanthanide complexes were first described in 1968.[18] They are generally prepared by treatment of anhydrous lanthanide trichlorides with sodium indenide in THF solution (Equation (117)). This method yields the mono-THF adducts $[Ln(THF)(C_9H_7)_3]$.[556,560–2] A pyridine adduct of tris(indenyl)cerium(III) has been synthesized by reacting $[Ce(OPr^i)_4(py)]$ with triethylaluminum in indene as solvent. In the course of this reaction the cerium(IV) is reduced to the +3 oxidation state. Unsolvated $[Sm(C_9H_7)_3]$ can be prepared from $SmCl_3$ and $Mg(C_9H_7)_2$ in benzene solution.

$$LnCl_3 + 3\ M(C_9H_7) \longrightarrow [Ln(THF)(C_9H_7)_3] + 3\ MCl \qquad (117)$$

Ln = Y, La, Ce, Pr, Nd, Sm, Gd, Tb, Dy, Yb
M = Na, K

$[Sm(THF)(C_9H_7)_3]$ was also obtained by treatment of bis(pentafluorophenyl)samarium(II) with indene in THF solution via an intermediate samarium(II) indenyl complex.[231] In addition, a small number of mixed indenyl cyclopentadienyl complexes of the type $[Ln(C_9H_7)(THF)(Cp)_2]$ (Ln = Sm, Dy, Ho, Er, Yb) have been reported.[563]

Structural investigations revealed the presence of monomeric molecules containing η^5-coordinated indenyl ligands. The sterically demanding indenyl ligand is large enough to prevent the formation of polymeric species. In unsolvated $[Sm(C_9H_7)_3]$ the centers of the five-membered rings surround the samarium atom in a trigonal planar fashion. A tetrahedral coordination geometry was established by x-ray crystallography for the pyridine adduct $[Ce(py)(C_9H_7)_3]$. Other structurally characterized derivatives are the THF adducts $[Nd(THF)(C_9H_7)_3]$ (Nd–C 0.2812 nm) and $[Gd(THF)(C_9H_7)_3]$ (Gd–C 0.2795 nm) as well as the diethyl etherates $[Ln(Et_2O)(C_9H_7)_3]$ (Ln = Nd, Gd, Er).[561,562] Binuclear anionic complexes of the type $[Na(THF)_6][Ln(\mu\text{-}Cl)(C_9H_7)_3Ln(C_9H_7)_3]$ (Ln = Nd, Sm) have been prepared and structurally characterized.[564,565]

The very bulky heptamethylindenyl ligand combines an effective shielding of the metal atoms with a better solubility of the products. Homoleptic $[Ln(C_9Me_7)_3]$ are known as well as THF solvates of the mono- and disubstitution products $LnCl_2(C_9Me_7)$ and $LnCl(C_9Me_7)_2$.[566] The fluorenyl ligand (= dibenzocyclopentadienyl) also provides sufficient steric bulk to give stable organolanthanide derivatives, but the chemistry of these compounds is sometimes hampered by solubility problems. $[Sm(THF)_2(C_{13}H_9)_2]$ was prepared from $[SmI_2(THF)_2]$ and two equivalents of potassium fluorenide. The molecule adopts a bent sandwich structure as in $[Sm(THF)_2(Cp^*)_2]$.[494] Tris(fluorenyl)lanthanides, $[Ln(C_{13}H_9)_3]$ (Ln = Y, La, Ce, Pr, Nd, Gd, Dy, Sm), have been isolated as unsolvated, presumably monomeric species.[560] Furthermore a small number of disubstituted complexes of the type $[Ln(\mu\text{-}Cl)_2Li(THF)_2(C_{13}H_9)_2]$ (Ln = La, Nd, Sm, Ho, Lu) have been obtained by reacting anhydrous lanthanide trichlorides with two equivalents of lithium fluorenide in THF solution.[567]

2.2.6.3 *Pentamethylcyclopentadienyls*

(i) *M(Cp*)X compounds*

Mono(pentamethylcyclopentadienyl) complexes of divalent lanthanides are exceedingly rare. The first such compound was described in 1986. Treatment of $[SmI_2(THF)_2]$ with $K(Cp^*)$ in a 1:1 molar ratio in THF solution produced dark green $[\{Sm(\mu\text{-}I)(THF)_2(Cp^*)\}_2]$. The dimeric nature of this complex was established by an x-ray diffraction study (Figure 35).[568] The analogous europium(II) complex $[\{Eu(\mu\text{-}I)(THF)_2(Cp^*)\}_2]$ has been prepared and studied by ^{151}Eu Mössbauer spectroscopy.[569] Although $[\{Sm(\mu\text{-}I)(THF)_2(Cp^*)\}_2]$ appears to be an ideal starting material for the synthesis of other mono(pentamethylcyclopentadienyl)lanthanide complexes, virtually no derivative chemistry has been reported.

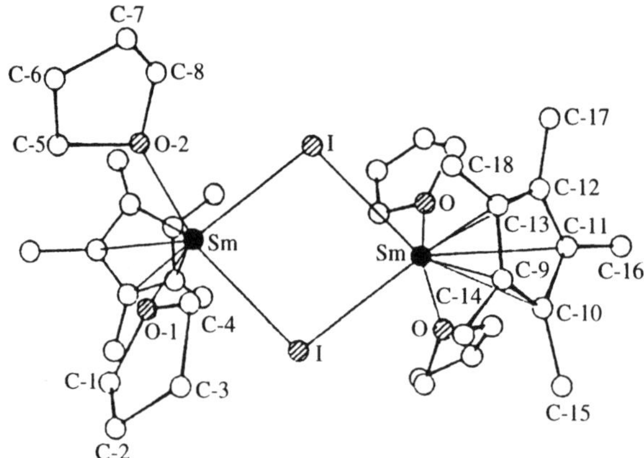

Figure 35   The molecular structure of $[\{Sm(\mu\text{-}I)(THF)_2(Cp^*)\}_2]$.[568]

Two ytterbium(II) complexes of this type were prepared by metathetical reactions using $YbCl_2$ (Equation (118)).[217]

$$YbCl_2 + NaCp^* \xrightarrow{\quad L \quad} [YbCl(L)_nCp^*] + NaCl \qquad (118)$$

$$L = THF, n = 2$$
$$L = DME, n = 1$$

The ytterbium(II) stannyl complex $[Yb\{Sn(CH_2Bu^t)_3\}(THF)_2(Cp^*)]$ was formed along with $[Yb(THF)_2(Cp^*)_2]$ when $[Yb\{Sn(CH_2Bu^t)_3\}_2(THF)_2]$ was reacted with one equivalent of pentamethylcyclopentadiene.[147]

(ii) *M(Cp*)₂ compounds*

Decamethylmetallocene(II) derivatives form a truly exciting class of low-oxidation-state organolanthanides. Compounds such as $[Sm(THF)_2(Cp^*)_2]$ or $[Sm(THF)_2(C_5Me_4Et)_2]$ were initially prepared by metal vapor syntheses.[570] Since the early 1980s this class of highly reactive lanthanide organometallics has become very important and well-investigated.

The successful preparation and characterization of unsolvated decamethylsamarocene by Evans *et al.* can be considered a landmark in organo-*f*-element chemistry. This compound is one of the most reactive organometallics ever made and has given rise to a large number of unusual reactions and unexpected products. It was first observed as a minor by-product in the metal vapor reaction between samarium and pentamethylcyclopentadiene at −120 °C. [580] When the reaction mixture was worked up with THF, the major product was the THF solvate $[Sm(THF)_2(Cp^*)_2]$. Later a more straightforward synthesis involving the reaction of samarium diiodide with a slight excess of two equivalents of $K(Cp^*)$ was developed (Scheme 22).[571] Unsolvated decamethylsamarocene can be obtained by desolvation and sublimation of the THF adduct.[572] Bis(pentamethylcyclopentadienyl)europium(II) is prepared analogously from $[EuI_2(THF)_2]$ and $K(Cp^*)$. In this case the resulting THF solvate is more difficult to desolvate than the samarium analogue. Pure $Eu(Cp^*)_2$ was obtained only after repeated vacuum sublimations.[573] Unsolvated $Yb(Cp^*)_2$ is accessible analogously.

$$[SmI_2(THF)_2] + 2\,KCp^* \xrightarrow{\text{THF}} [Sm(THF)_2Cp^*_2] + 2\,KI \xrightarrow{\Delta} SmCp^*_2$$

Scheme 22

Solvated derivatives of the type $[LnL(Cp^*)_2]$ (L = Et_2O, THF) can also be obtained by reacting $LnCl_2$ with two equivalents of $Na(Cp^*)$ in THF or diethyl ether (Equation (119)).[574] In the case of europium $EuCl_3$ may be employed as starting material because reduction to the +2 oxidation state takes place during the course of the reaction with three equivalents of $Na(Cp^*)$ (Equation (120)).[574] The europium(II) complexes $[Eu(THF)(Cp^*)_2]$ and $[Eu(THF)(Et_2O)(Cp^*)_2]$ have been investigated by [151]Eu Mössbauer spectroscopy.[569]

$$LnCl_2 + 2\,NaCp^* \xrightarrow{+L} [LnLCp^*_2] + 2\,NaCl \qquad (119)$$

$$Ln = Sm,\ Eu,\ Yb$$

$$EuCl_3 + 3\,NaCp^* \xrightarrow[\text{ii, Et}_2\text{O}]{\text{i, THF}} [Eu(Et_2O)(THF)Cp^*_2] + 3\,NaCl + 0.5\,Cp^*_2 \qquad (120)$$

Various solvates have been described for the ytterbium(II) derivative, which in some cases can be interconverted. Recrystallization of $[Yb(THF)(Cp^*)_2]$ from toluene yielded $[Yb(THF)(Cp^*)_2]\cdot 1/2$ toluene. Similar reactions of $YbBr_2$ or YbI_2 with $Na(Cp^*)$ afforded the bis-solvates $[LnL_2(Cp^*)_2]$ (L = Et_2O, THF, 1/2 DME). Simple heating reverts $[Yb(THF)_2(Cp^*)_2]$ back to $[Yb(THF)(Cp^*)_2]$.[18,574]

An alternative preparation of ytterbium(II) metallocenes involves the reaction of a liquid ammonia solution of ytterbium metal with pentamethylcyclopentadiene. Work-up with THF yielded the solvated complex $[Yb(NH_3)(THF)(Cp^*)_2]$, the structure of which was determined by x-ray diffraction (Yb–N 0.255(3) nm).[575] Several other adducts of $Yb(Cp^*)_2$ with nitrogen and even phosphorus donor ligands have been prepared from $[Yb(Et_2O)(Cp^*)_2]$ by ligand exchange (Equations (121) and (122), C_5H_5N = pyridine).[18] [171]Yb NMR spectroscopy has recently been shown to be a useful tool to study such organoytterbium(II) complexes.[576]

$$[Yb(Et_2O)Cp^*_2] + 2\,C_5H_5N \longrightarrow [Yb(C_5H_5N)_2Cp^*_2] + Et_2O \qquad (121)$$

$$[Yb(Et_2O)Cp^*_2] + Me_2P(CH_2)_nPMe_2 \longrightarrow [Yb\{Me_2P(CH_2)_nPMe_2\}Cp^*_2] + Et_2O \qquad (122)$$

$$n = 1,\ 2$$

Solvated decamethylsamarocene $[Sm(THF)_2(Cp^*)_2]$ **(10)** is obtained as a purple material, while $Sm(Cp^*)_2$ **(11)** forms dark green crystals. Both complexes are highly air-sensitive, turning yellow in the presence of traces of oxygen. The molecular and crystal structure of $[Sm(THF)_2(Cp^*)_2]$ was determined by x-ray diffraction. The molecule adopts a typical bent metallocene structure with a pseudotetrahedral coordination geometry around the divalent samarium.[18,572]

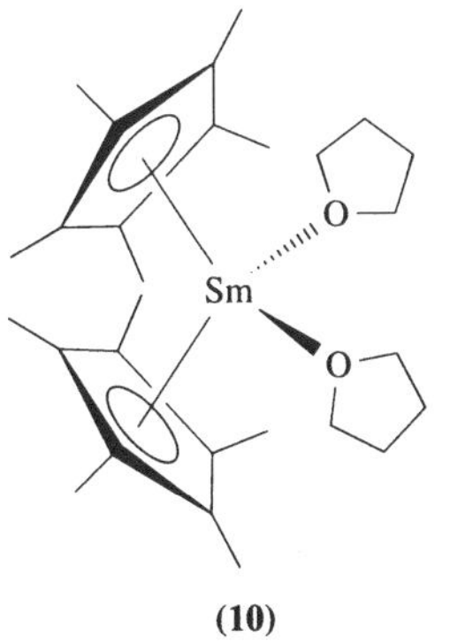

(10) 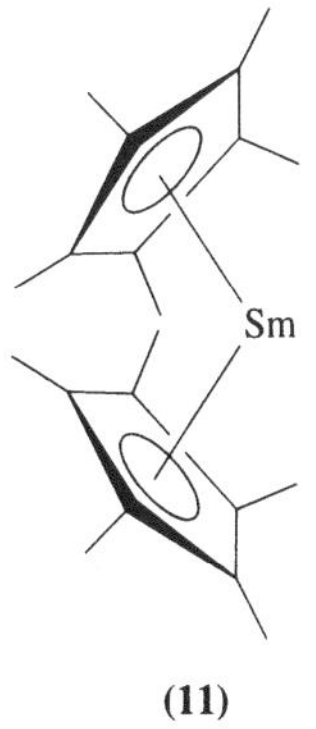   **(11)**

Other structurally characterized solvates of decamethylsamarocene are $[Sm(dihydropyran)_2(Cp^*)_2]$ (Sm–C(Cp^*) = 0.2842(4) nm),[577] $[Sm(tetrahydropyran)(Cp^*)_2]$ (Sm–C(Cp^*) = 0.2816(3) nm),[577] $[Sm(THF)(Cp^*)_2]$ (Sm–C(Cp^*) = 0.2816(4) nm),[244] and $[Sm(DME)(Cp^*)_2]$ (Sm–C(Cp^*) = 0.282(3) nm).[502]

The unusual molecular structure of $Sm(Cp^*)_2$ was reported in 1984. The structure consists of bent metallocene molecules with a centroid–Sm–centroid angle of 140.1°. The shortest separation between two monomeric units is 0.322(1) nm (Figure 36).[578]

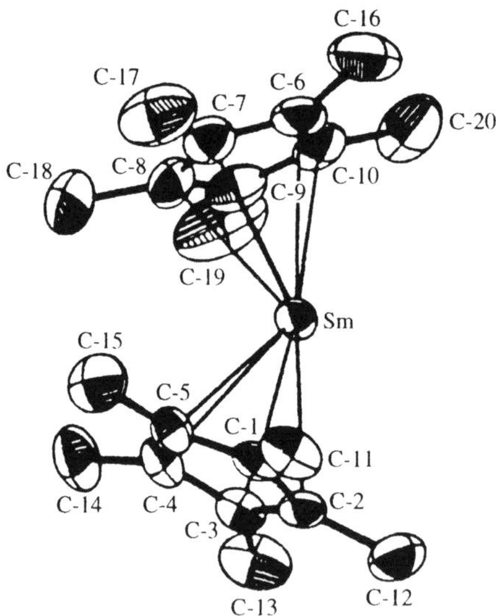

Figure 36 The molecular structure of $Sm(Cp^*)_2$.[578]

Base-free $Eu(Cp^*)_2$ and $Yb(Cp^*)_2$ too have been structurally characterized by x-ray methods,[578,579] showing once again a bent metallocene geometry similar to that in the samarium analogue or in $[M(Cp^*)_2]$ (M = Ca, Ba).[580,581] A bent metallocene structure similar to decamethylsamarocene was also deduced from electron diffraction studies in the gas phase.[582,583] Earlier calculations on alkaline earth and lanthanide metallocenes, $[M(Cp)_2]$ (M = Ca, Sr, Ba, Sm, Eu, Yb) were in favor of a linear structure, although the energy differences between bent and linear forms were found to be very small. Weak van der Waals attractive forces between Cp^* rings are made responsible for the bent structures in the solid state.[584,585] $Ln^{II}(Cp^*)_2$ complexes have also been investigated by gas phase He^I and He^{II} photoelectron spectroscopy[84,579] and photoluminescence studies.[586] Most results were in agreement with a predominantly ionic character of the metal–ligand bonding.

As stated above, decamethylsamarocene and its THF adduct are highly reactive species, which react with a great variety of substrates to give samarium(III) complexes containing one or more $Sm(Cp^*)_2$ units. The outcome of virtually all reactions of decamethylsamarocene and $[Sm(THF)_2(Cp^*)_2]$ is governed by the strong reducing power of divalent samarium. In that respect these complexes have been termed "dissolved alkali metals," as they too are powerful one-electron reducing agents. As compared to samarium(II) analogues, the reactivity of $Yb(Cp^*)_2$ and its solvates is significantly diminished, and only very little derivative chemistry has been described for bis(pentamethylcyclopenta-dienyl)europium(II) complexes.

Very few examples of π-adducts of lanthanide metallocenes have been described. Paramagnetic ^1H NMR spectroscopy has been used to study complex formation between unsolvated $Eu(Cp^*)_2$[573] and neutral substrates such as H_2, C_2H_4, or CH_4. In the case of methane no interaction was found, indicated by the absence of a paramagnetic shift. In the other cases formation of the adducts $[Eu(L)(Cp^*)_2]$ (L = H_2, C_2H_4) was made plausible by the occurrence of line-broadening as well as paramagnetically shifted resonances.[587] However, no simple lanthanide metallocene derivative involving η^2-H_2 or η^2-C_2H_4 coordination has so far been isolated. Their instability is in agreement with the results of SCF–X–SW calculations on the model compounds $YbCO(Cp)_2$ and $Yb(C_2H_4)(Cp)_2$.[588] In the case of π-bonded alkynes it was possible to isolate the adduct $[Yb(\eta^2\text{-}MeC\equiv CMe)(Cp^*)_2]$, made from $Yb(Cp^*)_2$[579] and the free alkyne ligand.[589] The structural data indicate a weak coordination of the alkyne ligand and little or negligible π-backbonding (Yb–C(Cp*) 0.2659(9) nm, Yb–C(η^2) 0.285(1) nm). The same is true for the first organolanthanide alkene complex. Bimetallic $[Yb(\mu\text{-}C_2H_4)(Cp^*)_2Pt(PPh_3)_2]$ was prepared by reacting $Yb(Cp^*)_2$ and $[Pt(PPh_3)_2(C_2H_4)]$ (Equation (123)), Figure 37, Yb–C(Cp*) 0.267(2) nm, Yb–C(η^2) 0.2781(6) nm).[589]

$$YbCp^*_2 + [Pt(PPh_3)_2(C_2H_4)] \longrightarrow [Yb(\mu\text{-}C_2H_4)Cp^*_2Pt(PPh_3)_2] \qquad (123)$$

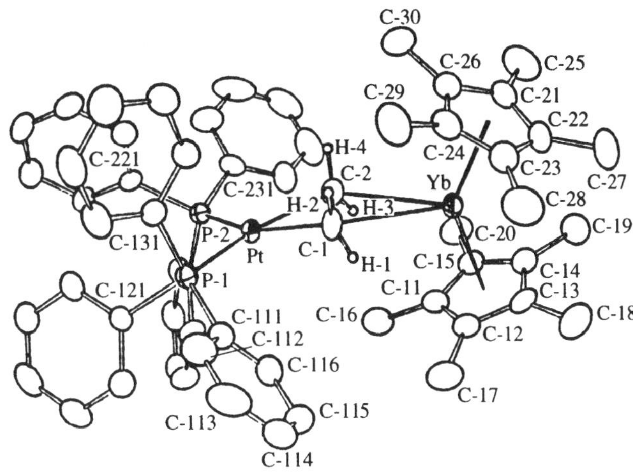

Figure 37 The molecular structure of [Yb(μ-C$_2$H$_4$)(Cp*)$_2$Pt(PPh$_3$)$_2$].[589]

In the platinum complex the coordinated ethylene is more electron-rich than the free alkene and can thus interact as a Lewis base with the Lewis acid Yb(Cp*)$_2$.[590] In fact, both ytterbium complexes are best described as Lewis acid–base adducts in which the lanthanide metallocenes do not act as π-donors.[579]

A truly remarkable compound is the substituted methyl complex [Yb(μ-Me)(Cp*)$_2$Be(Cp*)], which was prepared according to Equation (124) and structurally characterized (Figure 38).[591]

$$[YbCp*_2] + [BeMeCp*] \longrightarrow [Yb(\mu\text{-Me})Cp*_2BeCp*] \tag{124}$$

Figure 38 The molecular structure of [Yb(μ-Me)(Cp*)$_2$Be(Cp*)].[591]

The most remarkable structural feature is the nearly linear Yb–Me–Be unit (177.2°). Similar μ-methyl bridging was found in [Lu(μ-Me)(Cp*)$_2$Lu(Me)(Cp*)$_2$][187] and the anionic uranium complex [U(μ-Me)(MeC$_5$H$_4$)$_3$U(MeC$_5$H$_4$)$_3$]$^-$ (cf. Section 2.3.4).

An interesting application of the strong reducing power of [Sm(THF)$_2$(Cp*)$_2$] is the elegant preparation of decamethylsilicocene according to Equation (125). SiCl$_2$(Cp*)$_2$ reacts with decamethylsamarocene to give the divalent silicon species and the corresponding chloro complex of samarium(III). The outcome of this reaction shows that [Sm(THF)$_2$(Cp*)$_2$] is a stronger reducing agent than Si(Cp*)$_2$.[592]

$$2\,[Sm(THF)_2Cp*_2] + [SiCl_2Cp*_2] \xrightarrow{\text{THF}} SiCp*_2 + 2\,[SmCl(THF)Cp*_2] \tag{125}$$

Very few reactions of [Sm(THF)$_2$(Cp*)$_2$] have been reported in which the +2 oxidation state of samarium is retained or in which the oxidation state in the product is not well defined. One of the few examples is the adduct formation with triphenylphosphine oxide, which leads to [Sm(OPPh$_3$)(THF)(Cp*)$_2$].[593]

Apart from the three classical lanthanide(II) complexes samarium, europium, and ytterbium there has been evidence for the existence of "nonclassical" organolanthanide(II) complexes. Neodymium dichloride can be prepared in THF solution by reduction of $NdCl_3$ with lithium naphthalide. Treatment of the $NdCl_2(THF)_2$ thus obtained with two equivalents of K(Cp*) resulted in the formation of a dark purple, exceedingly air-sensitive compound characterized as $[K(THF)_n][NdCl_2(Cp*)_2]$ containing neodymium(II). The presence of neodymium in the +2 oxidation state was made plausible by the unambiguous redox reactions leading to the organoneodymium(III) complexes $[NdCl(THF)(Cp*)_2]$, $[Nd(S_2CNMe_2)(Cp*)_2]$, and $[Nd(SeMes)(THF)(Cp*)_2]$.[594]

(iii) M(Cp*)X₂ compounds

Single-ring organolanthanide cyclopentadienyl complexes are receiving considerable attention in recent years. In addition to increased coordinative unsaturation these compounds allow a greater variation of the ligand set as compared to the $Ln(Cp*)_2$ derivatives. However, the chemistry is often complicated by the pronounced tendency of the products to disproportionate as well as facile loss of coordinated solvent and/or formation of "ate" complexes. Mono(pentamethylcyclopentadienyl)-lanthanide halides cannot be isolated as unsolvated monomers.[595] This situation parallels the corresponding chemistry of the parent mono(cyclopentadienyl)lanthanide dihalides, which are isolable only as THF adducts.[18,256] 1:1-Reactions of lanthanide trihalides with M(Cp*) generally produce "ate" complexes in which retained alkali metal halide is incorporated, as for example in $[Ln(\mu\text{-}Cl)_3Li(Et_2O)_2(Cp*)]$ (Ln = Yb, Lu),[18,313] $[Yb(\mu\text{-}Cl)_3Li(THF)(Cp*)]$,[313] $[Yb(\mu\text{-}I)_3Li(Et_2O)_2(Cp*)]$,[18,313] and $[Ln(\mu\text{-}Cl)_3Na(Et_2O)_2(Cp*)]$ (Ln = Pr, Nd).[18,26,596] These compounds contain three halide bridges between lanthanide and alkali metal. $[YbCl_2(THF)_2(Cp*)]$ and $[YbCl_2(DME)(Cp*)]$ are examples of alkali-metal-free complexes of this type. These complexes are formed during electron transfer reactions between $[Yb(Et_2O)(Cp*)_2]$ and alkyl or aryl chlorides.[597]

Reactions of inorganic cerium(IV) compounds such as $[NH_4]_2[Ce(NO_3)_6]$ or $[pyH][CeCl_6]$ (pyH = pyridinium cation) with NaCp invariably lead to the formation of cerium(III) cyclopentadienyl complexes. Reduction to the +3 oxidation state was also observed with Cp* reagents. The mono(pentamethylcyclopentadienyl)cerium(III) complexes $[CeCl_2(py)_2(Cp*)]$, $[Li(py)_2][CeCl_3(Cp*)]$ and $[Li(py)_2][CeCl_3(C_5Me_4Pr)]$ have been prepared by reacting $[pyH][CeCl_6]$ with either M(Cp*) (M = Li, Na) or $Li(C_5Me_4Pr)$ under various reaction conditions. There was no indication of the formation of a stable organocerium(IV) complex.[598]

Even mono(pentamethylcyclopentadienyl)lanthanide complexes containing bulky alkoxide ligands have been found to be somewhat unstable and to undergo ligand redistribution reactions. The lanthanum derivative $[La(OAr)_2(Cp*)]$ was prepared by reacting $[La(OAr)_3]$ with one equivalent of LiCp* (OAr = 2,6-di-*t*-butylphenoxide). Addition of THF leads to the solvated species $[La(OAr)_2(THF)_2(Cp*)]$. In solution, however, disproportionation takes place and $[LaOAr(Cp*)_2]$, $[La(OAr)_2(Cp*)]$ and $[La(OAr)_3]$ are formed in equal amounts.[599]

In some cases the retention of alkali metal halide leads to the formation of mono(pentamethylcyclopentadienyl)lanthanide complexes exhibiting more complicated molecular structures. This is the case, for example, in the reaction of gadolinium trichloride with NaCp*. As a minor product the compound $[Na(\mu\text{-}THF)\{Gd(THF)Cp*\}_2(\mu_2\text{-}Cl)_3(\mu_3\text{-}Cl)_2]_2 \cdot 6THF$ was isolated and characterized by x-ray crystallography.[600] Other structurally characterized mono(Cp*) lanthanide cluster complexes are $[Yb_5(\mu_3\text{-}O)(\mu\text{-}Cl)_8(OEt)_2(Cp*)_5]$[601] and $[Yb_4(\mu\text{-}F)_4(Cp*)_6]$ (Figure 39). The latter is unusual as it contains two ytterbium(II) ions with one $\bar{C}_5Me_5$ ligand (Yb–C 0.265(2) nm) and two ytterbium(III) ions with two $\bar{C}_5Me_5$ ligands.[602]

A bulky ligand capable of stabilizing simple unsolvated mono(pentamethylcyclopentadienyl)-lanthanide derivatives is the bis(trimethylsilyl)amido anion. Monomeric $[Nd\{N(TMS)_2\}_2(Cp*)]$ was synthesized according to Equation (126).[18]

$$[Nd(\mu\text{-}Cl)_3Na(Et_2O)_2Cp*] + 2\,NaN(TMS)_2 \xrightarrow[-2\,Et_2O]{toluene} [Nd\{N(TMS)_2\}_2Cp*] + 3\,NaCl \qquad (126)$$

An interesting scandium compound involving nitrogen coordination is $[Sc(OEP)(Cp*)]$ (OEP = octaethylporphyrin dianion), which was prepared by reacting $[ScCl(OEP)]$ with Li(Cp*).[183]

$Sm(Cp*)_2$ reacts with excess hydrazine to form the tetranuclear hydrazido complex $[\{Sm(Cp*)\}_4(NHNH)_2(NHNH_2)_4(NH_3)_2]$, which was structurally characterized (Figure 40). The molecule consists of a distorted tetrahedral arrangement of four samarium atoms with bridging hydrazido anions on each edge of the tetrahedron.[603]

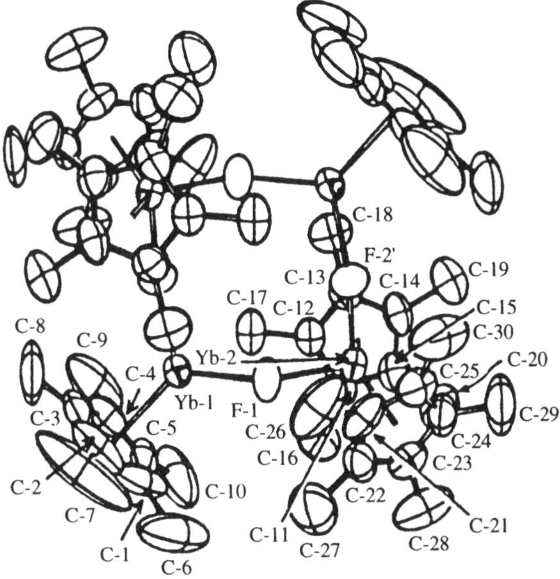

Figure 39 The molecular structure of [Yb$_4$(μ-F)$_4$(Cp*)$_6$].[602]

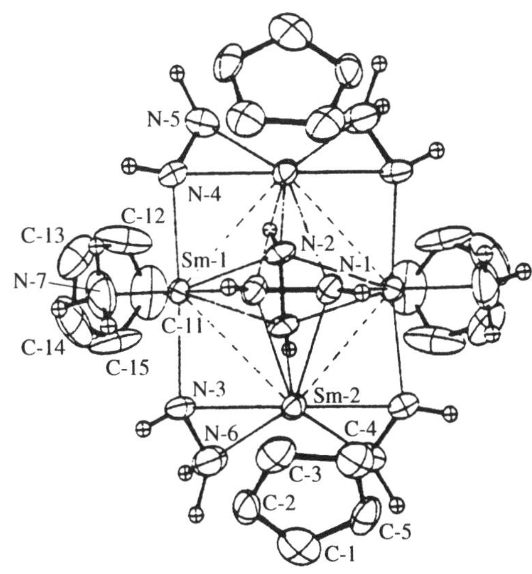

Figure 40 The molecular structure of [{Sm(Cp*)}$_4$(NHNH)$_2$(NHNH$_2$)$_4$(NH$_3$)$_2$].[603]

Information on mono(pentamethylcyclopentadienyl)lanthanide(III) hydrocarbyls was scant until recently. One of the first reported complexes of this type was made with the use of a chelating anionic phosphoylide ligand (Equation (127)).[18]

$$LuCl_3 + NaCp* + 2\,Li(CH_2)_2PMe_2 \longrightarrow [Lu\{(CH_2)_2PMe_2\}_2Cp*] + 2\,LiCl + NaCl \qquad (127)$$

Simple lanthanide hydrocarbyls containing a (Cp*)Ln unit were first isolated as anionic derivatives such as the structurally characterized (Ln = Lu) [Li(TMEDA)$_2$][LnMe$_3$(Cp*)] (Ln = Yb, Lu)[604,605] or [Li(TMEDA)$_2$][Lu(Bu^i)$_2$Cl(Cp*)] and [Li(THF)$_3$][Lu(Bu^i)$_2$Cl(Cp*)].[606] Recent studies in this area have focused on circumventing the problem of salt retention by using nonchloride precursors such as iodides or alkoxides. Interesting differences in chemical behavior have been observed depending on the size of the lanthanide ion.

A useful scandium precursor, polymeric [{ScCl$_2$(Cp*)}$_n$], has been prepared by the following two-step procedure (Equations (128) and (129)).[607]

$$Sc(acac)_3 + MgCl(THF)Cp* \longrightarrow [Sc(acac)_2Cp*] + MgCl(acac) \qquad (128)$$

$$[Sc(acac)_2Cp^*] + 2\,AlCl_3 \longrightarrow 1/n\,[\{ScCl_2Cp^*\}_n] + 2\,AlCl_2(acac) \qquad (129)$$

Reaction of [{ScCl$_2$(Cp*)}$_n$] with MeLi afforded oligomeric [{ScMe$_2$(Cp*)}$_n$] while treatment with trimethylphosphine afforded monomeric soluble [ScCl$_2$(PMe$_3$)(Cp*)].[607] Both compounds have been further used as precursors for new mono(pentamethylcyclopentadienyl)scandium alkoxides and hydrocarbyls. The LiCl adducts [ScCl(OR)(Cp*)·LiCl] have been prepared by reacting [ScCl$_2$(PMe$_3$)(Cp*)] with LiOR (R = C$_6$H$_2$(Bu^t)$_3$-2,4,6, C$_6$H$_3$(Bu^t)$_2$-3,5). Treatment of [ScCl(OC$_6$H$_3$(Bu^t)$_2$-3,5)(Cp*)·LiCl] with methyllithium afforded the mixed alkoxy-hydrocarbyl derivative [{Sc(OC$_6$H$_3$(Bu^t)$_2$-3,5)(μ-Me)(Cp*)}$_2$], which can alternatively be prepared by protonation of [{ScMe$_2$(Cp*)}$_n$] with the free phenol. For steric reasons the bridging methyl complex is inert toward various reagents such as H$_2$, ethylene, propylene, butyne, or donating solvents like THF or pyridine.[607] The structurally characterized anionic scandium hydrocarbyl complex [Li(THF)$_3$][Sc(C$_2$B$_9$H$_{11}$)-CH(TMS)$_2$(Cp*)] has been prepared by addition of LiCH(TMS)$_2$ to neutral [Sc(C$_2$B$_9$H$_{11}$)-(THF)(Cp*)].[608]

The search for novel catalytic systems has led to a modification of the pentamethylcyclopentadienyl ligand by pendant amido functions. Hydrogenation of the scandium hydrocarbyl [ScCH(TMS)$_2$(C$_5$Me$_4$SiMe$_2$NBu^t)] in the presence of trimethylphosphine afforded the dimeric hydride [{Sc(PMe$_3$)(μ-H)(C$_5$Me$_4$SiMe$_2$NBu^t)}$_2$], which was structurally characterized.[25,609] [{Sc(PMe$_3$)(μ-H)(C$_5$Me$_4$SiMe$_2$NBu^t)}$_2$] is a single-component alkene polymerization catalyst (cf. Section 2.2.13). Stoichiometric addition of two equivalents of propylene yielded the corresponding bis(μ-propyl) derivative (Equation (130)). This compound was structurally characterized by x-ray diffraction.[25]

$$(130)$$

(Cp*)Y hydrocarbyl and hydride complexes are now accessible by several synthetic routes. An effective preparation of such complexes starts from the monomeric unsolvated yttrium phenoxide [Y(OAr)$_3$] (Ar = C$_6$H$_3$(Bu^t)$_2$-2,6). One pentamethylcyclopentadienyl ligand is selectively introduced by reacting [Y(OAr)$_3$] with one equivalent of K(Cp*) in toluene. Established routes lead from the resulting phenoxide [Y(OAr)$_2$(Cp*)] to the corresponding hydrocarbyl and hydride derivatives. The only exception is that the hydrogenation step requires the use of hydrogen under pressure, whereas complexes of the type [LnCH(TMS)$_2$(Cp*)$_2$] normally undergo facile hydrogenation under normal pressure.[184-6] A major factor for the success of these syntheses is the excellent leaving group character of the phenoxide ligand which is easily eliminated as insoluble KOAr. The method thus circumvents problems associated with halide complexation or solvate formation (Scheme 23).[610]

The structurally characterized bis(phenoxide) [Y(OAr)$_2$(Cp*)] contains one almost linear Y–O–C unit (168°), while the bond to the second aryloxide ligand is significantly bent (129°) (Y–O 0.2096(4) nm and 0.2058(3) nm).[611] The dimeric hydrides [{Y(OR)(μ-H)(Cp*)}$_2$] are interesting precursors for reactions with alkenes and alkynes. Terminal olefins are regiospecifically added in a 1:1 molar ratio to afford the corresponding μ-η-alkyl complexes (Equation (131)). The products are stable toward β-hydride elimination.[610,612]

The terminal alkyne TMS-C≡CH reacts similarly with [{Y(OR)(μ-H)(Cp*)}$_2$] to give the bridging acetylide derivatives [(μ-H)(μ-C≡C-TMS){Y(OR)(Cp*)}$_2$].[380] Further reactions of the remaining hydride ligand have not been observed even under forcing reaction conditions. Dimeric complexes with bridging methyl ligands have been synthesized by reacting [Y(OAr)$_2$(Cp*)] with equimolar amounts of methyllithium (Equation (132)).[610,612]

The dimeric alkoxide [{Y(μ-OBu^t)(OBu^t)(Cp*)}$_2$] was prepared by reacting the trimetallic yttrium alkoxide [Y$_3$(OBu^t)$_7$Cl$_2$(THF)$_2$] with NaCp*. Disruption of the trimetallic structure was also observed in the presence of other alkali metal cyclopentadienyl reagents and several products have been structurally characterized.[279] A closely related reaction was used to prepare the first pentamethylcyclopentadienyl complex of europium(III). This is a remarkable result because the Cp*$^-$ anion had previously been shown to reduce Eu^{III} to Eu^{II}, for example in the preparation of [Eu(Et$_2$O)(THF)(Cp*)$_2$] from EuCl$_3$ and (Cp*)$^-$ (cf. Section 2.2.6.3). A two-step procedure was developed to prepare the organoeuropium(III) alkoxide [{Eu(μ-OBu^t)(OBu^t)(Cp*)}$_2$] (Scheme 24). From this result it was suggested that alkoxide

$$[Y(OAr)_3] + [KCp^*] \xrightarrow[100\,°C]{\text{toluene}} [Y(OAr)_2Cp^*] \xrightarrow{KCH(TMS)_2} \cdots \xrightarrow{H_2(20\ \text{bar})}$$

Scheme 23

$$[\{Y(OR)(\mu\text{-}H)Cp^*\}_2] + H_2C{=}CHR \longrightarrow [(\mu\text{-}H)(\mu\text{-}CH_2CH_2R)\{Y(OR)Cp^*\}_2] \qquad (131)$$

$$R = H,\ Me,\ Bu^n$$

$$2\,[Y(OAr)_2Cp^*] + 2\,MeLi \longrightarrow [\{Y(OAr)(\mu\text{-}Me)Cp^*\}_2] + 2\,LiOAr \qquad (132)$$

ligands can be used broadly in organolanthanide chemistry to stabilize higher oxidation states. This phenomenon has been observed before in cerium(IV) chemistry as alkoxide ligands have been found to stabilize cerium(IV) with respect to reduction by cyclopentadienyl ligands (cf. Section 2.2.5.8).[613]

$$3\,EuCl_3 + 7\,NaOBu^t \xrightarrow[-NaCl]{\text{THF}} [Eu_3(OBu^t)_7Cl_2(THF)_2] \xrightarrow{3\,KCp^*} 3/2\,[\{Eu(\mu\text{-}OBu^t)(OBu^t)Cp^*\}_3]$$

Scheme 24

A binuclear mono(Cp*)–samarium complex involving samarium–sulfur bonding was prepared by an unusual reaction of $[Sm(THF)_2(Cp^*)_2]$ with a bis(thiophosphoryl) disulfide (Equation (133)) and structurally characterized. Pentamethylcyclopentadienyl ligands are eliminated in the course of the reaction. A notable structural feature is the presence of triply bridging O,O'-dimethyldithiophosphate ligands in which one of the methoxy groups is involved in the coordination to samarium.[614]

$$2\,[Sm(THF)_2Cp^*_2] + 2\,[(MeO)_2P(S)S]_2 \longrightarrow [Sm\{S_2P(OMe)_2\}_2Cp^*]_2 + 4\,THF + Cp^*_2 \qquad (133)$$

Another suitable precursor for the preparation of mono(Cp*)–lanthanide complexes is the homoleptic hydrocarbyl $[Y(o\text{-}C_6H_4CH_2NMe_2)_3]$, which can be selectively protonated by one equivalent of pentamethylcyclopentadiene (Equation (134)). The preparation was not successful with the corresponding lanthanum and cerium compounds.[468] A similar reaction of $[Y\{N(TMS)_2\}_3]$ with Cp*H only afforded mixtures of mono- and bis(Cp*) complexes.[595]

$$[Y(o\text{-}C_6H_4CH_2NMe_2)_3] + Cp^*H \longrightarrow [Y(o\text{-}C_6H_4CH_2NMe_2)_2Cp^*] + C_6H_5CH_2NMe_2 \qquad (134)$$

An x-ray crystal structure determination of $[Y(o\text{-}C_6H_4CH_2NMe_2)_2(Cp^*)]$ revealed agostic $Y\cdots C\text{–}N$ interactions. Thermolysis of the compound produced $[Y(o\text{-}C_6H_4CH_2NMe)(\mu\text{-}CH_2)(Cp^*)\{\mu\text{-}o\text{-}C_6H_4CH_2NMe(\mu\text{-}CH_2)\}Y(THF)(Cp^*)]$, which was structurally characterized.[468] $[Y(acac)_2(Cp^*)]$ was prepared by selective replacement of a cyclooctatetraenyl ligand (Equation (135)).[615]

$$[YCp^*(C_8H_8)] + 2\,MeC(O)CH_2C(O)Me \longrightarrow [Y(acac)_2Cp^*] + C_8H_8 + H_2 \qquad (135)$$

Mono(pentamethylcyclopentadienyl)lanthanide diiodides have been prepared by a two-step procedure. Prolonged heating of lanthanide metals with EtI in THF solution produces the THF-soluble adducts $[LnI_3(THF)_3]$, which can subsequently be treated with $K(Cp^*)$ (Equation (136)).[616]

$$[LnI_3(THF)_3] + [KCp^*] \longrightarrow [LnI_2(THF)_3Cp^*] + KI \qquad (136)$$
$$Ln = Ce, La$$

The large iodide ions seem to be an important factor governing the stability of the products. In addition, alkali metal iodides show a reduced tendency to form "ate" complexes with the products than the chlorides. Diiodides of the type $[LnI_2(THF)_3(Cp^*)]$ have been successfully used as precursors for highly reactive and catalytically useful bis(hydrocarbyls). Treatment of the lanthanum derivative with two equivalents of $KCH(TMS)_2$ afforded the THF solvate $[La\{CH(TMS)_2\}_2(THF)(Cp^*)]$. The unsolvated bis(hydrocarbyl) can be prepared either by starting from unsolvated $[\{LaI_2(Cp^*)\}_n]$ or by elimination of the coordinated THF in the presence of TMS-I. The reaction sequence is summarized in Scheme 25.[145,617,618]

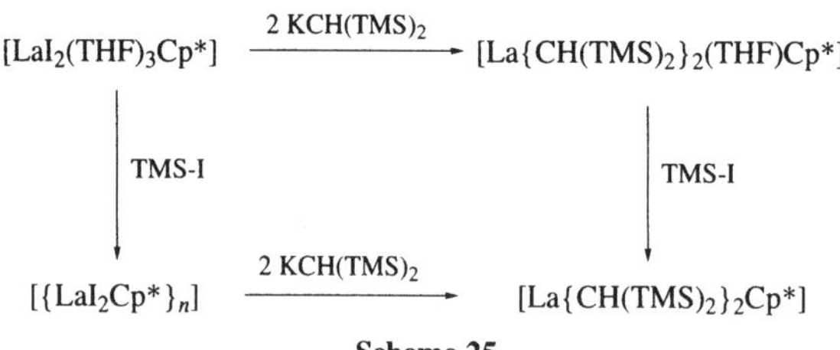

Scheme 25

Both $[La\{CH(TMS)_2\}_2(Cp^*)]$ and its THF solvate have been structurally characterized. In both cases the bis(trimethylsilyl)methyl ligands exhibit significant distortions, which are indicative of β-Si–Me–La interactions. Especially characteristic is the near planar arrangement of the $La-C_\alpha-Si-Me_\gamma$ units. A related planar, four-center transition state has been proposed for the β-hydride elimination.[617]

Unsolvated $[La\{CH(TMS)_2\}_2(Cp^*)]$ has been used as starting material in the preparation of the first cationic lanthanide hydrocarbyl complex containing a "noncoordinating" cation. Such species are important homogeneous catalysts in alkene polymerization processes. Structural investigations on related cationic zirconium species have revealed that the "noncoordinating" anion $[BPh_4]^-$ often interacts with the metal atom via phenyl π coordination. η^3-Phenyl coordination was proposed for $[Zr(Bz)_3(\eta^3-C_6H_5BPh_3)]$[619] and $[ZrMe(BPh_4)(Cp^*)_2]$[620] while η^6-coordination of a phenyl ring was found in the structurally characterized tribenzyl complex $[Zr(Bz)_3\{\eta^6-C_6H_5CH_2B(C_6F_5)_3\}]$.[621] The preparation of a cationic lanthanum hydrocarbyl complex is outlined in Scheme 26. Treatment of unsolvated $[La\{CH(TMS)_2\}_2(Cp^*)]$ with $[PhMe_2NH][BPh_4]$ yielded the zwitterionic monoalkyl derivative $[La\{CH(TMS)_2\}\{BPh_4\}(Cp^*)]$. Solid-state ^{13}C CP-MAS NMR data suggested an unusual π-coordination of two phenyl rings to lanthanum. This bonding situation reflects the severe electronic and coordinative unsaturation of the lanthanum center. Interestingly a similar coordination mode of tetraphenylborate was recently found in a number of structurally characterized niobium complexes of the type $[Nb(L)\{Ph_2B(C_6H_5-\eta^6)_2\}]$ (L = CO, MeC≡CMe).[622,623] The zwitterionic intermediate readily reacts with THF to give the first fully characterized cationic lanthanide hydrocarbyl complex, $[La\{CH(TMS)_2\}(THF)_3(Cp^*)][BPh_4]$.[624]

Mono(pentamethylcyclopentadienyl) chemistry has also been successfully extended to cerium.[625-7] The following reaction sequence allows the preparation of bis(alkoxides) and bis(hydrocarbyls) starting from $[Ce(OAr)_3]$ (Ar = $C_6H_3(Bu^t)_2$-2,6) (Equations 137 and 138).

The bis(amide) $[Ce\{N(TMS)_2\}_2(Cp^*)]$ has been prepared analogously by reacting $[Ce(OAr)_2(Cp^*)]$ with two equivalents of $NaN(TMS)_2$. The complexes $[Ce(OAr)_2(Cp^*)]$ (Ce–C 0.276(1) nm), $[Ce\{CH(TMS)_2\}_2(Cp^*)]$ (Figure 41, Ce–C(Cp^*) 0.279(3) nm, Ce–C(σ) 0.253 nm), and $[Ce\{N(TMS)_2\}_2(Cp^*)]$ (Ce–C 0.277(1) nm) have all been crystallographically characterized. Like its yttrium analogue the cerium bis(aryloxide) contains one nearly linear Ce–O–C unit (158°), while the other is significantly bent (105°).[625,626]

The reaction of $[CeCH(TMS)_2(Cp^*)_2]$ with excess t-butanol afforded a dimeric mono(Cp^*) cerium bis(alkoxide) (Equation (139)).[627]

When changing to the smaller lutetium, subtle but interesting differences in the chemistry of mono(pentamethylcyclopentadienyl) complexes are once again encountered. Some species are surprisingly difficult to synthesize or not accessible at all. For example, a base-free bis(alkyl) of the type $[Ln\{CH(TMS)_2\}_2(Cp^*)]$, which is well known for Ln = La, Ce[617,625,626] is not isolable with lutetium.

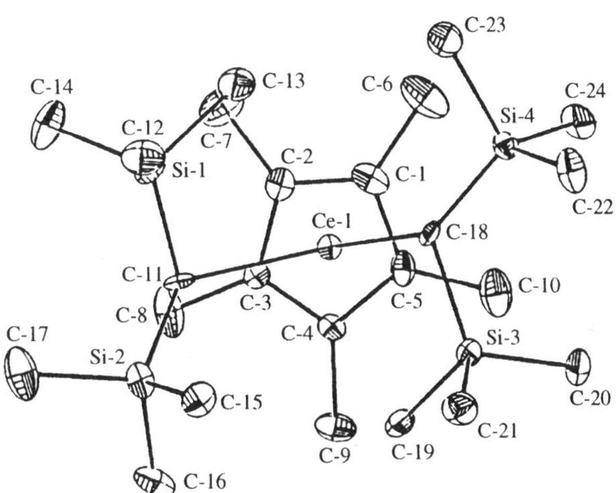

Scheme 26

$$[Ce(OAr)_3] + LiCp^* \longrightarrow [Ce(OAr)_2Cp^*] + LiOAr \qquad (137)$$

$$[Ce(OAr)_2Cp^*] + 2\,LiCH(TMS)_2 \longrightarrow [Ce\{CH(TMS)_2\}_2Cp^*] + 2\,LiOAr \qquad (138)$$

Figure 41 The molecular structure of $[Ce\{CH(TMS)_2\}_2(Cp^*)]$.[625]

$$2\,[CeCH(TMS)_2Cp^*_2] + 4\,Bu^tOH \longrightarrow [\{Ce(OBu^t)_2Cp^*\}_2] + 2\,CH_2(TMS)_2 + 2\,Cp^*H \qquad (139)$$

Steric reasons have been suggested for this different behavior. However, a remarkable preparation of a mixed bis(alkyl) species was achieved. The synthetic route is outlined in Scheme 27.[628]

The molecular structures of $[Li(TMEDA)][Lu\{CH(TMS)_2\}Cl_2(Cp^*)]$ and $[Li(THF)_3][Lu\{CH(TMS)_2\}(CH_2TMS)Cl(Cp^*)]$ have been determined by x-ray diffraction.[628] Most of the mono(Cp*)–lutetium complexes are thermally labile and slowly decompose above room temperature. A thermally more stable bis(alkyl), $[Lu(CH_2TMS)_2(THF)(C_5Me_3Ph_2)]$, was prepared by using the sterically more demanding 1,3-diphenyl-2,4,5-trimethylcyclopentadienyl anion as supporting ligand.[628] A mono(Cp*)–lutetium bis(thiolate) was isolated as the lithium thiolate adduct $[LuSBu^t(\mu\text{-}SBu^t)_2Li(TMEDA)(Cp^*)]$.[369]

(iv) MX(Cp*)₂ compounds

(a) Classical coligands. Especially in the area of bis(cyclopentadienyl)lanthanide(III) complexes the use of the bulky pentamethylcyclopentadienyl ligand has had a major impact on the development of organolanthanide chemistry. This particular ligand provides enhanced stability and better solubility to

$$[LuCl_3(THF)_3] \xrightarrow[Et_2O]{NaCp} [Na(THF)(OEt_2)][LuCl_3Cp^*] \xrightarrow{LiCH(TMS)_2}$$

Scheme 27

many such complexes and its use was crucial for the successful synthesis of many highly reactive and catalytically active lanthanide metallocene derivatives. The bond disruption enthalpies for various $Sm(Cp^*)_2$ complexes with classical coligands have been determined.[62] Such thermochemical investigations have been shown to be helpful for understanding known and designing new reactivity pathways for organolanthanide complexes.

Cations derived from decamethylmetallocenes have been prepared as disolvates of the type $[Ln(L)_2(Cp^*)_2][BPh_4]$ (Ln = Sm, Ce; L = THF, THT; THT = tetrahydrothiophene).[629,630] According to *ab initio* calculations on the parent cations $[Ln(Cp)_2]^+$ these species should adopt a bent metallocene structure. The preparation of cationic complexes involves protolysis of bis(Cp*)–lanthanide hydrocarbyls by suitable ammonium salts or redox reactions in the case of samarium (Equations (140) and (141)).

$$[CeCH(TMS)_2Cp^*_2] + [NHEt_3][BPh_4] \xrightarrow{L} [Ce(L)_2Cp^*_2][BPh_4] + CH_2(TMS)_2 + NEt_3 \quad (140)$$
$$L = THF, THT$$

$$[Sm(THF)_2Cp^*_2] + Ag[BPh_4] \longrightarrow [Sm(THF)_2Cp^*_2][BPh_4] + Ag \quad (141)$$

The crystal structures of $[Ce(THF)Cp^*_2][BPh_4]$ and $[Sm(THF)_2(Cp^*)_2][BPh_4]$ have been reported (Sm–C 0.269(2) nm, Sm–O 0.246(1) nm; Ce–C 0.274(3) nm, Ce–S 0.3058(1) nm and 0.3072(1) nm). The THF ligands in $[Sm(THF)_2(Cp^*)_2][BPh_4]$ are not readily substituted, but it does react with alkyl anions to provide a halide-free route to trivalent $(Cp^*)_2Sm$ hydrocarbyls.[629]

Bis(pentamethylcyclopentadienyl)lanthanide fluorides are not generally available via the normal metathetical routes because of the insolubility of lanthanide trifluorides. However, defluorination of perfluoroalkenes by the divalent organolanthanides $[Ln(L)(Cp^*)_2]$ (Ln = Sm, Eu, Yb; L = Et_2O, THF) has been found to be an elegant way to make the corresponding lanthanide(III) fluorides. The compounds $[LnF(L)(Cp^*)_2]$ (Ln = Sm, Eu, Yb; L = Et_2O, THF) have been prepared and fully characterized. Crystal structure determinations of $[YbF(Et_2O)(Cp^*)_2]$ and $[YbF(THF)(Cp^*)_2]$ revealed the first terminal lanthanide–fluorine bonds (Yb–F 0.2015(4) and 0.2026(2) nm respectively).[299]

Dimeric mixed-valence $[(\mu\text{-}F)\{Yb(Cp^*)_2\}_2]$ has been prepared and structurally characterized. The compound contains an asymmetric Yb^{2+}–F–Yb^{3+} bridge.[631] Purple $[Yb(DME)(Cp^*)_2][PF_6]$ was prepared by oxidation of the ytterbium(II) metallocene $[Yb(DME)(Cp^*)_2]$ with $[Fe(Cp)_2][PF_6]$.[18] The pentaytterbium(III) fluoride cluster $[Yb_5(\mu_4\text{-}F)(\mu_3\text{-}F)_2(\mu\text{-}F)_6(Cp^*)_6]$ resulted from further reaction of $[(\mu\text{-}F)\{Yb(Cp^*)_2\}_2]$ with perfluoroalkenes. Three different environments for bridging ytterbium–fluorine bonds have been found in the x-ray-determined molecular structure of this compound (Figure 42).[299]

Numerous reports on the synthesis and characterization of solvated and unsolvated bis(pentamethylcyclopentadienyl)lanthanide halides have been published. Complexes of this type have turned out to be the most important starting materials for the preparation of catalytically active organolanthanide

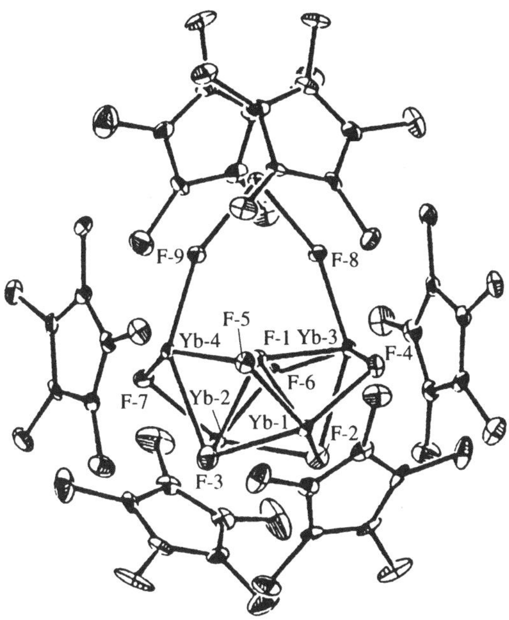

Figure 42 The molecular structure of $[Yb_5(\mu_4\text{-}F)(\mu_3\text{-}F)_2(\mu\text{-}F)_6(Cp^*)_6]$.[299]

hydrocarbyls and hydrides. Compounds containing a $LnX(Cp^*)_2$ (X = Cl, Br, I) fragment are easily available by reacting anhydrous lanthanide trihalides with two equivalents of a Cp* transfer reagent. These reactions are generally carried out in a coordinating solvent such as THF or DME. Meanwhile such complexes have been described for all members of the lanthanide series with the exception of promethium. Normally a monomeric $[LnX(Cp^*)_2]$ unit is coordinatively unsaturated. As a consequence, $[LnX(Cp^*)_2]$ complexes show a strong tendency either to form adducts with coordinating solvent molecules or to retain one equivalent of alkali metal halide and form "ate" complexes. This behavior has led to an interesting structural variety of these compounds. Not less than five different types of bis(pentamethylcyclopentadienyl)lanthanide halides have been characterized. (i) Monomeric $[LnCl(Cp^*)_2]$ is stable only in the case of the small scandium(3+) ion. $[ScCl(Cp^*)_2]$ was the starting point for an interesting class of novel organoscandium derivatives (see below). (ii) THF adducts of the type $[LnX(THF)(Cp^*)_2]$ have been prepared and structurally characterized for several lanthanide elements. (iii) In most cases $LnX(Cp^*)_2$ derivatives are initially isolated as "ate" complexes of the type $[Ln(\mu\text{-}X)_2ML_2(Cp^*)_2]$ (X = Cl, Br, I; M = Li, Na, K; L = THF, Et$_2$O, 1/2 DME, 1/2 TMEDA). In these materials one equivalent of alkali metal halide is retained to increase the formal coordination number around the lanthanide atom. Lanthanide and alkali metal are bridged by two halide ligands to give a central four-membered $[Ln(\mu\text{-}X)_2M]$ ring. Formally these compounds could also be described as "ate" complexes of the type $[ML_2][LnX_2(Cp^*)_2]$. One or two solvent molecules are needed to complete the coordination sphere around the alkali metal ion. (iv) Unsolvated species normally exist as symmetrical halide-bridged dimers of the type $[\{Ln(\mu\text{-}X)(Cp^*)_2\}_2]$. (v) The compound $[Y(\mu\text{-}Cl)(Cp^*)_2YCl(Cp^*)_2]$ is a rare example of an unsymmetrical dimer, while $[\{Sm(\mu\text{-}Cl)(Cp^*)_2\}_3]$ is trimeric in the solid state. These compounds will be discussed in the above order.

An elegant way of preparing salt-free THF adducts of the type $[LnX(THF)(Cp^*)_2]$ (Ln = Sm, Eu, Yb; X = Cl, I) is the controlled oxidation of lanthanide(II) metallocenes using alkyl or aryl halides.[388,632] This method is of course restricted to those lanthanide elements for which the +2 oxidation state is readily accessible. Monomeric $[SmCl(THF)(Cp^*)_2]$ was prepared by treatment of $[Sm(THF)_2(Cp^*)_2]$ with Bu^{t}Cl, while 1,2-diiodoethane was used for an analogous preparation of $[SmI(THF)(Cp^*)_2]$. Presumably these oxidation reactions proceed via a radical mechanism.[388,568] The molecular and crystal structures of both samarium complexes $[SmX(THF)(Cp^*)_2]$ (X = Cl, I) have been determined. A two-step procedure was reported for the preparation of $[YbCl(THF)(Cp^*)_2]$. Unsolvated $[Yb(\mu\text{-}Cl)_2Li(Cp^*)_2]$ reacts with aluminum trichloride to give the blue bimetallic complex $[Yb(\mu\text{-}Cl)_2AlCl_2(Cp^*)_2]$. Subsequent treatment of the intermediate tetrachoroaluminate with THF results in elimination of AlCl$_3$ and formation of $[YbCl(THF)(Cp^*)_2]$.[18] In some cases salt-free THF adducts are obtained directly from reactions of anhydrous lanthanide halides with Cp* reagents. For example, $[YCl(THF)(Cp^*)_2]$ was isolated as a by-product from the reaction of YCl$_3$ with two equivalents of K(Cp*) and its structure was determined.[568] All $[LnX(THF)(Cp^*)_2]$ derivatives were found to be isostructural exhibiting a pseudotetrahedral geometry around the metal atom. Besides the yttrium derivative the compounds [LnCl(THF)(Cp*)$_2$] with Ln = Ce,[633] Nd,[18] Sm,[314] Ho,[596,605] Yb,[314] Lu,[634] and [CeI(MeCN)$_2$(Cp*)$_2$][635] have been structurally characterized.

Synthesis and characterization of "ate" complexes of the type [Ln(μ-X)$_2$ML$_2$(Cp*)$_2$] (X = Cl, Br, I; M = Li, Na, K; L = THF, Et$_2$O, 1/2 DME) have been described in detail in a number of publications. These complexes are the most common type of Ln(Cp*)$_2$ derivatives. They are generally the initial products when anhydrous lanthanide trihalides are reacted with two equivalents of alkali metal pentamethylcyclopentadienides.[18,569,605,636,637] The first compound of this type to appear in the literature was [Nd(μ-Cl)$_2$Li(THF)$_2$(Cp*)$_2$] (**12**, Ln = Nd), made by reacting NdCl$_3$ with Li(Cp*) in THF solution.[638] A typical reaction sequence is given in Equation (142).

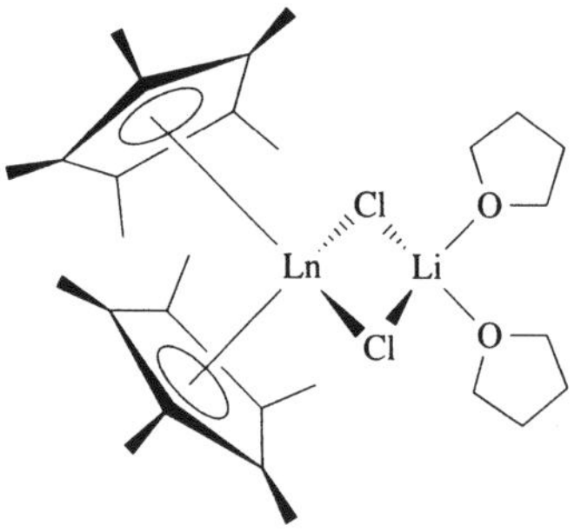

(12)

$$LnCl_3 + 2\,LiCp^* \xrightarrow{\text{THF}} [Ln(\mu\text{-Cl})_2Li(THF)_2Cp^*_2] + LiCl \qquad (142)$$

The vast majority of these reactions have been carried out with the lanthanide trichlorides. Li(Cp*), Na(Cp*), and K(Cp*) have been used as pentamethylcyclopentadienyl transfer reagents, and suitable coligands are diethyl ether, THF, DME, or TMEDA. In most cases the resulting "ate" complexes are well-crystallizing solids which are surprisingly soluble in organic solvents despite the presence of coordinated alkali metal halide. Complexes of this type have been synthesized for all lanthanide elements except promethium.[596,605] As so often, the cerium derivatives are the most air-sensitive members of the series. In the presence of traces of air, the bright yellow cerium complexes instantaneously turn purple or brown. However, it has so far not been possible to isolate a well-defined cerium(IV) species from these oxidized materials. Alternatively, treatment of inorganic cerium(IV) precursors with Li(Cp*) invariably produced [Ce(μ-Cl)$_2$LiL$_2$(Cp*)$_2$] (L = Et$_2$O, pyridine) species containing cerium(III).[639] The compounds have been characterized by various spectroscopic methods as well as x-ray crystal structure analyses.[616,636] Other structurally characterized complexes of this type are [Ce(μ-Cl)$_2$K(THF)$_n$(Cp*)$_2$],[633] [Pr(μ-Cl)$_2$Na(DME)$_2$(Cp*)$_2$],[605] and [Yb(μ-Cl)$_2$Li(Et$_2$O)$_2$(Cp*)$_2$].[18]

Similar "ate" complexes have been prepared with related pentaalkylcyclopentadienyl ligands. For example, the preparation and spectroscopic characterization of [Ln(μ-Cl)$_2$M(THF)$_2$(C$_5$Me$_4$Et)$_2$] (Ln = Nd, Gd, Yb; M = Li, K) have been reported. These complexes are obtained by procedures analogous to Equation (142).[640]

Finally, one case of an unsymmetrical dimer and a trimer have been described. The reaction of anhydrous YCl$_3$ with two equivalents of K(Cp*) in THF solution at room temperature was reported to yield [Y(μ-Cl)$_2$K(THF)$_2$(Cp*)$_2$] as the initial product. Upon heating the compound lost THF and KCl and sublimed as [Y(μ-Cl)(Cp*)$_2$YCl(Cp*)$_2$]. One bridging chloride ligand connects the two yttrium centers while the second chloride is terminally bonded to yttrium. Thus, one yttrium atom is formally seven-coordinate and the formal coordination number around the second yttrium center is 8. The Y–Cl–Y angle was determined to be 162.8(2)° (Figure 43).[641]

[SmCl(THF)(Cp*)$_2$] can be desolvated at 150 °C to form a material which was shown by x-ray structural analysis to be the cyclic trimer [{Sm(μ-Cl)(Cp*)$_2$}$_3$] (Sm–C 0.273(4) nm).[642] The molecular structure closely resembles that of the uranium(III) complex [{U(μ-Cl)(Cp*)$_2$}$_3$] (cf. Section 2.3.5.3). When the desolvation of [SmCl(THF)(Cp*)$_2$] is carried out in the presence of tetraglyme, the ionic complex [Sm(Cp*)$_2$Cl(μ-Cl)Sm(Cp*)$_2${μ-η^4-Me(OCH$_2$CH$_2$)$_4$OMe}Sm(Cp*)$_2$]$^+$[{SmCl(Cp*)$_2$}$_2$(μ-Cl)]$^-$ can be isolated. This material has also been structurally characterized. The structure contains four independent Sm(Cp*)$_2$ units and a variety of Sm–Cl bonds which allow a detailed analysis of trends in various structural features to be made.[642] Also related to the cyclotrimeric chloro complex [{Sm(μ-Cl)(Cp*)$_2$}$_3$] is the μ-cyanide trimer [{Sm(C$_6$H$_{11}$NC)(μ-CN)(Cp*)$_2$}$_3$], which was formed in an unusual

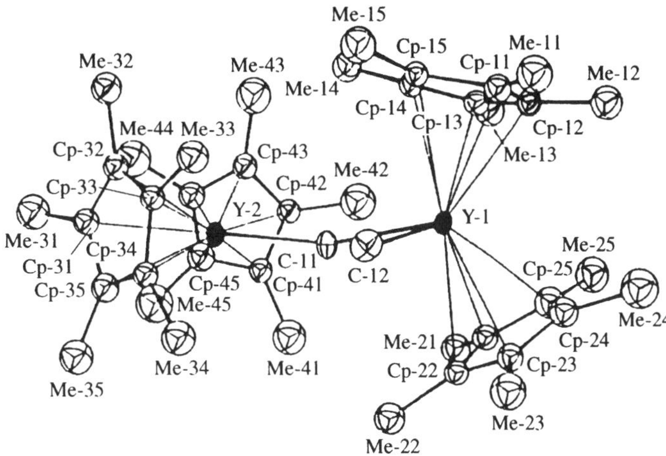

Figure 43 The molecular structure of [Y(μ-Cl)(Cp*)$_2$YCl(Cp*)$_2$].[641]

reaction between [Sm(THF)$_2$(Cp*)$_2$] and cyclohexyl isocyanide. The bridging cyanide ligands result from fragmentation of the organic isocyanide. The molecular structure of the trinuclear cyanide complex is depicted in Figure 44 (Sm–C(Cp*) 0.275(2) nm, Sm–C(CNR) 0.258(2) nm).[643,644]

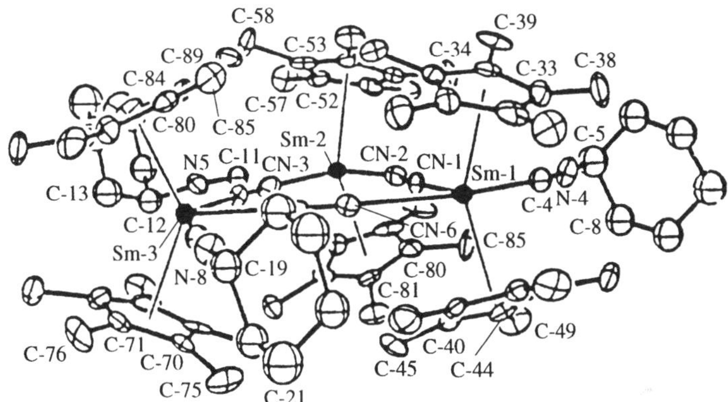

Figure 44 The molecular structure of [{Sm(C$_6$H$_{11}$NC)(μ-CN)(Cp*)$_2$}$_3$].[643,644]

A large number of Ln(Cp*)$_2$ derivatives containing additional group 16-centered ligands have been prepared and characterized. Many of them result from transformations of the corresponding samarium(II) or ytterbium(II) precursors. One typical reaction product is easily observed when [Sm(THF)$_2$(Cp*)$_2$] decomposes in the presence of traces of oxygen. The purple color of the samarium(II) complex then changes to orange-yellow, indicating the formation of [(μ-O){Sm(Cp*)$_2$}$_2$]. This oxygen-bridged dimer is also commonly the final product when [Sm(THF)$_2$(Cp*)$_2$] is reacted with various oxygen-containing substrates such as nitrosobenzene and pyridine-N-oxide.[397,593] The binuclear complex [(μ-O){Sm(Cp*)$_2$}$_2$] has been structurally characterized. As shown crystallographically, the compound contains an exactly linear Sm–O–Sm bridge. The two Sm(Cp*)$_2$ units are in a perpendicular arrangement relative to each other.[593] A structurally characterized derivative is the *t*-butyl cyanide adduct [(μ-O){Sm(Bu^{t}CN)(Cp*)$_2$}$_2$] (Sm–C 0.289(1) nm). Apparently the europium and ytterbium analogues of [(μ-O){Sm(Cp*)$_2$}$_2$] have not been prepared, although reactions of [Yb(THF)$_2$(Cp*)$_2$] and [YbCl(THF)(Cp*)$_2$] with O$_2$ and O$_2^-$ have been investigated and the chemiluminescence characteristics of the resulting ytterbium(III) species have been reported.[645]

Only a small number of alkoxide and aryloxide complexes of the type [LnOR(Cp*)$_2$] have been reported. An unsolvated samarium complex was prepared by reacting [Sm(THF)$_2$(Cp*)$_2$] with 2,3,5,6-tetramethylphenol in toluene solution (Equation (143)):[646]

$$[\text{Sm(THF)}_2\text{Cp*}_2] + \text{HOC}_6\text{HMe}_4\text{-2,3,5,6} \xrightarrow[-2\,\text{THF}]{} [\text{Sm(OC}_6\text{HMe}_4\text{-2,3,5,6)Cp*}_2] + 0.5\,\text{H}_2 \quad (143)$$

The x-ray crystal structure determination of [Sm(OC$_6$HMe$_4$-2,3,5,6)(Cp*)$_2$] revealed a nearly linear Sm–O–C unit.[646]

The reaction of $K(Cp^*)$ with the salt $[Sm(THF)_2(Cp^*)_2][BPh_4]$ unexpectedly produced the alkoxy derivative $[Sm\{O(CH_2)Cp^*\}(THF)(Cp^*)_2]$ via ring-opening of a THF ligand (Sm–O 0.2081(8) nm).[629] O,N-Coordination is found in iminoacetylacetonate chelate complexes of the type $[Ln\{O-C(Me)=CHC(Me)=NR\}(Cp^*)_2]$ (Ln = Y, La, Lu; R = Ph, C_6H_4Me-p). These compounds were made by metathetical reaction between $LnCl(Cp^*)_2$ precursors and sodium iminoacetylacetonates, and the molecular structure of $[La\{OC(Me)=CHC(Me)=NC_6H_4Me-p\}(Cp^*)_2]$ was reported.[647]

In some cases $Ln(Cp^*)_2$ derivatives containing sulfur ligands can be prepared in a conventional manner from the corresponding chloro complexes by salt-elimination reactions. Chelating dithiocarbamate ligands are suitable for such metathetical reactions; thus the first organolanthanide complexes with sulfur-coordinated ligands have been prepared according to Equation (144).[648]

$$LnCl(THF)Cp^*_2 + NaS_2CNEt_2 \xrightarrow{\text{THF}} [Ln(S_2CNEt_2)Cp^*_2] + NaCl \qquad (144)$$

$$Ln = Nd,\ Yb$$

The ytterbium derivative $[Yb(S_2CNEt_2)(Cp^*)_2]$ was structurally characterized by x-ray crystallography. The dithiocarbamate anion acts as chelating ligand and no additional solvent molecule is necessary to coordinatively saturate the central ytterbium atom.[648]

Another useful approach to bis(pentamethylcyclopentadienyl)lanthanide alkoxides and thiolates is protolysis of hydrocarbyls by free alcohols or thiols. For example, methane is eliminated upon treatment of $LuMe(Cp^*)_2$ with ethanol to afford the monomeric alkoxide $[LuOEt(Cp^*)_2]$.[649] One of the first organolanthanide complexes containing bridging thiolate ligands was prepared according to Equation (145).[650]

$$[Lu(\mu\text{-Me})_2Li(THF)_2Cp^*_2] + 2\ Bu^tSH \longrightarrow [Lu(\mu\text{-}SBu^t)_2Li(THF)_2Cp^*_2] + 2\ CH_4 \qquad (145)$$

Similar treatment of $[Lu(\mu\text{-Me})_2Li(THF)_2(Cp^*)_2]$ with Bu^tOH gave the THF solvate $[Lu(OBu^t)(THF)(Cp^*)_2]$. The molecular structure of $[Lu(\mu\text{-}SBu^t)_2Li(THF)_2(Cp^*)_2]$ was determined (Lu–C 0.266(2), Lu–S 0.2709(3) and 0.2723(3) nm).[369,370,650]

Various other complexes in which $Ln(Cp^*)_2$ units are bonded to organic ligands via group 15 or 16 atoms have been obtained starting from $[Sm(THF)_2(Cp^*)_2]$ or $[Yb(Et_2O)(Cp^*)_2]$. Reductive cleavage of E–E bonds by lanthanide(II) complexes probably represents the most elegant route to compounds containing Ln–E bonds (E = S, Se, Te). For example, monomeric organosamarium(III) chalcogenolates have been prepared according to Equations (146) and (147).[651]

$$2\ [Sm(THF)_2Cp^*_2] + [Me_2NC(S)S]_2 \longrightarrow 2\ [Sm(S_2CNMe_2)Cp^*_2] + 4\ THF \qquad (146)$$

$$2\ [Sm(THF)_2Cp^*_2] + REER \longrightarrow 2\ [Sm(ER)(THF)Cp^*_2] + 2\ THF \qquad (147)$$

$$E = S,\ Se,\ Te$$
$$R = Mes,\ C_6H_2(CF_3)_3\text{-}2,4,6$$

The molecular structures of $[Sm(S_2CNMe_2)(Cp^*)_2]$ (Sm–C 0.2711(7) nm, Sm–S 0.2808(2) nm), $[Sm\{SeC_6H_2(CF_3)_3\text{-}2,4,6\}(THF)(Cp^*)_2]$ (Figure 45, Sm–C 0.2734(4) nm, Sm–Se 0.2919(1) nm), and $[Sm(TeMes)(THF)(Cp^*)_2]$ (Sm–C 0.273(1) nm, Sm–Te 0.3088(2) nm) have been determined by x-ray crystallography.[651]

Monomeric and dimeric $Yb(Cp^*)_2$ complexes of the type $[Yb(ER)L(Cp^*)_2]$ (E = O, S, Se, Te; L = Lewis base), $[\{Yb(Cp^*)_2\}_2(\mu\text{-}E)]$ (E = S, Se, Te) and $[\{Yb(Cp^*)_2\}_2(\mu\text{-}Te_2)]$ have been investigated by Andersen et al.[652–4] Monomeric species are readily formed upon treatment of $[Yb(Et_2O)(Cp^*)_2]$ with dichalcogenides. The same ytterbium(II) precursor reacts with elemental sulfur, selenium, or tellurium to afford the $\mu\text{-}E$ or $\mu\text{-}E_2$-bridged complexes. The monomeric ytterbium(III) chalcogenolates $[Yb(SPh)(NH_3)(Cp^*)_2]$ (Yb–C 0.264(2) nm, Yb–N 0.2428(5) nm, Yb–S 0.2674(5) nm) and $[Yb(TePh)(NH_3)(Cp^*)_2]$ (Yb–C 0.263(3) nm, Yb–N 0.250(1) nm, Yb–Te 0.3039(1) nm) were crystallographically characterized.[653,655] In the binuclear complex $[\{Yb(Cp^*)_2\}_2(\mu\text{-}Te_2)]$ the two $Yb(Cp^*)_2$ fragments are bridged by a side-on coordinated $\mu\text{-}Te_2$ unit (Yb–C 0.263(2) nm, Yb–Te 0.3156(4) nm).[654]

Related scandium tellurides have been prepared either from $[ScR(Cp^*)_2]$ or $ScD(Cp^*)_2$. Elemental tellurium slowly inserts into the Sc–C bond of the scandium hydrocarbyls to afford the corresponding tellurolates. More rapid tellurium insertion can be achieved when Bu^n_3PTe is used as a soluble tellurium source. The tellurolates decompose upon heating in solution to afford the tellurium-bridged binuclear complex $[\{Sc(Cp^*)_2\}_2(\mu\text{-}Te)]$ (Scheme 28). Photochemically induced reversion of the thermal reaction

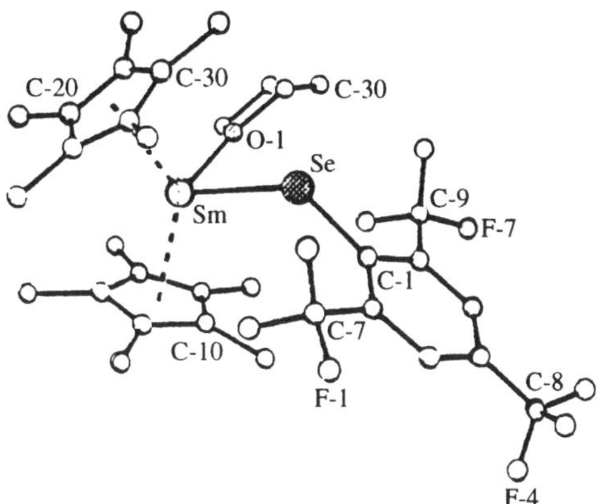

Figure 45 The molecular structure of $[Sm\{SeC_6H_2(CF_3)_3\text{-}2,4,6\}(THF)(Cp^*)_2]$.[651]

was observed upon UV irradiation of a mixture of $[\{Sc(Cp^*)_2\}_2(\mu\text{-Te})]$ and TeR_2.[656] The binuclear telluride can be obtained directly by treatment of $ScD(Cp^*)_2$ with $Bu^n{}_3PTe$. It was found to be isostructural with its ytterbium analogue $[\{Yb(Cp^*)_2\}_2(\mu\text{-Te})]$.[657]

$$2\,[ScRCp^*{}_2] + 2\,Te\ \text{or}\ 2\,Bu^n{}_3PTe \longrightarrow 2\,[ScTeRCp^*{}_2] \xrightarrow[-TeR_2]{\Delta} [\{ScCp^*{}_2\}_2(\mu\text{-Te})]$$

$$R = CH_2TMS,\ Bz$$

Scheme 28

Reactions of $[Sm(THF)_2(Cp^*)_2]$ with group 15 compounds are often quite unusual and in several cases further transformations of the organic ligands in the coordination sphere of samarium can be carried out. A typical example for such a reaction sequence is the samarium-mediated insertion of carbon monoxide into the N=N double bond of azobenzene.[568,658] Treatment of $[Sm(THF)_2(Cp^*)_2]$ with azobenzene in toluene solution (molar ratio 2:1) initially affords the binuclear complex $[\{Sm(Cp^*)_2\}_2(\mu\text{-}\eta^1{:}\eta^1\text{-PhNNPh})]$ in which the two decamethylsamarocene units are in a trans arrangement. Although samarium is coordinated primarily to the nitrogen atoms of the ligand, an x-ray structure determination of the product also revealed an agostic interaction between the samarium atoms and ortho hydrogens of the phenyl rings (Sm–C(Cp*) 0.274(3) nm, Sm–N 0.240(1) nm).

In THF solution the compound $[\{Sm(Cp^*)_2\}_2(\mu\text{-PhNNPh})]$ undergoes a ligand redistribution reaction to give the solvated complexes $[\{Sm(THF)(Cp^*)_2\}_2(\mu\text{-}\eta^2{:}\eta^2\text{-PhNNPh})_2]\cdot2THF$ (Sm–C(Cp*) 0.276(1) nm, Sm–N 0.2323(8) nm, 0.2559(7) nm) and $[Sm(\eta^2\text{-PhNNPh})(THF)(Cp^*)_2]\cdot0.5THF$ (Sm–C(Cp*) 0.276(2) nm, Sm–N 0.242(2) nm). A reaction of $[Yb(THF)(Cp^*)_2]$ with azobenzene produced the binuclear complex $[\{Yb(THF)(Cp^*)\}_2(\mu\text{-}\eta^2{:}\eta^2\text{-PhNNPh})_2]\cdot$toluene (Yb–C(Cp*) 0.263(2) nm, Yb–N 0.2197(9), 0.253(4) nm).[658] The unsolvated samarium azobenzene complex was subsequently treated with carbon monoxide, which resulted in a double insertion of CO into the nitrogen–nitrogen bond. The resulting binuclear complex $[\{Sm(Cp^*)_2\}_2\{\mu\text{-}\eta^4\text{-}(PhN)OCCO(NPh)\}]$ was structurally characterized. Not surprisingly, the oxygen atoms of the new ligand were involved in the coordination to samarium (Scheme 29, Sm–C 0.271(1) nm, Sm–N 0.249(1) nm, Sm–O 0.230(1) nm).[568,658,659]

$$[Sm(THF)_2Cp^*{}_2] + PhN=NPh \longrightarrow [\{SmCp^*{}_2\}_2(\mu\text{-PhNNPh})] \xrightarrow{+2\,CO}$$

Scheme 29

A similar reaction sequence was performed by reacting [Sm(THF)$_2$(Cp*)$_2$] first with diphenylacetylene and subsequently treating the initial product with carbon monoxide.[660] Formal addition of two carbon monoxide molecules to diphenylacetylene resulted in the formation of an indenoindene ring system. In the final product two Sm(Cp*)$_2$ units were coordinated only to oxygen atoms of the new ligand showing again the strong preference of lanthanides for hard donor ligands (Scheme 30).[660]

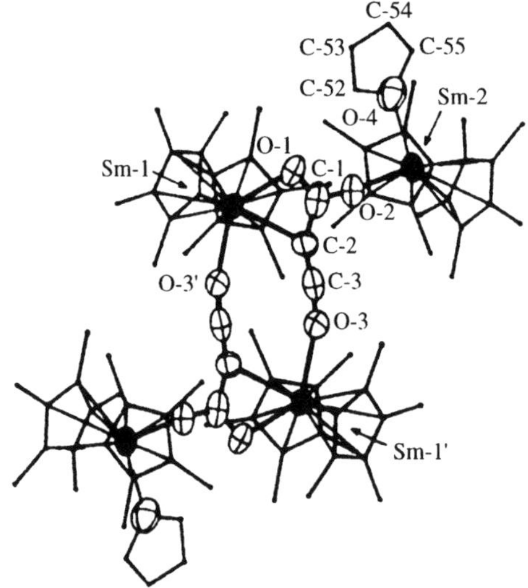

Scheme 30

5,10-Dihydroindeno[2,1,a]indene was isolated after reductive cleavage of [(μ-C$_{16}$H$_{10}$O$_2$){Sm(Cp*)$_2$}$_2$] with sodium in THF. Similar results have been reported for reactions [Sm(THF)$_2$(Cp*)$_2$] with bis(2-pyridyl)ethylene.[661] A 2:1 stoichiometry leads to the formation of [(μ-η^2:η^2-pyCHCHpy){Sm(Cp*)$_2$}$_2$], while combination of the same reagents in a 1:1 molar ratio produces [(μ-η^3:η^3-py$_4$C$_4$H$_4$){Sm(Cp*)$_2$}$_2$] via dimerization of the bis(2-pyridyl)ethylene ligand. Both products have been structurally characterized by x-ray crystallography.[661]

One of the most unusual organolanthanide complexes results from the reaction of [Sm(THF)$_2$(Cp*)$_2$] with carbon monoxide alone. Treatment of the samarium(II) precursor with CO under pressure gave a black crystalline material which was shown by x-ray structural analysis to be the tetranuclear samarium(III) complex [{Sm$_2$(μ-O$_2$CCCO)(THF)(Cp*)$_4$}$_2$] (Figure 46).[662]

Figure 46   The molecular structure of [{Sm$_2$(μ-O$_2$CCCO)(THF)(Cp*)$_4$}$_2$].[662]

The result of this unexpected reaction is a trimerization of carbon monoxide to give a ketenecarboxylate dianion, O$_2$CCCO^{2-}. In the product two of these ketenecarboxylate dianions act as bridging ligands between four Sm(Cp*)$_2$ units.

The use of the bulky bis(trimethylsilyl)amide ligands allows the synthesis of unsolvated organolanthanide amides of the type [LnN(TMS)$_2$(Cp*)$_2$] (Equation (148)). The molecular structure of [SmN(TMS)$_2$(Cp*)$_2$] has been determined.[18,185,663]

$$[Ln(\mu\text{-Cl})_2Li(Et_2O)_2Cp^*_2] + MN(TMS)_2 \longrightarrow [LnN(TMS)_2Cp^*_2] + MCl + LiCl \qquad (148)$$

Ln = Y, Ce, Nd, Sm
M = Li, Na, K

Adducts of Sm(Cp*)$_2$ with bipyridyl and diazadienes (Equation (149)) have been described which are readily formed by treatment of [Sm(THF)$_2$(Cp*)$_2$] with the corresponding nitrogen donor ligands.[664,665]

$$[Sm(THF)_2Cp^*_2] + RN=CH-CH=NR \longrightarrow [Sm(\eta^2\text{-}RN=CH-CH=NR)Cp^*_2] + 2\,THF \qquad (149)$$

$$R = Pr^i,\ Bu^t,\ c\text{-}C_6H_{11},\ p\text{-}C_6H_4Me$$

The N-*t*-butyl derivative [Sm(η^2-Bu^{t}N=CH–CH=NBu^t)(Cp*)$_2$] has been structurally characterized (Figure 47, Sm–C 0.2773(7) nm, Sm–N 0.2480(5) and 0.2489(5) nm).[665] The structural as well as the spectroscopic data are consistent with the presence of samarium(III) complexes of the bipyridyl or diazadiene radical anions. Similar reactions of [Sm(THF)$_2$(Cp*)$_2$] with pyridazine and benzaldazine afforded the reductively coupled products [{μ-η^4-(CH=NNCH=CHCH)$_2$}{Sm(THF)(Cp*)$_2$}$_2$] (Figure 48) and [{μ-η^4-(PhHC=NNCHPh)$_2$}{Sm(Cp*)$_2$}$_2$].[664]

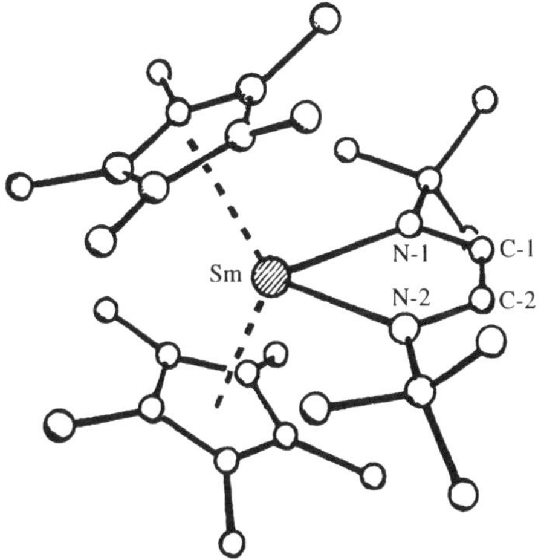

Figure 47　The molecular structure of [Sm(η^2-Bu^{t}N=CH–CH=NBu^t)(Cp*)$_2$].[665]

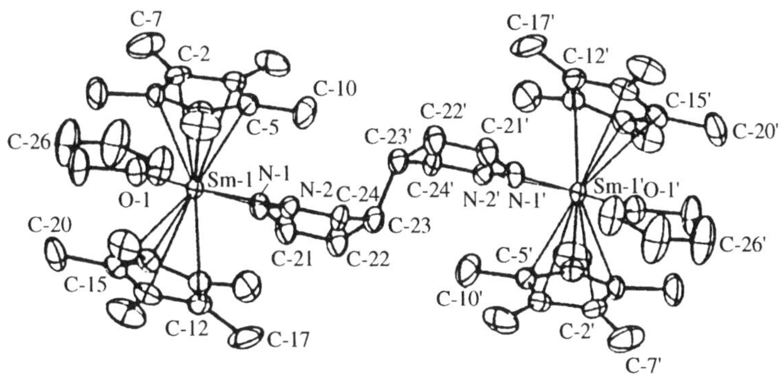

Figure 48　The molecular structure of [{μ-η^4-(CH=NNCH=CHCH)$_2$}{Sm(THF)(Cp*)$_2$}$_2$].[664]

Dianions of certain nitrogen heterocycles have been found to act as antiaromatic bridging ligands between decamethyllanthanocene units. The first complexes of this type were prepared by *in situ* preparation of the dianion followed by treatment with [La(μ-Cl)$_2$K(DME)$_2$(Cp*)$_2$]. Equation (150) shows the example of 2,3-dimethylquinoxaline. A related binuclear complex was synthesized using the dianion of phenazine. The molecular structure of 2,3-dimethylquinoxaline derivative was determined by x-ray methods.[666]

The first dinitrogen complex of an *f*-element, [{Sm(Cp*)$_2$}$_2$(μ-η^2:η^2-N$_2$)] (**13**) was obtained by slow evaporation of a dark green toluene solution of Sm(Cp*)$_2$ in a nitrogen atmosphere. The red-brown crystals were structurally characterized. In the unusual binuclear complex the dinitrogen ligand is side-on coordinated to two Sm(Cp*)$_2$ units. The central Sm(μ-η^2:η^2-N$_2$)Sm unit is planar. (Figure 49, Sm–C 0.273(2) nm, Sm–N 0.236(1) nm).[667]

Scandium hydrazido(1-) complexes were prepared according to Equation (151). The products undergo insertion reactions with acetonitrile.[668]

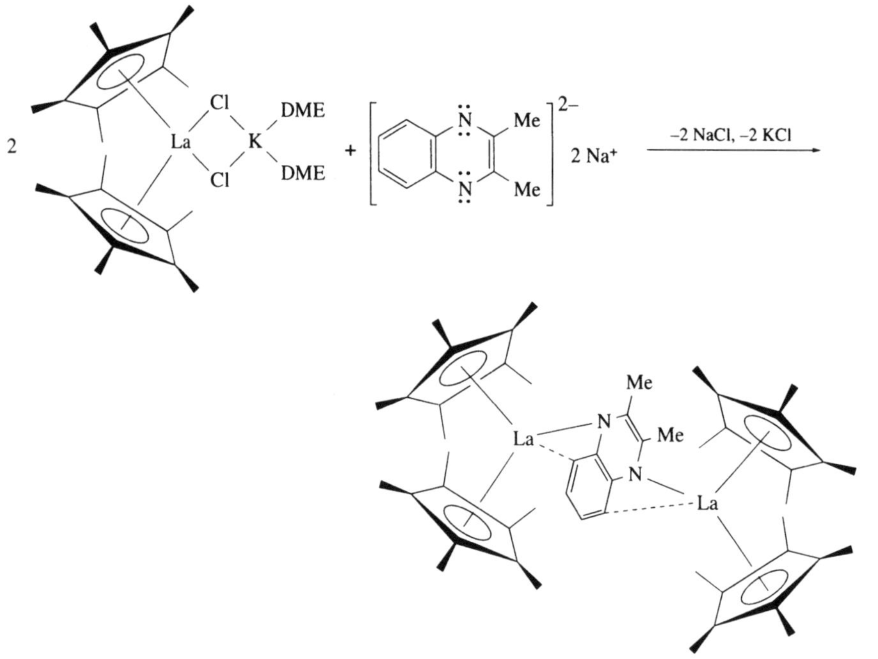

(150)

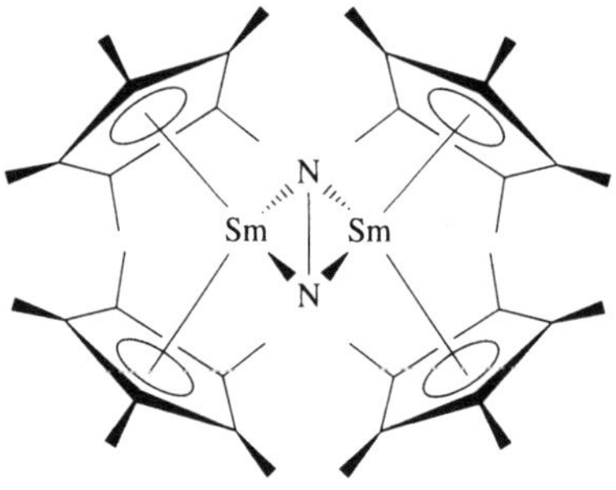

(13)

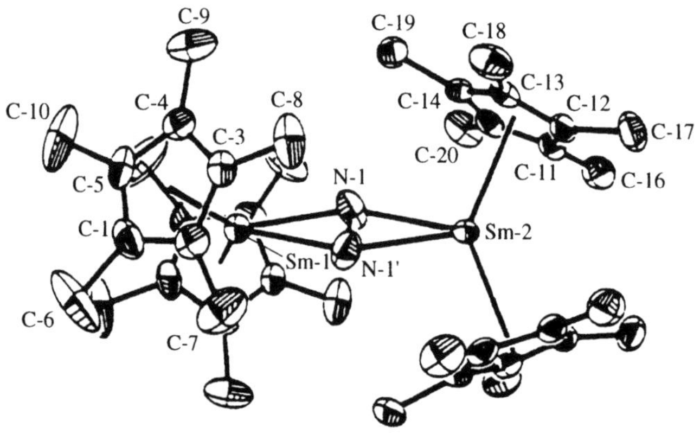

Figure 49 The molecular structure of [{Sm(Cp*)$_2$}$_2$(μ-η^2:η^2-N$_2$)].[667]

$$ScMeCp^*_2 + H_2NNR_2 \longrightarrow [ScN(H)NR_2Cp^*_2] + CH_4 \qquad (151)$$

$$R = H, Me$$

Hydrazine reacts with Sm(Cp*)$_2$ or [{Sm(μ-H)(Cp*)$_2$}$_2$] to afford the structurally characterized binuclear μ-hydrazido(2-) complex [(μ-η^2:η^2-NHNH){Sm(THF)(Cp*)$_2$}$_2$] which adopts a butterfly structure in contrast to the planar core in [(μ-η^2:η^2-N$_2$){Sm(Cp*)$_2$}$_2$]. The μ-hydrazido(2-) ligand can be protonated by treatment with [Et$_3$NH][BPh$_4$] to give the cationic hydrazine complex [{Sm(η^2-NH$_2$NH$_2$)(THF)(Cp*)$_2$}(BPh$_4$)] (Equation (152)).[669]

$$[\{Sm(THF)Cp^*_2\}_2(\mu\text{-}\eta^2\text{:}\eta^2\text{-NHNH})] + 2\,[Et_3NH][BPh_4] \longrightarrow [Sm(\mu^2\text{-NH}_2NH_2)(THF)Cp^*_2][BPh_4] +$$
$$[Sm(THF)_2Cp^*_2][BPh_4] \qquad (152)$$

With excess hydrazine, $Sm(Cp^*)_2$ reacts under formation of the tetranuclear hydrazido complex $[\{Sm(Cp^*)\}_4(NHNH)_2(NHNH_2)_4(NH_3)_2]$, which was structurally characterized (see above).[603]

A related reaction of $[Sm(THF)_2(Cp^*)_2]$ with N,N'-diphenylhydrazine takes a different pathway, as the N–N bond is cleaved to afford the mononuclear amido complex $[SmNHPh(THF)(Cp^*)_2]$.[670] Apparently no polyphosphide complex derived from $[Sm(THF)_2(Cp^*)_2]$ has been reported, although some interesting results have been obtained with antimony and bismuth. $Sb(Bu^n)_3$ reacts with unsolvated $Sm(Cp^*)_2$ to give the structurally characterized trinuclear product $[(\mu\text{-}\eta^2\text{:}\eta^2\text{:}\eta^1\text{-Sb}_3)\{Sm(Cp^*)_2\}_3(THF)]$, which can be described as a derivative of the triantimony Zintl ion (Sm–C 0.272(2) nm, Sm–Bi 0.329(2) nm).[671] The analogous reaction with $BiPh_3$ illustrates the fact that it is often difficult if not impossible to predict the outcome of reactions involving $[Sm(THF)_2(Cp^*)_2]$ or $Sm(Cp^*)_2$. In this case the bismuth analogue of the above-mentioned dinitrogen complex is formed among other products (Equation (153)). $[\{SmCp^*_2\}_2(\mu\text{-}\eta^2\text{:}\eta^2\text{-Bi}_2)]$ was structurally characterized.[672]

$$4\,[SmCp^*_2] + 2\,BiPh_3 \longrightarrow [\{SmCp^*_2\}_2(\mu\text{-}\eta^2\text{:}\eta^2\text{-Bi}_2)] + PhPh + 2\,[SmPhCp^*_2] \qquad (153)$$

(b) Hydrocarbyls. Organolanthanide hydrocarbyls of the type $LnR(Cp^*)_2$ are highly important as catalytically active species and precursors for catalytically active organolanthanide hydrides.[187,673-7] Typical preparations start with "ate" complexes of the type $[Ln(\mu\text{-Cl})_2ML_2(Cp^*)_2]$ (M = alkali metal, L = Et_2O, THF, $\tfrac{1}{2}$DME), which are treated with alkyllithium reagents. In the case of small alkyl groups such as methyl the products are again "ate" complexes in which the lanthanide and alkali metal are bridged by the hydrocarbyl ligands. Instead of tetrahydrofuran, DME or TMEDA have also been used to saturate the coordination sphere of the alkali metal. The stepwise formation of such complexes was demonstrated by the following reaction sequence (Scheme 31).[18,26,604,605,678]

$$[Ln(\mu\text{-Cl})_2Li(THF)_2Cp^*_2] + MeLi \xrightarrow[-LiCl]{} [Ln(\mu\text{-Cl})(\mu\text{-Me})Li(THF)_2Cp^*_2] \xrightarrow[-LiCl]{MeLi}$$

$$[Ln(\mu\text{-Me})_2Li(THF)_2Cp^*_2]$$

Ln = La, Pr, Yb, Lu

Scheme 31

The reaction of $[Sm(THF)_2Cp^*_2][BPh_4]$ with lithium alkyls provides a halide-free access to $(Cp^*)_2Sm$ hydrocarbyls (Equation (154)).[629]

$$[Sm(THF)_2Cp^*_2][BPh_4] + LiR \longrightarrow [SmR(THF)Cp^*_2] + Li[BPh_4] \qquad (154)$$

R = Me, Ph

The THF adducts $[LnMe(THF)(Cp^*)_2]$ (Ln = Sc, Y, Sm, Yb, Lu) have been isolated and all are thermally stable. Structurally characterized solvated organolanthanide methyl complexes are $[YMe(THF)(Cp^*)_2]$ (Y–C(Cp^*) 0.266(2) nm, Y–C(Me) 0.255(2) nm),[679] $[SmMe(THF)(Cp^*)_2]$ (Sm–C(Cp^*) 0.2711(6) nm, Sm–C(Me) 0.248(1) nm),[680] $[YMe(THF)(Cp^*)_2]$, and $[Yb\text{-}Me(Et_2O)(Cp^*)_2]$.[677] An excess of organolithium reagent can lead to the formation of anionic hydrocarbyls such as $[Li(DME)_2][Ln(CH_2TMS)_2(Cp^*)_2]$ (Ln = Pr, Lu).[604,605]

Monomeric organoscandium hydrocarbyls[681] are readily prepared from $ScCl(Cp^*)_2$ and organolithium reagents (Equation (155)).[673]

$$ScClCp^*_2 + LiR \longrightarrow [ScRCp^*_2] + LiCl \qquad (155)$$

R = Me, Ph, $C_6H_4Me(o,m,p)$, CH_2Ph

In some cases solvent-free hydrocarbyls have been prepared by adding stoichiometric amounts of alkenes to the corresponding hydrides (e.g. Equation (156)).[673]

$$1/n\,[\{ScHCp^*_2\}_n] + H_2C{=}CHR \longrightarrow [ScCH_2CH_2RCp^*_2] \qquad (156)$$

R = H, Me

The successful preparation of the base-free, highly reactive lanthanide methyls [ScMe(Cp*)$_2$][673] and [{LnMe(Cp*)$_2$}$_2$] (Ln = Y, Lu) can be considered a landmark in organo-*f*-element chemistry.[18,649,675] Base-free [{YbMe(Cp*)$_2$}$_2$] was recently prepared from Yb(Cp*)$_2$ and methylcopper. It forms an asymmetrically (μ-Me)$_2$ bridged dimer in the solid state.[682] The monomeric 14-electron species [ScMe(Cp*)$_2$] has been structurally characterized. Unexpectedly, α-agostic C–H–Sc interaction does not play any significant role in this compound.[673] A dimeric structure was found by x-ray crystallography for [{LuMe(Cp*)$_2$}$_2$]. One methyl group acts as a bridge between the two lutetium atoms, while the second methyl ligand is terminal.[187]

The methyl complexes are active catalysts for alkene polymerizations, and they activate C–H bonds. Hydrogenolysis of [ScMe(Cp*)$_2$] and [{LnMe(Cp*)$_2$}$_2$] (Ln = Y, Lu) yields the corresponding hydrides, [{Sc(μ-H)(Cp*)$_2$}$_n$] and [{Ln(μ-H)(Cp*)$_2$}$_2$] (Ln = Y, Lu).[649,673] All three methyl derivatives activate the C–H bonds of a range of hydrocarbons including ^{13}CH$_4$, arenes, styrene, and alkynes.[403,404,675,683] Facile elimination of methane is observed when the σ-methyl complexes are treated with benzene, tetramethylsilane, pyridine, or methylenetriphenylphosphorane (Scheme 32).[649,675]

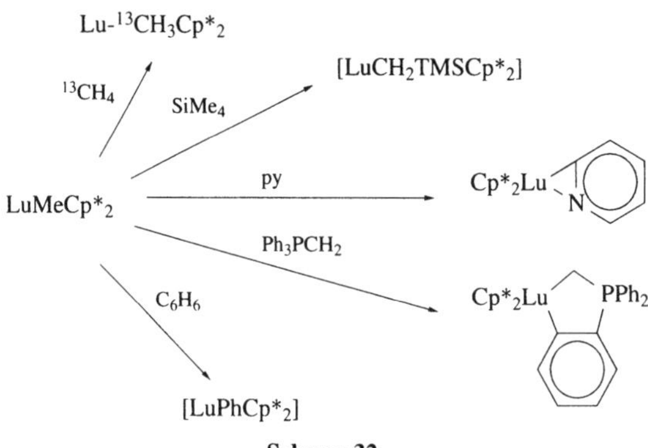

Scheme 32

The C–H activation of ^{13}CH$_4$ by LnMe(Cp*)$_2$ species in cyclohexane solution has been studied by a kinetic analysis. The relative rates of this reaction are in the order Y (250) > Lu (5) > Sc (1). A four-center transition state was formulated as an intermediate. Scheme 33 shows the proposed scenario. H–H and C–H activation reactions involving LnH(Cp*)$_2$ and LnR(Cp*)$_2$ derivatives have also been the subject of theoretical studies.[684,685]

Scheme 33

For [ScR(Cp*)$_2$] the relative reactivity toward various types of C–H bonds has been studied in detail (Equation (157)).[677] The reaction rate has been found to depend very much on the *s* character of the reacting bonds, that is, it follows the order R^1,R^2 = H > sp C > sp^2 C > sp^3 C.[25]

[ScMe(Cp*)$_2$] reacts with substituted alkenes (e.g., propylene) at the terminal *sp*^{2} C–H bond to afford the corresponding vinylic C–H activation products (Equation (158)).

Of great importance as precatalysts in organolanthanide-catalyzed processes are lanthanide hydrocarbyls containing the bulky bis(trimethylsilyl)methyl ligand. They are also the most suitable precursors for the preparation of the dimeric hydrides [{Ln(μ-H)(Cp*)$_2$}$_2$]. Such compounds can be isolated completely free of coordinating solvents and alkali metal halides (Equation (159)).[184–6,540,686–8]

In the case of yttrium, salt-free [YCl(THF)(Cp*)$_2$] is the preferred starting material.[185] The molecular structure of [YCH(TMS)$_2$(Cp*)$_2$] has been determined, showing α-C–H–Y and γ-Me⋯Y agostic interactions. Both types of secondary interactions relieve the electronic unsaturation in the formally 14-electron complex.[185] A somewhat different behavior was reported for the corresponding organocerium hydrocarbyl [CeCH(TMS)$_2$(Cp*)$_2$].[186,687] The original preparation called for treatment of the "ate" complex [Ce(μ-Cl)$_2$Li(THF)$_2$(Cp*)$_2$] with LiCH(TMS)$_2$. However, the yields were low because the LiCl

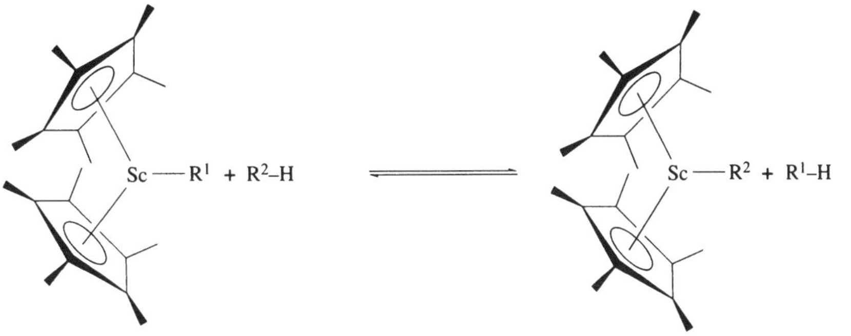

$$\text{[Sc]}-R^1 + R^2-H \rightleftharpoons \text{[Sc]}-R^2 + R^1-H \qquad (157)$$

$$[ScMeCp^*_2] + H_2C=CHR \longrightarrow [ScCH=CHRCp^*_2] + CH_4 \qquad (158)$$

$$[Ln(\mu\text{-}Cl)_2Li(Et_2O)_2Cp^*_2] + LiCH(TMS)_2 \longrightarrow [LnCH(TMS)_2Cp^*_2] + 2\,LiCl + 2\,Et_2O \qquad (159)$$

$$Ln = Y,\ La,\ Ce,\ Nd,\ Sm,\ Lu$$

by-product (two equivalents) reacts with $[CeCH(TMS)_2(Cp^*)_2]$. A more reliable preparation of $[CeCH(TMS)_2(Cp^*)_2]$ involves the reaction of $[CeCl(Cp^*)_2]$ with the organolithium reagent (Equation (160)).

$$CeClCp^*_2 + LiCH(TMS)_2 \longrightarrow [CeCH(TMS)_2Cp^*_2] + LiCl \qquad (160)$$

The molecular structures of both $[CeCH(TMS)_2(Cp^*)_2]$ and $[NdCH(TMS)_2(Cp^*)_2]$[686] have been determined by x-ray diffraction. Both complexes exhibit an agostic α-C–H–Ln interaction as well as a secondary β-Si–Me–Ln interaction.

Scandium hydrocarbyls of the type $[ScR^1(Cp^*)_2]$ (R^1 = Me, *p*-tolyl) react with nitriles R^2CN to afford the azomethine complexes $[ScNCR^1R^2(Cp^*)_2]$.[689] However, in the case of $[ScMe(Cp^*)_2]$ the reactivity of the scandium center is so high, that further addition of nitrile occurs (Scheme 34).[25]

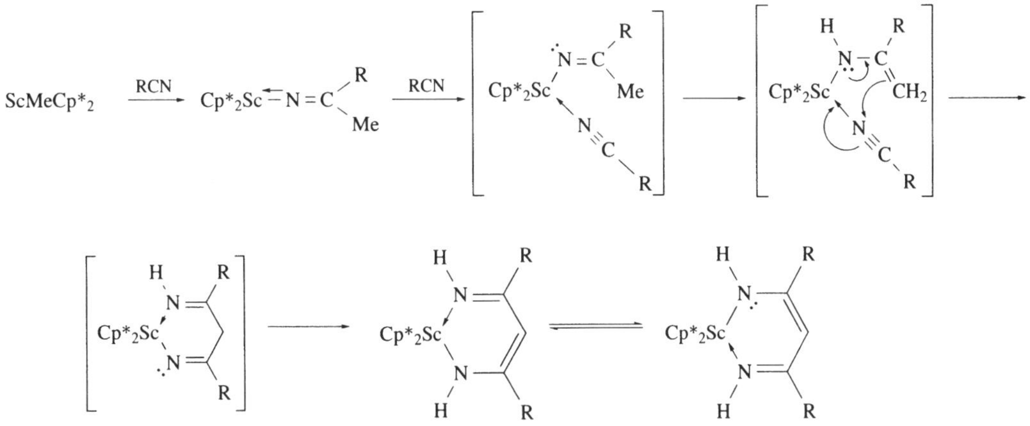

Scheme 34

In contrast, C–H activation takes place when the lanthanide hydrocarbyls $[LnCH(TMS)_2(Cp^*)_2]$ (Ln = La, Ce) are reacted with acetonitrile. In this case the binuclear μ-cyanomethyl complexes $[\{LnCH_2CN(Cp^*)_2\}_2]$ have been isolated. The lanthanum derivative was structurally characterized (La–C(σ) 0.2748(4) nm, La–N 0.2537(4) nm).[690] Monomeric $[ScR(Cp^*)_2]$ complexes also metallate pyridine to give the structurally characterized compound $[Sc(C,N\text{-}\eta^2\text{-}C_5H_4N)(Cp^*)_2]$. The thermal decomposition of $[ScMe(Cp^*)_2]$ in cyclohexane at 80 °C has been studied. The product, $[\{Sc(\eta^6\text{-}C_5Me_4CH_2)(Cp^*)\}_n]$, was shown to contain a "tucked-in" tetramethylfulvene ligand.[673] The same material was also formed besides methylcyclopentane during the slow decomposition of $[ScCH_2(c\text{-}C_5H_8)(Cp^*)_2]$. An x-ray structure determination revealed that the decomposition product was the fulvene-bridged dimer $[\{Sc(\mu\text{-}\eta^1,\eta^5\text{-}C_5Me_4CH_2)(Cp^*)\}_2]$ (Figure 50).[691]

CO and CO_2 readily insert into the Sc–C bond of $[ScR(Cp^*)_2]$ to afford the corresponding η^2-acyls and carboxylates, respectively (Equations (161) and (162)).[25]

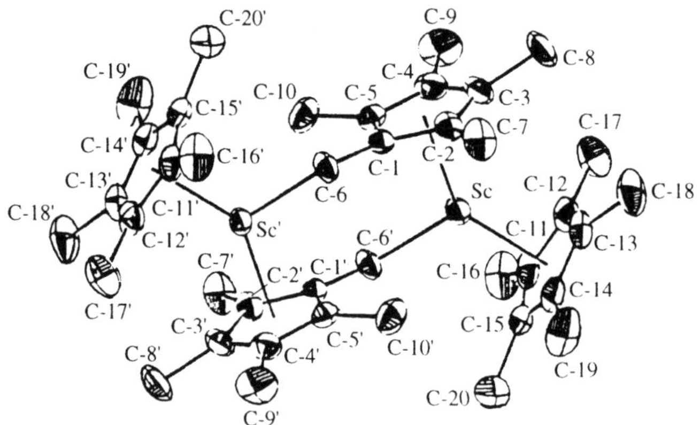

Figure 50   The molecular structure of [{Sc(μ-η^1,η^5-C$_5$Me$_4$CH$_2$)(Cp*)}$_2$].[691]

$$[\text{ScPhCp*}_2] + \text{CO} \longrightarrow [\text{Sc}(\eta^2\text{-COPh})\text{Cp*}_2] \tag{161}$$

$$[\text{ScC}_6\text{H}_4\text{MeCp*}_2] + \text{CO}_2 \longrightarrow [\text{Sc}(\eta^2\text{-O}_2\text{CC}_6\text{H}_4\text{Me})\text{Cp*}_2] \tag{162}$$

Coordinated carbon monoxide also reacts with the Sc–C bond in [ScR(Cp*)$_2$]. In this case scandiumoxycarbene complexes are obtained (Equation (163)).[25]

$$[\text{ScRCp*}_2] + [\text{M(CO}_2)\text{Cp}] \longrightarrow [\text{ScCp*}_2\text{OC(R)}{=}\text{M(CO)}_2\text{Cp}] \tag{163}$$

$$\text{M = Co, Rh; R = Me, CH}_2\text{CH}_2\text{Ph}$$

An extensive derivative chemistry based on the decamethylyttrocene hydrocarbyls [YCH(TMS)$_2$(Cp*)$_2$] and [YMe(THF)(Cp*)$_2$] has been developed by Teuben *et al.*[692,693] Typical reactions of [YCH(TMS)$_2$(Cp*)$_2$] are summarized in Scheme 35.

Scheme 35

As expected, facile cleavage of the Y–C bond occurs upon treatment of the hydrocarbyls with protic reagents such as alcohols or terminal alkynes.[692,694] Other typical reactions include insertions of CO$_2$ or Bu^{t}NC into the yttrium–carbon bond. Bu^{t}NC only forms a Lewis base adduct with [YCH(TMS)$_2$(Cp*)$_2$]. Xylyl isocanide did not react with [YCH(TMS)$_2$(Cp*)$_2$] at all.[693] In contrast, the less sterically encumbered [YCH$_2$C$_6$H$_3$Me$_2$-3,5(Cp*)$_2$] afforded an insertion product when reacted with xylyl isocyanide.[692] The resulting [Y{η^2-C(CH$_2$C$_6$H$_3$Me$_2$)=N(2,6-xylyl)}(THF)(Cp*)$_2$] was structurally characterized (Y–C(Cp*) 0.2728(4) nm, Y–C(η^2) 0.2392(3) nm, Y–N(η^2) 0.2407(3) nm). The starting material [YCH$_2$C$_6$H$_3$Me$_2$-3,5(Cp*)$_2$] was made by thermolysis of [YCH(TMS)$_2$(Cp*)$_2$] in mesitylene.[694] Like other reactive lanthanide hydrocarbyls [YCH(TMS)$_2$(Cp*)$_2$] metallates pyridine to afford the η^2-pyridine complex [Y(η^2-C$_5$H$_4$N)(Cp*)$_2$], which was also obtained from [YMe(THF)(Cp*)$_2$].[693]

The hydrocarbyls [LnCH(TMS)$_2$(Cp*)$_2$] (Ln = La, Ce) display a differentiated reactivity toward ketones.[695] It was proposed that the differences in reactivity are thermodynamic and not kinetic in origin. The early lanthanide hydrocarbyls did not react with the sterically hindered di-*t*-butyl ketone. Hydrogen transfer and formation of the lanthanide aldolates [LnOCMe$_2$CH$_2$C(=O)Me(Cp*)$_2$] (Ln = La, Ce) was observed when the hydrocarbyls were treated with acetone (Equation (164)). The cerium derivative was structurally characterized.

$$[LnCH(TMS)_2Cp^*_2] + 2 \; \overset{O}{\underset{}{\|}} \longrightarrow Cp^*_2Ln \overset{O=\!\!<}{\underset{O-\!\!<}{}} + CH_2(TMS)_2 \qquad (164)$$

In contrast, addition of diethyl ketone to $[LnCH(TMS)_2(Cp^*)_2]$ did not result in C–C coupling but afforded the enolate–ketone adducts $[Ln\{OC(Et)=C(H)Me\}(O=CEt_2)(Cp^*)_2]$.[695]

The thermolysis of the hydrocarbyls $[LnCH(TMS)_2(Cp^*)_2]$ (Ln = Y, La, Ce) in C_6D_{12} has been investigated. With Ln = Y a complex product mixture was obtained. In the other two cases novel tetranuclear products were isolated (Equation (165)). The reaction was proposed to proceed via an intermediate fulvene complex $Ln(C_5Me_4CH_2)(Cp^*)$ (Ln = La, Ce). The cerium derivative was structurally characterized. Three cerium atoms are bridged by $C_5Me_3(CH_2)_2$ ligand, which resulted from double metallation of a $\bar{C}_5Me_5$ ligand. The structure of the cerium complex is shown in Figure 51. When the thermolysis experiments were carried out in toluene, the benzyl derivatives $[LnBz(Cp^*)_2]$ (Ln = La, Ce) were formed through C–H activation. The cerium benzyl complex was crystallographically characterized.[696] Structurally characterized $[SmBz(THF)(Cp^*)_2]$ was obtained similarly by thermolysis of $[\{Sm(\mu\text{-}H)\}_2(Cp^*)_2]$ in toluene and isolated after recrystallization from THF.[697]

$$4\,[LnCH(TMS)_2Cp^*_2] \xrightarrow{\;\Delta\;} [\{Ln_2Cp^*_3\{\mu_3\text{-}\eta^5,\eta^1,\eta^1\text{-}C_5Me_3(CH_2)_2\}\}_2] + 4\,CH_2(TMS)_2 \qquad (165)$$

$$Ln = La,\ Ce$$

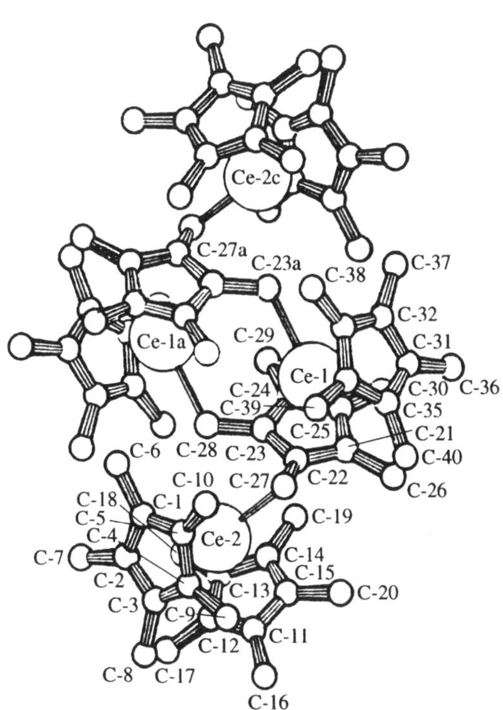

Figure 51　The molecular structure of $[\{Ce_2(Cp^*)_3\{\mu_3\text{-}\eta^5,\eta^1,\eta^1\text{-}C_5Me_3(CH_2)_2\}\}_2]$.[696]

Interesting bimetallic methyl and ethyl species have been prepared by reacting lanthanide metallocenes with alkylaluminum or gallium reagents. The bimetallic hydrocarbyls $[Ln(\mu\text{-}Me)_2MMe_2(Cp^*)_2]$ (Ln = Y, Lu; M = Al, Ga) are formed when the dimeric $[\{LnMe(Cp^*)_2\}_2]$ species are reacted with MMe_3.[187,680,698,699] In solution there is an equilibrium between the monomeric complexes and cyclic dimers. The latter contain eight-membered ring systems with methyl bridges between the different metal atoms (Equation (166)).

Similar bimetallic alkyls can also be prepared by using organolanthanide(II) precursors. For example, excess $[AlMe_3]$ reacts with $[Sm(THF)_2(Cp^*)_2]$ to produce the cyclic dimer $[SmCp^*_2\{\mu\text{-}MeAlMe_2(\mu\text{-}Me)\}_2SmCp^*_2]$.[680] Two derivatives of this type, $[(Cp^*)_2Ln\{(\mu\text{-}Me)_2Al(\mu\text{-}Me)_2\}_2Ln(Cp^*)_2]$ (Ln = Y, Sm)

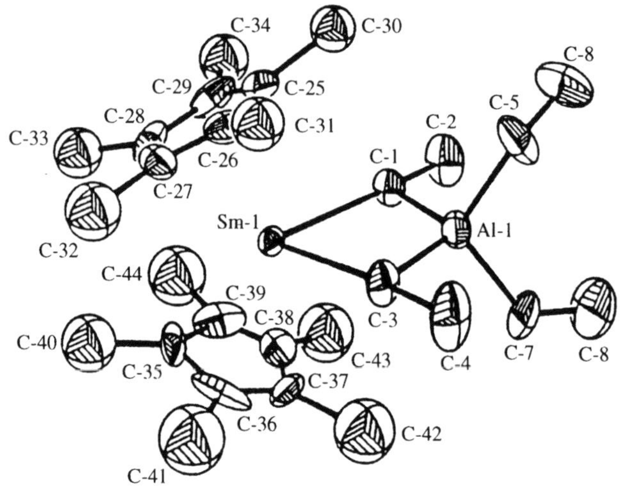

$$2\ Cp^*_2Ln\diagup\hspace{-1em}\diagdown Al \diagdown\hspace{-1em}\diagup \rightleftharpoons Cp^*_2Ln\cdots Al\cdots LnCp^*_2 \tag{166}$$

have been structurally characterized (Ln = Y: Y–C(Me) 0.266(2) nm, Y–Me–Al 176(1)°; Ln = Sm: Sm–C(Me) 0.275(2) nm, Sm–Me–Al 176(1)°).[680,698]

An unusual monomeric species with bridging μ-Ēt ligands can be isolated, when unsolvated Sm(Cp*)₂ is reacted with excess AlEt₃ (Equation (167)).[700]

$$3\ SmCp^*_2 + 4\ AlEt_3 \longrightarrow 3\ [Sm(\mu\text{-}Et)_2AlEt_2Cp^*_2] + Al \tag{167}$$

An x-ray crystal structure determination of [Sm(μ-Et)₂AlEt₂(Cp*)₂] revealed a monomeric complex with highly unsymmetrical ethyl bridges between samarium and aluminum (Figure 52, Sm–C(Cp*) 0.2712(2) nm, Sm–C(CH₂) 0.2662(4) nm).[700]

Figure 52 The molecular structure of [Sm(μ-Et)₂AlEt₂(Cp*)₂].[700]

Yet another coordination mode of a μ-Et group was found in the structurally characterized bimetallic ytterbium complex [Yb(μ-Et)AlEt₂(THF)(Cp*)₂] (**14**) (Equation (168)).[701] In this case ytterbium and aluminum are bridged by a single ethyl group but both carbon atoms are coordinated to the ytterbium atom (Yb–C(Cp*) 0.268(2) nm, Yb–C(1) 0.285(2) nm, Yb–C(2) 0.294(2) nm).

(14)

$$[Yb(THF)Cp^*_2] + AlEt_3 \longrightarrow [Yb(\mu\text{-}Et)AlEt_2(THF)Cp^*_2] \tag{168}$$

The first bis(pentamethylcyclopentadienyl)lanthanide allyl complex was [Lu{CH₂C(Me)CH₂}(Cp*)₂]. This compound was one of the products of the thermal decomposition of

[Lu(Bu^i)(Cp*)$_2$]. NMR spetral studies indicated the presence of a fluxional allyl ligand.[18] The reaction of allene with bis(pentamethylcyclopentadienyl)scandium hydride afforded [Sc(η^3-C$_3$H$_5$)(Cp*)$_2$].[673] Recently the first organolanthanide butadiene complex was synthesized by reacting [La(μ-Cl)$_2$K(DME)(Cp*)$_2$] with "magnesium butadiene," Mg(C$_4$H$_6$)(THF)$_2$. The product, binuclear [La(THF)(μ-η^1,η^3-CH$_2$CHCHCH$_2$)(Cp*)$_2$La(Cp*)$_2$], was structurally characterized (Figure 53, La–C(η^1) 0.2633(4) nm, La–C(η^3) 0.273(3) nm).[702] A mononuclear η^2-butadienyl complex, [Sm{η^2-Bu^{t}CH=CC(Bu^t)=CH$_2$}(Cp*)$_2$] was isolated from a reaction of [{Sm(μ-H)}$_2$(Cp*)$_2$] with the terminal alkyne Bu^{t}C≡CH and structurally characterized. The ligand can be formally viewed as a 1,3-di-*t*-butyl-1,3-butadiene metallated in the 2-position.[663]

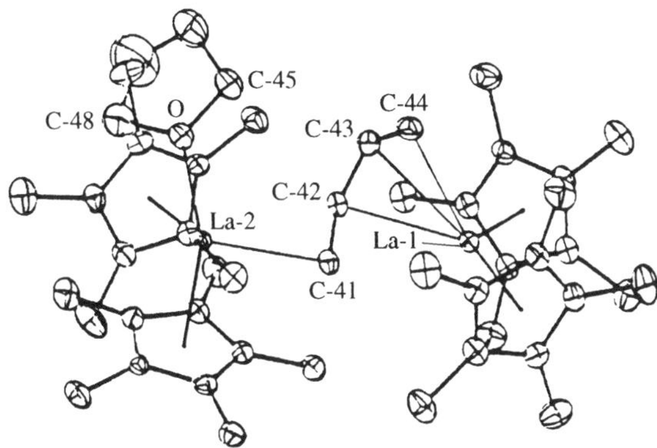

Figure 53 The molecular structure of [La(THF)(μ-η^1,η^3-CH$_2$CHCHCH$_2$)(Cp*)$_2$La(Cp*)$_2$].[702]

Numerous hydrocarbyls containing a Sm(Cp*)$_2$ unit have been prepared from decamethylsamarocene or its THF adduct [Sm(THF)$_2$(Cp*)$_2$]. For example, treatment of [Sm(THF)$_2$(Cp*)$_2$] with diphenylmercury (Equation (169)) gave the structurally characterized σ-phenyl complex (Sm–C(σ) 0.2511(8) nm). [SmPh(THF)(Cp*)$_2$] was also formed in a C–H activation reaction between [SmMe(THF)(Cp*)$_2$] and benzene.[680]

$$2\,[Sm(THF)_2Cp^*_2] + HgPh_2 \longrightarrow 2\,[SmPh(THF)Cp^*_2] + Hg + 2\,THF \qquad (169)$$

The samarium hydrocarbyls [SmR(THF)(Cp*)$_2$] (R = Me, Ph) have been alternatively prepared by treatment of the cationic complex [Sm(THF)$_2$(Cp*)$_2$][BPh$_4$] with MeLi or PhLi, respectively.[629]

The highly reactive lanthanide(II) complex also reacts with various hydrocarbons to give samarium(III) hydrocarbyls. In most cases the oxidation of samarium to the +3 oxidation state is accompanied by the formation of η^3-allyl units in the ligand system. A typical example is the reaction of unsolvated Sm(Cp*)$_2$ with styrene, which affords [{Sm(Cp*)$_2$}$_2$(μ-η^2:η^4-CH$_2$CHPh)]. Similarly *trans*-stilbene yields the binuclear complex [{Sm(Cp*)$_2$}$_2$(μ-η^2:η^4-PhCHCHPh)] while *cis*-stilbene is isomerized to the *trans* form in the presence of [Sm(THF)$_2$(Cp*)$_2$]. Figure 54 shows the bridging interaction of the η^2:η^4-CH$_2$CHPh ligand with the two samarium centers.[703]

Aliphatic alkenes generally react with Sm(Cp*)$_2$ to afford η^3-allyl complexes. One equivalent of alkene is hydrogenated during the course of the reaction (Equation (170)). The same complexes are accessible via addition of alkenes to the dimeric hydride [{Sm(μ-H)(Cp*)$_2$}$_2$].[704]

$$SmCp^*_2 + 2\,R^1CH=CH_2 \longrightarrow [Sm(\eta^3\text{-}CH_2CHCHR^2)Cp^*_2] + CH_3CH_2CH_2R^2 \qquad (170)$$

$$R^1 = Me,\ Et,\ Bz \qquad\qquad R^2 = H,\ Me,\ Ph$$

Sm(Cp*)$_2$ induces a dimerization of 1,3-butadiene to form the bis(η^3-allyl) complex [{Sm(Cp*)$_2$}$_2$(μ-η^3:η^3-CH$_2$CHCHCH$_2$CH$_2$CHCHCH$_2$)]. In extending this work, a detailed study on the reactivity of Sm(Cp*)$_2$ with polycyclic aromatic hydrocarbons has been carried out. Decamethylsamarocene reacts with hydrocarbons which have reduction potentials more positive than −2.22 V vs. SCE to form a series of binuclear complexes in high yield. The hydrocarbons included in this investigation were anthracene, 9-methylanthracene, pyrene, 2,3-benzanthracene, acenaphthylene, azulene, and the nitrogen heterocycles phenazine and acridine. In the case of acridine, reductive dimerization of the ligand is observed. Several products have been characterized by x-ray crystallography. In all cases the Sm(Cp*)$_2$ fragments are coordinated to η^3-allyl or η^3-azaallyl units.[705]

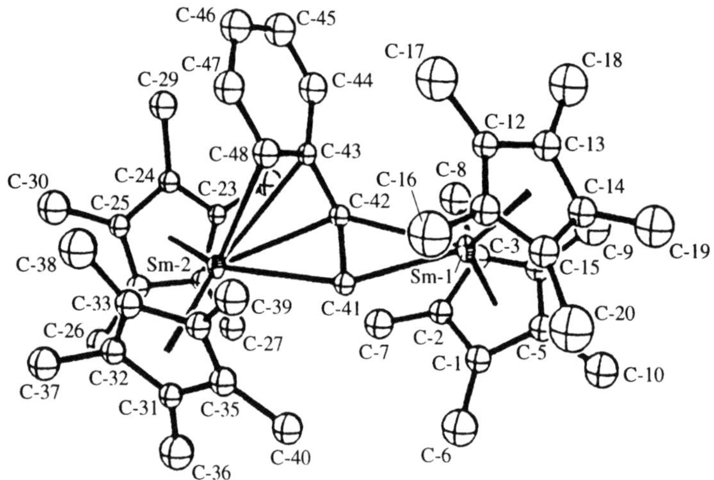

Figure 54　The molecular structure of [{Sm(Cp*)$_2$}$_2$(μ-η^2:η^4-CH$_2$CHPh)].[703]

Diphenylacetylene was found to add two equivalents of [Sm(THF)$_2$(Cp*)$_2$] to give a black compound formulated as [{Sm(Cp*)$_2$}$_2$(μ-PhC=CPh)].[416] It should be regarded as a samarium(III) complex of the dianion of *trans*-stilbene rather than a samarium(II) adduct of neutral diphenylacetylene. In accordance with this description of the bonding the complex yields pure *trans*-stilbene upon hydrolysis. [{Sm(Cp*)$_2$}$_2$(μ-PhC=CPh)] is a useful starting material for the preparation of the highly reactive hydride species [{Sm(μ-H)(Cp*)$_2$}$_2$], which is obtained in high yield by simple hydrogenolysis with molecular hydrogen (see below).[706]

Structurally different acetylides have been prepared by reacting [Ln(Et$_2$O)(Cp*)$_2$] (Ln = Eu, Yb) with phenylacetylene. With [Eu(Et$_2$O)(Cp*)$_2$] binuclear [{Eu(μ-C=CPh)(THF)$_2$(Cp*)$_2$}$_2$] was formed, while the ytterbium(II) complex yielded trinuclear [Yb$_3$(μ-C=CPh)$_4$Cp*$_4$].[707] Monomeric acetylides of the type [Ln(C=CR)(THF)(Cp*)$_2$] (R = Bu^t, Ph) were prepared by reacting the cationic species [Sm(THF)$_2$(Cp*)$_2$][BPh$_4$] with KC=CR or by protonation of [LnN(TMS)$_2$(Cp*)$_2$] (Ln = Ce, Nd, Sm) with phenylacetylene.[629,654] The samarium phenylacetylide was crystallographically characterized.[629] Acetylide bridging between yttrium and lithium was found by x-ray analysis in the organoyttrium acetylide [Y(μ-C=CBu^t)$_2$Li(THF)(Cp*)$_2$], which was made straightforwardly from [YCl(THF)(Cp*)$_2$] and LiC=CBu^t.[644] Mixed lanthanide potassium acetylides were prepared similarly according to Equation (171).[663]

$$[Ln(\mu\text{-Cl})_2K(THF)_2Cp*_2] + 2\,KC{\equiv}CPh \longrightarrow 1/n\,[\{Ln(\mu\text{-C}{\equiv}CPh)_2KCp*_2\}_n] + 2\,KCl \qquad (171)$$

$$Ln = Ce,\,Nd,\,Sm$$

An x-ray crystal structure determination of [{Sm(μ-C=CPh)$_2$K(Cp*)$_2$}$_n$] revealed a polymeric structure with the phenylacetylide ligands bridging samarium and potassium (Figure 55).

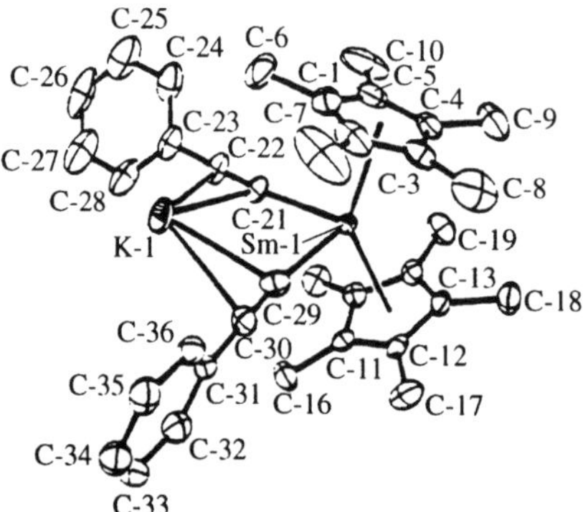

Figure 55　The molecular structure of [{Sm(μ-C=CPh)$_2$K(Cp*)$_2$}$_n$].[663]

The binuclear diacetylene complex [{Sm(Cp*)$_2$}$_2$(μ-η^2:η^2-PhC$_4$Ph)] was first obtained by treatment of [Sm(THF)$_2$(Cp*)$_2$] with PhC=CC=CPh and structurally characterized (Figure 56, Sm–C(Cp*) 0.271(2) nm; Sm–C(η^2) 0.248(1) nm for C-1 and 0.276(2) nm for C-2).[708]

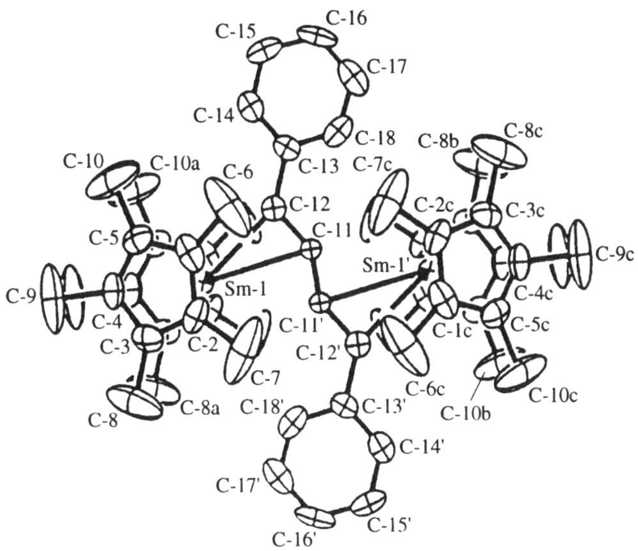

Figure 56 The molecular structure of [{Sm(Cp*)$_2$}$_2$(μ-η^2:η^2-PhC$_4$Ph)].[708]

Several other preparative routes leading to [{Sm(Cp*)$_2$}$_2$(μ-η^2:η^2-PhC$_4$Ph)] via coupling of PhC≡C units have been uncovered. The μ-η^2:η^2-diacetylene complex was formed during thermolysis of [Sm(C≡CPh)(THF)(Cp*)$_2$] as well as by reacting [SmCH(TMS)$_2$(Cp*)$_2$] or [{Sm(μ-H)(Cp*)$_2$}$_2$] with phenylacetylene.[709] Later it has been demonstrated that acetylide carbon–carbon coupling in the coordination sphere of lanthanide metallocenes is a very common reactivity pattern for unsolvated acetylides of the type [{Ln(μ-C≡CR)}$_n$(Cp*)$_2$] (Ln = La, Ce, Sm; R = Me, Bu^t, Ph).[663,710,711] Several binuclear μ-η^2:η^2-RC$_4$R products have been structurally characterized, including [{La(Cp*)$_2$}$_2$(μ-η^2:η^2-Bu^{t}C$_4$Bu^t)}$_2$],[711] [{La(Cp*)$_2$}$_2$(μ-η^2:η^2-PhC$_4$Ph)],[711] [{Ce(Cp*)$_2$}$_2$(μ-η^2:η^2-MeC$_4$Me)], and [{Ce(Cp*)$_2$}$_2$(μ-η^2:η^2-Bu^{t}C$_4$Bu^t)].[710] Reactions of Sm(Cp*)$_2$ with terminal alkynes in the absence of THF may even take a different course. The novel trienediyl complexes [{Sm(Cp*)$_2$}$_2$(μ-η^2:η^2-RCH$_2$CH$_2$C=C=C=CCH$_2$CH$_2$R)] (R = Pr^i, Ph) have been isolated from reactions of Sm(Cp*)$_2$ with HC≡CCH$_2$CH$_2$R (R = Pr^i, Ph). The phenyl-substituted derivative was characterized by x-ray crystallography.[663]

An unprecedented reductive dimerization of a stable phosphaalkyne occurred when [Sm(THF)$_2$(Cp*)$_2$] was treated with one equivalent of Bu^{t}C≡P (Equation (172)).[712]

$$2\,[\text{Sm(THF)}_2\text{Cp*}_2] + 2\,\text{Bu}^t\text{C≡P} \longrightarrow [\{\text{Sm(Bu}^t\text{C=P)Cp*}_2\}_2] + 4\,\text{THF} \tag{172}$$

The new ligand is the dianion of a 2,3-diphosphabutadiene. The resulting binuclear samarium complex was structurally characterized (Figure 57, Sm–C 0.2557(6) nm, Sm–P 0.2949(2) nm).

Formally the reductive dimerization of a phosphaalkyne closely resembles the above-mentioned dimerization of acetylide units in the coordination sphere of lanthanide elements.

Anionic chelating phosphoylide ligands have frequently been shown to be versatile ligands in organo-*f*-element chemistry. The combination of these ligands with lanthanide decamethylmetallocene fragments results in thermally highly stable rare-earth hydrocarbyl complexes. The first complex of this type was synthesized according to Equation (173).[18]

$$[\text{Lu}(\mu\text{-Cl})_2\text{Na(Et}_2\text{O)}_2\text{Cp*}_2] + \text{Li(CH}_2)_2\text{PMe}_2 \longrightarrow [\text{Lu}\{(\text{CH}_2)_2\text{PMe}_2\}\text{Cp*}_2] + \text{LiCl} + \text{NaCl} \tag{173}$$

Similarly the compounds [Ln{(CH$_2$)$_2$PR^{1}R^2}(Cp*)$_2$] (Ln = Nd, Sm; R^1 = Me, R^2 = Ph; R^1 = R^2 = Me, Bu^t, Ph) have been synthesized from the corresponding [Ln(μ-Cl)$_2$Li(Et$_2$O)$_2$(Cp*)$_2$] precursors.[713] The molecular structure of [Lu{(CH$_2$)$_2$PMe$_2$}(Cp*)$_2$] was established by an x-ray diffraction study. It shows a typical bent lanthanide(III) metallocene derivative with a pseudotetrahedral coordination around the central lutetium atom.[713]

Equation (174) shows how the unusual (trimethylsilyl)diazomethyl substituent has been introduced into organolanthanides. The products are thermally quite sensitive.[714]

Recently the first [Ln(Cp*)$_2$] derivatives containing Ln–Si bonds have been prepared and characterized. Treatment of [LnCH(TMS)$_2$(Cp*)$_2$] with H$_2$Si(TMS)$_2$ in the absence of solvent afforded the neutral silyl complexes [LnSiH(TMS)$_2$(Cp*)$_2$] (Ln = Y, Sm, Nd) (Equation (175)). An x-ray crystal

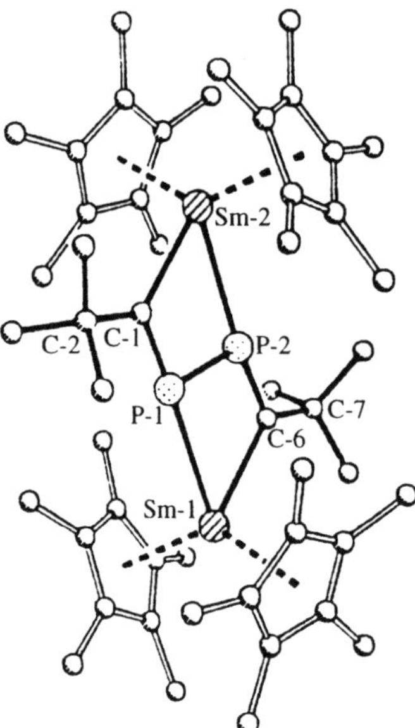

Figure 57   The molecular structure of [{Sm(Bu^{t}C=P)(Cp*)$_2$}$_2$].[712]

$$[\text{LuCl(THF)Cp*}_2] + \text{LiC(=N}_2)\text{TMS} \longrightarrow [\text{Ln}\{\text{C(=N}_2)\text{TMS}\}(\text{THF)Cp*}_2] + \text{LiCl} \qquad (174)$$

$$\text{Ln = Y, Yb, Lu}$$

structure analysis of the samarium derivative revealed the presence of a dimer stabilized by intermolecular Sm···Me–Si contacts (Figure 58).[715,716] The complexes [LnSiH$_2$Ph(Cp*)$_2$] (Ln = Y, Nd) were postulated as intermediates in the organolanthanide catalyzed dehydrogenation of PhSiH$_3$ and the hydrosilylation of alkenes.[717,718]

$$[\text{LnCH(TMS)}_2\text{Cp*}_2] + \text{H}_2\text{Si(TMS)}_2 \longrightarrow [\text{LnSiH(TMS)}_2\text{Cp*}_2] + \text{CH}_2(\text{TMS})_2 \qquad (175)$$

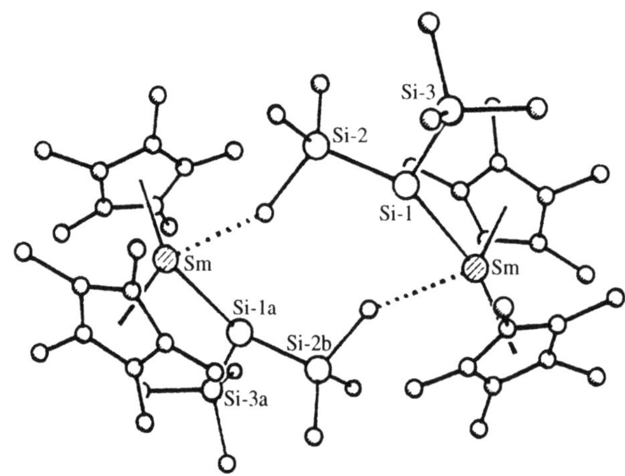

Figure 58   The molecular structure of [{SmSiH(TMS)$_2$(Cp*)$_2$}$_2$].[715]

(c) Hydrides. Organolanthanide hydrides of the type LnH(Cp*)$_2$ are of great importance as rare-earth-based homogeneous catalysts. Thus, this group of compounds has been thoroughly investigated and elaborate methods for their synthesis have been designed.[184–6,649,694,706,719] Since a [LnH(Cp*)$_2$] fragment is coordinatively unsaturated, these hydride species are usually isolated as dimers or solvates. Bis(pentamethylcyclopentadienyl)lanthanide hydrides are generally prepared by hydrogenolysis of the corresponding hydrocarbyls. Unsolvated hydrocarbyls containing the bulky –CH(TMS)$_2$ ligand have been shown to be especially useful for the preparation of hydrides. A typical synthesis starting from the THF adduct of yttrium trichloride is outlined below (Equations (176)–(178)).[720]

The dimeric cerium, neodymium, and samarium hydrides [{Ln(μ-H)(Cp*)$_2$}$_2$] (Ln = Ce, Nd, Sm) have been prepared analogously by hydrogenation of [LnCH(TMS)$_2$(Cp*)$_2$].[184,629,686]

$$[YCl_3(THF)_3] + 2\,NaCp^*_2 \xrightarrow{\ THF\ } [YCl(THF)Cp^*_2] + 2\,NaCl \qquad (176)$$

$$[YCl(THF)Cp^*_2] + LiCH(TMS)_2 \xrightarrow{\ Et_2O\ } [YCH(TMS)_2(THF)Cp^*_2] + LiCl + THF \qquad (177)$$

$$2\,[YCH(TMS)_2Cp^*_2] + H_2 \longrightarrow [\{Y(\mu\text{-}H)Cp^*_2\}_2] + 2\,CH_2(TMS)_2 \qquad (178)$$

Two factors are important for the success of this preparation. Due to the high steric demand of the $\overline{C}H(TMS)_2$ ligand the intermediate hydrocarbyl complex can be isolated completely free of donor solvents and alkali halides. Such base-free hydrocarbyls can be easily isolated by recrystallization from hydrocarbon solvents. The last step, the hydrogenolysis reaction, proceeds under very mild conditions at one atmosphere of hydrogen and ambient temperature. The facile conversion of organolanthanide hydrocarbyls into hydrides in the presence of molecular hydrogen appears to be generally applicable. The three-step procedure outlined above has been successfully carried out with other lanthanide elements.

Unsolvated $[\{Lu(\mu\text{-}H)(Cp^*)_2\}_2]$ was prepared by hydrogenolysis of base-free $[\{LuMe(Cp^*)_2\}_2]$ in hexane solution. For this complex, ^1H NMR spectral studies revealed an equilibrium at room temperature between the dimeric form and the monomer $[LuH(Cp^*)_2]$ (Scheme 36).[18,649]

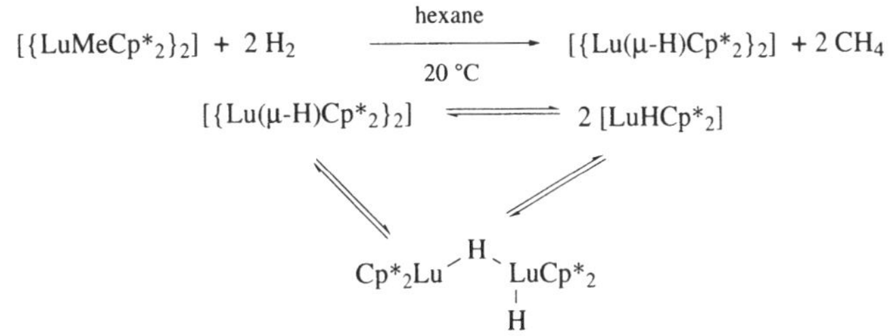

$$[\{LuMeCp^*_2\}_2] + 2\,H_2 \xrightarrow[20\,°C]{\ hexane\ } [\{Lu(\mu\text{-}H)Cp^*_2\}_2] + 2\,CH_4$$

$$[\{Lu(\mu\text{-}H)Cp^*_2\}_2] \rightleftharpoons 2\,[LuHCp^*_2]$$

$$Cp^*_2Lu\underset{H}{\overset{H}{\diagdown}}LuCp^*_2$$

Scheme 36

On the basis of its ^1H NMR spectral data $[\{Lu(\mu\text{-}H)(Cp^*)_2\}_2]$ was originally proposed to be dimeric with both hydride ligands asymmetrically bridging.[649] However, more recent studies revealed that the hydrides $[\{Ln(\mu\text{-}H)(Cp^*)_2\}_2]$ (Ln = Y, Lu) are dimers in which the lanthanide atoms are symmetrically bridged by μ-H ligands.[719] $[\{Sc(\mu\text{-}H)(Cp^*)_2\}_n]$ is quite unstable in solution unless a hydrogen atmosphere is provided. $[\{Lu(\mu\text{-}H)(Cp^*)_2\}_2]$ decomposes in diethyl ether solution to give $[LuOEt(Cp^*)_2]$ and ethane. Generally more stable are the THF solvates $[LnH(THF)(Cp^*)_2]$ (Sc, Y, Lu). The thermolysis of $[\{Y(\mu\text{-}H)(Cp^*)_2\}_2]$ in n-octane or cyclohexane solution has been reported to yield the structurally characterized bridging fulvene complex $[Y(\mu\text{-}\eta^1,\eta^5\text{-}CH_2C_5Me_4)(\mu\text{-}H)(Cp^*)_2Y(Cp^*)]$.[719,721] Thermolysis of $[\{Sm(\mu\text{-}H)(Cp^*)_2\}_2]$ has also been investigated. In alkanes or benzene the thermal decomposition leads to the binuclear tetramethylfulvene complex $[Sm(\mu\text{-}\eta^1,\eta^5\text{-}CH_2C_5Me_4)(\mu\text{-}H)(Cp^*)_2Sm(Cp^*)]$. When the thermolysis was carried out in toluene, the benzyl complex $[SmBz(Cp^*)_2]$ was formed via C–H activation and isolated as the THF adduct $[SmBz(THF)(Cp^*)_2]$.

Hydride complexes of the type $[\{Ln(\mu\text{-}H)(Cp^*)_2\}_2]$ are exceedingly reactive. A most astonishing reaction is the formation of $[(\mu\text{-}OSiMe_2OSiMe_2O)\{Sm(THF)(Cp^*)_2\}_2]$. This compound is formed when a THF solution of $[\{Sm(\mu\text{-}H)(Cp^*)_2\}_2]$ reacts with high-vacuum grease! An alternative preparation involves treatment of the dimeric hydride with hexamethylcyclotrisiloxane.[697]

The dimeric hydrides have also been found to polymerize alkenes and to activate small molecules such as carbon monoxide as well as sp^2 and sp^3 C–H bonds.[18,26,649,675] $[\{Sc(\mu\text{-}H)(Cp^*)_2\}_n]$ adds ethylene or propylene to afford the alkyls $[ScEt(Cp^*)_2]$ and $[Sc(Pr^n)(Cp^*)_2]$, respectively.[673] Treatment of $[\{Sc(\mu\text{-}H)(Cp^*)_2\}_n]$ with allene produced the addition product $[Sc(\eta^3\text{-}C_3H_4)(Cp^*)_2]$, while pyridine was metallated to give $[Sc(\eta^2\text{-}C_5H_4N)(Cp^*)_2]$. The methyl groups of the C_5Me_5 ligands in $[\{Sc(\mu\text{-}D)(Cp^*)_2\}_n]$ have been reported to undergo intramolecular H/D exchange.[673] Treatment of other hydrides of the type $[\{Ln(\mu\text{-}H)(Cp^*)_2\}_2]$ (Ln = La, Nd, Sm, Lu) with α-alkenes normally results in the formation of η^3-allyl complexes. As shown in Scheme 37 their formation involves insertion of the alkene into the Ln–H bond followed by hydrogen abstraction.

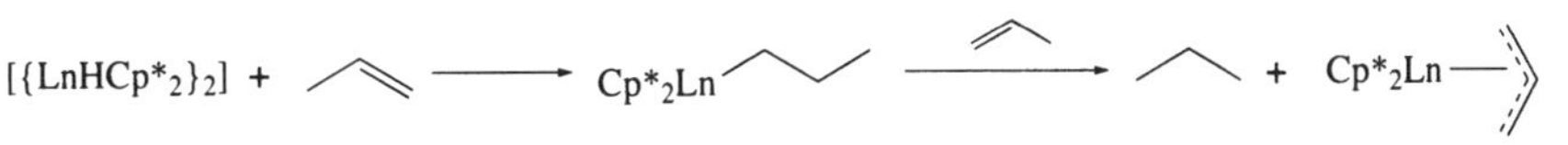

Scheme 37

Various C–H activation reactions have also been reported for $[\{Y(\mu\text{-}H)(Cp^*)_2\}_2]$. Metallation of benzene and toluene leads to the σ-aryl compounds $[YPh(Cp^*)_2]$ and $[YBz(Cp^*)_2]$, respectively. With substituted aromatic molecules PhX (X = OMe, SMe, NMe_2, CH_2NMe_2, PMe_2, $PPh_2{=}CH_2$, F, Cl, Br), *ortho*-metallation is the dominant reaction. The product resulting from *ortho*-metallation of $Ph_3P{=}CH_2$, $[Y(o\text{-}C_6H_4PPh_2CH_2)(Cp^*)_2]$, was structurally characterized.

An especially remarkable case of C–H activation is the reaction of $LuH(Cp^*)_2$ with tetramethylsilane, which yields $[LuCH_2TMS(Cp^*)_2]$ (Equation (179)).[649]

$$LuHCp^*_2 + SiMe_4 \longrightarrow [LuCH_2TMSCp^*_2] + H_2 \tag{179}$$

The C–H bonds in benzene are also readily attacked by $[\{LuH(Cp^*)_2\}_2]$ (dimer). In the first step, the σ-phenyl complex $[LuPh(Cp^*)_2]$ and hydrogen are formed. $[LuPh(Cp^*)_2]$ can further react with $LuH(Cp^*)_2$ yielding a phenylene bridged binuclear complex (Scheme 38):[224,649]

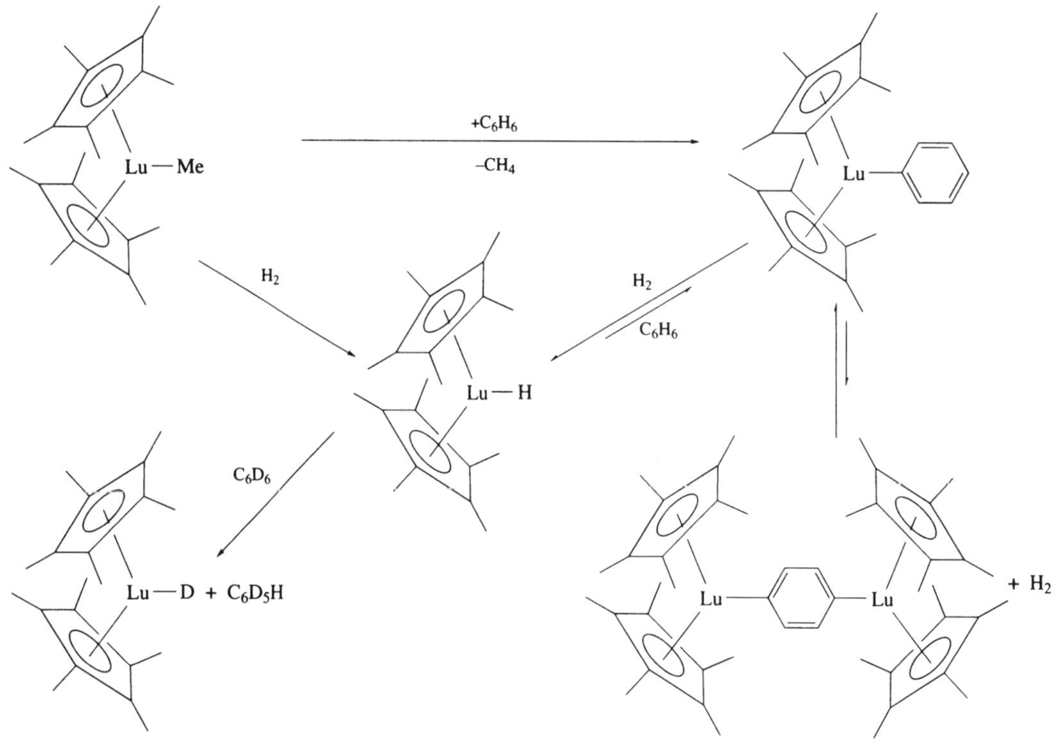

Scheme 38

The activation of carbon monoxide by $[\{Sm(\mu\text{-}H)(Cp^*)_2\}_2]$ was investigated by Evans *et al.*[722] $[\{Sm(\mu\text{-}H)(Cp^*)_2\}_2]$ reacts with CO in aromatic hydrocarbon solvents to give *cis*- and *trans*-$[(\mu\text{-}OCH{=}CHO)\{Sm(OPPh_3)(Cp^*)_2\}_2]$ after recrystallization in the presence of triphenylphosphine oxide as auxiliary ligand. The *cis*-isomer was found to isomerize at room temperature to give the *trans*-complex. The *cis*-isomer was structurally characterized by x-ray diffraction.

Facile addition of Ln–H bonds has also been reported for coordinated carbon monoxide. This reaction was investigated for $ScH(Cp^*)_2$ (Equation (180)).[25]

$$ScHCp^*_2 + [M(CO)_2Cp] \longrightarrow [Sc\{OC(H){=}MCp^*(CO)\}Cp_2] \tag{180}$$

$$M = Co, Rh$$

Novel $Ln(Cp^*)_2$ borohydride derivatives have been prepared by addition of $[Mes_2BH]_2$ to dimeric bis(pentamethylcyclopentadienyl)lanthanide hydrides (Equation (181)). Thermolyses of these complexes did not result in isolable boryl complexes of the type $LnBMes_2(Cp^*)_2$.[715]

$$[\{Ln(\mu\text{-}H)Cp^*_2\}_2] + [Mes_2BH]_2 \longrightarrow 2\,[Ln(\mu\text{-}H)_2BMes_2Cp^*_2] \tag{181}$$

(v) M(Cp)₃ compounds*

Homoleptic lanthanide(III) complexes bearing three pentahapto-coordinated pentamethylcyclopentadienyl ligands have long been elusive. Steric oversaturation was supposed to be the main reason for the failure of classical approaches to such compounds, for example by the metathetical reaction of lanthanide trichlorides with three equivalents of K(Cp*). The first well-characterized example was [Sm(Cp*)₃], which was unexpectedly prepared by reacting unsolvated decamethylsamarocene with cyclooctatetraene (Equation (182)).[723]

$$2\,SmCp^*_2 + C_8H_8 \longrightarrow [SmCp^*_3] + [Sm(C_8H_8)Cp^*] \tag{182}$$

According to the crystal structure determination the three $\overline{C}_5Me_5$ ligands are η^5-coordinated and surround the samarium atom in a pseudotrigonal planar fashion, thus minimizing interligand repulsion (Figure 59, Sm–C 0.282(5) nm).

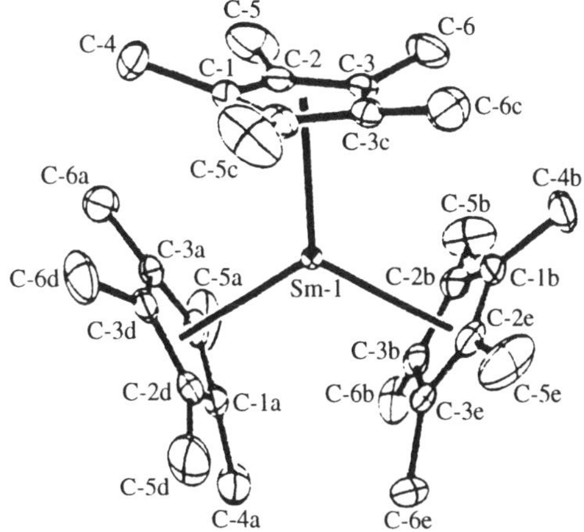

Figure 59 The molecular structure of [Sm(Cp*)₃].[723]

In contrast, the corresponding tris(tetramethylcyclopentadienyl)lanthanide complexes are easily accessible by classical metathesis reactions (Equation (183)). In this case the terbium derivative was crystallographically characterized.[724]

$$LnCl_3 + 3\,Na(C_5Me_4H) \longrightarrow [Ln(C_5Me_4H)_3] + 3\,NaCl \tag{183}$$

$$Ln = La,\ Sm,\ Tb$$

A related mixed-valence bridging cyclopentadienyl complex, [Sm(μ-η⁵:η²-Cp)(Cp*)₂Sm(Cp*)₂], was synthesized as shown in Equation (184). The μ-cyclopentadienyl ligand is η^5-coordinated to a $Sm^{III}(Cp^*)_2$ unit and η^2-bonded to $Sm^{II}(Cp^*)_2$. The starting material [Sm(η⁵-Cp)(Cp*)₂] was made from Sm(Cp*)₂ and excess cyclopentadiene. Its structure has also been determined by x-ray crystallography (Sm–C(Cp*) 0.2770(3) nm, Sm–C(Cp) 0.2738(4) nm).[725]

$$[Sm(\eta^5\text{-}Cp)Cp^*_2] + SmCp^*_2 \longrightarrow [Sm(\mu\text{-}\eta^5{:}\eta^2\text{-}Cp)Cp^*_2SmCp^*_2] \tag{184}$$

2.2.7 Cyclopentadienyl-like Compounds

2.2.7.1 Pentadienyl compounds

To some extent, the chemistry of metal complexes containing pentadienyl ligands resembles that of the corresponding cyclopentadienyl derivatives. This has been demonstrated in the past mostly for transition metal complexes. Due to this striking similarity the bis(pentadienyl)metal complexes have been called "open metallocenes" and accordingly the pentadienyl ligands can be regarded as open cyclopentadienyls. For many years tris(2,4-dimethylpentadienyl)neodymium, [Nd(C₇H₁₁)₃] (C₇H₁₁ = 2,4-dimethylpentadienyl), was the only lanthanide derivative of that type. The bright green material was prepared by treatment of anhydrous neodymium trichloride with three equivalents of 2,4-dimethylpentadienylpotassium in THF solution (Equation (185)).[726]

$$\text{NdCl}_3 + 3\,\text{KC}_7\text{H}_{11} \xrightarrow{\quad\text{THF}\quad} [\text{Nd}(\text{C}_7\text{H}_{11})_3] + 3\,\text{KCl} \qquad (185)$$

According to the crystal structure determination the homoleptic complex contains three equally bonded, nearly planar 2,4-dimethylpentadienyl ligands in a pentahapto fashion. The six methyl groups are slightly out of plane pointing away from the central neodymium atom. More recently several other series of lanthanide pentadienyls have been reported, including [LnCl$_2$(THF)$_3$(C$_7$H$_{11}$)] (Ln = Nd, Sm, Gd), [LnCl(THF)(C$_7$H$_{11}$)$_2$] (Ln = Nd, Sm, Gd), and [Ln(C$_7$H$_{11}$)$_3$] (Ln = La, Sm, Gd, Tb, Lu).[727,728] The tris(η^5-2,4-dimethylpentadienyl) complexes [Ln(C$_7$H$_{11}$)$_3$] (Ln = Gd, Tb, Lu) have been crystallographically characterized.[727,728] In [Gd(C$_7$H$_{11}$)$_3$] the Gd–C bond distances range from 0.2738 nm (C-1 and C-5 positions) to 0.2821 nm (C-2 and C-4 positions). A remarkable structural difference was found in the lutetium derivative. Due to the small ionic radius of lutetium this compound contains only two pentahapto-coordinated 2,4-dimethylpentadienyl ligands while the third one is η^3-coordinated (Figure 60, Lu–C(η^5) 0.264 nm, Lu–C(η^3) 0.259(1) nm). The hexanuclear cluster complex [Nd$_6$Cl$_{12}$(C$_7$H$_{11}$)$_6$(THF)$_2$] will be discussed in Section 2.2.13.[729]

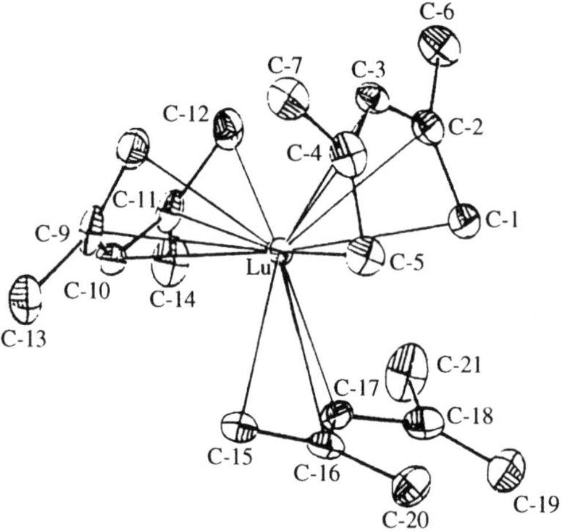

Figure 60   The molecular structure of [Lu(η^5-C$_7$H$_{11}$)$_2$(η^3-C$_7$H$_{11}$)].[728]

An interesting ligand modification is the linking of two pentadienyl units by a hydrocarbon bridge. A bridged bis(pentadienyl) dianion was generated by double deprotonation of two methyl groups in 2,4,7,9-tetramethyl-1,3,7,9-decatetraene. Subsequent treatment of the dipotassium salt with ytterbium diiodide afforded the "open" ytterbium(II) metallocene [Yb(THF){4,4'-(CH$_2$)$_2$(2-C$_6$H$_8$)$_2$}], which can be regarded as a formal analogue of [Yb(Et$_2$O)(Cp*)$_2$].[730]

2.2.7.2 *Heteroatom five-membered ring ligands*

One of the more recent developments in the field of organo-*f*-element chemistry is the increasing use of anionic heteroatom five-membered ring ligands. For example, substituted phospholyl and arsolyl ligands have been shown to be interesting alternatives to the pentamethylcyclopentadienyl ligand. Lanthanide(II) derivatives containing tetrasubstituted phospholyl and arsolyl ligands have been prepared by reacting lanthanide diiodides with two equivalents of K(C$_4$Me$_4$E) (E = P, As) (Equation (186)). An alternative preparation involves treatment of biphospholyl or biarsolyl derivatives with the metal powders of samarium or ytterbium. In this case the products are formed by reductive cleavage of P–P or As–As bonds.[731,732]

$$[\text{LnI}_2(\text{THF})_2] + 2\,\text{K}(\text{C}_4\text{Me}_4\text{E}) \longrightarrow [\text{Ln}(\text{THF})_2(\text{C}_4\text{Me}_4\text{E})_2] + 2\,\text{KI} \qquad (186)$$

$$\text{E = P, As; Ln = Sm, Yb}$$

The molecular structure of [Yb(THF)$_2$(2,5-Ph$_2$C$_4$H$_2$P)$_2$] has been determined by x-ray crystallography (Figure 61). The structure closely resembles those of substituted lanthanide(II) metallocenes such as [Sm(THF)$_2$(Cp*)$_2$].[731]

Phospholyl analogues of the well-known bis(pentamethylcyclopentadienyl)lanthanide chlorides [Ln(μ-Cl)$_2$LiS$_2$(Cp*)$_2$] (S = solvent) have also been prepared. Li(C$_4$Me$_4$P) was generated *in situ* from

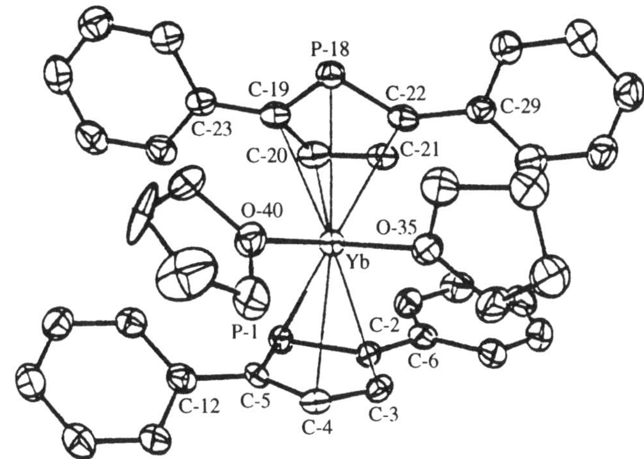

Figure 61　The molecular structure of $[Yb(THF)_2(2,5\text{-}Ph_2C_4H_2P)_2]$.[731]

bis(tetramethylphospholyl)butane and treated with anhydrous lanthanide trichlorides to afford the products in moderate yields (Equation (187)).[733]

$$LnCl_3 + 2\,Li(C_4Me_4P) \xrightarrow{\quad S \quad} [Ln(\mu\text{-}Cl)_2LiS_2(C_4Me_4P)_2] + LiCl \qquad (187)$$

$$Ln = Y,\ S = 1/2\ DME$$
$$Ln = Lu,\ S = Et_2O$$

2.2.8 Arene Complexes

The lanthanide elements Ce, Pr, Nd, Sm, Eu, Dy, Ho, Er, and Yb react with naphthalene[734-46] or anthracene[743,745] to form highly reactive complexes of the corresponding radical anions or dianions. These and other complexes obtained from lanthanide metals and unsaturated hydrocarbons[747] are usually formulated as lanthanide(0) complexes and thus their preparation and derivative chemistry[132,199,748-50] will not be further discussed in this section.

Examples of organolanthanide(III) complexes containing π-coordinated arenes are still rare, although a small number of complexes has been structurally characterized. The majority of the compounds described so far are of the type $[Ln(AlCl_4)_3(\eta^6\text{-arene})]$ (**15**). They are easily prepared by reacting anhydrous lanthanide trichlorides with $AlCl_3$ in the presence of the arene. Preferably the reaction is carried out in the appropriate arene solvent. Reactions with hexamethylbenzene can be carried out in toluene solution (Equation (188)).[751-7]

(15)

$$LnCl_3 + 3\,AlCl_3 + C_6R_6 \longrightarrow [Ln(AlCl_4)_3(\eta^6\text{-}C_6R_6)] \qquad (188)$$

$$R = H,\ Me$$
$$Ln = La,\ Nd,\ Sm,\ Gd,\ Yb$$

The electron-rich arene ligands benzene, m-xylene, and hexamethylbenzene have thus far been employed. Among the well-characterized complexes are $[Sm(AlCl_4)_3(\eta^6\text{-}m\text{-}C_6H_4Me_2)]$[754,755,757] and $[Ln(AlCl_4)_3(\eta^6\text{-}C_6Me_6)]$ (Ln = Sm, Er).[751,758,759] The bonding in these complexes has been described as an electrostatically induced dipole interaction between the Ln^{3+} cation and the π-electron system of the aromatic ring. The structurally characterized $[Sm(AlCl_4)_3(\eta^6\text{-}C_6Me_6)]$ is monomeric in the solid state (Sm–C 0.289(5) nm).[756] Another crystallographically established hexamethylbenzene derivative is $[Yb(AlCl_4)_3(\eta^6\text{-}C_6Me_6)]\cdot$toluene.[755] The complexes $[Sm(AlCl_4)_3(\eta^6\text{-}C_6H_6)]$ (Sm–C 0.289–0.293 nm) and $[Nd(AlCl_4)_3(\eta^6\text{-}C_6H_6)]$ (Nd–C 0.293 nm) are isostructural with the uranium(III) analogue, $[U(AlCl_4)_3(\eta^6\text{-}C_6H_6)]$.[752,760] The erbium complex $[Er(AlCl_4)_3(\eta^6\text{-}C_6Me_6)]_4$ has been found to be a cyclotetramer in the solid state with η^2-tetrachloroaluminate ligands bridging the erbium atoms.[761]

η^6-Coordination between lanthanide atoms and remote phenyl groups has been found in some special alkoxides. Thus far two different types of these bonding situations have been described. Lanthanide metal powders react with $Hg(C_6F_5)_2$ and 2,6-diphenylphenol in THF solution to produce the lanthanide(III) alkoxides $[Ln(OC_6H_3Ph_2\text{-}2,6)_3]$ (Ln = Nd, Sm, Er, Yb, Lu).[762] The structurally characterized THF solvate $[Yb(OC_6H_3Ph_2\text{-}2,6)_3(THF)_2]$ showed no unusual structural features. However, in the unsolvated tris(alkoxide) $[Yb(OC_6H_3Ph_2\text{-}2,6)_3]$ (**16**) one phenyl substituent is in a geometrically favorable position to engage in a novel intramolecular π-interaction with the central ytterbium ion (Yb–C 0.2978(6) nm).

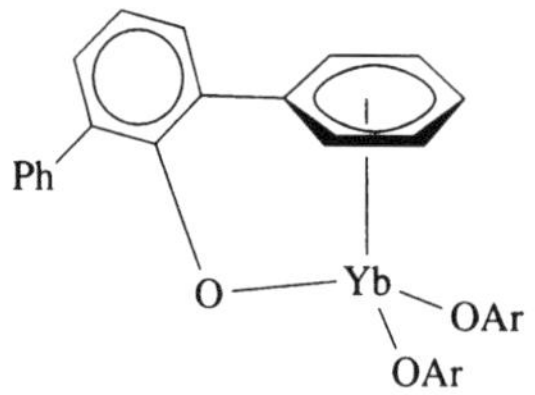

(16)

An interesting case of intermolecular $\eta^6\text{-}\pi$-interaction between a phenyl substituent and a lanthanide metal has been found in the structurally characterized dimeric alkoxides $[\{Ln(OC_6H_3(Pr^i)_2\text{-}2,6)_3\}_2]$ (**17**, Ln = Nd, Sm, Er, Yb; Nd–C 0.3035 nm, Sm–C 0.2986 and 0.3016 nm).[763] A very similar formation of a centrosymmetric dimer held together by $\eta^6\text{-}\pi$-interactions to phenyl ring had been reported earlier for a uranium(III) analogue.[764]

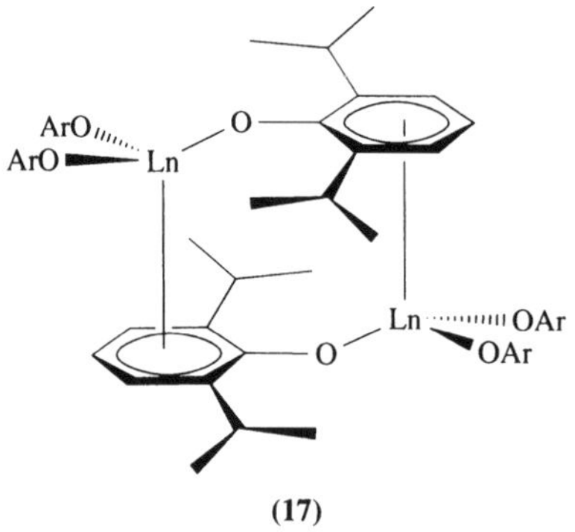

(17)

2.2.9 Cyclooctatetraenyl Compounds

2.2.9.1 $M(C_8H_8)$ compounds

The first (cyclooctatetraenyl)lanthanide(II) complexes were prepared in 1969 by reacting the blue solutions of metallic europium or ytterbium in liquid ammonia with cyclooctatetraene (Equation (189)).[18,26] The orange products $Eu(C_8H_8)$ and $Yb(C_8H_8)$ were found to be extremely air-sensitive and insoluble in organic solvents. So far no structural information is available on these presumably polymeric materials.

$$Ln + C_8H_8 \xrightarrow{\ NH_3(l.)\ } Ln(C_8H_8) \qquad (189)$$

Upon treatment of $Yb(C_8H_8)$ with pyridine the polymeric structure is broken up and soluble $[Yb(C_5H_5N)_3(C_8H_8)]$ is formed, which was structurally characterized (Yb–C 0.264(3) nm, Yb–N 0.258(2) nm).[26,765]

Cerocene(IV) was reported to react with metallic potassium in DME solution to give an anionic cerium(II) derivative as a green, microcrystalline solid (Equation (190)).[18] The product has not been further investigated and the molecular structure remains unknown. In view of the known solid-state properties of cerium(II) it seems highly unlikely that the reduction product contains cerium in this unusually low oxidation state. The green color of the product is more consistent with a saltlike compound containing the $[Ce(C_8H_8)_2]^-$ anion. Thus, the presence of cerium(II) in an organometallic compound remains questionable.

$$[Ce(C_8H_8)_2] + 2\,K \xrightarrow{\quad DME \quad} [K(DME)]_2[Ce(C_8H_8)_2] \qquad (190)$$

Such anionic lanthanide(II) complexes are, however, well-established for ytterbium[766] and samarium.[767] The potassium salts $[K(diglyme)_2][Yb(C_8H_8)]$ and $[\{K(DME)_2\}_2][Yb(C_8H_8)]$[768] (Figure 62) have been structurally characterized. Both are sandwich complexes in which the cyclooctatetraenyl dianions act as bridging ligands between ytterbium and potassium.

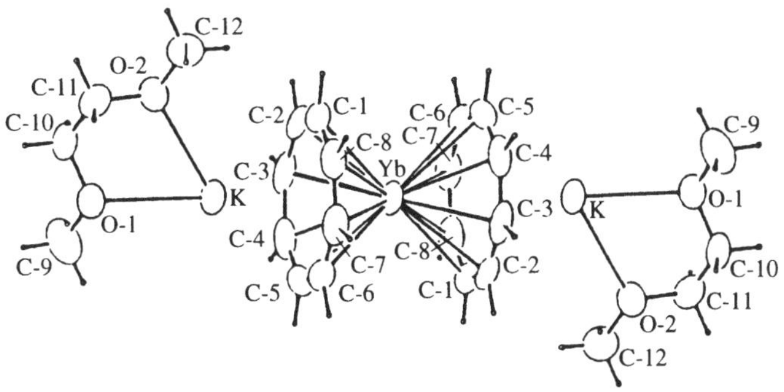

Figure 62 The molecular structure of $[\{K(DME)_2\}_2][Yb(C_8H_8)]$.[768]

Polymeric $Sm(C_8H_8)$ has been prepared in a straightforward manner from $[SmI_2(THF)_2]$ and $K_2C_8H_8$.[767]

2.2.9.2 $M(C_8H_8)X$ compounds

The first mono(cyclooctatetraenyl)lanthanide(III) complexes were reported as early as 1971. Until now, dimeric (cyclooctatetraenyl)lanthanide chlorides (**18**) have been the most important starting materials in this area. These compounds are generally made by reacting anhydrous lanthanide trichlorides with $K_2C_8H_8$ in a molar ratio of 1:1 (Equation (191)).[18,26,51,52,769,770]

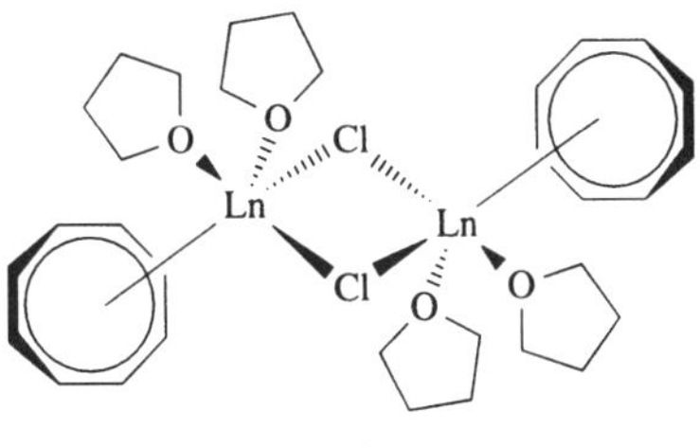

(18)

$$2\,LnCl_3 + 2\,K_2C_8H_8 \longrightarrow [\{Ln(\mu\text{-}Cl)(THF)_2(C_8H_8)\}_2] + 4\,KCl \qquad (191)$$

An x-ray structural analysis of the cerium derivative $[\{Ce(\mu\text{-}Cl)(THF)_2(C_8H_8)\}_2]$ showed a chloro-bridged dimeric molecule with an η^8-cyclooctatetraenyl ligand and two THF molecules coordinated to each cerium atom.[18,26] Dimeric mono(cyclooctatetraenyl)lanthanide chlorides are available for nearly all members of the lanthanide series. They form highly colored crystalline solids which are thermally not as robust as the $[Ln(C_8H_8)_2]^-$ sandwich complexes and highly oxygen-sensitive. Upon heating *in vacuo* above ca. 60 °C the compounds lose the coordinated THF, although well-defined unsolvated mono(cyclooctatetraenyl)lanthanide chlorides have never been isolated. The bridging chlorine atoms are readily replaced by other anionic ligands such as cyclopentadienyl, amide, alkoxide, and hydrocarbyl ligands. However, the synthetic value of these complexes is somewhat diminished by their low solubility even in THF. The isolation procedure for [{Ln(μ-Cl)(THF)$_2$(C$_8$H$_8$)}$_2$] derivatives requires prolonged Soxhlet extraction of the crude products with THF. Recently two other precursors have been used as preparative alternatives. Replacement of the starting lanthanide trichlorides by lanthanide triiodides afforded monomeric mono(cyclooctatetraenyl)lanthanide iodides (**19**) as THF solvates.[771] The same compounds are also accessible by reacting lanthanide metal powders with cyclooctatetraene and iodine (Equations (192) and (193)).[197,772]

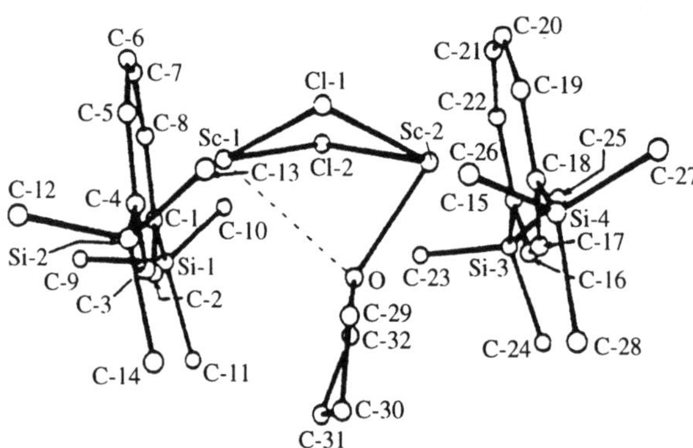

(**19**)

$$\text{LnI}_3 + \text{K}_2\text{C}_8\text{H}_8 \longrightarrow [\text{LnI(THF)}_n(\text{C}_8\text{H}_8)] + 2\,\text{KI} \qquad (192)$$

$$\text{Ln = Nd, Sm}$$

$$\text{Ln} + \text{C}_8\text{H}_8 + 1/2\,\text{I}_2 \xrightarrow{\text{THF}} [\text{LnI(THF)}_n(\text{C}_8\text{H}_8)] \qquad (193)$$

$$\text{Ln = La, Ce, Pr, Nd, Sm; } n = 1,2,3$$

The molecular structures of the cerium (Ce–I 0.3299(1) nm)[194] and neodymium derivatives have been determined. The monomeric molecules adopt a characteristic "piano stool" geometry with an η^8-coordinated cyclooctatetraenyl ligand.[771]

Interesting structural differences were found in the corresponding complexes containing the bulky 1,4-bis(trimethylsilyl)cyclooctatetraenyl ligand. The mono(cyclooctatetraenyl)lanthanide chlorides [{Ln(μ-Cl)(THF)1,4-(TMS)$_2$C$_8$H$_6$}$_2$] are isolated from equimolar reactions of [LnCl$_3$(THF)$_3$] (Ln = Sc, Y) with the dilithium salt [Li(THF)]$_2$[1,4-(TMS)$_2$C$_8$H$_6$]. A single-crystal x-ray structural analysis of the scandium derivative revealed the presence of an unusual semibridging THF ligand (Figure 63).[773]

Figure 63   The molecular structure of [{Sc(μ-Cl)(THF)1,4-(TMS)$_2$C$_8$H$_6$}$_2$].[773]

In contrast to the dimeric chloro derivatives the iodo complexes show a significantly higher solubility in THF. The same effect was found for the recently reported dimeric triflates of the type [{Ln(μ-O$_3$SCF$_3$)(THF)$_2$(C$_8$H$_8$)}$_2$]. As was shown earlier for cyclopentadienyl lanthanide chemistry (cf. Section 2.2.5.4), the use of anhydrous lanthanide triflates offers some important advantages over the previously used lanthanide trichlorides. Lanthanide triflates react with equimolar amounts of K$_2$C$_8$H$_8$ in THF solution to give the dimeric organolanthanide triflates [{Ln(μ-O$_3$SCF$_3$)(THF)$_2$(C$_8$H$_8$)}$_2$] (Equation (194)).[771,774]

The corresponding yttrium complex is isolated as the bis(THF) adduct [{Y(μ-O$_3$SCF$_3$)(THF)(C$_8$H$_8$)}$_2$]. The crystal and molecular structures of the neodymium complex (Figure 64)[774] has been determined by x-ray analysis. In addition to an η^8-coordinated cyclooctatetraenyl ligand two THF molecules are bonded to each lanthanide atom. The resulting Ln(THF)$_2$(C$_8$H$_8$) fragments are

$$2\,K_2C_8H_8 \;+\; 2\,Ln(O_3SCF_3)_3 \;\xrightarrow{\text{THF}}\; 2$$ 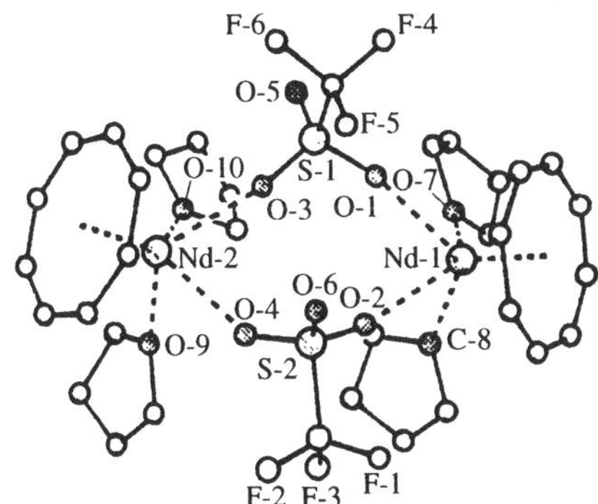

$$+\; 4\,KO_3SCF_3 \qquad (194)$$

Ln = Ce, Pr, Nd, Sm

connected via bridging triflate ligands to give an eight-membered $[Ln_2O_4S_2]$ ring as the central part of the molecule. A very similar situation was found in the dimeric bis(cyclopentadienyl)lanthanide(III) triflates $[\{Ln(\mu\text{-}O_3SCF_3)(Cp)_2\}_2]$ (cf. Section 2.2.5.4). From a synthetic standpoint these organolanthanide triflates offer several advantages: despite their dimeric constitution their solubility is dramatically increased as compared to the corresponding chlorides. Furthermore, the triflate anions are excellent leaving groups and thus facilitate substitution reactions with other anionic ligands. LiO_3SCF_3 or NaO_3SCF_3 are much more easily separated by filtration than finely divided LiCl or NaCl.

Figure 64   The molecular structure of $[\{Nd(\mu\text{-}O_3SCF_3)(THF)_2(C_8H_8)\}_2]$.[774]

Mono(cyclooctatetraenyl)lanthanide chlorides and triflates are useful precursors for the preparation of other lanthanide half-sandwich complexes containing the cyclooctatetraenyl ligand. Half-sandwich complexes prepared by metathetical reactions include $[Ln\{p\text{-}MeOC_6H_4C(N\text{-}TMS)_2\}(THF)(C_8H_8)]$ and $[Ln\{Ph_2P(N\text{-}TMS)_2\}(THF)(C_8H_8)]$ as well as the polypyrazolylborates $[Ln(HBpz_3)(THF)(C_8H_8)]$ and $[Ln\{HB(3,5\text{-}Me_2pz)_3\}(THF)(C_8H_8)]$ (Ln = Y, Ce, Pr, Nd, Sm).[774-6] An anionic lutetium bis(amide), $Li[Y(NMeCH_2CH_2NMe_2)_2(Bu^nC_8H_7)]$, was prepared by reacting $[LuCl(THF)(Bu^nC_8H_7)]$ with two equivalents of lithium N,N,N'-trimethylethylenediamide.[376] Several dimeric alkoxides have been prepared according to Equation (195).[615]

$$[\{Ln(\mu\text{-}Cl)(THF)_2(C_8H_8)\}_2] \;+\; 2\,NaOR \;\longrightarrow\; [\{Ln(\mu\text{-}OR)(THF)(C_8H_8)\}_2] \;+\; 2\,NaCl \qquad (195)$$

Ln = Y, Lu
R = Ph, $C_6H_3Me_2$-2,6

$[\{Y(\mu\text{-}OPh)(THF)(C_8H_8)\}_2]$[615] and $[\{Dy\{\mu\text{-}OCH_2(CH_2)_2CH{=}CH_2\}(THF)(C_8H_8)\}_2]$[777] have been structurally characterized by x-ray diffraction. The corresponding tri-t-butylmethoxides and triphenylsiloxides $[Ln(OCBu^t_3)(THF)(C_8H_8)]$ and $[Ln(OSiPh_3)(THF)(C_8H_8)]$ (Ln = Y, Lu) are monomeric in the solid state.[615]

The first cyclooctatetraenyl lanthanide thiolates and selenolates were prepared by the reaction of metallic samarium with cyclooctatetraene and diaryl disulfides or diselenides as oxidants (Equation (196)). Catalytic amounts of iodine accelerate the reactions. The molecular structure of dimeric $[\{Sm(\mu\text{-}SePh)(THF)_2(C_8H_8)\}_2]$ was determined by x-ray crystallography.[772]

$$2\,Sm \;+\; 2\,C_8H_8 \;+\; ArEEAr \;\xrightarrow{\text{THF}}\; [\{Sm(\mu\text{-}EAr)(THF)_n(C_8H_8)\}_2] \qquad (196)$$

$n = 1$; EAr = $SC_6H_2(Pr^i)_3$
$n = 2$; EAr = SPh, $SC_6H_2Me_3$, SePh

The lutetium derivative $[\{Lu(\mu\text{-}Cl)(THF)_2(C_8H_8)\}_2]$ was used to prepare the first mono(cyclo-octatetraenyl)lanthanide hydrocarbyls. Treatment of the chloro complex with $LiCH_2TMS$ or *o*-$Me_2NCH_2C_6H_4Li$ afforded the monomeric hydrocarbyls $[LuR(THF)(C_8H_8)]$ (Equation (197)). In these cases the successful isolation of mononuclear compounds was made possible by the comparatively small ionic radius of lutetium and the steric bulk of the organic ligands. $[Lu(o\text{-}Me_2NCH_2C_6H_4)(THF)(C_8H_8)]$ was structurally characterized. Further reaction of $[Lu(o\text{-}Me_2NCH_2C_6H_4)(THF)(C_8H_8)]$ with *t*-butylacetylene produced $[LuC{\equiv}CBu^t(C_8H_8)]$.[778,779]

$$[\{Lu(\mu\text{-}Cl)(THF)_2(C_8H_8)\}_2] + 2\,LiR \longrightarrow 2\,[LuR(THF)(C_8H_8)] + 2\,LiCl + 2\,THF \qquad (197)$$

$$R = CH_2TMS,\ o\text{-}Me_2NCH_2C_6H_4$$

Similar alkylation of $[\{Ln(\mu\text{-}Cl)(THF)_2(C_8H_8)\}_2]$ (Ln = Y, Sm, Lu) with two equivalents of $LiCH(TMS)_2$ did not lead to neutral hydrocarbyls of the type $[Ln\{CH(TMS)_2\}(THF)_n(C_8H_8)]$. Instead, the "inverse sandwich complexes" $[Li(THF)_2\{\mu\text{-}\eta^2,\eta^8\text{-}C_8H_8\}Ln\{CH(TMS)_2\}_2]$ were isolated. An x-ray crystal structure determination (Figure 65) revealed an unusual zwitterionic structure with η^2-coordination of a $Li(THF)_2$ unit to the bridging cyclooctatetraenyl ligand.[780]

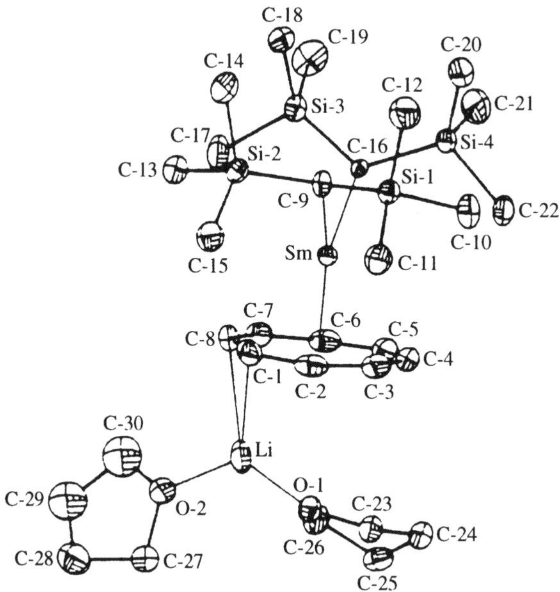

Figure 65   The molecular structure of $[Li(THF)_2\{\mu\text{-}\eta^2,\eta^8\text{-}C_8H_8\}Sm\{CH(TMS)_2\}_2]$.[780]

2.2.9.3 [M(C$_8$H$_8$)$_2$]$^-$ compounds

Anionic sandwich complexes of the type $[Ln(C_8H_8)_2]^-$ were the first lanthanide(III) cyclooctatetraenyl complexes reported in the literature. Potassium salts of these anions are easily available by reacting anhydrous lanthanide trichlorides with two equivalents of $K_2C_8H_8$ in THF solution (Equation (198)).[18,26,51,52]

$$LnCl_3 + 2\,K_2C_8H_8 \longrightarrow K[Ln(C_8H_8)_2] + 3\,KCl \qquad (198)$$

$$Ln = Sc,\ Y,\ La,\ Ce,\ Pr,\ Nd,\ Sm,\ Gd,\ Tb$$

In contrast to the chloro-bridged dimers $[\{Ln(\mu\text{-}Cl)(THF)_2(C_8H_8)\}_2]$ the anionic complexes are highly soluble in THF and can thus be easily separated by extraction. The potassium salts $K[Ln(C_8H_8)_2]$ are thermally highly stable, but extremely air-sensitive. Like $K_2C_8H_8$ they decompose almost explosively upon contact with air. The have been characterized by various spectroscopic methods (IR, Raman, ^{1}H NMR, UV–VIS)[18,78] and have also been the subject of theoretical studies. Early semiempirical LCAO calculations have indicated a predominantly ionic bonding between the lanthanide ions and the cyclooctatetraenyl ligands.[18] The structural chemistry of the cerium(III) anion $[Ce(C_8H_8)_2]^-$ has been investigated in detail. X-ray structural analyses have been reported for the lithium, sodium, and potassium salts. [Li(THF)$_4$][Ce(C$_8$H$_8$)$_2$] was obtained as a by-product when [{Ce(μ-Cl)(THF)$_2$(C$_8$H$_8$)}$_2$] was reacted with LiCH(TMS)$_2$. The bright green, highly air-sensitive compound consists of separated [Li(THF)$_4$]$^+$ cations and [Ce(C$_8$H$_8$)$_2$]$^-$ sandwich anions (Figure 66).[781]

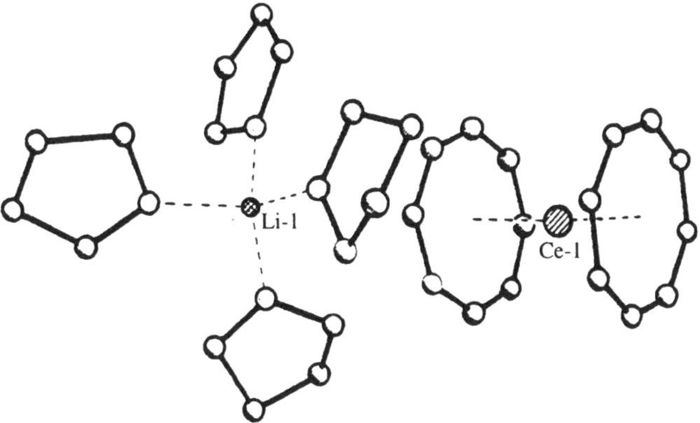

Figure 66 The molecular structure of [Li(THF)$_4$][Ce(C$_8$H$_8$)$_2$].[781]

A different situation is found in the sodium salt [Na(THF)$_3$][Ce(C$_8$H$_8$)$_2$] (Figure 67). In this heterobimetallic sandwich complex sodium and cerium are bridged by a cyclooctatetraenyl ring. The coordination sphere around sodium is filled up by three THF ligands. The "competition" between sodium and cerium for the π-electrons of the bridging cyclooctatetraenyl ring is clearly seen in the longer Ce–C bond distances to the bridging COT ring as compared to the "terminal" COT ligand. Potassium salts of the [Ce(C$_8$H$_8$)$_2$]$^-$ anion have been described as solvates with DME and diglyme.[18] The molecular structures of [K(diglyme)][Ce(C$_8$H$_8$)$_2$] and [K(diglyme)][Yb(C$_8$H$_8$)$_2$][782] were shown to be very similar to the corresponding sodium salt.

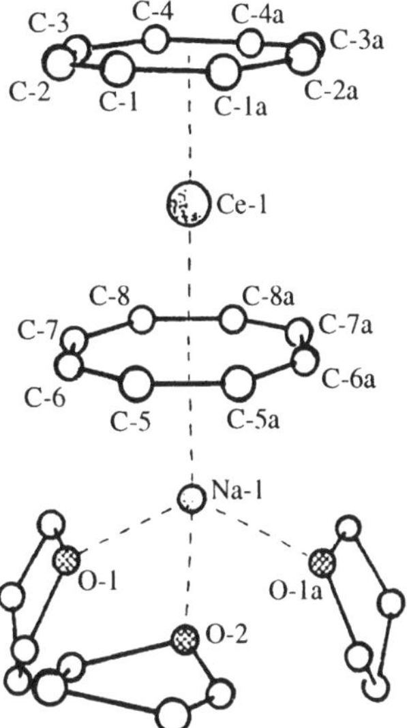

Figure 67 The molecular structure of [Na(THF)$_3$][Ce(C$_8$H$_8$)$_2$].[781]

Formal combination of cationic [Ln(THF)$_2$(C$_8$H$_8$)]$^+$ fragments with the [Ln(C$_8$H$_8$)$_2$]$^-$ anions results in the binuclear cycoocatetraenyl complexes [Ln(THF)$_2$(C$_8$H$_8$)][Ln(C$_8$H$_8$)$_2$]. Such compounds have not been prepared by metathetical reactions but can be made by metal-vapor synthesis. Cocondensation of lanthanide vapors with cyclooctatetraene at −196 °C followed by extraction of the crude materials with THF gave the binuclear complexes [Ln(THF)$_2$(C$_8$H$_8$)][Ln(C$_8$H$_8$)$_2$] (Ln = La, Ce, Nd, Er). An x-ray structure determination of the neodymium complex revealed the presence of an unsymmetrically bridging cyclooctatetraenyl ligand.[18]

A triple-decker sandwich structure was proposed for the binuclear cerium complex Ce$_2$(C$_8$H$_8$)$_3$, but no structural information is available on this unusual material. The microcrystalline complex was prepared by reacting Ce(OPr^i)$_4$ with cyclooctatetraene in the presence of triethylaluminum.[18]

An unusual tetranuclear erbium complex, [Er(μ-C_8H_8)(C_8H_8)K(μ-C_8H_8)Er(μ-C_8H_8)K(THF)$_4$] was obtained serendipitously by reacting benzylcyclopentadienylerbium dichloride, [ErCl$_2$(THF)$_3$-(PhCH$_2$C$_5$H$_4$)] with K$_2$C$_8$H$_8$ in a 1:1 molar ratio in THF solution. Figure 68 shows the molecular structure.[783]

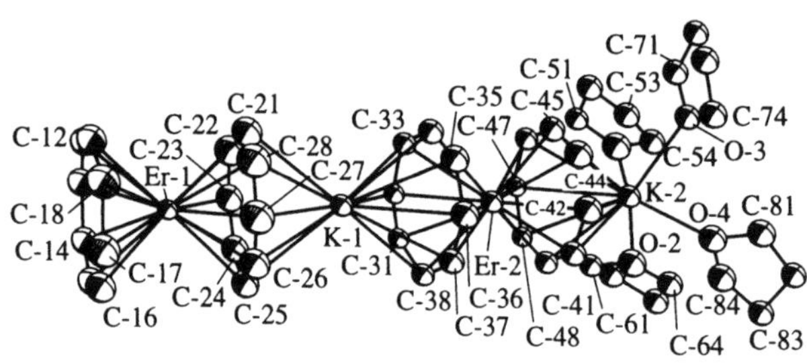

Figure 68   The molecular structure of [Er(μ-C_8H_8)(C_8H_8)K(μ-C_8H_8)Er(μ-C_8H_8)K(THF)$_4$].[783]

2.2.9.4 [M(C$_8$H$_8$)$_2$] compounds

Among the lanthanide elements only cerium can be expected to form stable organolanthanide(IV) complexes. For many years, the so-called cerocene, [Ce(C_8H_8)$_2$] (= bis([8]annulene)cerium(IV)), was the only well-characterized organocerium(IV) complex in which cerium could adopt a Ce^{4+} ground state configuration.[784,785] This cerium analogue of uranocene was first prepared by reacting Ce(OPr^i)$_4$ with cyclooctatetraene in the presence of triethylaluminum.[18] It was isolated as a black-red, pyrophoric solid which was surprisingly resistant toward hydrolysis. A symmetrical sandwich-type structure was indicated by the presence of a single resonance (δ 5.6) in the ^{1}H NMR spectrum. Reduction of [Ce(C_8H_8)$_2$] with alkali metals afforded the known anionic cerium(III) sandwich complex [Ce(C_8H_8)$_2$]$^-$.[18] Recent self-consistent field calculations led to the surprising conclusion that the ground state in cerocene is almost entirely $4f^1$. This would correspond to a formulation as [Ce^{3+}(C$_8$H$_8$$^{1.5-}$)$_2$].[786]

More recently cerocene(IV) chemistry has gained renewed interest through the use of substituted cyclooctatetraenyl ligands. As expected, ring substituted cerocenes are more soluble and less air-sensitive than the parent compound. The first structurally characterized derivative was the 1,1'-dimethyl complex [Ce(MeC$_8$H$_7$)$_2$] with an average Ce–C bond distance of 0.2692(6) nm and a ring centroid–Ce–ring centroid angle of 176.03°.[782]

Soluble cerocene(IV) derivatives have first become readily accessible through the use of trimethylsilyl-substituted cyclooctatetraenyl ligands (Schemes 39 and 40).[787]

$$CeCl_3 + Li(THF)_n\{1,4\text{-}(TMS)_2C_8H_6\} \xrightarrow[-3\,LiCl]{THF} Li[Ce\{1,4\text{-}(TMS)_2C_8H_6\}_2] \xrightarrow[\substack{-Ag \\ -LiI}]{+AgI} [Ce\{1,4\text{-}(TMS)_2C_8H_6\}_2]$$

Scheme 39

$$Ce(O_3SCF_3)_3 + K(THF)_n\{1,3,6\text{-}(TMS)_3C_8H_5\} \xrightarrow[-3\,KO_3SCF_3]{THF} K[Ce\{1,3,6\text{-}(TMS)_3C_8H_5\}_2] \xrightarrow[\substack{-Ag \\ -KI}]{+AgI}$$

$$[Ce\{1,3,6\text{-}(TMS)_3C_8H_5\}_2]$$

Scheme 40

The dark purple crystalline [Ce$\{$1,3,6-(TMS)$_3$C$_8$H$_5\}_2$] (**20**) has been structurally characterized (Figure 69, Ce–C 0.2670–0.2751 nm). The ring centroid–Ce–ring centroid angle is 176.1°.

2.2.9.5 Other cyclooctatetraenyl compounds

Mixed-ring cyclooctatetraenyl sandwich complexes of the trivalent rare-earth elements are well investigated. The COT ligand has been combined with the unsubstituted as well as several substituted cyclopentadienyl ligands. Solvated and unsolvated complexes of this type are known. The parent

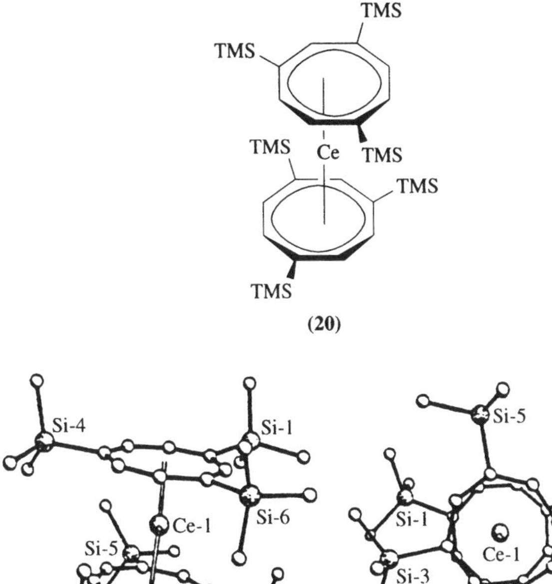

(20)

Figure 69 The molecular structure of $[Ce\{1,3,6\text{-}(TMS)_3C_8H_5\}_2]$.[787]

complexes were first described by Takats *et al.* They have been prepared as THF solvates by two different synthetic approaches. The two methods outlined below (Equations (199) and (200)) are equally useful and the choice of a particular method basically depends on the availability of the starting materials.[18,776,788]

$$[LnCl_2(THF)_3Cp] + K_2C_8H_8 \xrightarrow{\text{THF}} [Ln(THF)_n(C_8H_8)Cp] + 2\,KCl \qquad (199)$$

$$n = 1;\ Ln = Y,\ Sm,\ Ho,\ Er$$
$$n = 2;\ Ln = Pr,\ Nd,\ Gd$$

$$[\{Ln(\mu\text{-}Cl)(THF)_2(C_8H_8)\}_2] + 2\,NaCp \xrightarrow{\text{THF}} 2\,[Ln(THF)(C_8H_8)Cp] + 2\,KCl \qquad (200)$$

$$Ln = Y,\ Nd$$

Simultaneous treatment of anhydrous lanthanide trichlorides with equimolar amounts of NaCp and $K_2C_8H_8$ has been shown to be equally successful.[788,789] Yellow, unsolvated $[Sc(C_8H_8)(Cp)]$ was prepared similarly by reacting $[ScCl(THF)(C_8H_8)]$ with sodium cyclopentadienide.[18] The bis(THF) solvate $[Pr(THF)_2(C_8H_8)(Cp)]$ has been structurally characterized. The two ring centroids and the oxygen atoms of the THF ligands form a distorted tetrahedron around the central praseodymium atom.[788,789]

The mixed-ring cyclooctatetraenyl yttrium complex $[Y(THF)(C_8H_8)(MeC_5H_4)]$ was prepared according to Equation (201) and characterized by x-ray crystallography.[790]

$$[\{Y(\mu\text{-}Cl)(THF)(C_8H_8)\}_2] + 2\,Na(MeC_5H_4) \xrightarrow{\text{THF}} 2\,[Y(THF)(C_8H_8)(MeC_5H_4)] + 2\,NaCl \qquad (201)$$

The centroid–Y–centroid angle in $[Y(THF)(C_8H_8)(MeC_5H_4)]$ was found to be 149°.[790] Similar indenyl mixed-ligand sandwich complexes were prepared by simultaneous addition of KC_9H_7 and $K_2C_8H_8$ to anhydrous lanthanide trichlorides in THF (Equation (202)). The praseodymium complex was structurally characterized. The two ring centroids and the two THF oxygen atoms form a distorted tetrahedral coordination geometry around the metal (Pr–C(C_9H_7) 0.287 nm, Pr–C(C_8H_8) 0.272 nm).[788,791,792]

$$LnCl_3 + K_2C_8H_8 + KC_9H_7 \xrightarrow{\text{THF}} [Ln(THF)_2(C_8H_8)(C_9H_7)] + 3\,KCl \qquad (202)$$

$$Ln = Pr,\ Nd$$

Other ring-substituted mixed-sandwich complexes are $[Ln(THF)(C_8H_8)(C_5Me_4H)]$ (Ln = Y, La, Pr, Sm, Gd, Dy, Er, Lu),[793] $[Ln(THF)(C_8H_8)(Cp^*)]$ and $[Ln(Cp^*)(C_8H_8)]$ (Equation (203)),[665,790,794,795] $[Lu\{C_5(Bz)_5\}(C_8H_8)]$,[796] as well as $[Ln\{1,3\text{-}(TMS)_2C_5H_3\}\{1,4\text{-}(TMS)_2C_8H_6\}]$.[797]

$$[\{Ln(\mu\text{-}Cl)(THF)_2(C_8H_8)\}_2] + 2\,NaCp^* \xrightarrow{\text{THF}} 2\,[Ln(THF)(C_8H_8)Cp^*] + 2\,NaCl \qquad (203)$$

$$\text{Ln = Sc, Y, La, Pr, Nd, Sm, Gd, Tb, Dy, Er, Lu}$$

Structurally characterized derivatives are $[Lu(Cp^*)(C_8H_8)]$,[794] $[Tb(Bu^t_2C_5H_3)(C_8H_8)]$,[615] and $[LuC_5(Bz)_5(C_8H_8)]$.[796] The pentamethylcyclopentadienyl derivatives $[Ln(Cp^*)(C_8H_8)]$ (Ln = Sc, Y, La) have been studied by UPS spectroscopy and their ionization energies have been compared with the early transition metal analogues $[M(Cp^*)(C_8H_8)]$ (M = Ti, Zr).[795,798] From these studies an almost entirely ionic bonding was concluded for the lanthanide derivatives. The structurally characterized samarium complex $[Sm(THF)(2,4\text{-}C_7H_{11})(C_8H_8)]$, made from $[\{Sm(\mu\text{-}Cl)(THF)_2(C_8H_8)\}_2]$ and 2,4-dimethylpentadienyl potassium, is an interesting example of an "open" sandwich complex.[769]

A novel development in the chemistry of cyclooctatetraenyl lanthanide sandwich complexes is the introduction of donor function to make "metalloligands." The first examples contain phosphino substituents at the cyclopentadienyl ring. Three samarium sandwich complexes of this type have been prepared according to Equation (204). The products have been successfully employed as "metalloligands" in the preparation of heterobimetallic complexes (cf. Section 2.2.11.2).[799]

$$[\{Sm(\mu\text{-}Cl)(THF)_2(C_8H_8)\}_2] + 2\,M(C_5R^1_4PR^2_2) \longrightarrow 2\,[Sm(THF)_n(C_8H_8)(C_5R^1_4PR^2_2)] + 2\,MCl \qquad (204)$$

$$M = Li;\ R^1 = H;\ R^2 = Ph;\ n = 2$$
$$M = K;\ R^1 = Me;\ R^2 = Ph;\ n = 0$$
$$M = K;\ R^1 = R^2 = Me;\ n = 0$$

A heterobimetallic mono(cyclooctatetraenyl)cerium(III) complex was isolated, when cerium(IV) tetraisopropoxide was reacted with triethylaluminum in the presence of cyclooctatetraene. In the course of this reaction cerium(IV) is reduced to the trivalent oxidation state. Cerium and aluminum are bridged by isopropoxide ligands in the product (Equation (205)).[18]

$$Ce(OPr^i)_4 \xrightarrow{\text{AlEt}_3,\ C_8H_8} [Ce(\mu\text{-}OPr^i)_2AlEt_2(C_8H_8)] \qquad (205)$$

2.2.10 Fullerenes

The first communications on lanthanide elements intercalated in fullerenes appeared in 1985.[800,801] The first extraction and isolation of an endohedral lanthanum fullerene, La@C_{82}, was achieved in 1991.[802] By now the electronic properties of lanthanide metal-encapsulated fullerenes are well studied. It has been established that these compounds are best described as $Ln^{3+}@C_{2n}^{3-}$.[803-7] Lanthanide metallofullerenes are prepared by arc burning of a composite rod[803,808-10] made from a metal oxide and graphite in a depressurized helium atmosphere.[801-3,806,811] By using such composite rods, the endohedral metallofullerenes of scandium, yttrium,[803,811,812] and lanthanum[803,810-12] have been successfully produced, extracted, and characterized by electron spin resonance and mass spectrometry. However, the yield of metallofullerenes is very low, normally far below 1% of the original soot. Recently a high yield synthesis of lanthanofullerenes using a newly designed fullerene arc generator equipped with an anaerobic sampling apparatus and lanthanum carbide-enriched composite carbon rods was reported. The yield of the lanthanofullerene La@C_{82} increases by a factor of 10 or more, when LaC$_2$-enriched graphite rods are used for the generation of soot instead of hydrooxidized or oxidized lanthanum composite rods. Furthermore, the yields of other types of lanthanofullerenes LaC$_{2n}$ increase significantly under these conditions.[812] However, the isolation of pure endohedral La@C_{82} proved to be quite difficult. Only recently was the chromatographic separation of La@C_{82} reported.[813] For some time La@C_{82} appeared to be the only soluble endohedral lanthanum fullerene.[802] It was discovered that La@C_{82} is soluble in a wide range of solvents, including cold and boiling toluene, carbon disulfide, and pyridine.[802,803,808,811] In the meantime, other soluble endohedral lanthanum buckyballs have been detected and isolated. The use of greater La:C ratios in the soot production afforded the toluene extractable species La$_2C_{80}$.[809] La@C_{76} has become available through the use of anaerobic sampling methods.[810] A high-pressure toluene extraction device has been designed which allows efficient extraction of La@C_{2n} for n from 37 to 45.[814] In addition to the endohedral metallofullerenes, externally bound LaC$_{60}^+$ ions have also been prepared.[815]

2.2.11 Heterobimetallic Complexes

2.2.11.1 Metal–metal bonded complexes

The tendency of lanthanide ions to form stable unsupported metal–metal bonds to transition metals or other lanthanides is extremely low. No complex containing a direct lanthanide-to-lanthanide bond has ever been described. Metathetical reactions of, for example, bis(cyclopentadienyl)lanthanide chlorides with carbonyl metallates usually lead to the formation of complexes, in which the metals are linked via isocarbonyl bridges, that is, they contain M–CO–Ln units (M = transition metal). These compounds are discussed in Section 2.2.10.2. True metal–metal bonds may exist in some organomercury ytterbium halides, although these are not organolanthanide complexes in the strict sense, and structural information is lacking. Such compounds have been prepared according to Equation (206).[816]

$$\mathrm{Yb\ +\ RHgI\ \xrightarrow{\ THF\ }\ RHgYbI(THF)}_n \qquad (206)$$

A milestone in the chemistry of heterobimetallic organolanthanide complexes was the preparation of the first complex containing an unsupported lanthanide–transition metal bond. The compound [(Cp)$_2$(THF)LuRu(CO)$_2$(Cp)] was prepared from [LuCl(THF)(Cp)$_2$] and the sodium salt of the [Ru(CO)$_2$(Cp)]$^-$ anion. The crystal structure determination revealed a Lu–Ru bond length of 0.2955(2) nm and no bridging carbonyl ligands (Figure 70).[90,817]

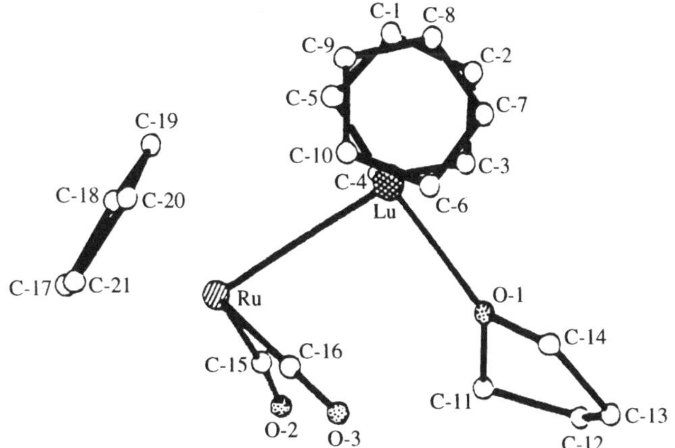

Figure 70　The molecular structure of [(Cp)$_2$(THF)LuRu(CO)$_2$(Cp)].[90,817]

The reaction has later been successfully extended to the preparation of [Lu(Cp*)$_2$Ru(CO)$_2$(Cp)] and [Lu(THF){1,3-(TMS)$_2$C$_5$H$_3$}$_2$Ru(CO)$_2$(Cp)]. A La–Ru bond is present in the bimetallic complexes [La(THF)$_3$I$_2$Ru(CO)$_2$(Cp)] and [La(THF)(Cp){Ru(CO)$_2$(Cp)}(µ-I)$_2$Na(THF)$_2$], which were prepared according to Equations (207) and (208) and characterized by ^{139}La NMR spectroscopy.[90]

$$\mathrm{[LaI_3(THF)_3]\ +\ Na[CpRu(CO)_2]\ \longrightarrow\ [La(THF)_3I_2Ru(CO)_2Cp]\ +\ NaI} \qquad (207)$$

$$\mathrm{[LaI_2(THF)_3Cp]\ +\ Na[CpRu(CO)_2]\ \xrightarrow{\ THF\ }\ [La(THF)Cp\{Ru(CO)_2Cp\}(\mu\text{-}I)_2Na(THF)_2]} \qquad (208)$$

Initial reactivity studies have been reported for the metal–metal bonded species. In all cases cleavage of the Lu–Ru bond occurs (Equations (209)–(211)).[90]

$$\mathrm{2\ [Lu(THF)\{1,3\text{-}(TMS)_2C_5H_3\}_2Ru(CO)_2Cp]\ +\ 2\ HX\ \longrightarrow\ [\{Lu(\mu\text{-}X)\{1,3\text{-}(TMS)_2C_5H_3\}_2\}_2]\ +\ 2\ [RuH(CO)_2Cp]} \qquad (209)$$

$$\mathrm{[Lu(THF)\{1,3\text{-}(TMS)_2C_5H_3\}_2Ru(CO)_2Cp]\ +\ HgCl_2\ \longrightarrow}$$
$$\mathrm{[\{Lu(\mu\text{-}Cl)\{1,3\text{-}(TMS)_2C_5H_3\}_2\}_2]\ +\ 2\ Hg[Ru(CO)_2Cp]_2} \qquad (210)$$

$$\mathrm{[Lu(THF)Cp_2Ru(CO)_2Cp]\ +\ (Pr^i)_2C{=}O\ \longrightarrow\ [Lu(THF)Cp_2OC(Pr^i){=}CMe_2]\ +\ [RuH(CO)_2Cp]} \qquad (211)$$

2.2.11.2 Heterobimetallic complexes without metal–metal bonds

Heterobimetallic organo-*f*-element complexes derived from transition metal carbonyls have been reviewed by Beletskaya *et al.*[818] Two fundamentally different synthetic routes to such heterobimetallics have been developed. The most elegant way involves the reduction of transition metal carbonyl complexes by lanthanide(II) complexes. This way a number of interesting new complexes have been prepared, although the method has so far been applied only to samarium and ytterbium derivatives. Similar reactions of europium(II) complexes have not yet been studied. The second, more generally applicable route is the metathetical reaction of organolanthanide halides with carbonyl metallates. These preparations can be carried out in principle with all elements of the lanthanide series. Furthermore, adduct formation of Ln(Cp)$_3$ with transition metal carbonyls or nitrosyls leads to similar compounds.[18] In all cases, however, the resulting anionic metal carbonyl fragments are normally bonded to the lanthanide ions via so-called isocarbonyl linkages, that is, transition metal-coordinated carbon monoxide ligands are linked to the lanthanide ions through the oxygen atoms. This situation once again reflects the highly oxophilic character of the lanthanide elements. It is also the main reason for the often encountered inaccessibility of metal–metal bonded species in organolanthanide chemistry. If oxygen-containing ligands are present, these elements will usually find a way to circumvent the formation of direct metal–metal bonds. A typical example is the transmetallation reaction of elemental lanthanides with mercury bis(tetracarbonylcobaltate) (Equation (212)). The resulting products were initially discussed as being metal–metal bonded species, although the carbonylmetallate anions are coordinated to the lanthanides via isocarbonyl bridges.[819]

$$\text{Ln} + \text{Hg}[\text{Co(CO)}_4]_2 \xrightarrow{\text{THF}} [\text{Ln}\{\text{Co(CO)}_4\}_2(\text{THF})_n] + \text{Hg} \qquad (212)$$

$$\text{Ln} = \text{Sm, Eu}$$
$$n = 3, 4$$

SmI$_2$(THF)$_x$ was reported to react with dicobalt octacarbonyl to give the saltlike species [SmI$_2$(THF)$_5$][Co(CO)$_4$]. Treatment of [Co$_2$(CO)$_8$] with the THF adduct of decamethylsamarocene resulted in a clean redox reaction and formation of the isocarbonyl-linked product (Equation (213)).[820]

$$2\,[\text{Sm(THF)}_2\text{Cp*}_2] + [\text{Co}_2(\text{CO})_8] \xrightarrow{-2\,\text{THF}} 2\,[\text{Sm(THF)}(\mu\text{-OC})\text{Co(CO)}_3\text{Cp*}] \qquad (213)$$

A mixture of the two samarium-cobalt complexes was obtained, when dicobalt octacarbonyl was reacted similarly with the mono(pentamethylcyclopentadienyl)samarium(II) complex [{Sm(μ-I)(THF)$_2$(Cp*)}$_2$].[820]

Andersen *et al.* have studied electron transfer reactions between [Yb(Et$_2$O)(Cp*)$_2$] and various transition metal carbonyls. Treatment of the ytterbium(II) complex with [Mn$_2$(CO)$_{10}$] gave dark blue polymeric YbMn(CO)$_5$(Cp*)$_2$·0.25toluene. The crystal structure of this material consists of polymeric sheets of the composition Yb(μ-OC)$_3$Mn(CO)$_2$(Cp*)$_2$ and dimeric Yb(μ-OC)$_2$Mn(CO)$_3$(Cp*)$_2$ units between the sheets.[821] A similar reaction of [Yb(Et$_2$O)(Cp*)$_2$] with [Co$_2$(CO)$_8$] produced the monomeric isocarbonyl-linked species [Yb(THF)(μ-OC)Co(CO)$_3$(Cp*)$_2$].[18]

Polynuclear heterobimetallic complexes were isolated from electron transfer reactions of [Yb(Et$_2$O)(Cp*)$_2$] and cyclopentadienyl cobaltdicarbonyl derivatives as well as [Fe$_2$(CO)$_9$] or [Fe$_3$(CO)$_{12}$] (Equation (214)).[822]

$$2\,[\text{Yb(Et}_2\text{O)Cp*}_2] + 3\,[\text{Co(CO)}_2(\text{RC}_5\text{H}_4)] \longrightarrow [\{\text{YbCp*}_2\}_2(\mu_3\text{-CO})_4\text{Co}_3(\text{RC}_5\text{H}_4)_2] \qquad (214)$$

$$\text{R} = \text{H, Me, TMS}$$

An x-ray structural analysis (Figure 71) of [{Yb(Cp*)$_2$}$_2$(μ$_3$-CO)$_4$Co$_3$(TMS-C$_5$H$_4$)$_2$] showed that each of the [Yb(Cp*)$_2$] fragments was coordinated to the oxygen atoms of two triply bridging isocarbonyl ligands.[822]

A related complex, [{Yb(Cp*)$_2$}$_2${Fe$_3$(CO)$_{11}$}], was obtained by reacting [Yb(Et$_2$O)(Cp*)$_2$] with [Fe$_3$(CO)$_{12}$].[18] The ytterbium–iron compound {[YbFe(CO)$_4$(MeCN)$_3$]$_2$·MeCN}∞ does not contain lanthanide–carbon bonds.[823]

Isocarbonyl bridges were also found in the structurally characterized compounds [Yb(THF)(μ-OC)Co(CO)$_3$(Cp)$_2$],[824,825] [Ce{1,3-(TMS)$_2$C$_5$H$_3$}$_2$(μ-OC)$_2$W(CO)(Cp)],[826] [Dy$_2$(μ-OC)(Cp*)$_4$Fe$_2$(CO)(μ-CO)$_2$(Cp)$_2$]$_2$,[827] and [{Sm(μ-OC)$_2$(Cp*)$_2$Fe(Cp*)}$_2$].[828] The latter two form twelve-membered ring systems as illustrated by the samarium–iron complex (Figure 72, Sm–C 0.2694(9) nm).

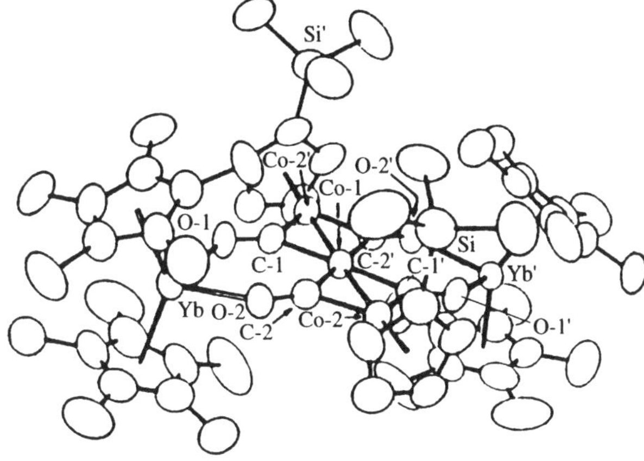

Figure 71 The molecular structure of [{Yb(Cp*)$_2$}$_2$(μ_3-CO)$_4$Co$_3$(TMS-C$_5$H$_4$)$_2$].[822]

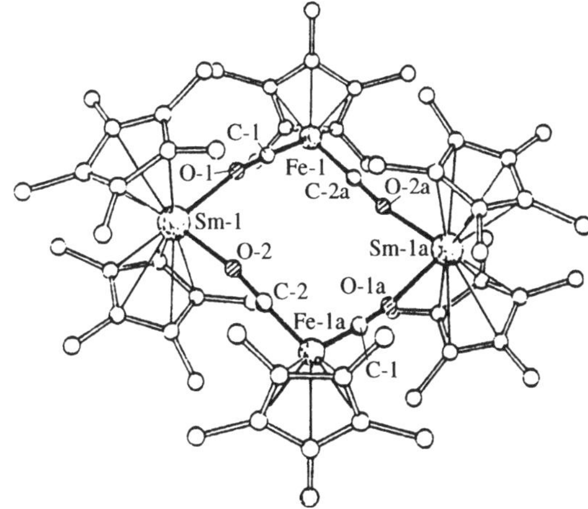

Figure 72 The molecular structure of [{Sm(μ-OC)$_2$(Cp*)$_2$Fe(Cp*)}$_2$].[828]

Rhenium hydride complexes have been found to give heterobimetallic species upon reaction with organolanthanides. The compounds [Y(THF)Re$_2$H$_7$(PMe$_2$Ph)$_4$(Cp)$_2$] (Y$\cdots$Re 0.3090(2) nm) and [Lu(THF)Re$_2$H$_7$(PMe$_2$Ph)$_4$(Cp)$_2$] (Lu$\cdots$Re 0.3068(1) and 0.3025(1) nm) have been prepared and structurally characterized.[829] The related bimetallic hydrides [Yb(diglyme){(PMe$_3$)$_3$WH$_5$}$_2$] and [{(Cp)$_2$NbH$_2$}Yb(diglyme)] have been prepared by reacting either K[WH$_5$(PMe$_3$)$_3$] or K[NbH$_2$(Cp)$_2$] with ytterbium diiodide.[830] However, these materials are not organolanthanide complexes in the strict sense and are thus not further discussed.

Reactions of the dimeric hydrides [{Ln(μ-H)(Cp*)$_2$}$_2$] with [WH$_2$(Cp)$_2$] afforded the σ-bond metathesis products [Ln(μ-η^1,η^5-C$_5$H$_4$)(μ-H)$_2$(Cp*)$_2$W(Cp)] via dehydrocoupling between the Ln–H bond and the C–H (but not the weaker W–H) bond of the organotungsten hydride. [Sm(μ-η^1,η^5-C$_5$H$_4$)(μ-H)$_2$(Cp*)$_2$W(Cp)] was structurally characterized. The yttrium compound [Y(μ-η^1,η^5-C$_5$H$_4$)(μ-H)$_2$(Cp*)$_2$W(Cp)] reacts with pyridine to give [Y(η^2-C$_5$H$_4$N)(Cp*)$_2$] and [WH$_2$(Cp)$_2$] (Scheme 41).[715,831]

Scheme 41

Numerous heterobimetallic complexes have been obtained by salt-elimination reactions between bis(cyclopentadienyl)lanthanide halides and carbonylmetallates (e.g., $[Mn(CO)_5]^-$, $[Re(CO)_5]^-$, or $[(Cp)W(CO)_3]^-$) (Equation (215)).[18,832-4] In some cases, especially with the manganese-containing anions, metal–metal bonded products have been discussed on the basis of IR spectroscopic studies. However, no conclusive proof for the existence of such metal–metal-bonded species has been obtained in these early publications. It seems most likely that all of the products described in these studies contains isocarbonyl linkages between the lanthanide and transition elements.

$$[LnCl(THF)Cp_2] + Na[W(CO)_3Cp] \xrightarrow{\text{THF}} [Ln(THF)(\mu\text{-OC})Cp_2\{W(CO)_2Cp\}] + CO \qquad (215)$$

$$Ln = Dy, Er, Yb$$

The first examples of cationic organolanthanide complexes with $[Co(CO)_4]^-$ as anion, $[Ln(THF)(C_5H_4CH_2CH_2OMe)_2][Co(CO)_4]$ (Ln = Sm, Yb), have been synthesized by metathesis of $[LnI(C_5H_4CH_2CH_2OMe)_2]$ with $K[Co(CO)_4]$ or by one-electron oxidation of the lanthanide(II) metallocenes $[Ln(THF)(C_5H_4CH_2CH_2OMe)_2]$ with $[Co_2(CO)_8]$. In this case the formation of Ln–OC–Co linkages is avoided because the lanthanide atoms are coordinatively saturated through chelate formation with the methoxyethyl substituents.[835] Unexpectedly, similar treatment of $[LuCl_2(THF)_3(Cp)]$ with an equimolar mixture of $Na_2[W(CO)_5]$ and $Na_2[W_2(CO)_{10}]$ in DME solution afforded only the trinuclear anionic tungsten cluster $[Na(DME)_3][W_3(CO)_{14}]$ and no isolable lanthanide-containing species.[836]

A novel type of heterobimetallic organolanthanide complex involves the use of lanthanide-containing "metalloligands." While such heterobimetallic compounds are well established in *d*-transition metal chemistry, they have only recently been described for the first time in organolanthanide chemistry. The first example of such ligand was the diphenylphosphinocyclopentadienylytterbium(II) complex $[Yb(THF)_2(C_5H_4PPh_2)_2]$. The "metalloligand" has been successfully utilized to prepare the structurally characterized heterobimetallic ytterbium–transition metal complexes $[Yb(THF)_2\{(C_5H_4PPh_2)_2PtMe_2\}]\cdot$toluene[837] (Figure 73) and $[Yb(THF)_2\{(C_5H_4PPh_2)_2Ni(CO)_2\}]$.[250] These complexes resulted from reactions of the ytterbium(II) precursor with $[Ni(CO)_2(PPh_3)_2]$ or $[PtMe_2(COD)]$ (COD = cycloocta-1,5-diene), respectively. The molybdenum carbonyl derivative $[Yb(THF)_2\{(C_5H_4PPh_2)_2Mo(CO)_4\}]$ was made analogously from $[Mo(CO)_4(norbornadiene)]$.

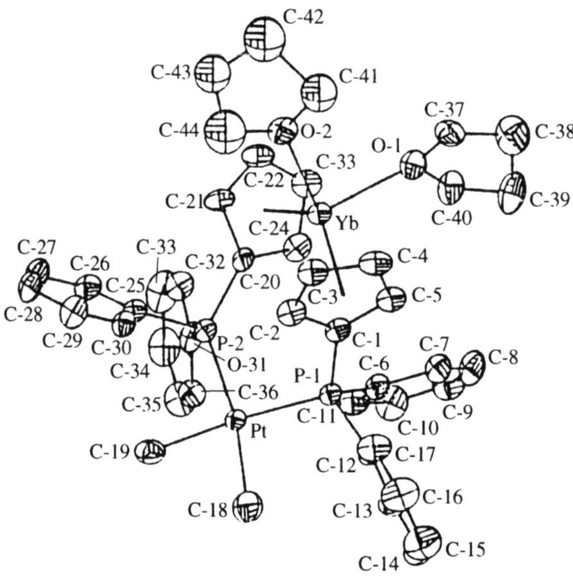

Figure 73   The molecular structure of $[Yb(THF)_2\{(C_5H_4PPh_2)_2PtMe_2\}]\cdot$toluene.[837]

Another potentially useful "metalloligand" is $[La(C_5H_4PPh_2)_2(THF)(Cp)]$.[472] This compound and the ytterbium complex $[Yb(THF)_2(C_5H_4PPh_2)_2]$ have both been designed to act as chelating ligands toward transition metals. Monodentate metallophosphines have been synthesized based on cyclooctatetraenyl-lanthanide mixed sandwich complexes (cf. Section 2.2.9.5). The metallophosphines $[Sm(C_5R^1_4PR^2_2)(THF)(C_8H_8)]$ react with cyclopentadienylrhodium dicarbonyl to provide heterobimetallic samarium–rhodium complexes (Equation (216)).[799]

Yet another type of heterobimetallic organolanthanide without isocarbonyl bridges involves coordination of transition metal sandwich complexes *via* σ-bonds to the rare-earth elements (Equations

$$[Sm(C_5R^1_4PR^2_2)(THF)(C_8H_8)] + [Rh(CO)_2Cp] \longrightarrow [Sm(C_5R^1_4PR^2_2)RhCp(CO)_2(C_8H_8)] + THF + CO \qquad (216)$$

$$R^1 = H, R^2 = Ph$$
$$R^1 = Me, R^2 = Me, Ph$$

(217) and (218)). Early examples of such complexes include bis(ferrocenyl)ytterbium(II) and bis(cymantrenyl)ytterbium(II), which have both been prepared by metal exchange reactions between ytterbium and the corresponding organomercury reagents.[838–42] Although these interesting compounds have not been structurally investigated by x-ray crystallography, coordinative stabilization by solvent molecules (THF) seems highly likely.[839]

$$Ln + Hg[\{Fe(C_5H_4)Cp\}_2] \xrightarrow{\text{THF}} [Ln(THF)_n\{Fe(C_5H_4)Cp\}_2] + Hg \qquad (217)$$
$$Ln = Sm, Eu, Yb$$

$$Ln + Hg[Mn(CO)_3(C_5H_4)]_2 \xrightarrow{\text{THF}} [Ln(THF)_n\{Mn(CO)_3(C_5H_4)\}_2] + Hg \qquad (218)$$
$$Ln = Sm, Eu, Yb$$

Other complexes of this type are $[\{Mn(CO)_3(C_5H_4)\}YbI_2(THF)_3]$[843] and the chromium derivatives $[Ln(THF)_n\{(C_6H_5)Cr(CO)_3\}_2]$ (Ln = Sm, Eu, Yb).[844] The complex $[Yb(DME)(Cp)_2][Hg\{Ge(C_6F_5)_3\}_2]$ was prepared by reaction of $Yb(Cp)_2$ with $Hg[Ge(C_6F_5)_3]_2$.[845]

Especially useful for the preparation of heterobimetallic organolanthanides is the chelating 2-(dimethylaminomethyl)ferrocenyl ligand (= FcN)⁻. Several classes of organolanthanides containing σ-bonded FcN⁻-ligands have been prepared by metathetical reactions of the corresponding halide precursors and stoichiometric amounts of Li(FcN). Among the fully characterized lanthanide FcN derivatives are Ln(FcN)Cl(Cp) (Ln = Dy, Ho),[846] [Ln(FcN)_2(Cp)] (Ln = Dy, Yb),[846,847] $[Ln(FcN)(THF)_n(Cp)_2]$ (Ln = Sc, Y, Sm, Dy, Ho, Er; n = 0, 1),[846–8] [Ln(FcN)Cl(Cp*)] (Ln = Y, Sm),[849] and $[Sm(FcN)(Cp^*)_2]$.[849] Another ferrocene-containing lanthanide organometallic is the alkoxide $[\{Sm(\mu\text{-}OFc)(Cp)_2\}_2]$ (Fc = ferrocenyl).[850] Lanthanide FcN complexes without supporting cyclopentadienyl ligands are available via direct treatment of lanthanide trichlorides with stoichiometric amounts of Li(FcN) in THF solution (Equations (219)–(221)). All these heterobimetallic lanthanide organometallics are thermally highly stable and contain chelating 2-(dimethylaminomethyl)ferrocenyl ligands.[847]

$$LaCl_3 + 2\,Li(FcN) \longrightarrow LaCl(FcN)_2 + 2\,LiCl \qquad (219)$$

$$LnCl_3 + 3\,Li(FcN) \longrightarrow [Ln(FcN)_3 \bullet LiCl \bullet THF] + 2\,LiCl \qquad (220)$$
$$Ln = Ce, Pr, Nd$$

$$YbCl_3 + 3\,Li(FcN) \longrightarrow [Yb(FcN)_3] + 3\,LiCl \qquad (221)$$

The structurally characterized ytterbium "ate" complex $[Yb(\mu\text{-}Cl)_2(FcN)_2Li(THF)_2]$ was prepared by reacting anhydrous $YbCl_3$ with two equivalents of Li(FcN) (Equation (222)).[851]

$$(222)$$

$[Yb(\mu\text{-}Cl)_2(FcN)_2Li(THF)_2]$ represents the first structurally characterized example of a diorganolanthanide(III) halide containing only σ-bonded organic ligands (Yb–C 0.2651, 0.2372 nm).[851]

An unusual trimetallic 2-(dimethylaminomethyl)ferrocenylyttrium complex (**21**) was obtained when $[\{Y(\mu\text{-}Cl)(Cp)_2\}_2]$ was reacted with two equivalents of Li(FcN) (Equation (223)).[852]

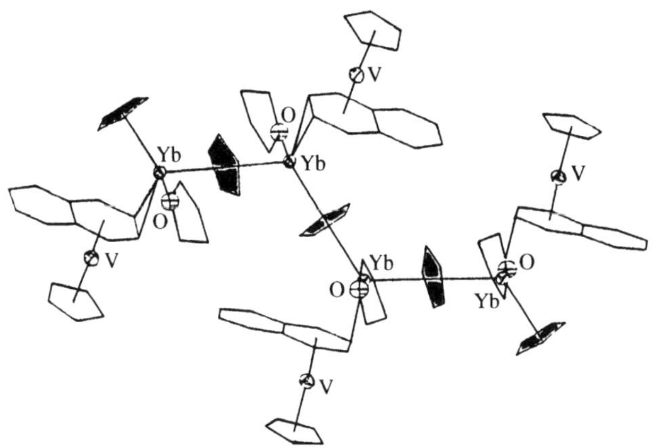

(21)

$$[\{Y(\mu\text{-Cl})Cp_2\}_2] + 4\,Li(FcN) \longrightarrow 2\,Li[Y(FcN)_2Cp_2] + 2\,LiCl \tag{223}$$

Formally $Li[Y(FcN)_2(Cp)_2]$ is the first structurally characterized derivative of the $[Ln(Cp)_4]^-$ anions (cf. Section 2.2.5.9). Two cyclopentadienyl rings in this compound exhibit a novel $\eta^5(M),\mu\text{-}\eta^1(M'),\mu\text{-}\eta^1(M'')$ coordination mode.[852]

Finally, a polymeric structure was found in the unusual ytterbium–vanadium heterobimetallic complex $[\{V(\mu\text{-}\eta^6{:}\eta^2\text{-}C_{10}H_8)(Cp)Yb(THF)(Cp)\}_n]$, which was synthesized by the reaction of $V(Cp)_2$ with an excess of naphthaleneytterbium, $[Yb(THF)_2(C_{10}H_8)]$ in THF. An x-ray structural investigation revealed that $[\{V(\mu\text{-}\eta^6{:}\eta^2\text{-}C_{10}H_8)(Cp)Yb(THF)(Cp)\}_n]$ is a polymeric two-dimensional multidecker complex. The molecular structure consists of an infinite zig-zag chain formed by Yb(Cp) moieties with one $V(\mu\text{-}C_{10}H_8)(Cp)$ unit coordinated η^2 via the naphthalene ligand to each ytterbium atom (Figure 74).[853]

Figure 74   The molecular structure of $[\{V(\mu\text{-}\eta^6{:}\eta^2\text{-}C_{10}H_8)(Cp)Yb(THF)(Cp)\}_n]$.[853]

The redox properties of cerocene(IV) derivatives can be utilized to prepare novel heterobimetallic complexes. Reduction of $[Ce\{1,4\text{-}(TMS)_2C_8H_6\}_2]$ with cobaltocene produced the saltlike compound **(22)**, in which both the cation and the anion are sandwich complexes.[787]

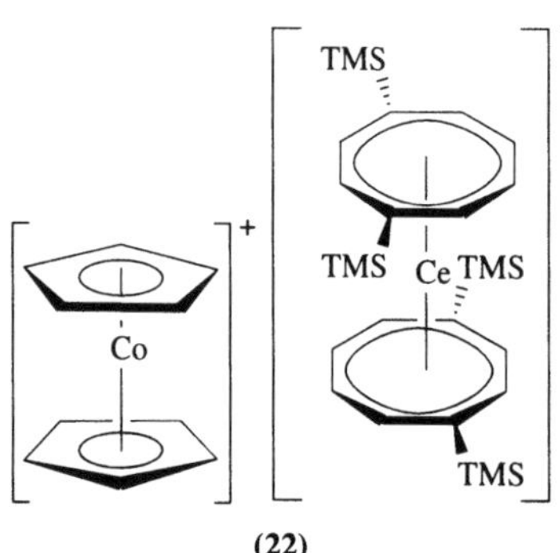

(22)

2.2.12 Heteronuclear NMR Spectroscopy

Only four lanthanide nuclei are accessible to direct observation by heteronuclear NMR spectroscopy, namely ^{45}Sc, ^{89}Y, ^{139}La, and ^{171}Yb.[854]

Only a few reports on the use of ^{45}Sc NMR spectroscopy in organoscandium chemistry have appeared, although scandium-45 is an excellent NMR nucleus. It is 100% abundant and has a relative sensitivity more than 1700 times that of ^{13}C, which makes ^{45}Sc the sixth most favorable NMR nucleus.[855] Chemical shifts are measured relative to saturated $ScCl_3$ in D_2O. Thus far, only the complexes [{Sc(μ-Cl)(Cp)$_2$}$_2$], [Sc(BH$_4$)(Cp)$_2$] and Sc(Cp)$_3$ have been investigated by ^{45}Sc NMR spectroscopy. For Sc(Cp)$_3$ the formation of a dimer at low temperature was studied by variable-temperature ^{45}Sc NMR.[855]

^{89}Y NMR spectroscopy is by now fairly well developed and has become a valuable diagnostic tool to characterize organoyttrium complexes.[320,611,856] In order to determine the influence of various ligand environments on the yttrium center, the compounds [Y(OAr)$_3$] (Ar = [C$_6$H$_3$(Bu^t)$_2$-2,6]), [Y{CH(TMS)$_2$}$_3$], [YCH(TMS)$_2$(Cp*)$_2$], [Y(OAr)$_2$(Cp*)], and [YOAr(Cp*)$_2$] have been studied. The ^{89}Y NMR chemical shifts vary significantly depending on the electron-donating properties of the different ligands. The Cp*, OAr, and CH(TMS)$_2$ group contributions to the ^{89}Y chemical shifts were calculated to be -100, $+56$, and -298 ppm.[611] Other organoyttrium complexes studied by ^{89}Y NMR spectroscopy include [Y(THF)(MeC$_5$H$_4$)$_3$], [{Y(μ-Cl)(MeC$_5$H$_4$)$_2$}$_2$], [YCl(THF)(MeC$_5$H$_4$)$_2$], [{Y(μ-Me)(MeC$_5$H$_4$)$_2$}$_2$], [{Y(μ-C≡CBu^t)(MeC$_5$H$_4$)$_2$}$_2$], [{Y(μ-H)(THF)(MeC$_5$H$_4$)$_2$}$_2$], [Li(THF)$_4$][(μ_3-H){(MeC$_5$H$_4$)$_2$Y(μ-H)}$_3$], [Y(μ-Cl)$_2$K(THF)$_2$(Cp*)$_2$],[856] [{Y(μ-H)(Cp*)$_2$}$_2$],[705] and [Y(μ-η^1:η^5-C$_5$H$_4$)(μ-H)$_2$(Cp*)$_2$W(Cp)][715,831] as well as several cyclooctatetraenyl complexes.[774] Solid-state CP/MAS ^{89}Y NMR spectroscopy has recently been extended to a series of air-sensitive yttrium complexes with a chemical shift range of over 1000 ppm. Previously this method had been limited to air-stable hydrated materials. It has now been demonstrated that ^{89}Y NMR CP/MAS spectra can be easily obtained for a broad range of complexes, including organometallics, in a short period of time.[857]

Several adducts of tris(cyclopentadienyl)lanthanum have been investigated by ^{139}La NMR spectroscopy.[506,511,854] Among the complexes studied are [La(DMF)(Cp)$_3$], [La(DMSO)(Cp)$_3$], [La{OP(OMe)$_3$}(Cp)$_3$], and [La(OCMe$_2$)(Cp)$_3$] as well as adducts with THF, nitriles, and isonitriles.[507,512,858] More recently complexes containing a direct La–Ru bond have been investigated by ^{139}La NMR spectroscopy.[90] Chemical shifts are measured relative to the resonance of La(ClO$_4$)$_3$ in H$_2$O/D$_2$O. The range of chemical shifts for the hitherto known ^{139}La NMR spectra is extremely wide. It ranges from $+1090$ ppm for [LaBr$_6$]$^{3-}$ to -578 ppm for [La(MeCN)(Cp)$_3$]. Thus, the method is a very sensitive tool for studying organolanthanum complexes.[854]

More recently ^{171}Yb NMR spectroscopy has been introduced as a useful method to study organoytterbium(II) complexes. High-resolution ^{171}Yb data have been published for the metallocenes [Yb(Et$_2$O)(Cp*)$_2$], [Yb(THF)$_2$(Cp*)$_2$], and [Yb(C$_5$H$_5$N)$_2$(Cp*)$_2$].[576] By now a large number of diamagnetic ytterbium(II) organometallics is available, and well-resolved ^{171}Yb NMR spectra are fairly easy to obtain. ^{171}Yb has a spin-1/2 nucleus, a natural abundance of 14.27% and a moderately sized, positive gyromagnetic ratio. The receptivity of ^{171}Yb is about four times greater than that of ^{13}C. The established ^{171}Yb chemical shift standard is [Yb(THF)$_2$(Cp*)$_2$]. Various Yb(Cp*)$_2$ derivatives as well as ytterbium(II) amide alkoxides, and stannyls have been studied, with observed chemical shifts ranging from $+36$ to $+1228$ ppm.[96,106,147,249,527,576]

2.2.13 Homogeneous Catalysis

Chemists first became interested in catalytic applications of organolanthanides when it turned out that some lanthanide organometallics are highly active as alkene polymerization catalysts.[559] The first indications of catalytic activity of organolanthanides came from the observation that active species were formed when lanthanide oxides, halides, or alkoxides were treated with cocatalysts such as lithium alkyls, aluminum alkyls, or other hydrocarbyl transfer reagents. The transient lanthanide σ-hydrocarbyl lanthanide species were found to catalyze the cracking of hydrocarbons as well as oligomerizations and polymerizations of alkenes.[18,859] Later, when well-defined organolanthanide complexes became available, it was found that some of them are also effective polymerization catalysts. For example, polymerization of ethylene or butadiene was observed in the presence of bis(cyclopentadienyl)-lanthanide hydrocarbyls[18] and (cyclooctatetraenyl)cerium complexes,[859] as well as lanthanide allyl complexes.[859] Butadiene polymerization was most effectively catalyzed by anionic tetra(allyl)-lanthanate(III) complexes.

A review article covering the use of organolanthanides in homogeneous catalysis was published by Watson and Parshall in 1985.[187] Other review articles on special aspects of organolanthanide catalysis (e.g., C–H activation) have also been published.[66–8,860] Most catalytic applications of organolanthanide complexes involve transformations of alkenes and, to a lesser extent, alkynes. Recent research in this area has focused on extending the scope of catalytic reactions involving organolanthanides and on varying the bis(cyclopentadienyl) ligand coordination sphere in order to achieve new catalytic applications in particular, variations of the cyclopentadienyl substituents or the use of ring-bridged cyclopentadienyl ligands have led to novel catalytic applications such as oligomerization and isospecific polymerization of α-alkenes,[168,172,173] hydrocyclization of α,ω-dienes, C–C σ-bond activation, and hydroamination/cyclization of alkenes. The following alkene or alkyne transformations have been shown to be effectively catalyzed by lanthanide organometallics:

   (i)   hydrogenation
  (ii)   oligomerization
 (iii)   cyclization
 (iv)   polymerization
  (v)   hydroamination
 (vi)   hydrosilylation
(vii)   hydroboration

These organolanthanide-catalyzed reactions will be discussed in the above order. This will be followed by a discussion of other catalytic reactions. Probably the most useful and active catalytic species in various reactions are lanthanide metallocenes of the type $LnR(Cp^*)_2$ (R = CH(TMS)$_2$ or H; Ln = Y, La, Ce, Nd, Sm, Lu). They may act as homogeneous catalysts or precatalysts. These two classes of compounds have been systematically investigated by Marks and Schumann et al.[540,689] The alkyls are prepared by a straightforward two-step synthesis. Subsequent hydrogenolysis affords the hydrides in almost quantitative yields (cf. Section 2.2.6.3) (Equations (224)–(226)):

$$LnCl_3 + 2\,LiCp^* \xrightarrow[\text{ii, Et}_2\text{O}]{\text{i, THF}} [Ln(\mu\text{-}Cl)_2Li(Et_2O)_2Cp^*_2] + LiCl \qquad (224)$$

$$[Ln(\mu\text{-}Cl)_2Li(Et_2O)_2Cp^*_2] + LiCH(TMS)_2 \longrightarrow [LnCH(TMS)_2Cp^*_2] + 2\,LiCl + 2\,Et_2O \qquad (225)$$

$$2\,[LnCH(TMS)_2Cp^*_2] + 2\,H_2 \longrightarrow [\{Ln(\mu\text{-}H)Cp^*_2\}_2] + 2\,CH_2(TMS)_2 \qquad (226)$$

$$Ln = Y, La, Ce, Nd, Sm, Lu$$

A catalyst for the hydrogenation of ethylene was prepared from a samarium–magnesium alloy. $SmMg_3$ reacts with anthracene in THF solution to give a complex of unknown structure. This intermediate absorbs hydrogen and is then able to catalyze the hydrogenation of ethylene.[18] Catalytic activity in hydrogenation reactions was also observed for the reaction products of lanthanide vapors and alkynes. Although incompletely characterized, these materials are believed to contain Ln–H bonds.[18,92,93] Hydrogenation of alkenes and alkynes is also effectively catalyzed by various organolanthanide hydrides[91,202,708,861] as well as cyclopentadienylytterbium(II) complexes. Extraordinarily high catalytic activity in alkene hydrogenation is displayed by the above-mentioned bis(pentamethyl-cyclopentadienyl)lanthanide hydrides, $[\{Ln(\mu\text{-}H)(Cp^*)_2\}_2]$. Terminal as well as internal alkenes are easily hydrogenated in the presence of these organolanthanide hydrides.[540,862] The catalytic hydrogenation of 1-hexene to *n*-hexane was studied in detail. Increasing catalytic activity was found for the series La < Nd < Sm < Lu and the highest turnover number was 120 000 h^{-1}. This enormous value can be favorably compared to the activities of well-known transition metal homogeneous catalysts. Turnover numbers are 3000 h^{-1} for $[RhCl(PPh_3)_3]$ and 6400 h^{-1} for $[Ir(cod)(PCy_3)][PF_6]$ under similar reaction conditions (25 °C, 1 atm of H$_2$). Even higher activities have been reported for "tied-back" metallocene hydrides $[\{Ln(\mu\text{-}H)\{Me_2Si(C_5Me_4)_2\}_2\}_2]$.[688] A plausible mechanism for organolanthanide-catalyzed α-alkene hydrogenation is shown in Scheme 42. Asymmetric alkene hydrogenation was recently achieved with the use of chiral organolanthanide catalysts. For example, hydrogenation of 2-phenyl-1-butene proceeded with high catalyst activity and moderate to excellent enantioselectivity.[717]

The same mechanism has been proposed for the selective hydrogenation of substituted dienes catalyzed by $[YMe(THF)(Cp^*)_2]$. The methyl precatalyst reacts with H$_2$ to yield the active hydride catalyst and methane.[858]

Catalytic hydrogenation of the cyanide function in *t*-butyl cyanide to afford neopentylamine has been achieved with $[\{ScH(Cp^*)_2\}_n]$ (Equation (227)).[25]

Various dimerization and oligomerization reactions catalyzed by organolanthanides have been studied. A "tied back" organoscandium hydride has been found to catalyze the dimerization of α-alkenes. Head-to-tail dimers are obtained in a highly regiospecific manner (Scheme 43).[25]

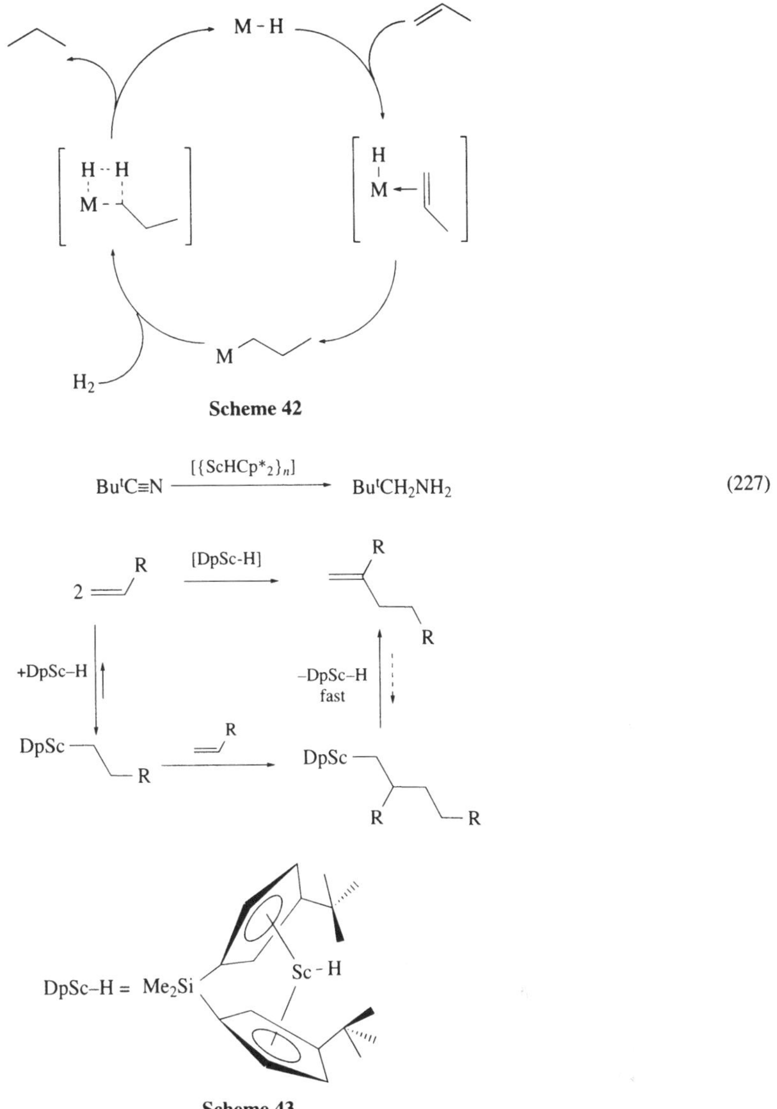

Scheme 42

$$Bu^tC\equiv N \xrightarrow{[\{ScHCp^*_2\}_n]} Bu^tCH_2NH_2 \tag{227}$$

Scheme 43

One of the first results reported in this area was the finding that diphenylacetylene is dimerized in the presence of tris(cyclopentadienyl)samarium.[540] Propylene oligomerization was reported for $[\{LuMe(Cp^*)_2\}_2]$.[187] The lanthanide hydrocarbyls $[LnCH(TMS)_2(Cp^*)_2]$ (Ln = Y, La, Ce) have been shown to be precatalysts in the oligomerization of terminal alkynes $RC\equiv CH$.[863] The catalytically active species are the acetylides $LnC\equiv CR(Cp^*)_2$. Solvated acetylides $[LnC\equiv CR(L)(Cp^*)_2]$ (L = THF, Et$_2$O) have been isolated for Ln = Y, Ce, Sm and the samarium complex $[SmC\equiv CPh(THF)(Cp^*)_2]$ has been structurally characterized.[629] The ionic radius of the lanthanide metal significantly influences the nature of the products. For example, with the larger lanthanides lanthanum and cerium higher oligomers (preferably trimers) are formed. In the case of Ln = Y the products are generally dimers, but the regioselectivity depends on the substituent R. The dimers $RHC=C(H)-C\equiv CR$ are obtained with R = Ph or TMS. With bulky substituents (R = Pr^i, Bu^t) the result is a head-to-tail dimerization of the terminal alkyne (Equation (228)).[863]

$$2\,RC\equiv CH \longrightarrow H_2C=C(R)-C\equiv CR \tag{228}$$

Alkynes of the type $MeC\equiv CR$ (R = Me, Et, Pr^n) are cyclodimerized in the presence of $[LnCH(TMS)_2(Cp^*)_2]$ (Ln = La, Ce) to afford 3-alkylidenecyclobutenes.[863] In this case the active catalyst is $[Ln\text{-}CH_2C\equiv CR(Cp^*)_2]$, which is formed by C–H activation upon reaction of the alkyne with the precatalyst $[LnCH(TMS)_2(Cp^*)_2]$. The proposed mechanism is illustrated in Scheme 44. Steric factors strongly influence the outcome of this reaction. Presumably for steric reasons, isolated $[Y\text{-}CH_2C\equiv CR(Cp^*)_2]$ has been found inactive towards $MeC\equiv CMe$. In the case of sterically more demanding

R substituents only C–H activation of the α-methyl group is observed and the reaction stops at the stage of precatalyst formation.[864] Similar results were recently obtained with MeYbI as catalyst. $MeC{\equiv}CC_5H_{11}$ and $TMS\text{-}C{\equiv}CR$ ($R = C_5H_{11}$, Ph) were dimerized to afford *cis*- and *trans*-cyclobutenes.[865]

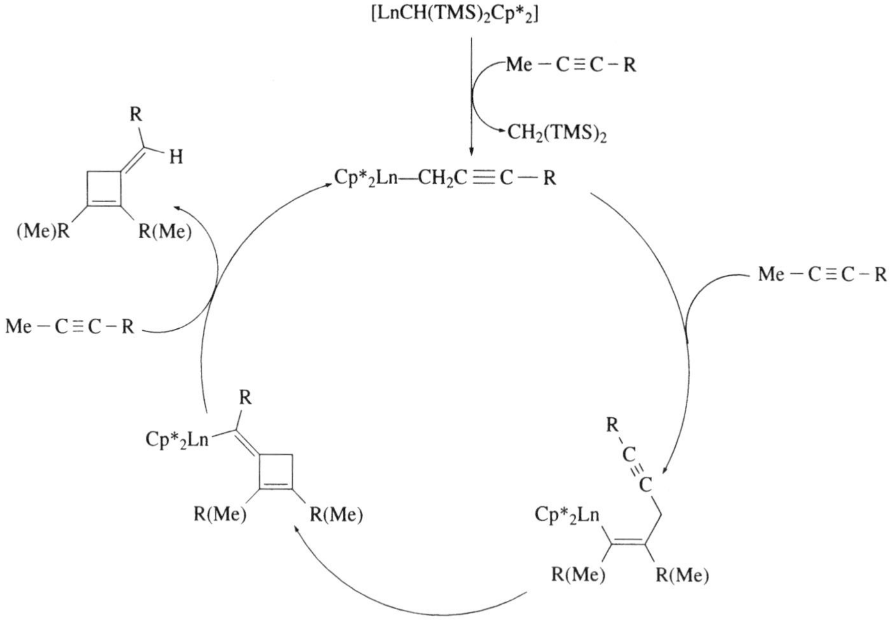

Scheme 44

The catalytic hydrocyclization of 1,5- and 1,6-dienes has been achieved by using $[YMe(THF)(Cp^*)_2]$ as precatalyst. The catalytic species is bis(pentamethylcyclopentadienyl)yttrium hydride, which is formed by C–H-activation and elimination of methane (Scheme 45)[866]

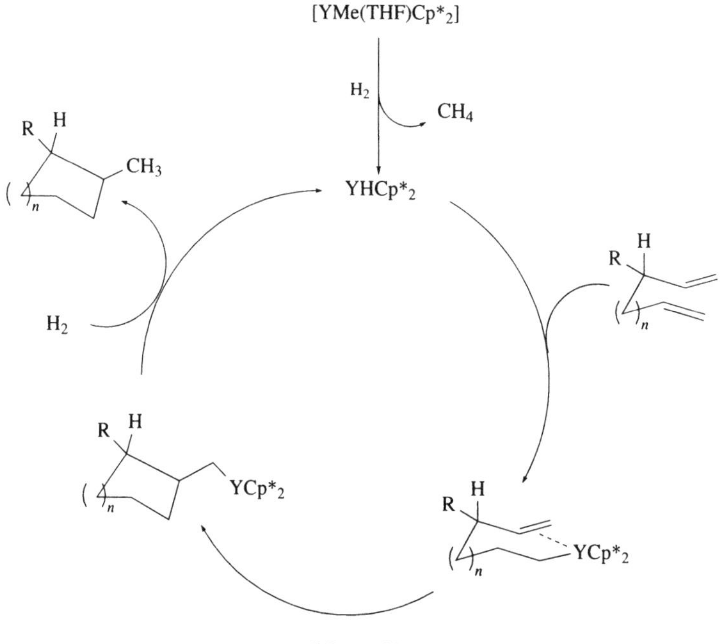

Scheme 45

Hydrocyclization of 1,5-hexadiene to methylcyclopentane is also effectively catalyzed by the scandium hydride $[ScH(PMe_3)\{Me_2Si(C_5Me_4)_2\}]$ (Equation (229)).[25] The mechanism of this process was investigated by deuteration experiments.[867] A model complex, the pendant phosphine-substituted cyclopentadienyl complex $[ScCH(TMS)_2\{Me_2Si(C_5Me_4)(C_5H_3CH_2CH_2P(Bu^i)_2)\}]$, was synthesized in order to study the influence of a high effective phosphine concentration on these scandium-catalyzed processes.[541]

$$ \ce{/\\=/\\=} \quad \xrightarrow[\substack{25\ °C,\ 99.2\ \%\\ 78\ \text{turnovers}}]{[ScH(PMe_3)\{Me_2Si(C_5Me_4)_2\}]} \quad \ce{=<} \qquad (229) $$

Early mechanistic studies concerning organolanthanide-catalyzed alkene polymerization reactions showed that insertion of the unsaturated hydrocarbon into the lanthanide–carbon σ-bond is a key step. It was first demonstrated for the reversible insertion of propylene into the Lu–C bond of methyl-bis(pentamethylcyclopentamethyl)lutetium (Equation (230)).[18]

$$ \text{LuMeCp*}_2 + \text{MeCH=CH}_2 \quad \rightleftharpoons \quad \text{LuCH}_2\text{CHMe}_2\text{Cp*}_2 \qquad (230) $$

In this sense [{LuMe(Cp*)$_2$}$_2$] can be regarded a model compound for Ziegler–Natta catalysis. The reverse reaction proceeds via β-alkyl elimination. A competing process is β-hydride elimination, which leads to LuH(Cp*)$_2$ dimer and isobutene. Alternatively, LuCH$_2$CHMe$_2$(Cp*)$_2$ can be prepared by insertion of isobutene into the lutetium–hydrogen bond.[18] More recent thermochemical studies on the basis of *f*-element bond enthalpies have shown that insertion of alkenes into an Ln–H bond is quite exothermic (ca. 15 kcal mol^{-1}), while β-hydride elimination is less favorable.[863]

Very high catalytic activities in the homogeneous polymerization of ethylene are also displayed by ScMe(Cp*)$_2$, NdR(Cp*)$_2$,[868] LaNMe$_2$(Cp*)$_2$[869] and the bis(pentamethylcyclopentadienyl)lanthanide hydrides of the type [{Ln(μ-H)(Cp*)$_2$}$_2$].[18,185,186,649,675-7,688,694,706] The highest turnover frequency (greater than 1800 s^{-1} at 25 °C and 1 atm of ethylene) was reported for [{Lu(μ-H)(Cp*)$_2$}$_2$]. In a series of important papers Marks and Schumann *et al.* have compared the behavior of [LnCH(TMS)$_2$(Cp*)$_2$] and [LnCH(TMS)$_2${Me$_2$Si(C$_5$Me$_4$)$_2$}] (Ln = La, Nd, Sm, Lu) as precatalysts in alkene polymerization and hydrogenation reactions.[184,540,686,862] The hydrides [{Ln(μ-H)(Cp*)$_2$}$_2$] were shown to be extremely active in alkene polymerization. An even higher activity was found for the "tied back" hydrides [{Ln(μ-H){Me$_2$Si(C$_5$Me$_4$)$_2$}$_2$]. In the case of ethylene polymerization, polyethylene with a high molecular weight (M_w = 3.8–15 × 10^5) and moderate polydisperity (M_w/M_n = 1.4–6.4) was produced. The activity of the ethylene polymerization catalysts was found to increase parallel with the ionic radius of the lanthanide metal (Lu ≪ Nd < La). Propylene and butadiene were not polymerized in the presence of [{Lu(μ-H)(Cp*)$_2$}$_2$] but afforded η^3-allyl complexes via stoichiometric reactions.[862]

Dimeric yttrium hydrides of the type [{Y(OR)(μ-H)(Cp*)}$_2$] have been found to catalyze the slow polymerization of 1-hexene to give poly(1-hexene) with M_w = 15 700 and M_w/M_n = 1.67. The active catalyst was proposed to be [(μ-H)(μ-hexyl){Y(OR)(Cp*)}$_2$]. The chain propagation proceeds via 1,2-addition and termination via β-hydride elimination.[537] Similar results were obtained with scandium hydrocarbyls and hydrides containing the internally chelating cyclopentadienyl-amide ligand $\bar{\text{C}}_5$Me$_4$SiMe$_2$$\bar{\text{N}}$-Bu^{t-}.[25,609] The dimeric hydride [{Sc(PMe$_3$)(μ-H)(C$_5$Me$_4$SiMe$_2$N-Bu^t)}$_2$] acts as a single-component, regiospecific polymerization catalyst. It catalyzes the polymerization of α-alkenes RCH=CH$_2$ (R = Me, Et, Pr^n) to polyalkenes with M_w = 4000–6000 and M_w/M_n = 1.7–2.1. As in the case of [{Y(OR)(μ-H)(Cp*)}$_2$] the polymerizations are slow as compared to single-component zirconium catalysts.[609] The dimeric yttrium hydride [(Y(μ-H){*rac*-Me$_2$Si(2-TMS-4-Bu^{t}C$_5$H$_2$)$_2$})$_2$] has recently been reported to be the first isospecific single-component α-alkene polymerization catalyst. Using this catalyst highly (97%) isotactic poly(1-alkenes) with moderate molecular weights have been manufactured (e.g., poly(1-hexene): M_w = 24 000 and M_w/M_n = 1.75).[869]

A well-defined, single component Ziegler–Natta catalyst is the amido-cyclopentadienyl scandium hydride [{Sc(PMe$_3$)(μ-H)}(C$_5$Me$_4$SiMe$_2$NBu^t)$_2$] (**23**) which was described by Bercaw *et al.*[870]

Me$_2$Si   PMe$_3$ H   Sc   Sc   N   SiMe$_2$ N   H   Bu^t Me$_3$P   Bu^t

(**23**)

This hydride cleanly catalyzes the polymerization of propylene, 1-butene, and 1-pentene. Atactic polyalkenes with low molecular weight are obtained. The proposed mechanism is illustrated in Scheme 46.[25,870,871]

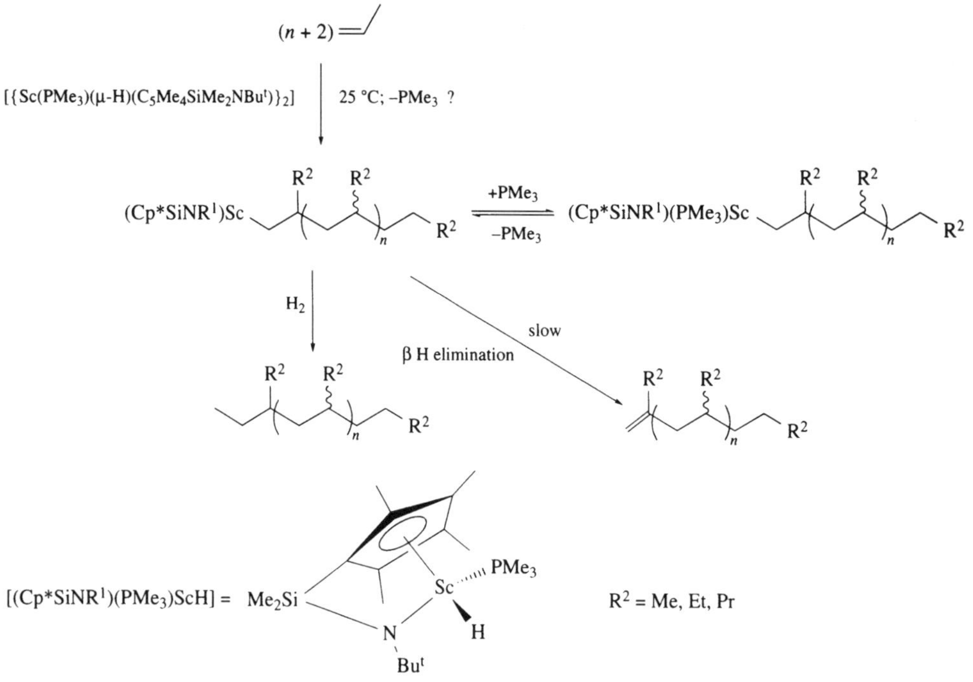

Scheme 46

Treatment of $[\{Sc(PMe_3)(\mu\text{-}H)(C_5Me_4SiMe_2NBu^t)\}_2]$ with two equivalents of propylene at low temperature yielded the structurally characterized phosphine-free di-μ-propyl complex $[\{Sc(\mu\text{-}Pr^n)(C_5Me_4SiMe_2NBu^t)\}_2]$ **(24)**. This dimeric organoscandium alkyl was found to be an even more active α-alkene polymerization catalyst than the hydride precursor.[25]

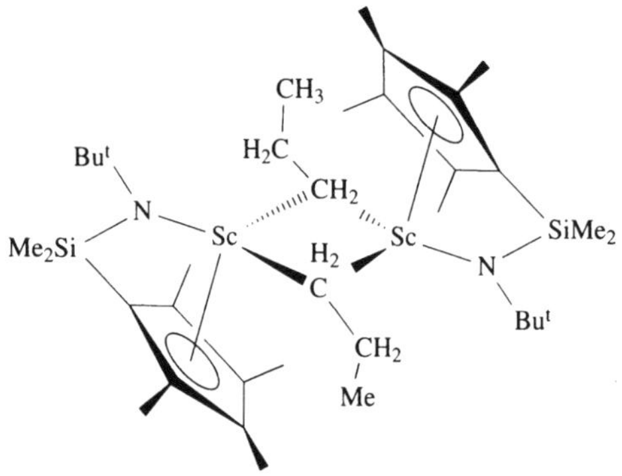

(24)

Mono(cyclopentadienyl)- and mono(indenyl)lanthanide dichlorides have been used as catalysts for the polymerization of butadiene, and the kinetics have been studied.[557,872,873] Aluminum alkyls have been used as cocatalysts. In the case of $LnCl_2(Cp)$ the activity in catalyzing the stereospecific polymerization of butadiene to *cis*-1,4-polybutadiene decreased in the order Nd > Pr > Y > Ce > Gd.[873] Butadiene polymerization was also reported for the catalyst systems "Ph_3Nd_2"/$Al_2Et_2Cl_3$,[874] Nd-(octoate)$_3$/$AlEt_2Cl$/$AlEt_3$,[875] and $LnCl_3$/Bu^nLi (Ln = Nd, Ho, Dy, Er).[876] The bimetallic complex $[\{Nd_6Al_3(\mu\text{-}Cl)_6(\mu\text{-}Et)_9Et_5(OPr^i)\}_2]$ has been isolated from the system $Nd(OPr^i)_3$/$AlEt_3$/$AlEt_2Cl$ and structurally characterized.[174,877] The novel mono(cyclopentadienyl)tris(allyl)lanthanate(III) complexes $[Li(dioxane)_2][Ln(\eta^3\text{-}C_3H_5)_3Cp']$ (Cp' = Cp, Cp*, C_9H_7, $C_{13}H_9$) catalyze the polymerization of butadiene in toluene under standard conditions with moderate activity and a high *trans* selectivity.[215] Highly active catalysts for the production of *cis*-1,4-polybutadiene are formed by combining tris(2,4-

dimethylpentadienyl)neodymium, $Nd(2,4-C_7H_{11})_3$, with Lewis acids such as $SnCl_4$, $SnPh_2Cl_2$, $AlEtCl_2$, or $AlBr_3$. An active precatalyst in this system, hexanuclear $[Nd_6Cl_{12}(2,4-C_7H_{11})_6(THF)_2]$ has been isolated and structurally characterized (Figure 75). In this unusual cluster compound two hexagonal bipyramidal Nd_3Cl_5 units are connected via two chloro bridges. The pentadienyl ligands are coordinated to neodymium in a pentahapto fashion.[729]

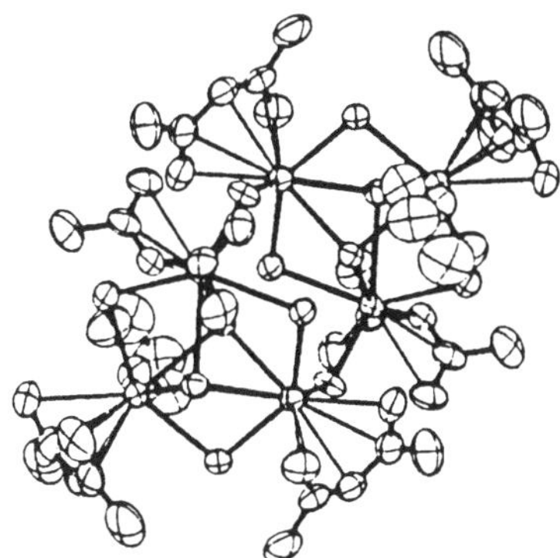

Figure 75 The molecular structure of $[Nd_6Cl_{12}(2,4-C_7H_{11})_6(THF)_2]$.[729]

Polymerization of methylmethacrylate (MMA) has been effected with the organolanthanide precatalysts $[\{Sm(\mu-H)(Cp^*)_2\}_2]$ and $[LnMe(THF)(Cp^*)_2]$. Both complexes have been found to initiate the living polymerization of methylmethacrylate.[878] Similar activity was reported for the divalent lanthanide organometallics $[Yb(DME)(MeC_5H_4)_2]$,[224] $[Yb(THF)_2(Cp^*)_2]$, $[Sm(THF)_2(Cp^*)_2]$, and $[Yb(THF)_2(C_9H_7)_2]$.[879] The catalytic activity was found to decrease with increasing steric bulk of the cyclopentadienyl ligands ($Cp > MeC_5H_4 > Cp^*$). In the case of samarium it was possible to determine the molecular structure of the active catalyst, $[Sm\{(MMA)_2H\}(Cp^*)_2]$. This intermediate contains an eight-membered ring formed by samarium and two MMA molecules. One MMA unit coordinates to the metal in an enolate form and the other one via the carbonyl group. MMA polymerization catalyzed by $[\{Sm(\mu-H)(Cp^*)_2\}_2]$ affords poly(methyl methacrylate) of high molecular weight ($M_w = 55$–563×10^3) and extremely narrow polydispersity ($M_w/M_n = 1.02$–1.04).[878,880]

The production of syndiotactic poly(methyl methacrylate) was also achieved by using the organolanthanide catalysts $[LnMe(THF)(Cp^*)_2]$ (Ln = Yb, Lu) and $[Ln(\mu-Me)_2AlMe_2(Cp^*)_2]$ (Ln = Y, Yb, Lu).[879] Lanthanide catalysts of the type $LnR(Cp^*)_2$ (Ln = Sm, Yb, Lu; R = H, Me) also allowed well-controlled block copolymerizations as shown in Equations (231) and (232)).[878]

$$\left[Cp^*_2Sm - (CH_2CH_2)_n - R \right] \xrightarrow{m\ MMA} \left[Cp^*_2Sm - O - \underset{OMe}{\overset{Me}{\underset{|}{C}}} = \underset{H_2}{\overset{Me}{\underset{|}{C}}} - \underset{CO_2Me}{C} - (C-CH_2)_{m-1} \longrightarrow (CH_2CH_2)_n - R \right] \quad (231)$$

$$\left[Cp^*_2Sm - (CH_2CH_2)_n - R \right] \xrightarrow{m\ \overset{(CH_2)_x}{\underset{O}{O}}} \left[Cp^*_2Sm \left[O - CH_2(CH_2)_x - \underset{O}{\overset{}{C}} \right]_m (CH_2CH_2)_n - R \right] \quad (232)$$

$$R = H, Me; x = 3, 4$$

$$MMA = methylmethacrylate$$

Ring-opening polymerization of lactones has been achieved with various bis(cyclopentadienyl)-lanthanide halides as catalysts. For example, polymerization of ε-caprolactone in the presence of $[YCl(THF)(Cp)_2]$ afforded polycaprolactone with $M_w = 125\,000$ and $M_w/M_n = 1.78$.[881] A variety of samarium(II) complexes have also been shown to catalyze the ring-opening polymerization of ε-caprolactone.[882] A wide range of aldehydes are efficiently dimerized to the corresponding esters by catalytic amounts of $[LnCH(TMS)_2(Cp^*)_2]$ (Ln = Nd, La). This reaction led to a new organolanthanide-catalyzed synthesis of polyesters from dialdehydes.[883]

Organolanthanides have been found to catalyze the dehydrogenation of organosilanes $RSiH_3$ to give polyorganosilanes.[683,715,717,718] The latter are of practical interest as precursors for silicon carbide ceramics. Scheme 47 illustrates the proposed catalytic cycle. Precatalysts are the lanthanide hydrocarbyls $[LnCH(TMS)_2(Cp^*)_2]$ (Ln = Y, La, Nd, Sm, Lu), while the active catalyst is thought to be the corresponding hydride $[\{Ln(\mu-H)(Cp^*)_2\}_2]$. The rate law for the dehydrogenative coupling of monoorganosilanes is $[LnCH(TMS)_2(Cp^*)_2]^1[RSiH_3]^1$. As found earlier for organolanthanide-catalyzed alkene hydrogenation,[718] the relative rates parallel the ionic radii of the lanthanide elements and increase in the order La < Nd < Sm < Y < Lu.[682]

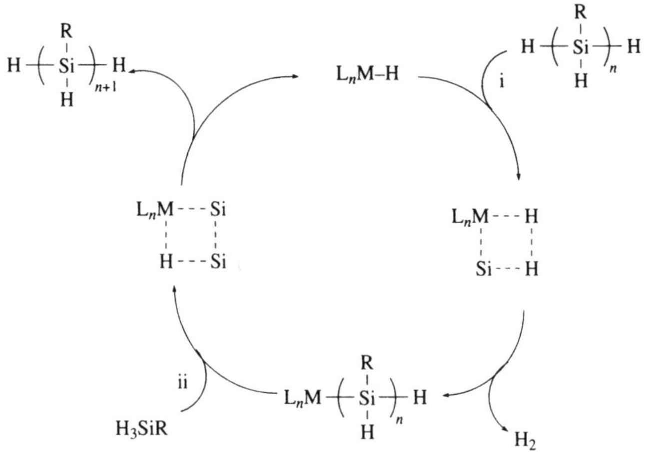

Scheme 47

The hydroamination/cyclization of aminoalkenes catalyzed by organolanthanides has been studied in great detail by Marks *et al.*[869,884-6] The catalytic cycle is initiated by formation of an amido complex from $[\{Ln(\mu\text{-}H)(Cp^*)_2\}_2]$ or $LnR(Cp^*)_2$ and the free amine, followed by intramolecular alkene insertion into the Ln–N bond. The latter step is irreversible and rate limiting. Subsequent protolysis of the resulting lanthanide hydrocarbyl by the free amine liberates the cyclization product and regenerates the precatalyst $[\{Ln(\mu\text{-}H)(Cp^*)_2\}_2]$ (Scheme 48). The active catalytic species was proposed to be $[LnNHR(H_2NR)(Cp^*)_2]$. A model compound, $[LaNHMe(H_2NMe)(Cp^*)_2]$ has been structurally characterized.[885]

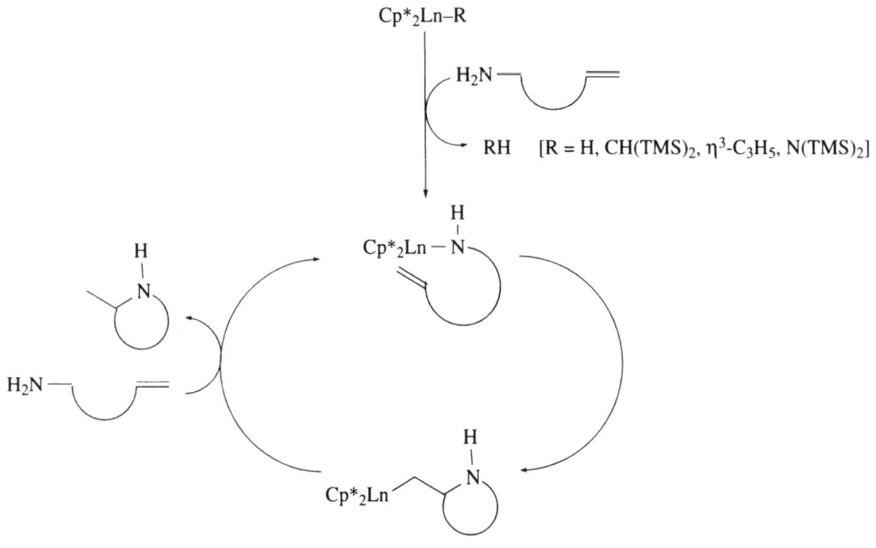

Scheme 48

More recently, chiral organolanthanide precatalysts have been ingeniously designed and used in diastereoselective hydroamination/cyclization processes. For example, a >95% diastereoselectivity was achieved with a samarium catalyst for the hydroamination/cyclization of $H_2C{=}CHCH_2CH_2CH(Me)NH_2$ to 2,5-dimethylpyrrolidine. The precatalysts are of the type $[LnCH(TMS)_2\{Me_2Si(C_5Me_4)(C_5H_3R^*)\}]$ where R^* is a chiral substituent such as (−)-menthyl or (+)-neomenthyl.[884,887]

The samarium(II) complexes $Sm(Cp^*)_2$ and $[Sm(THF)_2(Cp^*)_2]$ have also been found to be effective precatalysts in these hydroamination/cyclization reactions.[886] In this case the initial step is the formation of samarium(III) intermediates via allylic C–H activation (Equation (233)).

Both intermediates can then be protonated by a primary or secondary amine to generate the active catalyst $SmNR^1R^2(Cp^*)_2$ (Scheme 49).[886]

$$2 \, SmCp^*_2 + CH_2=CH(CH_2)_3NMe_2 \longrightarrow [Sm(\eta^3\text{-}H_2CCHCH(CH_2)_2NMe_2)Cp^*_2] + 1/2 \, [\{Sm(\mu\text{-}H)Cp^*_2\}_2] \quad (233)$$

Scheme 49

Another extension of this catalytic process is the use of amino-substituted alkynes as substrates. The organolanthanide complex $[SmCH(TMS)_2(Cp^*)_2]$ serves as precatalyst for the efficient and regiospecific hydroamination/cyclization of aliphatic and aromatic aminoalkynes $RC{\equiv}C(CH_2)_nNH_2$ to yield the corresponding heterocycles. The turnover-limiting step is the intramolecular alkyne insertion into the Sm–N bond, which is followed by rapid protolysis of the resulting Sm–C bond.[888]

Alkyl complexes of the type $[LnCH(TMS)_2(Cp^*)_2]$ (Ln = Y, Nd) have been reported to catalyze the hydrosilylation of various substituted and functionalized alkenes to a mixture of linear and branched products (Equation (234)).[718]

$$Ph\diagup\!\!\!\diagdown + PhSiH_3 \longrightarrow Ph\diagup\!\!\!\diagdown SiH_2Ph + \underset{Ph}{\overset{SiH_2Ph}{\diagup\!\!\!\diagdown}} \quad (234)$$

For yttrium, the hydrocarbyl $[YCH(TMS)_2(Cp^*)_2]$ has been proposed to react with $PhSiH_3$ to give the precatalyst. Further reaction of the intermediate silyl derivative with $PhSiH_3$ affords the active catalyst $[\{Y(\mu\text{-}H)(Cp^*)_2\}_2]$ besides the coupling product $PhSiH_2SiH_2Ph$.[718]

Even an organolanthanide-catalyzed variant of the (anti-Markownikoff) hydroboration of alkenes has been developed. The catalytic cycle is illustrated in Scheme 50. Reaction of the precatalyst $[LnCH(TMS)_2(Cp^*)_2]$ with catecholborane produces the active catalyst $[\{Ln(\mu\text{-}H)(Cp^*)_2\}_2]$, together with the corresponding alkylated catecholborane. This is followed by alkene insertion and hydroboration of the intermediate lanthanide alkyl. Conventional oxidative work-up of the resulting alkylated catecholborane affords the alcohol. $[Sm(THF)_2(Cp^*)_2]$ has been found to be another suitable precatalyst.[889]

Monoorganolanthanide dihalides of the type LnX_2R have been reported to catalyze the Tischenko reaction, that is, the conversion of aldehydes into the corresponding esters (Equation (235)).[890–2]

A typical organolanthanide catalyst was prepared for example from praseodymium, neodymium, or samarium metal and iodoethane in THF. It was supposed that the active catalytic species is actually $HLnI_2$, which is formed in addition to ethylene by β-hydride elimination from an intermediate $EtLnI_2$.[892] The activity of this catalytic system is low. A more active catalyst has been prepared by action of iodobenzene on metallic cerium.[893]

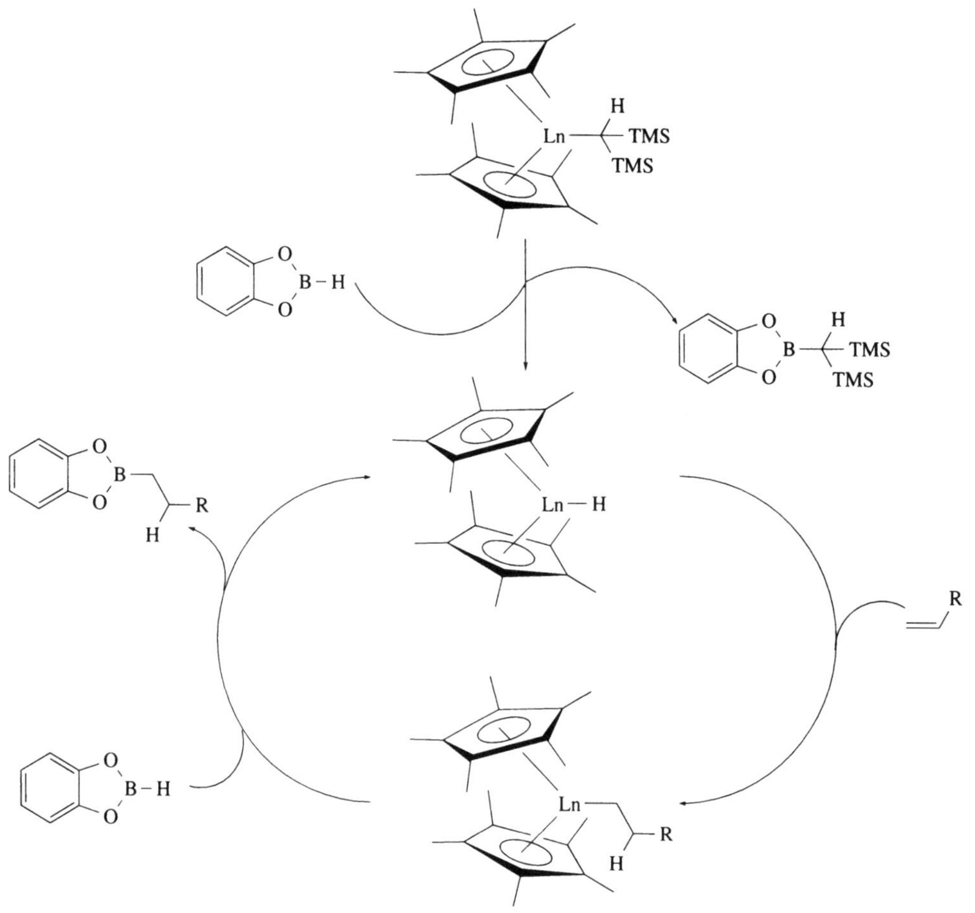

Scheme 50

$$2\ RCHO \xrightarrow{\text{Ln cat.}} R\text{–}C(O)\text{–}OCH_2R \qquad (235)$$

An interesting catalytic application of $[\{Sc(\mu\text{-}D)(Cp^*)_2\}_n]$ originated from the observation, that this compound undergoes intramolecular H/D exchange reaction with the Cp* methyl groups.[673] Subsequently, it was discovered that $[\{Sc(\mu\text{-}H)(Cp^*)_2\}_n]$ catalyzes the preparation of high isotopic purity $C_5(CD_3)_5H$ from pentamethylcyclopentadiene. The deuteration is conducted under a D_2 atmosphere in C_6D_6 at 145 °C.[894]

Isomerization of terminal alkenes was achieved with the system $Ln(Cp)_3/NaH$ (Ln = Y, Er, Lu). The reactions were carried out at 45 °C in THF and afforded *cis*- and *trans*-2-alkenes in very good yields. The proposed mechanism is depicted in Scheme 51.[895]

A single report on heterobimetallic organolanthanide catalysts has appeared. The catalytic hydroformylation of 1-octene (Equation (236)) has been achieved with heterobimetallic lanthanide catalysts such as $[Ln(\mu\text{-}CO)Co(CO)_3(Cp)_2]$ (Ln = Yb, Lu) or $[\{Lu(THF)(Cp)_2\}Re(CO)_5]$. Similar results have been obtained with the *in situ* prepared catalytic systems $YbCl(Cp)_2/Na[Co(CO)_4]$ and $LuCl(Cp)_2/Na[Co(CO)_4]$.[896]

2.2.14 Organic Synthesis

Since the early 1980s there has been an increasing interest among organic chemists in the rare-earth elements. The use of lanthanide compounds in organic synthesis is well established and has been documented in several review articles.[897-9] For example, the preparation and reactions of organosamarium complexes has been reviewed by Inagawa,[900] while the organic chemistry involving samarium(II) compounds has been discussed in a review article by Kagan.[901] The latter author has also published a comprehensive review on lanthanides in organic synthesis, which covers synthetic uses of organolanthanide complexes.[902] A review on highly selective syntheses using organolanthanide reagents

$$LnCp_3 \ + \ NaH$$

$$\downarrow \ THF$$

$$[\{Ln(\mu\text{-}H)(THF)Cp_2\}_2]$$

$$LnHCp_2$$

Scheme 51

$$n\text{-}C_6H_{13}CH=CH_2 \ \xrightarrow[\ H_2 \ + \ CO\ ]{catalyst} \ n\text{-}C_6H_{13}CH_2CH_2CHO \ + \ n\text{-}C_6H_{13}CH(Me)CHO \qquad (236)$$

was published by Utimoto *et al.*[903] This rapid development has been stimulated in part by the fact that the lanthanides exhibit a low toxicity and some are much less expensive than other synthetically useful metals such as palladium or rhodium. While the majority of the synthetic applications involve the use of inorganic lanthanide compounds (SmI_2, $LnCl_3$, $Ln(O_3SCF_3)_3$, $[NH_4]_2[Ce(NO_3)_6]$), an increasing number of organic reactions using organolanthanides have now been investigated. The possible uses of organolanthanide complexes in organic synthesis will be discussed in the order of increasing oxidation states. Divalent lanthanides are very effective one-electron reducing agents and, conversely lanthanum(IV) derivatives are powerful oxidants. The hardness of the lanthanide ions and their strongly oxophilic character also guide the reactivity of these metals.

Divalent organolanthanides offer a great potential for selective transformations of organic molecules. Grignard analogues of the type LnXR (R = alkyl, aryl; Ln = Sm, Eu, Yb; X = halide) have been studied since 1970 (Equation (237)).[18,892,904-9] YbIR species are most conveniently prepared from ytterbium metal and RI in THF. In some cases activated lanthanide powders are used for these preparations.[910] Although no structural information is available, these compounds can be regarded as Grignard analogues. They are soluble in THF and behave like typical Grignard reagents.[902] THF solutions of LnIR are highly colored.

$$Ln \ + \ RI \ \xrightarrow{\ THF\ } \ LnIR \qquad (237)$$

$$Ln = Sm, \ Eu, \ Yb$$
$$R = Me, \ Ph, \ p\text{-}MeC_6H_4, \ 2,6\text{-}Me_2C_6H_3, \ Mes$$

Such Grignard-type organolanthanides have not been isolated as crystalline materials nor have they ever been fully characterized either by spectroscopic methods or x-ray crystallography. However, the *in situ* prepared solutions are synthetically useful as they react like typical Grignard reagents. For example, alcohols are obtained when YbIPh (obtained from Yb and PhI in THF at 30 °C[902,911]) is treated with aldehydes or ketones.[902,912] Treatment of either esters or nitriles with YbIPh affords ketones.[912] Other reagents reported to react with phenylytterbium iodide include alkyl halides,[911] allyl bromide or iodide,[911] *cis*- or *trans*-styryl bromide,[911] amines and azoles,[913] as well as α,β-unsaturated carbonyl compounds such as acrolein, 2-cyclohexen-1-one, and chalcone.[912,914] Other organoytterbium(II) complexes have been prepared by reacting YbIMe or YbXPh with C–H-acidic hydrocarbons like phenyl-*o*-carborane, phenylacetylene, pentafluorobenzene, indene, or fluorene.[915] Phenyl isocyanate is converted into N-phenylbenzamide upon treatment with YbIPh (Equations (238)–(240)).[902]

$$ClSiMe_2Ph \ \xrightarrow{\ YbIMe\ } \ PhTMS \qquad (238)$$

$$Ph_2C=O \xrightarrow{SmIPh} Ph_3COH \tag{239}$$

$$PhN=C=O \xrightarrow{YbIPh} PhC(O)NHPh \tag{240}$$

Aromatic substitution has been reported for the reaction of YbIPh with hexafluorobenzene (Equation (241)). In this case the selectivity and the yields are higher than those obtained with organolithium or Grignard reagents.[99,916]

$$C_6F_6 + YbIPh \longrightarrow C_6F_5Ph + YbIF \tag{241}$$

Nucleophilic substitution at trifluorostyrene was achieved with EuIPh and YbIPh (Equation (242)).

$$PhCF=CF_2 + LnIPh \longrightarrow trans\text{-}PhCF=CFPh + LnIF \tag{242}$$

$$Ln = Eu, Yb$$

A remarkable difference from normal Grignard reagents is observed in the reaction of phenylytterbium iodide with benzoyl chloride, which afforded benzophenone as the only product.[913] Generally the organolanthanide halides differ from Grignard reagents as they are one-electron reducing agents. In addition, they can cause deoxygenation reactions due to the high oxophilicity of the lanthanide ions.[917,918] For example, deoxygenation of the intermediate carbinolate was observed when carbonyl compounds were treated with an excess of LnXR.[919,920] Dibenzyl was formed instead of the expected diphenylmethane when YbIPh and BzBr were reacted in the presence of copper bromide as catalyst.[911] Deoxygenation and substitution occurred upon treatment of anthraquinone with four equivalents of YbIPh (Equation (243)).[919,920]

$$\tag{243}$$

Ytterbium(II) hydrocarbyls such as $Yb(C_6F_5)_2$ have been prepared by Deacon *et al.* from ytterbium and organomercury derivatives in THF and have been used *in situ* for further reactions. They readily add to the carbonyl group of aliphatic and aromatic aldehydes and ketones to afford the corresponding alcohols in yields around 40–80%.[99,100,921] Carboxylation of $Yb(C_6F_5)_2$ afforded a mixture of the expected $C_6F_5CO_2H$ and $o\text{-}C_6F_4HCO_2H$.[100] Similar reactions have been carried out with lanthanide(II) acetylides, especially with the well-characterized compound bis(phenylethynyl)ytterbium(II).[204] This compound adds to aldehydes and ketones to give after work-up the corresponding secondary and tertiary alcohols containing a $PhC\equiv C$ substituent (Equations (244) and (245)). Towards benzophenone, bis-(phenylethynyl)ytterbium(II) simply acts as a reducing agent. In this case $Ph_2C(OH)C(OH)Ph_2$ is obtained after hydrolysis.[921]

$$RCHO \xrightarrow[ii, H^+]{i, Yb(PhC\equiv C)_2} (PhC\equiv C)RCHOH \tag{244}$$

$$R_2C=O \xrightarrow[ii, H^+]{i, Yb(PhC\equiv C)_2} (PhC\equiv C)R_2COH \tag{245}$$

Various lanthanides have been used in Reformatsky-type reactions. The metal powders of cerium, lanthanum, neodymium, and samarium have been found to react with ethyl-2-bromopropionate in THF to give an intermediate organolanthanide insertion product. This intermediate was proposed to be internally stabilized by coordination of the ester carbonyl group. Subsequent treatment with ketones at room temperature readily afforded γ-lactones in ca. 60–70% yield. In a side reaction the ketones were reduced to pinacols (Scheme 52).[922]

The vast majority of organic transformations involving lanthanide(II) reagents have been carried out using samarium diiodide. Since the introduction of samarium diiodide into preparative organic chemistry by Kagan *et al.* in 1977,[923] this compound has become an extremely popular and useful reagent. Its synthetic uses have already been compiled in several review articles,[901,902,923,924] and

$$Ln + Br(CH_2)_2CO_2Et \longrightarrow \underset{\text{OEt}}{BrLn\text{----}O} \xrightarrow[\text{ii, H}^+]{\text{i, R}^1\text{R}^2\text{C=O}} \underset{R^2}{\overset{R^1}{\diagdown}}O$$

Scheme 52

standardized THF solutions of samarium diiodide are now commercially available. Samarium diiodide and bis(cyclopentadienyl)samarium are easily prepared starting from samarium metal powder (Equations (246) and (247)).[230,923]

$$Sm + ICH_2CH_2I \xrightarrow{\text{THF}} SmI_2 + CH_2{=}CH_2 \qquad (246)$$

$$SmI_2 + 2\,NaCp \xrightarrow{\text{THF}} SmCp_2 + 2\,NaI \qquad (247)$$

In contrast to the ever increasing amount of work using samarium diiodide, similar reactions with bis(cyclopentadienyl)samarium(II) have only rarely been studied. However, in a few cases a comparison has been made between the reactivities of SmI_2 and $Sm(Cp)_2$.[254,255,902,925] For example, aromatic acid chlorides are efficiently reduced to α-diketones by samarium diiodide in THF. α-Ketols are obtained from aliphatic acid chlorides (Equations (248) and (249)).[926] For these transformations a mechanism has been suggested, which involves transient formation of an acylsamarium species $R{-}C(O)SmI_2$, which then reacts as a nucleophile with $RC(O)Cl$.[927] This mechanism is in agreement with the observation that mixed ketols are formed when equimolar mixtures of acid chlorides and aldehydes or ketones are treated with samarium diiodide. Similar reactions of samarium allyl complexes with acid chlorides produced β,γ-unsaturated ketones.[928]

$$2\,R{-}C(O)Cl + SmI_2 \xrightarrow{\text{THF}} R{-}C(O){-}C(O){-}R \qquad (248)$$

$$R^1{-}C(O)Cl + R^2R^3C{=}O + SmI_2 \xrightarrow{\text{THF}} R^1{-}C(O){-}C(OH)R^2R^3 \qquad (249)$$

The formation of an acylsamarium intermediate is not totally unusual, as acyllanthanide complexes with cyclopentadienyl auxiliary ligands have already been isolated by CO insertion into an Ln–C bond (cf. Scheme 11).[929,930] In a related study the reactivity of bis(cyclopentadienyl)samarium(II) towards $Bu^tC(O)Cl$ was investigated. At $-20\,°C$ in THF a dark solution of an acyl complex was formed, which was stable on storage and could be reacted with various electrophiles (Equation (250)). For example, treatment of the solution with heptanal afforded the corresponding α-ketol (cf. Equation (249), $R^1 = Bu^t$, $R^2 = H$, $R^3 = n\text{-}C_6H_{13}$). The initial formation of acylsamarium intermediates was also made plausible by deuterolysis experiments. $Sm(Cp)_2$ was reacted with $AdC(O)Cl$ ($Ad = 1$-adamantyl) to give an acylsamarium intermediate, which was soluble in THF. Subsequent deuterolysis produced AdCDO in good yield.[901]

$$2\,SmCp_2 + Bu^tC(O)Cl \xrightarrow{\text{THF}} Bu^tC(O)SmCp_2 + [SmCl(THF)Cp_2] \qquad (250)$$

Closely related to the dihapto acyls are η^2-iminoacylsamarium intermediates. These novel reagents are formed by treatment of organic halides and 2,6-xylyl isocyanide with samarium diiodide. The organosamarium(III) intermediates have been successfully utilized in new carbon–carbon bond forming reactions. For example, subsequent reactions with carbonyl compounds produce α-hydroxy-imines, which can be hydrolyzed to the corresponding α-hydroxy-ketones. Unsymmetrical α-diketones are synthesized by autoxidation of the α-hydroxy-imines. Other applications include the preparation of hydroxy-diketones and vicinal triketones. Scheme 53 outlines the typical three-component coupling reaction of organic halides and carbonyl compounds in the presence of η^2-iminoacylsamarium(III) intermediates.[931,932]

$$R^1{-}X + \text{xylyl–NC} \xrightarrow{SmI_2} SmI_2(\eta^2\text{-xylyl–N=C–R}^1) \xrightarrow{R^2R^3C{=}O} R^1{-}C({=}N\text{–xylyl}){-}C(OH)R^2R^3$$

Scheme 53

In situ formed lanthanide(II) metallocene derivatives have been employed as reducing agents. For example, the system $Ln(Cp)_3/NaH$ in THF has been used to reduce 1-hexene. Dienes too are regioselectively reduced.[300]

Organolanthanide(II) complexes have been shown to be effective reagents for dehalogenation and deoxygenation reactions. Bis(pentafluorophenyl)ytterbium has been employed to selectively remove *ortho* halogen substituents from polyhalogenated benzoic acids (Equations (251) and (252)).[933]

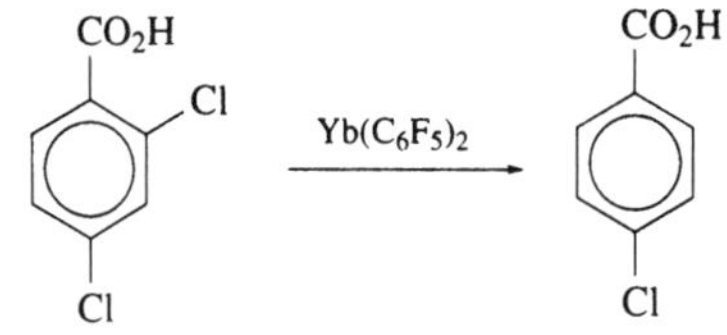

$$\text{(251)}$$

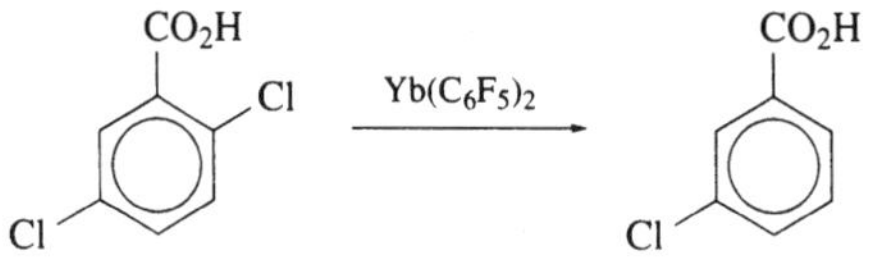

$$\text{(252)}$$

The ytterbium(II) reagent has to be used in excess and the yields of dehalogenated products are in the range of 76–84%. A mechanism for this reaction was proposed, which initially involves protolysis of the ytterbium(II) aryl to give a transient (pentafluorophenyl)ytterbium(II) carboxylate, $YbOC(O)Ar(C_6F_5)$. This is followed by electron transfer to the *ortho* halide substituent via a six-membered transition state. The final step is hydrogen abstraction from the solvent THF by the *ortho* aryl radical.[933] Epoxides such as epoxypropane are smoothly deoxygenated when treated with the THF adduct of decamethylsamarocene, $[Sm(THF)_2(Cp^*)_2]$.[593] The samarium(II) reagent is thereby transformed into the oxo-bridged binuclear organosamarium(III) complex $[(Cp^*)_2Sm–O–Sm(Cp^*)_2]$ (Equation (253)). Similarly pyridine-*N*-oxide is quantitatively deoxygenated in the presence of $[Sm(THF)_2(Cp^*)_2]$.

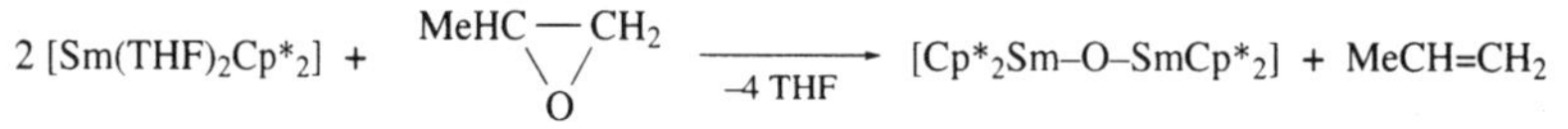

$$2\,[Sm(THF)_2Cp^*_2] + \underset{O}{MeHC\!-\!CH_2} \xrightarrow{-4\,THF} [Cp^*_2Sm–O–SmCp^*_2] + MeCH=CH_2 \qquad \text{(253)}$$

$[Sm(THF)_2(Cp^*)_2]$ has also been used in interesting reductive carbonylation reactions, for example of diphenylacetylene and 1,2-dipyridylethylene (cf. Section 2.2.6.3). These reactions are outlined in Schemes 54 and 55.[660,706] Stepwise reaction of $[Sm(THF)_2(Cp^*)_2]$ with diphenylacetylene and carbon monoxide produced an indenoindene complex in which the organic ligand was attached to samarium only via oxygen. The organic moiety was liberated upon protolysis. It was also possible to reductively cleave the complex with sodium in THF to afford 5,10-dihydroindeno[2,1,a]indene.[706]

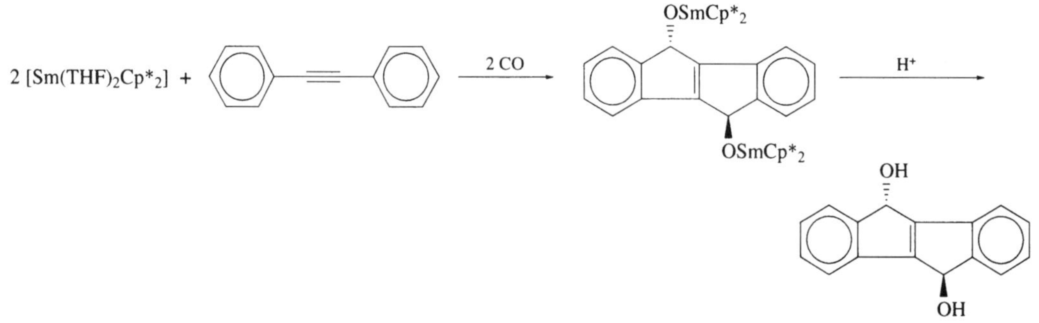

Scheme 54

Reductive carbonylation of 1,2-dipyridylethylene was achieved analogously by first adding the organic precursor to $[Sm(THF)_2(Cp^*)_2]$ followed by double CO insertion. Subsequent protolysis of the resulting bis(enolate) complex yielded the corresponding α-diketone.[661,934]

C–C-Coupling reactions have been observed when $[Sm(THF)_2(Cp^*)_2]$ was reacted with pyridazine or $PhCH=N–N=CHPh$.[664] One-electron reduction of pyridazine leads to coupling of two $C_4H_4N_2$ molecules in the 4-position (Equation (254)). The resulting binuclear samarium complex was structurally characterized.

The reaction of $[Sm(THF)_2(Cp^*)_2]$ with $PhCH=N–N=CHPh$ also resulted in reductive coupling of two substrate molecules to form $[\{\mu\text{-}\eta^4\text{-}PhCH=NNCHPh–CHPhNN=CHPh\}\{Sm(Cp^*)_2\}_2]$ (Equation (255)).[664]

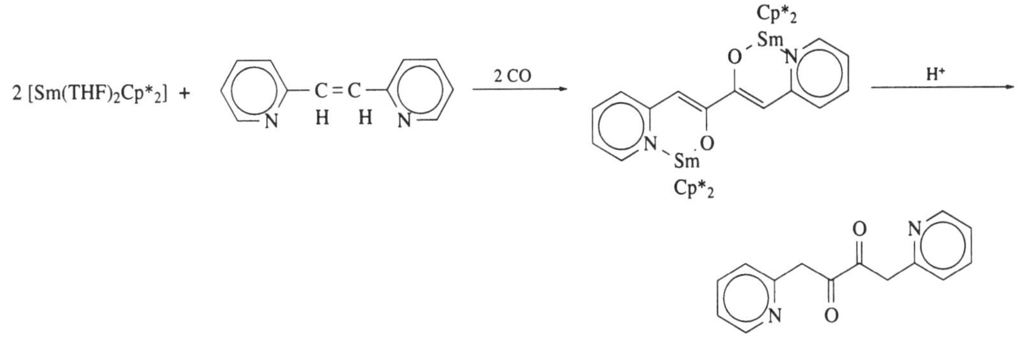

Scheme 55

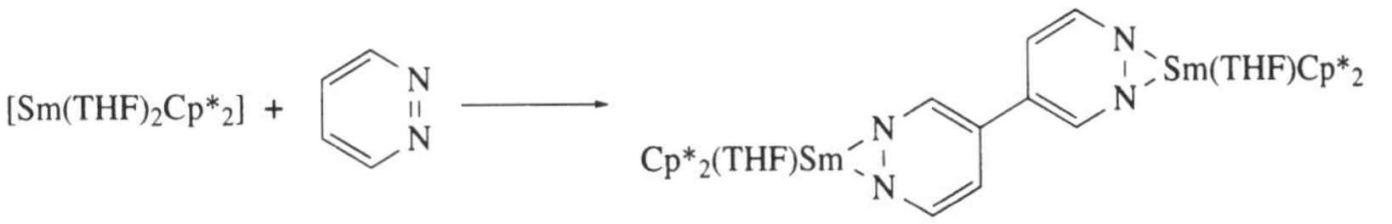

$$[Sm(THF)_2Cp^*_2] + \quad \longrightarrow \quad \tag{254}$$

$$[Sm(THF)_2Cp^*_2] + PhCH=NN=CHPh \quad \longrightarrow \quad \tag{255}$$

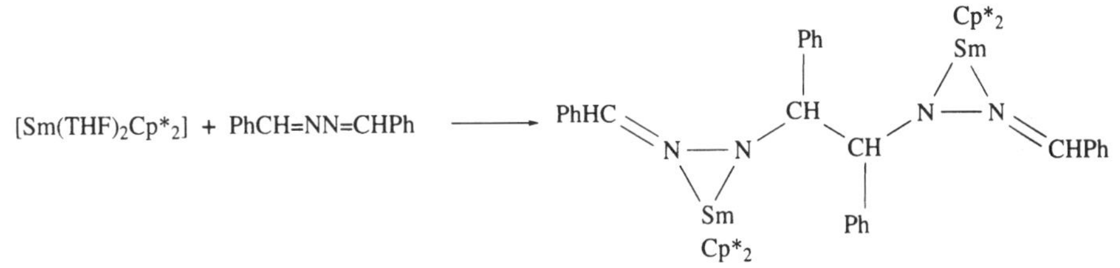

Due to their strong reducing power, organolanthanide(II) complexes have been employed more frequently in organic synthesis than related lanthanide(III) species. It has not been possible to isolate $RSmI_2$ species from samarium diiodide and RI (R = alkyl).[927,935] Such intermediates also seem to be absent in Barbier reactions, which have been successfully carried out with samarium diiodide (Equation (256)).[923,925,936] Radical intermediates have been postulated in this reaction instead of samarium hydrocarbyls.[935,937]

$$R^1\text{-C(O)-}R^2 + R^3X + SmI_2 \quad \xrightarrow{\text{THF}} \quad R^1R^2R^3C\text{-OH} \tag{256}$$

In the case of Barbier reactions (Equation (256)), it was found that $Sm(Cp)_2$ was superior to SmI_2, as the reactions proceeded much faster. In some cases replacement of SmI_2 by $Sm(Cp)_2$ even allowed the isolation of organosamarium intermediates, which were stable in THF solution. For example, addition of benzylic halides to slurries of insoluble bis(cyclopentadienyl)samarium in THF produced dark brown solutions that contained organosamarium species.[255] These solutions could be reacted with various electrophiles such as D_2O, aldehydes, or acid chlorides to afford the expected products (i.e., deuterated aldehydes, alcohols, and ketones). In a related investigation, Finke *et al.* have carefully studied electron transfer reactions between organic halides and $[Yb(Et_2O)(Cp^*)_2]$.[388,597]

Organolanthanide(III) complexes have also found various uses in synthetic organic chemistry, although very little is known about the structures of the reagents or possible intermediates. The homoleptic methyl complexes $[LnMe_6]^{3-}$ have been found to methylate α,β-unsaturated aldehydes and ketones, giving preferentially 1,2-adducts.[151] Anionic η^3-allyl complexes of the type $Li[Ln(C_3H_5)_4]$ (Ln = Ce, Nd, Sm) react with α,β-unsaturated carbonyl compounds to afford 3-hydroxy-1,5-dienes with 1,2-regioselectivity. The reagents can be prepared *in situ* from $LnCl_3$, $Sn(C_3H_5)_4$, and *n*-butyllithium in THF solution.[214] $YCl(Cp)_2$ has been shown to be a useful cyclopentadienyl transfer reagent. It reacts with aldehydes and ketones at 80 °C in DME to generate the corresponding fulvenes in excellent yields. At lower temperatures, cyclopentadienyl-substituted alcohols are isolated after hydrolytic work-up.[938] The ytterbium(III) complex $YbCl\{(TMS)_2C_5H_3\}_2$ has been used as a catalyst in a Mukaiyama addition reaction,[939] and the reactivity of benzylsamarium complexes of the type $SmCH_2R(Cp)_2$ (R = Ph, $C_6H_5Bu^t$-*p*, $C_6H_3Me_2$-2,5) toward organic reagents has been investigated.[255,940] The latter have been prepared *in situ* by treatment of $Sm(Cp)_2$ with benzylic chlorides and reacted with various aldehydes, ketones, and acid chlorides (Scheme 56).[940]

$$\text{ArCH}_2\text{Cl} + 2\,\text{SmCp}_2 \xrightarrow{\text{THF}} \text{SmClCp}_2 + \text{SmCH}_2\text{ArCp}$$

Ar = (phenyl)

Ar = Bu^t—(phenyl)—

Ar = (dimethylphenyl, Me, Me)

$$i,\ R^1\text{-}\underset{\underset{O}{\|}}{C}\text{-}R^2 \qquad \text{ii, }H_3O^+$$

$$D_2O$$

$$\text{ArH}_2C-\underset{\underset{OH}{|}}{\overset{\overset{R^1}{|}}{C}}-R^2 \qquad\qquad \text{ArCH}_2D$$

Scheme 56

A well investigated group of organolanthanide(III) reagents are monoorganolanthanide dihalides, which can be prepared *in situ*, for example from lanthanide triiodides or trichlorides and lithium alkyls.[902,941] The preparation of alkylsamarium reagents of the type SmI_2R by reduction of alkyl bromides or iodides offers a convenient and highly selective entry to a variety of functionalized organic compounds. Such RX/SmI_2 (X = Br, I) reagents have been utilized in 1,2-carbonyl addition reactions[942-5] and various natural product syntheses.[946-51] Experimental observations support the intermediacy of solution-stable organosamarium(III) species in many samarium-mediated processes.[952] Recently the first x-ray crystal structure determination of such a compound, $[\text{GdCl}_2(\text{THF})_4\text{Ph}]$, has been reported (cf. Section 2.2.2.3).[168] In a comparative study the compounds LnX_2R (Ln = Y, La, Ce, Pr, Nd, Sm, Gd; R = Me, Bu^n; X = F, Cl, I) were prepared *in situ* and reacted with aldehydes and ketones.[556] Especially useful are the derivatives of lanthanum and cerium. For example, reaction of cerium with iodobenzene affords a mixture of phenylcerium species, which behaves like a Grignard reagent.[953] Such reagents react with ketones at low temperatures (– 65 °C) to give alcohols.[954] 1,2-Addition is the preferred reaction pathway in the case of enones.[955] At 0 °C reductive coupling[954] or pinacol formation is observed.[956] Other synthetic uses of CeX_2R reagents include the preparation of allylic alcohols,[957] aldol condensation reactions,[958] and the reduction of alkenes and alkynes.[959,960] More recently, organocerium reagents of the type CeX_2R have been reported to react with 1-imino-*E,E*-butadieneiron tricarbonyl complexes in a stereoselective manner to afford single diastereomers in good yields.[961] Organocerium dihalides have been studied in detail in order to compare these compounds to Grignard reagents. The main difference is the reduced basicity of the lanthanide reagents, which allows reactions with enolizable ketones.[962] The following two examples may illustrate that this behavior has been utilized in some steps of natural products syntheses. In both cases ethynylcerium reagents have been employed, which are accessible from cerium trichloride and lithium acetylides (Equation (257)).[963,964]

$$\text{CeCl}_3 + \text{LiC}\equiv\text{CX} \longrightarrow \text{CeCl}_2\text{C}\equiv\text{CX} + \text{LiCl} \qquad (257)$$

$$X = H,\ \text{TMS}$$

Both ethynylcerium(III) reagents readily add to the carbonyl function of 1,2,3,4-tetrahydronaphthalene-2-ones to give high yields of carbinols, which in turn are useful intermediates in anthracyclinone syntheses (Scheme 57).[963]

$$\text{(OMe, OMe-tetralone, O)} + \text{CeCl}_2\text{C}\equiv\text{C-X} \xrightarrow[\text{1 h}]{\text{THF, }-78\ °\text{C}} \text{(OMe, OMe, OH, -C}\equiv\text{C-X)} \longrightarrow \text{(OMe, OMe, OH, C(=O)CH}_3)$$

X = TMS : 100%

X = H   : 85%

Scheme 57

Intermediates in the total synthesis of 11-deoxyanthracyclinones have been prepared in a similar manner using the 2-trimethylsilylethynylcerium(III) reagent (Equation (258)). The yield was only 11% when the corresponding ethynyllithium reagent was used, because the ketone was mainly enolized.[964]

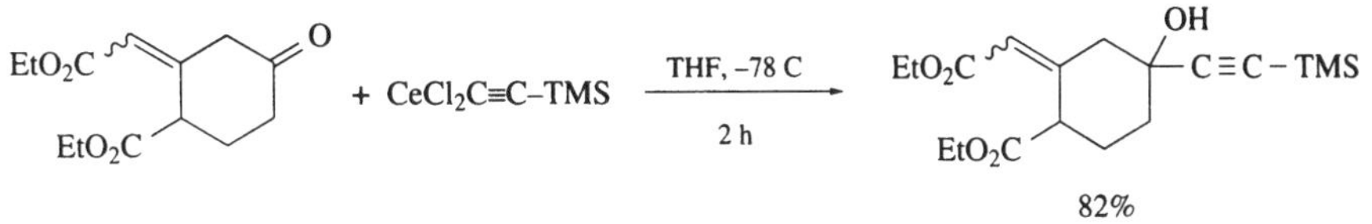

$$\text{(258)}$$

82%

In some cases the nucleophilic addition of organocerium reagents to chiral ketones can lead to a different stereochemistry as compared to addition of the corresponding organolithium reagents. In the following example the different results were explained by the different coordination environments of lithium and cerium. Cerium is thought to form a very stable chelate complex with the neighboring benzyl ether moiety while lithium is solvated by THF ligands (Equation (259)).[965]

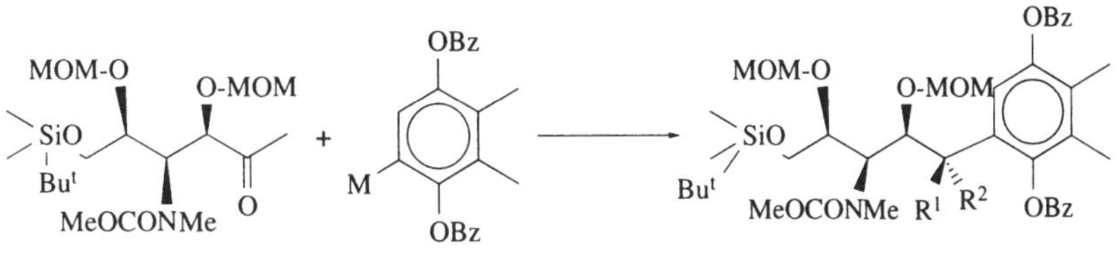

$$\text{(259)}$$

M = Li,   74%; R^1 = OH; R^2 = Me

M = Cl$_2$Ce, 95%; R^1 = Me; R^2 = OH

Reformatsky reactions have also been performed with organocerium reagents. In the preparation of substituted coumarins shown in Scheme 58 the cerium(III) reagent has been found to be superior over the classical zinc Reformatsky reagent.[966]

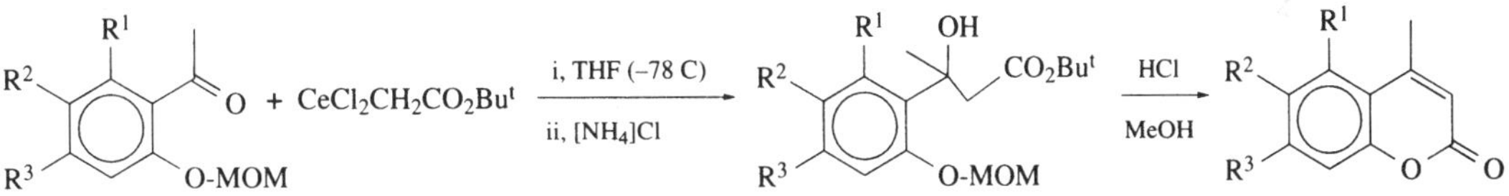

Scheme 58

More recently, soluble organolanthanide reagents have also been prepared by reacting Ln(OPr^i)$_3$ and [Ln{N(TMS)$_2$}$_3$] (Ln = Ce, Sm) with alkyllithium or Grignard reagents at low temperature. The resulting reagents were, for example, used for ring-opening alkylation reactions of epoxides (Equation (260)) and selective carbonyl addition reactions.[967,968]

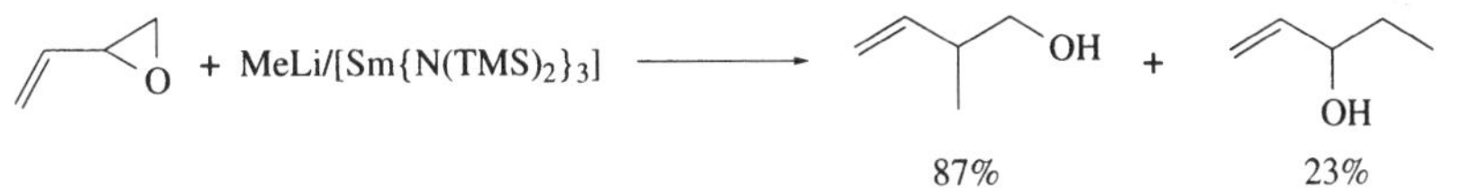

$$\text{(260)}$$

87%   23%

The reduction of alkenes has been achieved with the reagent Ln(Cp)$_3$/NaH. The reactivity of the system was found to parallel the ionic radii of the lanthanide metals. The same reagent has also been used for reductive dehalogenation of aryl and vinyl halides to produce the corresponding aromatic and alkane products.[443,969,970]

Rapid transcyanation from acetone cyanohydrin to several aldehydes is promoted by catalytic amounts of lanthanide(III) compounds, including the *in situ* prepared organometallic reagents Yb(Bu^n)$_3$ and La(O$_3$SCF$_3$)$_2$Me (Equation (261)).[971] Monoorganolanthanide triflates are readily prepared *in situ* by treatment of Ln(O$_3$SCF$_3$)$_3$ with LiR (R = alkyl, aryl).[972]

$$R\text{--}CHO + Me_2C(OH)CN \longrightarrow R\text{--}CH(OH)CN + Me_2C{=}O \qquad \text{(261)}$$

R = substituted aryl or cyclohexyl

2.2.15 Organolanthanides in Materials Science

Several homoleptic tris(cyclopentadienyl)lanthanide complexes have been investigated as molecular precursors in chemical vapor deposition (MOCVD) processes. Volatility is generally the main problem that has to be overcome when organolanthanides are to be employed. The low volatility of the parent compounds $Ln(Cp)_3$ limits their use as MOCVD precursors. However, the volatility of such complexes can be greatly improved by using certain substituted cyclopentadienyl ligands. In this respect the *i*-propylcyclopentadienyl ligand has been most intensively studied. Various complexes of the type $Ln(Pr^iC_5H_4)_3$ have been prepared by treatment of $LnCl_3$ with three equivalents of $[Na(Pr^iC_5H_4)]$.[18,238] The dysprosium derivative, $Dy(Pr^iC_5H_4)_3$, has been used as precursor for the deposition of high purity Gd_2O_3.[238] $Er(Pr^iC_5H_4)_3$ has been utilized to grow erbium-doped indium phosphide layers.[470] Similarly, $Yb(MeC_5H_4)_3$ was the source of ytterbium in the growth of highly doped InP:Yb layers by organometallic vapor phase epitaxy.[973] Erbium-doped GaAs has been manufactured using tris(*n*-butylcyclopentadienyl)erbium, $Er(Bu^nC_5H_4)_3$, as a liquid precursor. The gallium arsenide morphology was excellent at growth temperatures near 620 °C.[974,975] The parent tris(cyclopentadienyl)erbium has been utilized as a source material for doping AlGaAs.[976] In a comparative study the production of different III–V semiconductors doped with the rare-earth elements erbium, thulium, and ytterbium using atmospheric pressure MOVPE was investigated. Best results were obtained using tris(*i*-propylcyclopentadienyl)lanthanide precursors which have an acceptable vapor pressure and can be used as liquids at bubbler temperatures of 60–90 °C. Ytterbium was found to occupy a regular lattice site in InP, whereas the other lanthanides were incorporated as different centers and clusters.[977] An unusual application of cyclopentadienyllanthanide complexes is the manufacturing of lanthanide-doped quartz glass.[978]

Mixed-ring sandwich complexes of the type $[Ln(cot)(Cp^*)]$ have been used to deposit thin films of yttrium and other rare-earth elements using chemical vapor deposition (MOCVD) techniques.[977]

Organolanthanide-catalyzed dehydrogenative coupling of monoorganosilanes has been utilized to manufacture polyorganosilanes and finally silicon carbide ceramics. For example, the neodymium hydrocarbyl $[NdCH(TMS)_2(Cp^*)_2]$ catalytically reacts with methylsilane, $MeSiH_3$, to give poly(methylsilane). Subsequent pyrolysis of the polymers afforded β-SiC.[718]

2.3 ACTINIDES

The question of covalency in organoactinide complexes has been the subject of much debate since the early 1980s, and this discussion is expected to continue over the next few years. The nature of the bonding between π-donor ligands and *f*-elements has been the subject of several recent review articles.[38,64,88] Theoretical calculations on various levels of approximation (*ab initio*,[979] extended Hückel, relativistically parameterized extended Hückel,[980] quasirelativistic X_α scattered wave SCF,[981-3] and other calculations[984-6]) have been carried out on organouranium complexes with cyclopentadienyl or cyclooctatetraenyl ligands. A large amount of theoretical work has been devoted especially to the electronic structure of uranocene.[31,785,979,984,987,988] Some of the calculations came to the conclusion that the bonding between U^{4+} and these ligands is almost exclusively ionic.[989] Others were in favor of partial covalency involving the uranium $6d$ and to a lesser extent the uranium $5f$ orbitals.[785,987] Photoelectron spectroscopy provided experimental evidence for covalency in uranocene.[990,991] Attempts to detect covalency in actinide cyclopentadienyl and cyclooctatetraenyl complexes by careful analysis of structural parameters were equally inconsistent.[36,77]

Other articles of general interest have focused on geometric control in organoactinide complexes. For example, a packing saturation rule for organoactinides has been developed by Xing-Fu.[992-8] In this model the sum of the ligand cone angles determines the stability of complexes. Low stability can be predicted for sterically "oversaturated" coordination compounds. Other attempts to rationalize steric crowding around an actinide metal center led to a new definition of a "steric coordination number" by Pires de Matos *et al.*[75] These models assume, however, purely ionic bonding in organoactinide complexes. Steric similarities between chemically different ligands may be a helpful guide in the design of new classes of complexes. For example, the cone angles of chloride and BH_4 are similar, thus $[UCl_3(Cp)]$ and $[U(BH_4)_3(Cp)]$ are isosteric. The same considerations have been made for the cone angles of $Cp \approx$ tritox $\approx PhC(N\text{-}TMS)_2$ or $Cp^* \approx B_9C_2H_{11}$.[999]

Thermochemical studies are becoming increasingly important in the organometallic chemistry of the $5f$-elements. It has been clearly demonstrated in recent years that the knowledge of metal–ligand bond disruption enthapies is of fundamental importance, as it allows the prediction of novel organometallic reaction patterns and catalytic processes.[887,1000-4]

2.3.1 Carbonyls

In sharp contrast to the rich and extensive metal carbonyl chemistry of the *d*-block elements, there is virtually no such chemistry in the case of the actinides. Various attempts have been made in the past to prepare stable actinide carbonyl complexes. In their reluctance to form stable metal carbonyls in their +3 or +4 oxidation states the actinides very much resemble the lanthanide and early transition metal ions. All these ions have in common that they are good σ acceptors but very poor π-donors. Thus, in metal carbonyl derivatives of these elements π-back-donation is negligible. This is reflected, for example, in the IR spectroscopic data of a thermally labile carbon monoxide adduct of uranium tetrafluoride. The compound $UF_4(CO)$ was obtained analogous to the homoleptic uranium carbonyls by cocondensation of UF_4 with carbon monoxide in an argon matrix at 4 K. The high C–O stretching frequency of 2184 cm^{-1} indicated a predominant σ character of the U–CO bond.[32] These spectroscopic results were further confirmed by EHMO calculations, although the theoretical studies also provided some evidence that organoactinide carbonyls could be isolable.[1005] CU(O)CO and OCU(O)CO species have been identified as secondary reaction products during cocondensation reactions of uranium vapor with carbon monoxide in an argon matrix.[1006] In particular, some actinide hydrocarbyl complexes show interesting CO activation reactions (cf. Sections 2.3.4.5 and 2.3.5.3). Although carbonyl complexes have been postulated as intermediates in these reactions, no such species have ever been isolated. It can be anticipated that the uranium(III) ion should be a better candidate for preparing more stable uranium carbonyls than uranium(IV). Indeed, it has been demonstrated that dark green $[U(TMS\text{-}C_5H_4)_3]$ reversibly absorbs carbon monoxide to give red $[UCO(TMS\text{-}C_5H_4)_3]$, which is stable in solution for prolonged periods of time.[1007] The ν_{CO} frequency of 1976 cm^{-1} (1935 cm^{-1} for ^{13}CO) was interpreted in terms of a significant π-base character of the $U(TMS\text{-}C_5H_4)_3$ fragment. Solid $[U(TMS\text{-}C_5H_4)_3]$ also reversibly absorbs CO, but a pure crystalline uranium carbonyl could not be isolated. The corresponding isocyanide complex $[U(CNEt)(TMS\text{-}C_5H_4)_3]$, however, can be prepared as a crystalline solid.[1007]

2.3.2 Hydrocarbyls

2.3.2.1 Homoleptic compounds

The first attempts to synthesize homoleptic uranium hydrocarbyls date back to the 1940s, when Gilman *et al.* searched for volatile uranium compounds.[1008] However, it turned out that it was impossible to isolate simple alkyluranium compounds due to their extreme thermal instability. Reactions of uranium tetrachloride with lithium alkyls have been carefully reinvestigated by Marks and Seyam.[32,1009] The reaction shown in Equation (262) is quite complex and the resulting mixtures are thermally unstable. Homoleptic uranium tetraalkyls are coordinatively unsaturated and cannot be isolated. It has been shown that β-hydride elimination is the main decomposition pathway of intermediates containing uranium–carbon bonds. Similar reactions of thorium tetrachloride with lithium alkyls also failed to yield the desired thorium tetraalkyls. In this case too, only β-hydride elimination products were observed.[32]

$$UCl_4 + 4\,RLi \longrightarrow U + \text{organic products} + 4\,LiCl \qquad (262)$$

Early attempts to saturate the coordination sphere around uranium by forming "ate" complexes were only partially successful. Complexes of the type $[LiS_4]_2[UR_6]$ (R = Me, Ph, CH$_2$TMS; S = Et$_2$O, THF) and Li$_3$UR$_8$·3dioxane (R = Me, CH$_2$Bu^t, CH$_2$TMS) have been prepared by treating either UCl$_4$ or $[U_2(OEt)_{10}]$ with excess alkyllithium in the appropriate solvents. The species $[LiS_4]_2[UR_6]$ can be further reacted with tetramethylethylenediamine to afford Li$_2$UR$_6$·7TMEDA. However, all these products are thermally unstable above room temperature and have only partially been characterized. A general problem involved in the preparation of anionic uranium alkyls is the ease of reduction which occurs in the presence of excess alkyllithium. None of these "ate" complexes has ever been subjected to low-temperature x-ray diffraction studies. In the case of the Li$_3$UR$_8$·3dioxane species eight-coordinate $[UR_8]^{3-}$ anions have been proposed, in which three faces of the coordination polyhedra are capped by $[Li(dioxane)]^+$ cations.[32]

A better suited candidate for the preparation of homoleptic neutral and anionic hydrocarbyl complexes should be thorium, as thorium is much less susceptible to reduction processes than uranium. A stable homoleptic thorium hydrocarbyl is tetrabenzylthorium, which has been prepared according to Equation (263).[32]

Related complexes such as tetrabenzylzirconium or $[ThBz_3(Cp^*)]$ (cf. Section 2.3.5.3) have been shown to contain multihapto-coordinated benzyl ligands. Thus, it seems highly likely that allyl-like

$$ThCl_4 + 4\,LiBz \xrightarrow[-20\,°C]{THF} [Th(Bz)_4] + 4\,LiCl \tag{263}$$

multihapto-coordination also plays an important role in tetrabenzylthorium.[32] Recently a straightforward preparation of tetrakis(3,5-dimethylbenzyl)thorium(IV), [Th(CH$_2$C$_6$H$_3$Me$_2$-3,5)$_4$], was reported.[1010]

A major achievement in this field was the isolation of the first anionic thorium methyl complex by Marks *et al.* Yellow crystalline lithium heptamethylthorate was isolated when thorium tetrachloride was reacted with excess methyllithium in the presence of tetramethylethylenediamine in diethyl ether at −78 °C (Equation (264)).[1011]

$$ThCl_4 + 7\,MeLi + 4\,TMEDA \xrightarrow[-78\,°C]{Et_2O} [\{Li(TMEDA)\}_3\{ThMe_7\}\cdot TMEDA] + 4\,LiCl \tag{264}$$

In contrast to the thermally labile anionic uranium alkyls the thorium complex is stable at room temperature. Figure 76 shows the molecular structure as derived from an x-ray diffraction study. The coordination polyhedron around the central thorium atom can be described as a monocapped trigonal prism. Except for an additional σ-methyl group the structure resembles that of Schumann's homoleptic hexamethyllanthanidates [{Li(TMEDA)}$_3${LnMe$_6$}] (cf. Section 2.2.2.2). The Th–C bond distances to the bridging methyl groups range from 0.2667(8) to 0.2765(9) nm while the terminal Th–C(Me) distance is 0.2571(9) nm. Rapid reactions occur when the heptamethylthorate complex is treated with molecular hydrogen or carbon monoxide, but no well-defined hydrides or CO insertion products have been isolated.[1011]

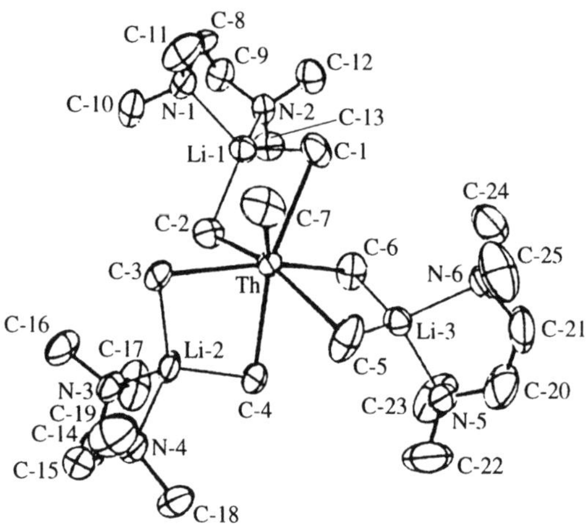

Figure 76 The molecular structure of [{Li(TMEDA)}$_3${ThMe$_7$}·TMEDA].[1011]

All attempts to synthesize diorganouranium dioxides of the type UO$_2$R$_2$ (R = alkyl, aryl) by treatment of uranyl chloride with various organolithium or Grignard reagents have failed. Once again the reaction products turned out to be thermally highly unstable. Either β-hydride elimination or reductive elimination in an intermediate UO$_2$R$_2$ have been discussed for the decomposition processes.[32,1012]

The first homoleptic uranium(III) alkyl was prepared by Sattelberger *et al.* from a monomeric uranium(III) phenoxide and isolated as a royal blue crystalline material (Equation (265)).[1013]

$$[U\{OC_6H_3(Bu^t)_2\text{-}2,6\}_3] + 3\,LiCH(TMS)_2 \longrightarrow [U\{CH(TMS)_2\}_3] + 3\,LiOC_6H_3(Bu^t)_2\text{-}2,6 \tag{265}$$

Thermally unstable [Np{CH(TMS)$_2$}$_3$] and [Pu{CH(TMS)$_2$}$_3$] have been prepared analogously.[1014] The molecular structure of [U{CH(TMS)$_2$}$_3$] consists of trigonal-pyramidal molecules which are stabilized by γ-agostic interactions (U–C 0.248(2) nm). A direct synthesis of the homoleptic alkyl from UCl$_3$ and LiCH(TMS)$_2$ in a 1:3 molar ratio failed because of LiCl retention and formation of the anionic complex [Li(THF)$_3$][UCl{CH(TMS)$_2$}$_3$].[1013]

2.3.2.2 Heteroleptic compounds

Base-free homoleptic actinide tetraalkyls are thermally highly unstable. However, it was found that certain additional ligands are capable of adding substantial thermal stability to such organoactinides. For example, highly basic chelating phosphines have been found to effectively stabilize tetrabenzyluranium and even tetramethyluranium.[32,1015] This is especially remarkable as actinide ions are normally very reluctant in forming stable coordination compounds with "soft" donor ligands such as phosphines or arsines. The phosphine-stabilized actinide hydrocarbyls have been prepared as shown in Equations (266)–(268).[1015]

$$AnCl_4 + 2\,dmpe \longrightarrow [AnCl_4(dmpe)_2] \tag{266}$$

$$[AnCl_4(dmpe)_2] + 4\,RLi \xrightarrow[-dmpe]{} [AnR_4(dmpe)] + 4\,LiCl \tag{267}$$
$$An = Th,\ U$$
$$R = Me,\ Bz$$
$$dmpe = Me_2PCH_2CH_2PMe_2$$

$$[AnCl_4(dmpe)_2] + 3\,[LiBz] + MeLi \xrightarrow[-dmpe]{} [AnMe(Bz)_3(dmpe)] + 4\,LiCl \tag{268}$$

Interestingly, the stability of the phosphine-stabilized actinide hydrocarbyls decreases significantly when bulky alkyl ligands such as CH_2Bu^t, CH_2TMS, or CH_2CPhMe_2 are employed. The same is true for substituents containing β-hydrogen atoms. In all these cases the products were found to decompose above ca. 0 °C. The stabilizing effect of benzyl groups, in particular, stems from the possibility of variable coordination modes. This was demonstrated by single-crystal x-ray structural analyses of $[ThBz_4(dmpe)]$ and $[UMeBz_3(dmpe)]$. The thorium derivative contains three monohapto-benzyl ligands and one polyhapto-coordinated benzyl while in the uranium complex two benzyl ligands are monohapto coordinated and the third one polyhapto. At room temperature the molecule is fluxional and shows three magnetically equivalent benzyl ligands in the ^1H NMR spectrum. At -60 °C the exchange process is slowed down and the spectrum becomes very complex.[1015]

The reactivity of $[U(Bz)_4MgCl_2]$ toward CO and CO_2 was investigated. Carbon monoxide yields the insertion product $[\{U(Bz)_3(COBz)MgCl_2\}_2]$, while reactions with CO_2 afforded the complexes $[\{UBz_3(CO_2Bz)MgCl_2\}_2]$, $[\{U(CO_2Bz)_4MgCl_2\}_2]$, and $[U_2(CO_2Bz)_4MgCl_2]$. However, none of these unusual materials has been structurally characterized.[1016] The homoleptic eight-coordinate thorium dialkylphosphide $[Th\{P(CH_2CH_2PMe_2)_2\}_4]$ reacts with carbon monoxide to give a double insertion product where CO is incorporated into a coordinated secondary alcohol derivative by coupling of two dialkyl phosphide groups at the same carbon atom. The structurally characterized product contains two η^2-phosphidoacyl units (25).[1017]

(25)

Bulky amido ligands have also been demonstrated to effectively stabilize the actinide–carbon σ-bond. In these cases, however, the products contain only one directly bonded hydrocarbyl ligand. Thermally stable amido-actinide methyl derivatives can be prepared by the reaction sequence shown below (Equations (269)–(271)).[32]

$$AnCl_4 + 3\,NaN(TMS)_2 \longrightarrow [AnCl\{N(TMS)_2\}_3] + 3\,NaCl \tag{269}$$
$$An = Th,\ U$$

$$2\,[ThCl\{N(TMS)_2\}_3] + MgMe_2 \xrightarrow{Et_2O} 2\,[ThMe\{N(TMS)_2\}_3] + MgCl_2 \qquad (270)$$

$$[UCl\{N(TMS)_2\}_3] + MeLi \xrightarrow{Et_2O} [UMe\{N(TMS)_2\}_3] + LiCl \qquad (271)$$

An x-ray diffraction study of the corresponding tetrahydroborate $[Th(BH_4)\{N(TMS)_2\}_3]$ revealed a distorted tetrahedral coordination geometry. Thus, the bulky bis(trimethylsilyl)amido ligand stabilizes the low coordination number 4 around actinide atoms. An interesting derivative chemistry has been developed starting from the methyl compounds $[AnMe\{N(TMS)_2\}_3]$.[1018–28] Insertion into the actinide–carbon bond was observed with aldehydes, ketones, nitriles, and isocyanides. The most remarkable derivatives are formed when $[ThMe\{N(TMS)_2\}_3]$ or $[UMe\{N(TMS)_2\}_3]$ are thermolyzed to give methane and novel four-membered metallacycles (Equation (272)).

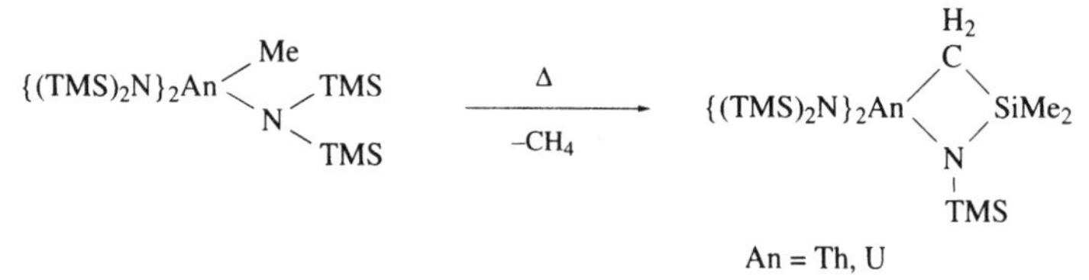

$$(272)$$

An = Th, U

The metallacyclic compounds display an interesting insertion chemistry. They have been found to react readily with carbon monoxide, ketones, nitriles, or *t*-butyl isocyanide to afford five- and six-membered metallacycles. Pyridine is metallated to give an η^2-C_5H_4N complex and alkoxides are formed upon treatment with alcohols. Scheme 59 summarizes the reactions of $[UMe\{N(TMS)_2\}_3]$ and the metallacycle.[1018–28]

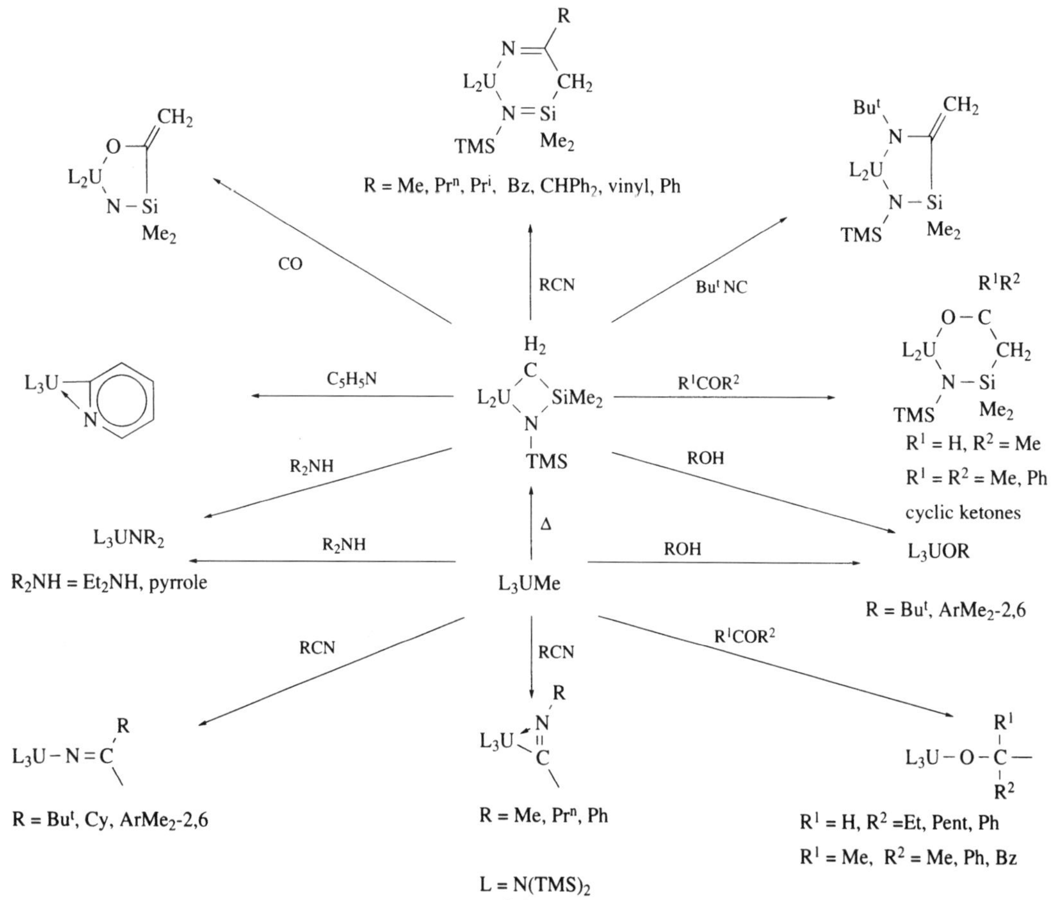

Scheme 59

L = N(TMS)$_2$

The initial step in the CO and Bu^{t}NC reactions is believed to be a migratory insertion which produces either a dihapto-acyl (CO) or a dihapto-iminoalkyl (Bu^{t}NC). Similar insertion reactions are well established in the chemistry of the cyclopentadienylactinide hydrocarbyls $[AnR(Cp)_3]$ and $[AnR_2(Cp^*)_2]$. The carbenelike acyl or iminoalkyl carbon atom then inserts into a silicon–carbon bond (Equation (273)).

$$X = O, Bu^tN \tag{273}$$

Treatment of the four-membered metallacycles with hydrogen produces amido-substituted actinide hydrides (Equation (274)).

$$\{(TMS)_2N\}_2An\;\text{(ring)}\;SiMe_2 + H_2 \longrightarrow [AnH\{N(TMS)_2\}_3] \tag{274}$$

The hydrides are also available by two alternative routes (Equations (275) and (276)).[32]

$$UCl_4 + 4\,NaN(TMS)_2 \xrightarrow{\;THF\;} [UH\{N(TMS)_2\}_3] \tag{275}$$

$$[AnCl\{N(TMS)_2\}_3] + NaN(TMS)_2 \xrightarrow{\;THF\;} [AnH\{N(TMS)_2\}_3] \tag{276}$$

Interestingly, the hydrides can be converted back to the σ-methyl compounds $[AnMe\{N(TMS)_2\}_3]$ by a two-step sequence involving deprotonation and subsequent alkylation (Equation (277)). Deprotonation may be effected by treatment with strong bases, although the complexes $[AnH\{N(TMS)_2\}_3]$ are considered to be mainly hydridic.[32]

$$[AnH\{N(TMS)_2\}_3] \xrightarrow[\;ii,\,MeBr\;]{\;i,\,Bu^nLi\;} [AnMe\{N(TMS)_2\}_3] \tag{277}$$

Bulky heteroallylic ligands are becoming increasingly popular in f-element chemistry and have recently been used to stabilize uranium alkyls. N,N'-Bis(trimethylsilyl)benzamidinate ligands have been shown to be especially useful for this purpose. Benzamidinate analogues of $[UMe(Cp)_3]$ have been prepared according to Equation (278).[129]

$$[UCl\{4\text{-}RC_6H_4C(N\text{-}TMS)_2\}_3] + MeLi \longrightarrow [UMe\{4\text{-}RC_6H_4C(N\text{-}TMS)_2\}_3] + 2\,LiCl \tag{278}$$
$$R = H, MeO, CF_3$$

A long uranium–carbon bond in $[UMe\{PhC(N\text{-}TMS)_2\}_3]$ (U–C 0.2498(5) nm) indicates severe steric crowding in these molecules (Figure 77).[1029]

Besides $\bar{C}_5Me_5$ the bulky benzamidinate ligand $[2,4,6\text{-}(CF_3)_3C_6H_2C(N\text{-}TMS)_2]^-$ is the only ancillary ligand which allows stabilization of a dimethyluranium fragment. The methyl derivative was prepared by alkylation of the chloro precursor (Equation (279)).[1029]

$$[UCl_2\{2,4,6\text{-}(CF_3)_3C_6H_2C(N\text{-}TMS)_2\}_2] + 2\,MeLi \longrightarrow [UMe_2\{2,4,6\text{-}(CF_3)_3C_6H_2C(N\text{-}TMS)_2\}_2] + 2\,LiCl \tag{279}$$

Polyprazolylborate ligands have also been found to stabilize heteroleptic uranium alkyls. The preparation of the complexes $[UCl_2\{HB(3,5\text{-}Me_2pz)_3\}\{CH(TMS)_2\}]$, $[UCl_{3-x}\{HB(3,5\text{-}Me_2pz)_3\}(CH_2TMS)_x]$ ($x = 1\text{–}3$), $[U(HBpz_3)_2(CH_2TMS)_2]$, and $[U\{HB(3,5\text{-}Me_2pz)_3\}\{CH(TMS)_2\}_2(THF)]$ by salt metathesis reactions has been reported.[54,1030] $[U\{HB(3,5\text{-}Me_2pz)_3\}\{CH(TMS)_2\}_2(THF)]$ was fully characterized by a single-crystal x-ray structural analysis (U–C 0.264(2) nm, U–N 0.260(3) nm, U–O 0.258(1) nm).[1031]

Intramolecular nitrogen coordination is a key factor in the stability of dialkylthorium compounds comprising the 2-(6-methylpyridyl) ligand ($= MeC_5H_3NCH_2$). An interesting precursor,

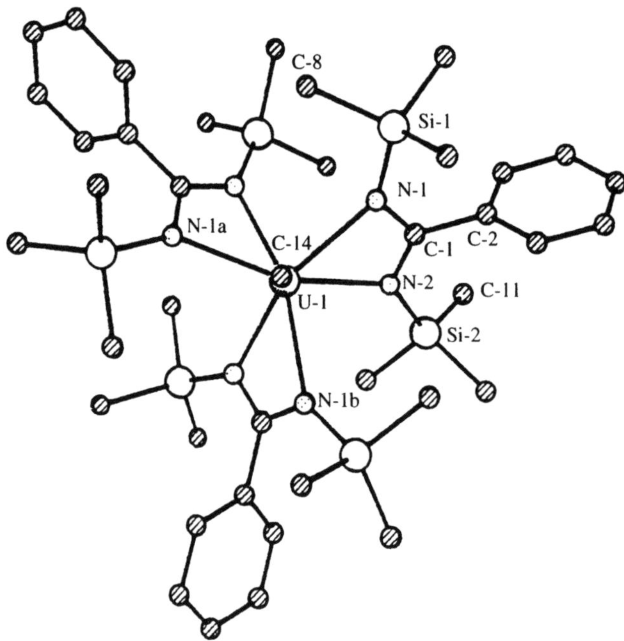

Figure 77   The molecular structure of [UMe{PhC(N-TMS)$_2$}$_3$].[1029]

[ThCl$_2$(MeC$_5$H$_3$NCH$_2$)$_2$], was prepared by reacting ThCl$_4$ with excess MeC$_5$H$_3$NCH$_2$Li. The dichloride was further reacted with two equivalents of LiOAr (OAr = 2,6-di-*t*-butylphenoxide) to afford [Th(MeC$_5$H$_3$NCH$_2$)$_2$(OAr)$_2$], which was structurally characterized. The six-coordinated thorium atom is surrounded by two chelating 2-(6-methylpyridyl) ligands and two bulky phenoxides (Th–C 0.255(1) nm, Th–N 0.261(1) nm, Th–O 0.2190(9) nm).[1032]

The bulky alkoxide ligand tri-*t*-butylmethoxide (tritox)$^-$ has also recently been shown to stabilize heteroleptic uranium dialkyls. This ligand can be regarded as a steric equivalent to cyclopentadienyl. The dibenzyl derivative [U(tritox)Bz$_2$] was prepared from the dichloride precursor and benzyllithium (Equation (280)). Unlike many [UX$_2$(Cp)$_2$] derivatives the tritox compound shows no tendency to disproportionate in solution.[999,1033]

$$[UCl_2(THF)_2(tritox)_2] + 2\,LiBz \longrightarrow [U(Bz)_2(tritox)_2] + 2\,LiCl + 2\,THF \qquad (280)$$

In a related study the homoleptic uranium(IV) alkoxide [U{OCH(Bu^t)$_2$}$_4$] was reacted with methyllithium to afford a crystallographically characterized methyl alkoxide derivative (Equation (281)).[1034]

$$[U\{OCH(Bu^t)_2\}_4] + MeLi \longrightarrow [U\{OCH(Bu^t)_2\}_4MeLi] \qquad (281)$$

Lithium and uranium are bridged by two bulky alkoxide ligands. The coordination geometry around uranium is square-pyramidal with the σ-methyl group in the apical position (U–C 0.2465(7) nm).[1034]

2.3.3 Allyls

Only a handful of actinide allyls without supporting cyclopentadienyl ligands has been reported in the literature. Homoleptic tetrakis(allyl) derivatives have been described for uranium and thorium. These two compounds as well as U(C$_3$H$_4$Me-2)$_4$ are obtained by reacting the metal tetrachlorides with the corresponding allyl-Grignard reagents (Equation (282)).[31,1010]

$$AnCl_4 + 4\,MgXC_3H_5 \longrightarrow An(C_3H_5)_4 + 4\,MgClX \qquad (282)$$

$$An = U, Th$$

All these compounds are thermally labile and decompose above ca. 0 °C. Apparently no attempts have been made to prepare thermally more robust actinide allyls by employing sterically demanding substituted allyl ligands. NMR spectral data of $U(C_3H_5)_4$ were in agreement with a rigid η^3-coordination of the allyl ligands, whereas the allyl units in $Th(C_3H_5)_4$ showed fluxionality at higher temperatures. A mechanism for the observed exchange of *syn-* and *anti-*η^3-allyl protons is shown in Scheme 60.[31,1010]

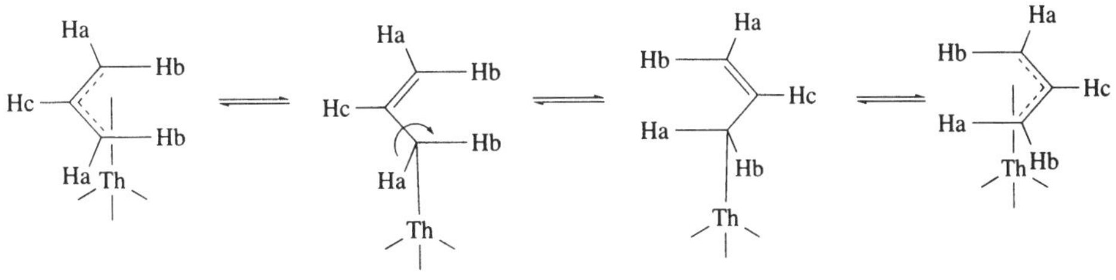

Scheme 60

Several protolytic reactions have been reported for tetrakis(allyl)uranium.[31] The allyl ligands are completely removed when the compound is treated with an excess of alcohols. Partial replacement of the allyl moieties was achieved by reacting $U(C_3H_5)_4$ with stoichiometric amounts of alcohols or HX (Equations (283) and (284)).

$$2\,U(C_3H_5)_4 + 4\,ROH \longrightarrow [\{U(OR)_2(C_3H_5)_2\}_2] + 4\,C_3H_6 \tag{283}$$

$$R = Et, Pr^i, Bu^t$$

$$U(C_3H_5)_4 + HX \longrightarrow UX(C_3H_5)_3 + C_3H_6 \tag{284}$$

$$X = Cl, Br, I$$

The *t*-butoxide derivative $[\{U(OBu^t)_2(C_3H_5)_2\}_2]$ has been structurally characterized by x-ray diffraction.[31] The molecule contains two bridging and two terminal alkoxide ligands, and two η^3-allyl ligands are attached to each uranium center with U–C distances ranging from 0.2644(16) to 0.2736(18) nm. With the very bulky tritox ligand (= tri-*t*-butylmethoxide) the monomeric allyl complex $[U(tritox)_2(C_3H_5)_2]$ has been isolated.[999]

A compound formulated as $[U(bipy)_3(C_3H_5)_4]$ has been obtained be reacting $U(C_3H_5)_4$ with 2,2-bipyridine.[31] This complex is not well characterized. It appears that two allyl ligands are monohapto coordinated, while the other two have been transferred to coordinated 2,2'-bipyridine. However, these rather speculative results still await further conformation by structural investigations. The reaction of $U(C_3H_5)_4$ with $(Bu^nO)_3PO$ as well as CO_2 insertion into the U-allyl bond have also been studied.[1035-7]

2.3.4 Cyclopentadienyl Compounds

2.3.4.1 M(Cp)₂X compounds

Bis(cyclopentadienyl)metal(III) complexes of the type $MX(Cp)_2$ (X = halide, hydride, hydrocarbyl, etc.) form a large and very common class of organometallic compounds in lanthanide chemistry (see above). In marked contrast, only a small number of such compounds has been described for the actinide elements.[31] Simple derivatives like $UCl(Cp)_2$ are not available, and in the case of uranium and thorium none of the known compounds has been prepared by standard salt-elimination reactions. The purple form of $Th(Cp)_3$ was reported to react with ammonium chloride in benzene to give a material formulated as $ThCl(Cp)_2$ (Equation (285)).[31] The exact (dimeric?) nature of this compound is not known as structural information is lacking.

$$ThCp_3 + [NH_4]Cl \xrightarrow{\;C_6H_6\;} ThClCp_2 + NH_3 + C_5H_6 \tag{285}$$

Similarly a cyano complex of uranium(II) was synthesized by protonation of tris(cyclopentadienyl)uranium with HCN (Equation (286)).[31]

$$UCp_3 + HCN \xrightarrow{\;C_6H_6\;} UCNCp_2 + C_5H_6 \tag{286}$$

For some transuranium elements the +3 oxidation state is significantly more stable than for uranium and thorium. In this case it is possible to use metal trichlorides as precursors. The microscale reaction of berkelium trichloride with molten bis(cyclopentadienyl)beryllium afforded $BkCl(Cp)_2$ (Equation (287)).[31]

$$BkCl_3 + BeCp_2 \longrightarrow BkClCp_2 + BeCl_2 \tag{287}$$

The molecular structure of bis(cyclopentadienyl)berkelium chloride has not been established by x-ray studies, but it seems highly likely that the compound is dimerized via chloride bridges in the solid state. This is the most common structural type for bis(cyclopentadienyl)lanthanide chlorides.

An anionic bis(cyclopentadienyl)uranium(III) complex, $[Na(18\text{-crown-}6)][U(BH_4)_2(Cp)_2]$ was prepared by sodium amalgam reduction of $[U(BH_4)_2(Cp)_2]$ in the presence of the crown ether. Rare examples of mono(cyclopentadienyl)uranate(III) complexes $[U(BH_4)_3(Cp)]^-$ and $[U(BH_4)_3(Cp^*)]^-$ have been made similarly.[36,1038,1039]

A much better understanding of this class of organoactinides was made possible through the use of substituted cyclopentadienyl ligands. Lappert *et al.* succeeded in preparing a complete series of bis(cyclopentadienyl)uranium halides by employing the bulky 1,3-bis(trimethylsilyl)cyclopentadienyl ligand.[1040–3] The compounds $[\{U(\mu\text{-}X)\{1,3\text{-}(TMS)_2C_5H_3\}_2\}_2]$ (X = Cl, Br, I) were prepared by Na/K reduction of the corresponding uranium(IV) dihalides in toluene solution. In the case of X = F the precursor was the tetrafluoroborate derivative $[\{U(\mu\text{-}F)(\mu\text{-}BF_4)\{1,3\text{-}(TMS)_2C_5H_3\}_2\}_2]$. X-ray analyses of the isostructural chloride and bromide revealed the dimeric nature of these compounds with symmetric halide bridges (U–Cl 0.2818(4) nm, U–Cl' 0.2802(4) nm, U···U 0.4357(1) nm).[1040]

The structurally characterized chloro-bridged "ate" complex $[U(\mu\text{-}Cl)_2Li(THF)_2\{1,3\text{-}(TMS)_2C_5H_3\}_2]$ (U–C 0.277(2) nm, U–Cl 0.2730(1) nm) was alternatively prepared by reacting $[UCl_2\{1,3\text{-}(TMS)_2C_5H_3\}_2]$ with $[Li\{P(TMS)_2\}(THF)_2]_2$ in an attempt to make the corresponding uranium diorganophosphide.[1043] Adduct formation with neutral ligands such as isocyanides is a typical reaction pattern. The crystal structures of the adducts $[UBr(Bu^tNC)_2\{1,3\text{-}(TMS)_2C_5H_3\}_2]$, $[UCl(TMS\text{-}CN)_2\{1,3\text{-}(TMS)_2C_5H_3\}_2]$, and $[UCl(2,6\text{-}Me_2C_6H_3NC)_2\{1,3\text{-}(TMS)_2C_5H_3\}_2]$ have been reported.[1044–6] The derivative chemistry of $[U(\mu\text{-}Cl)\{1,3\text{-}(TMS)_2C_5H_3\}_2]_2$ (Scheme 61) also includes the formation of the anionic complex $[PPh_4][UCl_2\{1,3\text{-}(TMS)_2C_5H_3\}_2]$ (U–Cl 0.2666(8) nm) as well as the preparation of the first uranium(III) aryloxides.[1041–3] Reactions of $[UCl(THF)\{1,3\text{-}(TMS)_2C_5H_3\}_2]$ with TMS-N_3 and $Sn\{N(TMS)_2\}_2$ produced the complexes $[UCl(=N\text{-}TMS)\{1,3\text{-}(TMS)_2C_5H_3\}_2]$ and $[UCl\{N(TMS)_2\}\{1,3\text{-}(TMS)_2C_5H_3\}]$, respectively.[1041]

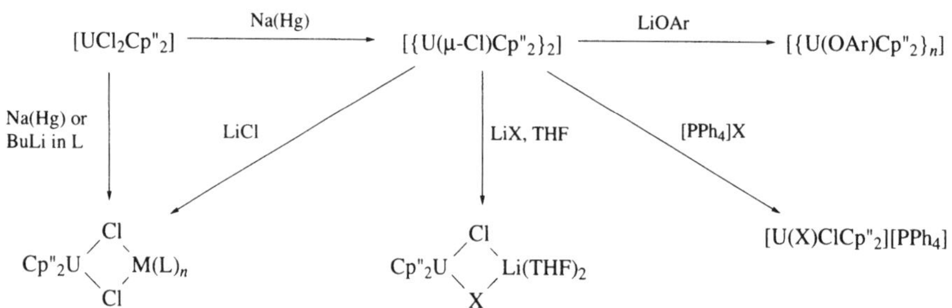

X = Cl, Br; Ar = $C_6H_3Pr^i\text{-}2,6$; M(L)$_n$ = Na(THF)$_2$, Li(THF)$_2$, Na(TMEDA), Li(TMEDA), Li(PMDETA) with TMEDA = $(Me_2NCH_2)_2$ and PMDETA = $(Me_2NCH_2CH_2)_2NMe$

Scheme 61

t-Butyl-substituted cyclopentadienyl ligands too have been successfully employed to stabilize bis(cyclopentadienyl)uranium(III) complexes. For example, UCl_3 reacts with two equivalents of $K\{(Bu^t)_2C_5H_3\}$ to afford the chloro-bridged dimer $[\{U(\mu\text{-}Cl)\{(Bu^t)_2C_5H_3\}_2\}_2]$ (U–C 0.279(4) nm, C–Cl 0.2856(4) nm).[1047] The dimeric bis(cyclopentadienyl)uranium(III) thiolate $[\{U(\mu\text{-}SPh)(Bu^tC_5H_4)_2\}_2]$ was prepared by protolysis of $[U(Bu^tC_5H_4)_3]$ with one equivalent of thiophenol. Similar reactions with $[U(THF)(MeC_5H_4)_3]$ and various alcohols and thiols produced only the uranium(IV) complexes $[U(ER)(MeC_5H_4)_3]$ (ER = OMe, OPr^i, OPh, SPr^i).[367] A rare example of a bis(cyclopentadienyl)uranium(III) alkyl was synthesized according to Equation (288).[1048]

$$2\,[U(Bu^tC_5H_4)_3] + 2\,MeLi \longrightarrow [\{U(\mu\text{-}Me)(Bu^tC_5H_4)_2\}_2] + 2\,Li(Bu^tC_5H_4) \tag{288}$$

2.3.4.2 *M(Cp)₃ compounds*

Tris(cyclopentadienyl) complexes of actinides(III) have been prepared for thorium, uranium, and neptunium as well as for several transplutonium elements. The best known uranium(III) cyclopentadienyl complex is U(Cp)₃, which can be prepared either by reducing uranium(IV) precursors such as [U(Cp)₄] or [UCl(Cp)₃] or by starting with uranium trichloride or even uranium metal. The most straightforward method appears to be the reduction of easily available [UCl(Cp)₃] with sodium hydride.[31] An electrochemical reduction of [UCl(Cp)₃] yielded the stable anionic complex [UCl(Cp)₃]⁻ as the initial product.[1049] A photochemical synthesis of the tris(cyclopentadienyl) derivatives of uranium(III) and thorium(III) has also been described.[31,1050]

U(Cp)₃ behaves like a strong Lewis acid. It readily forms stable 1:1 adducts with various donor ligands (THF, nitriles, isonitriles, nicotine, etc.).[31] Synthetically more important is the THF adduct [U(THF)(Cp)₃], because its solubility is significantly higher than that of unsolvated tris(cyclopentadienyl)uranium(III). The THF solvate can be prepared directly via the following routes (Equations (289)–(291)).[31,1051]

$$[UCl_3(THF)_n] + 3\,NaCp \xrightarrow{\text{THF}} [U(THF)Cp_3] + 3\,NaCl \tag{289}$$

$$[UClCp_3] + Na \xrightarrow[\text{THF}]{C_{10}H_8} [U(THF)Cp_3] + NaCl \tag{290}$$

$$[UCl_2\{N(Et_3Si)_2\}] + 2\,NaCp \xrightarrow{\text{THF}} [U(THF)Cp_3] + \text{other products} \tag{291}$$

Once again the most convenient synthetic route is that involving reduction of the readily prepared precursor [UCl(Cp)₃]. Sodium amalgam reduction gives a 70% yield.[1039,1052,1053] The crystal and molecular structure of [U(THF)(Cp)₃] has been reported.[1051] Three η⁵-coordinated cyclopentadienyl rings and a THF ligand describe a distorted tetrahedron around the central uranium atom (U–C 0.276(2)–0.282(2) nm, U–O 0.255(1) nm). The coordinated THF in [U(THF)(Cp)₃] is readily displaced by nitriles to afford [U(CNR)(Cp)₃] (R = Me, Prⁿ, Prⁱ, Buᵗ). [U(NCPrⁿ)(Cp)₃] (U–N 0.261(1) nm) and [U(NCPrⁱ)(Cp)₃] (U–N 0.2551(9) nm) have been crystallographically characterized.[1054] Similar reaction of [U(THF)(Cp)₃] with benzonitrile or thermolysis of [U(CNR)(Cp)₃] (R = Me, Prⁿ) resulted in formation of an equimolar mixture of the uranium(IV) complexes [UCN(Cp)₃] and [UR(Cp)₃] (Equation (292)).[1054]

$$2\,[U(CNR)Cp_3] \xrightarrow{\Delta} [UCNCp_3] + [URCp_3] + RCN \tag{292}$$

$$R = Me, Pr^n, Ph$$

Sodium amalgam reduction of [UCl(Cp)₃] in the presence of 18-crown-6 afforded [Na(18-crown-6)][UCl(Cp)₃].[1039,1052,1053] The anions [UX(Cp)₃]⁻ (X = BH₄, alkyl) have been prepared analogously or by addition of X⁻ to [U(THF)(Cp)₃]. A chloro-bridged binuclear anion was obtained by combining [U(THF)(Cp)₃] with [UCl(Cp)₃]⁻ according to Equation (293). The x-ray crystal structure revealed a symmetrical bent chloro-bridge.[1039,1053]

$$[U(THF)Cp_3] + [Na(18\text{-crown-}6)][UClCp_3] \xrightarrow{\text{THF}} [Na(18\text{-crown-}6)(THF)_2][U(\mu\text{-Cl})Cp_3UCp_3] \tag{293}$$

Due to their higher solubility, tris(cyclopentadienyl)uranium(III) complexes containing substituted cyclopentadienyl ligands are better suited for reactivity studies than the parent compound U(Cp)₃. For example, tris(methylcyclopentadienyl)uranium(III) forms adducts with triphenylphosphine oxide,[1055] tetrahydrothiophene,[1056] and nitrogen heterocycles. The molecular structures of [U(OPPh₃)(MeC₅H₄)₃][1055] and [U(SC₄H₈)(MeC₅H₄)₃][1056] very much resemble that of [U(THF)(Cp)₃]. The U–S distance in the latter complex is 0.2986(5) nm. Other structurally characterized adducts are [U(NH₃)(MeC₅H₄)₃],[1057] [U(NC₅H₄NMe₂)(MeC₅H₄)₃],[1058] [U{N(CH₂CH₂)₃CH}(MeC₅H₄)₃],[503] [U{P(OCH₂)₃CEt}(MeC₅H₄)₃],[1055] and [{U(Cp)₃}₂(μ-dmpe)]. The compound [U(PMe₃)(MeC₅H₄)₃] was obtained by *in situ* preparation of [U(THF)(MeC₅H₄)₃] from UCl₃ and Na(MeC₅H₄) followed by addition of trimethylphosphine. The U–P distance was found to be 0.2972(6) nm.[1059] A comparison has been made between the structurally characterized adducts [UL(MeC₅H₄)₃] and the cerium(III) analogues (L = PMe₃, N(CH₂CH₂)₃CH, P(OCH₂)₃CEt).[503,504] The results were interpreted in terms of a more pronounced π back-bonding in the uranium compounds involving a significant contribution of the uranium 5*f* orbitals. This back-bonding is negligible in the cerium complexes.[1060-2] [U(THF)(Cp)$_3$] is oxidized upon treatment with alkyl halides (Equation (294)).[1063]

$$2\,[\text{U(THF)Cp}_3] + RX \longrightarrow [\text{UXCp}_3] + [\text{URCp}_3] \tag{294}$$

$$X = Cl,\ Br,\ I$$

$$R = Me,\ Bu^n,\ Pr^i,\ allyl,\ Bz$$

Similar oxidation of [U(THF)(MeC$_5$H$_4$)$_3$] with alcohols and thiols produced the uranium(IV) complexes [UER(MeC$_5$H$_4$)$_3$] (ER = OMe, OPr^i, OPh, SPr^i).[367,1064]

Monomeric, base-free [U(RC$_5$H$_4$)$_3$] complexes can be synthesized by using cyclopentadienyl ligands with bulky substituents. The compounds [U(RC$_5$H$_4$)$_3$] (R = Bu^t, TMS) have been isolated and the trimethylsilyl derivative structurally characterized.[367,1064] A trigonal planar arrangement of the cyclopentadienyl ligands around uranium was found (U–C 0.278(4) nm). Addition of ethyl isocyanide afforded the pseudotetrahedral adduct [U(CNEt)(TMS-C$_5$H$_4$)$_3$].[1065] Nucleophilic addition of azide produced a binuclear μ-azido-bridged anion (Equation (295)).[1039]

$$2\,[\text{U(C}_5\text{H}_4\text{TMS)}_3] + \text{NaN}_3 + \text{18-crown-6} \xrightarrow{\text{THF}} [\text{Na(18-crown-6)(THF)}_2][\text{U(μ-N}_3)(\text{C}_5\text{H}_4\text{TMS})_3\text{U(C}_5\text{H}_4\text{TMS})_3] \tag{295}$$

A well-defined derivative of tris(cyclopentadienyl)thorium has been prepared by Lappert *et al.*[1066] The paramagnetic compound [Th{C$_5$H$_3$(TMS)$_2$-1,3}$_3$] has been prepared by reduction of [ThCl$_2${C$_5$H$_3$(TMS)$_2$-1,3}$_2$] with sodium–potassium alloy in toluene and isolated as a bright blue, exceedingly air-sensitive crystalline material. The x-ray crystal structure analysis revealed the presence of an authentic monomeric thorium(III) complex with pseudotrigonal planar coordination geometry (Figure 78, Th–C 0.280(2) nm). An EPR spectrum was interpreted with a $6d^1$ ground state of the molecule.[1067]

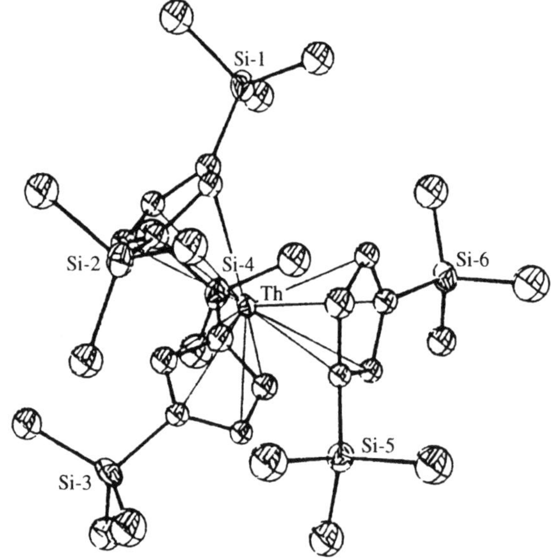

Figure 78   The molecular structure of [Th{C$_5$H$_3$(TMS)$_2$-1,3}$_3$].[1066]

A tris-THF adduct of tris(cyclopentadienyl)neptunium has been prepared by sodium naphthalide reduction of [NpCl(Cp)$_3$] (Equation (296)). The ^{237}Np Mössbauer data of [Np(THF)$_3$(Cp)$_3$] was consistent with a +3 oxidation state for neptunium and a largely ionic bonding between neptunium and the cyclopentadienyl ligands, because the isomer shift was almost identical with that found for NpCl$_3$.[31]

$$[\text{NpClCp}_3] + \text{Na} \xrightarrow{\text{C}_{10}\text{H}_8/\text{THF}} [\text{Np(THF)}_3\text{Cp}_3] + \text{NaCl} \tag{296}$$

Solvent-free microscale syntheses have been developed for the preparation of transplutonium tris(cyclopentadienyl) complexes. These compounds have been made by reacting the metal trichlorides with molten bis(cyclopentadienyl)beryllium (Equation (297)).[31] Unreacted [Be(Cp)$_2$] as well as BeCl$_2$ are easily removed by vacuum sublimation. Sublimation at significantly higher temperatures afforded the pure actinide(III) cyclopentadienyls. This method had already been successfully employed to the synthesis of tris(cyclopentadienyl)promethium. More recently the solvated triiodides have been used as very convenient precursors (Equation (298)).[1014]

$$2\,AnCl_3 + 3\,BeCp_2 \xrightarrow{65\,^\circ C} 2\,AnCp_3 + 3\,BeCl_2 \tag{297}$$

An = Pu, Am, Cm, Bk, Cf

$$[PuI_3(THF)_4] + 3\,LiCp \xrightarrow{THF} [Pu(THF)Cp_3] + 3\,LiI \tag{298}$$

According to x-ray powder diffraction studies the transplutonium tris(cyclopentadienides) are isostructural with the $Ln(Cp)_3$ derivatives of the early and middle lanthanides (Ln = Pr, Sm, Gd). Single-crystal optical and fluorescence spectroscopic data of $An(Cp)_3$ (An = Cm, Am) have been used to predict the degree of covalency in the metal–ligand bonding.[31] X_α-SW molecular orbital calculations have been carried out on $An(Cp)_3$ complexes. The most important result is that in the series An = U, Np, Pu, Am, Cm, Bk, Cf, the participation of the $5f$ orbitals in the metal–ligand bonding increases from left to right, while at the same time the participation of the $6d$ orbitals decreases.[86]

Reduction of $[U(Bu^n)(Cp)_3]$ has been reported to afford the anionic uranate(III) complex $[U(Bu^n)(Cp)_3]^-$. The latter was isolated as its $[Li(2.1.1\text{-cryptate})]^+$ salt and characterized by x-ray crystallography. The structure of the anion differs very little from that of the neutral uranium(IV) precursor (U–C(σ) 0.2557(9) nm, U–C(Cp) 0.283 nm).[1052,1068–70] Such anionic species are more conveniently prepared either by sodium amalgam reduction of $[UR(Cp)_3]$ or addition of RLi to $[U(THF)(Cp)_3]$ (Equations (299) and (300)).[1039,1070,1071]

$$[URCp_3] \xrightarrow{Na/Hg} Na[URCp_3] \tag{299}$$

R = Me, Bu^n

$$[U(THF)Cp_3] + LiR \longrightarrow Li[URCp_3] \tag{300}$$

R = Me, Pr^i, Pr^n, Bu^n, allyl, Bz

The anionic complexes react with terminal alkenes and hydrogen[1070] as well as alkyl halides. In the latter case the uranium(IV) alkyls $[UR(Cp)_3]$ are obtained.[1071,1072]

A methyl-bridged binuclear anion was prepared by addition of methyllithium to $[U(THF)(MeC_5H_4)_3]$ (Equation (301)).[1073]

$$[U(THF)(MeC_5H_4)_3] \xrightarrow[\text{TMEDA}]{MeLi} [Li(TMEDA)_2][(\mu\text{-}MeC_5H_4)\{Li(TMEDA)\}_2][(MeC_5H_4)_3U(\mu\text{-}Me)U(MeC_5H_4)_3]_2 \tag{301}$$

An x-ray crystal structure determination of the product revealed a U–C(Me)–U angle of 176.9(1)° and U–C bond distances of 0.271(3) nm and 0.274(2) nm.

Addition of H$^-$ to $[U(RC_5H_4)_3]$ (R = H, Me, TMS, Bu^t) or sodium amalgam reduction of the uranium(IV) hydrides $[UH(RC_5H_4)_3]$ (R = TMS, Bu^t) afforded the hydrido-bridged anions $[U(\mu\text{-}H)(RC_5H_4)_3U(C_5H_4R)_3]^-$ (R = H, Me) or the monomeric anions $[UH(RC_5H_4)_3]^-$ (R = TMS, Bu^t) (Scheme 62). The crystal structure of $[Na(18\text{-crown-6})(THF)_2][U(\mu\text{-}H)(TMS\text{-}C_5H_4)_3U(C_5H_4TMS)_3]$ was also discussed. This compound was obtained from an equimolar mixture of $[Na(18\text{-crown-6})][(TMS\text{-}C_5H_4)_3UH]$ and $[U(TMS\text{-}C_5H_4)_3]$.[1074]

2.3.4.3 *M(Cp)X$_3$ compounds*

Monocyclopentadienyl complexes of uranium and thorium are stable only as adducts of the type $[AnX_3L_n(Cp)]$ (X = Cl, Br, NCS) with one or two additional oxygen or nitrogen donor ligands L.[36,1075] Obviously the $AnX_3(Cp)$ fragment alone is coordinatively unsaturated and unstable. In the presence of oxygen donor ligands, however, the synthesis of such complexes is simple and straightforward. In this case thallium cyclopentadienides are the preferred cyclopentadienyl transfer reagents (Equations (302)–(304)).[31,1076]

The coordinated ether ligands can be replaced by phosphine oxides such as Ph_3PO or $(Me_2N)_3PO$. In an alternative synthetic approach the corresponding Lewis-base adducts of uranium tetrachloride can be used as precursors and treated with thallium cyclopentadienide (Equations (305) and (306)).[31]

$$[U(C_5H_4R)_3] \qquad\qquad [UH(C_5H_4R)_3]$$
$$R = TMS\ or\ Bu^t \qquad\qquad R = TMS\ or\ Bu^t$$

NaH | 18-crown-6   [Et$_3$NH][BPh$_4$]   Tl[BPh$_4$]   NaHg / 18-crown-6

$$[Na(18\text{-}crown\text{-}6)][U(C_5H_4R)_3(\mu\text{-}H)U(C_5H_4R)_3] \longrightarrow [Na(18\text{-}crown\text{-}6)][UH(C_5H_4R)_3]$$
$$R = TMS \qquad\qquad\qquad\qquad R = TMS,\ Bu^t$$

$$2\ [U(THF)(C_5H_4R)_3] \underset{[Et_3NH][BPh_4]}{\overset{NaH}{\rightleftharpoons}} [Na(THF)_2][(C_5H_4R)_3U(\mu\text{-}H)U(C_5H_4R)_3]$$
$$R = H,\ Me \qquad\qquad\qquad\qquad R = H,\ Me$$

Scheme 62

$$AnX_4 + TlCp \xrightarrow{THF} [AnX_3(THF)_2Cp] + TlX \qquad (302)$$

$$AnX_4 + TlCp \xrightarrow{DME} [AnX_3(DME)Cp] + TlX \qquad (303)$$
$$An = Th,\ U$$
$$X = Cl,\ Br$$

$$AnCl_4 + Tl(MeC_5H_4) \xrightarrow{THF} [AnCl_3(THF)_2(MeC_5H_4)] + TlCl \qquad (304)$$
$$An = Th,\ U$$

$$UX_4L_2 + TlCp \xrightarrow{THF} [UX_3L_2Cp] + TlX \qquad (305)$$
$$X = Cl,\ Br$$
$$L = Ph_3PO,\ (Me_2N)_3PO$$

$$[UCl_2(acac)_2(OPPh_3)_2] + TlCp \longrightarrow [UCl(acac)_2(OPPh_3)_2Cp] + TlCl \qquad (306)$$

Structurally all these complexes are very similar, as has been demonstrated by single-crystal x-ray analyses of [UCl$_3$(THF)$_2$(MeC$_5$H$_4$)], [UCl$_3$(Ph$_3$PO)$_2$(Cp)], and [UCl$_3${(Me$_2$N)$_3$PO}(Cp)].[31,1075] They exhibit a pseudooctahedral coordination geometry around the central uranium atom with the donor ligands in *cis*-positions, cyclopentadienyl *trans* to a donor ligand, and the chloride ions in a meridional arrangement (**26**). The dynamic behavior of these complexes in solution has been studied by NMR spectroscopy.[1077]

(**26**)

The majority of the work on mono(cyclopentadienyl)actinide organometallics has been carried out with uranium and thorium. However, the thallium cyclopentadienide route has been successfully extended to neptunium and plutonium (Equation (307)).[31]

$$NpCl_4 + Tl(MeC_5H_4) \xrightarrow{THF} [NpCl_3(THF)_2(MeC_5H_4)] + TlCl \qquad (307)$$

Although the [237]Np Mössbauer data for [NpCl$_3$(THF)$_2$(MeC$_5$H$_4$)] were interpreted in terms of η^1-coordination of the methylcyclopentadienyl ligand,[1058] η^5-bonding analogous to the related uranium and thorium complexes seems more likely. In fact, the molecular structure of [NpCl$_3$(MePh$_2$PO)$_2$(Cp)] features the same ligand arrangement as the above-mentioned uranium derivatives with an η^5-coordinated cyclopentadienyl ligand (Np–Cl 0.265(2) nm, Np–O 0.275(5) nm).[1078]

Numerous other adducts of the type $[AnX_3L_2(Cp)]$ have been reported mostly by Bagnall *et al.* The different compounds are summarized in Table 2.

Table 2 Mono(cyclopentadienyl)actinide complexes of the type $[AnX_3L_2(Cp)]$.

An	X	L	Ref.
Th	Cl	THF	1076
		$MeC(O)N(Pr^i)_2$, $MeC(O)N(c\text{-}C_6H_{11})_2$, $EtC(O)N(Pr^i)_2$, $Pr^iC(O)NMe_2$, $Pr^iC(O)N(Pr^i)_2$,	1079
		$Bu^tC(O)N(Pr^i)_2$, $OC(NMe_2)_2$, $Me_2NC(O)N(CH_2CHMe_2)_2$	
Th	Br	$Pr^iC(O)NMe_2$, $Bu^tC(O)N(Pr^i)_2$, $EtC(O)NEt_2$	1079
U	Cl	THF, Ph_3PO, $(Me_2N)_3PO$, $MeC(O)N(Pr^i)_2$, $EtC(O)N(Pr^i)_2$, $OC(NMe_2)_2$	1077,1080
		Me_3PO, Me_2PhPO, $MePh_2PO$, $Ph_2P(C_5H_4N)$	1076,1081
Np	Cl	$HC(O)NMe_2$, $MeC(O)N(Pr^i)_2$, $EtC(O)N(Pr^i)_2$, Me_3PO, Me_2PhPO, $MePh_2PO$	1078
Np	NCS	Me_2PhPO, $MePh_2PO$, Ph_3PO	1080
Pu	Cl	$HC(O)NMe_2$, $MeC(O)NMe_2$, $MeC(O)N(Pr^i)_2$, $EtC(O)N(Pr^i)_2$, $Bu^tC(O)NMe_2$, Me_3PO,	
		Me_2PhPO, $MePh_2PO$, Ph_3PO, $(Me_2N)_3PO$	1082
Pu	NCS	Me_3PO, Me_2PhPO, $MePh_2PO$, Ph_3PO	1080

Some tris-adducts have also been prepared including the complexes $[ThCl_3(MeCN)_3(Cp)]$,[1076] $[UCl_3\{MeC(O)N(c\text{-}C_6H_{11})_2\}_2(THF)(Cp)]$,[1079] $[U(NCS)_3L_3(Cp)]$ $(L = Me_2PhPO,\ MePh_2PO)$,[1076] and $[Np(NCS)_3(Me_3PO)_3(Cp)]$.[1080]

Mono(cyclopentadienyl)actinide complexes of the type $[AnX_3(THF)_2(Cp)]$ have been used as precursors in a variety of subsequent reactions, for example with oxygen,[1083] lithium alkyls,[1084] and β-diketonate and pyrazolylborate ligands (Equations (308)–(315)).[31,36,1038,1085-91]

$$[UCl_3(THF)_2Cp] + LiR \longrightarrow [UCl_2R(THF)_2Cp] + LiCl \quad (308)$$
$$R = Me,\ Et,\ Bu^n$$

$$[UCl_3(THF)_2Cp] + Na(acac) \xrightarrow{\text{THF}} [UCl_2(acac)Cp] + NaCl + 2\,THF \quad (309)$$

$$[UCl_2(acac)Cp] + 2\,Ph_3PO \xrightarrow{\text{THF}} [UCl_2(acac)(Ph_3PO)_2Cp] \quad (310)$$

$$[UCl_2(acac)_2(THF)_2] + TlCp + Ph_3PO \longrightarrow [UCl(acac)_2(Ph_3PO)Cp] + TlCl + 2\,THF \quad (311)$$

$$[UCl_3(THF)_2Cp] + 2\,Na(acac) \xrightarrow{\text{THF}} [UCl(acac)_2Cp] + 2\,NaCl + 2\,THF \quad (312)$$

$$[UCl_3(THF)_2Cp] + n\,Na(tropolonate) \xrightarrow{\text{THF}} [U(tropolonate)_nCl_{3-n}Cp] + 2\,NaCl + 2\,THF \quad (313)$$
$$n = 1\text{--}3$$

$$[AnX_3L_2Cp] + K[Bpz_3] \xrightarrow{\text{THF}} [AnX_2\{HBpz_3\}LCp] + KX + L \quad (314)$$
$$An = Th,\ U$$
$$X = Cl,\ Br$$
$$L = THF,\ Ph_3PO,\ (Me_2N)_3PO$$

$$[AnCl_3(THF)_2Cp] + K[HB(3,5\text{-}Me_2pz)_3] \xrightarrow{\text{THF}} [AnCl_2\{HB(3,5\text{-}Me_2pz)_3\}Cp] + KCl \quad (315)$$
$$An = Th,\ U$$

The acetylacetonate derivative $[UCl(acac)_2(Ph_3PO)(Cp)]\cdot THF$ was structurally characterized (U–C 0.279(1) nm, U–Cl 0.2662(8) nm, U–O(acac) 0.235(1) nm, U–O(Ph_3PO) 0.240(1) nm). The coordination polyhedron around the central uranium atom is a pentagonal bipyramid, in which five oxygen atoms define the equatorial plane.[1090]

Mixed cyclopentadienyl pyrazolylborate complexes $[AnCl_2\{HB(3,5\text{-}Me_2pz)_3\}(Cp)]$ have been synthesized and the uranium derivative structurally characterized (U–C 0.274(1) nm, U–N 0.251(2) nm, U–Cl 0.260(1) nm).[1030] The pyrazolylborate derivatives $[AnCl_2\{HB(3,5\text{-}Me_2pz)_3\}L(Cp)]$ served as starting materials for some alkoxides. The compounds $[AnCl\{HB(3,5\text{-}Me_2pz)_3\}(OR)(Cp)]$ (An = Th, U; R = Pr^i, Bu^t, $C_6H_2Me_3$-2,4,6) were prepared by metathetical reactions of the dichlorides with NaOR.[54] A different approach to mono(cyclopentadienyl)uranium alkoxides utilizes the facile protolytic elimination of cyclopentadienyl ligands from $[UCl(RC_5H_4)_3]$ in the presence of alcohols (Equation (316)).[1091]

$$[UCl(R^1C_5H_4)_3] + 2 R^2OH \longrightarrow [U(OR^2)_2Cl(R^1C_5H_4)] + 2 R^1Cp \qquad (316)$$

$$R^1 = H, Me$$
$$R^2 = Et, Pr^n, Pr^i, Bu^n, Bu^t$$

A monocyclopentadienyluranium(IV) complex containing the tripod ligand $[Co\{P(O)(OEt)_2\}_3(Cp)]^-$ was prepared by treatment of $[[Co\{P(O)(OEt)_2\}_3(Cp)]UCl_3]$ with thallium cyclopentadienide. The starting material was obtained by reacting UCl_4 with $Na[Co\{P(O)(OEt)_2\}_3(Cp)]$ in a 1:1 molar ratio (Equations (317) and (318)). The molecular and crystal structure of the bright green complex $[[Co\{P(O)(OEt)_2\}_3(Cp)]UCl_2(Cp)]$ has been determined.[1092,1093]

$$UCl_4 + Na[Co\{P(O)(OEt)_2\}_3Cp] \longrightarrow [[Co\{P(O)(OEt)_2\}_3Cp]UCl_3] + NaCl \qquad (317)$$

$$[[Co\{P(O)(OEt)_2\}_3UCl_3]Cp] + NaCp \longrightarrow [[Co\{P(O)(OEt)_2\}_3Cp]UCl_2Cp] + NaCl \qquad (318)$$

The cobalt-containing tripod ligand is considered a steric cyclopentadienyl equivalent. A monodentate steric cyclopentadienyl equivalent is the bulky alkoxide anion $Bu^t_3CO^-$ (= tritox). A mixed-ligand monocyclopentadienyluranium(IV) complex containing the tritox ligand was prepared according to Equation (319).[999]

$$[UCl_2(THF)_2(tritox)_2] + TlCp \longrightarrow [UCl(tritox)_2Cp] + TlCl \qquad (319)$$

One of the rare examples of mono(cyclopentadienyl)uranium complexes with nitrogen donor ligands is the tris-amide $U(NEt_2)_3(Cp)$.[1094] The thorium analogue has been prepared by protolytic cleavage of Th–N bonds in $Th(NEt_2)_4$ with a stoichiometric amount of cyclopentadiene (Equation (320)).[1095]

$$Th(NEt_2)_4 + C_5H_6 \longrightarrow Th(NEt_2)_3Cp + HNEt_2 \qquad (320)$$

Cyclic bis(cyclopentadienyl)actinide amides were obtained by protonation of the metallacycles $[\{N(TMS)_2\}_2AnCH_2SiMe_2N\text{-}TMS]$ with stoichiometric amounts of cyclopentadiene.[1019,1021]

No mono(cyclopentadienyl)actinide trihydride has been reported, but monomeric $[U(BH_4)_3(Cp)]$ and $[U(BH_4)_3(MeC_5H_4)]$ have been prepared and the former was structurally characterized.[1088,1096] Although the positions of the hydrogen atoms were not located, the U–B distances (0.246(4) nm and 0.257(5) nm) were in agreement with tridentate coordination of the borohydride ligands.[1038,1087] The volatile compound $[U(BH_4)_3(Cp)]$ can be conveniently prepared by reacting $[U(BH_4)_4]$ with thallium cyclopentadienide or free cyclopentadiene in a 1:1 molar ratio.[1038,1088] Like $UCl_3(Cp)$ the tris(borohydride) tends to disproportionate in the presence of coordinating solvents and forms adducts with various donor ligands.[1089] The compounds $[U(BH_4)_3L_2(Cp)]$ (L = THF, 1/2DME, $(Me_2N)_3PO$, Ph_3PO) have been prepared and characterized.[1089]

A remarkable mono(cyclopentadienyl)uranium hydrocarbyl containing six uranium–carbon bonds has been prepared and structurally characterized. The compound $[U\{(CH_2)_2PPh_2\}_3(Cp)]$ was obtained by reacting $[UCl_3(THF)_2(Cp)]$ with three equivalents of the lithiated phosphoylide $Li(CH_2)_2PPh_2$. The coordination geometry around the central uranium atom is pseudotrigonal with the cyclopentadienyl ligand and one methylene group occupying the *trans* positions (Figure 79). The average U–C(σ) bond length is 0.266(3) nm.[1097]

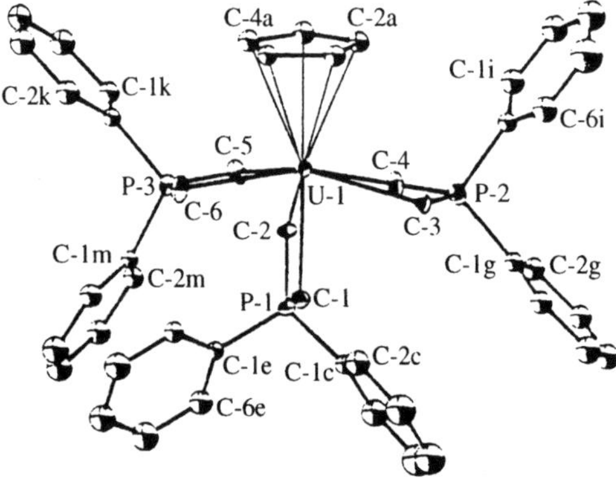

Figure 79 The molecular structure of $[U\{(CH_2)_2PPh_2\}_3(Cp)]$.[1097]

A tetrametallic complex containing two U(Cp) units was obtained serendipitously by reacting $[U(AlH_4)_2(Et_2O)_n(Cp)_2]$ with acetic acid. As shown by x-ray crystallography, the product $[\{U_2(MeCOO)_5O\}_2(Cp)]$ contains two organometallic and two purely inorganic uranium centers.[1098]

Among the numerous mono-, di-, and trisubstituted cyclopentadienyl ligands those with trimethylsilyl substituents are well suited to allow the isolation and characterization of mono(cyclopentadienyl)actinide complexes. Lappert *et al.* successfully employed the 1,3-bis(trimethylsilyl)cyclopentadienyl ligand in this area of organoactinide chemistry. Trivalent uranium forms a series of complexes of the type $UX\{1,3\text{-}(TMS)_2C_5H_3\}_2$ (X = Cl, Br, I, BH_4, CH_2TMS, Bz, $OC_6H_3Me_2\text{-}2,6$).[1041] Treatment of UCl_4 with one equivalent of $Li[1,2,4\text{-}(TMS)_3C_5H_2]$ afforded the structurally characterized lithium chloride adduct $[UCl_2(THF)\{1,2,4\text{-}(TMS)_3C_5H_2\}(\mu\text{-}Cl)_2Li(THF)_2]$ (U–Cl(terminal) 0.2606(4) nm, U–Cl(bridging) 0.2730(4) nm).[1099]

2.3.4.4 $M(Cp)_2X_2$ compounds

In sharp contrast to the well-developed chemistry of metallocene dihalides of titanium, zironum, and hafnium the development of the corresponding actinide chemistry was hampered by the non-availability of some key starting materials. Despite early reports on their synthesis the two basic starting materials $[UCl_2(Cp)_2]$ and $[ThCl_2(Cp)_2]$ are kinetically unstable and cannot be isolated. The reaction of uranium tetrachloride with two equivalents of thallium cyclopentadienide in DME was originally reported to yield "$[UCl_2(Cp)_2]$." This product was later shown to be a 1:1 mixture of $[UCl(Cp)_3]$ and $[UCl_3(DME)(Cp)]$.[31,36] Other cyclopentadienyl transfer reagents also failed to produce the elusive bis(cyclopentadienyl)uranium dichloride. It was only possible to stabilize the $UCl_2(Cp)_2$ unit by adduct formation with the bidentate phosphine oxide ligand $Ph_2P(O)CH_2P(O)Ph_2$. While bis(cyclopentadienyl)-thorium dichloride is also not available, the corresponding iodide has been prepared according to Equation (321).[31]

$$ThI_4 + MgCp_2 \longrightarrow [ThI_2Cp_2] + MgI_2 \tag{321}$$

Adduct formation with the chelating dmpe ligand (= 1,2-bis(dimethylphosphino)ethane) has been found to be an effective way to stabilize $ThCl_2(Cp)_2$ (Equation (322)).[31]

$$[ThCl_4(dmpe)] + 2\,NaCp \longrightarrow [ThCl_2(dmpe)Cp_2] + 2\,NaCl \tag{322}$$

Considerable efforts have been directed toward the development of other suitable precursors for $UX_2(Cp)_2$ complexes. One of the most successful approaches used the easily accessible bis(cyclopentadienyl)uranium(IV) bis(diethylamide). This compound is formed upon treatment of uranium tetrakis(diethylamide) with cyclopentadiene in a molar ratio of 1:2.[31] The corresponding thorium bis(diethylamide) has been prepared analogously (Equation (323)).[1094]

$$An(NEt_2)_4 + 2\,C_5H_6 \longrightarrow [An(NEt_2)_2Cp_2] + 2\,HNEt_2 \tag{323}$$
$$An = Th, U$$

Photoelectron spectra of $[U(NEt_2)_2(Cp)_2]$ revealed that the first ionization corresponds to removal of a uranium $5f$ electron, while the second ionization potential was N_{2p} lone pair in nature.[1100] The diethylamido ligands in $[U(NEt_2)_2(Cp)_2]$ are excellent leaving groups and thus are readily replaced when the compound is treated with a variety of reagents (Equations (324)–(327)). This way a number of alkoxides, phenoxides, thiolates, and dithiocarbamates containing a bis(cyclopentadienyl)uranium unit have been prepared.[31,1101,1102]

$$[U(NEt_2)_2Cp_2] + 2\,ROH \longrightarrow [U(OR)_2Cp_2] + 2\,HNEt_2 \tag{324}$$
$$R = \text{alkyl, aryl, 8-hydroxyquinolinate,}$$
$$\text{tropolonate}$$

$$[U(NEt_2)_2Cp_2] + 2\,RSH \longrightarrow [U(SR)_2Cp_2] + 2\,HNEt_2 \tag{325}$$
$$R = Et, Bu^t, Ph$$

Not all of the alkoxide derivatives were resistant to disproportionation reactions.[1101] An attempted preparation of $[U(8\text{-hydroxyquinolinate})_2(Cp)_2]$ from $[UCl_2(8\text{-hydroxyquinolinate})_2]$ and $Tl(Cp)$ yielded

$$[U(NEt_2)_2Cp_2] + LH_2 \longrightarrow ULCp_2 + 2\,HNEt_2 \qquad (326)$$

$$LH_2 = \text{catechol, 1,2-ethanedithiol,}$$
$$o\text{-mercaptophenol, toluene-3,4-dithiol}$$

$$[U(NEt_2)_2Cp_2] + 2\,CS_2 \longrightarrow [U(S_2CNEt_2)_2Cp_2] \qquad (327)$$

only $[UCl(Cp)_3]$.[1102] The products $[U(O_2CBu^t)_2(Cp)_2]$ and $[U(S_2CNEt_2)_2(Cp)_2]$ were studied by photoelectron spectroscopy.[1100] Isocyanides were found to readily insert into the U–N bonds of $[U(NEt_2)_2(Cp)_2]$ to stepwise give iminoalkylamido products (Scheme 63).[1103,1104]

$$[U(NEt_2)_2Cp_2] \xrightarrow{\ CNR\ } Cp_2U \xleftarrow[\substack{| \\ C \\ | \\ NEt_2}]{\substack{NEt_2 \\ /}} NR \xrightarrow{\ CNR\ } Cp_2U$$

$$R = c\text{-}C_6H_{11},\ C_6H_3Me_2\text{-}2,6$$

Scheme 63

$[U(BH_4)_2(Cp)_2]$ is yet another simple bis(cyclopentadienyl)uranium derivative that is accessible in a straightforward manner (Equations (328) and (329)).[51,1096,1100] The ring-substituted derivatives $[U(BH_4)_2(MeC_5H_4)_2]$, $[U(BH_4)_2(TMS\text{-}C_5H_4)_2]$, and $[U(BH_4)_2\{1,3\text{-}(TMS)_2C_5H_3\}_2]$ have been prepared in a similar way.[1041,1096] According to the IR spectroscopic data the borohydride ligands are coordinated to uranium in a tridentate fashion.

$$UCl_4 + 2\,Na[BH_4] \xrightarrow[\text{or DME}]{\text{THF}} UCl_2(BH_4)_2 \qquad (328)$$

$$UCl_2(BH_4)_2 + 2\,TlCp \xrightarrow[\text{or DME}]{\text{THF}} [U(BH_4)_2Cp_2] + 2\,TlCl \qquad (329)$$

Sodium amalgam reduction of $[U(BH_4)_2(Cp)_2]$ in the presence of 18-crown-6 was reported to yield the anionic uranate(III) complex $[Na(18\text{-crown-}6)][U(BH_4)_2(Cp)_2]$.[1038,1052] The same anion was generated electrochemically.[1105] Reactions of $[U(BH_4)_2(Cp)_2]$ with $Li[AlH_4]$ afforded the aluminohydride derivatives $Li[U(AlH_4)_2NEt_2(Cp)_2]$ and $U(AlH_4)_2(Cp)_2 \cdot n\,Et_2O$.[1038]

Among the anionic chelating and tridentate ligands which are able to form complexes of the type $UX_2(Cp)_2$ are acetylacetonate as well as certain poly(pyrazolyl)borate anions such as dihydrobis(pyrazolyl)borate and hydrotris(pyrazolyl)borate.[31] Bulky nitrogen ligands such as N,N-di-p-tolyltriazenide (= triaz) or N,N'-di-p-tolylformamidinate (= form) are also able to stabilize $An(Cp)_2$ units, for example, in the compounds $[An(triaz)_2(Cp)_2]$ (An = Th, U) or $[U(form)_2(Cp)_2]$.[1106]

Bis(cyclopentadienyl)thorium bis(hydrocarbyls) have been successfully stabilized with the use of the chelating 1,2-bis(dimethylphosphino)ethane ligand (= dmpe) (Equation (330)).[1107,1108]

$$[ThCl_2(dmpe)Cp_2] + 2\,RLi \longrightarrow [ThR_2(dmpe)Cp_2] + 2\,LiCl \qquad (330)$$

$$R = Me,\ Bz$$

The starting material $[ThCl_2(dmpe)(Cp)_2]$ (Th–C 0.280(2) nm, Th–Cl 0.2708(2) nm, Th–P 0.3122(2) nm) as well as the bis(alkyls) $[ThMe_2(dmpe)(Cp)_2]$ (Th–C(Cp) 0.284(3) nm, Th–C(Me) 0.257(1) nm, Th–P 0.3146(2) nm) and $[Th(Bz)_2(dmpe)(Cp)_2]$ (Th–C(Cp) 0.283(5) nm, Th–C(CH$_2$) 0.2657(9) nm, Th–P 0.3142(2) nm and 0.3237(2) nm) have been structurally characterized.[87,321,322]

Among the variety of mono- and disubstituted cyclopentadienyl ligands the 1,3-bis(trimethylsilyl)-cyclopentadienyl anion has been found particularly useful to stabilize disubstituted organoactinides. The dimeric fluoro complex $[U(\mu\text{-}F)(\mu\text{-}BF_4)\{1,3\text{-}(TMS)_2C_5H_3\}_2]_2$ was prepared by reacting either $[UCl_2\{1,3\text{-}(TMS)_2C_5H_3\}_2]$ or $[U(CH_2R)_2\{1,3\text{-}(TMS)_2C_5H_3\}_2]$ (R = TMS, Ph) with $Ag[BF_4]$ in diethyl ether. The crystal structure determination revealed unsymmetric fluoride bridges (U–F 0.2354(5) and 0.2260(5) nm) and bridging tetrafluoroborate ligands (Figure 80).[1108]

An interesting bis(μ-oxo) complex $[U(\mu\text{-}O)\{1,3\text{-}(TMS)_2C_5H_3\}_2]_2$, was prepared by controlled hydrolysis of $[U(NMe_2)_2\{1,3\text{-}(TMS)_2C_5H_3\}_2]$ (U–C 0.277(3) nm, U–O 0.211(2) nm).[1041] Several related

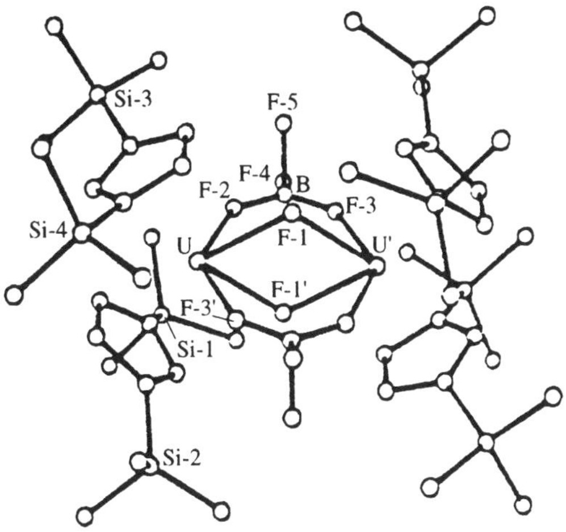

Figure 80   The molecular structure of $[U(\mu\text{-}F)(\mu\text{-}BF_4)\{1,3\text{-}(TMS)_2C_5H_3\}_2]_2$.[212]

cyclic imides containing four-membered U_2N_2 rings have been reported. Asymmetric imido bridges have been found in $[\{U(\mu\text{-}NPh)(MeC_5H_4)_2\}_2]$ (U–C 0.276(2) nm, U–N 0.2156(8) and 0.2315(8) nm), while the *N*-trimethylsilyl derivative $[\{U(\mu\text{-}N\text{-}TMS)(MeC_5H_4)_2\}_2]$ is symmetrically bridged (U–C 0.277(5) nm, U–N 0.2224(7) nm).[1109]

2.3.4.5 [M(Cp)₃X] compounds

(i) Classical co-ligands

As mentioned above, $[UCl(Cp)_3]$ was the first organoactinide complex ever made. The brown crystalline complex was first prepared in 1956 by Reynolds and Wilkinson via treatment of uranium tetrachloride with three equivalents of NaCp (Equation (331)).[2] A more convenient preparation of $[UCl(Cp)_3]$ involves treatment of UCl_4 with thallium cyclopentadienide (Equation (332)). This method gives high yields and has the advantage that the cyclopentadienyl reagent is air-stable.[31] A direct synthesis from uranium metal and cyclopentadienyl transfer reagents in the presence of mercury dichloride has also been described (Equation (333)).[1110] Tris(cyclopentadienyl)thorium chloride is most conveniently prepared by adopting the thallium cyclopentadienide method (Equation (332)).[31]

$$UCl_4 + 3\,NaCp \quad \xrightarrow{\text{THF}} \quad [UClCp_3] + 3\,NaCl \tag{331}$$

$$AnCl_4 + 3\,TlCp \quad \xrightarrow{\text{DME}} \quad [AnClCp_3] + 3\,TlCl \tag{332}$$
$$An = U, Th$$

$$U \quad \xrightarrow[\text{or TlCp/HgCl}_2]{\text{HgCp}_2/\text{HgCl}_2} \quad [UClCp_3] \tag{333}$$

$[UCl(Cp)_3]$ and $[ThCl(Cp)_3]$ have been shown by x-ray powder diffraction to be isomorphous. $[UCl(Cp)_3]$ has been thoroughly studied by various spectroscopic techniques including ^{13}C NMR and solid state ^1H NMR spectral studies.[31,1111,1112] He^I/He^{II} UV photoelectron spectra have been obtained for $[AnX(Cp)_3]$ (X = F, Cl, Br) and $[AnCl(MeC_5H_4)_3]$ (An = Th, U) as well as for $[UBr(MeC_5H_4)_3]$ and $[UBH_4(MeC_5H_4)_3]$.[31,981] The electronic structure of $[UX(Cp)_3]$ complexes (X = F, Cl, Br) has also been investigated by nonrelativistic DV-X_α first-principle MO calculations.[981] Absorption and magnetic circular dichroism spectra were used to determine the crystal field splitting pattern for $[UCl(Cp)_3]$ and $[UCl(C_5D_5)_3]$.[1113] Structurally, $[UCl(Cp)_3]$ has been characterized by an early x-ray diffraction study[31] and a more recent neutron diffraction investigation, both of which confirmed the expected overall structure with the ligands forming a distorted tetrahedron around the central uranium atom. This results in an approximate C_{3v} symmetry with the U–Cl bond along the molecular three-fold axis. The data were not very accurate due to disorder of cyclopentadienyl rings. More precise structural data were derived from an x-ray diffraction study of the ring-substituted derivative chloro-tris(η^5-benzylcyclopentadienyl)uranium(IV) (U–Cl 0.2627(2) nm, U–C 0.2733(1) nm). This compound too exhibited the distorted tetrahedral structure (Cl–U–ring centroid angles between 98.8° and 101.2°), which is common for all compounds of the type $[UX(Cp)_3]$.[31,36] For example, a very similar coordination geometry is displayed in $[UF(Cp)_3]$. This compound is remarkable as organouranium fluorides are very rare. Furthermore the U–F distance in $[UF(Cp)_3]$ (0.2196(12) nm) is shorter than any other previously reported uranium(IV)–fluorine bonds.[31] $[UF(MeC_5H_4)_3]$ has been obtained by treatment of $[U(Bu^t)(MeC_5H_4)_3]$ with various fluorocarbons.[1114] $[UBr(Cp)_3]$[1115] and $[UI(Cp)_3]$[1116] have also been structurally characterized (U–I 0.3059(2) nm, U–C 0.273(4) nm in the iodide).

It is interesting to note that $[UCl(Cp)_3]$ does not react with iron dichloride to yield ferrocene[2] as do the tris(cyclopentadienyl)lanthanide complexes. $[UCl(Cp)_3]$ is mildly air-sensitive.[31,1083] This different behavior suggests a higher degree of covalency in the metal–ligand bonding in the case of uranium. $[UCl(Cp)_3]$ can be reduced electrochemically with Na/Hg in the presence of $[N(Bu^n)_4][PF_6]$ in THF solution to afford the anionic uranate(III) $[N(Bu^n)_4][UCl(Cp)_3]$.[1117,1118] The reactive chlorine atom in $[UCl(Cp)_3]$ allows the preparation of a great variety of other $[UX(Cp)_3]$ derivatives by simple metathetical reactions. Despite its sensitivity to oxygen, the brown $[UCl(Cp)_3]$ slowly dissolves with a green coloration in deoxygenated water, thereby undergoing ionization to form the aquo cation $[U(H_2O)_2(Cp)_3]^+$.[1119] Salts containing the aquo cation can be precipitated by adding various anions to the resulting green solutions. By now the derivative chemistry of $[UCl(Cp)_3]$ is well investigated and $[UCl(Cp)_3]$ remains one of the most important starting materials in organoactinide chemistry. Typical preparations of other $[UX(Cp)_3]$ derivatives starting from either $[UCl(Cp)_3]$ or $[U(Cp)_4]$ are shown in Equations (334)–(336) and Scheme 64.[31,1120]

$$[UClCp_3] + X^- \longrightarrow [UXCp_3] + Cl^- \tag{334}$$

$$X = F, Br, I, OCN, SCN, [BH_4], [BPh_3CN], OR, C(CN)_3$$

$$2\,[UClCp_3] + [M(CN)_4]^{2-} \longrightarrow [UCp_3]_2[M(CN)_4] + 2\,Cl^- \tag{335}$$

$$M = Ni, Pt$$

$$[UCp_4] + [NH_4]X \longrightarrow [UXCp_3] + NH_3 + C_5H_6 \tag{336}$$

$$X = F, Cl, CN, NO_3, ClO_4, ReO_4, BPh_4, 1/2\,SO_4, 1/2\,C_2O_4$$

$$[UClCp_3] + AlCl_3 \longrightarrow [U(\mu\text{-}Cl)_2AlCl_2Cp_3] \xrightarrow{\text{THF}} [U(\mu\text{-}Cl)_2AlCl_2(THF)Cp_3]$$

Scheme 64

A series of azido complexes containing the trimethylsilylcyclopentadienyl ligand has been investigated. The tri- and tetravalent uranium derivatives include the compounds $[UN_3(TMS\text{-}C_5H_4)_3]$, $[\{U(TMS\text{-}C_5H_4)_3\}_2(\mu\text{-}N_3)]$, $[\{U(TMS\text{-}C_5H_4)_3\}_2(\mu\text{-}N_3)][BPh_4]$, $[Na(18\text{-crown-6})][UN_3(TMS\text{-}C_5H_4)_3]$, and $[Na(18\text{-crown-6})][\{U(TMS\text{-}C_5H_4)_3\}_2(\mu\text{-}N_3)]$.[1121]

While the parent hydroxide $[UOH(Cp)_3]$ appears to be ill-defined, the well-characterized hydroxo complexes $[UOH(RC_5H_4)_3]$ (R = Bu^t, TMS) have been prepared either by hydrolysis of the corresponding hydrides or by treatment of the cationic complexes $[U(RC_5H_4)_3][BPh_4]$ with sodium hydroxide (Equations (337) and (338)). The latter method was found to be more reliable.[1122]

$$[UH(RC_5H_4)_3] + H_2O \longrightarrow [UOH(RC_5H_4)_3] + H_2 \tag{337}$$
$$R = Bu^t, TMS$$

$$[U(RC_5H_4)_3][BPh_4] + NaOH \longrightarrow [UOH(RC_5H_4)_3] + Na[BPh_4] \tag{338}$$
$$R = Bu^t, TMS$$

Reaction of $[UOH(TMS\text{-}C_5H_4)_3]$ with $[UH(TMS\text{-}C_5H_4)_3]$ afforded the μ-oxo derivatives $[\{U(TMS\text{-}C_5H_4)_3\}_2(\mu\text{-}O)]$ and thermolysis of the hydroxo complexes produced the trinuclear oxo-bridged complexes $[\{U(\mu\text{-}O)(RC_5H_4)_2\}_3]$ (R = Bu^t, TMS) (Equation (339) and (340)). The trimethylsilyl-substituted trimer $[\{U(\mu\text{-}O)(TMS\text{-}C_5H_4)_2\}_3]$ was structurally characterized (U–O 0.208(4) nm).[1122]

$$[UOH(TMS\text{-}C_5H_4)_3] + [UH(TMS\text{-}C_5H_4)_3] \longrightarrow [\{U(\mu\text{-}O)(TMS\text{-}C_5H_4)_3\}_2] + H_2 \tag{339}$$

$$3\,[UOH(TMS\text{-}C_5H_4)_3] \longrightarrow [\{U(\mu\text{-}O)(TMS\text{-}C_5H_4)_2\}_3] + 3\,CpTMS \tag{340}$$

Various alkoxides of the type [UOR(Cp)$_3$] have been prepared by metathetical reactions starting from either [UCl(Cp)$_3$], [UNEt$_2$(Cp)$_3$], or [UPPh$_2$(Cp)$_3$] (Scheme 65). The ^{1}H-NMR spectra of several [UOR(Cp)$_3$] derivatives have been studied in detail.[31] Structurally characterized examples are [U{OC(CF$_3$)$_2$CCl$_3$}(Cp)$_3$] (U–O 0.223(1) nm),[1123] [UOPh(Cp)$_3$] (U–O 0.2119(7) nm, U–O–C 159.4(5)°),[1124] and [UOSiPh$_3$(Cp)$_3$] (U–O 0.2135(8) nm, U–O–Si 172.6(6)°).[1125] The short U–O distances and the nearly linear geometry around the oxygen atom have been interpreted in terms of strong π-bonding between uranium and oxygen. Various tris(cyclopentadienyl)thorium compounds containing other halide ligands, tetrahydroborate, or alkoxide ligands have been prepared in a similar manner to the corresponding uranium complexes.[31] [U(OBu^n)(Cp)$_3$] undergoes a migratory insertion reaction with CO$_2$.[1126]

Some tris(cyclopentadienyl)uranium(IV) complexes with additional group 15 ligands have been described. Dialkylamides of the type [AnNR$_2$(Cp)$_3$] can be prepared for example from homoleptic actinide dialkylamides via displacement of amido ligands upon treatment with stoichiometric amounts of cyclopentadiene (Scheme 65).[31,1095] [UNPh$_2$(Cp)$_3$] was structurally characterized (U–C 0.279(3) nm, U–N 0.229(1) nm).[1127] 2,6-Dimethylphenyl isocyanide adds to [UNEt$_2$(Cp)$_3$] to afford the structurally characterized insertion product [U{C(NEt$_2$)=NC$_6$H$_3$Me$_2$-2,6}(Cp)$_3$]. The resulting iminocarbamoyl ligand is η^2-coordinated to uranium (U–C(Cp) 0.2808 nm, U–C(η^2) 0.235(2) nm, U–N(η^2) 0.235(2) nm).[1128]

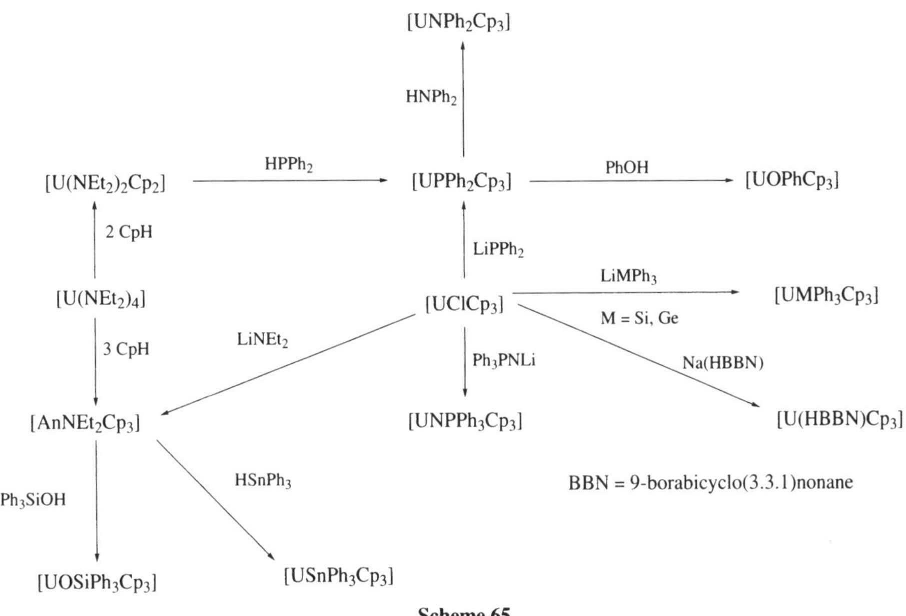

Scheme 65

Metathetical reactions (Scheme 65) have been used to prepare the first organoactinide phosphine imides [AnNPPh$_3$(Cp)$_3$]. The structurally characterized uranium derivative contains a very short uranium–nitrogen bond (U–C 0.278(2) nm, U–N 0.207(2) nm). An–N multiple bonding in [AnNPPh$_3$(Cp)$_3$] was further confirmed by an EHMO calculation on the model compound [UNPH$_3$(Cp)$_3$].[1129]

More recently, uranium(III) cyclopentadienyl complexes have been used as precursors for uranium(IV) and even uranium(V) organometallics. For reactivity studies ring-substituted derivatives such as [U(THF)(MeC$_5$H$_4$)$_3$] or [U(TMS-C$_5$H$_4$)$_3$] are much better suited than the parent tris(cyclopentadienyl)uranium(III) because of their greatly enhanced solubility. For example, oxidation of [U(THF)(MeC$_5$H$_4$)$_3$] with alcohols or thiols is a valuable route to the complexes [UER(MeC$_5$H$_4$)$_3$] (ER = OMe, OPr^i, OPh, SPr^i).[367,1064] The most interesting reactions of [U(THF)(MeC$_5$H$_4$)$_3$] lead to novel binuclear organouranium complexes. Scheme 66 summarizes this derivative chemistry of [U(THF)(MeC$_5$H$_4$)$_3$].

For example, the THF adduct [U(THF)(MeC$_5$H$_4$)$_3$] reacts with COS, Ph$_3$PSe, or (Bu^n)$_3$PTe to give the chalcogen-bridged dimers [{U(MeC$_5$H$_4$)$_3$}$_2$(μ-E)] (E = S, Se, Te).[1055] The symmetrically sulfur-bridged derivative was structurally characterized by x-ray diffraction (U–S 0.260(1) nm, U–S–U 164.9(5)°).

The oxo-bridged complex [{U(TMS-C$_5$H$_4$)$_3$}$_2$(μ-O)] was synthesized by reacting U(TMS-C$_5$H$_4$)$_3$ with CO$_2$ or N$_2$O. The crystal structure determination reveals a linear U–O–U unit (U–O 0.21053(2) nm,

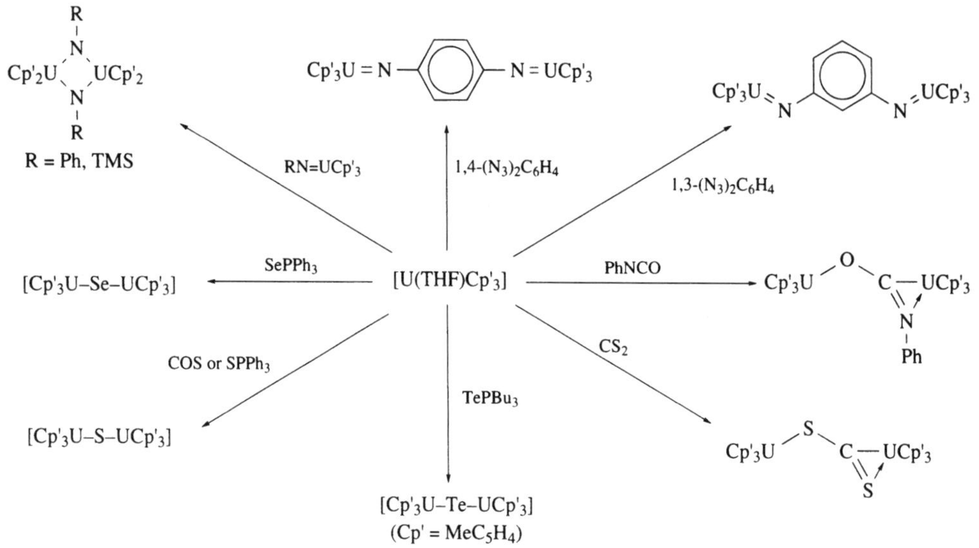

Scheme 66

U–C 0.280 nm). With its linear U–O–U bridge the compound closely resembles the structure of $[\{Sm(Cp^*)_2\}_2(\mu\text{-}O)]$ (cf. Section 2.2.6.3).[1130]

Reactions of $[U(THF)(MeC_5H_4)_3]$ with organic azides and with isocyanides leading to uranium(V) imido and uranium(IV) isocyanato complexes have also been investigated.[1131] The dinuclear complexes $[\{U(MeC_5H_4)_3\}_2(\mu\text{-}1,4\text{-}N_2C_6H_4)]$ and $[\{U(MeC_5H_4)_3\}_2(\mu\text{-}1,3\text{-}N_2C_6H_4)]$ have been synthesized, in which two uranium atoms are bridged by the diimide systems. Only the former complex showed antiferromagnetic coupling between the uranium centers (ordering temperature 20 K).[1132]

Related binuclear uranium complexes have been prepared by reacting $[U(THF)(RC_5H_4)_3]$ (R = Me, TMS) with carbon disulfide according to Equation (341). The x-ray crystal structure determination of the methylcyclopentadienyl derivative revealed an η^1,η^2-coordination of the bridging CS_2 ligand (U–S 0.2792(3) nm, U–C 0.252(2) nm).[1064]

$$2\,[U(THF)(RC_5H_4)_3] + CS_2 \longrightarrow [\{U(RC_5H_4)_3\}_2(\mu\text{-}\eta^1,\eta^2\text{-}CS_2)] + 2\,THF \qquad (341)$$

$$R = Me, TMS$$

Until recently organometallic complexes of high-valent uranium (formal oxidation states +5 or +6) were completely unknown. The $[UClCp_3]^+$ cation can be generated electrochemically but is highly susceptible to disproportionation.[1118] One of the first successful entries into this chemistry involved oxidation of cyclopentadienyluranium(III) complexes with organic azides. This reaction very much resembles the well-known Staudinger reaction in main group chemistry, where tertiary phosphines are converted to phosphine imines by means of organic azides (Equation (342)). Analogously the uranium(V) imides $[UNR(MeC_5H_4)_3]$ (R = Ph, TMS) were prepared from $[U(THF)(MeC_5H_4)_3]$ (Equation (343)).[1131]

$$PR^1_3 + R^2N_3 \longrightarrow R^1_3P{=}NR^2 + N_2 \qquad (342)$$

$$R^1 = \text{alkyl, aryl}$$
$$R^2 = \text{Ph, TMS, etc.}$$

$$(MeC_5H_4)_3U(THF) + RN_3 \longrightarrow (MeC_5H_4)_3UNR + N_2 + THF \qquad (343)$$

$$R = Ph, TMS$$

From an x-ray diffraction study of $[UNPh(MeC_5H_4)_3]$ a considerable degree of multiple bonding was deduced for the uranium–nitrogen bond in these imides (U–N 0.2019(6) nm, U–N–C 167.4(6)°).[1131]

Treatment of $[U(THF)(MeC_5H_4)_3]$ with phenyl isocyanate resulted in the formation of a binuclear complex with the two uranium atoms unsymmetrically bridged by the isocyanato ligand (Equation (344), U–N 0.236(2) nm, U–C 0.242(2) nm, U–O 0.211(1) nm).[1131]

$$2\,[U(THF)(MeC_5H_4)_3] \; + \; PhNCO \longrightarrow [\{U(MeC_5H_4)_3\}_2(\mu\text{-}\eta^2\text{-}PhNCO)] \; + \; 2\,THF \qquad (344)$$

The phosphido derivative $[UPPh_2(Cp)_3]$ was prepared by reacting $[U(NEt_2)_2(Cp)_2]$ with diphenylphosphine (Scheme 65).[1094] The reaction was assumed to proceed via an intermediate $[U(NEt_2)(PPh_2)(Cp)_2]$, which then disproportionates to give the tris(cyclopentadienyl)uranium phosphide. The complex may find some uses as a metallo-ligand as its spectroscopic data indicate the presence of a free electron pair on phosphorus.[1094] $[UPPh_2(Cp)_3]$ reacts with 2,6-dimethylphenyl isocyanide to give the insertion product $[U\{C(PPh_2)=NC_6H_3Me_2\text{-}2,6\}(Cp)_3]$.[1128]

In general, ring-substituted tris(cyclopentadienyl)uranium(IV) complexes are easily available via standard synthetic routes using ring-substituted cyclopentadienyl transfer reagents as illustrated in Equations (345)–(347).[31,1133,1134]

$$UCl_4 \; + \; 3\,Li(RC_5H_4) \xrightarrow{\;THF\;} [UCl(RC_5H_4)_3] \; + \; 3\,LiCl \qquad (345)$$

$$R = Me,\; Pr^i,\; Bu^t,\; TMS,\; PPh_2$$

$$[UCl_3(THF)_2Cp] \; + \; 2\,Li(RC_5H_4) \xrightarrow{\;THF\;} [UClCp(RC_5H_4)_2] \; + \; 2\,LiCl \qquad (346)$$

$$[UCl_3(THF)_2(RC_5H_4)] \; + \; 2\,LiCp \xrightarrow{\;THF\;} [UClCp_2(RC_5H_4)] \; + \; 2\,LiCl \qquad (347)$$

Tris(cyclopentadienyl)actinide complexes have also been described for neptunium. A radiochemical synthesis has been used to prepare $[NpCl(Cp)_3]$ (Equation (348)).[31]

$$[^{239}UXCp_3] \xrightarrow{\;\beta^-\;} [^{239}NpXCp_3] \qquad (348)$$

$$X = F,\; Cl$$

Either potassium cyclopentadienide or bis(cyclopentadienyl)beryllium can be used to convert neptunium tetrachloride to tris(cyclopentadienyl)neptunium(IV) chloride (Equations (349) and (350)). The latter reaction is carried out in the melt, that is, in the absence of solvent.[31]

$$NpCl_4 \; + \; 3\,KCp \xrightarrow{\;THF\;} [NpClCp_3] \; + \; 3\,KCl \qquad (349)$$

$$2\,NpCl_4 \; + \; 3\,BeCp_2 \longrightarrow 2\,[NpClCp_3] \; + \; 3\,BeCl_2 \qquad (350)$$

$[NpCl(Cp)_3]$ served as starting material for $[NpBH_4(Cp)_3]$ and several alkoxides of the type $[NpOR(Cp)_3]$. A number of these organoneptunium complexes have been investigated by ^{237}Np Mössbauer spectroscopy. Among the compounds studied are $[NpCl(Cp)_3]$, $[Np(Bu^n)(Cp)_3]$, $[Np(C_6H_4Et)(Cp)_3]$, $[NpOR(Cp)_3]$ $(R = Pr^i,\; Bu^t,\; CH(CF_3)_2)$, $[NpBH_4(MeC_5H_4)_3]$, $[NpOPr^i(MeC_5H_4)_3]$, $[NpL(Cp)_3]$ $(L = NCS,\; MeCN)$, $[Np(NC_4H_4)(Cp)_3]$, $[Nd(MeCN)_2(Cp)_3][AlCl_4]$ $[\{Np(Cp)_3\}_2(\mu\text{-}C_2)]$, and $[Np(\mu\text{-}Cl)_2AlCl_2(Cp)_3]$. In combination with magnetic susceptibility data the ^{237}Np Mössbauer results have been used to quantify the electron-donating capability and the relative crystal field strengths of the ligands at neptunium.[31,212,1058,1060,1061,1135,1136]

(ii) Hydrocarbyls

Among all ancillary ligands those of the cyclopentadienyl type appear to be best suited to stabilize actinide-carbon σ-bonds. Thus, actinide hydrocarbyls of the type $[AnR(Cp)_3]$ $(An = Th,\; U,\; Np)$ represent a large and well-investigated class of organoactinides. They are straightforwardly prepared by reacting the corresponding cyclopentadienylactinide chlorides with organolithium or Grignard reagents as outlined below (Equations (351)–(354)).[32,36]

$$[ThClCp_3] \xrightarrow[\text{or RMgX}]{\;RLi\;} [ThRCp_3] \qquad (351)$$

$$R = Pr^n,\; Pr^i,\; Bu^n,\; CH_2Bu^t,\; c\text{-}C_6H_{11},\; allyl,\; 2\text{-}cis\text{-}2\text{-butenyl},\; 2\text{-}trans\text{-}2\text{-butenyl}$$

$$[UClCp_3] \quad \xrightarrow{\begin{array}{c} RLi \\ \text{or } RMgX \end{array}} \quad [URCp_3] \tag{352}$$

R = Me, Pr^n, Pr^i, Bu^t, Bu^n, CH_2Bu^t, vinyl, allyl,

2-methylallyl, 2-*cis*-2-butenyl, 2-*trans*-2-butenyl, C≡CH,

C≡CPh, Ph, *p*-MeC_6H_4, Bz, C_6F_5, ferrocenyl

$$2\,[UClCp_3] + p\text{-}LiC_6H_4Li \quad \xrightarrow{\hspace{3cm}} \quad [Cp_3UC_6H_4UCp_3] + 2\,LiCl \tag{353}$$

$$[NpClCp_3] \quad \xrightarrow{RLi} \quad [NpRCp_3] + LiCl \tag{354}$$

R = Bu^n

Uranium alkyls of the type [$UR(Cp)_3$] are also formed upon reaction of [$UCl(Cp)_3$] with RX (X = Cl, Br, I) in the presence of Na/Hg.[1063] The thorium derivatives form colorless crystalline solids while the uranium complexes are mostly yellow-brown or brown in color. Ring-substituted derivatives have been prepared analogously.[1114] The vinyl complex [$U(CH{=}CH_2)(TMS\text{-}C_5H_4)_3$] was structurally characterized (U–C(Cp) 0.2759(4) nm, U–C(vinyl) 0.2436(4) nm).[1137] All tris(cyclopentadienyl)actinide hydrocarbyls are exceedingly air-sensitive. Theoretical calculations have been carried out on [$UMeCp_3$] and the model cationic complex [$UMe(Cp)_3$]$^{2+}$.[1004,1005] According to these studies the actinide–cyclopentadienyl bonding is rather ionic while large overlap populations are found for the U–CH_3 bond. The latter are similar to those found in σ-hydrocarbyl complexes of the *d*-transition metals. Marks *et al.* have thoroughly studied and interpreted the bond disruption enthalpies D(Th–R) for the complexes [$ThR(Cp)_3$] (Equation (355)).[1138–40] These are defined for the gas phase reaction as shown in Equation (356).

$$[ThRCp_3](g.) \quad \xrightarrow{\hspace{3cm}} \quad ThCp_3(g.) + R{\cdot}(g.) \tag{355}$$

$$D(Cp_3Th\text{-}R) = \Delta H^\circ_f(ThCp_3(g.)) + \Delta H^\circ_f(R{\cdot}(g.)) - \Delta H^\circ_f(Cp_3Th\text{-}R(g.)) \tag{356}$$

The process used to determine the bond disruption enthalpies is the alcoholysis of organothorium hydrocarbyls according to Equation (357). Trifluoroethanol, CF_3CH_2OH, is especially suited for these experiments.

$$[ThR^1Cp_3] + R^2OH \quad \xrightarrow{\hspace{3cm}} \quad [ThOR^2Cp_3] + R^1H \tag{357}$$

Alcoholysis enthalpies have been measured by anaerobic batch titration calorimetry.[1138] Iodinolysis has alternatively been used to measure absolute metal ligand bond disruption enthalpies.[1137,1141] The D(Th–R) gas phase values for [$ThR(Cp)_3$] range from 315 kJ mol^{-1} (R = Bz) to 375 kJ mol^{-1} (R = Me), thus showing that the Th–C σ-bonds are quite strong as compared to transition metal hydrocarbyl derivatives.[1142,1143] For corresponding Th/U pairs within a series of complexes the D(Th–R) values are generally ca. 10–30 kJ mol^{-1} higher than D(U–R). Other classes of organoactinide hydrocarbyls for which D(An–R) bond disruption enthalpies have been measured include [$UX(Cp)_3$] (X = Cl, OBu^n, CH_2Pr^i), [$AnR(C_9H_7)_3$], [$AnR_2(Cp^*)_2$], [$AnCl(R)(Cp^*)_2$] (An = Th, U), [$Th(OBu^t)(R)(Cp^*)_2$], and [$UR(TMS\text{-}C_5H_4)_3$] (R = Me, CH_2Bu^t, CH_2TMS, Bz, Ph).[1096,1137–41]

The decomposition of [$ThR(Cp)_3$] and [$UR(Cp)_3$] complexes under thermal or photochemical conditions has been studied in great detail.[32,1050,1144] The results consistently revealed that in the case of uranium free-radical pathways are predominant in these decomposition reactions whereas β-hydride elimination either plays a minor role or in some cases is completely absent. Reaction products may also result from hydrogen-abstraction from cyclopentadienyl ligands or the solvent. Photoinduced β-hydride elimination is the major pathway for the decomposition of thorium hydrocarbyls [$ThR(Cp)_3$]. As expected, tris(cyclopentadienyl)actinide hydrocarbyls of the type [$AnR(Cp)_3$] exhibit a high reactivity toward various reagents. In particular, migratory insertion reactions into the actinide carbon bond have been extensively studied. Suitable substrates for such insertion reactions are carbon monoxide and isocyanides as well as carbon dioxide and sulfur dioxide. Insertion of CO into the thorium–carbon bond in [$ThR(Cp)_3$] in most cases results in the formation of η^2-acyl complexes, which can be described by two resonance structures (Scheme 67, Equation (358)). The reaction is believed to be initiated by the formation of an intermediate metal carbonyl complex, but no simple adduct has ever been isolated.[1145–8]

The relative rates of CO migratory insertion have been correlated with steric and electronic factors as well as the [$ThR(Cp)_3$] bond disruption enthalpies. According to kinetic studies the reactions are first

$$[THRCp_3] + CO \longrightarrow \text{"}[Th(R)(CO)Cp_3]\text{"} \longrightarrow [Th(\eta^2\text{-}COR)Cp_3]$$

$$R = Me,\ Pr^n,\ Bu^n,\ CH_2Bu^t,\ Bz$$

Scheme 67

$$\tag{358}$$

order in $[ThR(Cp)_3]$ and first order in CO. Acceleration of the rate of CO insertion has been achieved by photochemical activation. In some cases the η^2-acyl complexes are unstable ($R = Pr^i$) or cannot be isolated ($R = CH_2TMS$) and undergo a 1,2-rearrangement to give enolate complexes (Equations (359) and (360)).[1145,1146]

$$[Th(\eta^2\text{-}COCHMe_2)Cp_3] \longrightarrow [ThOCH{=}CMe_2Cp_3] \tag{359}$$

$$[ThCH_2TMSCp_3] + CO \longrightarrow [ThOC(TMS){=}CH_2Cp_3] \tag{360}$$

Uranium hydrocarbyls of the type $[UR(Cp)_3]$ also undergo migratory CO insertion reactions (Equations (361) and (362)), but the products are less stable than the thorium analogues and some of the reactions are reversible. Recently an unusual decomposition pathway leading to alkylbenzene molecules has been investigated.[1149] Furthermore, the uranium acyls do not engage in 1,2-rearrangements to form enolate derivatives. CO also inserts into U–N and U–P bonds, although the products have mainly been studied in solution.[1150]

$$[URCp_3] + CO \longrightarrow [U(\eta^2\text{-}COR)Cp_3] \tag{361}$$

$$R = Me,\ Pr^n,\ Bu^n,\ Bu^t$$

$$[UXCp_3] + CO \longrightarrow [U(\eta^2\text{-}COX)Cp_3] \tag{362}$$

$$X = NEt_2,\ PPh_2,\ NCBH_3$$

In one case a uranium acyl complex was obtained by an alternative route. The compound $[U(COMe)Cl(Cp)_3]$ was isolated as an intermediate in the oxidative addition of acetyl chloride to $[U(Cp)_3]$.[1151]

Nitrogen analogues of the η^2-acyl complexes (η^2-iminoalkyls) have been prepared by insertion of isocyanides into the uranium–carbon bond of $[UR(Cp)_3]$ (Equations (363) and (364)). Like the η^2-acyls the isocyanide insertion products can be described by two resonance forms. The rotation of the η^2-$\bar{C}(R^1)NR^2$ ligand is sterically hindered as has been demonstrated by variable-temperature ^{1}H-NMR spectroscopy. In addition, the EI mass spectra of several $[U\{\eta^2\text{-}C(R^1)NR^2\}(Cp)_3]$ complexes have been reported and the fragmentation patterns of these and related compounds (e.g., $[UNEt_2(Cp)_3]$) analyzed.[1152–5]

$$[UR^1Cp_3] + R^2NC \longrightarrow [U\{\eta^2\text{-}C(R^1)NR^2\}Cp_3] \tag{363}$$

$$R^1 = Me,\ R^2 = c\text{-}C_6H_{11}$$

$$R^1 = Bu^n;\ R^2 = Bu^t,\ c\text{-}C_6H_{11},\ 2,6\text{-}Me_2C_6H_3$$

$$\tag{364}$$

The molecular and crystal structure of $[U\{\eta^2\text{-}C(Me)N(c\text{-}C_6H_{11})\}(Cp)_3]$ was reported. The entire U(CCNC) unit is planar and the C–N bond length of 0.125(2) nm corresponds to a double bond (U–N 0.240(2) nm, U–C 0.236(2) nm).[1156]

Heterocumulenes such as carbon dioxide and sulfur dioxide have also been found to insert into the actinide–carbon bond in $[AnR(Cp)_3]$. For example, migratory insertion reactions of $[ThR(Cp)_3]$ with

CO_2 afford the corresponding organothorium carboxylates (Equation (365)). Interestingly, insertion of carbon dioxide proceeds at a significantly slower rate than the analogous reactions with CO.[1145,1146,1157] η^3-Allyl is converted to butyrate upon insertion of CO_2 (Equations (366) and (367)).[1158]

$$[ThRCp_3] + CO_2 \longrightarrow [Th(O_2CR)Cp_3] \tag{365}$$

$$R = Me, Pr^i$$

$$[U(Bu^n)Cp_3] + CO_2 \longrightarrow [U(O_2CBu^n)Cp_3] \tag{366}$$

$$[U(\eta^3\text{-}C_3H_5)Cp_3] + CO_2 \longrightarrow [U(O_2CPr^n)Cp_3] \tag{367}$$

Significantly less stable are the insertion products of $[UR(Cp)_3]$ with sulfur dioxide (Equation (368)). Their formation has been demonstrated spectroscopically but the products readily decompose above room temperature. In the case of R = Me an insoluble decomposition product formulated as $[U_2Me_2(SO_2)_4(Cp)_3]$ has been isolated.[1159]

$$[URCp_3] + SO_2 \longrightarrow [U(O_2SCp_3R)] \tag{368}$$

$$R = Me, Bu^n$$

Lithium alkyls act as reducing agents toward $[UR(Cp)_3]$ derivatives and afford uranate(III) complexes of the type $[UR(Cp)_3]^-$ (Equation (369)).[1068] The product $[Li(crypt)][U(Bu^n)(Cp)_3]$ has been fully characterized by a single-crystal x-ray analysis.[1067,1071,1072]

$$[URCp_3] + LiR \xrightarrow{\text{THF}} [Li(THF)_n][URCp_3] + RH + \text{by-products} \tag{369}$$

$$R = Me, Pr^n, Pr^i, Bu^n, \text{allyl, Bz}$$

These reactions proceed via intermediate formation of anionic $[UMe_2(Cp)_3]^-$ and $[UR_3(Cp)_3]^{2-}$ species which subsequently undergo homolytic U–C bond cleavage to form the uranate(III) ions $[UR(Cp)_3]^-$.[1078,1079]

Oxidation of $[U(Bu^n)(Cp)_3]$ and $[U(Bu^i)(Cp)_3]$ with molecular oxygen resulted in the formation of insoluble uranium oxide products as well as oxidation of the organic ligands.[1083]

A truly remarkable class of organouranium hydrocarbyls are compounds of the type $[UCHPR_3(Cp)_3]$ (27). These uranium-substituted phosphoylides have been developed and investigated by Gilje and Cramer and co-workers. The synthesis involves treatment of tris(cyclopentadienyl)uranium chloride with lithiated phosphoylides (Equation (370)).[1160,1161]

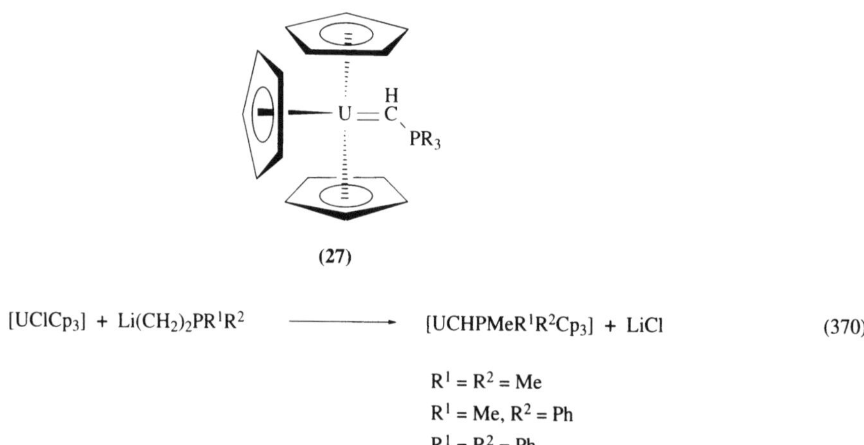

(27)

$$[UClCp_3] + Li(CH_2)_2PR^1R^2 \longrightarrow [UCHPMeR^1R^2Cp_3] + LiCl \tag{370}$$

$$R^1 = R^2 = Me$$
$$R^1 = Me, R^2 = Ph$$
$$R^1 = R^2 = Ph$$

The products are isolated as highly air-sensitive, dark green solids. Complexes of the type $[UCHPR_3(Cp)_3]$ were the first compounds exhibiting uranium–carbon multiple bond character. Alkylidene character of the uranium–carbon bond was supported by the very short U–C distance of 0.229(3) nm and the U–C–P angle of 142(1)° in the crystallographically characterized complex $[UCHPMe_2Ph(Cp)_3]$. Uranium–carbon distances that are considered typical σ-bonds are 0.243(2) nm in

[U(Bu^n)(Cp)$_3$] and 0.233(2) nm in [UC≡CPh(Cp)$_3$].[32] Uranium–carbon multiple bonding was further supported by extended Hückel MO calculations which also revealed significant multiple bond character for the uranium–carbon bond in [UCHPR$_3$(Cp)$_3$] complexes.[1005] [UCHPMe$_3$(Cp)$_3$] was the first organouranium complex to be structurally characterized by a low-temperature neutron diffraction analysis, which also gave strong evidence for U=C multiple bonding (Figure 81, U–C 0.2293(1) nm, U–C–P 141.49(7)°). Unexpectedly no agostic interaction between uranium and the α-hydrogen atom was found.[1162]

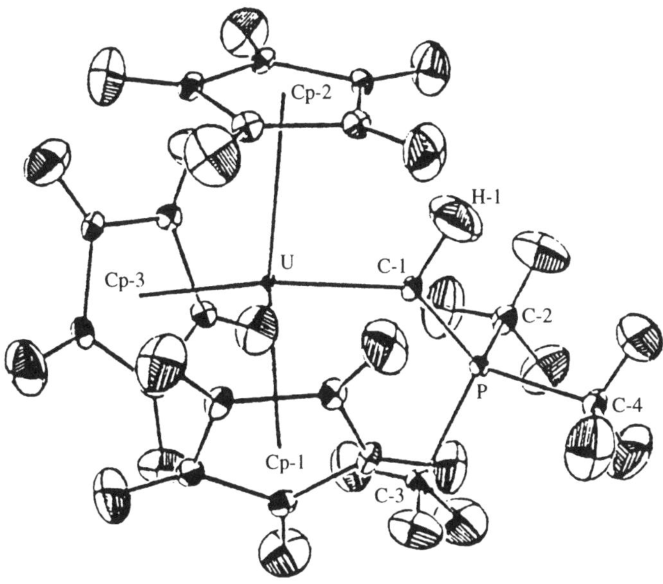

Figure 81 The molecular structure of [UCHPMe$_3$(Cp)$_3$].[1162]

The uranium phosphoylides [UCHPR$_3$(Cp)$_3$] have been termed a third form of metal–carbon doubly bonded species besides the Fischer carbene complexes and the Schrock alkylidene compounds. In [UCHPR$_3$(Cp)$_3$] the electron-withdrawing phosphorus substituents stabilize the formal negative charge on the α-carbon atom.

Various and unusual reactivity patterns have been uncovered for the organouranium phosphoylide complexes, including various unprecedented insertion reactions. Displacement of the phosphoylide unit and formation of [UC≡CPh(Cp)$_3$] was observed when [UCHPMePh$_2$(Cp)$_3$] was treated with phenylacetylene. Similarly the previously known[1094] amide [UNPh$_2$(Cp)$_3$] was obtained by reacting [UCHPMePh$_2$(Cp)$_3$] with diphenylamine.[1127] Reaction of carbon monoxide with the uranium–carbon multiple bond in [UCHPMe$_2$Ph(Cp)$_3$] occurs readily to afford [U(OCCH)PMe$_2$Ph(Cp)$_3$] (Scheme 68).[1163]

$$[\text{UCHPR}_3\text{Cp}_3] + \text{CO} \longrightarrow \quad \cdots \quad \longleftrightarrow \quad \cdots$$

Scheme 68

The x-ray structure determination of [U(OCCH)PMe$_2$Ph(Cp)$_3$] revealed a significant degree of ketene character in the OCCH unit (U–O 0.227(1) nm, U–C$_\alpha$ 0.237(2) nm, C$_\alpha$–O 0.127(3) nm, C$_\alpha$–C$_\beta$ 0.137(3) nm, O–C$_\alpha$–C$_\beta$ 128(2)°).[1163] PhNCO has been reported to insert into the U=C double bond to give the complexes [U{(NPh)(O)CCHP(Ph)(R)(Me)}(Cp)$_3$]. The derivative with R = Me was structurally characterized (U–C 0.284(2) nm, U–N 0.245(1) nm, U–O 0.234(1) nm).[1164] The uranium phosphoylide complexes also undergo interesting reactions with metal-coordinated carbon monoxide. Various transition metal carbonyls have been reacted with complexes of the type [UCHPR$_3$(Cp)$_3$] and the products have been structurally characterized by x-ray diffraction.[1165–8] A typical example is the reaction of [UCHPMePhR(Cp)$_3$] with dimeric [{Fe(CO)$_2$(Cp)}$_2$]. The pentanuclear reaction products formally result from insertion of a metal-coordinated carbon monoxide into the uranium–carbon multiple bond and subsequent coupling of two CO ligands (Equation (371)).[1165]

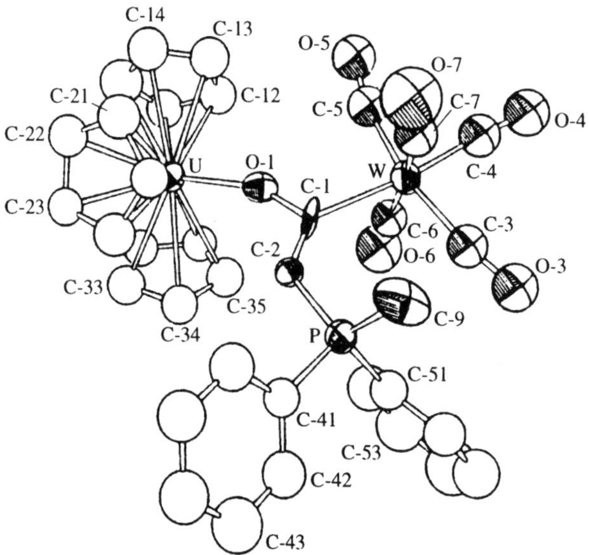

$$[UCHP(Me)(Ph)RCp_3] + [Fe_2(CO)_4Cp_2] \longrightarrow \qquad\qquad\qquad (371)$$

R = Me, Ph

Treatment of $[UCHPMe_2Ph(Cp)_3]$ with $[Mn(CO)_3(Cp)]$ initially affords a similar insertion product. Upon heating, this material eliminates poorly characterized "$UOH(Cp)_3$" to afford the manganese acetylide derivative $[Mn(CO)_2C{\equiv}CPMe_2Ph(Cp)]$ (Scheme 69).[1169]

$$[UCHPMe_2PhCp_3] + [Mn(CO)_3Cp] \longrightarrow \qquad \longrightarrow [Mn(CO)_2C{\equiv}CPMe_2PhCp] + \text{"}UOHCp_3\text{"}$$

Scheme 69

Similar insertion of $[Co(CO)_2(Cp)]$ into the U=C bond of $[UCHP(Me)(Ph)R(Cp)_3]$ (R = Me, Ph) produced the structurally characterized (R = Me) complexes $[UOC\{Co(CO)(Cp)\}CHP(Me)(Ph)R(Cp)_3]$. Heating of these materials resulted in formation of $[Co(CO)P(Me)(Ph)R(Cp)]$.[1168] Tungsten hexacarbonyl also adds to $[UCHP(Me)(Ph)R(Cp)_3]$ to yield the tungstaenolate complexes $[UOC[W(CO)_5]CHP(Me)(Ph)R(Cp)_3]$, which undergo an isomerization reaction upon heating to form the novel compounds $[UOCHCHP(Ph)(R)CH_2W(CO)_5(Cp)_3]$ (Scheme 70). The two products with R = Ph were structurally characterized (Figure 82).[1167]

$$[UCHP(Me)(Ph)RCp_3] + [W(CO)_6] \longrightarrow [UOC[W(CO)_5]CHP(Me)(Ph)RCp_3] \xrightarrow{\Delta} [UOCHCHP(Ph)(R)CH_2W(CO)_5Cp_3]$$

R = Me, Ph

Scheme 70

Figure 82 The molecular structure of $[UOC\{W(CO)_5\}CHP(Me)(Ph)_2(Cp)_3]$.[1167]

Nitriles such as MeCN add across the uranium–carbon multiple bond in $[UCHPR_3(Cp)_3]$ to give imido-uranium complexes of the type $[UNC(Me)CHPR_3(Cp)_3]$ (Equation (372)). The product $[UNC(Me)CHPMePh_2(Cp)_3]$ has been structurally characterized by x-ray diffraction. A short uranium–nitrogen distance of 0.206(1) nm (cf. U–N 0.229(1) nm in $[UNPh_2(Cp)_3]$ and a U–N–C bond angle of 163(1)° are consistent with significant uranium–nitrogen multiple bond character as well as a highly delocalized π-system in this compound.[1170]

$$[UCHPMePh_2Cp_3] + MeCN \longrightarrow [UNC(Me)CHPMePh_2Cp_3] \tag{372}$$

Isonitriles also readily insert into the uranium–carbon double bond as exemplified by the reaction of $[UCHPMePh_2(Cp)_3]$ with cyclohexylisonitrile.[1171] The product, $[U\{\eta^2\text{-}CN(C_6H_{11})CHPMePh_2\}(Cp)_3]$, was crystallographically characterized. The structural data indicate significant multiple bond character of the uranium–nitrogen bond (U–N 0.231(2) nm, U–C 0.244(3) nm, C–N 0.139(4) nm).

A few higher homologues of the hydrocarbyl derivatives $[UR(Cp)_3]$ have been prepared. Compounds containing direct U–Si, U–Ge, and U–Sn bond have been synthesized according to Equations (373)–(375) via metathetical reactions using either $[UCl(Cp)_3]$ or $[UNEt_2(Cp)_3]$.[1125,1172,1173]

$$[UClCp_3] + LiSiPh_3 \longrightarrow [USiPh_3Cp_3] + LiCl \tag{373}$$

$$[UClCp_3] + KGePh_3 \longrightarrow [UGePh_3Cp_3] + KCl \tag{374}$$

$$[UNEt_2Cp_3] + HSnPh_3 \longrightarrow [USnPh_3Cp_3] + HNEt_2 \tag{375}$$

In view of the high oxophilicity of both uranium and silicon the synthesis of a U–Si species was a remarkable achievement. Apparently the analogous thorium derivatives have not been reported. The uranium–tin distance in $[USnPh_3(Cp)_3]$ is 0.3166(1) nm.[1173] The reactivity of $[USiPh_3(Cp)_3]$ and $[UGePh_3(Cp)_3]$ toward xylyl isocyanide was investigated. Xylyl isocyanide is inserted into the uranium–element bonds to give the η^2-silyl or η^2-germyl carbamoyl derivatives $[U\{C(EPh_3)=NC_6H_3Me_2\text{-}2,6\}(Cp)_3]$ (E = Si, Ge).[1125,1172] In this respect $[UGePh_3(Cp)_3]$ behaves similarly to other organouranium complexes containing U–C or U–N bonds. Uranium–metal bond disruption enthalphies have recently been measured for the complexes $[UMPh_3(Cp)_3]$ (M = Si, Ge, Sn). These data were indicative for weak heterobimetallic bonding.[1174]

(iii) Hydrides

The parent tris(cyclopentadienyl)uranium hydride $UH(Cp)_3$ cannot be isolated as reduction of $[UCl(Cp)_3]$ with akali metal hydrides invariably leads to $[U(THF)(Cp)_3]$ containing trivalent uranium. This is in agreement with thermochemical data, which indicated that hydrogen elimination should be facile.[1000] The first stable organouranium(IV) hydride became accessible by employing the trimethylsilylcyclopentadienyl anion as ancillary ligand. The *t*-butyl and diphenylphosphino substituted derivatives were made analogously (Equation (376)).[1134,1175]

$$[UCl(RC_5H_4)_3] \xrightarrow{K[BHEt_3]} [UH(RC_5H_4)_3] + KCl \tag{376}$$

$$R = Bu^t, TMS, PPh_2$$

All three hydrides have been prepared as thermally stable materials and fully characterized. The greatly increased stability of these hydrides results from the fact, that the bulky substituted cyclopentadienyl ligands inhibit the bimolecular elimination of H_2. Reduction of the uranium(IV) hydride species with sodium amalgam in the presence of 18-crown-6 leads to the anionic uranate(III) hydrides $[Na(18\text{-crown-}6)][(RC_5H_4)_3UH]$ (R = Bu^t, TMS). The same products have been made by sodium hydride reduction of $[U(RC_5H_4)_3]$.[1175] An interesting difference was observed for the sodium hydride reduction of the parent $[U(THF)(Cp)_3]$, which did not afford the mononuclear anionic species $[UH(Cp)_3]^-$. Instead, the binuclear hydride-bridged anion $[U(\mu\text{-}H)(Cp)_3U(Cp)_3]^-$ was formed (Equation (377)). The sodium salt was structurally characterized (U–C 0.282 nm, U$\cdots$U 0.4403(2) nm).[36]

$$2\,[U(THF)Cp_3] + NaH \xrightarrow{\text{THF}} [Na(THF)_2][U(\mu\text{-}H)Cp_3UCp_3] \tag{377}$$

The uranium(IV) hydride $[UH(C_5H_4TMS)_3]$ undergoes an interesting addition reaction to afford the thermally stable binuclear μ-dioxymethylene complex $[UOCH_2O(TMS\text{-}C_5H_4)_3U(C_5H_4TMS)_3]$. This was the first example of the transformation: $2\,[M]\text{–}H + CO_2 \rightarrow [M]\text{–}OCH_2O\text{–}[M]$. UV photolysis of the dioxymethylene complex afforded $[\{U(C_5H_4TMS)_3\}_2(\mu\text{-}O)]$, while treatment with $[PPh_3H]Br$ quantitatively produced $[UBr(TMS\text{-}C_5H_4)_3]$.[1176]

Tris(cyclopentadienyl)uranium(IV) complexes with borohydride and aluminohydride ligands have received interest for their spectroscopic properties and as useful synthetic precursors. An interesting coordination behavior of substituted borohydride ligands was observed depending on the substituents on boron. The parent [U(BH$_4$)(Cp)$_3$] can be straightforwardly prepared from [UCl(Cp)$_3$] and Na[BH$_4$]. An alternative preparation involves treatment of various [UR(Cp)$_3$] derivatives (R = F, Me, Et, OMe, NEt$_2$, CONEt$_2$, COBu^n) with Lewis-base adducts of BH$_3$ (Equation (378)).[1154]

$$[URCp_3] + L \cdot BH_3 \xrightarrow{\text{toluene}} [U(BH_4)Cp_3] \tag{378}$$

$$L = BH_3, THF, Me_2S$$

[U(BH$_4$)(Cp)$_3$] served as starting material for organo-substituted borohydride derivatives (Equation (379)).[31]

$$[U(BH_4)Cp_3] + BR_3 \longrightarrow [U(BH_3R)Cp_3] + BHR_2 \tag{379}$$

$$R = Et, Ph$$

Infrared and Raman spectroscopic investigations have revealed tridentate coordination of the borohydride ligands in [U(BH$_4$)(Cp)$_3$] and [U(BH$_3$R)(Cp)$_3$], whereas the cyanoborohydride ligand in [U(NCBH$_3$)(Cp)$_3$] is coordinated via the nitrogen atom only. Rapid interchange between bridging and terminal hydrogen atoms in [U(BH$_4$)(Cp)$_3$] was detected by ^{1}H NMR spectroscopy. On the basis of its optical spectra, [U(BH$_4$)(Cp)$_3$] has been demonstrated to have a less distorted tetrahedral geometry than [UCl(Cp)$_3$]. A close similarity between the spectra of [U(BH$_4$)(Cp)$_3$] and [U(Cp)$_4$] led to the conclusion that tridentate tetrahydroborate and cyclopentadienyl have basically the same crystal field strengths.[31]

Sodium amalgam reduction of [U(BH$_4$)(Cp)$_3$] in the presence of 18-crown-6 led to the anionic uranium(III) complex [Na(18-crown-6)][U(BH$_4$)(Cp)$_3$].[1051] An interesting related complex is the structurally characterized compound [U(BBNH)(Cp)$_3$], in which the BBNH$^-$ ligand is attached to uranium via two hydrogen bridges (BBN = 9-borabicyclo(3.3.1)nonane; U–C 0.276(1) nm, U$\cdots$B 0.278(4) nm).[1177]

The easily accessible tetrahydroaluminate derivative [U(AlH$_4$)(Cp)$_3$] was shown to be a useful alternative as precursor for various other complexes containing the U(Cp)$_3$ unit. [U(AlH$_4$)(Cp)$_3$] can be prepared by the reaction of [U(BH$_4$)(Cp)$_3$] with Li[AlH$_4$]. It was proposed that the complex is polymeric with the tetrahydroaluminate ligands bridging pseudotrigonal planar U(Cp)$_3$ units.[1178] Typical reactions of [U(AlH$_4$)(Cp)$_3$] are summarized in Scheme 71.[1179]

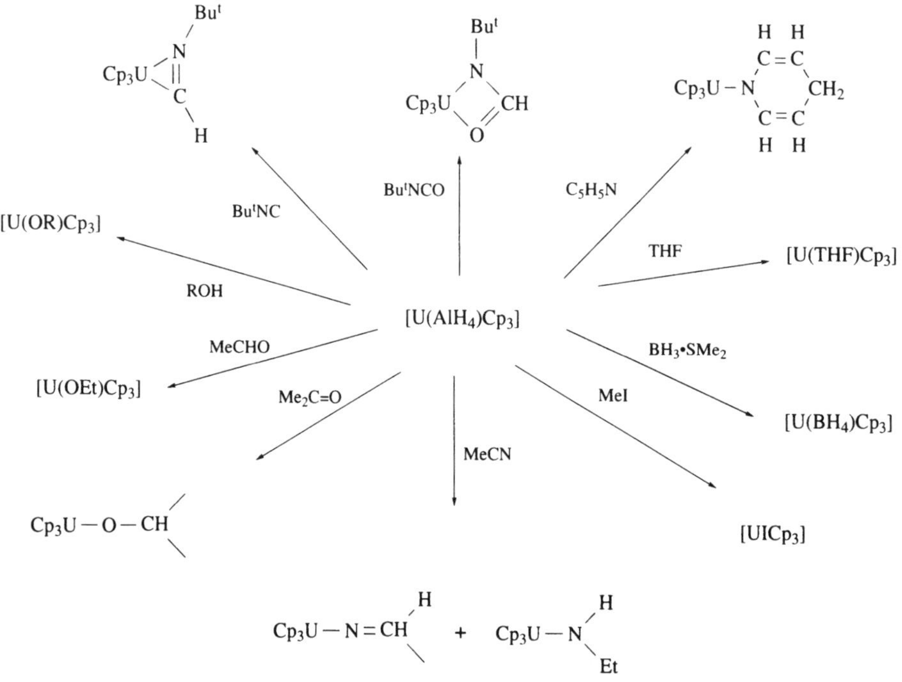

Scheme 71

The reactive $[AlH_4]^-$ ligand in $[U(AlH_4)(Cp)_3]$ is easily displaced by protic reagents such as alcohols, and in several cases the compound behaves like $[UH(Cp)_3]$ (e.g., insertion reactions with acetone, acetonitrile).[1179]

2.3.4.6 [MX(L)(Cp)₃] compounds

Trigonal-bipyramidal complexes of the type $[MX(L)(Cp)_3]$ (X = anionic ligand, L = neutral ligand) form an interesting and well-investigated class of organouranium compounds.[36,1180,1181] One suitable precursor for such complexes is the organouranium thiocyanate $[UNCS(Cp)_3]$, which can be made from $[UCl(Cp)_3]$ and potassium thiocyanate. $[UNCS(Cp)_3]$ adds Lewis bases such as water or acetonitrile to give pseudo-trigonal-bipyramidal adducts of the general formula $[U(NCS)(L)(Cp)_3]$ (Equation (380)). Anionic adducts of the type $[AnX_2(Cp)_3]^-$ (An = U, Np, Pu; X = OCN, SCN) have been prepared analogously by adding cyanates or thiocyanates to the $[An(NCS)(Cp)_3]$ species (Equations (380)–(382)).[1180–2]

$$[UNCSCp_3] + L \longrightarrow [U(NCS)(L)Cp_3] \tag{380}$$
$$L = H_2O, MeCN$$

$$[AnNCSCp_3] + MNCS \longrightarrow M[An(NCS)_2Cp_3] \tag{381}$$

$$[AnNCSCp_3] + MNCO \longrightarrow M[An(NCS)(OCN)Cp_3] \tag{382}$$
$$An = U, Np, Pu$$
$$M = K, K(crypt), AsPh_4$$

Absorption and MCD spectra of a number of $[MX(L)(Cp)_3]$ complexes have been reported.[1183] Several complexes of this type have also been characterized by x-ray methods. Typical for these structures is a pseudotrigonal bipyramidal coordination geometry around the central uranium atom with the three η^5-coordinated cyclopentadienyl ligands occupying the equatorial positions. The "slim" neutral or anionic coligands are arranged in the axial positions. In $[U(NCS)(MeCN)(Cp)_3]^{31}$ the average Cp–U–Cp angle (Cp = ring centroid) is 119.9°, while the average N–U–Cp angles approach the ideal value of 90° (87.8° and 92.2°, respectively). The U–N(NCS) distance was found to be 0.2407(15) nm and the U–N(MeCN) distance is 0.2678(16) nm. Very similar structural features were found for the anionic derivative $[AsPh_4][U(NCS)_2(Cp)_3]$ with U–N distances of 0.246(1) nm and 0.250(1) nm.[1180]

The neutral complex $[U(NCBH_3)(MeCN)(Cp)_3]$ was prepared by recrystallization of the salt $[NBu^n_4][U(NCBH_3)_2(Cp)_3]$ from acetonitrile and its crystal structure was determined.[1184] The anionic $[NCBH_3]$ ligand is nearly linear (172(1)°) with a U–N distance of 0.254(1) nm. As expected, the U–N bond to the neutral acetonitrile ligand is significantly longer (0.270(1) nm). Similar addition of alkyl isocyanides to the tetraphenylborates of either $[U(H_2O)_n(Cp)_3]^+$ or $[U(NCR^2)_2(Cp)_3]^+$ produced the cationic pseudotrigonal bipyramidal uranium compounds $[U(CNR^1)_2(Cp)_3][BPh_4]$ and $[U\text{-}(NCR^2)(CNR^1)Cp_3][BPh_4]$ (R^1 = Me, Pr^n, c-C$_6$H$_{11}$).[1184–7] The formation of cationic tris(cyclopentadienyl)uranium complexes is summarized in Scheme 72.

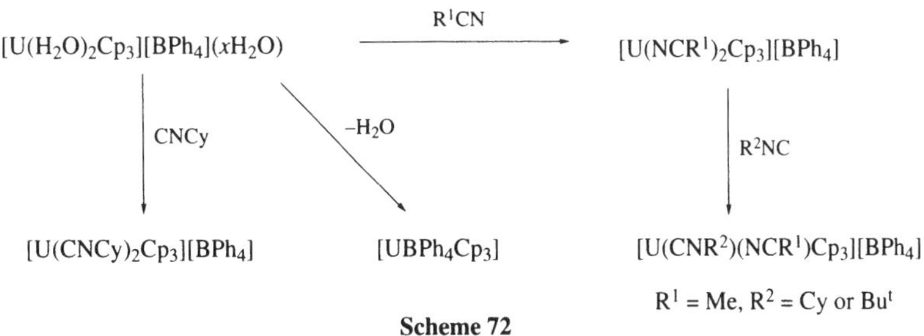

Scheme 72

The complex $[U(MeCN)_2(Cp)_3][BPh_4]$ was structurally characterized.[1186] In addition, the absorption and magnetic circular dichroism spectra have been measured in order to assign the crystal field levels.[1183,1188] The pseudo-salt $[UBPh_4(Cp)_3]$ was prepared by dehydration of the hydrated salt (Scheme 72). The related compound $[UMeBPh_3(Cp)_3]$ was obtained by addition of BPh₃ to $[UMe(Cp)_3]$. For both compounds cation/anion interactions (e.g., η^6-phenyl coordination and U–Me–B bridges) were deduced from spectroscopic data.[1187] Trigonal bipyramidal coordination was also discussed for the reaction products of $[UX(Cp)_3]$ (X = Cl, Me, Bu^n, Ph) with tetracyanoethylene and 7,7,8,8-tetracyanoquinodimethane.[1189]

Serendipitous formation of a cationic pseudo-trigonal bipyramidal uranium complex was observed when [UCl(Cp)$_3$] was treated in acetonitrile solution with butadiene containing traces of oxygen. The uranium–nitrogen distances in the product [U(MeCN)$_2$(Cp)$_3$][UO$_2$Cl$_2$] were found to be 0.262(2) nm and 0.258(2) nm.[1190] The same cation was also found in the structurally characterized salt [U(MeCN)$_2$(Cp)$_3$][ThCl$_4$(MeCN)(Cp)] (U–C 0.273(3) nm, U–N 0.254(4) nm).[1191]

2.3.4.7 *[M(Cp)$_4$] compounds*

Homoleptic tetrakis(cyclopentadienyl)metal complexes of uranium and thorium were first prepared by Fischer *et al.* The original synthesis involved treatment of the metal tetrachlorides with excess potassium cyclopentadienide in benzene solution (Equation (383)). Later this method was successfully employed to the synthesis of tetrakis(cyclopentadienyl)neptunium(IV).[31]

$$AnCl_4 + 4\,KCp \longrightarrow [AnCp_4] + 4\,KCl \qquad (383)$$

$$An = Th,\ U,\ Np$$

Alternatively the tetrafluorides of uranium and thorium have been reacted with molten bis(cyclopentadienyl)magnesium in the absence of solvent to afford the tetrakis(cyclopentadienyl) complexes (Equation (384)).[31]

$$AnF_4 + 2\,MgCp_2 \longrightarrow [AnCp_4] + 2\,MgF_2 \qquad (384)$$

Tetrakis(cyclopentadienyl)uranium has also been reported to be produced upon treatment of U(NEt$_2$)$_4$ with excess cyclopentadiene (Equation (385)).[1095]

$$U(NEt_2)_4 + 4\,C_5H_6 \longrightarrow [UCp_4] + 4\,HNEt_2 \qquad (385)$$

The only other actinide element for which a homoleptic [An(Cp)$_4$] complex is known as protactinium. This compound was made in two steps using a solvent-free approach as shown in Equations (386) and (387).[31]

$$Pa_2O_5 \xrightarrow[\;CCl_4\;]{\;Cl_2\;} PaCl_4 \qquad (386)$$

$$PaCl_4 + 2\,BeCp_2 \longrightarrow [PaCp_4] + 2\,BeCl_2 \qquad (387)$$

Theoretical calculations have been performed on the electronic structure of [An(Cp)$_4$] (An = Th–Np)[988] and the results of these calculations have been successfully correlated with the data obtained from optical spectroscopy and magnetic susceptibility measurements. From these studies it was concluded that the crystal field strength of cyclopentadienyl is comparable with that of a tridentate [BH$_4$]$^-$ ligand but significantly smaller than that of C$_8$H$_8^{2-}$.[31] Alternatively, cyclopentadienyl was found to be a better donor to uranium than Cl$^-$.[1192] Photoelectron spectra have been reported for [Th(Cp)$_4$] and [U(Cp)$_4$].[1193] The room temperature ^{1}H NMR spectrum of [U(Cp)$_4$] exhibits only one singlet at δ −13.1 (in C$_6$D$_6$). This indicated a tetrahedral coordination geometry around uranium with pentahapto coordination of all four cyclopentadienyl ligands. The structure was later confirmed by an x-ray diffraction study. With an average uranium–carbon distance of 0.281(2) nm these bonds are slightly longer than those normally found in other uranium(IV) cyclopentadienyl complexes and thus indicate steric crowding in [U(Cp)$_4$].[31]

Thermochemical combustion data have been used to estimate the mean uranium–ligand dissociation energy in the uranium derivative. At 247 kJ mol^{-1} the value found for [U(Cp)$_4$] was significantly smaller than those determined for both uranocene, [U(C$_8$H$_8$)$_2$], (347 kJ mol^{-1}) and ferrocene (297 kJ mol^{-1}).[31] The dipole moments of [An(Cp)$_4$] (An = Th, U, Np) have been measured in benzene solution. Surprisingly, the dipole moments were not zero although these complexes are known to have a tetrahedral coordination geometry in the solid state.[1194] As compared to the neptunium(III) cyclopentadienyl complex [Np(THF)$_3$(Cp)$_3$] the ^{237}Np Mössbauer spectrum of [Np(Cp)$_4$] was consistent with the assumption of a higher degree of covalency in the metal–ligand bonding. The ring-substituted derivative [Np(MeC$_5$H$_4$)$_4$] was also studied by ^{237}Np Mössbauer spectroscopy.[31,1195,1196]

The derivative chemistry of the tetracyclopentadienyls is very little developed, although various other (cyclopentadienyl)actinide complexes could be accessible via substitution of a cyclopentadienyl ligand.

CCl_4 was found to react with $[U(Cp)_4]$ to give a mixture of $[UCl(Cp)_3]$ and $[U(C_5H_4CCl_3)(Cp)_3]$. Similar reactions of tetrakis(cyclopentadienyl)uranium with MeCl, $CHCl_3$, Bu^nCl, Pr^iCH_2Cl, and Bu^iCl afforded $[UMe(Cp)_3]$, $[U(C_5H_4CHCl_2)(Cp)_3]$, $[U(Bu^n)(Cp)_3]$, $[UCH_2CHMe_2(Cp)_3]$, and $[UCH_2CMe_2Cl(Cp)_3]$, respectively.[1197,1198] In a related study $[U(Cp)_4]$ was reacted with PhBr, BzCl, and Ph_3CCl.[1199] One cyclopentadienyl ligand in $[U(Cp)_4]$ is selectively replaced by halide upon treatment with elemental bromine or iodine. However, this preparation of $[UX(Cp)_3]$ (X = Br, I) offers no advantage over the direct synthesis shown in Equation (334). The same is true for the reaction of $[U(Cp)_4]$ with ammonium salts, which has been used to prepare various $[UX(Cp)_3]$ derivatives. Both methods suffer from the fact that precursors like UX_4 (X = Br, I) or $[UClCp_3]$ are much more readily accessible than $[U(Cp)_4]$. Migratory insertion reactions of $[U(Cp)_4]$ with CO or CO_2 have been reported to yield η^2-acyl or carboxylato complexes, respectively. For example, the binuclear dihapto–acyl bridged complex $[\{U(Cp)_2\}_2(\mu\text{-}\eta^2\text{-}COC_5H_4)_2]$ was prepared by treatment of $[U(Cp)_4]$ with CO in benzene solution at 70 °C.[1200]

Oxidation of $[U(Cp)_4]$ (or $[UCl(Cp)_3]$) with molecular oxygen was reported to produce uranyl derivatives in addition to organic oxidation products such as cyclopentadiene, cyclopentanone, and mixtures of oxygen-containing oligomers. A cyclic voltammetry study of $[U(Cp)_4]$ and $[Np(Cp)_4]$ revealed a single reduction wave, thus indicating a reversible one electron An^{IV}/An^{III} process.[1063,1135,1201]

2.3.5 Modified Cyclopentadienyl Ligands

2.3.5.1 Ring-bridged cyclopentadienyls

Bridging of two cyclopentadienyl rings by an organic spacer group results in chelating ligands which impose interesting properties especially on *f*-element organometallics. In particular, ligand redistribution processes are effectively prevented by linking the cyclopentadienyl rings. Some of these ligands allow the preparation of $[AnCl_2Cp_2]$ derivatives which are inaccessible for the parent cyclopentadienyl systems.[31,32] The bifunctional cyclopentadienyl ligands (28) and (29) have been successfully employed in organoactinide chemistry.

$$X = CH_2, CH_2CH_2CH_2, SiMe_2$$

(28)   (29)

The molecular and crystal structure of $[U_2(\mu_3\text{-}Cl)_2(\mu\text{-}Cl)_3\{CH_2(C_5H_4)_2\}_2Li(THF)_2]$ has been determined by x-ray crystallography. The compound was prepared by reacting uranium tetrachloride with $Li_2\{CH_2(C_5H_4)_2\}$.[31] Systems in which unsubstituted cyclopentadienyl rings are bridged by alkylene groups are often hampered by the low solubility of the products and/or hydrogen abstraction from the cyclopentadienyl rings. Solubility is greatly improved by permethylation of the five-membered rings. Thus the bridged tetramethylcyclopentadienyl anion $[Me_2Si(C_5Me_4)_2]^{2-}$ appears to be most suitable for the preparation of soluble and thermally stable though coordinatively unsaturated and highly reactive organoactinide hydrocarbyls. Soluble thorium complexes containing the peralkylated chelating cyclopentadienyl ligand $[Me_2Si(C_5Me_4)_2]^{2-}$ have been prepared according to Equations (388) and (389).[1202,1203]

$$ThCl_4 + Li_2\{Me_2Si(C_5Me_4)_2\} \longrightarrow Li[ThCl_3\{Me_2Si(C_5Me_4)_2\}] + LiCl \qquad (388)$$

$$ThCl_4 + Li_2\{Me_2Si(C_5Me_4)_2\} \xrightarrow{DME} [ThCl_2\{Me_2Si(C_5Me_4)_2\}] + 2\,LiCl \qquad (389)$$

The compounds $Li[ThCl_3\{Me_2Si(C_5Me_4)_2\}]$ and $[ThCl_2\{Me_2Si(C_5Me_4)_2\}]$ served as starting materials for a series of hydrocarbyl derivatives (Equation (390)).[1202,1203]

$$[ThCl_2\{Me_2Si(C_5Me_4)_2\}] + 2\,RLi \xrightarrow{\text{DME}} [ThR_2\{Me_2Si(C_5Me_4)_2\}] + 2\,LiCl \qquad (390)$$

$$R = Bu^n,\ CH_2Bu^t,\ CH_2TMS,\ Bz,\ Ph$$

The molecular and crystal structure of $[Th(CH_2TMS)_2\{Me_2Si(C_5Me_4)_2\}]$ has been determined by x-ray crystallography (Figure 83). The "pulling back" of the cyclopentadienyl rings causes a significant opening of the coordination sphere around thorium. The ring centroid–Th–ring centroid angle in $[Th(CH_2TMS)_2\{Me_2Si(C_5Me_4)_2\}]$ is 118.4° as compared to values around 138° in the corresponding $[ThR_2(Cp^*)_2]$ complexes. This distortion of the bent metallocene geometry does not significantly affect the structural parameters of the hydrocarbyl ligands, but it has a strong effect on the catalytic activity of the hydride derived from the $Th\{Me_2Si(C_5Me_4)_2\}$ fragment.[1202]

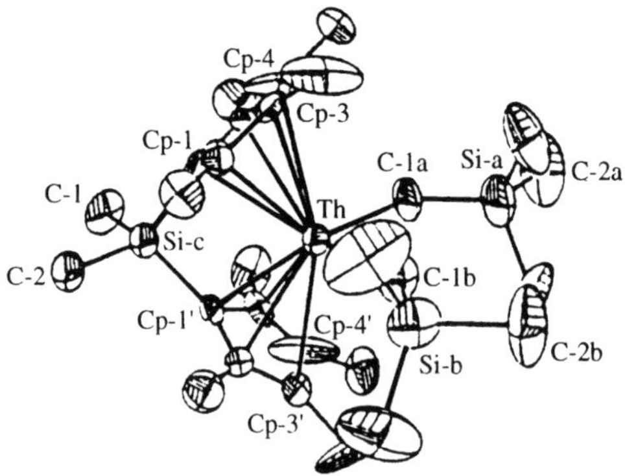

Figure 83　The molecular structure of $[Th(CH_2TMS)_2\{Me_2Si(C_5Me_4)_2\}]$.[1202]

The thorium dialkyls served as precursors for a highly reactive ring-bridged thorium hydride. Rapid formation of the hydride and elimination of RH is observed upon treatment with molecular hydrogen (Equation (391)).[1202,1203]

$$2\,[ThR_2\{Me_2Si(C_5Me_4)_2\}] + 2\,H_2 \longrightarrow [\{ThH_2\{Me_2Si(C_5Me_4)_2\}_2] + 4\,RH \qquad (391)$$

The dimeric molecular structure of $[\{ThH_2\{Me_2Si(C_5Me_4)_2\}\}_2]$ was determined by x-ray crystallography. The close Th–Th distance of 0.3632(2) nm (cf. 0.359 nm in thorium metal) indicated a structure with four hydride bridges instead of two in the case of $[\{ThH_2(Cp^*)_2\}_2]$ (see below). Accordingly, no terminal Th–H stretching bands were observed in the IR spectrum.[1202,1203]

Special ring-bridged cyclopentadienyl ligands have been developed, which comprise an additional donor function in the bridges which can participate in the coordinative saturation of the actinide metal. Ligands of this type, especially those with ether bridges, have been successfully employed in organolanthanide chemistry. A 2,6-pyridyl-bridged bis(cyclopentadienyl) dianion has been prepared and reacted with UCl_4 to afford the corresponding dichloro complex (Equation (392)).[1204]

$$UCl_4 + Na_2[2,6\text{-}C_5H_3N(CH_2C_5H_4)_2] \xrightarrow{\text{THF}} [UCl_2\{2,6\text{-}C_5H_3N(CH_2C_5H_4)_2\}] + 2\,NaCl \qquad (392)$$

A crystal structure determination revealed a five-coordinate uranium center with a strong U–N coordinative bond (U–N 0.262(1) nm, U–Cl 0.2615(3) and 0.2636(3) nm).[1204]

2.3.5.2　*Indenyl compounds*

The indenyl ligand (= benzocyclopentadienyl) has been shown to be a versatile ancillary ligand in organoactinide chemistry. Tris(indenyl)actinide(III) complexes, $[An(C_9H_7)_3]$, have been briefly described for uranium and thorium, although very little information about the synthesis and spectroscopic characterization of these compounds has been released.[31] $[U(C_9H_7)_3]$ was made from UCl_3 and NaC_9H_7. An attempted reduction of $[ThCl(C_9H_7)_3]$ with potassium produced only $[Th(C_9H_7)_4]$.[1205] The molecular structure of tris(indenyl)uranium(III) has been determined by x-ray crystallography. The molecule displays a nearly trigonal coordination with U–C bond lengths falling in the range between 0.275(2) and 0.281(2) nm. A THF adduct of tris(indenyl)uranium has been obtained by $Na[BH_4]$ reduction of $[UCl(C_9H_7)_3]$.[31] Monomeric $[U(TMS\text{-}C_9H_6)_3]$ was prepared in a straightforward manner from UCl_3 and $K(TMS\text{-}C_9H_6)$.[1206]

Monosubstituted indenyl derivatives of the actinide elements are stable only in the presence of additional Lewis bases. For example, THF adducts have been prepared via two different synthetic routes (Equations (393) and (394)). The coordinated THF molecules can be replaced by addition of triphenylphosphine oxide (Equation (395)).[31,36]

$$\text{AnX}_4 + \text{K}(\text{C}_9\text{H}_7) \xrightarrow{\text{THF}} [\text{AnX}_3(\text{THF})_2(\text{C}_9\text{H}_7)] + \text{KX} \qquad (393)$$

An = Th, U, Np
X = Cl, Br

$$[\text{UCl}(\text{C}_9\text{H}_7)_3] + 2\,\text{UCl}_4 \xrightarrow{\text{THF}} 3\,\text{UCl}_3(\text{THF})(\text{C}_9\text{H}_7) \qquad (394)$$

$$[\text{AnX}_3(\text{THF})(\text{C}_9\text{H}_7)] + \text{L} \longrightarrow [\text{AnX}_3\text{L}_2(\text{C}_9\text{H}_7)] + \text{THF} \qquad (395)$$

An = Th, U, Np
X = Cl, Br
L = Ph$_3$PO, Pr^{n}CN, PhCN

Trihapto coordination of indenyl ligands was observed in the saltlike complex [UBr$_2$(MeCN)$_4$(C$_9$H$_7$)]$_2$[UBr$_6$], which was made from UBr$_3$(C$_9$H$_7$)(THF) and acetonitrile. The coordination polyhedron in this compound is a distorted pentagonal bipyramid with four MeCN nitrogen atoms and one bromide ligand in the equatorial plane. The uranium–carbon distances to the coordinated carbon atoms range from 0.263(4) nm to 0.269(4) nm while those to the uncoordinated carbon atoms of the five-membered ring are 0.281(4) nm and 0.282(4) nm.[1207] In contrast, pentahapto coordination was found in the molecular structure of [UBr$_3$(THF)$_2$(C$_9$H$_7$)]. In this case the U–C bond lengths range from 0.268(2) nm to 0.279(2) nm.[1208]

Numerous complexes containing indenyl and several substituted indenyl ligands have been prepared by reacting [AnX$_3$(THF)$_2$(Ind)] with organic nitriles. The complexes prepared included [AnX$_3$(THF)$_2$(Ind)], [AnX$_3$(Pr^{n}CN)$_2$(Ind)], [AnX$_3$(PhCN)$_2$(Ind)], and [AnX$_2$(MeCN)$_4$(Ind)][AnX$_6$] (An = Th, U; X = Cl, Br; Ind = indenyl, 1-ethylindenyl, 1,4,7-trimethylindenyl, heptamethylindenyl).[1209] The oxygen-centered diuranium complex [{UBr(MeCN)$_4$(C$_9$H$_7$)}$_2$(μ-O)][UBr$_6$] was obtained by treatment of [UBr$_3$(THF)$_2$(C$_9$H$_7$)] with acetonitrile in the presence of dry oxygen. The coordination geometry around the two uranium centers is pseudopentagonal bipyramidal with the trihapto coordinated indenyl ligands and the bridging oxygen atom in *trans*-positions (U–O 0.2057(1) nm).[1209]

One of the few bis(indenyl)actinide complexes is [U(BH$_4$)$_2$(C$_9$H$_7$)$_2$]. The molecular structure of this compound was determined showing pentahapto indenyl (U–C 0.262(2)–0.282(2) nm) and tridentate borohydride ligands.[1210]

Straightforward substitution reactions have been used to prepare a series of tris(indenyl)actinide halides (Equation (396)).[31,1211] Ring-substituted indenyl ligands such as 1-ethylindenyl and 1,4,7-trimethylindenyl have also been used to prepare a series of tris(indenyl)actinide halides.[1212,1213] Recently the bulky trimethylsilylindenyl ligand allowed the preparation of a tris(indenyl)thorium hydride derivative (Scheme 73).[1001]

$$\text{AnX}_4 + 3\,\text{K}(\text{C}_9\text{H}_7) \xrightarrow{\text{THF}} [\text{AnX}(\text{C}_9\text{H}_7)_3] + 3\,\text{KX} \qquad (396)$$

An = Th, U
X = Cl, Br, I

$$\text{ThCl}_4 + 3\,\text{K}(\text{TMS-C}_9\text{H}_6) \xrightarrow[-3\,\text{KCl}]{} [\text{ThCl}(\text{TMS-C}_9\text{H}_6)_3] \xrightarrow{\text{K}[\text{BHEt}_3]} [\text{ThH}(\text{TMS-C}_9\text{H}_6)_3]$$

Scheme 73

[UCl(C$_9$H$_7$)$_3$] and [ThCl(C$_9$H$_7$)$_3$] are isomorphous according to x-ray diffraction studies. In these complexes as well as in the corresponding bromides the central uranium atom displays a distorted tetrahedral coordination geometry. The average ring centroid–U–ring centroid angle in [UCl(C$_9$H$_7$)$_3$] is 112° and the average ring centroid–U–Cl angle is 106.7°.[31] Due to stronger interligand nonbonded repulsion the tetrahedral geometry in the tris(indenyl)uranium halides is less distorted than in the corresponding tris(cyclopentadienyl)uranium halides. Trihapto coordinated indenyl ligands were observed in the uranium tris(indenyls) [UBr(C$_9$H$_7$)$_3$] (U–Br 0.2747(2) nm, U–C(η^3-Ind) 0.268(6)–0.291(3) nm) and [UI(C$_9$H$_7$)$_3$] (U–I 0.3041(1) nm, U–C(η^3-Ind) 0.257(7)–0.285(2) nm).[1214,1215]

Pentahapto coordination was found in the structure of [UCl(1,4,7-trimethylindenyl)$_3$].[1216] The molecular and crystal structure determination of [ThCl(1-EtC$_9$H$_6$)$_3$] revealed a trihapto coordination of the three ethylindenyl ligands (Th–Cl 0.2673(3) nm). The Th–C distances to the coordinated carbon atoms range from 0.271(1)nm to 0.286(1) nm and from 0.289(1)nm to 0.298(1) nm to the remaining carbon atoms of the five-membered ring.[1211]

Actinide–ligand bond disruption enthalpies have been determined for [U(C$_9$H$_7$)$_3$], [UI(C$_9$H$_7$)$_3$], [UMe(C$_9$H$_7$)$_3$], as well as [AnR(C$_9$H$_7$)$_3$] and [AnR(1-EtC$_9$H$_6$)$_3$] (An = Th, U; R = Me, Pr^i, CH$_2$TMS, Bz, OCH$_2$CF$_3$). Experimental values were obtained by iodinolysis batch titration calorimetry in toluene solution (U–I 267(3) kJ mol^{-1}, U–Me 196(7) kJ mol^{-1}). For [U(THF)(C$_9$H$_7$)$_3$] in toluene the value is 71(5) kJ mol^{-1}.[1217] The absolute uranium–ligand bond disruption enthalpy for [USEt(TMS-C$_9$H$_6$)$_3$] has been measured by using the oxidative addition reaction shown in Equation (397).[1206]

$$2\,[\text{U(TMS-C}_9\text{H}_6)_3] + \text{EtSSEt} \longrightarrow 2\,[\text{USEt(TMS-C}_9\text{H}_6)_3] \qquad (397)$$

Straightforward metathetical reactions have been used to prepare the corresponding alkoxides and tetrahydroborates from [AnX(C$_9$H$_7$)$_3$]. The only exception is the reaction of [UCl(C$_9$H$_7$)$_3$] with sodium borohydride, during which reduction to uranium(III) and formation of [U(THF)(C$_9$H$_7$)$_3$] occurs.[31]

Tetrakis(indenyl) complexes of some actinide elements can be prepared despite the increased steric demand of the indenyl ligand as compared to cyclopentadienyl. For example, [Th(C$_9$H$_7$)$_4$] has been obtained from a reaction of [ThCl(C$_9$H$_7$)$_3$] with potassium.[1205] Steric congestion is certainly the reason for different bonding modes in [An(Cp)$_4$] and [An(C$_9$H$_7$)$_4$]. The molecular structure of [Th(C$_9$H$_7$)$_4$] was determined (Figure 84) and the structural data were consistent with trihapto coordination of the indenyl ligands. In contrast, all four cyclopentadienyl ligands in [An(Cp)$_4$] are pentahapto coordinated.[1218]

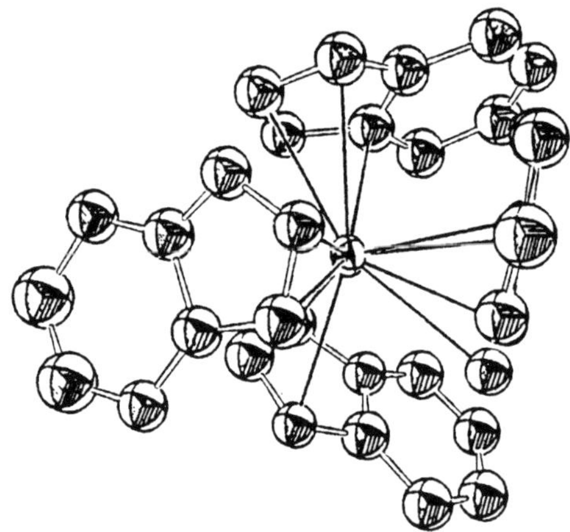

Figure 84 The molecular structure of [Th(C$_9$H$_7$)$_4$].[1218]

2.3.5.3 *Pentamethylcyclopentadienyls*

(i) *[MX$_2$(Cp*)] and MX(Cp*)$_2$ compounds*

Mono(pentamethylcyclopentadienyl) complexes of uranium(III) have become available through the use of [UI$_3$(THF)$_4$] as a valuable new starting material. This reagent is easily accessible by the reaction of clean uranium turnings with elemental iodine in THF.[1219] Treatment of [UI$_3$(THF)$_4$] with one equivalent of K(Cp*) afforded the diiodo complex [UI$_2$(THF)$_3$(Cp*)].[1220] The anionic complex [Na(THF)$_6$][U(BH$_4$)$_3$(Cp*)] was unexpectedly obtained via Cp* transfer when the metallo-ligand [Th(PPh$_2$)$_2$(Cp*)$_2$] was reacted with [U(BH$_4$)$_3$(THF)$_n$]. The compound contains tridentate borohydride ligands (U–C 0.274 nm, U···B 0.261 nm).[1221]

Uranium(III) complexes of the type UX(Cp*)$_2$ form a small but very interesting class of organoactinide compounds. Bis(pentamethylcyclopentadienyl)uranium(III) chloride is best prepared by hydrogenolysis of hydrocarbyl precursors according to Equation (398).[1222]

An x-ray structure determination showed the product to be a trimer having approximate D_{3h} symmetry in which three U(Cp*)$_2$ units are bridged by chloride ligands. The chloro bridges are readily cleaved upon treatment of the trimer with Lewis bases (Equation (399)).

$$3 \, [U(R)ClCp^*_2] + 3/2 \, H_2 \longrightarrow [\{U(\mu\text{-}Cl)Cp^*_2\}_3] + 3 \, RH \tag{398}$$

$$R = alkyl$$

$$[\{U(\mu\text{-}Cl)Cp^*_2\}_3] + 3 \, B \longrightarrow 3 \, [U(Cl)BCp^*_2] \tag{399}$$

$$B = THF, \, py, \, PMe_3$$

The monomeric THF adduct reacts with organic halides under halide abstraction to give the corresponding bis(pentamethylcyclopentadienyl)uranium(IV) dihalides.[1223,1224] A monomeric organouranium(III) amide derivative was obtained when [{U(μ-Cl)(Cp*)$_2$}$_3$] was reacted with sodium bis(trimethylsilyl)amide (Equation (400)).[1222]

$$[\{U(\mu\text{-}Cl)Cp^*_2\}_3] + 3 \, NaN(TMS)_2 \longrightarrow 3 \, [UN(TMS)_2Cp^*_2] + 3 \, NaCl \tag{400}$$

Although the monomeric complex [UN(TMS)$_2$(Cp*)$_2$] appears to be a highly useful synthetic precursor its derivative chemistry has not been further elucidated.

Uranium(III) hydrocarbyls are available via metathetical reactions between [{U(μ-Cl)(Cp*)$_2$}$_3$] and alkyllithium reagents (Equation (401)). However, with the exception of [UCH(TMS)$_2$(Cp*)$_2$] the products are thermally unstable above room temperature.[1222]

$$[\{U(\mu\text{-}Cl)Cp^*_2\}_3] + 3 \, LiR \longrightarrow 3 \, [URCp^*_2] + 3 \, LiCl \tag{401}$$

$$R = CH_2TMS, \, CH(TMS)_2, \, Ph$$

Reports on other organoactinide(III) complexes containing pentamethylcyclopentadienyl ligands are scarce. A reversible one-electron reduction ($E_{1/2} = -0.68$ V in acetonitrile) was observed for [NpCl$_2$(Cp*)$_2$] as well as for [PuCl$_2$(Cp*)$_2$].[1117,1225–7]

(ii) MX$_3$(Cp*) compounds

Only a handful of mono(pentamethylcyclopentadienyl)actinide(IV) complexes have been described in the literature. Such compounds have been reported so far only for uranium and thorium. The parent chloro derivatives have been prepared directly from the metal tetrachlorides by treatment with a stoichiometric amount of MgCl(Cp*) (Equation (402)). This Grignard-type reagent has the advantage that it is a nonreducing cyclopentadienyl precursor. Due to steric unsaturation the resulting complexes can only be isolated as THF adducts.[1228]

$$AnCl_4 + MgClCp^* \xrightarrow{\text{THF}} [AnCl_3(THF)_2Cp^*] + MgCl_2 \tag{402}$$

$$An = Th, \, U$$

A similar synthetic approach has been used to prepare Lewis-base adducts of single-ring complexes containing the ethyltetramethylcyclopentadienyl ligand (Equation (403)).[31]

$$AnCl_4L_2 + Li(C_5Me_4Et) \longrightarrow [AnCl_3L_2(C_5Me_4Et)] + LiCl \tag{403}$$

$$An = Th, \, U$$

$$L = Ph_3PO, \, MeC(O)NMe_2, \, Bu^tC(O)NMe_2$$

According to their spectroscopic data, complexes of the type [AnCl$_3$(THF)$_2$(Cp*)] display a typical pseudooctahedral structure. The three chlorine atoms in [AnCl$_3$(THF)$_2$(Cp*)] have been successfully replaced by allyl and 2-methallyl ligands (Equation (404)).[1203,1229,1230]

$$[AnCl_3(THF)_2Cp^*] + 3 \, RMgX \longrightarrow [AnR_3Cp^*_2] + 3 \, MgClX \tag{404}$$

$$An = Th, \, U, \, R = allyl, \, X = Br$$
$$An = U, \, R = 2\text{-methallyl}, \, X = Cl$$

The x-ray diffraction derived molecular structure of [U(2-methallyl)$_3$(Cp*)] exhibits average U–C distances of 0.266(1) nm (terminal) and 0.280(1) nm (central).[1229]

Information on other mono(pentamethylcyclopentadienyl)actinide hydrocarbyls of the type [AnR$_3$(Cp*)] is scarce. In the case of simple hydrocarbyl ligands the products appear to be thermally unstable. High thermal stability is only encountered in the case of benzylic or chelating ligands (Equation (405)).

$$[AnCl_3(THF)_2Cp^*] + 3\,RLi \longrightarrow [AnR_3Cp^*] + 3\,LiCl + 2\,THF \qquad (405)$$

$$An = Th,\ R = Bz,\ CH_2Bu^t,\ o\text{-}C_6H_4NMe_2$$
$$An = U,\ R = Bz$$

An x-ray crystal structure determination of [Th(Bz)$_3$(Cp*)] revealed multihapto coordination of the benzyl ligands similar to the allyl groups in [An(allyl)$_3$(Cp*)].[1228] On the NMR time-scale the tribenzyls are fluxional exhibiting a rapid η^3-η^1 equilibration. Steric saturation of the metal atom seems to be the main reason for the stability of [Th(CH$_2$Bu^t)$_3$(Cp*)] and [Th(o-C$_6$H$_4$NMe$_2$)$_3$(Cp*)]. The derivative chemistry of the tris(hydrocarbyls) is much less developed than that of the corresponding [AnR$_2$(Cp*)$_2$] complexes. In several cases rapid reactions with alcohols, formaldehyde, carbon monoxide, or hydrogen have been observed, but isolation of pure products proved to be rather difficult. For example, neopentane is eliminated when [Th(CH$_2$Bu^t)$_3$(Cp*)] is treated with molecular hydrogen, but a mixture of organothorium hydrides was isolated which could not be separated.[32]

The unusual bimetallic phosphoylide complex [{U{μ-(CH$_2$)PPh$_2$(CH$_2$)}}$_2$(Cp*)Mg-{CH$_2$PMePh$_2$}$_2$(μ_3-O)(μ-O)(μ-Cl)$_2$] was obtained serendipitously and structurally characterized.[1231]

(iii) [MX$_2$(Cp*)$_2$] compounds

(a) *Classical coligands.* The introduction of the bulky pentamethylcyclopentadienyl ligand into organoactinide chemistry by Marks *et al.* has had a strong impact on the development of this area. Peralkylation of the cyclopentadienyl ligand offers several advantages as compared to the unsubstituted ring. Besides a modification of the electronic properties it is basically the greatly increased steric bulk which makes pentamethylcyclopentadienyl an extremely useful ligand in organo-*f*-element chemistry. This ligand is ideally suited to coordinatively saturate the large lanthanide and actinide cations. Higher solubility and better crystallizing properties are also often encountered. The first actinide pentamethylcyclopentadienyl complexes were reported in 1978.[31,36,1227,1232] Treatment of uranium or thorium tetrachloride with two equivalents of MgCl(Et$_2$O)(Cp*) resulted in selective formation of the disubstitution products [AnCl$_2$(Cp*)$_2$] (Equation (406)). This result was in marked contrast to the chemistry of the parent cyclopentadienyl ligand, where [UCl$_2$(Cp)$_2$] and [ThCl$_2$(Cp)$_2$] are unstable and undergo ligand redistribution reactions. The Grignard reagent MgCl(Et$_2$O)(Cp*) is the reagent of choice as it does not reduce UCl$_4$ to uranium(III) species. Reduction was observed to occur when M(Cp*) (M = Li, Na, K) was reacted with uranium tetrachloride, although this problem can be circumvented by carrying out the reaction in pentane (Equation (407)).[1233] Another suitable reagent with nonreducing properties is (SnBu$^n_3$)C$_5$Me$_4$Et (Equation (408)).[31]

$$AnCl_4 + 2\,MgCl(Et_2O)Cp^* \longrightarrow [AnCl_2Cp^*_2] + 2\,MgCl_2 + 2\,Et_2O \qquad (406)$$
$$An = Th,\ U,\ Np$$

$$AnCl_4 + 2\,KCp^* \xrightarrow{\text{pentane}} [AnCl_2Cp^*_2] + 2\,KCl \qquad (407)$$
$$An = Th,\ U$$

$$UCl_4 + (SnBu^n_3)C_5Me_4Et \longrightarrow [UCl_2(C_5Me_4Et)_2] + 2\,[Bu^n_3SnCl] \qquad (408)$$

All products of the type [AnX$_2$(Cp*)$_2$] (X = halide, alkoxide, thiolate, amide, hydrocarbyl, etc.) are monomeric, pseudotetrahedral "bent-sandwich" complexes. This has been shown unambiguously by their spectroscopic data as well as numerous single-crystal x-ray analyses. The isostructural complexes [AnCl$_2$(Cp*)$_2$] (An = Th, U) have recently been structurally characterized.[1234] Despite the considerable steric bulk imposed by the two pentamethylcyclopentadienyl ligands, [UCl$_2$(Cp*)$_2$] is able to add an additional neutral ligand such as pyrazole. Due to the coordination of pyrazole the Cl–U–Cl is widened to 148.29(8)° whereas the ring centroid-U-ring centroid angle is 137.1°.[1235] Another structurally characterized complex of this type is [UCl$_2$(HNPPh$_3$)(Cp*)$_2$] (U–C 0.2773(1) nm, U–N 0.243(1) nm), in which U–H agostic interactions and N–H$\cdots$Cl hydrogen bonding were detected.[1236]

Reversible one-electron reduction and formation of $[AnCl_2(Cp^*)_2]^-$ occurs when $[AnCl_2(Cp^*)_2]$ (An = U, Np) is reduced electrochemically or by pulse radiolysis. As expected, $[ThCl_2(Cp^*)_2]$ was not reduced even at potentials as low as -2.7 V vs. SCE.[1117,1225-7] A rich derivative chemistry has been developed using $[AnCl_2(Cp^*)_2]$ as precursors. One or both chloride ligands are readily susceptible to various metathetical reactions as shown in Equations (409)–(413).[1237-9]

$$[AnCl_2Cp^*_2] + 2\,NaOR \longrightarrow [An(OR)_2Cp^*_2] + 2\,NaCl \tag{409}$$
$$M = Th,\ U$$
$$R = Me$$

$$[AnCl_2Cp^*_2] + LiNR_2 \longrightarrow [AnCl(NR_2)Cp^*_2] + LiCl \tag{410}$$
$$M = Th,\ U$$
$$R = Me,\ Et$$

$$[AnCl_2Cp^*_2] + 2\,LiNR_2 \longrightarrow [An(NR_2)_2Cp^*_2] + 2\,LiCl \tag{411}$$
$$M = Th,\ U$$
$$R = Me$$

$$[UCl_2Cp^*_2] + Na(pz) \longrightarrow [UCl(\eta^2\text{-}pz)Cp^*_2] + NaCl \tag{412}$$

$$[UCl_2Cp^*_2] + 2\,Na(pz) \longrightarrow [U(\eta^2\text{-}pz)_2Cp^*_2] + 2\,NaCl \tag{413}$$
$$pz = \text{pyrazolyl}$$

Organoactinide amides of the type $[AnCl(NR_2)(Cp^*)_2]$ and $[An(NR_2)_2(Cp^*)_2]$ have been found to readily insert carbon monoxide into the metal–nitrogen bond to give η^2-carbamoyl complexes (Equations (414) and (415)). In analogy to the η^2-acyl derivatives (see below) these compounds can be described by two resonance forms but in the case of the carbamoyls the carbenelike structure is less important.[1237]

$$[AnCl(NR_2)_2Cp^*_2] + CO \longrightarrow [AnCl(\eta^2\text{-}CONR_2)Cp^*_2] \tag{414}$$

$$[An(NR_2)_2Cp^*_2] + 2\,CO \longrightarrow [An(\eta^2\text{-}CONR_2)_2Cp^*_2] \tag{415}$$

Due to extensive electron delocalization, the entire carbamoyl ligands in $[U(\eta^2\text{-}CONMe_2)_2(Cp^*)_2]$ are essentially coplanar.[1237] Similarly, isocyanides can be inserted into the U–N bond of $[UCl(NEt_2)(Cp^*)_2]$ to give the corresponding iminoalkylamido products.[1103,1104]

Various $[AnX_2(Cp^*)_2]$ complexes containing alkoxide, thiolato, and amido ligands are also favorably prepared by protolysis of the corresponding hydrocarbyl precursors (see below).[1232,1238,1239]

The first organoactinide complex containing a polysulfide ligand was prepared by Ryan *et al.* Dilithium pentasulfide was generated from elemental sulfur and $Li[BHEt_3]$ in THF solution and treated with $[ThCl_2(Cp^*)_2]$ to give the polysulfide complex $[ThS_5(Cp^*)_2]$ (Equation (416)). The x-ray structure determination revealed an unusual twist–boat conformation of the six-membered ThS_5 ring as a result of maximizing the coordination number around the central thorium atom (Figure 85).[1240]

$$[ThCl_2Cp^*_2] + Li_2S_5 \longrightarrow [ThS_5Cp^*_2] + 2\,LiCl \tag{416}$$

The first organoactinide phosphides were prepared by reacting $[ThCl_2(Cp^*)_2]$ with lithium diorganophosphides (Equation (417)). The diphenylphosphido derivative was structurally characterized by x-ray crystallography and showed no evidence for Th–P multiple bonding (Th–P 0.287(2) nm).[1241] Bis(trimethylsilyl)phosphido complexes have been prepared with both thorium and uranium (Equation (418)).[1242]

$$[ThCl_2Cp^*_2] + 2\,LiPR_2 \longrightarrow [Th(PR_2)_2Cp^*_2] + 2\,LiCl \tag{417}$$

$$[An(X)ClCp^*_2] + KP(TMS)_2 \xrightarrow{\ \ THF\ \ } [An(X)P(TMS)_2Cp^*_2] + KCl \tag{418}$$
$$An = Th,\ U$$
$$X = Cl,\ Me$$

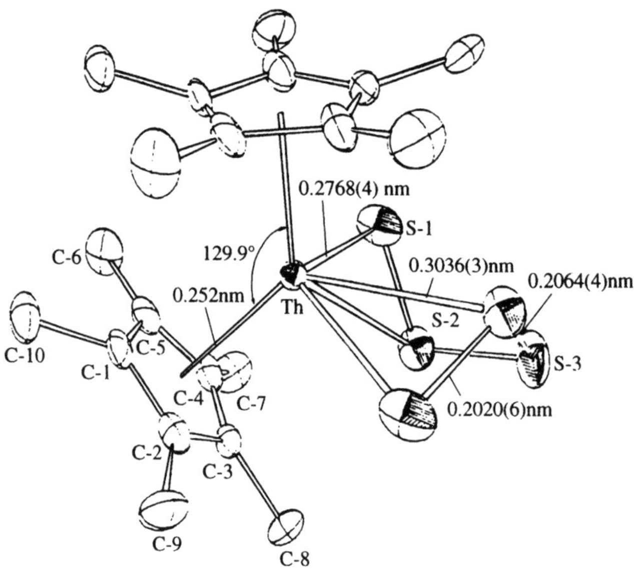

Figure 85 The molecular structure of $[ThS_5(Cp^*)_2]$.[1240]

Thermolytic decomposition of the silylphosphides $[An(Me)P(TMS)_2Cp^*_2]$ resulted in clean formation of novel metallacyclic complexes (Equation (419)).[1242]

$$[An(Me)P(TMS)_2Cp^*_2] \xrightarrow[120\ °C]{C_6D_6} Cp^*_2An\overset{\displaystyle TMS}{\underset{\displaystyle CH_2}{\diamond}}SiMe_2 + CH_4 \qquad (419)$$

An = Th, U

The complexes $[U(Cl)P(TMS)_2(Cp^*)_2]$, $[Th(Me)P(TMS)_2(Cp^*)_2]$ and the uranium metallacycle have been characterized by single crystal x-ray structural analyses.[1242]

Like several *d*-transition metals thorium has been found to form unusual complexes with polyphosphide ligands. Thermal reaction of $[Th(\eta^4\text{-}C_4H_6)\{C_5H_3(Bu^t)_2\}_2]$ with P_4 in toluene afforded $[Th(\mu\text{-}\eta^3{:}\eta^3\text{-}P_6)\{C_5H_3(Bu^t)_2\}_2Th\{C_5H_3(Bu^t)_2\}_2]$ while $[Th(\mu\text{-}\eta^3\text{-}P_3)\{C_5H_3(Bu^t)_2\}_2ThCl\{C_5H_3(Bu^t)_2\}_2]$ was isolated when the same reaction was carried out in the presence of $MgCl_2$. Both complexes were structurally characterized (Figure 86).[1243]

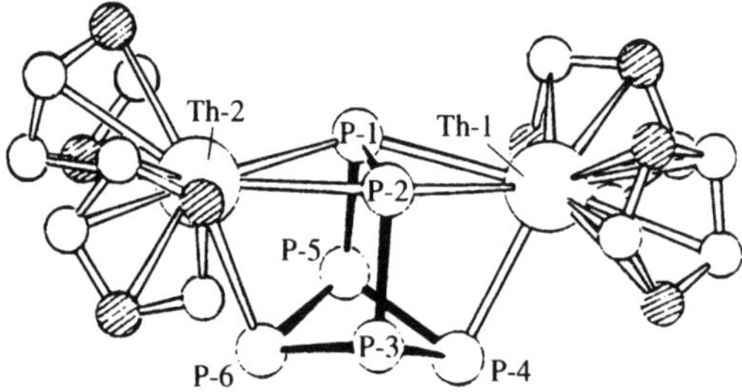

Figure 86 The molecular structure of $[Th\{C_5H_3(Bu^t)_2\}_2(\mu\text{-}\eta^3{:}\eta^3\text{-}P_6)Th\{C_5H_3(Bu^t)_2\}_2]$.[1243]

An elegant synthetic route has been designed to prepare the first organouranium(VI) complex. The structurally characterized bis(organoimido) complex $[U(NPh)_2(Cp^*)_2]$ was made in two steps (Equations (420) and (421)). The ring centroid–U–ring centroid angle is 141.9°.[1233]

$$[U(Me)ClCp^*_2] + LiNHPh + TMEDA \xrightarrow{Et_2O} [\{Li(TMEDA)\}\{U(NPh)ClCp^*_2\}] + CH_4 \qquad (420)$$

An alternative pathway to this exciting uranium(VI) complex involves treatment of $[U(Me)Cl(Cp^*)_2]$ with 1-lithio-1,2-diphenylhydrazine (Equation (422)).

$$[\{Li(TMEDA)\}\{U(NPh)ClCp^*_2\}] + PhN_3 \xrightarrow[-LiCl]{Et_2O} [U(NPh)_2Cp^*_2] + N_2 + TMEDA \quad (421)$$

$$[U(Me)ClCp^*_2] + LiN(Ph)NHPh \xrightarrow{Et_2O} [U(NPh)_2Cp^*_2] + LiCl + CH_4 \quad (422)$$

Oxidative atom-transfer chemistry was used to synthesize the first high-oxidation-state organouranium complexes containing a terminal oxo function (Equations (423)–(425)).[1244]

$$[UI(THF)Cp^*_2] + KOC_6H_3(Pr^i)_2\text{-}2,6 \xrightarrow{THF} [U\{OC_6H_3(Pr^i)_2\text{-}2,6\}(THF)Cp^*_2] + KI \quad (423)$$

$$[UMe_2Cp^*_2] + H_2NC_6H_3(Pr^i)_2\text{-}2,6 \xrightarrow{THF} [U\{NC_6H_3(Pr^i)_2\text{-}2,6\}(THF)Cp^*_2] + 2\,CH_4 \quad (424)$$

$$[U(EC_6H_3(Pr^i)_2\text{-}2,6)(THF)Cp^*_2] + C_5H_5NO \xrightarrow{THF} [U\{EC_6H_3(Pr^i)_2\text{-}2,6\}(O)Cp^*_2] + C_5H_5N \quad (425)$$

$$E = O, N$$
$$C_5H_5NO = \text{pyridine-N-oxide}$$

Both terminal oxo complexes $[U\{EC_6H_3(Pr^i)_2\text{-}2,6\}(O)(Cp^*)_2]$ (E = O, N) have been crystallographically characterized. The uranium oxo bond distances are 0.1859(6) nm (E = O) and 0.1844(4) nm (E = N), respectively.[1244]

(b) Hydrocarbyls. Probably no other ancillary ligand is better suited to stabilize monomeric actinide hydrocarbyls than pentamethylcyclopentadienyl. Thus, an important class of actinide hydrocarbyls of the type $[AnR_2(Cp^*)_2]$ has been prepared starting from the bis(pentamethylcyclopentadienyl)uranium and thorium dichlorides. These actinide hydrocarbyl complexes are of enormous interest due to their very high reactivity toward various reagents. Even C–H activation is a common phenomenon in the chemistry of uranium and thorium metallocenes containing alkyl ligands. The development of such complexes began with a series of papers by Marks and co-workers.[1232,1245] A large number of $[AnR_2(Cp^*)_2]$ derivatives has now been isolated. Today's research activities are mainly directed toward reactivity and catalytic studies. Other important topics of current interest are thermochemical and theoretical investigations. These are aimed to increase our understanding of this important class of compounds and eventually predict their reactivity and catalytic behavior.

The preparation of $[AnR_2(Cp^*)_2]$ derivatives is straightforward and involves treatment of the dichloride precursors with alkyllithium reagents (Equations (426) and (427)).[1232,1245]

$$[AnCl_2Cp^*_2] + LiR \longrightarrow [An(Cl)RCp^*_2] + LiCl \quad (426)$$
$$An = Th, U$$
$$R = Me, CH_2TMS, CH_2Bu^t, CH(TMS)_2, Ph,$$
$$Bz$$

$$[AnCl_2Cp^*_2] + 2\,LiR \longrightarrow [AnR_2Cp^*_2] + 2\,LiCl \quad (427)$$
$$An = Th, U, R = Me, CH_2TMS, CH_2Bu^t,$$
$$CH_2CMe_2Et, Ph, Bz, CH_2CMe_2Ph,$$
$$CH_2SiMe_2Ph$$

Marks *et al.* have pointed out that thermochemical data are a unique guide to understanding reactivity patterns in organoactinide chemistry.[63] Detailed thermochemical studies have been carried out on actinide hydrocarbyls of the type $[ThR_2(Cp^*)_2]$ and $[UR_2(Cp^*)_2]$ in order to determine actinide–ligand bond disruption enthalpies in these complexes. In general, it was concluded that β-hydride elimination is less favorable for the thorium hydrocarbyls than for first- and second-row *d*-transition metal hydrocarbyls. While the Th–R bonds in $[ThR_2(Cp^*)_2]$ appear to be rather strong, the values are lower than those found for $[ThR(Cp)_3]$. For the uranium derivatives the U–R bonds were found to be generally weaker than the corresponding Th–R bonds. This was attributed to the fact that uranium(IV) is much more easily reduced to the +3 oxidation state than thorium.[1246,1247] Ancillary alkoxide ligands appear to strengthen the An–R bond.[1136] The complexes $[MMe_2(Cp^*)_2]$ and their precursors $[MCl_2(Cp^*)_2]$ (M = Zr, Th, U) have also been studied by gas-phase He^I/He^{II} photoelectron spectroscopy. Surprisingly, the bonding was found to be quite similar for all three metals although in the case of thorium and uranium some participation of the 5*f*-orbitals in the bonding was deduced.[1246]

Agostic interactions play an important role in the stabilization of highly reactive uranium and thorium alkyls. The molecular structure of the compound $[Th(CH_2Bu^t)_2(Cp^*)_2]$ has been determined by neutron diffraction (Figure 87, Th–C_a(1) 0.2546(4) nm, Th–C_b(1) 0.2478(4) nm)[1245] These results in combination with theoretical calculations have revealed that an agostic interaction between thorium and the α-CH bond is responsible for the distortion of the alkyl groups and the large Th–C–C angles.[1247,1248] This structural investigation has revealed severe steric crowding in $[ThR_2(Cp^*)_2]$ derivatives containing bulky alkyl ligands as well as severe distortions of the alkyl groups. This appears to be the main reason why such complexes readily undergo cyclometallation reactions to form thoracyclobutanes (see below).[1245,1249-51]

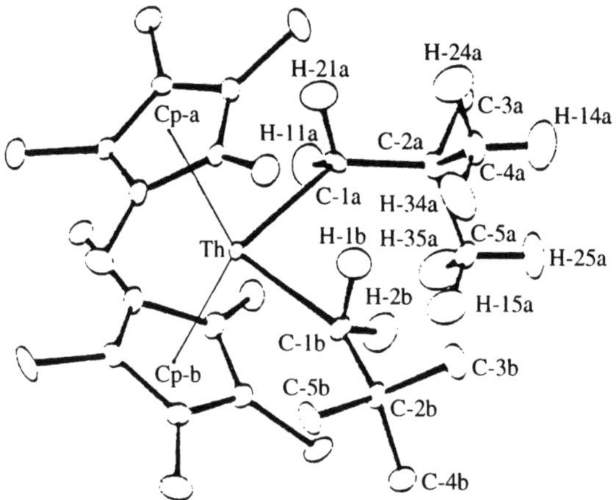

Figure 87   The molecular structure of $[Th(CH_2Bu^t)_2(Cp^*)_2]$.[1245,1249,1250]

X-ray crystal structure determinations have also been carried out on $[Th(CH_2Bu^t)_2(Cp^*)_2]$,[1245] $[Th(CH_2Bu^t)(CH_2TMS)(Cp^*)_2]$,[1245] and $[Th(CH_2TMS)_2(Cp^*)_2]$, all reflecting steric congestion. All bis(pentamethylcyclopentadienyl)actinide(IV) hydrocarbyls are exceedingly reactive and readily undergo protolytic cleavage of the actinide–carbon bonds as well as other transformations. Protolysis upon treatment with alcohols, thiols, or secondary amines represents an elegant approach to "bent-sandwich" $[AnX_2(Cp^*)_2]$ complexes containing alkoxide, thiolato, and amido ligands (Equations (428)–(431)).[1232,1238,1239]

$$[ThR^1_2Cp^*_2] + R^2OH \longrightarrow [Th(R^1)OR^2Cp^*_2] + R^1H \qquad (428)$$

$$R^1 = \text{hydrocarbyl}$$
$$R^2 = CH(Bu^t)_2$$

$$[ThR^1_2Cp^*_2] + 2 R^2SH \longrightarrow [Th(SR^2)_2Cp^*_2] + 2 R^1H \qquad (429)$$

$$R^1 = \text{hydrocarbyl}$$
$$R^2 = Pr^n$$

$$[ThR^1_2Cp^*_2] + 2 R^2OH \longrightarrow [Th(OR^2)_2Cp^*_2] + 2 R^1H \qquad (430)$$

$$R^1, R^2 = \text{hydrocarbyl}$$

$$[AnR^1_2Cp^*_2] + 2 R^2_2NH \longrightarrow [Th(NR^2_2)_2Cp^*_2] + 2 R^1H \qquad (431)$$

$$An = Th, U$$
$$R^1 = \text{hydrocarbyl}$$
$$R^2 = Et$$

The most unusual reactivity pattern of $[ThR_2(Cp^*)_2]$ complexes is C–H activating cyclometallation. These cyclometallation processes are thermally induced and lead to the formation of thoracyclobutane derivatives (Equations (432) and (433) and Scheme 74).[1245,1249-52] Thermochemical data revealed that the cyclization reactions are entropically driven. In the case of the silicon-containing four-membered ring the reaction is reversible, but the cyclic product is more stable. The molecular structure of $[Th(CH_2SiMe_2\text{-}o\text{-}C_6H_4)(Cp^*)_2]$ was determined by x-ray diffraction.[1245]

$$
Cp*_2Th\underset{CH_2Bu^t}{\overset{CH_2Bu^t}{<}} \quad \xrightarrow[-CMe_4]{\Delta} \quad Cp*_2Th\diamond CMe_2 \quad\quad (432)
$$

$$
Cp*_2Th\underset{CH_2TMS}{\overset{CH_2TMS}{<}} \quad \xrightarrow[-SiMe_4]{\Delta} \quad Cp*_2Th\diamond SiMe_2 \quad\quad (433)
$$

$$
Cp*_2Th\underset{CH_2SiMe_2Ph}{\overset{CH_2SiMe_2Ph}{<}} \quad \xrightarrow{\Delta} \quad Cp*_2Th\diamond SiMePh \quad \xrightarrow{\Delta\ \Delta} \quad Cp*_2Th\diamond SiMe_2
$$

Scheme 74

The thermochemically derived bond disruption enthalpies for $[Th(CH_2)_2CMe_2(Cp*)_2]$ and $[Th(CH_2)_2SiMe_2(Cp*)_2]$ are relatively low as compared to other thorium hydrocarbyls of the type $[ThR_2(Cp*)_2]$.[1138,1253] This finding is consistent with significant strain in the four-membered thoracyclobutane rings and explains the facile activation of saturated hydrocarbons by these unusual organoactinides. The derivative chemistry of thoracyclobutanes is particularly fascinating.[1203,1254] Treatment with molecular hydrogen affords the corresponding organothorium hydride. Facile ring-opening occurs with a variety of reagents. While the formation of alkoxides upon reaction with alcohols is quite normal, the thoracyclobutanes display a unique reactivity toward saturated and unsaturated hydrocarbons. Facile C–H activation is observed in reactions with benzene, tetramethylsilane, trimethylphosphine, and even methane. The relative rates for the C–H functionalities were measured to be $SnMe_4 \ge SiMe_4 >$ cyclopropane $\approx PMe_3 > C_6H_6 > CH_4 \ge C_2H_6 \gg C_6H_{12}$. The ring-opening reaction of $[Th(CH_2)_2CMe_2(Cp*)_2]$ with methane proceeds at normal pressure and slightly elevated temperatures(!) (Equation (434)).[1203,1254]

$$
\xrightarrow{\ +\ CH_4\ } \xrightarrow{60\ ^\circ C} \quad\quad (434)
$$

Interestingly, the example of propene shows that in this case alkene insertion is preferred over C–H activation. The derivative chemistry of thoracyclobutanes is summarized in Scheme 75.[1203,1254,1255]

The electronic structure of the phosphathoracyclobutane derivative $[Th(CH_2)_2PMe(Cp*)_2]$ was investigated by theoretical calculations. Metal–ligand bonding in this complex involves both $5f$ and $6d$ metal atomic orbitals with a major role of the latter.[1048]

In addition to C–H activation, CO migratory insertion reactions are also quite prominent in the chemistry of $[AnR_2(Cp*)_2]$ hydrocarbyls. Carbon monoxide readily inserts into the actinide–carbon bond to afford dihapto-acyl complexes (Equation 435).[980,1256]

Similar to the dihapto-acyls of the type $[U(\eta^2\text{-}COR)(Cp)_3]$ the bonding in $[An(\eta^2\text{-}COR)R(Cp*)_2]$ can be described by two resonance structures (Equation (436)).

According to theoretical calculations the dihapto-acyls behave like "anchored" Fischer-type carbene complexes.[983,1257–9] The localization of the LUMO on the acyl carbon atom is mainly responsible for the

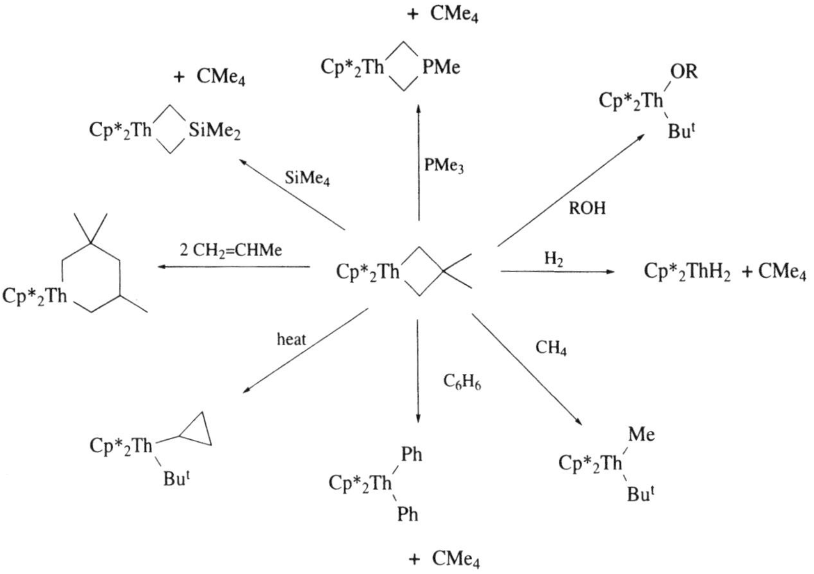

Scheme 75

$$[AnR_2Cp^*_2] + CO \longrightarrow [An(\eta^2\text{-}COR)RCp^*_2] \qquad (435)$$

$$An = Th,\ U$$

$$(436)$$

carbenelike reactivity of these organoactinides. Single-crystal x-ray structure determinations have been carried out on the two acyl complexes $[Th(\eta^2\text{-}COCH_2Bu^t)Cl(Cp^*)_2]$ and $[Th(\eta^2\text{-}COPh)Cl(Cp^*)_2]$.[32] The molecular structure of $[Th(\eta^2\text{-}COCH_2Bu^t)ClCp^*_2]$ has been determined (Th–C$_a$ 0.244(2) nm, Th–O 0.237(2) nm).

A rich derivative chemistry based on η^2-acyl complexes of the type $[An(\eta^2\text{-}COR)R(Cp^*)_2]$ has been developed. Thermal isomerization affords enolates.[32,1260] The greater thermodynamic stability of enolato derivatives had been noted before in the $[U(\eta^2\text{-}COR)(Cp)_3]$ series.[1145,1146] Further treatment with excess carbon monoxide results in CO tetramerization and formation of enedione diolato derivatives.[1260,1261] Catalytic hydrogenation occurs at the acyl carbon atom to give alkoxides of the type $[An(OR)R(Cp^*)_2]$.[1262] Other unusual reactions include the insertion of ketenes as well as the simultaneous addition of CO and phosphines.[1260,1261] Finally, C–C coupling results in the formation of monomeric or dimeric enediolates,[32,1256] which have been the subject of a theoretical study.[983] The characteristic reaction patterns of $[An(\eta^2\text{-}COR)R(Cp^*)_2]$ complexes are summarized in Scheme 76.[1260]

Similar insertion reactions have been carried out with isocyanides (Equation (437)). In analogy to the isocyanide adducts of $[UR(Cp)_3]$ the resulting products have been formulated as dihapto iminoacyl complexes (Equation (438)). Apparently no further reactions of these products have been reported.[1153]

Insertion reactions with CO_2 afforded the corresponding bis(acetates), which are also accessible directly from $[AnCl_2(Cp^*)_2]$ and NaOAc (Equation (439)).[1230]

The first butadiene complex of an actinide element was independently prepared by two research groups.[1263] "Magnesium butadiene" has been used to introduce the butadiene ligand (Equation (440)).

η^4-Butadienide coordination and the presence of thorium(IV) was established by an x-ray crystal structure determination of $[Th(\eta^4\text{-}C_4H_6)(Cp^*)_2]$ (Figure 88). The average Th–C(1,4) distance is 0.257(3) nm and the Th–C(2,3) distance is 0.274(3) nm. The carbon–carbon distances in the butadiene ligand are very similar with an average C-1–C-2 and C-3–C-4 distance of 0.146(4) nm and a C-2–C-3 bond length of 0.144(3) nm.[1263] $[Th(\eta^4\text{-}C_4H_6)\{(Bu^t)_2C_5H_3\}_2]$ and $[Th(\eta^4\text{-}CH_2CMeCMeCH_2)(Cp^*)_2]$ have been prepared analogously.[1243,1263]

Addition of pyridine to $[Th(\eta^4\text{-}C_4H_6)(Cp^*)_2]$ afforded an η^3-allyl complex with additional intramolecular nitrogen coordination (Equation (441)). A thorium oxycarbene–chromium complex was produced upon treatment with chromium hexacarbonyl.[1263]

Scheme 76

$$[U(R^1)XCp^*_2] + R^2NC \longrightarrow [U\{C(R^1)NR^2\}XCp^*_2] \tag{437}$$

$$R^1 = Me, X = Cl, R^2 = Bu^t$$
$$R^1 = Bu^n, X = Cl, R^2 = Bu^t$$
$$R^1 = Me, X = Me, R^2 = Bu^t$$

$$\tag{438}$$

$$[AnMe_2Cp^*_2] + 2\,CO_2 \longrightarrow [An(O_2CMe)_2Cp^*_2] \tag{439}$$

$$An = Th,\ U$$

$$[ThCl_2Cp^*_2] + MgC_4H_6 \longrightarrow [Th(\eta^4\text{-}C_4H_6)Cp^*_2] + MgCl_2 \tag{440}$$

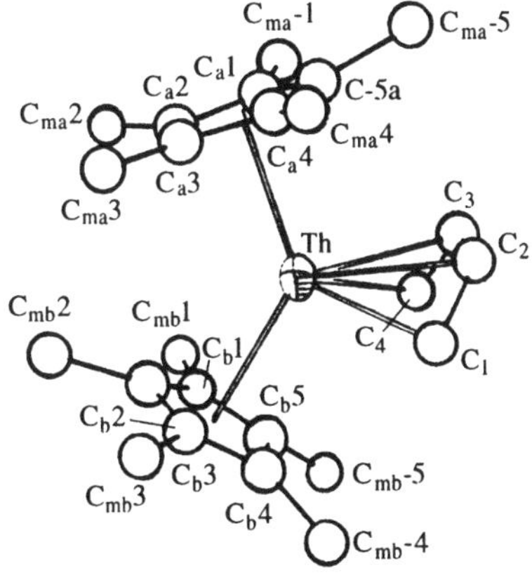

Figure 88 The molecular structure of $[Th(\eta^4\text{-}C_4H_6)(Cp^*)_2]$.[1263]

$$[Th(C_4H_6)Cp^*_2] \xrightarrow{\ py\ } \tag{441}$$

Other more special actinide hydrocarbyl complexes include vinyl and phosphoylide derivatives which have been prepared according to Equations (442) and (443)).[32,1264]

$$[AnCl_2Cp^*_2] + 2\,LiCH=CH_2 \xrightarrow[-2\,LiCl]{} [An(CH=CH_2)_2Cp^*_2] \qquad (442)$$

$$An = U, Th$$

$$[AnCl_2Cp^*_2] + Li(CH_2)_2PR^1R^2 \xrightarrow[-LiCl]{} [An\{(CH_2)_2PR^1R^2\}ClCp^*_2] \qquad (443)$$

$$An = U, Th$$

$$R^1 = Me, R^2 = Ph$$

$$R^1 = R^2 = Me \text{ or } Ph$$

The thermally highly stable phosphoylide complexes show fluxional behavior in solution, which has been explained by a series of bond breaking, rotation, and recombination processes. The average U–C(CH$_2$) bond length in the structurally characterized complexes [U$\{$(CH$_2$)$_2$PRPh$\}$Cl(Cp*)$_2$] (R = Me, Ph) is 0.260(1) nm.[1264]

The cationic species [ThMe(THF)$_2$(Cp*)$_2$][BPh$_4$] was prepared by reacting [ThMe$_2$(Cp*)$_2$] with [NHR$_3$][BPh$_4$] and structurally characterized (Th–C(Cp*) 0.280(1) nm, Th–C(Me) 0.249(1) nm). [Th(*o*-C$_6$H$_4$CH$_2$NMe$_2$)(Cp*)$_2$][BPh$_4$], stabilized by intramolecular nitrogen-coordination, was made analogously from [Th(*o*-C$_6$H$_4$CH$_2$NMe$_2$)Me(Cp*)$_2$].[1265] The search for cationic metallocene polymerization catalysts led to the investigation of the first unsolvated cationic bis(pentamethylcyclopentadienyl)thorium(IV) species. The saltlike compounds [ThMe(Cp*)$_2$][BPh$_4$] and [ThMe(Cp*)$_2$][B(C$_6$F$_5$)$_4$] were prepared free of coordinating bases according to Equations (444) and (445) by reacting [ThMe$_2$(Cp*)$_2$] with triorganoammonium tetraarylborates.[1265,1266]

$$[ThMe_2Cp^*_2] + [NHEt_3][BPh_4] \longrightarrow [ThMeCp^*_2][BPh_4] + Et_3N + CH_4 \qquad (444)$$

$$[ThMe_2Cp^*_2] + [NHBu^n][B(C_6F_5)_4] \longrightarrow [ThMeCp^*_2][B(C_6F_5)_4] + Bu^n_3N + CH_4 \qquad (445)$$

In analogy to comparable cationic metallocene complexes of zirconium or hafnium the bulky tetraarylborate anion is not completely inert with respect to the coordinatively unsaturated metal center. Thus, the crystal structure determination of [ThMe(Cp*)$_2$][B(C$_6$F$_5$)$_4$] revealed a weak coordination of the [B(C$_6$F$_5$)$_4$]$^-$ anion. The Th–C(Cp*)$_2$ bond distances averaged 0.2757(4) nm and the σ-bonded thorium–carbon distance was found to be 0.2399(8) nm, while the closest Th–F distances were 0.2757(4) nm and 0.2675(5) nm.[1266]

Metal bis(dicarbollides) of the type [M(B$_9$C$_2$H$_{11}$)$_2$]$^-$ (M = Co, Fe) have been combined with cationic thorium complexes [ThR(Cp*)$_2$]$^+$ as potential weakly coordinating anions (Equations (446) and (447)).[1267]

$$[ThR_2Cp^*_2] + [HNEt_3][Co(B_9C_2H_{11})_2] \longrightarrow [ThRCp^*_2][Co(B_9C_2H_{11})_2] + NEt_3 + RH \qquad (446)$$

$$R = Me, CH_2TMS$$

$$2\,[ThMe_2Cp^*_2] + [HNEt_3][Fe(B_9C_2H_{11})_2] \longrightarrow [ThMeCp^*_2]_2[Fe(B_9C_2H_{11})_2] + NEt_3 + CH_4 + 1/2\,C_2H_3 \qquad (447)$$

The structure of [ThMe(Cp*)$_2$]$_2$[Fe(B$_9$C$_2$H$_{11}$)$_2$] (Figure 89) revealed tight ion-pairing with three close Th$\cdots$H-B bridging interactions (0.242(3) nm, 0.250(3) nm, 0.267(4) nm).[1267]

(c) Hydrides. A rich and exciting derivative chemistry of soluble and highly reactive organoactinide hydrides has been made possible through the use of pentamethylcyclopentadienyl ancillary ligands. Hydrogenolysis of organoactinide hydrocarbyls represents the best access to the corresponding hydrido species (Equations (448) and (449)).[29,32,1232] These reactions have been kinetically and mechanistically studied in great detail. A heterolytic four-center transition state has been discussed. The rate law is first order in actinide complex and first order in hydrogen.[1230] The monohydrido species [$\{$Th(H)Cl(Cp*)$_2$$\}_2$] and [Th(H)(OR)(Cp*)$_2$] are more stable than the dihydrides.[1232,1253]

$$2\,[AnR_2Cp^*_2] + 4\,H_2 \longrightarrow [\{AnH_2Cp^*_2\}_2] + 4\,RH \qquad (448)$$

$$An = Th, U; R = alkyl$$

$$2\,[Th(R)ClCp^*_2] + 2\,H_2 \longrightarrow [\{Th(H)ClCp^*_2\}_2] + 2\,RH \qquad (449)$$

$$R = alkyl$$

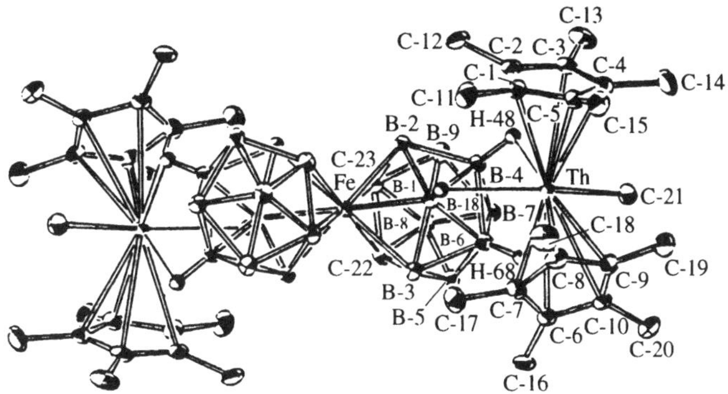

Figure 89 The molecular structure of $[ThMe(Cp^*)_2]_2[Fe(B_9C_2H_{11})_2]$. [1267]

The uranium monohydride $[\{U(H)Cl(Cp^*)_2\}_2]$ is not available via this route as treatment of the uranium hydrocarbyls $[U(R)Cl(Cp^*)_2]$ with hydrogen yields the trimeric uranium(III) species $[\{U(\mu\text{-}Cl)(Cp^*)_2\}_3]$ (see above). [32] Neutron diffraction has been employed to determine the molecular structure of $[\{ThH_2(Cp^*)_2\}_2]$. The dimeric molecule contains two bridging (Th–H 0.229(3) nm) and two terminal (Th–H 0.203(1) nm) hydrido ligands. The H–Th–H bond angle was found to be 58(1)° and the Th–H–Th angle is 122(4)°. The Th···Th distance is rather long (0.4007(8) nm), indicating the absence of metal–metal bonding interaction. Terminal and bridging hydrido ligands interchange at a rapid rate as shown by NMR spectral investigations. Rapid exchange with external H_2 is also observed.

Organothorium hydride species have also been obtained when either $ThMe_7^{3-}$ or $[Th(CH_2Bu^t)_3(Cp^*)]$ was reacted with molecular hydrogen. However, in these cases the products are mixtures which could not be fully characterized. [32]

Chemically the organoactinide complexes $[\{AnH_2(Cp^*)_2\}_2]$ behave like typical metal hydrides (Equations (450)–(453)): [29,1232]

$$An\text{–}H \ + \ RX \ \longrightarrow \ An\text{–}X \ + \ RH \tag{450}$$

$$X = Cl, Br, I$$

$$An\text{–}H \ + \ ROH \ \longrightarrow \ An\text{–}OR \ + \ H_2 \tag{451}$$

$$An\text{–}H \ + \ O{=}CR_2 \ \longrightarrow \ An\text{–}O\text{-}CHR_2 \tag{452}$$

$$An\text{–}H \ + \ CH_2{=}CHMe \ \longrightarrow \ An\text{–}Pr^n \tag{453}$$

The facile addition of terminal alkenes can be considered the reverse of the well-known β-hydride elimination. $[\{UH_2(Cp^*)_2\}_2]$ has been found to be an effective catalyst in homogeneous alkene hydrogenation (cf. Section 2.3.10). Alkenes also insert into the Th–H bond of $[Th(H)(OR)(Cp^*)_2]$. These reactions have been studied in great detail. [1231,1268]

Migratory insertion of carbon monoxide into the An–H bond of $[Th(H)(OR)(Cp^*)_2]$ has been shown to yield cis-enediolates as the final products (Equation (454)). [983,1269]

$$2\,[Th(H)(OR)Cp^*_2] \ + \ 2\,CO \ \longrightarrow \ [Cp^*_2Th(OR)\text{-}O\text{-}CH{=}CH\text{-}O\text{-}Th(OR)Cp^*_2] \tag{454}$$

$$R = Bu^t, CH(Bu^t)_2, 2,6\text{-}(Bu^t)_2C_6H_3$$

The mechanism of this fundamentally interesting reaction has been thoroughly studied using NMR spectroscopic kinetic and crossover experiments. At low temperatures (−40 °C) rapid though reversible formation of 1:1 adducts is observed, which have been formulated as dihapto-formyls (Equation (455)). Intermediate formation of thorium carbonyl species could not be detected.

$$[Th(H)(OR)Cp^*_2] \ + \ CO \ \longrightarrow \ [Th(\mu^2\text{-}CHO)(OR)Cp^*_2] \tag{455}$$

At room temperature dimerization occurs to afford the *cis*-enediolates. From kinetic data a plausible reaction mechanism for the formation of organoactinide *cis*-enediolates has been deduced (Scheme 77). [1270]

Simultaneous treatment of $[Th(H)(OR)(Cp^*)_2]$ with carbon monoxide and hydrogen affords bis-alkoxides. According to the proposed mechanism the products result from hydrogenolysis of μ-CH_2O-bridged binuclear intermediates (Equation (456)). [1270]

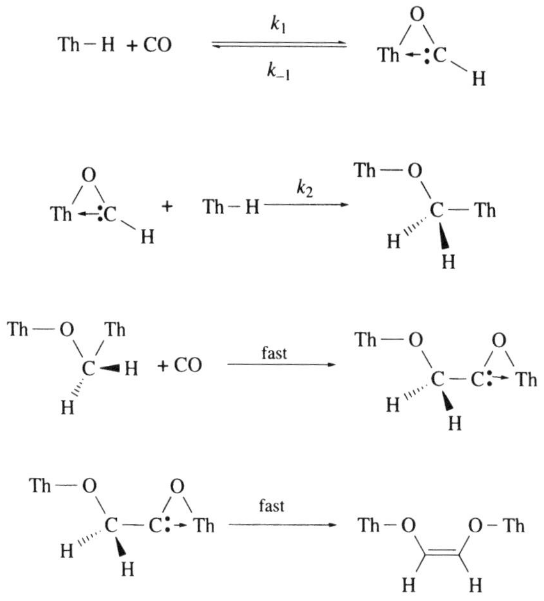

Scheme 77

$$[Th(H)(OR)Cp^*_2] + CO + H_2 \longrightarrow [Th(OMe)(OR)Cp^*_2] \tag{456}$$

Carbon dioxide also inserts into the thorium–hydrogen bond. The formate [Th(OC-HBu^i)(O$_2$CH)(Cp*)$_2$] was made by reaction of [Th(OCHBu^i)(H)(Cp*)$_2$] with CO$_2$.[982]

Yet another interesting class of organoactinide hydrides contains chelating phosphines as auxiliary ligands. The existence of such compounds is remarkable as actinide elements are normally very reluctant to form stable complexes with tertiary phosphine ligands. The phosphine-stabilized organouranium hydrides are accessible via two synthetic approaches (Equations (457) and (458)).[1271]

$$[UR_2Cp^*_2] + dmpe \xrightarrow{H_2} [U(H)dmpeCp^*_2] + 2\,RH \tag{457}$$

$$[\{UH_2Cp^*_2\}_2] + 2\,dmpe \longrightarrow 2\,[U(H)dmpeCp^*_2] + H_2 \tag{458}$$

Unexpectedly, an analogous complex containing TMEDA instead of the chelating phosphine is not accessible by a similar route. The compound [U(H)dmpe(Cp*)$_2$] exhibits a similar reactivity as [{UH$_2$(Cp*)$_2$}$_2$] combined with an improved solubility, which can be attributed to the phosphine ligand. The molecular structure of the organouranium phosphine hydride has been determined by x-ray diffraction, although the position of the hydrido ligand could not be accurately located. However, it is clearly detected in the low-temperature ^{1}H NMR spectra and in the IR spectrum ($\nu_{U-H} = 1219$ cm^{-1}).[1271]

An unusual reaction takes place when the organoactinide hydrides [{AnH$_2$(Cp*)$_2$}$_2$] are treated with trimethyl phosphite. Instead of simple addition as was the case with dmpe, a methoxy group is transferred to the metal atom while phosphorus is hydrogenated to give a PH bridge. This results in the formation of rather interesting dimetallophosphines of the type [{An(OMe)(Cp*)$_2$}$_2$PH] (Equation (459)).[1238]

$$5\,[\{AnH_2Cp^*_2\}_2] + 4\,P(OMe)_3 \xrightarrow[25\,°C]{toluene} [\{An(OMe)Cp^*_2\}_2PH] + 3\,[An(OMe)_2Cp^*_2] + 8\,H_2 \tag{459}$$

$$An = Th, U$$

The phosphinidene complexes have been characterized by spectroscopic methods as well as an x-ray structural investigation of the uranium derivative (Figure 90). The two U(OMe)(Cp*)$_2$ fragments are symmetrically bonded to the bridging PH ligand (U–P 0.2743(1) nm, U–P–U 157.7(2)°). A linear arrangement was found for the U–OMe units (U–O–C 178(1)°).[1238]

The cationic organothorium dicarbollide [ThCH$_2$TMS(Cp*)$_2$][Co(B$_9$C$_2$H$_{11}$)$_2$] slowly undergoes hydrogenolysis to yield [{ThH(Cp*)$_2$}{Co(B$_9$C$_2$H$_{11}$)$_2$}] (Equation (460)).[1267]

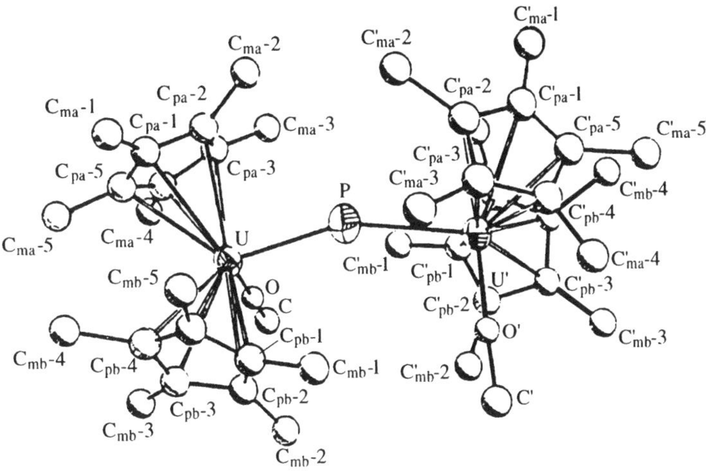

Figure 90 The molecular structure of [{U(OMe)(Cp*)$_2$}$_2$PH].[1238]

$$[ThCH_2TMSCp*_2][Co(B_9C_2H_{11})_2] + H_2 \longrightarrow [ThHCp*_2][Co(B_9C_2H_{11})_2] + SiMe_4 \qquad (460)$$

2.3.6 Cyclopentadienyl-like Compounds

2.3.6.1 *Pentadienyl and cyclohexadienyl compounds*

Pentadienyl and cyclohexadienyl ligands are being increasingly employed in *d*-transition metal chemistry. Due to their chemical similarity with the corresponding metallocene derivatives, pentadienyl complexes have been termed "open cyclopentadienyl complexes." Interestingly, the often used 2,4-dimethylpentadienyl ligand (dmpd)$^-$ was found to give a stable uranium(III) complex (Equation (461)).[1220,1272]

$$[UX_3(THF)_n] + 3\,K(dmpd) \longrightarrow [U(dmpd)_3] + 3\,KX \qquad (461)$$
$$X = Cl, I$$

Although the crystal structure [U(dmpd)$_3$] has not been determined, it can be anticipated that the molecular structure is analogous to that of the corresponding tris(2,4-dimethylpentadienyl)neodymium, in which all three pentadienyl ligands are η^5-coordinated.[726] The chemistry has recently been extended to various neutral, anionic, and cationic bis(pentadienyl)uranium(III) complexes as well as neutral pentadienyluranium(IV) borohydride complexes. Similar reactions have been reported with the related 6,6-dimethylcyclohexadienyl (dmch)$^-$ ligand. The results are summarized in Scheme 78.[1038,1273-5]

The uranium(IV) compounds are stable only in the absence of noncoordinating solvents. Addition of donor ligands causes immediate reduction to U(BH$_4$)$_3$ derivatives. Monomeric [U(BH$_4$)$_3$(dmpd)] has been structurally characterized (U–C 0.275(3) and 0.267(3) nm).[1038] [U(BH$_4$)$_3$(dmpd)] and [U(BH$_4$)$_3$(dmch)] are quite volatile having a vapor pressure which is comparable to that of [U(BH$_4$)$_4$]. Especially remarkable is the very stable [U(dmpd)$_2$]$^+$ cation, which was the first organometallic cation of an actinide(III) element.[1274]

Recently the first cycloheptatrienyl complexes of an *f*-element have been characterized. The anion [U(NEt$_2$)$_3$(μ-η^7,η^7-C$_7$H$_7$)U(NEt$_2$)$_3$]$^-$ was formed by treatment of U(NEt$_2$)$_4$ with the potassium salt of the cycloheptadienyl anion, K(C$_7$H$_9$). The potassium salt [K(18-crown-6)][U(NEt$_2$)$_3$(μ-η^7,η^7-C$_7$H$_7$)U(NEt$_2$)$_3$] was isolated after recrystallization in the presence of the crown ether. K[U(BH$_4$)$_3$(μ-η^7,η^7-C$_7$H$_7$)U(BH$_4$)$_3$] was prepared similarly from [U(BH$_4$)$_4$] and K(C$_7$H$_9$), but the product could not be separated from concomitant [U(BH$_4$)$_3$(THF)$_3$]. However, serendipitously isolated [{U(BH$_4$)$_2$(THF)$_5$}{U(BH$_4$)$_3$(μ-η^7,η^7-C$_7$H$_7$)U(BH$_4$)$_3$}] was fully characterized by an x-ray crystal structure analysis (Figure 91). The average U–C distance to the bridging cycloheptatrienyl ligand is 0.269(2) nm.[1276]

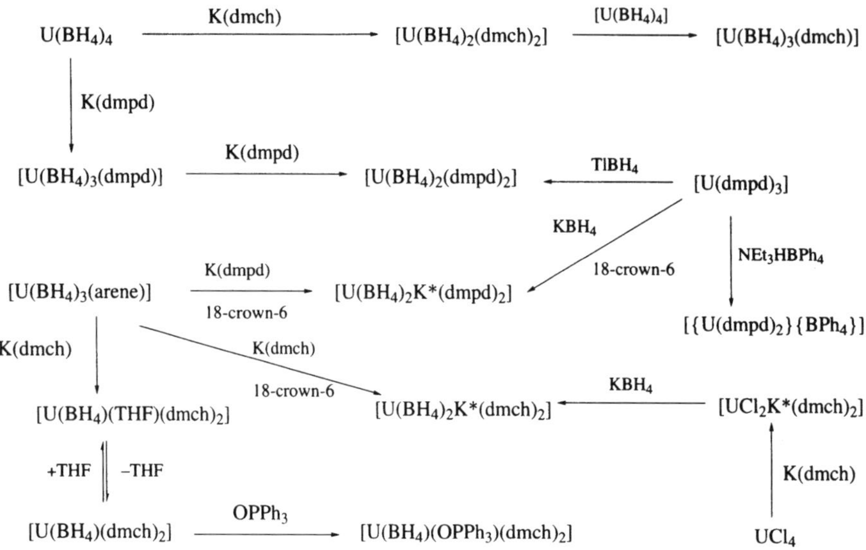

dmpd = 2,4-dimethylpentadienyl, dmch = 6,6-dimethylcyclohexadienyl, K* = [K(18-crown-6)]⁺

Scheme 78

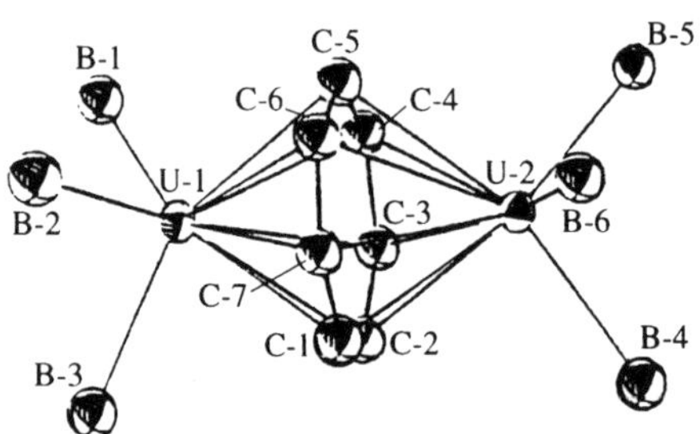

Figure 91   The molecular structure of the anion in $[\{U(BH_4)_2(THF)_5\}\{U(BH_4)_3(\mu\text{-}\eta^7,\eta^7\text{-}C_7H_7)U(BH_4)_3\}]$.[1276]

2.3.6.2 Heteroatom five-membered ring ligands

Recently the tetramethylphospholyl anion has been shown to be a suitable ligand in organoactinide chemistry.[1277] The synthetic procedures for the preparation of phospholyluranium complexes are summarized in Scheme 79.

An x-ray structure determination of $[U(BH_4)_2(C_4Me_4P)_2]$ revealed pentahapto coordination of the phospholyl rings and trihapto bonding of the borohydride ligands.[1277] Phospholyl analogues of $[UX(Cp)_3]$ (X = Cl, Me, OR) have also become available. $[UCl(C_4Me_4P)_3]$ was obtained by treatment of UCl_4 with three equivalents or an excess of KC_5Me_4P in toluene, while careful control of the stoichiometry allowed the preparation of $[UCl_2(C_4Me_4P)_2]$. All three phospholyl ligands in $[UCl(C_4Me_4P)_3]$ are pentahapto coordinated (U–Cl 0.2671 nm, U–P 0.2927 nm). Starting from tris(tetramethylphospholyl)uranium chloride the derivatives $[UMe(C_4Me_4P)_3]$, $[UH(C_4Me_4P)_3]$, and $[U(OPr^i)(C_4Me_4P)_3]$ were synthesized by metathetical reactions.[1277]

2.3.7 Arene Complexes

The number of well characterized actinide arene complexes is still fairly limited, although some recent progress has been made in this area. The discovery of the first η^6-arene complex of an actinide element dates back to 1971.[31,36] The compound $[U(AlCl_4)_3(C_6H_6)]$ was prepared by reacting anhydrous UCl_4 with aluminum trichloride and aluminum in benzene solution (Equation (462), which does not account for the exact stoichiometry).

The hexamethylbenzene derivative $[U(AlCl_4)_3(\eta^6\text{-}C_6Me_6)]$ was prepared analogously and its crystal structure was reported. The molecular structure closely resembles that of the analogous samarium complex (cf. Section 2.2.8) (U–C 0.292(2) nm).

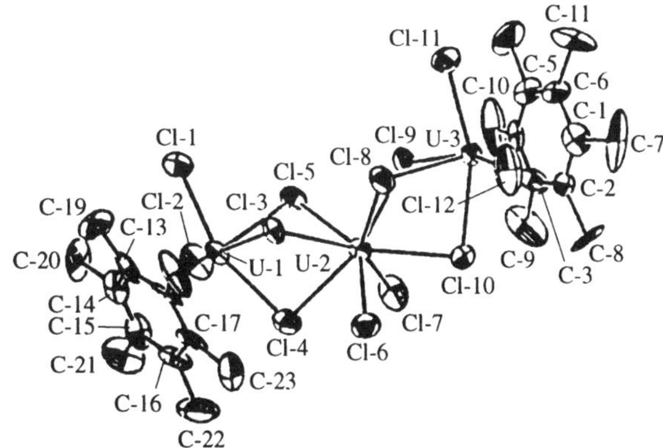

Scheme 79

$$UCl_4 + AlCl_3 + Al \xrightarrow{C_6H_6} [U(AlCl_4)_3(C_6H_6)] \qquad (462)$$

Several other complexes can be isolated when hexamethylbenzene is employed in this reducing Friedel–Crafts synthesis (Equation (463)), which does not account for the exact stoichiometry).[1278,1279]

$$UCl_4 + AlCl_3 + Al \xrightarrow{C_6Me_6} [U_2Cl_7(C_6Me_6)_2][AlCl_4] \qquad (463)$$

In the molecular structure of the $[U_2Cl_7(C_6Me_6)_2]^+$ cation the two uranium atoms are bridged by three chloride ligands. The uranium–uranium distance of 0.3937(1) nm indicates the absence of a direct U–U bond. Relatively long distances are also found between uranium and the η^6-coordinated hexamethylbenzene ligands (U–C 0.292(4) nm).[1278]

Other products isolated from this system are the trinuclear complexes $[(C_6Me_6)Cl_2U(\mu\text{-}Cl)_3UCl_2(\mu\text{-}Cl)_3UCl_2(C_6Me_6)]$ and $[\{(\eta^6\text{-}C_6Me_6)U_3(\mu\text{-}Cl)_3(\mu_3\text{-}Cl)_2(\mu,\eta^2\text{-}AlCl_4)_3\}\{AlCl_4\}]$. The former adopts a bent chain structure while the latter is a cyclic trimer (Figures 92 and 93).[1277,1279]

Figure 92 The molecular structure of $[(C_6Me_6)Cl_2U(\mu\text{-}Cl)_3UCl_2(\mu\text{-}Cl)_3UCl_2(C_6Me_6)]$.[1277,1279]

All the above-mentioned uranium complexes containing η^6-coordinated arene ligands cannot be dissolved without decomposition in donating solvents such as diethyl ether or THF because the arene ligands are easily displaced.[1277] Because of their insoluble nature the uranium arene complexes were the first organoactinide compounds to be investigated by CP MAS ^{13}C NMR spectroscopy.[1279]

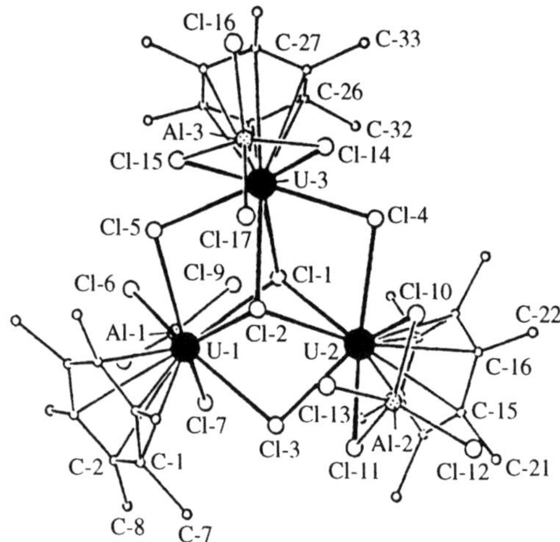

Figure 93 The molecular structure of $[\{(\eta^6\text{-}C_6Me_6)U_3(\mu\text{-}Cl)_3(\mu_3\text{-}Cl)_2(\mu,\eta^2\text{-}AlCl_4)_3\}\{AlCl_4\}]$.[1277,1279]

A novel development in this area was initiated by the preparation of $[U(BH_4)_3(\eta^6\text{-}C_6H_3Me_3)]$.[1273] This structurally characterized compound was prepared by a thermal reaction of $[U(BH_4)_4]$ with mesitylene (Equation (464)). The monomeric molecule adopts a "three-legged piano stool" geometry (U–C 0.293(2) nm). The aromatic ligands can be displaced in the following order: hexamethylbenzene > mesitylene > toluene > benzene. Complexes of the type $[U(BH_4)_3(\eta^6\text{-arene})]$ are promising precursors for the preparation of other organouranium(III) complexes.[1273,1280]

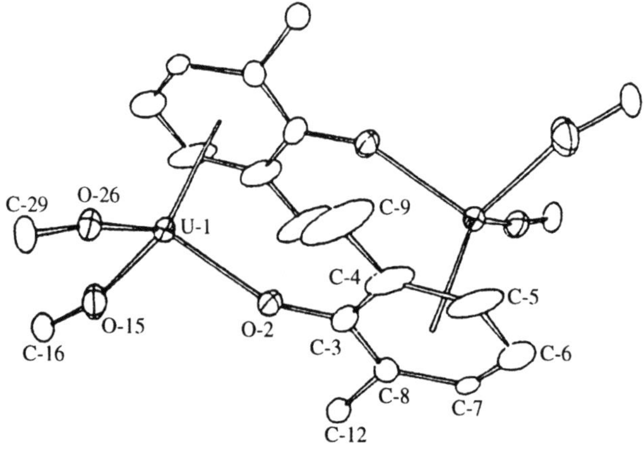

Uranium-η^6-arene interactions in the solid state were also observed unexpectedly in some phenoxide derivatives of uranium(III). The dimeric compounds $[\{U(OC_6H_3R_2\text{-}2,6)_3\}_2]$ (R = Pr^i, Bu^t) were prepared by treatment of $[U\{N(TMS)_2\}_3]$ with the corresponding phenols. The derivative with R = Pr^i was structurally characterized and shows η^6-arene coordination (Figure 94, U–C 0.292(2) nm).[764]

Figure 94 The molecular structure of $[\{U(OC_6H_3(Pr^i)_2\text{-}2,6)_3\}_2]$.[764]

2.3.8 Cyclooctatetraenyl Compounds

2.3.8.1 $[M(C_8H_8)_2]^-$ compounds

Reduction of neutral actinide bis(cyclooctatetraenyl) complexes formally leads to the sandwich anions $[An(C_8H_8)_2]^-$. Such anionic derivatives represent a well-known class of compounds in organolanthanide chemistry. The first indication for the formation of an anionic uranium(III) species was obtained in 1981.[1281] Products containing the $[U(C_8H_8)_2]^-$ anion were obtained either by lithium naphthalide reduction of uranocene (Equation (465)) or by treatment of uranium trichloride with two equivalents of $K_2C_8H_8$ (Equation (466)). In both cases, however, the resulting products were too unstable to be isolated.[1281]

$$[U(C_8H_8)_2] + LiC_{10}H_8 \xrightarrow{\text{THF}} Li[U(C_8H_8)_2] + C_{10}H_8 \tag{465}$$

$$[UCl_3(THF)_n] + 2\,K_2C_8H_8 \xrightarrow{\text{THF}} K[U(C_8H_8)_2] + 3\,KCl \tag{466}$$

Potassium reduction of uranocenes also afforded the potassium salts $K[U(RC_8H_7)_2]$ (R = H, Me).[782,1281,1282] The crystal and molecular structure of $[\{K(diglyme)\}\{U(MeC_8H_7)_2\}]$ has recently been determined. In this molecule uranium(III) and potassium are bridged by a methylcyclooctatetraenyl ligand to give a nearly linear $U(MeC_8H_7)(\mu\text{-}MeC_8H_7)K$ arrangement (ring centroid–U–ring centroid 176°). The coordination sphere of potassium is completed by coordination of the tridentate diglyme ligand. The average uranium–carbon distances have been reported to be 0.2707(7) nm and 0.2732(8) nm, respectively, with the longer distances corresponding to the bridging $MeC_8H_7^{2-}$ ligand. These differences in the bond lengths reflect the competition of the two metal atoms for the electronic π-system of the bridging methylcyclooctatetraenyl ligand.[782]

Anionic cyclooctatetraenyl sandwich complexes have also been reported for other actinide elements. The neptunium and plutonium derivatives can be prepared directly from the metal trihalides according to Equation (467).[31]

$$AnX_3 + 2\,K_2C_8H_8 \xrightarrow{\text{THF}} [K(THF)_2][An(C_8H_8)_2] + 3\,KX \tag{467}$$

$$An = Np, Pu$$
$$X = Br, I$$

According to their x-ray powder diffraction patterns the two THF-solvated potassium salts are isostructural. The salt $[K(diglyme)][Pu(C_8H_8)_2]$ has been prepared by carrying out the above synthesis in diglyme solution. In this case the x-ray powder diffraction data indicate a close similarity to the corresponding cerium derivative $[\{K(diglyme)\}\{Ce(C_8H_8)_2\}]$.[31] In this anionic sandwich complex of cerium(III) one cyclooctatetraenyl ring acts as a bridging ligand between cerium and potassium. The coordination sphere of potassium is saturated by addition of the tridentate diglyme ligand. A ^{237}Np Mössbauer study revealed that the degree of covalency in the metal–ligand bonding is smaller in the $[Np(C_8H_8)_2]^-$ anion than in the neptunium(IV) complex $[Np(C_8H_8)_2]$.[1057] The anionic derivatives are easily oxidized in the presence of air to give the neutral sandwich complexes $[An(C_8H_8)_2]$ (An = Np, Pu).[31] Rapid electron exchange was also described for the couple $[K\{An(Bu^tC_8H_7)_2\}]/[An(Bu^tC_8H_7)_2]$ (An = Np, Pu).[1282]

The cyclooctatetraenyl dianion has been used to prepare one of the rare examples of organometallic americium compounds. Treatment of americium triiodide with two equivalents of $K_2C_8H_8$ yielded the anionic sandwich complex $[Am(C_8H_8)_2]^-$ (Equation (468)). It was not possible to achieve further reduction to form an americium(II) derivative.[31]

$$AmI_3 + 2\,K_2C_8H_8 \longrightarrow K[Am(C_8H_8)_2] + 3\,KI \tag{468}$$

Neutral actinide sandwich complexes containing a cyclooctatetraenyl ligand have prepared by the following synthetic routes (Equations (469) and (470)). Analogous lanthanide mixed-sandwich complexes have been known for many years. The dimethylbipyridine adduct was structurally characterized.[1283]

$$[UI_2(THF)_3Cp^*] + K_2C_8H_8 \xrightarrow{\text{THF}} [U(THF)Cp^*(C_8H_8)] + 2\,KI \tag{469}$$

$$[U(THF)Cp^*(C_8H_8)] + Me_2bipy \longrightarrow [U(Me_2bipy)Cp^*(C_8H_8)] + THF \qquad (470)$$

$$Me_2bipy = 4,4'\text{-dimethyl-2,2'-bipyridine}$$

2.3.8.2 *[M(C$_8$H$_8$)$_2$] compounds*

Reduction of 1,3,5,7-cyclooctatetraene generates the Hückel $(4n + 2)$ π-electron dianion $C_8H_8^{2-}$, which has been shown to be a highly versatile ligand for early transition metals as well as the lanthanides and actinides. In organoactinide chemistry the neutral uranium sandwich complex $[U(C_8H_8)_2]$ (= bis([8]annulene)uranium(IV)), better known as "uranocene") is the most prominent representative of this class of compounds. Its first synthesis in 1968 was a major achievement in organoactinide chemistry. Uranocene chemistry was reviewed by Streitwieser[51,52,1284] and Sevastyanov *et al.*[1285] The original synthesis of uranocene involves treatment of uranium tetrachloride with two equivalents of $K_2C_8H_8$ (Equation (471)).[31,36]

$$UCl_4 + 2\ K_2C_8H_8 \longrightarrow [U(C_8H_8)_2] + 4\ KCl \qquad (471)$$

The reaction outlined in Equation (471) still represents the most convenient access to uranocene. A solvent-free approach uses the reaction of uranium tetrafluoride with cyclooctatetraenylmagnesium (Equation (472)).[31]

$$UF_4 + 2\ Mg(C_8H_8) \longrightarrow [U(C_8H_8)_2] + 2\ MgF_2 \qquad (472)$$

Finely divided uranium can also be reacted directly with cyclooctatetraene to yield uranocene. Such highly reactive, finely divided uranium may be obtained in different ways including the thermal decomposition of uranium trihydride or "$U(Bu^n)_4$," the reduction of uranium tetrachloride with sodium–potassium alloy, or electrolytic methods.[31]

Uranocene forms a dark green, pyrophoric solid that is only sparingly soluble in aliphatic hydrocarbons. A single-crystal x-ray analysis revealed that the uranocene sandwich exhibits rigorous D_{8h} symmetry. The planar eight-membered rings are arranged in an eclipsed conformation with average U–C bond lengths of 0.2647(4) nm and average C–C distances of 0.1392(13) nm.[31,51]

Numerous ring-substituted uranocenes have been prepared by reacting uranium tetrachloride with the corresponding substituted cyclooctatetraene dianions (Equation (473)).[31,51,52]

$$UCl_4 + 2\ K_2C_8H_7R^1 \longrightarrow [U(C_8H_7R^1)_2] + 4\ KCl \qquad (473)$$

$$R^1 = Et,\ CH=CH_2,\ Bu^n,\ Ph,\ o\text{-Tol, mesityl},\ C_8H_7,\ OMe,\ OEt,\ OBu^t, OCH_2CH=CH_2,\ NMe_2,\ CH_2NMe_2$$

$$(CH_2)_3NMe_2,\ p\text{-}C_6H_4NMe_2,\ PR^2_2\ (R^2 = Et,\ Bu^t,\ Ph)$$

Most of these ring-substituted uranocenes are significantly more soluble in nonpolar organic solvents than the parent compound. Uranocene derivatives bearing a larger number of substituents at the cyclooctatetraenyl rings have also been reported. Highly soluble 1,3,5,7,1',3',5',7'-octamethyluranocene was prepared according to Equation (474).[51,52]

$$UCl_4 + K_2(C_8H_4Me_4) \longrightarrow [U(C_8H_4Me_4)_2] + 4\ KCl \qquad (474)$$

The crystal structure of octamethyluranocene is remarkable as two crystallographically independent rotamers have been found in the unit cell. One rotamer has the eight methyl groups nearly staggered and the other one nearly eclipsed.[51,52] Other polysubstituted uranocenes and thorocenes with bulky substituents are 1,1',5,5'-tetrakis(*t*-butyl)uranocene[1286] and $[An\{1,4\text{-}(TMS)_2C_8H_6\}_2]$ (An = Th, U).[774] More recently, 1,3,6,1'3'6'-hexakis(trimethylsilyl)uranocene was synthesized by reacting UCl_4 with the dipotassium salt of 1,3,6-tris(trimethylsilyl)cyclooctatetraene.[787] Annulated uranocenes such as bis(cyclobuteno-cyclooctatetraenyl)uranium(IV) and bis(benzocyclooctatetraenyl)uranium(IV) have also been reported. The average U–C bond length in dibenzouranocene is 0.265(7) nm.[1287,1288] A truly remarkable species in the series of highly substituted uranocenes is 1,3,5,7,1',3',5',7'-octaphenyluranocene.[51,52] In contrast to the pyrophoric parent compound this material is completely air stable due to the effective shielding of the central uranium atom by eight phenyl substituents. Octaphenyluranocene can be sublimed at 400 °C/10^{-5} torr. The two cyclooctatetraenyl ligands are planar and arranged in an eclipsed conformation. The staggered phenyl rings are rotated out of the ligand plane by an average angle of 42°. The average uranium–carbon distances vary between 0.263(2) nm (U–CH) and 0.268(1) nm (U–CPh). Another especially remarkable uranocene derivative is biuranocene.[51,52,1289]

The preparation of uranocene as has been successfully adapted to thorium and other actinide elements. Thorocene, $[Th(C_8H_8)_2]$, has been prepared analogously to Equation (471) from thorium tetrachloride and $K_2C_8H_8$.[31,51,52] The bright yellow material is considered to be somewhat more ionic than uranocene. The crystal structure of thorocene has been found to be isomorphous with that of $[U(C_8H_8)_2]$ with average Th–C bond lengths of 0.2701(4) nm and average C–C distances of 0.1386(9) nm. The observed differences in the metal–carbon distances simply reflect the different ionic radii of the two actinide elements. Fully characterized substituted thorocenes include $[Th(C_8H_7R)_2]$ (R = Bu^t, *o*-tolyl, and mesityl[1290]).

Bis(cyclooctatetraenyl) complexes of neptunium(IV) and protactinium(IV) have been prepared in an analogous manner from the metal tetrachlorides and the cyclooocatetraene dianion (Equation (475)). Bis(tetraethylammonium) hexachloroplutonate(IV) has been found to be a suitable precursor for the preparation of plutonocene (Equation (476)). Alternatively, these bis(cyclooctatetraenyl) complexes can be prepared by reacting the finely divided actinide metals directly with cyclooctatetraene.[31,51,52]

$$AnCl_4 \ + \ 2\,K_2C_8H_8 \ \xrightarrow{\text{THF}} \ [An(C_8H_8)_2] \ + \ 4\,KCl \qquad\qquad (475)$$

$$[NEt_4]_2[PuCl_6] \ + \ 2\,K_2C_8H_8 \ \longrightarrow \ [Pu(C_8H_8)_2] \ + \ 4\,KCl \ + \ 2\,[NEt_4]Cl \qquad\qquad (476)$$

X-ray powder diffraction studies as well as IR data revealed that all these sandwich complexes are isomorphous with uranocene.

Similar synthetic routes as outlined below (Equations (477)–(479)) have been used to prepare ring-substituted bis(cyclooctatetraenyl) derivatives of protactinium, neptunium, and plutonium.[31,51,52,1282,1291]

$$NpCl_4 \ + \ 2\,K_2C_8H_7R \ \xrightarrow{\text{THF}} \ [Np(C_8H_7R)_2] \ + \ 4\,KCl \qquad\qquad (477)$$

$$[NEt_4]_2[PuCl_6] \ + \ 2\,K_2C_8H_7R \ \longrightarrow \ [Pu(C_8H_7R)_2] \ + \ 4\,KCl \ + \ 2\,[NEt_4]Cl \qquad\qquad (478)$$
$$R = Et, \ Bu^n$$

$$AnCl_4 \ + \ K_2(C_8H_4Me_4) \ \longrightarrow \ [An(C_8H_4Me_4)_2] \ + \ 4\,KCl \qquad\qquad (479)$$
$$An = Th, \ Pa, \ Np$$

A large body of spectroscopic data and theoretical calculations concerning the metal-ligand bonding in uranocene and related complexes has now been accumulated.[31,51,52] The main question involves the participation of the actinide 5*f* orbitals in the bonding. A detailed discussion can be found in the previous issue of this monograph. Theoretical calculations have been carried out at various levels of approximation.[785,979,984,987,988] Nonrelativistic and quasirelativistic SCF-X$_\alpha$ scattered wave calculations have been carried out on uranocene and thorocene.[784] The ionization fragmentation of uranocene was investigated in a flowing afterglow study.[1292] Volatilization studies have been carried out on uranocene[1293] and 1,1-di-*n*-butyluranocene[1294] and the heats of combustion, fusion, and evaporation have been determined for these compounds.[1295] Experimental methods which have been used to study $[An(C_8H_7R)_2]$ complexes include optical spectroscopy,[1148] electronic Raman spectroscopy, high-temperature gas-phase electron diffraction,[1296] He^I/He^{II} photoelectron spectroscopy,[1297] ^{237}Np Mössbauer spectroscopy,[1058,1196] as well as magnetic susceptibility measurements. The gas-phase x-ray photoelectron spectrum of uranocene was reported and the metal-ring covalency was compared to uranium hexafluoride, $[U(BH_4)_4]$, and $[U(BH_3Me)_4]$.[990] The overall results agree with the fact that in the case of uranocene the actinide 5*f* orbitals are involved in the metal–ligand bonding and that there is a significant degree of covalency in this bonding. The metal–ligand bonding in thorocene is very similar to that in cerocene, $[Ce(C_8H_8)_2]$, that is, the degree of metal–ligand bond covalency is smaller than in the uranium complex.[785] Detailed ^{1}H NMR spectral studies of uranocenes have mainly been concerned with separating the contact and dipolar contributions to the paramagnetically shifted spectra.[854,1111,1112,1287] In addition, specifically substituted uranocenes have been subjected to variable-temperature ^{1}H NMR investigations.[1298] These studies have revealed two different types of dynamic processes: (a) hindered rotation of bulky substituents on the cyclooctatetraenyl ring, and (b) hindered rotation about the metal–ring centroid axis. The former process was observed for example in 1,1'-dimesityluranocene in which the rotation about the ring-to-substituent C–C bond is hindered by the *ortho*-methyl groups of the mesityl substituents. The phenyl rotation barrier in 1,1'-diphenyluranocene was determined to be $\Delta H^{\ddagger} = 4.4$ kcal mol^{-1}. The second process was reported for 1,4,1',4'-tetrakis(*t*-butyl)uranocene[1299] and 1,3,6,1',3',6'-hexakis(trimethylsilyl)uranocene.[787] Calorimetric data have been reported for uranocene. The mean metal–ligand dissociation energy was found to be 347 kJ mol^{-1} as compared to the significantly smaller value (247 kJ mol^{-1}) reported for tetrakis(cyclopentadienyl)uranium(IV).[51,52]

Reactivity studies involving bis(cyclooctatetraenyl)actinide complexes are fairly limited. All complexes of the type $[An(C_8H_8)_2]$ are extremely air-sensitive. Uranocene itself is known for its pyrophoric nature. Some differences in hydrolytic behavior are notable, as thorocene is highly susceptible to hydrolysis whereas uranocene is only slowly hydrolyzed. Treatment of a THF solution of uranocene with water produces uranium dioxide and two isomers of cyclooctatriene. Some evidence for the formation of the cationic sandwich complex $[U(C_8H_8)_2]^+$ was obtained from an electrochemical investigation of uranocene, although this unstable species was observed only at low temperatures and no stable salt containing this uranium(V) cation has ever been reported. Organic nitro compounds have been found to react with uranocene to yield the corresponding azo compounds (cf. Section 2.3.11).[51,52]

2.3.8.3 *[MX₂(C₈H₈)] compounds*

Cyclooctatetraenyl half-sandwich complexes are a relatively new addition to organoactinide chemistry. Their synthesis proved to be significantly more difficult than the preparation of, for example, uranocene. Meanwhile, several synthetic approaches to monocyclooctatetraenyl complexes of uranium and thorium have been developed. Compounds of the type $[AnCl_2(THF)_2(C_8H_8)]$ (An = U, Th) are an interesting synthetic target, since they should be versatile precursors for a large number of other monocyclooctatetraenylactinide derivatives.

Early attempts to generate the compound $[UCl_2(THF)_2(C_8H_8)]$ by addition of one equivalent of the cyclooctatetraene dianion to uranium tetrachloride in THF failed, as the reaction was quite complex and appeared to involve uranium(III) intermediates. The only isolable product was uranocene (Equation (480)).[52]

$$UCl_4 + K_2C_8H_8 \xrightarrow{\text{THF}} 1/2\,[U(C_8H_8)_2] + 1/2\,UCl_4 + 2\,KCl \tag{480}$$

In marked contrast, the analogous reaction of thorium tetrachloride with one equivalent of the cyclooctatetraene dianion proceeded smoothly to give the thorium half-sandwich complex $[ThCl_2(THF)_2(C_8H_8)]$ in moderate yield (Equation (481)).[52]

$$ThCl_4 + K_2C_8H_8 \xrightarrow{\text{THF}} [ThCl_2(THF)_2(C_8H_8)] + 2\,KCl \tag{481}$$

In the case of thorium, ligand redistribution between thorocene and thorium tetrachloride also proved to be a suitable route to monocyclooctatetraenylthorium(IV) complexes. The reaction is preferably carried out in THF at ca. 100 °C in a pressure-bottle. Ring-substituted derivatives have been prepared accordingly (Equations (482) and (483)).[52,1300,1301] Once again the corresponding uranium system is more complicated. Refluxing of equimolar amounts of uranium tetrachloride and uranocene in THF produced only a small amount of the half-sandwich complex $[UCl_2(THF)_2(C_8H_8)]$ (Equation (484)).

$$[Th(RC_8H_7)_2] + ThCl_4 \xrightarrow[\text{reflux}]{\text{THF}} 2\,[ThCl_2(THF)_2(RC_8H_7)] \tag{482}$$
$$R = H,\ alkyl$$

$$[Th(1,3,5,7\text{-}Me_4C_8H_4)_2] + ThCl_4 \xrightarrow{\text{THF}} 2\,[ThCl_2(THF)_2(1,3,5,7\text{-}Me_4C_8H_4)] \tag{483}$$

$$[U(C_8H_8)_2] + UCl_4 \xrightarrow{\text{THF}} 2\,[UCl_2(THF)_2(C_8H_8)] \tag{484}$$

Another successful approach to the synthesis of monocyclooctatetraenylactinide half-sandwich complexes involves treatment of the corresponding bis(cyclooctatetraenyl)metal(IV) complexes with protic acids. While the reaction of thorocene with dry HCl generates the half-sandwich complex in good yield, only moderate yields of the uranium derivative were obtained when uranocene was treated with dry HCl in THF (Equation (485)). In this case the uranium reaction was complicated by the generation of large amounts of UCl_4. Nevertheless, the reaction was successfully employed in the preparation of uranium(IV) half-sandwich complexes containing alkyl-substituted cyclooctatetraenyl ligands.[51,52,1284] Selective replacement of one cyclooctatetraenyl ligand of uranocene was also achieved by treatment with iodine (Equation (486)).[1302]

The molecular structure of $[ThCl_2(THF)_2(C_8H_8)]$ has been determined by x-ray diffraction. The compound adopts the expected half-sandwich structure with a "four-legged piano stool" geometry.[52]

$$[An(RC_8H_7)_2] + 2\,HCl \xrightarrow{\ THF\ } [AnCl_2(THF)_2(RC_8H_7)] \qquad (485)$$

$$An = Th,\ U$$
$$R = H,\ alkyl$$

$$[U(C_8H_8)_2] + I_2 \xrightarrow{\ THF\ } [UI_2(THF)_2(C_8H_8)] + C_8H_8 \qquad (486)$$

The most successful synthetic route leading to monocyclooctatetraenyluranium(IV) complexes involves the use of uranium(III) compounds as one-electron reducing agents (Equation (487)). Uranium trichloride was generated *in situ* from UCl_4 and sodium hydride:[1303]

$$UCl_4 + C_8H_8 + 2\,NaH \xrightarrow{\ THF\ } [UCl_2(THF)_2(C_8H_8)] + 2\,NaCl + H_2 \qquad (487)$$

Yet another synthetic approach to monocyclooctatetraenylactinide(IV) complexes uses metal tetrahydroborate precursors. Typical preparations of thorium and uranium derivatives are shown below (Equations (488)–(490)). A tridentate coordination of the borohydride ligands was deduced from vibrational spectroscopy data. This is in fact the favored coordination mode for borohydride ligands in organoactinide complexes.[52,1304]

$$\text{"}[AnCl_2(BH_4)_2]\text{"} + K_2C_8H_8 \xrightarrow{\ THF\ } [An(BH_4)_2(THF)_2(C_8H_8)] + 2\,NaCl \qquad (488)$$

$$An = Th,\ U$$

$$[Th(BH_4)_4(THF)_2] + [Th(Bu^nC_8H_7)_2] \xrightarrow{\ THF\ } 2\,[\,Th(BH_4)_2(THF)_2(Bu^nC_8H_7)] \qquad (489)$$

$$[U(BH_4)_4] + C_8H_8 \xrightarrow{\hspace{2cm}} [U(BH_4)_2(C_8H_8)] \qquad (490)$$

A ring-substituted uranium analogue, $[U(BH_4)_2\{1,4\text{-}(TMS)_2C_8H_6\}]$, was prepared by reacting $[UCl_2(BH_4)_2]$ with the dilithium salt of the 1,4-bis(trimethylsilyl)cyclooctatetraene dianion.[773]

The anionic tetrahydroborate complexes $[U(BH_4)_3(C_8H_8)]^-$ and $[U(BH_4)_2(OEt)(C_8H_8)]^-$ have been prepared by addition of $[BH_4]^-$ to the neutral complexes $[U(BH_4)_2(C_8H_8)]$ and $[U(BH_4)(OEt)(C_8H_8)]$.[1305] Protonation of $[U(BH_4)_2(C_8H_8)]$ with alcohols ROH (R = Et, Pr^i, Bu^t) afforded the alkoxide derivatives $U(BH_4)(OR)(C_8H_8)$ and $U(OR)_2(C_8H_8)$. For R = Et and Pr^i the products are dimeric in THF solution and in the solid state, whereas the *t*-butoxides are monomers. The compounds $[\{U(BH_4)(\mu\text{-}OEt)(C_8H_8)\}_2]$ and $[\{U(OPr^i)(\mu\text{-}OPr^i)(C_8H_8)\}_2]$ have been structurally characterized.[1305,1306]

The mono(cyclooctatetraenyl)uranium bis(thiolate) complexes $[\{U(\mu\text{-}SR)_2(C_8H_8)\}_2]$ (R = Pr^i, Bu^n) were prepared by treating $[U(BH_4)_2(C_8H_8)]$ with the corresponding thiol or NaSR reagent. The structurally characterized isopropyl thiolate is dimeric in the solid state with four bridging $\overline{S}Pr^i$ ligands (Figure 95), and its structure is retained in solution. Reaction of $[U(BH_4)_2(C_8H_8)]$ with excess NaSBu^t afforded the anion $[U(SBu^t)_3(C_8H_8)]^-$, which adopts a three-legged piano-stool configuration.[1307]

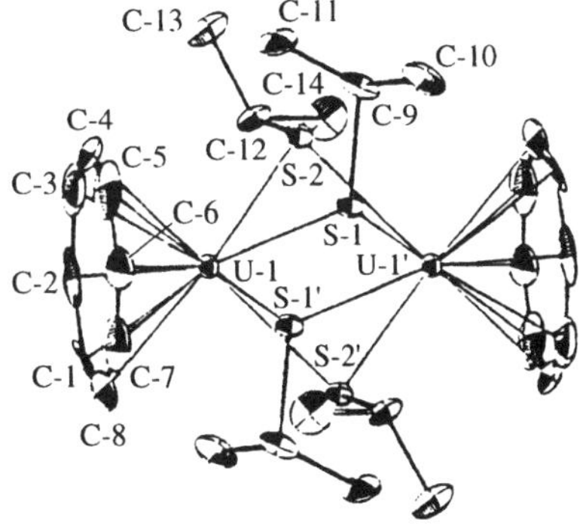

Figure 95 The molecular structure of $[\{U(\mu\text{-}SPr^i)_2(C_8H_8)\}_2]$.[1307]

Another recent development is the successful preparation of cyclooctatetraenylactinide bis(hydrocarbyls). Early attempts to generate such species by treatment of $[UCl_2(THF)_2(C_8H_8)]$ with alkyllithium or Grignard reagents had failed.[1303] However, the compound $[U(CH_2TMS)_2(HMPA)(C_8H_8)]$ was synthesized by reacting $[UI_2(HMPA)_2(C_8H_8)]$ with two equivalents of LiCH$_2$TMS. In the presence of excess lithium reagent the anionic complex $[Li(THF)_3][U(CH_2TMS)_3(C_8H_8)]$ could be isolated.[1304,1305]

An interesting derivative chemistry of mono(cyclooctatetraenyl) actinide complexes is currently being developed. Metathetical reactions using chloride or borohydride precursors led to the compounds $[U(acac)_2(C_8H_8)]$, $[U\{N(TMS)_2\}_2(C_8H_8)]$, $[U(BH_4)L(C_8H_8)(Cp)]$ (L = THF, Ph_3PO), and $[UI(C_8H_8)(Cp^*)]$. Adduct formation with Lewis bases afforded the complexes $[UCl_2(PMe_3)_2(C_8H_8)]$, $[UCl_2(C_5H_5N)_2(C_8H_8)]$ (U–N 0.2639(5), 0.2644(6) nm), $[UI_2(HMPA)_2(C_8H_8)]$, and $[U(BH_4)_2L(C_8H_8)]$ (L = THF, PPh_3, Ph_3PO). The molecular structure of $[U(BH_4)_2(Ph_3PO)(C_8H_8)]$ was determined (U–C 0.268(1) nm, U–O 0.227(1) nm).[1302-4]

The derivative chemistry of $[ThCl_2(THF)_2(C_8H_8)]$ is illustrated in Scheme 80.[1300,1301]

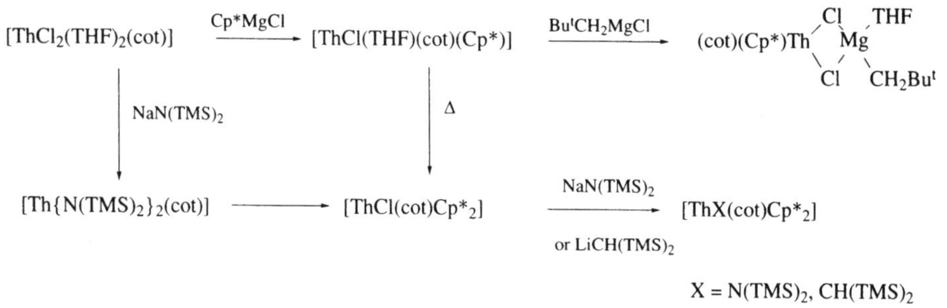

$[ThCl_2(THF)_2(cot)] \xrightarrow{Cp^*MgCl} [ThCl(THF)(cot)(Cp^*)] \xrightarrow{Bu^tCH_2MgCl} (cot)(Cp^*)Th\begin{smallmatrix} Cl \\ \diagup \quad \diagdown \\ \quad Mg \\ \diagdown \quad \diagup \\ Cl \end{smallmatrix}\begin{smallmatrix} THF \\ \\ \\ \\ CH_2Bu^t \end{smallmatrix}$

$\downarrow$ NaN(TMS)$_2$ $\downarrow$ Δ

$[Th\{N(TMS)_2\}_2(cot)] \longrightarrow [ThCl(cot)Cp^*_2] \xrightarrow[\text{or LiCH(TMS)}_2]{\text{NaN(TMS)}_2} [ThX(cot)Cp^*_2]$

$X = N(TMS)_2, CH(TMS)_2$

Scheme 80

Especially remarkable is the structurally characterized magnesium compound $[Th(\mu\text{-}Cl)_2(C_8H_8)(Cp^*)Mg(CH_2Bu^t)(THF)]$, which has been termed a "transition metal Grignard adduct." Due to coordinative unsaturation, agostic interactions are important structural features in the crystallographically characterized complexes $[ThCH(TMS)_2(C_8H_8)(Cp^*)]$ and $[Th\{N(TMS)_2\}_2(C_8H_8)]$.[1300,1301] A series of mono(cyclooctatetraenyl)uranium amides includes the compounds $[U(NEt_2)_2(C_8H_8)]$, $M[U(NEt_2)_3(C_8H_8)]$ (M = Li, Na, K), and $[(C_8H_8)U(NEt_2)(THF)_2][BPh_4]$. Most remarkable is the fact that the anionic species $M[U(NEt_2)_3(C_8H_8)]$ can be oxidized to the first uranium(V) cyclooctatetraenyl complexes (Scheme 81).[1308]

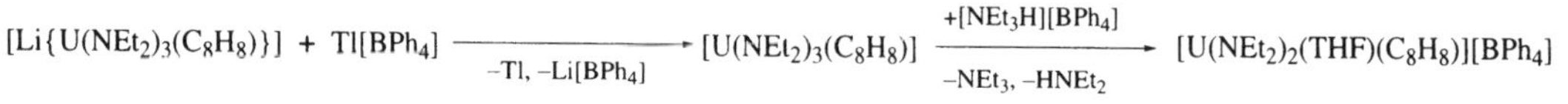

$[Li\{U(NEt_2)_3(C_8H_8)\}] + Tl[BPh_4] \xrightarrow[-Tl, -Li[BPh_4]]{} [U(NEt_2)_3(C_8H_8)] \xrightarrow[-NEt_3, -HNEt_2]{+[NEt_3H][BPh_4]} [U(NEt_2)_2(THF)(C_8H_8)][BPh_4]$

Scheme 81

2.3.9 Heterobimetallic Compounds

The synthesis and characterization of heterobimetallic organoactinides has attracted considerable interest. Interest in these compounds stems from the expectation that such heterobimetallics may exhibit unusual chemical and catalytic properties different from those of the individual mononuclear species.

2.3.9.1 Metal–metal bonded complexes

A fundamental difficulty implicated with the synthesis of mixed-metal complexes containing transition metals is the highly oxophilic character of the actinide ions. This is the main reason why early attempts to prepare such species have all failed. The first compound for which a uranium–transition metal bond was suggested was $[U\{Mn(CO)_5\}_4]$. This complex was prepared by treatment of uranium tetrachloride with four equivalents of $Na[Mn(CO)_5]$ (Equation (491)).[32]

$$UCl_4 + 4\,Na[Mn(CO)_5] \longrightarrow [U\{Mn(CO)_5\}_4] + 4\,NaCl \qquad (491)$$

The product $[U\{Mn(CO)_5\}_4]$, however, was not fully characterized. Its characterization was based only on elemental analysis, osmometric molecular weight determination, and IR and mass spectrometry. Presumably due to its extreme air-sensitivity, the bright orange crystalline material was not structurally characterized by x-ray diffraction. Generally, metal carbonyl fragments tend to coordinate to lanthanide or actinide ions via isocarbonyl linkages, that is, M-C≡O → Ln or M-C≡O → An, respectively. An isocarbonyl bridge between two metal atoms is easily detected by IR spectroscopy. Isocarbonyl linked species are characterized by low CO stretching frequencies in the region around 1600 cm^{-1}. Although the lowest CO band reported for $[U\{Mn(CO)_5\}_4]$ appeared at 1864 cm^{-1}, an isocarbonyl structure cannot definitively be ruled out. As depicted in Equations (492)–(495), most other related reactions of organouranium compounds with metal carbonyl derivatives did not lead to the formation of uranium–metal bonded products. Once again, isocarbonyl linked species were obtained.[32,1027,1309]

$$[U(NEt_2)_2Cp_2] + 2\,[Mo(CO)_3HCp] \longrightarrow [U(\mu\text{-}OC)Cp_2\{Mo(CO)_2Cp\}_2] + 2\,HNEt_2 \qquad (492)$$

$$[UClCp_3] + Na[Mo(CO)_3Cp] \longrightarrow [U(\mu\text{-}OC)\{Mo(CO)_2Cp\}Cp_3] + NaCl \qquad (493)$$

$$[UMeCp_3] + [M(CO)_3HCp'] \longrightarrow [U(\mu\text{-}OC)\{M(CO)_2Cp'\}Cp_3] + CH_4 \qquad (494)$$
$$M = Mo;\; Cp' = Cp \text{ or } Cp^*$$
$$M = W;\; Cp' = Cp$$

$$[UMe\{N(TMS)_2\}_3] + [M(CO)_3HCp] \longrightarrow [U(\mu\text{-}OC)\{N(TMS)_2\}_3\{M(CO)_2Cp\}] + CH_4 \qquad (495)$$
$$M = Mo,\, W$$

In all cases, low CO stretching frequencies showed the presence of isocarbonyl linkages between uranium and the transition metals.[1309]

In view of these results the successful synthesis of the first complexes with direct, unsupported actinide–transition metal bonds must be considered a major breakthrough in this area. The first successful approach involved the combination of the two fragments $ThX(Cp^*)_2$ (X = Cl, I) and $Ru(CO)_2(Cp)$ according to Equation (496):[1310]

$$[ThX_2Cp^*_2] + Na[Ru(CO)_2Cp] \xrightarrow[25\,^\circ C]{\text{toluene}} [ThX\{Ru(CO)_2Cp\}Cp^*_2] + NaX \qquad (496)$$
$$X = Cl,\, I$$

Crucial for the formation of a metal–metal bonded species is that the transition metal carbonyl fragment posesses an appropriately directed, high-lying, metal-centered HOMO (highest occupied molecular orbital). The molecular structure of $[ThI\{Ru(CO)_2Cp\}Cp^*_2]$ is shown in Figure 96. As compared to values available in the literature (i.e., calculated distances from metallic radii or from analogous Zr–Ru compounds[1311,1312]) the Th–U distance of 0.30277(6) nm is quite short.[1310]

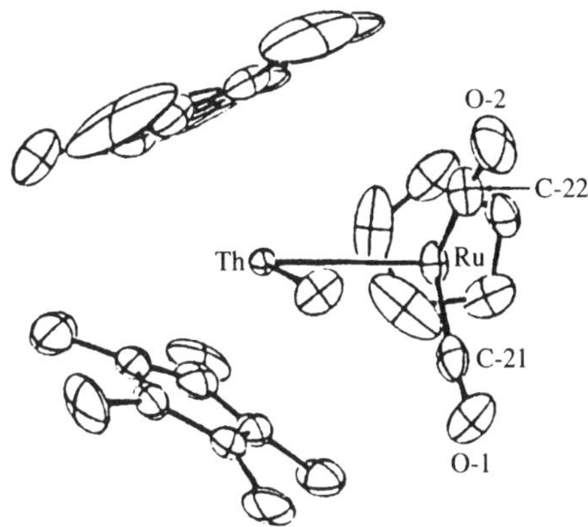

Figure 96 The molecular structure of $[ThI\{Ru(CO)_2Cp\}Cp^*_2]$.[1310]

Due to its highly polar character the thorium–ruthenium bond is easily cleaved upon treatment of the heterobimetallic compound with alcohols (Equation (497)).[1310]

$$[ThI\{Ru(CO)_2Cp\}Cp^*_2] + Bu^tOH \longrightarrow [ThI(OBu^t)Cp^*_2] + [Ru(CO)_2HCp] \qquad (497)$$

The heterobimetallic thorium–ruthenium complex was also reported to induce unusual oligomerization reactions with acetonitrile and acetone as illustrated in Equations (498) and (499). The thorium amidinate complex was structurally characterized (Th–C 0.280(2) nm, Th–N 0.246(1) nm).[1313]

Later it was discovered that metal–metal bonded species are also accessible by reacting tris(cyclopentadienyl)actinide chlorides with metal carbonyl anions of iron and ruthenium (Equation (500)).[1314]

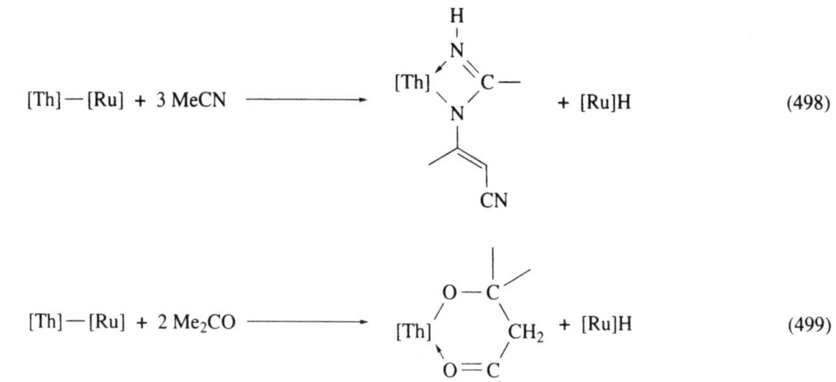

$$[Th]-[Ru] + 3\,MeCN \longrightarrow \quad + [Ru]H \tag{498}$$

$$[Th]-[Ru] + 2\,Me_2CO \longrightarrow \quad + [Ru]H \tag{499}$$

$$[Th] = Cp^*_2(Cl)Th$$
$$[Ru] = CpRu(CO)_2$$

$$[AnCl(RC_5H_4)_3] + Na[M(CO)_2Cp] \longrightarrow [An\{M(CO)_2Cp\}(RC_5H_4)_3] \tag{500}$$

$$R = H;\ An = Th,\ U$$
$$R = Me;\ An = Th$$
$$M = Fe,\ Ru$$

The compound $[Th\{Fe(CO)_2Cp\}Cp_3]$ was crystallographically characterized (Th–Fe 0.2940(5) nm). NMR spectral data revealed hindered rotation about the actinide–metal bonds in these complexes. Facile cleavage of the An–M bond was observed upon treatment with acohols or ketones.[1314] According to X_α-scattered wave MO calculations the $Ru(CO)_2(Cp)$ unit can be regarded as an organometallic pseudohalide.[1315] Uranium–metal bond disruption enthalphies have recently been measured for the complexes $[U\{M(CO)_2Cp\}Cp_3]$ (M = Fe, Ru). These data were indicative of weak heterobimetallic bonding.[1174]

2.3.9.2 Heterobimetallic complexes without direct metal–metal bonds

As mentioned above (cf. Section 2.3.9.1), reactions of organoactinide halides with anionic transition metal carbonyl reagents frequently lead to the formation of isocarbonyl-linked species.[32,36,1309] Other examples of heterobimetallic *f*-element complexes without direct metal–metal interactions include phosphido-bridged species. The following two phosphido-bridged complexes have been synthesized by reacting the "metallophosphine ligand" $[Th(PPh_2)_2(Cp^*)_2]$ with nickel(0) or platinum(0) reagents (Equations (501) and (502)).[1316,1317]

$$[Th(PPh_2)_2Cp^*_2] + [Ni(1,5\text{-cod})_2] + 2\,CO \xrightarrow{THF} [Th(\mu\text{-PPh}_2)_2Cp^*_2Ni(CO)_2] + 2\,1,5\text{-cod} \tag{501}$$

$$[Th(PPh_2)_2Cp^*_2] + [Pt(1,5\text{-cod})_2] + PMe_3 \xrightarrow{toluene} [Th(\mu\text{-PPh}_2)_2Cp^*_2Pt(PMe_3)] + 2\,1,5\text{-cod} \tag{502}$$

Both phosphido-bridged complexes have been structurally characterized. The Th–Pt distance in $[Th(\mu\text{-PPh}_2)_2(Cp^*)_2Pt(PMe_3)]$ (**31**) is 0.2984(1) nm.[1316,1317] With 0.3206(2) nm the Th–Ni separation in $[Th(\mu\text{-PPh}_2)_2(Cp^*)_2Ni(CO)_2]$ (**30**) is about 0.05 nm shorter than the expected nonbonding distance. MO calculations on the model complex $[Th(\mu\text{-PH}_2)_2(Cp)_2Ni(CO)_2]$ revealed the possibility of a weak donor–acceptor bond between nickel and thorium.[1318]

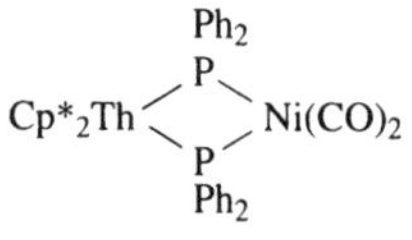

(30) 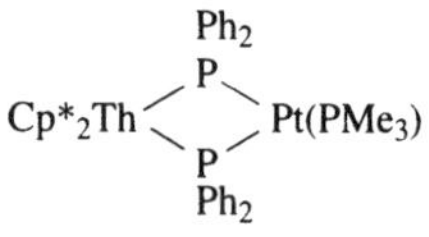   **(31)**

Bimetallic uranium complexes have been synthesized with the use of bridging diphenylphosphinocyclopentadienyl ligands (Equations (503) and (504)).[1028]

In a similar manner, the new metallo-ligands $[UX(C_5H_4PPh_2)_3]$ (X = Cl, NEt$_2$, BH$_4$) have been used to synthesize bimetallic complexes. Treatment of the diphenylphosphinocyclopentadienyl derivatives

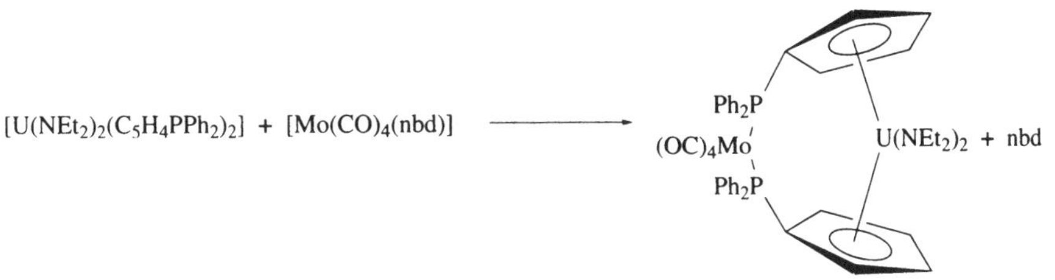

$$[U(NEt_2)_2(C_5H_4PPh_2)_2] + [Mo(CO)_4(nbd)] \longrightarrow (OC)_4Mo\langle{Ph_2P \atop Ph_2P}\rangle U(NEt_2)_2 + nbd \qquad (503)$$

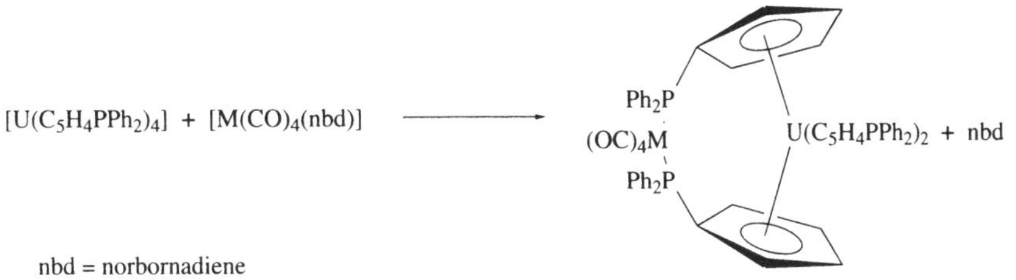

$$[U(C_5H_4PPh_2)_4] + [M(CO)_4(nbd)] \longrightarrow (OC)_4M\langle{Ph_2P \atop Ph_2P}\rangle U(C_5H_4PPh_2)_2 + nbd \qquad (504)$$

nbd = norbornadiene
M = Cr , Mo

with [Mo(CO)$_4$(nbd)] by analogy with Equation (503) afforded the compounds [{Mo(CO)$_4$(C$_5$H$_4$PPh$_2$)$_2$}(C$_5$H$_4$PPh$_2$)UX] (X = Cl, NEt$_2$, BH$_4$).[1134]

A hydrido-bridged bimetallic uranium–rhenium complex was synthesized by the reaction outlined in Equation (505). In analogy to [UBH$_4$(Cp)$_3$] it was concluded that the rhenium hydride unit acts as a tridentate ligand toward uranium.[1319]

$$[UClCp_3] + KReH_6L_2 \longrightarrow [UReH_6L_2Cp_3] + KCl \qquad (505)$$

$$L = PPh_3, P(C_6H_4F\text{-}p)_3$$

An unusual bimetallic μ-methylene complex containing a metallated cyclopentadienyl ring was prepared by thermolyzing mixtures of the thoracyclobutane [Th(CH$_2$)$_2$CMe$_2$(Cp*)$_2$] and dimethyl metallocenes of zirconium or hafnium (Equation (506)). The complex with M = Zr and R = Me was structurally characterized (Th–C(CH$_2$) 0.2377(8) nm).[1320]

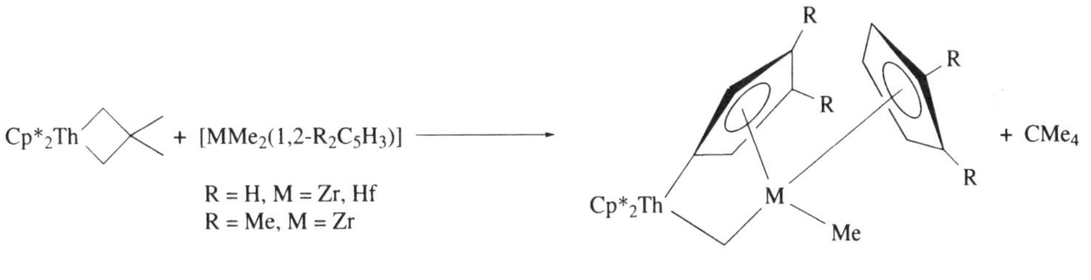

$$Cp^*_2Th\langle\rangle + [MMe_2(1,2\text{-}R_2C_5H_3)] \longrightarrow \qquad + CMe_4 \qquad (506)$$

R = H, M = Zr, Hf
R = Me, M = Zr

Another remarkable class of multimetallic compounds containing uranium or thorium are polyoxoanion-supported organoactinides which have been investigated by Klemperer *et al.* and prepared according to Equation (507).[1321]

$$[AnClCp_3] + 2\,[N(Bu^n)_4]_3[MW_5O_{19}] \longrightarrow [N(Bu^n)_4]_5[An(MW_5O_{19})_2Cp_3] + [N(Bu^n)_4]Cl \qquad (507)$$

An = Th, U
M = Nb, Ta

The compound [N(Bu^n)$_4$]$_5$[U(NbW$_5$O$_{19}$)$_2$(Cp)$_3$] was structurally characterized by x-ray diffraction, in which the uranium adopts a pseudotrigonal bipyramidal coordination geometry. Three η^5-coordinated cyclopentadienyl ligands are in the equatorial positions and the two polyoxoanions are axially coordinated via oxygen atoms.[1321] A smilar preparation afforded the dimeric bis(cyclopentadienyl)uraniumpolyoxometallate [N(Bu^n)$_4$][U(TiW$_5$O$_{19}$)(Cp)$_2$]. In this structurally characterized complex two U(Cp)$_2$ units are bridged by the oxygen-coordinated polyoxoanions (Figure 97).[1201]

Yet another type of bimetallic actinide complexes contains σ-bonded ferrocenyl ligands. Especially useful for the stabilization of such compounds is the chelating 2-(dimethylaminomethyl)ferrocenyl ligand (= FcN). Two actinide FcN complexes have been prepared and characterized (Equations (508) and (509)).[1322]

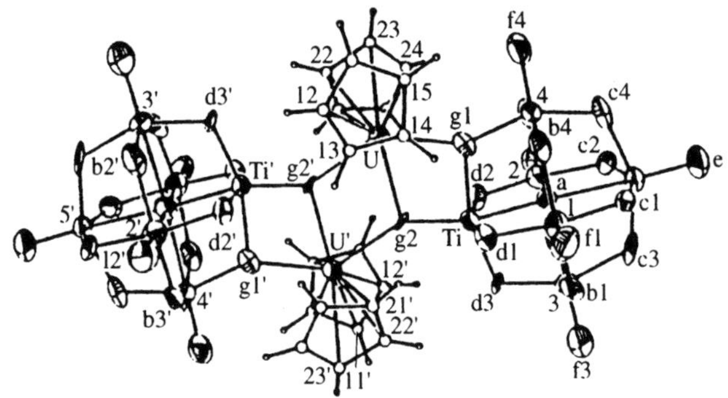

Figure 97 The molecular structure of the anion in $[N(Bu^n)_4][U(TiW_5O_{19})(Cp)_2]$.[1201]

$$[Th(acac)_4] + Li(FcN) \longrightarrow [Th(acac)_3(FcN)] + Li(acac) \tag{508}$$

$$[UO_2(acac)_2] + Li(FcN) \longrightarrow [UO_2(acac)(FcN)] + Li(acac) \tag{509}$$

Polymeric mixed-metal complexes were obtained by reacting $[UCl(Cp)_3]$ with $K_3[Co(CN)_6]$ to afford $[U(Cp)_3]_3[Co(CN)_6]$. From these experiments a similarity between the $[U(Cp)_3]$ unit and $[SnMe_3]$ was deduced and the structure of $[\{U(Cp)_3\}_3Co(CN)_6]$ was compared with that of polymeric $[Co(CN)_6(Me_3Sn)_3]$. From the spectroscopic data a pseudotrigonal bipyramidal geometry around uranium was proposed.[1323]

2.3.10 Homogeneous Catalysis

Catalytic aspects of organoactinide chemistry have been reviewed by Marks *et al.*[1251,1324] and Folcher.[859]

Organoactinide hydrides of the type $[\{AnH_2(Cp^*)_2\}_2]$ undergo a rapid and quantitative addition of terminal olefins (cf. Section 2.3.5.3).[1232] This process can be regarded as the reverse of β-hydride elimination. Consequently it was discovered that these hydrides effectively catalyze the hydrogenation of α-alkenes. In the case of $[\{UH_2(Cp^*)_2\}_2]$ N_t for the hydrogenation of 1-hexene at 25 °C and 1 atm H_2 in toluene is 63 000 h^{-1}. Scheme 82 shows the mechanism of homogeneous α-alkene hydrogenation using $[\{UH_2(Cp^*)_2\}_2]$ as catalyst. The mechanism was derived from detailed investigation analogous to reactions catalyzed by organolanthanide hydrides such as $[\{Lu(\mu\text{-}H)(Cp^*)_2\}_2]$.

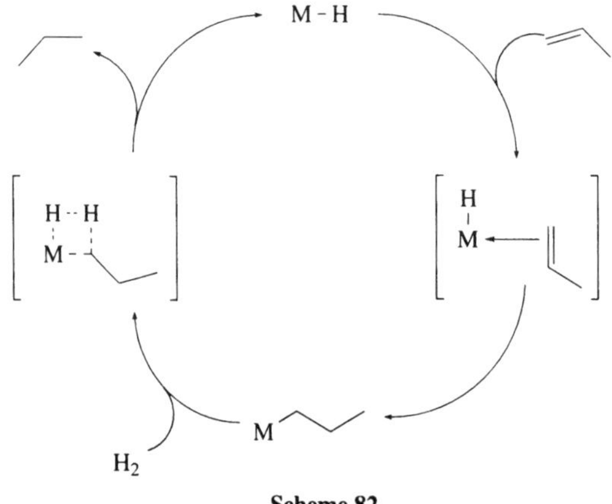

Scheme 82

Catalytic reactions can also be carried out heterogeneously when the organoactinide hydrides are absorbed on high-surface-area dehydroxylated alumina.[1103,1104] In this immobilized form the bis(pentamethylcyclopentadienyl)actinide hydrides are effective catalysts for alkene hydrogenation as well as ethylene polymerization.[1325-8] The "pulled-back" bis(cyclopentadienyl)thorium hydride $[(ThH_2\{Me_2Si(C_5Me_4)_2\})_2]$ exhibits a greatly enhanced activity (ca. 1000 times more active) in catalytic olefin hydrogenation as compared to $[\{UH_2(Cp^*)_2\}_2]$.[1202]

Catalytic hydrogenation of terminal alkenes was also observed in the presence of anionic uranate(III) complexes of the type $Li[UR(Cp)_3]$ (R = Me, Bu^n). The uranium(III) hydride $UH(Cp)_2$ was proposed to be an intermediate in the catalytic cycle.[1039,1070]

Several organoactinide hydrocarbyls such as $[ThMe(Cp)_3]$, $[ThR_2(Cp^*)_2]$ (R = Me, CD_3, Et), $[ThBz_3(Cp^*)]$, and $[ThBz_2(Cp^*)_2]$, and $[UMe_2(Cp^*)_2]$ have been immobilized on surfaces and the resulting potential catalysts studied by ^{13}C CPMAS NMR spectroscopy.[1329] Alkylation of the surfaces and formation of An–O-surface linkages was proposed. Catalytic activity was only observed for systems containing "cationlike" species of the type $[ThMe(Cp^*)_2]^+$ MeAl ~ surface$^-$ formed on dehydroxylated supports (e.g., $MgCl_2$, highly dehydroxylated silica or alumina), while no catalytic activity was displayed by systems with "μ-oxo-like" units of the type $[Th(Me)(Cp^*)_2–O–Al ~ surface]$ or $[Th(Me)(Cp^*)_2–O–Si ~ surface]$. The latter are formed on hydroxylated supports (MgO, Al_2O_3, SiO_2).[1327,1328,1330] $[ThBz_3(Cp^*)]$ on dehydroxylated alumina catalyzes ethylene polymerization as well as propylene and arene hydrogenation, but is inactive in propylene polymerization. Facile arene hydrogenation was also reported for the supported complexes $[Th(CH_2C_6H_3Me_2-3,5)_4/DA]$ and $[Th(\eta^3-C_3H_5)_4/DA]$ (DA = dehydroxylated alumina. The catalytic activity for benzene hydrogenation follows the order $[Th(\eta^3-C_3H_5)_4/DA] > [Th(CH_2C_6H_3Me_2-3,5)_4/DA] > [ThBz_3(Cp^*)/DA]$. The proposed mechanism for $[Th(\eta^3-C_3H_5)_4/DA]$-catalyzed arene hydrogenation is presented in Scheme 83.[1010]

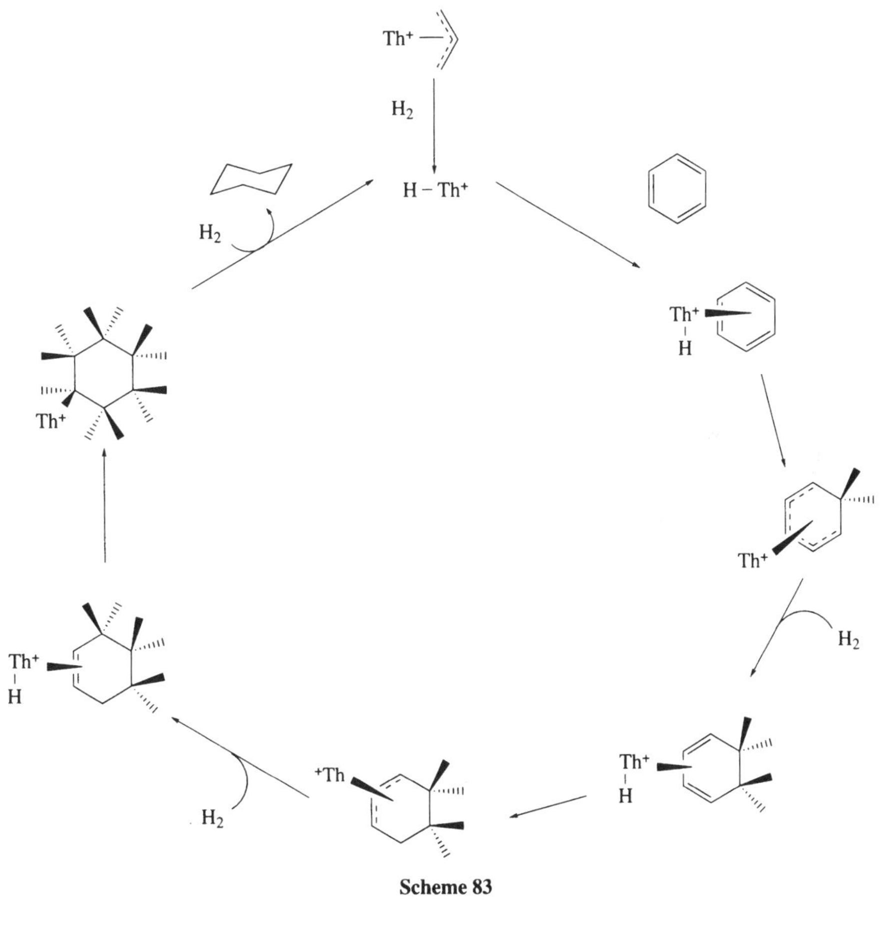

Scheme 83

The cationic complexes $[ThMe(Cp^*)_2][BPh_4]$ and $[ThMe(Cp^*)_2][B(C_6F_5)_4]$ have been investigated as alkene (ethylene, 1-hexene) polymerization catalysts. The latter has been found to be much more active because of the weakly coordinating anion, while the tetraphenylborate is deactivated due to π-arene coordination.[1266]

Dehydrocoupling of phenylsilane was studied in the presence of $[UMe_2(Cp^*)_2]$ and $[ThMe_2(Cp^*)_2]$ as catalysts. However, the practical use of these compounds as catalysts was ruled out.[1331]

2.3.11 Organic Synthesis

The silylamido substituted uranium hydrocarbyl [UMe{N(TMS)$_2$}$_3$] and the related metallacycle [UCH$_2$SiMe$_2$N-TMS{N(TMS)$_2$}$_2$] have been introduced as reagents in preparative organic chemistry by Dormond *et al.*[1018-28] The metallacycle is often more reactive than the uranium methyl complex.[1022,1027] Aldehydes and ketones can be transformed into alcohols or vinyl compounds upon treatment with the uranium reagents, and methyl ketones have been made similarly from nitriles (Equations (510)–(512)).[1023,1025,1026,1084]

$$R_2C{=}O \xrightarrow[\text{ii, } H_3O^+]{\text{i, [UMe\{N(TMS)}_2\}_3]} R_2C(Me)(OH) \tag{510}$$

$$RC{\equiv}N \xrightarrow[\text{ii, } H_3O^+]{\text{i, [UCH}_2\text{SiMe}_2\text{N-TMS\{N(TMS)}_2\}_2]} RC(O)Me \tag{511}$$

$$R_2C{=}O \xrightarrow[\text{ii, } H_3O^+]{\text{i, [UCH}_2\text{SiMe}_2\text{N-TMS\{N(TMS)}_2\}_2]} R_2C{=}CH_2 \tag{512}$$

In its capability to synthesize vinyl compounds from carbonyls the metallacycle resembles the well-known Tebbe reagent. The metallacycle also acts as a highly selective nucleophile in chemo- and stereoselective alkylation reactions of carbonyl compounds and even reacts with esters.[1022,1027,1084] Insertion reactions of aldehydes and ketones into the uranium–carbon bond have also been reported for the hydrotris(pyrazolyl)borate uranium hydrocarbyls [UCl$_2$(CH$_2$TMS){HB(3,5-Me$_2$pz)$_3$}] and [UCl$_2${CH(TMS)$_2$}{HB(3,5-Me$_2$pz)$_3$}].[1332]

Mono(cyclopentadienyl)uranium hydrocarbyls of the type UCl$_2$R(Cp) (R = Me, Et, Bu^n) have been investigated in alkylation reactions of carbonyl compounds.[1084]

Treatment of uranocenes with aliphatic or aromatic nitro compounds results in deoxygenation and formation of the corresponding azo derivatives (Equation (513)). Primary amines are often formed as by-products. Oxygen transfer from the nitro group to uranium seems to be the initial step in this reaction.[51,1333]

$$2\,[U(C_8H_7R^1)_2] + 2\,R^2NO_2 \longrightarrow R^2N{=}NR^2 + 4\,C_8H_7R^1 + 2\,UO_2 \tag{513}$$

$$R^1 = Bu^t, R^2 = Ph$$

2.4 REFERENCES

1.   G. Wilkinson and J. M. Birmingham, *J. Am. Chem. Soc.*, 1954, **76**, 6210.
2.   L. T. Reynolds and G. Wilkinson, *J. Inorg. Nucl. Chem.*, 1956, **2**, 246.
3.   R. D. Rogers and L. M. Rogers, *J. Organomet. Chem.*, 1991, **416**, 201.
4.   R. D. Rogers and L. M. Rogers, *J. Organomet. Chem.*, 1992, **442**, 83.
5.   R. D. Rogers and L. M. Rogers, *J. Organomet. Chem.*, 1992, **442**, 225.
6.   R. D. Rogers and L. M. Rogers, *J. Organomet. Chem.*, 1993, **457**, 41.
7.   U. Kilimann and F. T. Edelmann, *J. Organomet. Chem.*, 1994, in press.
8.   J. D. Miller, *Annu. Rep. Prog. Chem., Sect. A*, 1985, **81**, 325.
9.   J. D. Miller, *Annu. Rep. Prog. Chem., Sect. A*, 1986, **82**, 347.
10.   J. D. Miller, *Annu. Rep. Prog. Chem., Sect. A*, 1986, **83**, 357.
11.   J. D. Miller, *Annu. Rep. Prog. Chem., Sect. A*, 1987, **84**, 285.
12.   M. J. Winter, *Organomet. Chem.*, 1987, **15**, 229.
13.   M. J. Winter, *Organomet. Chem.*, 1989, **17**, 206.
14.   M. J. Winter, *Organomet. Chem.*, 1989, **18**, 200.
15.   C. J. Jones, *Annu. Rep. Prog. Chem., Sect. A*, 1991, **86**, 77.
16.   R. D. Ernst and T. J. Marks, *J. Organomet. Chem.*, 1987, **318**, 29.
17.   H. Schumann and W. Genthe, in 'Handbook of the Physics and Chemistry of Rare Earths', 1984, **7**, 445.
18.   H. Schumann, *Angew. Chem.*, 1984, **96**, 475; *Angew. Chem., Int. Ed. Engl.*, 1984, **23**, 474.
19.   H. Schumann, *J. Organomet. Chem.*, 1985, **281**, 95.
20.   H. Schumann, in 'Fundamental and Technological Aspects of Organo-*f*-Element Chemistry', NATO ASI Ser., ed. T. J. Marks and I. L. Fragalà, D. Reidel, Boston, MA, 1985, **155**, 1.
21.   J. A. McCleverty, *Trans. Met. Chem.*, 1985, **10**, 118.
22.   G. B. Deacon, *Chem. Aust.*, 1991, **58**, 162.
23.   M. N. Bochkarev, G. S. Kalinina and L. N. Bochkarev, *Usp. Khim.*, 1985, **54**, 1362.

24. Q. Shen, *Huaxue Tongbao*, 1985, 1.
25. W. E. Piers, P. J. Shapiro, E. E. Bunel and J. E. Bercaw, *Synlett*, 1990, 74.
26. C. J. Schaverien, *Adv. Organomet. Chem.*, 1994, **36**, 283.
27. W. J. Evans, *Adv. Organomet. Chem.*, 1985, **24**, 131.
28. W. J. Evans, *Polyhedron*, 1987, **6**, 803.
29. T. J. Marks, *Science*, 1982, **217**, 989.
30. P. J. Fagan, E. A. Maatta, J. M. Manriquez, K. G. Moloy, A. M. Seyam and T. J. Marks, in 'Actinides in Perspective', ed. N. Edelstein, Plenum Press, Oxford, 1992, 433.
31. T. J. Marks and A. Streitwieser, Jr., in 'The Chemistry of the Actinide Elements', ed. J. J. Katz, G. T. Seaborg and L. R. Morss, Chapman & Hall, London, New York, 1986, vol. 2, chap. 22.
32. T. J. Marks in 'The Chemistry of the Actinide Elements', ed. J. J. Katz, G. T. Seaborg and L. R. Morss, Chapman & Hall, London, New York, 1986, vol. 2, chap. 23.
33. T. J. Marks and V. W. Day, in 'Fundamental and Technological Aspects of Organo-*f*-Element Chemistry', NATO ASI Ser., ed. T. J. Marks and I. L. Fragalà, D. Reidel, Boston, 1985, **155**, 115.
34. J. Takats, in 'Fundamental and Technological Aspects of Organo-*f*-Element Chemistry', NATO ASI Ser., ed. T. J. Marks and I. L. Fragalà, D. Reidel, Boston, 1985, **155**, 159.
35. I. Santos, A. Pires de Matos and A. G. Maddock, *Adv. Organomet. Chem.*, 1989, **34**, 65.
36. M. Ephritikhine, *New J. Chem.*, 1992, **16**, 451.
37. M. D. Fryzuk, *Can. J. Chem.*, 1992, **70**, 2839.
38. R. Poli, *Chem. Rev.*, 1991, **91**, 509.
39. D. J. Cardin and R. J. Norton, *Organomet. Chem.*, 1984, **12**, 199.
40. K. J. Karel and P. L. Watson, *Organomet. Chem.*, 1985, **13**, 214.
41. M. J. Winter, *Organomet. Chem.*, 1990, **19**, 208.
42. M. D. Fryzuk, T. S. Haddad and D. J. Berg, *Coord. Chem. Rev.*, 1990, **99**, 137.
43. M. J. Winter, *Organomet. Chem.*, 1991, **20**, 205.
44. T. J. Marks and I. L. Fragalà (eds.) 'Fundamental and Technological Aspects of Organo-*f*-Element Chemistry', NATO ASI Ser., D. Reidel, Boston, MA, 1985, vol. 155.
45. A. P. Ginsberg (ed)., 'Inorganic Syntheses', Wiley (Interscience), New York, 1990, **27**, 136.
46. W. J. Evans, *ACS Symp. Ser.*, 1987, **333**, 278.
47. I. P. Beletskaya and G. Z. Suleimanov, *Metalloorg. Khim.*, 1988, **1**, 10.
48. S. J. Swamy, *Proc. Indian Natl. Sci. Acad., Part A*, 1989, **55**, 261.
49. Q. Shen, W. Chen, Y. Jin and C. Shan, *Pure Appl. Chem.*, 1988, **60**, 1251.
50. M. N. Bochkarev, *Proc. Indian Natl. Sci. Acad., Part A*, 1989, **55**, 170.
51. A. Streitwieser, Jr., in 'Fundamental and Technological Aspects of Organo-*f*-Element Chemistry', NATO ASI Ser., ed. T. J. Marks and I. L. Fragalà, D. Reidel, Boston, MA, 1985, **155**, 77.
52. A. Streitwieser and T. R. Boussie, *Eur. J. Solid State Inorg. Chem.*, 1991, **28**, 399.
53. J. W. Gilje, R. E. Cramer, M. A. Bruck, K. T. Higa and K. Panchanatheswaran, *Inorg. Chim. Acta*, 1985, **110**, 139.
54. A. Domingos *et al.*, *Eur. J. Solid State Inorg. Chem.*, 1991, **28**, 413.
55. V. N. Sokolov and V. K. Vasil'ev, *Radiokhim.*, 1985, **27**, 187.
56. V. Y. Mishin, G. V. Sidorenko and D. N. Suglobov, *Radiokhim.*, 1986, **28**, 293.
57. R. Bohlander, *KfK Report No. 4152*, Kernforschungszentrum Karlsruhe, 1986.
58. B. Kanellakopulos and R. Klenze, *KFK-Nachr.*, 1984, **16**, 119.
59. W. J. Evans and M. A. Hozbor, *J. Organomet. Chem.*, 1987, **326**, 299.
60. A. P. Sattelberger, *Chemtracts: Anal., Phys., Inorg. Chem.*, 1990, **2**, 38.
61. S. P. Nolan, D. Stern, D. Hedden and T. J. Marks, *ACS Symp. Ser.*, 1990, **428**, 159.
62. S. P. Nolan, C. L. Stern and T. J. Marks, *J. Am. Chem. Soc.*, 1989, **111**, 7844.
63. M. A. Giardello, W. A. King, S. P. Nolan, M. Porchia, C. Sishta and T. J. Marks, in 'Energetics of Organometallic Species', NATO ASI Ser., 1992, **367**, 35.
64. C. J. Burns and B. E. Bursten, *Comments Inorg. Chem.*, 1989, **9**, 61.
65. G.-X. Xu, J. Ren, C.-H. Huang and J.-G. Wu, *Pure Appl. Chem.*, 1988, **60**, 1145.
66. I. P. Rothwell, in 'Activation of Functionalized Alkanes', ed. C. L. Hill, Wiley, New York, 1989, 151.
67. P. L. Watson, in 'Selective Hydrocarbon Activation', ed. J. A. Davies, P. L. Watson, J. F. Liebman and A. Greenberg, VCH Publishers, New York, 1990, 79.
68. H. Raba, J.-Y. Saillard and R. Hoffmann, *J. Am. Chem. Soc.*, 1986, **108**, 4327.
69. V. M. Bulychev, G. L. Soloveichik, V. K. Bel'skii and A. I. Sizov, *Metalloorg. Khim.*, 1989, **2**, 82.
70. T. J. Marks, *Inorg. Chim. Acta*, 1987, **140**, 1.
71. V. V. Zagorskii and G. B. Sergeev, *Mol. Cryst. Liq. Cryst.*, 1990, **186**, 81.
72. G. B. Deacon, P. MacKinnon, R. S. Dickson, G. N. Pain and B. O. West, *Appl. Organomet. Chem.*, 1990, **4**, 439.
73. W. J. Evans and S. E. Foster, *J. Organomet. Chem.*, 1992, **433**, 79.
74. A. G. Orpen, L. Brammer, F. H. Allen, O. Kennard, D. G. Watson and R. Taylor, *J. Chem. Soc., Dalton Trans.*, 1989, S1.
75. J. Marçalo and A. Pires de Matos, *Polyhedron*, 1989, **8**, 2431.
76. X. F. Li, X. Z. Feng, Y. T. Xu, H. D. Wang, P. N. Sun and J. Shi, *Kexue Tongbao*, 1987, **32**, 1259.
77. S. C. Sockwell and T. P. Hanusa, *Inorg. Chem.*, 1990, **29**, 76.
78. D. W. Clark and K. D. Warren, *J. Organomet. Chem.*, 1976, **122**, C28.
79. G.-X. Xu and J. Ren, *Int. J. Quantum Chem.*, 1986, **29**, 1017.
80. J. Q. Ren and G. X. Xu, *Kexue Tongbao*, 1986, **31**, 882.
81. J. Q. Ren and G. X. Xu, *Gaodeng Xuexiao Huaxue Xuebao*, 1986, **7**, 441.
82. Y. Gao and G. Gao, *Wuji Huaxue*, 1986, **2**, 72.
83. P. N. Hazin, J. W. Bruno and H. G. Brittain, *Organometallics*, 1987, **6**, 913.
84. J. C. Green, D. Hohl and N. Rösch, *Organometallics*, 1987, **6**, 712.
85. Z. Li, W. Chen, G. X. Xu and J. Q. Ren, *Wuji Huaxue*, 1987, **3**, 111.
86. R. J. Strittmatter and B. E. Bursten, *J. Am. Chem. Soc.*, 1991, **113**, 552.
87. N. Rösch, *Inorg. Chim. Acta*, 1984, **94**, 297.

88. B. E. Bursten and R. J. Strittmatter, *Angew. Chem.*, 1991, **103**, 1085; *Angew. Chem., Int. Ed. Engl.*, 1991, **30**, 1069.
89. M. Pepper and B. E. Bursten, *Chem. Rev.*, 1991, **91**, 719.
90. I. P. Beletskaya *et al.*, *J. Am. Chem. Soc.*, 1993, **115**, 3156.
91. W. J. Evans, K. M. Coleson and S. C. Engerer, *Inorg. Chem.*, 1981, **20**, 4320.
92. W. J. Evans, S. C. Engerer, P. A. Piliero and A. L. Wayda, *Fundam. Res. Homogeneous Catal.*, 1979, **3**, 941.
93. W. J. Evans, S. C. Engerer, P. A. Piliero and A. L. Wayda, *J. Chem. Soc., Chem. Commun.*, 1979, 1007.
94. W. J. Evans, S. C. Engerer and A. C. Neville, *J. Am. Chem. Soc.*, 1978, **100**, 331.
95. S. Q. Sun and W. Q. Chen, *Kidorui*, 1989, **14**, 170.
96. P. B. Hitchcock, S. A. Holmes, M. F. Lappert and S. Tian, *J. Chem. Soc. Chem. Commun.*, 1994, 2691.
97. G. B. Deacon and D. G. Vince, *J. Organomet. Chem.*, 1976, **112**, C1.
98. G. B. Deacon, W. D. Raverty and D. G. Vince, *J. Organomet. Chem.*, 1977, **135**, 103.
99. G. B. Deacon, A. J. Koplick, W. D. Raverty and D. G. Vince, *J. Organomet. Chem.*, 1979, **182**, 121.
100. G. B. Deacon, P. I. MacKinnon and T. D. Tuong, *Aust. J. Chem.*, 1983, **36**, 43.
101. G. B. Deacon, C. M. Forsyth and R. H. Newnham, *Polyhedron*, 1987, 6, 1143.
102. G. Z. Suleimanov, V. I. Bregadze, N. A. Kovalchuk and I. P. Beletskaya, *J. Organomet. Chem.*, 1982, **235**, C17.
103. O. P. Syutkina, L. F. Rybakova, E. N. Egorova, A. B. Sigalov and I. P. Beletskaya, *Izv. Akad. Nauk SSSR*, 1983, 648; *Bull Acad. Sci. USSR*, 1983, **32**, 586.
104. G. Z. Suleimanov, V. I. Bregadze, N. A. Kovalchuk, K. S. Khalilov and I. P. Beletskaya, *J. Organomet. Chem.*, 1983, **255**, C5.
105. T. D. Tilley, Ph.D. Thesis, University of California, Berkeley, 1982; cited in ref. 188.
106. J. R. van den Hende, P. B. Hitchcock and M. F. Lappert, *J. Chem. Soc, Chem. Commun.*, 1994, 1413.
107. F. O. Rice and K. K. Rice, 'The Aliphatic Free Radicals', The Johns Hopkins Press, Baltimore, MD, 1935. p. 58.
108. W. M. Plets, *Dokl. Akad. Nauk SSSR*, 1938, **20**, 27.
109. B. N. Afanasev and P. A. Tsyganova, *Zh. Obshch. Khim.*, 1948, **18**, 306.
110. H. Gilman and R. G. Jones, *J. Org. Chem.*, 1945, **10**, 505.
111. F. A. Hart and M. S. Saran, *J. Chem. Soc., Chem. Commun.*, 1968, 1614.
112. F. A. Hart, A. G. Massey and M. S. Saran, *J. Organomet. Chem.*, 1970, **21**, 147.
113. T. A. Starostina, R. R. Shifrina, L. F. Rybakova and E. S. Petrov, *Zh. Obshch. Khim.*, 1987, **57**, 2402.
114. R. Cai, J. Wei and S. He, *Fudan Xuebao, Ziran Kexueban*, 1989, **28**, 111.
115. J. L. Atwood *et al.*, *J. Chem. Soc., Chem. Commun.*, 1978, 140.
116. H. Schumann and J. Müller, *J. Organomet. Chem.*, 1978, **146**, C5.
117. H. Schumann and J. Müller, *J. Organomet. Chem.*, 1979, **169**, C1.
118. M. F. Lappert and R. Pearce, *J. Chem. Soc., Chem. Commun.*, 1973, 126.
119. A. A. Trifonov, M. N. Bochkarev and G. A. Razuvaev, *Zh. Obshch. Khim.*, 1988, **58**, 931.
120. I. S. Guzman, N. N. Chigir, O. K. Sharaev, G. N. Bondarenko, E. I. Tinyakova and B. A. Dolgoplosk, *Dokl. Akad. Nauk SSSR*, 1979, **249**, 860; *Proc. Acad. Sci. USSR*, 1979, **249**, 519.
121. B. A. Dolgoplosk *et al.*, *J. Organomet. Chem.*, 1980, **201**, 249.
122. E. L. Vollershtein, V. A. Yakovlev, E. I. Tinyakova and B. A. Dolgoplosk, *Dokl. Akad. Nauk SSSR*, 1980, **250**, 365; *Proc. Acad. Sci. USSR*, 1980, **250**, 19.
123. V. A. Yakovlev, I. A. Oreshkin and B. A. Dolgoplosk, *Metalloorg. Khim.*, 1989, **2**, 93.
124. S. B. Gol'shtein, V. A. Yakovlev, A. I. Mikaya and B. A. Dolgoplosk, *Dokl. Akad. Nauk SSSR*, 1989, **309**, 1129.
125. T. A. Starostina, L. F. Rybakova and E. S. Petrov, *Metalloorg. Khim.*, 1989, **2**, 915.
126. J. Hu, E. Pan, Y. Na and Q. Shen, *Zhongguo Xitu Xuebao*, 1987, **5**, 949.
127. L. E. Manzer, *J. Am. Chem. Soc.*, 1978, **100**, 8068.
128. L. E. Manzer, *J. Organomet. Chem.*, 1977, **135**, C6.
129. A. L. Wayda, J. L. Atwood and W. E. Hunter, *Organometallics*, 1984, **3**, 939.
130. Z. Huang and W. Huang, *Gaodeng Xuexiao Huaxue Xuebao*, 1990, **11**, 1022.
131. Z. Huang and X. Wu, *Wuji Huaxue Xuebao*, 1990, **6**, 353.
132. M. N. Bochkarev, V. V. Khramenkov, Y. F. Rad'kov, L. N. Zakharov and Y. T. Struchkov, *Metalloorg. Khim.*, 1990, **3**, 1438.
133. M. N. Bochkarev, V. V. Khramenkov, Y. F. Rad'kov, L. N. Zakharov and Y. T. Struchkov, *J. Organomet. Chem.*, 1992, **429**, 27.
134. H. H. Karsch, A. Appelt and G. Müller, *Angew. Chem., Int. Ed. Engl.*, 1986, **25**, 823.
135. S. Hao, J.-I. Song, H. Aghabozorg and S. Gambarotta, *J. Chem. Soc., Chem. Commun.*, 1994, 157.
136. H. H. Karsch, G. Ferazin and P. Bissinger, *J. Chem. Soc., Chem. Commun.*, 1994, 505.
137. V. I. Bregadze, N. A. Kovalchuk, N. N. Godovikov, G. Z. Suleimanov and I. P. Beletskaya, *J. Organomet. Chem.*, 1983, **241**, C13.
138. H. Schumann and F. W. Reier, *J. Organomet. Chem.*, 1982, **235**, 287.
139. G. K. Barker and M. F. Lappert, *J. Organomet. Chem.*, 1974, **76**, C45.
140. P. B. Hitchcock, M. F. Lappert, R. G. Smith, R. A. Bartlett and P. P. Power, *J. Chem. Soc., Chem. Commun.*, 1988, 1007.
141. C. J. Schaverien and A. G. Orpen, *Inorg. Chem.*, 1991, **30**, 4968.
142. J. L. Atwood, M. F. Lappert, R. G. Smith and H. Zhang, *J. Chem. Soc., Chem. Commun.*, 1988, 1308.
143. F. T. Edelmann, A. Steiner, D. Stalke, J. W. Gilje, S. Jagner and M. Håkansson, *Polyhedron*, 1994, **13**, 539.
144. C. J. Schaverien and J. B. van Mechelen, *Organometallics*, 1991, **10**, 1704.
145. C. J. Schaverien and G. J. Nesbitt, *J. Chem. Soc., Dalton Trans.*, 1992, 157.
146. L. N. Bochkarev *et al.*, *J. Organomet. Chem.*, 1994, **467**, C3.
147. F. G. N. Cloke, C. I. Dalby, P. B. Hitchcock, H. Karamallakis and G. A. Lawless, *J. Chem. Soc., Chem. Commun.*, 1991, 779.
148. E. A. Fedorova, G. S. Kalinina, M. N. Bochkarev and G. A. Razuvaev, *Zh. Obshch. Khim.*, 1982, **52**, 1187; *J. Gen. Chem. USSR*, 1982, **52**, 1041.
149. M. N. Bochkarev, E. A. Fedorova, Y. F. Rad'kov, S. Y. Khorshev, G. S. Kalinina and G. A. Razuvaev, *J. Organomet. Chem.*, 1983, **258**, C29.
150. H. Schumann and G. Frisch, *Z. Naturforsch.*, 1982, **37**, 168.

151. H. Schumann *et al.*, *Organometallics*, 1984, **3**, 69.
152. H. Schumann, H. Lauke, E. Hahn and J. Pickardt, *J. Organomet. Chem.*, 1984, **263**, 29.
153. H. Schumann, J. Pickardt and N. Bruncks, *Angew. Chem.*, 1981, **93**, 127; *Angew. Chem., Int. Ed. Engl.*, 1981, **20**, 120.
154. H. Schumann, G. Kociok-Köhn, A. Dietrich and F. H. Görlitz, *Z. Naturforsch., Teil B*, 1991, **46**, 896.
155. A. Shakoor, K. Jacob and K.-H. Thiele, *Z. Anorg. Allg. Chem.*, 1985, **521**, 57.
156. A. L. Wayda and W. J. Evans, *J. Am. Chem. Soc.*, 1978, **100**, 7119.
157. H. Schumann, W. Genthe, E. Hahn, J. Pickardt, H. Schwarz and K. Eckart, *J. Organomet. Chem.*, 1986, **306**, 215.
158. A. B. Nikitin, V. Y. Yakovlev and B. A. Dolgoplosk, *Dokl. Akad. Nauk SSSR*, 1986, **291**, 393.
159. H. Schumann, in 'Organometallics of the *f*-Elements', eds. T. J. Marks and R. D. Fischer, Reidel, Dordrecht, 1979, 81.
160. S. A. Cotton, F. A. Hart, M. B. Hursthouse and A. J. Welch, *J. Chem. Soc., Chem. Commun.*, 1972, 1225.
161. N. S. Emel'yanova, A. A. Trifonov, L. N. Zakharov, M. N. Bochkarev, A. F. Shestakov and Y. T. Struchkov, *Metallorg. Khim.*, 1993, **6**, 363.
162. K.-H. Thiele, K. Unverhau, M. Geitner and K. Jacob, *Z. Anorg. Allg. Chem.*, 1987, **548**, 175.
163. D. J. Cardin and R. J. Norton, *J. Chem. Soc., Chem. Commun.*, 1979, 513.
164. W. Chen, J. Zheng, L. Bai and T. Gao, *Kexue Tongbao*, 1986, **31**, 1042.
165. K. Jacob and K.-H. Thiele, *Z. Anorg. Allg. Chem.*, 1986, **543**, 192.
166. G. Campari and F. A. Hart, *Inorg. Chim. Acta*, 1982, **65**, L217.
167. S. Wang and F. Wang, *Kexue Tongbao*, 1985, **30**, 1344.
168. G. Lin, Z. Jin, Y. Zhang and W. Chen, *J. Organomet. Chem.*, 1990, **396**, 307.
169. G. Lin, Z. Jin, Y. Zhang and W. Chen, *Jiegou Huaxue*, 1991, **10**, 192.
170. P. B. Hitchcock, M. F. Lappert and R. G. Smith, *J. Chem. Soc., Chem. Commun.*, 1989, 369.
171. X. Bao and W. Chen, *Chin. Sci. Bull.*, 1989, **34**, 207.
172. B. Boje and J. Magull, *Z. Anorg. Allg. Chem.*, 1994, **620**, 703.
173. C. J. Schaverien, N. Meijboom and A. G. Orpen, *J. Chem. Soc., Chem. Commun.*, 1992, 124.
174. Z. Shen, *Inorg. Chim. Acta*, 1987, **140**, 7.
175. P. Biagini, G. Lugli, L. Abis and R. Millini, *J. Organomet. Chem.*, 1994, **474**, C16.
176. J. W. Buchler, A. De Cian, J. Fischer, M. Kihn-Botulinski, H. Paulus and R. Weiss, *J. Am. Chem. Soc.*, 1986, **108**, 3652.
177. J. W. Buchler, A. De Cian, J. Fischer, M. Kihn-Botulinski and R. Weiss, *Inorg. Chem.*, 1988, **27**, 339.
178. J. Arnold, *J. Chem. Soc., Chem. Commun.*, 1990, 976.
179. J. W. Buchler and B. Scharbert, *J. Am. Chem. Soc.*, 1988, **110**, 4272.
180. J. W. Buchler, M. Kihn-Botulinski and B. Scharbert, *Z. Naturforsch., Teil B*, 1988, **43**, 1371.
181. J. W. Buchler, J. Hüttermann and J. Löffler, *Bull. Chem. Soc. Jpn.*, 1988, **61**, 71.
182. J. Arnold and C. G. Hoffmann, *J. Am. Chem. Soc.*, 1990, **112**, 8620.
183. J. Arnold, C. G. Hoffmann, D. Y. Dawson and F. J. Hollander, *Organometallics*, 1993, **12**, 3645.
184. P. J. Fagan, J. M. Manriquez, T. J. Marks, V. W. Day, S. H. Vollmer and C. S. Day, *J. Am. Chem. Soc.*, 1980, **102**, 5393.
185. K. H. den Haan *et al.*, *Organometallics*, 1986, **5**, 1726.
186. H. J. Heeres, J. Renkema, M. Booij, A. Meetsma and J. H. Teuben, *Organometallics*, 1988, **7**, 2495.
187. P. L. Watson and G. W. Parshall, *Acc. Chem. Res.*, 1985, **18**, 51.
188. J. M. Boncella and R. A. Andersen, *Organometallics*, 1985, **4**, 205.
189. M. D. Fryzuk and T. S. Haddad, *J. Chem. Soc., Chem. Commun.*, 1990, 1088.
190. M. D. Fryzuk, T. S. Haddad and S. J. Rettig, *Organometallics*, 1992, **11**, 2967.
191. M. D. Fryzuk, T. S. Haddad and S. J. Rettig, *Organometallics*, 1991, **10**, 2026.
192. L. Hasinoff, J. Takats, X. W. Zhang, A. H. Bond and R. D. Rogers, *J. Am. Chem. Soc.*, 1994, **116**, 8833.
193. R. Duchateau, C. T. van Wee, A. Meetsma and J. H. Teuben, *J. Am. Chem. Soc.*, 1993, **115**, 4931.
194. F. T. Edelmann, *J. Alloys Compd.*, 1994, **207/208**, 182.
195. F. T. Edelmann, *Coord. Chem. Rev.*, 1994, **137**, 403.
196. K. Mashima, H. Sugiyama and A. Nakamura, *J. Chem. Soc., Chem. Commun.*, 1994, 1581.
197. K. Mashima and H. Takaya, *Tetrahedron Lett.*, 1989, **30**, 3697.
198. K. Jacob and K.-H. Thiele, *Z. Anorg. Allg. Chem.*, 1986, **536**, 147.
199. M. N. Bochkarev, V. V. Khramenkov, Y. F. Rad'kov, L. N. Kakharov and Y. T. Struchkov, *Metalloorg. Khim.*, 1991, **4**, 1193.
200. S. Li, X. Yang, Y. Sun, P. Li, X. Wang and X. Zhou, *Gaodeng Xuexiao Huaxue Xuebao*, 1988, **9**, 1242.
201. G. Z. Suleimanov, Y. S. Bogachev, P. V. Petrovskii, Y. K. Grishin, I. L. Zhuravleva and I. P. Beletskaya, *Izv. Akad. Nauk SSSR, Ser. Khim.*, 1984, 471.
202. W. J. Evans, S. C. Engerer and K. M. Coleson, *J. Am. Chem. Soc.*, 1981, **103**, 6672.
203. G. B. Deacon, A. J. Koplick and T. D. Tuong, *Aust. J. Chem.*, 1982, **35**, 941.
204. G. B. Deacon and A. J. Koplick, *J. Organomet. Chem.*, 1978, **146**, C43.
205. L. N. Bochkarev, S. B. Shustov, T. V. Guseva and S. F. Shil'tsov, *Zh. Obshch. Khim.*, 1988, **58**, 923.
206. S. B. Shustov, L. N. Bochkarev and S. F. Zhil'tsov, *Metalloorg. Khim.*, 1990, **3**, 624.
207. F. Li, Y. Jin, F. Pei and F. Wang, *Yingyong Huaxue*, 1991, **8**, 81.
208. R. Taube, H. Windisch, F. H. Görlitz and H. Schumann, *J. Organomet. Chem.*, 1993, **445**, 85.
209. Z. Huang *et al.*, *Huaxue Xuebao*, 1986, **44**, 817.
210. W. Qiu, Z. Huang, S. Zhuang and W. Wu, *Wuji Huaxue*, 1985, **1**, 173.
211. W. Qiu, S. Zhuang, Z. Huang and W. Wu, *Fudan Xuebao Ziran Kexueban*, 1987, **26**, 113.
212. M. Brunelli, S. Poggio, U. Pedretti and G. Lugli, *Inorg. Chim. Acta*, 1987, **131**, 281.
213. Z. Huang, M. Chen, W. Qiu and W. Wu, *Inorg. Chim. Acta*, 1987, **139**, 203.
214. S. Fukuzawa, K. Sato, T. Fujinami and S. Sakai, *J. Chem. Soc., Chem. Commun.*, 1990, 939.
215. R. Taube and H. Windisch, *J. Organomet. Chem.*, 1994, **472**, 71.
216. W. Wu, M. Chen and P. Zhou, *Organometallics*, 1991, **10**, 98.
217. S. J. Swamy and H. Schumann, *J. Organomet. Chem.*, 1987, **334**, 1.
218. E. O. Fischer and H. Fischer, *Angew. Chem.*, 1964, **76**, 52; *Angew. Chem., Int. Ed. Engl.*, 1964, **3**, 132.
219. G. B. Deacon, P. I. MacKinnon, T. W. Hambley and J. C. Taylor, *J. Organomet. Chem.*, 1983, **259**, 91.
220. G. B. Deacon, C. M. Forsyth, R. H. Newnham and T. D. Tuong, *Aust. J. Chem.*, 1987, **40**, 895.

221. H. A. Zinnen, J. J. Pluth and W. J. Evans, *J. Chem. Soc., Chem. Commun.*, 1980, 810.
222. G. B. Deacon, A. J. Koplick and T. D. Tuong, *Aust. J. Chem.*, 1984, **37**, 517.
223. A. Hammel, W. Schwarz and J. Weidlein, *J. Organomet. Chem.*, 1989, **378**, 347.
224. T. Jiang, Q. Shen, Y. Lin and S. Jin, *J. Organomet. Chem.*, 1993, **450**, 121.
225. J. Jin, S. Jin and W. Chen, *J. Organomet. Chem.*, 1991, **412**, 71.
226. X. Jusong, G. Wei, Z. Jin and W. Chen, *Zhongguo Xitu Xuebao*, 1992, **3**, 203.
227. G. B. Deacon, B. M. Gatehouse and P. A. White, *Polyhedron*, 1989, **8**, 1983.
228. G. B. Deacon, G. N. Pain and T. D. Tuong, *Polyhedron*, 1985, **4**, 1149.
229. G. Z. Suleimanov and I. P. Beletskaya, *Metalloorg. Khim.*, 1989, **2**, 704.
230. J. L. Namy, P. Girard, H. B. Kagan and P. Caro, *Nouv. J. Chem.*, 1981, **5**, 479.
231. G. B. Deacon and R. H. Newnham, *Aust. J. Chem.*, 1985, **38**, 1757.
232. G. B. Deacon, D. L. Wilkinson and R. L. Davis, *Inorg. Chim. Acta*, 1988, **142**, 329.
233. G. Z. Suleimanov, T. K. Kurbanov, Y. A. Nuriev, L. F. Rybakova and I. P. Beletskaya, *Dokl. Akad. Nauk SSSR*, 1982, **265**, 896; *Proc. Acad. Sci. USSR*, 1982, **265**, 254.
234. G. Z. Suleimanov, L. F. Rybakova, Y. A. Nuriev, T. K. Kurbanov and I. P. Beletskaya, *J. Organomet. Chem.*, 1982, **235**, C19.
235. F. T. Edelmann, B. Kanellakopulos, D. Stalke and C. Apostolidis, unpublished results.
236. A. M. Bond, G. B. Deacon and R. H. Newnham, *Organometallics*, 1986, **5**, 2312.
237. L. Arnaudet and B. Ban, *New. J. Chem.*, 1988, **12**, 201.
238. V. D. Makhaev, Y. B. Zvedov, N. G. Chernorukov and V. I. Berestenko, *Vysokchist. Veshchestva*, 1990, 210.
239. A. L. Wayda, *J. Organomet. Chem.*, 1989, **361**, 73.
240. Q. Shen, D. Zheng, L. Lin and Y. Lin, *J. Organomet. Chem.*, 1990, **391**, 307.
241. V. K. Bel'skii, Y. K. Gun'ko, B. M. Bulychev, A. I. Sizov and G. L. Soloveichik, *J. Organomet. Chem.*, 1990, **390**, 35.
242. M. F. Lappert, P. I. W. Yarrow, J. L. Atwood and R. Shakir, *J. Chem. Soc., Chem. Commun.*, 1980, 987.
243. W. J. Evans, R. A. Keyer and J. W. Ziller, *J. Organomet. Chem.*, 1990, **394**, 87.
244. W. J. Evans, G. Kociok-Köhn, S. E. Foster, J. W. Ziller and R. J. Doedens, *J. Organomet. Chem.*, 1993, **444**, 61.
245. P. B. Hitchcock, J. A. K. Howard, M. F. Lappert and S. Prashar, *J. Organomet. Chem.*, 1992, **437**, 177.
246. D. Deng, C. Qian, F. Song, Z. Wang, G. Wu and P. Zheng, *J. Organomet. Chem.*, 1992, **443**, 79.
247. P. Jutzi, J. Dahlhaus and M. O. Kristen, *J. Organomet. Chem.*, 1993, **450**, C1.
248. P. Jutzi and J. Dahlhaus, *Phosphorus, Sulfur, and Silicon*, 1994, **87**, 73.
249. J. R. van den Hende, P. B. Hitchcock, M. F. Lappert and T. A. Nile, *J. Organomet. Chem.*, 1994, **472**, 79.
250. G. B. Deacon *et al.*, *Aust. J. Chem.*, 1992, **45**, 567.
251. G. B. Deacon, B. M. Gatehouse and P. A. White, *Aust. J. Chem.*, 1992, **45**, 1939.
252. G. B. Deacon *et al.*, *J. Organomet. Chem.*, 1984, **277**, C21.
253. G. B. Deacon, G. D. Fallon and C. M. Forsyth, *J. Organomet. Chem.*, 1993, **462**, 183.
254. J. L. Namy, J. Collin, J. Zhang and H. B. Kagan, *J. Organomet. Chem.*, 1987, **328**, 81.
255. J. Collin, J. L. Namy, C. Bied and H. B. Kagan, *Inorg. Chim. Acta*, 1987, **140**, 29.
256. C. S. Day, V. W. Day, R. D. Ernst and S. H. Vollmer, *Organometallics*, 1982, **1**, 998.
257. G. Z. Suleimanov, Y. A. Nutriev, O. P. Syutkina, T. K. Kurbanov and I. P. Beletskaya, *Izv. Akad. Nauk SSSR, Ser. Khim.*, 1982, 1671; *Bull. Acad. Sci. USSR*, 1982, **31**, 1490.
258. X. Zhou, Z. Wu, H. Ma, Z. Xu and X. You, *Polyhedron*, 1994, **13**, 375.
259. G. Yang, Y. Fan, Z. Jin, Y. Xing and W. Chen, *J. Organomet. Chem.*, 1987, **322**, 57.
260. Z. Z. Wu, Z. Xu, X. You, X. Zhou and L. Shi, *J. Coord. Chem.*, 1992, **26**, 329.
261. W. Ke, Z. Jin, G. Wei and W. Chen, *Jiegou Huaxue*, 1992, **11**, 244.
262. M. Adam, X.-F. Li, W. Oroschin and R. D. Fischer, *J. Organomet. Chem.*, 1985, **296**, C19.
263. G. B. Deacon, G. D. Fallon and D. L. Wilkinson, *J. Organomet. Chem.*, 1985, **293**, 45.
264. B. Zhang, Y. Wang, X. Bao and W. Chen, *Gaodeng Xuexiao Huaxue Xuebao*, 1990, **11**, 458.
265. V. K. Bel'skii, Y. K. Gun'ko, E. B. Lobkovskii and G. L. Soloveichik, *Metalloorg. Khim.*, 1991, **4**, 420.
266. G. Nie, J. Wang and W. Chen, *Yingyong Huaxue*, 1989, **6**, 60.
267. Y. Wang, W. Chen and B. Zhang, *Yingyong Huaxue*, 1989, **6**, 37.
268. R. Zhang, G. Yu and T. Ye, *Zhenkong Kexue Yu Jishu*, 1988, **8**, 89.
269. Z. Jin, H. Ninghai, L. Yi, X. Xiaolong and L. Guozhe, *Inorg. Chim. Acta*, 1988, **142**, 333.
270. Y. Li, X. Xu and G. Liu, *Youji Huaxue*, 1988, **8**, 43.
271. Y. Li, J. Liu and G. Liu, *Yingyong Huaxue*, 1989, **6**, 62.
272. X. Li, J. Liu, S. Jin, Y. Lin and G. Liu, *Jiegou Huaxue*, 1991, **10**, 60.
273. J. Guan, Q. Shen, J. Hu and Y. Lin, *Jiegou Huaxue*, 1990, **9**, 184.
274. J. Jizhu, G. Wei, Z. Jin and W. Chen, *Jiegou Huaxue*, 1992, **11**, 369.
275. J. Jin, S. Jin, Z. Jin and W. Chen, *Polyhedron*, 1992, **12**, 2873.
276. Z. Jin, J. Guan, G. Wei, J. Hu and Q. Shen, *Jiegou Huaxue*, 1990, **9**, 140.
277. H. Schumann, G. Kociok-Köhn and J. Loebel, *Z. Anorg. Allg. Chem.*, 1990, **581**, 69.
278. W. J. Evans and M. S. Sollberger, *J. Am. Chem. Soc.*, 1986, **108**, 6095.
279. W. J. Evans, T. J. Boyle and J. W. Ziller, *Organometallics*, 1993, **12**, 3998.
280. H. Schumann, J. A. Meese-Marktscheffel and A. Dietrich, *J. Organomet. Chem.*, 1989, **377**, C5.
281. H. Schumann, J. A. Meese-Marktscheffel, A. Dietrich and J. Pickardt, *J. Organomet. Chem.*, 1992, **433**, 241.
282. A. V. Protchenko, L. N. Zakharov, M. N. Bochkarev and Y. T. Struchkov, *J. Organomet. Chem.*, 1993, **447**, 209.
283. D. M. Roitershtein, A. M. Ellern, M. Y. Antipin, L. F. Rybakova, Y. T. Struchkov and E. S. Petrov, *Mendeleev Commun.*, 1992, 118.
284. X. Wang and Z. Ye, *Youji Huaxue*, 1985, 389.
285. H. Ma and X. Ye, *J. Organomet. Chem.*, 1987, **326**, 369.
286. L. Yang, L. Dai, H. Ma and Z. Ye, *Organometallics*, 1989, **8**, 1129.
287. Z. Ye, Y. Yu and H. Ma, *Polyhedron*, 1988, **7**, 1095.
288. L. Dai, Y. Cai, L. Yang, Y. Y. Yu, Z. Ye and H. Ma, *Jiegou Huaxue*, 1992, **11**, 325.
289. Y. Yu, S. Wang, H. Ma and Z. Ye, *Polyhedron*, 1992, **12**, 265.

290. K. Jacob, M. Glanz, K. Tittes and K.-H. Thiele, *Z. Anorg. Allg. Chem.*, 1988, **556**, 170.
291. Y. Yu, S. Wang, Z. Ye and H. Ma, *Polyhedron*, 1991, **10**, 1599.
292. Z. Ye and Z. Wu, *Synth. React. Inorg. Met.-Org. Chem.*, 1989, **19**, 157.
293. Z. Wu, Z. Xu and X. You, *Synth. React. Inorg. Met.-Org. Chem.*, 1993, **23**, 1155.
294. X. Shen, Y. Xie, H. Jiang and Q. Li, *Polish J. Chem.*, 1994, **68**, 1303.
295. Z. Wu and Z. Ye, *Polyhedron*, 1991, **10**, 27.
296. M. G. Sewchok, R. C. Haushalter and J. S. Merola, *Inorg. Chim. Acta*, 1988, **144**, 47.
297. G. Yu, W. Chen, D. Gong, M. Tong and J. Zheng, *Kexue Tongbao*, 1985, **30**, 1495.
298. P. S. Gradeff, K. Yünlü, T. J. Deeming, J. M. Olofson, J. W. Ziller and W. J. Evans, *Inorg. Chem.*, 1989, **28**, 2600.
299. P. L. Watson, T. H. Tulip and I. Williams, *Organometallics*, 1990, **9**, 1999.
300. Y. Zhongwen, M. Huaizhu and Y. Yongfei, *J. Less-Common Met.*, 1986, **126**, 405.
301. E. B. Lobkovskii, G. L. Soloveichik, B. M. Bulychev and A. B. Erofeev, *Zh. Strukt. Khim.*, 1984, **25**, 170.
302. H. Lueken, J. Schmitz, W. Lamberts, P. Hannibal and K. Handrick, *Inorg. Chim. Acta*, 1989, **156**, 119.
303. W. Lamberts, H. Lueken and B. Hessner, *Inorg. Chim. Acta*, 1987, **134**, 155.
304. T. Aknoukh, J. Müller, K. Qiao, X.-F. Li and R. D. Fischer, *J. Organomet. Chem.*, 1991, **408**, 47.
305. V. K. Bel'skii, S. Y. Knyazhanskii, Y. K. Gun'ko, B. M. Bulychev and G. L. Soloveichik, *Metalloorg. Khim.*, 1991, **4**, 1135.
306. W. Lamberts, H. Lueken and U. Elsenhans, *Inorg. Chim. Acta*, 1986, **121**, 81.
307. H. Lueken, W. Lamberts and P. Hannibal, *Inorg. Chim. Acta*, 1987, **132**, 111.
308. J. Schmitz, H. Schilder and H. Lueken, *J. Alloys Compd.*, 1993, **200**, 195.
309. W. Lamberts and H. Lueken, *Inorg. Chim. Acta*, 1987, **132**, 119.
310. W. Lamberts, B. Hessner and H. Lueken, *Inorg. Chim. Acta*, 1987, **139**, 215.
311. C. Qian, D. Deng, Z. Zhang and C. Ni, *Youji Huaxue*, 1985, 403.
312. C. Ni, Z. Zhang, D. Deng and C. Qian, *J. Organomet. Chem.*, 1986, **306**, 209.
313. Z. Lin, Y. Liu and W. Chen, *Sci. Sin., Ser. B*, 1988, **30**, 1136.
314. H. Yasuda, H. Yamamoto, K. Yokota and A. Nakamura, *Chem. Lett.*, 1989, 1309.
315. J. Jizhu, Z. Jin and W. Chen, *Jiegou Huaxue*, 1992, **11**, 204.
316. C. Qian *et al.*, *Synth. React. Inorg. Met.-Org. Chem.*, 1984, **14**, 663.
317. Z. Wu *et al.*, *J. Organomet. Chem.*, 1993, **455**, 93.
318. J. Stehr and R. D. Fischer, *J. Organomet. Chem.*, 1992, **430**, C1.
319. W. J. Evans, R. A. Keyer and J. W. Ziller, *J. Organomet. Chem.*, 1993, **450**, 115.
320. E. C. Baker, L. D. Brown and K. N. Raymond, *Inorg. Chem.*, 1975, **14**, 1376.
321. P. L. Watson, J. F. Whitney and R. L. Harlow, *Inorg. Chem.*, 1981, **20**, 3271.
322. Y. K. Gun'ko, V. K. Bel'skii, B. M. Bulychev and A. I. Sizov, *Metalloorg. Khim.*, 1990, **3**, 411.
323. Y. K. Gun'ko, V. K. Bel'skii, A. I. Sizov, B. M. Bulychev and G. L. Soloveichik, *Metalloorg. Khim.*, 1989, **2**, 1125.
324. V. K. Bel'skii, S. Y. Knyazhanskii, B. M. Bulychev and G. L. Soloveichik, *Metalloorg. Khim.*, 1989, **2**, 754.
325. V. K. Bel'skii, N. R. Strel'tsova, Y. K. Gun'ko, S. Y. Knyazhanskii, B. M. Bulychev and G. L. Soloveichik, *Metalloorg. Khim.*, 1991, **4**, 1139.
326. S. Song, Q. Shen, S. Jin, J. Guan and Y. Lin, *Polyhedron*, 1992, **12**, 2857.
327. V. K. Bel'skii, S. Y. Knyazhanskii, B. M. Bulychev and G. L. Soloveichik, *Metalloorg. Khim.*, 1989, **2**, 567.
328. A. Recknagel, F. Knösel, H. Gornitzka, M. Noltemeyer, F. T. Edelmann and U. Behrens, *J. Organomet. Chem.*, 1991, **417**, 363.
329. Q. Shen, M. Qi, J. Guan and Y. Lin, *J. Organomet. Chem.*, 1991, **406**, 353.
330. J. Guan, Y. Lin and Q. Shen, *Youji Huaxue*, 1991, **11**, 261.
331. M. F. Lappert, A. Singh, J. L. Atwood and W. E. Hunter, *J. Chem. Soc., Chem. Commun.*, 1981, 1191.
332. J. Jin, Z. Jin, G. Wei and W. Chen, *Chin. Sci. Bull.*, 1993, **38**, 526.
333. W. A. Herrmann, R. Anwander, H. Riepl, W. Scherer and C. R. Whitaker, *Organometallics*, 1993, **12**, 4342.
334. D. Deng, C. Qian, G. Wu and P. Zheng, *J. Chem. Soc., Chem. Commun.*, 1990, 880.
335. D. Deng, X. Zheng, C. Qian, S. Jin and Y. Lin, *Huaxue Xuebao*, 1992, **10**, 1024.
336. D. Deng, B. Li and C. Qian, *Polyhedron*, 1990, **9**, 1453.
337. C. Qian *et al.*, *Inorg. Chem.*, 1994, **33**, 3382.
338. W. A. Herrmann, R. Anwander, F. C. Munck and W. Scherer, *Chem. Ber.*, 1993, **126**, 331.
339. M. Kaupp, O. P. Charkin and P. R. von Schleyer, *Organometallics*, 1992, **11**, 2765.
340. W. J. Evans, M. A. Hozbor, S. G. Bott, G. H. Robinson and J. L. Atwood, *Inorg. Chem.*, 1988, **27**, 1990.
341. W. A. Herrmann, R. Anwander, M. Klein, K. Öfele, J. Riede and W. Scherer, *Chem. Ber.*, 1992, **125**, 2391.
342. D. Deng, F. Song, Z. Wang, C. Qian, G. Wu and P. Zheng, *Polyhedron*, 1992, **12**, 2883.
343. P. B. Hitchcock, M. F. Lappert and S. Prashar, *J. Organomet. Chem.*, 1991, **413**, 79.
344. J. W. Gilje and H. W. Roesky, *Chem. Rev.*, 1994, **94**, 895.
345. M. Adam, G. Massarweh and R. D. Fischer, *J. Organomet. Chem.*, 1991, **405**, C33.
346. H. Schumann, E. Palamidis and J. Loebel, *J. Organomet. Chem.*, 1990, **384**, C49.
347. H. Schumann, W. Genthe, N. Bruncks and J. Pickardt, *Organometallics*, 1982, **1**, 1194.
348. G. B. Deacon, S. Nickel and E. R. T. Tiekink, *J. Organomet. Chem.*, 1991, **409**, C1.
349. Z. Wu, Z. Xu, X. You, X. Zhou and Z. Jin, *Polyhedron*, 1992, **12**, 2673.
350. Z. Wu, Z. Xu, X. You, X. Zhou, Y. Xing and Z. Jin, *J. Chem. Soc., Chem. Commun.*, 1993, 1494.
351. J. Stehr and R. D. Fischer, *J. Organomet. Chem.*, 1993, **459**, 79.
352. G. Massarweh and R. D. Fischer, *J. Organomet. Chem.*, 1993, **444**, 67.
353. W. J. Evans, R. Dominguez and T. P. Hanusa, *Organometallics*, 1986, **5**, 1291.
354. Z. Ye, Z. Zhou, S. Yuan and Z. Luo, *Youji Huaxue*, 1984, 119.
355. X. Wang, B. Du and Z. Wu, *Youji Huaxue*, 1986, 383.
356. Z. Ye, Z. Zhou, Z. Lou, X. Wang and F. Shen, *Huaxue Xuebao*, 1986, **44**, 707.
357. F. Shen and Z. Ye, *Youji Huaxue*, 1987, 128.
358. R. Chen, Z. Gu, J. Zhu and Z. Ye, *Zhongguo Xita Xuebao*, 1988, **6**, 85.
359. S. Li, X. Yang, C. Sun, S. Liang and Z. Ye, *Gaodeng Xuexiao Huaxue Xuebao*, 1990, **11**, 367.

360. G. B. Deacon and D. L. Wilkinson, *Aust. J. Chem.*, 1989, **42**, 845.
361. H. Schumann, J. A. Meese-Marktscheffel, A. Dietrich and F. H. Görlitz, *J. Organomet. Chem.*, 1992, **430**, 299.
362. J. Zhou, Y. Ge and C. Qian, *Synth. React. Inorg. Met.-Org. Chem.*, 1984, **14**, 651.
363. L. Shi, H. Ma, Y. Yu and Z. Ye, *J. Organomet. Chem.*, 1988, **339**, 277.
364. Z. Ye, Y. Yu, S. Wang and X. Jin, *J. Organomet. Chem.*, 1993, **448**, 91.
365. X. Wang, X. Shen and B. Du, *Polyhedron*, 1992, **11**, 355.
366. X. Wang, X. Shen and B. Du, *Synth. React. Inorg. Met.-Org. Chem.*, 1993, **23**, 419.
367. S. D. Stults, R. A. Andersen and A. Zalkin, *Organometallics*, 1990, **9**, 1623.
368. I. P. Beletskaya *et al.*, *J. Organomet. Chem.*, 1994, **468**, 121.
369. H. Schumann *et al.*, *Polyhedron*, 1988, **7**, 2307.
370. H. Schumann, I. Albrecht, M. Gallagher, E. Hahn, C. Muchmore and J. Pickardt, *J. Organomet. Chem.*, 1988, **349**, 103.
371. P. N. Hazin, J. W. Bruno and G. K. Schulte, *Organometallics*, 1990, **9**, 416.
372. H. Schumann, P. R. Lee and A. Dietrich, *Chem. Ber.*, 1990, **123**, 1331.
373. H. Schumann, E. Palamidis and J. Loebel, *J. Organomet. Chem.*, 1990, **390**, 45.
374. J. Guan, Q. Shen, S. Jin and Y. Lin, *Polyhedron*, 1994, **13**, 1695.
375. I. P. Beletskaya, A. Z. Voskoboinikov and G. K.-I. Magomedov, *Metalloorg. Khim.*, 1989, **2**, 808.
376. H. Schumann, P. R. Lee and J. Loebel, *Chem. Ber.*, 1989, **122**, 1897.
377. Z. Wu, X. Zhou, W. Zhang, Z. Xu, X. You and X. Huang, *J. Chem. Soc., Chem. Commun.*, 1994, 813.
378. H. Schumann, E. Palamidis, G. Schmid and R. Boese, *Angew. Chem., Int. Ed. Engl.*, 1986, **25**, 718.
379. H. Schumann, E. Palamidis, J. Loebel and J. Pickardt, *Organometallics*, 1988, **7**, 1008.
380. W. J. Evans, I. Bloom, W. E. Hunter and J. L. Atwood, *Organometallics*, 1983, **2**, 709.
381. C. Qian *et al.*, *J. Organomet. Chem.*, 1983, **247**, 161.
382. W. J. Evans, J. H. Meadows, A. L. Wayda, W. E. Hunter and J. L. Atwood, *J. Am. Chem. Soc.*, 1982, **104**, 2008.
383. H. Schumann and G. Jeske, *Angew. Chem., Int. Ed. Engl.*, 1985, **24**, 255.
384. H. Schumann and G. Jeske, *Z. Naturforsch. Teil B*, 1985, **40**, 1490.
385. H. Schumann, F. W. Reier and E. Palamidis, *J. Organomet. Chem.*, 1985, **297**, C30.
386. G. B. Deacon and D. L. Wilkinson, *Inorg. Chim. Acta*, 1988, **142**, 155.
387. W. J. Evans, R. Dominguez and T. P. Hanusa, *Organometallics*, 1986, **5**, 263.
388. R. G. Finke, S. R. Keenan, D. A. Schiraldi and P. L. Watson, *Organometallics*, 1987, **6**, 1356.
389. H. Schumann, H. Lauke, E. Hahn, M. J. Heeg and D. Van der Helm, *Organometallics*, 1985, **4**, 321; H. Schumann, F. W. Reier and E. Hahn, *Z. Naturforsch., Teil B*, 1985, **40**, 1289.
390. J. Holton, M. F. Lappert, D. G. H. Ballard, R. Pearce, J. L. Atwood and W. E. Hunter, *J. Chem. Soc., Dalton Trans.*, 1979, **45**, 54.
391. Q. Shen, Y. Cheng and Y. Lin, *J. Organomet. Chem.*, 1991, **419**, 293.
392. W. J. Evans, D. K. Drummond, T. P. Hanusa and R. J. Doedens, *Organometallics*, 1987, **6**, 2279.
393. W. J. Evans, R. Dominguez, K. R. Levan and R. J. Doedens, *Organometallics*, 1985, **4**, 1836.
394. C. Qian and Y. Ge, *J. Organomet. Chem.*, 1986, **299**, 97.
395. H. Schumann, F. W. Reier and M. Dettlaff, *J. Organomet. Chem.*, 1983, **255**, 305.
396. H. Schumann and F. W. Reier, *Inorg. Chim. Acta*, 1984, **95**, 43.
397. W. J. Evans, D. K. Drummond, L. A. Hughes, H. Zhang and J. L. Atwood, *Polyhedron*, 1988, **7**, 1693.
398. Y. Li, X. Xu, L. Zhu and G. Liu, *Yingyong Huaxue*, 1987, **4**, 82.
399. S. Wang, Y. Yu, Z. Ye, C. Qian and X. Jin, *J. Chem. Soc., Chem. Commun.*, 1994, 1097.
400. Y. Chauvin, S. Heyworth, H. Olivier, F. Robert and L. Saussine, *J. Organomet. Chem.*, 1993, **455**, 89.
401. D. M. Roitershtein, L. F. Rybakova and E. S. Petrov, *Dokl. Akad. Nauk SSSR*, 1990, **315**, 1393.
402. D. M. Roitershtein, L. F. Rybakova, E. S. Petrov, A. M. Ellern, M. Y. Antipin and Y. T. Struchkov, *J. Organomet. Chem.*, 1993, **460**, 39.
403. H. Schumann, S. Nickel, E. Hahn and M. J. Heeg, *Organometallics*, 1985, **4**, 800.
404. H. Schumann, J. A. Meese-Marktscheffel and F. E. Hahn, *J. Organomet. Chem.*, 1990, **390**, 301.
405. H. Schumann, S. Nickel, J. Loebel and J. Pickardt, *Organometallics*, 1988, **7**, 2004.
406. B. K. Campion, R. H. Heyn and T. D. Tilley, *Organometallics*, 1993, **12**, 2584.
407. J. V. Ortiz and R. Hoffmann, *Inorg. Chem.*, 1985, **24**, 2095.
408. E. N. Zavadovskaya, O. K. Sharaev, G. K. Borisov, Y. P. Yampolskii, E. I. Tinyakova and B. A. Dolgoplosk, *Dokl. Akad. Nauk SSSR*, 1985, **284**, 143.
409. W. J. Evans, J. H. Meadows, W. E. Hunter and J. L. Atwood, *J. Am. Chem. Soc.*, 1984, **106**, 1291.
410. I. P. Beletskaya, A. Z. Voskoboinikov and G. K.-I. Magomedov, *Metalloorg. Khim.*, 1989, **2**, 810.
411. D. L. Deng, Y. Q. Jiang, C. T. Qian, G. Wu and P. J. Zheng, *J. Organomet. Chem.*, 1994, **470**, 99.
412. H. Schumann, W. Genthe, E. Hahn, M. B. Hossain and D. Van der Helm, *J. Organomet. Chem.*, 1986, **299**, 67.
413. C. Qian, D. Deng, C. Ni and Z. Zhang, *Inorg. Chim. Acta*, 1988, **146**, 129.
414. S. Y. Knyazhanskii, V. K. Bel'skii, B. M. Bulychev and G. L. Soloveichik, *Metalloorg. Khim.*, 1989, **2**, 570.
415. Y. K. Gun'ko, B. M. Bulychev, G. L. Soloveichik and V. K. Bel'skii, *J. Organomet. Chem.*, 1992, **424**, 289.
416. W. J. Evans, J. H. Meadows, W. E. Hunter and J. L. Atwood, *Organometallics*, 1983, **2**, 1252.
417. W. J. Evans, T. P. Hanusa, J. H. Meadows, W. E. Hunter and J. L. Atwood, *Organometallics*, 1987, **6**, 295.
418. W. J. Evans, J. H. Meadows, A. L. Wayda, W. E. Hunter and J. L. Atwood, *J. Am. Chem. Soc.*, 1982, **104**, 2015.
419. W. J. Evans, J. H. Meadows and T. P. Hanusa, *J. Am. Chem. Soc.*, 1984, **106**, 4454.
420. W. J. Evans, M. S. Sollberger, S. I. Khan and R. Bau, *J. Am. Chem. Soc.*, 1988, **110**, 439.
421. S. Y. Knyazhanskii, B. M. Bulychev, O. K. Kireeva, V. K. Bel'skii and G. L. Soloveichik, *J. Organomet. Chem.*, 1991, **414**, 11.
422. T. J. Marks and G. W. Grynkewich, *Inorg. Chem.*, 1976, **15**, 1302.
423. M. F. Lappert, A. Singh, J. L. Atwood and W. E. Hunter, *J. Chem. Soc., Chem. Commun.*, 1983, 206.
424. V. D. Makhaev and A. P. Borisov, *Koord. Khim.*, 1992, 466.
425. A. B. Erofeev, B. M. Bulychev, V. K. Bel'skii and G. L. Soloveichik, *J. Organomet. Chem.*, 1987, **335**, 189.
426. S. Y. Knjazhanskii, B. M. Bulychev, V. K. Bel'skii and G. L. Soloveichik, *J. Organomet. Chem.*, 1987, **327**, 173.
427. S. Y. Knyazhanskii, B. M. Bulychev and G. L. Soloveichik, *Metalloorg. Khim.*, 1989, **2**, 345.

428. V. K. Bel'skii, A. B. Erofeev, B. M. Bulychev and G. L. Soloveichik, *J. Organomet. Chem.*, 1984, **265**, 123.
429. E. B. Lobkovskii, G. L. Soloveichik, A. B. Erofeev, B. M. Bulychev and V. K. Bel'skii, *J. Organomet. Chem.*, 1982, **235**, 151.
430. E. B. Lobkovskii, G. L. Soloveichik, B. M. Bulychev, A. B. Erofeev, A. I. Gusev and N. I. Kirillova, *J. Organomet. Chem.*, 1983, **254**, 167.
431. E. B. Lobkovskii, Y. K. Gun'ko, B. M. Bulychev, V. K. Bel'skii, G. L. Soloveichik and M. Y. Antipin, *J. Organomet. Chem.*, 1991, **406**, 343.
432. V. K. Bel'skii, B. M. Bulychev, A. B. Erofeev and G. L. Soloveichik, *J. Organomet. Chem.*, 1984, **268**, 107.
433. V. K. Bel'skii, Y. K. Gun'ko, B. M. Bulychev and G. L. Soloveichik, *J. Organomet. Chem.*, 1991, **420**, 43.
434. E. B. Lobkovskii, A. N. Chekhlov, S. Y. Knyazhanskii, B. M. Bulychev and G. L. Soloveichik, *Metalloorg. Khim.*, 1989, **2**, 1190.
435. S. Y. Knyazhanskii, E. B. Lobkovskii, B. M. Bulychev, V. K. Bel'skii and G. L. Soloveichik, *J. Organomet. Chem.*, 1991, **419**, 311.
436. Y. B. Zverev, I. V. Runovskaya, S. G. Chesnokova, N. P. Chernyaev and E. F. Krupnova, *Khim. Elementoorg. Soedin.*, 1986, 50.
437. G. K. Borisov, A. I. Kuz'michev, N. N. Ryzhikova and V. N. Sinev, *Zh. Obshch. Khim.*, 1988, **58**, 71.
438. G. K. Borisov, S. G. Krasnova and G. G. Devyatykh, *Russ. J. Inorg. Chem.*, 1992, **18**, 346.
439. M. N. Bochkarev, A. A. Trifonov, G. A. Razuvaev, M. A. Ilatovskaya and V. B. Shur, *Izv. Akad. Nauk SSSR, Ser. Khim.*, 1986, 1898.
440. M. N. Bochkarev, A. A. Trifonov, G. A. Razuvaev, M. A. Ilatovskaya, V. B. Shur and M. E. Vol'pin, *Dokl. Akad. Nauk SSSR*, 1987, **295**, 1381.
441. H. G. Brittain, J. H. Meadows and W. J. Evans, *Organometallics*, 1985, **4**, 1585.
442. R. G. Bulgakov, S. P. Kuleshov, V. N. Khandozhko, I. P. Beletskaya, G. A. Tolstikov and V. P. Kazakov, *Izv. Akad. Nauk SSSR, Ser. Khim.*, 1988, **7**, 1689.
443. C. Qian, Y. Ge, D. Deng and Y. Gu, *Huaxue Xuebao*, 1987, **45**, 210.
444. G. Paolucci, R. D. Fischer, H. Breitbach, B. Pelli and P. Traldi, *Organometallics*, 1988, **7**, 1918.
445. W. Jahn, K. Yünlü, W. Oroschin, H.-D. Amberger and R. D. Fischer, *Inorg. Chim. Acta*, 1984, **95**, 85.
446. J. Li, J. Ren, G. Xu and C. Qian, *Inorg. Chim. Acta*, 1986, **122**, 255.
447. H.-D. Amberger and W. Jahn, *Spectrochim. Acta, Part A*, 1984, **40**, 1025.
448. H.-D. Amberger, W. Jahn and N. M. Edelstein, *Spectrochim. Acta, Part A*, 1985, **41**, 465.
449. H.-D. Amberger and W. Jahn, *Spectrochim. Acta, Part A*, 1985, **41**, 469.
450. H.-D. Amberger, H. Schultze and N. M. Edelstein, *Spectrochim. Acta, Part A*, 1985, **41**, 713.
451. H.-D. Amberger and N. M. Edelstein, *New Front. Rare Earths Sci. Appl., Proc. Int. Conf. Rare Earths Dev. Appl.*, 1985, **1**, 278.
452. H. Reddmann and H.-D. Amberger, *J. Less-Common Met.*, 1985, **112**, 297.
453. H.-D. Amberger, K. Yünlü and N. M. Edelstein, *Spectrochim. Acta, Part A*, 1986, **42**, 27.
454. H.-D. Amberger and K. Yünlü, *Spectrochim. Acta, Part A*, 1986, **42**, 393.
455. H.-D. Amberger, H. Schultze and N. M. Edelstein, *Spectrochim. Acta, Part A*, 1986, **42**, 657.
456. H.-D. Amberger and H. Schultze, *Spectrochim. Acta, Part A*, 1987, **43**, 1301.
457. H.-D. Amberger, H. Schultze, H. Reddmann, G. V. Shalimoff and N. M. Edelstein, *J. Less-Common. Met.*, 1989, **149**, 249.
458. C. Hagen, H. Reddmann, H.-D. Amberger, F. T. Edelmann, U. Pegelow, G. V. Shalimoff and N. M. Edelstein, *J. Organomet. Chem.*, 1993, **462**, 69.
459. S. H. Eggers, J. Kopf and R. D. Fischer, *Organometallics*, 1986, **5**, 383.
460. Z. Xie, F. E. Hahn and C. Qian, *J. Organomet. Chem.*, 1991, **414**, C12.
461. M. Adam, U. Behrens and R. D. Fischer, *Acta Crystallogr., Sect. C*, 1991, **47**, 968.
462. V. K. Bel'skii, Y. K. Gun'ko, G. L. Soloveichik and B. M. Bulychev, *Metalloorg. Khim.*, 1991, **4**, 577.
463. S. H. Eggers, W. Hinrichs, J. Kopf, W. Jahn and R. D. Fischer, *J. Organomet. Chem.*, 1986, **311**, 313.
464. S. H. Eggers, J. Kopf and R. D. Fischer, *Acta Crystallogr., Sect. C*, 1988, **43**, 2288.
465. S. H. Eggers, H. Schultze, J. Kopf and R. D. Fischer, *Angew. Chem., Int. Ed. Engl.*, 1986, **25**, 656.
466. S. D. Stults, R. A. Andersen and A. Zalkin, *Organometallics*, 1990, **9**, 115.
467. A. Hammel, W. Schwarz and J. Weidlein, *J. Organomet. Chem.*, 1989, **363**, C29.
468. M. Booij, N. H. Kiers, A. Meetsma, J. H. Teuben, W. J. J. Smeets and A. L. Spek, *Organometallics*, 1989, **8**, 2454.
469. Y. B. Zverev, I. V. Runovskaya, S. G. Chesnokova, N. P. Chernyaev and P. E. Gaivoronskii, *Vysokochist. Veshchestva*, 1988, **6**, 141.
470. J. Weber *et al.*, *J. Cryst. Growth*, 1990, **104**, 815.
471. C. Qiu and Z. Zhou, *Huaxue Xuebao*, 1986, **44**, 1058.
472. H. Schumann, J. A. Meese-Marktscheffel, B. Gorella and F. H. Görlitz, *J. Organomet. Chem.*, 1992, **428**, C27.
473. C. Qian, B. Wang, D. Deng, G. Wu and P. Zheng, *J. Organomet. Chem.*, 1992, **427**, C29.
474. C. Qian, B. Wang, D. Deng and X. Jin, *Polyhedron*, 1993, **12**, 2265.
475. G. B. Deacon, A. J. Koplick and T. D. Tuong, *Polyhedron*, 1982, **1**, 423.
476. H. D. Kaesz (ed)., 'Inorganic Syntheses', John Wiley & Sons (Interscience), New York, 1989, **26**, 17.
477. S. Song, Q. Shen and S. Jin, *Polyhedron*, 1992, **11**, 2863.
478. H. Reddmann, H. Schultze, H.-D. Amberger, G. V. Shalimoff and N. M. Edelstein, *J. Organomet. Chem.*, 1991, **411**, 331.
479. L. Shi, F. Shen, X. Zhou and Z. Ye, *Zhongguo Kexue Jishu Daxue Xuebao*, 1991, **21**, 109.
480. M. R. Spirlet, J. Rebizant, C. Apostolidis and B. Kanellakopulos, *Inorg. Chim. Acta*, 1987, **C43**, 2322.
481. M. R. Spirlet, J. Rebizant, C. Apostolidis and B. Kanellakopulos, *Inorg. Chim. Acta*, 1987, **139**, 211.
482. R. Maier, B. Kanellakopulos and C. Apostolidis, *J. Organomet. Chem.*, 1992, **427**, 33.
483. A. Domingos, N. Marques, A. Pires de Matos, M. G. Silva-Valenzuela and L. B. Zinner, *Polyhedron*, 1993, **12**, 2545.
484. H. Schumann and F. W. Reier, *J. Organomet. Chem.*, 1984, **269**, 21.
485. H. Schulz, H. Schultze, H. Reddmann, M. Link and H.-D. Amberger, *J. Organomet. Chem.*, 1992, **424**, 139.
486. H. Schulz and H.-D. Amberger, *J. Organomet. Chem.*, 1993, **443**, 71.
487. H.-D. Amberger and H. Schulz, *Spectrochim. Acta, Part A*, 1991, **47**, 233.
488. R. Maier, B. Kanellakopulos, C. Apostolidis and B. Nuber, *J. Organomet. Chem.*, 1992, **435**, 275.

489. H. Schulz, H. Reddmann and H.-D. Amberger, *J. Organomet. Chem.*, 1992, **440**, 317.
490. C. Apostolidis, B. Kanellakopulos, R. Klenze, H. Reddmann, H. Schulz and H.-D. Amberger, *J. Organomet. Chem.*, 1992, **426**, 307.
491. R. D. Rogers, J. L. Atwood, A. Eman, D. J. Sikora and M. D. Rausch, *J. Organomet. Chem.*, 1981, **216**, 383.
492. W. Q. Chen, G. Y. Lin, J. S. Xia, G. C. Wei, Y. Zhang and Z. S. Jin, *J. Organomet. Chem.*, 1994, **467**, 75.
493. F. Benetollo, G. Bombieri, C. B. Castellani, W. Jahn and R. D. Fischer, *Inorg. Chim. Acta*, 1984, **95**, L7.
494. S. Wang, Y. Yu, Z. Ye, C. Qian and X. Huang, *J. Organomet. Chem.*, 1994, **464**, 55.
495. Z. Wu, Z. Xu, X. You, X. Zhou, X. Huang and J. Chen, *Polyhedron*, 1994, **13**, 379.
496. R. D. Rogers, R. V. Bynum and J. L. Atwood, *J. Organomet. Chem.*, 1980, **192**, 65.
497. Z. Ye, S. Wang, Y. Lu and L. Shi, *Inorg. Chim. Acta*, 1990, **177**, 97.
498. C. Ni, D. Deng and C. Qian, *Inorg. Chim. Acta*, 1985, **110**, L7.
499. G. B. Deacon, B. M. Gatehouse, S. N. Platts and D. L. Wilkinson, *Aust. J. Chem.*, 1987, **40**, 907.
500. M. Adam, H. Schultze and R. D. Fischer, *J. Organomet. Chem.*, 1992, **429**, C1.
501. J. Rebizant, M. R. Spirlet, C. Apostolidis and B. Kanellakopulos, *Acta Crystallogr., Sect. C*, 1990, **46**, 2076.
502. S. J. Swamy, J. Loebel, J. Pickardt and H. Schumann, *J. Organomet. Chem.*, 1988, **353**, 27.
503. J. G. Brennan, S. D. Stults, R. A. Andersen and A. Zalkin, *Organometallics*, 1988, **7**, 1329.
504. S. Stults and A. Zalkin, *Acta Crystallogr., Sect. C*, 1987, **43**, 430.
505. H. Schumann, F. H. Görlitz, F. E. Hahn, J. Pickardt, C. Qian and Z. Xie, *Z. Anorg. Allg. Chem.*, 1992, **609**, 131.
506. S. H. Eggers, M. Adam, E. T. K. Haupt and R. D. Fischer, *Inorg. Chim. Acta*, 1987, **139**, 315.
507. M. Adam, E. T. K. Haupt and R. D. Fischer, *Bull. Magn. Reson.*, 1990, **12**, 101.
508. H. Schumann, C. Janiak and J. Pickardt, *J. Organomet. Chem.*, 1988, **349**, 117.
509. H. Gao, Q. Shen, J. Hu, S. Jin and Y. Lin, *Chin. Sci. Bull.*, 1989, **34**, 614.
510. H. Gao, Q. Shen, J. Hu, S. Jin and Y. Lin, *J. Organomet. Chem.*, 1992, **427**, 141.
511. L. Xing-Fu *et al.*, *Inorg. Chim. Acta*, 1985, **100**, 183.
512. S. H. Eggers and R. D. Fischer, *J. Organomet. Chem.*, 1986, **315**, C61.
513. H. Schulz, H. Reddmann and H.-D. Amberger, *J. Organomet. Chem.*, 1993, **461**, 69.
514. B. L. Kalsotra, S. P. Anand, R. K. Multani and B. D. Jain, *J. Organomet. Chem.*, 1971, **28**, 87.
515. G. B. Deacon, T. D. Tuong and D. G. Vince, *Polyhedron*, 1983, **2**, 969.
516. A. Gulino *et al.*, *Organometallics*, 1988, **7**, 2360.
517. W. J. Evans, T. J. Deming and J. W. Ziller, *Organometallics*, 1989, **8**, 1581.
518. K. Jacob, J. Glanz, K. Tittes, K.-H. Thiele, I. Pavlik and A. Lyçka, *Z. Anorg. Allg. Chem.*, 1989, **577**, 145.
519. K. Jacob, M. Glanz, J. Holeçek and A. Lyçka, *Z. Anorg. Allg. Chem.*, 1990, **581**, 33.
520. C. Qian, C. Ye, H. Lu, Y. Li and Y. Huang, *J. Organomet. Chem.*, 1984, **263**, 333.
521. C. Qian, C. Ye and Y. Li, *J. Organomet. Chem.*, 1986, **302**, 171.
522. C. Qian, C. Ye and Y. Li, *Youji Huaxue*, 1986, 130.
523. C. Ye, C. Qian and X. Yang, *J. Organomet. Chem.*, 1991, **407**, 329.
524. C. Qian, Z. Xie and Y. Huang, *Inorg. Chim. Acta*, 1987, **139**, 195.
525. C. Qian, X. Wang, Y. Li and C. Ye, *Polyhedron*, 1990, **9**, 479.
526. S. J. Swamy, J. Loebel and H. Schumann, *J. Organomet. Chem.*, 1989, **379**, 51.
527. A. Recknagel and F. T. Edelmann, *Angew. Chem., Int. Ed. Engl.*, 1991, **30**, 693.
528. F. T. Edelmann, M. Rieckhoff, I. Haiduc and I. Silaghi-Dumitrescu, *J. Organomet. Chem.*, 1993, **447**, 203.
529. P. Yan, N. Hu, Z. Jin and W. Chen, *J. Organomet. Chem.*, 1990, **391**, 313.
530. P. Yan and W. Chen, *Chin. Sci. Bull.*, 1991, **36**, 652.
531. C. Sun, G. Wei, Z. Jin and W. Chen, *J. Organomet. Chem.*, 1993, **453**, 61.
532. C. Sun, G. Wei, Z. Jin and W. Chen, *Polyhedron*, 1994, **13**, 1483.
533. S. Sun, Z. Jin, G. Wei and W. Chen, *Chin. J. Chem.*, 1992, 405.
534. N. Höck, W. Oroschin, G. Paolucci and R. D. Fischer, *Angew. Chem., Int. Ed. Engl.*, 1986, **25**, 738.
535. K. Qiao, R. D. Fischer, G. Paolucci, P. Traldi and E. Celon, *Organometallics*, 1990, **9**, 1361.
536. K. Qiao, R. D. Fischer and G. Paolucci, *J. Organomet. Chem.*, 1993, **456**, 185.
537. D. Stern, M. Sabat and T. J. Marks, *J. Am. Chem. Soc.*, 1990, **112**, 9558.
538. C. M. Fendrick, L. D. Schertz, V. W. Day and T. J. Marks, *Organometallics*, 1988, **7**, 1828.
539. H. Schumann, L. Esser, J. Loebel, A. Dietrich, D. van der Helm and X. Ji, *Organometallics*, 1991, **10**, 2585.
540. G. Jeske, L. E. Schock, P. N. Swepston, H. Schumann and T. J. Marks, *J. Am. Chem. Soc.*, 1985, **107**, 8103.
541. W. P. Schaefer, R. D. Köhn and J. E. Bercaw, *Acta Crystallogr., Sect. C*, 1992, **48**, 251.
542. R. E. Marsh, W. P. Schaefer, E. B. Coughlin and J. E. Bercaw, *Acta Crystallogr., Sect. C*, 1992, **48**, 1773.
543. N. Koga and K. Morokuma, *J. Am. Chem. Soc.*, 1988, **110**, 108.
544. E. Bunel, B. J. Burger and J. E. Bercaw, *J. Am. Chem. Soc.*, 1988, **110**, 976.
545. G. Fu, Y. Xu, Z. Xie and C. Qian, *Acta Chim. Sin.*, 1989, 431.
546. C. Qian, Z. Xie and Y. Huang, *J. Organomet. Chem.*, 1987, **323**, 285.
547. C. Chen, X. Zhong, C. Qian, Z. Xie and Y. Huang, *Youji Huaxue*, 1988, **8**, 235.
548. Z. Xie, C. Qian and Y. Huang, *J. Organomet. Chem.*, 1991, **412**, 61.
549. H. Schumann, J. Loebel, J. Pickardt, C. Qian and Z. Xie, *Organometallics*, 1991, **10**, 215.
550. C. Qian, Z. Xie and Y. Huang, *J. Organomet. Chem.*, 1990, **398**, 251.
551. C. Qian and D. Zhu, *J. Chem. Soc., Dalton Trans.*, 1994, 1599.
552. J. Gräper, R. D. Fischer and G. Paolucci, *J. Organomet. Chem.*, 1994, **471**, 87.
553. C. Qian and D. Zhu, *J. Organomet. Chem.*, 1993, **445**, 79.
554. G. Paolucci, R. D'Ippolito, C. Ye, C. Qian, J. Gräper and R. D. Fischer, *J. Organomet. Chem.*, 1994, **471**, 97.
555. W. J. Evans, T. S. Gummersheimer, T. J. Boyle and J. W. Ziller, *Organometallics*, 1994, **13**, 1281.
556. W. Chen and X. Wang, *Huaxue Xuebao*, 1985, **43**, 295.
557. W. Q. Chen, S. X. Xiao, Y. L. Wang and G. Q. Yu, *Kexue Tongbao*, 1984, **29**, 892.
558. G. Fuxing, G. Wei, Z. Jin and W. Chen, *J. Organomet. Chem.*, 1992, **438**, 289.
559. A. Zhou, Z. Wu, F. Zhao and X. Song, *Youji Huaxue*, 1989, **9**, 230.
560. R. K. Sharma and C. P. Sharma, *J. Indian Chem. Soc.*, 1987, **64**, 506.

561.   J. Xia, Z. Jin, G. Lin and W. Chen, *J. Organomet. Chem.*, 1991, **408**, 173.
562.   X. Jusong, Z. Jin, L. Guanyang and W. Chen, *Zhongguo Xitu Xuebao*, 1992, 1.
563.   Z. Zhennan, W. Zhongzhi, D. Baohu and Y. Zhongwen, *Polyhedron*, 1989, **8**, 17.
564.   Y. Su, Z. Jin, N. Hu and W. Chen, *Yingyong Huaxue*, 1990, **7**, 23.
565.   M. Chen, G. Wu, W. Wu, S. Zhuang and Z. Huang, *Organometallics*, 1988, **7**, 802.
566.   L. Chen, M. Tsutsui and D. E. Bergbreiter, *Youji Huaxue*, 1985, 223.
567.   A. B. Sigalov *et al.*, *Izv. Akad. Nauk SSSR, Ser. Khim.*, 1983, 918; *Bull. Acad. Sci. USSR*, 1983, **32**, 833.
568.   W. J. Evans *et al.*, *Inorg. Chem.*, 1986, **25**, 3614.
569.   A. F. Williams, F. Grandjean, G. J. Long, T. A. Ulibarri and W. J. Evans, *Inorg. Chem.*, 1989, **28**, 4584.
570.   W. J. Evans, I. Bloom, W. E. Hunter and J. L. Atwood, *Organometallics*, 1985, **4**, 112.
571.   W. J. Evans, *Inorg. Chim. Acta*, 1987, **139**, 169.
572.   W. J. Evans, J. W. Grate, H. W. Choi, I. Bloom, W. E. Hunter and J. L. Atwood, *J. Am. Chem. Soc.*, 1985, **107**, 941.
573.   W. J. Evans, L. A. Hughes and T. P. Hanusa, *Organometallics*, 1986, **5**, 1285.
574.   P. L. Watson, *J. Chem. Soc., Chem. Commun.*, 1980, 652.
575.   A. L. Wayda, J. L. Dye and R. D. Rogers, *Organometallics*, 1984, **3**, 1605.
576.   A. G. Avent, M. A. Edelman, M. F. Lappert and G. A. Lawless, *J. Am. Chem. Soc.*, 1989, **111**, 3423.
577.   W. J. Evans and T. A. Ulibarri, *Polyhedron*, 1989, **8**, 1007.
578.   W. J. Evans, L. A. Hughes and T. P. Hanusa, *J. Am. Chem. Soc.*, 1984, **106**, 4270.
579.   R. A. Andersen, J. M. Boncella, C. J. Burns, J. C. Green, D. Hohl and N. Rösch, *J. Chem. Soc., Chem. Commun.*, 1986, 405.
580.   R. A. Williams, T. P. Hanusa and J. C. Huffman, *Organometallics*, 1990, **9**, 1128.
581.   R. A. Williams, T. P. Hanusa and J. C. Huffman, *J. Chem. Soc., Chem. Commun.*, 1988, 1045.
582.   R. A. Andersen, J. M. Boncella, C. J. Burns, R. Blom, A. Haaland and H. V. Volden, *J. Organomet. Chem.*, 1986, **312**, C49.
583.   R. A. Andersen, R. Blom, J. M. Boncella, C. J. Burns and H. V. Volden, *Acta Chem. Scand.*, 1987, **A41**, 24.
584.   T. K. Hollis, J. K. Burdett and B. Bosnich, *Organometallics*, 1993, **12**, 3385.
585.   M. Kaupp, P. R. von Schleyer, M. Dolg and H. Stoll, *J. Am. Chem. Soc.*, 1992, **114**, 8202.
586.   A. C. Thomas and A. B. Ellis, *Organometallics*, 1985, **4**, 2223.
587.   S. P. Nolan and T. J. Marks, *J. Am. Chem. Soc.*, 1989, **111**, 8538.
588.   X.-M. Min, *Huaxue Xuebao*, 1992, **11**, 1098.
589.   C. J. Burns and R. A. Andersen, *J. Am. Chem. Soc.*, 1987, **109**, 941.
590.   C. J. Burns and R. A. Andersen, *J. Am. Chem. Soc.*, 1987, **109**, 915.
591.   C. J. Burns and R. A. Andersen, *J. Am. Chem. Soc.*, 1987, **109**, 5853.
592.   W. J. Evans, T. A. Ulibarri and P. Jutzi, *Inorg. Chim. Acta*, 1990, **168**, 5.
593.   W. J. Evans, J. W. Grate, I. Bloom, W. E. Hunter and J. L. Atwood, *J. Am. Chem. Soc.*, 1985, **107**, 405.
594.   M. Wedler, A. Recknagel and F. T. Edelmann, *J. Organomet. Chem.*, 1990, **395**, C26.
595.   M. Booij, N. H. Kiers, H. J. Heeres and J. H. Teuben, *J. Organomet. Chem.*, 1989, **364**, 79.
596.   I. Albrecht, E. Hahn, J. Pickardt and H. Schumann, *Inorg. Chim. Acta*, 1985, **110**, 145.
597.   R. G. Finke, S. R. Keenan, D. A. Schiraldi and P. L. Watson, *Organometallics*, 1986, **5**, 598.
598.   Z. Peng, H. Tian and Q. Shen, *Youji Huaxue*, 1985, **5**, 385.
599.   H. J. Heeres and J. H. Teuben, *Recl. Trav. Chim. Pays-Bas*, 1990, **109**, 226.
600.   Q. Shen, M. Qi and Y. Lin, *J. Organomet. Chem.*, 1990, **399**, 247.
601.   A. Zalkin and D. Berg, *Acta Crystallogr., Sect. C*, 1989, **45**, 1630.
602.   C. J. Burns, D. J. Berg and R. A. Andersen, *J. Chem. Soc., Chem. Commun.*, 1987, 272.
603.   K.-G. Wang, E. D. Stevens and S. P. Nolan, *Organometallics*, 1992, **11**, 1011.
604.   H. Schumann, I. Albrecht, J. Pickardt and E. Hahn, *J. Organomet. Chem.*, 1984, **276**, C5.
605.   H. Schumann, I. Albrecht, J. Loebel, E. Hahn, M. B. Hossain and D. van der Helm, *Organometallics*, 1986, **5**, 1296.
606.   I. Albrecht and H. Schumann, *J. Organomet. Chem.*, 1986, **310**, C29.
607.   W. E. Piers, E. E. Bunel and J. E. Bercaw, *J. Organomet. Chem.*, 1991, **407**, 51.
608.   R. E. Marsh, W. P. Schaefer, G. C. Bazan and J. E. Bercaw, *Acta Crystallogr., Sect C*, 1992, **48**, 1416.
609.   P. J. Shapiro, E. Bunel, W. P. Schaefer and J. E. Bercaw, *Organometallics*, 1990, **9**, 867.
610.   C. J. Schaverien, *Organometallics*, 1994, **13**, 69.
611.   C. J. Schaverien, J. H. G. Frijns, H. J. Heeres, J. R. van den Hende, J. H. Teuben and A. L. Spek, *J. Chem. Soc., Chem. Commun.*, 1991, 642.
612.   C. J. Schaverien, *J. Chem. Soc., Chem. Commun.*, 1992, 11.
613.   W. J. Evans, J. L. Shreeve and J. W. Ziller, *Organometallics*, 1994, **13**, 731.
614.   M. Rieckhoff, M. Noltemeyer, F. T. Edelmann, I. Haiduc and I. Silaghi-Dumitrescu, *J. Organomet. Chem.*, 1994, **469**, C19.
615.   H. Schumann, J. Winterfeld, R. D. Köhn, L. Esser, J. Sun and A. Dietrich, *Chem. Ber.*, 1993, **126**, 907.
616.   P. N. Hazin, J. C. Huffman and J. W. Bruno, *Organometallics*, 1987, **6**, 23.
617.   H. Van der Heijden, C. J. Schaverien and A. G. Orpen, *Organometallics*, 1989, **8**, 255.
618.   C. J. Schaverien, H. Van der Heijden and A. G. Orpen, *Polyhedron*, 1989, **8**, 1850.
619.   M. Bochmann, M. Kargar and A. J. Jaggar, *J. Chem. Soc., Chem. Commun.*, 1990, 1038.
620.   A. D. Horton and J. H. G. Frijns, *Angew. Chem., Int. Ed. Engl.*, 1991, **30**, 1152.
621.   C. Pellecchia, A. Grassi and A. Immirzi, *J. Am. Chem. Soc.*, 1993, **115**, 1160.
622.   F. Calderazzo, U. Englert, G. Pampaloni and L. Rocchi, *Angew. Chem., Int. Ed. Engl.*, 1992, **31**, 1235.
623.   F. Calderazzo, G. Pampaloni, L. Rocchi and U. Englert, *Organometallics*, 1994, **13**, 2592.
624.   C. J. Schaverien, *Organometallics*, 1992, **11**, 3476.
625.   H. J. Heeres, A. Meetsma and J. H. Teuben, *J. Chem. Soc., Chem. Commun.*, 1988, 962.
626.   H. J. Heeres, A. Meetsma, J. H. Teuben and R. D. Rogers, *Organometallics*, 1989, **8**, 2637.
627.   H. J. Heeres, J. H. Teuben and R. D. Rogers, *J. Organomet. Chem.*, 1989, **364**, 87.
628.   H. Van der Heijden, P. Pasman, E. J. M. de Boer, C. J. Schaverien and A. G. Orpen, *Organometallics*, 1989, **8**, 1459.
629.   W. J. Evans, T. A. Ulibarri, L. R. Chamberlain, J. W. Ziller and D. Alvarez, *Organometallics*, 1990, **9**, 2124.

630. H. J. Heeres, A. Meetsma and J. H. Teuben, *J. Organomet. Chem.*, 1991, **414**, 351.
631. C. J. Burns and R. A. Andersen, *J. Chem. Soc., Chem. Commun.*, 1989, 136.
632. R. G. Finke, S. R. Keenan and P. L. Watson, *Organometallics*, 1989, **8**, 263.
633. W. J. Evans, J. M. Olofson, H. Zhang and J. L. Atwood, *Organometallics*, 1988, **7**, 629.
634. L. Gong, A. Streitwieser, Jr. and A. Zalkin, *J. Chem. Soc., Chem. Commun.*, 1987, 460.
635. P. N. Hazin *et al.*, *Inorg. Chem.*, 1988, **27**, 1393.
636. M. D. Rausch, K. J. Moriarty, J. L. Atwood, J. A. Weeks, W. E. Hunter and H. G. Brittain, *Organometallics*, 1986, **5**, 1281.
637. M. Visseaux, A. Dormond and D. Baudry, *Bull. Soc. Chim. Fr.*, 1993, **130**, 173.
638. A. L. Wayda and W. J. Evans, *Inorg. Chem.*, 1980, **19**, 2190.
639. Q. Shen and P. Zhou, *Kexue Tongbao*, 1987, **32**, 28.
640. Q. Shen, H. Tian, Z. Sun and R. Shi, *Youji Huaxue*, 1985, 241.
641. W. J. Evans, T. T. Peterson, M. D. Rausch, W. E. Hunter and J. L. Atwood, *Organometallics*, 1985, **3**, 554.
642. W. J. Evans, D. K. Drummond, J. W. Grate, T. Zhang and J. L. Atwood, *J. Am. Chem. Soc.*, 1987, **109**, 3928.
643. W. J. Evans and D. K. Drummond, *Organometallics*, 1988, **7**, 797.
644. W. J. Evans, D. K. Drummond, T. P. Hanusa and J. M. Olofson, *J. Organomet. Chem.*, 1989, **376**, 311.
645. A. C. Thomas and A. B. Ellis, *J. Chem. Soc., Chem. Commun.*, 1984, 1270.
646. W. J. Evans, T. P. Hanusa and K. R. Levan, *Inorg. Chim. Acta*, 1985, **110**, 191.
647. K.-H. Thiele, A. Scholz, J. Scholz, U. Böhme, R. Kempe and J. Sieler, *Z. Naturforsch., Teil B*, 1993, **48**, 1753.
648. T. D. Tilley, R. A. Andersen, A. Zalkin and D. H. Templeton, *Inorg. Chem.*, 1982, **21**, 2644.
649. P. L. Watson, *J. Chem. Soc., Chem. Commun.*, 1983, 276.
650. H. Schumann, I. Albrecht and E. Hahn, *Angew. Chem., Int. Ed. Engl.*, 1985, **24**, 985.
651. A. Recknagel, M. Noltemeyer, D. Stalke, U. Pieper, H.-G. Schmidt and F. T. Edelmann, *J. Organomet. Chem.*, 1991, **411**, 347.
652. D. J. Berg, C. J. Burns, R. A. Andersen and A. Zalkin, *Organometallics*, 1989, **8**, 1865.
653. D. J. Berg, R. A. Andersen and A. Zalkin, *Organometallics*, 1988, **7**, 1858.
654. A. Zalkin and D. J. Berg, *Acta Crystallogr., Sect. C*, 1988, **44**, 1488.
655. A. Zalkin, T. J. Henly and R. A. Andersen, *Acta Crystallogr., Sect C*, 1987, **43**, 233.
656. W. E. Piers, *J. Chem. Soc., Chem. Commun.*, 1994, 309.
657. W. E. Piers, L. R. MacGillivray and M. Zaworotko, *Organometallics*, 1993, **12**, 4723.
658. W. J. Evans, D. K. Drummond, S. G. Bott and J. L. Atwood, *Organometallics*, 1986, **5**, 2389.
659. W. J. Evans *et al.*, *J. Am. Chem. Soc.*, 1988, **110**, 4983.
660. W. J. Evans, L. A. Hughes, D. K. Drummond, H. Zhang and J. L. Atwood, *J. Am. Chem. Soc.*, 1986, **108**, 1722.
661. W. J. Evans, R. A. Keyer, G. W. Rabe, D. K. Drummond and J. W. Ziller, *Organometallics*, 1993, **12**, 4664.
662. W. J. Evans, J. W. Grate, L. A. Hughes, H. Zhang and J. L. Atwood, *J. Am. Chem. Soc.*, 1985, **107**, 3728.
663. W. J. Evans, R. A. Keyer and J. W. Ziller, *Organometallics*, 1993, **12**, 2618.
664. W. J. Evans and D. K. Drummond, *J. Am. Chem. Soc.*, 1989, **111**, 3329.
665. A. Recknagel, M. Noltemeyer and F. T. Edelmann, *J. Organomet. Chem.*, 1991, **410**, 53.
666. J. Scholz, A. Scholz, R. Weimann, C. Janiak and H. Schumann, *Angew. Chem., Int. Ed. Engl.*, 1994, **33**, 1171.
667. W. J. Evans, T. A. Ulibarri and J. W. Ziller, *J. Am. Chem. Soc.*, 1988, **110**, 6877.
668. P. J. Shapiro, L. M. Henling, R. E. Marsh and J. E. Bercaw, *Inorg. Chem.*, 1990, **29**, 4560.
669. W. J. Evans, G. Kociok-Köhn and J. W. Ziller, *Angew. Chem., Int. Ed. Engl.*, 1992, **31**, 1081.
670. W. J. Evans, G. Kociok-Köhn, V. S. Leong and J. W. Ziller, *Inorg. Chem.*, 1992, **31**, 3592.
671. W. J. Evans, S. L. Gonzales and J. W. Ziller, *J. Chem. Soc., Chem. Commun.*, 1992, 1138.
672. W. J. Evans, S. L. Gonzales and J. W. Ziller, *J. Am. Chem. Soc.*, 1991, **113**, 9880.
673. M. E. Thompson *et al.*, *J. Am. Chem. Soc.*, 1987, **109**, 203.
674. P. L. Watson, *J. Am. Chem. Soc.*, 1982, **104**, 337.
675. P. L. Watson, *J. Am. Chem. Soc.*, 1983, **105**, 6491.
676. P. L. Watson and D. C. Roe, *J. Am. Chem. Soc.*, 1982, **104**, 6471.
677. P. L. Watson and T. Herskovitz, *ACS Symp. Ser (Initiation Polym.)*, 1983, **212**, 459.
678. M. L. H. Green, A. K. Hughes, N. A. Popham, A. H. H. Stephens and L.-L. Wong, *J. Chem. Soc., Dalton Trans.*, 1992, 3077.
679. K. H. den Haan, J. L. de Boer, J. H. Teuben, W. J. J. Smeets and A. L. Spek, *J. Organomet. Chem.*, 1987, **327**, 31.
680. W. J. Evans, L. R. Chamberlain, T. A. Ulibarri and J. W. Ziller, *J. Am. Chem. Soc.*, 1988, **110**, 6423.
681. B. W. Pfenning, M. E. Thompson and A. B. Bocarsly, *Organometallics*, 1993, **12**, 649.
682. P. T. Matsunaga, *Energy Res. Abstr.*, 1992, 17(6).
683. C. M. Forsyth, S. P. Nolan and T. J. Marks, *Organometallics*, 1991, **10**, 2543.
684. T. Ziegler, E. Folga and A. Berces, *J. Am. Chem. Soc.*, 1993, **115**, 636.
685. E. Folga, T. Ziegler and L. Fan, *New. J. Chem.*, 1992, **15**, 741.
686. H. Mauermann, P. N. Swepston and T. J. Marks, *Organometallics*, 1985, 4, 200.
687. J. Renkema and J. H. Teuben, *Recl. Trav. Chim. Pays-Bas*, 1986, **105**, 241.
688. G. Jeske, H. Lauke, H. Mauermann, P. N. Swepston H. Schumann and T. J. Marks, *J. Am. Chem. Soc.*, 1985, **107**, 8091.
689. J. E. Bercaw, D. L. Davies and P. T. Wolczanski, *Organometallics*, 1986, **5**, 443.
690. H. J. Heeres, A. Meetsma and J. H. Teuben, *Angew. Chem., Int. Ed. Engl.*, 1990, **29**, 420.
691. S. Hajela, W. P. Schaefer and J. E. Bercaw, *Acta Crystallogr., Sect. C*, 1992, **48**, 1771.
692. J. H. Teuben, in 'Fundamental and Technological Aspects of Organo-*f*-Element Chemistry', NATO ASI Ser., eds. T. J. Marks and I. L. Fragalà, D. Reidel, Boston, MA, 1985, **155**, 195.
693. K. H. den Haan, G. A. Luinstra, A. Meetsma and J. H. Teuben, *Organometallics*, 1987, **6**, 1509.
694. K. H. den Haan, Y. Wielstra and J. H. Teuben, *Organometallics*, 1987, **6**, 2053.
695. H. J. Heeres, M. Maters and J. H. Teuben, *Organometallics*, 1992, **11**, 350.
696. M. Booij, A. Meetsma and J. H. Teuben, *Organometallics*, 1991, **10**, 3246.
697. W. J. Evans, T. A. Ulibarri and J. W. Ziller, *Organometallics*, 1991, **10**, 134.
698. M. A. Busch, R. Harlow and P. L. Watson, *Inorg. Chim. Acta*, 1987, **140**, 15.

699. K. H. den Haan, Y. Wielstra, J. J. W. Eshuis and J. H. Teuben, *J. Organomet. Chem.*, 1987, **323**, 181.
700. W. J. Evans, L. R. Chamberlain and J. W. Ziller, *J. Am. Chem. Soc.*, 1987, **109**, 7209.
701. H. Yamamoto, H. Yasuda, K. Yokota, A. Nakamura, Y. Kai and N. Kasai, *Chem. Lett.*, 1988, 1963.
702. A. Scholz, A. Smola, J. Scholz, J. Loebel, H. Schumann and K.-H. Thiele, *Angew. Chem., Int. Ed. Engl.*, 1991, **30**, 435.
703. W. J. Evans, T. A. Ulibarri and J. W. Ziller, *J. Am. Chem. Soc.*, 1990, **112**, 219.
704. W. J. Evans, T. A. Ulibarri and J. W. Ziller, *J. Am. Chem. Soc.*, 1990, **112**, 2314.
705. W. J. Evans, S. L. Gonzales and J. W. Ziller, *J. Am. Chem. Soc.*, 1994, **116**, 2600.
706. W. J. Evans, I. Bloom, W. E. Hunter and J. L. Atwood, *J. Am. Chem. Soc.*, 1983, **105**, 1401.
707. J. M. Boncella, T. D. Tilley and R. A. Andersen, *J. Chem. Soc., Chem. Commun.*, 1984, 710.
708. W. J. Evans, R. A. Keyer, H. Zhang and J. L. Atwood, *J. Chem. Soc., Chem. Commun.*, 1987, 837.
709. W. J. Evans, R. A. Keyer and J. W. Ziller, *Organometallics*, 1990, **9**, 2628.
710. H. J. Heeres, J. Nijhoff, J. H. Teuben and R. D. Rogers, *Organometallics*, 1993, **12**, 2609.
711. C. M. Forsyth, S. P. Nolan, C. S. Stern, T. J. Marks and A. L. Rheingold, *Organometallics*, 1993, **12**, 3618.
712. A. Recknagel, D. Stalke, H. W. Roesky and F. T. Edelmann, *Angew. Chem., Int. Ed. Engl.*, 1989, **28**, 445.
713. W.-K. Wong, H. Chen and F.-L. Chow, *Polyhedron*, 1990, **9**, 875.
714. H. Siebald, M. Dartiguenave and Y. Dartiguenave, *J. Organomet. Chem.*, 1992, **438**, 83.
715. N. S. Radu and T. D. Tilley, *Phosphorus, Sulfur, and Silicon*, 1994, **87**, 209.
716. N. S. Radu, T. D. Tilley and A. L. Rheingold, *J. Am. Chem. Soc.*, 1992, **114**, 8293.
717. T. Kobayashi, T. Sakakura, T. Hayashi, M. Yumura and M. Tanaka, *Chem. Lett.*, 1992, 1157.
718. T. Sakakura, H.-J. Lautenschlager and M. Tanaka, *J. Chem. Soc., Chem. Commun.*, 1991, 40.
719. M. Booij, B.-J. Deelman, R. Duchateau, D. S. Postma, A. Meetsma and J. H. Teuben, *Organometallics*, 1993, **12**, 3531.
720. K. H. den Haan and J. H. Teuben, *Recl. Trav. Chim. Pays-Bas*, 1984, **103**, 333.
721. K. H. den Haan and J. H. Teuben, *J. Chem. Soc., Chem. Commun.*, 1986, 682.
722. W. J. Evans, J. W. Grate and R. J. Doedens, *J. Am. Chem. Soc.*, 1985, **107**, 1671.
723. W. J. Evans, S. L. Gonzales and J. W. Ziller, *J. Am. Chem. Soc.*, 1991, **113**, 7423.
724. H. Schumann, M. Glanz and H. Hemling, *J. Organomet. Chem.*, 1993, **445**, C1.
725. W. J. Evans and T. A. Ulibarri, *J. Am. Chem. Soc.*, 1987, **109**, 4292.
726. R. D. Ernst and T. H. Cymbaluk, *Organometallics*, 1982, **1**, 708.
727. N. Hu, L. Gong, Z. Jin and W. Chen, *Wuji Huaxue Xuebao*, 1989, **5**, 107; see also: X. Qiu and J. Liu, *Chin. J. Chem.*, 1991, **9**, 10.
728. H. Schumann and A. Dietrich, *J. Organomet. Chem.*, 1991, **401**, C33.
729. J. Sieler, A. Simon, K. Peters, R. Taube and M. Geitner, *J. Organomet. Chem.*, 1989, **362**, 297.
730. W. Weng, K. Kunze, A. M. Arif and R. D. Ernst, *Organometallics*, 1991, 10, 3643.
731. F. Nief, L. Ricard and F. Mathey, *Polyhedron*, 1993, **12**, 19.
732. F. Nief and F. Mathey, *Synlett*, 1991, 745.
733. F. Nief and F. Mathey, *J. Chem. Soc., Chem. Commun.*, 1989, 800.
734. M. N. Bochkarev, A. A. Trifonov, V. K. Cherkasov and G. A. Razuvaev, *Zh. Obshch. Khim.*, 1988, **58**, 719.
735. M. N. Bochkarev, A. A. Trifonov, V. K. Cherkasov and G. A. Razuvaev, *Metalloorg. Khim.*, 1988, **1**, 392.
736. M. N. Bochkarev, E. A. Fedorova, I. M. Penyagina, O. A. Vasina, S. Y. Khorshev and A. V. Protchenko, *Metalloorg. Khim.*, 1989, **2**, 703.
737. E. A. Fedorova and O. A. Vasina, *Metalloorg. Khim.*, 1989, **2**, 392.
738. M. N. Bochkarev *et al.*, *J. Organomet. Chem.*, 1989, **378**, 363.
739. M. N. Bochkarev, E. A. Fedorova, I. M. Penyagina, O. A. Vasina, A. V. Protchenko and S. Y. Khorshev, *Metalloorg. Khim.*, 1989, **2**, 1317.
740. Y. F. Rad'kov, V. V. Khramenkov, L. N. Zakharov, Y. T. Struchkov, S. Y. Khorshev and M. N. Bochkarev, *Metalloorg. Khim.*, 1989, **2**, 1422.
741. A. A. Trifonov, A. V. Protchenko, V. K. Cherkasov and M. N. Bochkarev, *Metalloorg. Khim.*, 1989, **2**, 471.
742. T. A. Basalgina, G. S. Kalinina and M. N. Bochkarev, *Metalloorg. Khim.*, 1989, **2**, 1142.
743. M. N. Bochkarev *et al.*, *J. Organomet. Chem.*, 1989, **372**, 217.
744. T. Arakawa, S. Shimada, G.-Y. Adachi and J. Shiokawa, *Inorg. Chim. Acta*, 1988, **145**, 327.
745. D. M. Roitershtein, L. F. Rybakova and E. S. Petrov, *Metalloorg. Khim.*, 1990, **3**, 559.
746. N. S. Emel'yanova, A. V. Protchenko, E. A. Fedorova, O. A. Yasina and M. N. Bochkarev, *Metalloorg. Khim.*, 1991, **4**, 895.
747. L. Saussine, H. Olivier, D. Commereuc and Y. Chauvin, *New. J. Chem.*, 1988, **12**, 13.
748. M. N. Bochkarev, V. V. Khramenkov, Y. F. Rad'kov, L. N. Zakharov and Y. T. Struchkov, *Metalloorg. Khim.*, 1990, **3**, 1439.
749. M. N. Bochkarev, I. L. Fedushkin, H. Schumann and J. Loebel, *J. Organomet. Chem.*, 1991, **410**, 321.
750. A. A. Trifonov, M. N. Bochkarev, H. Schumann and J. Loebel, *Angew. Chem., Int. Ed. Engl.*, 1991, **30**, 1149.
751. B. Fan, Q. Shen and Y. Lin, *J. Organomet. Chem.*, 1989, **377**, 51.
752. B. Fan, Y. Lin and Q. Shen, *Yingyong Huaxue*, 1990, **7**, 23.
753. B. Fan, S. Jin, Q. Shen and Y. Lin, *Chin. Sci. Bull.*, 1991, **36**, 84.
754. B. Fan, Q. Shen and Y. Lin, *Wuji Huaxue Xuebao*, 1991, **7**, 143.
755. H. Liang, Q. Shen, J. Guan and Y. Lin, *J. Organomet. Chem.*, 1994, **474**, 113.
756. F. A. Cotton and W. Schwotzer, *J. Am. Chem. Soc.*, 1986, **108**, 4657.
757. B. Fan, Q. Shen and Y. Lin, *J. Organomet. Chem.*, 1989, **376**, 61.
758. F. A. Cotton and W. Schwotzer, *Organometallics*, 1987, **6**, 1275.
759. B. Fan, Q. Shen and Y. Lin, *Youji Huaxue*, 1989, **9**, 414.
760. M. Cesari, U. Pedretti, A. Zazetta, G. Lugli and N. Marconi, *Inorg. Chim. Acta*, 1971, **5**, 439.
761. H. Liang, Q. Shen, S. Jin and Y. Lin, *J. Chem. Soc., Chem. Commun.*, 1992, 480.
762. G. B. Deacon, S. Nickel, P. MacKinnon and E. R. T. Tiekink, *Aust. J. Chem.*, 1990, **43**, 1245.
763. D. M. Barnhart *et al.*, *Inorg. Chem.*, 1994, **33**, 3487.
764. W. G. Van der Sluys, C. J. Burns, J. C. Huffman and A. P. Sattelberger, *J. Am. Chem. Soc.*, 1988, **110**, 5924.
765. A. L. Wayda, I. Mukerji, J. L. Dye and R. D. Rogers, *Organometallics*, 1987, **6**, 1328.

766. D. C. Eisenberg, S. A. Kinsley and A. Streitwieser, Jr., *J. Am. Chem. Soc.*, 1989, **111**, 5769.
767. A. L. Wayda, S. Cheng and I. Mukerji, *J. Organomet. Chem.*, 1987, **330**, C17.
768. S. A. Kinsley, A. Streitwieser, Jr. and A. Zalkin, *Organometallics*, 1985, **4**, 52.
769. J. Jin, S. Jin, Z. Jin and W. Chen, *J. Chem. Soc., Chem. Commun.*, 1991, 1328.
770. G. Qi and Q. Shen, *Yingyong Huaxue*, 1988, **5**, 61.
771. U. Kilimann, M. Schäfer, R. Herbst-Irmer and F. T. Edelmann, *J. Organomet. Chem.*, 1994, **469**, C10.
772. K. Mashima, Y. Nakayama, A. Nakamura, N. Kanehisa, Y. Kai and H. Takaya, *J. Organomet. Chem.*, 1994, **473**, 85.
773. N. C. Burton, F. G. N. Cloke, P. B. Hitchcock, H. C. de Lemos and A. A. Sameh, *J. Chem. Soc., Chem. Commun.*, 1989, 1462.
774. U. Kilimann and F. T. Edelmann, *J. Organomet. Chem.*, 1994, **469**, C5.
775. U. Kilimann and F. T. Edelmann, *J. Organomet. Chem.*, 1993, **444**, C15.
776. A. Masino, Ph. D. Thesis, University of Alberta, 1978.
777. S. Zhang, G. Wei, W. Chen and J. Liu, *Polyhedron*, 1994, **13**, 1927.
778. A. L. Wayda, *Organometallics*, 1983, **2**, 565.
779. A. L. Wayda and R. D. Rogers, *Organometallics*, 1985, **4**, 1440.
780. H. Schumann, J. Winterfeld, F. H. Görlitz and J. Pickardt, *Organometallics*, 1987, **6**, 1328.
781. U. Kilimann, M. Schäfer, R. Herbst-Irmer and F. T. Edelmann, *J. Organomet. Chem.*, 1994, **469**, C15.
782. T. R. Boussie, D. C. Eisenberg, J. Rigsbee, A. Streitwieser, Jr., and A. Zalkin, *Organometallics*, 1991, **10**, 1922.
783. J. Xia, Z. Jin and W. Chen, *J. Chem. Soc., Chem. Commun.*, 1991, 1214.
784. N. Rösch, *Inorg. Chim. Acta*, 1984, **94**, 297.
785. A. Streitwieser, Jr., S. A. Kinsley, J. T. Rigsbee, I. L. Fragalà, E. Ciliberto and N. Rösch, *J. Am. Chem. Soc.*, 1985, **107**, 7786.
786. M. Dolg, P. Fulde, W. Küchle, C.-S. Neumann and H. Stoll, *J. Chem. Phys.*, 1991, **94**, 3011.
787. U. Kilimann, R. Herbst-Irmer, D. Stalke and F. T. Edelmann, *Angew. Chem., Int. Ed. Engl.*, 1994, **33**, 1618.
788. K. Wen, Z. Jin and W. Chen, *J. Chem. Soc., Chem. Commun.*, 1991, 680.
789. K. Wen, Z. Jin and W. Chen, *Yingyong Huaxue*, 1992, **9**, 98.
790. H. Schumann, J. Sun and A. Dietrich, *Monatsh. Chem.*, 1990, **121**, 747.
791. K. Wen, Z. Jin and W. Chen, *Chin. Chem. Lett.*, 1991, **2**, 693.
792. K. Wen, Z. Jin, G. Wei and W. Chen, *Chin. J. Chem.*, 1992, **10**, 331.
793. H. Schumann, M. Glanz, J. Winterfeld and H. Hemling, *J. Organomet. Chem.*, 1993, **456**, 77.
794. H. Schumann, R. D. Koehn, F. W. Reier, A. Dietrich and J. Pickardt, *Organometallics*, 1989, **8**, 1388.
795. P. Bruin, M. Booij, J. H. Teuben and A. Oskam, *J. Organomet. Chem.*, 1988, **350**, 17.
796. H. Schumann, C. Janiak, R. D. Koehn, J. Loebel and A. Dietrich, *J. Organomet. Chem.*, 1989, **365**, 137.
797. P. Poremba and F. T. Edelmann, unpublished results.
798. R. R. Andréa, A. Terpstra, A. Oskam, P. Bruin and J. H. Teuben, *J. Organomet. Chem.*, 1986, **307**, 307.
799. A. Visseaux, A. Dormond, M. M. Kubicki, C. Moise, D. Baudry and M. Ephritikhine, *J. Organomet. Chem.*, 1992, **433**, 95.
800. J. R. Heath *et al.*, *J. Am. Chem. Soc.*, 1985, **107**, 7779.
801. H. W. Kroto, J. R. Heath, S. C. O'Brien, R. F. Curl and R. E. Smalley, *Nature*, 1985, **318**, 162.
802. Y. Chai *et al.*, *J. Phys. Chem.*, 1991, **95**, 7564.
803. R. D. Johnson, M. S. de Vries, J. Salem, D. S. Bethune and C. S. Yannoni, *Nature*, 1992, **355**, 239.
804. H. Shinohara, H. Sato, Y. Saito, M. Ohkohchi and Y. Ando, *J. Phys. Chem.*, 1992, **96**, 3571.
805. H. Shinohara *et al.*, *Nature*, 1992, **357**, 52.
806. C. S. Yannoni *et al.*, *Science*, 1992, **256**, 52.
807. J. H. Weaver *et al.*, *Chem. Phys. Lett.*, 1992, **190**, 460.
808. M. M. Ross, H. H. Nelson, J. H. Callahan and S. W. McElvany, *J. Phys. Chem.*, 1992, **96**, 5231.
809. E. G. Gillan, C. Yeretzian, K. S. Min, M. M. Alvarez, R. L. Whetten and R. B. Kaner, *J. Phys. Chem.*, 1992, **96**, 6869.
810. S. Bandow *et al.*, *J. Phys. Chem.*, 1992, **96**, 9609.
811. S. Suzuki *et al.*, *J. Phys. Chem.*, 1992, **96**, 7159; see also: M. Hoinkis *et al.*, *Chem. Phys. Lett.*, 1992, **198**, 461.
812. S. Bandow, H. Shinohara, Y. Saito, M. Ohkohchi and Y. Ando, *J. Phys. Chem.*, 1993, **97**, 6101.
813. K. Kikuchi *et al.*, *Chem. Phys. Lett.*, 1993, **216**, 67.
814. C. Capp, T. D. Wood, A. G. Marshall and J. V. Coe, *J. Am. Chem. Soc.*, 1994, **116**, 4987.
815. Y. Huang and B. S. Freiser, *Tetrahedron Lett.*, 191, **32**, 629.
816. G. Z. Suleimanov, L. F. Rybakova, Y. A. Nuriev, T. K. Kurbanov and I. P. Beletskaya, *Izv. Akad. Nauk SSSR, Ser. Khim.*, 1983, 190; *Bull. Acad. Sci. USSR*, 1983, **32**, 165.
817. G. K. Magomedov, A. Z. Voskoboinikov E. B. Chuklanova, A. I. Gusev and I. P. Beletskaya, *Metalloorg. Khim.*, 1990, **3**, 706.
818. I. P. Beletskaya, A. Z. Voskoboinikov and G. K. I. Magomedov, *Metalloorg. Khim.*, 1990, **3**, 516.
819. G. Z. Suleimanov *et al.*, *J. Chem. Soc., Chem. Commun.*, 1984, 191.
820. W. J. Evans, I. Bloom, J. W. Grate, L. A. Hughes, W. E. Hunter and J. L. Atwood, *Inorg. Chem.*, 1985, **24**, 4620.
821. J. M. Boncella and R. A. Andersen, *Inorg. Chem.*, 1984, **23**, 432.
822. J. M. Boncella and R. A. Andersen, *J. Chem. Soc., Chem. Commun.*, 1984, 809.
823. H. Deng and S. G. Shore, *J. Am. Chem. Soc.*, 1991, **113**, 8538.
824. I. P. Beletskaya, A. Z. Voskoboinikov, E. B. Chuklanova, A. I. Gusev and G. K.-I. Magomedov, *Metalloorg. Khim.*, 1988, **1**, 1383.
825. I. P. Beletskaya, A. Z. Voskoboinikov and G. K.-I. Magomedov, *Dokl. Akad. Nauk SSSR*, 1989, **306**, 108; see also: G. K.-I. Magomedov, A. Z. Voskoboinikov and I. P. Beletskaya, *Metalloorg. Khim.*, 1989, **2**, 823.
826. P. N. Hazin, J. C. Huffman and J. W. Bruno, *J. Chem. Soc., Chem. Commun.*, 1988, 1473.
827. L. Wu, Y. Fan, J. Gao and B. Li, *Chin. Sci. Bull.*, 1991, **36**, 648.
828. A. Recknagel, A. Steiner, S. Brooker, D. Stalke and F. T. Edelmann, *Chem. Ber.*, 1991, **124**, 1373.
829. D. Alvarez, Jr., K. G. Caulton, W. J. Evans and J. W. Ziller, *J. Am. Chem. Soc.*, 1990, **112**, 5674.
830. M. L. H. Green, A. K. Hughes, D. P. Michaelidou and P. Mountford, *J. Chem. Soc., Chem. Commun.*, 1993, 591.
831. N. S. Radu, P. K. Gantzel and T. D. Tilley, *J. Chem. Soc., Chem. Commun.*, 1994, 1175.

832. I. P. Beletskaya *et al.*, *J. Organomet. Chem.*, 1986, **299**, 239.
833. X. Wang, X. Zhou, J. Zhang, Y. Xia, R. Liu and S. Wang, *Kexue Tongbao*, 1985, **30**, 351.
834. G. Z. Suleimanov, V. N. Kandozhko, R. Y. Mekhdiev, P. V. Petrovskii, N. E. Kolobova and I. P. Beletskaya, *Izv. Akad. Nauk SSSR, Ser. Khim.*, 1986, 1210.
835. D. Deng *et al.*, *J. Chem. Soc., Dalton Trans.*, 1994, 1665.
836. I. P. Beletskaya, A. Z. Voskoboinikov, E. B. Chuklanova, A. I. Gusev and A. V. Kisin, *J. Organomet. Chem.*, 1993, **454**, 1.
837. G. B. Deacon, A. Dietrich, C. M. Forsyth and H. Schumann, *Angew. Chem.*, 1989, **101**, 1374.
838. G. Z. Suleimanov *et al.*, *Dokl. Akad. Nauk SSSR*, 1984, **276**, 378.
839. G. Z. Suleimanov, P. V. Petrovskii, Y. S. Bogachev, I. L. Zhuravleva, E. I. Fedin and I. P. Beletskaya, *J. Organomet. Chem.*, 1984, **262**, C35.
840. G. Z. Suleimanov *et al.*, *Polyhedron*, 1985, **4**, 29.
841. A. B. Sigalov and I. P. Beletskaya, *Izv. Akad. Nauk SSSR, Ser. Khim.*, 1988, 445.
842. N. E. Kolobova *et al.*, *Zh. Org. Khim.*, 1987, **23**, 1699.
843. G. Z. Suleimanov *et al.*, *Zh. Obshch. Khim.*, 1986, **56**, 1205.
844. G. Z. Suleimanov, V. N. Kandozhko, P. V. Petrovskii, R. Y. Mekhdiev, N. E. Kolobova and I. P. Beletskaya, *J. Chem. Soc., Chem. Commun.*, 1985, 596.
845. L. V. Pankratov *et al.*, *Izv. Akad. Nauk SSSR, Ser. Khim.*, 1986, 2832.
846. K. Jacob *et al.*, *Z. Anorg. Allg. Chem.*, 1992, **618**, 163.
847. K. Jacob *et al.*, *J. Organomet. Chem.*, 1992, **436**, 231.
848. K.-H. Thiele and H. Baumann, *Z. Anorg. Allg. Chem.*, 1993, **619**, 1111.
849. K. Jacob, I. Pavlik and F. T. Edelmann, *Z. Anorg. Allg. Chem.*, 1993, **619**, 1957.
850. H. Gornitzka, F. T. Edelmann and K. Jacob, *J. Organomet. Chem.*, 1992, **436**, 325.
851. H. Gornitzka *et al.*, *J. Organomet. Chem.*, 1992, **439**, C6.
852. K. Jacob, M. Schäfer, A. Steiner, G. M. Sheldrick and F. T. Edelmann, *J. Organomet. Chem.*, in press.
853. M. N. Bochkarev, I. L. Fedushkin, V. K. Cherkasov, H. Schumann and F. H. Görlitz, *Inorg. Chim. Acta*, 1992, **201**, 69.
854. R. D. Fischer, in 'Fundamental and Technological Aspects of Organo-*f*-Element Chemistry', NATO ASI Ser., eds. T. J. Marks and I. L. Fragalà, D. Reidel, Boston, MA, 1985, **155**, 277.
855. M. Mancini *et al.*, *Inorg. Chem.*, 1984, **23**, 1072; P. Bougeard, M. Mancini, B. G. Sayer and M. J. McGlinchey, *Inorg. Chem.*, 1985, **24**, 93.
856. W. J. Evans, J. H. Meadows, A. G. Kostka and G. L. Closs, *Organometallics*, 1985, **4**, 324.
857. J. Wu, T. J. Boyle, J. L. Shreeve, J. W. Ziller and W. J. Evans, *Inorg. Chem.*, 1993, **32**, 1130.
858. G. A. Molander and J. O. Hoberg, *J. Org. Chem.*, 1992, **57**, 3266.
859. M. Bruzzone, in 'Fundamental and Technological Aspects of Organo-*f*-Element Chemistry', NATO ASI Ser., eds. T. J. Marks and I. L. Fragalà, D. Reidel, Boston, MA, 1985, **155**, 387.
860. G. Folcher, *J. Less-Common Met.*, 1986, **122**, 139.
861. W. J. Evans, *J. Organomet. Chem.*, 1983, **250**, 217.
862. G. Jeske, H. Lauke, H. Mauermann, H. Schumann and T. J. Marks, *J. Am. Chem. Soc.*, 1985, **107**, 8111.
863. H. J. Heeres and J. H. Teuben, *Organometallics*, 1991, **10**, 1980.
864. H. J. Heeres, A. Meetsma and J. H. Teuben, *Organometallics*, 1990, **9**, 1508.
865. V. Kalinin, E. I. Kazimirchuk, S. V. Vitt, V. N. Khandozhko and I. P. Beletskaya, *Dokl. Akad. Nauk*, 1993, **329**, 449.
866. G. A. Molander and J. O. Hoberg, *J. Am. Chem. Soc.*, 1992, **114**, 3123.
867. W. E. Piers and J. E. Bercaw, *J. Am. Chem. Soc.*, 1990, **112**, 9406.
868. X. Olonde, A. Mortreux, F. Petit and K. Bujadoux, *J. Mol. Catal.*, 1993, **82**, 75.
869. M. R. Gagné, C. L. Stern and T. J. Marks, *J. Am. Chem. Soc.*, 1992, **114**, 275.
870. P. J. Shapiro, W. D. Cotter, W. P. Schaefer, J. A. Labinger and J. E. Bercaw, *J. Am. Chem. Soc.*, 1994, **116**, 4623.
871. S. Hajela and J. E. Bercaw, *Organometallics*, 1994, **13**, 1147.
872. H. Qian, G. Yu and W. Chen, *Gaofenzi Tongxun*, 1984, **3**, 226.
873. G. G. Yu, W. Q. Chen and Y. L. Wang, *Kexue Tongbao*, 1984, **29**, 421.
874. S. Qi, X. Gao, S. Xiao and W. Chen, *Yingyong Huaxue*, 1986, **3**, 63.
875. A. Oehme, U. Gebauer and K. Gehrke, *J. Mol. Catal.*, 1993, **82**, 83.
876. S. B. Gol'shtein, V. A. Yakovlev, G. N. Bondarenko, Y. P. Yampol'skii and B. A. Dolgoplosk, *Dokl. Akad. Nauk SSSR*, 1986, **289**, 657.
877. C. Shan, Y. Lin, J. Ouyang, Y. Fan and G. Yang, *Makromol. Chem.*, 1987, **188**, 629.
878. H. Yasuda, M. Furo and H. Yamamoto, *Macromolecules*, 1992, **25**, 5115.
879. H. Yasuda *et al.*, *Macromolecules*, 1993, **26**, 7134.
880. H. Yasuda, H. Yamamoto, K. Yokota, S. Miyake and A. Nakamura, *J. Am. Chem. Soc.*, 1992, **114**, 4908.
881. H. Yasuda, *Chem. Abstr.*, 1994, **120**, 135 475a.
882. W. J. Evans and H. Katsumata, *Macromolecules*, 1994, **27**, 2330.
883. S. Onozawa, T. Sakakura and M. Tanaka, *Chem. Lett.*, 1994, 531.
884. M. R. Gagné, L. Brard, V. P. Conticello, M. A. Giardello, C. L. Stern and T. J. Marks, *Organometallics*, 1992, **11**, 2003.
885. M. R. Gagné and T. J. Marks, *J. Am. Chem. Soc.*, 1989, **111**, 4108.
886. M. R. Gagné, S. P. Nolan and T. J. Marks, *Organometallics*, 1990, **9**, 1716.
887. V. P. Conticello *et al.*, *J. Am. Chem. Soc.*, 1992, **114**, 2761.
888. Y. Li, P.-F. Fu and T. J. Marks, *Organometallics*, 1994, **13**, 439.
889. K. N. Harrison and T. J. Marks, *J. Am. Chem. Soc.*, 1992, **114**, 9220.
890. T. Imamoto *et al.*, *J. Org. Chem.*, 1984, **49**, 3904.
891. J. Collin, J. L. Namy and H. B. Kagan, *Nouv. J. Chim.*, 1986, **10**, 229.
892. K. Yokoo, N. Mine, H. Taniguchi and Y. Fujiwara, *J. Organomet. Chem.*, 1985, **279**, C19.
893. J. D. White and G. L. Larson, *J. Org. Chem.*, 1978, **43**, 4555.
894. D. O'Hare, J. M. Manriquez and J. S. Miller, *J. Chem. Soc., Chem. Commun.*, 1988, 491.
895. C. Qian, D. Zhu and D. Li, *J. Organomet. Chem.*, 1992, **430**, 175.
896. I. P. Beletskaya, G. K.-I. Magomedov and A. Z. Voskoboinikov, *J. Organomet. Chem.*, 1990, **385**, 289.

897. H. B. Kagan, *Inorg. Chim. Acta*, 1987, **140**, 3.
898. H. B. Kagan, M. Sasaki and J. Collin, *Pure Appl. Chem.*, 1988, **60**, 1725.
899. B. E. Kahn and R. D. Rieke, *Chem. Rev.*, 1988, **88**, 733.
900. J. Inanaga, *Kidorui*, 1990, **17**, 49.
901. H. B. Kagan, *New. J. Chem.*, 1990, **14**, 453.
902. D. F. Evans, G. V. Fazakerley and R. F. Phillips, *J. Chem. Soc., Chem. Commun.*, 1970, 244.
903. K. Utimoto, K. Oshima, K. Takai and S. Matsubara, *Yuki Gosei Kagaku Kenkyusho Koenshu*, 1991, **5**, 101.
904. L. F. Rybakova, M. I. Terekhova, O. P. Syutkina, A. V. Garbar and E. S. Petrov, *Zh. Obshch. Khim.*, 1986, **56**, 2162.
905. O. P. Syutkina, L. F. Rybakova, E. S. Petrov and I. P. Beletskaya, *J. Organomet. Chem.*, 1985, **280**, C67.
906. L. F. Rybakova, A. V. Garbar and E. S. Petrov, *Dokl. Akad. Nauk SSSR*, 1986, **291**, 1386.
907. Z. Hou, Y. Fujiwara, T. Jintoku, N. Mine, K. Yokoo and H. Taniguchi, *J. Org. Chem.*, 1987, **52**, 3524.
908. L. F. Rybakova, O. P. Syutkina, A. V. Garbar and E. S. Petrov, *Zh. Obshch. Khim.*, 1988, **58**, 1053.
909. O. P. Syutkina, L. F. Rybakova and E. S. Petrov, *Metalloorg. Khim.*, 1989, **2**, 1146.
910. M. I. Terekhova, A. V. Garbar, L. F. Rybakova and E. S. Petrov, *Zh. Obshch. Khim.*, 1986, **56**, 1419.
911. K. Yokoo, T. Fukagawa, Y. Yamanaka, H. Taniguchi and Y. Fujiwara, *J. Org. Chem.*, 1984, **49**, 3237.
912. K. Yokoo, Y. Yamanaka, T. Fukagawa, H. Taniguchi and Y. Fujiwara, *Chem. Lett.*, 1983, 1301; see also: T. Fukagawa, Y. Fujiwara and H. Taniguchi, *Chem. Lett.*, 1982, 601.
913. A. V. Garbar, L. F. Rybakova, E. S. Petrov and I. P. Beletskaya, *Izv. Akad. Nauk SSSR, Ser. Khim.*, 1986, 923.
914. A. B. Sigalov, L. F. Rybakova and I. P. Beletskaya, *Izv. Akad. Nauk SSSR, Ser. Khim.*, 1983, 1690.
915. L. F. Rybakova, O. P. Syutkina, E. S. Petrov, R. R. Shifrina and I. P. Beletskaya, *Izv. Akad. Nauk SSSR, Ser. Khim.*, 1984, 1413.
916. A. B. Sigalov, E. S. Petrov and I. P. Beletskaya, *Izv. Akad. Nauk SSSR, Ser. Khim.*, 1984, 2386.
917. O. P. Syutkina, L. F. Rybakova, E. S. Petrov and I. P. Beletskaya, *Izv. Akad. Nauk SSSR, Ser. Khim.*, 1986, 2143.
918. Z. Hou, N. Mine, Y. Fujiwara and H. Taniguchi, *J. Chem. Soc., Chem. Commun.*, 1985, 1700.
919. A. B. Sigalov, E. S. Petrov, L. F. Rybakova and I. P. Beletskaya, *Izv. Akad. Nauk SSSR, Ser. Khim.*, 1983, 2615.
920. A. B. Sigalov, E. S. Petrov and I. P. Beletskaya, *Izv. Akad. Nauk SSSR, Ser. Khim.*, 1985, 2181.
921. G. B. Deacon and T. D. Tuong, *J. Organomet. Chem.*, 1981, **205**, C4.
922. S. Fukuzawa, T. Fuyinami and S. Sakai, *J. Chem. Soc., Chem. Commun.*, 1986, 475.
923. J. L. Namy, P. Girard and H. B. Kagan, *Nouv. J. Chem.*, 1981, **5**, 479.
924. G. A. Molander, *Chem. Rev.*, 1992, **92**, 29.
925. J. L. Namy, J. Collin, C. Bied and H. B. Kagan, *Synlett*, 1992, **9**, 733.
926. P. Girard, R. Couffignal and H. B. Kagan, *Tetrahedron Lett.*, 1981, **22**, 3959.
927. J. Souppe, J. L. Namy and H. B. Kagan, *Tetrahedron Lett.*, 1984, **25**, 2869.
928. J. Collin, C. Bied and H. B. Kagan, *Tetrahedron Lett.*, 1991, **32**, 629.
929. W. J. Evans, A. L. Wayda, W. E. Hunter and J. L. Atwood, *J. Chem. Soc., Chem. Commun.*, 1981, 706.
930. I. P. Beletskaya, G. Z. Suleimanov, E. I. Kazimirchuk, S. G. Mamedova, V. N. Kandozhko and N. E. Kolobova, *Izv. Akad. Nauk SSSR, Ser. Khim.*, 1985, 2832.
931. M. Murakami, T. Kawano and Y. Ito, *J. Am. Chem. Soc.*, 1990, **112**, 2437.
932. M. Murakami and Y. Ito, *J. Organomet. Chem.*, 1994, **473**, 93.
933. G. B. Deacon and P. I. MacKinnon, *Tetrahedron Lett.*, 1984, **25**, 783.
934. W. J. Evans and D. K. Drummond, *J. Am. Chem. Soc.*, 1988, **110**, 2772.
935. H. B. Kagan, J. L. Namy and P. Girard, *Tetrahedron*, 1981, **37**, (Suppl. 1), 175.
936. P. Girard, H. B. Kagan and J. L. Namy, *J. Am. Chem. Soc.*, 1980, **102**, 2693.
937. E. Hasegawa and D. P. Curran, *Tetrahedron Lett.*, 1993, **34**, 1717.
938. C. Qian and A. Qiu, *Tetrahedron Lett.*, 1988, **29**, 6931.
939. L. Gong and A. Streitwieser, Jr., *J. Org. Chem.*, 1990, **55**, 6235.
940. C. Bied, J. Collin and H. B. Kagan, *Tetrahedron*, 1992, **48**, 3877.
941. T. Kauffmann, C. Pahde, A. Tannert and D. Wingbermühle, *Tetrahedron Lett.*, 1985, **26**, 4063.
942. P. Girard, J. L. Namy and H. B. Kagan, *J. Am. Chem. Soc.*, 1980, **102**, 2693.
943. D. P. Curran, T. L. Fevig and M. J. Totleben, *Synlett*, 1990, 773.
944. G. A. Molander and C. A. Kenny, *J. Org. Chem.*, 1991, **56**, 1439.
945. D. P. Curran, M. Totleben, C. Jasperse and T. L. Fevig, *Synlett*, 1992, 973.
946. E. J. Enholm, H. Satici and A. Trivellas, *J. Org. Chem.*, 1989, **54**, 5841.
947. E. Vedejs and S. Ahmad, *Tetrahedron Lett.*, 1988, **29**, 2291.
948. T. L. Fevig, R. L. Elliott and D. P. Curran, *J. Am. Chem. Soc.*, 1988, **110**, 5064.
949. J. D. White and T. C. Somers, *J. Am. Chem. Soc.*, 1987, **109**, 4424.
950. D. P. Curran and M. J. Totleben, *J. Am. Chem. Soc.*, 1992, **114**, 6050.
951. M. J. Totleben, D. P. Curran and P. Wipf, *J. Org. Chem.*, 1992, **57**, 1740.
952. P. Wipf and S. Venkatraman, *J. Org. Chem.*, 1993, **58**, 3455.
953. S.-I. Fukuzawa, T. Fujinami and S. Sakai, *J. Organomet. Chem.*, 1986, **299**, 179.
954. T. Imamoto, T. Kusumoto and M. Yokoyama, *J. Chem. Soc., Chem. Commun.*, 1982, 1042.
955. T. Imamoto and Y. Sugiura, *J. Organomet. Chem.*, 1985, **285**, C21.
956. T. Imamoto, T. Kusumoto, Y. Hatanaka and M. Yokoyama, *Tetrahedron Lett.*, 1983, **24**, 1553.
957. T. Imamoto, Y. Hatanaka, T. Tawarayama and M. Yokoyama, *Tetrahedron Lett.*, 1981, **22**, 4987.
958. T. Imamoto, T. Kusumoto and M. Yokoyama, *Tetrahedron Lett.*, 1983, **24**, 5233.
959. T. Imamoto, T. Mita and M. Yokoyama, *J. Chem. Soc., Chem. Commun.*, 1984, 163.
960. T. Imamoto, Y. Tawarayama, T. Kusumoto and M. Yokoyama, *Yuki Gosei Kagaku Kyokaishi*, 1984, **42**, 143.
961. Y. Takemoto, J. Takeuchi and C. Iwata, *Tetrahedron Lett.*, 1993, **34**, 6069.
962. T. Imamoto, Y. Sugiura and N. Takiyama, *Tetrahedron Lett.*, 1984, **25**, 4233.
963. M. Suzuki, Y. Kimura and S. Terashima, *Chem. Lett.*, 1984, 1543.
964. Y. Tamura, M. Sasho, H. Ohe, S. Akai and Y. Kita, *Tetrahedron Lett.*, 1985, **26**, 1549.
965. M. Kawasaki, F. Matsuda and S. Terashima, *Tetrahedron Lett.*, 1985, **26**, 2693.
966. K. Nagasawa, H. Kanbara, K. Matsushita and K. Ito, *Tetrahedron Lett.*, 1985, **26**, 6477.

967. I. Mukerji, A. L. Wayda, G. Dabbagh and S. H. Bertz, *Angew. Chem., Int. Ed. Engl.*, 1986, **25**, 760.
968. M. T. Reetz, H. Haning and S. Stanchev, *Tetrahedron Lett.*, 1992, **33**, 6963.
969. C. Qian, D. Zhu and Y. Gu, *J. Organomet. Chem.*, 1991, **401**, 23.
970. C. Qian, A. Qiu, Y. Huang and W. Chen, *J. Organomet. Chem.*, 1991, **412**, 53.
971. H. Ohno, A. Mori and S. Inoue, *Chem. Lett.*, 1993, 375.
972. S. Collins, Y. Hong, G. J. Hoover and J. R. Veit, *J. Org. Chem.*, 1990, **55**, 3565.
973. J. Weber *et al.*, *Appl. Phys. Lett.*, 1988, **53**, 2525.
974. X. M. Fang, Y. Li and D. W. Langer, *J. Appl. Phys.*, 1993, **74**, 6990.
975. D. W. Langer, Y. Li, X. M. Fang and V. Coon, *Mater. Res. Soc. Symp. Proc.*, 1993, **301**, 15.
976. A. C. Greenwald, W. S. Rees and U. W. Lay, *Mater. Res. Soc. Symp. Proc.*, 1993, **301**, 21.
977. F. Scholz, J. Weber, K. Pressel and A. Doernen, *Mater. Res. Soc. Symp. Proc.*, 1993, **301**, 3.
978. N. Edakawa, H. Yoshida, K. Kamya and M. Taya, *Chem. Abstr.*, 1994, **120**, 142 056t.
979. A. H. H. Chang and R. M. Pitzer, *J. Am. Chem. Soc.*, 1989, **111**, 2500.
980. K. G. Moloy and T. J. Marks, *J. Am. Chem. Soc.*, 1984, **106**, 7051.
981. A. Vittadini, M. Casarin, D. Ajò, R. Bertoncello, E. Ciliberto, A. Gulino and I. L. Fragalà, *Inorg. Chim. Acta*, 1986, **121**, L23.
982. K. G. Moloy and T. J. Marks, *Inorg. Chim. Acta*, 1985, **110**, 127.
983. K. Tatsumi, A. Nakamura, P. Hofmann, R. Hoffmann, K. G. Moloy and T. J. Marks, *J. Am. Chem. Soc.*, 1986, **108**, 4467.
984. P. M. Boerrigter, E. J. Baerends and J. G. Snijders, *Chem. Phys.*, 1988, **122**, 357.
985. S. Di Bella, A. Gulino, G. Lanza, I. L. Fragalà and T. J. Marks, *J. Phys. Chem.*, 1993, **97**, 11 673.
986. A. Gulino, S. Di Bella, I. L. Fragalà, M. Casarin, A. M. Seyam and T. J. Marks, *Inorg. Chem.*, 1993, **32**, 3873.
987. N. Rösch and A. Streitwieser, Jr., *J. Am. Chem. Soc.*, 1983, **105**, 7237.
988. P. Pyykkö, L. J. Laakkonen and K. Tatsumi, *Inorg. Chem.*, 1989, **28**, 1801.
989. P. Pyykkö and L. L. Lohr, *Inorg. Chem.*, 1981, **20**, 1950.
990. D. B. Beach *et al.*, *Inorg. Chem.*, 1986, **25**, 1735.
991. J. G. Brennan, J. C. Green and C. M. Redfern, *J. Am. Chem. Soc.*, 1989, **111**, 2373.
992. L. Xing-Fu *et al.*, *Inorg. Chim. Acta*, 1986, **116**, 1185.
993. R. D. Fischer and L. Xing-Fu, *J. Less-Common Met.*, 1985, **112**, 303.
994. P. Sun, X. Feng, Y. Xu, A. Guo and X. Li, *Zhongguo Kexue Jishu Daxue Xuebao*, 1986, **16**, 455.
995. X. Li, P. Sun and X. Feng, *Zhongguo Kexue Jishu Daxue Xuebao*, 1986, **16**, 105.
996. X. Li, P. Sun and X. Feng, *Zhongguo Kexue Jishu Daxue Xuebao*, 1986, **16**, 210.
997. K. W. Bagnall and L. Xing-Fu, *J. Chem. Soc., Dalton Trans.*, 1982, 1365.
998. L. Xing-Fu and G. Ao-Ling, *Inorg. Chim. Acta*, 1987, **134**, 143.
999. C. Baudin *et al.*, *J. Organomet. Chem.*, 1991, **415**, 59.
1000. T. J. Marks, M. R. Gagné, S. P. Nolan, L. E. Schock, A. M. Seyam and D. Stern, *Pure Appl. Chem.*, 1989, **61**, 1665.
1001. X. Jemine, J. Goffart, M. Ephritikhine and J. Fuger, *J. Organomet. Chem.*, 1993, **448**, 95.
1002. W. A. King, T. J. Marks, D. M. Anderson, D. J. Duncalf and F. G. N. Cloke, *J. Am. Chem. Soc.*, 1992, **114**, 9221.
1003. J.-C. Berthet, J.-F. Le Maréchal, M. Lance, M. Nierlich, J. Vigner and M. Ephritikhine, *J. Chem. Soc., Dalton Trans.*, 1992, 1573; X. Jemine, J. Goffart, J.-C. Berthet and M. Ephritikhine, *J. Chem. Soc., Dalton Trans.*, 1992, 2439.
1004. J. P. Leal, N. Marques, A. Pires de Matos, M. J. Calhorda, A. M. Galvão and J. A. Martinho Simões, *Organometallics*, 1992, **11**, 1632.
1005. K. Tatsumi and A. Nakamura, *J. Organomet. Chem.*, 1984, **272**, 141; K. Tatsumi and R. Hoffmann, *Inorg. Chem.*, 1984, **23**, 1633.
1006. T. J. Tague, Jr., L. Andrews and R. D. Hunt, *J. Phys. Chem.*, 1993, **97**, 10920.
1007. J. G. Brennan, R. A. Andersen and J. L. Robbins, *J. Am. Chem. Soc.*, 1986, **108**, 335.
1008. H. Gilman, *Adv. Organomet. Chem.*, 1968, **7**, 33.
1009. A. M. Seyam, *Inorg. Chim. Acta*, 1983, **77**, L123.
1010. M. S. Eisen and T. J. Marks, *J. Am. Chem. Soc.*, 1992, **114**, 10358.
1011. H. Lauke, P. N. Swepston and T. J. Marks, *J. Am. Chem. Soc.*, 1984, **106**, 6841.
1012. A. M. Seyam, *Inorg. Chim. Acta*, 1985, **110**, 123.
1013. W. G. Van der Sluys, C. J. Burns and A. P. Sattelberger, *Organometallics*, 1989, **8**, 855.
1014. B. D. Zwick, A. P. Sattelberger and L. R. Avens, *Proc. Transuranium Elem. Symp.*, eds. L. R. Morss and J. Fuger, Washington, D. C., 1992, p. 239.
1015. P. C. Edwards, R. A. Andersen and A. Zalkin, *Organometallics*, 1984, **3**, 293.
1016. M. R. Leonov and G. V. Solov'eva, *Metalloorg. Khim.*, 1991, **4**, 323.
1017. P. G. Edwards, M. B. Hursthouse, K. M. Abdul Malik and J. S. Parry, *J. Chem. Soc., Chem. Commun.*, 1994, 1249.
1018. A. Dormond, A. El Bouadili, A. Aaliti and C. Moise, *J. Organomet. Chem.*, 1985, **288**, C1.
1019. A. Dormond, A. A. El Bouadili and C. Moise, *J. Chem. Soc., Chem. Commun.*, 1985, 914.
1020. A. Dormond, A. Aaliti and C. Moise, *Tetrahedron Lett.*, 1986, **27**, 1497.
1021. A. Dormond, A. A. El Bouadili and C. Moise, *J. Less-Common Met.*, 1986, **122**, 159.
1022. A. Dormond, A. Aaliti, A. A. El Bouadili and C. Moise, *Inorg. Chim. Acta*, 1987, **139**, 171.
1023. A. Dormond, A. A. El Bouadili and C. Moise, *J. Org. Chem.*, 1989, **54**, 3747.
1024. A. Dormond, A. A. El Bouadili and C. Moise, *J. Org. Chem.*, 1987, **52**, 688.
1025. A. Dormond, C. Moise, A. El Bouadili and H. Bitar, *J. Organomet. Chem.*, 1989, **371**, 175.
1026. A. Dormond, A. El Bouadili and C. Moise, *J. Organomet. Chem.*, 1989, **369**, 171.
1027. A. Dormond, A. Aaliti, A. El Bouadili and C. Moise, *J. Organomet. Chem.*, 1987, **329**, 187.
1028. A. Dormond, P. Hepiègne, A. Hafid and C. Moise, *J. Organomet. Chem.*, 1990, **398**, C1.
1029. M. Wedler, F. Knösel, F. T. Edelmann and U. Behrens, *Chem. Ber.*, 1992, **125**, 1313.
1030. J. Takats, X. W. Zhang and R. McDonald, 77th CSC Conference, Winnipeg, Manitoba, 1994, Abstract 538.
1031. A. Domongs, N. Marques and A. Pires de Matos, *Polyhedron*, 1990, **9**, 69.
1032. S. M. Beshouri, P. E. Fanwick, I. P. Rothwell and J. C. Huffman, *Organometallics*, 1987, **6**, 2498.
1033. C. Baudin and M. Ephritikhine, *J. Organomet. Chem.*, 1989, **364**, C1.
1034. J. L. Stewart and R. A. Andersen, *J. Chem. Soc., Chem. Commun.*, 1987, 1846.

1035. S. A. Toropov, V. K. Vasil'ev and V. N. Sokolov, *Radiokhim.*, 1986, **28**, 712.
1036. S. A. Toropov, V. K. Vasil'ev, V. N. Sokolov and S. P. Evdokimova, *Radiokhim.*, 1986, **28**, 300.
1037. S. A. Toropov, V. K. Vasil'ev and V. N. Sokolov, *Radiokhim.*, 1987, **29**, 150.
1038. D. Baudry *et al.*, *J. Organomet. Chem.*, 1989, **371**, 163.
1039. J.-F. Le Maréchal *et al.*, *J. Chem. Soc., Chem. Commun.*, 1989, 308.
1040. P. C. Blake, M. F. Lappert, R. G. Taylor, J. L. Atwood, W. E. Hunter and H. Zhang, *J. Chem. Soc., Chem. Commun.*, 1986, 1394.
1041. P. C. Blake, M. F. Lappert, R. G. Taylor, J. L. Atwood and H. Zhang, *Inorg. Chim. Acta*, 1987, **139**, 13.
1042. P. C. Blake, M. F. Lappert, J. L. Atwood and H. Zhang, *J. Chem. Soc., Chem. Commun.*, 1988, 1436.
1043. P. C. Blake, E. Hey, M. F. Lappert, J. L. Atwood and H. Zhang, *J. Organomet. Chem.*, 1988, **353**, 307.
1044. A. Zalkin and S. M. Beshouri, *Acta Crystallogr., Sect. C*, 1989, **45**, 1219.
1045. S. M. Beshouri and A. Zalkin, *Acta Crystallogr., Sect. C*, 1989, **45**, 1221.
1046. A. Zalkin and S. M. Beshouri, *Acta Crystallogr., Sect. C*, 1989, **45**, 1080.
1047. A. Zalkin, A. L. Stuart and R. A. Andersen, *Acta Crystallogr., Sect. C*, 1988, **44**, 2106.
1048. S. D. Stults, R. A. Andersen and A. Zalkin, *J. Organomet. Chem.*, 1993, **462**, 175.
1049. Y. Mugnier, A. Dormond and E. Laviron, *J. Chem. Soc., Chem. Commun.*, 1982, 257.
1050. J. W. Bruno, D. G. Kalina, E. A. Mintz and T. J. Marks, *J. Am. Chem. Soc.*, 1982, **104**, 1860.
1051. H. J. Wasserman, A. J. Zozulin, D. C. Moody, R. R. Ryan and K. V. Salazar, *J. Organomet. Chem.*, 1983, **254**, 305.
1052. J.-F. Le Maréchal, E. Bulot, D. Baudry, M. Ephritikhine, D. Hauchard and R. Godard, *J. Organomet. Chem.*, 1988, **354**, C17.
1053. J.-F. Le Maréchal *et al.*, *J. Organomet. Chem.*, 1989, **379**, 259.
1054. R. Adam, C. Villiers, M. Ephritikhine, M. Lance, M. Nierlich and J. Vigner, *J. Organomet. Chem.*, 1993, **445**, 99.
1055. J. G. Brennan, R. A. Andersen and A. Zalkin, *Inorg. Chem.*, 1986, **25**, 1761.
1056. A. Zalkin and J. G. Brennan, *Acta Crystallogr., Sect. C*, 1985, **41**, 1295.
1057. R. K. Rosen and A. Zalkin, *Acta Crystallogr., Sect. C*, 1989, **45**, 1139.
1058. V. I. Spitsyn and G. V. Ionova, *Dokl. Akad. Nauk SSSR*, 1987, **292**, 921.
1059. J. G. Brennan and A. Zalkin, *Acta Crystallogr., Sect. C*, 1985, **41**, 1038.
1060. B. E. Bursten, L. F. Rhodes and R. J. Strittmatter, *J. Am. Chem. Soc.*, 1989, **111**, 2756.
1061. B. E. Bursten, L. F. Rhodes and R. J. Strittmatter, *J. Am. Chem. Soc.*, 1989, **111**, 2758.
1062. B. E. Bursten and R. J. Strittmatter, *J. Am. Chem. Soc.*, 1987, **109**, 6606.
1063. C. Villiers and M. Ephritikhine, *J. Organomet. Chem.*, 1990, **393**, 339.
1064. J. G. Brennan, R. A. Andersen and A. Zalkin, *Inorg. Chem.*, 1986, **25**, 1756.
1065. A. Zalkin, J. G. Brennan and R. A. Andersen, *Acta Crystallogr., Sect. C*, 1988, **44**, 1553.
1066. P. C. Blake, M. F. Lappert, J. L. Atwood and H. Zhang, *J. Chem. Soc., Chem. Commun.*, 1986, 1148.
1067. W. K. Kot, G. V. Shalimoff, N. M. Edelstein, M. A. Edelman and M. F. Lappert, *J. Am. Chem. Soc.*, 1988, **110**, 986.
1068. L. Arnaudet, G. Folcher, H. Marquet-Ellis, E. Klähne, K. Yünlü and R. D. Fischer, *Organometallics*, 1983, **2**, 344.
1069. M. Foyentin, G. Folcher and M. Ephritikhine, *J. Chem. Soc., Chem. Commun.*, 1987, 494.
1070. M. Foyentin, G. Folcher and M. Ephritikhine, *J. Organomet. Chem.*, 1987, **335**, 201.
1071. C. Villiers and M. Ephritikhine, *New. J. Chem.*, 1991, **15**, 559.
1072. C. Villiers and M. Ephritikhine, *New. J. Chem.*, 1991, **15**, 565.
1073. S. D. Stults, R. A. Andersen and A. Zalkin, *J. Am. Chem. Soc.*, 1989, **111**, 4507.
1074. J.-C. Berthet *et al.*, *J. Organomet. Chem.*, 1992, **440**, 53.
1075. K. W. Bagnall, F. Benetollo, G. Bombieri and G. De Paoli, *J. Chem. Soc., Dalton Trans.*, 1984, 67.
1076. I. Ahmed and K. W. Bagnall, *J. Less-Common Met.*, 1984, **99**, 285.
1077. J.-F. Le Maréchal, M. Ephritikhine and G. Folcher, *J. Organomet. Chem.*, 1986, **299**, 85.
1078. K. W. Bagnall, G. F. Payne, N. W. Alcock, D. J. Flanders and D. Brown, *J. Chem. Soc., Dalton Trans.*, 1986, 783.
1079. A. G. M. Al-Daher and K. W. Bagnall, *J. Less-Common Met.*, 1986, **116**, 351.
1080. K. W. Bagnall, G. F. Payne and D. Brown, *J. Less-Common Met.*, 1986, **116**, 333.
1081. H. J. Wasserman, D. C. Moody, R. T. Payne, R. R. Ryan and K. V. Salazar, *J. Chem. Soc., Chem. Commun.*, 1984, 533.
1082. K. W. Bagnall, G. F. Payne and D. Brown, *J. Less-Common Met.*, 1985, **113**, 325.
1083. M. R. Leonov, V. A. Il'yushenko, V. M. Fomin and N. V. Il'yushenkova, *Radiokhim.*, 1988, **30**, 324.
1084. A. Dormond, A. Aaliti and C. Moise, *J. Org. Chem.*, 1988, **53**, 1034.
1085. P. Zanella, G. Rossetto, A. Berton and G. Paolucci, *Inorg. Chim. Acta*, 1984, **95**, 263.
1086. J. Marçalo, N. Marques, A. Pires de Matos and K. W. Bagnall, *J. Less-Common Met.*, 1986, **122**, 219.
1087. D. Baudry *et al.*, *J. Chem. Soc., Chem. Commun.*, 1985, 1553.
1088. D. Baudry and M. Ephritikhine, *J. Organomet. Chem.*, 1988, **349**, 123.
1089. D. Baudry, P. Dorion and M. Ephritikhine, *J. Organomet. Chem.*, 1988, **356**, 165.
1090. C. Baudin, P. Charpin, M. Ephritikhine, M. Lance, M. Nierlich and J. Vigner, *J. Organomet. Chem.*, 1988, **345**, 263.
1091. B. Delavaux-Nicot and M. Ephritikhine, *J. Organomet. Chem.*, 1990, **399**, 77.
1092. D. Baudry, M. Ephritikhine, W. Kläui, M. Lance, M. Nierlich and J. Vigner, *Inorg. Chem.*, 1991, **30**, 2333.
1093. M. Wedler, J. W. Gilje, M. Noltemeyer and F. T. Edelmann, *J. Organomet. Chem.*, 1991, **411**, 271.
1094. G. Paolucci, G. Rossetto, P. Zanella and R. D. Fischer, *J. Organomet. Chem.*, 1985, **284**, 213.
1095. F. Ossola, G. Rossetto, P. Zanella, G. Paolucci and R. D. Fischer, *J. Organomet. Chem.*, 1986, **309**, 55.
1096. V. I. Tel'noi, I. B. Rabinovich, V. N. Larina, R. M. Leonov and G. V. Solov'eva, *Radiokhim.*, 1989, **31**, 38.
1097. R. E. Cramer, A. L. Mori, R. B. Maynard, J. W. Gilje, K. Tatsumi and A. Nakamura, *J. Am. Chem. Soc.*, 1984, **106**, 5920.
1098. N. Brianese *et al.*, *J. Organomet. Chem.*, 1989, **365**, 223.
1099. M. A. Edelman, M. F. Lappert, J. L. Atwood and H. Zhang, *Inorg. Chim. Acta*, 1987, **139**, 185.
1100. A. L. Arduini, J. Malito, J. Takats, E. Ciliberto, I. L. Fragalà and P. Zanella, *J. Organomet. Chem.*, 1987, **326**, 49.
1101. A. Berton, M. Porchia, G. Rossetto and P. Zanella, *J. Organomet. Chem.*, 1986, **302**, 351.
1102. P. Zanella, G. Rosetto and G. Paolucci, *Inorg. Chim. Acta*, 1984, **82**, 227.
1103. A. Dormond, A. Aaliti and C. Moise, *J. Chem. Soc., Chem. Commun.*, 1985, 1231.
1104. A. Dormond, A. Aaliti and C. Moise, *J. Less-Common Met.*, 1986, **122**, 153.
1105. D. Hauchard, M. Cassir, J. Chivot, D. Baudry and M. Ephritikhine, *J. Electroanal. Chem.*, 1993, **347**, 399.

1106. G. Paolucci, P. Zanella and A. Berton, *J. Organomet. Chem.*, 1985, **295**, 317.
1107. A. Zalkin, J. G. Brennan and R. A. Andersen, *Acta Crystallogr., Sect. C*, 1987, **43**, 418; 421.
1108. P. B. Hitchcock, M. F. Lappert and R. G. Taylor, *J. Chem. Soc., Chem. Commun.*, 1984, 1082.
1109. J. G. Brennan, R. A. Andersen and A. Zalkin, *J. Am. Chem. Soc.*, 1988, **110**, 4554.
1110. G. B. Deacon and T. D. Tuong, *Polyhedron*, 1988, **7**, 249.
1111. B. R. McGarvey and S. Nagy, *Inorg. Chem.*, 1987, **26**, 4198.
1112. B. R. McGarvey and S. Nagy, *Inorg. Chim. Acta*, 1987, **139**, 319.
1113. H.-D. Amberger, H. Reddmann and N. M. Edelstein, *Inorg. Chim. Acta*, 1988, **141**, 313.
1114. M. Weydert, R. A. Andersen and R. G. Bergman, *J. Am. Chem. Soc.*, 1993, **115**, 8837.
1115. M. R. Spirlet, J. Rebizant, C. Apostolidis, G. D. Andreetti and B. Kanellakopulos, *Acta Crystallogr., Sect. C.*, 1989, **45**, 739.
1116. J. Rebizant, M. R. Spirlet, C. Apostolidis and B. Kanellakopulos, *Acta Crystallogr., Sect. C.*, 1991, **47**, 854.
1117. D. Hauchard, M. Cassir, J. Chivot and M. Ephritikhine, *Electroanal. Chem., Interfacial Electrochem.*, 1991, **313**, 227.
1118. D. C. Sonnenberger and J. G. Gaudiello, *Inorg. Chem.*, 1988, **27**, 2747.
1119. R. D. Fischer, E. Klähne and G. R. Sienel, *J. Organomet. Chem.*, 1982, **238**, 99.
1120. C. C. Chang, N. K. Sung-Yu, C. W. Kang and C. T. Chang, *Inorg. Chim. Acta*, 1987, **141**, 119.
1121. J.-C. Berthet, M. Lance, M. Nierlich, J. Vigner and M. Ephritikhine, *J. Organomet. Chem.*, 1991, **420**, C9.
1122. J.-C. Berthet, M. Ephritikhine, M. Lance, M. Nierlich and J. Vigner, *J. Organomet. Chem.*, 1993, **460**, 47.
1123. F. Knösel, H. W. Roesky and F. Edelmann, *Inorg. Chim. Acta*, 1987, **139**, 187.
1124. M. R. Spirlet, J. Rebizant, C. Apostolidis, G. van den Bossche and B. Kanellakopulos, *Acta Crystallogr., Sect. C.*, 1990, **46**, 2318.
1125. M. Porchia *et al.*, *J. Chem. Soc., Dalton Trans.*, 1989, 677.
1126. M. R. Leonov, G. V. Solov'eva and S. V. Patrikeev, *Metalloorg. Khim.*, 1990, **3**, 343.
1127. R. E. Cramer, U. Engelhardt, K. T. Higa and J. W. Gilje, *Organometallics*, 1987, **6**, 41.
1128. P. Zanella *et al.*, *J. Chem. Soc., Dalton Trans.*, 1987, 2039.
1129. R. E. Cramer *et al.*, *Organometallics*, 1988, **7**, 841.
1130. J.-C. Berthet, J.-F. Le Maréchal, M. Nierlich, M. Lance, J. Vigner and M. Ephritikhine, *J. Organomet. Chem.*, 1991, **408**, 335.
1131. J. G. Brennan and R. A. Andersen, *J. Am. Chem. Soc.*, 1985, **107**, 514.
1132. R. K. Rosen, R. A. Andersen and N. M. Edelstein, *J. Am. Chem. Soc.*, 1990, **112**, 4588.
1133. A. Dormond, *J. Organomet. Chem.*, 1983, **256**, 47.
1134. D. Baudry, A. Dormond and A. Hafid, *New. J. Chem.*, 1993, **17**, 465.
1135. S. P. Kuleshov, R. G. Bulgakov, G. A. Tostikov and V. P. Kazakov, *Izv. Akad. Nauk SSSR, Ser. Khim.*, 1989, 1207.
1136. G. Adrian, H. Appel, R. Bohlander, H. Haffner and B. Kanellakopulos, *Hyperfine Interact.*, 1988, **40**, 275.
1137. L. E. Schock, A. M. Seyam, M. Sabat and T. J. Marks, *Polyhedron*, 1988, **7**, 1517.
1138. D. C. Sonnenberger, L. R. Morss and T. J. Marks, *Organometallics*, 1985, **4**, 352.
1139. J. W. Bruno, H. A. Stecher, L. R. Morss, D. C. Sonnenberger and T. J. Marks, *J. Am. Chem. Soc.*, 1986, **108**, 7275.
1140. S. Bettonville, J. Goffart and J. Fuger, *J. Organomet. Chem.*, 1989, **377**, 59.
1141. S. Bettonville, J. Goffart and J. Fuger, *J. Organomet. Chem.*, 1990, **393**, 205.
1142. G. Pilcher and J. A. Skinner, in 'The Chemistry of the Metal–Carbon Bond', eds. F. R. Hartley and S. Patai, Wiley, New York, 1982, 43.
1143. J. A. Connor, *Top. Curr. Chem.*, 1977, **71**, 71.
1144. M. Burton, H. Marquet-Ellis, G. Folcher and C. Giannotti, *J. Organomet. Chem.*, 1982, **229**, 21.
1145. D. C. Sonnenberger, E. A. Mintz and T. J. Marks, *Proc. 10th Conf. on Organometallic Chemistry*, Toronto, 1981, Abstract 3D05.
1146. D. C. Sonnenberger, E. A. Mintz and T. J. Marks, *J. Am. Chem. Soc.*, 1984, **106**, 3484.
1147. M. R. Leonov and G. V. Solov'eva, *Zh. Obshch. Khim.*, 1987, **57**, 2522.
1148. M. R. Leonov, I. I. Grinval'd, G. V. Solov'eva, V. A. Il'yushenko and N. K. Rudnevskii, *Radiokhim.*, 1987, **29**, 107.
1149. C. Villiers, R. Adam and M. Ephritikhine, *J. Chem. Soc., Chem. Commun.*, 1992, 1555.
1150. G. Paolucci, G. Rossetto, P. Zanella, K. Yünlü and R. D. Fischer, *J. Organomet. Chem.*, 1984, **272**, 363.
1151. C. C. Chang and N. K. Sung-Yu, *Inorg. Chim. Acta*, 1986, **119**, 107.
1152. G. Paolucci, S. Daolio and B. Traldi, *J. Organomet. Chem.*, 1986, **309**, 283.
1153. A. Dormond, A. A. El Bouadili and C. Moise, *J. Chem. Soc., Chem. Commun.*, 1984, 749.
1154. G. Rossetto, M. Porchia, F. Ossola, P. Zanella and R. D. Fischer, *J. Chem. Soc., Chem. Commun.*, 1985, 1460.
1155. M. Porchia, N. Brianese, F. Ossola, G. Rossetto and P. Zanella, *J. Chem. Soc., Dalton Trans.*, 1987, 691.
1156. P. Zanella *et al.*, *J. Chem. Soc., Chem. Commun.*, 1985, 96.
1157. M. R. Leonov, G. V. Solov'eva and I. Z. Kozina, *Khim. Elementoorgan. Soedin., Gor'kii*, 1984, 75.
1158. S. A. Karova, V. K. Vasil'ev and V. N. Sokolov, *Radiokhim.*, 1985, **27**, 723.
1159. L. Arnaudet, G. Folcher, and H. Marquet-Ellis, *J. Organomet. Chem.*, 1981, **214**, 215.
1160. R. E. Cramer, R. B. Maynard, J. C. Paw and J. W. Gilje, *J. Am. Chem. Soc.*, 1981, **103**, 3589.
1161. R. E. Cramer *et al.*, *Chem. Ber.*, 1988, **121**, 417.
1162. R. C. Stevens, R. Bau, R. E. Cramer, D. Afzal, J. W. Gilje and T. F. Koetzle, *Organometallics*, 1990, **9**, 694.
1163. R. E. Cramer, R. B. Maynard, J. C. Paw and J. W. Gilje, *Organometallics*, 1982, **1**, 869.
1164. R. E. Cramer, J. H. Jeong and J. W. Gilje, *Organometallics*, 1987, **6**, 2010.
1165. R. E. Cramer, K. T. Higa, S. L. Pruskin and J. W. Gilje, *J. Am. Chem. Soc.*, 1983, **105**, 6749.
1166. R. E. Cramer, K. T. Higa and J. W. Gilje, *J. Am. Chem. Soc.*, 1984, **106**, 7245.
1167. R. E. Cramer, J. H. Jeong and J. W. Gilje, *Organometallics*, 1986, **5**, 2555.
1168. R. E. Cramer, J. H. Jeong, P. N. Richman and J. W. Gilje, *Organometallics*, 1990, **9**, 1141.
1169. R. E. Cramer, K. T. Higa and J. W. Gilje, *Organometallics*, 1985, **4**, 1140.
1170. R. E. Cramer, K. Panchanatheswaran and J. W. Gilje, *J. Am. Chem. Soc.*, 1984, **106**, 1853.
1171. R. E. Cramer, K. Panchanatheswaran and J. W. Gilje, *Angew. Chem., Int. Ed. Engl.*, 1984, **23**, 912.
1172. M. Porchia, F. Ossola, G. Rossetto, P. Zanella and N. Brianese, *J. Chem. Soc., Chem. Commun.*, 1987, 550.
1173. M. Porchia, U. Casellato, F. Ossola, G. Rossetto, P. Zanella and R. Graziani, *J. Chem. Soc., Chem. Commun.*, 1986, 1034.

1174.   S. P. Nolan, M. Porchia and T. J. Marks, *Organometallics*, 1991, **10**, 1450.
1175.   J.-C. Berthet, J.-F. Le Maréchal and M. Ephritikhine, *J. Chem. Soc., Chem. Commun.*, 1991, 360.
1176.   J.-C. Berthet and M. Ephritikhine, *New J. Chem.*, 1992, **16**, 767.
1177.   P. Zanella, F. Ossola, M. Porchia, G. Rossetto, A. C. Villa and C. Guastini, *J. Organomet. Chem.*, 1987, **323**, 295.
1178.   F. Ossola, N. Brianese, M. Porchia, G. Rossetto and P. Zanella, *J. Organomet. Chem.*, 1986, **310**, C1.
1179.   F. Ossola, N. Brianese, M. Porchia, G. Rossetto and P. Zanella, *J. Chem. Soc., Dalton Trans.*, 1990, 877.
1180.   G. Bombieri, F. Benetollo, K. W. Bagnall, M. J. Plews and D. Brown, *J. Chem. Soc., Dalton Trans.*, 1983, 45.
1181.   F. Benetollo, E. Klähne and R. D. Fischer, *J. Chem. Soc., Dalton Trans.*, 1983, 115.
1182.   K. W. Bagnall *et al.*, *J. Chem. Soc., Dalton Trans.*, 1982, 1999.
1183.   H.-D. Amberger, R. D. Fischer and K. Yünlü, *Organometallics*, 1986, **5**, 2109.
1184.   M. Adam, K. Yünlü and R. D. Fischer, *J. Organomet. Chem.*, 1990, **387**, C13.
1185.   H. Aslan and R. D. Fischer, *J. Organomet. Chem.*, 1986, **315**, C64.
1186.   H. Aslan, K. Yünlü, R. D. Fischer, G. Bombieri and F. Benetollo, *J. Organomet. Chem.*, 1988, **354**, 63.
1187.   H. Aslan, J. Förster, K. Yünlü and R. D. Fischer, *J. Organomet. Chem.*, 1988, **355**, 79.
1188.   H.-D. Amberger, H. Reddmann, G. Shalimoff and N. M. Edelstein, *Inorg. Chim. Acta*, 1987, **139**, 335.
1189.   G. Rossetto, N. Brianese, F. Ossola, M. Porchia, P. Zanella and S. Sostero, *Inorg. Chim. Acta*, 1987, **132**, 275.
1190.   G. Bombieri, F. Benetollo, E. Klähne and R. D. Fischer, *J. Chem. Soc., Dalton Trans.*, 1983, 1115.
1191.   J. Rebizant, C. Apostolidis, M. R. Spirlet and B. Kanellakopulos, *Inorg. Chim. Acta*, 1987, **139**, 209.
1192.   B. E. Bursten and A. Fang, *Inorg. Chim. Acta*, 1985, **110**, 153.
1193.   B. E. Bursten, M. Casarin, S. Di Bella, A. Fang and I. L. Fragalà, *Inorg. Chem.*, 1985, **24**, 2169.
1194.   B. Kanellakopulos, R. Maier and J. Heuser, *J. Alloys Compd.*, 1991, **176**, 89.
1195.   G. Adrian, *Inorg. Chim. Acta*, 1987, **139**, 323.
1196.   G. Adrian, *KfK Report*, Kernforschungszentrum Karlsruhe, 1988, 4356.
1197.   N. I. Gramoteeva and M. R. Leonov, *Metalloorg. Khim.*, 1989, **2**, 1430.
1198.   M. R. Leonov, N. I. Gramoteeva, I. Z. Kozina and N. V. Il'yushenkova, *Metalloorg. Khim.*, 1990, **3**, 37.
1199.   N. I. Gramoteeva, M. R. Leonov and N. V. Il'yushenkova, *Metalloorg. Khim.*, 1992, **4**, 842.
1200.   M. R. Leonov and G. V. Solov'yova, *Metalloorg. Khim.*, 1993, **6**, 198.
1201.   M. R. Leonov, V. A. Il'yushenko, V. M. Fomin, N. V. Il'yushenkova and Y. A. Alexandrov, *Radiokhim.*, 1986, **28**, 12.
1202.   C. M. Fendrick, E. A. Mintz, L. D. Schertz, T. J. Marks and V. W. Day, *Organometallics*, 1984, **3**, 819.
1203.   C. M. Fendrick and T. J. Marks, *J. Am. Chem. Soc.*, 1986, **108**, 425.
1204.   G. Paolucci, R. D. Fischer, F. Benetollo, R. Seraglia and G. Bombieri, *J. Organomet. Chem.*, 1991, **412**, 327.
1205.   J. Goffart and S. Bettonville, *J. Organomet. Chem.*, 1989, **361**, 17.
1206.   X. Jemine, J. Goffart, P. C. Leverd and M. Ephritikhine, *J. Organomet. Chem.*, 1994, **469**, 55.
1207.   M. R. Spirlet, J. Rebizant and J. Goffart, *Lanth. Act. Res.*, 1986, **1**, 273.
1208.   J. Rebizant, M. R. Spirlet and J. Goffart, *Acta Crystallogr., Sect. C*, 1985, **41**, 334.
1209.   W. Beeckman, J. Goffart, J. Rebizant and M. R. Spirlet, *J. Organomet. Chem.*, 1986, **307**, 23.
1210.   J. Rebizant, M. R. Spirlet, S. Bettonville and J. Goffart, *Acta Crystallogr., Sect. C*, 1989, **45**, 1509.
1211.   M. R. Spirlet, J. Rebizant, S. Bettonville and J. Goffart, *Acta Crystallogr., Sect. C*, 1990, **46**, 1234.
1212.   J. Goffart, J. F. Desreux, B. P. Gilbert, J. L. Delsa, J. M. Renkin and G. Duyckaerts, *J. Organomet. Chem.*, 1981, **209**, 281.
1213.   M. R. Spirlet, J. Rebizant and J. Goffart, *Acta Crystallogr., Sect. B*, 1982, **38**, 2400.
1214.   M. R. Spirlet, J. Rebizant and J. Goffart, *Acta Crystallogr., Sect. C*, 1987, **43**, 354.
1215.   J. Rebizant, M. R. Spirlet, G. van den Bossche and J. Goffart, *Acta Crystallogr., Sect. C*, 1988, **44**, 1710.
1216.   J. Meunier-Piret and M. van Meerssche, *Bull. Soc. Chim. Belg.*, 1984, **93**, 299.
1217.   X. Jemine, J. Goffart, S. Bettonville and J. Fuger, *J. Organomet. Chem.*, 1991, **415**, 363.
1218.   J. Rebizant, M. R. Spirlet, B. Kanellakopulos and E. Dornberger, *J. Less-Common Met.*, 1986, **122**, 211.
1219.   L. R. Avens, S. G. Bott, D. L. Clark, A. P. Sattelberger, J. G. Watkin and B. D. Zwick, *Inorg. Chem.*, 1994, **33**, 2248.
1220.   D. L. Clark, A. P. Sattelberger, S. G. Bott and R. N. Vrtis, *Inorg. Chem.*, 1989, **28**, 1771.
1221.   R. R. Ryan, K. V. Salazar, N. N. Sauer and J. M. Ritchey, *Inorg. Chim. Acta*, 1989, **162**, 221.
1222.   P. J. Fagan, J. M. Manriquez, T. J. Marks, C. S. Day, S. H. Vollmer and V. W. Day, *Organometallics*, 1982, **1**, 170.
1223.   R. G. Finke, Y. Hirose and G. Gaughan, *J. Chem. Soc., Chem. Commun.*, 1981, 232.
1224.   R. G. Finke, D. A. Shiraldi and Y. Hirose, *J. Am. Chem. Soc.*, 1981, **103**, 1875.
1225.   R. G. Finke, G. Gaughan and R. Voegeli, *J. Organomet. Chem.*, 1982, **229**, 179.
1226.   P. Reeb, Y. Mugnier, A. Dormond and E. Laviron, *J. Organomet. Chem.*, 1982, **239**, C1.
1227.   D. C. Sonnenberger and J. Gaudiello, *J. Less-Common Met.*, 1986, **126**, 411.
1228.   E. A. Mintz, K. G. Moloy, T. J. Marks and V. W. Day, *J. Am. Chem. Soc.*, 1982, **104**, 4692.
1229.   T. H. Cymbaluk, R. D. Ernst and V. W. Day, *Organometallics*, 1983, **2**, 963.
1230.   Z. Lin and T. J. Marks, *J. Am. Chem. Soc.*, 1987, **109**, 7979.
1231.   R. E. Cramer, M. A. Bruck and J. W. Gilje, *Organometallics*, 1988, **7**, 1465.
1232.   P. J. Fagan, J. M. Manriquez, E. A. Maatta, A. M. Seyam and T. J. Marks, *J. Am. Chem. Soc.*, 1981, **103**, 6650.
1233.   D. S. Arney, C. J. Burns and D. C. Smith, *J. Am. Chem. Soc.*, 1992, **114**, 10068.
1234.   M. R. Spirlet, J. Rebizant, C. Apostolidis and B. Kanellakopulos, *Acta Crystallogr., Sect. C*, 1992, **48**, 2135.
1235.   C. W. Eigenbrot, Jr. and K. N. Raymond, *Inorg. Chem.*, 1982, **21**, 2653.
1236.   R. E. Cramer, S. Roth and J. W. Gilje, *Organometallics*, 1989, **8**, 2327.
1237.   P. J. Fagan, J. M. Manriquez, S. H. Vollmer, C. S. Day, V. W. Day and T. J. Marks, *J. Am. Chem. Soc.*, 1981, **103**, 2206.
1238.   M. R. Duttera, V. W. Day and T. J. Marks, *J. Am. Chem. Soc.*, 1984, **106**, 2907.
1239.   Z. Lin, C. P. Brock and T. J. Marks, *Inorg. Chim. Acta*, 1988, **141**, 145.
1240.   D. A. Wrobleski, D. T. Cromer, J. V. Ortiz, T. B. Rauchfuss, R. R. Ryan A. P. Sattelberger, *J. Am. Chem. Soc.*, 1986, **108**, 174.
1241.   D. A. Wrobleski, R. R. Ryan, H. J. Wasserman, K. V. Salazar, R. T. Paine and D. C. Moody, *Organometallics*, 1986, **5**, 90.
1242.   S. W. Hall *et al.*, *Organometallics*, 1993, **12**, 752.
1243.   O. J. Scherer, B. Werner, G. Heckmann and G. Wolmershäuser, *Angew. Chem., Int. Ed. Engl.*, 1991, **30**, 553.

1244. D. S. J. Arney and C. J. Burns, *J. Am. Chem. Soc.*, 1993, **115**, 9840.
1245. J. W. Bruno, G. M. Smith, T. J. Marks, C. K. Fair, A. J. Schultz and J. M. Williams, *J. Am. Chem. Soc.*, 1986, **108**, 40.
1246. E. Ciliberto, G. Condorelli, P. J. Fagan, J. M. Manriquez, I. L. Fragalà and T. J. Marks, *J. Am. Chem. Soc.*, 1981, **103**, 4755.
1247. K. Tatsumi and A. Nakamura, *Organometallics*, 1987, **6**, 427.
1248. K. Tatsumi and A. Nakamura, *J. Am. Chem. Soc.*, 1987, **109**, 3195.
1249. J. W. Bruno, T. J. Marks and V. W. Day, *J. Am. Chem. Soc.*, 1982, **104**, 7357.
1250. J. W. Bruno, V. W. Day and T. J. Marks, *J. Organomet. Chem.*, 1983, **250**, 237.
1251. J. W. Bruno, M. R. Duttera, C. M. Fendrick, G. M. Smith and T. J. Marks, *Inorg. Chim. Acta*, 1984, **94**, 271.
1252. G. M. Smith, J. D. Carpenter and T. J. Marks, *J. Am. Chem. Soc.*, 1986, **108**, 6805.
1253. J. W. Bruno, T. J. Marks and L. R. Morss, *J. Am. Chem. Soc.*, 1983, **105**, 6824.
1254. C. M. Fendrick and T. J. Marks, *J. Am. Chem. Soc.*, 1984, **106**, 2214.
1255. S. Di Bella, A. Gulino, G. Lanza, I. L. Fragalà and T. J. Marks, *Organometallics*, 1993, **12**, 3326.
1256. P. J. Fagan, E. A. Maatta and T. J. Marks, in 'Catalytic Activation of Carbon Monoxide', *ACS Symp. Ser.*, 1981, **152**, 53.
1257. K. Tatsumi, A. Nakamura, P. Hofmann, P. Stauffert and R. Hoffmann, *J. Am. Chem. Soc.*, 1985, **107**, 4440.
1258. P. Hofmann, P. Stauffert, K. Tatsumi, A. Nakamura and R. Hoffmann, *Organometallics*, 1985, **4**, 404.
1259. K. Tatsumi and A. Nakamura, *Inorg. Chim. Acta*, 1987, **139**, 247.
1260. K. G. Moloy, P. J. Fagan, J. M. Manriquez and T. J. Marks, *J. Am. Chem. Soc.*, 1986, **108**, 56.
1261. K. G. Moloy, T. J. Marks and V. W. Day, *J. Am. Chem. Soc.*, 1983, **105**, 5696.
1262. E. A. Maatta and T. J. Marks, *J. Am. Chem. Soc.*, 1981, **103**, 3576.
1263. (a) G. Erker, T. Mühlenbernd, R. Benn and A. Rufinska, *Organometallics*, 1986, **5**, 402; (b) G. M. Smith, H. Suzuki, D. C. Sonnenberger, V. W. Day and T. J. Marks, *Organometallics*, 1986, **5**, 549.
1264. R. E. Cramer, S. Roth, F. Edelmann, M. A. Bruck, K. C. Cohn and J. W. Gilje, *Organometallics*, 1989, **8**, 1192.
1265. Z. Lin, J.-F. Le Maréchal, M. Sabat and T. J. Marks, *J. Am. Chem. Soc.*, 1987, **109**, 4127.
1266. X. Yang, C. L. Stern and T. J. Marks, *Organometallics*, 1991, **10**, 840.
1267. X. Yang, W. A. King, M. Sabat and T. J. Marks, *Organometallics*, 1993, **12**, 4254.
1268. Z. Lin and T. J. Marks, *J. Am. Chem. Soc.*, 1990, **112**, 5515.
1269. P. J. Fagan, K. G. Moloy and T. J. Marks, *J. Am. Chem. Soc.*, 1981, **103**, 6959.
1270. D. A. Katahira, K. G. Moloy and T. J. Marks, *Organometallics*, 1982, **1**, 1723.
1271. M. R. Duttera, P. J. Fagan, T. J. Marks and V. W. Day, *J. Am. Chem. Soc.*, 1982, **104**, 865.
1272. T. H. Cymbaluk, J.-Z. Liu and R. D. Ernst, *J. Organomet. Chem.*, 1983, **255**, 311.
1273. D. Baudry *et al.*, *J. Organomet. Chem.*, 1989, **371**, 155.
1274. D. Baudry, E. Bulot and M. Ephritikhine, *J. Chem. Soc., Chem. Commun.*, 1989, 1316.
1275. D. Baudry, E. Bulot and M. Ephritikhine, *J. Organomet. Chem.*, 1990, **397**, 169.
1276. T. Arliguie, M. Lance, M. Nierlich, J. Vigner and M. Ephritikhine, *J. Chem. Soc. Chem., Commun.*, 1994, 847.
1277. D. Baudry, M. Ephritikhine, F. Nief, L. Ricard and F. Mathey, *Angew. Chem., Int. Ed. Engl.*, 1990, **29**, 1485; see also: P. Gradoz *et al.*, *J. Chem. Soc., Chem. Commun.*, 1992, 1720.
1278. F. A. Cotton and W. Schwotzer, *Organometallics*, 1985, **4**, 942; see also: F. A. Cotton, W. Schwotzer and C. Q. Simpson, *Angew. Chem., Int. Ed. Engl.*, 1986, **25**, 637.
1279. G. C. Campbell, F. A. Cotton, J. F. Haw and W. Schwotzer, *Organometallics*, 1986, **5**, 274.
1280. D. Baudry, E. Bulot and M. Ephritikhine, *J. Chem. Soc., Chem. Commun.*, 1988, 1369.
1281. F. Billiau, G. Folcher, H. Marquet-Ellis, P. Rigny and E. Saito, *J. Am. Chem. Soc.*, 1981, **103**, 5603.
1282. D. C. Eisenberg, A. Streitwieser, Jr. and W. K. Kot, *Inorg. Chem.*, 1990, **29**, 10.
1283. A. R. Schake, L. R. Avens, C. J. Burns, D. L. Clark, A. P. Sattelberger and W. H. Smith, *Organometallics*, 1993, **12**, 1497.
1284. A. Streitwieser, Jr., *Inorg. Chim. Acta*, 1984, **94**, 171.
1285. V. G. Sevast'yanov, V. A. Mitin and V. P. Solov'ev, *Primen. Metalloorg. Soedin. Poluch. Neorg. Pokrytii Mater.*, 1986, 104.
1286. M. H. Lyttle, A. Streitwieser, Jr. and M. J. Miller, *J. Org. Chem.*, 1989, **54**, 2331.
1287. W. D. Luke, S. R. Berryhill and A. Streitwieser, Jr., *Inorg. Chem.*, 1981, **20**, 3086.
1288. A. Zalkin, D. H. Templeton, R. Kluttz and A. Streitwieser, Jr., *Acta Crystallogr. Sect. C*, 1985, **41**, 327.
1289. X. Wang, A. Streitwieser, Jr., J. P. Solar and M. H. Lyttle, *Gaodeng Xuexiao Huaxue Xuebao*, 1985, **6**, 699.
1290. X. Wang and A. Streitwieser, Jr., *Gaodeng Xuexiao Huaxue Xuebao*, 1985, **6**, 613.
1291. C. LeVanda and A. Streitwieser, Jr., *Inorg. Chem.*, 1981, **20**, 656.
1292. L. M. Babcock, C. R. Herd and G. E. Streit, *Chem. Phys. Lett.*, 1984, **112**, 173.
1293. J. Yan, Y. Yan, Z. Li, J. Shi and J. Yao, *Yuanzineng Kexue Jishu*, 1986, **20**, 156.
1294. V. G. Sevast'yanov and V. A. Mitin, *Zh. Neorg. Khim.*, 1985, **30**, 3161.
1295. N. T. Kuznetsov, V. A. Mitin, K. V. Kir'yanov, V. G. Sevast'yanov and V. A. Bogdanov, *Radiokhim.*, 1987, **29**, 109.
1296. S. A. Komarov, V. G. Sevast'yanov, N. T. Kuznetsov and Y. S. Ezhov, *Vysokochist. Veshchestva*, 1990, 106.
1297. J. C. Green, M. P. Payne and A. Streitwieser, Jr., *Organometallics*, 1983, **2**, 1707.
1298. X. Wang, A. Streitwieser, Jr., M. H. Lyttle and J. P. Solar, *Gaodeng Xuexiao Huaxue Xuebao*, 1985, **6**, 412.
1299. R. M. Moore, Jr., A. Streitwieser, Jr. and H.-K. Wang, *Organometallics*, 1986, **5**, 1418.
1300. T. M. Gilbert, R. R. Ryan and A. P. Sattelberger, *Organometallics*, 1988, **7**, 2514.
1301. T. M. Gilbert, R. R. Ryan and A. P. Sattelberger, *Organometallics*, 1989, **8**, 857.
1302. D. Baudry, E. Bulot, M. Ephritikhine, M. Nierlich, M. Lance and J. Vigner, *J. Organomet. Chem.*, 1990, **388**, 279.
1303. T. R. Boussie, R. M. Moore, A. Streitwieser, Jr., A. Zalkin, J. Brennan and K. A. Smith, *Organometallics*, 1990, **9**, 2010.
1304. J.-C. Berthet, J.-F. Le Maréchal and M. Ephritikhine, *J. Organomet. Chem.*, 1990, **393**, C47.
1305. T. Arliguie, D. Baudry, J.-C. Berthet, M. Ephritikhine and J.-F. Le Maréchal, *New J. Chem.*, 1991, **15**, 569.
1306. T. Arliguie, D. Baudry, M. Ephritikhine, M. Nierlich, M. Lance and J. Vigner, *J. Chem. Soc., Dalton Trans.*, 1992, 1019.
1307. P. C. Leverd, T. Arliguie, M. Lance, M. Nierlich, J. Vigner and M. Ephritikhine, *J. Chem. Soc., Dalton Trans.*, 1994, 501.
1308. J.-C. Berthet and M. Ephritikhine, *J. Chem. Soc., Chem. Commun.*, 1993, 1566.
1309. A. Dormond and C. Moise, *Polyhedron*, 1985, **4**, 595.
1310. R. S. Sternal, C. P. Brock and T. J. Marks, *J. Am. Chem. Soc.*, 1985, **107**, 8270.
1311. A. F. Wells, 'Structural Inorganic Chemistry', 5th edn., Clarendon Press, Oxford, 1984, 1286.

1312. C. P. Casey, R. F. Jordan and A. L. Rheingold, *Organometallics*, 1984, **3**, 504.
1313. R. S. Sternal, M. Sabat and T. J. Marks, *J. Am. Chem. Soc.*, 1987, **109**, 7920.
1314. R. S. Sternal and T. J. Marks, *Organometallics*, 1987, **6**, 2621.
1315. B. E. Bursten and K. J. Novo-Gradac, *J. Am. Chem. Soc.*, 1987, **109**, 904.
1316. P. J. Hay, R. R. Ryan, K. V. Salazar, D. A. Wrobleski and A. P. Sattelberger, *J. Am. Chem. Soc.*, 1986, **108**, 313.
1317. J. M. Ritchey *et al.*, *J. Am. Chem. Soc.*, 1985, **107**, 501.
1318. J. V. Ortiz, *J. Am. Chem. Soc.*, 1986, **108**, 550.
1319. D. Baudry and M. Ephritikhine, *J. Organomet. Chem.*, 1986, **311**, 189.
1320. G. M. Smith, M. Sabat and T. J. Marks, *J. Am. Chem. Soc.*, 1987, **109**, 1854.
1321. V. W. Day, W. G. Klemperer and D. J. Maltbie, *Organometallics*, 1985, **4**, 104.
1322. K. Jacob, W. Kretschmer, K.-H. Thiele, I. Pavlik, A. Lyçka and J. Holeçek, *Z. Anorg. Allg. Chem.*, 1992, **613**, 88.
1323. K. Yünlü, N. Höck and R. D. Fischer, *Angew. Chem., Int. Ed. Engl.*, 1985, **24**, 879.
1324. R. L. Burwell, Jr. and T. J. Marks, *Chem. Ind.*, (*Catal. Org. React.*), 1985, **22**, 207.
1325. R. G. Bowman, A. Nakamura, P. J. Fagan, R. L. Burwell, Jr. and T. J. Marks, *J. Chem. Soc., Chem. Commun.*, 1981, 257.
1326. M.-Y. He, R. L. Burwell, Jr. and T. J. Marks, *Organometallics*, 1983, **2**, 566.
1327. M.-Y. He, G. Xiong, P. J. Toscano, R. L. Burwell, Jr. and T. J. Marks, *J. Am. Chem. Soc.*, 1985, **107**, 641.
1328. P. J. Toscano and T. J. Marks, *J. Am. Chem. Soc.*, 1985, **107**, 653.
1329. D. Hedden and T. J. Marks, *J. Am. Chem. Soc.*, 1988, **110**, 1647.
1330. W. C. Finch, R. D. Gillespie, D. Hedden and T. J. Marks, *J. Am. Chem. Soc.*, 1990, **112**, 6221.
1331. C. Aitken, J.-P. Barry, F. Gauvin, J. F. Harrod, A. Malek and D. Rousseau, *Organometallics*, 1989, **8**, 1732.
1332. A. Domingos, N. Marques, A. Pires de Matos, I. Santos and M. Silva, *Organometallics*, 1994, **13**, 654.
1333. Z. V. Todres, K. I. Dyusengaliev and V. G. Sevast'yanov, *Izv. Akad. Nauk SSSR, Ser. Khim.*, 1985, 1416.

3

Titanium Complexes in Oxidation States Zero and Below

F. GEOFFREY N. CLOKE
University of Sussex, Brighton, UK

3.1 (η-ARENE) COMPLEXES

3.1.1 Bis(η-arene) Compounds

Matrix isolation studies on the products of co-condensation of titanium vapour with naphthalene or 1-methylnaphthalene has provided UV–visible spectroscopic evidence for the existence of bis(η^6-naphthalene)titanium complexes.[1] The latter decompose above 190 K, thus confirming the earlier observations of thermally unstable red solutions obtained from macroscale co-condensation experiments of titanium vapour with alkylnaphthalenes.[2]

Co-condensation of titanium vapour with 2,6-dimethylpyridine (Equation (1)) affords very low yields (ca. 1%) of the heteroatom-substituted sandwich compound $[Ti(\eta\text{-}C_5H_3NMe_2\text{-}2,6)_2]$.[3] The latter has been structurally characterized by low-temperature x-ray diffraction, and exists in the solid state as rotamer (**1**) in which the two rings are rotated by 20° relative to one another.[4] In solution, $[Ti(\eta\text{-}C_5H_3NMe_2\text{-}2,6)_2]$ is fluxional down to 180 K on the NMR timescale. Use of the more sterically demanding 2,4,6-tri-*t*-butylpyridine affords $[Ti(\eta\text{-}C_5H_2NBu^t_3\text{-}2,4,6)_2]$ in over 30% yield (Equation (2));[5] the significantly higher yield of the latter is attributed to the bulk of the substituents, which prevents σ-coordination of the pyridine ligand. In solution, at room temperature $[Ti(\eta\text{-}C_5H_2NBu^t_3\text{-}2,4,6)_2]$ adopts the conformation shown, but above 340 K is fluxional and a barrier to ring rotation of 68 kJ mol^{-1} may be determined from NMR coalescence experiments.[5]

While bis(η-arene)titanium complexes (e.g., $[Ti(\eta\text{-}C_6H_3Bu^t_3\text{-}1,3,5)_2]$)[6] have usually been prepared by metal vapour synthesis techniques, 'conventional' methods have been developed for their preparation. Thus, ultrasonically assisted reduction of $TiCl_4$ with $K[BEt_3H]$ in the presence of the arene (benzene, toluene, *p*-xylene or mesitylene) affords the appropriate bis(η-arene)titanium complex in modest (5–22%) yield (Equation (3)).[7]

$$\text{Ti (at.)} + \quad [\text{2,6-lutidine}] \quad \xrightarrow{\;77\text{ K}\;} \quad [\text{Ti(bis-2,6-lutidine) sandwich}] \tag{1}$$

(1)

$$\text{Ti (at.)} + \quad [\text{2,4,6-tri-}t\text{-butylpyridine}] \quad \xrightarrow{\;77\text{ K}\;} \quad [\text{Ti(bis-tri-}t\text{-butylpyridine) sandwich}] \tag{2}$$

$$[\text{TiCl}_4] + \text{arene} + \text{K}[\text{Et}_3\text{BH}] \;\xrightarrow{\;(((\;}\; [\text{Ti}(\eta\text{-arene})_2] \tag{3}$$

arene = benzene, toluene, *p*-xylene, mesitylene

Reduction of $[\text{Ti}(\eta\text{-C}_6\text{H}_6)_2]$ or $[\text{Ti}(\eta\text{-PhMe})_2]$ with potassium metal or potassium hydride in terahydrofuran affords the blue, 17-electron anionic sandwich compounds $\text{K}[\text{Ti}(\eta\text{-C}_6\text{H}_5\text{R})_2]$ (R = H, Me) (Equation (4)).[8] The ESR spectrum of $[\text{Ti}(\eta\text{-C}_6\text{H}_6)_2]^-$ exhibits hyperfine coupling to the ring protons, and ^{47}Ti and ^{49}Ti satellites.

$$[\text{Ti}(\eta\text{-arene})_2] \;\xrightarrow{\;\text{K or KH/THF}\;}\; \text{K}[\text{Ti}(\eta\text{-arene})_2] \tag{4}$$

arene = benzene, toluene

Bis(η-arene)titanium anions in which the arene is a biphenyl are also directly accessible from titanium tetrachloride by treatment with an excess of the potassium salt of the biphenyl: the crystallographically characterized salts $[\text{K}(2.2.2\ \text{cryptand})][\text{Ti}(\eta^6\text{-C}_{12}\text{H}_{10})]$ (**2**) and $[\text{K}(2.2.2\ \text{cryptand})][\text{Ti}(\eta^6\text{-C}_{12}\text{H}_8\text{Bu}^t_2\text{-4,4'})]$ (**3**) have been prepared by this method. The K(2.2.2 cryptand) salt of (**2**) reacts readily with CO to afford the known $[\text{K}(2.2.2\ \text{cryptand})]_2[\text{Ti}(\text{CO})_6]$ (see Section 3.2.2), and both (**2**) and (**3**) may be oxidized with iodine to yield the corresponding neutral bis(η-arene)titanium complexes, $[\text{Ti}(\eta^6\text{-C}_{12}\text{H}_{10})]$ and $[\text{Ti}(\eta^6\text{-C}_{12}\text{H}_8\text{Bu}^t_2\text{-4,4'})]$ in ca. 30% yield (Scheme 1).[9]

The molecular structures of $[\text{Ti}(\eta\text{-C}_6\text{H}_6)_2]$ and $[\text{Ti}(\eta\text{-C}_6\text{H}_5\text{Me})_2]$ show the anticipated parallel ring sandwich structures, and extended Hückel MO calculations on these molecules yield metal–ligand and ligand–metal charge-transfer band energies which are in good agreement with their experimental absorption spectra.[10]

3.1.2 Mono(η-arene) Compounds

Reduction of $[\text{TiCl}_3\{\text{Bu}^t\text{Si}(\text{CH}_2\text{PMe}_2)_3\}(\text{THF})]$ with sodium naphthalenide gives the deep-red, thermally stable 16-electron naphthalene complex $[\text{Ti}(\eta^6\text{-C}_{10}\text{H}_8)\{\text{Bu}^t\text{Si}(\text{CH}_2\text{PMe}_2)_3\}]$ (**4**), which has been crystallographically characterized (Scheme 2).[11] Unlike the 16-electron bis(η-arene)titanium compounds, which are diamagnetic, (**4**) is paramagnetic. The latter feature may be associated with unusually strong metal-to-ligand backbonding into a naphthalene δ-acceptor orbital due to the tripodal phosphine ligand, reflected in a folding (dihedral angle 12.4°) of the coordinated naphthalene ring which is thus bonded in a diene–diyl mode. $[\text{Ti}(\eta^6\text{-C}_{10}\text{H}_8)\{\text{Bu}^t\text{Si}(\text{CH}_2\text{PMe}_2)_3\}]$ reacts rapidly with CO to afford $[\text{Ti}\{\text{Bu}^t\text{Si}(\text{CH}_2\text{PMe}_2)_3\}\text{CO}_4]$ (Scheme 2) and is thus a likely intermediate in the synthesis of the latter (Section 3.2.1).

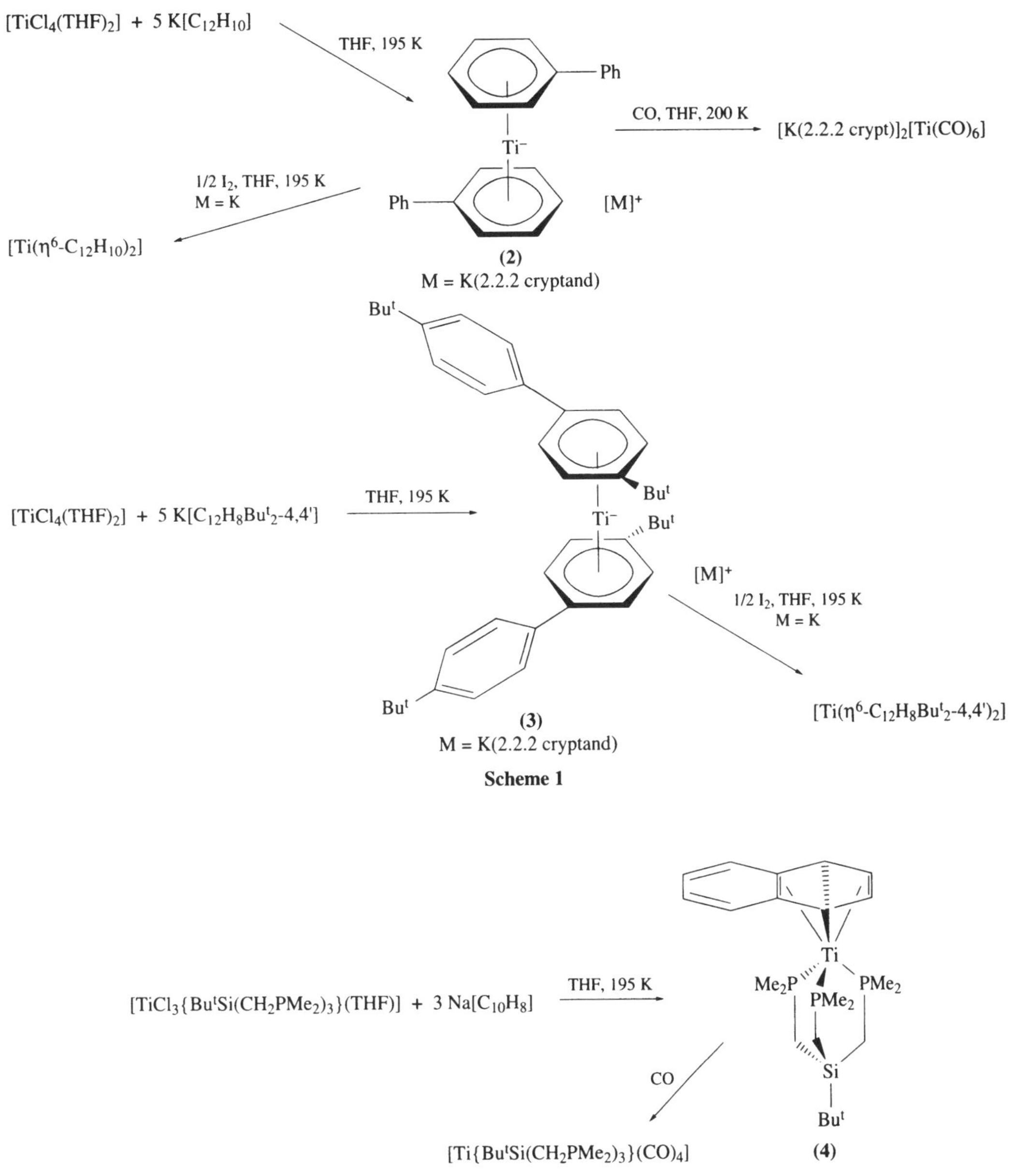

3.2 CARBONYL COMPLEXES

3.2.1 Heteroleptic Carbonyl Complexes

Treatment of [Ti(η-Cp)Cl$_3$] with sodium naphthalenide yields a thermally unstable product, which reacts as a monocyclopentadienylmetallate with CO to afford the tetracarbonylmetallate, isolated as the thermally robust tetraalkylammonium salt [Et$_4$N][Ti(η-Cp)(CO)$_4$] or the structurally characterized tetraphenylarsonium salt [Ph$_4$As][Ti(η-Cp)(CO)$_4$] (5) (Equation (5)).[12] Similarly, [Ti(η-Cp*)Cl$_3$] affords [Et$_4$N][Ti(η-Cp*)(CO)$_4$] on reduction and treatment with CO (Equation (6)); the latter exhibits an extremely low-field resonance at δ = 293.4 for the carbonyl carbon in the ^{13}C NMR spectrum.[13]

Naphthalenide reduction of [Ti(η-Cp)$_2$(CO)$_2$] in the presence of dmpe leads to loss of one cyclopentadienyl ligand (as the alkali metal cyclopentadienide) and formation of the thermally unstable (decomposes above 233 K) anion [Ti(η-Cp)(dmpe)(CO)$_2$]$^-$; the latter may be protonated to afford the structurally characterized hydride [Ti(η-Cp)H(dmpe)(CO)$_2$] (6), and the cryptand complexed salt [K(2.2.2 crypt)][Ti(η-Cp)(dmpe)(CO)$_2$], which is thermally robust (Scheme 3).[14]

$$[Ti(Cp)Cl_3] + 4\,Na[C_{10}H_8] \xrightarrow[\substack{ii,\ CO \\ iii,\ [R_4E]Br}]{i,\ THF,\ Ar,\ 203\ K} \left[\text{Cp–Ti(CO)}_4 \right]^- [R_4E]^+ \qquad (5)$$

R = Et, E = N; R = Ph, E = As

(5)
R = Ph, E = As

$$[Ti(Cp^*)Cl_3] + 4\,Na[C_{10}H_8] \xrightarrow[\substack{ii,\ CO \\ iii,\ [Et_4N]Br}]{i,\ THF,\ Ar,\ 203\ K} [Et_4N][Ti(Cp^*)(CO)_4] \qquad (6)$$

$$[Ti(Cp)_2(CO)_2] + 2\,K[C_{10}H_8] + dmpe \xrightarrow[\substack{200\ K}]{THF\ or\ DME} [Ti(Cp)(dmpe)(CO)_2]^-$$

[K(2.2.2 crypt)][Ti(Cp)(dmpe)(CO)_2]

THF, cryptand 2.2.2
200 K

DME, MeCO$_2$H
215 K

(6)

Scheme 3

The complex originally formulated as $[Ti(CO)_3(dmpe)_{3/2}]_n$,[15] prepared by reductive carbonylation of $TiCl_4$ or carbonylation of $[Ti(\eta\text{-}C_4H_6)_2(dmpe)]$ in the presence of excess dmpe (Equation (7)), has been structurally characterized and shown to be $[Ti(CO)_3(dmpe)_2]$ (7).[16] Complex (7) is fluxional on the NMR timescale and variable-temperature ^{13}C and ^{31}P studies show two distinguishable processes. The low-temperature process, $\Delta G^{\ddagger} = 32.5$ kJ mol^{-1}, is consistent with rotation of a P–P–P triangular face; the high-temperature process, $\Delta G^{\ddagger} = 48.4$ kJ mol^{-1}, equilibrates all phosphorus and carbonyl sites by an unknown mechanism.[16]

$$[Ti(\eta\text{-}C_4H_6)_2(dmpe)] \xrightarrow[\substack{CO,\ 70\ bar}]{THF,\ dmpe,\ 273\ K} (7) \xleftarrow[\substack{Na\text{–}Hg,\ CO,\ 70\ bar}]{THF,\ dmpe,\ 273\ K} [TiCl_4(THF)_2] \qquad (7)$$

(7)

The very reactive $[Ti(CO)_5(dmpe)]$ may be obtained in 50% yield by the synthetic procedure shown in Scheme 4:[17] $[Ti(CO)_3(dmpe)_2]$ (prepared by sodium naphthalenide reduction of titanium tetrachloride in the presence of dmpe and CO which affords a higher (70%) yield than the original method described above) reacts with CO to afford a mixture of starting material, free dmpe and $[Ti(CO)_5(dmpe)]$. Treatment of this mixture with $BH_3 \cdot THF$ removes the free dmpe as its borane adduct and slowly causes the decomposition of $[Ti(CO)_3(dmpe)_2]$, leaving essentially pure $[Ti(CO)_5(dmpe)]$. The latter is thermally unstable at room temperature, and exhibits ^{31}P and ^{13}C NMR spectra down to 180 K characteristic of a fluxional seven-coordinate molecule. $[Ti(CO)_5(dmpe)]$ reacts with a wide range of main-group nucleophiles to afford new Ti^0 carbonyl derivatives (Scheme 4).[17]

The thermal stability of the tripodal phosphine derivative $[Ti(CO)_4(trmpe)]$ (decomposes above 420 K; trmpe = $MeC\{CH_2PMe_2\}_3$) is particularly noteworthy; the latter is also directly accessible from titanium tetrachloride via naphthalenide reduction (Equation (8)),[18] as is the closely related, structurally characterized analogue $[Ti(CO)_4(trimpsi)]$ (trimpsi = $Bu^tSi\{CH_2PMe_2\}_3$) (8) (Equation (9)).[19]

$$[TiCl_4(THF)_2] + trmpe + 4\,K[C_{10}H_8] \xrightarrow{THF,\ CO,\ 210\ K} [Ti(CO)_4(trmpe)] \qquad (8)$$

The monodentate phosphine analogue of $[Ti(CO)_5(dmpe)]$, $[Ti(CO)_5(PMe_3)_2]$, is believed to be present in solutions obtained by low-temperature naphthalenide reduction of $[TiCl_4(DME)_2]$ in the presence of trimethylphosphine under a CO atmosphere (Scheme 5); a species formulated as

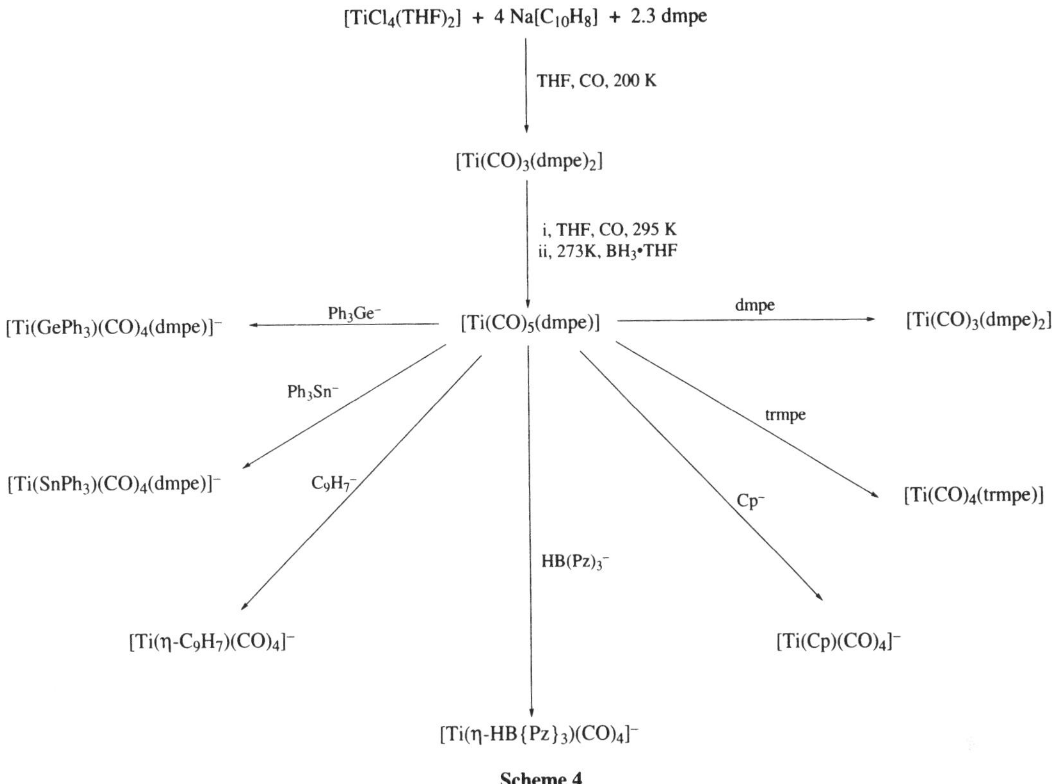

Scheme 4

(9)

[Ti(CO)$_4$(PMe$_3$)$_3$] is also proposed as a minor component of the products of this reaction.[20] [Ti(CO)$_5$(PMe$_3$)$_2$] is too thermally unstable to be isolated but displays ^{31}P and ^{13}C NMR spectra very similar to those of the known [Ti(CO)$_5$(dmpe)]. Solutions containing [Ti(CO)$_5$(PMe$_3$)$_2$] react rapidly with main-group nucleophiles to afford [Ti(CO)$_4$(triphos)] (triphos = MeC(CH$_2$PPh$_2$)$_3$), [Ti(CO)$_3$(diars)$_2$] (diars = o-C$_6$H$_4$(AsMe$_2$)$_2$), [Et$_4$N]$_2$[Ti(CO)$_5$(GePh$_3$)$_2$], or the structurally characterized [Et$_4$N]$_2$-[Ti(CO)$_5$(SnPh$_3$)$_2$] (**9**) (Scheme 5); the structure of the latter is described as an approximately monocapped octahedron where one triphenyltin group is the capping ligand. It is noteworthy that none of the latter compounds can be prepared from [Ti(CO)$_5$(dmpe)] or from reductive carbonylation of titanium tetrahalides in the absence of trimethylphosphine.

While acyclic amines such as R$_2$NCH$_2$CH$_2$NR$_2$ or RN(CH$_2$CH$_2$NR$_2$)$_2$ appear not to form isolable group 4 metal carbonyl derivatives, the macrocyclic amines tacn (tacn = 1,4,7-triazacyclononane) and Me$_3$tacn (Me$_3$tacn = 1,4,7-trimethyl-1,4,7-triazacyclononane) react with [Ti(CO)$_5$(dmpe)] to afford the thermally stable tetracarbonyl complexes [Ti(CO)$_4$(tacn)] and structurally characterized [Ti(CO)$_4$(Me$_3$tacn)] (**10**) (Equations (10) and (11)).[21] IR data indicate that tacn is the strongest neutral donor ligand to a zero oxidation state group 4 metal carbonyl currently known.

$$[\text{Ti(CO)}_5(\text{dmpe})] + \text{tacn} \xrightarrow[\text{THF, 295 K}]{} [\text{Ti(CO)}_4(\text{tacn})] \tag{10}$$

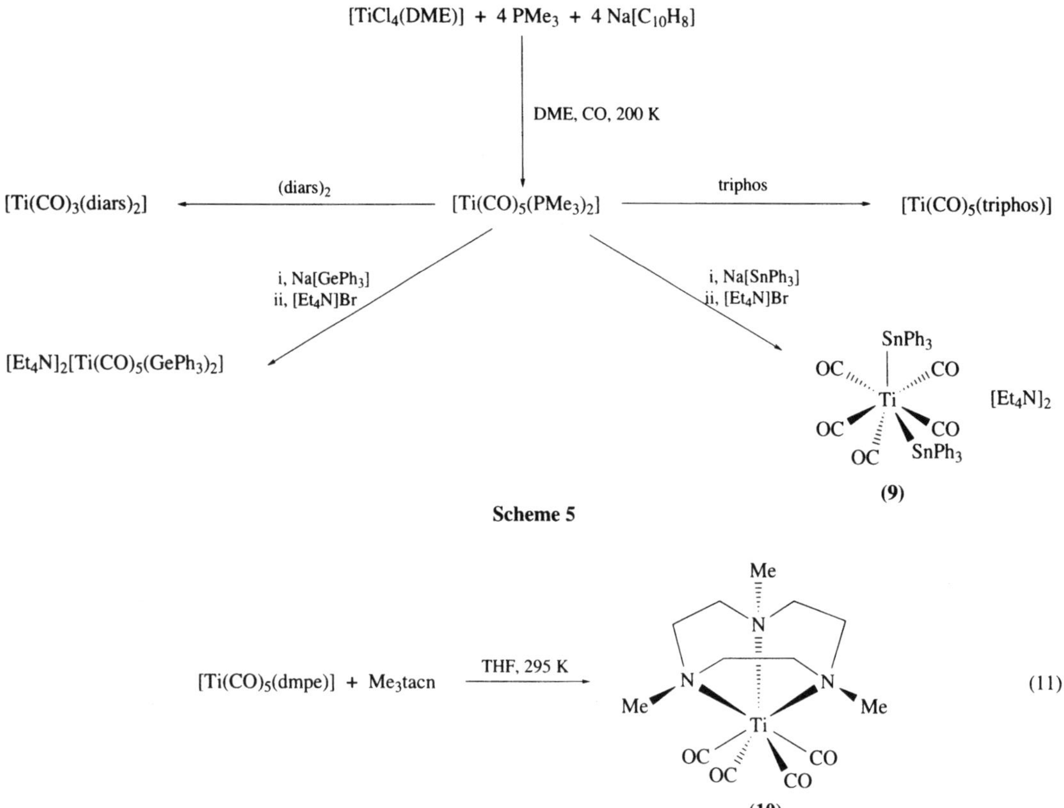

Scheme 5

$$[\text{Ti(CO)}_5(\text{dmpe})] + \text{Me}_3\text{tacn} \xrightarrow{\text{THF, 295 K}} \qquad (11)$$

3.2.2 Homoleptic Carbonyl Complexes

Reductive carbonylation of either [Ti(CO)$_3$(dmpe)$_2$], [Ti(CO)$_5$(dmpe)] or [TiCl$_4$(DME)] yields isolable salts of [Ti(CO)$_6$]$^{2-}$ provided that an appropriate crown ether or cryptand ligand is present to complex the alkali metal cations effectively (Equations (12) and (13)).[22,23] Solution IR spectral studies (ν_{CO} = 1745–1750 cm^{-1}) are consistent with the presence of discrete octahedral [Ti(CO)$_6$]$^{2-}$ units, as are the very narrow ^{47}Ti and ^{49}Ti NMR resonances which display a J_{Ti-C} of 23 Hz in ^{13}CO enriched samples.[22] Essentially unperturbed octahedral geometry is confirmed by the x-ray structure of [K(cryptand 2.2.2)]$_2$[Ti(CO)$_6$]·pyridine, which shows discrete anions and cations and one pyridine of crystallization.

$$\begin{array}{c}[\text{Ti(CO)}_4(\text{trmpe})] \\ \text{or} \\ [\text{Ti(CO)}_5(\text{dmpe})]\end{array} + 2\,\text{K}[\text{C}_{10}\text{H}_{18}] + 2\,\text{crypt 2.2.2} \xrightarrow[\text{200 K}]{\text{THF, CO}} [\text{K(crypt 2.2.2)}]_2[\text{Ti(CO)}_6] \qquad (12)$$

$$[\text{TiCl}_4(\text{DME})] + 6\,\text{K}[\text{C}_{10}\text{H}_8] + 4\,\text{15-crown-5} \xrightarrow[\text{200 K}]{\text{DME, CO}} [\text{K(15-crown-5)}]_2[\text{Ti(CO)}_6] \qquad (13)$$

3.3 HOMOLEPTIC DIENE COMPLEXES

Co-condensation of titanium vapour with 2,3-di-*t*-butylbuta-1,3-diene affords low (5–10%) yields of a deep-purple product formulated as the diamagnetic, 12-electron Ti^0 complex [Ti(η-C$_4$H$_4$Bu$^t_2$-2,3)$_2$] (Equation (14)).[24] The latter slowly decomposes at room temperature, but has been characterized by detailed NMR spectral studies which confirm the presence of two isomers (differing in the relative positions of the *t*-butyl substituents) in which the diene retains its *trans–trans* configuration.

Co-condensation of titanium vapour with the 1,4-isomer of di-*t*-butylbuta-1,3-diene gives low yields of the well-characterized, dark red 'sandwich' compound [Ti(η-C$_4$H$_4$Bu$^t_2$-1,4)$_2$] (Equation (15)).[25] The ligands in the latter adopt the expected *cis–cis* configuration, and NMR spectral studies show that the two dienes are mutually orthogonal, leading to pseudotetrahedral geometry about titanium. There is no NMR spectral evidence for agostic interactions, despite the compound being formally 12-electron Ti^0.

$$\text{(14)}$$

$$\text{(15)}$$

3.4 REFERENCES

1. P. D. Morand and C. G. Francis, *Inorg. Chem.*, 1985, **24**, 56.
2. E. P. Kundig and P. L. Timms, *J. Chem. Soc., Chem. Commun.*, 1977, 912.
3. E. J. Wucherer and E. L. Muetterties, *Organometallics*, 1987, **6**, 1691.
4. E. J. Wucherer and E. L. Muetterties, *Organometallics*, 1987, **6**, 1696.
5. K. Khan, D. Phil. Thesis, University of Sussex, 1991.
6. F. G. N. Cloke, K. A. E. Courtney, A. A. Sameh and A. C. Swain, *Polyhedron*, 1989, **8**, 1641.
7. H. Bönnemann and B. Korall, *Angew. Chem., Int. Ed. Engl.*, 1992, **31**, 1490.
8. J. A. Bandy, A. Berry, M. L. H. Green, R. N. Perutz, K. Prout and J.-N. Verpeaux, *J. Chem. Soc., Chem. Commun.*, 1984, 729.
9. D. W. Blackburn, D. Britton and J. E. Ellis, *Angew. Chem., Int. Ed. Engl.*, 1992, **31**, 1495.
10. O. N. Krasochka *et al.*, *Khim. Fiz.*, 1983, **11**, 1459.
11. T. G. Gardner and G. S. Girolami, *Angew. Chem., Int. Ed. Engl.*, 1988, **27**, 1693.
12. B. A. Kelsey and J. E. Ellis, *J. Am. Chem. Soc.*, 1986, **108**, 1344.
13. B. A. Kelsey and J. E. Ellis, *J. Chem. Soc., Chem. Commun.*, 1986, 331.
14. S. R. Frerichs, B. A. Kelsey and J. E. Ellis, *J. Am. Chem. Soc.*, 1987, **109**, 5559.
15. S. Wreford, M. B. Fischer, J.-S. Lee, E. J. James and S. C. Nyburg, *J. Chem. Soc., Chem. Commun.*, 1981, 458.
16. P. J. Domaille, R. L. Harlow and S. S. Wreford, *Organometallics*, 1982, **1**, 935.
17. K. M. Chi, S. R. Frerichs, B. A. Kelsey-Stein, D. W. Blackburn and J. E. Ellis, *J. Am. Chem. Soc.*, 1988, **110**, 163.
18. D. W. Blackburn, K. M. Chi, S. R. Frerichs, M. L. Tinkham and J. E. Ellis, *Angew. Chem., Int. Ed. Engl.*, 1988, **27**, 437.
19. T. G. Gardner and G. S. Girolami, *Organometallics*, 1987, **6**, 2551.
20. K. M. Chi, S. R. Frerichs and J. E. Ellis, *J. Chem. Soc., Chem. Commun.*, 1988, 1013.
21. J. E. Ellis, A. J. DiMaio, A. L. Rheingold and B. S. Haggerty, *J. Am. Chem. Soc.*, 1992, **114**, 10676.
22. K. M. Chi, S. R. Frerichs, S. B. Philson and J. E. Ellis, *J. Am. Chem. Soc.*, 1988, **110**, 303.
23. K. M. Chi and J. E. Ellis, *J. Am. Chem. Soc.*, 1990, **112**, 6022.
24. H. de Lemos, D. Phil. Thesis, University of Sussex, 1989.
25. F. G. N. Cloke and A. McCamley, *J. Chem. Soc., Chem. Commun.*, 1991, 1470.

4

Titanium Complexes in Oxidation States +2 and +3

MANFRED BOCHMANN
University of East Anglia, Norwich, UK

4.1 COMPLEXES OF TITANIUM(II)

4.1.1 Alkyl Complexes

Whereas the alkylation of $[TiCl_2(dmpe)_2]$ leads to impure chloride-containing products, $[Ti(BH_4)_2(dmpe)_2]$ reacts cleanly with methyllithium to give the diamagnetic titanium(II) complex (**1**). The resonances of the methyl ligands in the ^1H NMR spectrum occur at unusually high field (δ −5.72), with a relatively small value for the C–H coupling constant ($J_{CH} = 109.5$ Hz). However, structure determinations by both x-ray and neutron diffraction confirm normal Ti–C–H angles of 109°, with no indication of agostic metal–CH interactions.[1] Complex (**1**) reacts with ethene above −40 °C to give the mono- and dinuclear complexes (**2**) and (**3**) (Equation (1)). The alkene ligands are oriented 'in-plane' to maximize the extent of backbonding. There is no evidence for ethene rotation up to −10 °C. Above −40 °C (**2**) and (**3**) dimerize ethene to 1-butene via a titanium(IV) metallacyclopentane intermediate; significantly, there is no insertion of the alkene into the Ti–Me bond which would lead to a loss of backbonding stabilization of the product (Scheme 1).[2] At 25 °C and 4 bar pressure ethene is polymerized.[1] Dibenzyltitanium has been isolated as a fairly pure black solid from the reaction of $[Ti(CH_2Ph)_4]$ with butyllithium.[3]

$$(1)$$

Scheme 1

4.1.2 Monocyclopentadienyl Complexes

The reduction of $[TiCl_3Cp^*]$ by sodium naphthalenide in the presence of $Me_2PCH_2CH_2PMe_2$ followed by treatment with CO (1 bar) gives $[TiClCp^*(CO)_2(dmpe)]$. The compound was characterized by IR and NMR spectral methods and is rather unstable, while the structure of the analogous hafnium complex could be confirmed by x-ray diffraction.[4]

There are numerous complexes of the type $[TiClCp(diene)]$, and their chemistry has been briefly reviewed.[5] The reaction of $[TiClCp^*(diene)]$ with MeMgI gives $[TiMeCp^*(diene)]$ (diene = 2,3-dimethylbutadiene). The air-sensitive solid is thermally moderately stable, although solutions decompose at room temperature.[6] The reaction of $[TiCl_3Cp^*]$ with three equivalents of 1-methallylmagnesium bromide gives the brown allyl complex $[Ti(\eta^3\text{-}CH_2CHCHMe)Cp^*(\eta^4\text{-}C_4H_6)]$ in 30% yield, probably via $[Ti(\eta^1\text{-}CH_2CHCHMe)_3Cp^*]$ as the intermediate. The hafnium analogue of the latter is isolable.[7] The analogous treatment of $[TiCl_3Cp^*]$ with (dimethylbutadiene)magnesium leads to (**4**). The diene ligand is nonfluxional, and the bonding is best described by η^4 (**4a**) and $\eta^1{:}\eta^1{:}\eta^2$ (**4b**) resonance forms (Equation (2)).[6] The bonding modus of the diene ligand depends on its substitution pattern. With 1,3-butadiene and its 1,4-disubstituted derivatives RCH=CHCH=CHR (R = Me or Ph) the '*prone*' or *endo* conformation (**5**) is preferred, while with 2,3-disubstituted dienes $H_2C=CR^1CR^2=CH_2$ (R^1 = Me, R^2 = H; $R^1 = R^2$ = Me or Ph) the '*supine*' or *exo* isomer (**4**) results.[8] A series of such complexes $[TiXCp^*(diene)]$ (X = Cl, Br or I) has been isolated; the blue compounds are highly air sensitive. The $^{13}C{-}^{13}C$ coupling constants suggest pronounced η^4 diene–metal bonding for the *prone* conformers and a substantial contribution of the bent metallacyclopentene structure of type (**4b**) in the case of the *supine* complexes. This is borne out by the x-ray structures of the butadiene and the 1,4-diphenylbutadiene complexes, both of which are *prone* conformers, and the *supine* 2,3-diphenylbutadiene analogue. The C–C bond lengths in the *prone* complexes are fairly similar to one another, as are the metal-to-carbon distances, while in the *supine* isomers the C-2–C-3 bond is short. Pertinent structural parameters are given in Structures (**6**)–(**8**).[8]

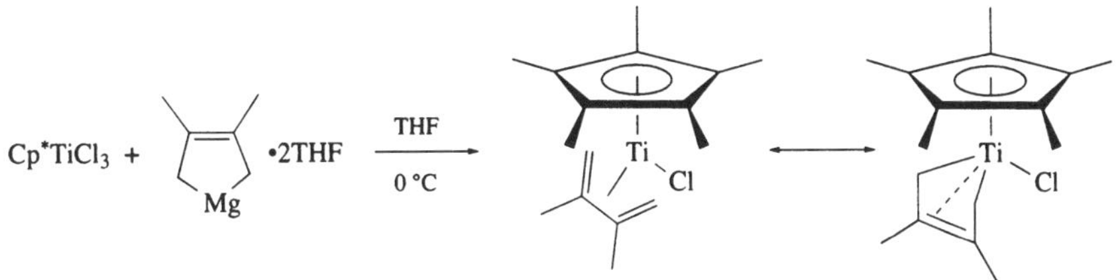

Cp*TiCl$_3$ + [diene]·2THF $\xrightarrow[\text{0 °C}]{\text{THF}}$ (**4a**) $\longleftrightarrow$ (**4b**) (2)

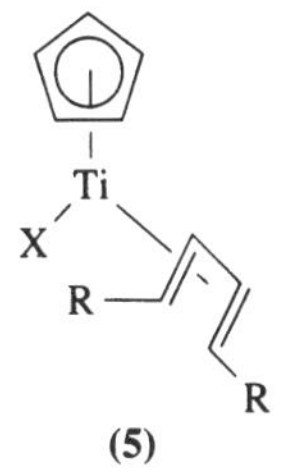

(**5**)

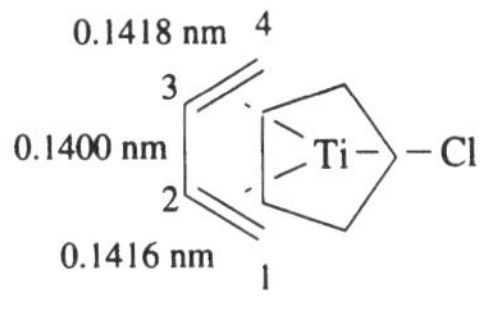

0.1418 nm 4
0.1400 nm 3
0.1416 nm 2
1
Ti–C-1 0.2178(10) nm
Ti–C-2 0.2288(9) nm
(**6**)

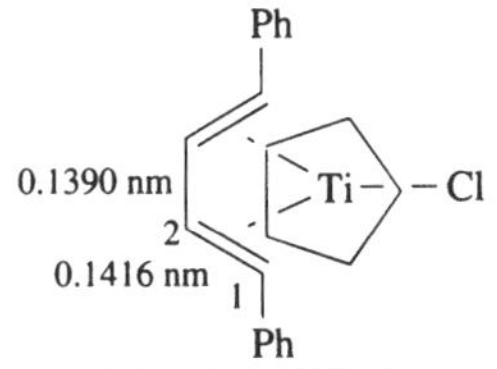

0.1390 nm
0.1416 nm
Ti–C-1 0.2233(6) nm
Ti–C-2 0.2293(5) nm
(**7**)

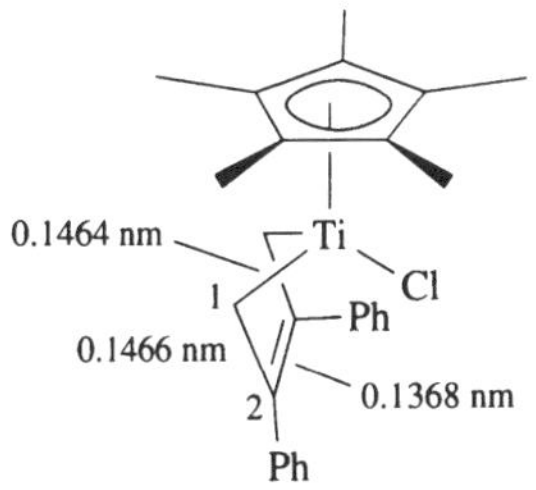

0.1464 nm
0.1466 nm
0.1368 nm
Ti–C-1 0.2150(6) nm
Ti–C-2 0.2405(6) nm
(**8**)

Structurally closely related compounds are obtained from [TiCl$_3$Cp*] and Mg[1,2-(CH$_2$)$_2$C$_6$H$_4$]. If this reaction is carried out in toluene the red xylenediyl complex (**9**) is formed which has a pronounced metallacyclopentene character. With THF as the solvent a dimeric product (**10**) was isolated as green crystals which contain a bent, Dewar-benzene-like arene ring and essentially isolated double bonds (Scheme 2).[9]

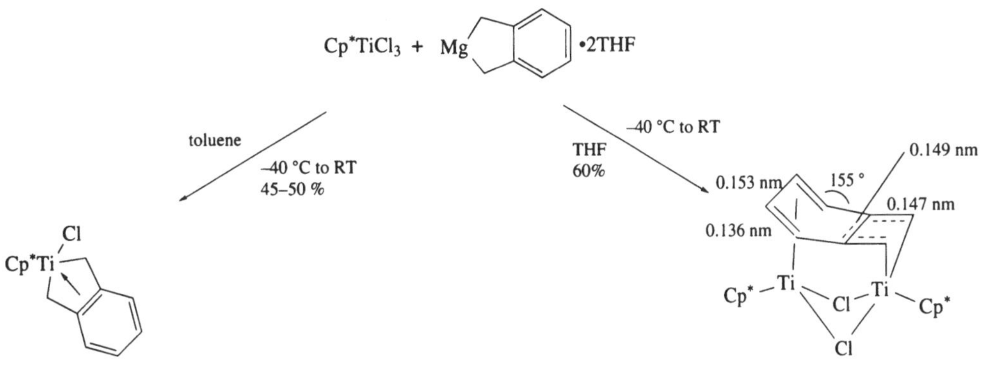

Cp*TiCl$_3$ + Mg[diene]·2THF

toluene
–40 °C to RT
45–50 %

THF
60%
–40 °C to RT

0.153 nm 155° 0.149 nm
0.136 nm 0.147 nm

(**9**)

(**10**)

Scheme 2

The diene ligand in butadiene and isoprene complexes of type (**6**) can be exchanged for ligands which are better π-acceptors. Thus, [TiClCp*(C$_4$H$_6$)] reacts with 1,4-diphenylbutadiene in hexane at 60 °C to give [TiClCp*(1,4-Ph$_2$C$_4$H$_4$)]. In contrast with the zirconium and hafnium analogues, the titanium diene complexes do not form stable adducts with pyridine, THF or PMe$_3$. They do not exhibit catalytic activity towards dienes by themselves but, on activation with Grignard reagents or magnesium

     Titanium Complexes in Oxidation States +2 and +3

in the presence of excess isoprene catalyse the dimerization of isoprene at 60 °C to give a 98:2 mixture of linear head-to-tail and head-to-head products. The Cp* complexes are more selective than Cp analogues.[8] Complex (**4**) reacts with xylyl isocyanide predominantly under displacement of the diene ligand and reductive coupling of the isocyanides to give the crystallographically characterized titanium(IV) complex (**11**) which contains an essentially planar $Ti_2C_2N_2$ core; a by-product (**12**) was identified spectroscopically (Equation (3)).[10] The reaction of (**4**) with CO leads to the the brown titanaoxirane complex (**13**) in 32% yield. As the x-ray structure shows, the carbon of CO has become part of a cyclopentene ring system and the C–O bond order has been reduced to a single bond ($r_{CO} = 0.1408(4)$ nm). By contrast, the reaction with the zirconium and hafnium analogues of (**4**) leads to the formation $[M(O)ClCp^*]_n$, with liberation of dimethylcyclopentadiene.[11]

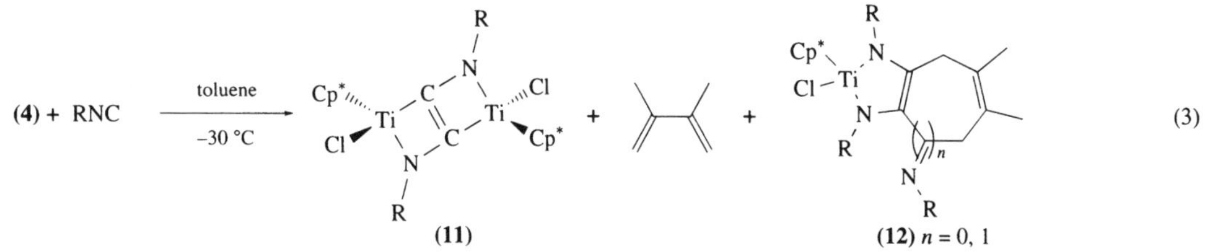

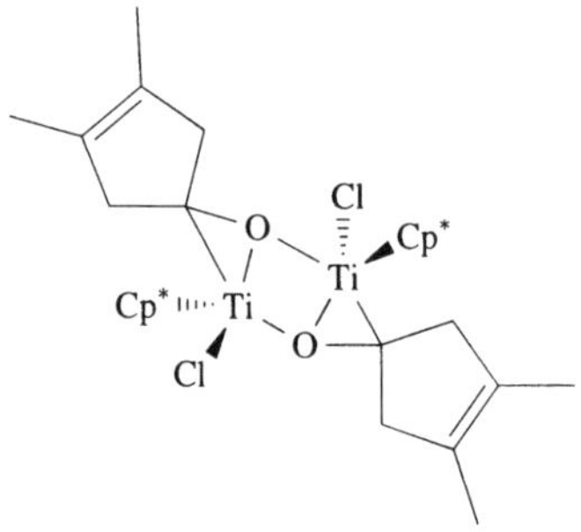

$$(\mathbf{4}) + RNC \xrightarrow[-30\,°C]{toluene} (\mathbf{11}) + \diagup\diagdown + (\mathbf{12})\ n = 0, 1 \tag{3}$$

(**13**)

4.1.3 Carbonyl Complexes of Titanium(II)

4.1.3.1 Synthesis of [TiCp$_2$(CO)$_2$] and [TiCp$_2$(CNR)$_2$] complexes

The synthesis and reactivity of carbonyl complexes of titanium, zirconium and hafnium has been reviewed extensively.[12] Detailed syntheses of $[TiCp_2(CO)_2]$ and $[TiCp^*_2(CO)_2]$ have been described.[13,14] $[Ti(Ind)_2(CO)_2]$ (Ind = η^5-indenyl) is obtained in 47% yield by reducing $[TiCl_2(Ind)_2]$ with aluminum powder in THF under a CO atmosphere. The compound forms dark green crystals which were characterized by x-ray diffraction.[15] The mixed-ligand complex $[TiCpCp^*(CO)_2]$ is similarly prepared by reduction of the corresponding titanocene dichloride with zinc powder.[16] Two equivalents of methyllithium in the presence of CO were used as the reducing agent for the synthesis of the *ansa*-metallocene complex $[Ti(\eta^5-C_5H_4)_2(CO)_2C_2H_4]$. The same product is obtained in high yield by treating $[Ti(\eta^5-C_5H_4)_2Me_2C_2H_4]$ with CO; the by-product is acetone.[17] Titanocene carbonyl complexes with phosphido-substituted cyclopentadienyl ligands are obtained from $[TiCl_2Cp(C_5Me_4PPh_2)]$ or $[TiCl_2(C_5Me_4PPh_2)_2]$ with aluminum powder activated by $HgCl_2$ under a CO atmosphere.[18] The reduction of $[TiCl_2Cp_2]$ with magnesium in THF at room temperature in the presence of an excess of RNC gives $[TiCp_2(CNR)_2]$ as black–violet crystals in 93% yield (R = 2,6-Me$_2$C$_6$H$_3$). The ν_{CN} bands occur in the IR spectrum at 2038 and 1937 cm^{-1} (free RNC: 2115 cm^{-1}).[19]

4.1.3.2 Properties of [TiCp$_2$(CO)$_2$] complexes

The electronic structure of $[TiCp_2(CO)_2]$ has been probed by gas-phase He^I and He^{II} photoelectron spectroscopy and Xα molecular orbital calculations. The bonding of CO to the metal centre relies predominantly on π^* backbonding. Contributions from σ and π donation from CO to the metal are unimportant.[20]

The mean metal–CO bond dissociation enthalpies $\overline{D}$, mean Ti–C bond energies $\overline{E}$ and the heat of formation of $[TiCp_2(CO)_2]$, ΔH_f^0, have been determined. For the first and second bond dissociation enthalpies D_1 and D_2, values of 150 (D_1) and 154 (D_2) kJ mol^{-1} were found;[21] later determinations gave values of 170 and 174 kJ mol^{-1}, respectively.[22] The mean Ti–CO bond energy is 174 ± 8 kJ mol^{-1}.[22,23] ΔH_f^0 for $[TiCp_2(CO)_2]$ (cryst.) was determined as -371.5 ± 12.9 kJ mol^{-1}.[24]

The diagnostically useful ν_{CO} frequencies of some titanium carbonyl derivatives are collected in Table 1.

Table 1 Infrared spectroscopic data for bis(cyclopentadienyl)titanium carbonyl complexes.

Compound	ν_{CO} (cm^{-1})	Medium	Ref.
$[TiCp_2(CO)_2]$	1977, 1899	Hexane	14
	1965, 1883	THF	14
	1982, 1903	Argon matrix	26
$[TiCp_2(CO)]$	1890	Argon matrix	26
$[TiCp_2(^{13}CO)_2]$	1928, 1850	Decalin	27
$[TiCp_2(C^{18}O)_2]$	1934, 1856	Hexane	25
$[TiCp_2(C^{16}O)(C^{18}O)]$	1959, 1870	Heptane	25
$[TiCp_2(C^{16}O)_2]$	1977, 1899	Heptane	25
$[Ti(\eta\text{-indenyl})_2(CO)_2]$	1987, 1913, 1906	Pentane	15
	1980, 1911, 1902	Decalin	27
$[Ti(\eta\text{-indenyl})_2(^{13}CO)_2]$	1936, 1868, 1860	Decalin	27
$[TiCp(\eta\text{-}C_5Me_5)(CO)_2]$	1956, 1875	Hexane	16
$[Ti(\eta\text{-}C_5HMe_4)_2(CO)_2]$	1946, 1862	Hexane	276
$[Ti(\eta\text{-}C_5Me_5)_2(CO)_2]$	1940, 1858		14
$[Ti(\eta\text{-}C_5Me_5)_2(^{13}CO)_2]$	1890, 1801	Decalin	27
$[TiCp_2(CO)(PF_3)]$	1932	Hexane	25
$[TiCp_2(CO)(PMe_3)]$	1864	Hexane	29
$[TiCp_2(CO)(PEt_3)]$	1864	Hexane	25
$[TiCp_2(CO)(PBu^n_3)]$	1868	Decalin	27
$[TiCp_2(CO)\{P(OEt)_3\}]$	1874	Decalin	27
$[TiCp_2(CO)\{P(OPr^i)_3\}]$	1869	Decalin	27
$[TiCp_2(CO)(PMe_2Ph)]$	1860	Decalin	27
$[TiCp_2(CO)(PMePh_2)]$	1860	Decalin	27
$[TiCp_2(CO)(PPh_3)]$	1850	THF	25
$[TiCp_2(CO)\{P(OPh)_3\}]$	1900	Hexane	25
$[TiCp_2(CO)(MeO_2CCH=CHCO_2Me)]$	2010; 1675 (CO_2Me)	Benzene-d_6	25
$[TiCp_2(CO)(C_2Ph_2)]$	2000; 1785 $(\nu_{C\equiv C})$	Benzene-d_6	25
$[TiCp_2(CO)\{C_2(C_6F_5)_2\}]$	2020; 1770 $(\nu_{C\equiv C})$	Benzene-d_6	25
$[TiCp(\eta\text{-}C_5Me_5)(CO)(C_2Ph_2)]$	1980	Hexane	16
$[TiCp_2(CO)(2\text{-}MeC_4H_3O)]$	1822	2-MeC$_4$H$_3$O	26
$[Ti(\eta\text{-indenyl})_2(CO)(PMe_3)]$	1861	Hexane	29
$[Ti\{C_2H_4(\eta\text{-}C_5H_4)_2\}(CO)_2]$	1980, 1900		17
$[Ti\{C_2Me_4(C_5H_4)_2\}(CO)_2]$	1965, 1888	Pentane	162
$[Ti(\eta\text{-}C_5Me_4PPh_2)_2(CO)_2]$	1955, 1882	THF	18
$[TiCp(\eta\text{-}C_5Me_4PPh_2)(CO)_2]$	1952, 1882	THF	18
$[Ti\{(C_5H_4)_2(SiMe_2)O\}(CO)_2]$	1960, 1855	Hexane	277
$[TiCp_2(CO)(CNBu^t)]$	1850; 2170 $(\nu_{C\equiv N})$	KBr pellet	28
$[TiCp_2(CO)(CNC_6H_3Me_2\text{-}2,6)]$	1862; 2020 $(\nu_{C\equiv N})$	KBr pellet	28
$[TiCp_2(CO)(CNMe)]$	1840; 2093 $(\nu_{C\equiv N})$	KBr pellet	28
$[Ti(\eta\text{-indenyl})_2(CO)(CNBu^t)]$	1855; 2173 $(\nu_{C\equiv N})$	KBr pellet	28
$[TiCp_2(CNC_6H_3Me_2\text{-}2,6)_2]$	2038, 1937 $(\nu_{C\equiv N})$	Pentane	19

4.1.3.3 Ligand exchange reactions of [TiCp$_2$(CO)$_2$] complexes

The ligand exchange reaction of $[TiCp_2(C^{18}O)_2]$ with $C^{16}O$ in hexane solution proceeds only in the presence of light and gives first $[TiCp_2(C^{16}O)(C^{18}O)]$, followed by $[TiCp_2(C^{16}O)_2]$. The likely intermediate is $[TiCp_2(CO)]$.[25] This intermediate was identified in solid argon or methane matrices by its single ν_{CO} band at 1890 cm^{-1} by irradiation of $[TiCp_2(CO)_2]$ with UV light ($\lambda = 313$ nm) close to the absorption maximum of $[TiCp_2(CO)_2]$ ($\lambda_{max} = 308$ nm, $\varepsilon = 8000$). There is no further decarbonylation to 'TiCp$_2$'.[26] The photolability of $[TiCp_2(CO)_2]$ is used for the preparation of a large number of derivatives $[TiCp_2(CO)(L)]$ (L = PF$_3$, PEt$_3$,[25] PBu$_3$, PMe$_2$Ph, PMePh$_2$, PPh$_3$, P(OMe)$_3$, P(OEt)$_3$, P(OPr^i)$_3$ or P(OPh)$_3$.[27] $[TiCp_2(CO)(PF_3)]$ is formed photolytically at 15 °C. Prolonged irradiation leads to $[TiCp_2(PF_3)_2]$ as sublimable yellow crystals in 86% yield. Irradiation of $[TiCp_2(CO)_2]$ in hexane in the presence of PEt_3 gives $[TiCp_2(CO)(PEt_3)]$, which is also obtained thermally in refluxing hexane in 60% yield. The thermal reaction with PPh_3 gives $[TiCp_2(CO)(PPh_3)]$. $[TiCp_2(CO)(PEt_3)]$ is a useful starting material because the PEt_3 ligand is readily displaced by PF_3, $P(OPh)_3$, *trans*-XCH=CHX ($X = CO_2R$; $R = Me$ or Et) or alkynes such as C_2R_2 ($R = Ph$ or C_6F_5) to give the corresponding complexes $[TiCp_2(CO)(L)]$. The green alkene complexes $[TiCp_2(CO)(XCH=CHX)]$ are thermally unstable above $0\,°C$. The strong π-acceptor properties of alkyne ligands is evident in $[TiCp_2(CO)(C_2R_2)]$ ($R = Ph$ or C_6F_5) by an increased ν_{CO} band to over $2000\ cm^{-1}$ and a decrease in the $\nu_{C\equiv C}$ frequency to $1770–1785$ cm^{-1}.[25] Mixed CO–RNC complexes are similarly accessible, often at low temperature even without irradiation. Thus, treatment of $[TiCp_2(CO)_2]$ with RNC ($R = Bu^t$ or $2,6\text{-}Me_2C_6H_3$) in THF at $-78\,°C$ to room temperature gives $[TiCp_2(CO)(CNR)]$. The PMe_3 ligand in $[TiCp'_2(CO)(PMe_3)]$ ($Cp' = Cp$ or η-indenyl) is very readily displaced to give $[TiCp'_2(CO)(NCR)]$ in high yield ($R = Me$ or Bu^t). $[TiCp^*_2(CO)_2]$ fails to react with RNC, even under photolytic conditions.[28]

The crystal structures of $[TiCp_2(PF_3)_2]$[25] $[TiCp_2(CO)(PMe_3)]$,[29] $[TiCp_2(CO)(PEt_3)]$[25] and $[TiCp_2(CO)(CNBu^t)]$[28] have been determined. In all cases titanium is tetrahedrally coordinated. The Ti–P distances increase in the order $PF_3 < PMe_3 \leq PEt_3$ from 0.2345 (average) and $0.2544(1)$ to $0.2585(1)\ nm$. In the isocyanide complex the Ti–CO bond length ($0.1993(10)\ nm$) is significantly shorter than the Ti–CN distance ($0.2112(9)\ nm$), illustrating the differences in backbonding.

The kinetics of CO substitution in $[TiCp'_2(CO)_2]$ ($Cp' = Cp$, $\eta\text{-}C_5Me_5$ or η-indenyl) by a variety of phosphines is first order in [Ti] and zero order in the entering nucleophile,[27] indicative of a dissociative mechanism and in agreement with the isotopic labelling[25] and matrix isolation[26] studies discussed above. The equilibrium constants for Equation (4) decrease in the order $L = PMe_3 > P(OEt)_3 \approx PMe_2Ph > PMePh_2 > P(OPr^i)_3 > PBu^n_3 > PPh_3$; in all cases CO as the better π-acceptor is strongly preferred.[27]

$$TiCp_2(CO)_2 + L \ \rightleftharpoons\ TiCp_2(CO)(L) + CO \tag{4}$$

Whereas the photolysis of $[TiCp_2(CO)_2]$ does not generate 'TiCp$_2$', this intermediate is apparently formed during the irradiation of $[Ti(N_3)_2Cp_2]$ deposited in a CO-doped argon matrix. The azide complex has absorption at $\lambda = 291$ and $450\ nm$ due to ligand-to-metal charge-transfer bands, and irradiation near these wavelengths leads to the formation of $[TiCp_2(CO)]$ and $[TiCp_2(CO)_2]$, evidently via $[TiCp_2]$, together with larger amounts of $[Ti(NCO)_2Cp_2]$. Evidence for $[TiCp_2(N_2)_2]$ was also obtained ($\nu_{NN} = 2140$ and $2080\ cm^{-1}$). By contrast, the photolysis of $[Ti(N_3)_2Cp_2]$ in solution under 50 bar CO gives $[Ti(NCO)_2Cp_2]$ but no $[Ti(CO)_2Cp_2]$. No evidence was found for Cp ring slippage and the formation of $[Ti(\eta^3\text{-}Cp)(\eta^5\text{-}Cp)(CO)_3]$ in a matrix of pure CO.[26]

The reaction of $[TiCpCp^*(CO)_2]$ with diphenylacetylene gives $[TiCpCp^*(CO)(C_2Ph_2)]$; the compound is unstable and could not be isolated.[16] The reaction of $[TiCp_2(CO)_2]$ with the tungsten carbyne complex $[W(\equiv CR)Cp(CO)_2]$ ($R = p$-tolyl) as a 'heteroalkyne' gives the carbyne-bridged bimetallic complex (**14**) (Equation (5)); the crystal structure confirmed the $\sigma{:}\eta^2$ bonding of the bridging CO ligand.[30] The same product is obtained by the reduction of $[TiCl_2Cp_2]$ with magnesium amalgam in the presence of $[W(\equiv CR)Cp(CO)_2]$.[31]

$$\tag{5}$$

Whereas the reactions discussed above have resulted in exchange of the CO ligand, a cyclopentadienyl ligand of $[TiCp_2(CO)_2]$ is displaced if the complex is reduced with sodium naphthalenide in the presence of dmpe at $-70\,°C$. The resulting anion $[TiCp(CO)_2(dmpe)]^-$ is protonated by acetic acid to give $[Ti(H)Cp(CO)_2(dmpe)]$, the first carbonyl hydride of titanium, as deep violet crystals in 15% yield. The ν_{CO} frequencies of the hydride complex (1933 and $1855\ cm^{-1}$) are comparable with those of $[TiCp_2(CO)_2]$. X-ray diffraction gave a terminal Ti–H distance of $0.174(6)\ nm$.[32]

4.1.3.4 Oxidative reactions with formation of Ti–O bonds

The hydrolysis of $[TiCp_2(CO)_2]$ in toluene at $80\,°C$ results in the loss of one cyclopentadienyl ligand per titanium to give $[(TiCp)_6(\mu\text{-}O)_8]$ (Equation (6)). The analogous reaction with zirconium gives intractable zirconium oxides. By contrast, $[TiCp^*_2(CO)_2]$ is not attacked by water.[33]

$$6\,[TiCp_2(CO)_2] + 8\,H_2O \xrightarrow[\text{toluene}]{80\,^\circ C} [Ti_6Cp_6(\mu\text{-}O)_8] + 5\,H_2 + 12\,CO + 6\,C_5H_6 \qquad (6)$$

The reaction of $[Ti(\eta\text{-}C_5Me_4R)_2(CO)_2]$ (R = Me or Et) with arsenic trioxide in refluxing xylene gives (**15**) as yellow crystals in 36–44% yield. The adamantoid structure of the $TiAs_3O_6$ cage was confirmed by an x-ray structure (R = Et).[34] Unsaturated dicarboxylic acids are reduced to saturated acids in reaction with $[TiCp_2(CO)_2]$, with formation of titanium(III) carboxylates; for example, maleic acid gives (**16**).[35] Azo compounds such as RC(O)N=NC(O)R react with $[TiCp_2(CO)_2]$ to give the chelate complex (**17**). This product is stable if R = *p*-tolyl; esters (R = OEt, OBu^t or MeC_6H_4O) lead to a further reaction with $[TiCp_2(CO)_2]$ to give a mixture of $[Ti(NCO)_2Cp_2]$, $[Ti(NCO)(OR)Cp_2]$ and $[Ti(OR)_2Cp_2]$. R_2CN_2 reacts with $[TiCp_2(CO)_2]$ under cleavage of the N=N bond and formation of the red-purple ketiminato complex $[Ti(NCO)(N=CR_2)Cp_2]$ (R_2 = Me_2 or $(CH_2)_5$).[36]

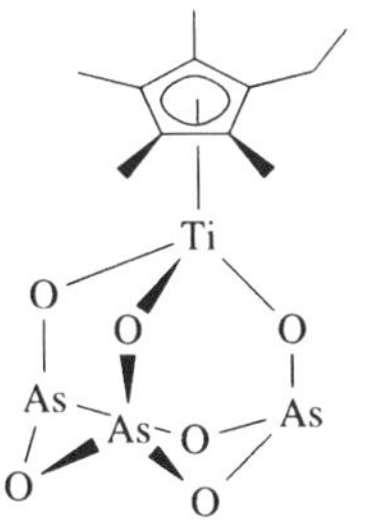

(**15**) 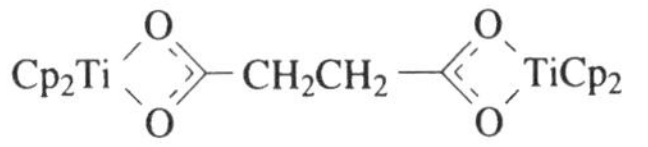　　　　(**16**) 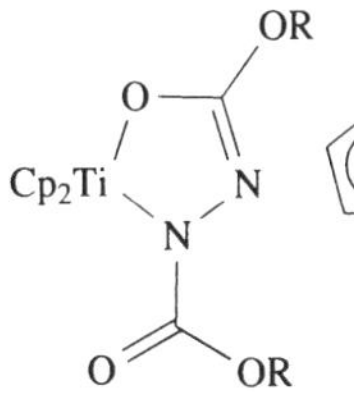　　　　(**17**) 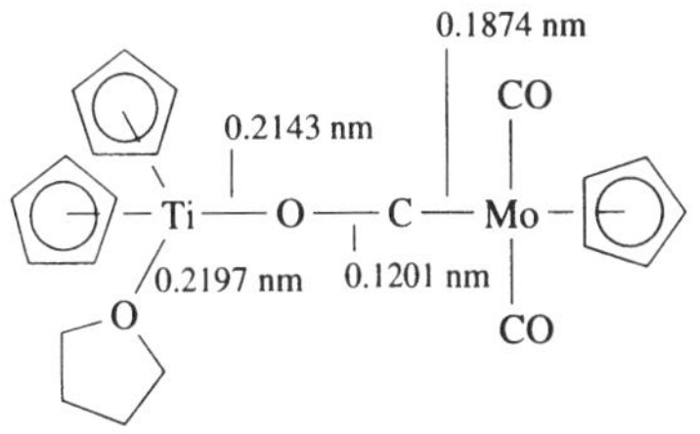　　　　(**18**)

Titanocene carbonyl compounds react with reducible metal carbonyls under oxidation to titanium(III) or titanium(IV) and formation of isocarbonyl complexes; that is, coordination from the oxygen of coordinated CO ligands. Thus, $[TiCp_2(CO)_2]$ and $[\{MoCp(CO)_2\}_2]$ in THF give (**18**) as bright green crystals whose structure was determined. Three ν_{CO} bands are seen in the IR spectrum at 1920, 1830 and 1650 cm^{-1}, the latter being indicative of the isocarbonyl moiety.[37] The analogous reaction with $[Co_2(CO)_8]$ in hexane leads to black crystalline $[Ti^{IV}Cp_2\{OCCo_3(CO)_9\}_2]$. $[TiCp^*_2(CO)_2]$ and $[Co_2(CO)_8]$ in hexane form $[Ti^{III}Cp^*_2\{Co(CO)_4\}]$ (**19**), while the use of toluene as the solvent results in $[Ti^{IV}Cp^*_2\{(\mu\text{-}OC)Co(CO)_3\}_2]$.[38] In a noncoordinating solvent such as toluene $[TiCp_2(CO)_2]$ reacts with $[\{Mo(\eta\text{-}C_5H_4Me)(CO)_2\}_2]$ to give (**20**) as green crystals in 72% yield (ν_{CO} = 1920, 1755 and 1710 cm^{-1}). The Ti–O–C moieties are markedly bent (141.2–143.4°). In THF (**20**) in converted to (**18**). Exposure of (**20**) to hydrogen leads to $[Ti(H)Cp_2]_x$ and $[Mo(H)Cp(CO)_3]$.[39]

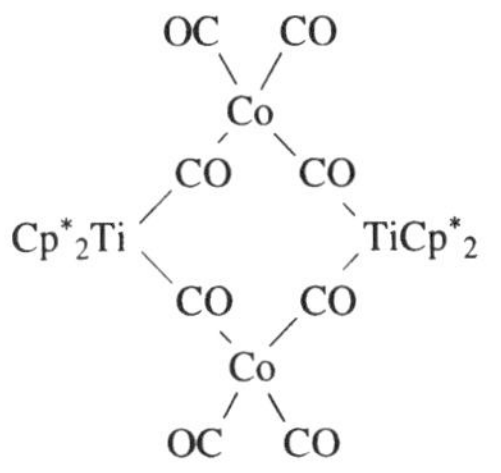

(**19**) 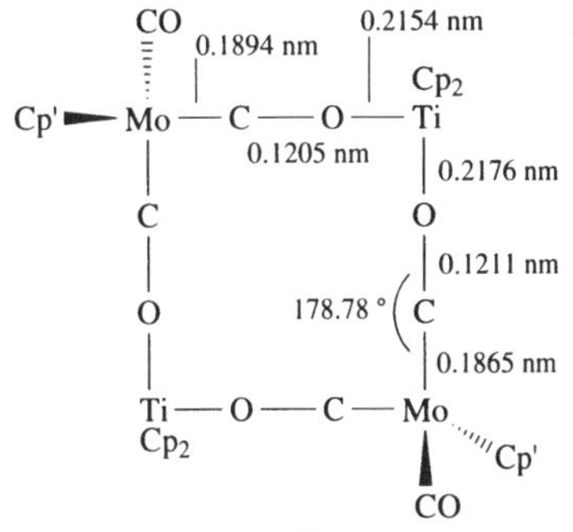　　　　(**20**)

Sulfoxides $R^1R^2S(O)$ are deoxygenated by $[TiCp_2(CO)_2]$ to give R^1R^2S and uncharacterized titanium oxo species, with partial loss of Cp.[40]

4.1.3.5 Oxidative reactions with formation of Ti–S and Ti–Se bonds

The reaction of $[TiCp_2(CO)_2]$ with H_2Se is very dependent on the reaction conditions. In hexane at 0 °C $[Ti_2Cp_4(\mu\text{-}Se)_2]$ is formed as a purple solid in 49% yield; reaction at 20 °C gives (**21**), and heating to 80 °C gives mainly the cluster $[Ti_6Cp_6(\mu\text{-}Se)_8]$ (Scheme 3). Treatment of $[TiCp^*_2(CO)_2]$ with H_2Se in a Ti:Se ratio of 1:4 leads to $[TiCp^*_2(SeH)_2]$, while lower hydrogen selenide concentrations also give $[TiCp^*_2(SeH)_2]$ together with $[Ti_2Cp^*_4(\mu\text{-}Se)_2]$.[41]

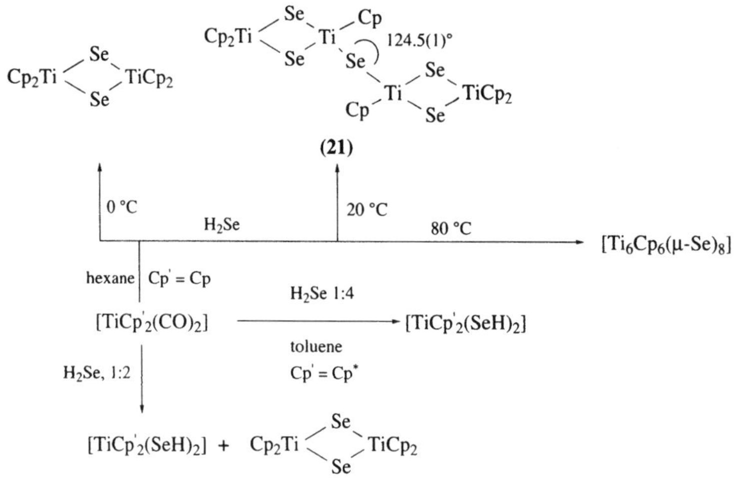

Scheme 3

Trisulfides RSSSR (R = *p*-tol, Pr^i, Bu^t or CH$_2$Ph) in toluene at $-20\,°C$ oxidatively add to [TiCp$_2$(CO)$_2$] to give [TiCp$_2$(SR)(SSR)]. The compound is desulfurized by PPh$_3$ to give Ph$_3$PS and [TiCp$_2$(SR)$_2$]; the latter is also produced from [TiCp$_2$(CO)$_2$] and RSSSR at 110 °C. The oxidative addition of Ph$_3$CSSX (X = Cl or phthalimide) gives [TiCp$_2$X(SSCPh$_3$)].[42]

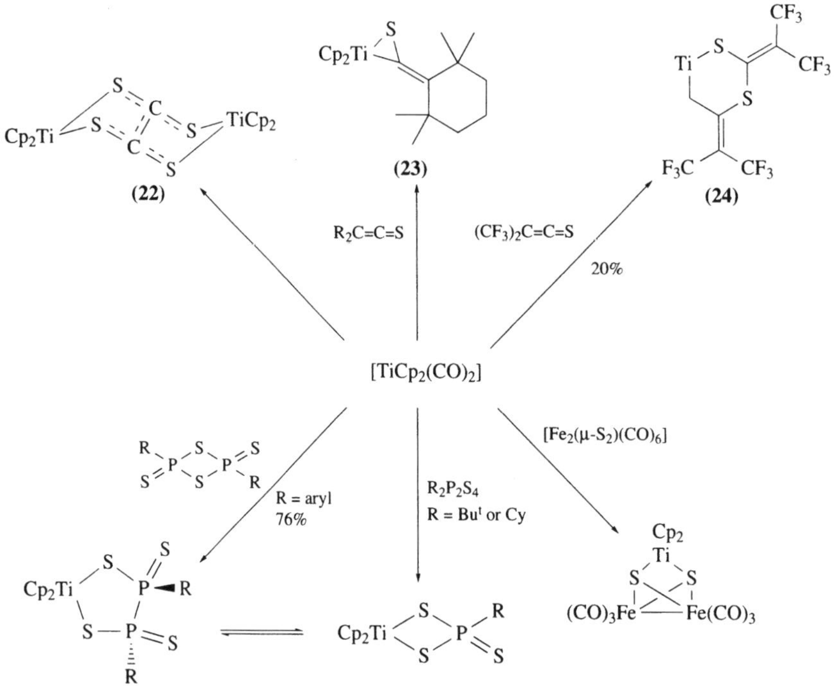

Scheme 4

Carbon disulfide is reductively coupled by [TiCp$_2$(CO)$_2$] to give the dark purple tetrathiooxalato complex (**22**) in 80% yield. Unlike the oxalato analogue the C–C bond possesses partial double-bond character in line with its better π-acceptor properties; that is, tetrathiooxalate acts as a [S$_2$C=CS$_2$]$^{4-}$ ligand rather than as [S$_2$C–CS$_2$]$^{2-}$. Hence (**22**) is best described as a diamagnetic titanium(IV) compound (Scheme 4).[43] The products of the reaction of thioketenes with [TiCp$_2$(CO)$_2$] depend on steric requirements: with bulky compounds, simple 1:1 adducts such as dark-green air-sensitive (**23**) is formed,[44] while (CF$_3$)$_2$C=C=S gives a more stable metallacycle incorporating two C–S moieties (**24**).[45] The crystal structures of both complexes were determined. The oxidation of [TiCp'$_2$(CO)$_2$] (Cp' = Cp or η-C$_5$H$_4$Me) with R$_2$P$_2$S$_4$ (R = *p*-C$_6$H$_4$OMe or *p*-C$_6$H$_4$OEt) gives red-purple air-sensitive (**25**). The complex is also obtained by treating R$_2$P$_2$S$_4$ first with Li[BHEt$_3$] followed by [TiCl$_2$Cp$_2$]. R$_2$P$_2$S$_4$

($R = Bu^t$ or cyclohexenyl), however, gives $[RPS]_n$ and (**26**). A more general route to these thiophosphoryl complexes is the reaction of $[TiCl_2Cp_2]$ with $Li_2[S_3PR]$. The complex $[Ti(\eta\text{-}C_5H_4Me)_2(\mu\text{-}S)_2P(S)(C_6H_4OMe)]$ was crystallographically characterized. The TiS_2 and PS_2 planes form a dihedral angle of 167.1°, rather wider than the comparable dihedral angle (131°) in the related complex $[TiCp^*_2S_3]$. Phosphorus NMR spectral data suggest that in solution compounds of type (**25**) and (**26**) are in equilibrium with each other ($R = $ anisyl, $K_{eq} = 0.22$) (Scheme 4).[46]

Titanium-containing sulfur–nitrogen heterocycles have been synthesized from $[TiCp_2(CO)_2]$ and S_4N_4 (Equation (7)), to give (**27**) and (**28**), the structures of which were determined.[47]

$$TiCp_2(CO)_2 + \text{(structure)} \longrightarrow \text{(27)} + \text{(28)} \tag{7}$$

The S–S bond of the dinuclear iron carbonyl $[Fe_2(\mu\text{-}S_2)(CO)_6]$ is cleaved by $[TiCp_2(CO)_2]$ to give a product for which structure (**29**) has been suggested (Scheme 4).[48]

4.1.3.6 Oxidative reactions with formation of Ti–N and Ti–P bonds

Azo compounds react with $[TiCp_2(CO)_2]$ effectively with oxidation of the metal centre to titanium(IV). Thus azoarenes RN=NR ($R = Ph$ or p-Tol; or $R_2N_2 = $ benzo[c]cinnoline) give (**30**); the phenyl derivative was structurally characterized. The Ti–N bonds are relatively short, while the N–N bond has been significantly elongated compared with uncoordinated azobenzene. *Ab initio* MO calculations support the notion that the metal centre has been oxidized.[49] Whereas diazoalkanes with ester substituents such as $(EtO_2C)_2CN_2$ react with $[TiCp_2(CO)_2]$ under CO displacement to give (**31**), Ph_2CN_2 leads to a coupling between CO and the diazoalkane to give (**32**). Oxidation of (**32**) with quinones leads to removal of the central titanium centre in the trimetallic unit and formation of (**33**) (Equation (8)).[50]

$$\text{(30)} \qquad \text{(31)}$$

$$\text{(32)} \qquad \longrightarrow \qquad \text{(33)} \tag{8}$$

The reduction of nitrogen-containing heterocycles with $[TiCp_2(CO)_2]$ in THF proceeds via radical intermediates which can dimerize or abstract hydrogen from the solvent.[51] 2-Vinylpyridine undergoes cyclometallation, with partial hydrogenation of the vinyl group (Scheme 5).[52]

The reaction of $[TiCp^*_2(CO)_2]$ with yellow phosphorus in refluxing xylene gives the red-brown dinuclear phosphido cluster $[Ti_2Cp^*_2(\mu\text{-}P_6)]$ (**34**). Unlike $[Mo_2Cp^*_2(\mu\text{-}P_6)]$, which possesses a tripledecker structure with a planar P_6 ring, (**34**) is a distorted cube. The compound is irreversibly

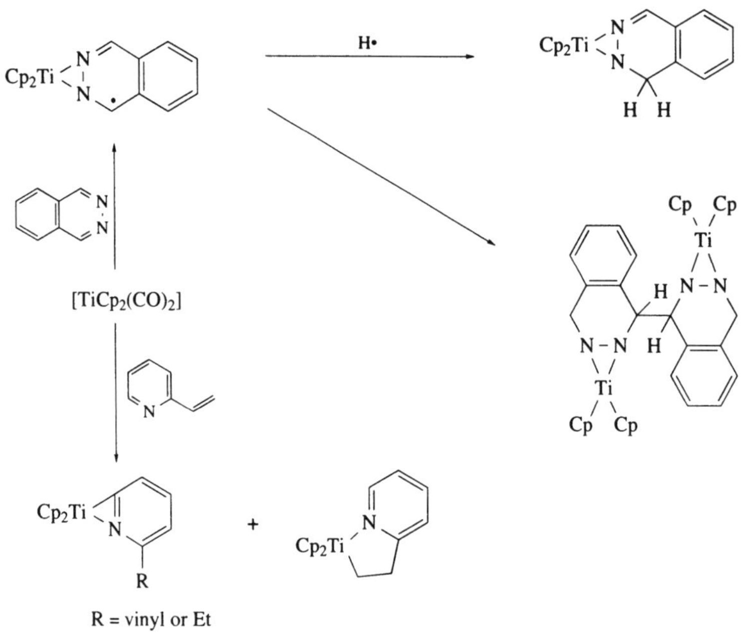

Scheme 5

oxidized ($E_{pa} = 0.70$ V vs. SCE) and reversibly reduced ($E_{red} = -1.31$ V) to give a titanium(III) anion the EPR spectrum of which shows a small coupling (3.8×10^{-4} T) of the unpaired electron to all six phosphorus atoms (Equation (9)).[53]

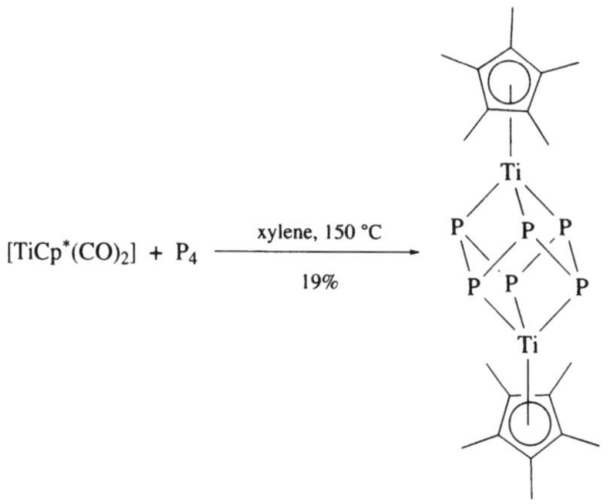

$$[TiCp^*(CO)_2] + P_4 \xrightarrow[\text{19\%}]{\text{xylene, 150 °C}} \quad \textbf{(34)} \qquad (9)$$

4.1.4 Phosphine Complexes of Titanium(II)

4.1.4.1 Synthesis

[TiCp$_2$(PMe$_3$)$_2$] is prepared in high yield by reduction of [TiCl$_2$Cp$_2$] with magnesium in THF in the presence of PMe$_3$. It has proved to be a very versatile starting material,[54,55] and a detailed synthetic procedure has been published.[56] The reaction of [TiCl$_2$(dmpe)$_2$] with NaCp' gives the analogous complex [TiCp'$_2$(dmpe)] as dark-orange diamagnetic crystals (Cp' = Cp or η-C$_5$H$_4$Me). The molecular structure of the C$_5$H$_4$Me complex has been determined.[57]

4.1.4.2 Reactions

The PMe_3 ligands in $[TiCp_2(PMe_3)_2]$ are easily displaced under milder conditions than from $[TiCp_2(CO)_2]$. Both ligand exchange and oxidative addition reactions are common (Scheme 6). Reaction with CO gives, successively, $[TiCp_2(PR_3)(CO)]$ and $[TiCp_2(CO)_2]$; similarly, one equivalent of $RC{\equiv}CR$ (R = H or Ph) displaces one phosphine to give $[TiCp_2(PMe_3)(C_2R_2)]$ (**35**), while an excess of alkyne leads to the well-known metallacyclopentadiene (**36**). Acetonitrile substitutes both phosphines to give $[TiCp_2(NCMe)_2]$.[54] Electrophiles such as diphenyl disulfide, iodomethane or acetyl chloride give the titanium(IV) complexes $[TiCp_2(SPh)_2]$, $[TiCp_2(Me)(I)]$ or $[TiCp_2(COMe)(Cl)]$, respectively, while the reaction with isocyanides gives access to $[TiCp_2(CNR)_2]$ (R = Me or Bu^i).[55] Carbon dioxide reacts with $[TiCp_2(PMe_3)_2]$ to give (**37**) as a yellow solid in high yield. The CO_2 ligand is η^2-bonded (Scheme 6). The tan-coloured CS_2 complex has an analogous structure. The CO_2 complex decomposes in solution above 0 °C to give, *inter alia*, $[TiCp_2(CO)_2]$ and $[TiCp_2(CO)(PMe_3)_2]$.[58] The IR spectrum of (**37**) has been investigated with the aid of ^{13}C and ^{18}O isotopic labelling and the assignment confirmed by a normal-coordinate analysis. The C–O stretching frequencies are found at 1671 and 1187 cm^{-1}.[59] By contrast, diphenyldiazomethane gives a complex with an η^1 coordinated diazo ligand (**38**) (Scheme 6) the structure of which was determined by x-ray diffraction. The short Ti–N distance (0.1829(11) nm) indicates partial Ti–N double-bond character, and the Ti–N–N and N–N–C fragments are significantly bent, with angles of 156.8(9)° and 126.0(9)°, respectively.[60]

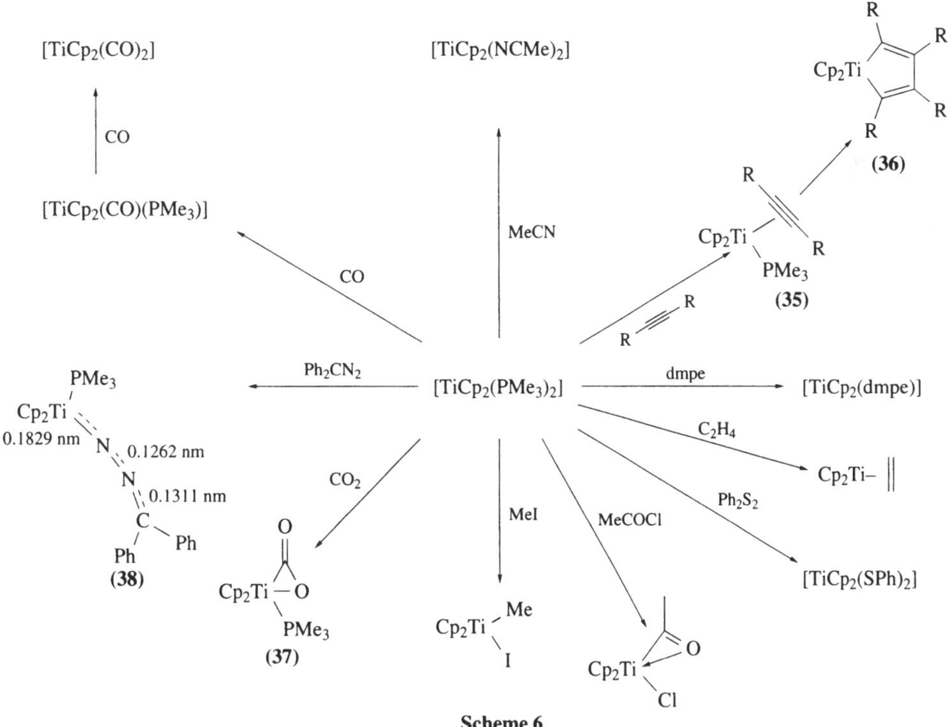

Scheme 6

Even dinitrogen is capable of displacing a PMe_3 ligand from $[TiCp_2(PMe_3)_2]$. Under 1 bar N_2, the solution equilibrium contains 30% $[Ti_2Cp_4(PMe_3)_2(\mu\text{-}N_2)]$ (**39**), the crystal structure of which was determined. The Ti–N bond is longer and the N–N bond shorter than in (**38**), indicating less backbonding to dinitrogen. At room temperature (**39**) is slowly converted to the titanium(III) complex (**40**) (Scheme 7).[61]

The interaction of ethene with $[TiCp_2(PMe_3)_2]$ leads to the displacement of both PMe_3 ligands and the formation of $[TiCp_2(\eta^2\text{-}C_2H_4)]$, which could not be isolated but was characterized by ^1H and ^{13}C NMR and mass spectra. Under ethene pressure (>15 bar) $[TiCp_2(\eta^2\text{-}C_2H_4)]$ catalyses the dimerization of ethene to give a 1:1 mixture of 1- and 2-butene. A likely mechanism for this reaction involves the addition of a second ethene molecule to give a titanacyclopentane which decomposes with elimination of butene. As expected, $[TiCp_2(\eta^2\text{-}C_2H_4)]$ is highly reactive (see Section 4.1.5).[62]

Sterically strained alkenes such as cyclopropenes react with $[TiCp_2(PMe_3)_2]$ under ring opening to give vinylcarbene complexes. Thus, 2,2-dialkylcyclopropenes $C_3H_2R^2R^1$ (R^2 = Me or R^1 = Ph; $R^2 = R^1$ = Ph) give the vinylcarbene complexes (**41**). The carbene character of the ligand is indicated by

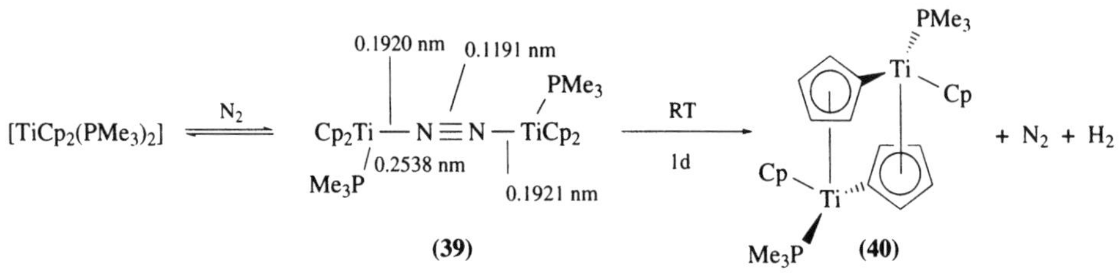

Scheme 7

the characteristically low ^{1}H NMR chemical shift of the hydrogen on the α-carbon atom (δ 12.5–12.9). For electronic reasons the carbene ligand is oriented perpendicular to the P–Ti–C plane. If $R^2 = R^1 = $ Me the reaction mixture at −30 °C also contains the alkene adduct (**42**) which on warming to 0 °C gives the coupled product (**43**).[63] A complex related to (**41**), the dark purple allenylidene compound (**44**), was obtained by an alternative method, the reaction of [TiCl$_2$Cp$_2$] with Li$_2$C=C=CPh$_2$ at 0 °C in the presence of PMe$_3$ in 70% yield.[64]

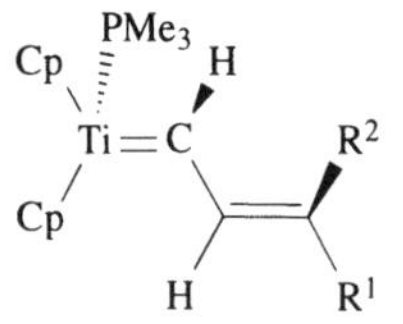

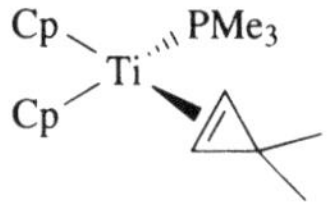

The reaction of [TiCp$_2$(PMe$_3$)$_2$] with methylenecyclopropane gives the corresponding red crystalline adduct (**45**) which undergoes an unusual rearrangement with elimination of 1,1-diphenylethene and 1,1-diphenyl-2-methylcyclopropane to give the dark green dinuclear complex (**46**) in good yield (Scheme 8). A more rational synthesis of (**46**) is the reduction of [Ti$_2$Cp$_4$Cl$_2$(μ-C$_2$)] with magnesium and PMe$_3$. The crystal structure of (**46**) revealed that the compound is best described as intermediate between a heteroallene and an alkynediyl formulation.[65] Alkene complexes are also obtained from [TiCp$_2$(PMe$_3$)$_2$] and 1,2-diphenylcyclopropene. The light-brown product slowly rearranges (faster on heating to 40 °C) to give the metallacyclobutene (**47**) (Scheme 9).[66]

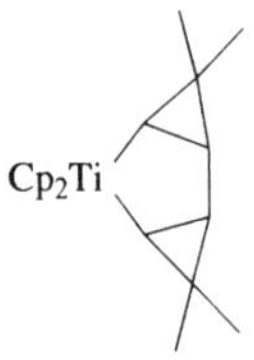

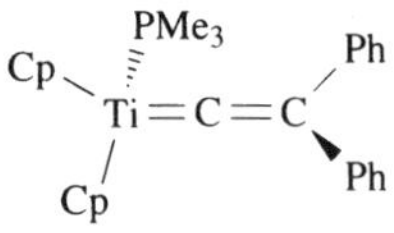

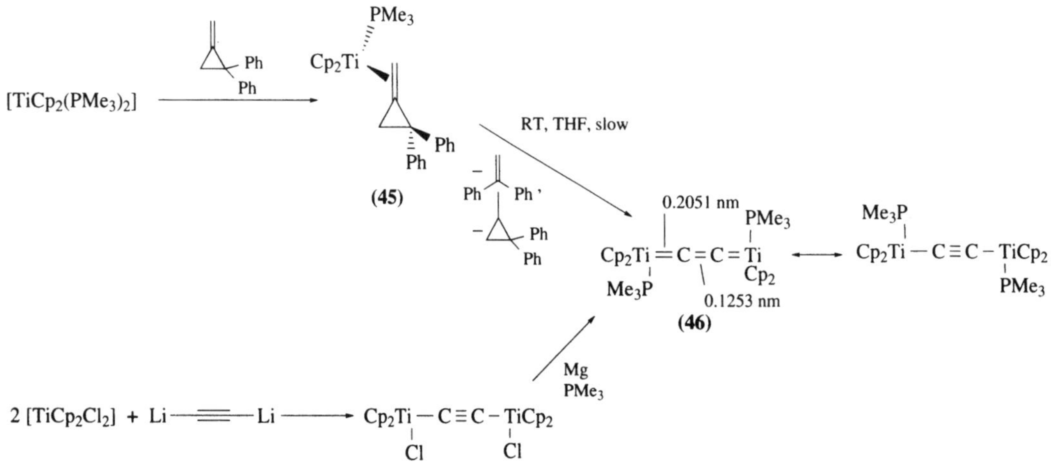

Scheme 8

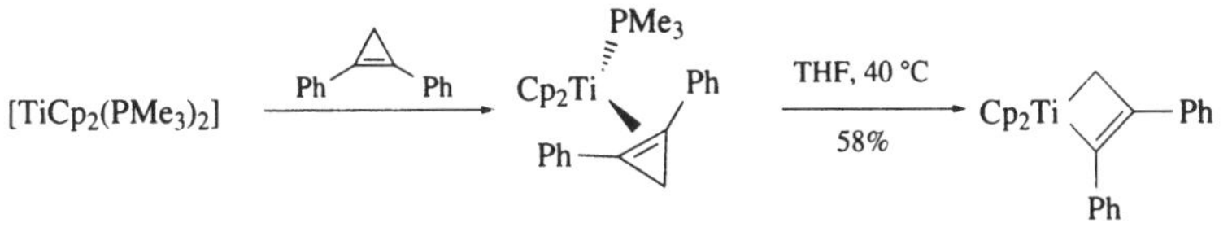

Scheme 9   (47)

The reaction of [TiCp$_2$(PMe$_3$)$_2$] with ethyne in a ratio of 1:1 gives purple solutions of TiCp$_2$(HC≡CH)(PMe$_3$) (**48**). The complex could not be isolated but was characterized spectroscopically. There is extensive backbonding to the C$_2$H$_2$ ligand, indicated by the low ν_{CC} band at 1618 cm^{-1} and the absence of any evidence for the rotation of the ethyne ligand in the variable-temperature NMR spectrum. [TiCp$_2$(HC≡CH)(PMe$_3$)] reacts with HCl to give [TiCl$_2$Cp$_2$] and ethene. In the presence of an excess of ethyne the titanacyclopentadiene complex (**49**) is formed as a light-brown air- and light-sensitive solid, accompanied by the precipitation of black *trans*-polyacetylene. Other terminal and disubstituted alkynes give similar alkyne complexes or metallacyclic products (Scheme 10). By contrast, with C$_2$H$_2$ terminal alkynes are not polymerized.[67,68] The polyalkyne produced with (**48**) has similar spectroscopic and doping properties to material produced with [Ti(OBu)$_4$]/AlEt$_3$ catalysts.[69] The PMe$_3$ ligand in (**48**) is labile; the complex reacts readily with carbonyl compounds such as CO$_2$ or RC(O)Me (R = H or Me) and with ethene to give the coupled products (**50**)–(**52**), respectively.[70]

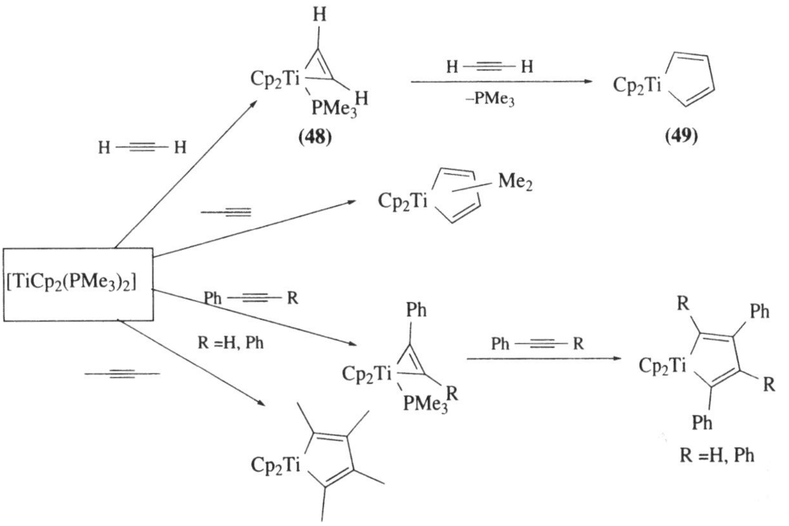

Scheme 10

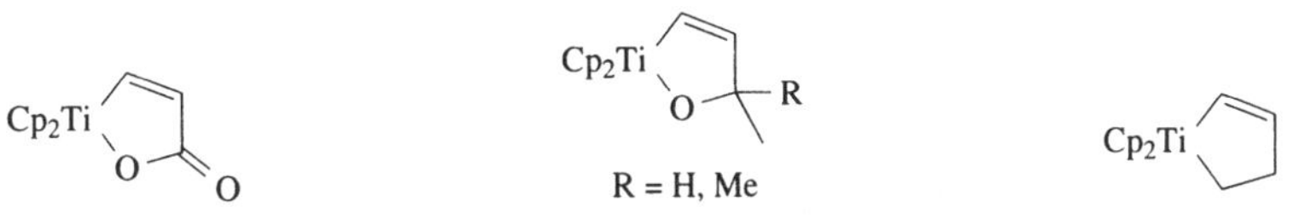

(50)   (51)   (52)

The alkyne ligand in (**48**) is readily protonated by HX (X = OMe, OEt, OPh, Cl or CF$_3$CO$_2$) to give the vinyl complexes [Ti(CH=CH$_2$)(X)Cp$_2$]. [TiCp$_2$(MeC≡CMe)] reacts similarly to give the corresponding methyl-substituted vinyl derivatives.[71,72] The trifluoroacetato complexes are unstable and react in solution with traces of water to give the oxo-bridged dimers [Ti$_2$Cp$_4$(CF$_3$CO$_2$)$_2$(μ-O)]. Similar binuclear complexes are obtained from (**48**) or [TiCp$_2$(MeC≡CMe)] and water; for example, (**53**), the structure of which shows a typical, near-linear Ti–O–Ti arrangement (angle 171.6(3)°) with short Ti–O bond distances.[72]

The ability of titanium phosphine complexes to induce C–C coupling reactions with alkenes, alkynes and carbonyl compounds has been applied to the synthesis of metallacycles (Scheme 11).[73]

Complexes analogous to (**48**) are produced from [TiCp$_2$(PMe$_3$)$_2$] and the phosphaalkyne Bu^{t}C≡P; for example, the crystallographically characterized complex (**54**) is obtained in 85% yield as dark-red crystals. The π-acceptor properties of the phosphaalkyne in (**54**) are evident from the elongation of the C–P bond (0.1636(2) nm) compared with that of the free ligand (0.1542 nm), and from the reduction

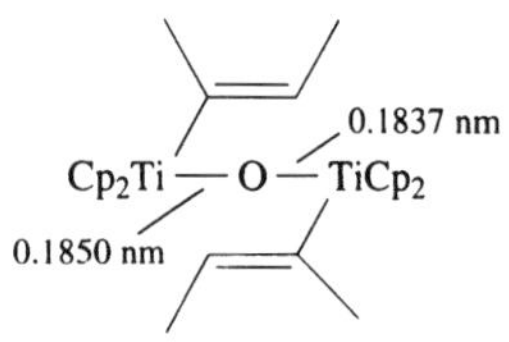

(53)

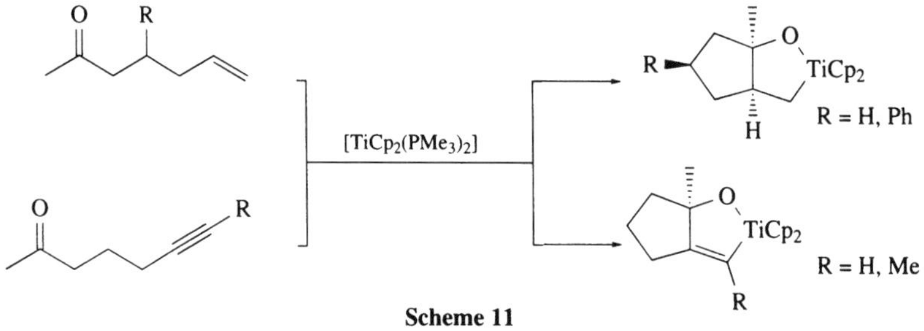

Scheme 11

of the P–C–C angle from 180° to 133.4(2)°. The PMe$_3$ ligand in **(54)** can be removed by the addition of one equivalent of BEt$_3$, to give the monomeric unstable compound [TiCp$_2$(Bu^{t}C≡P)] **(55)**. The structure of the stable dimer of **(53)** has been determined. As would be expected of ordinary alkyne complexes, **(54)** and **(55)** react with unsaturated substrates under metallacycle formation (Scheme 12).[74]

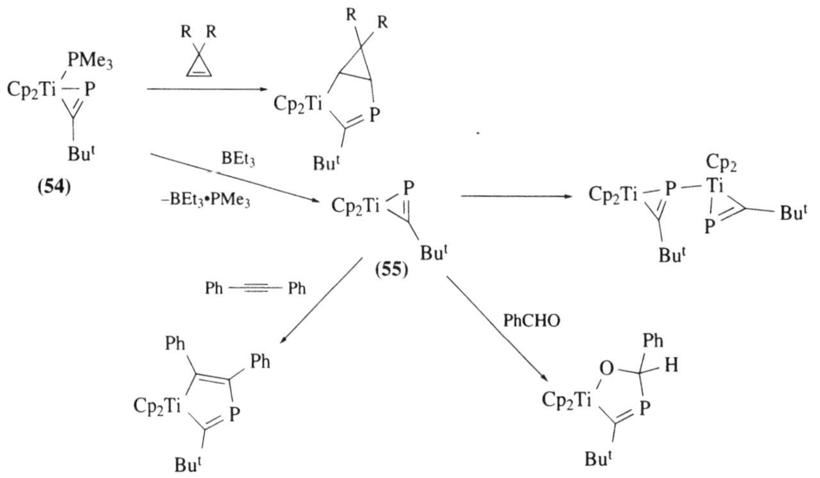

Scheme 12

4.1.5 Alkene Complexes of Titanium(II)

Bis(cyclopentadienyl)(ethene)titanium is prepared *in situ* from the reaction of [TiCp$_2$(PMe$_3$)$_2$] with ethene.[62] The same complex is generated from [TiCp$_2$Cl$_2$] with magnesium/dibromoethene in the presence of ethene.[75] As expected, it is highly reactive and reacts with unsaturated organic molecules, electrophiles and protic reagents in a manner similar to the alkyne trimethylphosphine complexes discussed in Sections 4.1.4 and 4.1.6. Some reactions are summarized in Scheme 13.[62,72,75]

Bis(pentamethylcyclopentadienyl)(ethene)titanium **(56)** has been prepared by reduction of [TiCl$_2$Cp*$_2$] with sodium amalgam in toluene at 25 °C under an ethene atmosphere, or by treatment of [{TiCp*$_2$}$_2$(μ-N$_2$)] with ethene. It forms lime-green crystals in 80% yield; its structure has been determined by x-ray crystallography.[76] The bonding of the ethene ligand is characterized by extensive backbonding. The C–C bond of 0.1438(5) nm is long compared with that of free ethene (0.1337 nm), and the hydrogen atoms are bent away from the metal so that the CH$_2$ plane forms an angle of 35° with the C–C axis (Figure 1). Under an excess of ethene **(56)** is in equilibrium with the red-orange metallacyclopentane complex [Ti(C$_4$H$_8$)Cp*$_2$] (Scheme 14).[76,77]

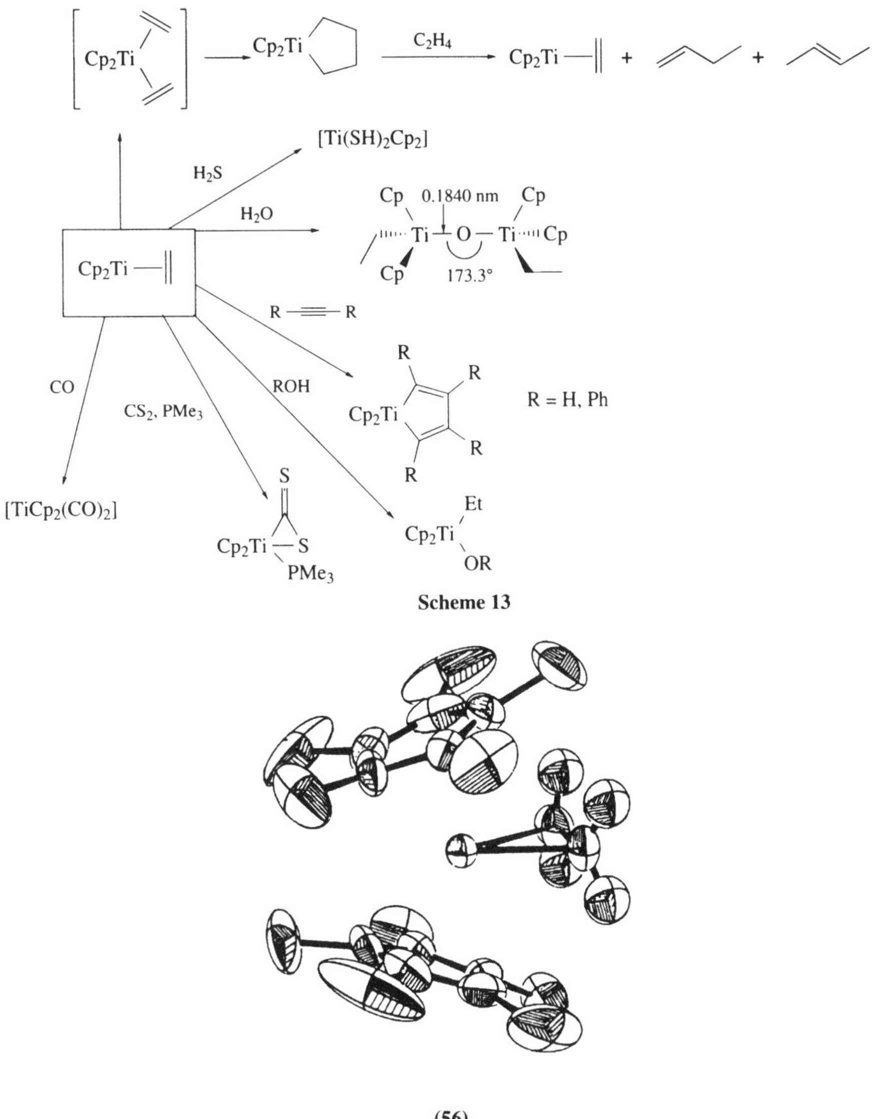

Scheme 13

(56)

Figure 1 Structure of [Ti(C$_2$H$_4$)Cp*$_2$] (56) (reproduced by permission of the American Chemical Society from *J. Am. Chem. Soc.*, 1983, **105**, 1136. Copyright (1983) American Chemical Society).

The reaction of (56) with nitriles RCN has been studied in detail. If R = Et, Bu^t or *p*-MeC$_6$H$_4$ the expected azametallacyclopentenes (57) are formed. The same product is obtained with acetonitrile at −50 °C (v_{CN} 1638 cm^{-1}). Above −10 °C, however, a 1,3-hydrogen shift takes place (Scheme 15). Treatment of (57) with ^{13}C$_2$H$_4$ shows that the C–C bond formation is reversible and leads to the incorporation of the ^{13}C label into the metallacyclic products. The 1,3-hydrogen shift is an intra- rather than an intermolecular reaction.[77]

Metal carbonyl complexes [M(CO)L$_n$] react with (56) to give the titanoxycarbenes (58) (ML$_n$ = Cr(CO)$_5$, Mo(CO)$_5$, W(CO)$_5$, Mn$_2$(CO)$_9$ or Re$_2$(CO)$_9$) (Scheme 15). The crystal structure of the rhenium complex has been determined. The five-membered ring is nonplanar.[78]

Methylenecyclopropenes oxidatively add to (56) to give the orange to red spirocyclic complexes (59) (R = H or Ph) in good yields. The phenyl derivative was structurally characterized. The compounds decompose in solution on heating to 95 °C to give open and cyclic organic products, derived from β-hydride and reductive elimination processes. Rapid heating to 200 °C, however, gives products indicative of the fast opening of the titanacyclopentane ring (Scheme 16).[79]

Titanocene alkene complexes are also thought to be involved in the isomerization of alkenes with [TiCl$_2$Cp$_2$]/Mg catalysts, although the nature of the active species in these mixtures has not been elucidated (Scheme 17). (*E*)-1-Phenyl-2-butene is the predominant initial product which is converted to

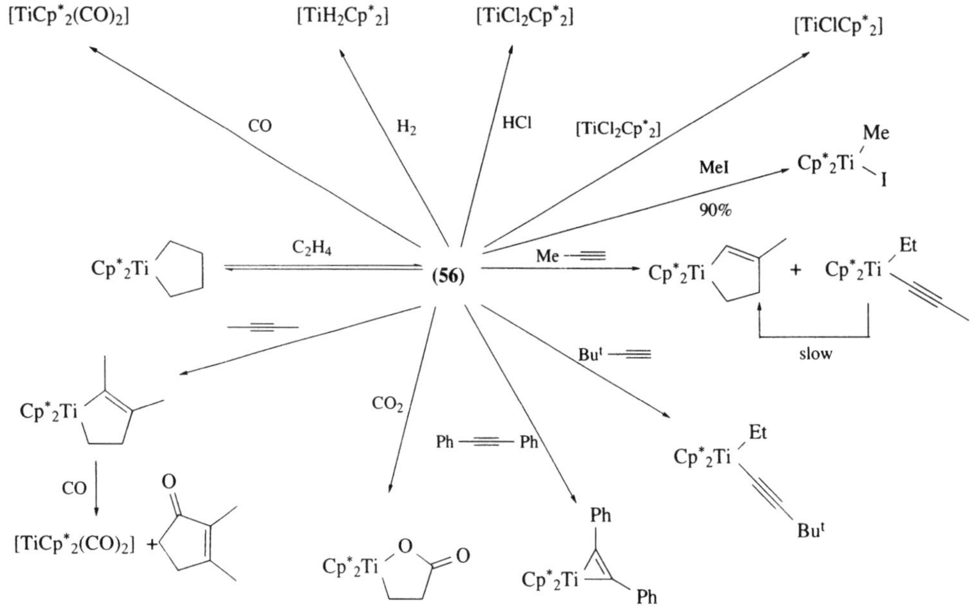

Scheme 14

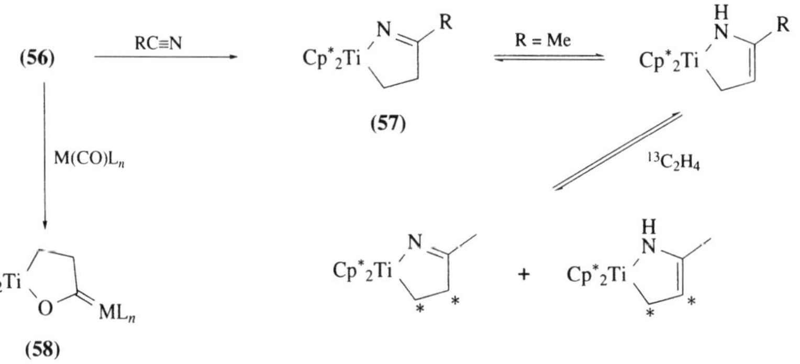

Scheme 15

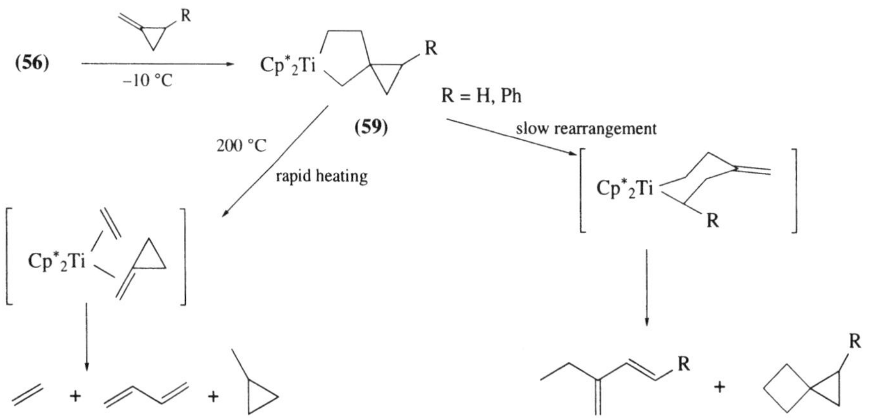

Scheme 16

the corresponding 1-phenyl-1-butene, while the (Z) isomer is isomerized only slowly. 2-Methyl-1-pentene is isomerized to 2-methyl-2-pentene. This reaction indicates that the active species in the $[TiCl_2Cp_2]/Mg$ catalyst differs from that in $[TiCl_2Cp_2]/Pr^iMgBr$ mixtures (which are generally thought to produce $[TiH_2Cp_2]$), as the latter is unable to catalyse this isomerization.[80] Similar complexes are presumably involved in the hydrogenation of alkenes and alkynes catalysed by $[TiCl_2Cp_2]/Mg$ in THF. The addition of PPh_3 facilitates the hydrogenation of 1-hexyne and of methyl oleate.[81] The reduction of

[TiCl$_2$Cp*$_2$] with sodium naphthalenide in the presence of 1-hexene gives complexes that catalyse the isomerization of the 1-alkene to an isomeric mixture of internal alkenes. A mechanism involving alkene complexes of TiCp*$_2$ has been suggested.[82] A diene complex of titanium(II) appears to be present in the dark-green complex [TiCp$_2$(anthracene)] which results from the reduction of [TiCl$_2$Cp$_2$] with magnesium in the presence of anthracene. Whereas magnesium and aluminum anthracene compounds contain two metal–carbon σ-bonds, the ^{13}C NMR spectrum of the titanium complex is best understood in terms of η^4-coordination to the central arene ring.[83]

Scheme 17

4.1.6 Alkyne Complexes of Titanium(II)

4.1.6.1 Synthesis and structure

As excellent π-acceptors, alkynes readily react with electron-rich titanium(II) complexes such as [TiCp$_2$(PMe$_3$)$_2$]. 2-Butyne displaces both trimethylphosphine ligands to give [TiCp$_2$(C$_2$Me$_2$)] (**60**) as a thermally unstable yellow-green solid. In solution in the presence of PMe$_3$ the compound is probably in equilibrium with [TiCp$_2$(C$_2$Me$_2$)(PMe$_3$)].[84] The displacement of two phosphines by one alkyne ligand suggests that the alkyne can act as a four- as well as a two-electron ligand. While the shift in the $v_{C≡C}$ frequency on coordination of the alkyne proved to be rather insensitive to the coordination mode of the alkyne, the ^{13}C NMR chemical shift is apparently a better guide, and as a general rule $\delta_{C≡C}$ values of *ca* 100–150 are taken to indicate that the alkyne acts as a two-electron donor, for example towards nickel, whereas alkynes as four-electron ligands give δ values of 190–270 ppm, typically if coordinated to Ti, V, W, etc. In agreement with this differing π-acceptor behaviour, the chemical shift differences between free and coordinated disubstituted alkynes C$_2$R$_2$ are much higher for early transition metals ($\Delta\delta_{C≡C}$ = 106–145 ppm) than for nickel ($\Delta\delta_{C≡C}$ = 38–55 ppm).[85] The complex [TiCp$_2$(C$_2$Ph$_2$)] has been isolated by the reduction of [TiCl$_2$Cp$_2$] with magnesium in the presence of diphenylacetylene in hexane at −45 °C as a brown microcrystalline solid which decomposes at 177–119 °C. The extensive backbonding contribution to the bonding of the alkyne in this complex is confirmed by the low-field ^{13}C chemical shift ($\delta_{C≡C}$ = 196 ppm) as well as the low $v_{C≡C}$ IR frequency of 1712 cm^{-1} (511 cm^{-1} lower than NMR for uncoordinated C$_2$Ph$_2$), which has led to the formulation of a metallacyclopropenylium resonance structure.[86]

The reaction of [TiCpCp*(CO)$_2$] with diphenylacetylene in the presence of PMe$_3$ gives the red-brown [TiCpCp*(C$_2$Ph$_2$)], while the addition of one equivalent of C$_2$Ph$_2$ in the absence of PMe$_3$ gives the unstable compound [TiCpCp*(CO)(C$_2$Ph$_2$)] which has been identified by its IR spectrum.[16] An alkyne complex without cyclopentadienyl ligands is generated by reducing [TiCl$_2$(oep)] (oep = octaethyl-porphyrinato) with LAH in the presence of C$_2$Ph$_2$ (Equation (10)). The product, the ruby-red diamagnetic complex [Ti(oep)(C$_2$Ph$_2$)] has been characterized crystallographically. The C≡C bond is elongated (0.130(1) nm), with a C–C–Ph angle of 142.9(6)°. This and the $v_{C≡C}$ frequency of 1701 cm^{-1} have again led to the suggestion that the alkyne acts as a four-electron ligand, although the rotation of the alkyne was found to be fast. The tetrakis(*p*-tolyl)porphyrinato analogue was also prepared.[87]

4.1.6.2 Reactions of alkyne complexes

The reactivity pattern of alkyne complexes resembles that of [TiCp*$_2$(C$_2$H$_4$)] (**56**), and formation of metallacycles is prevalent. Some reactions are shown in Scheme 18.[16,84,86,88–90] The metallacycle (**60**) is surprisingly stable and can obtained without prior isolation of the titanium alkyne complex.[88] Ketones

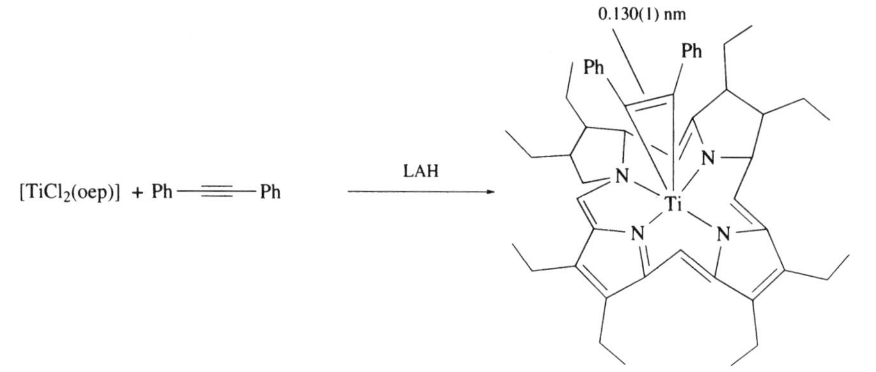

$$[TiCl_2(oep)] + Ph{-}{\equiv}{-}Ph \xrightarrow{\text{LAH}}$$

(10)

form red, air-stable, generally high melting compounds of type (**61**). The reaction product with acetone ($R^1 = R^2 = Me$) has been characterized by x-ray diffraction; it contains a rather short Ti–O bond of 0.1830(2) nm which indicates partial Ti–O double-bond character, a feature that no doubt contributes to the stability of these metallacycles.[90]

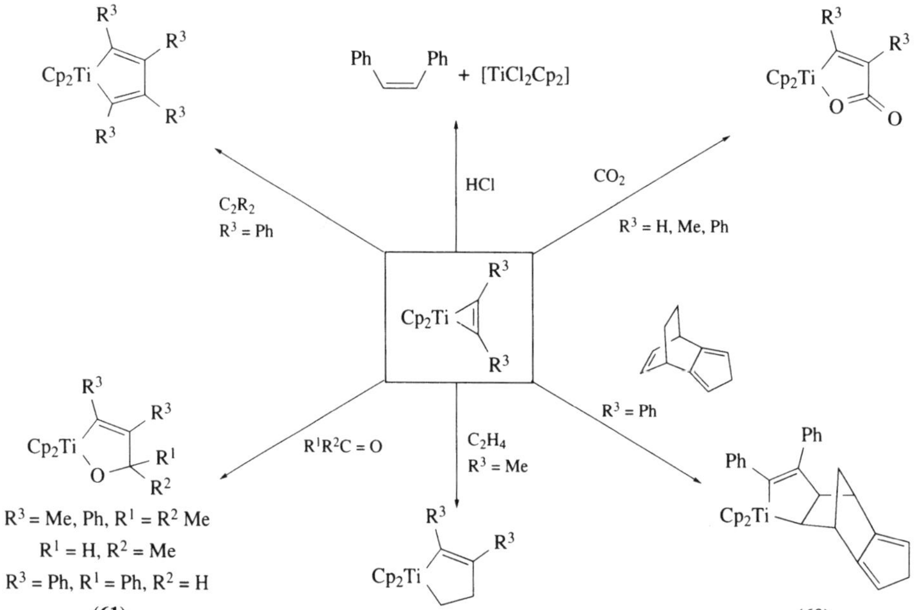

Scheme 18

An unusual metallacycle is obtained from the reaction of $[TiCp^*_2(C_2Ph_2)]$ with N_2O. There is no coordination to oxygen in this case since the formation of a five-membered ring is preferred (Equation (11)). The structure was confirmed by x-ray diffraction.[91]

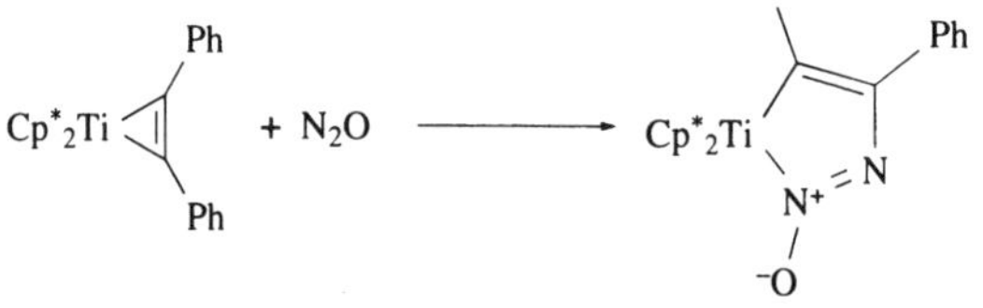

(11)

The complex $[TiCp_2(C_2Ph_2)]$ is a good hydrogenation catalyst. In the presence of dihydrogen (1 bar) at room temperature and substrate:titanium ratios of 40:1 to 100:1 diphenylacetylene is converted (100%) to 1,2-diphenylethane within minutes; stilbene, 1,4-diphenylbutadiene or cyclohexene rapidly give the corresponding alkanes, while 3-hexyne reacts more slowly to give hexane. The reaction conditions are much smoother than is required for hydrogenation reactions with $[TiCp_2(CO)_2]$.[92] The hydrogenation of diphenylacetylene is inhibited by CO and retarded by excess PMe_3 so that the reduction of the C≡C bond can be arrested at the C=C stage. The activity of the catalysts decreases in the order [TiCpCp*(C$_2$Ph$_2$)] ≫ [Ti(C$_5$H$_4$Me)$_2$(C$_2$Ph$_2$)(PMe$_3$)] > [TiCp$_2$(C$_2$Ph$_2$)(PMe$_3$)].[93]

The reduction of [TiCl$_2$Cp$_2$] with magnesium in the presence of 1,2-C$_6$H$_4$FBr generates an unstable titanium benzyne complex which behaves in a similar fashion to alkyne complexes; for example, with diphenylacetylene a dark-green titanaindene complex is formed.[94] The same benzyne intermediate is also thought to be formed during the thermolysis of [TiPh$_2$Cp$_2$] at 80 °C and can be trapped with Bu^{t}C≡P (Scheme 19).[95]

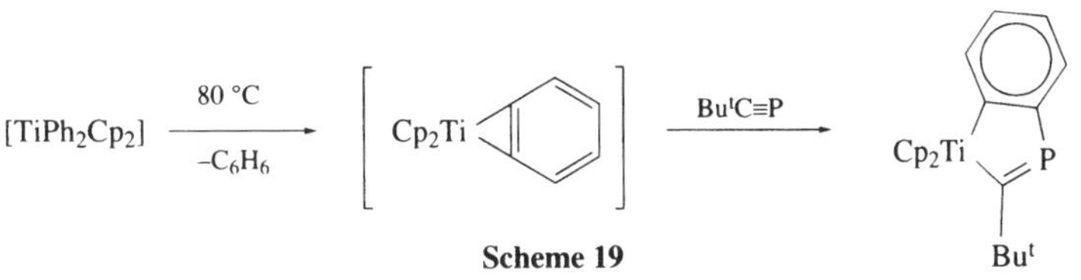

Scheme 19

4.1.7 Oxidation Reactions of Titanium(II)

The oxidation of titanium(II) to titanium(IV) complexes is the driving force in many of the reactions described above, and in the case of alkene and alkyne complexes leads C–C coupling and the formation of metallacycles; these reactions are discussed in Sections 4.1.5 and 4.1.6. This section contains reactions which do not fall into these categories.

[TiCp$_2$(bipy)] (bipy = 2,2'-bipyridyl) is reversibly oxidized electrochemically in a one-electron step on the cyclic voltammetry timescale, $E_{1/2}$ = 0.66 V vs. a platinum wire reference electrode in THF. On longer timescales chemical reactions follow the oxidation. By contrast, three successive reversible reduction steps have been observed.[96]

The oxidation of TiCp*$_2$ with N$_2$O in toluene at 0 °C gives the crystallographically characterized oxygen-bridged dimer (**62**) as lime-green crystals. The same compound is obtained from [TiCp*$_2$(N$_2$)].[97] By contrast, the oxidation of TiCp*$_2$, or better [TiCp*$_2$(C$_2$H$_4$)], by N$_2$O in the presence of a Lewis base such as pyridine gives the diamagnetic orange complexes (**63**). The Ti–O stretching frequency is observed at 852 cm^{-1} which shifts to 818 cm^{-1} on labelling with ^{18}O. The crystal structure confirms the presence of a terminal, very short Ti=O bond of 0.1665(3) nm with substantial multiple-bond character (Scheme 20).[98]

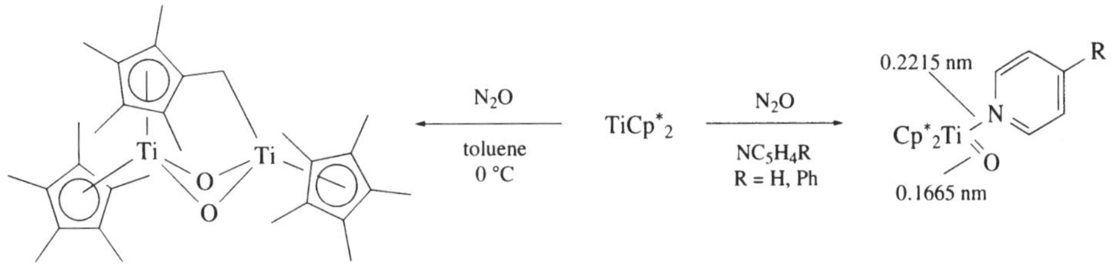

Scheme 20

(**62**)

(**63**)

4.1.8 Pentadienyltitanium Complexes

Titanium dichloride, generated *in situ* by the reduction of TiCl$_4$ with magnesium, reacts with potassium 2,4-dimethylpentadienide in THF at −78 °C to give the deep-green, diamagnetic 14-electron complex [Ti(η^5-C$_5$H$_5$Me$_2$-2,4)$_2$] (**64**). The compound is pyrophoric but can be purified by sublimation.[99] A detailed synthesis of (**64**) has been published.[100] The analogous reaction of TiCl$_2$ with K[C$_5$H$_5$Bu$^t_2$-2,4] gives the more stable complex [Ti(η^5-C$_5$H$_5$Bu$^t_2$-2,4)$_2$]. The NMR spectrum suggests that the two pentadienyl ligands are oriented at an angle of ca. 90° with respect to each other.[101] The 14-electron bis(dienyl) complexes readily form 16-electron adducts [Ti(η^5-C$_5$H$_5$Bu$^t_2$-2,4)$_2$(L)] (L = CO or a phosphine). The PF$_3$ adduct is more stable than the CO complex. The adduct has a bent sandwich structure of C_2 symmetry, that is, the pentadienyl ligands adopt an approximately eclipsed orientation.[102] This orientation was confirmed by the crystal structure of [Ti(η^5-C$_5$H$_6$Me-3)$_2$(CO)] (Figure 2).[103] A series of related pentadienyl carbonyl complexes has been prepared (Table 2).

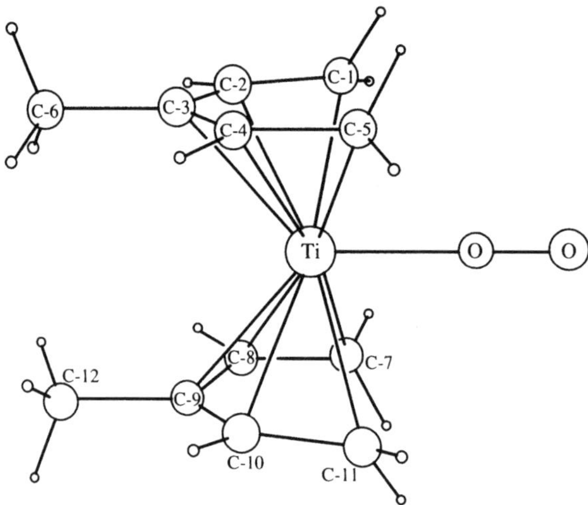

Figure 2   Structure of [Ti(η^5-C$_5$H$_6$Me-3)$_2$(CO)] (reproduced by permission of Elsevier Sequoia S. A. from *J. Organomet. Chem.*, 1993, **443**, 185).

Table 2   Bis(η^5-pentadienyl)titanium carbonyl complexes.

Ligand					Ref.
v_{co}(cm^{-1}) 1942	1946	1953 1957	1950	1960 1964	103 105

The green complex (**64**) reacts with a phosphine or a phosphite to give the orange to red compounds [Ti(η^5-C$_5$H$_5$Me$_2$-2,4)$_2$(L)]. The metal–ligand dissociation enthalpy (ΔH_d) correlates with the cone angle of L and ranges from 72.8 kJ mol^{-1} for L = PF$_3$ to 41.86 kJ mol^{-1} for L = PEt$_3$. However, the cone angles of P(OMe)$_3$, P(OEt)$_3$ and PEt$_3$ had to be revised because the alkyl or alkoxy substituents of the ligands adopt a more space-filling conformation than was envisaged when these angles were originally determined; the new (old) values are θ = 128° (107°) for P(OMe)$_3$, 134° (109°) for P(OEt)$_3$, and 137° (132°) for PEt$_3$.[104] [Ti(η^5-C$_5$H$_6$R-3)$_2$(PEt$_3$)] (R = H, Me) was obtained directly from [TiCl$_2$(THF)$_2$]; the phosphine ligand is readily displaced by CO.[105]

Mixed-ligand complexes are similarly prepared. Thus [TiCl$_2$Cp(THF)$_2$] reacts with K(C$_5$H$_5$R$_2$-2,4) in the presence of PEt$_3$ to give red (**65**), the structure of which was determined by x-ray diffraction (R = Me). The distances from the metal to the pentadienyl ligand are, on average, 0.01 nm shorter than to the Cp carbons and may indicate more effective bonding to the open dienyl ligand. The dienyl moiety is nevertheless quite reactive and undergoes coupling reactions with nitriles, isocyanides, ketones, imines and alkynes (Scheme 21).[106–8]

Bis(dimethylcyclohexadienyl)titanium(II) is obtained from potassion dimethylcyclohexadienide and either [TiCl$_3$(THF)$_3$] or [TiCl$_4$(THF)$_2$] as a very soluble, air-sensitive green product which readily takes up one CO ligand to give the crystallographically characterized red-orange complex (**66**) (Equation (12)).[109] PF$_3$, PMe$_3$ and P(OMe)$_3$ form more labile adducts. The CO ligand is lost on heating *in vacuo*.[109]

4.1.9   Arene Complexes of Titanium(II)

Treatment of TiX$_4$ with aluminum powder and AlX$_3$ in an arene solvent gives the well-known η^6-arene complexes (**67**) (X = Br, I; arene = benzene, toluene, *p*-xylene or mesitylene). These react with 1 bar CO at −20 °C to give CO adducts, probably of structure (**68**) (Equation (13)). The CO complexes decompose quickly at room temperature and more slowly at −40 °C. The v_{CO} values are in the range 2105–2040 cm^{-1} and vary with X. (**68**) reacts with TlCp to give [TiCp$_2$(CO)$_2$], and with L = *c*-C$_6$H$_{11}$NC or 1/2 Ph$_2$PCH$_2$CH$_2$PPh$_2$ to give products of the composition [Ti(arene)$_2$(L)$_2$(AlX$_4$)$_2$].[110,111]

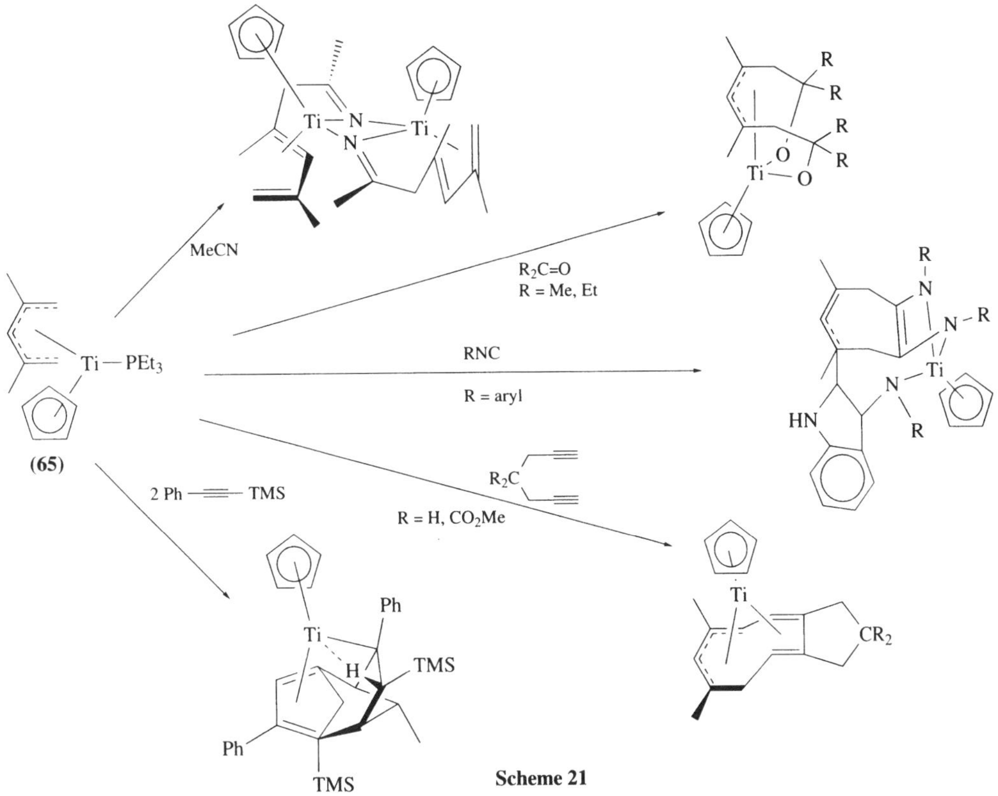

Scheme 21

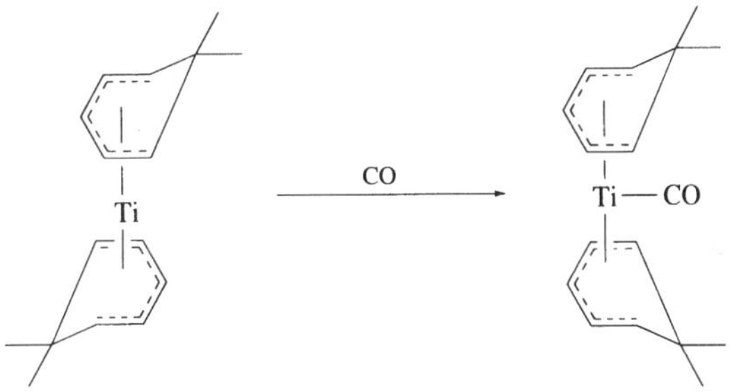

$$(12)$$

(66)

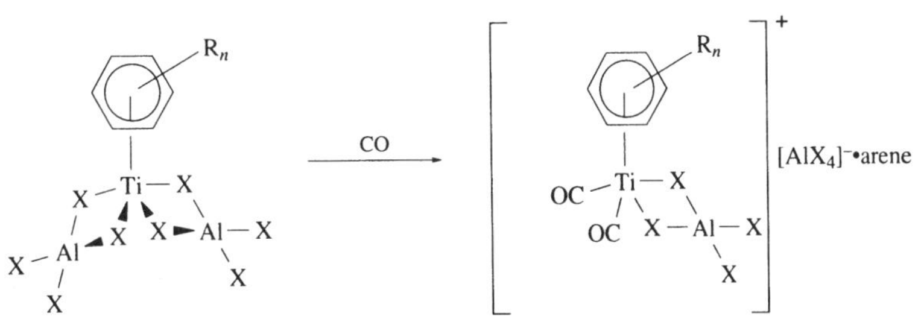

$$(13)$$

(67) **(68)**

The crystal structures of (**67**) (arene = C_6H_6 or $1,2,4,5\text{-}Me_4C_6H_2$) have been determined and confirm the suggested structures.[112,113] The durene complex contains both chlorine and iodine, with chlorine preferring the bridge positions. The structure of $[TiAl_2Cl_{8-x}Et_x(\eta\text{-}C_6Me_6)]$ has also been determined; the compound catalyses the polymerization of butadiene.[114] The reaction of (**67**) ($R_n = Me$) with disubstituted alkynes gives the cyclobutadienyl complexes (**69**) (R = Me, X = Br; R = Ph, X = Cl, Br). The crystal structure of one of these was determined (R = Ph). Compound (**69**) catalyses the cyclotrimerization of alkynes at 80 °C much more efficiently than $AlBr_3$ alone; the reaction with $K_2[C_8H_8]$ gives $[Ti(\eta\text{-}C_4Ph_4)(\eta\text{-}C_8H_8)]$, and with TlCp the titanacyclopentadiene complex $[Ti(\eta^2\text{-}$

$C_4Ph_4)Cp_2]$ is formed.[115] In contrast with the formation of (**69**), the reaction of $[Ti(AlBr_3Et)_2(\eta\text{-}C_6Me_6)]$ with diphenylacetylene in benzene leads quantitatively to the structurally characterized titanacycloheptatrienyl complex (**70**). The ring is distinctly nonplanar. All the ring carbon atoms are within π-bonding distance to titanium; there is no bond-length alternation within the ring. The compound can be regarded as a model for the intermediate during the catalytic cyclotrimerization of alkynes.[116]

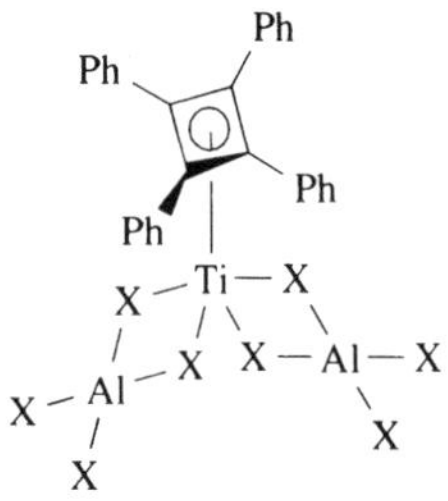

4.1.10 Cycloheptatrienyl Complexes

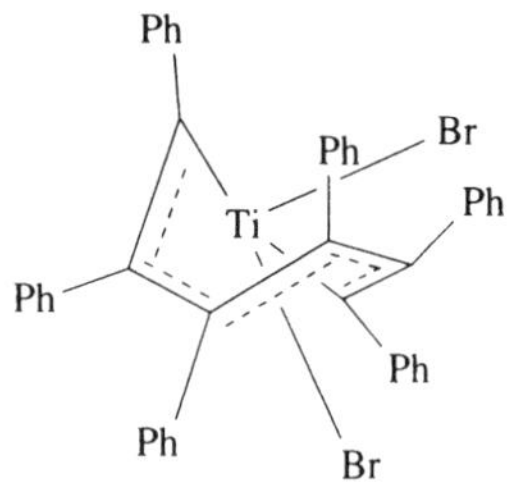

Although the assignment of a formal oxidation state in the case of cycloheptatrienyl complexes is somewhat ambiguous (see discussion below), there is evidence that complexes of the type $[TiCp(\eta\text{-}C_7H_7)]$ and their analogues should be regarded as titanium(II) complexes and they are therefore included in this section. The reaction of $[TiCl_3Cp^*]$ with cycloheptatriene in the presence of magnesium turnings in THF at room temperature gives blue crystals of $[TiCp^*(\eta\text{-}C_7H_7)]$; the compound sublimes above 100 °C/0.01 torr and has an absorption maximum at $\lambda = 675$ nm.[117] An x-ray diffraction study confirms the expected sandwich structure.[118] The helium(I) and helium(II) photoelectron spectra of the C_7H_7 complexes have been reported. The ionization energy of $[TiCp^*(\eta\text{-}C_7H_7)]$ (6.70 eV) is 0.13 eV less than that of $[TiCp(\eta\text{-}C_7H_7)]$, while the ionization energy of the C_7H_7 ligand shows hardly any dependence on the metal (Ti, Zr > Hf). The Cp^* e_1 orbital ionization energy is raised by 0.95 eV with respect to Cp. *Ab initio* MO calculations and XPS measurements suggest that cycloheptatrienyl is not present as $C_7H_7^+$ but as $C_7H_7^-$; the charge on C_7H_7 is more negative than on Cp. Charges for titanium of +1.75 (*ab initio*) or +1.0 (XPS) have been determined.[119]

$[TiCp(\eta^7\text{-}C_7H_7)]$ undergoes a one-electron reduction process at -2.0 V (vs. Ag/AgCl/KCl) to give the paramagnetic 17-electron radical anion $[TiCp(\eta^7\text{-}C_7H_7)]^-$. EPR spectral confirm that the geometry of the parent molecule is preserved on reduction.[120] In THF a one-electron oxidation step is also observed at 0.76 V (vs. SCE). $[TiCp^*(\eta^7\text{-}C_7H_7)]$ is reduced and oxidized at more negative potentials (-2.25 V and 0.13 V, respectively).[121]

Both rings in $[TiCp(\eta^7\text{-}C_7H_7)]$ can be functionalized, although the C_7H_7 reacts preferentially. Mono- and disubstituted derivatives containing CO_2H,[122] CO_2R,[122] TMS,[122] PPh_2[123-5] and PMe_2[126] substituents have been prepared as green to blue-green solids, generally in very good yields. Crystal structures show that the substituents in (**71**) are *trans* to each other for steric reasons, whereas they are almost eclipsed in (**72**). The PR_2-substituted complexes act as mono- and bidentate ligands and give heterobimetallic complexes (Scheme 22). $[Ti(\eta\text{-}C_5H_4PMe_2)(\eta\text{-}C_7H_6PMe_2)]$ (**73**) also reacts with $[Fe_2(CO)_9]$ to give the trimetallic complex $[(\textbf{73})\{Fe(CO)_4\}_2]$ and with $[Ni(cod)_2]$ to give the blue-black, very air-sensitive tetrahedral nickel phosphine complex $[Ni(\textbf{73})_2]$.[126]

Bis(toluene)titanium and cycloheptatriene in the presence of $AlClEt_2$ react under hydrogen transfer to give the cycloheptatrienyl–cycloheptadienyl complex $[Ti(\eta\text{-}C_7H_7)(\eta^5\text{-}C_7H_9)]$. In the presence of $AlEtCl_2$, however, dark-red, crystallographically characterized $[Ti(\mu\text{-}Cl)(\eta\text{-}C_7H_7)(THF)]_2$ (**74**) is formed. (**74**) reacts with donor ligands to give complexes of the type $[TiCl(\eta\text{-}C_7H_7)(L)_2]$ (**75**); these can be alkylated with Grignard reagents to give $[TiR(\eta\text{-}C_7H_7)(L)_2]$ ($R = Me$, Et; $L_2 = dmpe$, 1,2-bis(dimethylphosphino)cyclopentane) (Scheme 23). The crystal structure of $[TiEt(\eta\text{-}C_7H_7)(dmpe)]$ gives no indication of an agostic Ti–Et interaction.[127] Bis(diphenylphosphino)ethane binds only weakly to (**74**), and the adduct formation is reversible. Mixtures of (**74**) and $AlEt_2Cl$ catalyse the slow polymerization of ethene. Photoelectron spectroscopy suggests that the C_7H_7 ligand carries significant negative charge in these complexes and is best regarded as $C_7H_7^{3-}$, rather than $C_7H_7^+$, a situation similar to that in $[TiCp(\eta\text{-}C_7H_7)]$.[128] Simplified syntheses for complexes of type (**75**) have been reported which avoid the use of $[Ti(toluene)_2]$ as a starting material. Thus, sodium amalgam reduction of $TiCl_4$ in toluene in the presence of more than five equivalents of cycloheptatriene and L gives (**75**) in moderate yields ($L_2 = 2PMe_3$, 10%; $Me_2NCH_2CH_2NMe_2$, 27%; dmpe, 44%).[129,130]

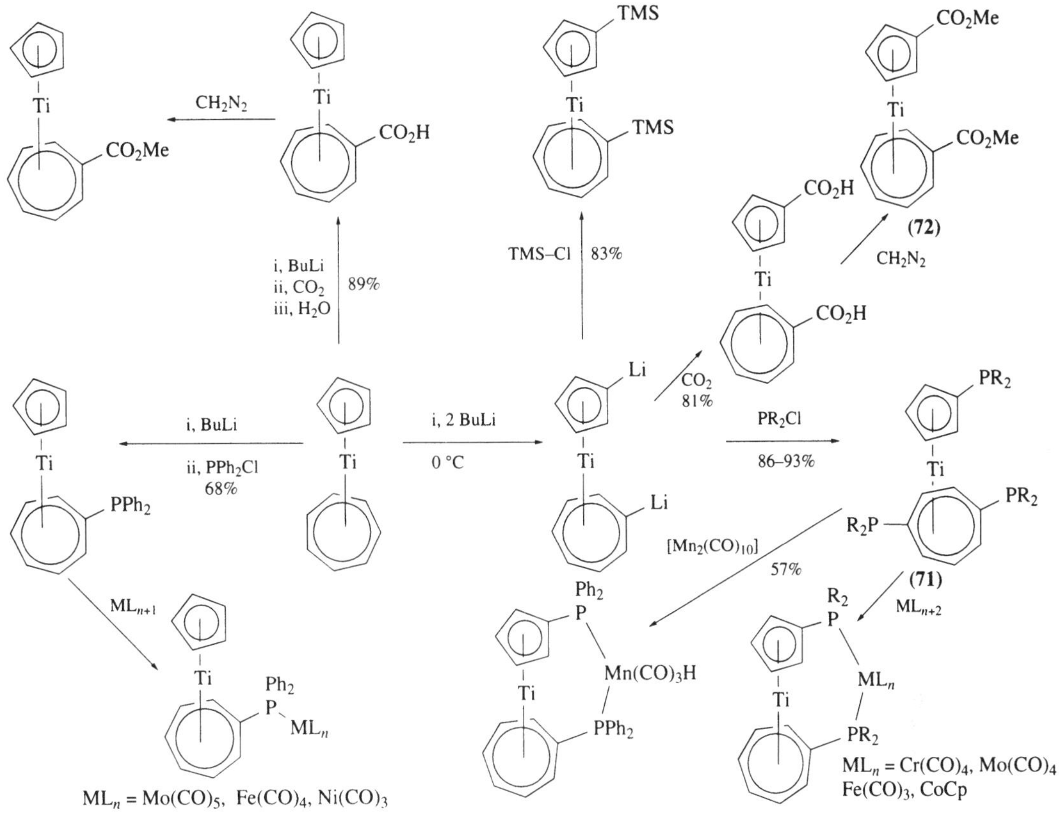

Scheme 22

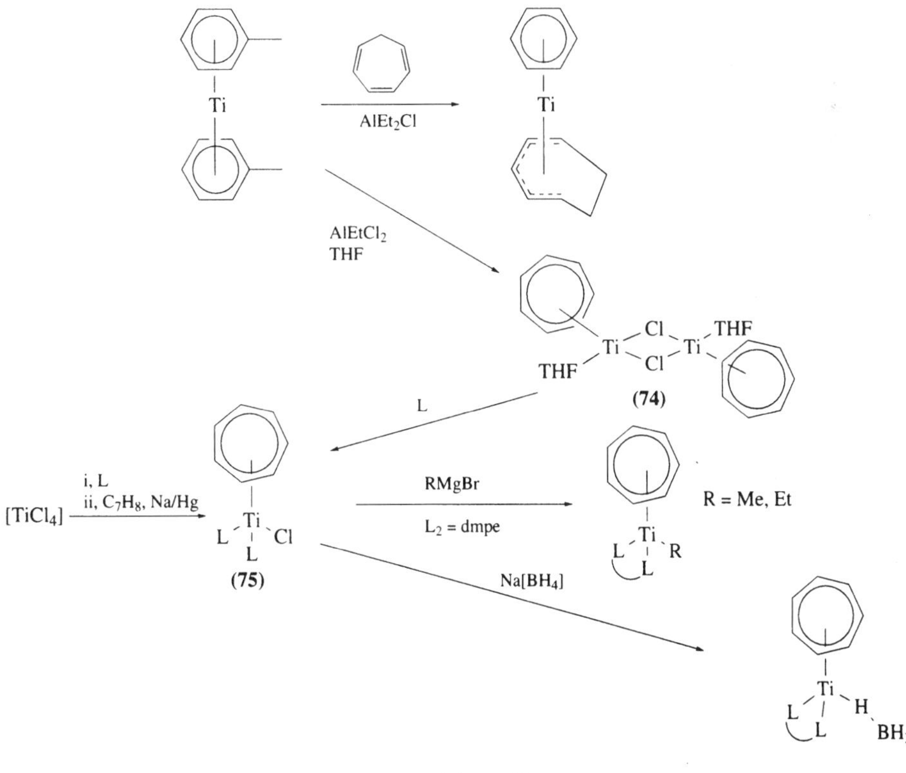

Scheme 23

4.2 COMPLEXES OF TITANIUM(III)

4.2.1 Titanium(III) Alkyls without Anionic π-Ligands

There are conflicting reports concerning the existence of tribenzyltitanium. This compound has been described as the product of the reaction of $[Ti(CH_2Ph)_4]$ with LiEt at temperatures below 0 °C, followed by the destruction of excess lithium reagent with CO_2 at −78 °C. The product can be extracted with toluene in high yield. It is also obtained from β-$TiCl_3$ and $LiCH_2Ph$ in diethyl ether at −30 °C in 50% yield. $Ti(CH_2Ph)_3$ disproportionates at 20 °C to titanium(II) and titanium(IV) and reacts rapidly with iodine at −35 °C to give, successively, $TiI(CH_2Ph)_2$ and $TiI_2(CH_2Ph)$. The alkyl content of these compounds was determined iodometrically and by hydrolysis. The reaction of β-$TiCl_3$ with $LiCH_2TMS$ gave $Ti(CH_2TMS)_3$ as a yellow viscous oil.[131] However, the reaction of $[Ti(CH_2Ph)_4]$ with ethyl- or butyllithium has been shown to produce the isolable adducts $Li[Ti(CH_2Ph)_4(R)]$ which decompose between −30 °C and 0 °C to give a product with a composition between $Li_2[Ti(CH_2Ph)_4]$ and $Li[Ti(CH_2Ph)_3]$ from which benzyllithium can be removed with toluene. The black residue is thought to consist of reasonably pure $[Ti^{II}(CH_2Ph)_2]$, whereas there was no evidence for an isolable benzyl complex of titanium(III).[3]

The reaction of the pentacoordinate chelate complex $[TiCl_2\{N(CH_2CH_2NEt_2)_2\}]$ with LiR gives the monomeric complexes $[TiR_2\{N(CH_2CH_2NEt_2)_2\}]$ in 60% (R = Me) and 83% (R = Ph) yield. The products were characterized by EPR spectroscopy; the magnetic moments of the methyl and phenyl complexes are 1.719 and 1.414 BM, respectively.[132] The Schiff base complex $[TiCl_2(salen)]$ reacts with RMgBr (R = Ph, mesityl) in THF under reduction of titanium(III) to give the octahedral monoalkyl titanium(III) complexes $[TiR(salen)(THF)]$ (Equation (14)). The x-ray crystal structure of the phenyl derivative was determined. By contrast, the alkylation with mesitylmagnesium bromide in toluene leads to the monoalkylation of the salen ligand.[133,134] These complexes have also been described in a brief review.[135]

$$\text{(14)}$$

4.2.2 Mono(cyclopentadienyl)titanium(III) Complexes

The reaction of $[TiCl_3Cp]$ with lithium nitride in THF in a 1:1 ratio leads to reduction and formation of sky-blue crystals $[TiCl_2Cp(THF)]$ from which solvent-free $[\{TiCl_2Cp\}_n]$ is obtained in toluene in 67% yield.[136,137] A nitrido complex is not formed. Although the THF complex is often formulated as $[TiCl_2Cp(THF)]$, the crystal structure of this compound shows that it actually consists of a mixture of the mono- and bis(THF) adducts in a 1:1 ratio (Figure 3); that is, the stoichiometry is $[TiCl_2Cp(THF)_{1.5}]$.[138] The reduction of $[TiCl_3Cp]$ with lithium nitride in a Ti:N ratio of 3:2 gives the green, diamagnetic cluster $[\{TiClCp\}_4]$ for which a cubane structure has been suggested.[136,137] Similarly, the reaction of $[TiCl_3Cp]$ with $[Li\{\mu\text{-}P(TMS)_2\}(THF)_2]_2$ in THF fails to give a phosphido complex but leads to $[TiCl_2Cp(THF)]$ in high yield.[139] The same product is obtained from the reduction of $[TiCl_2Cp_2]$ with magnesium in THF,[140] and by zinc reduction of $[TiCl_3Cp]$.[138]

Loss of Cp also occurs on photolysis of $[TiCl_2Cp_2]$; the resulting $TiCl_2Cp$ radical gives only a broad signal in the ESR spectrum but forms stable paramagnetic adducts with quinones such as 3,5-di-t-butyl-1,2-benzoquinone or 9,10-phenanthrenequinone, with transfer of the unpaired electron to the quinone ligand.[141] The reduction of $[TiCl_3Cp]$ with magnesium in THF in the presence of PMe_3 or 2,6-dimethylphenyl isocyanide at −30 °C gives the compounds $[TiCl_2Cp(L)_2]$ in good yields (Equation (15)); reduction at higher temperatures leads to titanium(II). The ν_{CN} vibrations of the isocyanide ligands are observed at 2165(s) and 2265(w) cm^{-1}. The piano-stool geometry of the complexes was confirmed by the crystal structure of the PMe_3 complex; the phosphine ligands are approximately *trans* to one another, with a Ti–P bond length of 0.2599(2) nm.[142] The addition of pentamethylcyclopentadiene to Ziegler catalysts prepared from $TiCl_4/AlEt_xCl_{3-x}$ mixtures gives green halide-bridged species such as $[TiAl_2Cl_8Cp^*]$ and $[TiAl_2Cl_7EtCp^*]$; these were investigated by ESR spectroscopy. If triphenylphosphine is added to these solutions, $TiCl_2Cp^*$ and phosphine complexes of the aluminum component are formed.[143]

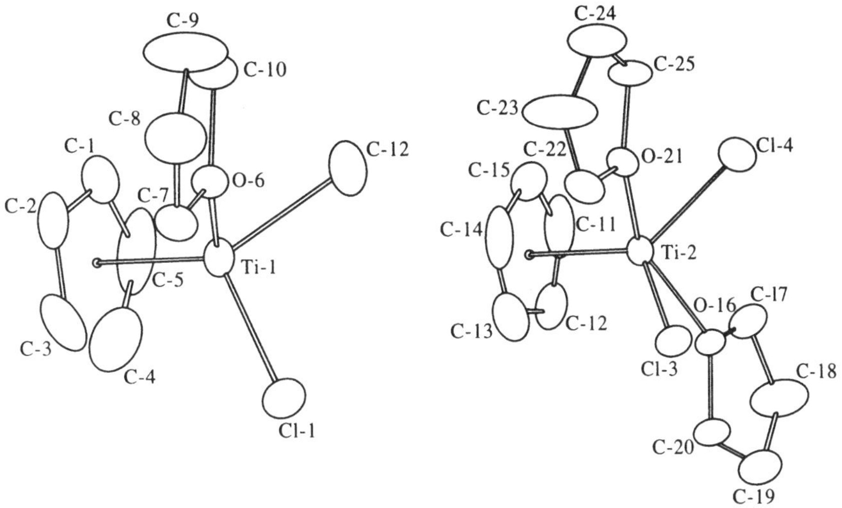

Figure 3 Structures of $[TiCl_2Cp(THF)]$ and $[TiCl_2Cp(THF)_2]$ (reproduced by permission of the American Chemical Society from *J. Am. Chem. Soc.*, 1983, **105**, 7295. Copyright (1983) American Chemical Society.)

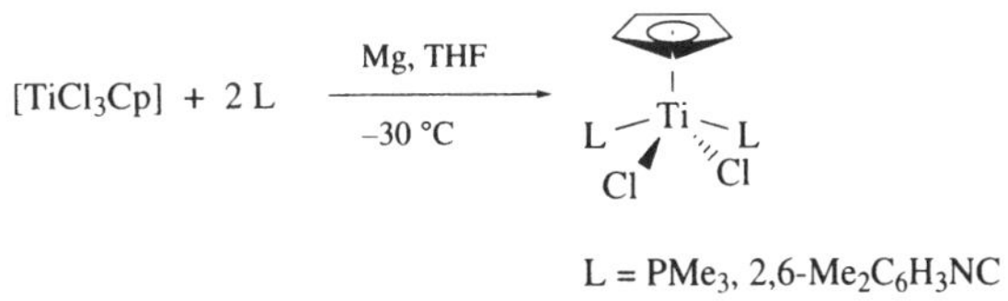

(15)

Treatment of $[TiCl_3Cp^*]$ with zinc gives $[TiCl_2Cp^*(THF)]$ in 30% yield which reacts with allylmagnesium chloride to give $[Ti(\eta^3\text{-allyl})_2Cp^*]$; the paramagnetic complex is thermally stable and melts at 78 °C. Similarly, $TiCl_2Cp$ reacts with benzyl Grignard reagents to give $[Ti(CH_2Ph)_2Cp]$, and with $CH_2{=}CHCH_2MgCl$ to give $[Ti(\eta^3\text{-allyl})_2Cp]$ which melts at 46 °C. By contrast, 3-methallyl complexes can only be generated at low temperature and decompose at 0 °C to give diene complexes (Scheme 24). $[Ti(\eta^3\text{-allyl})_2Cp]$ reacts with CO to unidentified products but fails to react with alkenes under isomerization or dimerization.[144]

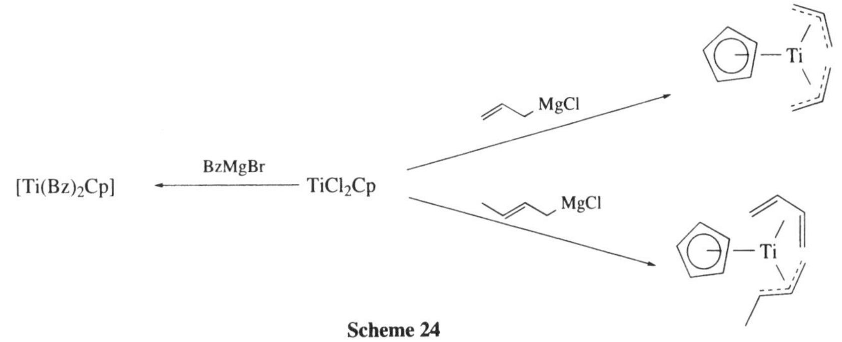

Scheme 24

Sodium cyclopentadienide reacts with the macrocylic complex $[TiCl_2(C_{22}H_{22}N_4)]$ ($C_{22}H_{22}N_4$ = tetramethyldibenzocyclotetrazatetradecane dianion) or with the titanium(III) complex $[TiCl(C_{22}H_{22}N_4)]\cdot[Li_2Cl_2(THF)_4]_{0.5}$ to give (**76**), the crystal structure of which was determined independently by two groups[145,146] (Figure 4). The Cp ring is parallel to the N_4 plane; the titanium atom is raised 0.090 nm above this plane. The magnetic moment of the compound is close to the spin-only value for a single unpaired electron (μ_{eff} = 1.81 BM).[146]

The reaction of diphenyldiazomethane with $[\{TiCl_2Cp\}_n]$ in THF gives the dinuclear, diazo-bridged compound (**77**). On heating, (**77**) eliminates $[TiCl_3Cp]$ and forms orange-red (**78**). Another dinuclear compound (**79**), in this case with a Ti_2N_3 five-membered ring, is isolated from the reaction of $[\{TiCl_2Cp\}_n]$ with azobenzene (Scheme 25). The structures of these products were confirmed by x-ray diffraction.[138,147]

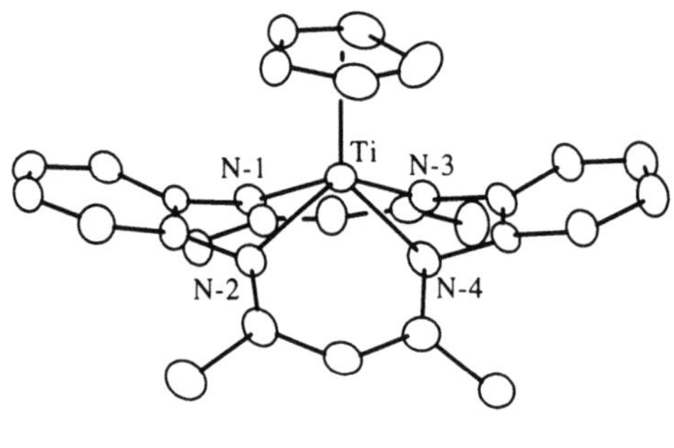

(76)

Figure 4 Structure of [TiCp(C$_{22}$H$_{22}$N$_4$)] (76). (reproduced by permission of the Royal Society of Chemistry from *J. Chem. Soc., Chem. Commun.*, 1986, 1101).

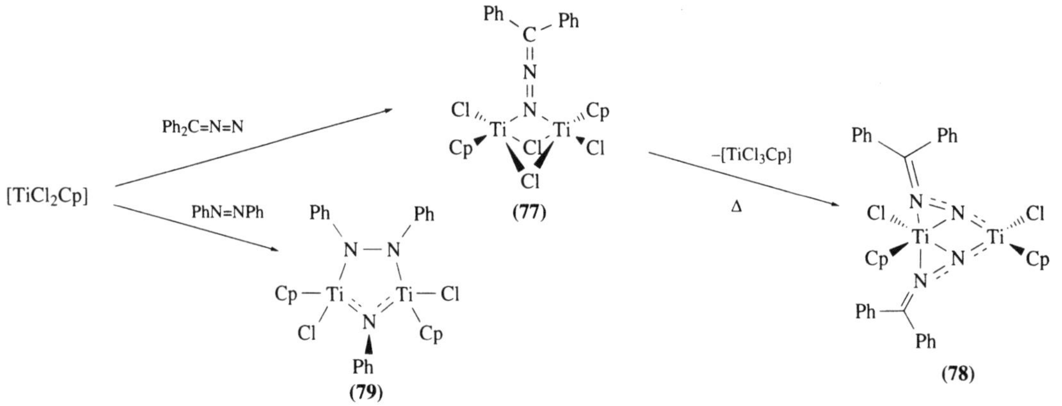

Scheme 25

The reaction between [TiCl$_2$Cp(THF)$_{1.5}$] and formaldehyde leads to C–O bond scission and the elimination of ethene to give the oxo-bridged compound [(TiCl$_2$Cp)$_2$(μ-O)]. Some [TiClCp(μ-O)]$_4$ is also produced due to hydrolysis.[148] The thermal decomposition of [TiCl(CH$_2$OMe)Cp'$_2$] (Cp' = Cp, C$_5$H$_4$Me) results in the loss of one Cp ligand per titanium and gives the tetranuclear oxo cluster (80) as dark red crystals (Equation (16)). The compound contains two significantly different Ti–Ti distances of 0.3236(1) and 0.3772(1) nm. Extensive spin-pairing is observed; the magnetic moment at 293 K is μ_{eff} = 1.40 BM.[149]

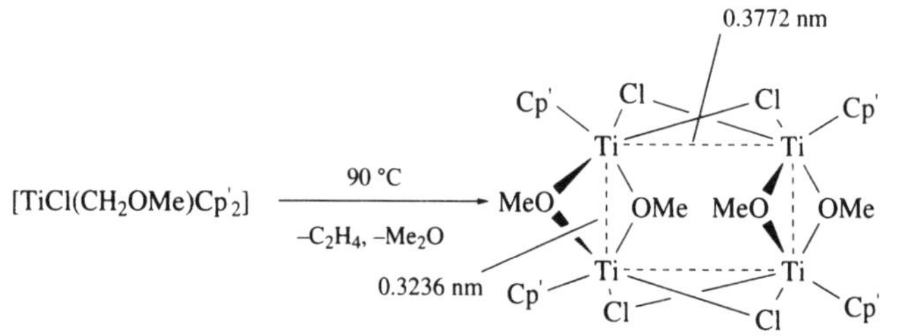

$$[\text{TiCl(CH}_2\text{OMe)Cp'}_2] \xrightarrow[-\text{C}_2\text{H}_4,\ -\text{Me}_2\text{O}]{90\ °\text{C}} \textbf{(80)} \qquad (16)$$

The reaction of [TiCl$_2$Cp(OC$_6$H$_4$X)] (X = *m*- or *p*-Me, H, *p*-Cl, *m*-, *p*-NO$_2$) with butyllithium gives dark brown to black solutions the EPR spectra of which suggest the presence of titanium(III), possibly [TiClCp(μ-OC$_6$H$_4$X)]$_2$. This product slowly hydrogenates cyclopentene and polymerizes ethene. The catalytic activity correlates with the Hammett σ constant; increased electron density on titanium increases the activity.[150] Photolysis, even exposure to sunlight, of [TiMe(O$_2$CR)$_2$Cp*] leads to Ti–Me bond homolysis and formation of [{TiCp*(μ,η^2-O$_2$CR)$_2$}$_2$]. This dimeric compound contains four carboxylato bridges and a relatively long Ti–Ti distance of 0.3660(7) nm. Nevertheless, it is only weakly paramagnetic ($\mu_{eff} \cong$ 0.6 BM), probably as the result of magnetic superexchange through the π-systems of the bridging ligands.[151]

The sulfido clusters [Ti(μ-S)(C₅H₄R)]₄ are obtained from the reaction of [TiCl₂(C₅H₄R)(THF)₂] (R = H, Me, Prⁱ) with (TMS)₂S. The 52-electron cluster compounds are diamagnetic; they do not show any reversible electrochemical behaviour. The structure of the methyl derivative was determined. The Ti₄S₄ is a distorted cube, with the Ti–Ti distances being slightly shorter than the S–S distances.[152]

4.2.3 Bis(cyclopentadienyl)titanium(III) Complexes

4.2.3.1 Halide complexes

Details of the synthesis of [{Ti(μ-Cl)Cp₂}₂] from TiCl₃ and TlCp have been published.[153] The comproportionation of [TiBr₂Cp₂] with [TiCp₂(CO)₂] in benzene produces [{Ti(μ-Br)Cp₂}₂] in 90% yield. ESR spectral investigations suggest that the dimer is in equilibrium with the monomeric complex. There is also evidence for the unstable intermediate TiBrCp₂(CO). The analogous chloro complexes are similarly obtained.[154] The reduction of [TiCl₂{C₂H₄(η-C₅H₄)₂}] with sodium sand in toluene gives the orange-brown complex [TiCl{C₂H₄(η-C₅H₄)₂}] which can be purified by sublimation at 130 °C/10⁻³ torr.[17] The reaction of [TiCl₃(THF)₃] with [MgClCp*(THF)] in THF at room temperature gives the blue, highly air-sensitive monomeric complex [TiClCp*₂] in 90% yield. The crystal structure is shown in Figure 5. The centroids of the cyclopentadienyl groups and the chloride constitute a trigonal-planar coordination geometry around the metal. The Ti–Cl bond length is 0.2363(1) nm, very similar to [TiCl₂Cp*₂]. The compound undergoes halide exchange with BBr₃ or LiBr to give blue-green [TiBrCp*₂], while treatment with LiI gives dark-green [TiICp*₂]. Other derivatives [TiXCp*₂] with X = BH₄, NMe₂, OBuᵗ or HCO₂ are synthesized by similar exchange reactions.[155] The reduction of [TiCl₂(η-C₅HMe₄)₂] with LAH gives [TiCl(C₅HMe₄)₂] as blue crystals. Steric interactions are less important here than with [TiClCp*₂], and consequently the cntr–Ti–cntr angle is 4° smaller. [TiCl(η-C₅HMe₄)₂] and LiI give [TiI(η-C₅HMe₄)₂] (cntr = centroid of the cyclopentadienyl ring).[156]

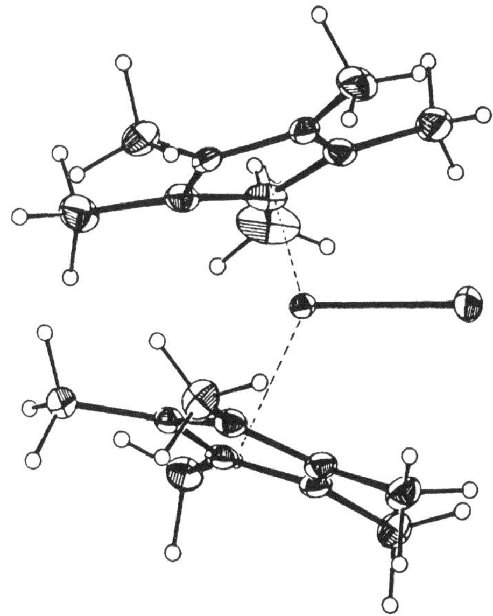

Figure 5 Structure of [TiClCp*₂] (reproduced by permission of the American Chemical Society from *Organometallics*, 1987, **6**, 1009. Copyright (1987) American Chemical Society.)

Sterically hindered disubstituted cyclopentadienyl ligands also lead to monomeric titanium(III) complexes. Thus, the compounds [TiCl(C₅H₃R₂)₂] are produced by the reaction of Li[C₅H₃R₂] with [TiCl₃(THF)₃] as black-purple (R = TMS)[157] or green (R = Buⁱ)[158] solids. [TiCl(C₅H₃Buⁱ₂)₂] was also obtained in lower overall yield by the reduction of [TiCl₂(C₅H₃Buⁱ₂)₂] with potassium in pentane; it has a temperature-independent magnetic moment of μ_eff = 1.65 BM.[159] The reaction of TiCl₂ or TiCl₃ with K[C₅HPh₄] in THF gives [TiCl(C₅HPh₄)₂]·1.5 THF as green crystals in 68% yield. The structure of a dichloromethane solvate ([TiCl(C₅HPh₄)₂]·CH₂Cl₂·0.5 THF) was determined. The cyclopentadienyl ligands adopt a staggered conformation to allow for the bulky phenyl substituents; the Ti–Cl distance is 0.2312(2) nm, slightly shorter than in [TiClCp*₂]. Cyclic voltammetry shows an irreversible oxidation at 0.14 V and a partly reversible reduction wave at −1.92 V.[160]

A number of titanium(III) complexes of *ansa*-metallocenes have been prepared. The reduction of [TiCl$_2${Me$_2$Si(C$_5$H$_4$)$_2$}] with 1.1 equivalents of sodium amalgam in THF or toluene gives [(TiCl{Me$_2$Si(C$_5$H$_4$)$_2$})$_2$] as a brownish-pink solid in 65% yield. In contrast with the zirconium analogue the compound is paramagnetic; an EPR spectral signal is observed at $g = 1.9809$ which shows [47]Ti and [49]Ti satellites.[161] The electrochemical reduction of [TiCl$_2${C$_2$Me$_4$(C$_5$H$_4$)$_2$}] under argon generates the anionic complex [TiCl$_2${C$_2$Me$_4$(C$_5$H$_4$)$_2$}]$^-$ ($E_p = -0.92$ V) which undergoes rapid ligand exchange with THF to give [TiCl{C$_2$Me$_4$(C$_5$H$_4$)$_2$}(THF)]. This complex may be further reduced ($E_p = -2.5$ V) to [TiCl{C$_2$Me$_4$(C$_5$H$_4$)$_2$}(THF)]$^-$. The reduction potential for [TiCl$_2${C$_2$Me$_4$(C$_5$H$_4$)$_2$}] to titanium(III) is very close to that of [TiCl$_2$Cp$_2$], but the potential for further reduction to titanium(II) is ca. 0.3 V more negative. This titanium(II) product decays considerably faster than the nonbridged analogue. The electrochemical reduction of [TiCl$_2${C$_2$Me$_4$(C$_5$H$_4$)$_2$}] under an atmosphere of CO gives first [TiCl{C$_2$Me$_4$(C$_5$H$_4$)$_2$}(CO)] followed by [Ti{C$_2$Me$_4$(C$_5$H$_4$)$_2$}(CO)$_2$].[162] The reduction of [TiCl$_2${(CH$_2$)$_n$(C$_5$H$_4$)$_2$}] ($n = 2$ or 3) with AlHCl$_2$ leads to [TiCl{(CH$_2$)$_n$(C$_5$H$_4$)$_2$}] in 64–85% yield.[163]

Whereas anions of the type [TiCl$_2$Cp$_2$]$^-$ have mainly been observed as unstable intermediates during the reduction of titanium(IV) in solution, the reaction of [TiX$_2$Cp$_2$] (X = Cl, Br or I) with [CoCp$_2$] in toluene has for the first time allowed the isolation of stable salts such as [CoCp$_2$][TiX$_2$Cp$_2$] as deep-green crystals. They possess a magnetic moment of 1.7–1.8 BM. Whereas the same product is obtained in THF if X = Cl, by contrast the reaction of [CoCp$_2$] with [TiX$_2$Cp$_2$] (X = Br or I) in THF gives [{TiXCp$_2$}$_2$]. While the reduction of [TiCl$_2$Cp$_2$] with [CoCp$_2$] under a CO atmosphere gives a mixture of [TiCp$_2$(CO)$_2$] and [CoCp$_2$][TiX$_2$Cp$_2$], the more labile bromo and iodo complexes react cleanly under these conditions to give [TiCp$_2$(CO)$_2$] and [CoCp$_2$]X in excellent yield.[164] The reduction of [TiCl$_2$Cp$_2$] with 1.5 equivalents of sodium amalgam leads to a mixture of the purple fulvalene complex (81) and the dark green hydrido complex (82) in ca. 40% each (Scheme 26). Compound (81) reacts with dry oxygen gas to give (83).[165]

[TiCl$_2$Cp$_2$] $\xrightarrow[\text{toluene}]{\text{Na/Hg}}$ (81) + (82)

(81) $\xrightarrow[85\%]{\text{O}_2}$ (83)

Scheme 26

The structures of the primary species formed during the reduction of [TiCl$_2$Cp$_2$] with magnesium have been established. During attempts to generate phosphorus derivatives by reacting [TiCl$_2$Cp$_2$] with magnesium in the presence of P$_4$ in THF, light-green crystals of (84) were obtained and characterized by x-ray diffraction. P$_4$ is not necessary for the synthesis. The same compound is formed in 25% yield from [TiClCp$_2$]$_2$ and MgCl$_2$ in THF. Similarly, the reaction of [TiCl$_2$Cp$_2$] with magnesium in the presence of RPH$_2$ (R = Ph or c-C$_6$H$_{11}$) gave aquamarine crystals of (85) in low yield. Both (84) and (85) react with PMe$_3$ to give [TiClCp$_2$(PMe$_3$)] which can be further reduced with magnesium in the presence of PMe$_3$ to give [TiCp$_2$(PMe$_3$)$_2$].[166]

A number of related halide-bridged heteronuclear compounds have been observed. The reaction of [TiCl$_2$Cp$_2$] with LAH in the presence of 1,5-hexadiene gives [TiCp$_2$(μ-Cl)$_2$Al(hexyl)$_2$].[167] Mixtures of [TiCl$_2$Cp'$_2$] and AlEt$_3$, AlEt$_2$Cl or AlEtCl$_2$ give [TiCp'$_2$(μ-Cl)$_2$AlEt$_2$], [TiCp'$_2$(μ-Cl)$_2$Al(Cl)Et] and [TiCp'$_2$(μ-Cl)$_2$AlCl$_2$], respectively. If Cp' = Cp or C$_5$H$_4$Me the reaction gives the blue products quantitatively. For the more sterically hindered ligands Cp' = C$_5$H$_2$Me$_3$, C$_5$HMe$_4$ or Cp*, the reduction method does not give quantitative yields, and the compounds are better prepared from TiClCp'$_2$ and AlCl$_{3-x}$Et$_x$. The system [TiCl$_2$(C$_5$H$_2$Me$_3$)$_2$]/AlEt$_3$ was slow enough to allow the determination of the rate of the reaction ($k = 0.14$ L mol^{-1} s^{-1}). The rates of reduction decrease in the order

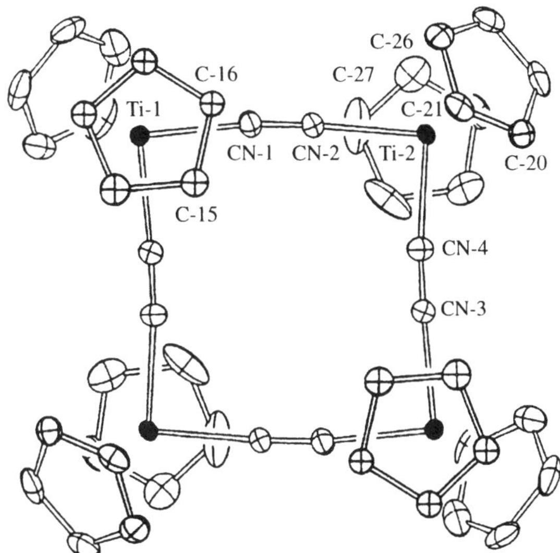

(84)   (85)

Cp > C_5H_4Me > $C_5H_2Me_3$ ≈ C_5HMe_4, and $AlEt_3$ > $AlEt_2Cl$ > $AlEtCl_2$; the latter requires heating to 70 °C for up to 20 h to go to completion. Yields of the Cp* compounds were low, with diamagnetic by-products being formed. Whereas [$TiCp*_2(\mu-Cl)_2AlCl_2$] is stable in the presence of excess $AlCl_3$, [Ti(η-C_5HMe_4)$_2(\mu-Cl)_2AlCl_2$] reacts with $AlCl_3$ with loss of one C_5HMe_4 ligand to give [Ti(η-C_5HMe_4){(μ-Cl)$_2AlCl_2$}$_2$].[168]

The ^{1}H NMR spectra of the paramagnetic compounds [$TiCp_2(\mu-Cl)_2Al(Cl)(X)$] (X = Cl or Et) have been recorded. With decreasing temperature the Al–CH_2 signal is shifted to lower frequency, while the Me signal moves in the opposite direction. The ^{13}C NMR signals follow the opposite trend. Reaction of [{$TiClCp_2$}$_2$] with [$AlHEt_2$]$_2$ gives a hydrido intermediate which was partially characterized.[169] [$TiCp_2(\mu-Cl)_2AlEt_2$] reacts with less than 0.75 equivalents of methanol per titanium under cleavage of an Al–Et bond. With higher methanol concentrations solvated [$TiClCp_2$] is obtained. Aldehydes, formate esters, formamide and dimethyl carbonate react with the Al–Et bond independent of the concentration.[170]

A new crystal modification of [{$TiClCp_2$}$_2$] has been characterized by x-ray diffraction. This β-form has longer Ti–Ti distances than the known α-form.[171] The structure of the pseudohalide complex [{$Ti(CN)Cp_2$}$_4$] has been determined. The tetrameric compound is crystallized as red-brown air-stable needles by the slow diffusion of pentane into THF and is quite air-stable, in contrast with the powdered material. The 12-membered ring is a nonplanar. The carbon and nitrogen positions could not be distinguished (Figure 6). Ti–CN distances range from 0.2127(4) to 0.2139(4) nm.[172]

Figure 6   Structure of [{$Ti(CN)Cp_2$}$_4$] (reproduced by permission of Elsevier Sequoia S. A. from *J. Organomet. Chem.*, 1992, **431**, 41).

The mean bond dissociation enthalpy of the Ti–Cl bond in [$TiClCp*_2$], as determined by reaction-solution calorimetry, is $\overline{D}$ = 481 ± 21 kJ mol^{-1}.[173] The helium(I) and helium(II) photoelectron spectra of [{$TiXCp'_2$}$_2$] have been measured (Cp' = Cp, C_5H_4Me, $C_5H_2Me_3$, C_5HMe_4 or C_5Me_5; X = Cl or Br). All the compounds are monomeric in the gas phase. The order of ionization is d(Ti) < π(Cp') < p(X) > σ(Ti–X). The data suggest that a strong interaction exists between the orbitals of the TiCp'$_2$ moiety and the halogen p orbitals. The effect of methylation of the Cp ring is additive within the series. All ionization potentials decrease with the introduction of a further methyl group, with the largest increment of 0.30 eV being found for $\pi(Cp')$, less for $d(Ti)$ (0.22 eV) and for $\sigma(Ti-Cl)$ (0.10 eV).[174] The ESR spectra of $TiX(\eta-C_5H_{5-n}Me_n)_2$ in toluene and 2-methyltetrahydrofuran have been recorded. Compounds with $n = 3-5$ are monomeric in toluene, those with $n = 0$ or 1 are dimers. These dissociate in THF with solvent coordination. $TiX(\eta-C_5H_2Me_3)_2$ is present in 2-MeTHF solution as a mixture of solvated and unsolvated species, $[TiX(\eta-C_5HMe_4)_2]$ solvates partially only on cooling to 77 K, while $[TiXCp^*_2]$ does not form solvent adducts.[175] The ESR spectra of the adducts $[TiXCp_2(L)]$ (L = PR_3, RNC, pyridine, amine or CO; X = F, Cl, Br or I) show a relationship between the g value and the hyperfine coupling constant $a(Ti)$ for X = Cl, Br or I, which is a reflection of the electronic character of L.[176] The ESR spectra of the ZnX_2 adducts $[Ti_2Zn(\mu-X)_4Cp_4]$ (X = Cl or Br) are typical of a triplet state, $S = 1$. For X = Cl the magnetic parameters derived from the analysis of the ESR spectrum are in good agreement with magnetic and structural measurements on the solid compound. Electron nuclear double resonance (ENDOR) measurements show that dissolution in THF ruptures the trimetallic structure and forms $S = 1/2$ species, most probably $[TiCp_2(THF)_2]^{+}$.[177]

The compounds $[TiXCp^*_2]$ (X = Cl, OMe or N = CHBu^t) react with dimethylcadmium under oxidation to titanium(IV) (Equation (17)). Alkyl complexes of titanium(III) are not formed. $[TiClCp^*_2]$ and $CdEt_2$ gives $[TiCl(Et)Cp^*_2]$ in 51% yield. The reaction with $ZnMe_2$ is very much slower, and there is no reaction with $HgMe_2$ (X = Cl or Me).[178]

$$2\,[TiXCp^*_2] + CdMe_2 \xrightarrow[\substack{-30\,°C \\ 55-73\%}]{} 2\,[Ti(Me)(X)Cp^*_2] + Cd \qquad (17)$$

Carbon monoxide induces the disproportionation of $TiClCp_2$ into $[TiCl_2Cp_2]$ and $[TiCp_2(CO)_2]$.[179] The complexes $[TiXCp^*_2]$ (X = Cl, Br or I) undergo a similar reaction; in contrast with the Cp analogue this disproportionation is irreversible. With over 2 bar CO pressure at $-30\,°C$ the intermediate complex $[TiXCp^*_2(CO)]$ is observed (Scheme 27); its concentration is dependent on the CO pressure between 2 and 90 bar, although at $-30\,°C$ the compound is stable even in the absence of excess CO. At 50 bar CO pressure the temperature dependence of the rate k_{obs} gives a linear Eyring plot with $\Delta H^{\ddagger}; = 80 \pm 12$ kJ mol^{-1} and $\Delta S^{\ddagger}; = 105 \pm 3$ J mol^{-1} K^{-1}. At $5\,°C$ at 20 bar CO, $[TiBrCp^*_2(CO)]$ ($\nu_{CO} = 2014$ cm^{-1}) disproportionates ca. 45 times slower that the chloro analogue, while under these conditions the disproportionation of the iodo complex is very fast.[180]

$$[TiXCp^*_2] \xrightarrow[\substack{>2\ bar \\ -30\,°C}]{CO} Cp^*_2Ti\underset{CO}{\overset{X}{<}} \xrightarrow{CO} [TiX_2Cp^*_2] + [TiCp^*_2(CO)_2]$$

Scheme 27

Benzo[c]cinnoline reacts with electrochemically generated $TiClCp_2$ at $-35\,°C$ in THF to give $[TiCp_2(Cl)N(R)-N(R)Ti(Cl)Cp_2]$. Electrochemical evidence, in particular the absence of the reduction wave for the coordinated azobenzene moiety, suggests that this compound should be formulated as **(86)** rather than **(87)**.[181] Benzyl chloride is reduced by $TiClCp_2$ to give the C–C coupled product, 1,2-diphenylethane. Bromocyclohexene gives a similar coupled product, and α,α,α-trichlorotoluene gives diphenylacetylene. The reaction is thought to involve radicals.[182] $TiClCp_2$ reduces nitrobenzene to azobenzene in moderate yield. A mixture of aluminum powder and $[TiCl_2Cp_2]$ converts benzylic or allylic chlorides and vicinal dibromides to the corresponding coupled products and to alkenes.[183]

(86)   **(87)**

4.2.3.2 Alkyl complexes

The reaction of [TiCl$_2$Cp$_2$] with butyllithium in the presence of tertiary allylamines gives titanium(III) azametallacycles, most probably via TiHCp$_2$ as intermediate. Allyl(butyl)ethylenediamine gives the corresponding dinuclear complex (**88**) as green crystals (Scheme 28).[184] Substituted cyclopropyl complexes are obtained from the reaction of TiClCp$_2$ with the corresponding Grignard reagents at −30 °C as dark red solids; there is evidence that the vinylic double bond is coordinated if R = H but not if R = Me (Equation (18)).[185]

Scheme 28

(18)

The reaction of [TiClCp*$_2$] with lithium, potassium or magnesium reagents gives a range of titanium(III) alkyl, benzyl, vinyl, allyl or alkynyl complexes (Table 3). All these compounds are deeply coloured, very air-sensitive materials. The crystal structure of [Ti(CH$_2$Bu^t)Cp*$_2$] was determined. The complex is sterically congested, as indicated by the wide Ti–C–C angle of 131.1(3)°. [TiMeCp*$_2$] shows normal Curie behaviour. Proton NMR spectra are observed for [TiMeCp*$_2$] and [Ti(C≡CMe)Cp*$_2$] only. At 30 °C the Cp* resonance of [TiMeCp*$_2$] is observed in the ^{1}H NMR spectrum as a broad singlet at δ = 18.38 and the Me group at δ = 22.74; the signals show the expected temperature dependence. The Me, Et, Pr^n, vinyl and Ph complexes were also prepared as perdeutero compounds, and all the expected ^{2}H signals could be observed in the ^{2}H NMR spectrum. The EPR spectra are less informative and consist of single lines without any hyperfine coupling to the hydrogen atoms of the alkyl ligands.[186]

Table 3 Synthesis of alkyl complexes [TiRCp*$_2$] from the reaction of [TiClCp*$_2$] with M–R.[186]

R	M	Appearance	Yield (%)	Comment
Me	MgI	Green crystals	88	
Et	MgBr	Brown crystals	87	Thermally unstable
Pr^n	MgBr	Green crystals	62	Stable below −20 °C
CH$_2$Bu^t	Li	Brown-green crystals	74	X ray
CH$_2$Ph	K	Brown crystals	76	
η^3-C$_3$H$_5$	MgBr	Dark green crystals	36	
η^3-C$_4$H$_7$	MgBr	Black-purple crystals	47	
CH=CH$_2$	MgCl	Dark green crystals	85	
C≡CMe	Li	Green solid	56	IR, $v_{C≡C}$ = 2080 cm^{-1}
Ph	MgBr	Dark green crystals	90	

The alkylation of [TiCl{Me$_2$Si(η-C$_5$H$_4$)$_2$}(L)] (L = PMe$_2$Ph) with LiR (R = Me or CH$_2$TMS) gives the complexes [TiR{Me$_2$Si(η-C$_5$H$_4$)$_2$}(L)] as air-sensitive, dark-red crystals in 58% and 47% yield, respectively. The methyl complex is also obtained by the reduction of [TiMeCl{Me$_2$Si(η-C$_5$H$_4$)$_2$}] with sodium amalgam in the presence of PMe$_2$Ph.[161]

The Ti–C bond dissociation enthalpy of [TiMeCp*$_2$] was determined by solution-reaction calorimetry and gave a value of 300 ± 29 kJ mol^{-1}, which is comparable with the values obtained for [TiMe$_2$Cp*$_2$].[173]

The complexes [TiRCp*$_2$] (R = Me, Et or Pr^n) undergo facile thermolysis to give the fulvene complex (**89**). This can react further with loss of hydrogen to give the titanium(IV) complex (**90a**) for which a contribution by the η^3:η^4 structure (**90b**) is discussed (Scheme 29).[187] The heat of formation of (**89**) is 620.0 ± 4.8 kJ mol^{-1}, while for (**90**) a value of 552.1 ± 4.3 kJ mol^{-1} has been measured. Although the reaction enthalpy ΔH for the conversion of [TiMeCp*$_2$] into (**89**) is positive (29.5 ± 7.3 kJ mol^{-1}), there is a significant entropy contribution to give $\Delta G = -26$ kJ mol^{-1} at 25 °C, rising to -34 kJ mol^{-1} at 70 °C.[188]

(89)   (90a)   (90b)

Scheme 29

Protolysis of [TiRCp*$_2$] with alcohols, carboxylic acids or hydrogen halides gives the corresponding alkoxo, carboxylato or halide compounds. Propyne reacts with [TiRCp*$_2$] to give [Ti(C≡CMe)Cp*$_2$]; excess propyne is catalytically dimerized to CH$_2$=C(Me)C≡CMe. The alkyl complexes [TiRCp*$_2$] have low Lewis acidity and show little tendency to coordinate ligands such as THF, in contrast with the isostructural scandium alkyls. [TiEtCp*$_2$] undergoes ligand exchange reactions with aluminum alkyls; it reacts with AlMe$_3$ to give [TiMeCp*$_2$], and with AlMeEtCl to give [TiClCp*$_2$]. However, the reaction of [TiEtCp*$_2$] with MeMgCl leads to methane elimination and formation of [TiCp*$_2$(η-C$_2$H$_4$)]. Some reactions of [TiRCp*$_2$] are shown in Scheme 30.[186,189] Unusually, the insertion products of [TiRCp*$_2$] with CO and Bu^{t}NC are unstable; and although at 0 °C a compound of the probable composition [Ti(η^2-COMe)Cp*$_2$] is observed in the IR spectrum ($\nu_{C=O} = 1530$ cm^{-1}), this decomposes on warming to [TiCp*$_2$(CO)$_2$]. [Ti(η^2-COEt)Cp*$_2$] is more stable and was characterized by PbCl$_2$ oxidation as [Ti(η^2-COEt)(Cl)Cp*$_2$] which forms two isomers in a 93:7 ratio, most probably conformers with 'O-inside' and the less favourable 'O-outside' acyl ligands. The iminoacyl analogue was also identified by PbCl$_2$ oxidation as [TiCl{η^2-C(Me)=NBu^t}Cp*$_2$]; this oxidation with PbCl$_2$ is a general very smooth route to mixed-ligand titanium(IV) complexes of the type [TiR(Cl)Cp*$_2$].[189] [TiMeCp*$_2$] forms a green-brown crystalline adduct with acetone which dissociates readily in solution. There is no insertion, probably for steric reasons. The adduct of [TiMeCp*$_2$] with Bu^{t}CN is unstable but isolable as brown crystals in 64% yield. [TiMeCp*$_2$] reacts with paraformaldehyde to give the CH$_2$O insertion product [Ti(OEt)Cp*$_2$]. Whereas CO$_2$ reacts with [TiMeCp*$_2$] to give the acetato complex, [TiEtCp*$_2$] affords the formate [Ti(O$_2$CH)Cp*$_2$], apparently as the result of very facile β-hydrogen elimination and formation of TiHCp*$_2$ as the intermediate. This process is probably facilitated by the β-agostic structure of [TiEtCp*$_2$]. Similarly, ethene extrusion reactions are observed when [TiEtCp*$_2$] is treated with CO$_2$ or RC≡CR. [TiEtCp*$_2$] reacts with 1-alkenes to give ethene and the corresponding titanium alkyl; this reaction is reversible under ethene pressure. There is no evidence for the formation of alkene complexes during this reaction. The equilibrium constants for the alkene exchange reaction indicate that the titanium ethyl complex is the most stable of the higher alkyls; the synthesis of [Ti(CH$_2$CH$_2$Ph)Cp*$_2$] from styrene could only be achieved by removal of the liberated ethene from the reaction mixture by evacuation. Selective ^{13}C and deuterium labelling revealed that the two ethyl carbon atoms in [TiEtCp*$_2$] exchange rapidly (Scheme 31), apparently via [TiH(C$_2$H$_4$)Cp*$_2$] as a short-lived intermediate. This process would also provide a mechanism for the observed exchange reaction with terminal alkenes and the formation of higher alkyl derivatives. It also explains why [TiEtCp*$_2$], in contrast with [TiMeCp*$_2$], inserts acetone to give the isopropoxy complex [Ti(OCHMe$_2$)Cp*$_2$].[190]

Treatment of [TiRCp*$_2$] (R = Me, CH$_2$Bu^t or tolyl) with CO at 0 °C followed by the addition of [Mo$_2$Cp$_2$(CO)$_6$] affords a yellow diamagnetic product (**91**) which decomposes to give, surprisingly, C$_6$Me$_6$ and an oxo cluster of the probable structure (**92**). A possible mechanism for the ring expansion of Cp* to C$_6$Me$_6$ is outlined in (Scheme 32).[191] The reaction of (**88**) with CO$_2$ gives [TiCp$_2${O$_2$C(CH$_2$)$_3$N(Bu)CH$_2$CH$_2$N(Bu)(CH$_2$)$_3$CO$_2$}TiCp$_2$] from which the dicarboxylic acid can be liberated on hydrolysis.[192]

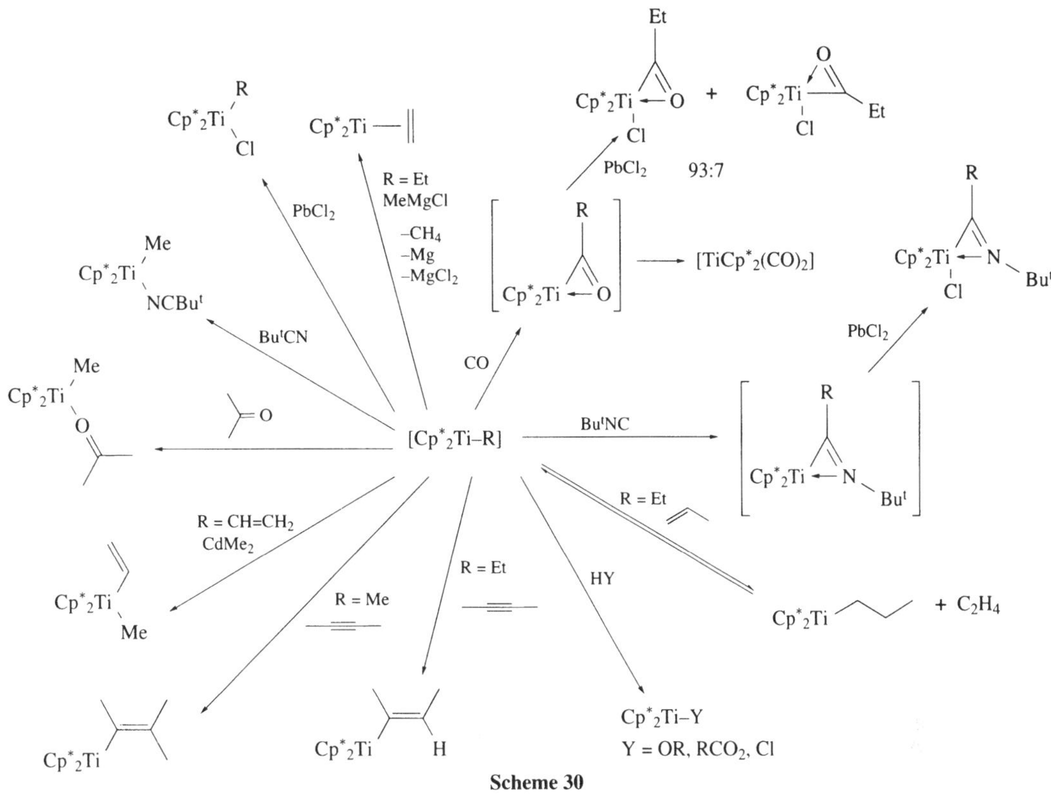

Scheme 30

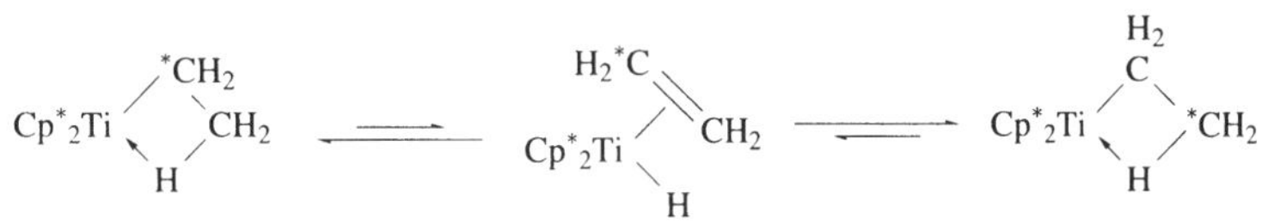

Scheme 31

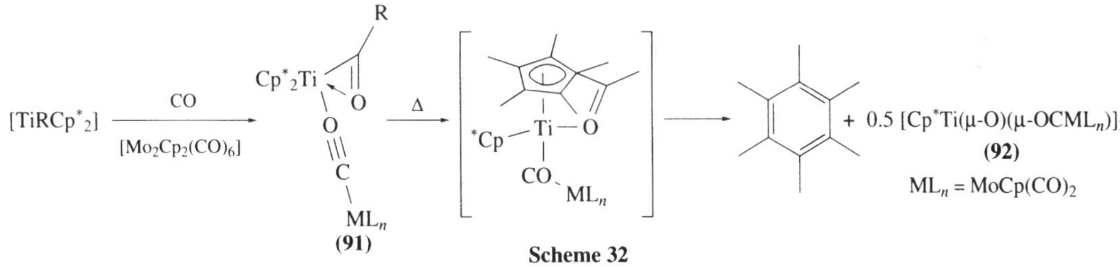

Scheme 32

4.2.3.3 σ-Bonded alkenyl and alkynyl complexes

The reaction of 2-butenyllithium with $TiClCp_2$ produces $[Ti\{C(Me)=CHMe\}Cp_2]$ (**93**) as green crystals in 55% yield. The compound decomposes at 60 °C and forms adducts with dinitrogen reversibly below −60 °C. Some of the reactions of (**93**) are outlined in Scheme 33.[193]

Treatment of $TiClCp_2$ with $NaC{\equiv}C(TMS)$ gives the acetylide complex (**94**) as dark-burgundy crystals;[194] this compound was also obtained by the oxidative addition of $(TMS)C{\equiv}C-C{\equiv}C(TMS)$ to 'TiCp$_2$' generated *in situ* from $TiCl_2Cp_2$ and magnesium.[195] The compound is diamagnetic at room temperature; the IR spectrum shows a low ν_{CC} vibration at 1789 cm^{-1}. The crystal structure of (**94**) was determined. The reaction of $[TiCl\{Me_2Si(\eta-C_5H_4)_2\}]$ with $LiC{\equiv}CPh$ at −40 to −10 °C gives a red acetylide complex with a high $\nu_{C{\equiv}C}$ stretching frequency of 2068 cm^{-1}. Stirring this compound for 12 h at −10 °C gives a new product with $\nu_{C{\equiv}C} = 1769$ cm^{-1} which is converted at room temperature to a green complex for which no $\nu_{C{\equiv}C}$ band was observed but which, on treatment with HCl gas, liberates a mixture of *cis*- and *trans*-PhCH=CH−C≡CPh. This sequence of events suggests the conversion of the originally formed μ,σ-acetylide complex of type (**95**) via a σ,π intermediate (**96**) to the final C–C coupled product

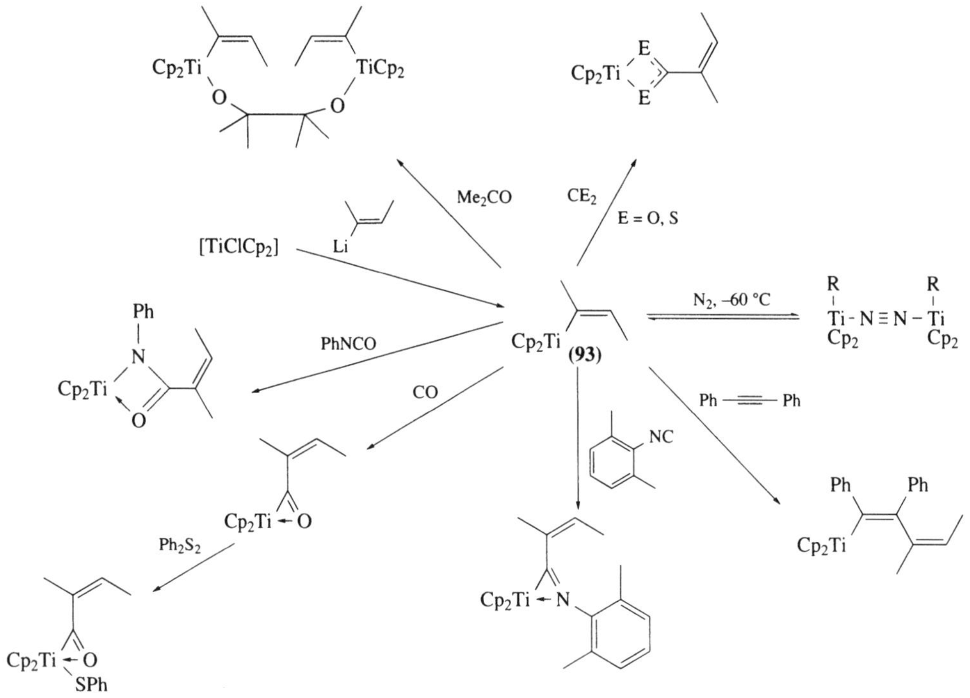

Scheme 33

(**97**) (Scheme 34).[196] EHMO and MNDO calculations on a range of compounds [M(C≡CR)]$_2$ have been carried out. If M = Li, AlH$_2$ or BeH(NH$_3$), the π-orbitals of C≡CR do not contribute to bonding, and structures such as (**95**) are preferred. With M = TiCp$_2$ or ZrCp$_2$ they do, and the C≡CR moiety bends towards the metal to increase overlap with the metal orbitals which leads to a lengthening of the C≡C bond. This also initiates weak bonding interactions between opposite alkynic carbon atoms which can lead to bond formation, as in (**97**).[197]

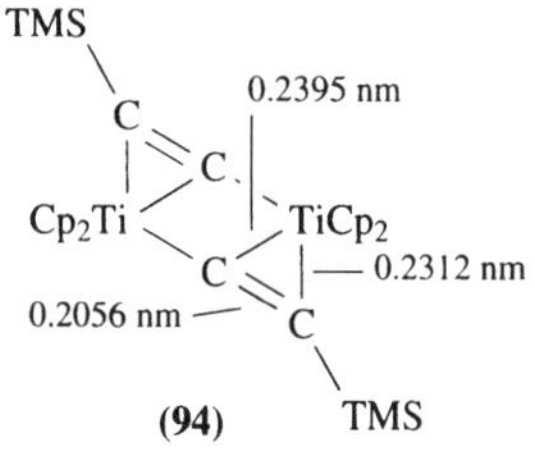

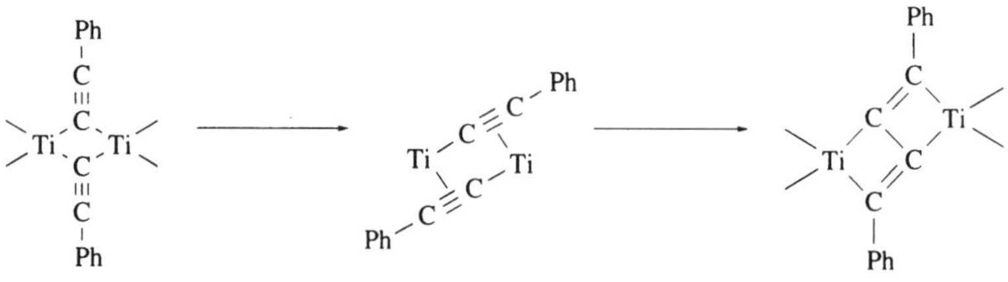

Scheme 34

The anion [TiCp$_2$(C≡CPh)$_2$]$^-$ has been electrochemically generated by the exclusion of light and the reduction of [TiCp$_2$(C≡CPh)$_2$] in THF at a potential of −1.10 V vs. SCE. The process is reversible at −30 °C where the anion is quite stable; there is no loss of Cp$^-$.[198] A complex of such an anion has been synthesized by reducing [TiCl$_2$(η-C$_5$HMe$_4$)$_2$] with magnesium in the presence of (TMS)C≡C–C≡C(TMS) to give (**98**); the complex was characterized by x-ray diffraction. The Ti–C≡C arrangement is close to linear (178.3°), whereas the C≡C–Si angle is 165.4°.[199]

The reduction of [TiCl$_2$Cp$_2$] with magnesium in the presence of only one equivalent of PMe$_3$ gives a black-violet diamagnetic η^1:η^5-Cp bridged complex (Equation (19)), the crystal structure of which was

(98)

determined. The *ipso* carbon atoms of the bridging cyclopentadienyl ring are apparently bonded to both metal centres, with a σ-bond length of 0.221 nm to one titanium (η^1 coordination), and 0.228 nm to the other. The Ti–Ti distance is 0.322 nm.[200] The deep-green mono-Cp bridged complex $[Ti_2(\mu,\eta^1{:}\eta^5{-}C_5H_4)Cp_3]$ reacts with dinitrogen in nonpolar solvents to give a nitrogen adduct, $[\{Ti_2(\mu,\eta^1{:}\eta^5{-}C_5H_4)Cp_3\}_2(N_2)]$.[201]

$$[TiCl_2Cp_2] + Mg + PMe_3 \longrightarrow 1/2 \qquad\qquad\qquad\qquad (19)$$

4.2.3.4 Allyl complexes

A detailed synthetic procedure for the preparation of $[Ti(\eta^3{-}C_3H_5)Cp_2]$ from $TiClCp_2$ and allylmagnesium bromide has been reported.[202] 1-Methallyltitanium complexes are formed when $TiClCp_2$ is treated with vinyl Grignard reagents in the presence of ethene at 0 °C. The expected vinyl(ethene) complex is apparently unstable and cannot be detected; instead, rapid rearrangement to the allyl complex takes place (Scheme 35). The same compound is obtained from $TiClCp_2$, alkyl Grignard reagents and ethene.[203]

$$TiClCp_2 + XMgCH{=}CH_2 \xrightarrow[0\,°C]{C_2H_4}$$

Scheme 35

A range of titanium allyl complexes of the general structure **(99a)**–**(99f)** were synthesized either from $TiClCp_2$ and the corresponding allyl Grignard reagent, or from $[TiCl_2Cp_2]$, Pr^iMgBr and a diene. The crystal structures of **(99a)**–**(99c)** were determined in order to evaluate steric effects which were quantified by a number of structural parameters including the the the bite angle β and the fold angle θ, and the structures of $[Ti(\eta^3{-}C_3H_4Me{-}2)Cp_2]$ and $[Ti(\eta^3{-}C_3H_3Me_2{-}1,3)Cp_2]$ are shown in Figures 7 and 8, respectively. Steric repulsion leads to an increase in the Ti–C-2 bond distance from 0.2376 nm in **(99a)** to 0.2464 nm in **(99b)** and consequently to an increase of θ from 112.3° to 121.5° for **(99b)**, significantly wider than the angles observed for sterically nonrestrained titanium allyls and for allyl compounds of other metals such as platinum (108°). Methyl substitution on the terminal allyl carbons leads to an elongation of the Ti–C bond lengths.[204] The reaction of $[TiCl_2\{C_2H_4(\eta{-}C_5H_4)_2\}]$ with allyllithium gives $[Ti(\eta^3{-}C_3H_5)\{C_2H_4(\eta{-}C_5H_4)_2\}]$.[17]

The reaction of $[Ti(\eta^3{-}C_4H_7)Cp_2]$ at 80 °C under 20 bar ethene pressure gives *cis*- and *trans*-2-butenes and, presumably, titanium vinyl complexes via a σ-bond metathesis process. The titanium allyl complex also catalyses the codimerization of ethene and butadiene at 80 °C to give a mixture of 2,4-hexadiene isomers, although the reaction is slow. An increase in temperature accelerates the reaction and favours the formation of the thermodynamically preferred *trans*,*trans*-hexadiene isomer.[203] $[Ti(\eta^3{-}C_3H_3Me_2{-}1,2)Cp_2]$, prepared from $[TiCl_2Cp_2]$, Pr^nMgBr and isoprene in diethyl ether, reacts with acyl chlorides RCOCl to give mono- and bis(allylation) products (Equation (20)). The proportion of the latter decreases with increasing steric hindrance of R in the order R = Me ≈ PhCH=CH > Me(CH₂)₄ > 2-furyl > Bu ≈ PhCH₂ > Ph > Pr^i. A range of other titanium allyls $[Ti\{\eta^3{-}MeCHC(R^1)CHR^2\}Cp_2]$ was

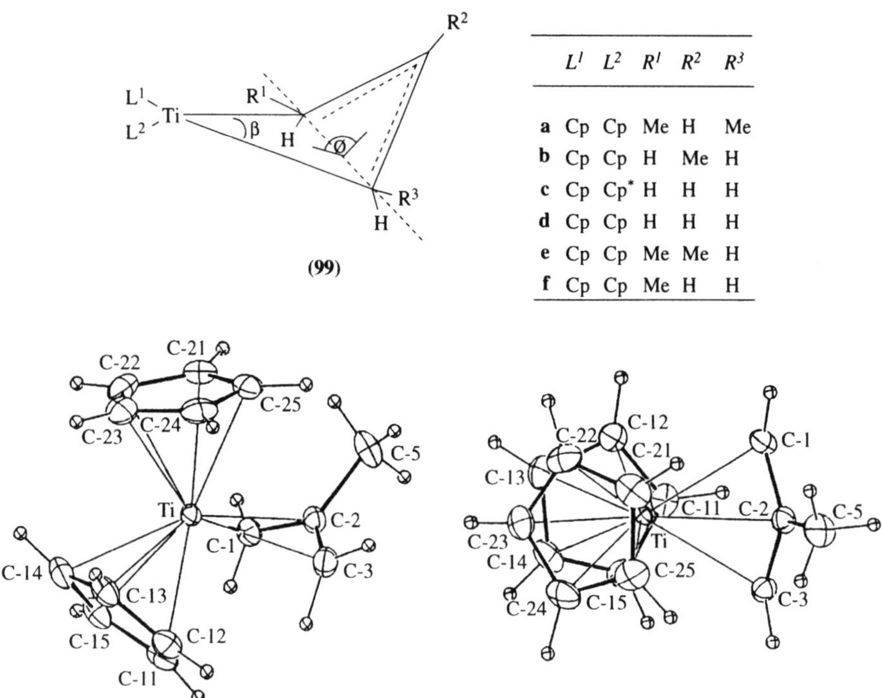

	L¹	L²	R¹	R²	R³
a	Cp	Cp	Me	H	Me
b	Cp	Cp	H	Me	H
c	Cp	Cp*	H	H	H
d	Cp	Cp	H	H	H
e	Cp	Cp	Me	Me	H
f	Cp	Cp	Me	H	H

Figure 7 Structure of [Ti(3-methylallyl)Cp₂] (reproduced by permission of Elsevier Sequoia S. A. from *J. Organomet. Chem.*, 1991, **407**, 191).

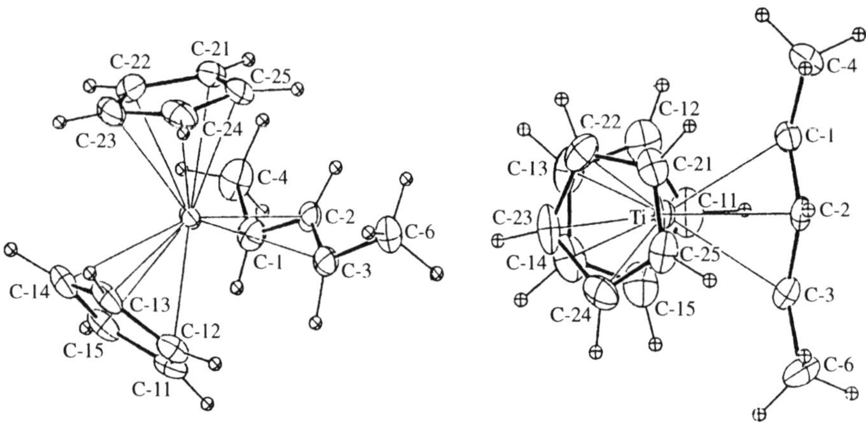

Figure 8 Structure of [Ti(1,3-dimethylallyl)Cp₂] (reproduced by permission of Elsevier Sequoia S. A. from *J. Organomet. Chem.*, 1991, **407**, 191).

prepared from the corresponding dienes ($R^1 = H$, CH_2Ph, $CH_2CH=CH_2$, $SiEt_2Me$ or CH_2TMS, and $R^2 = H$; or $R^1 = H$ and $R^2 = Me$); they generally give high yields of ketones on treatment with acyl halides. The reaction proceeds regiospecifically with attack on the more highly substituted allylic carbon atom, even at 100 °C. Diketones can be similarly prepared.[205] The reaction of [Ti(η^3-C₃H₅)Cp₂] with PhCH=CHCH₂Br gives a mixture of homo- and cross-coupled products; the coupling reaction with other allylic substrates such as allyl acetates or allyl ethers can be catalysed with palladium phosphine complexes.[206]

$$Cp_2Ti \quad + \quad R-\overset{O}{\underset{}{C}}-Cl \quad \longrightarrow \quad R-\overset{O}{\underset{}{C}} \quad + \quad \left(R-\overset{OH}{\underset{}{C}} \right)_2 \tag{20}$$

Refluxing the heterobinuclear compound [TiCp₂(μ-Cl)₂Al(C₆H₁₁)₂], prepared from [TiCl₂Cp₂], LAH and hexadiene, in the presence of hexadiene at 80 °C gives [Ti(η^3-MeCHCHCHEt)Cp₂]. The allyl complexes catalyse the isomerization of alkenes and dienes, probably via $\eta^3 \to \eta^1$ rearrangement to give [Ti(η^1-allyl)(alkene)Cp$_2$] intermediates.[167] Isomerization was also observed on hydrolysis of titanium allyl compounds with long alkyl substituents, generated by the isopropyl Grignard method from $CH_2=CH(CH_2)_nCH=CH_2$ ($n = 2$–10). With $n = 4$ a mixture of predominantly internal octenes was obtained.[185] The reactions of titanium allyls with electrophiles resemble those of Grignard reagents and generally result in allyl transfer reactions, although displacement of an allyl radical is sometimes observed, for example, in the reaction with CS_2 (Scheme 36). The corresponding organic products can be isolated on hydrolysis.[207] The reaction of aldehydes with variously substituted titanium allyl complexes as allyl anion equivalents proceeds with a high degree of regio- and stereoselectivity.[208]

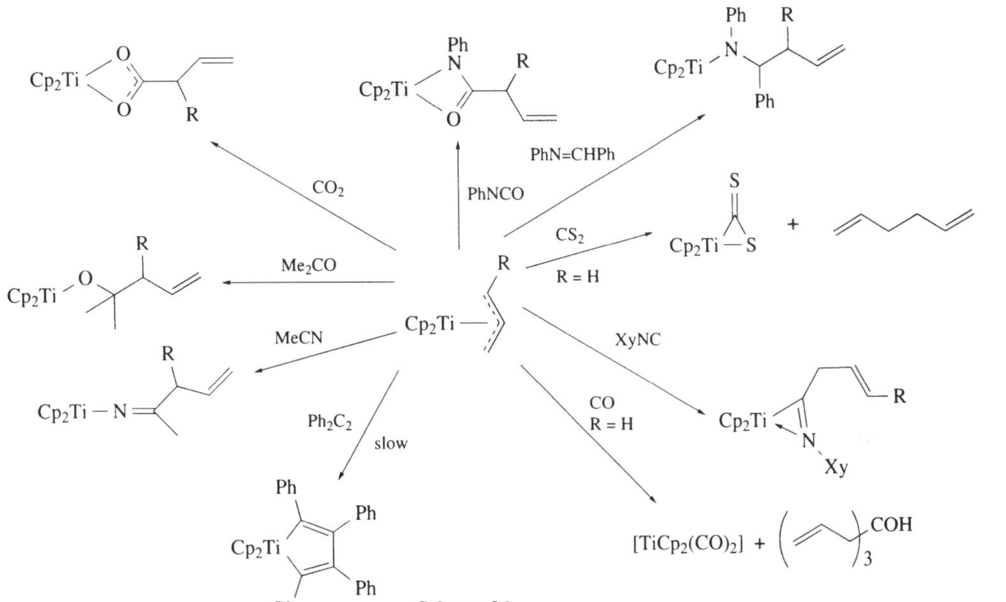

Scheme 36

Treatment of a solution of [Ti(η^3-C$_3$H$_5$)Cp$_2$] with more than six equivalents of AlClMe$_2$ gives a blue solution which smoothly catalyses the polymerization of ethene. [Ti(η^3-C$_3$H$_5$)Cp$_2$] is inactive on its own. There is no polymerization if AlMe$_3$ or AlMeCl$_2$ is used. Although neither the titanium nor the aluminum component reacts with PhC≡C(TMS), the mixture catalyses the carbotitanation of the alkyne to give, after hydrolysis, 1-trimethylsilyl-2-phenyl-1,4-pentadiene.[209,210]

4.2.3.5 Complexes with Ti–N bonds

The nitrogen ligand in the titanium(III) amide complex [Ti(NMePh)Cp*$_2$] is oriented parallel to the plane bisecting the Cp–Ti–Cp angle. The nitrogen atom is trigonal planar, and there is no donation of the nitrogen lone pair to the metal centre. The Ti–N bond is correspondingly long (0.2054(2) nm). The complex forms an adduct with Bu^{n}NC (Equation (21)) in which titanium is tetrahedrally coordinated, with an even longer Ti–N distance (0.2157(5) nm).[211] The titanium ketimido complex [Ti{N=C(H)Bu^t}Cp*$_2$] reacts with CdMe$_2$ to give the titanium(IV) complex [TiMe{N=C(H)Bu^t}Cp*$_2$]. Since the ketimido ligand lies in the Me–Ti–N plane and rotation about the Ti–N bond is slow, the compound exists as a mixture of two stereoisomers in a ratio of 6:1; the isomer with the *t*-butyl substituent pointing outwards is thought to predominate.[178]

4.2.3.6 Complexes with Ti–O bonds

The compound [TiCp$_2$(Me$_2$NCH$_2$CH$_2$NMe$_2$)$_2$] reacts with traces of oxygen to give green crystals of the titanium(III) oxobridged complex. [Ti$_2$Cp$_4$(μ-O)] (**100**). The coordination geometry of the metal is trigonal planar (angle sum 360°). The Ti–O–Ti bridge is almost linear, 170.9°, and the Ti–O bond distances of 0.1838(1) nm are characteristic of some π-bonding contribution of the oxygen lone pairs. By contrast, the reaction of [TiCp$_2$(PMe$_3$)$_2$] with oxygen gives [{Ti(O)Cp$_2$}$_n$] as a yellow polymeric

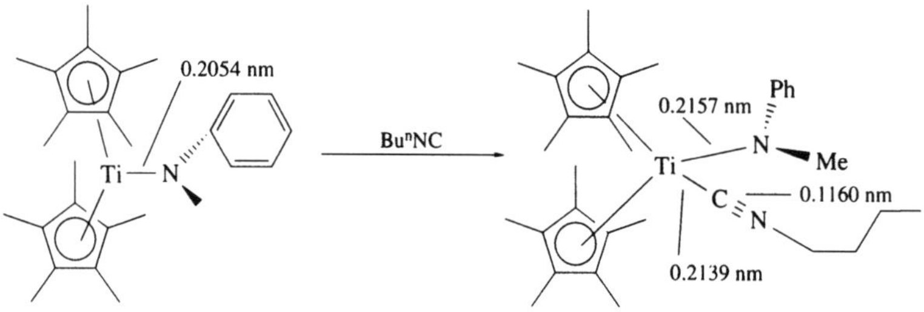

$$ (21) $$

solid.[212] The reaction of [TiMe$_2$Cp$_2$] with SiH(OR)$_3$ leads to the elimination of methane and formation of [TiCp$_2$(μ-OR)]$_2$ as dark-green crystals (R = Me or Et) the structures of which were determined. The Ti–O distances are significantly longer than in (**100**): 0.2065(2) nm (R = Me) and 0.2076(3) nm (R = Et). The Ti–Ti distance of only 0.3358(2) nm is significantly shorter than in [{Ti(μ-Cl)Cp$_2$}$_2$] (0.3955 nm). The unpaired electrons in the alkoxo complex occupy predominantly d_{z^2} orbitals.[213]

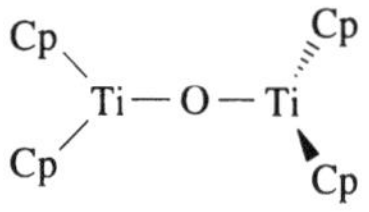

(100)

The reaction of [{TiClCp$_2$}$_2$] with the manganese carbamato complex [Mn$_6$(O$_2$CNEt$_2$)$_{12}$] in toluene gives MnCl$_2$ and blue needles of [TiCp$_2$(O$_2$CNEt$_2$)] in 61% yield. The compound is monomeric and is also accessible from [(TiClCp$_2$)$_2$], HNEt$_2$ and CO$_2$.[214] The reaction of titanium(III) alkyls bearing β-hydrogen atoms with CO$_2$ leads to alkene elimination and formation of titanium formate complexes. Thus [Ti(CH$_2$CH$_2$R)Cp*$_2$] reacts with CO$_2$ to give [TiCp*$_2$(O$_2$CH)] and RCH=CH$_2$. By contrast, [Ti(CH$_2$R)Cp*$_2$] gives [TiCp*$_2$(O$_2$CCH$_2$R)] (R = H, Bu^t or Ph).[215]

For carboxylato complexes it is usually assumed that the difference between the asymmetric and symmetric OCO vibrational frequencies can be taken as a probe for the coordination mode of the carboxylato group, with $\Delta = v_{as} - v_{sym} > 200$ cm^{-1} indicating η^1-coordination, while a value of <105 cm^{-1} is taken to imply η^2-bonding. However, the crystal structures of [TiCp$_2$(O$_2$CBu^t)] ($\Delta = 110$ cm^{-1}) and [NbCp$_2$(O$_2$CBu^t)] ($\Delta = 347$ cm^{-1}) demonstrate that both complexes contain η^2-coordinated carboxylato ligands, and the assignment of coordination modes on the basis of vibrational data must be regarded as doubtful.[216]

[TiCp$_2$(O$_2$CPh)] has been prepared either from [(TiClCp$_2$)$_2$[and PhCO$_2$K in water or from [TiCp$_2$(CO)$_2$] and PhCO$_2$H in THF. The complex contains a bidentate benzoato ligand, with the phenyl substituent only 12.8° inclined with respect to the TiO$_2$C plane. Although [TiCp$_2$(O$_2$CPh)] is monomeric in the solid state, it shows evidence for significant antiferromagnetic exchange interaction ($J = -4.0$ cm^{-1}). This can be traced to the the chain arrangement of the molecules in the crystal which show Ti–Ti long-range contacts of 0.593 nm. Magnetic exchange is not observed for compounds with different crystal packing, such as [TiCp$_2$(O$_2$CR)] (R = 2-naphthyl, p-C$_6$H$_4$Bu^t or p-C$_6$H$_4$Ph), or with [Ti(η-C$_5$Me$_5$)$_2$(O$_2$CPh)]. The dimers (**101**) and (**102**) were also characterized crystallographically; they have Ti–Ti distances of 0.985 nm and 0.783 nm, respectively, and show no magnetic exchange.[217] In contrast with the formation of these titanium(III) carboxylates, the reaction of [(TiClCp$_2$)$_2$] with KO$_2$C–C$\equiv$C–CO$_2$K in water gives the titanium(IV) complex (**103**).[35]

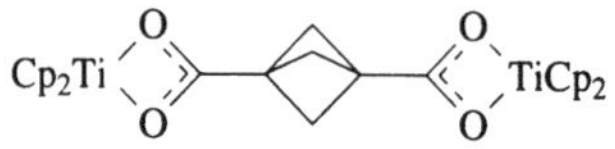

(101)

The reaction of [(TiClCp$_2$)$_2$] with [Co(CO)$_4$]$^-$ in toluene at room temperature over 2 h gives the air-sensitive emerald complex [TiCp$_2${OC–Co(CO)$_3$}] in 37% yield. Dissociation takes place in THF, and the IR spectrum shows the bands of the free [Co(CO)$_4$]$^-$ anion.[218] On prolonged reaction in toluene the compound [TiCp$_2${OCCo$_3$(CO)$_9$}] is formed. The analogous derivatives [TiCp$'_2${OCCo$_3$(CO)$_9$}] have been prepared from [(TiClCp$'_2$)$_2$] and [Co(CO)$_4$]$^-$, or alternatively from [TiCp$'_2$(CO)$_2$] and [Co$_2$(CO)$_8$] (Cp$'$ = C$_5$H$_4$Bu^t, C$_5$H$_4$CMe$_2$Ph, C$_5$H$_4$TMS or C$_5$H$_4$CO$_2$Me) in 49–62% yield.[218] The ionic THF complex,

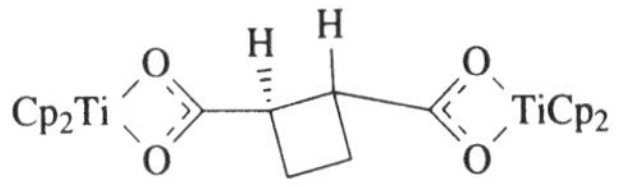

(102)

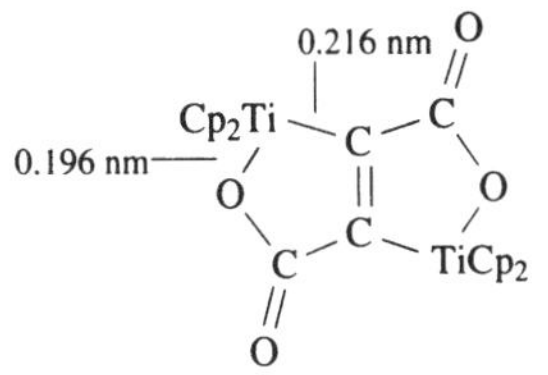

(103)

[TiCp$_2$(THF)$_2$][Co(CO)$_4$], is isolated from the reaction of [(TiClCp$_2$)$_2$] with Tl[Co(CO)$_4$] in THF.[38] Its crystal structure shows the presence of three independent [TiCp$_2$(THF)$_2$]$^+$ cations in the unit cell. The THF ligands are puckered and twisted by 35–50° relative to the TiO$_2$ plane. While the Ti–O(THF) distances are fairly similar, ranging from 0.2190(5) nm to 0.2219(6) nm, the O–Ti–O angles differ by almost 6° between the three cations, with values of 77.2°, 80.0° and 82.9°. Although structural parameters are often rationalized in terms of differences in d^n configuration, such a wide angle variation cautions against such practice.[219]

The reaction of [Ti(CH$_2$Bu^t)Cp*$_2$] with [Mo$_2$Cp$_2$(CO)$_6$[in benzene at room temperature gives the paramagnetic tetranuclear complex (104). The IR spectrum shows C–O stretching frequencies at 1872 cm^{-1} and 1351 cm^{-1}. The crystal structure confirms strong isocarbonyl bonding to Ti (r_{Ti-O} 0.1929(4) nm) and the presence of a Mo–Mo bond (0.2647(1) nm). The C–O bond length of the carbonyl ligands coordinated to titanium is 0.1271(7) nm, approximately a double bond. Extended Hückel calculations suggest a four-electron interaction between oxygen and titanium, in line with π-donation of the oxygen lone pairs to titanium.[220]

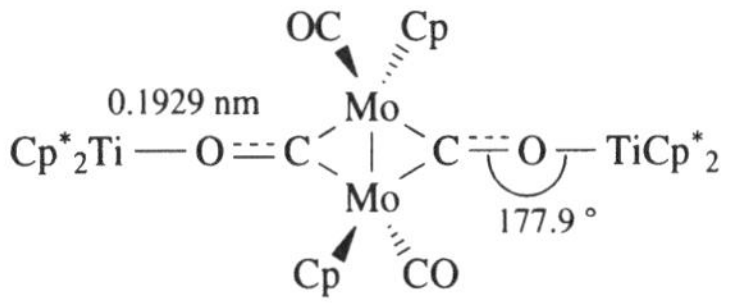

(104)

4.2.3.7 Bis(cyclopentadienyl)titanium(III) hydrides

(i) Homonuclear hydrides

The reaction of the titanium fulvene complex (89) (Scheme 29) with 2-methylpyridine leads to activation of the *o*-CH bond of the pyridine to give the first structurally characterized terminal hydrido complex of titanium(III) (Equation (22)). The v_{Ti-H} vibration is observed at 1495 cm^{-1} (v_{Ti-D} = 1065 cm^{-1}). The Ti–H distance is 0.170(4) nm, and the Ti–N bond is relatively long (0.2340(3) nm).[221] The reduction of [TiCl$_2$Cp$_2$] with potassium metal in toluene at 20–100 °C in a K:Ti ratio of 1.5:1 gives the fulvalene hydrido chloride complex (105; X = H, Y = Cl) in 50–70% yield. The crystal structure of the compound was determined. The Ti–Ti distance is 0.3124(2) nm, the Ti–H bond lengths are 0.214(6) nm and 0.196(5) nm.[222] Similarly, the dihydride (105; X = Y = H) was obtained from the reduction of [TiCl$_2$Cp$_2$] with 2.1 equivalents of Na/Hg in toluene as green crystals in 80% yield. The complex reacts with chlorine gas in dichloromethane at room temperature[223] or with HCl[224] to give the chloro-bridged compound (105; X = Y = Cl). The bromo analogue is obtained from the dihydride and Br$_2$.[224]

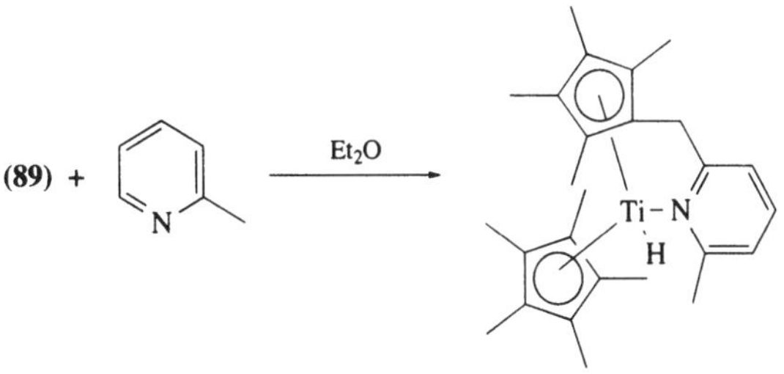

(22)

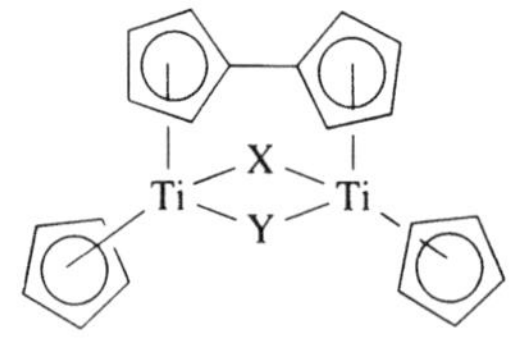

(105)

The reaction of [TiMe$_2$Cp$_2$] or [Ti(CH$_2$Ph)$_2$Cp$_2$] with silanes SiHR$_3$ (SiHR$_3$ = SiH$_3$Ph, SiH(OR)$_3$ or SiHMe(OR)$_2$) gives [(TiHCp$_2$)$_2$(μ-H)] as dark-blue crystals (Equation (23)). The reaction does not proceed with alkyl- or halosilanes, nor with [TiPh$_2$Cp$_2$] as the starting material. The Ti–H stretching vibrations are found at 2000(s) cm^{-1} and 1240–1220(br) cm^{-1}. The EPR spectrum shows a multiplet at g = 1.9937.[225]

$$4\,[\text{TiMe}_2\text{Cp}_2] + 8\,\text{SiHR}_3 \longrightarrow 2\,[(\text{TiHCp}_2)_2(\mu\text{-H})] + 6\,\text{SiMeR}_3 + \text{Si}_2\text{R}_6 + 2\,\text{CH}_4 \tag{23}$$

(ii) Heteronuclear hydrides

The reaction of LAH with [(TiClCp$_2$)$_2$] in aromatic solvents was studied by EPR spectroscopy and was thought to give a linear polynuclear form of [TiCp$_2$(AlH$_4$)] bearing TiHCp$_2$ side groups. Binuclear complexes such as [TiCp$_2$(μ-H$_2$)AlCl$_2$], [TiCp$_2$(μ-H$_2$)AlHCl] and [TiCp$_2$(μ-H$_2$)AlH$_2$] are successively formed as minor components.[226] Since this investigation many such products have been isolated and structurally characterized. [(TiClCp$'_2$)$_2$] and M[AlH$_4$] (M = Li or Na) in diethyl ether or THF give [TiCp$'_2$(μ-H$_2$)AlH$_2$] (Cp$'$ = Cp, C$_5$H$_4$Me or C$_5$H$_4$Bu^t). The complexes [Ti{(CH$_2$)$_n$(η-C$_5$H$_4$)$_2$}(μ-H$_2$)AlH$_2$] (n = 1–3) were obtained similarly. The compounds were characterized by EPR and IR spectroscopy.[227] The system [(TiClCp$'_2$)$_2$]/LAH is complex and may lead to several different products, depending on the reaction conditions. For example, (**106**) was isolated from the reaction in diethyl ether and characterized by x-ray diffraction. The aluminum centre has a distorted trigonal-bipyramidal coordination geometry. The Ti–Al distances are 0.2772(3) nm and 0.2770(3) nm. The compound is a catalyst for the hydrogenation and isomerization of alkenes, and it is thought that partial dissociation of TiHCp$_2$ takes place which is responsible for the catalytic activity.[228] The reaction of [TiCl$_2$Cp$_2$] with [AlH$_3$(OEt$_2$)[in diethyl ether gives purple crystals of (**107**). There are two independent molecules in the unit cell. The average Ti–H and Al–H distances were 0.180 nm and 0.173 nm, respectively. Again the aluminum geometry is trigonal-bipyramidal, and the complex is more appropriately described as an AlH$_3$ adduct of TiHCp$_2$ than a salt of the [AlH$_2$Cl$_2$]$^-$ anion.[229] The reaction of [(TiClCp$_2$)$_2$] with LAH in diethyl ether/ benzene followed by the addition of tetramethylethylenediamine leads to (**108**) the composition of which was confirmed by a crystal structure determination. Ti–H bond lengths of 0.160(11) nm and 0.163(8) nm and a Ti–Al distance of 0.2788 nm were found. The trigonal-bipyramidal geometry of aluminum is maintained.[230] The thermal decomposition of (**108**) in toluene leads to the loss of the amine and attack on the Cp rings to give a complex with Al–Cp σ-bonds (**109**). The compound crystallizes as the toluene solvate.[231,232]

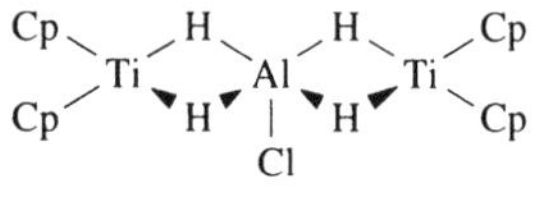

(106) 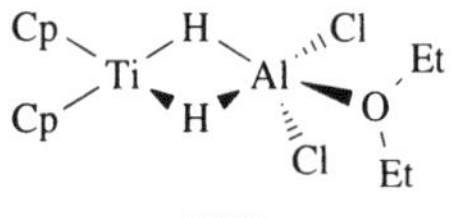   **(107)**

The reaction of [(TiCp$_2$)$_2$(AlH$_4$X)] (X = Cl or [BH$_4$]) with LiMe or LAH in diethyl ether/benzene gives black crystals of [(TiCp$_2$)$_2$(AlH$_4$Me)], whose molecular structure resembles that of (**106**), with a terminal Al–Me bond. During the attempted isolation of this product, crystals of (**109**) (as benzene solvate) were also obtained.[233] [TiCp$_2$(μ-H$_2$)AlH$_2$] in diethyl ether reacts with alcohol vapour with partial alcoholysis of a terminal Al–H bond to give the dark-purple to grey complexes [{TiCp$_2$(μ-H$_2$)AlH(OR)}$_2$] (R = Me, Et or Bu^i) (**110**). The addition of NHEt$_2$ to [TiCp$_2$(μ-H$_2$)AlH$_2$] in ethyl acetate gives the dark-purple amide [{TiCp$_2$(μ-H$_2$)AlH(NEt$_2$)}$_2$] in 68% yield. The crystal structures of the ethoxo and diethylamido complexes show that these compounds are hetereoatom-bridged dimers.[234] Dimethoxyethane is cleaved by [TiCp$_2$(μ-H$_2$)AlH$_2$] to give the methoxy-bridged complex (**110**; X = OMe) which was characterized by x-ray diffraction.[235]

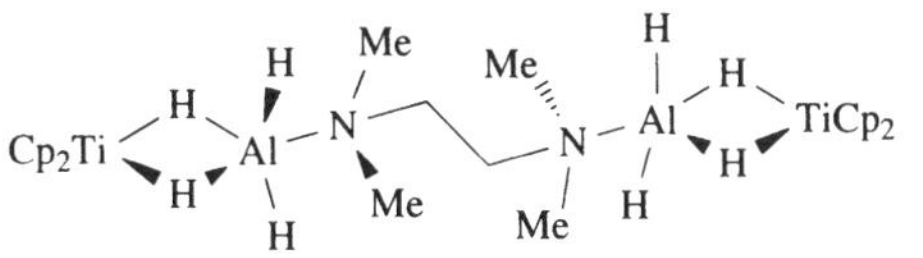

(108)

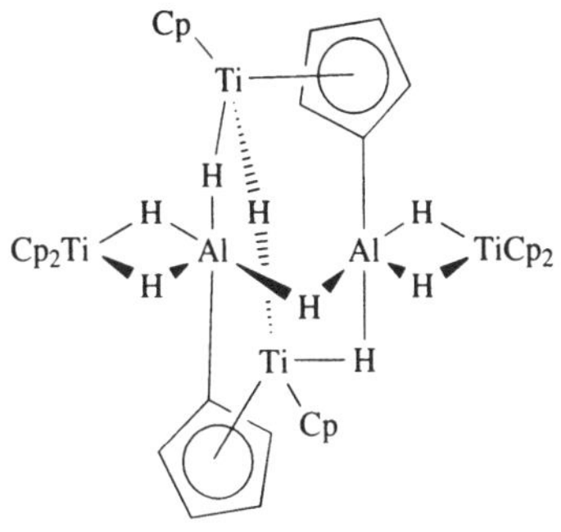

(109)

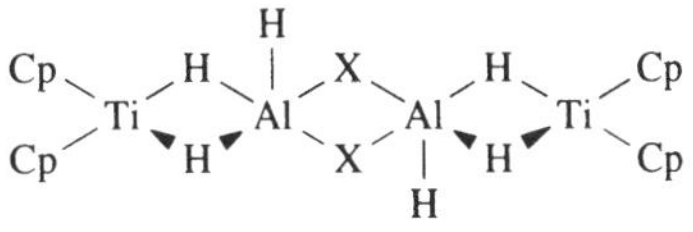

(110) X = OMe, NEt$_2$

The complexes [(TiCp$_2$)$_2$(AlH$_4$Cl)], [(TiCp$_2$)$_2$(AlH$_3$Cl)(Et$_2$O)] and [(TiCp$_2$)$_2$(AlH$_2$Cl$_2$)(Et$_2$O)] all react with Li[BH$_4$] to give the same product (**111**), the molecular structure of which was determined. The aluminum atom is in the centre of a distorted octahedron of hydrogen atoms. The same product is obtained from [(TiCp$_2$)(BH$_4$)] and [(TiCp$_2$)(AlH$_4$)] or Li[AlBH$_4$].[236]

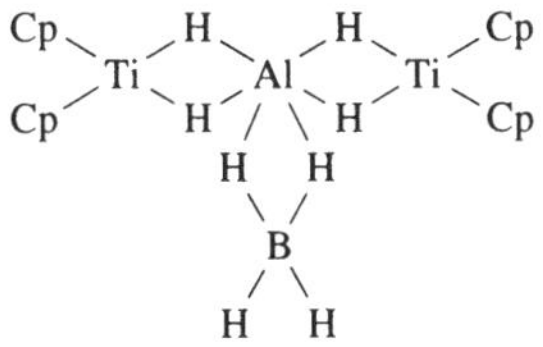

(111)

The complexes [(TiCp'$_2$)(AlH$_4$)] with substituted Cp ligands (Cp' = Cp, C$_5$H$_4$Me, C$_5$H$_2$Me$_3$, C$_5$HMe$_4$, Cp* or C$_5$Me$_4$Et) were synthesized from [(TiClCp'$_2$)$_2$] and LAH in toluene. All except the peralkylated compounds give paramagnetic clusters on ageing which react with butadiene to give [(TiCp'$_2$)(AlH$_4$)] and [Ti(η^3-C$_5$H$_7$)Cp'$_2$].[237] [{TiCp*$_2$}(BH$_4$)] and Li[AlBH$_4$] in diethyl ether give the green, crystallographically characterized tetranuclear compound (**112**). The Al–H–Al bridges are apparently unsymmetric, with Al–H bond distances of 0.16 nm and 0.19 nm. Complex (**112**) slowly decomposes in solution to give the purple trinuclear complex [{TiCp*$_2$}$_2$(μ-H)$_4$AlH], the structure of which resembles that of (**105**), although in this case the hydrogen atoms could not be located.[238] The reaction of [TiClCp*$_2$] with LAH in a 2:1 ratio gives green-black crystals of (**113**) in low yield. The pentamethylcyclopentadienyl ligand has been metallated to form C$_5$Me$_4$CH$_2$–Al moieties. The oxygen probably originates from traces of water. The structure of (**113**) is shown in Figure 9.[239]

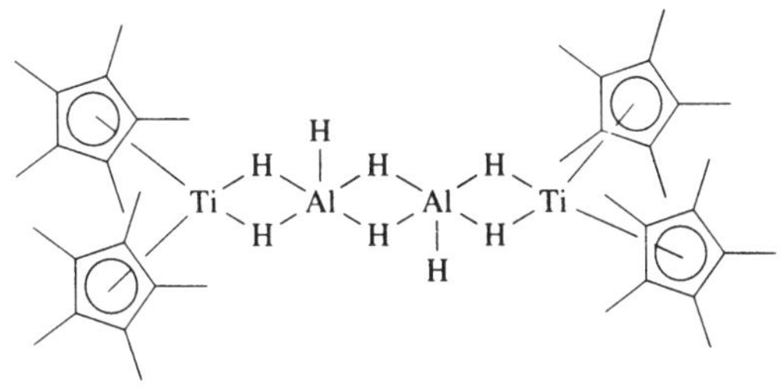

(112)

Terminal alkenes RCH=CH$_2$ react with [TiCp$_2$(μ-H$_2$)AlH$_2$] to give [TiCp$_2$(μ-H)(μ-CH$_2$CH$_2$R)Al(CH$_2$CH$_2$R)$_2$] which eliminates alkanes REt, leading to complexes with Cp–Al σ-bonds, formulated as (**114**) and (**115**) (Scheme 37).[167]

The reaction of [TiCl$_2$(η-C$_5$H$_3$Bu$^t_2$)$_2$] with Li[BH$_4$] gives [Ti(η-C$_5$H$_3$Bu$^t_2$)$_2$(μ-H)$_2$BH$_2$]. The molecular structure of the compound shows a H–Ti–H angle of 58.2° and a Ti–B distance of 0.241 nm.[240]

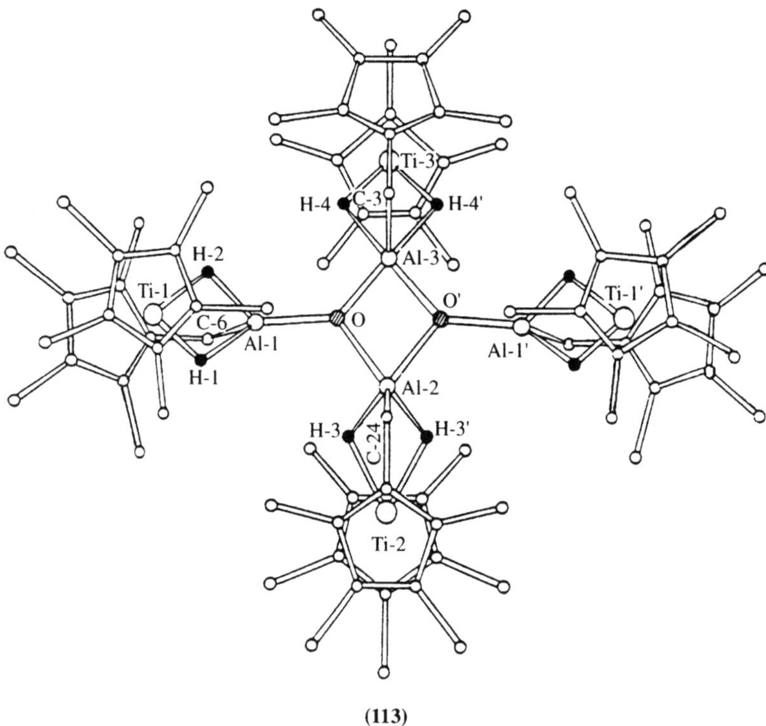

(113)

Figure 9 Structure of [{TiAl(μ_2-H)$_2$(η^1:η^5-C$_5$Me$_4$CH$_2$)(η^5-Cp*)}$_2$(μ_3-O)]$_2$ **(113)** (reproduced by permission of Elsevier Sequoia S. A. from *J. Organomet. Chem.*, 1987, **319**, 69).

Scheme 37

$R^1 = CH_2CH_2R^2$

(114)   **(115)**

(iii) Titanium(III) hydrides as hydrogenation catalysts

The complex [(TiCp$_2$)$_2$(AlH$_4$Me)] catalyses the hydrogenation of 1-hexene more efficiently than [(TiCp$_2$)$_2$(AlH$_4$Cl)] and with significantly higher activity than **(111)**.[233] Most hydrogenation catalysts are generated *in situ* without characterization of the titanium component, although the active species is generally thought to be TiHCp$_2$. Mixtures of [TiCl$_2$Cp$_2$] and sodium or potassium hydride, which have been generated freshly from the metal and hydrogen in the presence of naphthalene and a [TiCl$_4$] catalyst, catalyse the hydrogenation of 1-hexene with an initial activity of over 2000 turnovers per minute. The catalysts decay quickly with time. Cyclopentadienyl ligands (C$_5$H$_4$R) with small substituents give high initial activity but decay rapidly, while large substituents (R) give lower activity but better stability. It is argued that there may be two active species in this system: one, for example, Ti^{IV}(H)ClCp$_2$, which may be very active initially but deteriorates quickly; and Ti^{III}HCp$_2$ which has better long-term stability. Commercially available NaH or KH do not work as activators, probably because of their low surface areas, and LiH is not very reactive. The hydrogenation activity increases in the order [TiCl$_4$] $\ll$ TiCl$_3$Cp < [TiCl$_2$Cp$_2$].[241] [TiCl$_2$Cp$_2$]/Pr^{i}MgBr mixtures are also thought to give TiHCp$_2$. These mixtures catalyse the hydrogenation of simple alkenes and dienes, while α-alkenes with substituents in the 2-position are inert to hydrogenation.[242] [TiCl$_2$Cp$_2$] catalyses the hydrogenation of phenylacetylene with LAH to give, after hydrolysis, a 85:15 mixture of styrene and ethylbenzene. Evidently vinylaluminum reagents are formed; iodine addition prior to hydrolytic workup gives 1- and

2-iodostyrene. A mixture of the *ansa*-metallocene $[TiCl_2\{C_2H_4(\eta\text{-}C_5H_4)_2\}]$ and ethyllithium under a hydrogen atmosphere catalyses the hydrogenation of cyclohexene.[17] Diphenylacetylene is reduced to *cis*- and *trans*-styrene and 1,2-diphenylethane.[243] $[TiCl_2Cp_2]/Bu^iMgBr$ catalyse the hydromagnesation of alkynes. Presumably $(TiBu^iCp_2)_2$ is formed which readily decomposes to $TiHCp_2$, followed by the insertion of alkyne into the Ti–H bond and transmetallation to give a vinylmagnesium reagent (Scheme 38). These reaction steps were shown to be reversible by the addition of a second alkyne $R^1C\equiv CR^1$ which is hydromagnesated with regeneration of the first alkyne $R^2C\equiv CR^2$ (R^2, R^1 = TMS or alkyl).[244]

$$[TiCl_2Cp_2] \xrightarrow{Bu^iMgBr} Cp_2Ti\text{-}CH_2CHMe_2 \longrightarrow TiHCp_2 \xrightarrow{R^2 \equiv R^2}$$

Scheme 38

Aromatic and α,β-unsaturated ketones are reduced to alcohols by $[TiCl_2(C_5H_4R^*)_2]/Pr^iMgBr$ catalysts (R^* = neomenthyl), with elimination of propene. Similar unsubstituted Cp complexes are ineffective.[245] Polymer-attached titanocene dichloride (**116**) is activated by butyllithium to give a catalyst for the hydrogenation of cyclohexene under 1 bar hydrogen. The activity is ca. 10–70 times higher than with comparable homogeneous systems, probably because of better site isolation.[246] The activity of heterogenized catalysts tethered to supports via a silyl linkage (**117**) (n = 0, 1, 3) activated with BuLi has been compared with that of homogeneous analogues. The rate of hydrogenation of 1-octene is similar for the heterogeneous and homogeneous systems. No EPR spectral signals have been observed. The activation parameters for the $[TiCl_2Cp_2]/BuLi$ system are $\Delta H^{\ddagger} = 23.8$ kJ mol^{-1} and $\Delta S^{\ddagger} = -127.8$ J mol^{-1} K^{-1}, and for the heterogeneous catalyst $\Delta H^{\ddagger} = 26.8$ kJ mol^{-1} and $\Delta S^{\ddagger} = -112.9$ J mol^{-1} K^{-1}. These systems are more active than rhodium phosphine catalysts.[247]

(**116**)   (**117**)

(iv) Alkene isomerizations

The catalytic activity for 1-hexene isomerization decreases in the order $[TiCp_2(AlH_4)] \gg [TiCp_2(AlH_3Cl)] > [TiCp_2(AlH_3Br)]$. The CH_2-bridged complexes $[Ti\{(CH_2)_n(C_5H_4)_2\}(AlH_4)]$ (n = 1–3) show little activity. 2-Hexene is produced with a *trans/cis* ratio of 4.7–5.8 which exceeds the equilibrium value of 3.4.[248] Mixtures of $[TiCl_2Cp_2]$ or $[(TiClCp_2)_2]$ with Pr^iMgBr catalyse the isomerization of 1,5-hexadiene to linear and cyclic products, such as methylenecyclopentane and methylcyclopentene, the relative ratios of which depend on the solvent and the reaction conditions. The catalyst is deactivated slowly due to the formation of titanium allyl complexes.[249] The effect of substuted cyclopentadienyl ligands C_5H_4R has been studied for R = H, Me, Et, Pr^n, Bu^n, *n*-pentyl, cyclo-C_5H_9, cyclo-C_6H_{11}, 1-(*m*- or *p*-tolyl)cyclohexyl, C_2H_4OMe, $CH(Me)CH_2OMe$ and $CH_2CH(Me)OMe$, as well as for a wide range of vinyl-, allyl- and benzyl-substituted titanocene derivatives, some 75 titanium catalysts in total. The cyclic:linear ratio varies from 16:84 for R = 1-(*p*-tolyl)cyclohexyl to 90:10 for R = C_2H_4OMe; the latter gives 90% selectivity for methylenecyclopentane at 100% 1,5-hexadiene conversion. Most unsaturated substituents have little effect on selectivity, although R = allyl gives a 95.5 preference for linear hexadiene isomers with internal *cis* and *trans* bonds. Mixed-ligand systems $[TiCl_2CpCp']$ have also been used. Pentamethylcyclopentadienyl complexes have poor activity.[250] $[TiCl_2Cp_2]/Pr^iMgBr$ catalysts are also active for the transformation of 1,5-cyclooctadiene into the

1,3-isomer with excellent activity at room temperature. The effects of solvent, the Ti:Mg ratio, the Cp substituents and the nature of the alkyl groups of the Grignard reagent have been investigated.[251] The isomerization of nonconjugated linear dienes catalysed by TiHCp$_2$ has been studied with the aid of molecular mechanics (MM2) calculations. The ratio of linear to cyclic alkenes produced is, *inter alia*, governed by the cyclic-like conformation of the diene.[252] [TiCl$_2$Cp$_2$]/RMgBr (R = Me$_2$CHCH$_2$ or Bu^t) or [TiCl$_2$Cp$_2$]/C$_2$H$_4$Br$_2$/Mg catalysts convert 1-alkenes to *trans*-2-alkenes. For example, allylbenzene gives 77% *trans*-β-methylstyrene.[253] The isomerization of vinylcyclohexene with [TiCl$_2$Cp$_2$]/Pr^{i}MgBr proceeds stepwise, and good selectivity is achieved by controlling the reaction temperature (Scheme 39).[254]

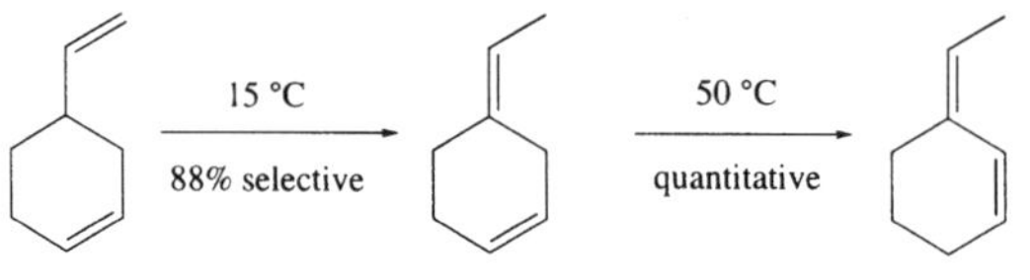

Scheme 39

The reaction of Pr^{i}MgBr with phenyl isocyanate in the presence of [TiCl$_2$Cp$_2$] gives only the *n*-propyl product, presumably because of the rapid isomerization of the Ti–Pr^i intermediate to the sterically less hindered Ti–Pr^n. The more reactive PhNCS gives a mixture of both possible products (Scheme 40).[255]

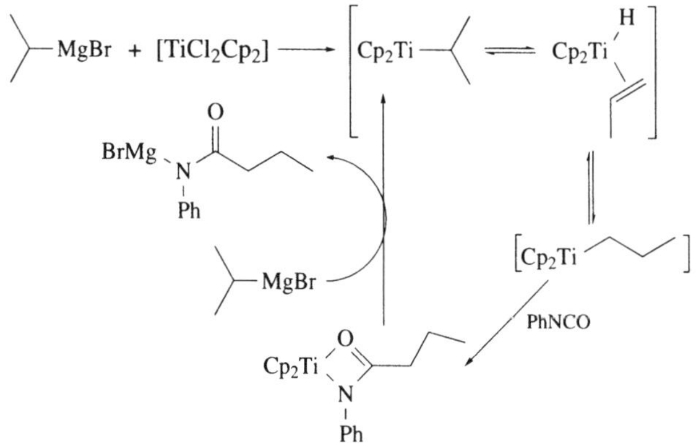

Scheme 40

(v) Silane polymerizations

Silanes react with titanocene dialkyls to give titanium(III) hydrido complexes (see Equation (23)); these are active catalysts for the oligomerization and polymerization of phenylsilane (Equation (24)). The polymer is a brittle white glass if R = Ph but a viscous oil if R = cyclohexyl. The average molecular weight is 900–1400. Dialkylsilanes react much more slowly to give oligomers, while trialkylsilanes are unreactive.[256]

$$n\ RSiH_3 \quad \xrightarrow{[(TiHCp_2)_2(\mu\text{-}H)]} \quad \left[\begin{array}{c} R \\ | \\ -Si- \\ | \\ H \end{array}\right]_n + n\,H_2 \tag{24}$$

Some of the organometallic species involved in this process have been isolated. [TiMe$_2$Cp$_2$] reacts with a moderate excess (ca. 1:3) of SiH$_3$Ph to give blue-black crystals of the silyl complex (**118**), the crystals structure of which was determined. The complex is unstable in solution and is converted to the μ-hydrido complex (**119**); this, too, was identified by x-ray diffraction. The reaction is reversible, and the addition of SiH$_3$Ph to (**119**) regenerates (**118**), while treatment of (**119**) with CO gives SiH$_3$Ph and [TiCp$_2$(CO)$_2$].[257]

[TiMe$_3$Cp], [TiMe$_2$(η-C$_5$H$_4$Me)$_2$] and their zirconium analogues also catalyse the polymerization of SiH$_3$Ph to give products with a degree of polymerization of 10–20. [TiMe$_3$Cp] and [TiMe$_3$Cp*] are, however, much less effective than bis(cyclopentadienyl) complexes. [TiMe$_2$Cp*$_2$] does not catalyse the

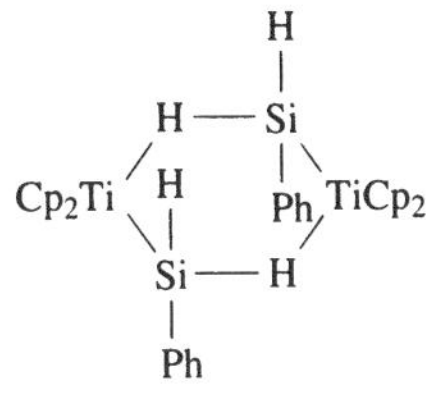

$$(118) \qquad (119)$$

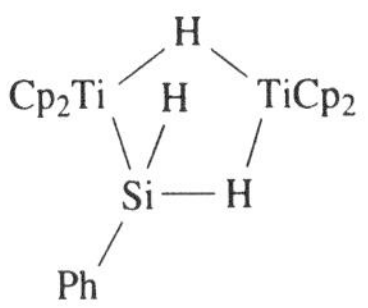

reaction, although line broadening is observed in the variable-temperature NMR spectrum; it is thought that TiCp*$_2$ is formed which rapidly undergoes reversible oxidative addition of SiH$_3$Ph. Other metals are poor catalysts; [UMe$_2$Cp*$_2$] gives oligomers, while [ThMe$_2$Cp*$_2$] produces dimers.[258] [TiMe$_2$Cp$_2$] catalyses the linear oligomerization of Si$_2$H$_2$Me$_4$ in benzene at room temperature to give products with up to five silicon atoms; the reaction involves Si–Si bond cleavage. Higher oligomers are obtained when the temperature is raised to 130 °C. Si$_2$H$_4$Me$_2$ is more reactive; it gives linear and branched polymers and forms a solid product after several weeks which is pyrophoric in air.[259] GeH$_3$Ph in the presence of [TiMe$_2$Cp$_2$] produces a three-dimensional gel, while GeH$_2$Ph$_2$ is selectively dimerized.[260] The silanes SiH$_2$R^{1}R^2 (R^1 = R^2 = Ph; or R^1 = Me, R^2 = Ph) in the presence of [TiPh$_2$Cp$_2$] react with ketones to give high yields of hydrosilation products. It is suggested that [(TiHCp$_2$)$_2$(μ-H)] is produced as the catalytically active species.[261]

4.2.3.8 Silyl complexes

The complexes (118) and (119) were mentioned in the previous section.[257] The reaction of [TiMe$_2$Cp$_2$] with SiH$_2$(R)Ph (R = H, Me or Ph) in the presence of phosphines gives the titanium(III) silyl complexes [Ti(SiHRPh)Cp$_2$(L)] (L = PMe$_3$ or PEt$_3$). The structures of [Ti(SiH$_2$Ph)Cp$_2$(PEt$_3$)] and [Ti(SiHPh$_2$)Cp$_2$(PMe$_3$)] were determined by x-ray diffraction; they show Ti–Si bond lengths of 0.2634(5) nm and 0.2652(1) nm, respectively. The EPR spectra display hyperfine coupling with the phosphorus ligands and the α-hydrogen atoms of the silyl groups, as well as with ^{47}Ti and ^{49}Ti.[262] The complexes [Ti(SiH$_2$Ph)Cp$_2$(PMe$_3$)] and [Ti(SiHMePh)Cp$_2$(PMe$_3$)] were prepared similarly; they have closely related structures. Molecular mechanics calculations suggest that distortions in the structures are the result of intramolecular steric factors.[263]

4.2.3.9 Titanium(III) complexes with bonds to phosphorus or arsenic

The complex [(TiClCp$_2$)$_2$] reacts with LiP(TMS)$_2$ to give the phosphido complex (120). The TiPSi$_2$ core is planar. The Cp–Ti–Cp and Si–P–Si planes form a dihedral angle of only 12.04°, indicative of extensive π-donation of the phosphorus lone pair to titanium. In consequence, the observed Ti–P bond of 0.2468(1) nm is unusually short. By contrast, the reaction between [(TiClCp$_2$)$_2$] and LiAs(TMS)$_2$ gives the paramagnetic dimer (121) with a long Ti–Ti distance of 0.422 nm.[264] The assumption of extensive π-contribution to the Ti–P bonding in (120) is supported by *ab initio* MO calculations.[265] [TiCl$_2$Cp$_2$] is reduced by magnesium in the presence of PMe$_3$ to give [TiClCp$_2$(PMe$_3$)]. The x-ray structure shows a much longer Ti–P bond length of 0.2599(2) nm, typically of a simple adduct without involvement of π-bonding.[266] Methods of generating titanium phosphido complexes and their reactions are shown in Scheme 41. [(TiClCp$_2$)$_2$] reacts with LiPR$_2$ (R = Me or Ph) to give the dimers (122). The PMe$_2$ complex is pyrophoric and reacts violently with chlorocarbons, while the PPh$_2$ compound is more stable. The complexes are paramagnetic.[267] Complexes (122) (R = Et or Ph) are also obtained from [TiCl$_2$Cp$_2$] and LiPR$_2$ which acts as the reducing agent. If the reaction is carried out in the presence of PMe$_3$, complex (123) is isolated as green crystals in 82% yield. The crystal structure of (123) reveals relatively long Ti–PPh$_2$ and Ti–PMe$_3$ distances of 0.2681(3) nm and 0.2636(3) nm, respectively; the lone pair on the PPh$_2$ ligand is approximately perpendicular to the TiP$_2$ plane, with little π-bonding. Even longer Ti–P distances of 0.267(1) nm and 0.271(1) nm are observed for the anion (124) which is formed when [TiCl$_2$Cp$_2$] is treated with an excess of LiPPh$_2$. The lone pairs of both PPh$_2$ groups are oriented out of the TiP$_2$ plane, and the phosphorus atoms are pyramidal.[268] Two equivalents of PMe$_2$TMS react with [TiCl$_2$Cp$_2$] under reduction to give [TiClCp$_2${PMe$_2$(TMS)}] (125) as green crystals. The x-ray structure shows Ti–P and Ti–Cl distances of 0.2639(1) nm and 0.2486(1) nm, respectively. While heating of (125) to 100 °C does not lead to the expected elimination of TMS-Cl and formation of a phosphido complex, reduction with sodium amalgam in THF gives (122) (R = Me) as blue crystals, the molecular structure of which was determined. The Ti_2P_2 ring is planar, with an average Ti–P bond length of 0.2627 nm.[269] Complex (122) (R = Et) reacts with $[\{MCp(CO)_3\}_2]$ (M = Mo or W) in benzene to give the μ-phosphido complex (126); the molybdenum compound was characterized by x-ray diffraction.[270] By contrast, the same reaction in THF gives a phosphorus-free product, $[TiCp_2(THF)\{MoCp(\mu\text{-}CO)(CO)_2\}]$ (18) (*cf.* Merola *et al.*[37]).[271] Cyclic voltammetry measurements on (126) show a one-electron reduction wave at -1.47 V vs. Ag/AgCl typical for reduction of the titanium centre. Similarly, reduction with sodium amalgam gives $[(126)]^-$. The reduced species could not be isolated and was identified by its IR spectrum ($\nu_{CO} = 1742$, 1671 cm^{-1}) and by regeneration of (126) on air oxidation.[270]

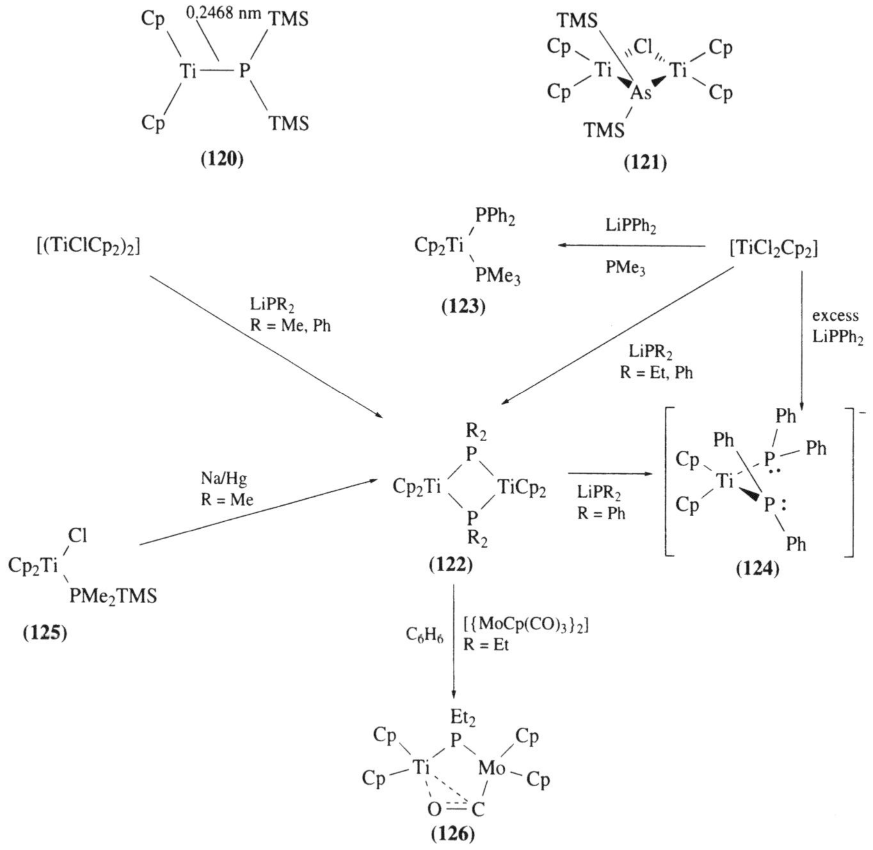

Scheme 41

4.2.4 Cyclooctatetraenyl Complexes of Titanium(III) (by F. G. N. Cloke)

The mixed ring sandwich compound $[TiCp^*(\eta\text{-}C_8H_8)]$ may be obtained from the reaction of $[TiCl_3Cp^*]$ with $K_2[C_8H_8]$ in THF in 34% yield[118] or by the reduction of $[TiCl_3Cp^*]$ with magnesium in the presence of cyclooctatetraene in 31–75% yield (Equation (25)).[117,118]

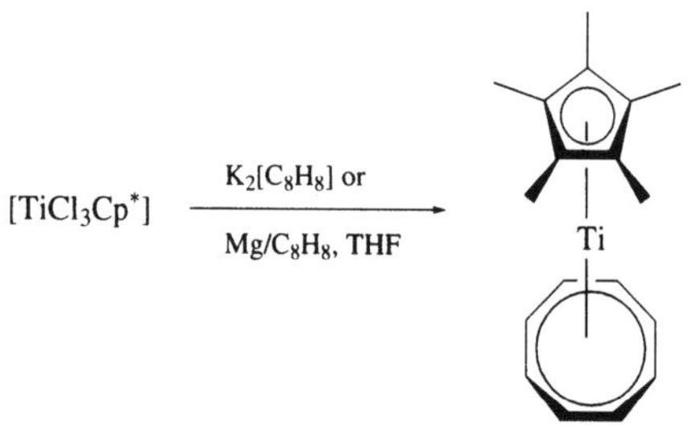

$$[TiCl_3Cp^*] \xrightarrow[\text{Mg/C}_8\text{H}_8,\ \text{THF}]{K_2[C_8H_8]\ \text{or}} \qquad (25)$$

The x-ray crystal structure of $[TiCp^*(\eta\text{-}C_8H_8)]$ shows the expected parallel ring sandwich structure, very similar to that of the unsubstituted analogue $[TiCp(\eta\text{-}C_8H_8)]$.[272] The electronic structures of

[TiCp(η-C$_8$H$_8$)] and [TiCp*(η-C$_8$H$_8$)] have been examined by ENDOR[273] and photoelectron spectroscopy,[119] respectively, and are fully consistent with the molecular orbital scheme for a parallel ring sandwich molecule in which the HOMO is the partially occupied d_{z^2} orbital.

Electrochemical studies in THF on a variety of monocyclooctatetraenyltitanium(III) complexes of the general class [TiCp'(η-C$_8$H$_8$)] (Cp' = Cp, η-C$_5$H$_4$Me, Cp*, η^5-indenyl or fluorenyl) have been carried out and show fully reversible one-electron oxidations to the [TiCp(η-C$_8$H$_8$)]$^+$ cations;[121,274] in the case of [TiCp*(η-C$_8$H$_8$)], bulk electrolysis in THF/[NBu$_4$][ClO$_4$] affords the orange, stable diamagnetic salt [Ti(Cp*(η-C$_8$H$_8$))][ClO$_4$].[121] In no case was one-electron reduction to the 18-electron anion [TiCp'(η-C$_8$H$_8$)]$^-$ observed.

The reaction of [TiCl$_3$(THF)$_3$] with cyclooctatetraene and sodium in the presence of 2.2'-bipyridyl (bipy) affords the formally titanium(III) complex [TiCl(η-C$_8$H$_8$)(bipy)] which has been structurally authenticated (Equation (26)).[275]

$$[\text{TiCl}_3(\text{THF})_3] + \qquad \xrightarrow[\text{THF}]{\text{C}_8\text{H}_8/\text{Na}} \qquad \qquad (26)$$

4.3 REFERENCES

1. J. A. Jensen, S. R. Wilson, A. J. Schultz and G. S. Girolami, *J. Am. Chem. Soc.*, 1987, **109**, 8094.
2. M. D. Spencer, P. M. Morse, S. R. Wilson and G. S. Girolami, *J. Am. Chem. Soc.*, 1993, **115**, 2057.
3. A. Roeder, J. Scholz and K. H. Thiele, *Z. Anorg. Allg. Chem.*, 1983, **505**, 121.
4. B. K. Stein, S. R. Frerichs and J. E. Ellis, *Organometallics*, 1987, **6**, 2017.
5. B. Hessen and J. H. Teuben, *J. Organomet. Chem.*, 1988, **358**, 135.
6. J. Blenkers, B. Hessen, F. van Bolhuis, A. J. Wagner and J. H. Teuben, *Organometallics*, 1987, **6**, 459.
7. J. Blenkers, H. J. de Liefde Meijer and J. H. Teuben, *J. Organomet. Chem.*, 1981, **218**, 383.
8. (a) J. Chen, Y. Kai, N. Kasai, H. Yamamoto, H. Yasuda and N. Nakamura, *Chem. Lett.*, 1987, 1545; (b) H. Yamamoto *et al.*, *Organometallics*, 1989, **8**, 105.
9. M. Mena, P. Royo, R. Serrano, M. A. Pellinghelli and A. Tiripicchio, *Organometallics*, 1988, **7**, 258.
10. B. Hessen, J. Blenkers, J. H. Teuben, G. Helgesson and S. Jagner, *Organometallics*, 1989, **8**, 830.
11. B. Hessen, J. Blenkers, J. H. Teuben, G. Helgesson and S. Jagner, *Organometallics*, 1989, **8**, 2809.
12. D. J. Sikora, D. W. Macomber and M. D. Rausch, *Adv. Organomet. Chem.*, 1986, **25**, 317.
13. B. Demerseman, in 'Organometallic Syntheses', eds. R. B. King and J. J. Eisch, Elsevier, Amsterdam, 1986, vol. 3, p. 22.
14. (a) D. J. Sikora, K. J. Moriarty and M. D. Rausch, *Inorg. Synth.*, 1986, **24**, 147; (b) D. J. Sikora, K. J. Moriarty and M. D. Rausch, *Inorg. Synth.*, 1990, **28**, 248.
15. M. D. Rausch, K. J. Moriarty, J. L. Atwood, W. E. Hunter and E. Samuel, *J. Organomet. Chem.*, 1987, **327**, 39.
16. B. Demerseman, R. Mahe and P. H. Dixneuf, *J. Chem. Soc., Chem. Commun.*, 1984, 1394.
17. J. A. Smith and H. H. Brintzinger, *J. Organomet. Chem.*, 1981, **218**, 159.
18. J. Szymonicek, M. M. Kubicki, J. Besançon and C. Moise, *Inorg. Chim. Acta*, 1991, **180**, 153.
19. L. B. Kool, M. D. Rausch, H. G. Alt, H. E. Engelhardt and M. Herberhold, *J. Organomet. Chem.*, 1986, **317**, C38.
20. M. Casarin, E. Ciliberto, A. Gulino and I. Fragala, *Organometallics*, 1989, **8**, 900.
21. M. J. Calhorda, A. R. Dias, A. M. Galvao and J. A. M. Simoes, *J. Organomet. Chem.*, 1986, **307**, 167.
22. A. R. Dias and J. A. M. Simoes, *Polyhedron*, 1988, **7**, 1531.
23. A. R. Dias, P. B. Dias, H. P. Diogo, A. M. Galvao, M. E. Minas da Piedade and J. A. M. Simoes, *Organometallics*, 1987, **6**, 1427.
24. M. A. V. Ribeiro da Silva and M. D. M. C. Ribeiro da Silva, *J. Organomet. Chem.*, 1990, **393**, 359.
25. B. H. Edwards, R. D. Rogers, D. J. Sikora, J. L. Atwood and M. D. Rausch, *J. Am. Chem. Soc.*, 1983, **105**, 416.
26. M. Tacke C. Klein, D. J. Stufkens and A. Oskam, *J. Organomet. Chem.*, 1993, **444**, 75.
27. G. T. Palmer, F. Basolo, L. B. Kool and M. D. Rausch, *J. Am. Chem. Soc.*, 1986, **108**, 4417.
28. L. B. Kool, M. D. Rausch, M. Herberhold, H. G. Alt, U. Thewalt and B. Honold, *Organometallics*, 1986, **5**, 2465.
29. L. B. Kool, M. D. Rausch, H. G. Alt, M. Herberhold, B. Wolf and U. Thewalt, *J. Organomet. Chem.*, 1985, **297**, 159.
30. G. M. Dawkins, M. Green, K. A. Mead, J.-Y. Salaün, F. G. A. Stone and P. Woodward, *J. Chem. Soc., Dalton Trans.*, 1983, 527.
31. A. F. Hill, H. D. Hönig and F. G. A. Stone, *J. Chem. Soc., Dalton Trans.*, 1988, 3031.
32. S. R. Frerichs, B. K. Stein and J. E. Ellis, *J. Am. Chem. Soc.*, 1987, **109**, 5558.
33. F. Bottomley, D. F. Drummond, G. O. Egharevba and P. S. White, *Organometallics*, 1986, **5**, 1620.
34. B. Bettinger, O. J. Scherer and G. Heckmann, *J. Organomet. Chem.*, 1991, **405**, C19.
35. D. R. Corbin, J. L. Atwood and G. D. Stucky, *Inorg. Chem.*, 1986, **25**, 98.
36. G. Avar, W. Rüsseler and H. Kisch, *Z. Naturforsch., Teil B*, 1987, **42**, 1441.
37. J. S. Merola, R. A. Gentile, B. Ansell, M. Modrick and S. Zentz, *Organometallics*, 1982, **1**, 1731.
38. J. S. Merola, K. S. Campo and R. A. Gentile, *Inorg. Chem.*, 1989, **28**, 2950.
39. J. S. Merola, K. S. Campo, R. A. Gentile, M. A. Modrick and S. Zentz, *Organometallics*, 1984, **3**, 334.

40. T. L. Chen, A. Shaver and T. H. Chan, *J. Organomet. Chem.*, 1989, **367**, C5.
41. F. Bottomley, T. T. Chin, G. O. Egharevba, L. M. Kane, D. A. Pataki and P. S. White, *Organometallics*, 1988, **7**, 1214.
42. A. Shaver and S. Morris, *Inorg. Chem.*, 1991, **30**, 1926.
43. H. A. Harris, A. D. Rae and L. F. Dahl, *J. Am. Chem. Soc.*, 1987, **109**, 4739.
44. K. Seitz and U. Behrens, *J. Organomet. Chem.*, 1985, **288**, C47.
45. J. Benecke and U. Behrens, *J. Organomet. Chem.*, 1989, **363**, C15.
46. G. A. Zank and T. B. Rauchfuss, *Organometallics*, 1984, **3**, 1191.
47. C. G. Marcellus, R. T. Oakley, W. T. Pennington and A. W. Cordes, *Organometallics*, 1986, **5**, 1395.
48. M. Cowie, R. L. DeKock, T. R. Wagenmaker, D. Seyferth, R. S. Henderson and M. K. Gallagher, *Organometallics*, 1989, **8**, 119.
49. G. Fochi, C. Floriani, J. C. Bart and G. Giunchi, *J. Chem. Soc., Dalton Trans.*, 1983, 1515.
50. S. Gambarotta, C. Floriani, A. Chiesi-Villa and C. Guastini, *J. Am. Chem. Soc.*, 1982, **104**, 1918.
51. D. R. Corbin, G. D. Stucky, W. S. Willis and E. G. Sherry, *J. Am. Chem. Soc.*, 1982, **104**, 4298.
52. G. Fochi, *Congr. Naz. Chim. Inorg. 15th*, 1982, 19.
53. O. J. Scherer, H. Swarowsky, G. Wolmershäuser, W. Kaim and S. Kohlmann, *Angew. Chem.*, 1987, **99**, 1178; *Angew. Chem., Int. Ed. Engl.* 1987, **26**, 1153.
54. L. B. Kool, M. D. Rausch, H. G. Alt, M. Herberhold, U. Thewalt and B. Wolf, *Angew. Chem.*, 1985, **97**, 425; *Angew. Chem., Int. Ed. Engl.*, 1985, **24**, 394.
55. L. B. Kool, M. D. Rausch, H. G. Alt, M. Herberhold, B. Honold and U. Thewalt, *J. Organomet. Chem.*, 1987, **320**, 37.
56. L. B. Kool, M. D. Rausch, H. G. Alt and M. Herberhold, in 'Organometallic Syntheses', eds. R. B. King and J. J. Eisch, Elsevier, Amsterdam, 1986, vol. 3, p. 13.
57. G. S. Girolami, G. Wilkinson, M. Thornton-Pett and M. B. Hursthouse, *J. Chem. Soc., Dalton Trans.*, 1984, 2347.
58. H. G. Alt, K. H. Schwind and M. D. Rausch, *J. Organomet. Chem.*, 1987, **321**, C9.
59. C. Jegat, M. Fouassier, M. Tranquille and J. Mascetti, *Inorg. Chem.*, 1991, **30**, 1529.
60. L. B. Kool *et al.*, *J. Chem. Soc., Chem. Commun.*, 1986, 408.
61. D. H. Berry, L. J. Procopio and P. J. Carroll, *Organometallics*, 1988, **7**, 570.
62. H. G. Alt, K. H. Schwind, M. D. Rausch and U. Thewalt, *J. Organomet. Chem.*, 1988, **349**, C7.
63. P. Binger, P. Müller, R. Benn and R. Mynott, *Angew. Chem.*, 1989, **101**, 647; *Angew. Chem., Int. Ed. Engl.*, 1989, **28**, 610.
64. P. Binger, P. Müller, R. Wenz and R. Mynott, *Angew. Chem.*, 1990, **102**, 1070; *Angew. Chem., Int. Ed. Engl.*, 1990, **29**, 1037.
65. P. Binger *et al.*, *Chem. Ber.*, 1992, **125**, 2209.
66. P. Binger *et al.*, *Chem. Ber.*, 1991, **124**, 2165.
67. H. G. Alt, H. E. Engelhardt, M. D. Rausch and L. B. Kool, *J. Am. Chem. Soc.*, 1985, **105**, 3717.
68. G. Alt, H. E. Engelhardt, M. D. Rausch and L. B. Kool, *J. Organomet. Chem.*, 1987, **329**, 61.
69. J. R. Martinez, M. D. Rausch, J. C. W. Chien and H. G. Alt, *Makromol. Chem.*, 1989, **190**, 1309.
70. H. G. Alt, G. S. Herrmann, M. D. Rausch and D. T. Mallin, *J. Organomet. Chem.*, 1988, **356**, C53.
71. H. G. Alt, G. S. Herrmann and M. D. Rausch, *J. Organomet. Chem.*, 1988, **356**, C50.
72. G. S. Herrmann, H. G. Alt and U. Thewalt, *J. Organomet. Chem.*, 1990, **393**, 83.
73. D. F. Hewlett and R. J. Whitby, *J. Chem. Soc., Chem. Commun.*, 1990, 1684.
74. P. Binger *et. al.*, *Chem. Ber.*, 1990, **123**, 1617.
75. S. A. Rao and M. Periasamy, *J. Organomet. Chem.*, 1988, **352**, 125.
76. S. A. Cohen, P. R. Auburn and J. E. Bercaw, *J. Am. Chem. Soc.*, 1983, **105**, 1136.
77. S. A. Cohen and J. E. Bercaw, *Organometallics*, 1985, **4**, 1006.
78. K. Mashima, K. Jyodoi, A. Ohyoshi and H. Takaya, *J. Chem. Soc., Chem. Commun.*, 1986, 1145.
79. M. Mashima and H. Takaya, *Organometallics*, 1985, **4**, 1464.
80. Y. Qian, G. Li, X. Zheng and Y. Z. Huang, *J. Mol. Catal.*, 1993, **78**, L31.
81. F. Scott, H. G. Raubenheimer, G. Pretorius and A. M. Hamese, *J. Organomet. Chem.*, 1990, **384**, C17.
82. M. Akita, H. Yasuda, K. Nagasuna and A. Nakamura, *Bull. Chem. Soc. Jpn.*, 1983, **56**, 554.
83. J. Scholz and K. H. Thiele, *J. Organomet. Chem.*, 1986, **314**, 7.
84. H. G. Alt and G. S. Herrmann, *J. Organomet. Chem.*, 1990, **390**, 159.
85. U. Rosenthal, G. Oehme, V. V. Burlakov, P. V. Petrovskii, V. B. Shur and M. E. Vol'pin, *J. Organomet. Chem.*, 1990, **391**, 119.
86. V. B. Shur, V. V. Burlakov and M. E. Vol'pin, *J. Organomet. Chem.*, 1988, **347**, 77.
87. L. K. Woo, J. A. Hays, R. A. Jacobson and C. L. Day, *Organometallics*, 1991, **10**, 2102.
88. V. V. Bhide, M. F. Farona, A. Djebli and W. J. Youngs, *Organometallics*, 1990, **9**, 1766.
89. V. B. Shur, V. V. Burlakov, A. I. Yanovskii, Y. T. Struchkov and M. E. Vol'pin, *Izv. Akad. Nauk SSSR, Ser. Khim.*, 1983, 1212.
90. V. B. Shur, V. V. Burlakov, A. I. Yanovskii, P. V. Petrovsky, Y. T. Struchkov and M. E. Vol'pin, *J. Organomet. Chem.*, 1985, **297**, 51.
91. G. A. Vaughan, C. D. Sofield, G. L. Hillhouse and A. L. Rheingold, *J. Am. Chem. Soc.*, 1989, **111**, 5491.
92. V. B. Shur, V. V. Burlakov and M. E. Vol'pin, *J. Organomet. Chem.*, 1992, **439**, 303.
93. B. Demerseman, P. Le Coupanec and P. H. Dixneuf, *J. Organomet. Chem.*, 1985, **287**, C35.
94. V. B. Shur, E. G. Berkovich, M. E. Vol'pin, B. Lorenz and M. Wahren, *J. Organomet. Chem.*, 1982, **228**, C36.
95. P. Binger, B. Biedenbach, R. Mynott and M. Regitz, *Chem. Ber.*, 1988, **121**, 1455.
96. J. E. Anderson, T. P. Gregory, C. M. McAndrews and L. B. Kool, *Organometallics*, 1990, **9**, 1702.
97. F. Bottomley, G. O. Egharevba, I. J. B. Lin and P. S. White, *Organometallics*, 1985, **4**, 550.
98. M. R. Smith, P. T. Matsunaga and R. A. Anderson, *J. Am. Chem. Soc.*, 1993, **115**, 7049.
99. J. Z. Liu and R. D. Ernst, *J. Am. Chem. Soc.*, 1982, **104**, 3737.
100. D. R. Wilson, L. Strahl and R. D. Ernst, in 'Organometallic Synthesis', eds. R. B. King and J. J. Eisch, Elsevier, Amsterdam, 1986, vol. 3, p. 136.
101. R. D. Ernst, J. W. Freeman, P. N. Swepston and D. R. Wilson, *J. Organomet. Chem.*, 1991, **402**, 17.
102. R. D. Ernst, J. Z. Liu and D. R. Wilson, *J. Organomet. Chem.*, 1983, **250**, 257.
103. Y. Wang, Y. Q. Liang, Y. J. Liu and J. Z. Liu, *J. Organomet. Chem.*, 1993, **443**, 185.

104. L. Stahl and R. D. Ernst, *J. Am. Chem. Soc.*, 1987, **109**, 5673.
105. T. D. Newbound, A. L. Rheingold and R. D. Ernst, *Organometallics*, 1992, **11**, 1693.
106. E. Melendez, A. M. Arif, M. L. Ziegler and R. D. Ernst, *Angew. Chem.*, 1988, **100**, 1132; *Angew. Chem., Int. Ed. Engl.*, 1988, **27**, 1099.
107. T. E. Waldman, A. M. Wilson, A. L. Rheingold, E. Melendez and R. D. Ernst, *Organometallics*, 1992, **11**, 3201.
108. A. M. Wilson, T. E. Waldman, A. L. Rheingold and R. D. Ernst, *J. Am. Chem. Soc.*, 1992, **114**, 6252.
109. P. T. Di Mauro and P. T. Wolczanski, *Organometallics*, 1987, **6**, 1947.
110. P. Biagini, F. Calderazzo and G. Pampaloni, *J. Chem. Soc., Chem. Commun.*, 1987, 1015.
111. P. Biagini, F. Calderazzo and G. Pampaloni, *J. Organomet. Chem.*, 1988, **355**, 99.
112. U. Thewalt and F. Stollmaier, *J. Organomet. Chem.*, 1982, **228**, 149.
113. S. I. Troyanov and K. Mach, *J. Organomet. Chem.*, 1990, **389**, 41.
114. S. I. Troyanov, J. Polácek, H. Antropiusová and K. Mach, *J. Organomet. Chem.*, 1992, **430**, 317.
115. F. Calderazzo, F. Marchetti, G. Pampaloni, W. Hiller, H. Antropiusová and K. Mach, *Chem. Ber.*, 1989, **122**, 2229.
116. K. Mach and S. I. Troyanov, *J. Organomet. Chem.*, 1991, **414**, C15.
117. J. Blenkers, P. Bruin and J. H. Teuben, *J. Organomet. Chem.*, 1985, **297**, 61.
118. L. B. Kool, M. D. Rausch and R. D. Rogers, *J. Organomet. Chem.*, 1985, **297**, 289.
119. R. R. Andrea, A. Terpstra, A. Oskam, P. Bruin and J. H. Teuben, *J. Organomet. Chem.*, 1986, **307**, 307.
120. D. Gourier and E. Samuel, *Inorg. Chem.*, 1988, **27**, 3018.
121. J. E. Anderson, E. T. Maher and L. B. Kool, *Organometallics*, 1991, **10**, 2084.
122. M. D. Rausch, M. Ogasa, R. D. Rogers and A. N. Rollins, *Organometallics*, 1991, **10**, 2084.
123. B. Demerseman, P. H. Dixneuf, J. Douglade and R. Mercier, *Inorg. Chem.*, 1982, **21**, 3942.
124. L. B. Kool, M. Ogasa, M. D. Rausch and R. D. Rogers, *Organometallics*, 1989, **8**, 1785.
125. M. D. Rausch, M. Ogasa, M. A. Ayers, R. D. Rogers and A. N. Rollins, *Organometallics*, 1991, **10**, 2481.
126. M. Ogasa, M. D. Rausch and R. D. Rogers, *J. Organomet. Chem.*, 1991, **403**, 279.
127. M. L. H. Green, N. J. Hazel, P. D. Grebenik, V. S. B. Mtetwa and K. Prout, *J. Chem. Soc., Chem. Commun.*, 1983, 356.
128. C. E. Davies *et. al.*, *J. Chem. Soc., Dalton Trans.*, 1985, 669.
129. G. M. Diamond, M. L. H. Green and N. M. Walker, *J. Organomet. Chem.*, 1991, **413**, C1.
130. G. M. Diamond, M. L. H. Green, P. Mountford and N. M. Walker, *J. Chem. Soc., Dalton Trans.*, 1992, 2259.
131. B. A. Dolgoplosk, E. I. Tinyakova, I. S. Gusman and L. L. Afinogenova, *J. Organomet. Chem.*, 1983, **244**, 137.
132. A. R. Wills, P. G. Edwards, M. Harman and M. B. Hursthouse, *Polyhedron*, 1989, **8**, 1457.
133. C. Floriani, E. Solari, F. Corazza, A. Chiesi-Villa and C. Guastini, *Angew. Chem.*, 1989, **101**, 91; *Angew. Chem., Int. Ed. Engl.*, 1989, **28**, 64.
134. E. Solari, C. Floriani, A. Chiesi-Villa and C. Rizzoli, *J. Chem. Soc., Dalton Trans.*, 1992, 367.
135. C. Floriani, *Polyhedron*, 1989, **8**, 1717.
136. M. Kilner, G. Parkin and A. G. Talbot, *J. Chem. Soc., Chem. Commun.*, 1985, 34.
137. M. Kilner and G. Parkin, *J. Organomet. Chem.*, 1986, **302**, 181.
138. S. Gambarotta, C. Floriani, A. Chiesi-Villa and C. Guestini, *J. Am. Chem. Soc.*, 1983, **105**, 7295.
139. P. C. Blake, E. Hey, M. F. Lappert, J. L. Atwood and H. Zhang, *J. Organomet. Chem.*, 1988, **353**, 307.
140. W. E. Lindsele and R. A. Parr, *Polyhedron*, 1986, **5**, 1259.
141. A. Vlcek, *J. Organomet. Chem.*, 1985, **297**, 43.
142. A. Razavi, D. T. Mallin, R. O. Day, M. D. Rausch and H. G. Alt, *J. Organomet. Chem.*, 1987, **333**, C48.
143. K. Mach, H. Antropiusová and J. Polácek, *J. Organomet. Chem.*, 1990, **385**, 35.
144. J. Nieman, J. W. Pattiasina and J. H. Teuben, *J. Organomet. Chem.*, 1984, **262**, 157.
145. C. H. Yang and V. L. Goedken, *J. Chem. Soc., Chem. Commun.*, 1986, 1101.
146. S. Cuirli, C. Floriani, A. Chiesi-Villa and C. Guastini, *J. Chem. Soc., Chem. Commun.*, 1986, 1401.
147. S. Gambarotta, C. Floriani, A. Chiesi-Villa and C. Guastini, *J. Chem. Soc., Chem. Commun.*, 1982, 1015.
148. S. Gambarotta, C. Floriani, A. Chiesi-Villa and C. Guastini, *Organometallics*, 1986, **5**, 2425.
149. G. Erker, C. Krüger and R. Schlund, *Z. Naturforsch., Teil B*, 1987, **42**, 1009.
150. W. Skupinski and A. Wasilewski, *J. Organomet. Chem.*, 1985, **282**, 69.
151. P. Gómez-Sal, B. Royo, P. Royo, R. Serrano, I. Saez and S. Martinez-Carreras, *J. Chem. Soc., Dalton Trans.*, 1991, 1575.
152. J. Darkwa, J. R. Lockemeyer, T. B. Rauchfuss and A. L. Rheingold, *J. Am. Chem. Soc.*, 1988, **110**, 141.
153. L. E. Manzer, *Inorg. Synth.*, 1982, **21**, 84; L. E. Manzer, *Inorg. Synth.*, 1990, **28**, 260.
154. J. Wang, F. Liu and L. Bai, *Org. React. Mech.*, 1991, 47.
155. J. W. Pattiasina, H. J. Heeres, F. van Bolhuis, A. Meetsma, J. H. Teuben and A. L. Spek, *Organometallics*, 1987, **6**, 1004.
156. S. I. Troyanov, V. B. Rybakov, U. Thewalt, V. Varga and K. Mach, *J. Organomet. Chem.*, 1993, **447**, 221.
157. J. Okuda, *J. Organomet. Chem.*, 1988, **356**, C43.
158. J. Okuda, *J. Organomet. Chem.*, 1990, **385**, C39.
159. I. F. Urazowski, V. I. Ponomaryov, O. G. Ellert, I. E. Nifant'ev and D. A. Lemenovskii, *J. Organomet. Chem.*, 1988, **356**, 181.
160. M. P. Castellani, S. J. Geib, A. T. Rheingold and W. C. Trogler, *Organometallics*, 1987, **6**, 2524.
161. R. Gómez, T. Cuenca, P. Royo, M. A. Pellinghelli and A. Tiripicchio, *Organometallics*, 1991, **10**, 1505.
162. H. Schwemlein, W. Tritschler, H. Kiesele and H. H. Brintzinger, *J. Organomet. Chem.*, 1985, **293**, 353.
163. A. I. Sizov, M. K. Alekseev, G. L. Soloveichik and B. M. Bulychev, *Koord. Khim.*, 1983, **9**, 492.
164. P. Biagini, F. Calderazzo, G. Pampaloni and P. Franceso-Zanazzi, *Gazz. Chim. Ital.*, 1987, **117**, 27.
165. T. Cuenca, W. A. Herrmann and T. V. Ashworth, *Organometallics*, 1986, **5**, 2514.
166. D. W. Stephan, *Organometallics*, 1992, **11**, 996.
167. K. Mach, H. Antropiusová, V. Hanus, F. Turecek and P. Sedmera, *J. Organomet. Chem.*, 1984, **269**, 39.
168. K. Mach and V. Varga, *J. Organomet. Chem.*, 1988, **347**, 85.
169. D. Cozak and M. Chouinard, *Inorg. Chim. Acta*, 1987, **131**, 75.
170. S. K. Tyrlik, A. Korda, L. Poppe, A. Rockenbauer and M. Gyoer, *J. Organomet. Chem.*, 1987, **336**, 343.
171. V. K. Bel'skii, I. V. Sokolova, B. M. Bulychev and A. I. Sizov, *Zh. Strukt. Khim.*, 1986, **28**, 187.
172. P. Schinnerling and U. Thewalt, *J. Organomet. Chem.*, 1992, **431**, 41.
173. A. R. Dias, M. S. Salema, J. A. M. Simoes, J. W. Pattiasina and J. H. Teuben, *J. Organomet. Chem.*, 1988, **346**, C4.

174. T. Vondrák, K. Mach and V. Varga, *Organometallics*, 1992, **11**, 2030.
175. K. Mach and J. B. Raynor, *J. Chem. Soc., Dalton Trans.*, 1992, 683.
176. L. Bai, F. Lui and J. Wang, *Sci. China, Ser. B*, 1991, **34**, 257.
177. D. Gourier, D. Vivien and E. Samuel, *J. Am. Chem. Soc.*, 1985, **107**, 7418.
178. G. A. Luinstra and J. H. Teuben, *J. Organomet. Chem.*, 1991, **420**, 337.
179. L. P. Battaglia, M. Nardelli, C. Pelizzi, G. Predieri and G. P. Chiusoli, *J. Organomet. Chem.*, 1983, **259**, 301.
180. G. A. Luinstra, J. H. Teuben and H. H. Brintzinger, *J. Organomet. Chem.*, 1989, **375**, 183.
181. S. Renaud, Y. Mugnier, L. Roullier and E. Laviron, *J. Organomet. Chem.*, 1986, **309**, C11.
182. Q. Yanlong, L. Guisheng and Y. Z. Huang, *J. Organomet. Chem.*, 1990, **381**, 29.
183. Y. Qian, Z. Guisheng, H. Xiaofan and Y. Z. Huang, *Synlett*, 1991, 489.
184. H. Schreer and H. O. Fröhlich, *Z. Chem.*, 1983, **23**, 347.
185. H. Lehmkuhl, Y. L. Tsien, E. Janssen and R. Mynott, *Chem. Ber.*, 1983, **116**, 2426.
186. G. A. Luinstra, L. C. ten Cate, H. J. Heeres, J. W. Pattiasina, A. Meetsma and J. H. Teuben, *Organometallics*, 1991, **10**, 3227.
187. J. W. Pattiasina, C. E. Hissink, J. L. de Boer, A. Meetsma, J. H. Teuben and A. L. Spek, *J. Am. Chem. Soc.*, 1985, **107**, 7758.
188. A. R. Dias, M. S. Salema, J. A. M. Simoes, J. W. Pattiasina and J. H. Teuben, *J. Organomet. Chem.*, 1989, **364**, 97.
189. G. A. Luinstra and J. H. Teuben, *J. Chem. Soc., Chem. Commun.*, 1990, 1470.
190. G. A. Luinstra, J. Vogelzang and J. H. Teuben, *Organometallics*, 1992, **11**, 2273.
191. E. J. M. de Boer and J. de With, *J. Organomet. Chem.*, 1987, **320**, 289.
192. H. O. Fröhlich and H. Schreer, *Z. Chem.*, 1983, **23**, 348.
193. E. Klei and J. H. Teuben, *J. Organomet. Chem.*, 1981, **222**, 79.
194. G. L. Wood, C. B. Knobler and M. F. Hawthorne, *Inorg. Chem.*, 1989, **28**, 382.
195. U. Rosenthal and H. Görls, *J. Organomet. Chem.*, 1992, **439**, C36.
196. T. Cuenca, R. Gómez, P. Gómez-Sal, G. M. Rodriguez and P. Royo, *Organometallics*, 1992, **11**, 1229.
197. P. N. V. P. Kumar and E. D. Jemmis, *J. Am. Chem. Soc.*, 1988, **110**, 125.
198. L. Koch, A. Fakhr, V. Mugnier, L. Roullier, C. Moise and E. Laviron, *J. Organomet. Chem.*, 1986, **314**, C17.
199. S. I. Troyanov, V. Varga and K. Mach, *Organometallics*, 1993, **12**, 2820.
200. L. B. Kool, M. D. Rausch, H. G. Alt, M. Herberhold, U. Thewalt and B. Honold, *J. Organomet. Chem.*, 1986, **310**, 27.
201. G. P. Pez, P. Apgar and R. K. Crissey, *J. Am. Chem. Soc.*, 1982, **104**, 482.
202. M. P. Boleslawski and J. J. Eisch, in 'Organometallic Synthesis', eds. R. B. King and J. J. Eisch, Elsevier, Amsterdam, 1986, vol. 3, p. 19.
203. H. Lehmkuhl, E. Janssen and R. Schwickardi, *J. Organomet. Chem.*, 1983, **258**, 171.
204. J. Chen, Y. Kai, N. Kasai, H. Yasuda, H. Yamamoto and A. Nakamura, *J. Organomet. Chem.*, 1991, **407**, 191.
205. A. N. Kasatkin, A. N. Kulak and G. A. Tolstikov, *J. Organomet. Chem.*, 1988, **346**, 23.
206. A. N. Kasatkin, A. N. Kulak, G. A. Tolstikov and S. I. Lomakina, *Zh. Org. Khim.*, 1988, **24**, 2080.
207. E. Klei, J. H. Teuben, H. J. de Liefde Meijer, E. J. Kwak and A. P. Bruins, *J. Organomet. Chem.*, 1982, **224**, 327.
208. F. Sato, U. Uchiyama, K. Iida, Y. Kobayashi and M. Sato, *J. Chem. Soc., Chem. Commun.*, 1983, 921.
209. J. J. Eisch and M. P. Boleslawski, *J. Organomet. Chem.*, 1987, **334**, C1.
210. J. J. Eisch, M. P. Boleslawski and A. M. Piotrowski, in 'Proceedings of the International Symposium on Transition Metals as Organometallic Catalysts for Olefin Polymerization', eds. W. Kaminsky and H. Sinn, Springer-Verlag, Berlin, 1988, p. 371.
211. J. Feldman and J. C. Calabrese, *J. Chem. Soc., Chem. Commun.*, 1991, 1042.
212. B. Honold, U. Thewalt, M. Herberhold, H. G. Alt, L. B. Kool and M. D. Rausch, *J. Organomet. Chem.*, 1986, **314**, 105.
213. E. Samuel, J. F. Harrod, D. Gourier, Y. Dromzee, F. Robert and Y. Jeannin, *Inorg. Chem.*, 1992, **31**, 3252.
214. A. Belforte, F. Calderazzo and P. F. Zanazzi, *J. Chem. Soc., Dalton Trans.*, 1988, 2921.
215. G. A. Luinstra and J. H. Teuben, *J. Chem. Soc., Chem. Commun.*, 1987, 849.
216. I. L. Eremenko, Y. V. Skripkin, A. A. Pasynskii, V. T. Kalinnikov, Y. T. Struchkov and G. G. Aleksandrov, *J. Organomet. Chem.*, 1981, **220**, 159.
217. A. W. Clauss, S. R. Wilson, R. M. Buchanan, C. G. Pierpont and D. N. Hendrickson, *Inorg. Chem.*, 1983, **22**, 628.
218. J. Martin, M. Fauconet and C. Moise, *J. Organomet. Chem.*, 1989, **371**, 87.
219. J. S. Merola, K. S. Campo, R. A. Gentile and M. A. Modrick, *Inorg. Chim. Acta.*, 1989, **165**, 87.
220. E. J. M. de Boer, J. de With and A. G. Orpen, *J. Chem. Soc., Chem. Commun.*, 1985, 1666.
221. J. W. Pattiasina, F. van Bolhuis and J. H. Teuben, *Angew. Chem.*, 1987, **99**, 342; *Angew. Chem., Int. Ed. Engl.*, 1987, **26**, 330.
222. E. G. Perevalova, I. F. Urazowski, D. A. Lemenovskii, Y. L. Slovokhotov and Y. T. Struchkov, *J. Organomet. Chem.*, 1985, **289**, 319.
223. A. Cano, T. Cuenca, G. Rodriguez, P. Royo, C. Cardin and D. J. Wilcock, *J. Organomet. Chem.*, 1993, **447**, 51.
224. T. Wöhrle and U. Thewalt, *J. Organomet. Chem.*, 1993, **447**, 45.
225. E. Samuel and J. F. Harrod, *J. Am. Chem. Soc.*, 1984, **106**, 1859.
226. K. Mach and H. Antropiusová, *J. Organomet. Chem.*, 1983, **248**, 287.
227. A. I. Sizov, G. L. Sohoveichik and B. M. Bulychev, *Koord. Khim.*, 1985, **11**, 339.
228. A. I. Sizov, *et al.*, *J. Organomet. Chem.*, 1987, **335**, 323.
229. E. B. Lobkovskii, G. L. Soloveichik, B. M. Bulychev, R. G. Gerr and Y. T. Struchkov, *J. Organomet. Chem.*, 1984, **270**, 45.
230. E. B. Lobkovskii, G. L. Soloveichik, A. I. Sizov, B. M. Bulychev, A. I. Gusev, and N. I. Kirillova, *J. Organomet. Chem.*, 1984, **265**, 167.
231. E. B. Lobkovskii, G. L. Soloveichik, A. I. Sizov and B. M. Bulychev, *J. Organomet. Chem.*, 1985, **280**, 53.
232. E. B. Lobkovskii, A. I. Sizov, G. L. Soloveichik and B. M. Bulychev, *Isv. Akad. Nauk SSSR, Ser. Khim.*, 1985, 86.
233. A. I. Sizov, I. V. Molodnitskaya, B. M. Bulychev, E. V. Evdokimova, V. K. Bel'skii and G. L. Soloveichik, *J. Organomet. Chem.*, 1988, **344**, 293.
234. T. Y. Sokolova, A. I. Sizov, B. M. Bulychev, E. A. Rozova, V. K. Belskii and G. L. Soloveichik, *J. Organomet. Chem.*, 1990, **388**, 11.

235. V. K. Bel'skii, A. I. Sizov, B. M. Bulychev and G. L. Soloveichik, *Koord. Khim.*, 1985, **11**, 1003.
236. A. I. Sizov, I. V. Molodnitskya, B. M. Bulychev, V. K. Bel'skii and G. L. Soloveichik, *J. Organomet. Chem.*, 1988, **344**, 185.
237. K. Mach, H. Antropiusová, V. Varga and V. Hanus, *J. Organomet. Chem.*, 1988, **358**, 123.
238. V. K. Bel'skii, A. I. Sizov, B. M. Bulychev and G. L. Soloveichik, *J. Organomet. Chem.*, 1985, **280**, 67.
239. E. B. Lobkovskii, A. I. Sizov, B. M. Bulychev, I. V. Sokolova and G. L. Soloveichik, *J. Organomet. Chem.*, 1987, **319**, 69.
240. A. I. Sizov, G. L. Soloveichik, I. F. Urazovskii, V. K. Bel'skii and B. M. Bulychev, *Metalloorg. Khim.*, 1988, **1**, 793.
241. Y. Zhang, S. Liao, Y. Xu and S. Chen, *J. Organomet. Chem.*, 1990, **382**, 69.
242. Y. Qian, G. Li and Y. Huang, *J. Mol. Catal.*, 1989, **54**, L19.
243. H. S. Lee, *Bull. Korean Chem. Soc.*, 1987, **8**, 484.
244. M. A. Djadchenko, K. K. Pivnitsky, J. Spanig and H. Schick, *J. Organomet. Chem.*, 1991, **401**, 1.
245. Y. Zhang and Z. Hu, *Tetrahedron Lett.*, 1988, **29**, 4113.
246. L. Verdet and J. K. Stille, *Organometallics*, 1982, **1**, 380.
247. M. Capka and A. Reissova, *Collect. Czech. Chem. Commun.*, 1989, **54**, 1760.
248. B. M. Bulychev, E. V. Evdokimova, A. I. Sizov and G. L. Soloveichik, *J. Organomet. Chem.*, 1982, **239**, 313.
249. H. Lehmkuhl and Y. L. Tsien, *Chem. Ber.*, 1983, **116**, 2437.
250. Y. Qian, W. Chen, B. Li and S. Chen, *J. Mol. Catal.*, 1990, **60**, 19.
251. Q. Yanlong, L. Jiaqui and X. Weihua, *J. Mol. Catal.*, 1986, **34**, 31.
252. S. Li and Y. Qian, *J. Mol. Sci. (Int. Ed.)*, 1987, **5**, 73.
253. S. A. Rao and M. Periasamy, *J. Organomet. Chem.*, 1988, **342**, 15.
254. J. Lu, Y. Qian, B. He, C. Nie and C. Zhang, *Huaxue Xuebao*, 1988, **46**, 711.
255. Y. Zhang, J. Jiang and Y. Chen, *Tetrahedron Lett.*, 1987, **28**, 3815.
256. C. Aitken, J. F. Harrod and E. Samuel, *J. Organomet. Chem.*, 1985, **279**, C11.
257. C. T. Aitken, J. F. Harrod and E. Samuel, *J. Am. Chem. Soc.*, 1986, **108**, 4059.
258. C. T. Aitken, J. P. Barry, F. Gauvin, J. F. Harrod, A. Malek and D. Rousseau, *Organometallics*, 1989, **8**, 1732.
259. E. Hengge, M. Weinberger and C. Jammegg, *J. Organomet. Chem.*, 1991, **410**, C1.
260. C. T. Aitken, J. F. Harrod, A. Malek and E. Samuel, *J. Organomet. Chem.*, 1988, **349**, 285.
261. T. Nakano and Y. Nagai, *Chem. Lett.*, 1988, 481.
262. E. Samuel, Y. Mu, J. F. Harrod, Y. Dromzee and Y. Jeannin, *J. Am. Chem. Soc.*, 1990, **112**, 3435.
263. J. Britten, Y. Mu, J. F. Harrod, J. Polowin, M. C. Baird and E. Samuel, *Organometallics*, 1993, **12**, 2672.
264. D. Fenske, A. Grissinger, E. M. Hey-Hawkins and J. Magull, *Z. Anorg. Allg. Chem.*, 1991, **595**, 57.
265. M. Ehrig, W. Koch and R. Ahlrichs, *Chem. Phys. Lett.* 1991, **180**, 109.
266. L. B. Kool, M. D. Rausch, H. G. Alt, M. Herberholt, B. Wolf and U. Thewalt, *J. Organomet. Chem.*, 1985, **297**, 159.
267. S. R. Wade, M. G. H. Wallbridge and G. R. Willey, *J. Chem. Soc., Dalton Trans.*, 1983, 2555.
268. D. G. Dick and D. W. Stephan, *Organometallics*, 1991, **10**, 2811.
269. R. Payne, J. Hachgenei, G. Fritz and D. Fenske, *Z. Naturforsch., Teil B*, 1986, **41**, 1535.
270. D. G. Dick, Z. Hou and D. W. Stephan, *Organometallics*, 1992, **11**, 2378.
271. D. G. Dick and D. W. Stephan, *Organometallics*, 1990, **9**, 1910.
272. P. A. Kroon and R. B. Helmholdt, *J. Organomet. Chem.*, 1970, **25**, 451.
273. D. Gourier and E. Samuel, *J. Am. Chem. Soc.*, 1987, **109**, 4571.
274. E. Samuel, D. Guery and J. Vedel, *J. Organomet. Chem.*, 1984, **263**, C43.
275. R. Kempe and J. Sieler, *Z. Kristallogr.*, 1992, **201**, 290.
276. P. Courtot, R. Pichon, J. Y. Salaün and L. Toupet, *Can. J. Chem.*, 1991, **69**, 661.
277. D. M. Curtis, J. J. D'Errico, D. N. Duffy, P. S. Epstein and L. G. Bell, *Organometallics*, 1983, **2**, 1808.

5
Titanium Complexes in Oxidation State +4

MANFRED BOCHMANN
University of East Anglia, Norwich, UK

5.1 INTRODUCTION

Since the mid-1980s the organometallic chemistry of titanium has grown in breadth, and although some new bonding concepts such as alkyl group coordination via metal–H–C ('agostic') interactions were discovered, there is a growing emphasis on applications, particularly in organic synthesis and catalysis, with the development of titanium alkyls as selective alkylating agents and the stereoselective elaboration of bis(cyclopentadienyl) complexes (Section 5.5.1) being particularly noteworthy.

A number of reference works and reviews have appeared. Additional volumes have been added to the *Gmelin Handbook of Inorganic Chemistry* covering mononuclear and dinuclear organometallic compounds of titanium.[1] Annual surveys on titanium, zirconium and hafnium have appeared,[2] and the crystal structures of organometallic titanium complexes have been reviewed.[3]

The extensive use of titanium complexes in the development of organic synthetic methodology is beyond the scope of this chapter. No attempt has been made to cover this aspect comprehensively, particularly since the role of organotitanium reagents in organic synthesis, with an emphasis on the use of titanium alkyls, titanium allyls and related compounds as selective alkylating agents, is the subject of a recent monograph.[4] The use of low oxidation state titanium complexes in organic synthesis, such as reductive coupling of carbonyl compounds, epoxide deoxygenations, coupling reactions with 'Cp$_2$Ti=CH$_2$', and carbonyl alkenations, have been reviewed.[5] The following discussion gives some representative recent examples of these various reaction types.

The reaction of [TiCl$_3$Cp] with chiral alcohols R*OH derived from carbohydrates, such as (1), gives [TiCl(OR*)$_2$Cp] (2) which reacts with allylmagnesium chloride[6] or enolates[7] to give reagents for enantioselective C–C coupling reactions with aldehydes with (for C–C bond formation) unusually high optical yields (Scheme 1). The crystal structure of (2) has been determined;[8] it confirms that R*O is monodentate, without coordination of an ether function to the metal, and with normal Ti–O σ bond distances (0.178–0.181 nm). The synthetic applications of these complexes have been surveyed.[9]

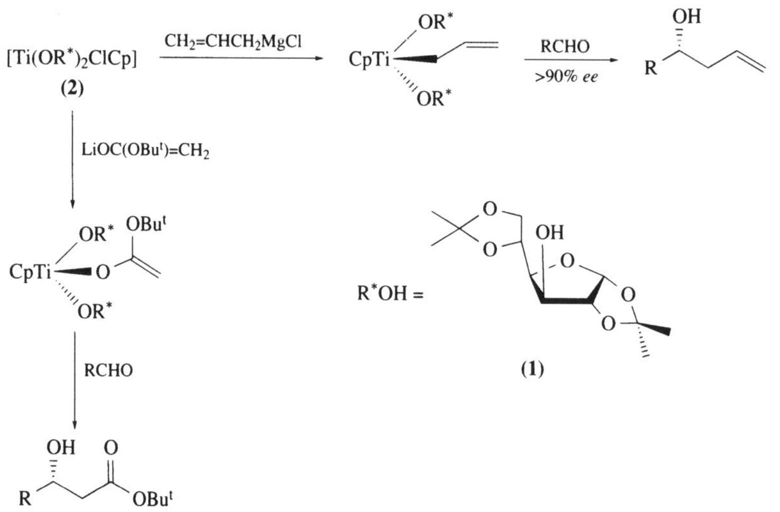

Scheme 1

The chiral complex [Ti(η^1-allyl)(C$_6$F$_5$)Cp(η-C$_5$H$_4$Bu^t)], prepared from allylmagnesium chloride and [TiCl(C$_6$F$_5$)Cp(η-C$_5$H$_4$Bu^t)], reacts with aldehydes less selectively under allyl transfer to give titanium alkoxides in 12–40% *de*. It is likely that there is a loss of chiral information due to ligand rearrangement during the allylation step.[10] Oxiranes react with [Ti(η^1-allyl)(OPh)$_3$] with preferential attack at the more substituted carbon atom to give a mixture of alcohols in which (3) predominates (Equation (1)).[11] Titanium alkyls, such as [TiMeCl$_3$] prepared *in situ*, induce the stereoselective nucleophilic cleavage of acetals (Equation (2)).[12]

[(TiClCp$_2$)$_2$] and MgClBu^i in the presence of dienes, such as cyclopentadiene, gives [Ti(η^3-allyl)Cp$_2$] complexes *in situ*; these react with aldehydes stereoselectively to give the *erythro* isomer only, for steric reasons (Equation (3)).[13] Titanium(III) allyls are also obtained from titanocene dichloride, MgClPr^n and

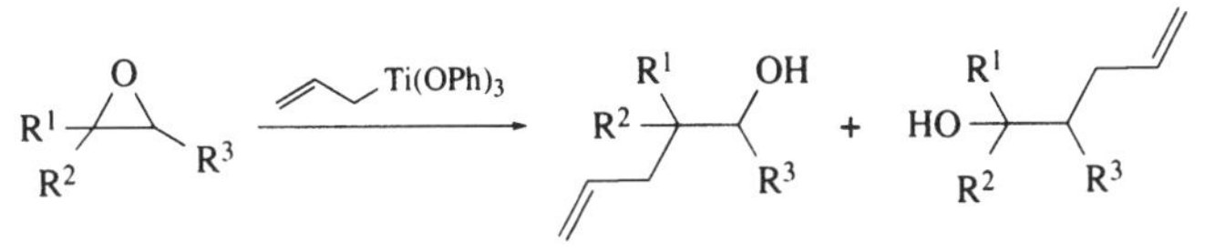

$$\text{(1)}$$

(3)

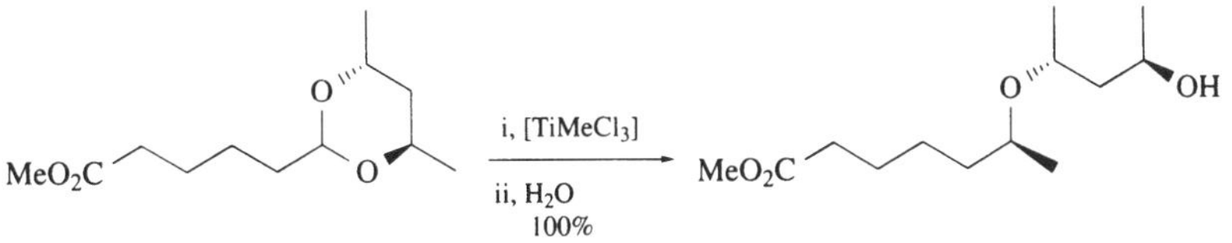

$$\text{(2)}$$

butadiene; allylation of aldehydes followed by hydrolysis with HCl in THF gives the corresponding *syn* and *anti* homoallylic alcohols (Scheme 2). The *anti:syn* ratio increases with increasing steric hindrance of R^1 and R^2, reflecting steric crowding in the transition state.[14]

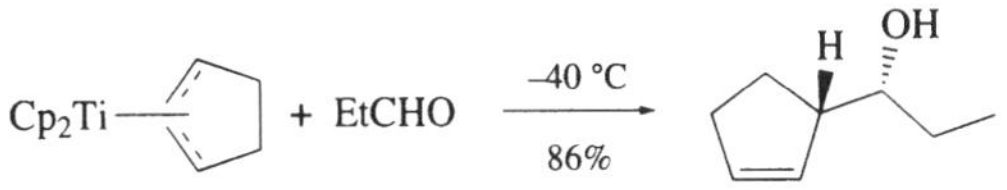

$$\text{(3)}$$

erythro

Favoured   Disfavoured

anti   *syn*

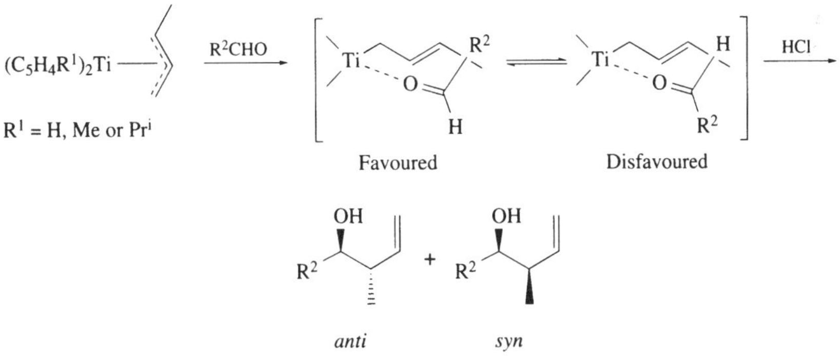

Scheme 2

Electrophilic titanium complexes, such as $[\text{Ti}(O_3SCF_3)_2Cp_2]$, are highly effective catalysts for Diels–Alder reactions; for example, the reaction of 3-pentenone with isoprene in the presence of the catalyst proceeds 10^3–10^5 times faster than the thermal reaction (Equation (4)).[15] A similarly high activity was found for the air-stable and readily isolable ionic complex $[\text{Ti}(Cp^*)_2(H_2O)_2][CF_3SO_3]_2$.[16]

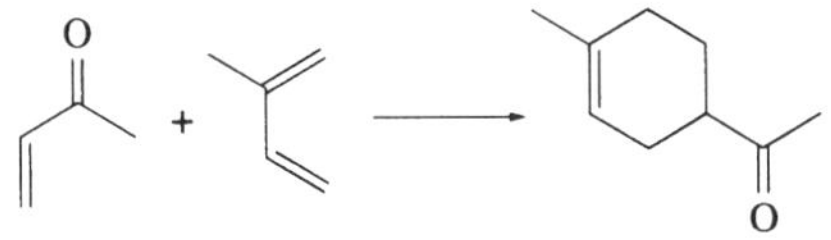

$$\text{(4)}$$

A number of C–C coupling reactions and reductions mediated by low oxidation state titanium compounds are thought to proceed via titanium alkoxide radical intermediates. Thus, the reaction of diaryl ketones with $[\text{TiCp}_2(CO)_2]$ leads to tetraarylethene as the final coupling product, while aliphatic ketones are reduced to alcohols (Scheme 3).[17]

Unsaturated epoxides react with $[(\text{TiClCp}_2)_2]$ to give titanium(IV) radical intermediates which immediately cyclize to give cyclic alcohols after hydrolytic work-up.[18] In a similar way, $[(\text{TiClCp}_2)_2]$ induces the coupling between epoxides and α,β-unsaturated carbonyl compounds. Two equivalents of the titanium(III) complex are necessary for these transformations, and dimetallic titanium alkoxide intermediates are thought to be involved (Scheme 4).[19] By contrast, the reaction of epoxides with $[\text{TiClCp}_2]_2$ in the presence of 1,4-cyclohexadiene leads to reduction to the alcohol, without coupling. Since the regioselectivity in the titanium reaction is given by the ease of titanium attack, it contrasts with that obtainable by epoxide reduction with $[\text{BHEt}_3]^-$, which is controlled by the greatest ease of H^-

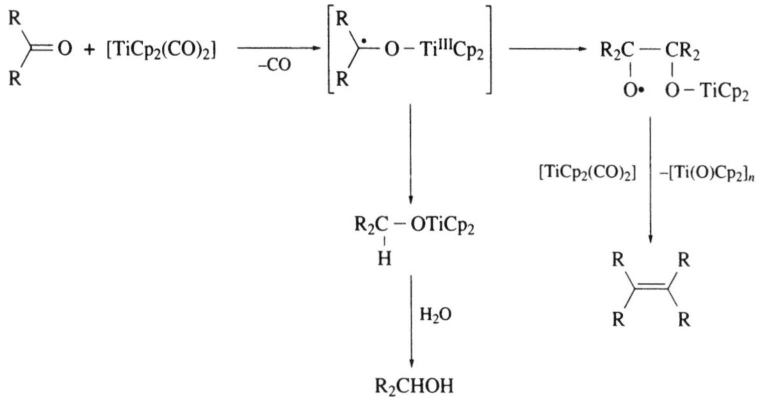

 Titanium Complexes in Oxidation State +4

Scheme 3

approach; hence the two methods are complementary (Scheme 5). In the absence of the diene as hydrogen donor the epoxide is deoxygenated to the corresponding alkene.[20] Similarly, hydroxymethyloxiranes react with two equivalents of $[TiClCp_2]_2$ under opening of the oxirane ring to give a wide range of allylic alcohols. Related epoxyesters are deoxygenated under similar conditions to give α,β-unsaturated esters.[21]

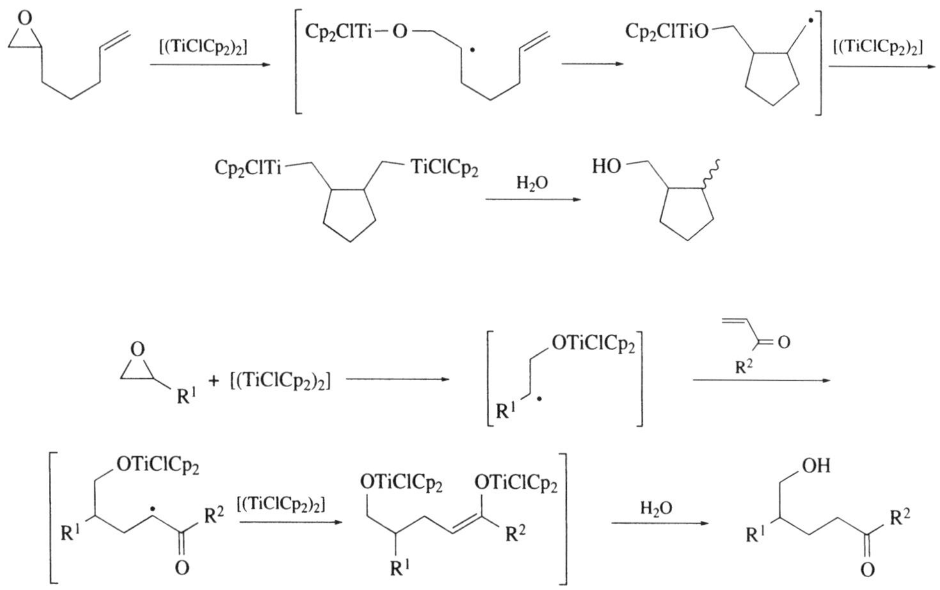

Scheme 4

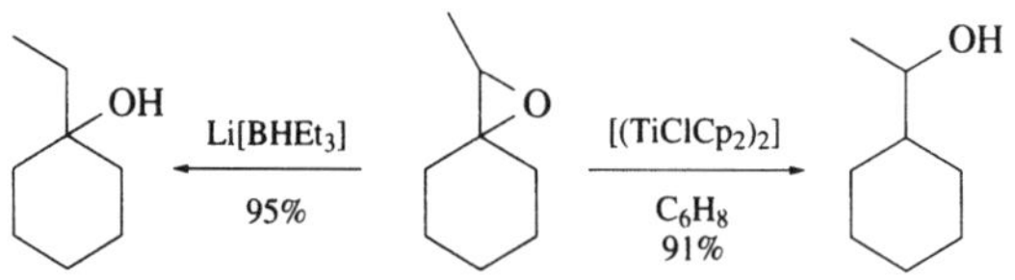

Scheme 5

Titanocene dichloride in the presence of a reducing agent is effective for the diastereoselective coupling of aldehydes to give *threo* and *erythro* isomers of diols (Equation (5)). The selectivity is dependent on the nature of the reducing agent; zinc gives a *threo:erythro* ratio of 11:1, which increases to 80:1 if $MgIPr^i$ is used. Even higher ratios are achieved with $[TiCl_2Cp_2]/MgClBu^s$ in THF. The reason for this effect is unknown.[22]

A mixture of $[TiCl_3Cp]$ and $LiAlH_4$ deoxygenates 1,4-endoxines to arenes; for example, oxabicycloheptadiene dicarboxylate derivatives are converted into dimethyl phthalates.[23] The reaction

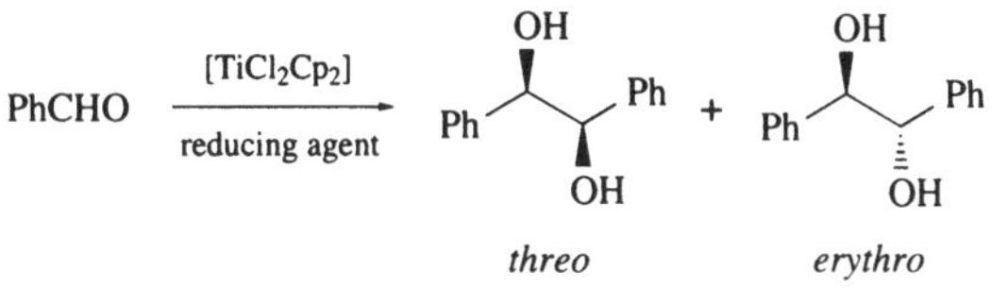

$$PhCHO \xrightarrow[\text{reducing agent}]{[TiCl_2Cp_2]} \quad \text{(threo + erythro products shown)} \tag{5}$$

products of [TiCl$_2$Cp$_2$] with BuLi catalyse the reduction of esters R^{1}C(O)OR^2 by SiH(OEt)$_3$ to give, after hydrolysis, primary alcohols R^{1}CH$_2$OH under mild conditions. The reaction is thought to proceed via titanium(II) intermediates, which would be able to oxidatively add silanes to give species such as [Ti^{IV}(H){Si(OEt)$_3$}Cp$_2$] as reactive intermediates.[24]

The titanacyclobutane complex (4) converts esters into enol ethers and enolizable ketones into alkenes since it is not a strong base abstracting acidic hydrogen, unlike Wittig reagents. However, with sterically hindered ketones such as Bu^{t}C(O)Me titanium enolates result, evidently because the formation of a metallaoxacyclobutane intermediate is disfavoured on steric grounds.[25] Acid chlorides react with (4) under methylene transfer to give enolate complexes which can either be hydrolysed to give ketones, or coupled with benzaldehyde (Scheme 6).[26] The chemistry of (4) and related metallacycles is discussed in Section 5.5.2.4.

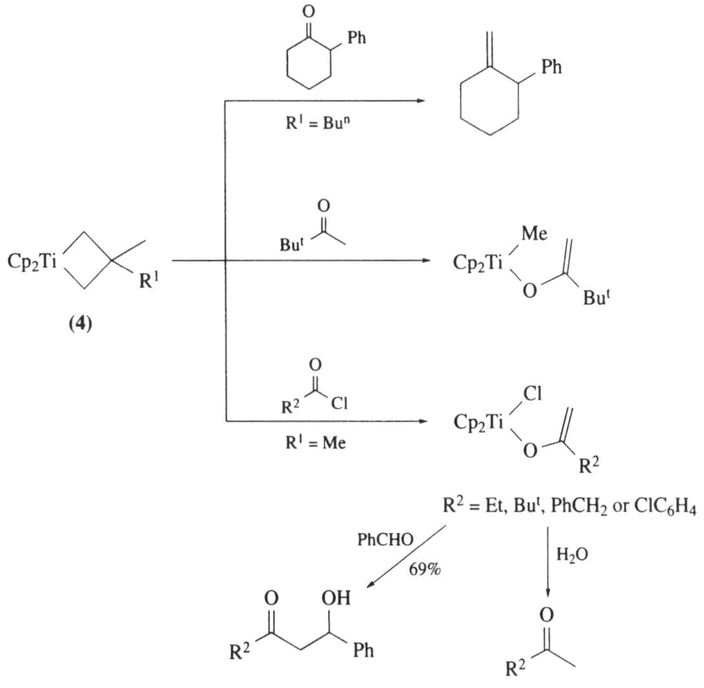

Scheme 6

The reaction of [TiClMe$_2$Cp] with unsaturated primary amines is thought to form (unobserved) titanium imidio intermediates which undergo cyclization, as suggested by the trapping of products with MeOH(D) and Pr^{i}CN (Scheme 7).[27] Titanacycles are also postulated as intermediates in the 'CpTi'-catalysed coupling of enynes with isocyanides to give bicyclic cyclopentenones (Scheme 8).[28]

A number of reactions proceed via titanium(III) intermediates, examples being the silyltitanations and hydromagnesations of alkynes and titanium(III)-catalysed cyclizations. For example, [TiCl$_2$Cp$_2$] reacts with two equivalents of LiSiMe$_2$Ph and 5-decyne, and hydrolysis gives BuC(H)=C(Bu)SiMe$_2$Ph, most probably via [Ti^{III}(SiMe$_2$Ph)Cp$_2$] generated *in situ*.[29] Similarly, Bu^{i}MgBr reacts with RC≡C(TMS) in the presence of catalytic amounts of [TiCl$_2$Cp$_2$] under elimination of isobutene to give (Z)-RCH=C(TMS)MgBr, that is, the product of a *cis* addition of 'MgHBr' to the alkyne. The reaction of the magnesium compound with iodoalkanes gives selectively trisubstituted alkenes in high yields.[30] The hydromagnesation of 1-alkenes with Grignard reagents carrying β-hydrogens follows a similar pathway, most probably involving [Ti^{III}RCp$_2$].[31] The fulvalene hydride complex (5) catalyses the ring closure of 1,2-divinylcyclohexanes at 135–190 °C; the bicyclic compound (6) is thought to be an intermediate in the formation of (7) and (8) (Equation (6)).[32]

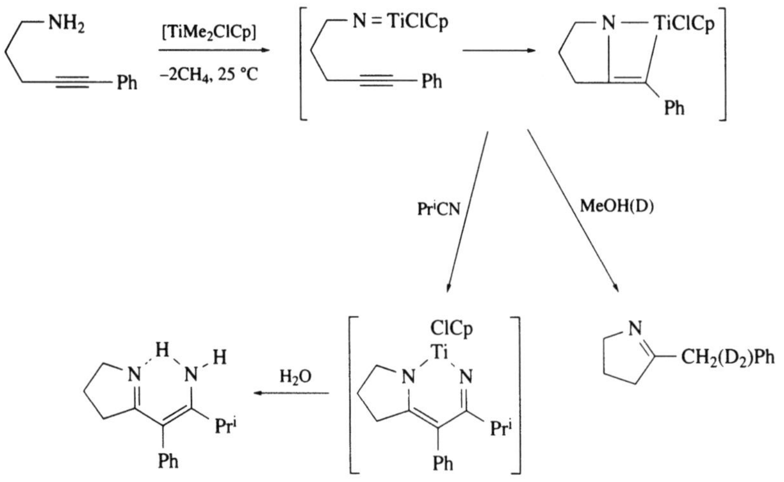

Scheme 7

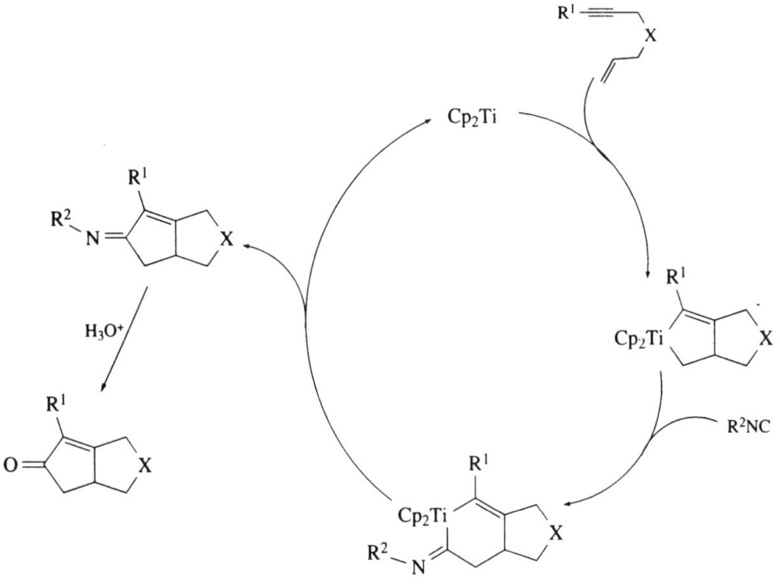

$X = O, NPh, CH_2$ or $C(CO_2R^3)_2$

Scheme 8

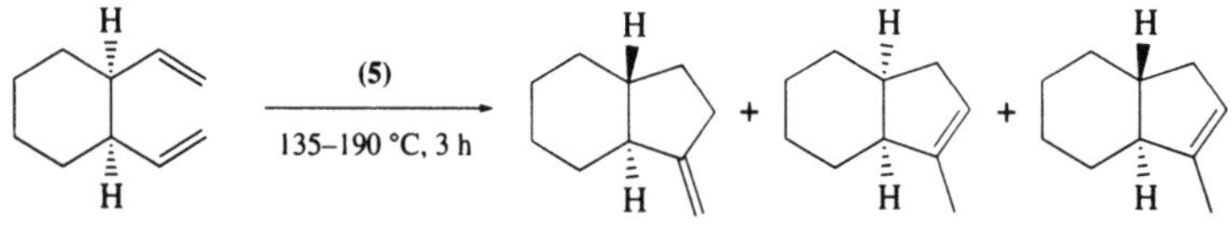

(6)

(6)   **(7)**   **(8)**

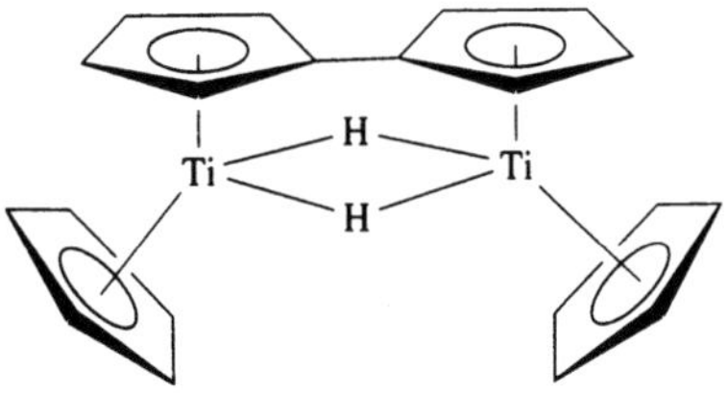

(5)

In a number of cases the nature of the titanium intermediate in reactions of this type is uncertain. The addition of highly active MgH_2 prepared *in situ* with 1,3-dienes to give a mixture of allylic magnesium compounds is catalysed by $[TiCl_4]$ or $[TiCl_2Cp_2]$; zirconium, hafnium and chromium complexes are less effective.[33] $[TiCl_4]/AlClEt_2$ mixtures codimerize dienes and alkynes to give substituted 1,4-cyclohexadienes. Although the presence of titanium(III) complexes can be demonstrated by EPR spectroscopy, most of the titanium in this mixture is apparently diamagnetic, possibly titanium(II).[34] Titanium(II) intermediates have also been postulated in the cycloaddition of norbornadiene with bis(trimethylsilyl)acetylene mediated by $[TiCl_4]/AlCl_2Et$ (Scheme 9).[35] Even less certain are the intermediates in the polymerization of alkynes by $[Ti(OBu)_4]/AlEt_3$ catalysts, although elegant experiments have been able to confirm that chain growth proceeds via alkyne insertion into Ti–C σ bonds. The copolymerization of doubly ^{13}C-labelled and ordinary phenylacetylene in the presence of the titanium catalyst results in a polymer in which the carbon atoms of the original alkyne remain connected via C=C double bonds, whereas ^{13}C NMR spectral measurements of the same polymer prepared with a $MoCl_5/SnMe_4$ catalyst indicate that the labelled atoms are now connected via single bonds. The results are compatible with a conventional, Cossée-type insertion mechanism in the case of titanium, and chain growth via metal alkylidene intermediates in the case of molybdenum (Scheme 10).[36]

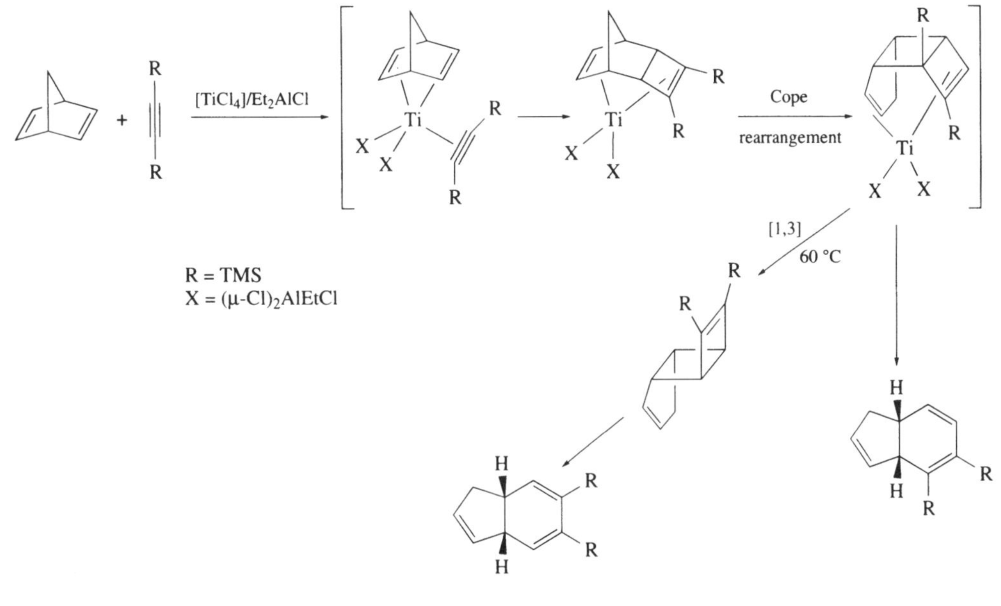

Scheme 9

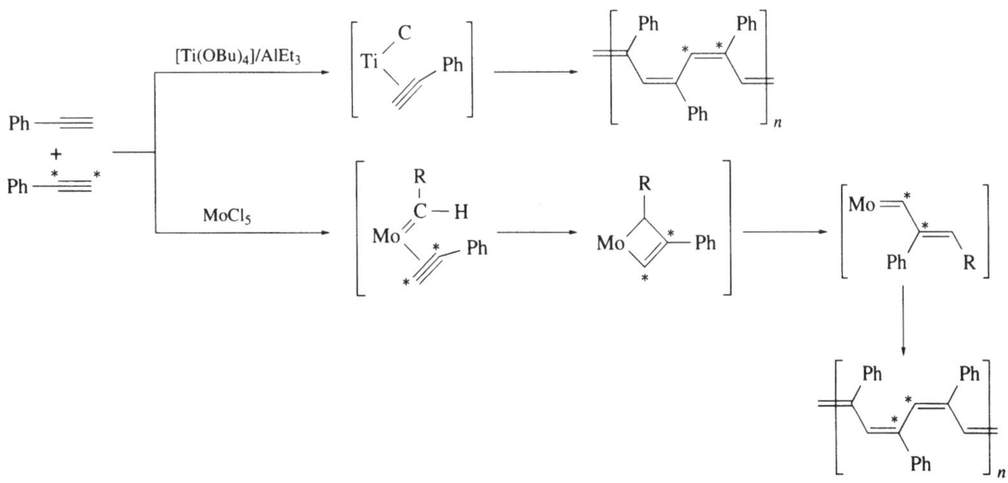

Scheme 10

5.2 TITANIUM ALKYL COMPLEXES WITHOUT ANIONIC π LIGANDS

5.2.1 Synthesis of Titanium(IV) Alkyl Complexes

The monoalkyl complexes [TiEtCl$_3$] and [TiEtBr$_3$] are obtained in high purity by reacting [TiCl$_4$] with PbEt$_4$ in CFCl$_3$ at −80 °C. [TiEtCl$_3$] can be trapped at −25 °C as blue-purple needles which melt at 10 ± 5 °C to a wine-red liquid which decomposes rapidly. The compound is volatile and forms blue adducts with dioxane or THF. Bubbling ethylene through a solution of [TiEtCl$_3$] in C$_2$Cl$_4$ at room temperature leads to TiCl$_3$ and the formation of polyethene.[37] The conproportionation of [TiMe$_4$] with three equivalents of [TiX$_4$] in diethyl ether at −78 °C affords [TiMeX$_3$] (X = Cl, Br or I). The iodide complex is obtained ether-free, while the other compounds form the purple crystalline adducts [TiMeX$_3$(Et$_2$O)$_2$] (X = Cl or Br) in 85% and 67% yields, respectively, which are freed from diethyl ether on recrystallization from pentane at −40 °C. Iodoalkyl complexes are also obtained according to Equation (7). Solutions of [TiMeX$_3$] are yellow to orange in hydrocarbons but purple in diethyl ether and decompose at ~40 °C. Purple [TiMe$_2$Br$_2$] decomposes at 10 °C, the less stable compounds [TiMe$_2$Cl$_2$] and [TiMe$_3$I] at −10 °C, and [TiMe$_3$X] (X = Cl or Br) at −40 °C. The addition of THF to [TiMeX$_3$] leads to reduction and formation of [TiX$_3$(THF)$_3$] (X = Cl, Br or I), while heating [TiMe$_2$X$_2$] in diethyl ether gives black, pyrophoric 'TiX$_2$·nEt$_2$O' (n = 0.1–0.5).[38]

$$[\text{TiMe}_4] + n\,\text{I}_2 \longrightarrow [\text{TiMe}_{4-n}\text{I}_n] + n\,\text{MeI} \qquad (7)$$

The alkylation of [TiCl$_4$] with Grignard reagents in THF gives [MgCl$_2$(THF)$_2$], which can react with [TiCl$_4$] to give the compounds [Mg(THF)$_6$][TiCl$_5$(THF)], [Mg(THF)$_4$Ti(μ-Cl)$_2$Cl$_4$] and [Mg$_2$(μ-Cl)$_2$(THF)$_6$][TiCl$_5$(THF)], probable intermediates in the formation of titanium alkyls.[39] The addition of dmpe to [TiRCl$_3$], generated *in situ* from [TiCl$_4$] and PbR$_4$, gives the deep-red, very sensitive complexes [TiRCl$_3$(dmpe)] (9) (R = Me)[40] and (10) (R = Et)[41] (dmpe = Me$_2$PCH$_2$CH$_2$PMe$_2$). As Figure 1 shows, these complexes contain bonding interactions between the metal centres and hydrogen atoms bonded to the α or β carbon of the alkyl ligand and provide textbook examples for what has become known as 'α-agostic' or 'β-agostic' alkyl bonding modes. This interaction results in significant distortions of the alkyl ligands (see Section 5.2.2). The compounds are seen as possible models for the bonding of the alkyl chain in Ziegler–Natta alkene polymerization catalysts.[40,41]

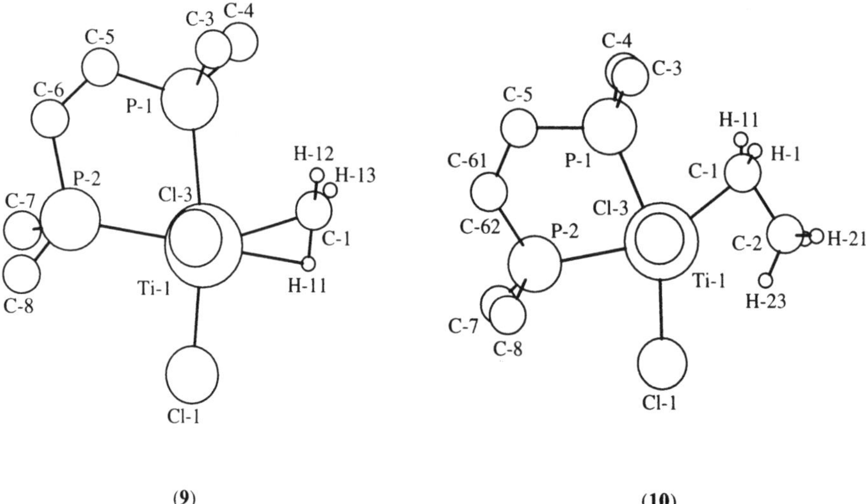

Figure 1   Molecular structures of [TiMeCl$_3$(dmpe)] (9) and [TiEtCl$_3$(dmpe)] (10) (reproduced by permission of the Royal Society of Chemistry from *J. Chem. Soc., Chem. Commun.*, 1982, 802, 1410).

The bulky titanium aryloxo complex [TiCl$_2$(OR)$_2$] reacts with Mg(CH$_2$TMS)$_2$ in petroleum to give the yellow–tan monoalkyl complex [TiCl(CH$_2$TMS)(OR)$_2$] in 61% yield. The compound is thermally stable and melts at 169–170 °C (R = 2,6-Bu$^t_2$-4-MeC$_6$H$_2$).[42]

The reaction of the amido complex [TiCl{N(TMS)$_2$}$_3$] with MeLi gives [Ti(Me){N(TMS)$_2$}$_3$], whose crystal structure was determined.[43] The structurally related complex [TiCl{N(C$_2$H$_4$N(TMS)$_3$}] reacts with lithium alkyls to give the alkyl complexes (11) (R = Bu^n or Bu^s). The tetradentate ligand provides significant chelate stabilization, and unlike most other titanium alkyl complexes containing β-hydrogen

$$\text{Scheme 11}$$

Scheme 11

atoms these complexes are thermally stable at room temperature and can be purified by flash chromatography. On heating to 66 °C, however, alkane is lost, and one of the silyl groups is metallated (Equation (8)).[44] The diazadiene complex $[TiCl_2(Bu^tNCHCHNBu^t)]$ is best considered as a titanium(IV) complex because of the pronounced π acceptor capacity of the diazadiene ligand. The compound is alkylated by $MgBr(CH_2Bu^t)$ to give $[TiCl(CH_2Bu^t)(Bu^tNCHCHNBu^t)]$ as red crystals in 35% yield. The compound sublimes at >120 °C/0.1 torr.[45] Titanium methyl and even t-butyl complexes are prepared by the alkylation of the imido complex $[TiCl(NSiBu^t_3)(NHSiBu^t_3)(THF)]$ with LiR. The yellow t-butyl complex is thermally unstable but can be stored at −20 °C, while the methyl complex thermolyses in benzene in the presence of hydrogen to give a dinuclear imido compound (Scheme 11).[46] Treatment of $[TiCl_2(NMe_2)_2]$ with an excess of the phosphorus ylide $P(CH_2)(NMe_2)_3$ gives low yields of (**12**) as very air-sensitive red crystals, whose structure was determined by x-ray diffraction (Equation (9)). Compound (**12**) reacts with excess $[TiCl_2(NMe_2)_2]$ to give orange crystals of $[TiCl_2\{C=P(NMe_2)_3\}_2]$ in 17% yield, whose reactions with aldehydes lead to C–C coupled products such as allenes and dienes.[47]

$$(Me_2N)_3P{=}CH_2 + [TiCl_2(NMe_2)_2] \longrightarrow (Me_2N)_3P{=}\!\!\!\!\diamondsuit\!\!\!\!{=}P(NMe_2)_3 \qquad (9)$$

(**12**)

The bulky alkoxo ligand $Bu^t_3CO^-$ can be thought of as a steric cyclopentadienyl equivalent and provides an even greater degree of stabilization. Thus, the dimethyl complex $[TiMe_2(OR)_2]$ is readily obtained from $[TiCl_2(OR)_2]$ and MeLi as a thermally stable, white, crystalline solid in 76% yield ($R = CBu^t_3$). Less than 10% decomposition is observed on heating benzene solutions of the compound for 2 weeks to 100 °C. The white trimethyl complex $[TiMe_3(OR)]$ is prepared similarly; it is thermally sensitive.[48] Analogous aryloxide compounds are prepared from the reaction of titanium alkyls with phenols. Thus $[TiMe_4]$, prepared *in situ* from $[TiCl_4]$ and MgMeI, reacts with ArOH in THF at −78 °C to give the lemon-yellow compound $[TiMe_2(OAr)_2]$ in 42% yield ($Ar = 2,6\text{-}Ph_2C_6H_3$). The compounds $[TiR_2(OAr)_2]$ ($R = CH_2Ph$, CH_2TMS or Ph) were obtained similarly as red (benzyl) or yellow (alkyl or phenyl) solids. The crystal structures of the latter two complexes were determined; the compounds are approximately tetrahedral and show characteristically short Ti–O bond distances of ~0.18 nm, indicative of a π-bonding contribution of the oxygen lone pairs to the metal centre. The dialkyls $[TiR_2(OAr)_2]$ ($R = Me$ or CH_2TMS) react with further ArOH to give the yellow complex $[TiR(OAr)_3]$.[49] The related complexes (**13**) were prepared by the alkylation method from the corresponding dichloride and obtained as yellow (R = Me) and green (R = Ph) crystals. The crystal structure of the methyl complex again shows a short Ti–O bond (0.179 9(3) nm).[50]

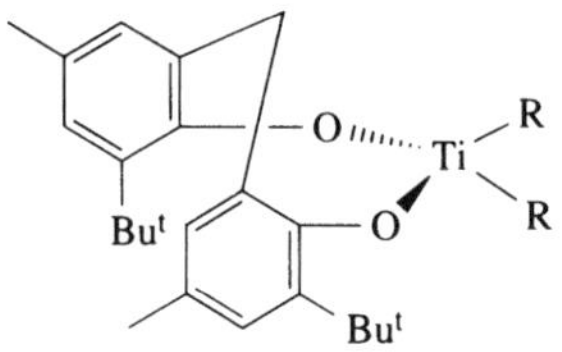

(**13**)

R = Me or Ph

The Schiff base complex *trans*-[TiCl$_2$(salen)] (salen = N,N'-bis(salicylaldehydo)ethylenediamine) reacts with methyllithium in benzene at 6–10 °C to give the expected red–brown *trans*-dimethyl complex. On heating, however, one of the methyl groups migrates to the macrocyclic ligand. The reaction of *trans*-[TiCl$_2$(salen)] with mesitylmagnesium bromide in toluene/dioxane leads to a similar ring-alkylated titanium(IV) product, whereas the same reaction in THF affords the titanium(III) alkyl *trans*-[Ti(C$_6$H$_2$Me$_3$)(salen)(THF)] (Scheme 12).[51]

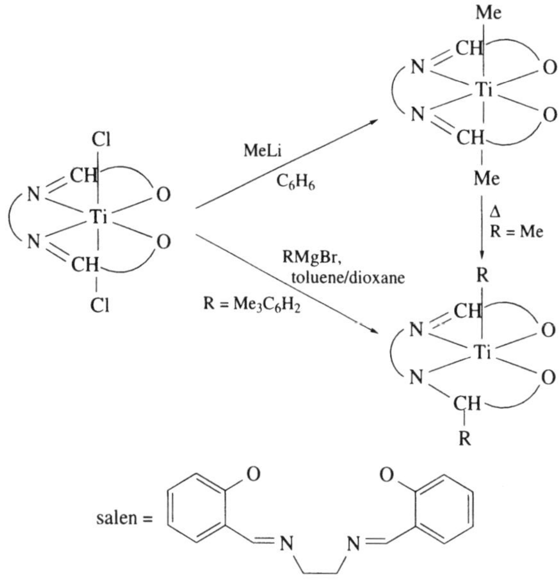

Scheme 12

The cyclopropyl complex (**14**) was prepared by alkylation of [TiCl(OPr^i)$_3$] and was crystallographically characterized (Equation (10)). One oxygen atom of each sulfonyl group is coordinated to titanium, with Ti–O bond lengths of 0.2314(3) nm and 0.2243(2) nm, whereas the Ti–OPr^i bond lengths of 0.1763 nm are very short. The Ti–C bond length is 0.2177 nm.[52]

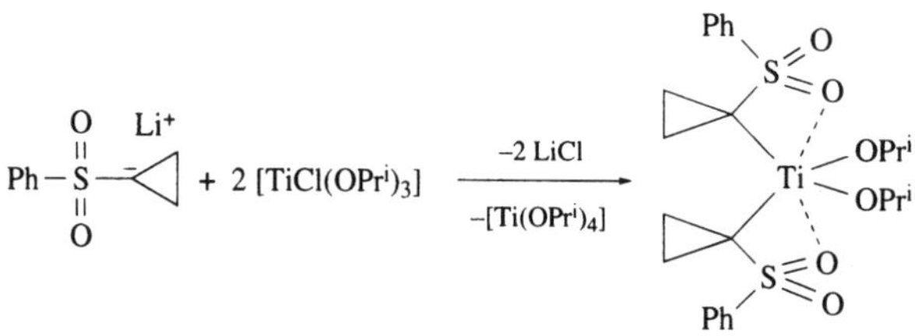

(**14**)

$$\qquad\qquad\qquad (10)$$

The reduction of [TiCl$_2$(OAr)$_2$] (Ar = 2,6-Ph$_2$C$_6$H$_3$) with sodium in the presence of 3-hexyne affords the reactive metallacyclopentadiene (**15**).[53] The alkylation of [TiCl$_4$(THF)$_2$] with 1-camphenyllithium at temperatures below $-70\ °C$ gives the yellow tetraalkyl complex (**16**). The compound is thermally quite stable (decomposes at 120–130 °C) and can be briefly handled in air. It is easier to reduce than tetrakis(1-norbornyl)titanium and shows a reduction wave at $E_{1/2} = -1.32$ V.[54] Attempts to obtain a titanium vinyl complex from $TiCl_4$ and vinyllithium failed. Reduction and vinyl coupling takes place instead, and the addition of dmpe to the reaction mixture gives the butadiene complex $[Ti(C_4H_6)_2(dmpe)]$.[55]

(**15**) (**16**)

5.2.2 Structures and Properties of Titanium(IV) Alkyl Complexes

The agostic alkyl coordination in (**9**) and (**10**) was mentioned in the previous section. In this context the gasphase structure of $[TiMeCl_3]$ was determined by electron diffraction. Early results indicated a flattened tetrahedral geometry for the Me^- ligand, with a Ti–C–H angle of $101.0 \pm 2.2°$, which could be interpreted as an agostic interaction averaged over all three hydrogen atoms, although a nonbonding $Ti\cdots H$ distance of 0.253 nm was found.[56] A redetermination of the $[TiMeCl_3]$ structure in an all-glass apparatus which eliminated the decomposition of $[TiMeCl_3]$ on metal surfaces gave more precise data and established an undistorted geometry for Me, with a Ti–C–H angle of 109.0(17)°. The Ti–C and C–H distances were 0.2047(6) nm and 0.1098(6) nm, respectively.[57] These results are in agreement with MO calculations.[58] The sign of $^2J_{HD}$, which was originally thought to be positive, is in fact negative, -11.3 Hz, similar to most methyl compounds.[59] A theoretical analysis of the bonding in $[TiMeCl_3]$ by the paired interacting orbitals (PIO) method showed that tilting of the methyl ligand towards the metal is favoured if there is a ligand (such as a phosphine) *trans* to Me.[60] Similar calculations of the geometries and vibrational frequencies of $[TiXCl_3]$ and $[TiXCl_5]^{2-}$ (X = Me or Cl) give good agreement with experimental values and show an undistorted methyl group.[61]

The x-ray diffraction structures of (**9**) and (**10**) were compared with the neutron diffraction structure of (**9**) at 20 K. Table 1 gives a comparison of the structural parameters for the methyl ligand obtained by the two methods. Neutron diffraction gives the geometry of the Me ligand as a regular pyramid, not a distorted flattened pyramid found in the x-ray structure. The position of the agostically bonded hydrogen atom is subject to considerable error. The ^{13}C NMR spectral resonance of the methyl ligand in $[Ti(CH_2D)Cl_3(dmpe)]$ at $-90\ °C$ is observed at δ 87.52 ($J_{CH} = 130.5$ Hz). The Et group in (**10**) gives J_{CH} values of 150.2 Hz and 126.8 Hz for the α and β carbons, respectively.[62] *Ab initio* MO studies on $[Ti(Me)Cl_3(PH_3)_2]$ show a Ti–C–H angle surprisingly close to that found by the neutron diffraction study (99.6° vs. 93.5°). The distortion of the methyl ligand was found to be sensitive to the P–Ti–P angle.[63] An *ab initio* geometry optimization on $[Ti(Et)Cl_2H(PH_3)_2]$ reproduces the β-agostic bonding, whereas $[Ti(Et)H_3(PH_3)_2]$ does not; an axial chloride ligand is obviously required.[64]

Table 1 Structural parameters of the Me^- ligand in (**9**) obtained by x-ray and neutron diffraction.

	X-ray diffraction	Neutron diffraction
Bond 1	0.2149(5) nm	0.2122(2) nm
Bond 2	0.100(2) nm	0.1095(3) nm
Bond 3	0.203(3) nm	0.2447(3) nm
Angle α	70(2)°	93.5(2)°
Angle β	105(4)°	118.4(2)°
Angle γ	105(5)°	111.6(3)°

Ab initio MO calculations on $[Ti\{C(TMS){=}CH_2\}X_2]^+$ find a small Ti–C–Si angle, a long Si–C_γ bond and a short $Ti{\cdots}C_\gamma$ distance, which suggests a dative interaction from the C–Si σ bond to a titanium vacant orbital. The geometry optimizations show a γ-agostic $CH{\cdots}Ti$ interaction ($Ti{\cdots}H$ 0.2104 nm). The data reproduce the bonding in $[Ti\{C(TMS){=}CH_2\}Cp_2]^+$.[65]

The ^1H and ^{13}C NMR spectral data of $[TiR_4]$, $[TiR_3X]$ and $[TiR_2X_2]$ (R = CH_2Ph or CH_2-1-naphthyl) have been recorded and confirm an η^2-type bonding of the benzyl ligand. This bonding mode involves acute Ti–C–C angles and consequently increased J_{CH}. Thus, while $[Ti(CH_2Ph)_4]$ gives J_{CH} = 136 Hz, the value decreases to 126 Hz and 124 Hz for $[Ti(CH_2Ph)_3Cp]$ and $[Ti(CH_2Ph)_2Cp_2]$, respectively, as the coordination mode changes from η^2 to η^1.[66] Tetrakis(1-norbornyl)titanium shows absorptions in the electronic spectrum at 245 nm, 286(sh) nm, 367 nm and 412(sh) nm, which were assigned to ligand-to-metal charge transfer bands. Irradiation gives Ti–C bond homolysis and leads to the formation of 1,1'-binorbornyl via norbornyl radicals.[67,68]

5.2.3 Reactions of Titanium(IV) Alkyl Complexes

Tetrabenzyltitanium behaves towards lithium alkyls as a Lewis acid and gives the salts $Li[Ti(CH_2Ph)_4(R)]$ (R = Me, Et or Bu), which can be isolated as red–brown crystalline solids. The compounds decompose between $-30\,°C$ and $0\,°C$ with formation of $Li_2[Ti(CH_2Ph)_4]$ and $Li[Ti(CH_2Ph)_3]$. The formation of $Ti(CH_2Ph)_3$ could not be confirmed.[69] The protolysis of $[Ti(CH_2Ph)_4]$ with 2,6-di-t-butylphenol leads to metallation of one of the t-butyl groups to give (17). $[Ti(CH_2TMS)_4]$ reacts similarly. Compound (17) can add bases such as pyridine to give (18) and undergoes further protolysis to give (19) (Scheme 13).[70] Protolysis of $[Ti(CH_2Ph)_4]$ with $[CoH(CO)_4]$ in toluene at $-40\,°C$ generates $[Ti(PhCH_2)_3\{Co(CO)_4\}]$ as unstable red–brown crystals that decompose on removal of the solvent at low temperature. No significant stabilization is achieved if one of the CO ligands is substituted by PPh_3. A further equivalent of $[CoH(CO)_4]$ gives $[Ti(PhCH_2)_2\{Co(CO)_4\}_2]$. The compounds were characterized spectroscopically and are thought to contain Ti–Co σ bonds.[71] Similarly, $[TiMe(OBu^t)_3]$ reacts with $[CoH(CO)_4]$ to afford $[Ti(OBu^t)_3\{Co(CO)_4\}]$. The red crystalline compound is thermally more stable and decomposes at $90\,°C$.[72]

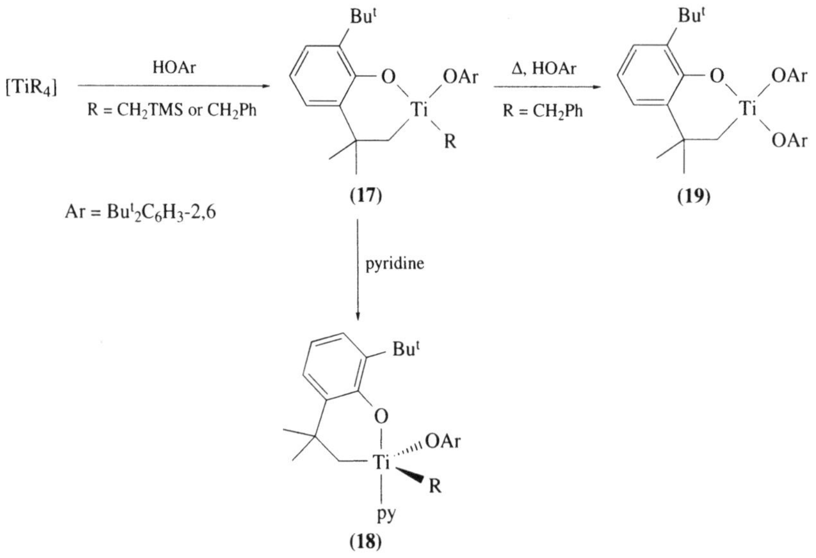

Scheme 13

The alkyl amido complex $[TiMe_2\{N(TMS)_2\}_2]$ is thermally stable up to 190 °C, in contrast with its zirconium and hafnium analogues, and does not give carbene complexes. However, reduction of $[TiCl_2\{N(TMS)_2\}_2]$ with sodium amalgam leads to metallation of TMS groups with formation of the μ-carbene compound (20), which was characterized by x-ray diffraction.[73] The thermolysis of $[Ti(CH_2Bu^t)_4]$ at 150 °C / 10^{-5} torr leads to the deposition of amorphous silvery titanium carbide films on pyrex substrates. Increasing the pressure to 10^{-4} torr gave films of crystalline TiC.[74]

Dry oxygen gas inserts cleanly into the Ti–C bonds of $[TiMe_2(OCBu^t_3)_2]$ to give $[Ti(OMe)_2(OCBu^t_3)_2]$ as white crystals in over 87% yield. Treatment of $[TiMe_3(OCBu^t_3)]$ with O_2 gives

(20)

[TiMe$_{3-x}$(OMe)$_x$(OCBu$^t_3$)]. Bimolecular ligand exchange processes are facile routes to mixed alkyl–alkoxo compounds (Equation (11)). Peroxo intermediates such as [TiMe(η^2-O$_2$Me)X$_2$] are thought to be involved, which undergo rearrangement to [Ti(OMe)$_2$X$_2$].[75] [TiMe$_2$(OMe)(OCBu$^t_3$)] is monomeric in solution but dimeric in the solid state, with five-coordinate titanium. The crystal structure of the compound shows two bridging methoxide ligands with relatively long Ti–O bonds of 0.2015(4) nm, while the bonds to the terminal ⁻OCBu$^t_3$ ligands contain π contributions (Ti–O 0.1752(6) nm). [TiMe(OMe)$_2$(OCBu$^t_3$)] and [Ti(OMe)$_3$(OCBu$^t_3$)] also give monomer:dimer equilibria in solution.[76]

$$[TiMe_3(OCBu^t_3)] + [TiMe(OMe)_2(OCBu^t_3)] \longrightarrow 2\,[TiMe_2(OMe)(OCBu^t_3)] \qquad (11)$$

The relative rates of the reaction of [TiMeX$_3$] with cyclic ketones according to Equation (12) were measured at −18 °C (X = Cl) and 22 °C (X = OPr^i) and found to vary systematically with the internal ring strain of the carbonyl compounds. A ring size with $n = 10$ gives the slowest reaction, and when $n = 6$ the fastest reaction.[77] A mixture of [TiMe(OCPr^i)$_3$] with two different ketones and aldehydes as trapping agents in a 1:1:1 ratio allows a series of relative reactivities to be established; the relative rate constants are given in Scheme 14. Heptanal reacts 680 times faster than 3-heptanone. Steric factors are more important than electronic influences. An oxygen function in the α position enhances reactivity, no doubt due to coordination to the metal centre. In line with this, the kinetic activation parameters show that in THF displacement of the solvent by the carbonyl compound precedes the alkyl transfer step.[78]

$$(CH_2)_n\ C{=}O \xrightarrow[\substack{Et_2O \\ X = Cl\ or\ OPr^i}]{[TiMeX_3]} (CH_2)_n\ C{\overset{\displaystyle}{\underset{OH}{\big\langle}}} \qquad (12)$$

k_{rel}	1	2.2	13	44	105	680	1087

Scheme 14

The reaction of [TiR$_4$] (R = CH$_2$Ph or CH$_2$-1-naphthyl) with CO gives [Ti(COR)R$_3$] as an unstable intermediate that readily decomposes to R$_2$CO and 'TiR$_2$'; the ketone is the main product if R = CH$_2$-1-naphthyl. In the presence of excess [TiR$_4$], insertion of the ketone occurs to give [TiR$_3$(OCR$_3$)], which on hydrolysis gives mainly R$_2$C=CHR (R = CH$_2$Ph). On the basis of an analysis of the organic hydrolysis products, minor by-products are derived from the reaction of [TiR$_4$] with the acyl intermediate [Ti(COR)R$_3$].[79]

Aryl isocyanides react with titanium dialkyl aryloxo compounds [TiR$_2$(OAr)$_2$] (R = CH$_2$Ph or CH$_2$TMS; Ar = 2,6-Pr$^i_2$C$_6$H$_3$ or 2,6-Ph$_2$C$_6$H$_3$) to give stepwise the monoiminoacyl complexes (**21**) followed by the bis(insertion) products (**22**), which undergo quantitative thermal rearrangement to (**23**) (Scheme 15). The latter contains a fluxional, nonplanar Ti$_2$N$_2$C$_2$ ring.[80,81] The intermediate complexes (**21**) were detected in all cases when less than two equivalents of isocyanide were used. The presence of the iminoacyl moieties is indicated in the IR spectra of the complexes by strong bands at 1560–1595 cm^{-1}. The crystal structures of (**21a**) and (**21c**) have been determined by x-ray diffraction.[81] In some cases, multiple isocyanide insertion is observed, as shown in Equations (13)[80] and (14).[82] The addition of CO under pressure to (**21**) affords the TiONC$_2$ metallacycle (**24**) (Scheme 16, R^1 = CH$_2$Ph), presumably via an unobservable acyl–iminoacyl intermediate that undergoes rapid rearrangement.[82] Since isocyanide insertion into [TiR$_2$(OAr)$_2$] proceeds sequentially, the mixed-ligand complexes (**22**) can be prepared (Scheme 15; R^2 = Bu^t; R^3 = Ph or 2,6-Me$_2$C$_6$H$_3$).

(21)	Ar	R^1	R^2
(a)	2,6-Pr$^i_2$C$_6$H$_3$	CH$_2$Ph	Bu^t
(b)	2,6-Pr$^i_2$C$_6$H$_3$	CH$_2$Ph	2,6-Me$_2$C$_6$H$_3$
(c)	2,6-Ph$_2$C$_6$H$_3$	CH$_2$TMS	Ph
(d)	2,6-Ph$_2$C$_6$H$_3$	CH$_2$TMS	2,6-Me$_2$C$_6$H$_3$
(e)	2,6-Ph$_2$C$_6$H$_3$	CH$_2$Ph	2,6-Me$_2$C$_6$H$_3$

Scheme 15

(17)
R = CH$_2$TMS
R^1 = 2,6-Me$_2$C$_6$H$_3$
$\xrightarrow{R^1NC}$
    (13)

[TiR$_2$(OAr)$_2$]
R = CH$_2$TMS
Ar = 2,6-Ph$_2$C$_6$H$_3$
$\xrightarrow{PhNC}$
    (14)

The conversion of (22) to (23) follows first-order kinetics, with rates decreasing for Ti > Zr > Hf. Electron-withdrawing substituents on the aryl ring of the isocyanide accelerate the reaction.[83] The addition of a donor ligand L to (21) induces the transfer of a second alkyl ligand to the iminoacyl carbon to give the η^2-imine complexes (25) (L = NC$_5$H$_4$X-4, X = H, Ph, Et or pyrrolyl). The crystal structures of two of these (X = Et or Ph) have been determined; the complex with X = Ph, for example, shows a relatively short Ti–N bond of 0.184 6(4) nm compared with rather long Ti–C and N–C bonds of 0.2158(5) nm and 0.1421(7) nm, respectively, suggesting dianionic rather than η^2-imine character for the [R$_2$C–NBu^t] moiety. The imine ligand is displaced by a 1,4-diazadiene to give (26), and by azobenzene to afford [Ti(OAr)$_2$(η^2-N$_2$Ph$_2$)(L)$_2$].[84] Treatment of the red–orange compound (25) (R = CH$_2$Ph, Ar = 2,6-Pr$^i_2$C$_6$H$_3$) with 4-phenylpyridine leads to the reductive elimination of R$_2$C=NBu^t and formation of the deep purple, paramagnetic titanium(II) complex (27) (Scheme 16).[85] Similarly, the addition of excess pyridine to (25) induces reductive elimination to give the deep purple compound (28), while 2,2'-bipyridyl leads to the blue–green complex [Ti(OAr)$_2$(bipy)$_2$]. The structures of both of these products were confirmed by x-ray diffraction.[86] (25) reacts with moisture in the presence of pyridine ligands to give a rare example of a titanium complex with a terminal oxo ligand, (29) (Scheme 16). The molecular structure of (29) shows a very short Ti=O bond length of 0.1657(6) nm, compared with the Ti–OAr bond lengths of 0.1863(6) nm and 0.1879(6) nm.[87]

The metallacyclopentadiene complex (15) (Ar = 2,6-Ph$_2$C$_6$H$_3$) reacts with acetonitrile to give peralkylated pyridine derivatives, with pyridine in the presence of ButNC to give (30), with diazadienes to afford (26), and with azobenzene to give (31) (Scheme 17). The formation of the latter requires the breaking of the N=N bond of azobenzene, and a mechanism for this reaction has been suggested.[53] Benzo[c]cinnoline reacts with (15) to give the light-yellow crystalline compound (32), while excess

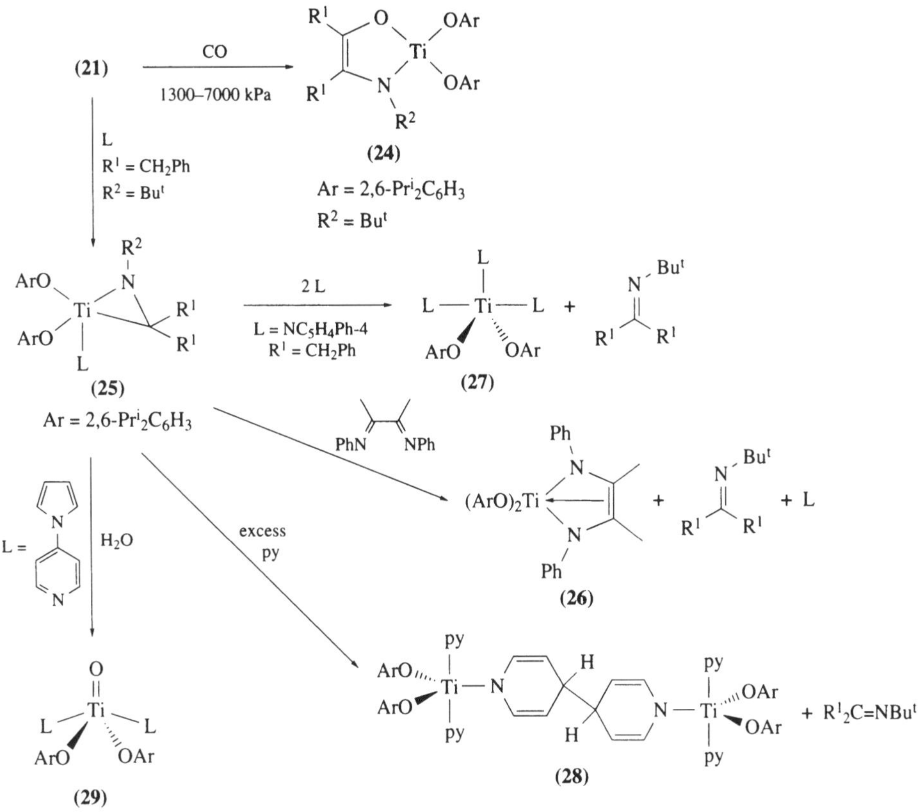

Scheme 16

benzo[*c*]cinnoline results in the unexpected formation of (**33**), whose crystal structure confirms the presence of a terminal Ti=N bond. Compound (**33**) is also obtained on heating (**32**) in hydrocarbon solution to 100 °C.[88] Ethene and (**15**) produce the crystallographically characterized and thermally comparatively stable titanacyclopentane complex (**34**); the by-product is tetraethylcyclohexadiene, which is rapidly formed catalytically if an excess of 3-hexyne and ethene are employed. The substituted metallacycles (**35**) (R = Me or Et) are prepared similarly. Compound (**34**) is fluxional since the cycloaddition of the two ethene units is reversible; consequently, there is an exchange of CH_2 groups, and coalescence is reached on heating to 55 °C ($\Delta G^{\ddagger}$ = 66.5 kJ mol^{-1}).[89] The codimerization of ethene and Bu^{t}C≡CH gives mainly (**36**), with minor amounts of other isomers (Equation (15)). Styrene instead of ethene reacts similarly. Diynes and ethene result in a mixture of isomeric bicyclic products.[90]

The addition of 2,3-dimethylbutadiene to (**34**) leads to the release of two ethene molecules and formation of (**37**) as dark purple crystals, which, according to the x-ray structure, contain a bent metallacyclopentene arrangement with a fold angle of 75°. Under a pressure of 1 bar of ethene, (**37**) reacts to give (**38**), which is present as a mixture of two isomers in an 8:2 ratio (Scheme 18). Butadiene or isoprene instead of 2,3-dimethylbutadiene give only the titanium η^3-allyl complex. Under these conditions 2,3-dimethylbutadiene and ethene are codimerized catalytically; a mechanism is outlined in Scheme 18. The deep green metallacycle (**39**) could also be observed.[91]

Trimethylphosphine reacts with (**34**) with the displacement of one ethene ligand to give the dark red titanium(II) complex (**40**). As is common for alkene complexes of early transition metals, the C–C bond of the C_2H_4 ligand is considerably elongated (to 0.1425(3) nm), indicative of significant backbonding. The compound reacts with ketones to give five- and seven-membered metallacycles (Scheme 19).[92]

Titanium alkyl complexes in the presence of aluminum alkyl activators are catalysts for the polymerization of alkenes and dienes. Thus, [Ti(CH$_2$Ph)$_4$] in the presence of a large excess of methylaluminoxane catalyses the syndiotactic polymerization of styrene. The nature of the active species is unknown; titanium complexes in the oxidation states +2, +3 and +4 can give active catalysts.[93] The [Ti(CH$_2$Ph)$_4$]/methylaluminoxane system also catalyses the syndiotactic 1,2-polymerization of 4-methyl-1,3-pentadiene, although the reaction is slow.[94] Titanium alkyl alkoxo intermediates have been postulated as active species in the polymerization of ethyne with [Ti(OBu^n)$_4$]/AlEt$_3$ catalysts. At low aluminum:titanium ratios (~ 3:1) the *trans* polymer is produced predominantly.[95]

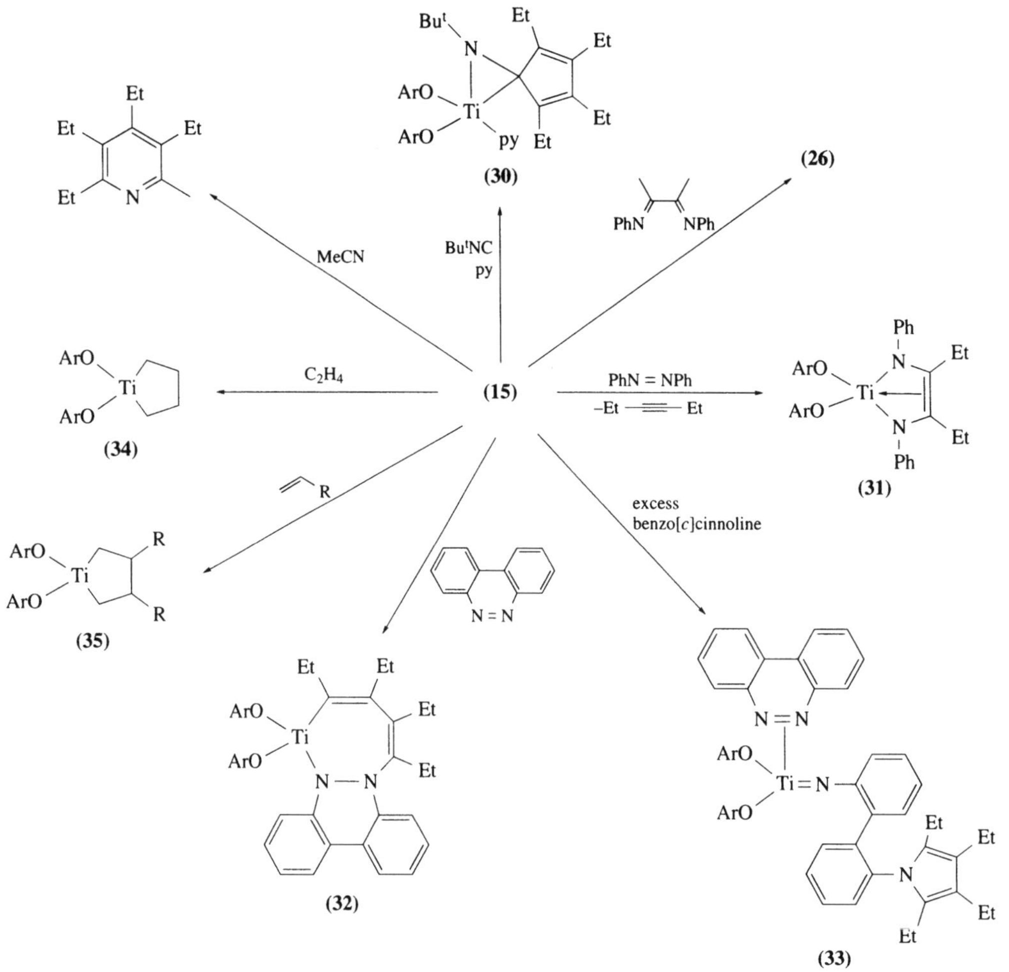

Scheme 17

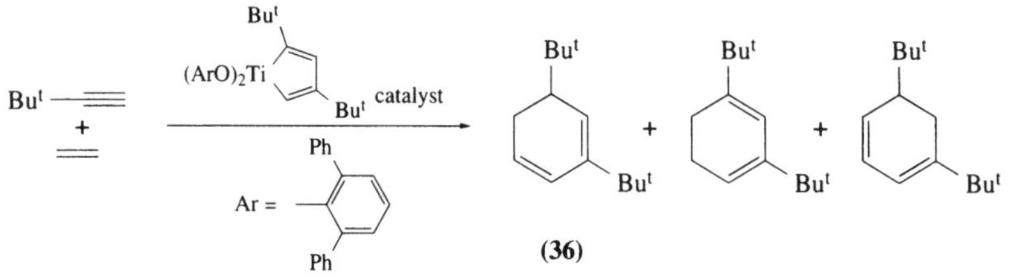

$$(15)$$

5.3 TITANIUM(IV) ISOCYANIDE COMPLEXES

It used to be thought that the reaction of $TiCl_4$ with isocyanides proceeded with insertion of the isocyanide into the Ti–Cl bond and formation of iminoacyl complexes.[96,97] It has now been shown that this is not the case: $TiCl_4$ and Bu^tNC give *cis*-$[TiCl_4(CNBu^t)_2]$, independent of the titanium:isocyanide ratio. The complex shows ν_{CN} bands at $2226\ cm^{-1}$ (Nujol) and $2210\ cm^{-1}$ (CH_2Cl_2 solution); the structure was confirmed by x-ray diffraction (Figure 2). The Ti–C bond lengths are $0.2240(8)$ nm and $0.2256(6)$ nm. The dimeric compounds $[TiCl_4(CNC_6H_3Me_2\text{-}2,6)_2]$ and $[TiCl_4\{CNCH_2P(O)(OEt)_2\}_2]$ were prepared similarly. The titanium halide catalyses the hydrolysis of isocyanides in the presence of traces of moisture, and it was shown that the bands originally assigned to C=N double bond vibrations are due to hydrolysis.[98]

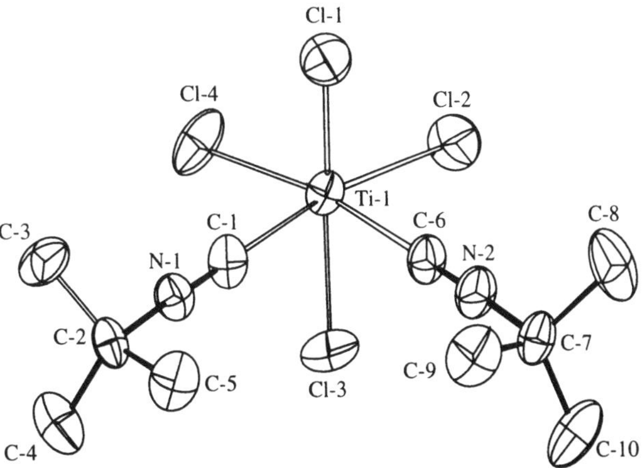

Figure 2 Molecular structure of *cis*-[TiCl₄(CNBuᵗ)₂] (reproduced by permission of the American Chemical Society. Copyright (1989) from *Inorg. Chem.*, 1989, **28**, 4417).

5.4 MONO(CYCLOPENTADIENYL)TITANIUM(IV) COMPLEXES

5.4.1 Halides

5.4.1.1 Synthesis

A number of new substituted cyclopentadienyl derivatives have been prepared, either from the cyclopentadienyl anion and [TiCl$_4$], by the oxidation of a titanium(III) precursor, or by removal of a cyclopentadienyl ligand from a titanocene complex. Thus, the reaction of Li[C$_5$H$_3$Bu$^t_2$-1,3] with [TiCl$_4$] in a xylene/hexane mixture gives orange–red plates of [TiCl$_3$(η-C$_5$H$_3$Bu$^t_2$-1,3)] in 70% yield. The crystal structure of the monomeric compound was determined.[99] By contrast, the reaction of Li[C$_5$H$_2$(TMS)$_3$-1,2,4] with [TiCl$_4$] leads to intractable products, while [TiCl$_3$(THF)$_3$] in THF affords the turquoise complex [TiCl$_2${η-C$_5$H$_2$(TMS)$_3$}], which is oxidized by HCl to give orange crystals of [TiCl$_3${η-C$_5$H$_2$(TMS)$_3$}] in 60% yield.[100,101] Refluxing the 2-methoxyethyl-substituted titanocene (**41**) in SOCl$_2$/SO$_2$Cl$_2$ gives the orange compound (**42**) (Equation (16)), whose crystal structure shows the coordination of the ether side-chain to the metal centre.[102] Complex (**43**) was obtained similarly.[103] In both cases the Ti–O dative bonds are relatively long, 0.2214 nm and 0.2165 nm, respectively. The Ti–Cl bonds *trans* to Ti–O are noticeably elongated. The chelating titanium–arsenic complex (**44**) is prepared according to Equation (17), while (**45**) is produced from [TiCl$_4$] and the corresponding lithium reagent.[104] The compounds are yellow to orange oils. The chelating cyclopentadienyl amido complex (**46**) is formed from [TiCl$_4$], and the dianion in THF at −78 °C as a waxy brown solid in 35% yield (Equation 18).[105]

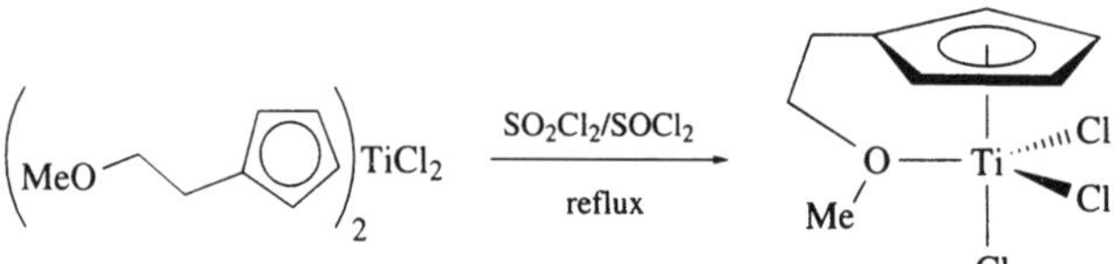

$$\text{(41)} \xrightarrow[\text{reflux}]{\text{SO}_2\text{Cl}_2/\text{SOCl}_2} \text{(42)} \tag{16}$$

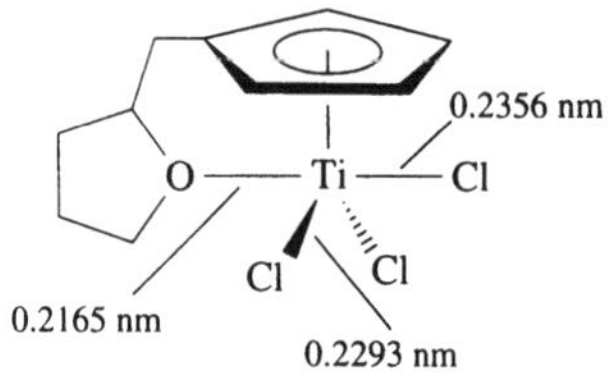

(43)

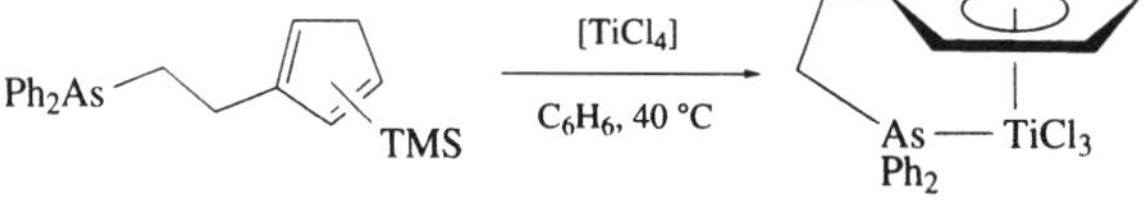

$$\text{(Ph}_2\text{As-TMS ligand)} \xrightarrow[\text{C}_6\text{H}_6,\ 40\ ^\circ\text{C}]{\text{[TiCl}_4\text{]}} \text{(44)} \tag{17}$$

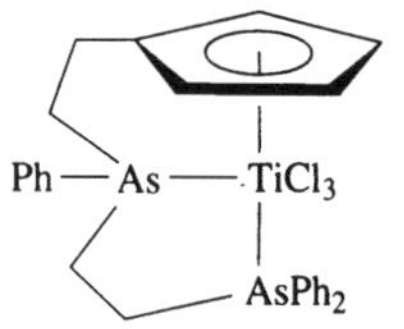

(45)

The highly substituted compound [TiCl$_3$(η-C$_5$HMe$_4$)] is prepared from C$_5$HMe$_4$TMS and [TiCl$_4$] in 65% yield.[106] The pentamethylcyclopentadienyl complexes [TiX$_3$(Cp*)] (X = Cl, Br or I) are similarly obtained from [TiX$_4$] and Cp*TMS in almost quantitative yield without the need for further purification. The zirconium and hafnium analogues were also isolated.[107] The reaction of Li[C$_5$Me$_4$Pr^i] with [TiCl$_3$(THF)$_3$] followed by oxidation with HCl gives [TiCl$_3$(η-C$_5$Me$_4$Pr^i)] as a red solid in 66% yield, which can be purified by sublimation at 140–150 °C/0.001 torr.[108] The thallium reagent

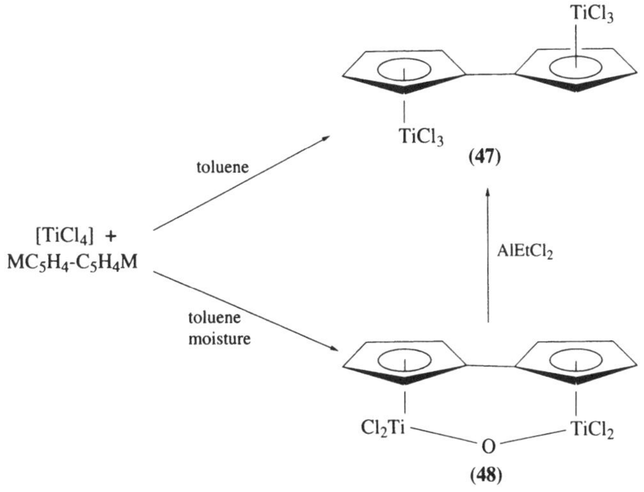

TI[$C_5Me_4CHCH_2$] reacts with [$TiCl_4$] to give the vinyl-substituted complex [$TiCl_3(\eta\text{-}C_5Me_4CH\text{=}CH_2)$] in 82% yield as a deep-red solid which sublimes at 130–140 °C.[109] The optically active complex [$TiCl_3\{\eta\text{-}C_5Me_4C(H)(Et)Ph\}$] is prepared from the lithium cyclopentadienide as yellow crystals.[110]

The red–brown fulvalene complex (**47**) is prepared in toluene in 15–30% yield according to Scheme 20. In the presence of traces of moisture the oxo-bridged complex (**48**) is formed, which can be converted into (**47**) on treatment with $AlEtCl_2$.[111]

Scheme 20

A titanium complex of an aminofluorenylideneborane (**49**) was obtained following Equation (19). The crystal structure shows that the coordination of the fluorenyl moiety to titanium is distorted in the direction of η^3-allylic bonding. The B–N bond is short, 0.137(2) nm.[112] The phospholyl complexes (**50**) are obtained from the corresponding phosphorus–tin reagents, whereas the reaction of [$TiCl_4$] with the lithium reagent leads to reduction of the transition metal and formation of the diphosphine (Scheme 21). The complexes form red, moderately air-sensitive solids. The tin reagent does not lead to bis(cyclopentadienyl) complexes.[113]

5.4.1.2 Structural aspects

The crystal structure of [$TiCl_3Cp$] has been determined. The compound is monomeric in the solid state and possesses the familiar piano-stool geometry. The Ti–Cl bond lengths range from 0.2201(5) nm to 0.2248(5) nm, appreciably longer than in [$TiCl_4$] (0.2170(2) nm).[114] A slightly longer Ti–Cl distance of 0.2303(3) nm has also been quoted.[115] By contrast, [$ZrCl_3Cp$] is a chloride-bridged polymer.[114] [$TiCl_3(\eta\text{-}C_5H_4Me)$] is also monomeric, with a Ti–Cl distance of 0.2229 nm. The methyl group and one chloride ligand are eclipsed.[116] The structures of [$TiCl_3(\eta\text{-}C_5Me_4Et)$][115] and [$TiCl_3\{\eta\text{-}C_5H_3(TMS)_2\text{-}1,3\}$][117] are similar; there are no unusual features resulting from the increased bulk of the cyclopentadienyl ligands.

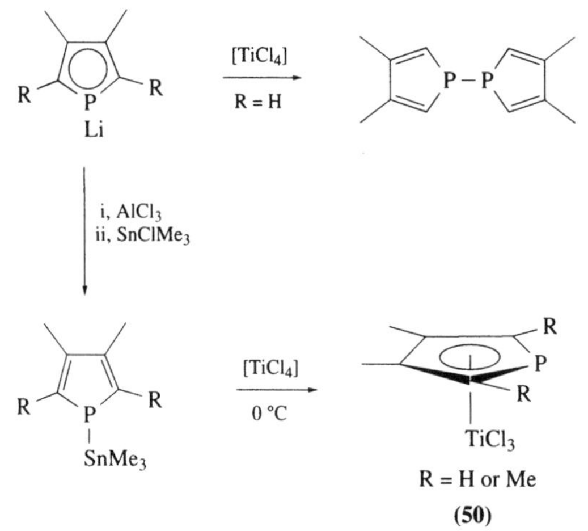

Scheme 21

Proton NMR spin–lattice relaxation time measurements have been used to determine the barrier of Cp ring rotation of [TiCl$_3$Cp] in the solid state. A value of 8.8 ± 0.8 kJ mol^{-1} was found, similar to that in ferrocene.[118] The ^{47}Ti and ^{49}Ti NMR chemical shifts of a series of titanium complexes including [TiCl$_3$Cp] and [TiCl$_3$(η-C$_5$Me$_4$Et)] have been determined at natural abundance.[119,120] Increased shielding was found with decreased electronegativity of the halide ligands. The assignment to ^{47}Ti and ^{49}Ti given earlier[119] should be reversed.[121]

The helium(I) and helium(II) photoelectron spectra of a series of complexes [TiX$_3$Cp'] (Cp' = Cp, C$_5$H$_4$Me or Cp*; X = Cl or Br) show a decrease of the ionization energies with increasing methyl substitution or replacement of chlorine with bromine.[122] Increasing methyl substitution in [TiCl$_3$Cp'] (Cp' = Cp, C$_5$H$_4$Me, C$_5$H$_2$Me$_3$, C$_5$HMe$_4$, Cp* or C$_5$Me$_4$Et) shifts the charge transfer band in the electronic spectrum to longer wavelengths (from 384 nm to 438 nm).[123]

5.4.1.3 Reactions

Redox and catalytic reactions of monocyclopentadienyltitanium trihalides are covered here. Reactions leading to specific complexes with Ti–C, Ti–N and Ti–O bonds are described in subsequent sections.

The reduction of [TiCl$_3$Cp] with cobaltocene gives dark-green crystals of the titanium(III) salt [CoCp$_2$][TiCl$_3$Cp], whose crystal structure has been determined. The mean Ti–Cl bond distance is 0.2349(8) nm, slightly longer than in [TiCl$_3$Cp]. Although air-sensitive, the [TiCl$_3$Cp]$^-$ anion is stable in solution.[124,125] [TiCl$_3$(η-C$_5$H$_4$Me)] undergoes a reversible reduction process at -0.355 V vs. Ag/AgCl, and is reduced by cobaltocene to give [CoCp$_2$][TiCl$_3$(η-C$_5$H$_4$Me)]. Proton NMR spectroscopy has proved useful for the characterization of this paramagnetic compound.[126] Substituted cyclopentadienyl complexes [TiCl$_3$Cp'] (Cp' = Cp, C$_5$H$_4$Me, C$_5$H$_2$Me$_3$, C$_5$HMe$_4$, C$_5$HMe$_5$ or C$_5$Me$_4$Et) react with aluminum alkyls such as AlEtCl$_2$ and are reduced to titanium(III) to give the compounds [TiAl$_2$Cp'Cl$_{8-x}$Et$_x$] ($x = 0$–4). The rate of reaction can be followed by electronic spectroscopy, and decreases with increasing methyl substitution of the Cp ring. The thermal stability of the products increases with increasing bulk of Cp'; the compounds [TiAl$_2$Cp'Cl$_4$Et$_4$] are stable at room temperature for Cp' = C$_5$H$_2$Me$_3$, C$_5$HMe$_4$ and Cp*.[123]

[TiCl$_3$Cp] forms adducts with Lewis bases and reacts with Me$_2$PCH$_2$CH$_2$PMe$_2$ (dmpe) to give the pseudo-octahedral, crystallographically characterized complex (51).[127] Due to the oxophilic character of titanium, the compound can be used as a mild chlorinating agent and reacts, for example, with [Re(O)(C$_2$Me$_2$)(Cp*)] or [Re(O)Me$_2$(Cp*)] to give [ReCl$_2$(C$_2$Me$_2$)(Cp*)] and [ReCl$_2$Me$_2$(Cp*)], respectively, together with [(TiCl(O)Cp)$_4$].[128,129]

Monocyclopentadienyltitanium complexes have been used as catalysts in a number of reactions. The addition of excess AlClEt$_2$ to the arsenic complex (44) gives a catalyst for the cyclotrimerization of butadiene, and mixtures of (45) and AlClEt$_2$ catalyse the dimerization of isoprene; only 10–15% higher oligomers are formed.[104] The compounds [TiCl$_2$XCp] (X = OCH$_2$CF$_3$, NMe$_2$ or Me) initiate the polymerization of isocyanates according to Equation (20). [TiCl$_2$(OCH$_2$CF$_3$)(Cp*)] can be used similarly. Due to the reduced Lewis acidity of these initiators the polymerization of functionalized isocyanates is also possible.[130]

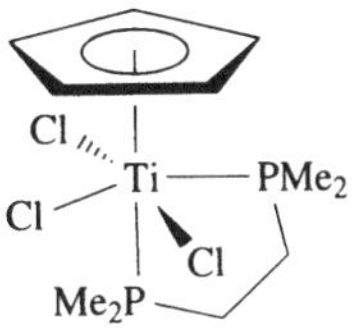

$$
\textbf{(51)}
$$

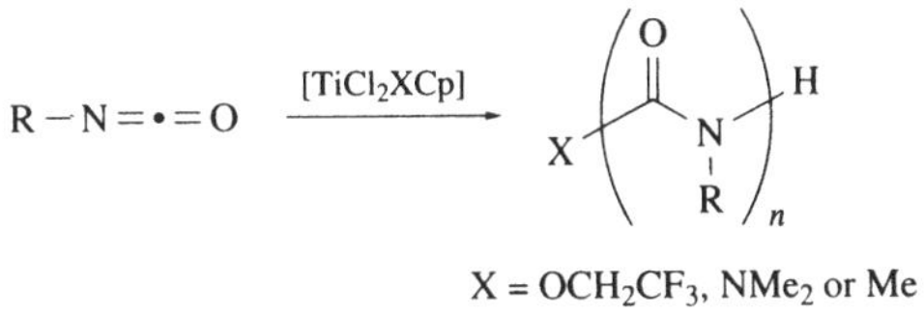

$$
R-N=\bullet=O \xrightarrow{\text{[TiCl}_2\text{XCp]}} \left(X-\!\!\!\!\!\begin{array}{c} O \\ \| \\ C \end{array}\!\!\!\!\!-N\!\!\begin{array}{c} H \\ | \\ R \end{array} \right)_n \tag{20}
$$

$$
X = OCH_2CF_3,\ NMe_2\ \text{or}\ Me
$$

The reaction of [TiCl$_3$(Cp*)] with methylaluminoxane (MAO) leads initially to a multitude of products. After long reaction times the ESR spectrum is consistent with the formation of a hydrido species, possibly a dimethylenecyclopentadienyl complex [TiH(Cl){η^3:η^4-C$_5$Me$_3$(CH$_2$)$_2$}(MAO)]; this compound exchanges hydrogens with D$_2$. The hydrido species is not formed if AlMe$_3$ is used instead of MAO.[131] Mixtures of [TiCl$_3$Cp] or [TiCl$_3$(Cp*)] with MAO catalyse the polymerization of polystyrene to a highly syndiotactic polymer. Analogous Zr, Hf, V, Nb, Cr, Co and Ni complexes give only atactic polymers.[132] [TiCl$_3$Cp] can be attached to a SiO$_2$/Al$_2$O$_3$ gel; only one chlorine atom per titanium is substituted. Reduction with butyllithium generates a catalyst for the hydrogenation of alkenes.[133] Similar supported and unsupported catalysts obtained from the reduction of [TiCl$_2$(OC$_6$H$_4$R)Cp] (R = H, Me, NO$_2$ or Cl) and [TiCl$_3$Cp]–SiO$_2$/Al$_2$O$_3$ gels give hydrogenation catalysts for simple alkenes such as ethene and propene. Cyclopentene is hydrogenated by the homogeneous system. The activity increases with increasing donor properties of the aryloxide ligand or the support.[134]

5.4.2 Complexes with Ti–C σ Bonds

5.4.2.1 Synthesis

Some syntheses of mono(cyclopentadienyl)titanium alkyls and their properties are summarized in Table 2. The alkylation of [TiCl$_3$Cp] with ZnMe$_2$ in benzene over 10–15 min gives the thermally stable monoalkyl complex [TiMeCl$_2$Cp] as a dark orange solid. Decomposition in the dark is very slow, even at 100 °C. The analogous CD$_3$ and C$_5$D$_5$ compounds were also obtained. The reaction with diethylzinc gives the less stable complex [TiEtCl$_2$Cp] as red crystals which melt and decompose at ~20 °C. [TiMeCl$_2$Cp] in benzene reacts with dry oxygen gas to give [Ti(OMe)Cl$_2$Cp] but is hydrolysed by traces of water to [{TiCl$_2$Cp}$_2$(μ-O)].[135] The alkyls [TiMeCl$_2$Cp] are conveniently isolated from [TiCl$_3$Cp]/ZnR$_2$ reaction mixtures by sublimation in good yields (R = Me, 71%; R = Et, 65%).[136] The homoenolate complex [Ti(CH$_2$CH$_2$CO$_2$Et)Cl$_2$Cp] is obtained from the reaction of [TiMeCl$_2$Cp] with Zn(I)(CH$_2$CH$_2$CO$_2$Et); the alkyl ligand acts as a chelate via the ester C=O group (Equation (21)).[137] The alkylation of [TiCl$_3$Cp] with 1-ferrocenyllithium gives [Ti(1-ferrocenyl)$_3$Cp] as a very sensitive solid. With [TiCl$_3$(Cp*)] only two of the chloride ligands could be exchanged.[138]

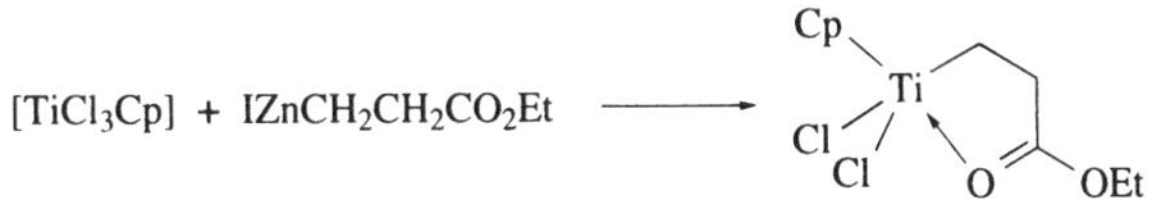

$$
\text{[TiCl}_3\text{Cp]} + \text{IZnCH}_2\text{CH}_2\text{CO}_2\text{Et} \longrightarrow \tag{21}
$$

A series of (pentamethylcyclopentadienyl)titanium alkyl complexes has been prepared, and the reactions have been summarized in a brief review.[139] Except for the pentafluorophenyl complexes these are thermally stable substances (Table 2), significantly more stable than their Cp analogues.[140–3] Trialkyls of the type [TiR$_3$Cp] are formally 12-electron complexes; hence they are electronically unsaturated. The solid-state structure of the benzyl derivative [Ti(CH$_2$Ph)$_3$(Cp*)] deserves particular attention in this respect. Whereas two of the CH$_2$Ph ligands are unremarkable, the third is positioned in such a way that the hydrogens of the benzylic CH$_2$ group can interact with vacant metal orbitals, resulting in small Ti–C$_\alpha$–H angles of 93.6 (7)° and 96.9(7)°, consistent with a double agostic interaction. As a result, the

Table 2 Synthesis of mono(cyclopentadienyl)titanium alkyl complexes.

Compound	Synthetic method[a]	Colour	m.p. (°C)	Yield (%)	Ref.
$[TiCl_2MeCp]$	A	Dark orange solid	20 (sublimes)	60	^{1}H NMR, analysis[135]
$[TiCl_2EtCp]$	A	Dark red crystals/liquid	~20		^{1}H NMR[135]
$[TiCl_2(CH_2CH_2CO_2Et)Cp]$	B	Red solid		56	^{1}H and ^{13}C NMR, IR, analysis[137]
$[Ti\{(\eta^1:\eta^5\text{-}C_5H_4)FeCp\}_3Cp]$	C	Blue–black solid	180 (decomposes)	24	^{1}H NMR, IR, MS, analysis[138]
$[TiCl\{(\eta^1:\eta^5\text{-}C_5H_4)FeCp\}_2(\eta\text{-}C_5Me_5)]$	C	Deep purple powder	240 (decomposes)	6	^{1}H NMR, IR, MS, analysis[138]
$[TiMe_3\{\eta\text{-}C_5H_2(TMS)_3\}]$	C	Yellow crystals	58, >110 (decomposes)	80	100
$[TiMe_3Cp^*]$	C	Yellow crystals	75–76	98	^{1}H and ^{13}C NMR, analysis[140,141] Electron diffraction[142]
$[TiClMe_2Cp^*]$	C	Light yellow solid		85	^{1}H and ^{13}C NMR, analysis[140,141]
$[Ti(CH_2SiMe_3)_3Cp^*]$	C	Yellow crystals	83–84	60	^{1}H and ^{13}C NMR, analysis[140,141]
$[TiCl(CH_2SiMe_3)_2Cp^*]$	C	Yellow solid		66	^{1}H and ^{13}C NMR, analysis[141]
$[Ti(CH_2Ph)_3Cp^*]$	D	Orange–red crystals		90	^{1}H and ^{13}C NMR, x-ray analysis[140,141]
$[TiCl(CH_2Ph)_2Cp^*]$	E	Red solid		85	^{1}H and ^{13}C NMR, analysis[141]
$[TiBr(CH_2Ph)_2Cp^*]$	E	Red solid		85	^{1}H and ^{13}C NMR, analysis[141]
$[Ti_2\{1,2\text{-}(CH_2)_2C_6H_4\}_3Cp^*_2]$	E	Brown–red crystals		50	^{1}H and ^{13}C NMR, x-ray analysis[141]
$[Ti(C_6F_5)_3Cp^*]$	C	Dark red crystals		70	^{1}H and ^{13}C NMR, analysis[141]
$[Ti_2(CH_2Ph)_6(\mu\text{-}\eta^5,\eta^5\text{-}C_{10}H_8)]$	E	Red crystals		97	^{1}H NMR, analysis[111]
$[\{TiMe_2Cp^*\}_2(\mu\text{-}O)]$	C	Yellow		95	^{1}H and ^{13}C NMR, analysis[143]
$[\{Ti(CH_2TMS)_2Cp^*\}_2(\mu\text{-}O)]$	C	Yellow		90–95	^{1}H and ^{13}C NMR, x-ray analysis[143]
$[\{Ti(CH_2Ph)_2Cp^*\}_2(\mu\text{-}O)]$	E	Red		85	^{1}H and ^{13}C NMR, analysis[143]

[a] Method A: titanium halide + $ZnMe_2$, benzene, room temperature. Method B: titanium halide + $IZnCH_2CH_2CO_2Et$, THF/benzene. Method C: titanium halide + lithium alkyl, hexane. Method D: titanium halide + RMgBr. Method E: titanium halide + magnesium dialkyl.

Ti–C_α–C_β angle of this benzyl moiety is widened to 139.0(7)°, and the Ti–C distance is short, 0.2081(10) nm, compared with Ti–C bond lengths to the other two benzyl ligands of 0.2140(10) nm and 0.2156(10) nm.[140] By contrast, however, the gas-phase electron diffraction structure of $[TiMe_3(Cp^*)]$ does not show any unusual features; there is a near tetrahedral Ti–C–H angle of 103.80(12)°, a normal Ti–C bond distance of 0.2107(5) nm, and a nonbonding Ti···H distance of 0.2604 nm.[142]

The reaction of $[TiCl_3(Cp^*)]$ with the *o*-xylenediyl magnesium reagent $[Mg\{1,2\text{-}(CH_2)_2C_6H_4\}(THF)_2]$ gives (**52**) (Equation (22)), which contains three nearly planar *o*-xylenediyl ligands. The six-membered rings of the two nonbridging xylenediyl ligands are bent towards the metal, with short Ti–C_β distances (0.245–0.247 nm) and narrow Ti–C_α–C_β angles (84.7–85.0°) suggestive of η^4-type bonding interactions (Figure 3). In agreement with this, the ^{13}C NMR spectrum of (**52**) shows $^1J_{CH}$ for the xylene CH_2 groups of 121 Hz for the bridging moiety but only 76.4 Hz for the two η^4-bonded ligands.[141]

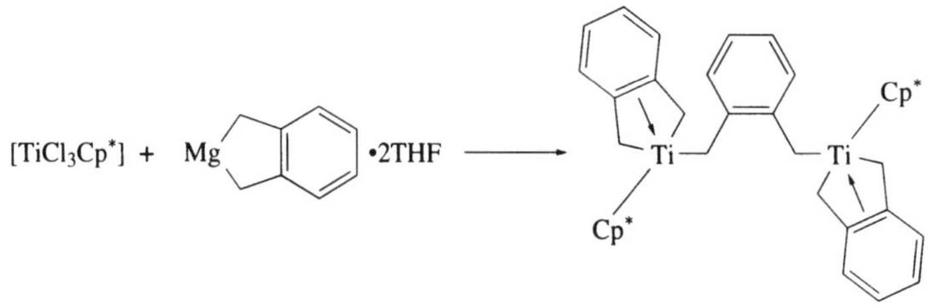

(**52**) (22)

The reaction of the fulvalene complex $[Ti_2Cl_6(\eta^5:\eta^5\text{-}C_{10}H_8)]$ with $Mg(CH_2Ph)_2$ gives $[Ti_2(CH_2Ph)_6(\eta^5:\eta^5\text{-}C_{10}H_8)]$, in which three different coordination modes of the benzyl ligand are realised: one (Ti-C-20) is η^1-coordinate, with a Ti–C_α–C_β angle of 118.5(3)°, one (Ti-C-6) contains an α-agostically bonded benzylic hydrogen atom and a widened Ti–C_α–C_β angle of 128.0(3)°, and the third

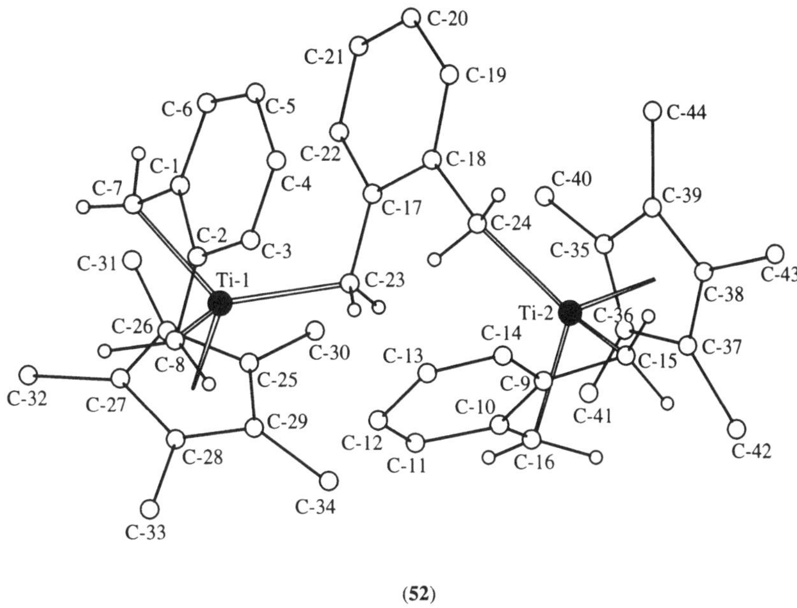

(52)

Figure 3   Molecular structure of (52) (reproduced by permission of the American Chemical Society. Copyright (1989) from *Organometallics*, 1989, **8**, 476).

Ti–C-13) is η^2-bonded, with a narrow Ti–C_α–C_β angle of only 90.9(3)° and a short Ti–C_β-14 distance of 0.2611(4) nm. Figure 4 shows the molecular structure of the compound, while Figure 5 illustrates the coordination geometries of the benzyl ligands around one of the titanium centres.[111]

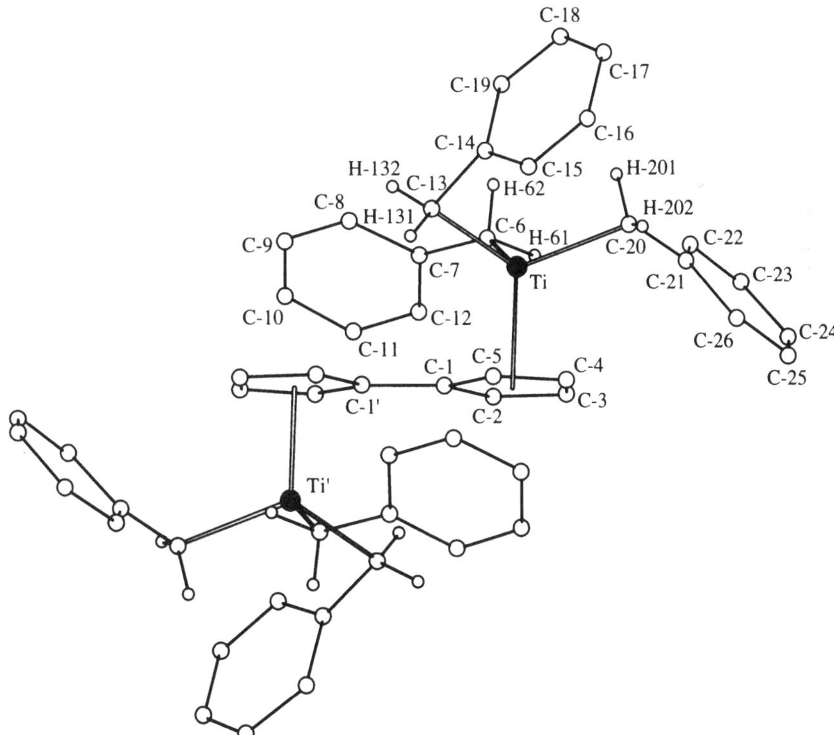

Figure 4   Molecular structure of the [Ti$_2$(CH$_2$Ph)$_6$(η^5:η^5-C$_{10}$H$_8$)] (reproduced by permission of the American Chemical Society. Copyright (1992) from *Organometallics*, 1992, **11**, 3301).

The reaction of the μ-imido complex [{TiCl(μ-NPh)Cp}$_2$] with methyllithium gives [{TiMe(μ-NPh)Cp}$_2$].[144] Oxo-bridged dimeric alkyls complexes are obtained by the alkylation of [{TiCl$_2$(Cp*)}$_2$(μ-O)] with lithium reagents (Table 2). The structure of the trimethylsilylmethyl complex

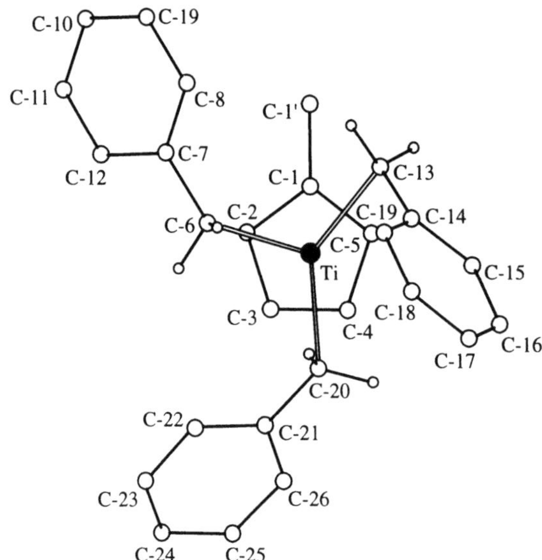

Figure 5 Projection of $[Ti_2(H_2Ph)_6(\eta^5:\eta^5-C_{10}H_8)]$ along an axis perpendicular to the fulvalene ligand, illustrating the different bonding modes of the benzyl ligands (reproduced by permission of the American Chemical Society. Copyright (1992) from *Organometallics*, 1992, **11**, 3301).

was determined and shows a bent Ti–O–Ti bridge of 155.9° with a typically short Ti–O distance of 0.1842(1) nm.[143]

The reaction of $[TiCl_3\{\eta-C_5Me_4(CH_2)_3OMe\}]$ with $Ph_3P=CH_2$ gives the very air-sensitive titanium ylide complex (**53**) (Scheme 22). Heating (**53**) in benzene solution to 155 °C leads to the formation of a titanium alkoxide and unidentified by-products.[145]

$$[TiCl_3Cp'] \xrightarrow[-HCl]{Ph_3P=CH_2} [TiCl_2(CH=PPh_3)Cp'] \xrightarrow{155\,°C}$$

(**53**)

$$Cp' = C_5Me_4(CH_2)_3OMe$$

Scheme 22

5.4.2.2 Reactions

The hydrolysis of $[TiMe_3Cp']$ with one equivalent of water in diethyl ether at room temperature gives a 90% yield of the pale yellow complex $[\{TiMe_2Cp'\}_2(\mu-O)]$ ($Cp' = \eta-C_5H_2(TMS)_3-1,2,4$).[100] The hydrolysis of $[TiMe_3(Cp^*)]$ gives dimeric or trimeric oxo-bridged alkyl complexes, depending on the reaction conditions (Scheme 23). The structure of (**54**) was determined by x-ray crystallography; it shows a nearly planar Ti_3O_3 ring with Ti–O bond lengths of 0.1816–0.1830 nm and Ti–O–Ti angles of 131.6–133.5°.[146] The protolysis of $[TiMe_3(Cp^*)]$ with an excess of ammonia at room temperature gives (**55**) as yellow crystals which decompose above 200 °C.[147]

Protolysis of $[TiMe_3Cp]$ with anilinium salts such as $[NHMe_2Ph][BPh_4]$ in THF or dimethoxyethane gives the cationic alkyl complexes $[TiMe_2Cp(L)_2][BPh_4]$ (**56**) as thermally stable orange–red solids ($L = THF$ or $\frac{1}{2}$ $MeOCH_2CH_2OMe$) (Equation (23)).[148]

The analogous reaction between $[TiMe_3(Cp^*)]$ and $[NHEt_3][B(C_6F_5)_4]$ in toluene is thought to give $[TiMe_2(Cp^*)][B(C_6F_5)_4]$, described in a patent as a black solid which catalyses the polymerization of styrene to a syndiotactic polymer. Similar cationic species are thought to be formed in the reactions of $[Ti(CH_2Ph)_3Cp]$ with $[NHEt_3][B(C_6F_5)_4]$ or $[FeCp_2][B(C_6F_5)_4]$.[149] The zwitterionic compound $[TiMe_2(Cp^*)(\mu-Me)B(C_6F_5)_3]$ (**57**) is formed when $[TiMe_3(Cp^*)]$ is treated with one equivalent of $B(C_6F_5)_3$ in dichloromethane at −50 °C (Scheme 24).[150] The complex is fluxional and dissociates partly in solution into $[TiMe_2(Cp^*)(CH_2Cl_2)_n][MeB(C_6F_5)_3]$. In the presence of donor ligands such as PMe_3, NEt_3 or dmpe, adducts analogous to (**56**) are formed. The zwitterionic complex is an active catalyst for the polymerization of ethene and styrene.[151] Whereas mixtures of $[Ti(CH_2Ph)_3(Cp^*)]$ and $B(C_6F_5)_3$ are

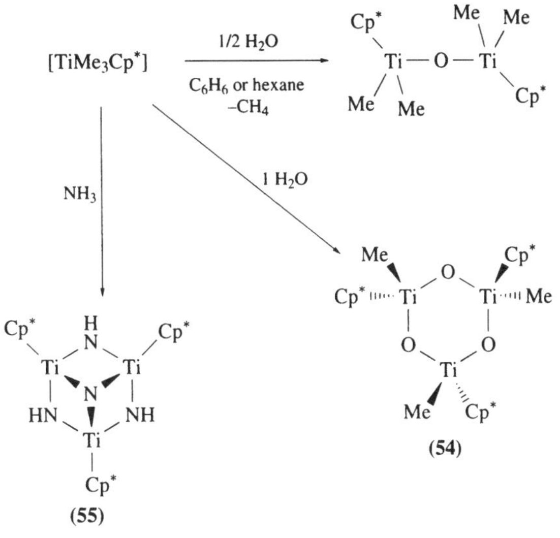

Scheme 23

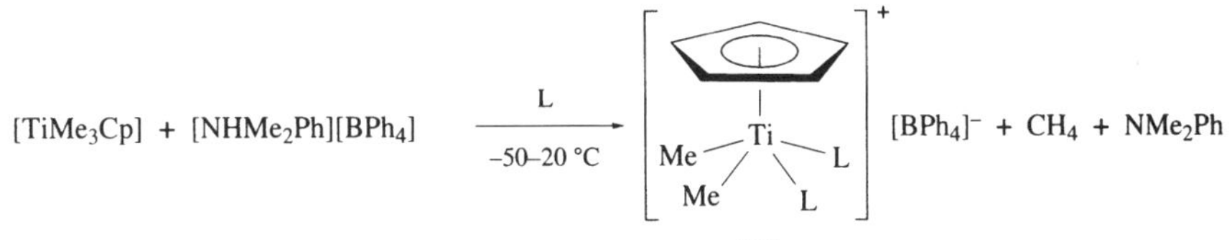

$$[TiMe_3Cp] + [NHMe_2Ph][BPh_4] \xrightarrow[-50–20\ °C]{L} \left[\ \underset{Me}{\overset{}{\underset{}{Ti}}}\ \right]^{+} [BPh_4]^- + CH_4 + NMe_2Ph \qquad (23)$$

(56)

good catalysts for the syndiotactic polymerization of styrene, mixtures of $[Ti(CH_2Ph)_3(Cp^*)]$ and $[NHMe_2Ph][B(C_6F_5)_4]$ give only atactic material with low activity. The catalytic activity for styrene polymerization of metal alkyls activated with $B(C_6F_5)_3$ decreases in the order $[Ti(CH_2Ph)_3(Cp^*)] > [TiMe_3(Cp^*)] \gg [Zr(CH_2Ph)_3(Cp^*)]$.[152] Similar cationic metal alkyls, for example, **(58)** and **(59)**, are obtained *in situ* from the reaction of the corresponding neutral titanium dimethyl complexes and $[NHEt_3][B(C_6F_5)_4]$ in benzene; they are active catalysts for the polymerization of ethene.[153]

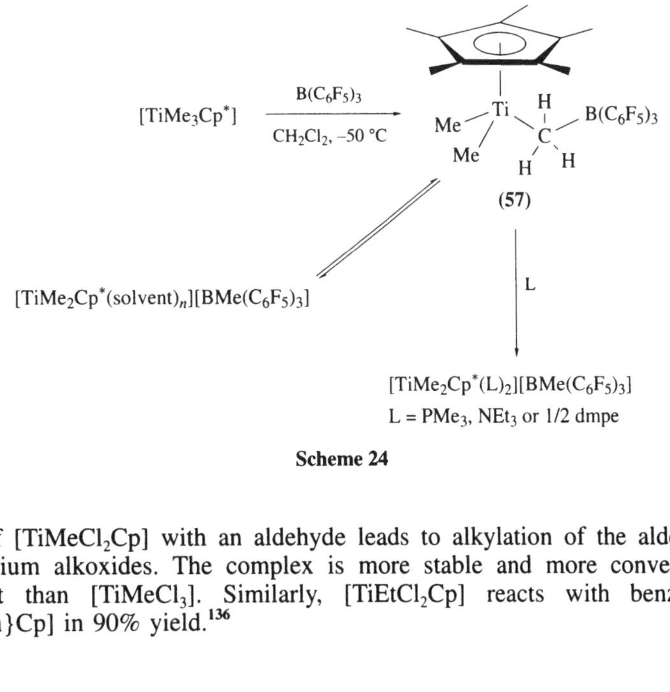

Scheme 24

The reaction of $[TiMeCl_2Cp]$ with an aldehyde leads to alkylation of the aldehyde function and formation of titanium alkoxides. The complex is more stable and more convenient to use as an alkylating reagent than $[TiMeCl_3]$. Similarly, $[TiEtCl_2Cp]$ reacts with benzaldehyde to give $[TiCl_2\{OCH(Et)Ph\}Cp]$ in 90% yield.[136]

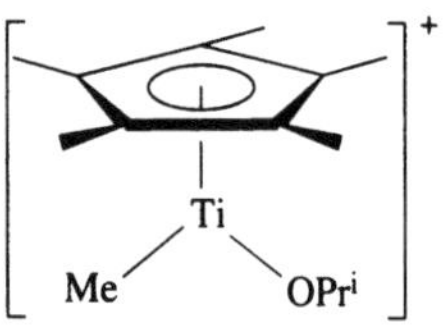

(58) 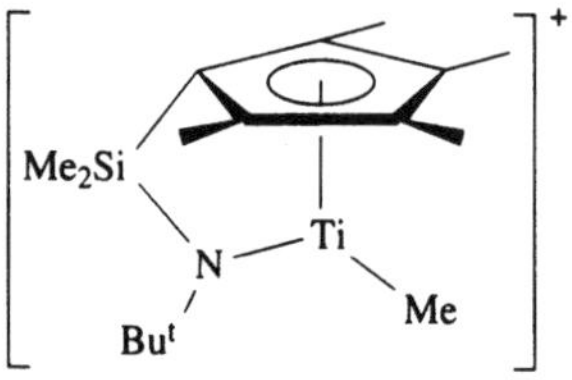   **(59)**

Treatment of the oxo-bridged complexes [{TiMe(Y)(Cp*)}$_2$(μ-O)] (Y = Cl or Me) (**60**) with 1 bar of carbon monoxide leads to a double methyl migration to the CO carbon and formation of the μ,η^2-acetone complex (**61**). Both the chloro and the methyl product are thermally unstable and decompose at room temperature within a few days. The ν_{CO} stretching mode of the acetone ligand is observed at 1190 cm^{-1}. Acetone is released on oxidation, while hydrolysis gives isopropanol. Thermolysis of (**61**) gives oxo-bridged clusters (Scheme 25).[154] The reaction of (**60**) with diphenyldiazomethane leads to mono- and di-insertion products (**62**) and (**63**) in a stepwise manner, depending on the reaction temperature (Scheme 25). The molecular structure of (**62**) was determined by x-ray diffraction. The N–N bond length of 0.1326(10) nm is intermediate between an N–N single and a double bond, while the N–C distance of 0.1298(10) nm is similar to that of the free diazoalkane. The trimer [{TiMe(Cp*)(μ-O)}$_3$] is less reactive than (**60**) and reacts with Ph$_2$CN$_2$ to give only a monoinsertion product, even on heating.[155]

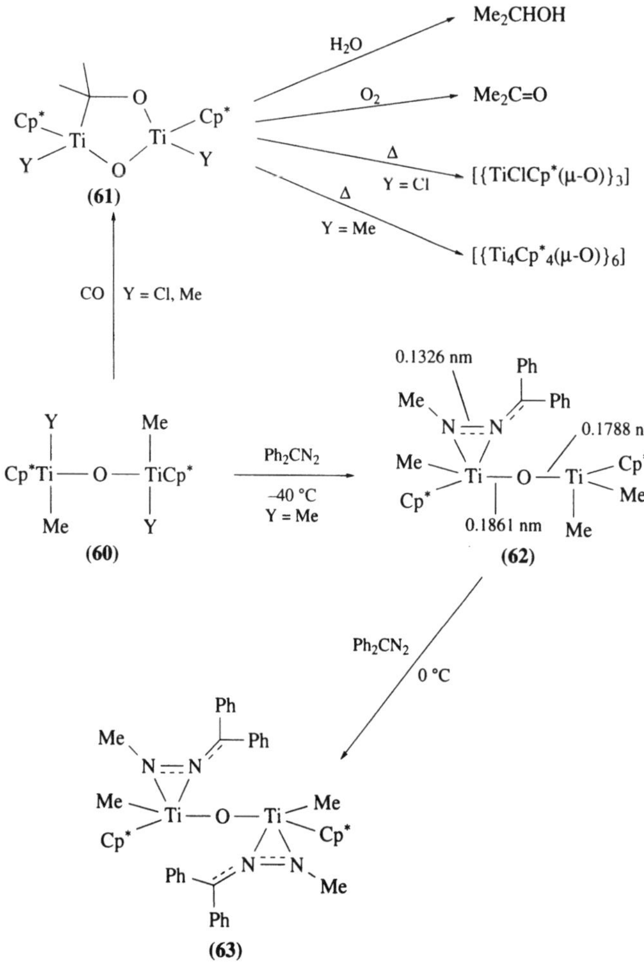

Scheme 25

5.4.3 Complexes with Ti–N σ Bonds

Treatment of [TiCl$_3$Cp] with NHR(TMS) in THF at room temperature leads to the yellow to red–brown monoamido complexes [TiCl$_2$(NHR)Cp] (R = Et, Pr^i, Bu^t or Ph). Heating these compounds *in vacuo*, or treatment with MeLi, generates the dimeric imido compounds [{TiCl(NR)Cp}$_2$]; the structure of the complex (R = Ph) was determined. Electrochemical studies suggest that a stable one-electron reduction product is formed.[144,156] [TiCl$_2$(NHBu^t)Cp] is an exception; it does not liberate HCl on heating and sublimes without decomposition.[156] The amido complexes [TiCl$_2$(NHBu^t)(η-C$_5$H$_4$R)] were obtained from [TiCl$_3$(η-C$_5$H$_4$R)] and LiNHBu^t in 80–90% yield. The crystal structures give Ti–N bond lengths of 0.1879(3) nm (R = H) and 0.1871(5) nm (R = Me), indicative of π-donation of the nitrogen lone pair to titanium. Nevertheless, a very low barrier to rotation (≤21 kJ mol^{-1}) of the amido ligand is suggested.[157] The reaction of [TiCl$_3$Cp] with Me$_2$Si(NRLi)$_2$ gives the bisamido complexes (**64**) (R = Bu^t or TMS) (Scheme 26). The third chloride ligand cannot be substituted, even with the potentially tridentate trianion PhSi(NRLi)$_3$, which leads to (**65**), whose structure was confirmed by x-ray diffraction (Figure 6). The lithium ion is coordinated to one amido nitrogen, one diethyl ether, and weakly to the chloride ligand. The TiN$_2$Si ring is folded by 11.5° along the N–N vector.[158]

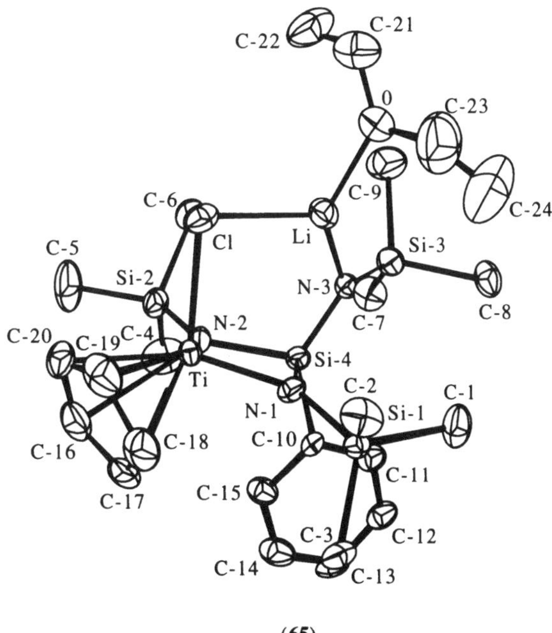

Scheme 26

(**65**)

Figure 6 Molecular structure of complex (**65**) (reproduced by permission of Elsevier from *J. Organomet. Chem.*, 1986, **307**, 177).

The reaction of $[TiCl_3Cp']$ with $LiN(TMS)_2(Et_2O)$ in toluene affords $[TiCl_2\{N(TMS)_2\}Cp']$ [Cp' = Cp, C_5H_4TMS, $C_5H_3(TMS)_2$ or Cp*] in 69–90% yield. In the case of Cp' = Cp the bis(amido) complex $[TiCl\{N(TMS)_2\}_2Cp]$ was also obtained and crystallographically characterized.[159] $[TiX_3Cp']$ and $N(SnMe_3)_3$ in toluene or pyridine give $[\{TiXCp'(\mu\text{-}NSnMe_3)\}_2]$ (X = F or Cl; Cp' = C_5H_4TMS or Cp*). The structures of the Cp* derivative and of $[\{TiCl(\eta\text{-}C_5H_4TMS)(\mu\text{-}NBu^t)\}_2]$ were determined.[160] The diazadiene salt $Na_2[dad]$ reacts with $[TiCl_3Cp]$ to give $[TiClCp(dad)]$, which contains a titanadiazametallacyclopentene unit folded along the N–N vector by 60.5° (dad = ArN=C-(Me)C(Me)=NAr; Ar = $p\text{-}C_6H_4OMe$). It adopts a 'supine' orientation, with an average titanium–nitrogen distance of 0.192 1 nm. The structure is maintained in solution. The compound reacts with benzylmagnesium chloride to give $[Ti(\eta^1\text{-}CH_2Ph)Cp(dad)]$.[161]

The phosphorus imide $(TMS)N=PPh_2CH_2PPh_2$ reacts with $[TiCl_3Cp]$ to give (66), which can act as a bidentate ligand, for example, towards $PdCl_2$ (Equation (24)).[162,163] The related compound $Me_3SiN=PPh_2OSiMe_3$ reacts with $[TiCl_3Cp]$ to give the cyclotrimer $[\{TiClCp\{OP(Ph)_2=N\}\}_3$ as a yellow solid in almost quantitative yield.[164]

$$\underset{CpCl_2Ti}{\overset{Ph_2P\frown PPh_2}{\underset{\diagdown N \diagup}{\|}}} \quad \xrightarrow{PdCl_2} \quad \underset{CpCl_2Ti}{\overset{Ph_2P\frown PPh_2}{\underset{N-PdCl_2}{\backslash\!\backslash \quad /}}} \qquad (24)$$

$$(66)$$

The chemistry of hydrazido and diazenido ligands has attracted attention because of the possible involvement of these entities in the reduction of dinitrogen. $[TiCl_3Cp]$ reacts with $(TMS)N(Me)NMe_2$ to give the hydrazido complex $[TiCl_2Cp\{\eta^2\text{-}N(Me)NMe_2\}]$ (67).[165,166] The compounds $[TiCl_2Cp\{\eta^2\text{-}N(H)NMe_2\}]$ and $[TiCl_2Cp\{\eta^2\text{-}N(Ph)NH_2\}]$ (68) are prepared similarly. The crystal structures of (67)[167] and (68)[165,168] have been determined, and the bonding parameters for the nitrogen ligands are given in Figure 7. The molecular structures of (67) and (68) are shown in Figure 8. There is considerable π-contribution from the σ-bonded amido nitrogen, whereas the bond distance to the second nitrogen is much longer, indicating a simple n-donor interaction. The crystal packing of (68) shows that the molecules form hydrogen-bonded pairs, while $[TiCl_2Cp\{\eta^2\text{-}N(H)NMe_2\}]$ forms hydrogen-bonded one-dimensional chains in the solid state. A series of such complexes has been prepared from $[TiCl_3Cp]$ and $MN(R^1)NR^2R^3$ (M = Li^{168} or TMS^{170}) (Equation (25)). By contrast, the reaction between $[TiCl_3Cp]$ and $(TMS)NHNH(TMS)$ or $LiNHNH_2$ in THF does not give a hydrazido complex but leads to reduction, with formation of $[TiCl_2Cp(THF)]$.[171]

$$[TiCl_3Cp] \; + \; \underset{M}{\overset{R^1}{\underset{\diagup}{\diagdown}}} N - N \overset{R^2}{\underset{R^3}{\diagdown}} \quad \xrightarrow{M = Li \text{ or } TMS} \qquad (25)$$

R^1	R^2	R^3
Ph	H	H
H	Me	Me
Me	Me	Me
H	H	Bu^t
H	H	TMS
H	Ph	Ph

The reaction of $[TiCl_3Cp]$ with $(TMS)N=NPh$ affords the red, air-stable diazenido compound $[TiCl_2Cp\{\eta^2\text{-}NNPh\}]$ (69).[165,169] Vibrations attributed to the diazenido moiety are observed at 1593 cm^{-1} and 1632 cm^{-1}. Investigations by x-ray diffraction of (69) (Figure 9) show that the N_2 ligand is oriented perpendicular to those in (67) and (68) (Figures 7 and 8). Although the unsubstituted nitrogen in (69) should be basic, the protonated form, (69)·HCl, is unstable and decomposes on attempted recrystallization.[169] The hydrazido complex $[TiCl_2Cp\{\eta^2\text{-}NHN(H)Bu^t\}]$ reacts with $EtO_2CN=NCO_2Et$ under hydrogen transfer to give $[TiCl_2Cp\{\eta^2\text{-}NNBu^t\}]$.

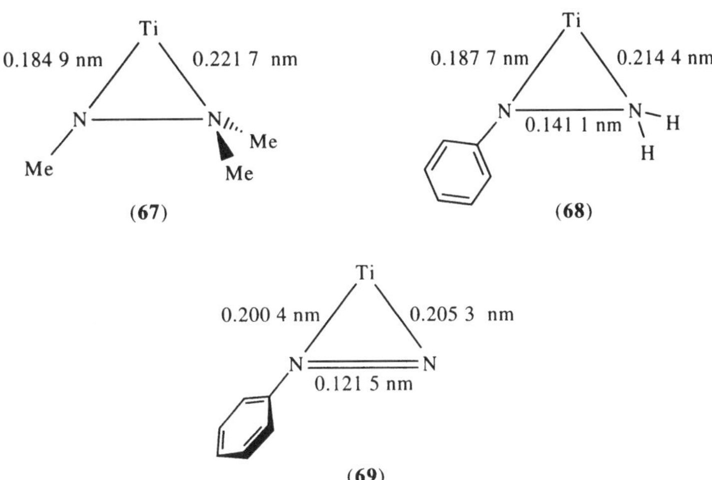

Figure 7 Bonding parameters of the hydrazido and diazenido ligands in (67), (68) and (69).

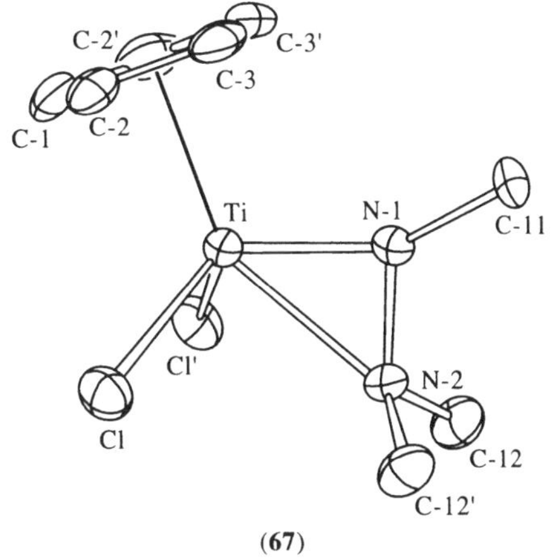

Figure 8 Molecular structures of the hydrazido complexes (67) and (68) (reproduced by permission of the Royal Society of Chemistry from *J. Chem. Soc., Chem. Commun.*, 1983, 1368 and reproduced by permission of Elsevier from *J. Organomet. Chem.*, 1987, 323, C29).

(69)

Figure 9 Molecular structure of (**69**) (reproduced by permission of the Royal Society of Chemistry from *J. Chem. Soc., Dalton Trans.*, 1986, 377).

The nitrogen-, oxygen-bonded compound $[TiCl_2Cp\{\eta^2\text{-}ONMe_2\}]$ is obtained from $[TiCl_3Cp]$ and TMS–O–NMe$_2$; its structure is analogous to (**67**), with substantial π-interaction between the titanium and oxygen atoms (Ti–O 0.1866 nm).[170]

The treatment of the hydrazido(1−) complex (**68**) with Bu^{t}NH$_2$ in dichloromethane at room temperature leads to deprotonation to the dark brown hydrazido(2−) complex $[TiClCp\{NN(H)Ph\}]_2$. The analogous NNMe$_2$ and NNPh$_2$ compounds were also obtained. As the crystal structure of the latter shows, one hydrazido ligand is side-on bonded, while the other forms a simple μ_2-bridge (Figure 10).[172]

Figure 10 Molecular structure of $[(TiClCp)_2(\mu\text{-}NNPh_2)_2]$ (reproduced by permission of the Royal Society of Chemistry from *J. Chem. Soc., Dalton Trans.*, 1986, 393).

The red ketimido and yellow phosphorus imido complexes (**70**) and (**71**) with η^1-bonded nitrogen ligands are obtained by reacting $[TiCl_3Cp]$ with $LiN=C(Bu^n)Bu^t$ and $(TMS)N=PPh_3$, respectively. The analogous reaction with PhNNN(Ph)Li gives a green–black 1,3-diphenyltriazenido complex.[169]

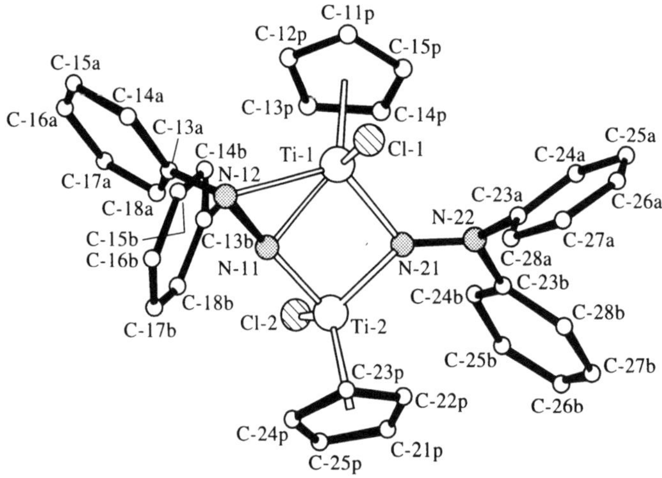

5.4.4 Monocyclopentadienyltitanium Complexes with Schiff Bases

The reactions of [TiCl$_3$Cp] with Schiff bases and heterocycles have been investigated in depth and have been summarized in a review.[173] The complexes are usually of the stoichiometry [TiCl(L)Cp] or [TiCl$_2$(L)Cp] and adopt pseudo-octahedral coordination geometries. Thus, benzoylhydrazone derivatives LH = (72) give [TiCl$_2$(L)Cp] or [TiCl(L)$_2$Cp], depending on the stoichiometry.[174] Dibasic Schiff bases such as (73) act as N,O-chelates and give compounds of the composition [TiCl(L)Cp].[175] Schiff bases derived from S-methyldithiocarbazate (LH = (74)) react with [TiCl$_3$Cp] in dichloromethane to give a series of compounds [TiCl$_m$(L)$_n$Cp] (m,n = 1 or 2).[176] Complexes of Schiff bases derived from carbohydrazide and carbonyl compounds such as salicyladehyde or MeCOC$_6$H$_4$X (X = H, p-OH, m-NO$_2$ p-NO$_2$ or p-OMe) have also been described.[177] The ligands (75a) act as neutral donors and form the adducts [TiCl$_3$(L)Cp], while LH = (75b) act as dianions to give [TiCl(L)Cp].[178] 1,5-Diarylthiocarbazones form chelates of the type (76) or (77), depending on the reagent ratio, where the N–H unit is hydrogen-bonded to the coordinated sulfur.[179] Silylated Schiff bases such as (78) act as tetradentate ligands and react with [TiCl$_3$Cp] or [TiCl$_2$(OMe)Cp] to give the complexes [TiX(L)Cp] (79) (X = Cl and OMe, respectively). These undergo ligand exchange with a variety of anions to give (79) (X = SMe, NMe$_2$, SnPh$_3$ or N$_3$), while treatment of (79) (X = Cl) with thallium pentanedionate [Tl(acac)] gives a titanium complex which is thought to contain a carbon-bonded η^1-pentanedionato ligand.[180]

(72)

(73)

(74)

R^1 = H, Me, Ph or C$_6$H$_4$OMe

R^2 = Me, Et, Bu, Ph, PhCH=CH or furyl

R^1, R^2 = –(CH$_2$)$_n$– (n = 4 or 5)

R^1 = H, R^2 =

R^1, R^2 = –(CH$_2$)$_4$–

(75)

(76)

(a) R^1 = H, Me or Ph

R^2 = Ph, C$_6$H$_4$X (X = OH, NO$_2$ or OMe), CH=CHPh or C$_6$H$_3$(OH)(OMe)

(b) R^1 = H or Me

R^2 = o-C$_6$H$_4$(OH)

R = Ph, o,p-C$_6$H$_4$X (X = Me or Cl) or 3,5-Me$_2$C$_6$H$_3$

(77)

(78)

(79)

R = –CH$_2$CH$_2$– or o-C$_6$H$_4$

Heterocyclic thiones such as LH = (80), (81) and (82) react with [TiCl$_3$Cp] to give yellow to brown products [TiCl$_n$(L)$_{3-n}$Cp] where L is bound via sulfur and nitrogen atoms but whose structures are unknown.[181,182] Similar thioketonate complexes are obtained from the hydantoin and tetrazoline derivatives (**83**) and (**84**).[183] The formulation of these compounds is primarily based on NMR and IR spectroscopic data. The complexes are nonelectrolytes.

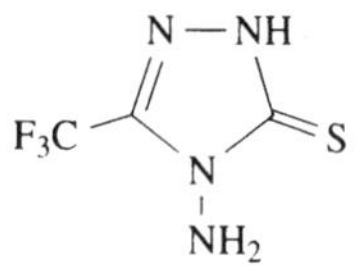

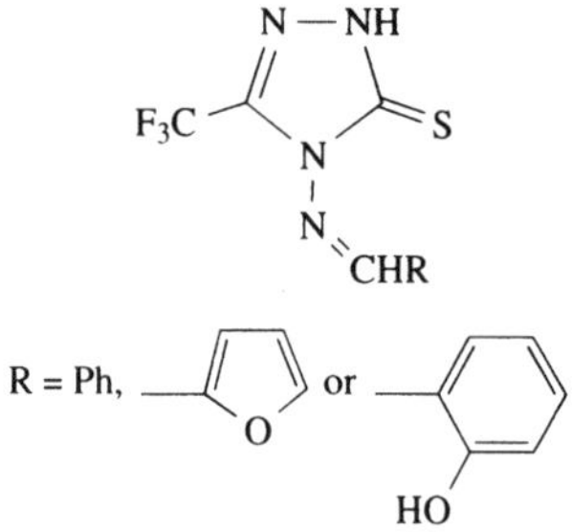

R = Ph, $\quad$ or $\quad$

(**80**)   (**81**)   (**82**)

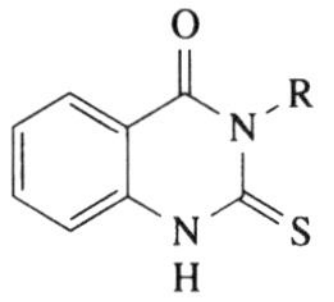

R = Me, Ph or *o*-tolyl

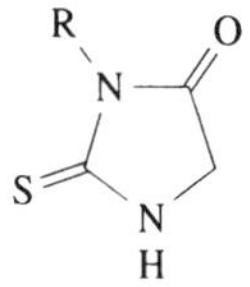

R = Ph, C_6H_4Me, Bz, *o,p*-OC_6H_4Me or C_{10}H_7

(**83**)

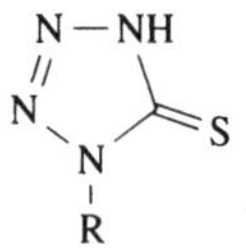

R = Ph, *p*-C_6H_4X (X = Me or Cl) or *p*-OC_6H_4Me

(**84**)

5.4.5 Complexes with Ti–O Bonds

The hydrolysis of [TiCl_3Cp'] can lead to dimeric, trimeric and tetrameric oxo-bridged titanium complexes, depending on the water concentration and the reaction conditions. The controlled hydrolysis of [TiCl_3Cp'] gives (**85**) (Cp' = Cp), which can react further to give (**86**) (X = Cl) (Scheme 27).[184] The dimer (**85**) (Cp' = Cp) was also obtained as a hydrolysis product during the attempted slow recrystallization of [Ti(SeMe)_3Cp] in dichloromethane/pentane. The x-ray structure of the compound gives Ti–Cl bond lengths of 0.2333 nm and 0.2256 nm, short Ti–O bonds of 0.181 nm and a Ti–O–Ti angle of 167.5(6)°, indicative of substantial π-bonding contributions between titanium and oxygen.[185] A similar dimer (**85**) (Cp' = η-C_5H_2(TMS)_3) was also obtained by hydrolysing [TiCl_3{η-C_5H_2(TMS)_3-1,2,4}] with 0.5 equivalent of H_2O in THF or acetone. A diagnostic feature of this dimer is the v_{as}(Ti–O) frequency of 762 cm^{-1}.[101,186] If one equivalent of water is present as well as base, the bis(oxo)-bridged lemon-yellow complex [(**87**), (Cp' = η-C_5H_2(TMS)_3] is formed, which possesses a planar Ti_2O_2 ring. The crystal structure gives Ti–O bond lengths of 0.1814(1) nm and 0.1835(1) nm and a Ti–Ti distance of 0.2707 nm.[101] A similar doubly bridged compound (**88**) is obtained when the product of the reaction of the pyridyl-substituted cyclopentadienyl anion (**89**) and TiCl_4 is recrystallized from wet dichloromethane (Equation (26)). In this case the oxo bridge is asymmetric, with Ti–O bonds lengths of 0.1791(3) nm and 0.1910(2) nm. The Ti–N distance of (**88**) is typical for a simple donor bond, 0.2297(3) nm.[187]

The hydrolysis of [TiX_3(Cp*)] with one equivalent of water in the presence of diethylamine gives (**85**) (Cp' = Cp*). The reaction rate increases in the order X = I < Br < Cl, opposite to what would have been expected on the basis of bond strengths and covalent character, and it is possible that steric factors are important.[188]

The hydrolysis of [TiCl_3(Cp')] with aqueous ammonia solution leads to (**90**), whose adamantoid cage structure was confirmed by x-ray diffraction (Scheme 27). The average Ti–O distance is 0.1837 nm, and the Ti–O–Ti angles are 123°. Refluxing [TiCl_3(Cp*)] in acetone leads to the trimer (**91**) (Cp' = Cp*).[189,190] The latter compound seems to be formed only with pentamethylcyclopentadienyl complexes, while smaller Cp'$^-$ ligands lead to (**86**).[184] Apart from the nature of the cyclopentadienyl ligand, the outcome of the hydrolysis reaction is also dependent on the halide and the presence of base. For example, [TiCl_3(Cp*)] in the presence of diethylamine gives (**91**), while [TiBr_3(Cp*)] under these conditions affords (**86**) (X = Br). In the presence of larger concentrations of water and base the tetramers (**90**) (Cp' = Cp*) are formed from either halide complex.[188] In the absence of base, on the other hand, hydrolysis of [TiBr_3(Cp')] gives the trimer (**91**) (X = Br, Cp' = Cp*), whose structure was confirmed by x-ray diffraction. The sterically slightly less hindered complex [TiBr_3(η-C_5HMe_4)] is hydrolysed under base-free conditions to give the tetramer (**86**) (X = Br, Cp' = C_5HMe_4).[191]

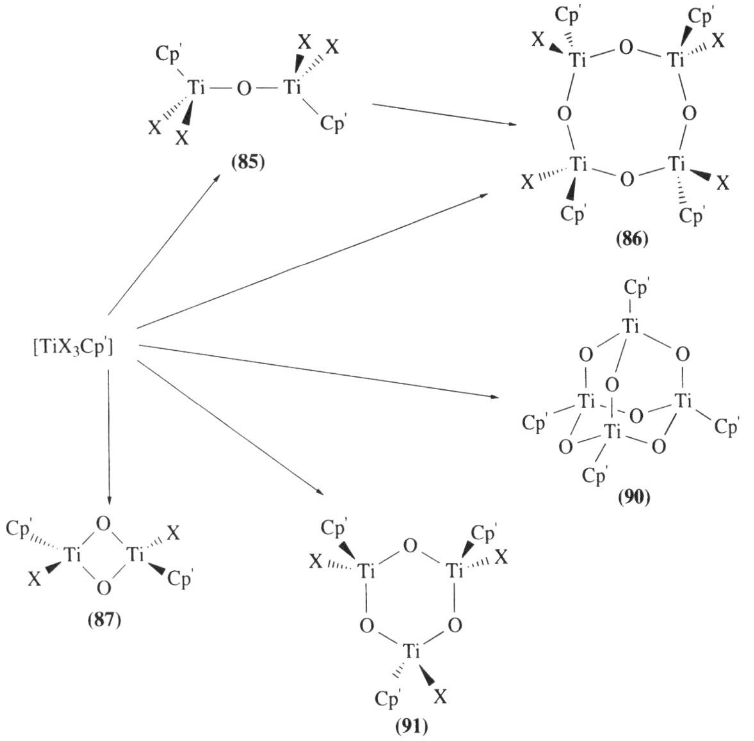

Scheme 27

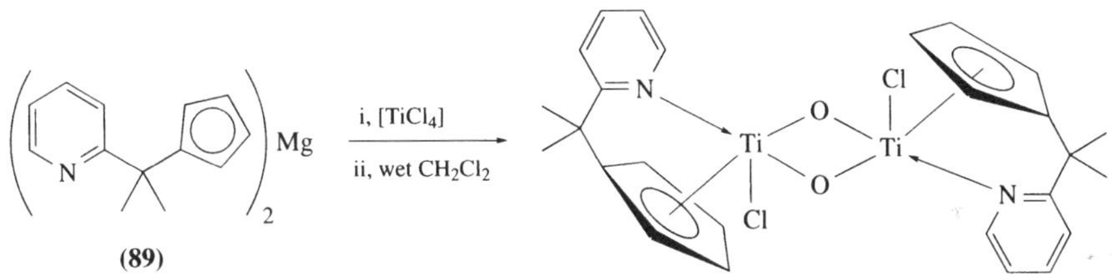

$$[\text{Ti}_4(\eta\text{-Cp}^*)_4(\mu\text{-O})_6] + 1/2\ [\text{TiCl}_4] \xrightarrow[-\text{TiO}_2]{} \qquad (27)$$

Complex (**86**) (X = Cl, Cp' = C_5H_4TMS) reacts with [NH$_4$]SCN to give the crystallographically characterized tetramer (**86**) (X = SCN).[184] The reaction of (**90**) (Cp' = Cp*) with 0.5 equivalents of [TiCl$_4$] leads to the substitution of one oxygen atom by two chloride ligands and the formation of a tetramer with a butterfly structure (**92**) (Equation (27)). While treatment of (**90**) with more than 0.75 equivalents of [TiCl$_4$] gives mixtures of compounds, (**91**) reacts with [TiCl$_4$] to give pure (**85**) (Cp' = Cp*).[190]

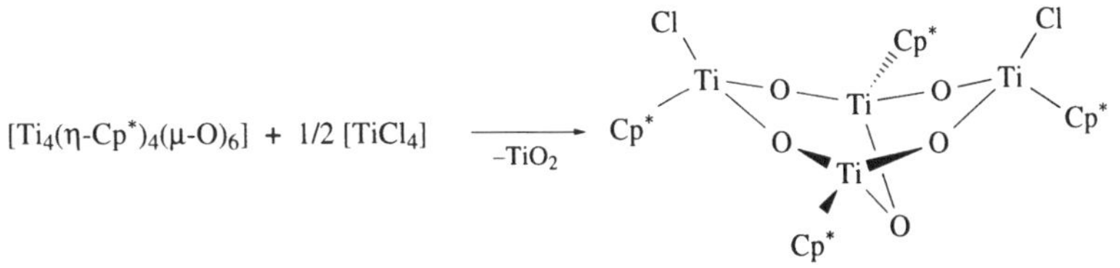

The complexes (**86**) (X = Cl; Cp' = Cp or C_5H_4Me) are reduced by aluminum powder to give the titanium(III) clusters [Ti$_6$Cp'$_6$(μ_3-Cl)$_4$(μ_3-O)$_4$] (**93**), which contain two unpaired electrons. The structure of (**93**) is shown in Figure 11. The reduction of (**86**) with zinc or SnHBu$_3$ leads to only partial reduction and formation of the diamagnetic $\text{Ti}^{\text{III}}_4\text{Ti}^{\text{IV}}_2$ compound [Ti$_6$Cp'$_6$(μ_3-Cl)$_2$(μ_3-O)$_6$]; it is isostructural with (**93**), with two oxygens taking the place of two chloride ligands.[192] The diamagnetism of the $\text{Ti}^{\text{III}}_2\text{Ti}^{\text{IV}}_4$ cluster [Ti$_6$Cp$_6$(μ_3-O)$_8$] is explained by extended Hückel molecular orbital (EHMO) calculations, which show that the two excess electrons are placed in a metal-centred a_{1g} orbital.[193]

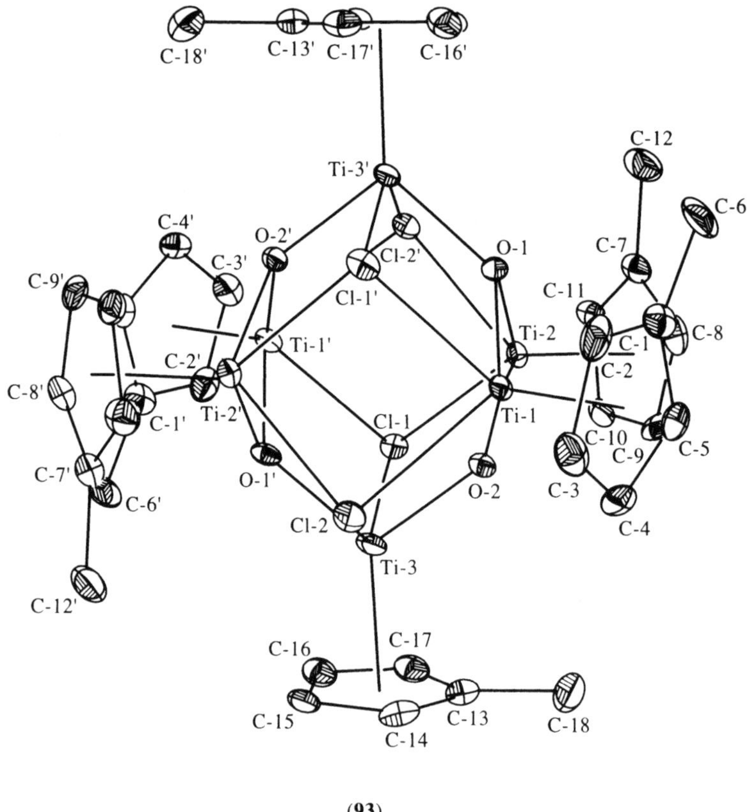

(93)

Figure 11 Molecular structure of $[Ti_6O_4Cl_4Cp'_6]$ (**93**) (reproduced by permission of the American Chemical Society. Copyright (1986) from *J. Am. Chem. Soc.*, 1986, **108**, 6823).

The structure of $[TiCp(NO_3)_3]$ (**94**) has been determined (Figure 12). The compound has a pentagonal-bipyramidal structure, with bidentate nitrato ligands. The average Ti–O distance is quite long, 0.2134 nm. The structure can be considered as a prototype for many mono(cyclopentadienyl)-titanium complexes with chelating ligands.[194] Protolysis of $[TiMe_3(Cp*)]$ with benzoic acid affords $[Ti(Cp*)(\eta^2\text{-}O_2CPh)_3]$. Investigation by x-ray diffraction confirms that the compound has a pentagonal-bipyramidal structure analogous to (**94**).[195] The reaction of $[TiCl_3(\eta\text{-}C_5H_3Bu^t_2\text{-}1,3)]$ with methanol in refluxing benzene gives the yellow complex $[TiCl_2(OMe)(\eta\text{-}C_5H_3Bu^t_2\text{-}1,3)]$, while treatment with sodium oxalate in THF leads to the orange binuclear oxalato complex $[\{TiCl_2(\eta\text{-}C_5H_3Bu^t_2\text{-}1,3)\}_2(\mu\text{-}C_2O_4)]$ in 43% yield. Each $-CO_2-$ group of the oxalato ligand is η^1-bonded. The reaction of $[TiCl_3(\eta\text{-}C_5H_3Bu^t_2\text{-}1,3)]$ with 2-pyrazinecarboxylic acid in benzene affords $[TiCl(\eta\text{-}C_5H_3Bu^t_2\text{-}1,3)(O_2CC_4H_3N_2)_2]$ in 84% yield. Here, too, the carboxylate acts as a monodentate ligand.[99]

The reaction of $[TiCl_3\{\eta\text{-}C_5H_{5-n}(TMS)_n\}]$ (n = 2 or 3) with acetylacetone derivatives $CH_2(COR^2)_2$ in the presence of $Ag[O_3SCF_3]$ gives the ionic compounds (**95**) (R^2 = Me or Bu^t). The crystal structures of (**95**) show no close contacts between the metal cation and the anion (Scheme 28). The highly TMS-substituted cyclopentadienyl ligand imparts high solubility, and in spite of their ionic character these compounds are noticeably soluble in cyclohexane. If R^2 = CF_3, high yields of the neutral complexes (**96**) are obtained.[196]

The reaction of acetylacetone (acacH) with $[TiCl_2(\eta\text{-}C_5H_4R)_2]$ (R = H, Me or CHPh_2) in isobutyronitrile in the presence of triethylamine leads to loss of one Cp$^-$ ligand and the formation of the six-coordinate complex $[TiCl(acac)_2(\eta\text{-}C_5H_4R)]$, while irradiation of the reaction mixture in THF gives $[TiCl_2(acac)(\eta\text{-}C_5H_4R)]$. Similarly, treatment of $[TiCl_2(\eta\text{-}C_5H_4R)_2]$ with LH = 8-hydroxyquinoline affords $[TiClL_2(\eta\text{-}C_5H_4R)]$.[197] The reaction of $[TiCl_3\{\eta\text{-}C_5H_2(TMS)_3\}]$ with one equivalent of acacH leads to the orange–red complex $[TiCl_2(acac)\{\eta\text{-}C_5H_2(TMS)_3\}]$.[101]

The cationic oxo-centred trinuclear methoxo complex (**97**) was isolated from the reaction of $[TiCp_2(H_2O)_2][BPh_4]$ with methanol and characterized by x-ray diffraction.[198] The diphenol $Me_2C(C_6H_4OH\text{-}p)_2$ reacts with $[TiCl_4]$ to give a polymeric bis(phenoxo) complex. Treatment of this product with TlCp affords $[TiCl\{Me_2C(C_6H_4O)_2\}Cp]$. Similarly, $[TiCl_2(1,2\text{-}O_2C_6H_3R\text{-}3)]$ and TlCp give $[TiCl(1,2\text{-}O_2C_6H_3R\text{-}3)Cp]$.[199] The hydrolysis of $[TiCl_3(Cp*)]$ with $Bu^t_2Si(OH)_2$ proceeds with

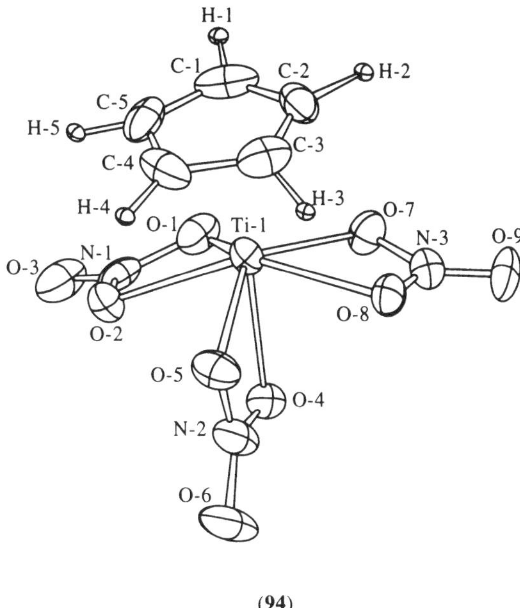

(94)

Figure 12 Molecular structure of [Ti(NO$_3$)$_3$Cp] (**94**) (reproduced from *J. Organomet. Chem.*, 1982, **232**, 41).

(95)

R^2 = Me, n = 2
R^2 = Bu^t, n = 2 or 3

(96)

R^2 = CF$_3$, n = 2 or 3

R^1 = TMS, X = CF$_3$SO$_3$

Scheme 28

substitution of one chloride ligand to give (**98**). The product of further hydrolysis of this complex by water in the presence of triethylamine was identified crystallographically as the yellow, hydrogen-bonded hydroxo complex (**99**) (Equation (28)). Surprisingly, (**99**) does not eliminate water on heating to give a μ-oxo complex.[200]

Cyclohexanol and [TiCl$_3$\{η-C$_5$H$_3$(TMS)$_2$\}] form the alkoxo complex [Ti(OC$_6$H$_{11}$)$_3$\{η-C$_5$H$_3$(TMS)$_2$\}], while 1,3,5-cyclohexanetriol in the presence of NEt$_3$ as a base acts as a tridentate ligand to give (**100**), whose structure was determined. The Ti–O bond length is 0.1833 nm. Both compounds react rapidly with ethanol with complete exchange of the alkoxo ligands.[201] The metallatrane compound [Ti\{(OCH$_2$CH$_2$)$_3$N\}Cp] is obtained from [TiCl\{(OCH$_2$CH$_2$)$_3$N\}] and TlCp in THF as colourless crystals in 90% yield, or in lower yield from [TiCl$_3$Cp] and N(CH$_2$CH$_2$OH)$_3$.[202] [TiCl$_3$Cp] and phenol in the presence of NEt$_3$ give the bright-yellow complex [Ti(OPh)$_3$Cp], which reacts with further [TiCl$_3$Cp] in

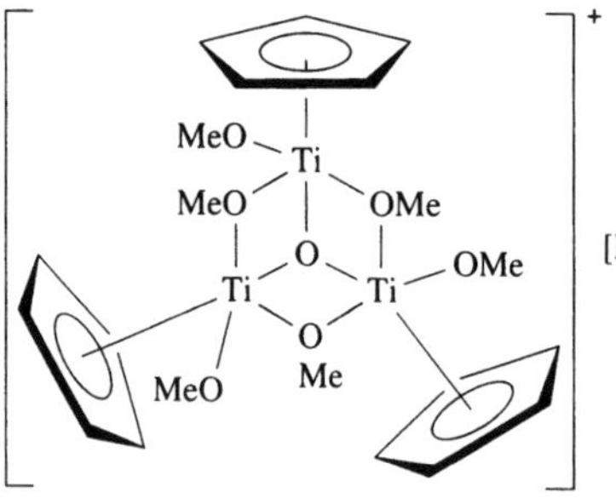

(97)

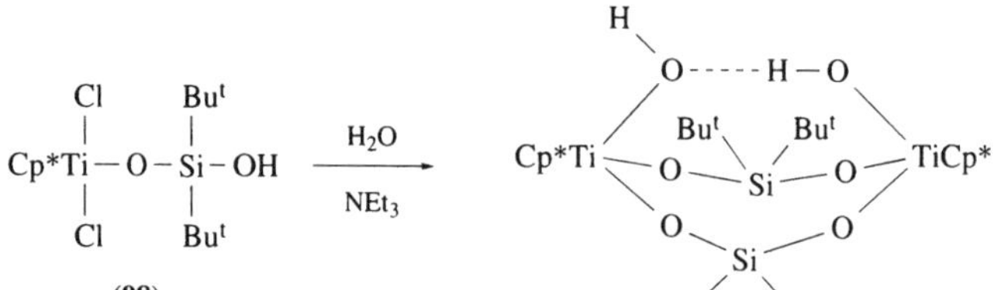

$$2\ [\text{TiCl}_2\text{Cp}_2] + 5\ [\text{Mo}_2\text{O}_7]^{2-} \xrightarrow{\text{CH}_2\text{Cl}_2} 2\ [\text{Ti(Cp)(Mo}_5\text{O}_{18})]^{3-} + 2\ \text{C}_5\text{H}_6 + 4\ \text{Cl}^- \tag{28}$$

(98)   **(99)**

n-hexane to give yellow needles of [TiCl(OPh)$_2$Cp]. Both compounds are monomeric.[203] A series of titanium complexes [TiCl(O–R–O)Cp] of dihydridic phenols has been prepared by related methods from variously substituted phenols and binaphthols.[204] The solvolysis of (**86**) (X = Cl) with 2,4,6-Me$_3$C$_6$H$_2$OH proceeds without disruption of the Ti$_4$O$_4$ framework to give (**86**) (X = 2,4,6-Me$_3$C$_6$H$_2$O), whose crystal structure was determined.[205] The reaction of [TiCl$_3$Cp] with XC$_6$H$_4$OH (X = H, *o*-, *m*- or *p*-NO$_2$, *o*-, *m*- or *p*-Me, or *o*- or *p*-Cl) affords [TiCl$_2$(XC$_6$H$_4$O)Cp] as light yellow to orange solids in moderate yields. The frequencies of the C–O and Ti–Cl bands in the IR spectra depend on the positions of the substituents X on the aryl ring. A linear correlation was found between the Hammett σ factors of X and the wavenumbers of the IR bands.[206] Mixtures of [Ti(OBu)$_3$Cp] and methylaluminoxane are active catalysts for the syndiotactic polymerization of styrene. Redox titration indicates that reduction to titanium(III) occurs under these conditions, and the oxidation state of the active species is thought to be titanium(III).[207]

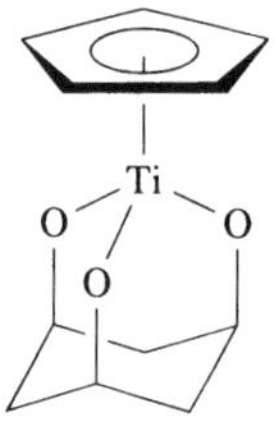

(100)

A detailed method for the synthesis of the heterosilicate compound [NBu$_4$]$_4$[TiCp(SiW$_9$V$_3$O$_{40}$)] has been published.[208,209] The reaction of [TiCl$_2$Cp$_2$] with [Mo$_2$O$_7$]$^{2-}$ leads to [TiCp(Mo$_5$O$_{18}$)]$^{3-}$ (Equation (29)). The analogous reaction with [WO$_4$]$^{2-}$ gives the isostructural tungsten complex. The structure of the molybdenum compound was determined by x-ray diffraction. The compounds were isolated as [NBu$_4$]$^+$ salts.[210] The Cp* complex [NBu$_4$]$_3$[Ti(Cp*)(W$_5$O$_{18}$)] was similarly obtained from [TiCl$_3$(Cp*)] and [NBu$_4$]$_2$[WO$_4$] in acetonitrile in the presence of trichloroacetic acid. [TiCp(W$_5$O$_{18}$)]$^{3-}$ is protonated by Cl$_3$CCO$_2$H to give [TiCp(W$_5$O$_{18}$)H]$^{2-}$, while HCl in MeCN leads to [TiCp(W$_5$O$_{18}$)H$_2$]$^-$. The crystal structures of both protonated complexes were determined.[211]

$$2\ [\text{TiCl}_2\text{Cp}_2] + 5\ [\text{Mo}_2\text{O}_7]^{2-} \xrightarrow{\text{CH}_2\text{Cl}_2} 2\ [\text{Ti(Cp)(Mo}_5\text{O}_{18})]^{3-} + 2\ \text{C}_5\text{H}_6 + 4\ \text{Cl}^- \tag{29}$$

A series of complexes of the type (**101**) with long functionalized alkyl chain substituents on the cyclopentadienyl ligands have been prepared following Equation (30) (R = (CH$_2$)$_5$CO$_2$Bu^t, (CH$_2$)$_6$O-THP, (CH$_2$)$_5$CH(OMe)$_2$, (CH$_2$)$_6$NH$_2$ or (CH$_2$)$_2$CH=CHCH=CH$_2$).[212] Complexes with aldehyde functionalities on the side-chain, such as (**101a**), can be aminated, for example with

N6-[{(aminohexyl)carbamoyl}methyl]adenosine-5'-triphosphate to give ATP derivatives designed for the location by electron microscopy of ATP binding sites in some proteins.[213] Compounds (**101**) with R = $(CH_2)_2CH=CHCH=CH_2$ undergo Diels–Alder reactions as shown in Equation (31) to give the products (**102**), which are suitable for attachment to proteins.[214]

$$[Ti(NMe_2)_3(C_5H_4R)] \xrightarrow[\text{DMF, pH 3–4}]{K_{10}[P_2W_{17}O_{61}]} \left[\begin{array}{c} R \\ TiP_2W_{17}O_{61} \end{array} \right]^{7-} \quad (30)$$

(**101**)

R = $-(CH_2)_3O$—⬡—$CH(OMe)_2$

(**101a**) R = $-(CH_2)_3O$—⬡—CHO

(**101b**) R = $-(CH_2)_3O$—⬡—$CH_2NH(CH_2)_6NHC(O)CH_2NH$ ...

$$\left[\begin{array}{c} TiP_2W_{17}O_{61} \end{array} \right] K_7 \xrightarrow{} K_7 \left[\begin{array}{c} TiP_2W_{17}O_{61} \end{array} \right] \quad (31)$$

(**102**)

5.4.6 Complexes with Ti–S Bonds

[TiCl$_3$Cp] reacts with $(NH_4)_2S_x$ in acetonitrile in the presence of [PPh$_4$]Cl to give olive–brown crystals of (**103**), whose crystal structure was determined.[215] The oxidation of [TiCp$_2$(CO)$_2$] with H$_2$S leads to the dark-green–brown paramagnetic mixed-valence Ti$^{III}_3$Ti$^{IV}_2$ cluster [Ti$_5$Cp$_5$(μ_3-S)$_6$].[216] The electronic structure of cubane clusters including [{Ti(η-C$_5$H$_4$Me)(μ_3-S)}$_4$] has been investigated by helium(I) and helium(II) photoelectron spectroscopy.[217] The reaction of [TiCl$_3$Cp'] (Cp' = Cp or C$_5$H$_4$Me) with 1.5 equivalents of Li$_2$S followed by treatment with oxygen gas gives the clusters [(TiCp')$_4$S$_8$O] and [(TiCp')$_4$S$_8$O$_2$]; the latter was characterized by x-ray diffraction (Figure 13).[218] The reaction of [TiCl$_2$Cp$_2$] with H$_2$Se proceeds according to Equation (32), with loss of cyclopentadiene to give the crystallographically characterized selenide cluster (**104**).[219]

$$\left[\begin{array}{c} \text{0.2073 nm} \\ \text{0.2050 nm} \\ S \\ S-S-S-Ti-S \\ S \quad S \\ \text{0.2391 nm} \end{array} \right]^{-} [PPh_4]^+$$

(**103**)

The reaction of [TiCl$_3$Cp] with RSLi gives the yellow thiolato complexes [TiCl$_2$(SR)Cp] (R = Et, c-C$_6$H$_{11}$, or *o*- or *p*-tolyl). Analogous C$_5$H$_4$Me complexes were obtained similarly. The reaction of

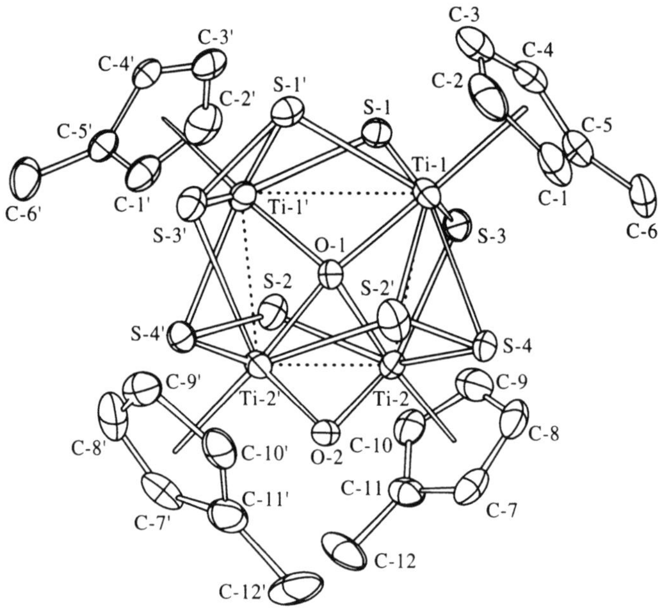

Figure 13 Molecular structure of [Ti$_4$O$_2$S$_8$Cp'$_4$] (reproduced by permission of the American Chemical Society. Copyright (1986) from *Inorg. Chem.*, 1986, **25**, 1886).

$$4\,[TiCl_2Cp_2] + 6\,H_2Se + 8\,NEt_3 \xrightarrow{THF} 4\,C_5H_6 + 8\,NEt_3{\cdot}HCl + \quad (\mathbf{104}) \qquad (32)$$

[TiCl$_3$Cp] with thiophenols affords [TiCl$_2$(SAr)Cp] (Ar = p-ClC$_6$H$_4$, p-BrC$_6$H$_4$ or C$_6$F$_5$), while treatment of [TiCl(SPh)$_2$Cp] with LiSPh generates the purple complex [Ti(SPh)$_3$Cp].[220] [TiBr$_3$Cp] reacts with NaSMe or LiSeMe to give [Ti(EMe)$_3$Cp] (E = S or Se). The analogous tellurium complex could not be obtained.[221] A 1:1 mixture of 1,2-(LiS)$_2$C$_6$H$_3$Me-4 and [TiCl$_3$Cp] at −45 °C affords the red–black complex [TiCl(S$_2$C$_6$H$_3$Me-4)Cp$_2$], while the reaction with two equivalents of the lithium thiolate leads to the deep-purple ionic complex [NEt$_4$][Ti(S$_2$C$_6$H$_3$Me-4)$_2$Cp].[222] Treatment of [TiCl$_3$Cp] with 1,3-propanedithiol in the presence of imidazole as base gives (**105**) (Scheme 29). The potentially tridentate ligand S(CH$_2$CH$_2$S(TMS))$_2$ affords (**106**). Complex (**105**) reacts with NaSPh to give (**107**), and with bis(dimethylphosphino)ethane to give the adduct (**108**). Hydrolysis of (**105**) in the presence of base leads to (**109**). The crystal structures of complexes (**105**)–(**109**) have been determined, and the molecular structures were rationalized with the aid of Fenske–Hall MO calculations.[223]

The interaction of [TiCl$_2$Cp$_2$] with the dithiocarbamates NaS$_2$CNR^{1}R^2 in refluxing dichloromethane gives a mixture of the five-coordinate bis-Cp and seven-coordinate mono-Cp complexes [TiCl(S$_2$CNR^{1}R^2)Cp$_2$] and [Ti(S$_2$CNR^{1}R^2)$_3$Cp] (R^1 = H or Et; R^2 = cyclopentyl, cycloheptyl or m-tolyl).[224] Analogous complexes with Cp = C$_5$H$_4$Me and R^2 = CH$_2$Ph were similarly obtained.[225] Treatment of [TiCl$_3$Cp] with potassium alkylxanthate (K[S$_2$COR]) in dichloromethane gives [TiCl$_2$(S$_2$COR)Cp] and [TiCl(S$_2$COR)$_2$Cp] (R = Me, Et, Pr, Bu or C$_5$H$_{11}$). The compounds are non-electrolytes, and the xanthate ligands are bidentate.[226] According to NMR spectra the complex [Ti(S$_2$CNMe$_2$)$_3$Cp] has a pentagonal-bipyramidal structure and is fluxional. The methyl substituents of the equatorial ligands interchange with $\Delta G^{\ddagger}$(70 °C) ~59 kJ mol^{-1} and 102–103 times faster than the methyl exchange within the unique (axial) ligand. A third pathway concerns the interchange between equatorial and the unique ligand, for which a $\Delta G^{\ddagger}$ of ~75 kJ mol^{-1} was estimated.[227]

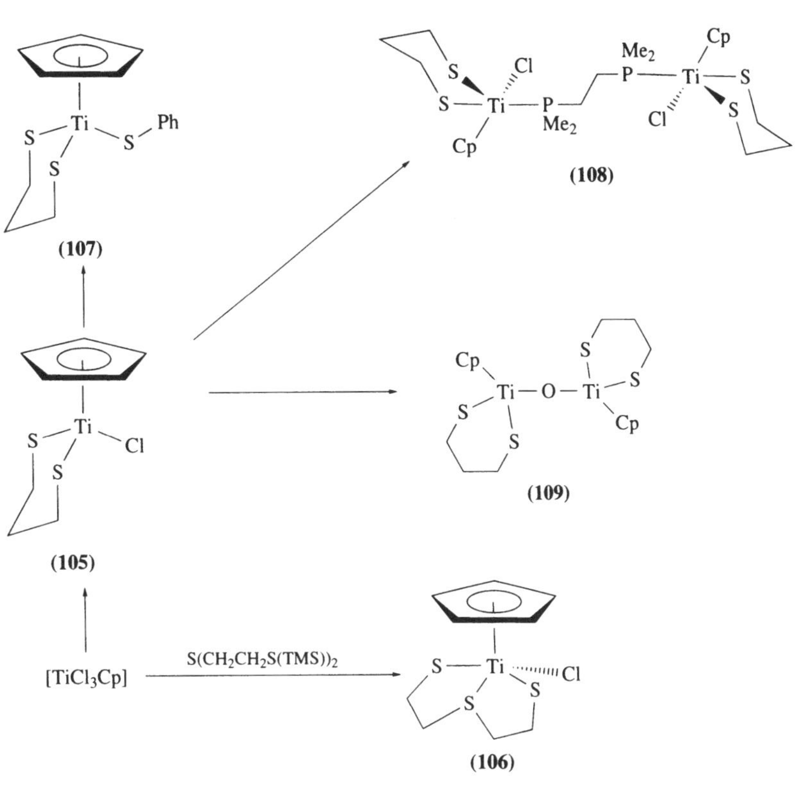

Scheme 29

5.5 BIS(CYCLOPENTADIENYL)TITANIUM COMPLEXES

5.5.1 Halide Complexes

5.5.1.1 Synthesis

The synthetic routes to titanocene dichlorides are now well established and involve the reaction of lithium, sodium, potassium or thallium cyclopentadienides with either [TiCl$_4$] or TiCl$_3$, followed by oxidation in the presence of aqueous HCl. The adducts [TiCl$_4$(THF)$_2$] and [TiCl$_3$(THF)$_3$] are commonly used precursors. The reaction with titanium(III) halides is preferred in cases where the cyclopentadienyl ligand is bulky or where the use of titanium(IV) gives unacceptably low yields. Titanium(III) intermediates of the type TiClCp$'_2$ are only isolated in exceptional cases. Most of the syntheses will not be described here in detail; a summary is given in Tables 3–5.

Bimetallic titanocene derivatives are readily accessible via the 4-stannatetrahydro-*s*-indacene (**110**). Thus, reaction with [TiCl$_3$Cp] in THF gives a mixture of (**111**) and (**112**) (Scheme 30), while the reaction in toluene at room temperature leads only to (**111**) in 82% yield.[279] Refluxing (**112**) in toluene for several hours gives the *ansa*-metallocene (**113**), which was also obtained by more conventional routes.[312,315] Analogues of (**112**) with a CMe$_2$ group instead of SiMe$_2$ have also been prepared. Compound (**111**) reacts with [{RhCl(1,5-C$_8$H$_{12}$)}$_2$] to give the mixed-metal complex (**114**). Mixed titanium–zirconium species are similarly accessible. The reduction of (**112**) with zinc powder affords the titanium(III) complex (**115**) (Scheme 30). The dinuclear complexes (**116**) are prepared in a similar manner following Equation (33). Compound (**116**) is reduced by zinc to the expected dinuclear titanium(III) product, and decomposes on sublimation *in vacuo* to give (**113**).[279] A heterobinuclear manganese–titanium complex related to (**116**) was obtained according to Equation (34).[280] Steric hindrance limits the range of titanocenes that can be prepared. For example, the reaction of [TiCl$_3$Cp] with K[C$_5$Ph$_4$R] gives [TiCl$_2$Cp(η-C$_5$Ph$_5$R)] (R = H or Ph) but fails for R = Me.[281] The reaction of K[C$_5$HPh$_4$] with [TiCl$_4$] produces brown intractable materials, although [TiCl$_2$(η-C$_5$HPh$_4$)] is accessible via oxidation of [TiCl(η-C$_5$HPh$_4$)] with silver chloride in 73% yield.[267]

Table 3 Preparation and properties of bis(cyclopentadienyl)titanium dichloride complexes, [TiCl₂L₂].

L	Preparative method[a]	Colour	m.p. (°C)	Yield (%)	Ref.
C₅H₄Me	A	Red crystals	215–219	60	228
C₅H₄-c-C₅H₉	A		197–198		x-ray[229,230]
C₅H₄-c-C₆H₁₁	A		252–253		x-ray[229,230]
C₅H₄Pr^i					x-ray[231]
C₅H₄Bu^t	A		50–60		synthesis[232] x-ray[233]
C₅H₄CH₂CH₂OMe	B	Violet–red	96–97	60	NMR, x-ray[102,234]
(R)-C₅H₄CH(Me)CH₂OMe	C	Red	136–137	61	x-ray[235]
(S)-C₅H₄CH₂CH(Me)OMe	C	Red	127–128	55	235
(S)-C₅H₄CH₂CH(OMe)Pr^i	C	Red	178–182	30	236
C₅H₄CH₂CH(OMe)Ph	C	Red		9.6	236
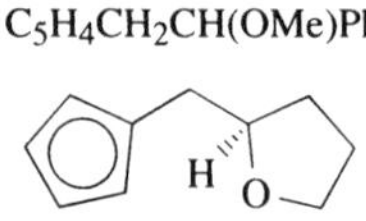 (structure: cyclopentadienylmethyl-tetrahydrofuran, H and O labels)	C	Red	112–113	79	236
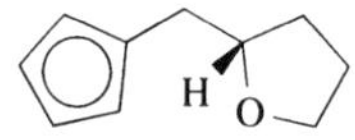 (structure: cyclopentadienylmethyl-tetrahydrofuran, H and O labels)	C	Red			x-ray[236]
C₅H₄CH=CH₂	D	Red–black		75	x-ray[109]
C₅H₄CH=CHPh	C	Green	244–246	86	236
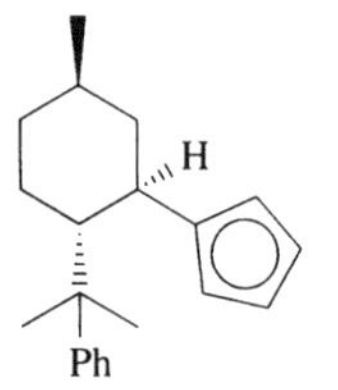 (structure: substituted cyclohexyl with phenyl-cyclopentadienyl, Ph, H labels)	E	Red	283–284	45.6	237
C₅H₄CO₂Me	D	Light red	193–194 (dec.)	62	238
C₅H₄(CH₂)₃Si(OMe)₃	B	Red oil		55	239
C₅H₄CH₂PPh₂	B	Maroon		85	240
C₅H₄(CH₂)₂PPh₂	B	Red oil			240
C₅H₄(CH₂)₂AsPh₂	A			32	104
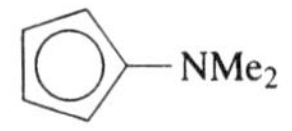 (structure: cyclopentadienyl–NMe₂)	A	Green–black	225 (dec.)	30	x-ray[241]
C₅H₄Cl	A	Deep red	254–256 (dec.) sublimes 220/0.005 torr[242]	67	242, 243
C₅H₄TMS					x-ray[244]
C₅H₃Bu^{t}₂-1,3	A	Orange,[245] red needles[99]	223–225	60	synthesis[245,246,247] x-ray[246]
C₅H₃MeBu^t (isomers)	A	Red			248
C₅H₃(TMS)₂-1,3	E	Dark red	182[249] 192–193[250]	85[249] 46[250]	249, 250
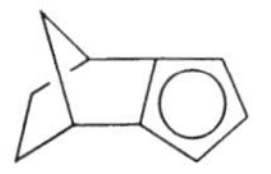 (structure: norbornane-fused cyclopentadienyl)	D	Red	235 (dec.)	70	251
exo,exo- 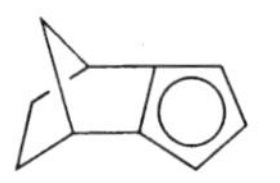 (structure: norbornane-fused cyclopentadienyl)	E	Red		49	252, 253
endo,endo- 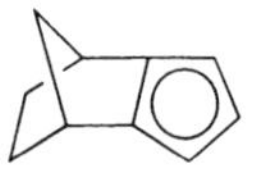 (structure: norbornane-fused cyclopentadienyl)	E	Red			252

Table 3 (continued)

L	Preparative method[a]	Colour	m.p. (°C)	Yield (%)	Ref.
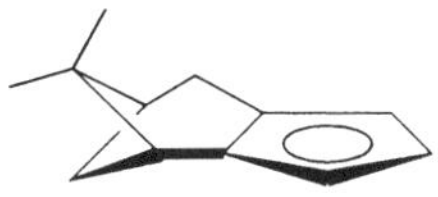	E	Red	240 (dec.)[254]	71	254, 255, x-ray[256]
endo,endo- 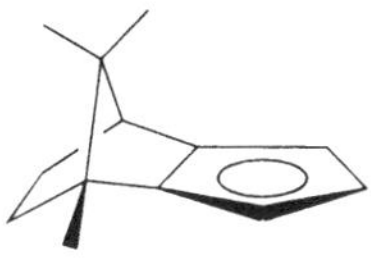	A[256]	Purple–brown	175–178[256]	23[256]	237, 256, 257
	E[237]		178–179[237] 179–180[257]	43[257] 11[237]	
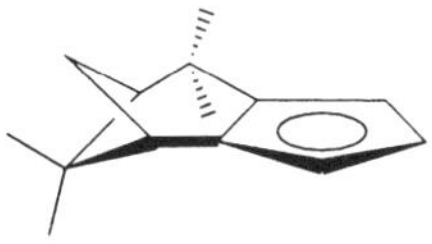	E	Purple–brown		45	258, 259
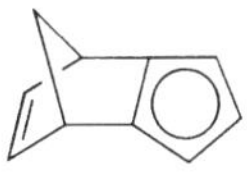	A	Red powder		54	260
		4.3% *exo,exo* 78.3% *exo,endo* 17.4% *endo,endo*			
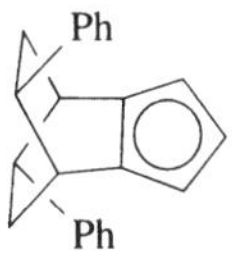	E			53	x-ray[261]
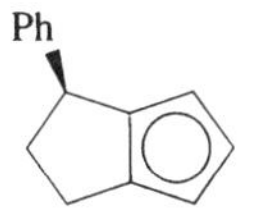	F	Red		6	262
	E	Dark red	165–166 $([\alpha]_D^{23} = +482°)$	45	263
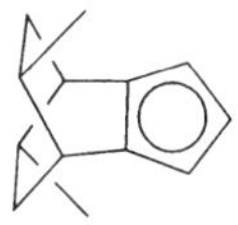	E	Dark red	189–190 $([\alpha]_D^{23} = -534°)$	54.5	x-ray[263]
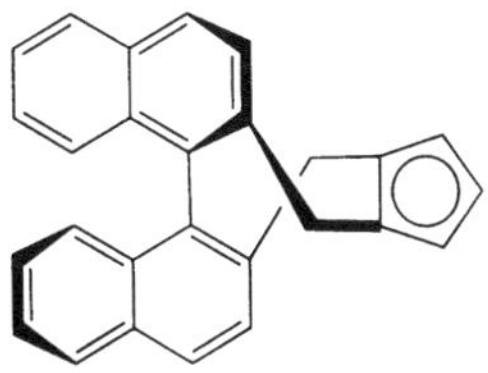	E	Red (*R,R*)-(+)	265	83	x-ray[264]

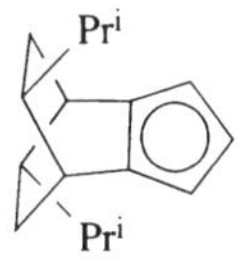

Table 3 (continued)

L	Preparative method[a]	Colour	m.p. (°C)	Yield (%)	Ref.
C_5HMe_4	E	Red		70	106, 265
$C_5Me_4Pr^i$	E	Purple		73	266
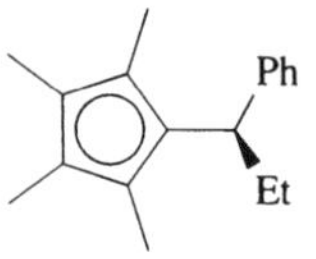 (structure: C_5Me_4 ring with CH(Ph)(Et) substituent)	E	Partially crystalline red oil (*rac* and *meso*)			110
C_5HPh_4	G	Burgundy (with 3/2 THF of crystallization)		73	267
$C_5Me_4PPh_2$	E	Brown powder		90	268
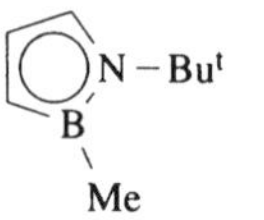 (structure: azaborolinyl, $N-Bu^t$, B, Me)	A	Light red (with [TiBr$_4$])		14	269
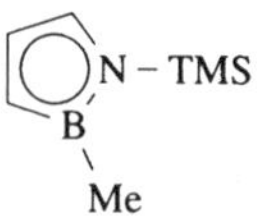 (structure: azaborolinyl, $N-TMS$, B, Me)	A			33	270
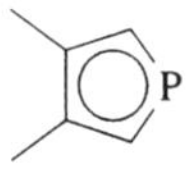 (structure: phospholyl ring, P)	H	Green–brown	160 (dec.)	64	271
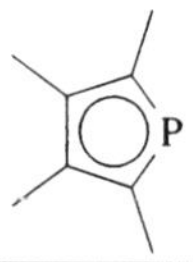 (structure: tetramethylphospholyl ring, P)	H	Green–black powder	85	39	271

[a] (Cp' = cyclopentadienyl derivative) Method **A**: LiCp', [TiCl$_4$], Et$_2$O. Method **B**: NaCp', TiCl$_4$, THF. Method **C**: Cp'H, potassium sand, [TiCl$_4$], THF. Method **D**: 2 TlCp', [TiCl$_4$(THF)$_2$]. Method **E**: LiCp', [TiCl$_3$] or [TiCl$_3$(THF)$_3$], followed by oxidation with conc. aqueous HCl. Method **F**: Cp'MgCl, [TiCl$_3$(THF)$_3$], THF, followed by oxidation with HCl or HCl/O$_2$. Method **G**: Oxidation of [TiClCp'$_2$] with AgCl, THF. Method **H**: Cp'SnMe$_3$, [TiCl$_4$], CH$_2$Cl$_2$/hexane.

The oxidation of the fulvalene complex (**117**) (X = Cl) with chlorine gas in dichloromethane[277] or with HCl in THF[278] leads to the green, poorly soluble compound (**118**) in 70–74% yield (Equation (35)). The oxidation of (**117**) (X = Br) with Br$_2$ gives the corresponding light-green bromo complex.[278]

Various heterocyclic titanocenes have been prepared. The reaction of azaborolinyl lithium (**119**) (R = Bu') with [TiBr$_4$] leads to (**120**) (X = Br) in 14% yield (Equation (36)).[269] The analogous complex (**120**) (R = TMS, X = Cl) is obtained similarly and has been characterized crystallographically (Figure 14). The dimensions are in principle similar to [TiCl$_2$Cp$_2$]. The Ti–B distances are rather long (0.257 nm), which leads to a slight folding of the five-membered ring. The boron atoms are oriented so that they can form weak interactions with the chloro ligands (B–Cl distances 0.2954 nm and 0.2978 nm).[270] The 1-trimethylstannylphosphole (**121**) is a convenient starting material for early transition metal phospholyl complexes and reacts with [TiCl$_3$Cp] and [TiCl$_4$] to give the corresponding phosphacyclopentadienyl compounds (**122**) and (**123**), respectively (R^1 = H or Me). The crystal structure of (**123**) (R^1 = H) was determined. Surprisingly, (**122**) on reaction with [W(CO)$_5$(MeCN)], does not form a simple titanium–tungsten binuclear complex, but instead decomposes to give a tungsten biphospholyl product (Scheme 31).[271]

Titanocene dichlorides are readily converted into the fluorides or bromides. Thus, treatment of [TiCl$_2$(Cp*)$_2$] with sodium fluoride in methanol gives the yellow complex [TiF$_2$(Cp*)$_2$] in 41% yield. The orange complex [TiF$_2$Cp(Cp*)] was obtained similarly in 25% yield. Treatment of [TiCl$_2$Cp(Cp*)] or [TiCl$_2$(Cp*)$_2$] with BBr$_3$ affords the corresponding bromo complexes as black crystals in 54% and 58% yields, respectively.[321] The reaction of titanocene dichlorides with two to three equivalents of Ag[BF$_4$] in dichloromethane gives the bright yellow difluoro complexes. This method was used to prepare a series of sterically hindered silyl-substituted complexes, such as [TiF$_2$CpCp''], [TiF$_2$Cp'Cp''], [TiF$_2$CpCp'''], [TiF$_2$Cp'''$_2$] and [TiF$_2$Cp'Cp'''] [Cp' = η-C$_5$H$_4$(TMS), Cp'' = η-C$_5$H$_3$(TMS)$_2$, Cp''' = η-C$_5$H$_2$(TMS)$_3$].[250]

Table 4 Preparation of mixed-ligand bis(cyclopentadienyl)titanium dichloride complexes, $L^1L^2TiCl_2$.

L^1	L^2	Preparative method[a]	Colour	m.p. (°C)	Yield (%)	Ref.
Cp	C_5H_4Cl	A	Red plates	Sublimes at 125/0.01 torr, 218–222(dec.)	82	243
Cp	C_5H_4TMS	B	Deep red needles		76	228
Cp	η^5-indenyl	B	Red–brown powder (with 3% $[TiCl_2Cp_2]$)			228
Cp	$C_5H_4CF_3$	A		141–142	84	272
Cp	$C_5H_4(CH_2)_2OMe$	B		98–100	79	102
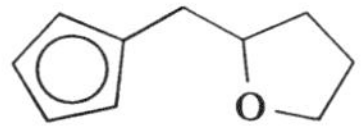	$C_5H_4(CH_2CH{=}CH_2)$	C	Red	71–72	48	103
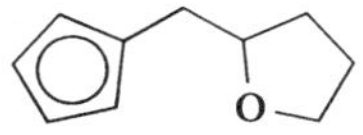	$C_5H_4CH_2Ph$	C	Red	121–122	39	103
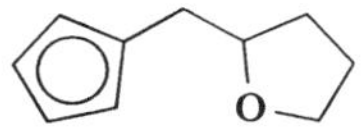	$C_5H_4CPhEt_2$	C	Red	166–167	43	103
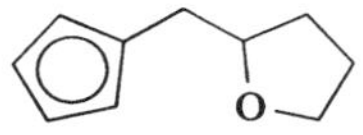	$C_5H_4CPhMe_2$	C	Red	106–107	74	103
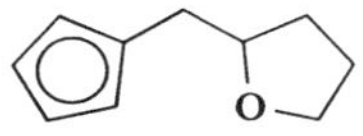	$C_5H_4CPh(Me)Et$	C	Red	125	52	103
Cp	$C_5H_4CH_2CH(OMe)Pr^i$	C	Red	174–177	49	236
Cp	(*R*)- 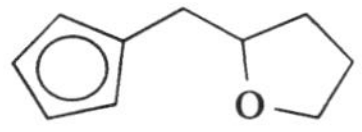	C	Red	108–109	70	236
Cp	(*S*)- 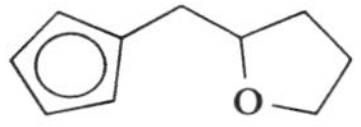	C	Red			236
Cp	$C_5H_4CH_2CH(Me)OMe$	C	Red	88–90	52	235
Cp	$C_5H_4CH(Me)CH_2OMe$	C	Red	134–135	98	235
Cp	$C_5H_4CH_2CH_2PPh_2$	B	Red	130		240
Cp	$C_5H_4CH{=}CH_2$	A	Dark red	154.5–157	44	109

Table 4 (continued)

L^1	L^2	Preparative method[a]	Colour	m.p. (°C)	Yield (%)	Ref.
$C_5Me_4CH=CH_2$	Cp	A	Red–purple		89	109
Cp	$C_5H_4CO_2Me$	A	Red	191–193 (decomposes), sublimes at 135–140/0.001 torr	57	238
Cp	$C_5H_4PPh_2$	A	Orange	130–132	95	273
Cp*	$C_5H_4PPh_2$	A	Red–brown	174–177 (decomposes)	62	274
Cp*	$C_5H_4PMe_2$	A	Red–purple	131–133		274
Cp	$C_5H_4(CH_2)Si(OEt)_3$	B				275
Cp	C_4H_4N	A		186–188		276
Cp	η^5-Fluorenyl	A		190–192 (decomposes)		276
Cp	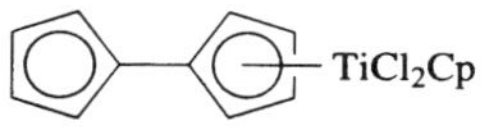 (biphenyl)–$TiCl_2Cp$	D	Green		70^{277}, 74^{278}	277, 278
Cp	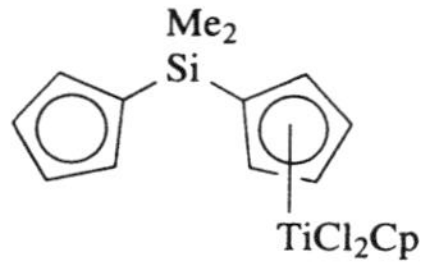 Me_2Si-bridged bis(cyclopentadienyl)–$TiCl_2Cp$	E			81	279
Cp	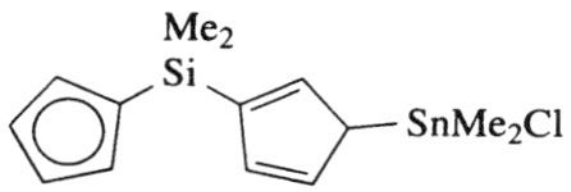 Me_2Si-bridged cyclopentadienyl–$SnMe_2Cl$	F			82	279
Cp	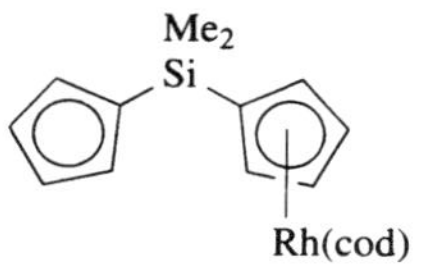 Me_2Si-bridged bis(cyclopentadienyl)–$Rh(cod)$	G	Violet		53	279
Cp	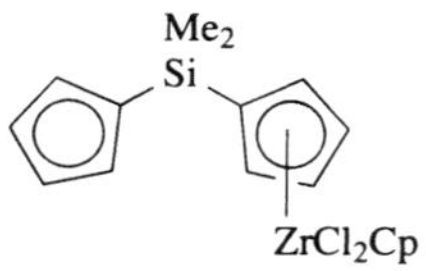 Me_2Si-bridged bis(cyclopentadienyl)–$ZrCl_2Cp$	H	Red		50	279

Cp	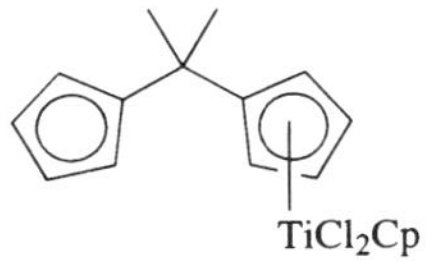 (structure: $TiCl_2Cp$)	E			89	279
Cp	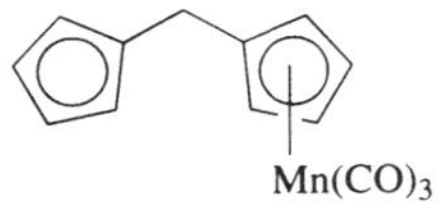 (structure: $Mn(CO)_3$)	B	Peach-coloured microcrystals		68	280
Cp	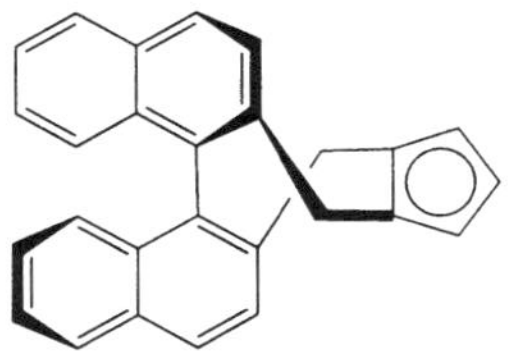 (binaphthyl structure)	B	Red (*S*)-(−)	312–314	24,[181] 28[264]	264, 281
Cp	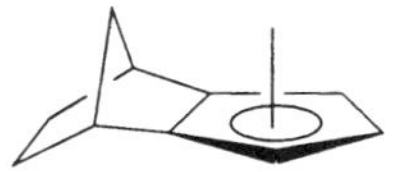 (*exo*)	B	Red	172–173	40	282, x-ray[283]
Cp	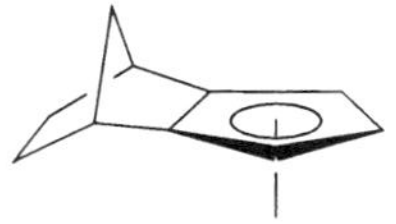 (*endo*)	I	Red	141–142	21	282
C_5H_4Me	 (*exo*)	B	Black	177	40	282, 283

Table 4 (continued)

L^1	L^2	Preparative method[a]		Colour	m.p. (°C)	Yield (%)	Ref.
C_5H_4Me		(endo)	I	Red	151	65	282, x-ray[283]
Cp*		(exo)	A	Red	185–186 (decomposes)	50	282, x-ray[283]
$C_5H_4Bu^t$		(exo)	B	Red	191	55	283
$C_5H_4Bu^t$		(endo)	I	Red	169	45	283
Cp			B	Red	181–182	58	257
Cp			B	Red	201–202	45	255

Cp	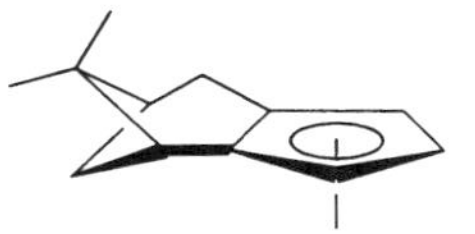	J		$[\alpha]^{19}{}_D = +817°$	69	284
Cp	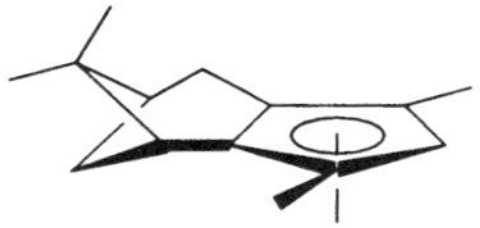		Red–black	$[\alpha]^{20}{}_D = -83.3°$		284
Cp	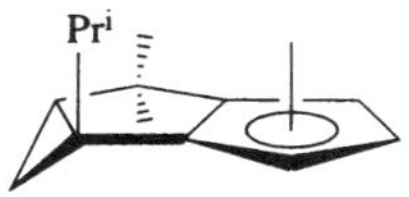 Pr^i	I			63	284
Cp	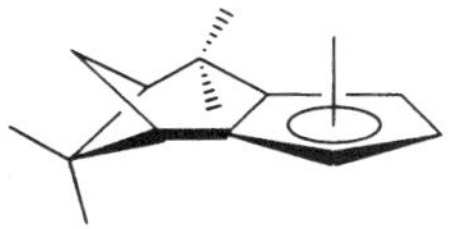	B	Purple–brown	201–202	45	258
Cp	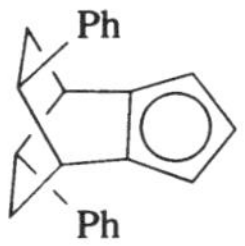 Ph … Ph	B			73	261
Cp	$C_5H_2Me_3$-1,2,4	B	Scarlet plates		48–56	228
Cp	$C_5H_3Ph_2$-1,3	B	Dark purple–red		20	228
Cp*	Cp	A,[285] B,[227] K,[286]	Deep red		46^{227}, 100^{285}, 82^{286}	218, 286, 287, x-ray[285]
$C_5H_3(TMS)_2$-1,3	Cp	B		177–179	82	250
$C_5H_3(TMS)_2$-1,3	C_5H_4TMS	B	Red	200–201	68	250

Table 4 (continued)

L^1	L^2	Preparative method[a]	Colour	m.p. (°C)	Yield (%)	Ref.
Cp	$C_5H_2(TMS)_3$-1,2,4	B	Deep red	122–123	61	250, x-ray[288]
$C_5H_4(TMS)$	$C_5H_2(TMS)_3$-1,2,4	B	Deep red	159–160	45	250
$C_5H_3(TMS)_2$-1,3	$C_5H_2(TMS)_3$-1,2,4	B	Deep red	140–141	31	250
C_5HMe_4	Cp	K			90	265
$C_5Me_4Pr^i$	Cp	A	Violet		95	266
(ligand structure with Ph, Et)	C_5Me_5	B	Red	186		110
Cp	C_5HPh_4	C	Dark red	218–221	70	289
Cp	C_5Ph_5	C	Black–purple		60	x-ray[289]
Cp	$C_5Me_4PPh_2$	B	Brick-red powder		85	268
Cp	(phospholyl structure)	L	Dark red	130 (decomposes)	47	271
Cp	(phospholyl structure)	L	Red	135 (decomposes)	45	271

[a] Method A: $[TiCl_3L^1]$ + TlL^2, THF or benzene, room temperature or reflux. Method B: $[TiCl_3L^1]$ + LiL^2, THF, room temperature or reflux. Method C: $[TiCl_3L^1]$ + KL^2, THF. Method D: oxidation of $[(TiClCp)_2(\eta^5,\eta^5\text{-}C_{10}H_8)]$ with Cl_2 or HCl in THF. Method E: $[TiCl_3L^1]$ + $(C_5H_3)_2(\mu\text{-}SiMe_2)(\mu\text{-}SnMe_2)$ in THF. Method F: $[TiCl_3L^1]$ + $(C_5H_3)_2(\mu\text{-}SiMe_2)(\mu\text{-}SnMe_2)$ in toluene. Method G: $[TiCl_2Cp(\eta^5\text{-}C_5H_4SiMe_2C_5H_3SnMe_2Cl)]$, $[\{RhCl(cod)\}_2]$, THF. Method H: $[ZrCl_2Cp(\eta^5\text{-}C_5H_4SiMe_2C_5H_3SnMe_2Cl)]$, $[TiCl_3Cp]$, THF. Method I: $[TiCl_3L^1]$ + LiL^2 in THF, $-78\,°C$, 12 h. Method J: solid $[TiCl_3L^1]$ + LiL^2, THF, 50 °C. Method K: $[TiCl_3L^1]$ + NaL^2, THF. Method L: $[TiCl_3L^1]$ + L^2–$SnMe_3$ in toluene.

Table 5 Preparation of *ansa*-titanocene dichloride complexes, [TiCl$_2$L].

L	Preparative method[a]	Colour	m.p. (°C)	Yield (%)	Ref.
	A	Dark red	Sublimes at 130–140/0.01 torr	91	290, x-ray[291]
	B	Green–brown powder (mixture of *rac* and *meso*)		22 (*rac*),[292] 60 (*rac* and *meso*)[292,293]	
	C	Red		70–90	x-ray[292,293]
(*S*)-	D	Red	$[\alpha]^{25}_{435} = -3300°$		292
	A	Red		86 (*rac/meso* =1.3)	294
Et ... Et	A	Red		84 (*rac/meso* = 1.5)	294
Pr^i ... Pr^i	A	Red		85 (*rac/meso* = 1.8)	294

Table 5 (continued)

L	Preparative method[a]	Colour	m.p. (°C)	Yield (%)	Ref.
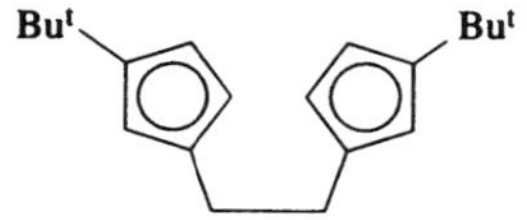	A	Red		86 (*rac/meso* = 2.0)	x-ray[294]
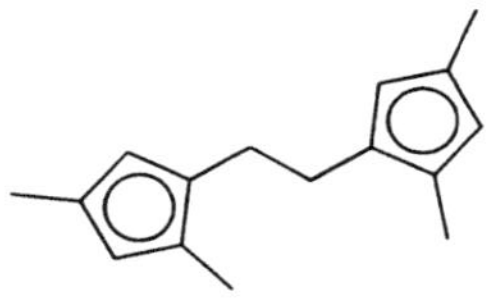	A	Deep red		75 (*rac/meso* = 2.0)	295
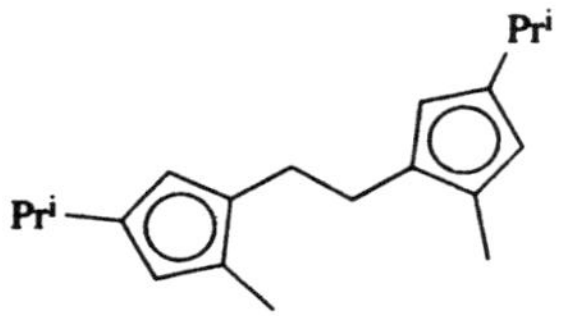	A	Red		83 (*rac/meso* = 2.6)	295
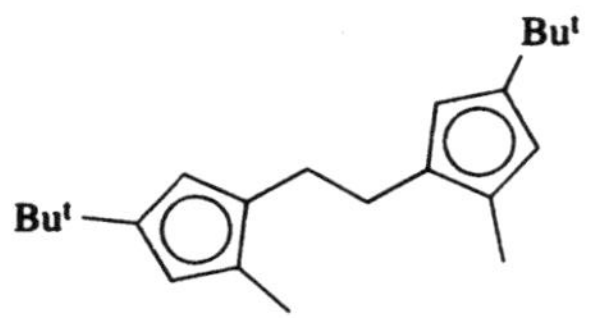				85 (*rac/meso* = 1.6)	x-ray[295]
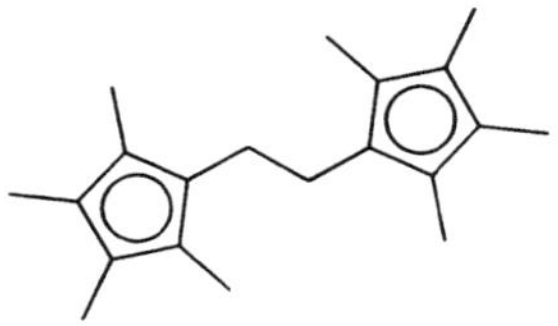	A	Dark red–brown		15	x-ray[296]

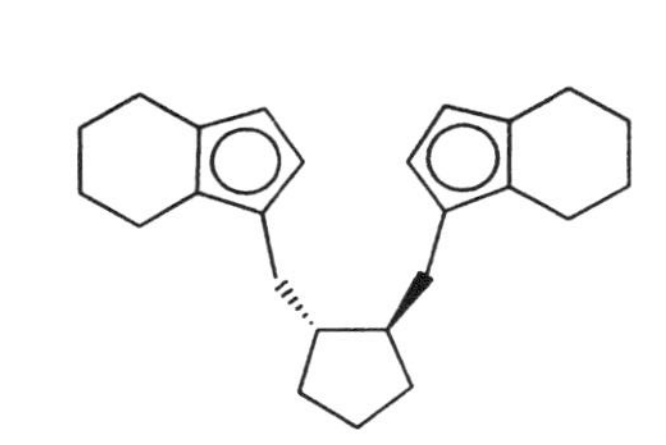	A	Red powder	63	synthesis,[297] NMR[298]
	A, B, D	Red	40 (R,S):(R,R):(S,S) = 4.2:2.5:1	x-ray of (R,R)[299]
	B	Dark green	34	300
	C	Red	58	x-ray of (S,S)[300]
	A	Red solid (2.6:1:1 mixture of isomers)	98	x-ray of major

Table 5 (continued)

L	Preparative method[a]	Colour	m.p. (°C)	Yield (%)	Ref.
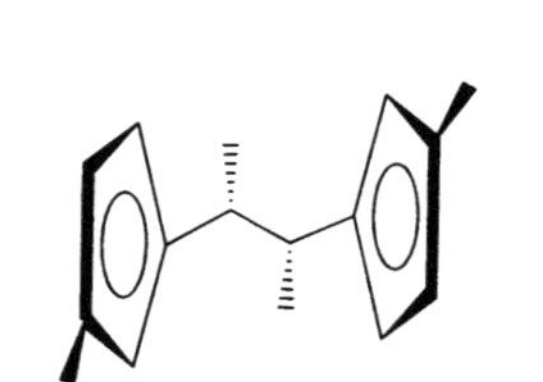 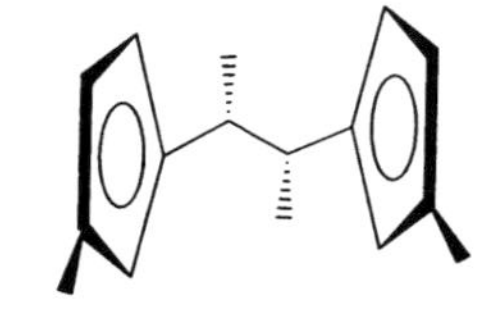	A	Red oil (4.6:2.8:1.0 mixture of isomers)		83	301

Structure	Method	Colour	m.p. (°C) / properties	Yield (%)	Structure determination
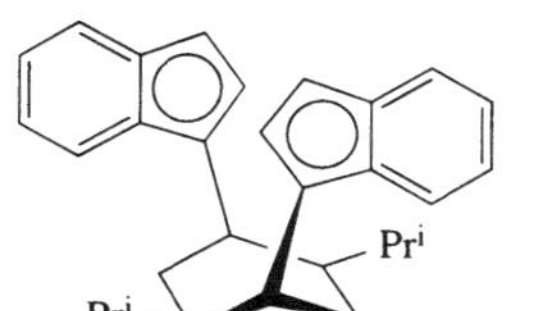	A	Dark green	264–265, $[\alpha]_D^{23} = +2900°$	80	x-ray[202]
	A, E A (*rac/meso* = 2:1), E (*rac/meso* = 6:1)	Red solid		15	x-ray of *rac* isomer[303]
	A (rac only) Red			8	x-ray [303]
	A E (1:1)[305] A (1:4)[304]	Red (2.5:1 mixture of *rac* and *meso* isomers)	193–194 (*meso*)[304] 248 (*rac*)[304]	46[305]	304, 305, x-ray of *meso*[306]
	A E(1.8:1)	Red (0.6:1 mixture of *rac* and *meso* isomers)		45–50	x-ray of *rac* and *meso* isomers[305]

Table 5 (continued)

L	Preparative method[a]	Colour		m.p. (°C)	Yield (%)	Ref.
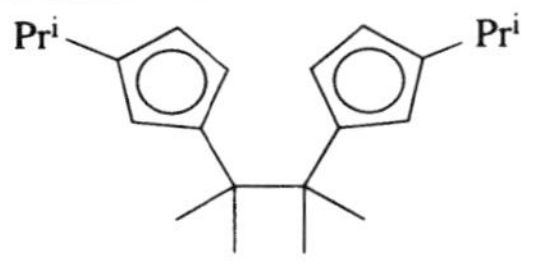 Pr^i ... Pr^i (structure)	E	Red	(ca. 1:1 mixture of *rac* and *meso* isomers)		40–50	305
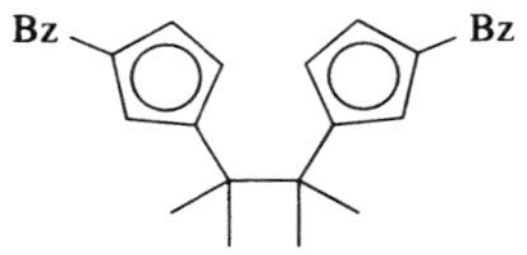 Bz ... Bz (structure)	E	Red	(ca. 1:1 mixture of *rac* and *meso* isomers)		40–45	305
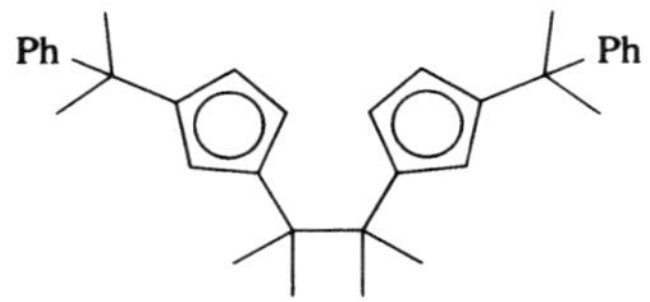 Ph ... Ph (structure)	E	Red	(1.2:1 mixture of *rac* and *meso* isomers)			305
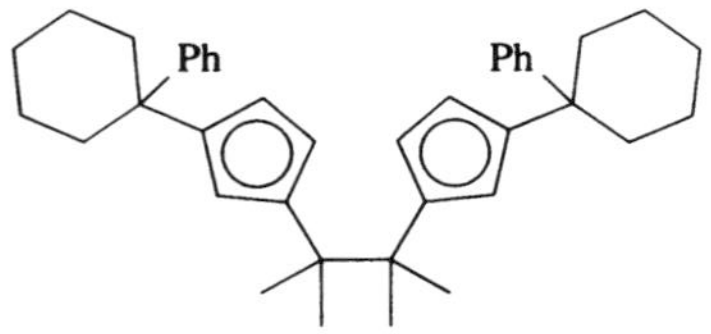 Ph ... Ph (structure)	E	Red	(1.2:1 mixture of *rac* and *meso* isomers)			x-ray of *rac* isomer[305]
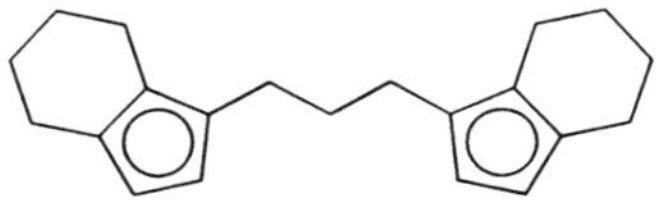 (structure)	B, C	Red			11	x-ray[307]

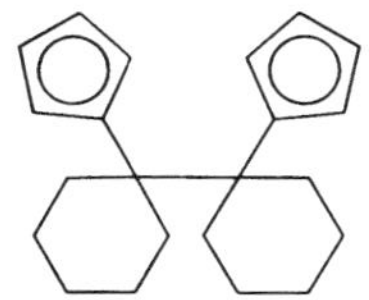	A			x-ray[291]
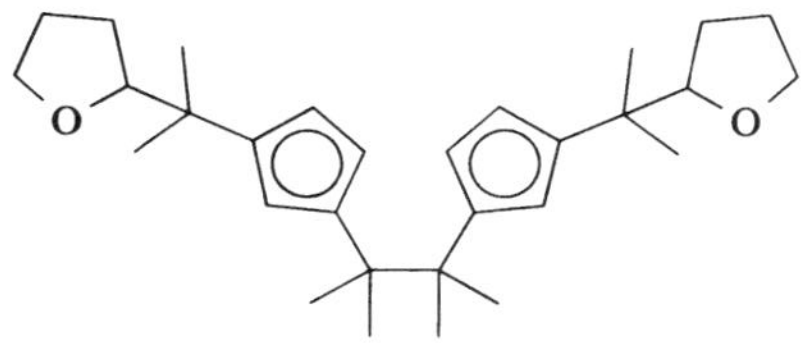	A			x-ray[291]
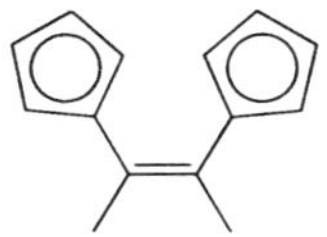	F		8	Synthesis,[308] x-ray[291]
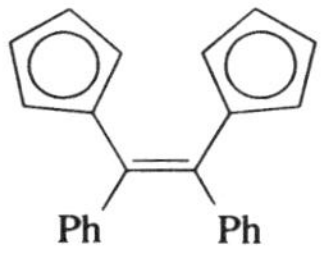	F		25	Synthesis,[308] x-ray[291]
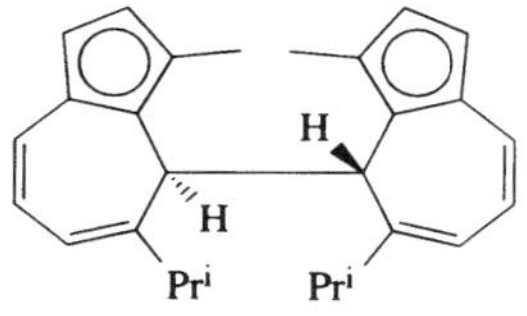	E	Black–purple	9.9	x-ray[262]

Table 5 (continued)

L	Preparative method[a]	Colour	m.p. (°C)	Yield (%)	Ref.
	A	Green (R)-(+) isomer only	270, $[\alpha]_D^{23} = -1800°$	42	x-ray[309]
	G	Dark red		18	x-ray[310]
	G	Red		4	x-ray[310]

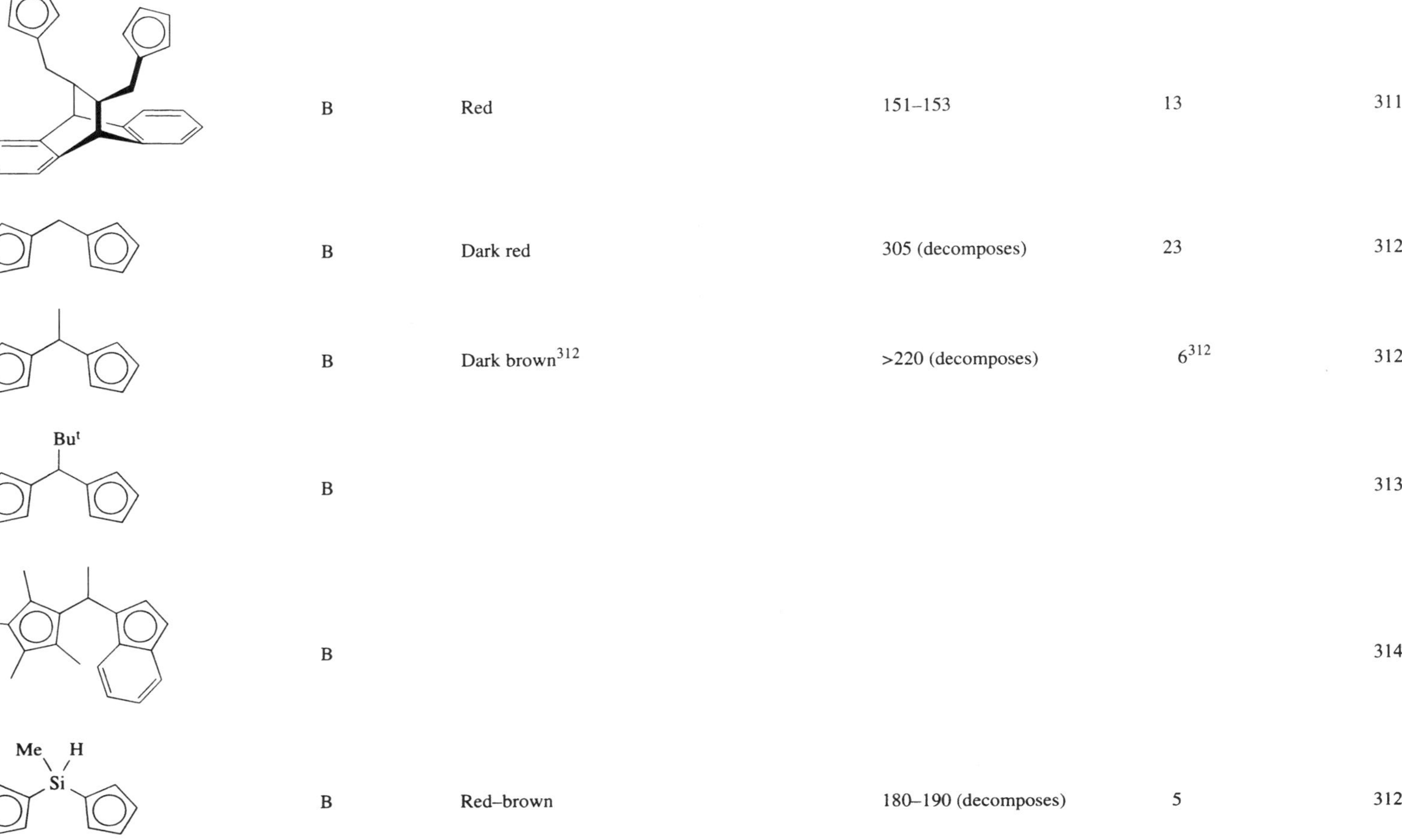

	B	Red	151–153	13	311
	B	Dark red	305 (decomposes)	23	312
	B	Dark brown[312]	>220 (decomposes)	6^{312}	312, 313
	B				313
	B				314
	B	Red–brown	180–190 (decomposes)	5	312

Table 5 (continued)

L	Preparative method[a]	Colour	m.p. (°C)	Yield (%)	Ref.
(Me)(Me)Si(C5H4)2	B	Dark brown,[312] red–brown[315]	250–270 (decomposes)[312]	7,[312] 69.3[315]	312, x-ray[315]
(Et)(Et)Si(C5H4)2	B	Red–purple	190	10	312
(Me)(Me)Ge(C5H4)2	B	Light brown	185–186 (decomposes)	2	312
(Me2)Si–CH2CH2–Si(Me2)(C5H4)2	B	Red	253	28	316
(Me2)Si–O–Si(Me2)(C5H4)2	B	Red	188–190	80	x-ray[317]

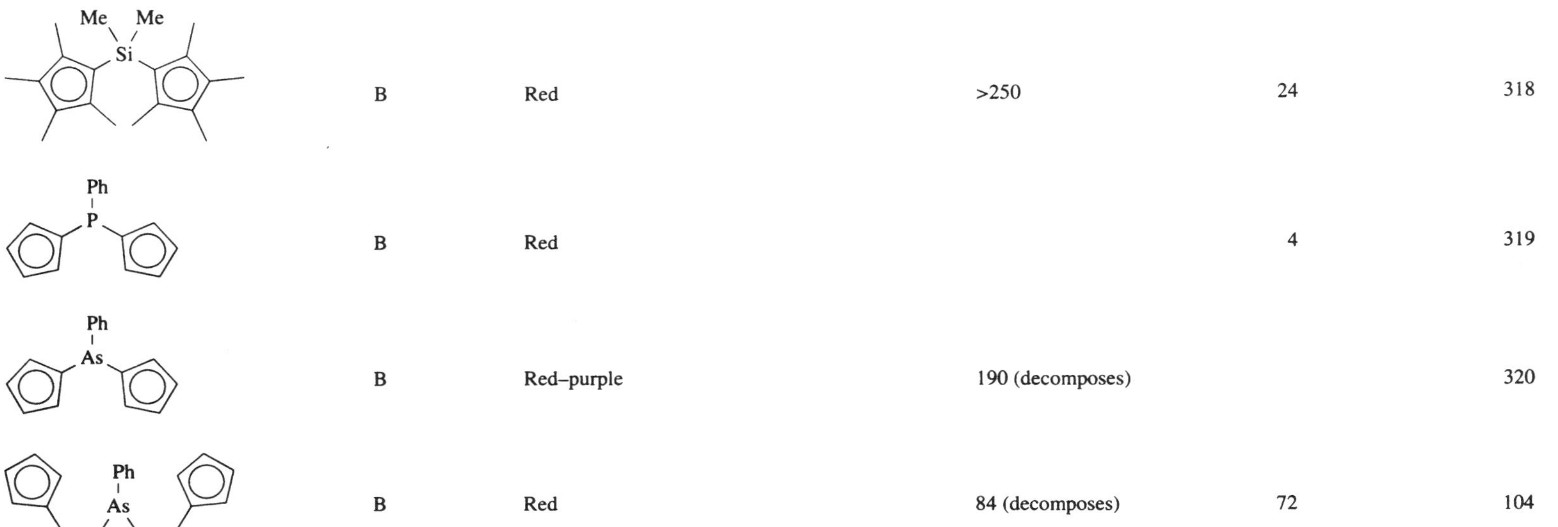

(Me)(Me)Si-bridged tetramethylcyclopentadienyl ligand	B	Red	>250	24	318
Ph-P(cyclopentadienyl)$_2$	B	Red		4	319
Ph-As(cyclopentadienyl)$_2$	B	Red–purple	190 (decomposes)		320
Ph-As-bis(ethylcyclopentadienyl)	B	Red	84 (decomposes)	72	104

^a Method A: Li$_2$L and [TiCl$_3$(THF)$_3$] in THF, followed by oxidation with concentrated aqueous HCl. Method B: Li$_2$L and TiCl$_4$ in THF. Method C: hydrogenation of the indenyl complex, PtO$_2$ catalyst, dimethoxymethane, hydrogen under pressure. Method D:Enantiomer separation by reaction of [TiCl$_2$L] with enantiomerically pure binaphthol, followed by conversion of the titanium naphthalato complex into enantiomerically pure [TiCl$_2$L]. Method E: L(MgBr)$_2$ and [TiCl$_3$(THF)$_3$] in THF, followed by oxidation. Method F: Na$_2$L and [TiCl$_3$(THF)$_3$] in THF, followed by oxidation and sublimation *in vacuo*. Method G: K$_2$L and [TiCl$_3$(THF)$_3$] in THF, followed by oxidation.

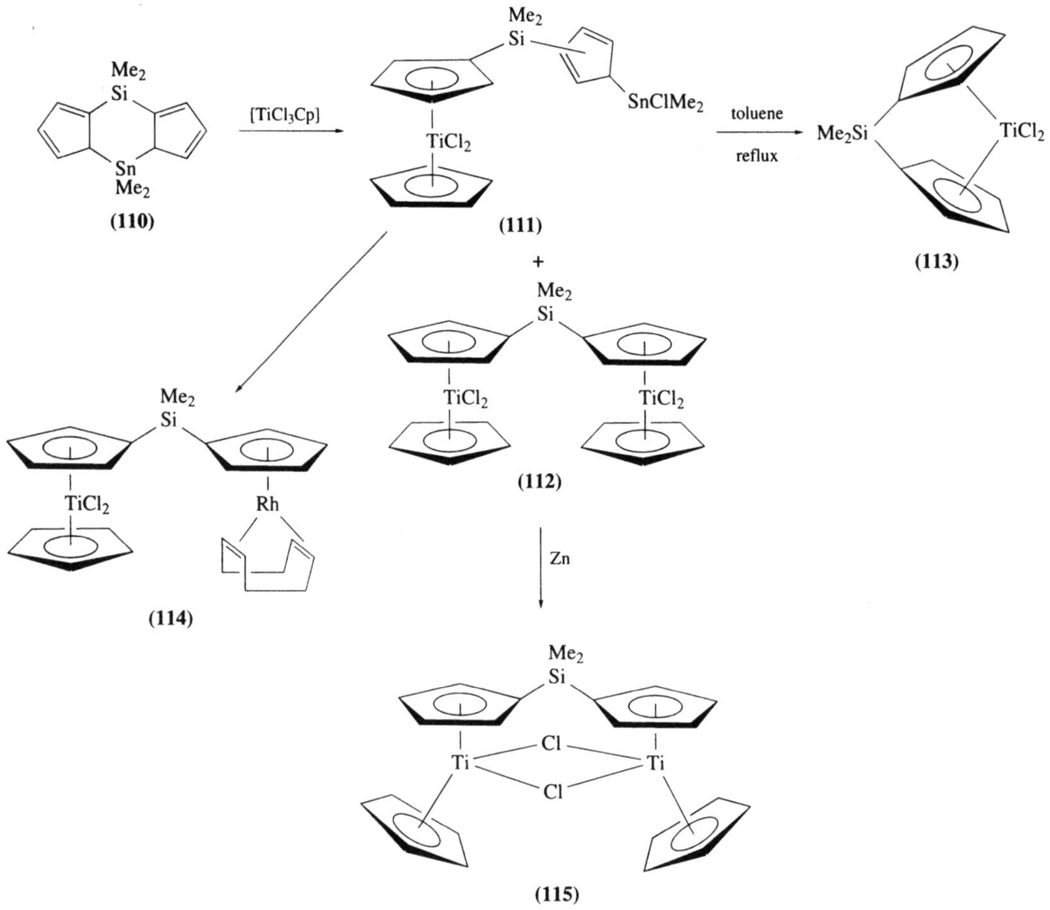

Scheme 30

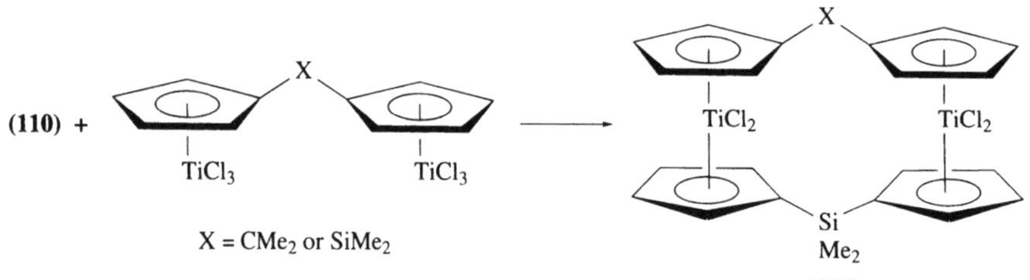

$$X = CMe_2 \text{ or } SiMe_2 \tag{33}$$

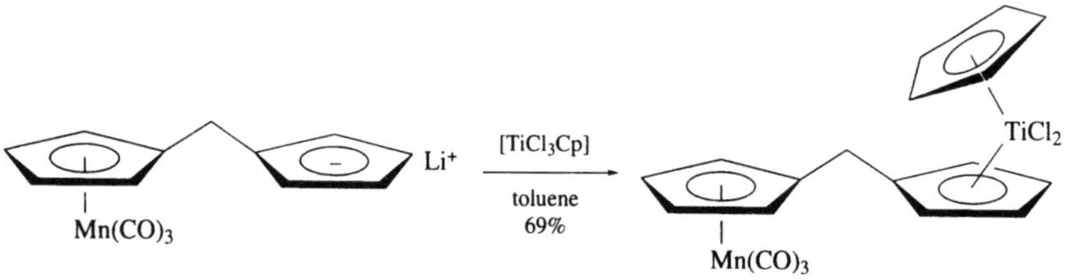

A very significant synthetic effort has been directed in recent years towards stereodifferentiating titanocene derivatives, ultimately in order to develop the use of these compounds in stereoselective synthesis and catalysis. For example, 3-phenyl-1,2-dihydropentalenylmagnesium chloride reacts with [TiCl$_3$(THF)$_3$] and oxidation with HCl/O$_2$ in THF gives the chiral *rac* complex (**124**) (Equation (37)). Reduction of guaiazulene with magnesium followed by treatment with [TiCl$_3$(THF)$_3$] gives the *ansa*-metallocene (**125**) in 10% yield (Equation (38)). The structures of both titanium complexes were confirmed by x-ray diffraction.[262]

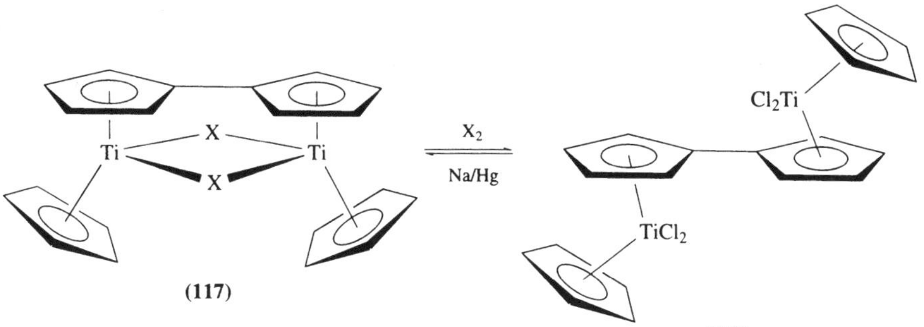

(35)

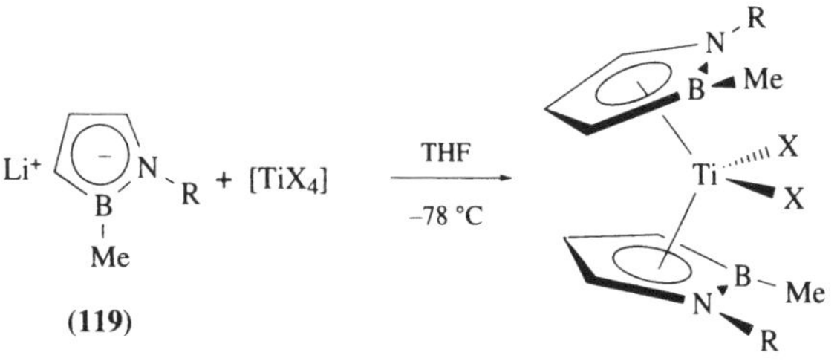

(36)

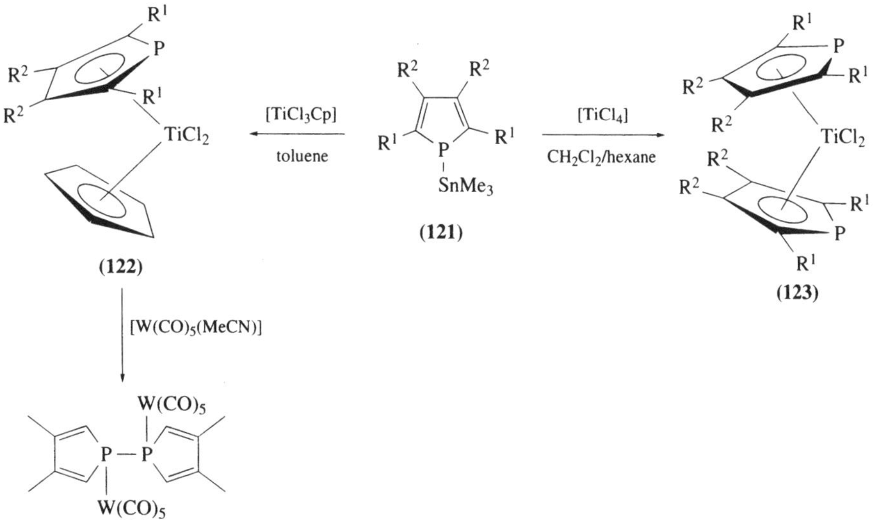

Scheme 31

Isodicyclopentadienyl ligands (IDCp⁻) (**126**) possess two possible ways of attachment to a metal centre, via the *exo* or the *endo* face. The stereochemical course of attaching this ligand to titanium is highly sensitive to the reaction conditions, in particular dilution, temperature and duration of the reaction. The reaction of [Li(**126**)] with [TiCl₃Cp] in THF at room temperature gives exclusively the *exo* product (**127**) in 40% yield (Scheme 32). The thallium reagent requires some heat and gives an identical titanium product. By contrast, the same reaction mixture kept at −78 °C for 12 h leads to the *endo* isomer in 21% yield. The reaction of [Li(**126**)] with [TiCl₃(η-C₅H₄Me)] proceeds in an analogous fashion and gives the *exo* isomer at room temperature as black crystals in 40% yield, while at −78 °C the red *endo* product is stereoselectively obtained in 65% yield. The configurations of *exo*-(**127**) and *endo*-[TiCl₂(η-C₅H₄Me)(**126**)] have been confirmed by x-ray diffraction. No crossover is observed when mixtures of *exo*-(**127**)- and *endo*-(**127**)-d₃ are refluxed for prolonged periods of time. The reaction of [Li(**126**)] with [TiCl₃(η-C₅H₄Buⁱ)] gives only *endo*-(**128**) in 45% yield if the temperature of −78 °C is maintained throughout the reaction. If the mixture is allowed to warm from −78 °C to 20 °C during the course of the reaction, a 1:9 *endo:exo* mixture is isolated, from which the pure *exo* isomer can be obtained by fractional crystallization. The reaction of [TiCl₃(Cp*] requires the use of the thallium reagent [Tl(**126**)] in refluxing benzene to avoid side reactions. The *exo* product [TiCl₂(Cp*)(**126**)] is observed exclusively in 50% yield and was characterized by x-ray diffraction (Figure 15). The

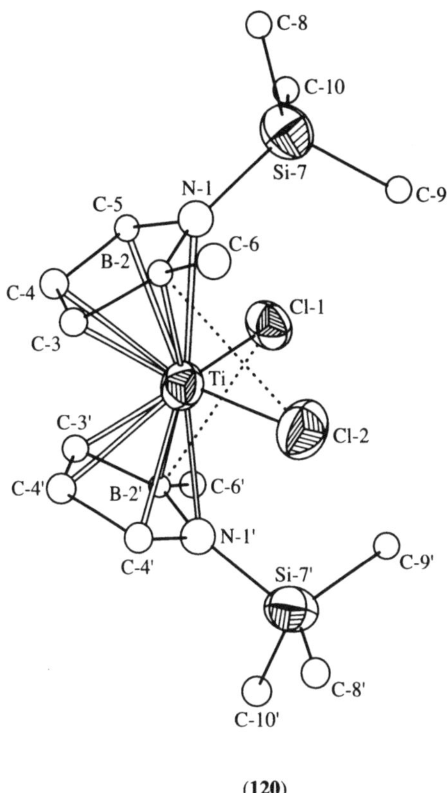

(120)

Figure 14   Molecular structure of the complex **(120)** (reproduced by permission of VCH from *Chem. Ber.*, 1985, **118**, 2418).

$$(37)$$

(124)

$$(38)$$

(125)

orientation adopted in the solid state minimizes the steric interaction by placing the CH_2 bridge of the IDCp moiety between the two chloride ligands. It is evident that an *endo* orientation would result in significantly stronger repulsive interaction between a chlorine atom and the hydrocarbon framework.[282,283]

If [Li(**126**)] is treated with [TiCl$_3$(THF)$_3$] at −78 °C and the mixture brought quickly to 25 °C, followed by stirring for 14 h, [TiCl$_2$(IDCp)$_2$] (**129**) is formed (Scheme 32), which is 98% *exo,exo*, with 2% of the *exo,endo* and traces of the *endo,endo* isomers. The molecular structure of (**129**) is shown in

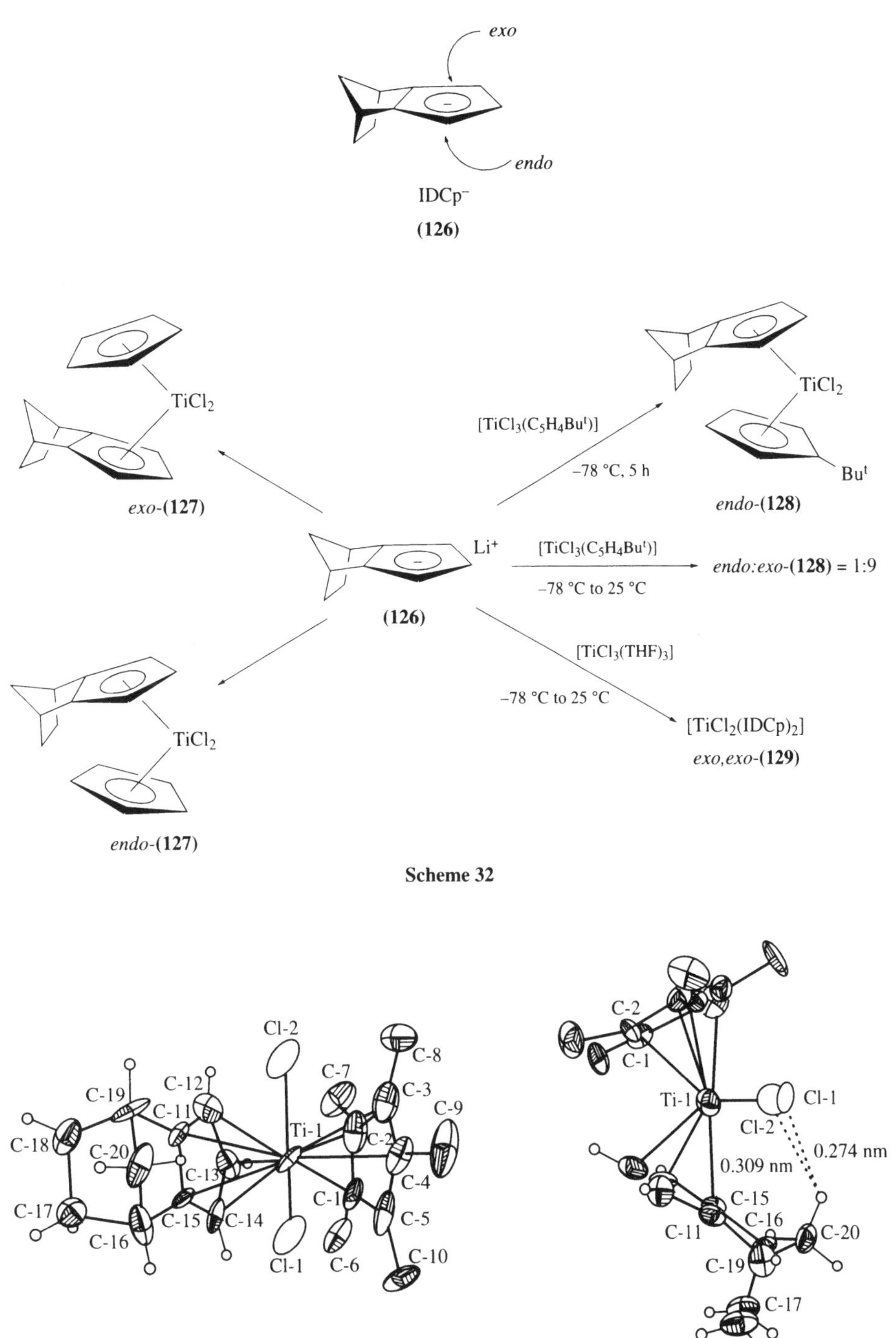

Scheme 32

Figure 15 Two views of the molecular structure of [TiCl₂Cp*(**126**)] (reproduced by permission of the American Chemical Society. Copyright (1989) from *Organometallics*, 1989, **8**, 2159).

Figure 16. The stereoselectivity appears to be concentration-dependent, with high concentrations of the lithium reagent favouring the *exo,endo* and *endo,endo* components. As was observed before, low temperatures (−64 °C, 5 h) also give the sterically less favourable *endo,endo* isomer preferentially in 91% selectivity, together with 9% of the *endo,exo* product.[252,253] The pure *endo,exo* isomer is obtained via IDCp(TMS) in a sequential reaction, as outlined in Scheme 33.[252] With the closely related ligand (**130**) other authors observed less stereoselective reactions and obtained the corresponding titanocene [TiCl₂(**130**)₂] in reactions between −78 °C and room temperature as a 4.3:78.3:17.4 *exo,exo/exo,endo/ endo,endo* mixture. The reason for this behaviour may be the shorter reaction times and different concentrations of the reagents.[260]

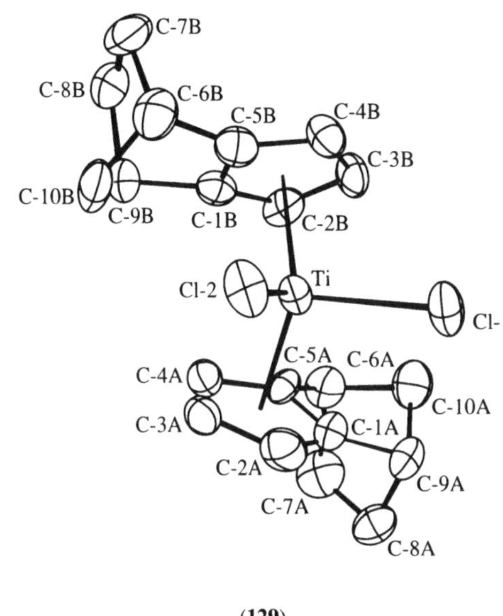

Scheme 33

(129)

Figure 16   Molecular structure of the bis(isodicyclopentadienyl) complex [TiCl₂(**129**)] (reproduced by permission of the American Chemical Society. Copyright (1987) from *Organometallics*, 1987, **6**, 15).

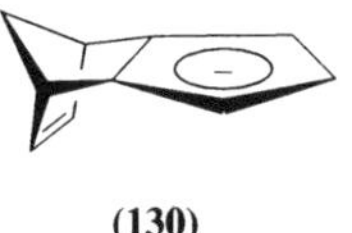

(**130**)

The pinane-fused cyclopentadienyl ligand system PCp⁻ (**131**) has two possible orientations for complexation; above plane and below plane. Treatment of [Li(**131**)] with [TiCl₃(THF)₃] in dimethoxyethane gives the sterically less encumbered above-plane isomer (**132**) in 71% yield (Scheme 34).[254-6] The analogous reaction with [TiCl₃Cp] in THF between −78 °C and room temperature over a period of 18 h gives a 5:1 mixture of (**133**) and (**134**) (Scheme 34). In this case the main isomer is the more hindered below-plane product (**133**), which can be isolated in 45% yield. If the reaction is performed at room temperature or in refluxing THF, the isomer ratio is reversed, to give a 1:3 mixture of (**133**) and (**134**). Once formed, there is no detectable interconversion between these stereoisomers.[255] The reaction is even more selective if solid [TiCl₃Cp] is added to a solution of (*R*)-(−)-(**131**) at 50 °C. Enantiomerically pure (**134**) is obtained after a single recrystallization, as indicated by the optical rotation ([α]¹⁹_D = +817°, *c* = 0.05 in toluene).[284]

The reaction of the norpinone-derived ligand (**135**) with solid [TiCl₃Cp] leads to the red-black compound (**136**) ([α]²⁰_D = −83.3°), where the titanium is attached to the less hindered π-face (Scheme 35). As in the case of (**129**), the reversion of stereoselectivity was achieved by a suitable choice of the reagents. Treatment of (**135**) with TMS-Cl, followed by [TiCl₄] in toluene at −78 °C affords the red–orange complex (**137**). A similar reaction sequence with the ligand (**138**) gives [TiCl₂Cp(**138**)] as an 'above-plane' isomer, and [TiCl₃(**138**)] with 'below-plane' coordination.[284]

Ligand (**139**), derived from (1*S*,5*S*)-(−)-verbenone, coordinates via the less encumbered π-face to give predominantly (**140**) (Equation (39)). The isomer (**141**) is formed as a consequence of optical impurities in the starting material. Both complexes have been characterized crystallographically. The

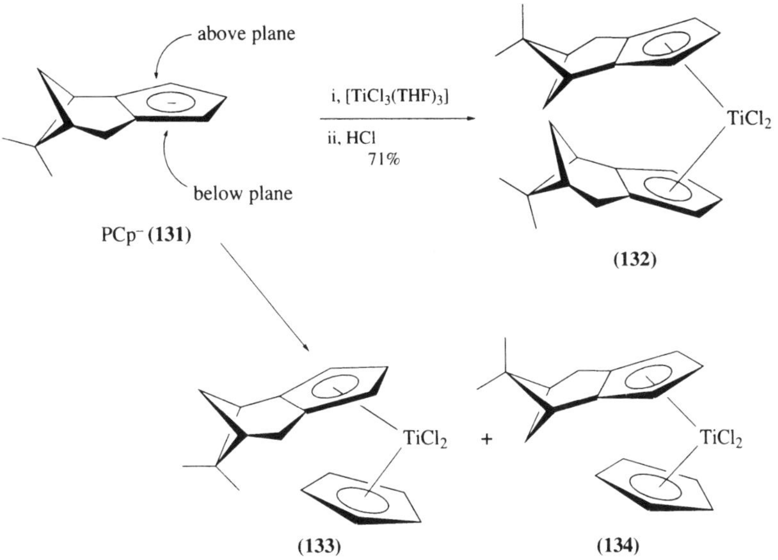

Scheme 34

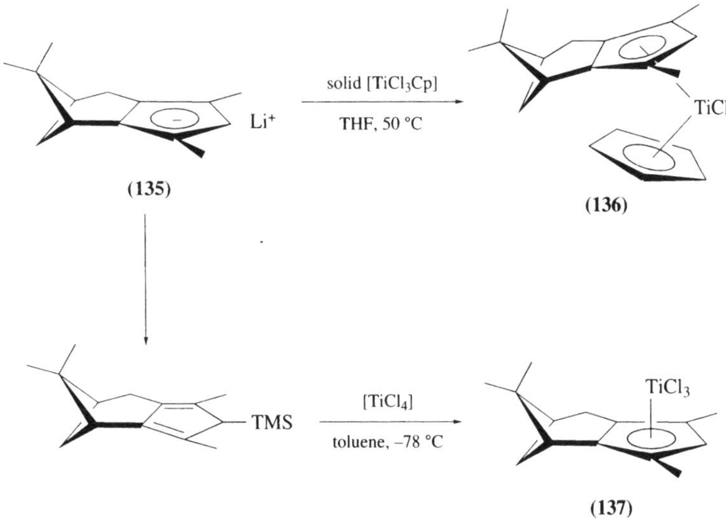

Scheme 35

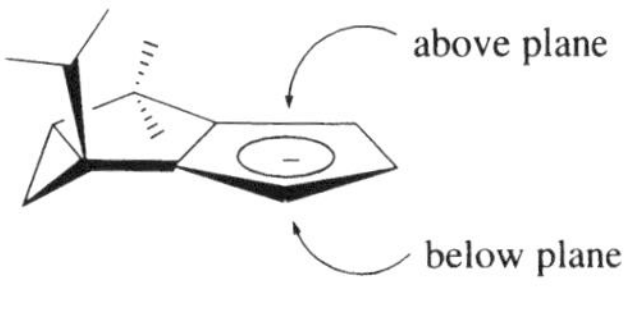

structures indicate a significant degree of steric crowding, which leads to an elongation of the bond distances between the metal centre and the cyclopentadienyl ring.[258,259]

The camphor derivative (**142**) reacts with [TiCl$_3$(THF)$_3$] to give a 9:1 mixture of *endo,endo-* and *endo,exo*-[TiCl$_2$(**142**)$_2$]. The red–purple major isomer was obtained pure. The reaction with [TiCl$_3$Cp] gives a 7:1 mixture of *endo-* and *exo*-[TiCl$_2$Cp(**142**)].[256,257] Titanocene derivatives with C_2 symmetry are obtained according to Equation (40) from (**143**). The complex (**144a**) was isolated in 53% yield and characterized by x-ray diffraction.[261] The complexes (R = Ph, Pr^i or Me) are isolated enantiomerically pure in 45–55% yield. The molecular structure of (1*S*,7*S*,8*R*,10*R*)-(–)-(**144c**) is shown in Figure 17.[262]

The binaphthyl-annulated ligand (**145**) reacts with [TiCl$_3$(THF)$_3$] to give the complex (**146**) in high yield (Equation (41)). The same product is formed from [TiCl$_4$] but only in 8% yield. If the racemic

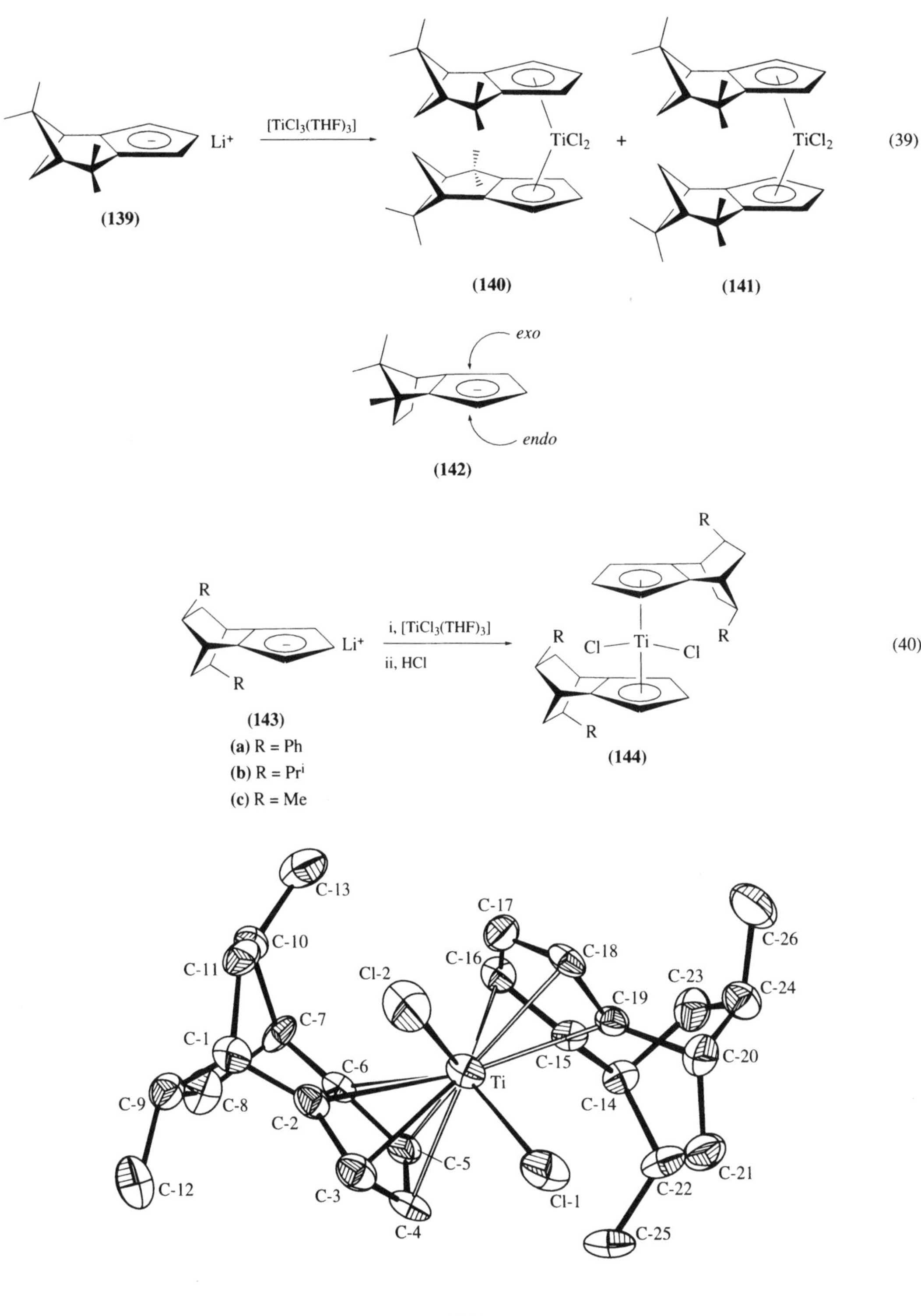

(144c)

Figure 17 Molecular structure of the complex (**144c**) (reproduced by permission of the American Chemical Society. Copyright (1991) from *Organometallics*, 1991, **10**, 3449).

ligand is used, both *rac*- and *meso*-(**146**) can be formed, while only DL-(**146**) results from enantiomerically pure (**145**). The structure of *rac*-(**146**) was determined by x-ray diffraction. The mixed-ligand complex (*S*)-(−)-[TiCl$_2$Cp(**145**)] was similarly obtained from (*S*)-(−)-(**145**) and [TiCl$_3$Cp].[264,265] The binaphthyl-bridged ligand (**147**) gives the *ansa*-titanocene (**148**) (Equation (42)). Using the (*R*)-(+) enantiomer of (**147**) gives the dark-green (*R*)-(−) complex as a single stereoisomer in 42% yield. The molecular structure of (**148**) is shown in Figure 18. The compound is not C_2-symmetric in the crystal. In solution two rapidly interchanging conformations with C_1 symmetry exist and can be frozen out by cooling to −60 °C. A spectrum typical of C_2 symmetry is observed above +60 °C.[309]

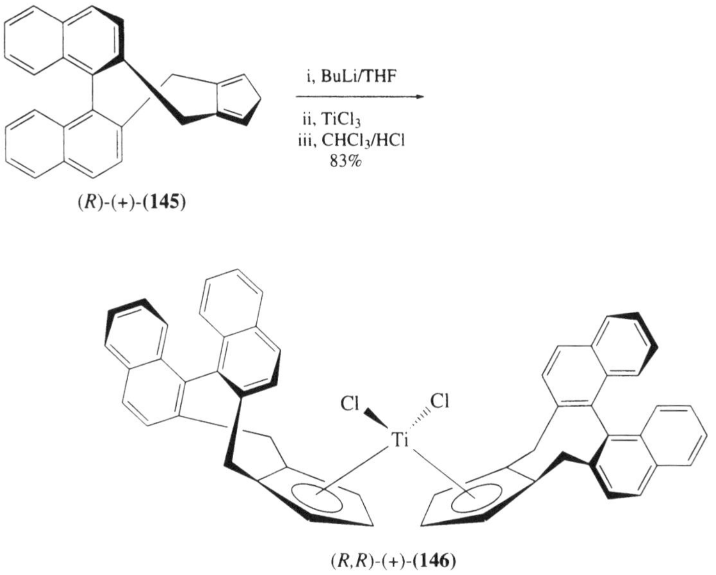

$$(R)\text{-}(+)\text{-}(\mathbf{145}) \quad \xrightarrow[\substack{\text{ii, TiCl}_3 \\ \text{iii, CHCl}_3/\text{HCl} \\ 83\%}]{\text{i, BuLi/THF}} \quad (R,R)\text{-}(+)\text{-}(\mathbf{146})$$

(41)

$$(R)\text{-}(+)\text{-}(\mathbf{147}) \quad \xrightarrow[\text{ii, HCl, CHCl}_3, \text{ air}]{\text{i, [TiCl}_3(\text{THF})_3]} \quad (\mathbf{148})$$

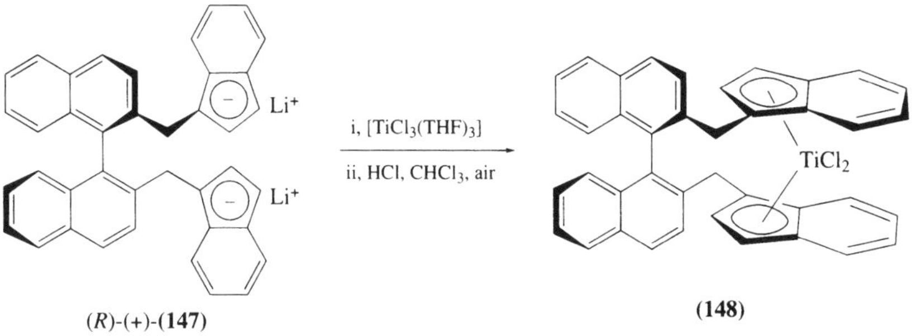

(42)

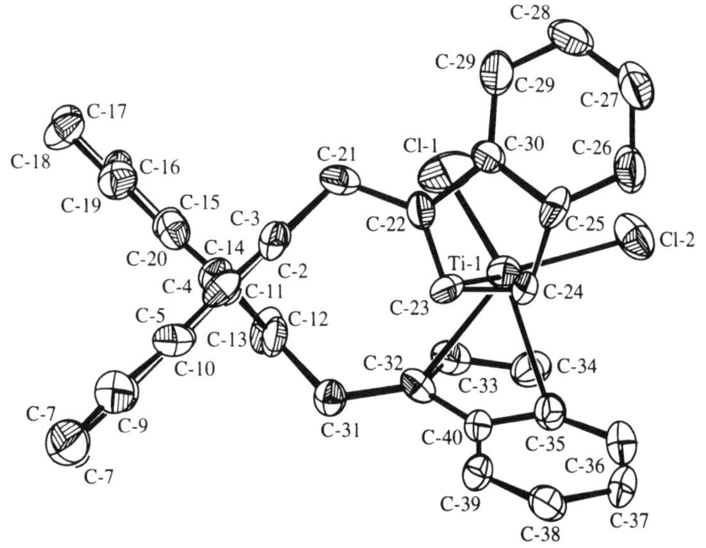

(148)

Figure 18 Molecular structure of the 1,1'-binaphthyl-bridged *ansa*-titanocene complex (**148**) (reproduced by permission of the American Chemical Society. Copyright (1991) from *Organometallics*, 1991, **10**, 2998).

A particularly significant development in the chemistry of early transition metal sandwich compounds was the synthesis of complexes where the two cyclopentadienyl ligands are tethered to each other by a bridge, the so-called *ansa*-metallocenes (Latin: *ansa*, handle). Complexes of this structural type have become important as highly stereoselective homogeneous alkene polymerization catalysts. Much research effort has been directed towards improvements in activity and stereoselectivity by suitable tailoring of the cyclopentadienyl ligand system, particularly in the case of zirconocenes. An overview of *ansa*-titanocenes and the preparative methods used is given in Table 5. The following discussion describes synthetic details for a representative number of these compounds, without being comprehensive.

The first examples of this kind, $[TiCl_2\{(CH_2)_n(C_5H_4)_2\}]$ (n = 1 or 2) were reported by Brintzinger and co-workers in 1979.[322] The standard method is the reaction of the dilithium salt of the bridged bis(cyclopentadienyl) dianion with $TiCl_3$, or better $[TiCl_3(THF)_3]$, followed by oxidation with 6 M or 12 M aqueous HCl, as outlined for the reaction of (149) in Equation (43).[290] The yields can sometimes be small because of the formation of polymeric by-products. Complex (150) is readily reduced by sodium sand in toluene to the corresponding titanium(III) complex. The permethylated complex, $[TiCl_2\{C_2H_4(C_5Me_4)_2\}]$, is prepared by this method as dark red crystals in 15% yield and was structurally characterized.[296]

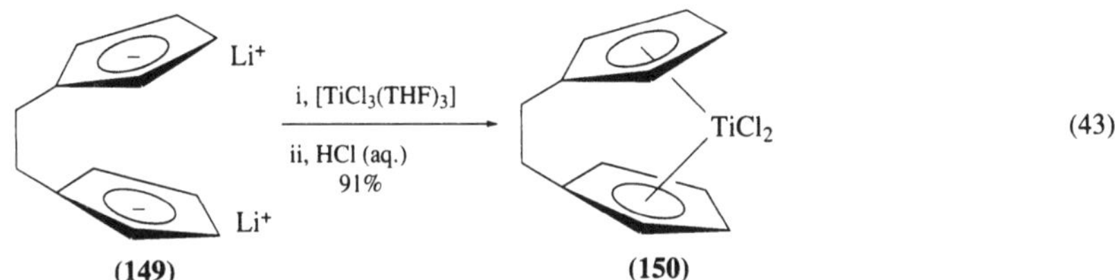

$$\text{(43)}$$

The reaction of the bis(indenyl) dianion with $[TiCl_4(THF)_2]$ in THF followed by quenching with HCl leads to two products, a dark green major fraction and an olive–brown compound, which were partially separated by column chromatography. Separation is, however, much facilitated if the indenyl six-membered rings of the crude product mixture are hydrogenated to give the red tetrahydroindenyl complex (151) as a mixture of the *rac* and the *meso* isomers (Scheme 36), which can be readily separated. Higher reaction temperatures favour the formation of the desired *rac* isomer. It is possible to convert the *meso* isomer into the *rac* product by irradiation of the product mixture with a mercury pressure lamp and a dichromate solution as light filter. The identity of these isomers was confirmed by x-ray diffraction (Section 5.5.1.2). The *rac* isomer is a mixture of two enantiomers, which can be separated by converting the titanocene dichlorides into the binaphtholato compounds with (*R*)-(+)- or (*S*)-(−)-1,1'-bi-2-naphthol. This method is now commonly adopted for the enantiomer separation of chiral metallocenes. In order to obtain the enantiomerically pure titanocene dichloride the binaphtholato complex is protolysed with dry HCl in toluene. In the present case the removal of the binaphthol can, however, be difficult and is facilitated by alkylating the titanocene complex to give the soluble dimethyl complex $[TiMe_2\{C_2H_4(Ind)_2\}]$, which is separated from the binaphthol and reconverted into the chloro complex by treatment with HCl gas. The optical rotation of pure (*S*,*S*)-(151) is $[\alpha]^{25}_{435} = -3300°$.[292]

Rac and *meso* isomers also result if a substituent R is introduced in cyclopentadienyl complexes of the type (150). Thus, a series of complexes (152) is formed according to Equation (44) in high yields, whose *rac:meso* ratio varies in the order 1.3:1 (R = Me), 1.5:1 (R = Et), 1.8:1 (R = Pr^i) and 2.0:1 (R = Bu^t). Only the last of these could be obtained pure and characterized by x-ray diffraction of the *rac* isomer. Several conformations are energetically accessible in the solid state, and two crystal modifications are observed, depending on the method of crystallization.[294] Disubstituted cyclopentadienyl derivatives similarly lead to *rac* and *meso* titanocenes (153) in a ratio of 2:1 to 3:1. Yields of 75–85% are recorded. The *rac:meso* ratio of (153) (R = Bu^t) was increased by irradiation to up to 15:1. The molecular structure of *rac*-(153) (R = Bu^t) possesses C_2 symmetry in the solid state; there is significant steric interaction between the ethene bridge and the methyl substituents and between the Bu^t groups and the chloride ligands.[295]

Several *ansa*-metallocences have been made with substituents on the bridge, either in order to influence the *rac:meso* ratio or as a means of introducing chirality. The tetramethyl ligand $[C_2Me_4(C_5H_4)_2]^{2-}$ is obtained by reducing dimethylfulvene with sodium amalgam. The subsequent reaction with $[TiCl_3(THF)_3]$ affords $[TiCl_2\{C_2Me_4(C_5H_4)_2\}]$ in 35% yield as a red powder. The ligand is also obtained if the reduction step is carried out with magnesium in the presence of CCl_4 to give the Grignard reagent.[297] The C_2Me_4-bridged complexes (154)[303] and (155)[304] have been prepared using the

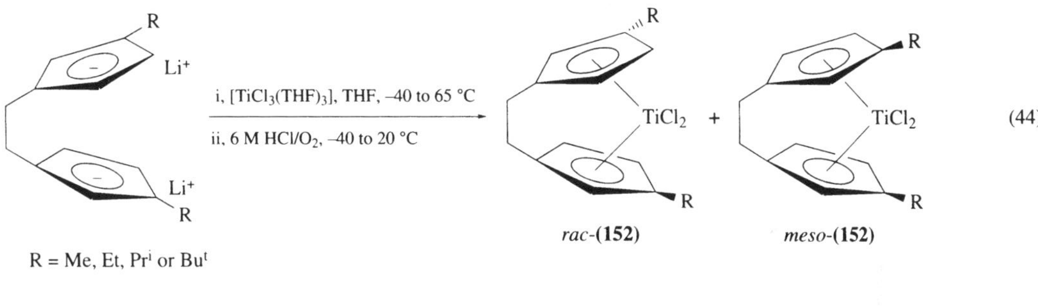

Scheme 36

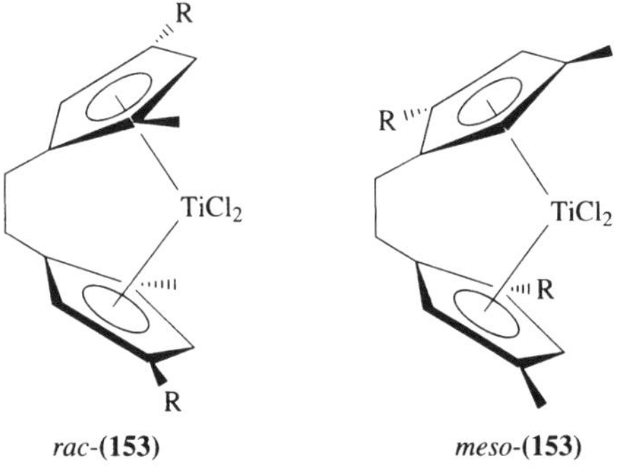

method given in Equation (44). The *rac:meso* ratio proved to be influenced by the nature of the cyclopentadienide reagent and the temperature. Thus, (**154**) is formed as a 6:1 *rac:meso* product if the cyclopentadienyl Grignard is used, whereas the lithium reagent gives a 2:1 mixture (Ar = Ph). If R = 1-naphthyl the Grignard reagent produces the *rac* isomer only, although the yield is low (8%). Generally the lithium reagent gives higher overall yields, 30% vs. 15% if R = Ph. The use of [TiCl$_4$] in toluene instead of [TiCl$_3$(THF)$_3$] affords 40% of a product with a *rac:meso* ratio of 1:1.[303] A series of complexes of the general structure (**155**) is obtained from the corresponding dimagnesium reagents and [TiCl$_3$(THF)$_3$] (R = Bu^t, TMS, Pr^i, CH$_2$Ph, CMe$_2$Ph or 1-phenylcyclohexyl) in 40–50% yield. The compounds are isolated as mixtures with *rac:meso* ratios of 0.6:1 to 2.5:1, depending on R. Isomer separation of compounds (**155**) (R = Bu^t, CMe$_2$Ph or 1-phenylcyclohexyl) was achieved by fractional crystallization or flash chromatography. The crystal structures of *meso*-(**155**) (R = TMS) and *rac*-(**155**)

(R = 1-phenylcyclohexyl) were determined.[305] A significantly higher *meso* content was found if the dilithium salt of the cyclopentadienide ligand was used to give the complex (**155**) (R = Bu^i) in 20–44% yield. A *rac:meso* ratio of 1:4 is measured if [TiCl$_3$(THF)$_3$] or [TiCl$_4$] is added at −78 °C; this increases to 1:2 if the addition is carried out at room temperature. Recrystallization of the product from toluene affords the pure *meso* isomer,[304] whose crystal structure has been reported.[306] The *meso* compound could not be photochemically converted to the *rac* isomer. The reaction of the dilithium reagent with [TiCl$_2$Cp$_2$] leads to an exchange of cyclopentadienyl ligands to give (**155**) in 30% yield with a *rac:meso* ratio of 1:1. The *rac* product was obtained in isomerically pure form by reacting the *in situ* generated binaphtholato complex [TiCl$_2$(1,1'-bi-2-naphtholate)] with the ligand dianion, followed by conversion of the resulting *ansa*-titanocene binaphtholato complex into *rac*-(**155**) by treatment with HCl in hexane.[304]

        rac-(**154**)   *rac*-(**155**)

Complexes with two chiral centres on the bridge are prepared as outlined in Scheme 37. The name 'chiracene' was suggested for the hydrogenated (*S,S*)-bis(tetrahydroindenyl) ligand system. Lithium or potassium salts of the ligand precursor can be used for the titanocene synthesis, and little difference in reactivity is observed between [TiCl$_3$(THF)$_3$] and [TiCl$_4$(THF)$_2$] as the source for the metal. The highest yields were obtained at 40 °C by adding THF solutions of the ligand dianion and [TiCl$_4$(THF)$_2$] from separate dropping funnels simultaneously to a large volume of THF over a period of 4.5 h. The brown crude product is then hydrogenated using a PtO$_2$ catalyst under a pressure of 30 bar of hydrogen gas for 20 h. The red 'chiracene' complex (**156**) is obtained as a 4.3:2.5:1 mixture of the (*R,S*), (*R,R*) and (*S,S*) diastereomers. The isomers could not be separated by fractional crystallization. Irradiation of the mixture leads to extensive photoisomerization, to give up to 85% of the (*R,R*) isomer, which can be separated by reaction with 1,1'-bi-2-naphthol. The crystal structure of (*R,R*)-(**156**) is shown in Figure 19.[299] A similar reaction sequence to that given in Scheme 37 leads to the 'cyclacene' complex (*S,S*)-[TiCl$_2${(*R,R*)-cyclacene}] (**157**) in 20% overall yield. The structure of (**157**) was determined by x-ray crystallography. No other isomer is detected. Although the compound has an unsymmetrical structure in the solid state, it possesses C_2 symmetry in solution.[300]

The biphenyl-bridged titanocenes (**158**) (R = H or Me) were prepared following Equation (45). No titanocene products are isolated with nonsubstituted cyclopentadienyl groups. The compounds are configurationally stable and free from diastereomers.[310] The synthesis and crystal structure of the related binaphthyl-bridged complex (**148**) has been mentioned (Equation (42) and Figure 18).[309]

Trimethylene-bridged titanocenes are obtained from the bis(1-indenyl)propane dianion and [TiCl$_4$(THF)$_2$] followed by hydrogenation under a pressure of 100 bar H$_2$ in the presence of a PtO$_2$ catalyst, which gives (**159**) as red crystals in 11% yield. The etheno-bridged compounds (**160**) and (**161**) are obtained as outlined in Scheme 38 in 8% and 25% yields, respectively.[308]

While bridged bis(cyclopentadienyl) ligands are usually prepared from cyclopentadienenes or fulvenes, that is, moieties with preformed C$_5$ rings, an alternative route is possible using the double Skattebøl rearrangement of (**162**). Three isomeric *ansa*-titanocenes (**163**) are obtained (Scheme 39), of which (**163a**) was isolated by fractional crystallization as red prisms. The others were separated after conversion to the 2-naphtholato complexes. The structures of (**163a**) and (**163b**) were confirmed by x-ray diffraction.[301]

5.5.1.2 Properties and structures

The stepwise metal–chloride bond dissociation enthalpies D_1 and D_2 of [TiCl$_2$Cp$_2$] were estimated from thermochemical data and EHMO calculations as 390 kJ mol^{-1} and 471 kJ mol^{-1}, respectively.[323] The first and second bond dissociation enthalpies of the Ti–Cl bonds in [TiCl$_2$Cp*$_2$] are 380 ± 20 kJ

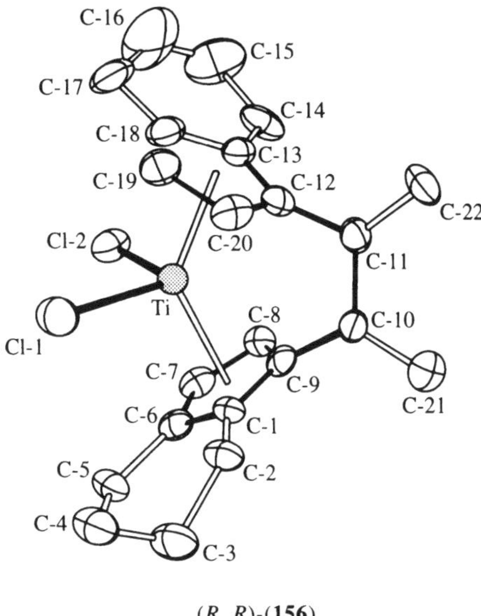

Figure 19 Molecular structure of (*R,R*)-[TiCl$_2${(*S,S*)-chiracene}] (**156**) (reproduced by permission of the American Chemical Society. Copyright (1992) from *Organometallics*, 1992, **11**, 1869).

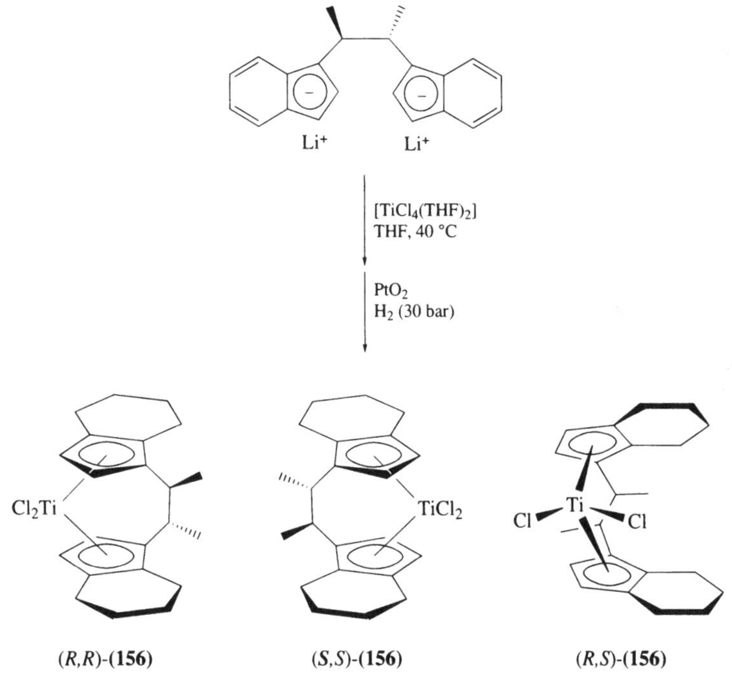

Scheme 37

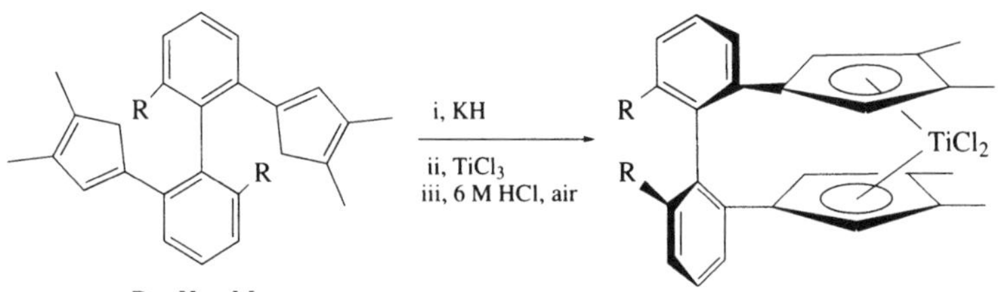

$$\text{(45)}$$

(158)

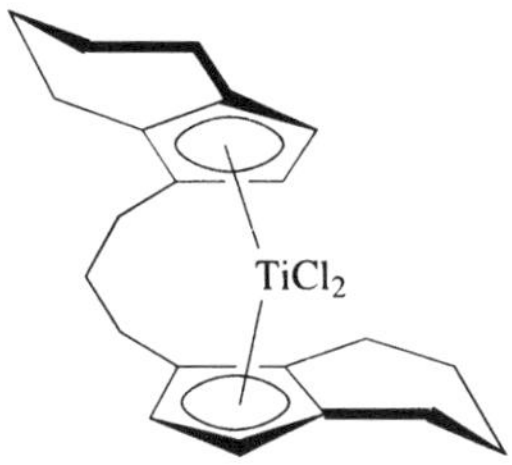

(159)

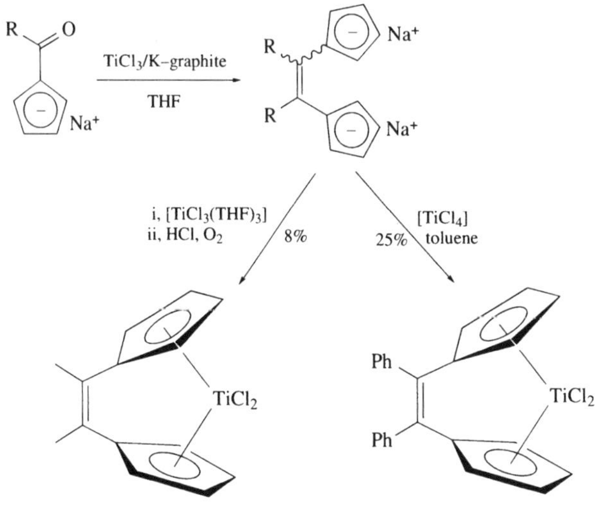

(160)   **(161)**

Scheme 38

mol^{-1} and 481 ± 21 kJ mol^{-1}, respectively, the latter being identical to D(Ti–Cl) of [TiCl(Cp*)$_2$].[324] The bond enthalpies E(Ti–X) of [TiX$_2$Cp$_2$] are 329 ± 10 kJ mol^{-1} for X = N$_3$ and 298 ± 9 kJ mol^{-1} for X = I. The latter is in good agreement with values found for [TiI$_4$].[325]

SCF-X$_\alpha$-scattered wave MO calculations have been carried out on [TiX$_2$Cp$_2$] (X = F, Cl, Br, I or Me). For X = F, Cl or Br, the lowest-energy transition is predicted to be the Cp $\rightarrow$ Ti charge-transfer, while for X = I the transitions Cp $\rightarrow$ Ti and I $\rightarrow$ Ti are very close.[326] The electronic spectrum of [TiI$_2$Cp$_2$] shows a low-energy I $\rightarrow$ Ti charge-transfer absorption. Accordingly, low-energy irradiation ($\lambda > 600$ nm) of [TiI$_2$Cp$_2$] in benzene gives [(TiICp$_2$)$_2$] and I$_2$. Cyclopentadienyl radicals are not formed. In polar solvents such as dichloromethane or acetonitrile, equilibria such as shown in Equation (46) exist.[327]

The relative ordering of low-energy charge-transfer transitions in titanocene diiodies can be reversed by introducing electron-donating substituents on the cyclopentadienyl ligand. Thus, in the photoelectron spectrum of [TiI$_2$(Cp*)$_2$] the lowest-energy band is no longer the I $\rightarrow$ Ti but the Cp* $\rightarrow$ Ti transition. As a consequence, while irradiation of [TiI$_2$Cp$_2$] leads to Ti–I bond cleavage, the cyclopentadienyl ligand is removed in [TiI$_2$(Cp*)$_2$]. Methyl- and trimethylsilyl-substituted cyclopentadienyl ligands behave like Cp^{-}.[328]

The helium(I) photoelectron spectra of the complexes [TiCl$_2$(η-C$_5$H$_{5-n}$Me$_n$)$_2$] (n = 0, 1, 3, 4 or 5) were measured. Each successive methyl substituent lowers the ionization potential by 0.61–1.11 eV with respect to [TiCl$_2$Cp$_2$]. The methyl effect is additive but does not extend to the very crowded Cp*

(162)   → MeLi, THF, −78 °C

i, BuLi
ii, [TiCl₃(THF)₃]
iii, HCl, O₂

(163a)   +   **(163b)**   +   **(163c)**

Scheme 39

$$[TiI_2Cp_2] + n\,\text{solvent} \rightleftharpoons [Ti(I)_{2-n}(\text{solvent})_nCp_2]^{n+}(nI)^{n-} \tag{46}$$

complex, possibly because of a change in the Cp–Ti–Cp angle. The chloro- and the bromotitanocene complexes show the same sequence of MO levels.[329]

ESCA data for the core electron bonding energies have been correlated with the electrochemical oxidation potentials of a series of titanocene difluorides, dichlorides and dibromides. The data are collected in Table 6. The core electron energies are influenced by the electron-donating effect of the methyl substituents of the cyclopentadienyl ligands. Substituting Cp by Cp* shifts the binding energy by ~0.4 eV, that is, two pentamethylcyclopentadienyl substituents have an electronic effect approaching that of a one-electron reduction of the metal.[330] The ⁴⁹Ti NMR chemical shifts have an inverse relationship with the Ti($2p_{3/2}$) core electron binding energies: substitution of Cp by Cp* leads to a high-frequency shift of the ⁴⁹Ti resonance of 148 ppm (Table 6).[321]

Table 6   Binding energies and oxidation potentials for a series of titanocene halides.

	Binding energy (± 0.1 eV)		*Oxidation potential,*	*⁴⁹Ti NMR chemical*
Compound	$Ti(2p_{3/2})^a$	$Ti(2p_{1/2})$	$E_{1/2}^b$ (± 0.02 V)^c	*shift (ref. [TiCl₄])^e*
[TiF₂Cp₂]	457.5	463.6	1.62	−1051
[TiF₂CpCp*]	457.0	463.1	1.24	−964
[TiF₂Cp*₂]	456.7	462.7	1.02	−823
[TiCl₂Cp₂]	456.9	463.0	1.75	−773
[TiCl₂CpCp*]	456.6	462.7	1.46	−601
[TiCl₂Cp*₂]	456.1	462.2	1.22^d	−443
[TiBr₂Cp₂]	456.8	462.9	1.70	−671
[TiBr₂CpCp*]	456.5	462.6	1.44	−486
[TiBr₂Cp*₂]	455.9	462.0	1.21	−293

[a] At 23 °C. Binding energies are calibrated against C($1s$) of polyethene. [b] $E_{1/2}$ values for the irreversible oxidation in acetonitrile. [c] Single-sweep cyclic voltammetry, 22 °C, ~10^{-3} M concentration in [Ti]. Scan rates 100 mVs⁻¹, saturated NaCl–SCE reference electrode and platinum bead working electrode. Supporting electrolyte [NBu₄][ClO₄]. [d] Oxidation is reversible at 0 °C in CH₂Cl₂ and at −40 °C in MeCN but irreversible at 22 °C in MeCN. [e] [TiCl₄] as external standard. Spectra were measured in CDCl₃.

The ion decay pathways in the mass spectra of [TiX₂{Me₂Si(C₅H₄)₂}] (X = Cl, Br or I) and [TiCl₂(η-C₅H₄EMe₃)₂] (E = C, Si or Ge) have been studied[331] and the electron-impact (EI) and negative-ion chemical ionization (NCI) mass spectra of some 24 titanocene dichlorides have been reported. Molecular ions are usually not found with the EI method but are readily detected by NCI.[332]

The ^{1}H NMR resonances for the cyclopentadienyl ligands of a series of *ansa*-titanocenes [TiX$_2${C$_2$Me$_4$(C$_5$H$_4$)$_2$}] (X = F, Cl, Br, I, CO or PMe$_3$) have been assigned by means of nuclear Overhauser and selective decoupling experiments. For the 16-electron metal halides the signal for the hydrogen in the β position to the bridge is observed at higher frequency than α-H, with the chemical shift difference Δ(β − α) decreasing in the order X = I > Br > Cl > F. The 18-electron titanium(II) complexes have β-H at lower frequency than α-H.[298] The rotational barriers in sterically hindered silyl-substituted titanocene dihalides have been determined. A value of 37 ± 2 kJ mol^{-1} is found for [TiX$_2${η-C$_5$H$_3$(TMS)$_2$}$_2$] (X = F or Cl) in CD$_2$Cl$_2$, while there is no indication for cyclopentadienyl rotation in [TiX$_2${η-C$_5$H$_4$(TMS)}{η-C$_5$H$_2$(TMS)$_3$}], even on heating to 60 °C; that is, the rotational barrier must be ≥71 kJ mol^{-1}.[250] The phenyl substituents in [TiCl$_2$(η-C$_5$HPh$_4$)$_2$] form a propeller arrangement and rotate with an energy barrier $\Delta G^{\ddagger}_{rot}$ ~41 ± 0.8 kJ mol^{-1}. The motions within one ring are independent of the other, as is the case in [Fe(η-C$_5$HPh$_4$)$_2$].[267]

The crystal structures of a series of titanocene dihalides have been reported. References are given in Tables 3–5, and most will not be discussed here in detail. Features of interest are the conformation of the cyclopentadienyl ligands as a function of repulsive steric interactions either with each other or the chloro ligands. This aspect has attracted particular attention in relation to the role of titanocenes in stereoselective synthesis, 1-alkene polymerizations and hydrogenations.

An important consequence of the introduction of substituents on or bridges between cyclopentadienyl ligands is their effect on the coordination gap aperture, defined as the largest possible angle α between two planes through the metal centre which touch van der Waals surfaces of each of the cyclopentadienyl ligands, as illustrated in Figure 20.[266] Values for α decrease with increasing bulk of the cyclopentadienyl ligand and are usually in the region of 60–90° but may be increased to 100–110° by small bridges such as CH$_2$.

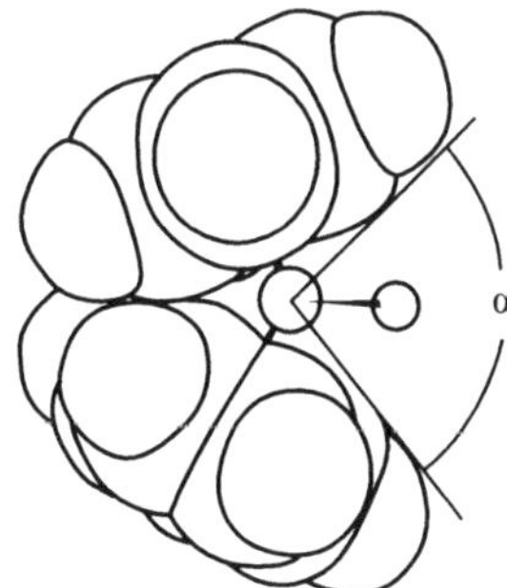

Figure 20 Space-filling model of [MCl$_2$Cp$_2$] indicating the van der Waals outlines and the coordination gap aperture α (reproduced by permission of Hüthig and Wepf from *Macromol. Chem., Macromol. Symp.*, 1993, **66**, 127).

The isopropyl substituents in [TiCl$_2$(η-C$_5$H$_4$Pr^i)$_2$] (**164**) are located directly above and below the two chloride ligands; the molecule possesses C_{2v} symmetry.[231] Sterically more hindered substituents, such as R = Bu^t or TMS, enforce a transoid conformation, as shown in (**165**).[233,244] Although bis- and tristrimethylsilyl-substituted cyclopentadienyl ligands are well known for their steric hindrance, the introduction of only one of these, as in [TiCl$_2$Cp{η-C$_5$H$_2$(TMS)$_3$}], imposes only moderate steric constraints, although the C$_5$H$_2$(TMS)$_3$ ligand is distorted toward an η^3-bonding mode.[288] Rather subtle steric effects seem to dictate the conformation of the *rac* and *meso* isomers of complexes with chiral cyclopentadienyl ligands such as (**140**) and (**141**) (Equation (39)). Whereas in (**140**) the bicyclic substituents point approximately in the same direction toward the open wedge of the bent sandwich above and below the TiCl$_2$ plane, they are directed away from each other in the *meso* compound (**141**), as shown in Figure 21. The steric strain imposed by these substituents is reflected in the bond distances between the titanium atom and the individual cyclopentadienyl carbon atoms, which differ by up to 0.031 nm.[259]

In the sterically hindered mixed-ligand complex [TiCl$_2$Cp(C$_5$Ph$_5$)] the Ti–C distances to the Cp ring range from 0.235 nm to 0.241 nm and are significantly shorter than the bonds to the C$_5$Ph$_5$ ligand (0.249–0.255 nm). The phenyl groups are bent away from the metal centre. Due to the propeller arrangement of the phenyl substituents, the complex is chiral and crystallizes as the racemate (Figure 22).[289]

The structures of *ansa*-titanocenes have attracted detailed attention because of the existence of configurational isomers and the ligand control these reagents can exert in stereoselective hydrogenations, isomerizations, C–C bond formations and 1-alkene polymerizations. The prototype of

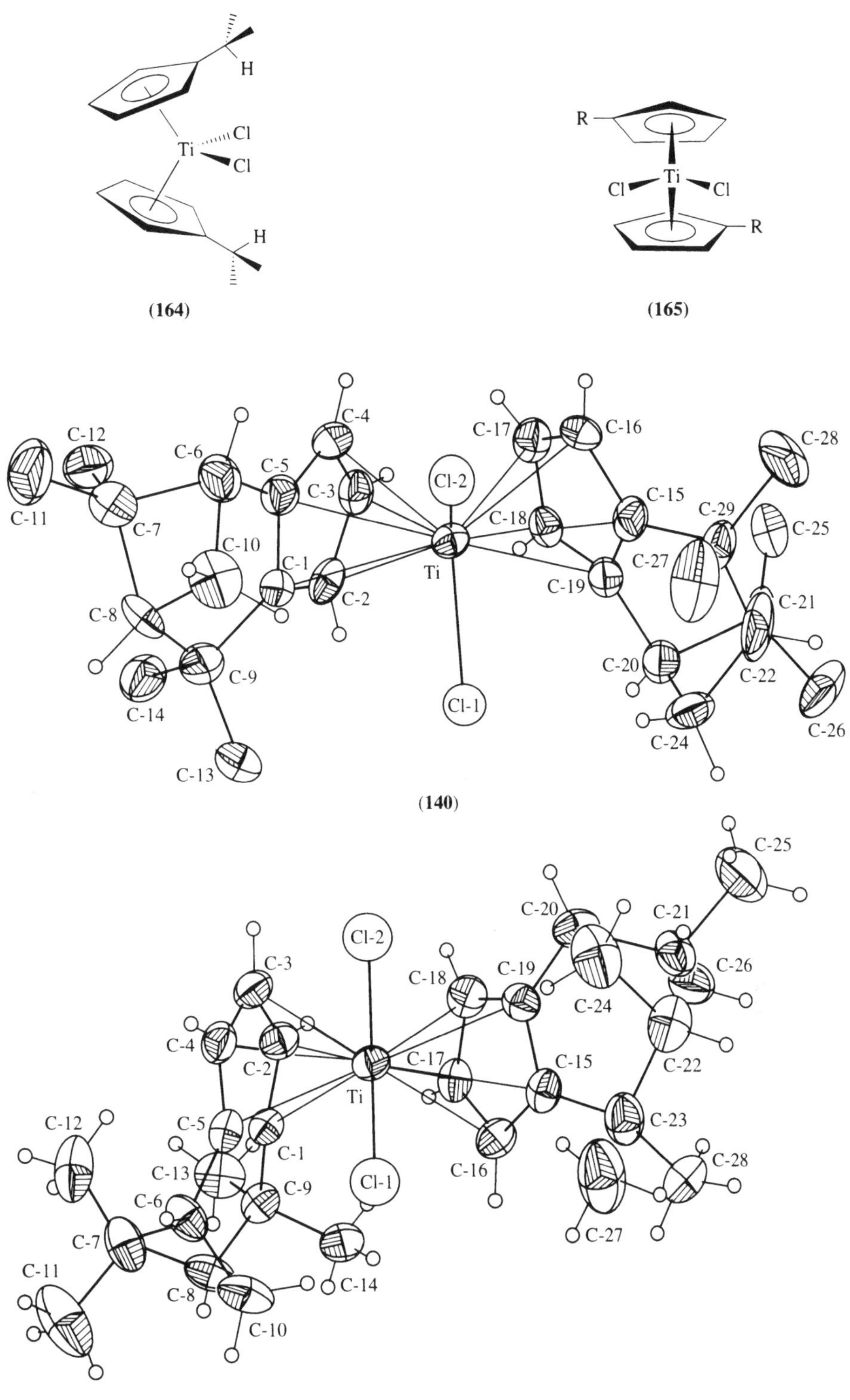

Figure 21 Molecular structures of (**140**) and (**141**) (reproduced by permission of Elsevier from *J. Organomet. Chem.*, 1990, **397**, 177).

the stereoselective *ansa*-titanocenes is (**151**); the molecular structures of *rac*-(**151**) and *meso*-(**151**) are shown in Figures 23 and 24, respectively.[292] In *rac*-(**151**) the methylene-substituted edges of the cyclopentadienyl rings are bent away from the TiCl$_2$ moiety, presumably as the result of steric repulsion between the CH$_2$ groups and the chloro ligands, and the axis normal to a C$_5$ ring plane forms an angle of 7.5 ± 1.5° with the plane bisecting the Cl–Ti–Cl angle. In *meso*-(**151**) the ethylene bridge is twisted

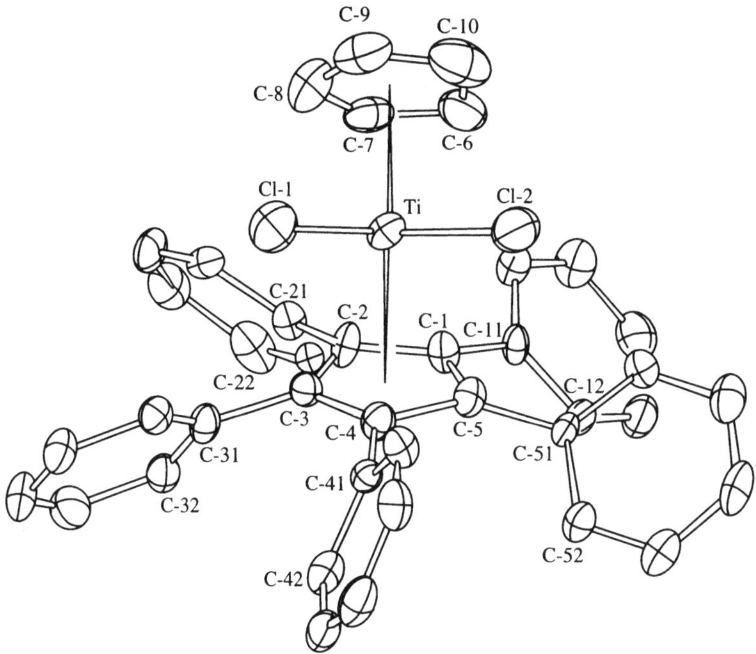

Figure 22   Molecular structure of [TiCl₂Cp(η-C₅Ph₅) (reproduced by permission of Elsevier from *J. Organomet. Chem.*, 1991, **412**, 343).

to one side of the molecule, and although the two tetrahydroindenyl units have slightly different positions in the solid state, it is evident they can interconvert by very minor conformational changes.

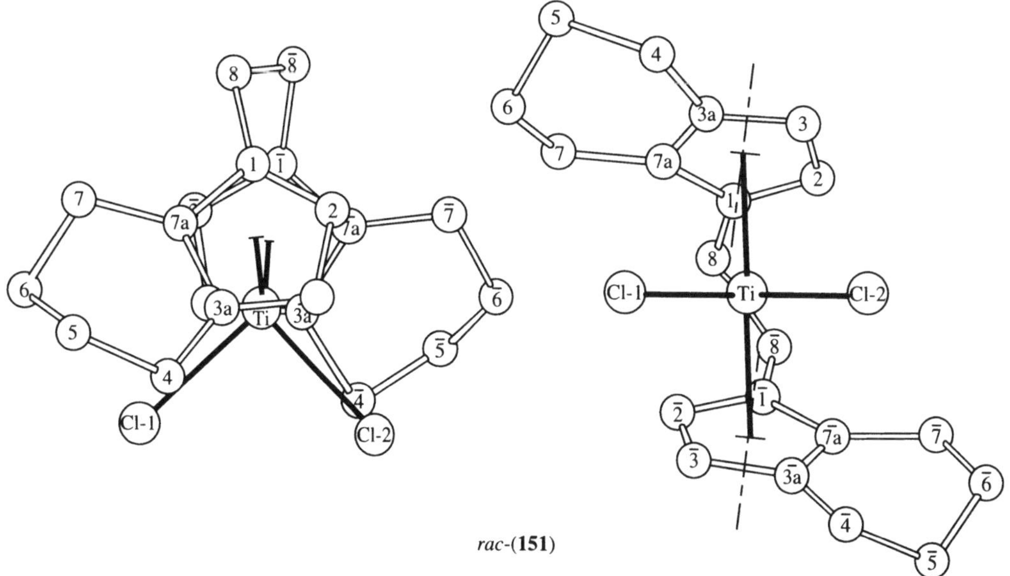

Figure 23   Top and front views of racemic ethylenebis(4,5,6,7-tetrahydro-1-indenyl)titanium dichloride, *rac*-(**151**) (reproduced by permission of Elsevier from *J. Organomet. Chem.*, 1982, **232**, 233).

The racemic isomers of *ansa*-titanocenes consist of two enantiomers which can often be resolved by converting the chloro complex into the binaphtholato derivative using enantiomerically pure 1,1'-bi-2-naphthol. An early application of this method was the synthesis of enantiomerically pure (*S,S*)-*rac*-(**151**). The structure of one such intermediate, ethylenebis(tetrahydro-(*S*)-1-indenyl)titanium-(*S*)-1',1"-bi-2-naphthotate, is shown in Figure 25.[292]

As was described above, the 'coordination gap aperture' is defined by two intersecting planes tangential to the van der Waals surfaces of the cyclopentadienyl ligands (see Figure 20). In simple

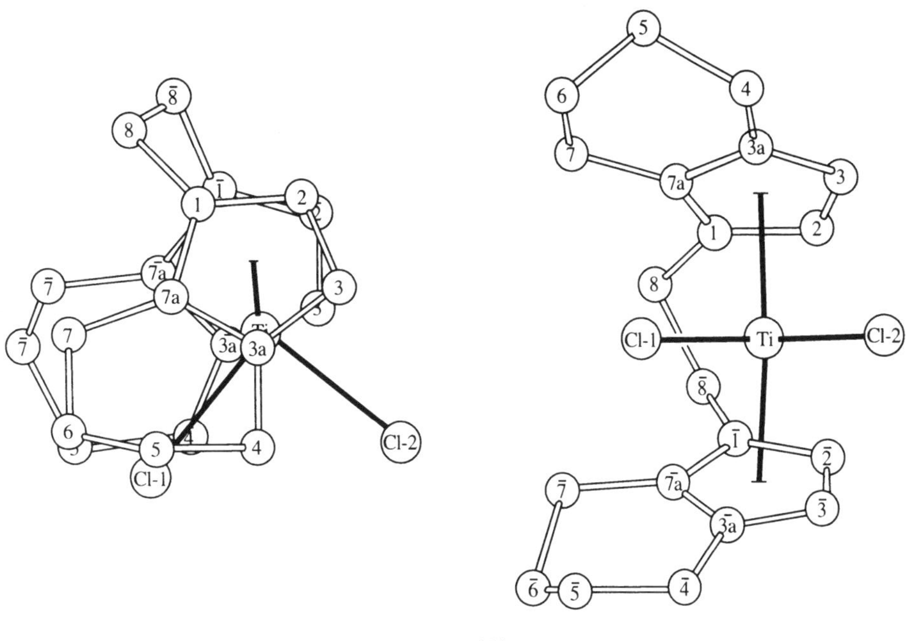

meso-(**151**)

Figure 24 Top and front views of *meso*-(**151**) (reproduced by permission of Elsevier from *J. Organomet. Chem.*, 1982, **232**, 233).

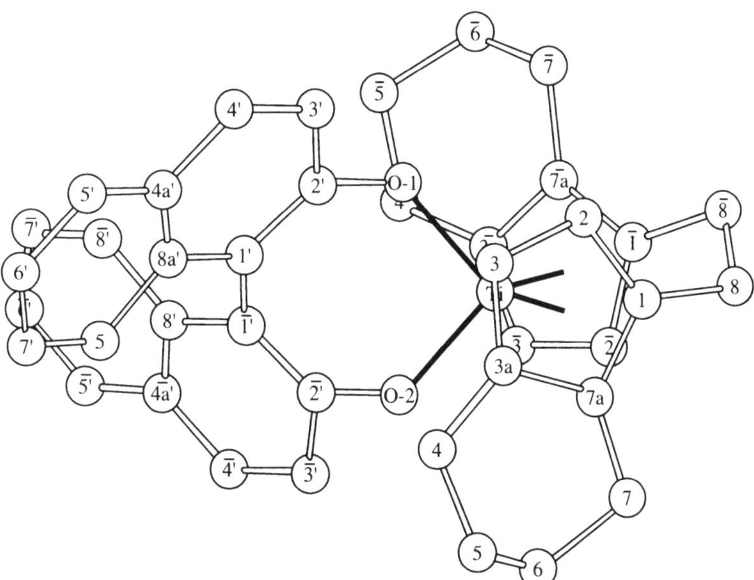

Figure 25 Molecular structure of ethylenebis[4,5,6,7-tetrahydro-(*S*)-1-indenyl]titanium-(*S*)-1',1"-bi-2-naphtholate (reproduced by permission of Elsevier from *J. Organomet. Chem.*, 1982, **232**, 233).

metallocenes, these planes are perpendicular to the plane bisecting the Cl–Ti–Cl angle. In *ansa*-metallocenes, however, the conformation of the bridge and the presence of substituents in α or β positions to the bridge result in a slanting of the tangential planes with respect to the MCl$_2$ plane. This resulting angle between the intersection of the planes defining the coordination gap aperture and the MCl$_2$ plane has been termed the *coordination gap obliquity*. The principle is illustrated in Figure 26 for the complex [ZrCl$_2$\{Me$_2$Si(C$_5$H$_3$Ph-3)$_2$\}], which shows an obliquity angle of 21°.[266,333] The same principle applies to titanocene complexes. Since the enantiomers of C_2-symmetric *ansa*-metallocenes show obliquity angles of opposite sign, these can be used to describe the chirality of the metal centre. Parameters such as coordination gap obliquity are relevant to the enantiofacial preference of binding a prochiral substrate molecule to the metal centre, for example, in 1-alkene polymerizations and asymmetric hydrogenations with chiral titanocenes.

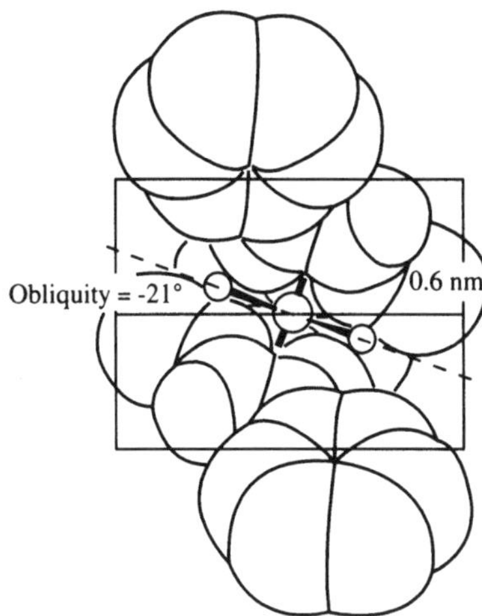

Figure 26   Obliquity of the coordination gap aperture in *ansa*-metallocene dihalides (reproduced by permission of Hüthig and Wepf, *Macromol. Chem., Macromol. Symp.*, 1993, **66**, 127).

ansa-Titanocenes frequently exhibit deviations from C_2 symmetry in the solid state. The position of the bridge often does not coincide with the plane bisecting the Cl–Ti–Cl angle, and the whole ligand framework is shifted to one side of it. This conformation minimizes repulsive steric interactions within the ligand framework. These intramolecular interactions are present even in nonsubstituted Cp derivatives such as (**160**) and (**161**), and less so in (**150**), where adherence to C_2 symmetry would result in an ecliptic conformation of the CH_2 hydrogens of the ethano bridge and lead to close contacts between the cyclopentadienyl hydrogen atoms in the β position to the bridge and the chlorine atoms (Figures 27 and 28). Two shallow conformational energy minima exist which are located about ± 20° away from the $TiCl_2$ bisector axis. In addition, close scrutiny of the crystal structures of (**150**) and of the cyclohexyl-substituted derivative $[TiCl_2\{R_2C_2(C_5H_4)_2\}]$ (R = $(CH_2)_5$) (**166**) reveals a number of close Cl–H contacts below 0.29 nm which help to stabilize the dissymmmetric conformation in the solid state (Figure 29).[291]

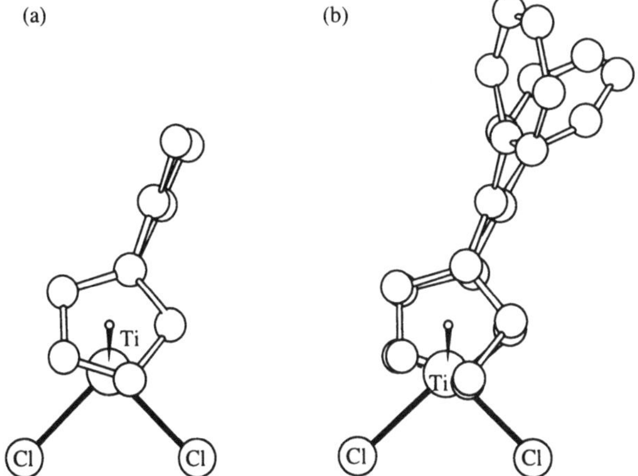

Figure 27   Top views of etheno-bridged titanocenes (a) $[TiCl_2\{Me_2C_2(C_5H_4)_2\}]$ and (b) $[TiCl_2\{Ph_2C_2(C_5H_4)_2\}]$, illustrating the deviation of the cyclopentadienyl ligand framework from the plane bisecting the Cl–Ti–Cl angle in the solid state (reproduced by permission of the American Chemical Society. Copyright (1992) from *Organometallics*, 1992, **11**, 1319).

Substituents on the cyclopentadienyl ring in the β-position deepen the conformational energy minima and stabilize the ligand conformation away from the Cl–Ti–Cl bisector axis, whereas α substituents have the opposite effect. Figures 30 and 31 illustrate the conformational preferences for a doubly substituted ethano-bridged and a propano-bridged complex ((**163a**) and (**159**), respectively).

Two examples of deviation from axial symmetry for tetramethylethano-bridged complexes are shown in Figure 32. In spite of a similar substitution pattern, the conformations adopted by the bridges differ

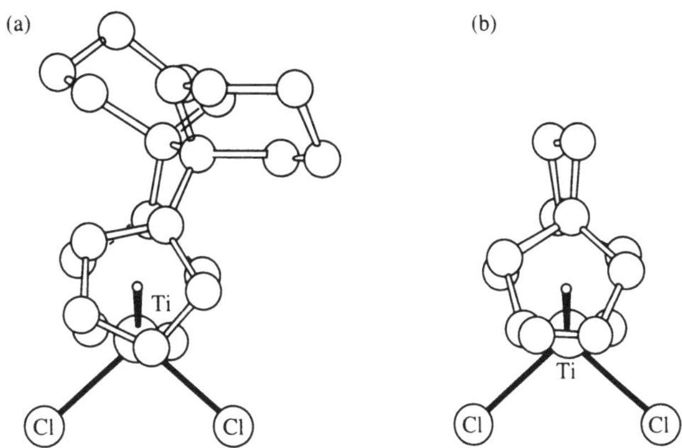

Figure 28   Top views of the solid-state conformations in ethano-bridged titanocenes [TiCl$_2${R$_4$C$_2$(C$_5$H$_4$)$_2$}]: (a) R$_2$ = (CH$_2$)$_5$; (b) R = H (reproduced by permission of the American Chemical Society. Copyright (1992) from *Organometallics*, 1992, **11**, 1319).

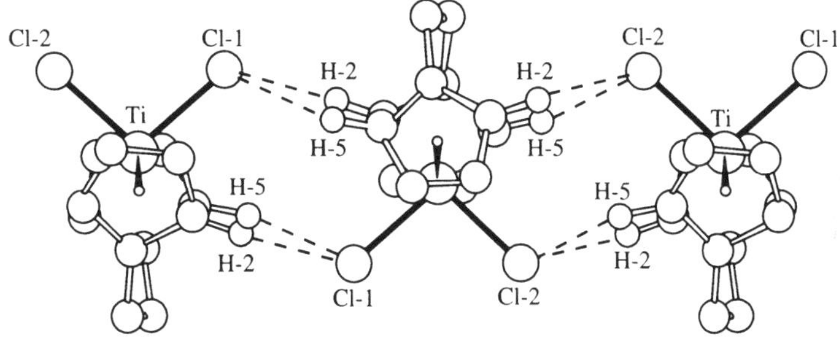

Figure 29   Intermolecular Cl$\cdots$H contacts in crystalline [TiCl$_2${H$_4$C$_2$(C$_5$H$_4$)$_2$}] (reproduced by permission of the American Chemical Society. Copyright (1992) from *Organometallics*, 1992, **11**, 1319).

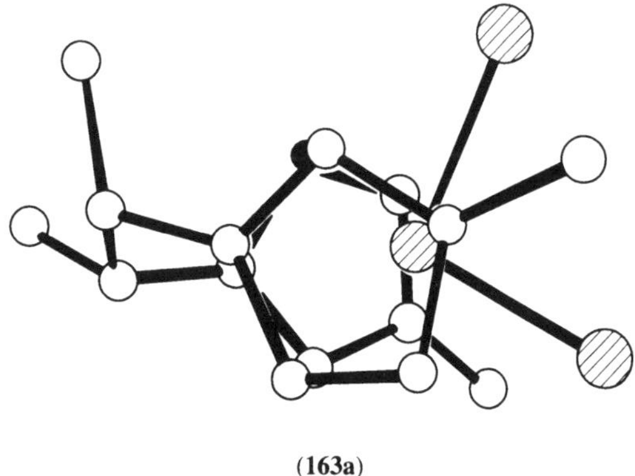

(**163a**)

Figure 30   Solid-state conformation of the complex (**163a**) (reproduced by permission of the American Chemical Society. Copyright (1993) from *Organometallics*, 1993, **12**, 2248).

in each case.[291,305] The energy barrier between the alternative geometries is quite low, and in solution there is rapid interchange between these minima, resulting in a time-averaged C_2 symmetry of the molecule.

5.5.1.3 Reactions

The thermolysis of titanocene dihalides and pseudohalides [TiX$_2$Cp$_2$] (X = Cl, Br, NCS, NCO or N$_3$) produces C$_5$H$_6$ and (Cp)$_2$ in a single sharp step. Pseudohalides, too, give rise to breakdown products at that temperature; for example, isocyanate gives HCN, and azide decomposes to N$_2$. The activation enthalpies for the thermolysis of [TiX$_2$Cp$_2$] (X = NCO and N$_3$) are 397 kJ mol^{-1} and 230 kJ mol^{-1}, respectively. The compounds have been used to deposit a titanium-containing film on CaF$_2$ or quartz

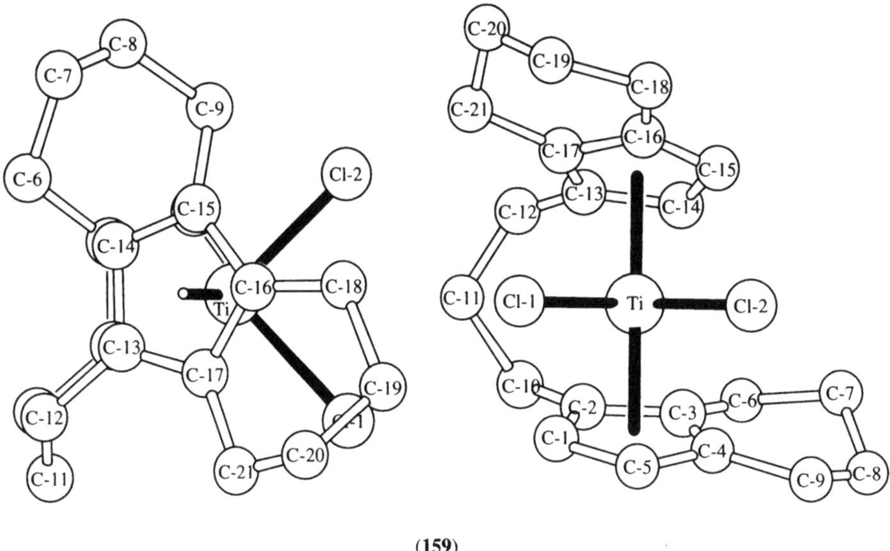

(159)

Figure 31 (a) Top and (b) side views of the propano-bridged complex (**159**) in the solid state (reproduced by permission of the American Chemical Society. Copyright (1992) from *Organometallics*, 1992, **11**, 1319).

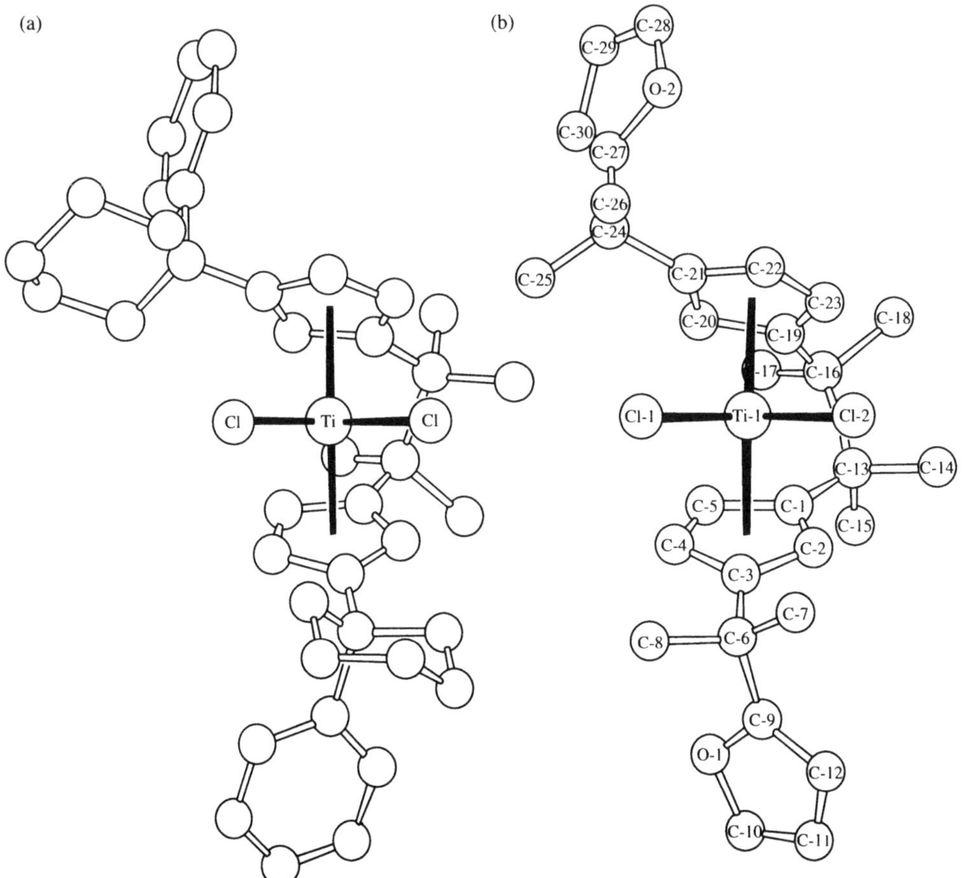

Figure 32 Variation in bridge conformation and deviation from axial symmetry in tetramethylethano-bridged *ansa*-titanocenes. The cyclopentadienyl rings carry (a) 1'-phenylcyclohexyl and (b) α,α-dimethylfurfuryl substituents (reproduced by permission of the American Chemical Society. Copyright (1992) from *Organometallics*, 1992, **11**, 1319).

substrates at 400 °C. The films are rich in carbon.[334] The photolysis of $[TiCl_2Cp_2]$ in toluene gives cyclopentadienyl radicals and $TiCl_2Cp$. Both photoproducts were detected by ESR spectroscopy. In the presence of oxygen gas, a signal for $[TiCl_2Cp(O-O)\cdot]$ is observed. The titanium(III) species reacts with Me_3CNO_2 to give an adduct whose hyperfine coupling to two chlorides can be resolved. Other spin traps such as diones also lead to relatively stable adducts.[335] The photolysis of $[TiCl_2Cp_2]$ induces the polymerization of methyl methacrylate. The resulting polymer has an almost entirely isotactic structure.[336]

The reaction of $[TiF_2Cp_2]$ with HCl in an argon matrix gives a distinct hydrogen-bonded species (Equation (47)), which was identified by IR spectroscopy.[337]

$$[TiF_2Cp_2] + HCl \xrightarrow{\text{argon matrix}} Cp_2Ti \overset{F}{\underset{F}{<}} H-Cl \qquad (47)$$

The kinetics of halide exchange of $[TiCl_2Cp_2]$ with X^- (X = Br, I, SCN or CN) in acetonitrile under pseudo-first-order conditions, that is, in the presence of a large excess of X^-, has been studied. Both chloride ligands are replaced simultaneously, without evidence for the intermediate $[TiCl(X)Cp_2]$. The rate is independent of [X]. The rate-limiting step is the dissociation of Cl^-; that is, a dissociative rather than associative pathway predominates. The substitution rate is slowed down if Cp is replaced by Cp* but significantly accelerated if up to 5% water is present because of the more favourable solvation of the Cl^- leaving group.[338] A series of substituted titanocene dichlorides was converted into the corresponding fluorides, bromides and iodides by the reaction with BX_3 (X = F, Br or I); the fluorides are also accessible by the reaction with ammonium fluoride.[339] The reaction of titanocene dichlorides with $[NH_4]_2[SiF_6]$ also leads to titanocene difluorides.[340]

Titanocene difluoride reacts with MF_5 in liquid sulfur dioxide to give $[TiCp_2\{(\mu\text{-}F)MF_5\}_2]$ (167) (Scheme 40). The structures of the arsenic and antimony compounds were confirmed by x-ray diffraction.[341,342] $[TiCp_2\{(\mu\text{-}F)SbF_5\}_2]$ reacts with further SbF_5 in SO_2 at 20 °C to give $[TiCp_2(Sb_2F_{11})_2]$; the compound is prone to loss of SbF_5 on recrystallization.[342] $[TiCp_2\{(\mu\text{-}F)AsF_5\}_2]$ exists in solution as the SO_2 solvate, $[TiCp_2(SO_2)_2][AsF_6]_2$.[343] The IR spectrum of the $[BiF_6]^-$ ligands in the analogous bismuth complex suggests a distorted octahedral geometry.[344] According to ^{14}N and ^{19}F NMR spectroscopy compound (167) (M = As) reacts with S_4N_4 in SO_2 to give the adduct $[TiCp_2(S_4N_4)][AsF_6]_2$ in solution, which decomposes to $[TiF(AsF_6)Cp_2]$ and $S_4N_4\cdot AsF_5$. The analogous reaction with Se_4N_4 gives $[Se_6N_4][AsF_6]_2$ and $[TiF_2Cp_2]$, besides N_2 and AsF_3.[345] Exposure of (167) (M = As) to acetonitrile affords the adduct $[TiCp_2(MeCN)_3][AsF_6]_2$ as red microcrystals, which slowly lose MeCN *in vacuo* to give $[TiCp_2(MeCN)_2][AsF_6]_2$. No adduct is formed with CF_3CN.[346] Ligands such as HCN, ICN and $AsMe_3$ give the expected adducts, while treatment with As_2Me_4 leads to (168) (Scheme 40). The Ti_2As_4 ring is fluxional, with $\Delta G^{\ddagger}$ ~54 kJ mol^{-1} for the ring inversion.[347]

Neutral and ionic metallocenes, notably those of titanium, have been thoroughly investigated as antitumour agents. Significant activity against diverse forms of animal and human tumours has been demonstrated.[348] A series of complexes $[Ti(X)(Y)Cp_2]$ (X = Cl; Y = 1,2-MeC_6H_4S or 1,2-$(NMe_2)(S)C_6H_4$) and $[Ti(X)_2(Y)Cp]$ (X = Cl or NCS; Y = Cl, SPh or NCS) has been tested against Ehrlich ascites tumour in mice. The compounds induced cures in 13–31% of the animals. Cures were, however, sporadic, and a good dose–activity relationship could not be established. The hydrolysis products of the titanium complexes, such as $[(TiClCp_2)_2(O)]$ and $[\{TiCl(O)Cp\}_4]$, were also active.[349] $[TiCl_2Cp_2]$ and *cis*-$[PtCl_2(NH_3)_2]$ were compared in rats against acute anti-inflammatory activity, antiarthritic activity against polyarthritis, immunosuppressant activity and irritant effects at the site of administration. Both titanium and platinum have similar effects, though titanium is less nephro- and gastrotoxic. *In vitro*, the titanium complex inhibits thymidine incorporation by isolated thymocytes and prevents the germination of radish seeds. There is no skin irritation.[350] The chemistry underlying some of these physiological effects has been investigated. In pure water, $[TiCl_2Cp_2]$ hydrolyses very fast to $[TiClCp_2(H_2O)]^+$ and more slowly to $[TiCp_2(H_2O)_2]^{2+}$. The half-life for the loss of the second chloride ligand is ~50 min. Under physiological conditions the $[TiCp_2]^{2+}$ framework does not remain intact, in contrast with the corresponding vanadium analogue.[351] While TiCp and $TiCp_2$ moieties are reasonably stable under acidic aqueous conditions, at pH 5.5 loss of Cp from $[TiCl_2Cp_2]$ and formation of $[Ti_4Cp_4O_6]$ is observed, which is poorly soluble in water. In the presence of carboxylate functions of the kind commonly found *in vivo*, titanocene complexes do not survive, and titanium cyclopentadienyl complexes do not apparently exist in blood or tissue. It is speculated that the carcinostatic activity may be caused not by the metal but by the slow release of cyclopentadiene or one of its products. The fast release and higher concentrations of cyclopentadiene can lead to toxic side effects.[352,353]

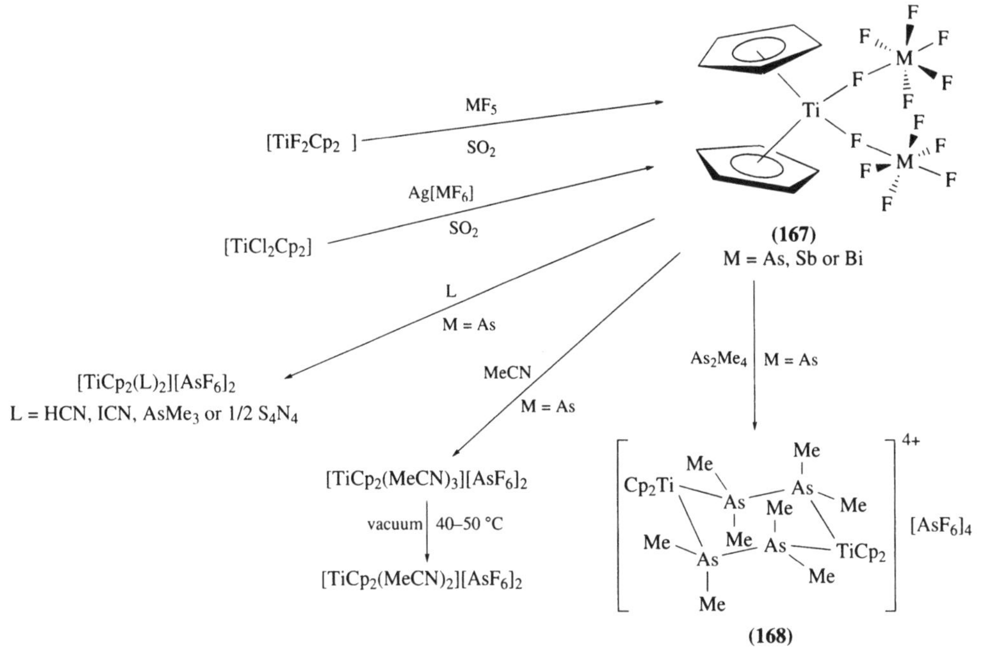

(167)
M = As, Sb or Bi

$[TiF_2Cp_2]$ — MF$_5$ / SO$_2$ →

$[TiCl_2Cp_2]$ — Ag[MF$_6$] / SO$_2$ →

$[TiCp_2(L)_2][AsF_6]_2$
L = HCN, ICN, AsMe$_3$ or 1/2 S$_4$N$_4$

L / M = As

MeCN / M = As

$[TiCp_2(MeCN)_3][AsF_6]_2$

vacuum | 40–50 °C

$[TiCp_2(MeCN)_2][AsF_6]_2$

As$_2$Me$_4$ | M = As

(168)

Scheme 40

Biological silica surfaces can be stained with [TiCl$_2$Cp$_2$] for investigation by electron microscopy. In heterogeneous biological silicas, different structural motifs could be identified by the levels of titanium incorporation.[354]

The reduction of titanocene dichloride is achieved with a variety of reagents. For example, treatment with [YbCp*$_2$(THF)$_2$] gives [(TiClCp$_2$)$_2$] and [YbClCp*$_2$(THF)],[355] while reduction with magnesium under an atmosphere of dinitrogen affords a mixture of titanium complexes with η^5-fulvalene ligands, accompanied by the reduction of N$_2$ to N^{3-} in the form of titanium–magnesium nitrido complexes.[356] The conproportionation of [TiCl$_2$Cp$_2$] and [TiCp$_2$(CO)$_2$] is reversible, and *in situ* IR spectroscopy under CO pressure allows the detection of the intermediate, [TiClCp$_2$(CO)], which exhibits a ν_{CO} band at 2068 cm^{-1}. At 30 °C and 1 bar pressure, about equal amounts of [TiCp$_2$(CO)$_2$], [(TiClCp$_2$)$_2$] and [TiCl$_2$Cp$_2$] are present in the mixture. At lower CO pressures the conproportionation dominates. The mechanism and the rate laws for this reaction were investigated.[357]

Electrochemically, [TiCl$_2$Cp$_2$] is reduced under a CO pressure of 3 bar to the radical anion [TiCl$_2$Cp$_2$(CO)]$^{\cdot-}$ (ν_{CO}1950 cm^{-1}), which takes up a second electron with loss of Cl$^-$ and formation of [TiCp$_2$(CO)$_2$].[358] The effects of cyclopentadienyl substituents on the electroreduction of a series of titanocene dichlorides have been evaluated by cyclic voltammetry. The reduction potentials can be described by a two-parameter Taft equation. Electron donors such as alkyl substituents push the reduction to more negative potentials.[359] If the cyclic voltammetry experiments are conducted in THF over a temperature range of −100 °C to +20 °C all main intermediates can be observed, such as [TiCl$_2$Cp$_2$], [TiCl$_2$Cp$_2$]$^-$ and [(TiClCp$_2$)$_2$].[360] On the other hand, following this reaction by ESR spectra in THF between −50 °C and room temperature did not give any evidence for the existence of [TiCl$_2$Cp$_2$]$^-$, rapid loss of Cl$^-$ and formation of [TiClCp$_2$(THF)] being observed instead. The bromo and iodo complexes behave similarly. Although the THF was distilled from LiAlH$_4$, traces of water still appeared to be present and gave rise to small amounts of [TiClCp$_2$(H$_2$O)] alongside the THF adduct. Halide-free species such as [TiCp$_2$(THF)$_2$]$^+$ were also found.[361] The dimethylamino substituent in [TiCl$_2$(η-C$_5$H$_4$NMe$_2$)$_2$] cannot be quaternized with iodomethane. Electrochemical reduction of this complex to titanium(III) is observed at $E^0 = -1.11$ V vs. SCE; irreversible reduction takes place at −1.84 V. The dimethylamino substituent stabilizes the higher oxidation state and makes the compound more difficult to reduce compared with the parent complex.[241] By contrast, the silyl-bridged *ansa*-titanocene [TiCl$_2${Me$_2$Si(C$_5$H$_4$)$_2$}] is reduced in THF reversibly at $E^0 = -0.793$ V vs. SCE to give [TiCl$_2${Me$_2$Si(C$_5$H$_4$)$_2$}]$^-$; this potential is very similar to that of [TiCl$_2$Cp$_2$] (0.80 V).[315]

Potassium metal in pentane reduces [TiCl$_2$(η-C$_5$H$_3$Bu$^t_2$)$_2$] to the deep-blue complex [TiCl(η-C$_5$H$_3$Bu$^t_2$)$_2$].[247] The reduction of the *ansa*-titanocene complex **(150)** proceeds with sodium sand in toluene to give the orange–brown compound [TiCl{C$_2$H$_4$(C$_5$H$_4$)$_2$}].[290] On the other hand, the reduction of [TiCl$_2$(η-C$_5$H$_4$Me)$_2$] with potassium–naphthalene in THF is thought to give rise to pyrophoric 'Ti(η-C$_5$H$_4$Me)$_2$'.[362]

The electrochemical oxidation of $[TiCl_2Cp_2]$ in acetonitrile with $[NBu_4][ClO_4]$ as the supporting electrolyte is found at $+1.90$ V vs. SCE. The solvent-stabilized cation $[TiClCp_2(MeCN)]^+$ is formed, which gives rise to a Cp signal in the ^{1}H NMR spectrum at δ 6.88, downfield from $[TiCl_2Cp_2]$ (δ 6.57). In the presence of CD_2Cl_2, another species is observed, thought to be $[TiClCp_2(CD_2Cl_2)]^+$ (δ 7.08), which reacts with the chlorinated solvent to regenerate $[TiCl_2Cp_2]$.[363]

Mixtures of $[TiCl_2Cp_2]$, aluminum triethyl and either NaH, CaH_2 or methanol catalyse the disproportionation of formaldehyde into methyl formate, methanol and dimethoxymethane. The methanol-containing catalyst is the most active. Methyl formate is obtained in 38.5% molar conversion, up to 7.7 mol per titanium atom.[364] Titanocenes with long-chain silyl side chains, such as $[TiCl_2\{\eta\text{-}C_5H_4(CH_2)_3Si(OMe)_3\}_2]$, can be attached to silica supports. After reduction with butyllithium they give efficient catalysts for the selective hydrogenation of 1-alkenes. Thus, 1-octene was completely hydrogenated after 3 h, while with *trans*-2-octene only 9% hydrogenation had taken place after 48 h.[239] The asymmetric hydrogenation of 2-phenyl-1-butene is catalysed by the chiral titanocenes (**169**) and (**170**) in the presence of butyllithium (Equation (48)). The complex (**169**) requires an induction period of 1 h at room temperature and gives the reduction product in 33% *ee* in a 40 h reaction. The catalyst (**170**) is very much faster. The induction time is only 2 min, and the reaction proceeds between 20 °C and -20 °C in 22% and 34% *ee*, respectively.[237] The same reaction catalysed with (**144a**) (cf. Equation (40)) is more enantioselective, with 68% *ee* at 25 °C and 100% conversion, rising to 95% *ee* at -75 °C, although the conversion decreases under these conditions to 25%. Both (*S*)- and (*R*)-2-phenylbutane can be prepared selectively.[261]

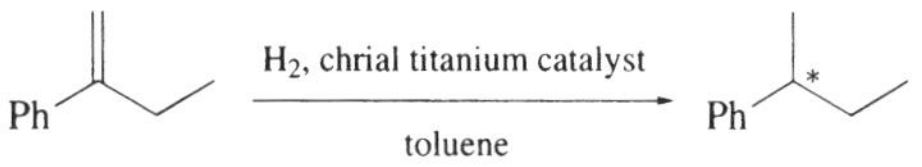

$$Ph\diagup\diagdown \xrightarrow[\text{toluene}]{H_2,\ \text{chrial titanium catalyst}} Ph\diagup\overset{*}{\diagdown} \qquad (48)$$

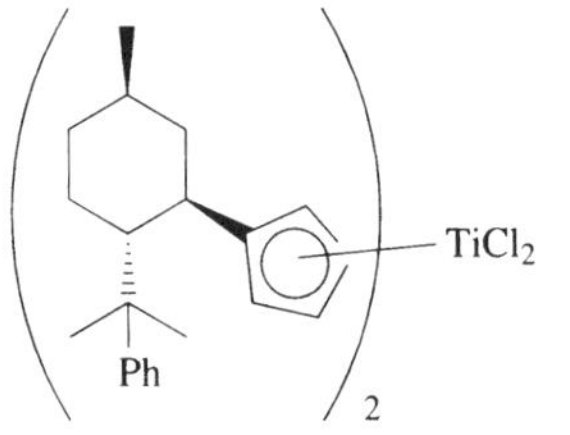

(169) 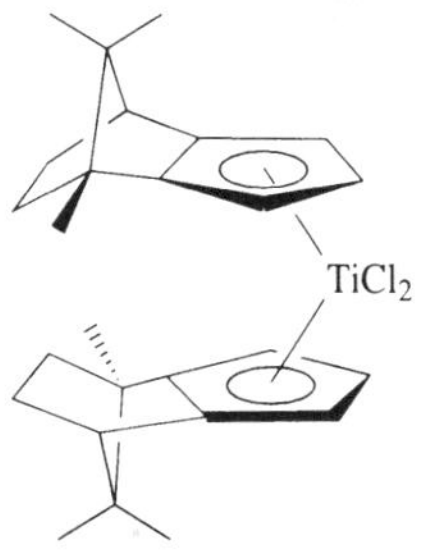   **(170)**

The isomerization of the vinylcyclohexane derivative (**171**) is catalysed by 2 mol.% of the chiral complex (1*R*,2*R*,4*R*,5*R*)-(**172**) in the presence of 8 mol.% of $LiAlH_4$, to give (**173**) (Equation (49)). At 180 °C, 44% *ee* is found at 100% conversion after 2 h, while at 23 °C, 76% *ee* is obtained, although at this temperature the reaction time for quantitative conversion had to be extended to 120 h.[302]

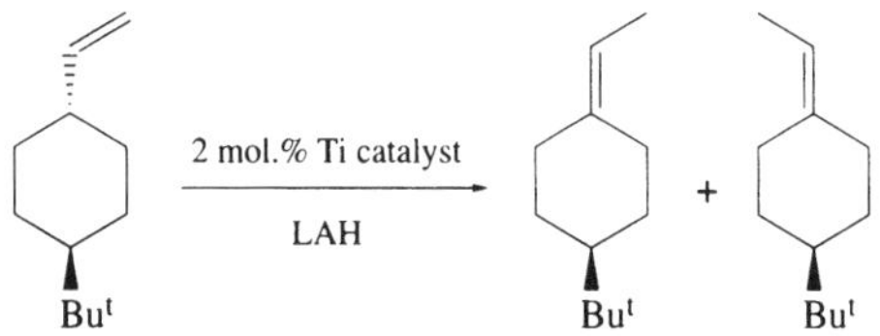

$$\text{(171)} \xrightarrow[\text{LAH}]{2\ \text{mol.\% Ti catalyst}} \text{(S)} + \text{(R)} \qquad (49)$$

(171)   (*S*)   (*R*)

(173)

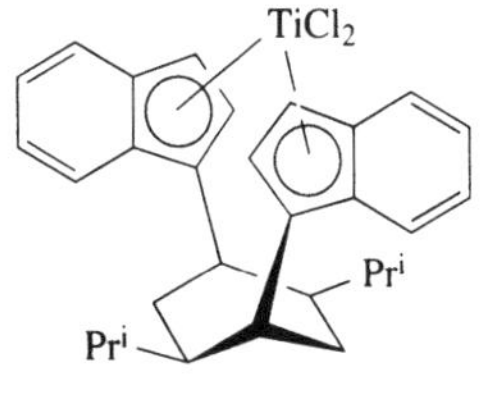

(172)

Bis(cyclopentadienyl) complexes are significantly less Lewis acidic than analogous mono(cyclopentadienyl) compounds and, hence, while the heteroatom substituents in the mono(cyclopentadienyl) complexes (**43**) and (**45**) are coordinated to the metal, no such coordination is found in [TiCl$_2$(η-C$_5$H$_4$CH$_2$OR)$_2$] and [TiCl$_2$\{PhAs(CH$_2$CH$_2$C$_5$H$_4$)$_2$\}].[102–4] On the other hand, coordination of a heteroatom substituent to the metal centre can be found if a chloride ligand is removed by reduction, for example, treatment of [TiCl$_2$\{η-C$_5$H$_4$(CH$_2$)$_2$PPh$_2$\}$_2$] with aluminum powder gives (**174**). The phosphino substituent can coordinate to late transition metal fragments to give the heterodinuclear complexes (**175**) (Scheme 41). Monosubstituted titanocenes have a similar chemistry and give a series of mixed-metal compounds (**176**) (ML$_n$ = Cr(CO)$_5$, Mo(CO)$_5$ or Fe(CO)$_4$), which can be reduced to the corresponding titanium(III) or titanium(II) compounds, as shown in Scheme 42.[240]

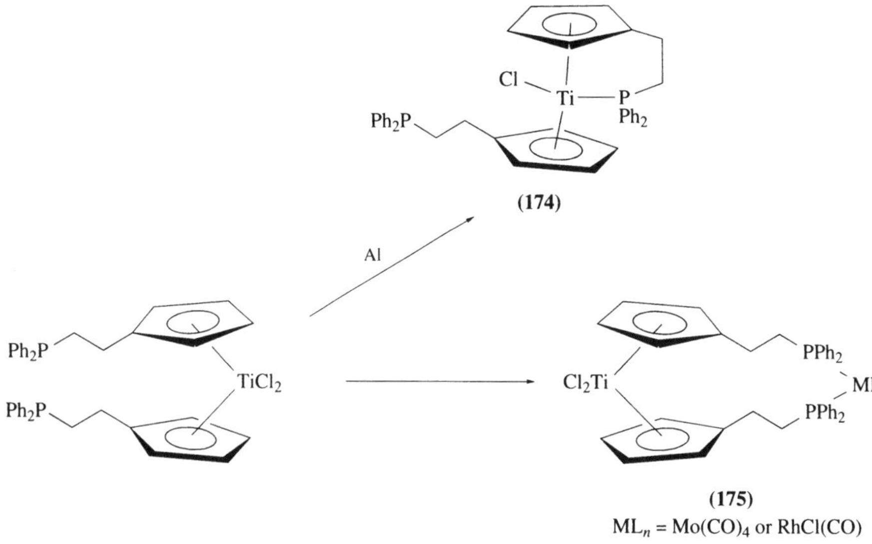

(**174**)

(**175**)
ML$_n$ = Mo(CO)$_4$ or RhCl(CO)

Scheme 41

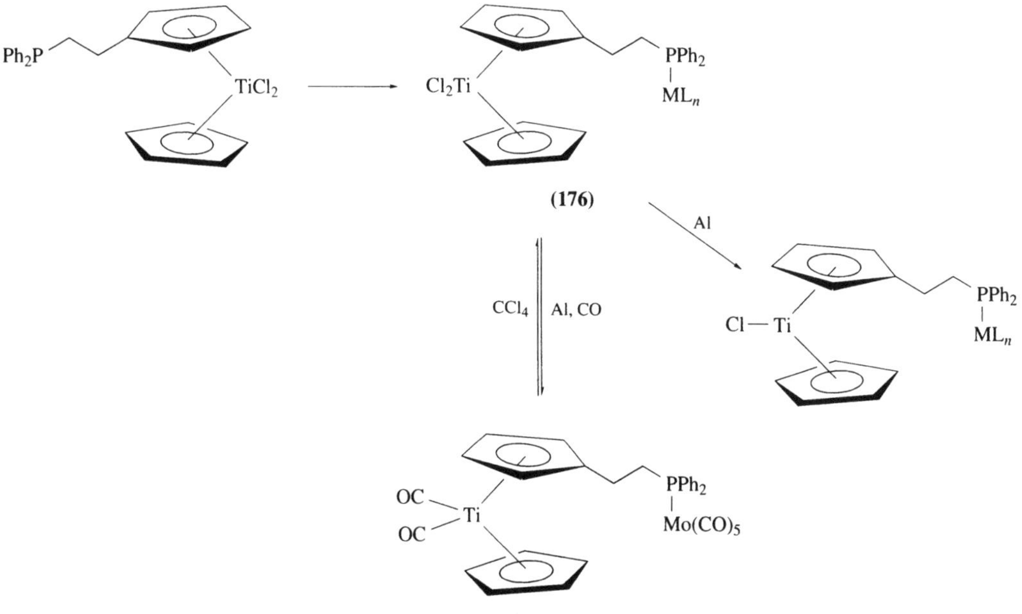

(**176**)

Scheme 42

The PPh$_2$-substituted complexes [TiCl$_2$Cp(η-C$_5$R$_4$PPh$_2$)] (**177**) (R = H or Me) also act as potential ligands. Compound (**177**) (R = H) forms the crystallographically characterized adduct [TiCl$_2$Cp(η-C$_5$R$_4$PPh$_2$ML$_n$)] (ML$_n$ = MnCp(CO)$_2$),[273] while (**177**) (R = Me) on treatment with [Cr(CO)$_5$(THF)] gives the analogous compound with ML$_n$ = Cr(CO)$_5$ as a grey-green powder. [TiCl$_2$(η-C$_5$Me$_4$PPh$_2$)$_2$] acts as a chelating ligand towards the $Mo(CO)_4$ fragment; the structure of the product (**178**) was determined by x-ray diffraction (Figure 33). As was discussed in the case of *ansa*-titanocenes, the P_2Mo moiety is tilted away from the plane bisecting the Cl–Ti–Cl angle in order to minimize steric interactions between the methyl groups and the chloride ligands. The cyclopentadienyl rings are staggered, and the Me groups are bent out of the C_5 plane as a result of steric hindrance.[268]

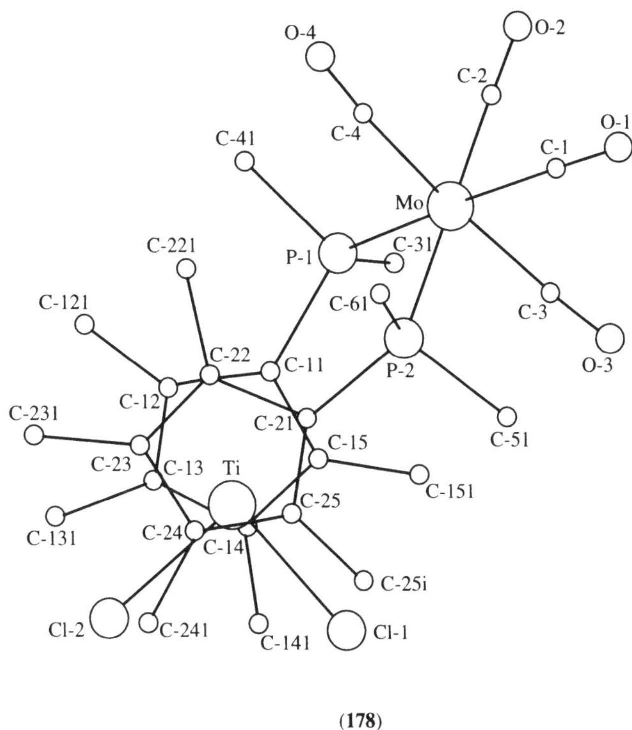

(178)

Figure 33 Top view of the molecular structure of $[TiCl_2\{\eta\text{-}C_5Me_4PPh_2\}_2\{Mo(CO)_4\}]$ (**178**). The phenyl rings have been omitted for clarity (reproduced by permission of Elsevier from *Inorg. Chim. Acta*, 1991, **180**, 153).

5.5.2 Complexes with Ti–C σ Bonds

5.5.2.1 Synthesis

The most frequently used method for synthesizing titanium alkyl, alkenyl, benzyl and aryl complexes is the metathetical ligand exchange between a titanocene dihalide and a lithium or magnesium reagent. In some cases, however, alternative routes have proved to be advantageous because of higher yields or greater selectivity. For example, the titanium(III) complexes $[TiX(Cp^*)_2]$ (X = Cl, OMe or N=C(H)Bu^i) react with $CdMe_2$ at −30 °C under oxidation to titanium(IV), to give selectively the mixed-ligand complexes $[Ti(X)Me(Cp^*)_2]$ and cadmium metal in 55–73% yield (Equation (50)). Similarly, $[Ti(Et)Cl(Cp^*)_2]$ is obtained from $[TiCl(Cp^*)_2]$ and $CdEt_2$ in 51% yield. The reactions in the cases of X = ethyl or vinyl proceed similarly but the products are thermally unstable. $[Ti(Me)(Et)(Cp^*)_2]$ is formed at −60 °C and decomposes to $[TiMe_2(Cp^*)_2]$ and $[Ti(\eta^2\text{-}C_2H_4)(Cp^*)_2]$. $[TiPh(Cp^*)_2]$ reacts with $CdMe_2$ only slowly at room temperature to give a mixture of products. $ZnMe_2$ instead of $CdMe_2$ reacts only slowly, and there is no reaction with $HgMe_2$.[365] Similarly, titanium(III) alkyls, vinyls, aryls and alkoxides are oxidized by $PbCl_2$ to give $[Ti(X)Cl(Cp^*)_2]$ (X = Me, Et, Pr^n, CH=CH$_2$, Ph or OPr^n) as red or orange crystalline products in high yields. The reaction also proceeds with $PbBr_2$. This reaction is cleaner than the alkylation of $[TiCl_2(Cp^*)_2]$ with one equivalent of a lithium reagent, and there is no tendency for the mixed-ligand products to disproportionate. The rotation of the phenyl ligand in $[Ti(Ph)Cl(Cp^*)_2]$ is hindered, and the *ortho* and *meta* carbon atoms are inequivalent in the ^{13}C NMR spectrum, even on heating to 90 °C. However, the reaction of $[TiX(Cp^*)_2]$ (X = CH$_2$Bu^t or CH$_2$Ph) with $PbCl_2$ only leads to Ti–C bond homolysis and formation of $[TiCl(Cp^*)_2]$.[366]

A series of sterically hindered titanocene dimethyl complexes $[TiMe_2Cp'Cp'']$ has been prepared by alkylation of the dichlorides with methyllithium (Cp' = Cp, Cp'' = $C_5H_3(TMS)_2$ or $C_5H_2(TMS)_3$;

$$[TiXCp^*_2] + 1/2\,CdMe_2 \xrightarrow{-30\,^\circ C} Cp^*_2Ti\begin{smallmatrix}Me\\X\end{smallmatrix} \tag{50}$$

$Cp' = C_5H_4(TMS)$, $Cp'' = C_5H_3(TMS)_2$ or $C_5H_2(TMS)_3$; $Cp' = C_5H_3(TMS)_2$, $Cp'' = C_5H_2(TMS)_3$; $Cp' = Cp' = C_5H_3(TMS)_2$.[250] The treatment of $[TiCl_2Cp'_2]$ with ArLi gives the aryl complexes $[TiAr_2Cp'_2]$ ($Cp' = Cp$ or C_5H_4Me; Ar = substituted phenyl).[367] The alkylation of $[TiCl_2(\eta\text{-}C_5HMe_4)_2]$ affords $[TiR_2(\eta\text{-}C_5HMe_4)_2]$ and the monoalkyls $[TiCl(R)(\eta\text{-}C_5HMe_4)_2]$, depending on the reactant ratio (R = Me, Ph or p-tolyl).[106] Ferrocenyllithium reacts with $[TiCl_2(\eta\text{-}C_5H_4Me)_2]$ to give $[Ti(Fc)_2(\eta\text{-}C_5H_4Me)_2]$ (Fc = 1-ferrocenyl) as dark green microcrystals.[138] The alkylation of the fulvalene complex (**118**) (Equation (35)) affords the compounds $[(TiR_2Cp)_2(\eta^5{:}\eta^5\text{-}C_{10}H_8)]$ (R = Me and CH_2TMS) in 75% and 60% yields, respectively.[277,278] The complexes are surprisingly insoluble in hexane and only slightly soluble in toluene, THF or dichloromethane. The solids are stable in air for several days. Alkylation attempts of (**118**) with $Mg(CH_2Ph)_2$ gave only dibenzyl, while no reaction was observed with R = p-tolyl, CH_2Bu^t or $N(TMS)_2$.[277]

Methylation of the *ansa*-titanocene complex (**150**) gives $[TiMe_2\{C_2H_4(C_5H_4)_2\}]$. In contrast with $[TiMe_2Cp_2]$ the complex is not hydrogenolysed on treatment with H_2. Exposure to CO leads to reduction to $[Ti(CO)_2\{C_2H_4(C_5H_4)_2\}]$ and the formation of acetone.[290] The reaction of MeMgI with the dimethylmethylene-bridged complex $[TiCl_2\{Me_2C(C_5H_4)_2\}]$ leads to the dark yellow complex $[TiMe_2\{Me_2C(C_5H_4)_2\}]$, which is considerably more stable than $[TiMe_2Cp_2]$. The crystal structure shows that the *ansa* ligand framework is strained, and the Cp–C–Cp angle is only 99°. Although the distances between the titanium atom and the cyclopentadienyl carbons increase towards the open end of the coordination wedge, there is no indication for η^3-bonding of the cyclopentadienyl rings.[368] The reaction of the related silyl-bridged complex $[TiCl_2\{Me_2Si(C_5H_4)_2\}]$ with MeLi gives $[TiMe_2\{Me_2Si(C_5H_4)_2\}]$ as red-orange crystals in 65% yield, whereas the alkylation with AlR_3 affords $[TiCl(R)\{Me_2Si(C_5H_4)_2\}]$ as red (R = Me, 54% yield) to dark red (R = CH_2TMS, 33%) solids. By treating $[TiCl_2\{Me_2Si(C_5H_4)_2\}]$ with magnesium dialkyls, benzylic complexes $[TiR_2\{Me_2Si(C_5H_4)_2\}]$ are obtained as violet (R = CH_2Ph) or dark green (R = $1,2\text{-}C_6H_4(CH_2)_2$) crystals in 76% and 44% yields, respectively.[369] The chiral *ansa*-titanocene dimethyl derivative (**179**) (X = Me) was obtained by alkylation of the dichloride with methyllithium. The crystal structure of the compound shows two distinctly different Ti–Me bond lengths. Mixtures of the parent dichloride with methylaluminoxane catalyse the polymerization of propene to give a polymer with thermoplastic properties. The structure of the polymer is thought to indicate the existence of two different propagation sites, for which the structure of the dimethyl complex is regarded as a model (see Section 5.5.2.5).[370,371]

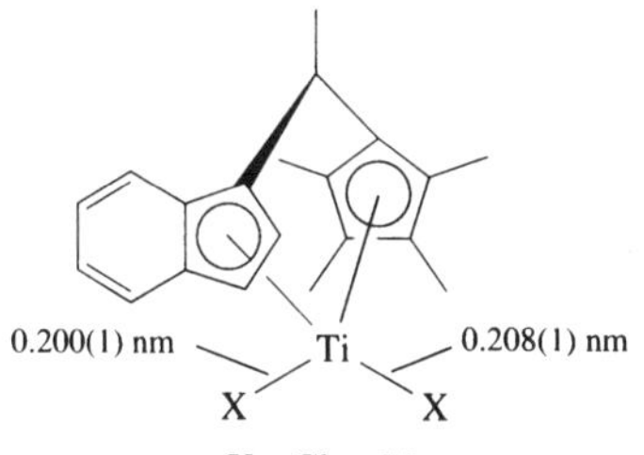

0.200(1) nm   Ti   0.208(1) nm

X   X

X = Cl or Me

(**179**)

The reaction of $[TiCl_2Cp_2]$ with one equivalent of 1-norbornyllithium at $-30\,^\circ$C in diethyl ether gives the red-brown complex $[Ti(R)ClCp_2]$ in 71% yield (R = 1-norbornyl). The addition of two equivalents of lithium reagent affords the red complex $[TiR_2Cp_2]$, which reacts with further LiR at $-90\,^\circ$C to give orange crystals of $Li[TiR_3Cp_2]$. The ate complex is thermally unstable and decomposes on warming to $0\,^\circ$C to the yellow compound $[TiR_3Cp]$ and LiCp. The solid titanium alkyls are air-stable, but solutions are slowly oxidized. The complexes react only slowly with water or alcohols.[372]

Complexes of alkyl ligands carrying functional groups can be prepared in a variety of ways. The reaction of $[TiCl_2Cp_2]$ with the dimagnesium reagent $CH_2(MgBr)_2$ at $-20\,^\circ$C affords $[Ti(CH_2MgBr)_2Cp_2]$ (**180**), which reacts with $[MCl_2Cp_2]$ to give the dark purple to brown bimetallic complexes (**181**) (M = Ti, Zr or Hf) in 31–36% yield. The complexes are characterized by ^1H NMR resonances for the μ-CH_2 ligands at δ 8.72 (M = Ti), 7.63 (M = Zr) and 6.26 (M = Hf).[373] Compound (**180**) reacts with $SiCl_4$ to give the red-purple, only slightly air-sensitive spiro compound (**182**) in 48% yield (Scheme 43). The complex possesses an unusually shielded silicon atom, which gives rise to a ^{29}Si NMR spectral signal at δ -145.3.[374]

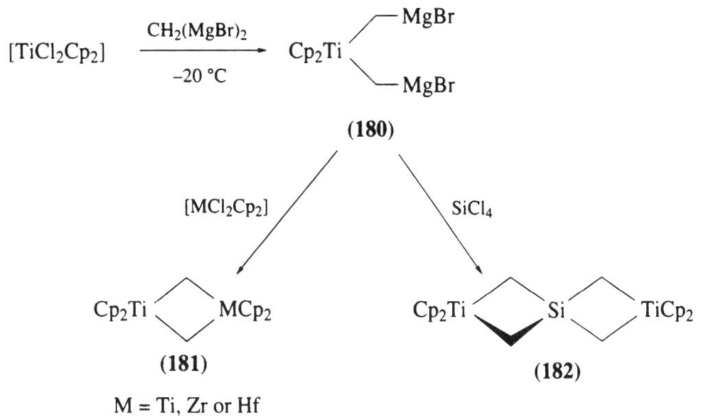

Scheme 43

Methoxy-substituted alkyl complexes are accessible by reacting $[TiCl_2(\eta\text{-}C_5H_4Me)_2]$ with $MgCl(CH_2OMe)$. Even in the presence of an excess of Grignard reagent only one of the chloride ligands is substituted. The crystal structure shows that in contrast with the analogous zirconium complex there is no coordination of the methoxy oxygen to the metal centre.[375] By contrast, the reaction of $[TiCl_2Cp_2]$ with $ICH_2CH_2CO_2Et$ in the presence of zinc powder gives the cationic homoenolate complex (**183**) as red crystals in 62% yield according to Equation (51). The structure was determined by x-ray diffraction. The alkyl group acts as a chelating ligand via the C=O group of the ester function.[128]

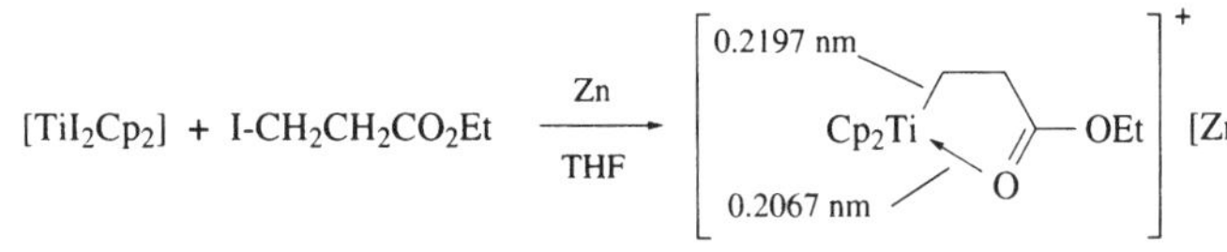

$$[TiI_2Cp_2] + I\text{-}CH_2CH_2CO_2Et \xrightarrow[\text{THF}]{Zn} \left[\begin{array}{c} 0.2197\ nm \\ Cp_2Ti \\ 0.2067\ nm \end{array} \quad OEt \right]^+ [ZnI_3]^- \qquad (51)$$

(**183**)

The reaction of $[TiCl_2Cp_2]$ with phosphorus ylids proceeds in two stages, first giving the ionic adduct $[TiCl(CH_2PMe_3)Cp_2]Cl$, which reacts with further $Me_3P=CH_2$ to give the red complex $[TiCl(CH=PMe_3)Cp_2]$.[376] The alkylation of $[TiCl_2Cp(Cp^*)]$ with two equivalents of $LiCH_2PPh_2$ leads to isolation of the red phosphinomethyl complex $[Ti(CH_2PPh_2)_2Cp(Cp^*)]$. The crystal structure confirms that the alkyl ligands are η^1-coordinate, without Ti–P interactions. The CH_2 hydrogens are inequivalent, and there is some indication of weak α-agostic bonding to one of them (Ti⋯H 0.246 nm).[377] The oxo-bridged complex (**184**) is similarly obtained and reacts with $[\{RhCl(CO)_2\}_2]$ to give the binuclear complex (**185**) (Scheme 44), whose *trans* structure is suggested on the basis of NMR spectral studies.[378]

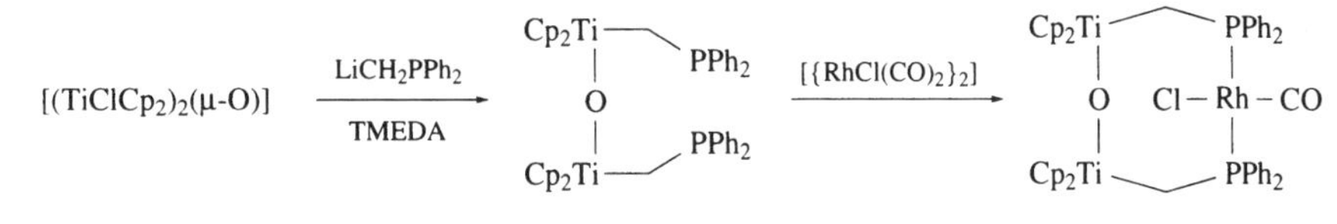

Scheme 44

Sulfur-containing alkyl complexes such as $[Ti(CH_2SPh)_2Cp_2]$ are accessible from $[TiCl_2Cp_2]$ and $LiCH_2SPh$. The moderately sensitive yellow-orange complex decomposes on heating to $100\,^\circ C$ in toluene, to give a mixture of the thiolato compounds $[Ti(SPh)_2Cp_2]$ and $[Ti(SPh)Cp_2]_2$.[379] $[TiCl_2Cp_2]$ reacts with 2-thienyllithium to give $[Ti(2\text{-}C_4H_3S)_2Cp_2]$.[380]

The reaction of $[TiCl_2(Cp^*)_2]$ with vinyllithium in toluene at $-30\,^\circ C$ gives the highly unstable $[Ti(CH=CH_2)_2(Cp^*)_2]$, which was spectroscopically characterized. The zirconium and hafnium analogues are more stable. The reaction in THF produces the thermally stable metallacycle (**186**) (Scheme 45). The crystal structure of (**186**) shows significant differences in the two Ti–C bond lengths and a very wide Ti–C=C angle α of $152.4(6)^\circ$. The compound decomposes on treatment with HCl in CCl_4 to liberate ethene and *cis*-2-butene, and with *n*-propanol to give 83% ethene and 17% 1-butene.[381,382]

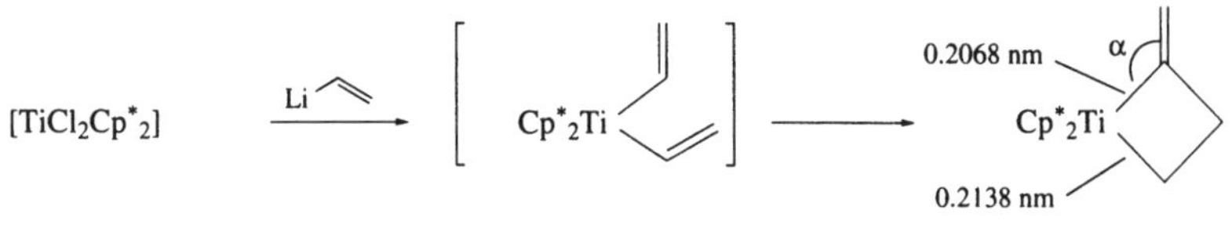

(186)

Scheme 45

The reaction of lithium acetylides LiC≡CR[2] with $[TiCl_2(\eta\text{-}C_5H_4R^1)_2]$ in diethyl ether gives first the orange complex $[TiCl(C_2R^2)(\eta\text{-}C_5H_4R^1)_2]$, which reacts further to give $[Ti(C_2R^2)_2(\eta\text{-}C_5H_4R^1)_2]$ (R^1 = H, R^2 = TMS). A series of acetylide complexes has been prepared (R^1 = H, Me or TMS; R^2 = Ph, Et, Pr^n, Bu^t or TMS). The addition of excess LiC≡CR[2] can lead to reduction to titanium(III). The $\nu_{C\equiv C}$ frequencies range from 2009 cm^{-1} (R^2 = TMS) to 2082 cm^{-1} (R^2 = Bu^t).[383]

A number of compounds containing metal complexes as σ-ligands have been prepared. The interaction of $[TiCl_2Cp_2]$ with the ferrocenyllithium derivative (**187**) gives (**188**) as thermally stable, dark green needles, while the alkylation of $[TiCl_2(Cp^*)_2]$ only leads to the substitution of one of the chloride ligands to give the green compound (**189**) (Scheme 46). The compounds are more stable than the simple ferrocene derivatives.[384] The titanium–chromium complexes (**190**) and (**191**) are accessible from $[Cr(C_6H_5RLi)(CO)_3]$ and $[TiCl_2Cp_2]$. Disubstitution could not be achieved in the case of R = *o*-F (Scheme 47). The more bulky complex $[Cr(C_6Me_5CH_2Li)(CO)_3]$ only gives the orange monoalkyl product (**192**) in low yield.[385]

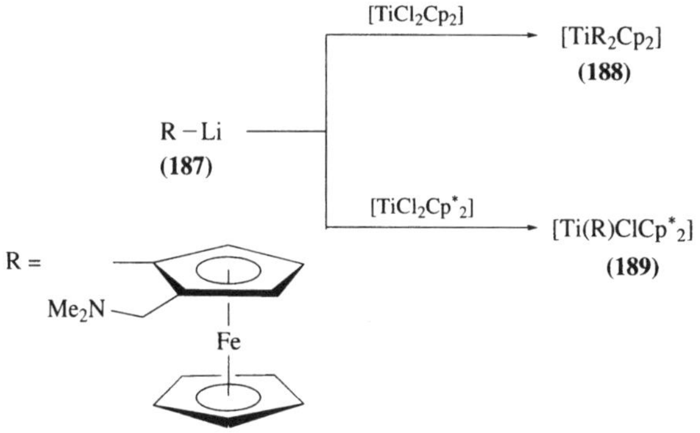

Scheme 46

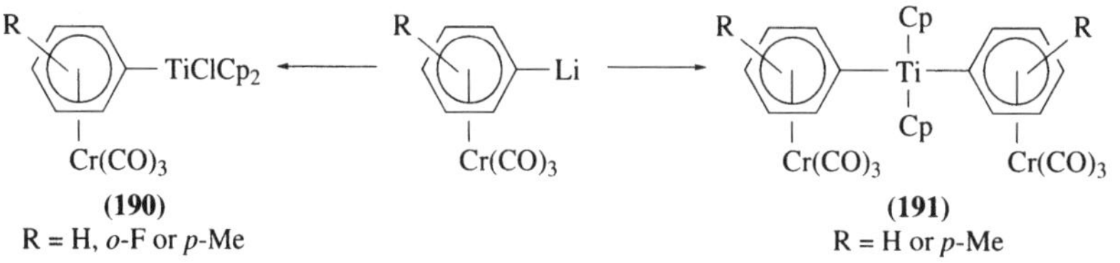

Scheme 47

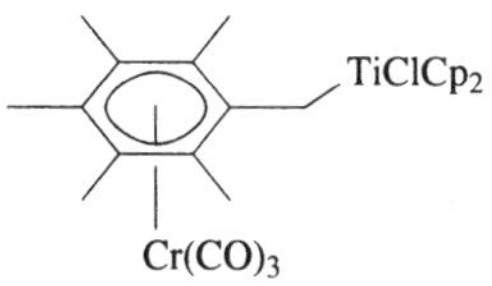

(192)

The reaction of $[TiCl_2Cp_2]$ with the lithiated allene $Li_2C{=}C{=}CPh_2$ in the presence of PMe_3 at 0 °C generates the dark purple allenylidene complex (**193**) in 71% yield (Scheme 48). Complex (**193**) is identified by $\nu_{C=C}$ at 1870 cm^{-1} and high-frequency ^{13}C NMR spectral resonances of δ 264.9, 168.3 and 94.0 for the α, β and γ carbons, respectively.[386] The reaction of $[TiCl_2Cp_2]$ with LiC≡CLi leads to the binuclear μ-acetylide complex (**194**), which can be reduced by magnesium in the presence of PMe_3 to give (**195**), whose molecular structure is best rationalized by a mixture of allenylidene and acetylide resonance forms (Scheme 48).[387]

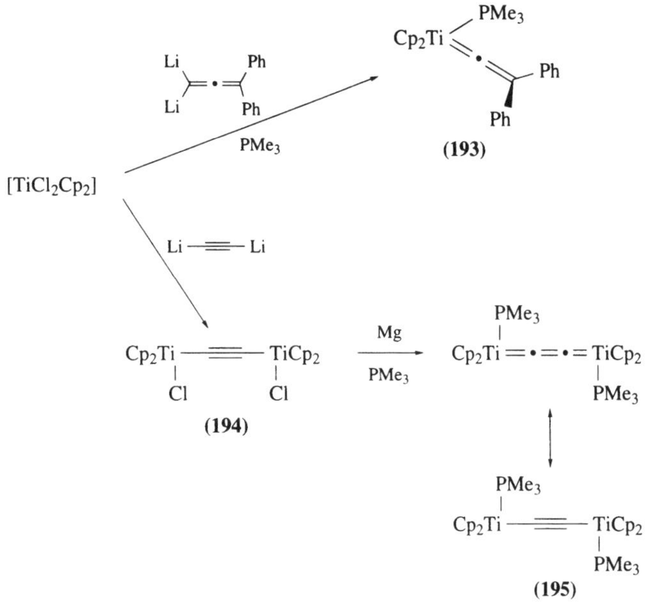

Scheme 48

An alternative method to complexes with titanium–carbon σ-bonds is the hydrolysis of alkene or alkyne complexes. The formation of the titanium alkyl and vinyl complexes **(196)**–**(198)** by this route is shown in Equations (52)–(54).[388–90] Complexes **(196)** and **(198)** were characterized by x-ray diffraction. The compounds contain almost linear Ti–O–Ti bridges with short (~0.184 nm) Ti–O bonds. The molecular structure of **(196)**[388] is shown in Figure 34.

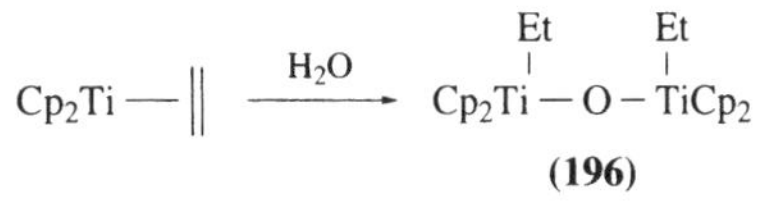

$$\text{Cp}_2\text{Ti}\!-\!\| \quad \xrightarrow{\;\text{H}_2\text{O}\;} \quad \underset{(196)}{\text{Cp}_2\text{Ti}\!-\!\text{O}\!-\!\text{TiCp}_2} \qquad (52)$$

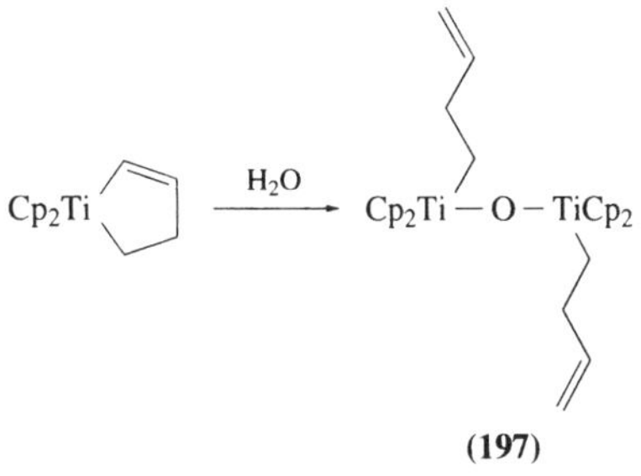

$$(53)$$

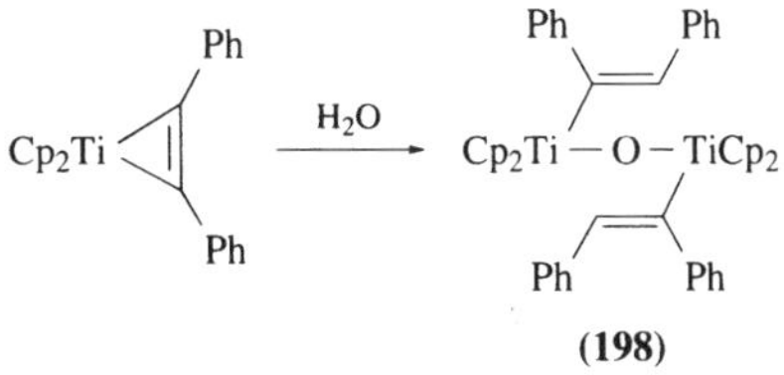

$$(54)$$

5.5.2.2 Structures and properties

A comprehensive review summarizes the M–H and M–C bond enthalpies of a large number of early transition metal complexes including alkyls of titanium. Most of these data are based on reaction solution calorimetry.[391] The mean bond enthalpies E(Ti–R) of [TiR$_2$Cp$_2$] for R = Ph and (C$_5$H$_4$)FeCp are 271 ± 9 kJ mol^{-1} and 262 ± 11 kJ mol^{-1}, based on E(Ti–Cl) for [TiCl$_2$Cp$_2$] of 430.5 ± 1.3 kJ mol^{-1}.[392] The first and the second Ti–Me bond dissociation enthalpies of [TiMe$_2$Cp$_2$] have been determined as

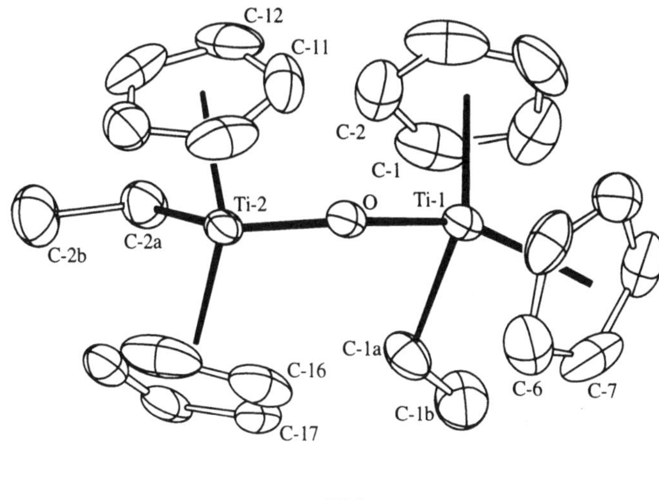

(196)

Figure 34   Molecular structure of [(TiEtCp₂)₂(μ-O)] (**196**) (reproduced by permission of Elsevier from *J. Organomet. Chem.*, 1988, **349**, C7).

$D_1 = 287$ kJ mol^{-1} and $D_2 = 309$ kJ mol^{-1}.[323] Similar values were found for [TiMe₂(Cp*)₂] ($D_1 = 271 \pm 21$ kJ mol^{-1}, $D_2 = 300 \pm 29$ kJ mol^{-1}) and [TiPh₂(Cp*)₂] ($D_1 = 269 \pm 28$ kJ mol^{-1}, $D_2 = 366 \pm 31$ kJ mol^{-1}).[324] The nature and position of the substituents (*o*-Me, *p*-Me, *p*-OMe or *p*-CF₃) in a series of titanocene diaryl complexes has little influence on the heats of formation and the mean Ti–C bond dissociation enthalpies.[393]

The J_{CH} coupling constants of the methyl ligands in [Ti(Me)ClCp₂], [Ti(Me)Cl(η-C₅HMe₄)₂], [Ti(Me)Cl(Cp*)₂] and a number of mixed-cyclopentadienyl complexes are almost identical (126–129 Hz), indicating little influence of the more or less electron-donating Cp ligands on the bonding of the methyl group.[106] The IR spectra of [TiR₂Cp₂] (R = Me, CHD₂ or CD₃) suggest that two C–H bonds in each Me group are stronger than the third, possibly as a result of steric hindrance between the ligands. This effect is absent in the zirconium and hafnium analogues. The v_{C-H} frequency can be correlated with the bond dissociation energy, and an increase of 76–143 kJ mol^{-1} is predicted on going from [TiMe₂Cp₂] to [HfMe₂Cp₂].[394]

The x-ray crystal structure of [Ti(CH₂Ph)₂Cp₂] has been determined. The benzyl ligand is η¹-bonded, with Ti–C bond lengths of 0.221–0.224 nm and Ti–C–C angles of 120.8° and 123.1°. In agreement with this, the *ipso* carbon of the phenyl group gives rise to a typical high-frequency ¹³C NMR spectral signal at δ 153.7.[395] An η² coordination would have shifted this value by ~30 ppm to lower frequency. Slightly shorter Ti–C bonds of 0.218 nm are found in the structure of [Ti(CH₂TMS)₂{Me₂Si(C₅H₄)₂}]. Although it was argued[315] that the parent compound [TiCl₂{Me₂Si(C₅H₄)₂}] possesses η³-bonded Cp⁻ ligands, no evidence for this was found in the alkyl complex.[396] The styryl complex (**199**) (X = 0.5 Cl or 0.5 Br) possesses a relatively small Ti–C–C angle of 105.7° (100.3° for the zirconium analogue), suggestive of a bonding interaction between the metal centre and the *ipso* carbon of the phenyl ring, similar to an η²-coordinate benzyl ligand.[397]

(199)

The ¹³C NMR spectral chemical shifts of a series of acetylide complexes [Ti(C≡CR)₂Cp₂] (R = Me, Ph or TMS) have been reported. Whereas the methyl and phenyl derivatives behave normally, the TMS compound shows a significant deshielding of the α carbon. This is thought to be the result of the stabilization of the metal-centred LUMO, which increases the paramagnetic shielding contribution and enhances ligand-to-metal charge transfer.[398] *Ab initio* (STO-3G basis set) calculations on [Ti(C≡CR)₂Cp₂] (R = H or Me) and the titanacyclopentadiene complex [Ti(C₄H₄)Cp₂] show extensive orbital mixing between filled acetylide π and bonding Cp orbitals but little evidence for π-donation from filled acetylide to vacant metal orbitals. In the metallacycle there is very little mixing between the Cp_2Ti bonding orbitals and the π-orbitals of the TiC_4H_4 ring.[399]

Solid-state ^{13}C cross-polarization magic-angle spinning (Cp–MAS) NMR spectra of $[Ti(\eta^1\text{-}Cp)_2(\eta^5\text{-}Cp)_2]$ show that sigmatropic rearrangements occur in the solid state over a temperature range of 182–347 K. The main pathway is a [1,2] shift, with [1,3] shifts possibly as minor pathways. Lineshape analysis gives an Arrhenius activation energy for the [1,2] shift of 33.2 ± 1.0 kJ mol^{-1}. The activation barrier is similar to that suggested by solution NMR spectral. However, in contrast with the behaviour in solution, interconversion between the η^1- and η^5-coordinate Cp$^-$ ligands seems to occur only at the highest temperatures possible before the onset of decomposition.[400] MO calculations on $[Ti(\eta^1\text{-}Cp)Cl(\eta^5\text{-}Cp)_2]$ suggest that the fluxional process involves η^3 coordination of two Cp$^-$ ligands in a transition state which is destabilized by 78.6 kJ mol^{-1} (Scheme 49).[401] Geometry optimizations confirm that the 16-electron structure $[Ti(\eta^1\text{-}Cp)Cl(\eta^5\text{-}Cp)_2]$ is 182 kJ mol^{-1} more stable than a 20-electron formulation $[TiCl(\eta^5\text{-}Cp)_3]$. In $[TiCp_4]$ also, electronic factors strongly favour $[Ti(\eta^1\text{-}Cp)_2(\eta^5\text{-}Cp)_2]$ over $[Ti(\eta^1\text{-}Cp)(\eta^5\text{-}Cp)_3]$. For $[TiCp_3]$ the structure $[TiCp_2(\eta^2\text{-}Cp)]$ is predicted for the ground state, as was found experimentally .[402]

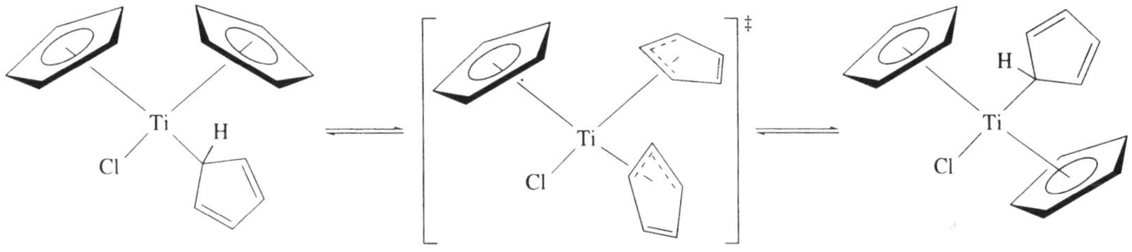

Scheme 49

5.5.2.3 Reactions

Titanium alkyls and aryls $[TiR_2Cp_2]$ (R = Me, Ph or p-tolyl) catalyse the hydrogen–deuterium exchange of arenes C_6H_5X (X = H, Me, OMe or F) at 130 °C. Deuteration of toluene occurs at the α position as well on the ring. The relative rate of exchange depends only slightly on R.[403] The thermal decomposition of $[TiMe_2(Cp^*)_2]$ proceeds according to Equation (55) to give the fulvene complex (**200**). The decomposition follows first-order kinetics, and experiments with the Me, CD_3 and $C_5(CD_3)_5$ complexes confirm that the reaction is intramolecular and that the methane is produced by the coupling of an Me$^-$ ligand with a hydrogen or deuterium atom from the other methyl group, generating a methylidene intermediate, $[Ti(=CH_2)(Cp^*)_2]$. The decomposition of $[Ti(Me)_2(Cp^*)_2]$ is 2.9 times faster at 98 °C than the thermolysis of $[Ti(CD_3)_2(Cp^*)_2]$, but no isotope effect is observed with $[Ti(Me)_2\{\eta\text{-}C_5(CD_3)_5\}_2]$. The measured activation enthalpy of $\Delta H^{\ddagger} \sim 117$ kJ mol^{-1} compares with a Ti–C bond enthalpy of ~ 251 kJ mol^{-1} and indicates that no Ti–C bond is completely broken during the rate-determining step.[404]

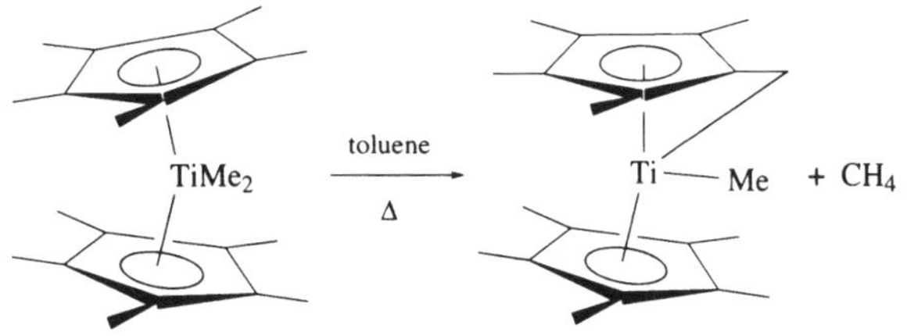

(55)

Heating the complex $[TiMe_2(\eta\text{-}C_5HMe_4)_2]$ to 130 °C induces a similar thermolysis process and forms a fulvene complex as a mixture of four isomers. Complexes with less highly substituted Cp$^-$ ligands do not give fulvene derivatives, possibly because the titanium methylidene intermediates are less stable and decompose reductively. The treatment of $[TiCl_2Cp'_2]$ (Cp' = C_5HMe_4 or Cp*) with LiAlH$_4$ followed by thermolysis in toluene gives the dimethylenecyclopentadienyl complex $[Ti\{\eta^5\text{-}C_5HMe_2(CH_2)_2\}\{\eta\text{-}C_5HMe_4\}]$ and its pentamethylcyclopentadienyl analogue.[405]

The processes of decomposition of bis(pentamethylcyclopentadienyl)titanium alkyls have been studied in detail.[366] The reaction of [TiCl(CD$_2$Me)$_2$\{η-C$_5$(CD$_3$)$_5$\}$_2$] with MeLi does not give [Ti(Me)(CD$_2$Me)$_2$\{η-C$_5$(CD$_3$)$_5$\}$_2$] but produces the ethene complex [Ti(η^2-CD$_2$CH$_2$)\{η-C$_5$(CD$_3$)$_5$\}$_2$] and CH$_4$. No MeD could be detected, that is, an α-H(D) activation process is not involved. The vinyl complex [TiCl(CH=CH$_2$)(Cp*)$_2$] reacts with lithium reagents LiR under elimination of RH to give presumably the allenylidene complex (201) as a transient species which readily reacts with the RH = ethene or propyne to give the metallacycle (186) or (202), respectively. The comparatively stable fulvene complex (203) is formed from (201) by hydrogen abstraction from a pentamethylcyclopentadienyl ligand, and on treatment of [TiCl(CH=CH$_2$)(Cp*)$_2$] with potassium benzyl (Scheme 50). Only in the case of R = Me can the formation of an alkyl–vinyl complex be observed, although at room temperature in benzene solution decomposition to (203) is facile. Deuteration studies show that this reaction proceeds according to Equation (56), evidently via [Ti(=C=CD$_2$)(Cp*)$_2$] as the intermediate. The decomposition of [Ti(Me)(Ph)(Cp*)$_2$] requires heating to 80 °C to give the phenyl analogue of (203) and methane. Deuteration of the phenyl group suggests that the intermediate in this case is a benzyne complex, which can be trapped by CO$_2$ (Scheme 51). The reaction of [TiCl(Me)(Cp*)$_2$] with KCH$_2$Ph leads to the only isolable mixed-ligand benzyl complex in this system, [Ti(Me)(CH$_2$Ph)(Cp*)$_2$], which decomposes on heating to 80 °C to an 85:15 mixture of [Ti^{III}(Me)(Cp*)$_2$] and (200).[360]

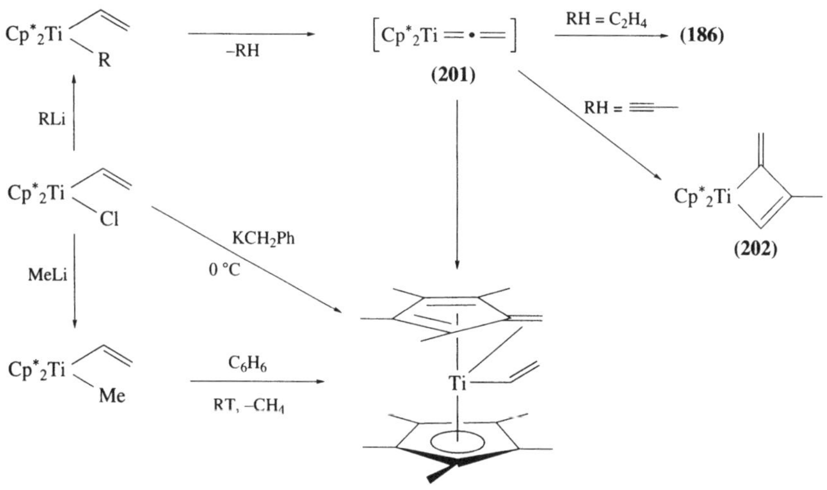

Scheme 50

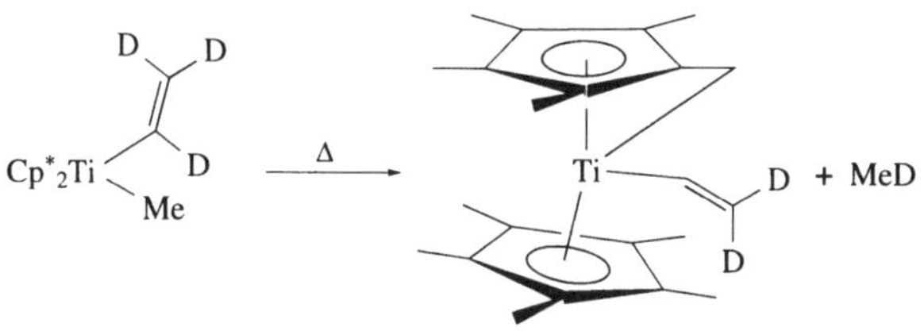

$$(56)$$

The photolysis of [TiMe$_2$Cp$_2$] or [TiMeClCp$_2$] generates methyl radicals which react with halide complexes such as [MnBr(CO)$_5$] under halide–methyl exchange to give [MnMe(CO)$_5$]. [ReBr(CO)$_5$], [MnBr(CO)$_3$(PEt$_3$)$_2$] and [FeCl$_2$(CO)$_2$(PMe$_3$)$_2$] react similarly. The metal–metal bonds, for example, in [Mn$_2$(CO)$_{10}$] and [Fe$_2$Cp$_2$(CO)$_4$], are cleaved to give the corresponding methyl complexes, while oxidative addition is observed with iron(0) and iridium(I) compounds.[406] The photolysis of [TiMe$_2$Cp$_2$] by tungsten light (λ = 300–700 nm) in the presence of CO$_2$ affords the acetato complex [Ti(Me)(O$_2$CMe)Cp$_2$] in 80% yield after 20 h of irradiation. There is no reaction with CO$_2$ in the dark.[407] On irradiation, [TiMe$_2$(η-C$_5$HMe$_4$)$_2$] catalyses the isomerization of 1-alkenes to (*E*)-2-alkenes. Prolonged irradiation gives only *cis–trans* isomerization but no other products. The reaction also proceeds thermally but gives (*E*)/(*Z*) mixtures. ESR spectroscopy suggests that titanium(III) species are formed during this process.[408] On flash photolysis of [Ti(C$_6$F$_5$)$_2$Cp$_2$] in a cold matrix the orange compound is reversibly converted into a blue isomer that is highly reactive towards CO, O$_2$, N$_2$ and MeCN. It is thought that an η^5 → η^1 rearrangement opens up the coordination sphere (Equation (57)).[409]

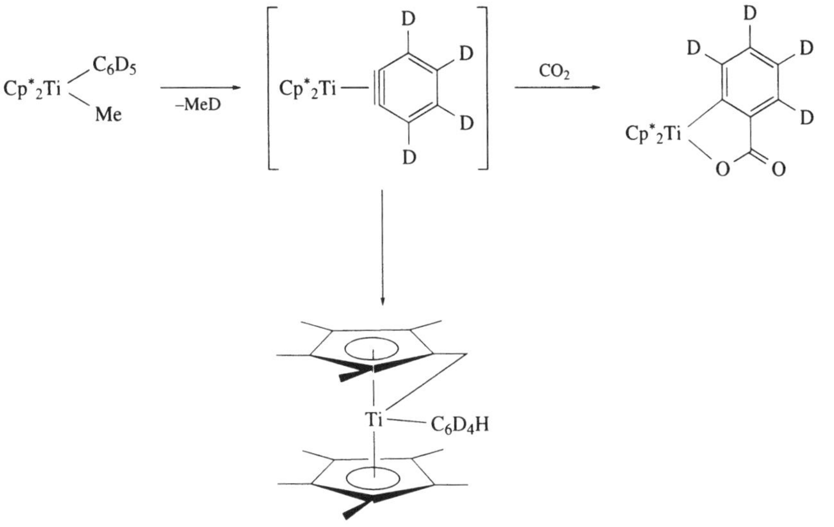

Scheme 51

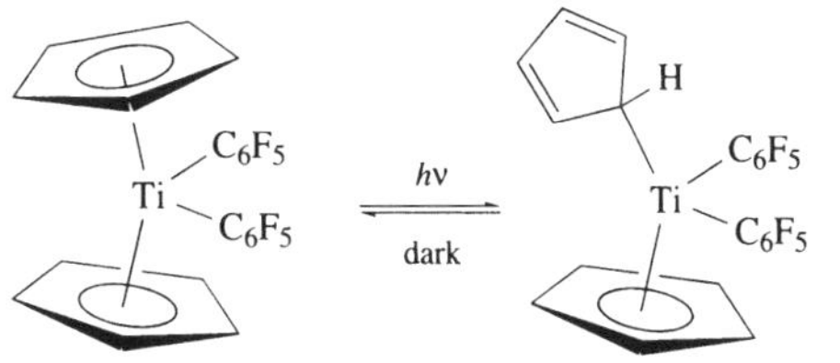

$$[\text{TiMe}_2\text{Cp}_2] + cis\text{-}[\text{PtCl}_2(\text{PMePh}_2)_2] \longrightarrow [\text{TiMeClCp}_2] + trans\text{-}[\text{Pt(Me)Cl}(\text{PMePh}_2)_2] \qquad (58)$$

(57)

The chemical or electrochemical reduction of $[\text{Ti}(\text{CH}_2\text{PPh}_2)_2\text{Cp}_2]$ leads to the liberation of a cyclopentadienyl anion and the formation of the titanium(III) complex $[\text{Ti}(\text{CH}_2\text{PPh}_2)_2\text{Cp}]$ in which the alkyl ligands are η^2-coordinate via the phophorus atoms.[410] The thermolysis of $[\text{Ti}(\text{CH}_2\text{Ph})_2\text{Cp}_2]$ in CCl_4 produces $[\text{TiCl}_2\text{Cp}_2]$, $\text{PhCH}_2\text{CCl}_3$ and $\text{PhCH}_2\text{CH}_2\text{Ph}$. The NMR spectral signals for the titanium benzyl and $\text{PhCH}_2\text{CCl}_3$ show a CIDNP effect consistent with radical formation and a cage recombination reaction. $\text{PhCH}_2\text{CCl}_3$ is formed by diffusive reaction of benzyl and CCl_3 radicals. The initiating role of CCl_4 is not clear.[411]

The complexes $[\text{TiR}_2\text{Cp}_2]$ (R = Me or Ph) react with $[\text{TiCl}_2\text{Cp}_2]$ to give the conproportionation products $[\text{Ti(R)ClCp}_2]$. There is little difference in the rates between the two. The mixed-ligand complex $[\text{Ti(Me)(Ph)Cp}_2]$ reacts with electrophiles such as HCl, HOAc, $[\text{HgCl}_2]$ and $[\text{HgMeCl}]$ under Ti–C bond cleavage; there is little selectivity. The mechanism does not seem to follow a classical S_E2 type mechanism but probably involves electron transfer. $[\text{TiMe}_2\text{Cp}_2]$ acts as a methylating agent towards noble metal complexes (cf. Equation (58)).[412]

The hexenyl complex (**204**) in the presence of AlEtCl_2 at $-78\,^{\circ}\text{C}$ cyclizes under insertion of the C=C bond into the Ti–C bond, possibly via a zwitterionic intermediate such as (**205**), to give the cyclopentylmethyl complex (**206**). The alkyl ligand can be detached from the metal centre by hydrolysis to give the alkane, or by treatment with *N*-bromosuccinimide to give the organic bromide (Scheme 52). This reaction has been applied to the synthesis of a number of alkyl-substituted cyclopentane derivatives.[413] The stereochemistry of the cyclic products provides information about the conformation of the ligand in the transition state (Scheme 53). The reaction proceeds with complete regioselectivity to give five-membered rings; cyclohexane derivatives are not formed.[414]

The protolysis of $[\text{TiMe}_2\text{Cp}_2]$ with one equivalent of $\text{H}[\text{BF}_4]\cdot\text{Et}_2\text{O}$ leads to the liberation of CH_4 and the precipitation of $[\text{Ti(Me)Cp}_2(\text{BF}_4)]$ as an insoluble orange solid which contains coordinated $[\text{BF}_4]^-$ anions. Treatment of this solid with THF or pyridine leads to decomposition and the formation of $[\text{TiF}_2\text{Cp}_2]$. The protolysis of $[\text{TiMe}_2\text{Cp}_2]$ with $[\text{NH}_4]\text{X}$ affords the ammine complexes $[\text{TiMe}(\text{NH}_3)\text{Cp}_2]^+\text{X}^-$ (X = PF_6 or ClO_4), which dissolve in acetonitrile to give $[\text{TiMe}(\text{MeCN})\text{Cp}_2]^+\text{X}^-$ (**207**) (Scheme 54). The analogous tetraphenylborate salt (**207**) (X = BPh_4) is accessible from $[\text{TiMeClCp}_2]$ and $\text{Na}[\text{BPh}_4]$. Related phosphine, nitrile and pyridine complexes (**208**) are prepared

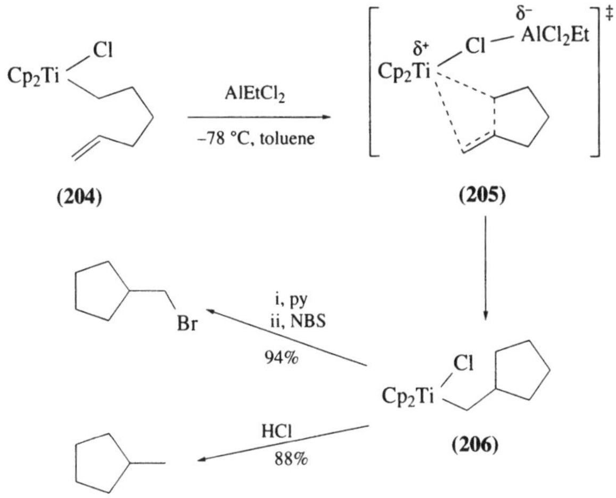

(204)   (205)

Scheme 52

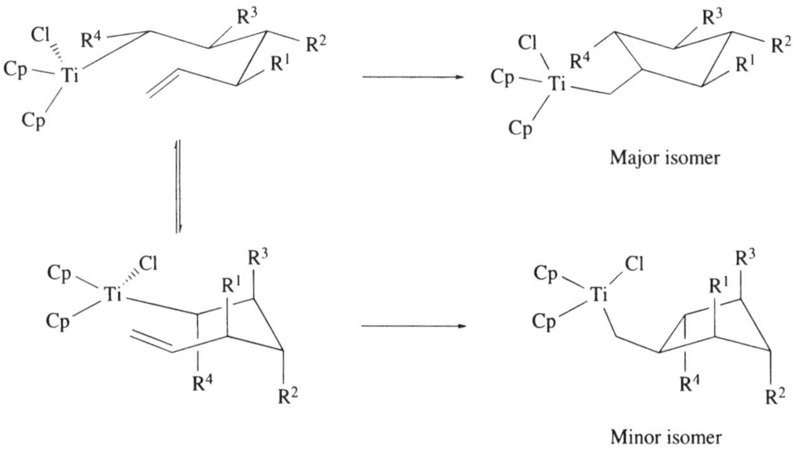

Major isomer

Minor isomer

Scheme 53

similarly. In contrast with the zirconium analogue, $[TiMe(pyridine)Cp_2]^+$ does not undergo *ortho*-metallation of the pyridine ligand. Compound (**207**) reacts with CO and Bu^tNC to give η^2-acyl and iminoacyl complexes, respectively. The structure of the iminoacyl complex (**209**), a rare case of an isocyanide complex of a d^0 metal, was determined by x-ray crystallography. A number of bis(indenyl) complexes of the type (**208**) ($Cp = \eta^5$-indenyl; $L = MeCN$, $PhCN$, Bu^tCN or PMe_3) were obtained in the same way. The phosphine complex $[TiMe(PMe_2Ph)Cp_2]^+$ fails to react with CO to give an acyl compound, although the expected product, $[Ti(\eta^2\text{-}COMe)(PMe_2Ph)Cp_2]^+$, is accessible by reacting $[Ti(\eta^2\text{-}COMe)(MeCN)Cp_2]^+$ with PMe_2Ph.[415,416] Compound (**207**) reacts with TMS-CN, which is in equilibrium with TMS-NC, to give $[Ti\{\eta^2\text{-}C(Me)=N(TMS)\}(CN(TMS))Cp_2]^+$, whose structure resembles that of (**209**).[417]

Cationic nitrile complexes of the general formula (**210**) undergo alkyl group migrations to give the azavinylidene complexes (**211**). The kinetics of the reaction depend on the substituent R ($R = Ph > Pr^n > Bu^t > Me$) and on the nature of the cyclopentadienyl ligand, with indenyl complexes being more reactive than Cp compounds, but are independent of [RCN]. The data suggest the formation of a 16-electron intermediate in the rate-limiting step, which is stabilized by coordination of another nitrile ligand to give the 18-electron product (**211**) (Scheme 55). The molecular structure of the benzonitrile complex $[Ti\{N=C(Me)Ph\}(\eta\text{-}C_9H_7)_2(PhCN)][BPh_4]$ has been determined (Figure 35). The Ti=N=C arrangement of the azavinylidene unit is linear. The short Ti–N bond of 0.1872(7) nm indicates double bond character through interaction of the nitrogen lone pair with a vacant metal-centred orbital; the whole arrangement is best described as a heteroallene structure. By contrast, the Ti–N distance to the nitrile donor ligand is rather long, 0.2175(7) nm.[417]

An alternative route to cationic nitrile complexes and their insertion products is the oxidation of the benzyl complexes $[Ti(CH_2Ph)_2Cp_2]$ with $Ag[BPh_4]$ in acetonitrile. The reaction produces a mixture of

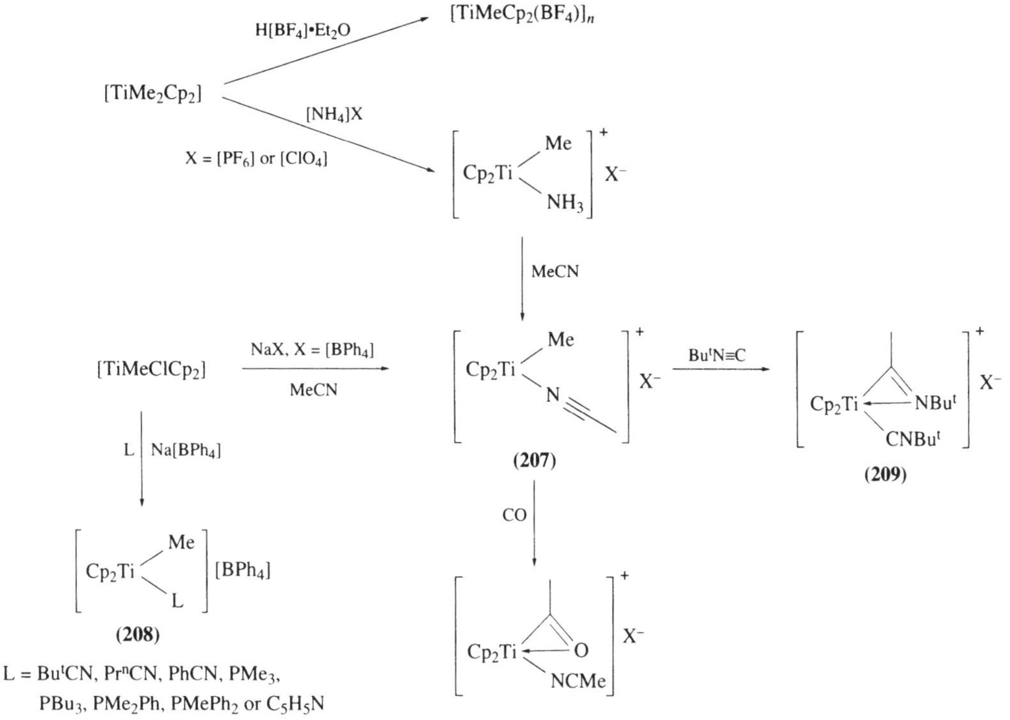

$[TiMeCp_2(BF_4)]_n$

$H[BF_4] \cdot Et_2O$

$[TiMe_2Cp_2]$

$[NH_4]X$

$X = [PF_6]$ or $[ClO_4]$

$\left[Cp_2Ti\begin{smallmatrix}Me\\NH_3\end{smallmatrix} \right]^+ X^-$

MeCN

$[TiMeClCp_2]$

$NaX, X = [BPh_4]$

MeCN

$\left[Cp_2Ti\begin{smallmatrix}Me\\N\end{smallmatrix} \right]^+ X^-$

(207)

$Bu^tN\equiv C$

$\left[Cp_2Ti\begin{smallmatrix}NBu^t\\CNBu^t\end{smallmatrix} \right]^+ X^-$

(209)

$L \mid Na[BPh_4]$

$\left[Cp_2Ti\begin{smallmatrix}Me\\L\end{smallmatrix} \right][BPh_4]$

(208)

$L = Bu^tCN, Pr^nCN, PhCN, PMe_3,$
$PBu_3, PMe_2Ph, PMePh_2$ or C_5H_5N

CO

$\left[Cp_2Ti\begin{smallmatrix}O\\NCMe\end{smallmatrix} \right]^+ X^-$

Scheme 54

$\left[Cp'_2Ti\begin{smallmatrix}Me\\N\end{smallmatrix} \right]^+$

(210)

$\left[Cp'_2Ti=N\begin{smallmatrix}\\R\end{smallmatrix} \right]^+$

$R-\equiv N$

$\left[Cp'_2Ti\begin{smallmatrix}N\\N\end{smallmatrix}\begin{smallmatrix}R\\R\end{smallmatrix} \right]^+$

(211)

$Cp' = Cp$ or Ind
$R = Me, Pr^n, Bu^t$ or Ph

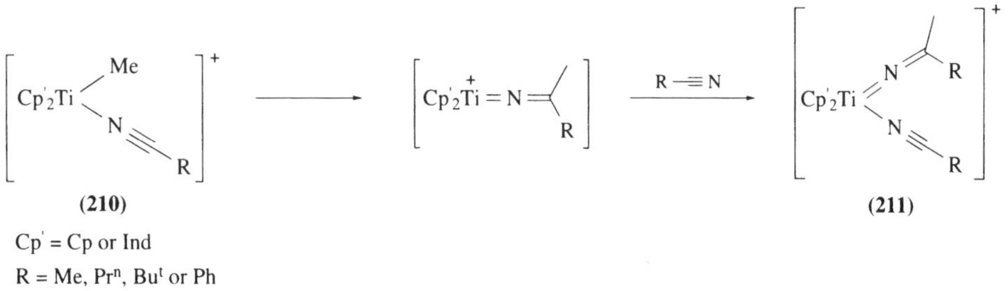

Scheme 55

two isomeric azavinylidene complexes (**212**) in low yield, besides the titanium(III) complex $[TiCp_2(MeCN)_2]^+$, which was isolated in 44% yield (Scheme 56). The latter can be recrystallized from THF to give $[TiCp_2(THF)(MeCN)]^+$. If the oxidation with $Ag[BPh_4]$ is carried out with mixtures of $[Ti(CH_2Ph)_2Cp_2]$ and $[Ti(CD_2C_6D_5)_2Cp_2]$, statistical mixtures of all possible dibenzyl by-products are obtained, confirming the involvement of free benzyl radicals. The reaction between $[Ti(CH_2Ph)_2Cp_2]$ and $[Fe(\eta\text{-}C_5H_4Me)_2][BPh_4]$ in MeCN is slower than the oxidation with Ag^+, but leads to a similar reaction mixture. In THF the reaction affords $[TiCp_2(THF)_2]^+$; unlike the zirconium analogue the intermediate $[Ti(CH_2Ph)Cp_2(THF)]^+$ is not isolable. The same result is obtained if $[Ti(CH_2Ph)_2Cp_2]$ is protolysed with $[NHMe_3][BPh_4]$ in THF. The oxidation of $[Ti(CH_2Ph)_2Cp_2]$ with $Ag[CB_{11}H_{12}]$ leads to the thermally labile complex $[Ti(CH_2Ph)(CB_{11}H_{12})Cp_2]$ in which the benzyl ligand is probably η^1-coordinate. The nature and extent of the anion coordination is uncertain.[418]

Cationic alkyl complexes of group 4 metals such as (**207**) have been discussed as models for the catalytically active species in metallocene-based Ziegler–Natta-type alkene polymerization reactions. This species is now thought to be a 14-electron cation such as $[TiRCp_2]^+$.[415-17] However, because of the strong coordination of donor ligands such as nitriles, pyridine or phosphines, ligand-stabilized complexes of the type (**207**) fail to undergo insertion reactions with alkenes and alkynes. More labile complexes are obtained by the protolysis of $[TiMe_2Cp_2]$ with $[NHMe_2Ph][BPh_4]$ in ethers to give $[TiMeCp_2(L)][BPh_4]$ (L = THF, Et_2O or PhOMe) as red-orange to brown solids (Scheme 57 and Equation 59). The complexes polymerize ethylene at room temperature with moderate activity (see Section 5.5.2.5).[419] The thermally labile and catalytically inactive donor-free complex

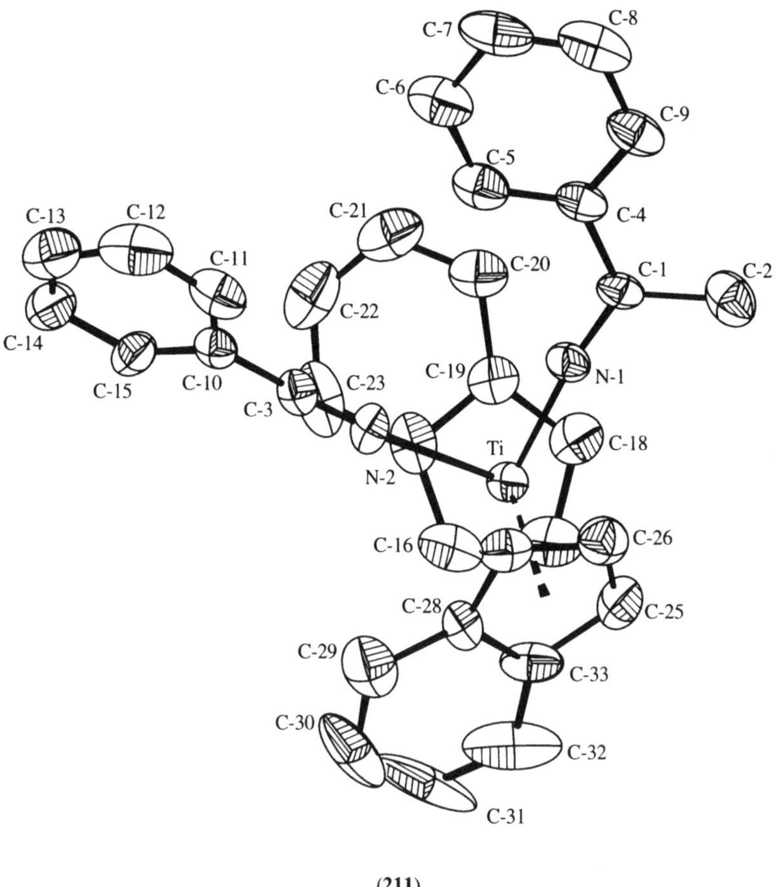

(211)

Figure 35 Molecular structure of [Ti(N=CMePh)(η-Ind)$_2$(NCPh)][BPh$_4$] **(211)** (reproduced by permission of the American Chemical Society. Copyright (1988) *Organometallics*, 1988, **7**, 1148).

Scheme 56

[TiMe(Cp*)$_2$][BPh$_4$] is obtained similarly as an orange-brown solid from dichloromethane solution (Scheme 57),[420] whereas the THF adduct [TiMe(Cp*)$_2$(THF)][BPh$_4$] is obtained by oxidizing [TiMe(Cp*)$_2$] with Ag[BPh$_4$] in THF (Equation (59)).[421] The crystal structure of this compound was determined. In contrast with the related zirconium complex [ZrMeCp$_2$(THF)][BPh$_4$][422] the THF ligand in [TiMe(Cp*)$_2$(THF)][BPh$_4$] lies in the C–Ti–O plane (Figure 36), no doubt as the result of steric hindrance from the Cp* ligands. In this conformation π donation of an oxygen lone pair to a vacant titanium orbital is not possible, and as a result the Ti–O bond (0.2154(6) nm) is longer than the Ti–C bond (0.1988(10) nm), whereas the opposite is the case for the zirconium complex.

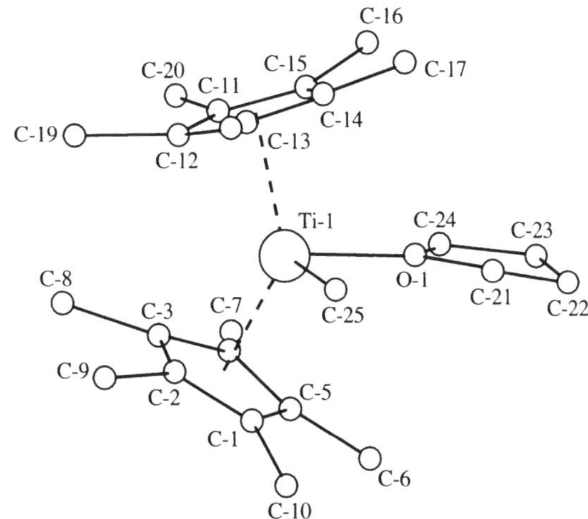

Scheme 57

$$[\text{TiMeCp}^*_2] \xrightarrow[\text{THF}]{\text{Ag[BPh}_4]} \left[\text{Cp}^*_2\text{Ti} \diagdown \begin{array}{c} \text{Me} \\ \text{THF} \end{array} \right] [\text{BPh}_4] \tag{59}$$

Figure 36 Molecular structure of [TiMe(Cp*)$_2$(THF)][BPh$_4$] (reproduced by permission of Elsevier from *Polyhedron*, 1989, **8**, 1838).

The protolysis of the metallacyclic complex (**213**) according to Equation (60) affords a cationic trimethylsilylmethyl complex. An analogous compound is obtained with Cp$_2$ = Me$_2$Si(C$_5$H$_4$)$_2$. The tetraphenylborate salts are isolated as orange solids; they are very slow catalysts for the polymerization of ethene.[423]

$$\text{Cp}_2\text{Ti} \diamondsuit \text{SiMe}_2 \xrightarrow[-\text{NEt}_3]{[\text{NHEt}_3][\text{BPh}_4],\ \text{THF}} \left[\text{Cp}_2\text{Ti} \diagdown \begin{array}{c} \text{TMS} \\ \text{THF} \end{array} \right] [\text{BPh}_4] \tag{60}$$
$$\textbf{(213)}$$

The carbene complex (**214**) reacts with excess Bu^{t}NC between −35 °C and room temperature under insertion into the Ti–C bond to the η^2-iminoacyl compound (**215**), as shown in Equation (61). The structure was confirmed by x-ray diffraction. An analogous complex was obtained with W(CO)$_5$ in place of Re$_2$(CO)$_9$.[424] The fulvene complex (**216**), obtained as green crystals by oxidation of [Ti^{III}(η-C$_5$Me$_4$CH$_2$)(Cp*)] with PbCl$_2$ in 92% yield, inserts one or two equivalents of xylyl isocyanide into the strained Ti–C bond. The structure of (**217**) was determined by x-ray diffraction (Scheme 58).[425] By contrast, benzaldehyde does not react with the Ti–C bond of (**216**) but gives the bis(alkoxo) complex (**218**) instead. The complex reacts with [TiCl$_4$] to give crystallographically characterized (**219**), and with HCl with release of the alcohol. The reaction is best understood by considering an σ,σ,η^2 formulation for the fulvene complex.[426]

The alkenation of ketones and esters R^{1}C(O)R^2 (R^1 = alkyl; R^2 = H, alkyl, aryl, vinyl or OR^3) can be carried out with [TiMe$_2$Cp$_2$] directly instead of with the Tebbe reagent [TiCp$_2$(μ-CH$_2$)(μ-Cl)AlMe$_2$]. The reaction proceeds smoothly with a wide variety of substrates in toluene or THF at 60–65 °C in good yields. At least two equivalents of [TiMe$_2$Cp$_2$] per C=O function are required.[427] [TiMe$_2$(Cp*)$_2$] reacts with substituted cyclohexanones to give titanium enolate complexes. Deuteration studies show that

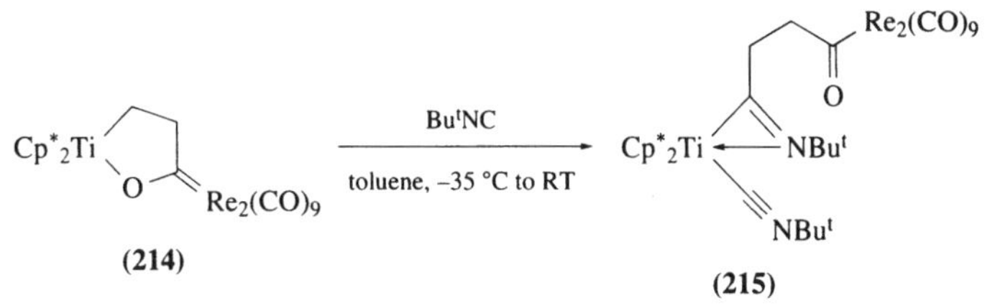

$$(61)$$

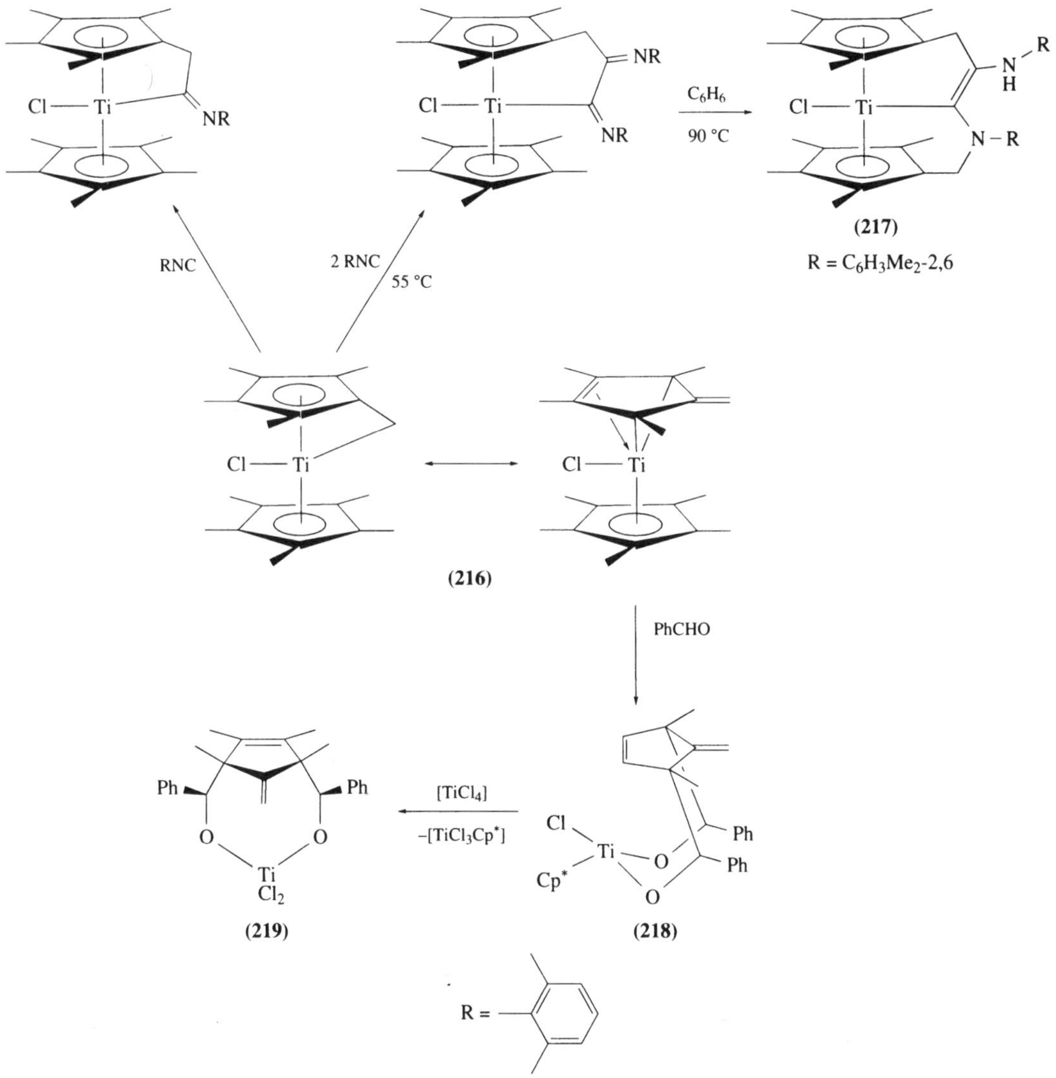

(214)   **(215)**

(217)

R = C$_6$H$_3$Me$_2$-2,6

(216)

(219)   **(218)**

$$R = $$

Scheme 58

[Ti(=CH$_2$)(Cp*)$_2$] is formed as an intermediate. The reaction is highly regioselective and produces the less highly substituted enolate, which can undergo quantitative thermal isomerization to the more highly substituted thermodynamic product (Scheme 59).[428] Epoxides give similar enolate products.[429] The enolate complex of acetone reacts with electrophiles in an unusual way; for example, Br$_2$ and HCl cleave the Ti–O bond rather than the Ti–Me bond, whereas I$_2$ and HgCl$_2$ leave the enolate ligand intact (Scheme 59).[430]

The protolysis of [TiPh$_2$Cp$_2$] with 2,2-dimethyl-1,3-propanediol gives the alkoxo compound (**220**), whose structure was determined by x-ray diffraction.[431]

The thermolysis of [TiPh$_2$(η-C$_5$H$_4$Me)$_2$] in toluene at 80 °C gives a benzyne species as a highly reactive intermediate which stabilizes itself by reaction with a methyl C–H bond to give (**221**) as a viscous dark green oil in nearly quantitative yield. The compound readily inserts ketones (Scheme 60).[432] Similarly, the thermolysis of [TiPh$_2$Cp$_2$] is thought to involve a benzyne intermediate which can

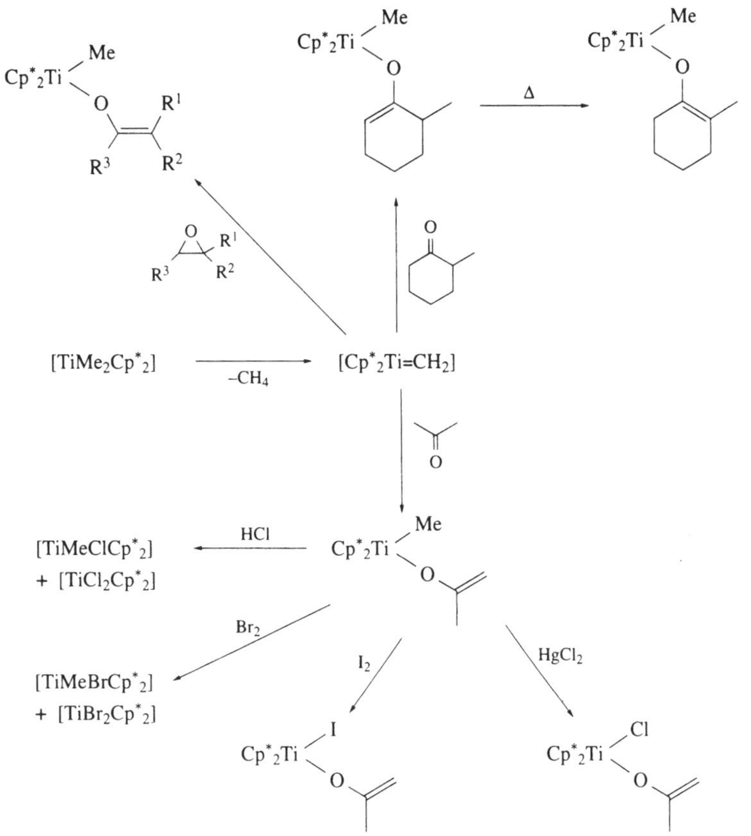

Scheme 59

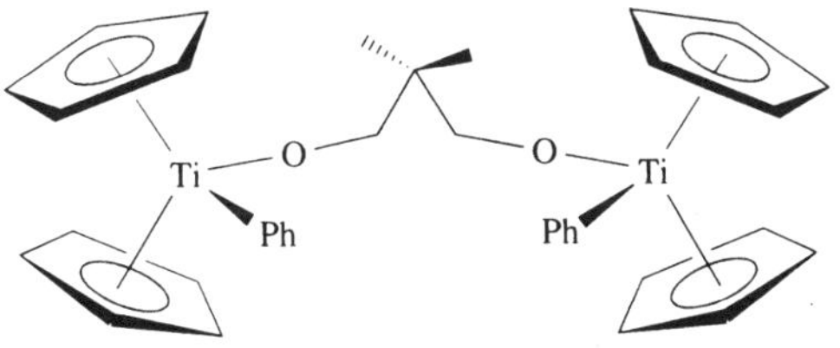

(220)

react with phosphorus ylides either via C–H activation to give (222), or by CH_2 transfer (Scheme 61). The high rotational barrier of the $^-CH=PPh_3$ ligand in (222) ($\Delta G^{\ddagger}_{rot} = 50.2 \pm 0.8$ kJ mol^{-1} at -34 °C) indicates a significant Ti–C π interaction, greater than for the zirconium and hafnium analogues.[433] The crystal structure of (222) confirms that the adopted conformation maximizes the π-interaction between titanium and the ylidic carbon atom.[434] The compound does not react with ketones. The reaction of the benzyne intermediate with metal carbonyls leads to titanium-substituted Fischer carbene complexes such as (223). The zirconium analogues are similarly accessible.[435]

The phosphinomethyl complex $[Ti(CH_2PPh_2)_2Cp_2]$ substitutes the PPh_3 ligands in $[Rh(1,5-C_8H_{12})(PPh_3)_2][BPh_4]$ in toluene at 0 °C to give the dimetallic complex $[TiCp_2(CH_2PPh_2)_2Rh(1,5-C_8H_{12})][BPh_4]$. The presence of such a chelating titanium-containing ligand has, however, little influence on the selectivity of the rhodium catalyst in the hydroformylation of 1-hexene, compared with the triphenylphosphine compound, although the reactivity is somewhat reduced.[436]

Vinylaluminum compounds generated *in situ* from terminal alkynes, butyllithium and $[AlMe_2Cl]_2$ react with $[TiMeClCp_2]$ to give titanium vinyl complexes such as (224) (Scheme 62). Treatment with hexamethylphosphoric triamide (HMPA) leads to the removal of dimethylaluminum chloride and the generation of an unstable titanium allenylidene species, which is trapped by ethyne to give a metallacycle. Although this intermediate is isolable in some cases, protolysis with HCl *in situ* liberates dienes. The overall process corresponds to a reductive ethyne codimerization. The same reaction principle can be applied to enynes and diynes, as shown in Scheme 63, to give the metallacycles (225)

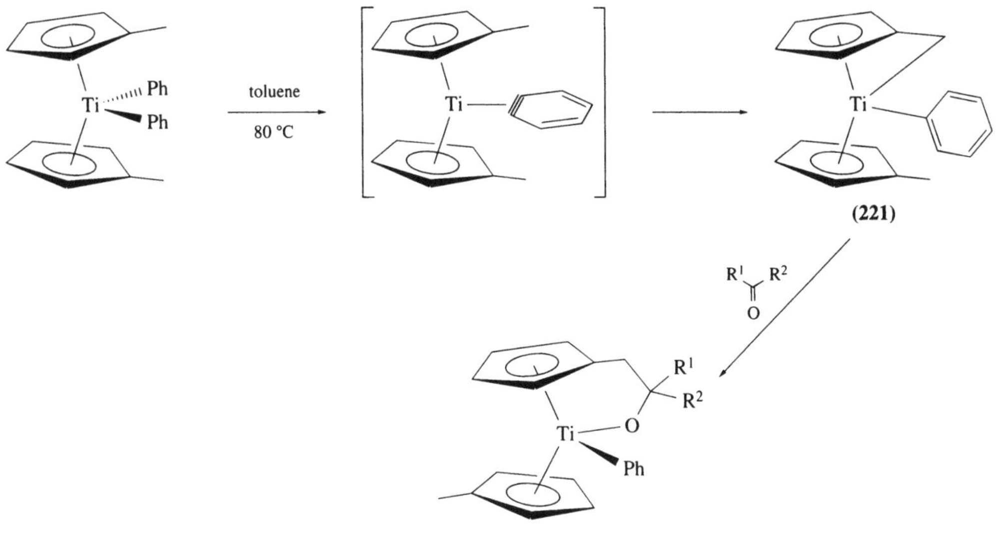

Scheme 60

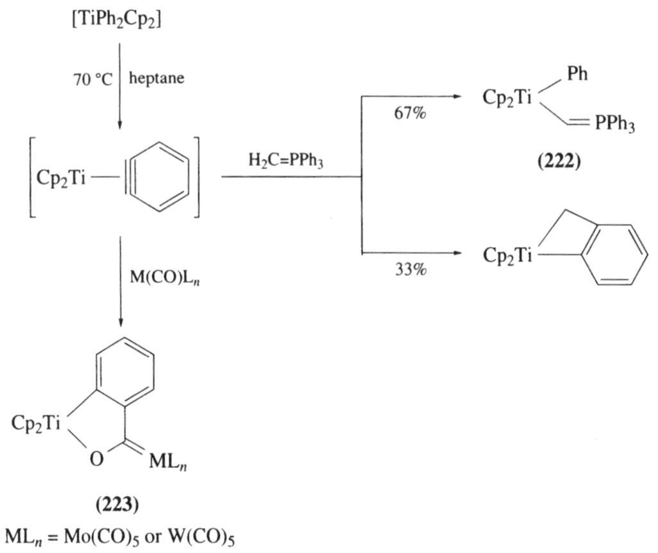

Scheme 61

and (**226**) in 64% and 89% yields, respectively. The formation and the reactivity of these complexes is illustrated for (**227**). The allenylidene compound is a postulated intermediate and could not be observed directly. Compound (**227**) is isolated as a dark red complex in 64% yield. It inserts isocyanides under ring expansion to give (**228**). The reaction product with CO could only be identified after hydrolysis; evidently two consecutive CO insertions have taken place. Bromination with *N*-bromosuccinimide leads to the expected organic dibromo compound.[437]

Iminoboranes react with [TiMe$_2$Cp$_2$] with elimination of methane via a methylidene intermediate in a manner similar to alkynes, to give the metallacycle (**229**) as a blue-black powder (Scheme 64). The compound has been characterized spectroscopically.[438]

Titanocene acetylide complexes can act as ligands towards a number of metal fragments in high and low oxidation states. For example, [Ti(C≡CR^2)$_2$(η-C$_5$H$_4$R^1)$_2$] (R^1 = TMS; R^2 = TMS or Ph) reacts with FeCl$_2$ or CuCl to give (**230**). The molecular structure of the iron complex was determined by x-ray diffraction and is shown in Figure 37.[439] The reaction of the titanium acetylide with [Co$_2$(CO)$_8$] or [Ni(CO)$_4$] in toluene at room temperature affords the paramagnetic, crystallographically characterized cobalt complex and the diamagnetic nickel analogue (**231**) as dark brown and deep green crystals, respectively (Scheme 65).[440,441] A magnetic moment of μ_{eff} = 4.2 BM was measured for the cobalt compound. The CO ligand of the nickel complex can be substituted by phosphine or phosphite ligands, and the compound undergoes a reversible one-electron reduction process. The geometric differences in

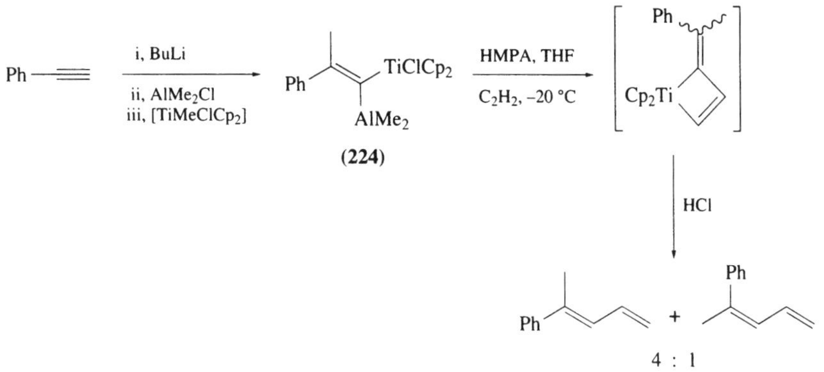

Scheme 62

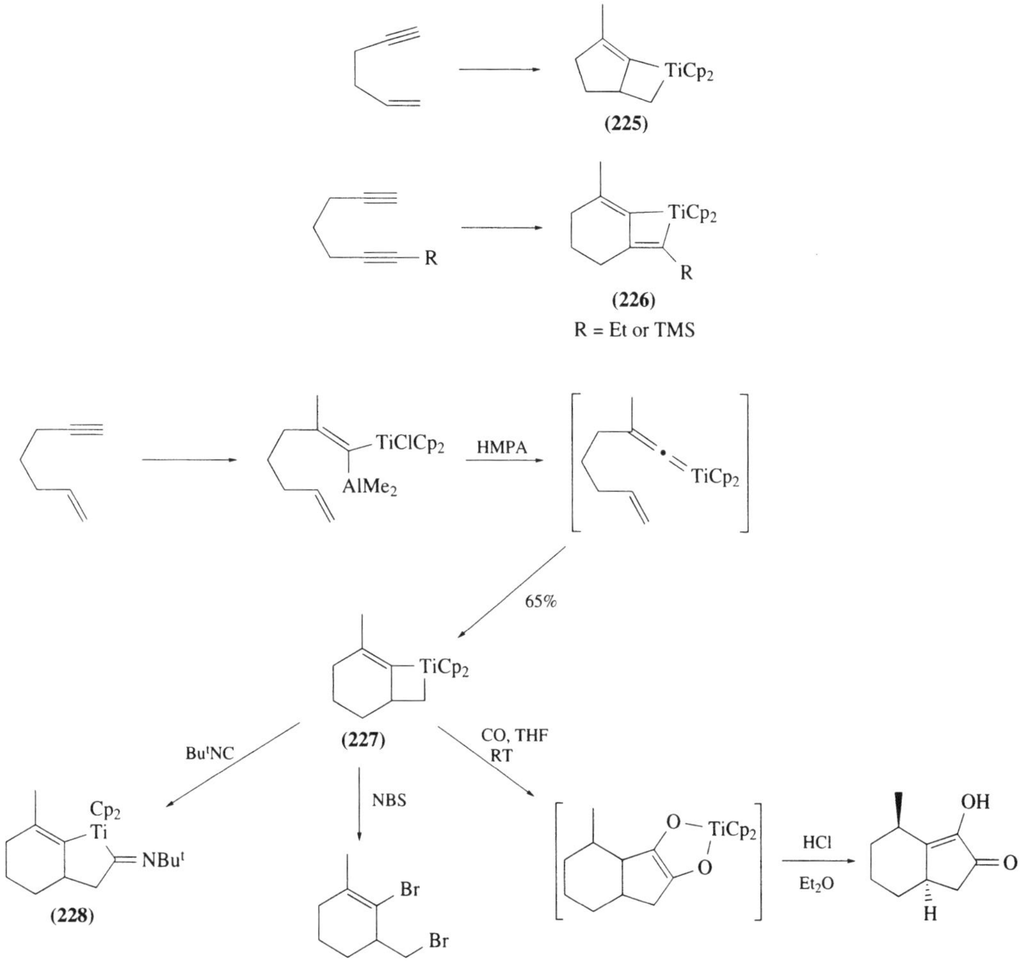

(225)

(226)

R = Et or TMS

(227)

(228)

Scheme 63

(229)

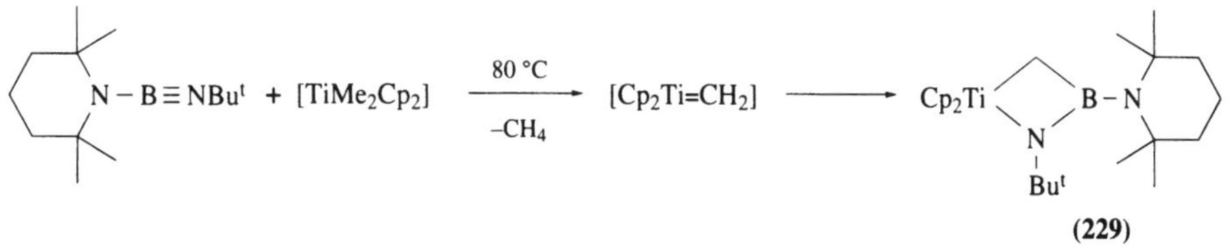

Scheme 64

                    Titanium Complexes in Oxidation State +4

the metal-coordinated alkynyl groups between (230) and (231) reflect the differences in the extent of backbonding. Thus, in (230) (MX_n = $FeCl_2$), the angle α is large (102.8°) (cf. Scheme 65) and the angles β and γ indicate close adherence to linearity for the Ti–C≡C–R moiety (β = 177°, γ = 176.5°). This may be compared with the corresponding values for (231) (M = Co), where α is much smaller (89.2°) and β and γ deviate significantly from 180° (β = 160.4°, γ = 152.1°).[440] Quite a different type of alkynyl bridge is found in the platinum complex (232), which is accessible from [Ti(C≡CBu^i)$_2$Cp$_2$] and *cis*-[Pt(C$_6$F$_5$)$_2$(THF)$_2$].[442] In this case both metals only interact with the alkyne α carbons. The molecular structure of (232) is shown in Figure 38.

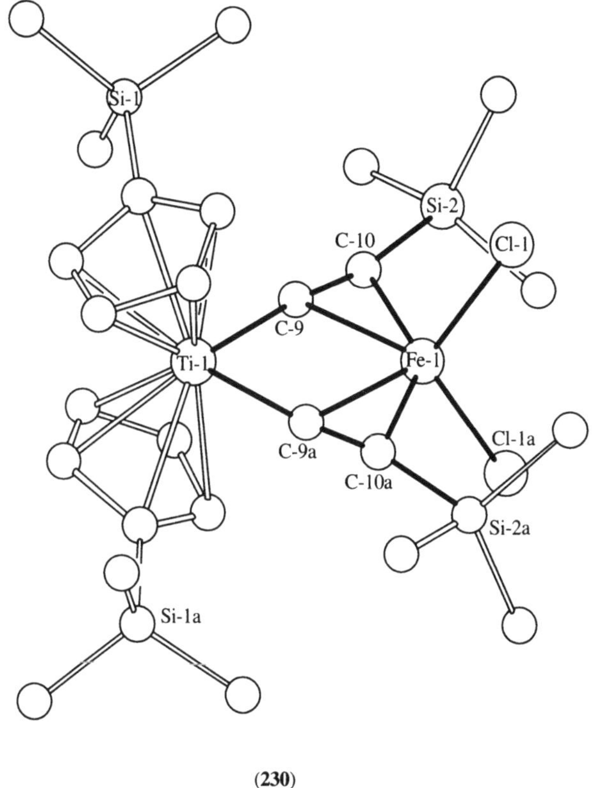

(230)

Figure 37   Molecular structure of [TiCp$_2$(μ_2-η^1-C≡C(TMS))$_2$FeCl$_2$] (230) (reproduced by permission of Elsevier from *J. Organomet. Chem.*, 1991, **409**, C7).

Titanium alkynyl complexes can be used for the gas-phase deposition of solid state materials. For example, [Ti(C≡CPh)$_2$Cp$_2$] reacts with H_2 and NH_3 at 500 °C to deposit titanium nitride films with low-grain boundaries oriented [111] with grains up to 2 μm long parallel to the substrate surface.[443]

5.5.2.4 *Titanium alkylidene complexes and metallacycles*

(i) *Stoichiometric reactions*

Titanium alkylidenes are usually highly unstable and require the addition of donor ligands to be observed and isolated. Some compounds belonging to this class have been mentioned before (cf., for example, Schemes 43 and 48). One of the first examples of a stabilized titanium methylidene compound is the 'Tebbe reagent', [TiCp$_2$(μ-CH$_2$)(μ-Cl)AlMe$_2$] (233). The aluminum alkyl can be replaced reversibly by alkenes, alkynes, donor ligands or other metal complexes, as shown in Scheme 66. The structure of the paramagnetic mixed-valence titanium dimer (234) is based on its ESR spectrum, which shows the unpaired electron to be equally distributed between the two metal centres. The presumed intermediate [Ti(=CH$_2$)Cp$_2$(THF)] reacts reversibly with methylenecyclohexane. Isotopic labelling of the exocyclic CH$_2$ group has been used to demonstrate that methylene exchange takes place in this reaction via the adduct (235). This process amounts to a nonproductive alkene metathesis.[444] A range of metallacycles is accessible by this route (Equation (62)). The relative stability, expressed by the

Scheme 65

(230)

$MX_n = FeCl_2$ or $CuCl$

$R^1 = R^2 = TMS$

(231)

$M = Co$ or Ni

$R^1 = TMS, R^2 = Ph$

$[Ti(C{\equiv}CR^2)_2(\eta\text{-}C_5H_4R^1)_2]$

$[Pt(C_6F_5)_2(THF)_2]$
$R^1 = H, R^2 = Bu^t$

(232)

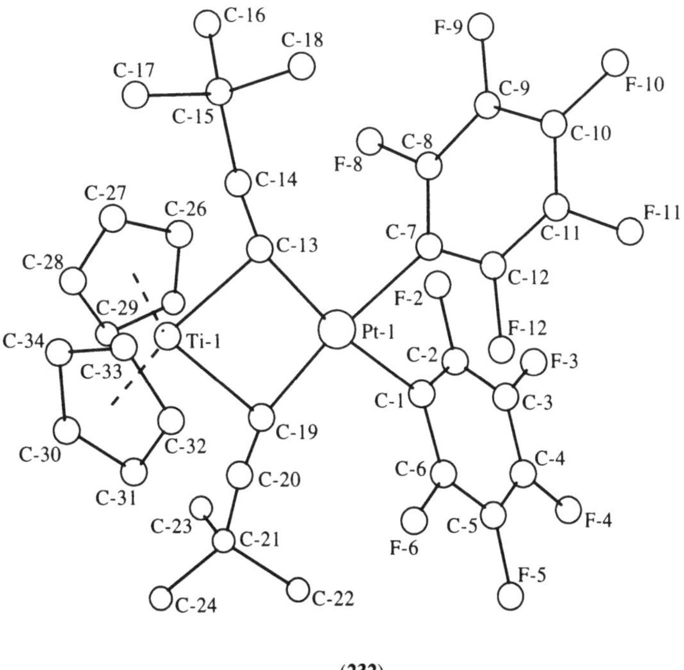

(232)

Figure 38 Molecular structure of $[TiCp_2(\mu_2\text{-}\eta^1\text{-}C{\equiv}CBu^t)_2Pt(C_6F_5)_2]$ (232) (reproduced by permission of the American Chemical Society. Copyright (1993) from *Organometallics*, 1993, **12**, 6).

equilibrium constant K, depends on the degree of steric hindrance, and increases for different titanacyclobutanes in the order shown in Scheme 67. The energy differences quoted are the differences in ΔG between pairs of metallacycles calculated from the K values.[445]

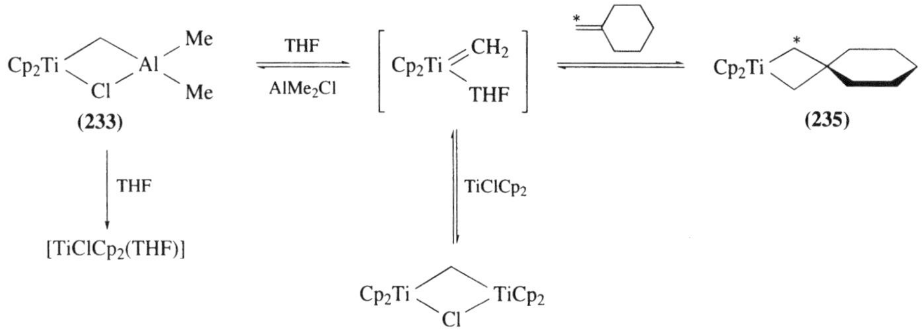

Scheme 66

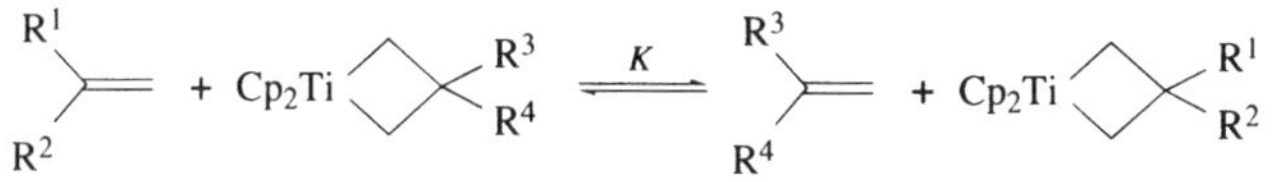

$$\text{(62)}$$

Increased stability $\longrightarrow$

3.4(3)   6.3(1)   11.3(3)   0.8(7)   3 .3(8)

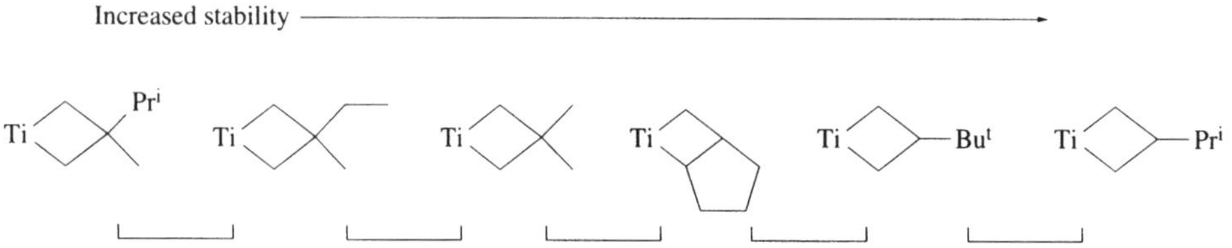

Difference in ΔG (kJ mol^{-1})

Scheme 67

The product of the reaction of $[TiCl_2Cp_2]$ with $CH_2(MgBr)_2$ was first formulated as (236) by analogy with (233).[446] However, in contrast with the Tebbe reagents, which exhibit high-frequency ^1H NMR spectral signals for the methylene group at δ 7–9, the magnesium complex shows a signal at δ 2.93, and is therefore suspected of having a different structure, possibly (237) or (238).[447] In a mixture of diethyl ether and benzene, $MgCl_2$ separates from the reaction mixture, with formation of the red-brown compound (239), whose μ-CH_2 groups now give rise to a singlet in the ^1H NMR spectrum at δ 9.87. Toluene solutions of (239) decompose slowly with further loss of magnesium halide to give the red-purple compound (181) (Equation (63)); the same reaction occurs instantaneously when THF is added to these solutions. The addition of benzophenone to solutions of (239) leads to alkenation of the ketone and produces 1,1-diphenylethene in 80% yield.[447] The addition of a substituted alkene $R^1CH=CR^2R^3$ or diphenylacetylene gives the metallacycles (240) and (241) ($R = Ph$), respectively (R^1, R^2, $R^3 = H$, H, Bu^n; H, H, Bu^t; H, Me, Me; or Me, Me, H).[446] The reaction of $[TiCl_2Cp_2]$ with the $CH_2(ZnI)_2$ gives $[TiCp_2(\mu\text{-}CH_2)ZnClI]$.[448]

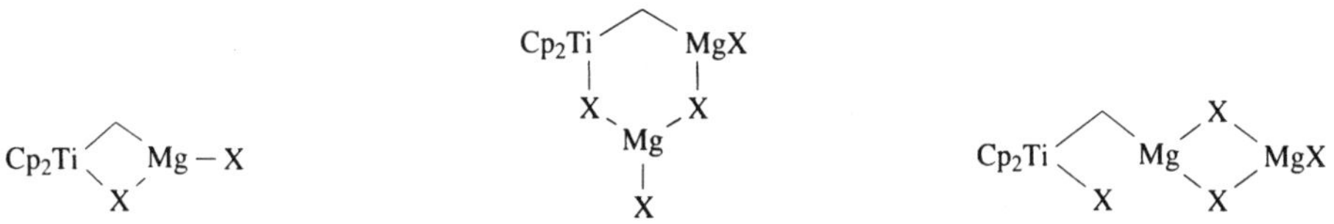

Complex (239) reacts rapidly with PMe_3 at 5 °C to give the methylidene complex $[Ti(CH_2)Cp_2(PMe_3)]$ in quantitative yield, whereas (181) (M = Ti) reacts with PMe_3 only slowly to give

$$\text{(239)} \xrightarrow{-\text{MgBr}_2} \text{(181)} \tag{63}$$

(242) (Scheme 68).[446] No reaction is observed between **(181)** and PEt_3, PBu^n_3 or PPh_3, or with the mixed-metal compound $[TiCp_2(\mu\text{-}CH_2)_2ZrCp_2]$.[449]

$$\text{(239) or (181)} \rightleftharpoons [Cp_2Ti=CH_2] \xrightarrow{PMe_3} Cp_2Ti\!\!\begin{array}{c}CH_2\\PMe_3\end{array}$$

(242)

Scheme 68

The metallacycles **(240)** ($R^1 = H$; $R^2 = R^3 = Me$ or $R^2 = Me$, $R^3 = Pr$) react with phosphines in a similar way to give $[Ti(CH_2)Cp_2(L)]$ as thermally unstable yellow-brown powders. The stability decreases in the order $L = PMe_3 > PMe_2Ph \gg PEt_3$. Compound **(240)**, the alkene and **(242)** are in equilibrium with each other; for PMe_3, $K_{eq} = 45 \pm 10$. Compound **(242)** reacts with ethene to give $[TiCp_2(CH_2)_3]$ **(243)** and with CO to give the η^2-ketene complex **(244)**.[450] The rate-limiting step in the conversion of the titanacyclobutane complex **(243)** into **(242)** is the rearrangement to the (methylene)(ethene) intermediate, $[TiCp_2(=CH_2)(C_2H_4)]$ **(245)**, from which ethene is displaced by the donor ligand L in a fast S_N2 process, although an S_N1 pathway could not be ruled out (Scheme 69).[451] Hartree–Fock and generalized valence bond calculations for the interconversion between **(243)** and **(245)**[452] and for the related structures $[Cl_2Ti(CH_2)_3]$ and $[Cl_2Ti(=CH_2)(C_2H_4)]$[453] show that the metallacycle is 46–50 kJ mol^{-1} more stable than the methylidene complex. In neither case is there a measurable activation barrier for this formally 'forbidden' $[2+2]$ cycloaddition process. The alkene coordination energy for the chloro complex was calculated as 43.1 kJ mol^{-1}, to give an activation enthalpy for the alkene metathesis with the titanacyclobutane of 90.8 kJ mol^{-1}.[453]

$$\text{(243)} \rightleftharpoons Cp_2Ti\!\!\begin{array}{c}CH_2\\\end{array}\text{(245)} \xrightarrow[-C_2H_4]{L} Cp_2Ti\!\!\begin{array}{c}CH_2\\L\end{array}$$

Scheme 69

(244)

The rates of cleavage of the titanacyclobutane ring are a function of the electronic nature of substituents on the cyclopentadienyl rings and were measured for a series of complexes $[Ti(CH_2CHBu^tCH_2)CpCp']$ ($Cp' = C_5H_4Me$, $C_5H_4CF_3$, C_5H_4Cl, C_5H_4TMS, $C_5H_2Me_3$ or Cp^*). The rates have been correlated with the lowest-energy transitions in the UV–visible spectra of $[TiCl_2CpCp']$ and with the ^{47}Ti and ^{49}Ti NMR spectral chemical shifts.[454] The metallacyclobutene complexes **(241)** ($R^1 = R^2 = Me$; $R^1 = Ph$, $R^2 = TMS$) react with CO in the presence of PMe_3 to give the crystallographically characterized vinylketene complexes **(246)** (Scheme 70). An intermediate, probably **(247)**, can be observed below $-30\,°C$; its IR spectrum shows a ν_{CO} band at 1615 cm^{-1}. The mechanism of rearrangement of **(247)** to **(246)** is not clear.[455]

Phosphorus ylides abstract HCl from $[TiCl(\eta^2\text{-}COMe)Cp_2]$ to give the thermally stable ketene complex **(248)** as red microcrystals. The compound is polymeric and can be recrystallized from benzene

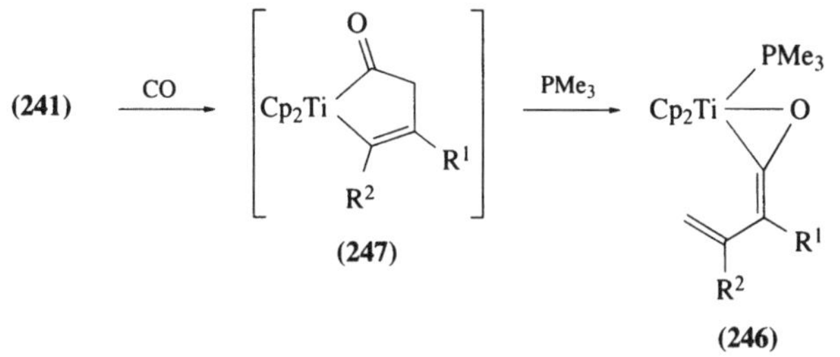

Scheme 70

to give a yellow crystalline isomer. Compound (**248**) reacts with ethene and ethyne under ring expansion, and with PMe$_2$Ph to give the monomeric phosphine adduct (Scheme 71).[456] The reaction of [TiCl(η^2-COMe)Cp$_2$] with sulfur ylids leads to methylene insertion into a ketene intermediate and formation of (**249**). The metallacycle is fluxional ($\Delta G^{\ddagger} \sim 54.4 \pm 0.8$ kJ mol^{-1}). A diphenyl-substituted analogue of (**249**) is accessible from [Ti(CH$_2$CHBu^{t}CH$_2$)Cp$_2$] and Ph$_2$C=C=O at 80 °C, with elimination of Bu^{t}CH=CH$_2$. This compound has a higher activation barrier for the conformational interchange of the ring ($\Delta G^{\ddagger} \sim 79.5 \pm 0.8$ kJ mol^{-1}) and forms two distinguishable isomers at room temperature.[457]

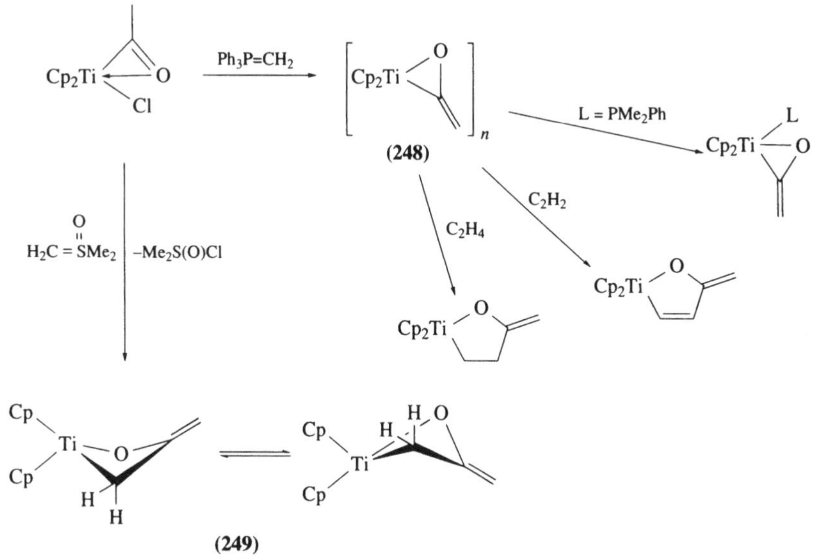

Scheme 71

Sulfur-containing metallacycles are obtained by reacting (**242**) with elemental sulfur or with propene sulfide to give the thioformaldehyde complex (**250**), or from trimethene sulfide to give (**251**). There is no reaction between (**242**) and less ring-strained sulfur heterocycles such as thiophene or tetrahydrothiophene. The structure of the thioformaldehyde complex was confirmed by x-ray diffraction. Some of its reactions are outlined in Scheme 72. Compound (**250**) is sensitive to electrophiles and reacts with methyl iodide by *S*-methylation to give (**252**) (X = I) and with acetyl chloride to give (**253**). The PMe$_3$ ligand in (**252**) (X = BF$_4$ or BPh$_4$) can be removed with CuCl. Hydrogenation of (**250**) gives [Ti(SMe)$_2$Cp$_2$] and some unidentified highly sensitive material.[458]

Apart from the formation of metallacycles via titanium methylidene and titanacyclobutane intermediates, several more 'conventional' syntheses of metallacyclic complexes have been reported. The complex (**254**) is prepared from the di-Grignard reagent (Equation (64)),[434] as is the red-orange *ansa*-titanocene (**255**) (Equation (65)). The crystal structure of (**255**) was determined; the molecule adopts the 'sideways' conformation of the *ansa* ligand framework frequently found in complexes of this kind to minimize steric interactions, as discussed in Section 5.5.1.2 for titanocene dichlorides.[459] The bridgehead alkyl complex (**256**) is obtained from [TiCl$_2$Cp$_2$], and the corresponding dilithium reagent in diethyl ether at 0 °C in low yield.[460] The fluxional complex (**257**) is obtained from [TiCl$_2$(η-C$_5$H$_4$TMS)$_2$] and the dimagnesium reagent; the structure was confirmed by x-ray diffraction. The five-membered ring is folded. Electrochemical reduction to an anionic titanium(III) complex probably involves cleavage of one Ti–C bond.[461] The analogous blue silyl-substituted metallacycle (**258**)[462] and the biphenylene-

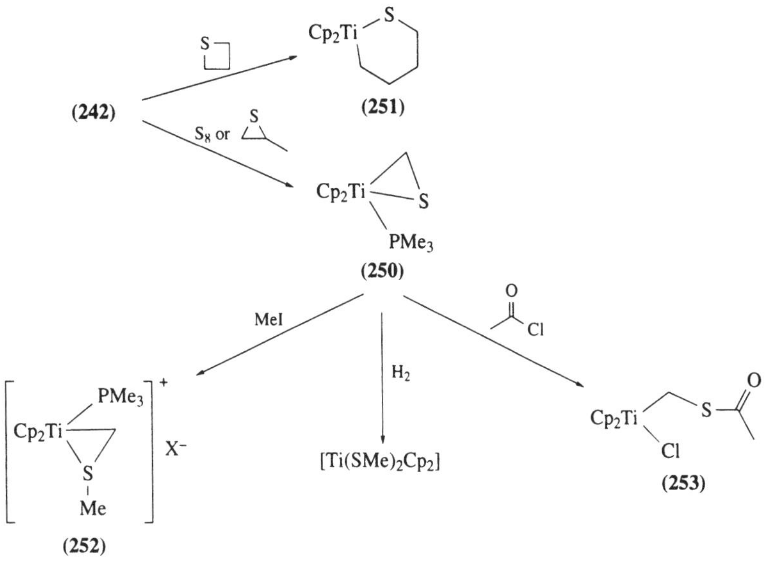

Scheme 72

bridged compound (**259**)[463,464] were similarly obtained from the dilithium reagents (Scheme 73). Compound (**258**) is thermally very stable; it appears to lose Cp^- on reduction.[462] The stable metallacycloheptadiene (**259**) sublimes at ~160 °C / 0.1 torr; its crystal structure has been determined. The complex undergoes a reversible one-electron reduction process, possibly with cleavage of one Ti–C bond, as was observed with (**257**).[464] The chlorothiophenyl complexes (**260**) are obtained as poorly characterized red solids.[465]

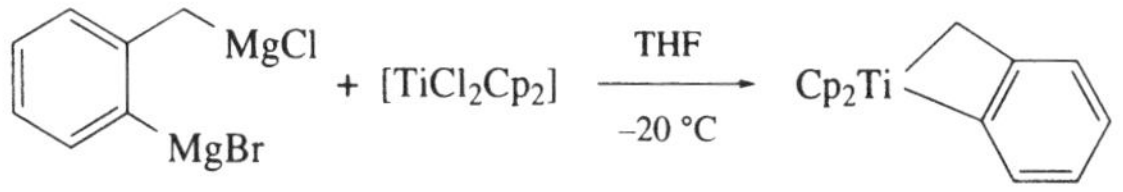

$$\text{(64)}$$

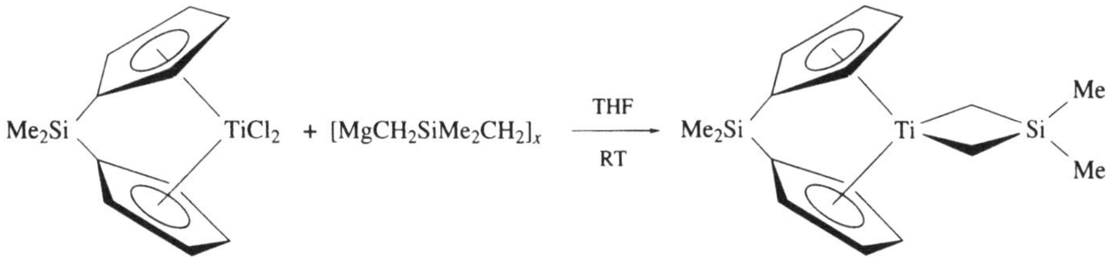

$$\text{(65)}$$

The reaction of $PhS(O)CH_2Li$ with $[(TiClCp_2)_2]$ followed by chromatography over activated alumina affords compound (**261**) as diamagnetic green crystals in 31% yield which sublime at 130 °C / 0.1 torr with partial decomposition to $[Ti(SPh)_2Cp_2]$. This decomposition product is also obtained from $PhS(O)CH_2Li$ and $[TiCl_2Cp_2]$. The sulfur atom in (**261**) can be methylated (Scheme 74).[466] The macrocyclic complex (**262**) is prepared from $[TiCl_2Cp_2]$ and the corresponding aryl dilithium reagent in benzene at room temperature as purple diamagnetic crystals in 55% yield.[467]

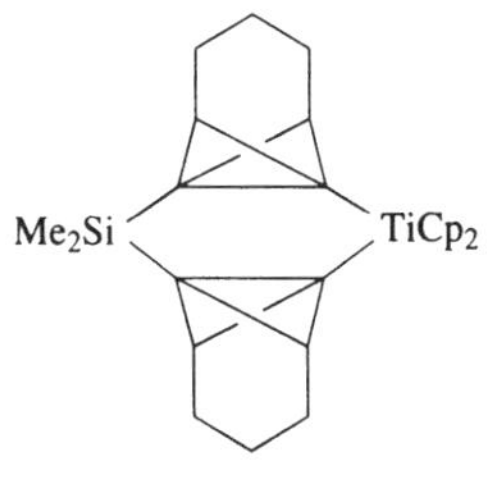

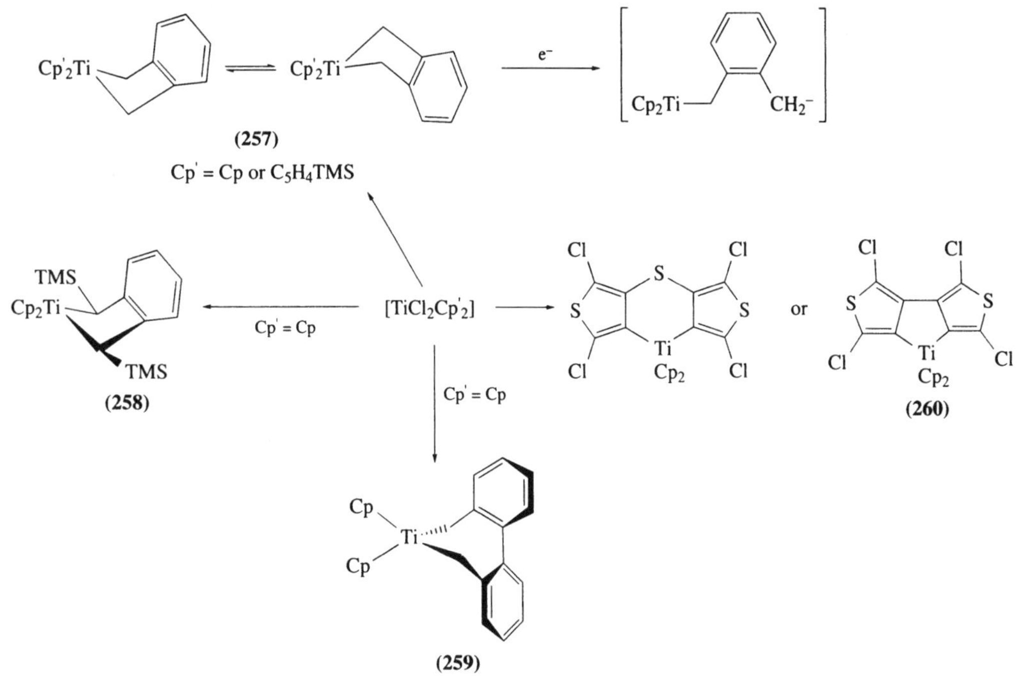

Scheme 73

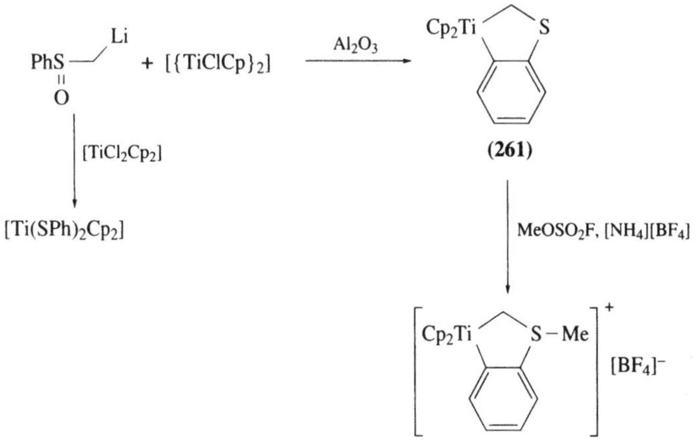

Scheme 74

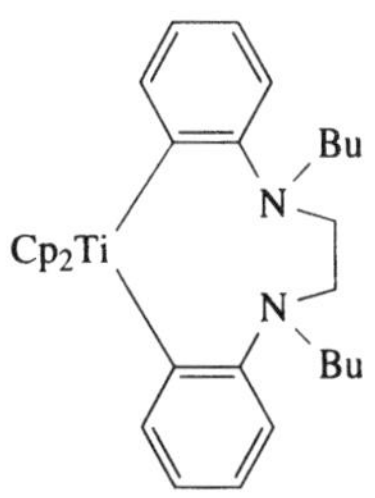

Titanacyclopentanes are formed when titanium(II) intermediates are exposed to ethene. Thus, the reaction of [{TiCl(Ind)$_2$}$_2$] with MeMgBr in the presence of ethene in THF at −30 °C gives a 53:6:40 mixture of (**263**), [Ti(η^2-C$_2$H$_4$)(Ind)$_2$] and [TiMe$_2$(Ind)$_2$], as shown in Equation (66). Recrystallization of the mixture from ethene-saturated pentane at −45 °C gives 80% of (**263**). The titanacyclopentane ring dissociates readily above −80 °C into [Ti(η^2-C$_2$H$_4$)(Ind)$_2$] and ethene.[468] Titanacyclopentadienes are formed by reducing [TiCl$_2$Cp$_2$] with magnesium in the presence of diynes. For example, the bicyclic compound (**264**) is isolated as a red solid in 54% yield (Equation (67)). Treatment with HCl liberates the diene.[469] Large metallacycles are accessible by reacting titanacyclopentadienes such as (**265**) with disubstituted alkynes. With two to four equivalents of the alkyne, titanacycloheptatrienes are obtained which undergo ring expansion to give larger rings (Scheme 75). The ring size depends to some extent on the substituents. With 3-hexyne, for example, the 17-membered metallacycle (**266**) is formed (Equation (68)). The method is useful for the synthesis of polyenes, which are released on addition of HCl.[470]

$$[\{TiCl(Ind)_2\}_2] \xrightarrow[C_2H_4]{MeMgBr} Ind_2Ti\diagdown\diagup + Ind_2Ti\!-\!\| + [TiMe_2(Ind)_2] \qquad (66)$$

(**263**)

$$[TiCl_2Cp_2] + R\!-\!\!\equiv\!\!-(CH_2)_n\!-\!\!\equiv\!\!-R \xrightarrow[THF]{Mg/HgCl_2} Cp_2Ti\diagup\!\diagdown(CH_2)_n \qquad (67)$$

(**264**)

$$Cp_2Ti \xrightarrow[R = Et\ or\ Ph]{2\text{-}4\ equiv.\ C_2R_2} Cp_2Ti \xrightarrow{excess\ C_2R_2} Cp_2Ti \qquad n = 2, 3\ or\ 5$$

(**265**)

Scheme 75

$$Cp_2Ti \xrightarrow{Et\!-\!\!\equiv\!\!-Et} Cp_2Ti \qquad (68)$$

(**266**)

The reactivity of titanacyclobutanes has been studied in detail, not least because of the involvement of these compounds as intermediates in alkene metathesis processes. The deuterium label in the complex (**267**) scrambles rapidly in the presence of catalytic amounts of AlMe$_2$Cl. A mechanism for this process involving β-hydride abstraction and formation of a titanium allyl intermediate was ruled out by selective deuteration studies, and a process involving titanium–aluminum bimetallic intermediates was suggested (Scheme 76).[471] The treatment of titanacyclobutanes such as (**268**) with iodine results in the reductive elimination of substituted cyclopropanes. The stereochemistry of this process has been probed with deuterium labelling and is thought to proceed via iodoalkyl intermediates such as (**269**), to give a 1:1 mixture of the expected deuterocyclopropanes on warming (Scheme 77). The dimethyl-substituted derivative (**270**), on the other hand, leads to a 9:1 mixture of *cis*- and *trans*-dimethylcyclopropanes, probably because iodine attack on the less hindered Ti–C bond to give the intermediate (**271a**) is favoured over the formation of (**271b**).[472] The electrochemical oxidation of (**272**) gives cyclopropanes with retention of configuration (Equation (69)).[473]

The photolysis of (**272**) in the presence of alkynes, ethene or PMe$_3$ affords cyclopropanes and titanium(II) or titanium(IV) products (Scheme 78), independent on the wavelength of light used (λ = 250–550 nm). Labelling studies suggest that the reaction involves biradicals such as (**273a**) (Scheme 79). The cyclopropanes *trans*- and *cis*-(**274**) are formed in 5:1 to 3:1 ratios, whereas (**273b**) leads to *trans*-(**274**) with retention of configuration.[474] Titanacyclobutanes with β-phenyl substituents (**275**) readily rearrange into their α-phenyl isomers (**276**), whereas the reverse reaction does not take place. The mechanism involves the dissociation of the metallacycle into titanium methylene intermediates and the alkene, which reassociates with scrambling of the deuterium label (Scheme 80).[475] The complexes (**272**) (R^1, R^2 = Me; R^1 = Me, R^2 = Pr^n) react with disubstituted allenes to give (**277**). By further treatment with ketones (R^5 = R^6 = Ph; R^5 = Ph, R^6 = Me) tetrasubstituted allenes are prepared (Scheme 81).[476] The complex (**277**) (R^3 = Me, R^4 = Pr^i) reacts with 3-hexyne via a titanium allenylidene

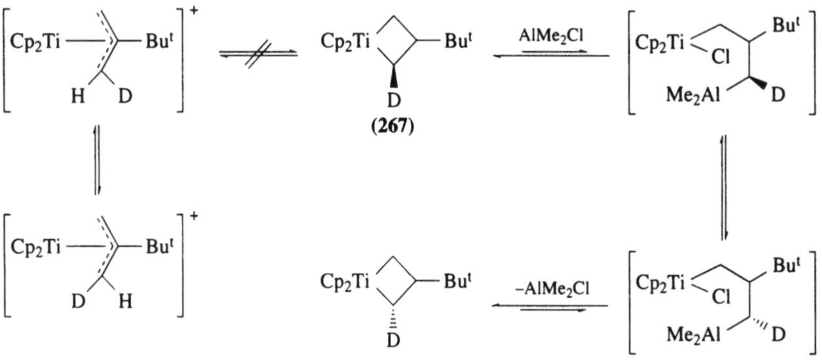

Scheme 76

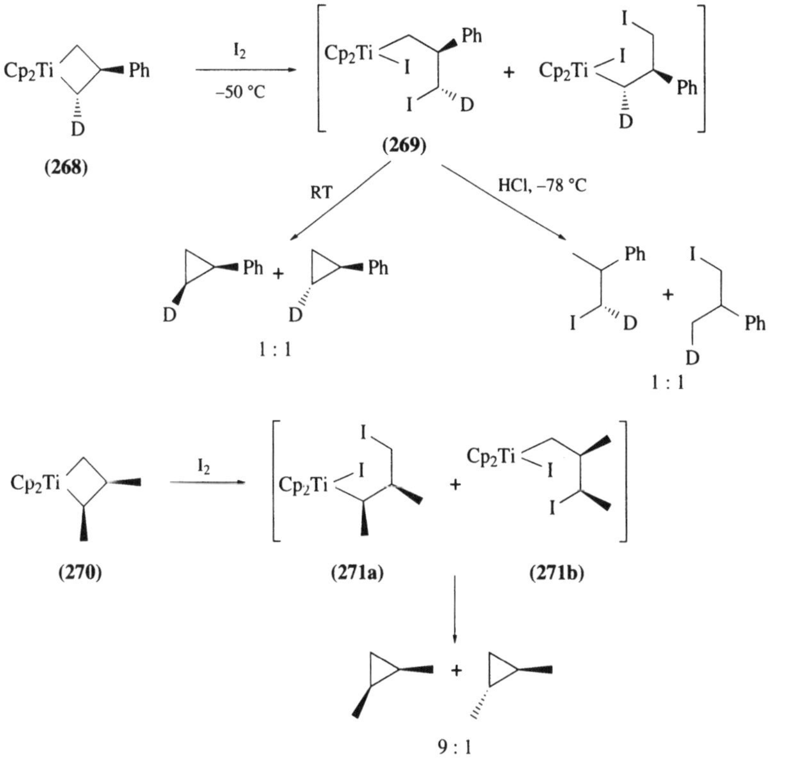

(268) → **(269)**

(270) → **(271a)** + **(271b)**

Scheme 77

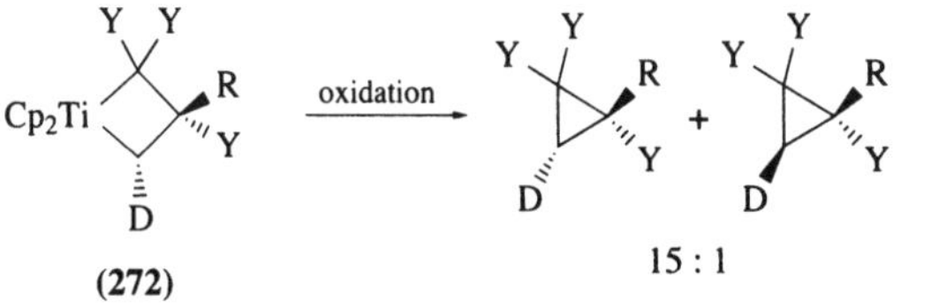

$$R = Bu^t \text{ or } Pr^i$$
$$Y = H \text{ or } D$$

(69)

intermediate to give the titanacyclobutenes (**278**) as a mixture of isomers. The same isomer ratio is obtained if the positions of R^3 and R^4 are reversed.[477]

The complex (**272**) ($R^1 = R^2 = Me$) reacts with allyl chloride to give the 3-butenyl complex (**279**), and with benzyl chloride to give (**280**). The formation of these products is explained if [Ti(CH$_2$)Cp$_2$] is postulated as the intermediate; this assumption is supported by the formation of (**280**) from benzyl chloride and [Ti(CH$_2$)Cp$_2$(PMe$_3$)] (Scheme 82). The kinetics of the benzyl chloride reaction show little sensitivity to the nature of the *para* substituents on the phenyl ring, and the reaction of (**272**) with

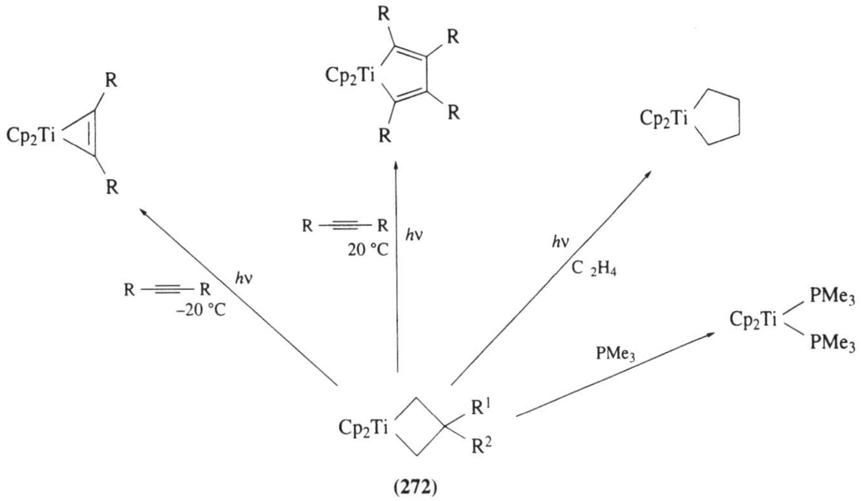

Scheme 78

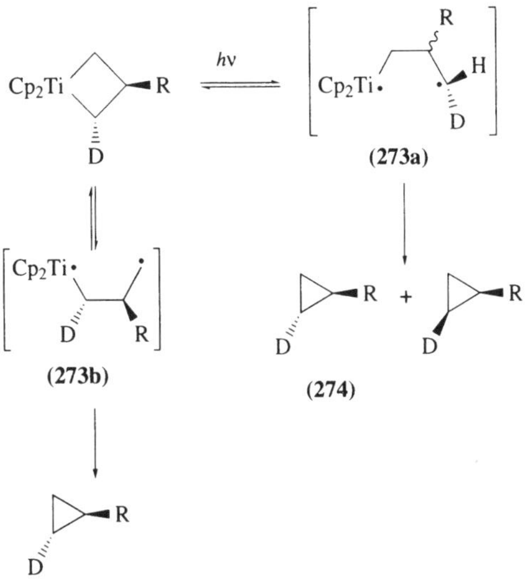

Scheme 79

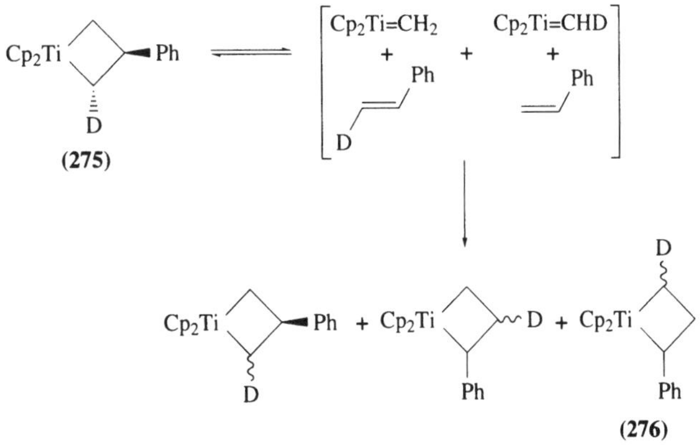

Scheme 80

optically active PhCH(D)Cl leads to (**280**) with complete loss of stereochemistry. These findings suggest that the reaction proceeds via radical intermediates.[478] Radicals are also involved in the reaction of (**272**) (R^1, $R^2 = Bu^i$, H; Me, Me; Pr^i, H) with tetrakis(trifluoromethyl)cyclopentadienone (Scheme 82). The electron transfer process induces the reductive elimination of the disubstituted cyclopropane, and the

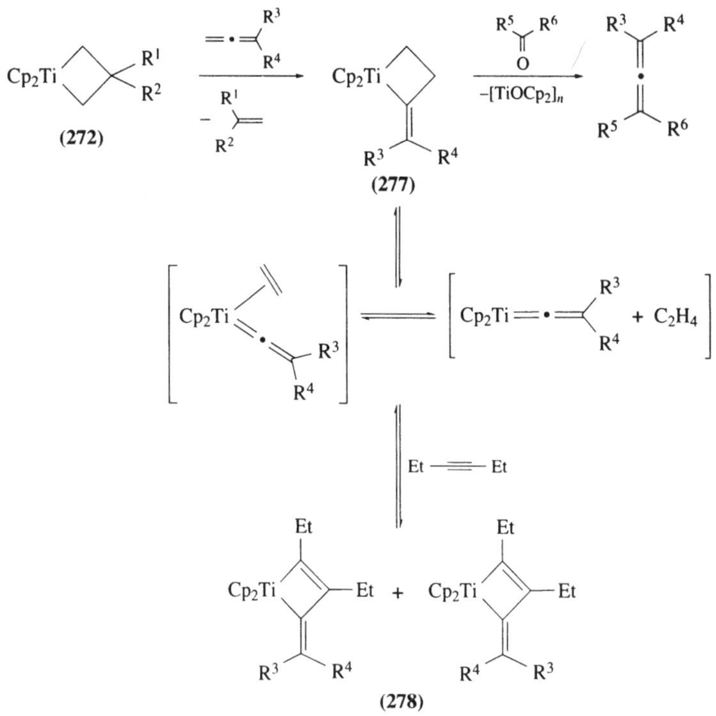

Scheme 81

generated 'TiCp$_2$' fragment is reactive enough to activate one of the C–F bonds of the cyclopentadienone to give (**281**), whose structure was confirmed by x-ray diffraction.[479]

The metallacycle (**282**), isolated as thermally relatively stable red needles in 33% yield by treating the Tebbe reagent with dimethylcyclopropene, reacts with benzophenone and HCl to give the expected alkenation and hydrolysis products, and with phosphines and AlMe$_2$Cl to give alkyl-substituted titanium carbene complexes, as shown in Scheme 83.[480] Compound (**272**) (R = Me, R^1 = Pr^i) reacts with iminoboranes to give the heterocyclic complexes (**283**) (Equation (70)). The products were characterized by NMR spectra. The product is stable for R^1 = Bu^n and R^2 = Bu^t, while others (R^1 = Pr^i or C$_6$F$_5$; R^2 = Pr^i or Bu^t) cyclize in solution to dimers and trimers.[481]

The titanacyclobutene complexes (**284**) (R^1 = R^2 = Ph; R^1 = Ph, R^2 = Me) react cleanly with acetone to give (**285**), and with aldehydes to give a mixture of (**286**) and (**287**) in a 1:1 (R^3 = Me) or 2:1 (R^3 = Ph or p-tolyl) ratio. Tetramethylguanidine leads to (**288**). Nitriles generally afford mono-insertion products such as (**289**) (R^1 = R^2 = Ph; R^4 = p-tolyl, Pr^i, CH(Me)Ph or C(Me)=CH$_2$) (Scheme 84). Acetonitrile does not lead to a stable product.[482] A series of compounds of the type (**289**) has been described (R^1 = R^2 = Ph; R^4 = Me, Et, Pr^i, Bu^t, Ph, p-C$_6$H$_4$X, o-C$_6$H$_4$Me or C$_6$F$_5$; X = F, OMe, Me or CF$_3$). Treatment of (**284**) (R^1 = R^2 = Me or Et) with Pr^{i}CN gives products such as (**290**), arising from the insertion of two nitriles. Hydrolysis of (**290**) with dry HCl gives the tetrasubstituted pyridine. Mixtures of (**289**) and (**290**) are formed if R^1 = R^2 = Me or Et and R^4 = Bu^t. The conversion of (**289**) into (**290**) by refluxing with excess nitrile did not prove possible. Evidently, the second insertion can only take place if the first nitrile molecule inserts into the titanium–vinyl and not the titanium–alkyl bond of (**284**).[483] Esters, carbonates, CO$_2$, CS$_2$ and imines do not react with (**284**).[482] Titanacyclobutenes of the type (**291**) are obtained by treating (**224**) (cf. Scheme 62) with alkynes (R^2 = Pr^n) in the presence of HMPA as a dark-purple mixture of configurational isomers (Scheme 85). Exposure of (**291**) to Bu^{t}NC results in the formation and isolation of the allenyl ketimine (**292**), presumably via isocyanide insertion into a titanium–vinyl bond of the metallacycle.[484] The complex (**284**) (R^1 = R^2 = Ph) reacts with RECl$_2$ (E = P, R = Ph, Et, Bu^t or OEt; E = As, R = Ph) to give phosphorus and arsenic heterocycles (Equation (71)).[485,486]

The Tebbe reagent (**233**) or titanacyclobutanes have been used in the synthesis of a series of heterobinuclear complexes. The reaction of (**233**) with the tungsten carbyne complex [W(≡CR)Cp(CO)$_2$] affords (**293**) via the intermediate (**294**) in 50–60% yield if an excess of (**233**) is employed (Scheme 86). The crystal structure of (**293**) has been determined (Figure 39).[487,488]

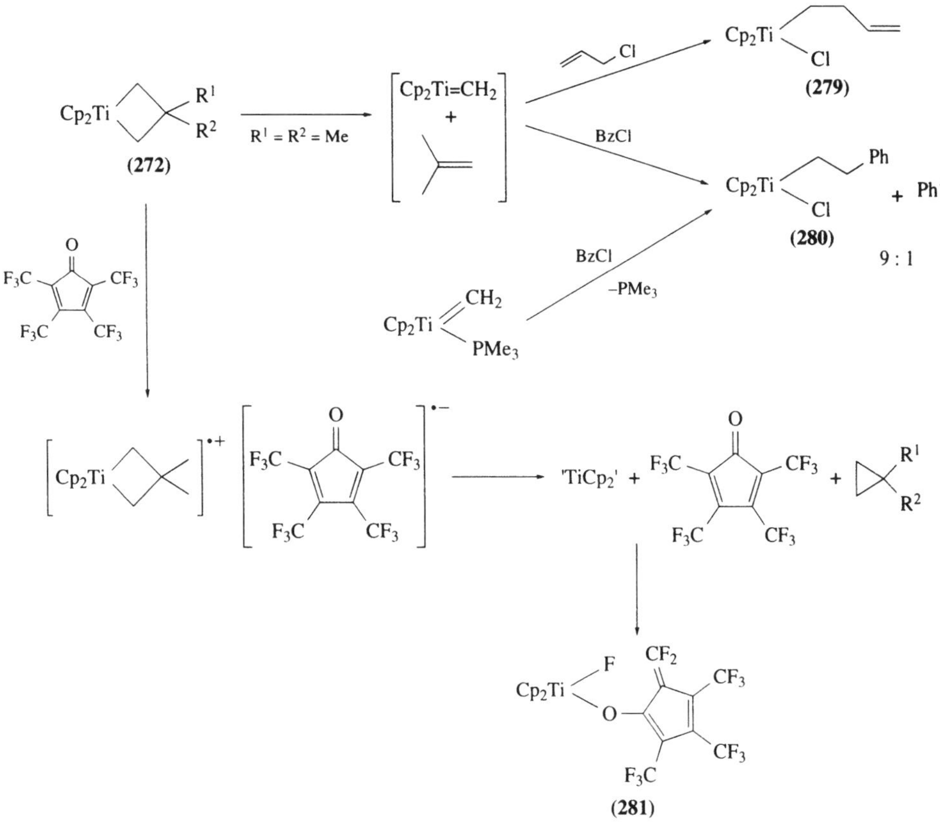

Scheme 82

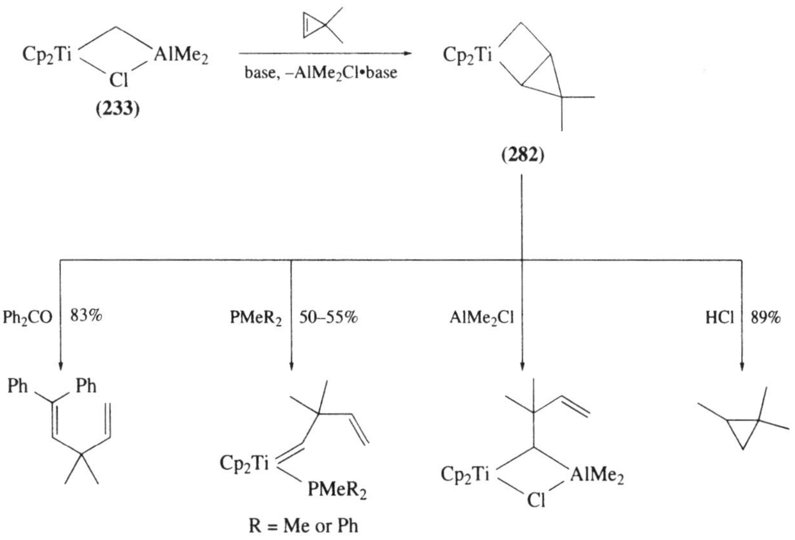

Scheme 83

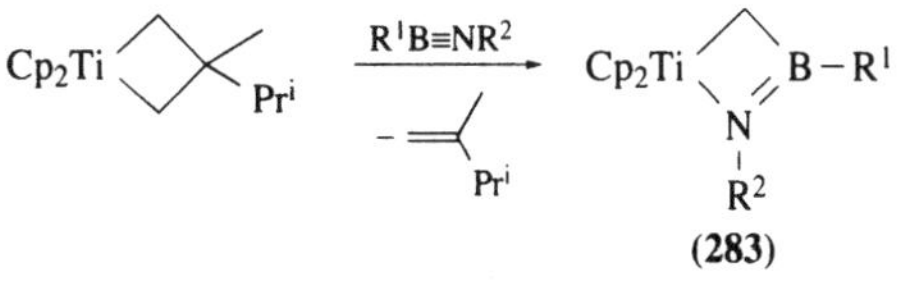

(70)

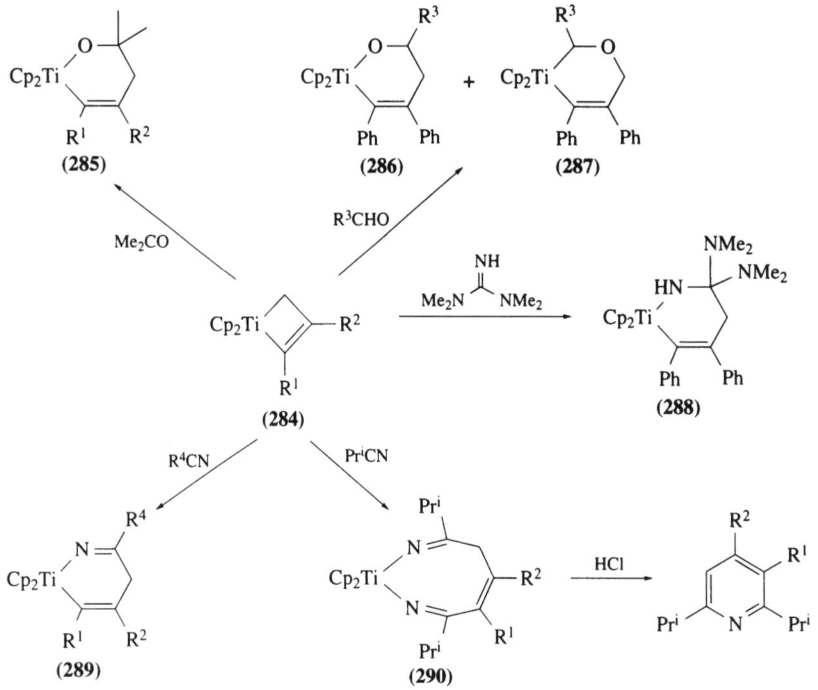

(285) (286) (287)

Me$_2$CO

R^{3}CHO

(284)

Me$_2$N$\overset{NH}{\diagup}$NMe$_2$

(288)

R^{4}CN

Pr^{i}CN

(289)

(290)

HCl

Scheme 84

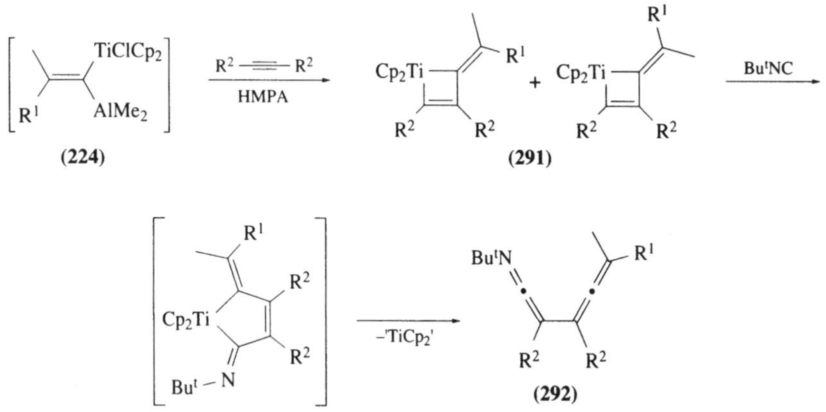

(224)

$R^2 \!\!\equiv\!\! R^2$

HMPA

(291)

Bu^{t}NC

$-$'TiCp$_2$'

(292)

Scheme 85

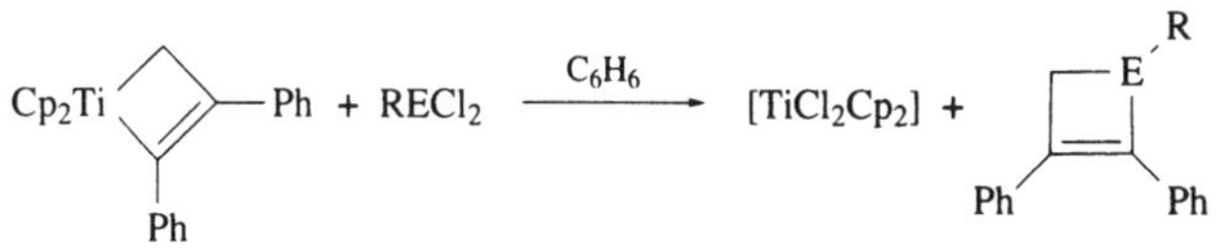

Cp$_2$Ti$\diagdown$—Ph + RECl$_2$ $\xrightarrow{C_6H_6}$ [TiCl$_2$Cp$_2$] +

(71)

(233) +

W$\equiv$CR

(294)

(293)

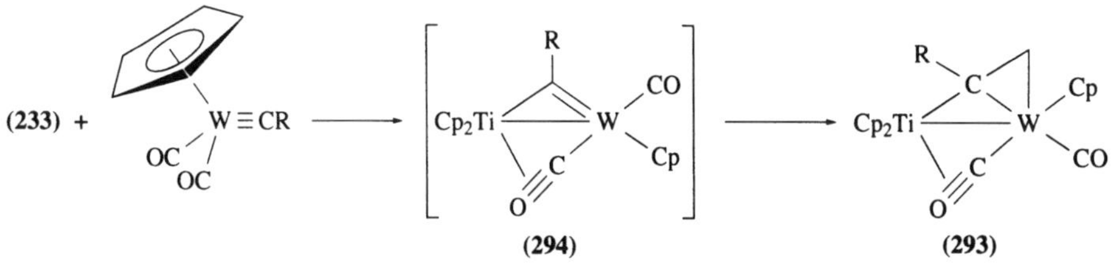

Scheme 86

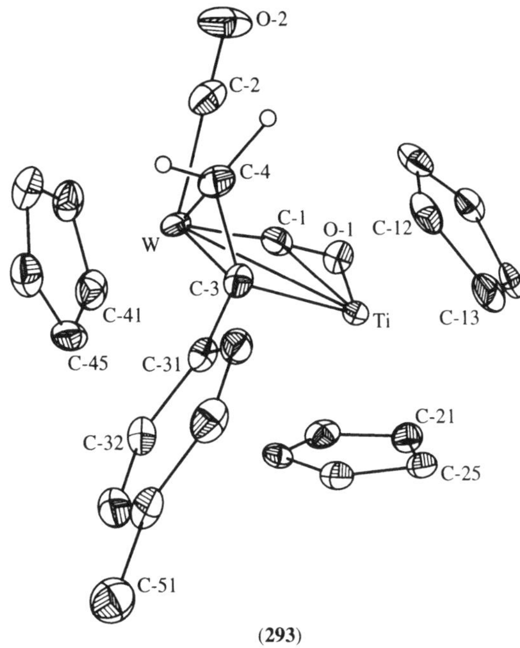

(**293**)

Figure 39 Molecular structure of the alkylidene-bridged titanium–tungsten complex (**293**) (reproduced by permission of the Royal Society of Chemistry from *J. Chem. Soc., Chem. Commun.*, 1983, 746).

Titanacyclobutanes react with noble metal halide complexes [MXL$_n$] to give the methylene-bridged complexes (**295**) (MXL$_n$ = RhCl(1,5-C$_8$H$_{12}$), Rh(OMe)(1,5-C$_8$H$_{12}$), IrCl(1,5-C$_8$H$_{12}$), PdCl(η^3-C$_4$H$_7$), PtCl(Me)(SMe$_2$), AuCl(PMe$_3$)), as shown in Equation (72). The rhodium, iridium and palladium complexes form crystalline products in 80–90% yield; the platinum and gold complexes were observed spectroscopically.[489,490] The structure of [TiCp$_2$(μ-CH$_2$)(μ-Cl)Rh(1,5-C$_8$H$_{12}$)] was determined by x-ray diffraction. The Ti–Rh distance is only 0.2986(2) nm, short enough for some Ti$\cdots$Rh interaction.[490] The molecular structure of [TiCp$_2$(μ-CH$_2$)(μ-Cl)PtMe(PMe$_2$Ph)] shows a Ti$\cdots$Pt distance of 0.296(2) nm.[491] Treatment of [TiCp$_2$(μ-CH$_2$)(μ-Cl)Rh(1,5-C$_8$H$_{12}$)] with methyllithium affords the methyl-bridged derivative (**296**). The crystal structure of the complex (Figure 40) indicates an even shorter Ti$\cdots$Rh distance of 0.2835(1) nm and the presence of an agostically bonded methyl group, with a short Ti$\cdots$H bond of only 0.202(5) nm. This structure is maintained in solution; at –90 °C the ^{1}H NMR spectral signal of the μ-Me group is split into a doublet at δ 1.28 and a triplet at δ –12.15 (J = 12.8 Hz). The signals coalesce at –40 °C.[492] The alkylation of [TiCp$_2$(μ-CH$_2$)(μ-Cl)PtMe(PMe$_2$Ph)] with MeMgBr produces the structurally related complex [TiCp$_2$(μ-CH$_2$)(μ-CH$_3$)PtMe(PMe$_2$Ph)] as a crystallographically characterized red-orange compound. The μ-CH$_2$ group and the terminal Pt–Me ligand are mutually *cis*. In the presence of the Grignard reagent, slow isomerization to the *trans* complex occurs. The Ti$\cdots$Pt distance is only 0.2776(1) nm, with a narrow Ti–CH$_2$–Pt angle of 82.9°. The structural parameters are suggestive of a Pt → Ti dative interaction.[491] The reaction of [TiCp$_2$(μ-CH$_2$)(μ-Cl)Rh(1,5-C$_8$H$_{12}$)] with substituted phenyllithium reagents affords the phenyl-bridged complexes (**297**) (Equation (73)). The compound with R = H is thermally unstable and could not be isolated; (**297**) (R = OMe or NMe$_2$) form dark green crystals in 18% and 10% yields, respectively. The structure of the *p*-NMe$_2$ derivative was determined by x-ray diffraction; it shows a Ti$\cdots$Rh distance of 0.2827(1) nm.[493]

$$\text{Cp}_2\text{Ti} \diamond \!\!<\! {\overset{R^1}{\underset{R^2}{}}} + [\text{MXL}_n] \longrightarrow \underset{-\,\,{\overset{R^1}{\underset{R^2}{}}}\!\!=\!\!=}{} \quad \text{Cp}_2\text{Ti} \diamond\!\! {\underset{X}{}}\! \text{ML}_n \qquad (72)$$

(**295**)

The complex [TiCp$_2$(μ-CH$_2$)(μ-Cl)PtMe(PMe$_2$Ph)] reacts with two equivalents of PMe$_2$Ph, with cleavage of the binuclear framework to give unstable [Ti(CH$_2$)Cp$_2$(PMe$_2$Ph)] and *trans*-[PtMeCl(PMe$_2$Ph)$_2$], but with one equivalent of PMe$_2$Ph at –70 °C only the chloride bridge is cleaved to give (**298**) (Scheme 87). The addition of CO at low temperature leads to a very sensitive enolate complex. The analogous palladium complex [TiCp$_2$(μ-CH$_2$)(μ-Cl)PdMe(PMe$_2$Ph)] is thermally unstable and decomposes under methylene–methyl coupling to give, *inter alia*, [TiEtClCp$_2$].[491]

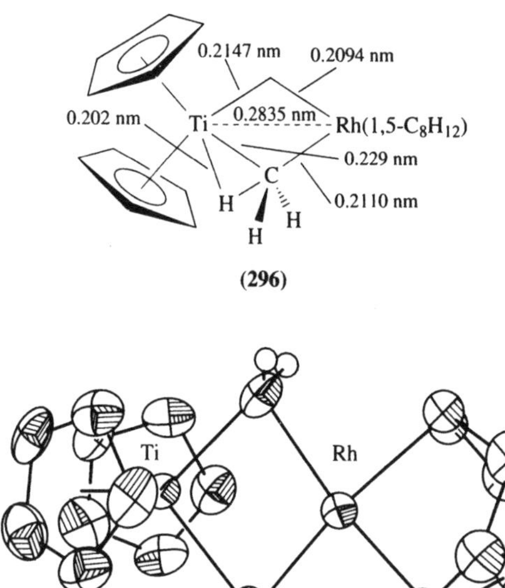

(296)

(296)

Figure 40   Molecular structure of [TiCp$_2$(μ-CH$_2$)(μ-Me)Rh(1,5-C$_8$H$_{12}$)] **(296)** (reproduced by permission of the American Chemical Society. Copyright (1986) from *J. Am. Chem. Soc.*, 1986, **108**, 6402).

(73)

(297)

Scheme 87

Titanacyclobutanes react with [M(CO)$_6$] (M = Cr or Mo) to give the ketene complex **(299)** (Scheme 88). The intermediates **(300)** and **(301)** are thermally unstable and decompose above 15 °C, but can be prepared independently from [TiCl$_2$Cp$_2$] and [Cr{=C(Me)OLi}(CO)$_5$] followed by HCl abstraction with NaN(TMS)$_2$. The complex **(300)** (M = Cr) is fluxional; the slow exchange limit is reached at −100 °C. Using [Mo(CO)$_5${P(OPh)$_3$}] instead of [Mo(CO)$_6$] gives the corresponding phosphite-substituted titanium–molybdenum complexes as a mixture of isomers.[494]

(ii) *Titanium-catalyzed metathesis polymerization*

Titanium methylidene complexes, and titanacyclobutanes as their precursors, catalyse alkene metathesis and the ring-opening metathesis polymerization of strained cyclic alkenes. The mechanistic principles and applications to the synthesis, for example, of conducting and functionalized polymers,

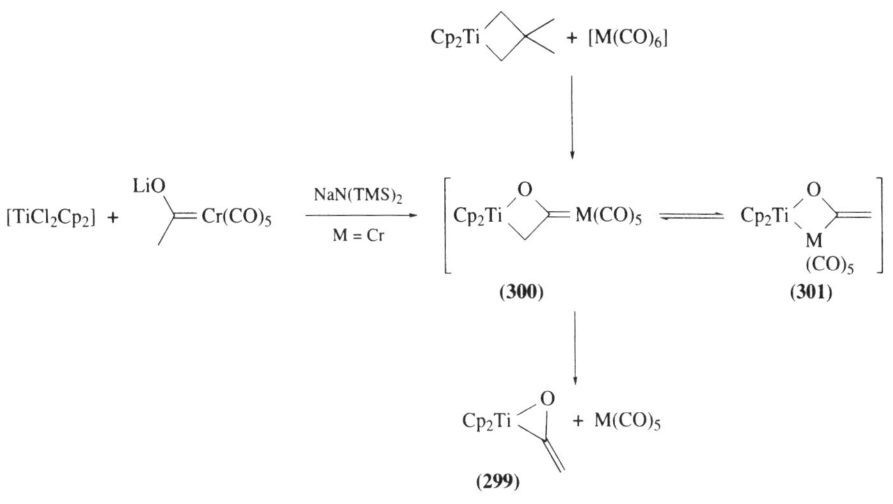

Scheme 88

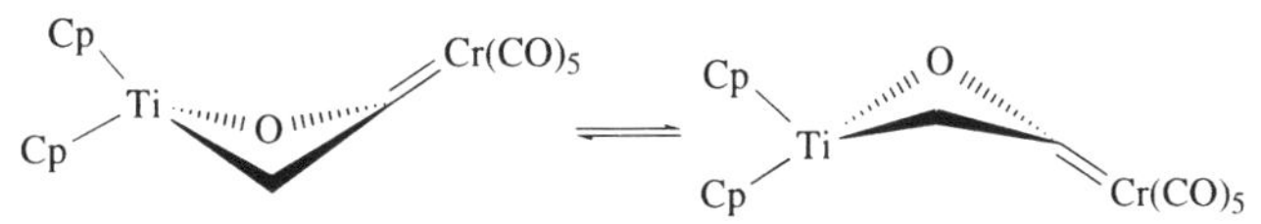

have been summarized in a review.[495] The degenerate metathesis of neohexene is catalyzed by (**272**) (R^1 = H, R^2 = Bu^t), as shown in Equation (74). The complexes (**272**) (R^1 = Me, R^2 = Pr^i) and (**272**) (R^1 = Me, R^2 = Ph) react more rapidly and at lower temperatures. The kinetics of the reaction of (**272**) (R^1 = H, R^2 = Bu^t) with diphenylacetylene to give (**284**) (R^1 = R^2 = Ph) are first order in [Ti] and independent of [C_2Ph_2]; the addition of excess neohexene has a slight inhibiting effect. These findings are consistent with a mechanism involving rearrangement of the titanacyclobutane into [Ti(CH_2)(alkene)Cp_2] in the rate-limiting step, followed by fast dissociation of the alkene and subsequent coordination of the alkyne. If one of the α-CH_2 groups of (**272**) is deuterated, a large secondary isotope effect is observed, consistent with the rehybridization of an sp^3 carbon to an sp^2 carbon, as required for the conversion of a metallacycle into a (methylidene)(alkene) complex.[496]

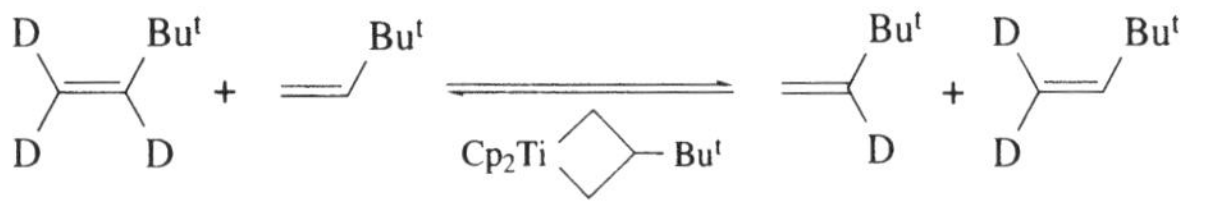

$$(74)$$

The reaction of (**233**) with norbornene in the presence of *p*-dimethylaminopyridine (dmap) affords the metallacycle (**302**), the product of '*exo*' addition of norbornene to [Ti(=CH_2)Cp_2] (Scheme 89). The compound can be isolated as red crystals in 58% yield. The addition of benzophenone leads to the diphenylmethylene end-capped hydrocarbon. The addition of an excess of norbornene at 65 °C leads to the ring-opening metathesis polymerization (ROMP) of the alkene to give poly(norbornene) with a *trans*:*cis* ratio of 1.5:1 and the narrow polydispersity (M_w/M_n = 1.13) typical of polymers produced by homogeneous metathesis processes. The complex (**303**) reacts with a 500-fold excess of norbornene to give a polymer with M_n = 92 000 and M_w/M_n = 1.08. The polymerization reaction can be followed by NMR spectroscopy. Signals for the titanium carbene intermediates are not found, but the signals of the α- and β-hydrogens of the metallacycles are observed throughout. The reaction is zero order in [norbornene].[497] The successive addition of different norbornene derivatives to catalyst precursors such as (**302**) allows the synthesis of di-block and triblock polymers.[498]

1-Methylnorbornene reacts with titanacycles following Equation (75) to give (**304**) as the only product. Under catalytic conditions, poly(1-methylnorbornene) is produced. The microstructure of the polymer is revealed by NMR spectra and shows a *trans*:*cis* ratio for the double bonds of 95:5 to 90:10, much higher than with poly(norbornene). The *syn* isomer of (**305**) does not give titanacycles for steric reasons, whereas *anti*-(**305**) is polymerized (Scheme 90).[499] Titanacyclobutanes rearrange on irradiation to titanium alkylidenes, which react rapidly with alkenes. This method has been applied to the synthesis

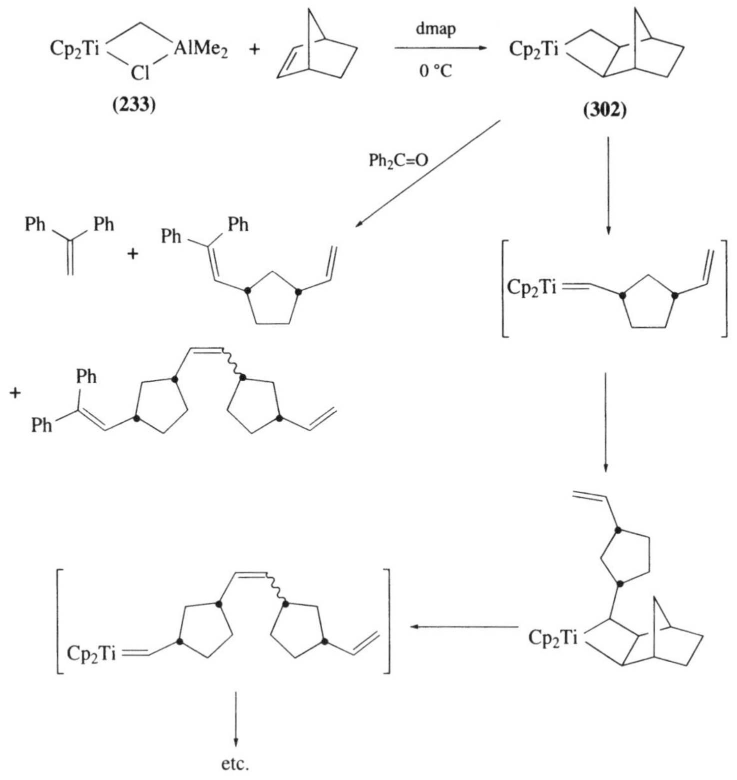

$$(233) \qquad\qquad (302)$$

Ph$_2$C=O

Scheme 89

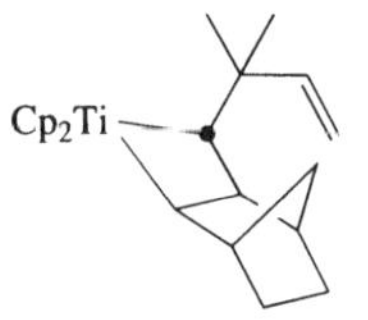

$$(303)$$

of polyfunctional initiators, using, for example, the diene (**306**) to give (**307**) as a mixture of two isomers (Scheme 91). Another example of a titanacycle which allows the growth of polymer chains linked by a siloxy bridge is (**308**), while the reaction with the cyclic norbornene derivative (**309**) leads to tetrameric catalyst precursors for the synthesis of starlike polymers Equation (76).[500]

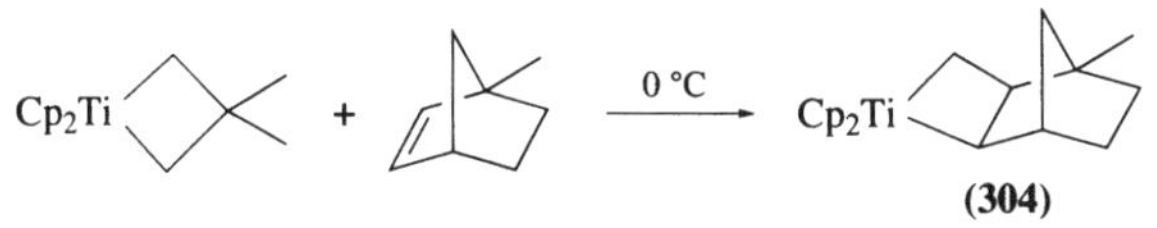

$$(304) \qquad\qquad\qquad (75)$$

5.5.2.5 *Ziegler–Natta polymerizations*

This section summarizes recent advances in the elucidation of the mechanism of the polymerization of 1-alkenes with homogeneous polymerization catalysts. Heterogeneous polymerization catalysts and systems where distinct species with titanium–carbon bonds have not been identified will not be covered.

A model of the polymerization process is the cyclization of terminally unsaturated titanium alkyl complexes. Thus, treatment of $[\text{TiCl}\{(\text{CH}_2)_n\text{CH=CH}_2\}\text{Cp}_2]$ ($n = 4$ or 5) with 1–10 equivalents of AlEtCl$_2$ in toluene at -100 °C leads to the cycloalkylmethyl complexes (**310**) as a mixture of *cis* and *trans* isomers (Scheme 92). Deuteration of the starting material produces pairs of diastereomers. A *cis:trans* ratio of 1.00 ± 0.05 was found by ^{2}H NMR spectroscopy. The results indicate that, at least in the titanium-catalysed system, α-C–H activation or agostic interactions between the α-methylene group and the metal, as suggested by Ivin *et al.*,[501] are not important.[502]

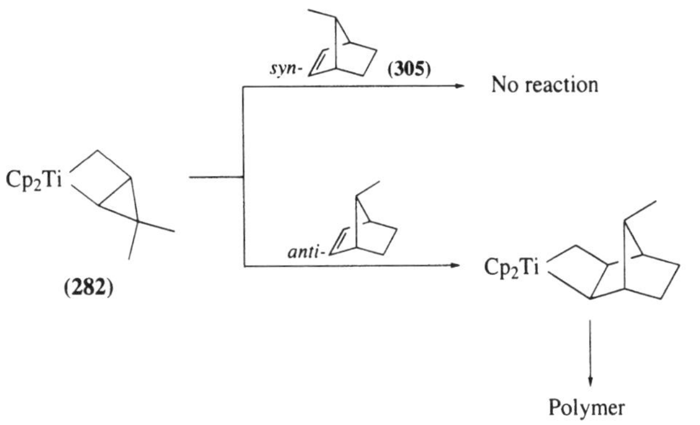

Scheme 90

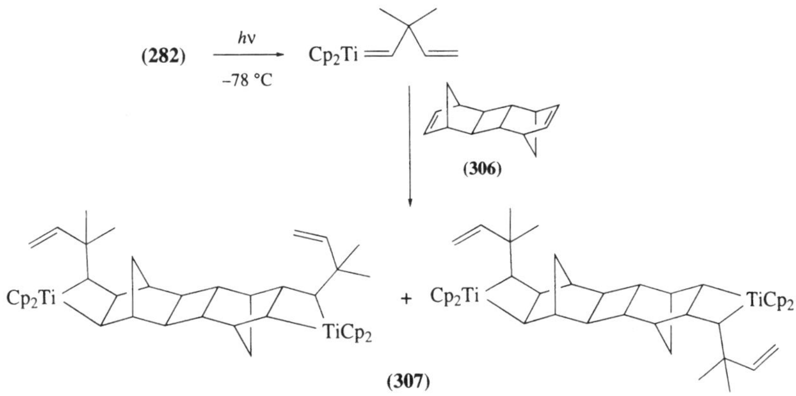

Scheme 91

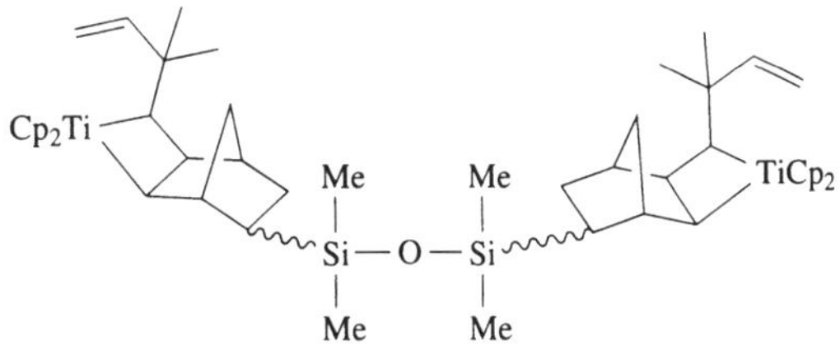

The determination of the oligomer distribution obtained from the oligomerization of ethene catalysed by [Ti(Et)ClCp$_2$]/[AlEtCl$_2$]$_2$ under fast flow conditions with different titanium:aluminum ratios and different reaction times leads to a kinetic 'intermittent ethene insertion' model of the polymerization process, as shown in Scheme 93. Here Al$_2$ is [AlEtCl$_2$]$_2$, Al$_2$' an unspecified aluminum dimer species,

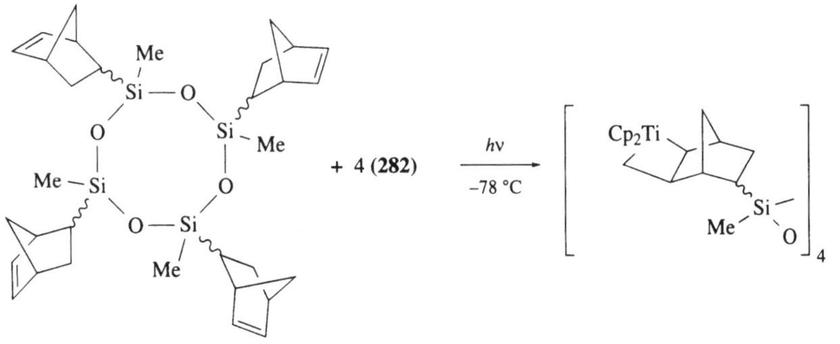

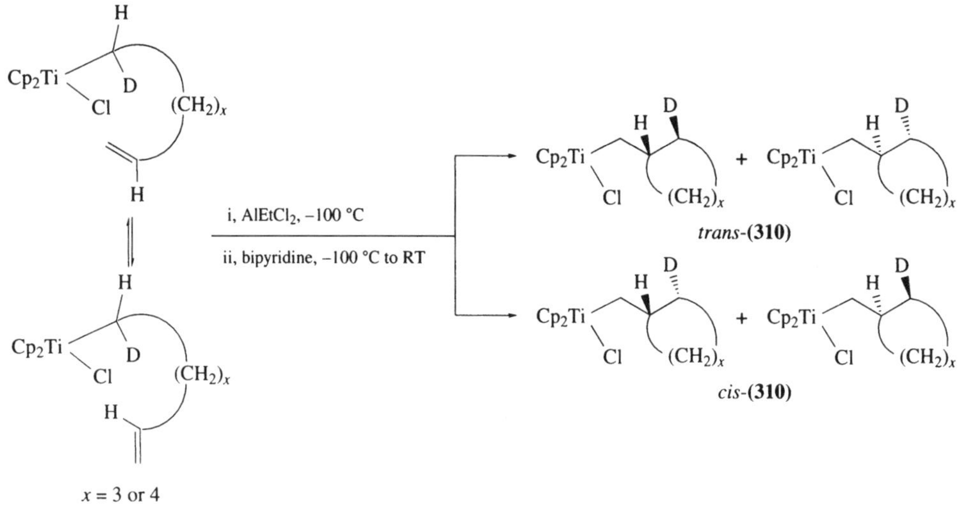

Scheme 92

C the titanium–aluminum 1:1 complex, and C* the catalytically active species of unknown structure. C_{n-1}, C_n and C_{n+1} are the observable resting stages of the catalyst. If titanium alkyls with different alkyl chain lengths are used to initiate the polymerization, for example, [TiRClCp$_2$] (R = n-propyl to n-hexyl), the growth rate k_w can be shown to decrease with increasing chain length. Alkyl transfer from titanium to aluminum also takes place, and at 283 K amounts to ~10% of total oligomer (Al:Ti = 10).[503]

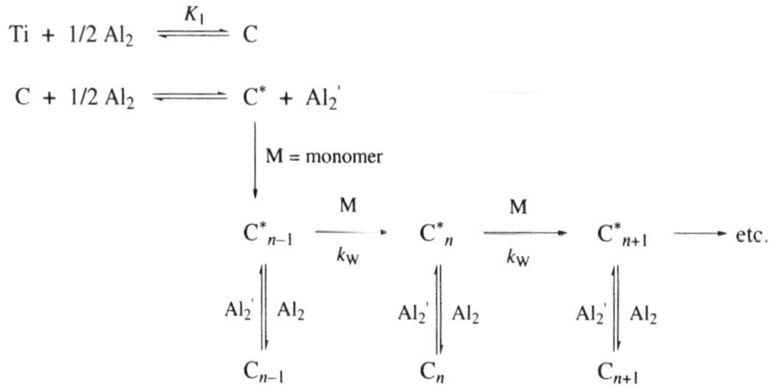

Scheme 93

Although the concentrations of the alkyl chains with maximum length lie below the detection limits of gel permeation chromatography (GPC), they can be calculated using a statistical model and related to the rate of chain growth k_w.[504] The average polymerization speed v_p depends linearly on the ethene concentration and increases with shorter reaction times since chain growth for shorter alkyl chain lengths C_nH_{2n+1} ($n \geq 2$) is faster than for longer alkyl chain lengths.[505] On the other hand, following the polymerization of ^{13}C-labelled ethene catalysed by a slow catalyst such as [Ti(Me)ClCp$_2$]/AlMe$_2$Cl shows that the rate of C_2H_4 insertion into a Ti–Me bond is 96 times slower than the insertion into a Ti–Pr bond. The relative rates for the insertion of ethene into a Ti–C_nH_{2n+1} bond are 1 (n = 1), 120 (n = 2), 96 (n = 3), 62 (n = 4), 48 (n = 5) and 47 (n = 6).[506] The determination of v_p via the oligomer distributions leads to an estimation of the concentration of active species [C*] which at Al:Ti = 2:1 is only 5% of the initial titanium concentration [Ti]$_0$, rising to 10% if the Ti:Al ratio is increased to 4:1. Alkyl chain transfer to aluminum is independent of chain length; the temperature dependence gives $\Delta G^{\ddagger}$ = 58 kJ mol^{-1}.[505] The presence of low concentrations of [C*] is confirmed by ^{13}C NMR spectra; even at a Ti:C$_2$H$_4$ ratio of 1:0.7 titanium heptyl and nonyl complexes are seen, while most of the initially employed catalyst precursor [Ti(Me)ClCp$_2$] remains unchanged. There is no evidence for detectable concentrations of a π-complex between ethene and the titanium catalyst.[507]

Although a number of chemical methods for the determination of the number of active sites [C*] in polymerization catalysts have been devised, none was found to be satisfactory. For example, tritiation with BuOT gives much lower values than the calculation of the number of polymer chains from the determination of the number-average molecular weight by GPC, while quenching with CO leads to side reactions. Quenching with SO$_2$ is unsuitable, and CO$_2$ is inert.[508]

Although kinetic data of the kind discussed above do not allow firm conclusions to be drawn about the structure of the active species C* in homogenously catalysed polymerization reactions, general mechanistic principles would suggest that the species must possess a vacant site for the coordination of the alkene in the *cis* position to the metal–alkyl bond so that migratory alkyl group transfer can occur. Since Lewis acids such as alkylaluminum halides will readily accept halide anions, the involvement of 14-electron cationic alkyl complexes of the type $[M(R)Cp_2]^+$ (M = Ti or Zr) has been suggested,[415,417,422,509,510] and the polymerization mechanism is thought to proceed in essence as depicted in Scheme 94. These ideas are borne out by MO calculations. The optimized geometries for the coordination of ethene to the $[Ti(Me)Cp_2]^+$ cation at various stages during the methyl migration process to give a titanium propyl product are shown in Figure 41. A reaction barrier for the methyl migration step of 41 kJ mol^{-1} was calculated, in agreement with experimental $\Delta G^{\ddagger}$ values of 25–50 kJ mol^{-1}. Changing the conformation of the methyl group from (**311a**) to (**311b**) stabilizes the transition state by 29 kJ mol^{-1} of which 24 kJ mol^{-1} are due to a reduction in van der Waals interactions with the cyclopentadienyl ligands. There is no evidence for a significant gain in energy as a result of agostic bonding of the methyl group, at least not in the case of titanium.[511]

(**311a**)   (**311b**)

$$Cp_2Ti\overset{R^1}{\underset{Cl}{<}} + AlR^2{}_nCl_{3-n} \;\rightleftharpoons\; Cp_2Ti\overset{R^1}{\underset{Cl}{<}}AlR^2{}_nCl_{3-n} \;\rightleftharpoons\; [Cp_2TiR^1]^+[AlR^1R^2{}_nCl_{3-n}]^- \;\xrightarrow{C_2H_4}$$

$$\left[Cp_2Ti\overset{R^1}{\underset{/\!/}{<}}\right]^+ X^- \;\longrightarrow\; [Cp_2Ti\text{-}CH_2CH_2R^1]^+X^- \;\xrightarrow{C_2H_4}\; \left[Cp_2Ti\overset{R^1}{\underset{/\!/}{<\!\!\sim}}\right]^+ X^- \;\longrightarrow\;\longrightarrow$$

$$[Cp_2Ti(CH_2CH_2)_xR^1]^+X^- \;\rightleftharpoons\; Cp_2Ti\overset{(CH_2CH_2)_xR^1}{\underset{Cl}{<}}AlR^2{}_nCl_{3-n}$$

Scheme 94

The notion of cationic titanium alkyl complexes as intermediates in insertion reactions of unsaturated organic substrates has received support in a number of stoichiometric transformations. For example, a mixture of $[TiCl_2Cp_2]$, $AlMeCl_2$ and $(TMS)C{\equiv}CPh$ in chloroform at $-20\,°C$ gives the cationic titanium vinyl complex (**312**), whose crystal structure confirmed the absence of coordination to the $[AlCl_4]^-$ anion; the narrow Ti–C–Si angle of 88.9° was explained on the basis of a Ti$\cdots$Si interaction (Scheme 95). $[TiMeCp_2]^+$ was postulated as the intermediate but could not be detected.[510] There is ^{1}H, ^{13}C and ^{27}Al NMR spectral evidence for the existence of neutral adducts and solvent-separated and contact ion pairs in the system $[Ti(CH_2TMS)ClCp_2]/AlXCl_2$ (X = Me or Cl) between $-60\,°C$ and $20\,°C$. Low temperatures favour the formation of $[Ti(CH_2TMS)Cp_2]^+[AlXCl_3]^-$, and equilibrium constants K_{eq} for the equilibrium shown in Equation (77) vary from $K_{eq} = 1.99$ at $-40\,°C$ to 0.20 at $20\,°C$. The end products of the reaction are $[TiCl_2Cp_2]$ and $Al(CH_2TMS)Cl_2$. Mixtures of $[Ti(CH_2TMS)ClCp_2]$ and $AlCl_3$ in dichloroethane at $0\,°C$ catalyse the polymerization of ethene; some of the polymer contains TMS end groups.[512,513]

In an attempt to synthesize isolable cationic titanium alkyl complexes in order to confirm their role in polymerization catalysis, various complexes with donor ligands have been prepared, such as $[TiMeCp_2(L)]^+$ (L = a nitrile, pyridine, or a phosphine).[415,416] Due to the electrophilic character of these compounds, they are only stable if weakly coordinating counteranions are present, such as

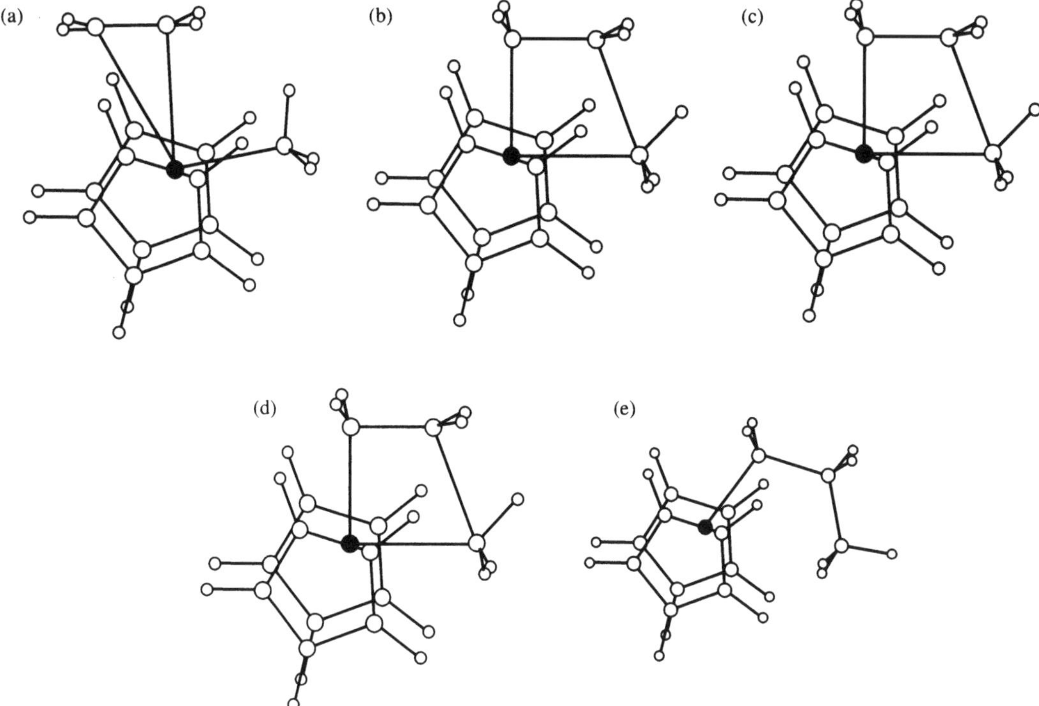

Figure 41 Calculated geometries for points along the potential energy surface for the insertion of ethene into the Ti–Me bond of a [Ti(Me)Cp$_2$]$^+$ cation. (a) Initial ethene adduct; (b) intermediate structure between reactant and transition state; (c) estimated transition state; (d) intermediate structure between transition state and product; and (e) product structure (reproduced by permission of the American Chemical Society. Copyright (1989) from *J. Am. Chem. Soc.*, 1989, **111**, 7968).

Scheme 95

$$[Cp_2Ti-R]^+[AlCl_4]^-$$

(77)

R = Cl or CH$_2$TMS

tetraphenylborate. The synthesis and reactivity of these complexes has been described (Section 5.5.2.3, cf. Schemes 54–57). In contrast with the neutral dialkyl precursors [TiMe$_2$Cp$_2$], ligand-stabilized cationic alkyl complexes undergo insertion reactions with unsaturated organic substrates such as nitriles, using the same metal orbitals that are involved during the migratory insertion of alkenes into the titanium–alkyl bond, and may hence be regarded as functional models for the chain growth step.[415,417] Ligands such as nitriles or phosphines are, however, too strongly bound to be displaced by alkenes or alkynes, and the compounds do not possess catalytic activity. By contrast, ether-stabilized complexes [TiMeCp$_2$(OR$_2$)]$^+$ (OR$_2$ = THF, Et$_2$O or PhOMe) (**313**), obtained by protolysis of [TiMe$_2$Cp$_2$] with [NHMe$_2$Ph][BPh$_4$] in the corresponding ether as isolable red–brown solids, dissociate in dichloromethane to a sufficient extent to catalyse the polymerization of ethene at 1 bar pressure (Scheme 96). The activity is modest; at 25 °C the THF complex converts ~50 mol of ethene per mole of titanium complex, while the less stable Et$_2$O complex is considerably more active and consumes 700 mole of ethene at −15 °C.[419] Similarly, the reaction of [MMe$_2$(Cp*)$_2$] (M = Ti, Zr or Hf) with [NHEt$_3$][BPh$_4$] in the presence of L = THF, tetrahydrothiophene (THT), dimethylaniline or anisole affords

[MMe(Cp*)$_2$(L)][BPh$_4$]. [TiMe(Cp*)$_2$(THT)][BPh$_4$] (**314**) is isolated as a brown solid in 61% yield. The titanium complexes are inactive, presumably as a result of steric hindrance, while the zirconium and hafnium compounds polymerize ethene and oligomerize propene. The tetrahydrothiophene ligands are rather labile and exchange rapidly with THF; consequently the THT complexes are more active catalysts than the THF compounds.[514] Base-free cationic complexes [TiMeCp'$_2$][BPh$_4$] (Cp' = Cp or Ind; Ind = η^5-indenyl) (**315**) are obtained according to Scheme 96 in dichloromethane at −40 °C. The [TiMeCp'$_2$]$^+$ cations were characterized in solution; they are most probably solvated and exist as a mixture of solvent-separated and tight ion pairs. There is no evidence for agostic bonding of the methyl ligand. In contrast with similar zirconium complexes there is no evidence for coordination to one of the phenyl rings of the tetraphenylborate anion. After injection of five equivalents of ethene into solutions of (**315**) (Cp' = Ind), the ^{1}H NMR spectrum shows the formation of polyethene, besides unreacted [TiMe(Ind)$_2$]$^+$. Exposure of these solutions to an excess of ethene at a pressure of 1 bar between −60 °C and 10 °C leads to rapid precipitation of polyethene. Propene is slowly polymerized to an atactic product.[515] While it cannot be ruled out in principle that the catalytically active species is not a cationic complex such as (**315**) but one of its decomposition or rearrangement products, the characterization of titanium alkyl cations in the absence of stabilizing donor ligands lends support to the mechanism outlined in Scheme 94.

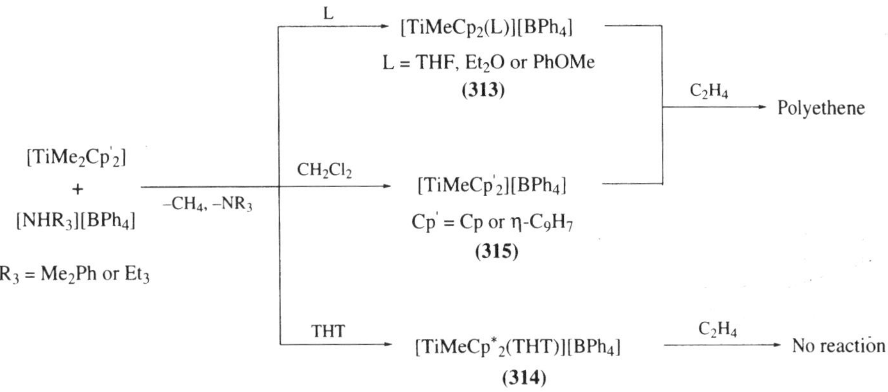

Scheme 96

The nature of the solvent and of the anion are important for the rate of formation, the stability and the catalytic activity of the cationic titanium alkyl complex. The reaction of [TiMe$_2$(Ind)$_2$] with [NHMe$_2$Ph][BR$_4$] to give [TiMe(Ind)$_2$][BR$_4$] is very much faster in more polar solvents such as dichloromethane and 1,2-dichloroethane than in bromo- and chlorobenzene. The polymerization activity is highest in CH$_2$Cl$_2$ and with less coordinating anions (R = Ph ≪ 3,5-C$_6$H$_3$(CF$_3$)$_2$).[420] Very much higher catalyst productivities are achieved with the very weakly basic anion [B(C$_6$F$_5$)$_4$]$^-$, using, for example, its triphenylcarbenium salt to generate the cationic metal alkyl complex (Scheme 97). At −60 °C a dinuclear methyl-bridged intermediate is observed. The polymerization activity of a series of complexes [MMe(η-C$_5$H$_4$TMS)$_2$][B(C$_6$F$_5$)$_4$] generated *in situ* in this way increases in the order M = Ti ≪ Hf < Zr. The productivity of the titanium catalyst at 21 °C and 1 bar of ethene, 82 kg PE ((mol Ti) bar h)$^{-1}$, is approximately two orders of magnitude less than that of zirconium.[516]

$$2\ [\text{TiMe}_2(\text{Ind})_2] \xrightarrow[\text{toluene, }-60\,°\text{C}]{[\text{CPh}_3]^+\text{X}^-} \left[\begin{array}{ccc} \text{Ind}_2\text{Ti} & -\text{Me}- & \text{TiInd}_2 \\ | & & | \\ \text{Me} & & \text{Me} \end{array}\right]^+ \text{X}^- \xrightarrow[-40\,°\text{C}]{[\text{CPh}_3]^+\text{X}^-} 2\ [\text{TiMe}(\text{Ind})_2]^+\text{X}^-$$

$$\text{X} = [\text{B}(\text{C}_6\text{F}_5)_4]$$

Scheme 97

While cationic 14-electron titanium alkyl complexes and their more extensively studied zirconium analogues can be regarded as 'well-defined' polymerization catalysts, most work on alkene polymerizations is carried out using metallocenes activated by aluminum alkyls, in particular, methylaluminoxanes (MAO). MAO are poorly defined polymeric hydrolysis products of trimethylaluminum, isolated as a toluene-soluble colourless viscous oil. Although often formulated as [AlMeO]$_n$, it is usually of the composition [AlMe$_{1.3-1.4}$O]$_n$. Used at Al:M ratios of 10^2–10^4:1 (M = Ti, Zr or Hf) it is a highly effective activator for metallocene-based polymerizations. For example, mixtures of [TiPh$_2$Cp$_2$] and MAO in toluene at −60 °C are highly active for the polymerization of propene; unusually for a metallocene without a rigid stereoselective ligand framework (cf. *ansa*-metallocenes) the resulting polypropene is highly isotactic. Maximum polymerization rates are achieved at $-45\,°C$; the catalyst decomposes at higher temperatures. The reaction is first order in $[C_3H_6]$, [Ti] and [Al]. The deactivation of the catalyst involves reduction to titanium(III). The stereoselectivity of polymerization is due to the chirality of the β-carbon of the growing polymer chain (chain-end control). By contrast, analysis of the microstructure of polypropene produced with a chiral *ansa*-titanocene catalyst, $[TiCl_2\{C_2H_4(Ind)_2\}]$/MAO, at $-60\,°C$ indicates that the stereochemistry of chain growth is subject to enantiomorphic site (as opposed to chain-end) control.[517] Structural analysis of isotactic polypropene with ^{13}C-enriched end groups synthesized with $[TiPh_2Cp_2]$/MAO/AlR$_3$ catalysts (R = Me or Et) confirms the chain-end control mechanism of isotactic-specific propagation.[518]

If the two cyclopentadienyl ligands in a metallocene catalyst are not linked as in *ansa*-metallocenes but contain substituents bulky enough to prevent free rotation, the stereoselectivity may be temperature dependent. Thus, a $[TiCl_2(\eta\text{-}C_5H_4Pr^i)_2]$/MAO catalyst polymerizes propene at $-50\,°C$ to give a predominantly isotactic polymer. At $-10\,°C$ the product is essentially atactic, and at $+10\,°C$ the polymer has an increased content of syndiotactic segments.[519]

The analysis of polypropene and poly(1-butene) with ^{13}C-enriched end groups prepared using $[TiMe_2\{C_2H_4(Ind)_2\}]$/MAO catalysts reveals a polymer structure similar to that obtained with δ-TiCl$_3$ activated by $Al(^{13}CH_3)_3$ or $Al(^{13}CH_2Me)_3$. In either case a chiral metal environment seems to determine the enantioselective monomer insertion. The insertion of 1-butene into a Ti–Me bond is partially enantioselective, while the insertion of propene into a Ti–Me bond is not selective at all; stereoselectivity is only observed once the alkyl group is longer. The insertion of propene into a Ti–Et bond shows much higher enantioselectivity.[520]

Mixtures of MAO and $[TiX_2\{Me(H)C(C_5Me_4)(\eta\text{-}Ind)\}]$ (**179**) (X = Cl) polymerize propene with a productivity of 2.5×10^5 g PP ((mol Ti) h atm)$^{-1}$ at $50\,°C$. The polymer has thermoplastic behaviour; it shows melt endotherms at $51.2\,°C$ and $66.0\,°C$, and seems to consist of isotactic and atactic blocks.[314] The crystal structure of (**179**) (X = Me) shows two distinctly different Ti–C bond lengths, and it is argued that this supports the notion that polymerization can take place at two different sites at the same metal, giving rise to atactic and isotactic polymer sections.[370,371] Similar polymers are obtained with (**179**) (X = Me) activated with $[CPh_3][B(C_6F_5)_4]$, although the organometallic species could not be identified.[371] This catalyst system is a rare case where the titanium complex is considerably more active than the zirconium analogue.[521]

Some investigations have been reported which have to be revised in the light of later results. For example, the activity of ethene polymerization by $[TiEtClCp_2]$ in the presence of excess AlCl$_3$ increases sharply on the addition of water; this has been explained assuming that protonation gives ion pairs such as $[TiEtCp_2(\mu\text{-}Cl)AlCl_3H]^+[AlCl_3(OH)]^-$.[522]

Mixtures of $[TiEtClCp_2]$ and MCl$_4$ (M = Ti or Sn) polymerize ethene to a waxy product.[523] The introduction of increasing amounts of CO into the ethene polymerization system $[TiCl_2Cp_2]$/AlEt$_2$Cl results in a decrease of activity and molecular weight. At low CO concentrations catalyst deactivation is reversible.[524] The polymerization activity of titanocene dichloride/MAO catalysts decreases with steric hindrance of the cyclopentadienyl ligand in the order $[TiCl_2Cp_2] > [TiCl_2Cp(Cp^*)] > [TiCl_2(Cp^*)_2]$, from 220 kg PE ((mol Ti) h atm)$^{-1}$ to 14 kg PE ((mol Ti) h atm)$^{-1}$.[108]

Numerous catalyst mixtures exist containing titanocene dihalides and preformed or *in situ* generated titanocene dialkyls, particularly in the patent literature. No attempt has been made to cover these comprehensively, but some representative examples are discussed below.

Exposure of $[TiCl_2Cp_2]$ to AlMe$_2$Cl at $80\,°C$ produces a green solid, which on addition of MAO gives an ethene polymerization catalyst several times more active than if pure $[TiCl_2Cp_2]$ is used.[525] Mixtures of $[TiMeClCp_2]$ and MAO in toluene at $-30\,°C$ produce a glassy solid with an Al:Ti ratio of 20:1, which on heating in solution to $80\,°C$ produces high-density polyethene.[526] Mixtures of $[TiMe_2Cp_2]$, AlMe$_3$ and water copolymerize ethene and propene to give a polymer with a nearly alternating structure. The reactivity ratios of the two alkenes have been determined by ^{13}C NMR spectroscopy. Yields and polymer molecular weights sharply decrease with increasing mole fraction of propene.[527]

Titanium alkyls such as $[TiMe_2Cp_2]$ are highly effective activators for heterogeneous polymerization catalysts, and numerous applications have been patented. Mixtures of $[TiCl_4]$, MgCl$_2$ and ethyl benzoate ground together and treated with $[TiMe_2(\eta\text{-}C_5H_4R)_2]$ (R = H or alkyl) give highly efficient catalysts for the polymerization of propene, with a high degree of stereospecificity, although the catalyst deactivates rapidly.[528] Mixtures of TiCl$_3$ and $[TiMe_2Cp_2]$ polymerize ethene to linear polyethene and higher 1-alkenes to extremely isotactic products. The catalyst is one of the most specific systems for isotactic polymerizations.[529] Mixtures of $[TiCl_4]$, MgCl$_2$ and $[TiMe_2Cp_2]$ without donor additives produce a polypropene with an isotacticity index of <97%. The addition of ethyl benzoate either as an external or internal donor caused a marked decrease of activity. It is thought that PhCO$_2$Et coordinates to active Ti^{3+} sites and cannot be removed by $[TiMe_2Cp_2]$, leading to deactivation. Catalysis in the absence of a donor is explained by assuming two types of active sites on the surface of the [TiCl$_4$]/MgCl$_2$ particles.[530,531] Block copolymers arise when TiCl$_3$/[TiMe$_2$Cp$_2$] catalysts are first exposed to propene, followed by ethene. The polymers have thermoplastic properties.[532] While TiCl$_3$/AlEt$_3$ catalysts react with propene–styrene mixtures to give blends of polypropene, polystyrene and copolymer, TiCl$_3$/[TiMe$_2$Cp$_2$] catalysts give a random styrene–propene copolymer. Propene reacts over 700 times faster than styrene.[533,534]

Catalysts prepared from [Ti(acac)$_3$], MgCl$_2$ and Al$_2$Cl$_3$Et$_3$ and activated by [TiMe$_2$Cp$_2$] in heptane polymerize styrene to give a strictly isotactic polymer. The use of AlEt$_3$ instead of [TiMe$_2$Cp$_2$] gives a catalyst with reduced activity.[535] Mixtures of [TiCl$_2$Cp$_2$] or [TiCl$_2$(Cp*)$_2$] and MAO react with styrene to give a highly syndiotactic polymer, while similar complexes of Zr, Hf, V, Nb, Cr, Co or Ni give atactic polystyrene.[132] Syndiotactic polystyrene is also produced from [TiCl$_2$Cp$_2$]/MAO/AlEt$_3$ catalysts in toluene at 50 °C.[536] Metallocene dihalides [MCl$_2$Cp$_2$] activated by aluminum alkyls in dichloromethane at 25 °C catalyse the transannular polymerization of 1,5-cyclooctadiene (Equation (78)), although the structure of the polymer remains somewhat uncertain. The activity decreases in the sequence M = Ti > Zr > Hf and Al$_2$Cl$_3$Et$_3$ > AlClEt$_2$ > AlEt$_3$. The molecular weight is low, 860–1760, with a very narrow polydispersity of only 1.2.[537]

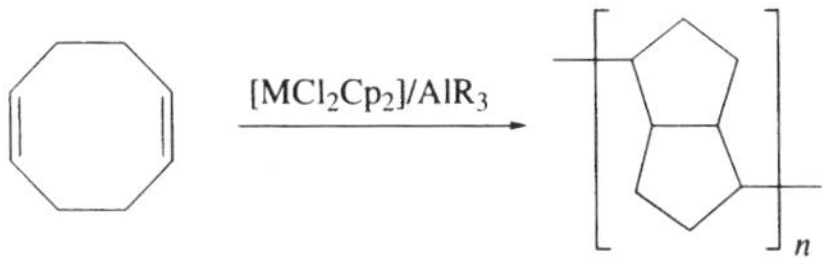

$$ \text{(78)} $$

5.5.3 Complexes with Ti–Si, Ti–Ge and Ti–Sn Bonds

The reaction of [TiCl$_2${Me$_2$Si(C$_5$H$_4$)$_2$}] (**108**) with Al(TMS)$_3$ affords the silyl complex [TiCl(TMS){Me$_2$Si(C$_5$H$_4$)$_2$}] as light green crystals in 43% yield.[369] On exposure to carbon monoxide the complex [TiCl(TMS)Cp$_2$] reductively eliminates TMS-Cl to give [TiCp$_2$(CO)$_2$], in contrast with the zirconium analogue, which undergoes CO insertion into the Zr–Si bond.[538] The crystal structure of [TiSi$_5$Ph$_{10}$Cp$_2$] (**316**) has been determined. The compound forms dark green crystals. The TiSi$_5$ six-membered ring adopts a chair conformation, with Ti–Si bond distances of 0.2765(8) nm and 0.2755(7) nm, significantly longer than the Ti–Si bond of 0.267(1) nm in [TiCl(TMS)Cp$_2$].[539]

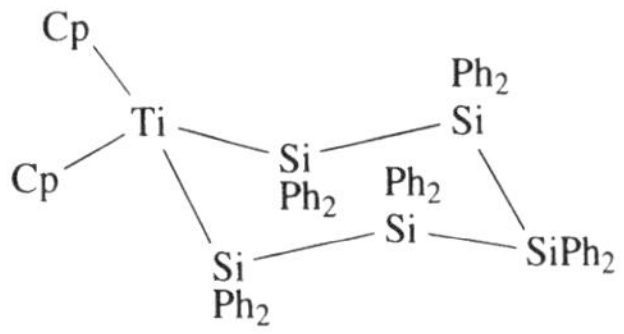

(**316**)

Titanocene dimethyl reacts with GeHPh$_3$ in hexane at 75 °C under protolysis of one Ti–Me bond to give [TiMe(GePh$_3$)Cp$_2$] as moderately air-sensitive dark violet crystals in 84% yield. No product was isolated from the reaction with SiHR$_3$ or SnHR$_3$ (R = Ph or Et), reduced titanium species being obtained instead. The germyl complex reacts with CO in diethyl ether/hexane at room temperature under insertion into the Ti–Me bond to give the black-purple acyl complex [Ti(η^2-COMe)(GePh$_3$)Cp$_2$] in 92% yield. The compound exhibits ν_{CO} at 1600 cm^{-1}; the crystal structure shows a Ti–Ge distance of 0.2710(2) nm.[540] Titanocene dichloride reacts with GeI$_2$(CF$_3$)$_2$ in dimethoxyethane in the presence of tin as the reducing agent to give [TiCl{Ge(CF$_3$)$_2$Cl}Cp$_2$]; free Ge(CF$_3$)$_2$ is probably not present under these conditions.[541] A conceptually similar reaction is the treatment of SnCl$_2$Ph$_2$ with lithium, followed by the addition of [TiCl$_2$Cp$_2$] to give [TiCl(SnPh$_3$)Cp$_2$] in 40% yield. The structure of the complex was confirmed by x-ray diffraction and shows a Ti–Sn bond length of 0.2843(1) nm.[542]

5.5.4 Complexes with Ti–N Bonds

This section describes complexes with simple donor and anionic nitrogen ligands. Reactions of nitrogen ligands with titanium alkyls are described in Section 5.5.2.3, and reactions with nitrogen-containing heterocycles in Section 5.5.5.

The mean titanium–nitrogen bond enthalpies E(Ti–N) of some titanocene derivatives have been determined by reaction solution calorimetry, together with corrections from EHMO calculations to evaluate the reorganizations enthalpies of the titanocene fragments. For $[Ti(N_3)_2Cp_2]$ and $[Ti(NC_8H_6)_2Cp_2]$ (NC_8H_6 = indolate), E(Ti–N) values of 376 ± 10 kJ mol^{-1} and 334 ± 10 kJ mol^{-1}, respectively, were determined,[543] while for the azobenzene complex $[TiCp_2(N_2Ph_2)]$ a value of 290 ± 12 kJ mol^{-1} was found.[544]

Cationic acetonitrile complexes are obtained when $[TiCl_2Cp_2]$ is treated with Lewis acids such as $SbCl_5$ or $SnCl_4$ in acetonitrile solution. Depending on the stoichiometry, the olive-green complex $[TiCl(MeCN)Cp_2][SbCl_6]$ (**317**) or the mauve complex $[Ti(MeCN)_2Cp_2][SbCl_6]_2$ (**318**) can be isolated (Scheme 98).[545] Dinitriles act only as monodentate ligands, as in the crystallographically characterized red to purple complexes (**319**) and (**320**). $[TiCl_2Cp_2]$ reacts with $PhCH=CHCN$ in the presence of $FeCl_3$ and traces of water to give $[\{TiCp_2(PhCH=CHCN)\}_2(\mu\text{-O})][FeCl_4]$ as orange–red crystals. The crystal structure of the complex has been determined; the two halves of the dimer are twisted by about 90° with respect to each other. The Ti–N bond lengths are 0.212–0.215 nm, as expected for nitrogen-donor ligands.[546] The reduction of $[TiCl_2Cp_2]$ by zinc in acetonitrile affords blue crystals of $[Ti(MeCN)_2Cp_2]_2[ZnCl_4]$; the average Ti–N distance in this complex is 0.216 nm. Similar titanium(III) complexes are accessible from $[(TiClCp_2)_2]$.[547] The primary product of the zinc reduction of titanocene dichloride is probably $[TiClCp_2 \cdot ZnCl_2]$. On addition of $PhCH_2CN$, an unstable deep-blue adduct is formed, which rearranges to give the orange titanium(IV) imido complex (**321**) (Scheme 99), whose molecular structure has been determined. The distances to the σ-bonded imido ligands are short, 0.1896(3) nm, indicative of partial double bond character and the participation of the nitrogen lone pair.[548]

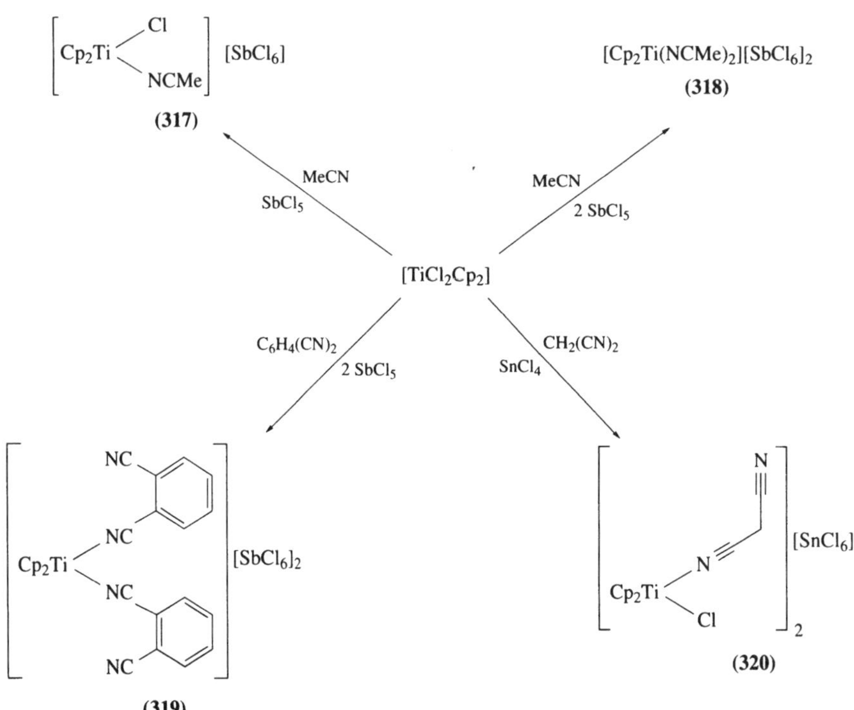

Scheme 98

[TiCl_2Cp_2] $\xrightarrow{\text{Zn}}$ [TiClCp_2·ZnCl_2] $\xrightarrow{\text{PhCH_2CN}}$ [TiCp_2(NCCH_2Ph)_2]_2[ZnCl_4] $\longrightarrow$
                                    green   blue

Scheme 99

Bidentate ligands L–L such as 2,2'-bipyridyl and 1,10-phenanthroline react with $[Ti(O_3SCF_3)_2Cp_2]$ under displacement of the triflate anions to give red-brown crystals of $[TiCp_2(L-L)][CF_3SO_3]_2$, whose crystal structures were determined.[549]

Titanocene dichloride reacts with two equivalents of purine in the presence of triethylamine as base to give the purinato complex (322) as dark-red crystals (Equation (79)). The compound is regarded as a possible model for the interaction of $[TiCl_2Cp_2]$ with DNA. The purinato ligand is almost coplanar with the TiClN plane; the dihedral angle is 7.2°.[550] Diazadienes react with $[TiCl_2Cp_2]$ in the presence of magnesium metal to give the folded diazametallacycles (323) (Equation (80)). The complexes are fluxional; the inversion barriers of the five-membered rings are $54–57 \pm 1.5$ kJ mol^{-1}.[551]

$$[TiCl_2Cp_2] + \text{purine} \xrightarrow{\text{NEt}_3} \text{(322)} \tag{79}$$

$$[TiCl_2Cp_2] + \text{diazadiene} \longrightarrow \text{(323)} \tag{80}$$

$R^1 = R^2 = Ph$

$R^1 = Ph, R^2 = Me$

Titanocene dichloride reacts with $1,2\text{-}(LiNH)_2C_6H_4$ to give the black diazametallacycle $[Ti\{(NH)_2C_6H_4\}Cp_2]$. The compound is light-sensitive in solution.[552] The Tebbe complex (233) in the presence of base reacts with nitriles with insertion and elimination of $AlClMe_2$, as outlined in Scheme 100. With two equivalents of pivalonitrile the diazametallacycle (324) is obtained, which rearranges slowly to the conjugated isomer (325). The structure of (324) (R = 1-adamantyl) was confirmed by x-ray diffraction. With only one equivalent of nitrile and an excess of base (dmap = 4-dimethylaminopyridine) the monomeric titanium imide complex (326) is obtained. If trimethylphosphine is present, the analogous phosphine complex (327) is isolated. The imido ligand can be liberated by hydrolysis. The complex (327) undergoes cyclizations, for example, with acetone to give (328), or with pivalonitrile to give (325).[553,554]

The addition of $LiN=CPh_2$ to $[TiCl(Cp^*)_2]$ leads to a brown oil, probably $[Ti(N=CPh_2)(Cp^*)_2]$, which is oxidized by $Ag[BPh_4]$ in Et_2O/THF in the presence of traces of moisture to give orange crystals of $[Ti(OH)(HN=CPh_2)(Cp^*)_2][BPh_4]\cdot Et_2O$. The molecular structure of the compound shows Ti–N and Ti–O distances of 0.2170(9) nm and 0.1853(7) nm, respectively, consistent with a partial double bond character of the Ti–O bond and a simple donor interaction to the imine ligand.[555]

Titanocene dichloride reacts with $[NH_4][NSO]$ to give the yellow complex $[Ti(NSO)_2Cp_2]\cdot 0.5CH_2Cl_2$.[556] The same product is obtained from $[TiCl_2Cp_2]$ and $K[NSO]$ in acetonitrile, or from $[TiCl_2Cp_2]$ and $[Hg(NSO)_2]$. The complex is air-stable in the solid state. The crystal structure demonstrates the bent geometry of the NSO ligand. $[Ti(NSO)_2Cp_2]$ reacts with two equivalents of $LiN(TMS)_2$ to give $[Ti(NSN(TMS))_2Cp_2]$ as yellow plates in 65% yield.[557,558] The closely related yellow sulfur diimido complexes $[TiCl(NSNBu^t)Cp_2]$ and $[Ti(NSNBu^t)_2Cp_2]$ are prepared from $[TiCl_2Cp_2]$ and one or two equivalents of $K[NSNBu^t]$ in 24% and 70% yields, respectively. On contact with silica, $[Ti(NSNBu^t)_2Cp_2]$ hydrolyses to give $[Ti(NSO)_2Cp_2]$. With methylcyclopentadienyl ligands only the monochloride $[TiCl(NSNBu^t)(\eta\text{-}C_5H_4Me)_2]$ was isolated, whereas from $[TiCl_2(Cp^*)_2]$ the red 1:2 complex $[Ti(NSNBu^t)_2(Cp^*)_2]$ is produced; in contrast with its Cp analogue it is resistant to hydrolysis.[559]

The isothiocyanato complex $[Ti(NCS)_2Cp_2]$ is electrochemically reduced in THF or dichloromethane to $[Ti(NCS)_2Cp_2]^-$. In DMF the reduction is followed by irreversible dimerization.[560] The photolysis of $[Ti(N_3)_2Cp_2]$ in a CO-doped argon matrix leads to a mixture of $TiCp_2(CO)$, $[TiCp_2(CO)_2]$ and larger amounts of $[Ti(NCO)_2Cp_2]$, presumably via titanocene as the intermediate. The same reaction in solution under a pressure of 50 bar of CO affords predominantly the isocyanato complex $[Ti(NCO)_2Cp_2]$.[561] $[Ti(NCO)_2Cp_2]$ reacts with two equivalents of $LiN(TMS)_2$ to give $[Ti(NCN(TMS))_2Cp_2]$, while

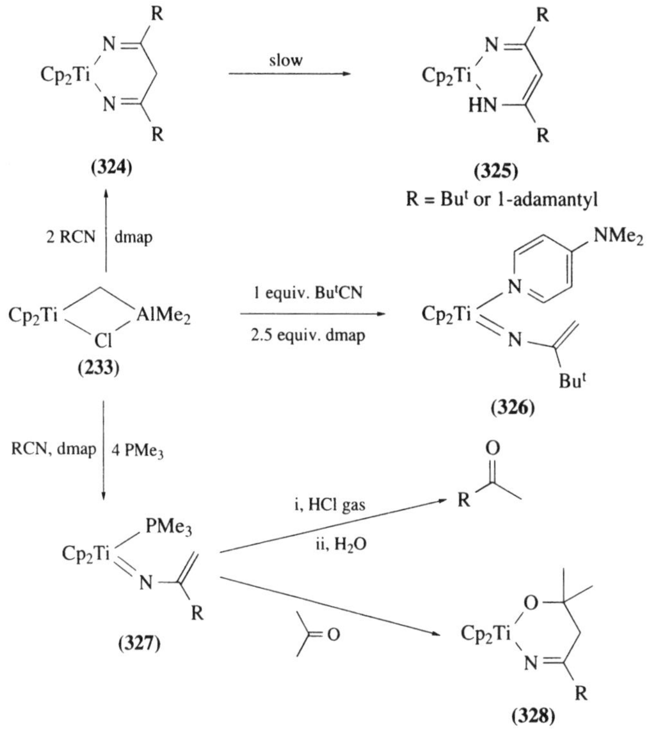

Scheme 100

[Ti(NCO)$_2$(η-C$_5$H$_4$Me)$_2$] and LiN(TMS)$_2$ give red crystals of [Ti(NCO)(NCN(TMS))(η-C$_5$H$_4$Me)$_2$]. The products are thermally sensitive.[562] The reaction of [TiCp$_2$(CO)$_2$] with S$_4$N$_4$ in THF followed by chromatography and recrystallization from acetonitrile affords the green–black air-stable metallacycle [Ti(N$_4$S$_3$)Cp$_2$]. The N$_4$S$_3$ ring is bound to titanium via two nitrogens, with Ti–N distances of 0.195(3) nm, and is essentially planar.[563]

5.5.5 Complexes with Schiff Bases and Heterocyclic Compounds

Whereas the mono(cyclopentadienyl) complexes with Schiff bases and polyfunctional heterocyclic ligands tend to adopt a pentagonal-bipyramidal geometry (cf. Section 5.4.4), bis(cyclopentadienyl) complexes retain the tetrahedral geometry of the metal centre and form either neutral or cationic compounds. Numerous derivatives have been described. Bidentate Schiff bases such as salicylimides HL = (329) react with [TiCl$_2$Cp$'_2$] to give [TiCl(L)Cp$'_2$]. According to the IR spectrum the ligand is bidentate (Cp' = Cp or η-Ind).[564-6] With HL = (330) (R = Ph, o-C$_6$H$_4$OMe or m-C$_6$H$_4$Cl) or (331), similar compounds [TiCl(L)Cp$_2$] with presumably bidentate ligands L are obtained,[567,568] while with H$_2$L = (332) a monomeric product [Ti(L)Cp$_2$] has been formulated.[568] Titanocene dichloride reacts with diarylthiocarbazones H$_2$L = (333) in THF in the presence of triethylamine to give compounds of the stoichiometries [Ti(LH)$_2$Cp$_2$], [TiCl(LH)Cp$_2$] or [Ti(L)Cp$_2$], depending on the reactant ratios.[569,570] Similar products are derived from H$_2$L = (334) and (335). Titanocene complexes with (334) (X = SH) have antibacterial activity.[571] The product of the reaction of [TiCl$_2$Cp$_2$] with H$_2$L = (336) has the composition [(TiCp$_2$)$_3$Cl(HL)].[572] The product of the reaction of [TiCl$_2$Cp$_2$] with 6-aminopenicillinic acid in the presence of NEt$_3$ has the structure (337). Coordination of the lactam C=O function in the water-soluble complex is indicated by a lowering of ν_{CO} by 115 cm^{-1} to 1650 cm^{-1}.[573]

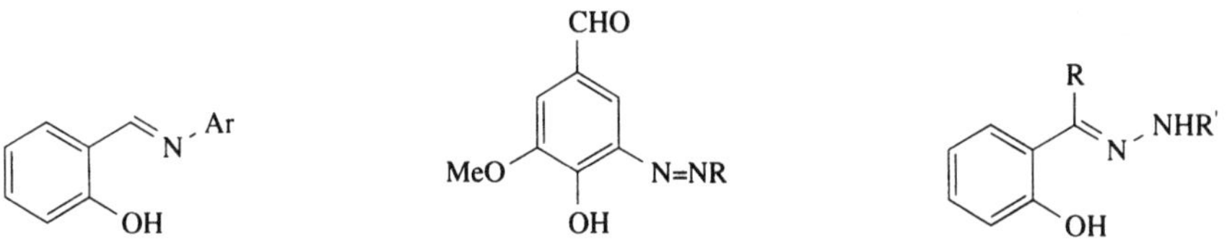

(332)

(333)

X = OH or SH
(334)

(335)

(336)

(337)

The reaction of [TiCl$_2$Cp$_2$] with salicylaldehyde in the presence of NaX or KX affords the 1:1 electrolyte [Ti(L)Cp$_2$]X (L is bidentate). A series of salts with X = variously substituted ROCS$_2^-$ and R^{1}R^{2}NCS$_2^-$ anions has been prepared. Similar products are obtained from **(338)**.[574,575] The reaction of [TiCl$_2$CpCp'] and 8-hydroxyquinoline proceeds in water to give [Ti(L)CpCp']X (Cp' = Cp, η-Ind or η-pyrrolyl). A series of salts with X = Cl, Br, I, CuCl$_3$, ZnCl$_3$(H$_2$O), $\frac{1}{2}$ CdCl$_4$, HgCl$_3$ or R^{1}R^{2}NCS$_2$ has been prepared. All these complexes are electrolytes.[576–80] The reaction of [TiCl$_2$Cp$_2$] with HL = 8-quinilinol-*N*-oxide in acetonitrile in the presence of NEt$_3$ as the base affords [TiL$_2$Cp$_2$], while the reaction with NaNH$_2$ in toluene gives [TiCl(L)Cp$_2$].[581] Ionic complexes with a wide range of anions are also isolated from the reaction of [TiCl$_2$Cp$_2$] with 3-hydroxy-2-methyl-1,4-naphthaquinone.[582] Nucleosides with ionizable protons react with [TiCl$_2$Cp$_2$] in methanol to give 1:2 electrolytes of the composition [TiCp$_2$(NuclH)(MeOH)]Cl$_2$, which on addition of OH$^-$ afford [TiCp$_2$(Nucl)(MeOH)]Cl (Nucl = adenosine, guanosine, cytidine or inosine). No bis(nucleoside) complexes have been obtained.[583]

X = OH or NH$_2$
(338)

5.5.6 Complexes with Ti–O Bonds

5.5.6.1 Titanium oxo, hydroxo and alkoxo complexes

The oxophilic character of titanium is well known. Thermochemical measurements confirm the strength of Ti–O bonds. For [Ti(OPh)$_2$Cp$_2$], mean Ti–O bond disruption enthalpies of 366 ± 11 kJ mol^{-1} and mean bond enthalpies of 455–462 kJ mol^{-1} have been measured. These values are typical for titanium alkoxides and phenoxides.[544,584,585]

The oxidation of [Ti(Cp*)$_2$] or [Ti(Cp*)$_2$(η^2-C$_2$H$_4$)] with N$_2$O in THF in the presence of L = pyridine or 4-phenylpyridine gives the unusual monomeric titanium oxo complexes **(339)** (Scheme 101). The compounds are diamagnetic and $v_{Ti=O}$ occurs at 852 cm^{-1}, in agreement with the short crystallographically determined Ti=O bond length of 0.1665(3) nm.[586] A similar reaction in toluene produces the lime-green dimeric μ-oxo complex **(340)**.[587]

The reaction of [TiCl$_2$Cp$_2$] with aqueous Na[HCO$_3$] at pH 4–4.5 produces a mixture of [TiCl$_2$Cp$_2$], [Ti$_2$Cl$_2$(μ-O)Cp$_4$] and [Ti$_3$Cl$_3$(μ-O)$_2$Cp$_5$] **(341)**. Only the trimer **(341)** is formed if [TiCl$_2$Cp$_2$] is treated with Ag$_2$O in chloroform in the presence of water. The crystal structure of the chloroform solvate has been determined and shows almost linear Ti–O–Ti arrangements with Ti–O bond lengths ranging from 0.1759 nm to 0.1880 nm.[588] The reaction of [Ti(O$_2$CPh)$_2$Cp$_2$] with sulfuric acid in acetic anhydride affords the dicationic aquo complex [TiCp$_2$(H$_2$O)$_2$][SO$_4$] in 85% yield. A similar reaction with H$_2$SeO$_4$ leads to [Ti(SeO$_4$)Cp$_2$], while on addition of K$_2$[Cr$_2$O$_7$] to the acidified reaction mixture the dichromate [TiCp$_2$(H$_2$O)$_2$][Cr$_2$O$_7$] is obtained. Both the selenate and the dichromate compounds are explosive on heating or mechanical agitation.[589] The reaction of [TiCl$_2$(Cp*)$_2$] with Ag[CF$_3$SO$_3$] in aqueous THF affords the dark-purple complex [Ti(Cp*)$_2$(H$_2$O)$_2$][CF$_3$SO$_3$]$_2$, whose crystal structure shows the presence of hydrogen bonds to two oxygens of the anion. From the hot reaction mixture the orange-red aquohydroxo complex α-[Ti(OH)(Cp*)$_2$(H$_2$O)][CF$_3$SO$_3$]·H$_2$O is crystallized. Another crystal

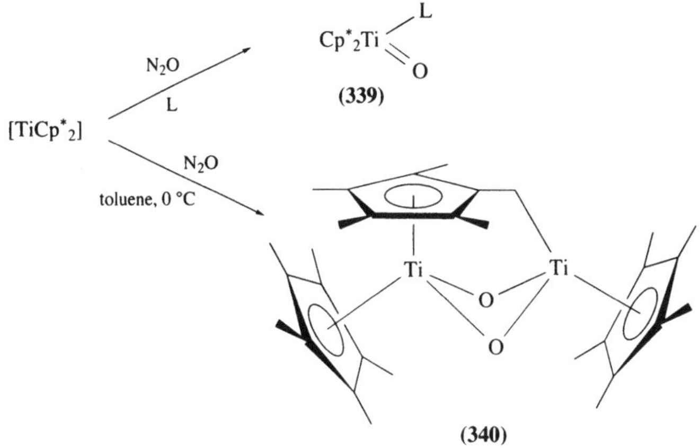

Scheme 101

modification, β-[Ti(OH)(Cp*)$_2$(H$_2$O)][CF$_3$SO$_3$]·H$_2$O, has also been identified. From the supernatant of this fraction cooled to 4 °C another hydrate, [Ti(OH)(Cp*)$_2$(H$_2$O)][CF$_3$SO$_3$]·2H$_2$O (**342**), can be isolated. The structures differ in the degree of hydrogen bonding.[590] A related complex with hydrogen bonds to THF, [Ti(OH)(Cp*)$_2$(H$_2$O)][BPh$_4$]·2THF (**343**), has been isolated as the product of the reaction of [TiMe(Cp*)$_2$] with Ag[BPh$_4$] in THF in the presence of traces of moisture. As with (**342**), the OH ligand in (**343**) is not involved in hydrogen bonding; it is oriented in such a way as to maximize the π-interaction between the oxygen lone pair and the metal centre (Figure 42). As a result, the Ti–OH bond length is 0.1853(5) nm, while the Ti–OH$_2$ distance is 0.2080(5) nm.[421] The reaction of [TiCl$_2$(Cp*)$_2$] with Ag[CF$_3$SO$_3$] in dry dimethylformamide affords the purple complex [Ti(Cp*)$_2$(DMF)$_2$][CF$_3$SO$_3$]$_2$ together with low yields of the red complex [TiCl(Cp*)$_2$(DMF)][CF$_3$SO$_3$]. Both compounds were characterized by x-ray diffraction; the Ti–O distances in the [Ti(Cp*)$_2$(DMF)$_2$]$^{2+}$ dication (**344**) are 0.1994(4) nm and 0.2007(4) nm.[591]

Titanocene phenoxides [Ti(OAr)$_2$Cp'$_2$] have been prepared from [TiCl$_2$Cp'$_2$] and substituted phenols ArOH using NaH or NaNH$_2$ as a base (Cp' = Cp or C$_5$H$_4$Me; Ar = C$_6$H$_4$Bu^t-4, C$_6$H$_3$Cl$_2$-2,4, C$_6$H$_2$Cl$_3$-2,4,6, C$_6$H$_3$Me$_2$-2,4, etc.).[592] Similar complexes are obtained in chloroform/water mixtures using Na[HCO$_3$] as the base. Depending on the reactant ratio, [TiCl(OAr)Cp$_2$] or [Ti(OAr)$_2$Cp$_2$] can be prepared (Ar = C$_6$H$_2$Cl$_3$-2,4,6).[593] The structure of [Ti(OC$_6$H$_2$Cl$_3$-2,4,6)$_2$Cp$_2$] has been determined by x-ray crystallography.[594] The reaction of [TiCl$_2$Cp'$_2$] with sodium α- or β-naphtholate or α- or β-naphthol in aqueous media leads to [TiCl(OAr)Cp'$_2$] or [Ti(OAr)$_2$Cp'$_2$] (Cp' = Cp or C$_5$H$_4$Me; Ar = naphthyl).[595,596] The reaction of [TiCl$_2$Cp$_2$] with H$_2$L = phenolphthalein gives a product of the stoichiometry [Ti(L)Cp$_2$]; IR and NMR spectral data suggest that the compound is monomeric.[597] By contrast, the product of the reaction of [TiCl$_2$Cp$_2$] with lithium catecholate gives oligomeric [{Ti(1,2-O$_2$C$_6$H$_4$)Cp$_2$}$_n$] (2 ≤ n ≤ 4).[598] Titanocene dichloride and sodium ascorbate afford the bis(ascorbato) complex (**345**) with monodentate ascorbato ligands in solution, although intermolecular interactions are possible in the solid state.[599] Sodium 5,6-isopropylidene-L-ascorbate (iasc) and [TiCl$_2$Cp$_2$] give [Ti(iasc)Cp$_2$].[600]

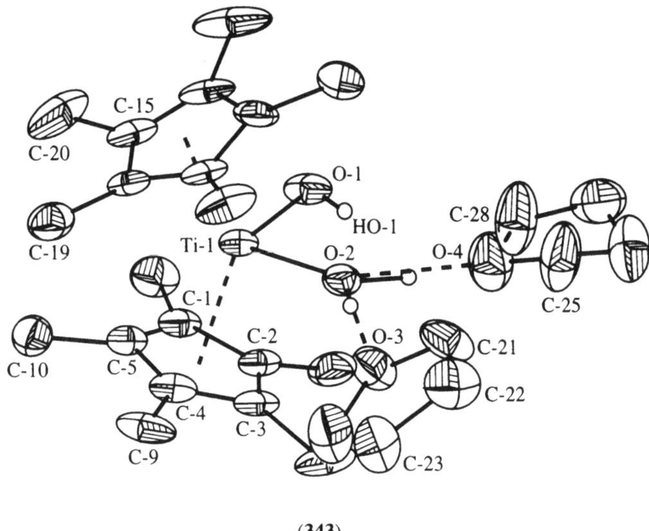

(343)

Figure 42　Molecular structure of [Ti(OH)(Cp*)$_2$(H$_2$O)][BPh$_4$]·2THF (**343**) (reproduced by permission of Elsevier from *Polyhedron*, 1989, **8**, 1838).

(344)

(345)

The protolysis of [TiMe$_2$Cp$_2$] with an alcohol R^{1}OH affords [TiMe(OR^1)Cp$_2$] (R^1 = CHMeC$_8$H$_{17}$, CMePh$_2$, Ch(Ph)C$_3$H$_7$, CH(Ph)CH$_2$Ph or CH$_2$CH=CHC$_3$H$_7$). Treatment of [TiMe(OCMePh$_2$)Cp$_2$] with benzoic acid leads to [Ti(OCMePh$_2$)(O$_2$CPh)Cp$_2$].[601] In other cases, further reaction with the carboxylic acid R^{2}CO$_2$H does not lead to [Ti(OR^1)(O$_2$CR^2)Cp$_2$] as expected, but produces [TiMe(O$_2$CR^2)Cp$_2$] (R^1 = PhCH$_2$, menthyl or bornyl; R^2 = Ph$_2$CH, Ph or PhCH=CH). In these cases the alkoxocarboxylato complexes are accessible in modest yields by refluxing [TiMe(O$_2$CR^2)Cp$_2$] with R^{1}OH in toluene.[602] The reaction of [TiMe$_2$Cp'$_2$] with squaric acid affords the squarate complexes (**346**) (Cp' = Cp or Cp*).[601] The squarate complex was also obtained from disodium squarate and [TiCl$_2$Cp$_2$] in the presence of water.[603]

(346)

The substitution reaction of [Ti(NCX)(OC$_6$H$_2$Me$_3$-2,4,6)Cp$_2$] with HBr affords [Ti(NCX)BrCp$_2$] with retention of configuration (X = O or S). By contrast, the subsequent substitution of [Ti(NCX)BrCp$_2$] with PhS$^-$ gives [Ti(NCX)(SPh)Cp$_2$] with retention of configuration if X = S but inversion if X = O. EHMO calculations and photoelectron spectroscopic results suggest strong through-bond interactions between the orbitals of the σ ligands (Br, NCX or OAr) and the Cp orbitals, which influence the localization of vacant orbitals; this in turn influences the site of nucleophilic attack and hence the stereochemistry.[604]

Titanocene enolate complexes $[Ti(OCH=CR_2)_2Cp_2]$ are prepared from $[TiCl_2Cp_2]$ and the lithium enolates $LiOCH=CR_2$ in THF as orange (R = H) or red (R = Me) crystals. $[TiMeClCp_2]$ and $LiOCH=CR_2$ give $[TiMe(OCH=CR_2)Cp_2]$. The molecular structure of $[Ti(OCH=CH_2)_2Cp_2]$ is shown in Figure 43. The Ti–O distances are normal, 0.1903(2) nm, while the C=C bonds are very short (0.1306(5) nm).[605] The structurally related ketene acetal complex (347) is obtained from the metallated isobutyric methyl ester as black crystals according to Equation (81); the structure of the compound was confirmed crystallographically (Ti-O 0.1888(7) nm).[606] The iron–titanium enolato complex (348) reacts stereospecifically with benzaldehyde to give (349), which is converted to the cationic complex (350) on chloride abstraction with sodium tetraphenylborate (Scheme 102). The structures of (349) and (350) were confirmed by x-ray diffraction.[607]

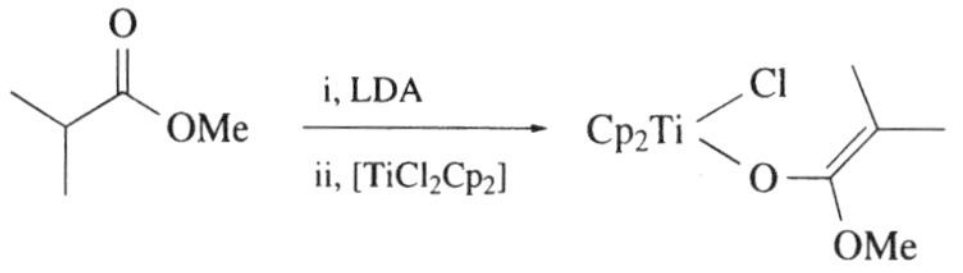

$$\text{(81)}$$

(347)

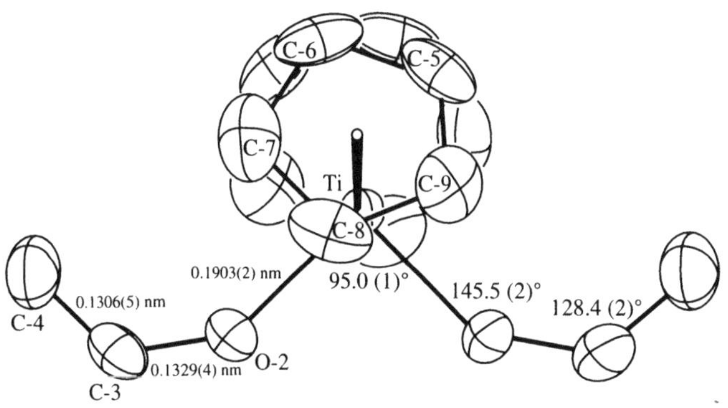

Figure 43 Molecular structure of the enolate complex $[Ti(OCH=CH_2)_2Cp_2]$ (reproduced by permission of the American Chemical Society. Copyright (1984) from *Organometallics*, 1984, **3**, 1855).

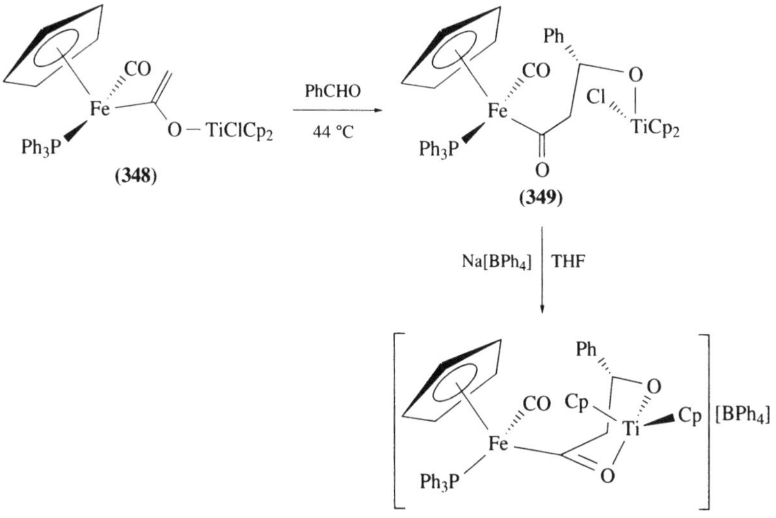

(348)　　(349)

(350)

Scheme 102

Several titanocene complexes with heteroatomic anions have been reported. Titanocene complexes with boron-containing oxygen ligands are accessible as shown in Equation (82). Due to the high steric hindrance of the $OB[CH(TMS)_2]_2$ ligand, only the orange monosubstitution product is formed.[608]

Examples for phosphate and molybdate complexes are the red compound (**351**) (Equation (83))[609] and the orange dimer [{TiMoO$_4$(η-C$_5$H$_4$Me)$_2$}$_2$] (**352**), prepared from [TiCl$_2$(η-C$_5$H$_4$Me)$_2$] and Na$_2$[MoO$_4$].[610] The structures of both complexes were determined by x-ray crystallography.

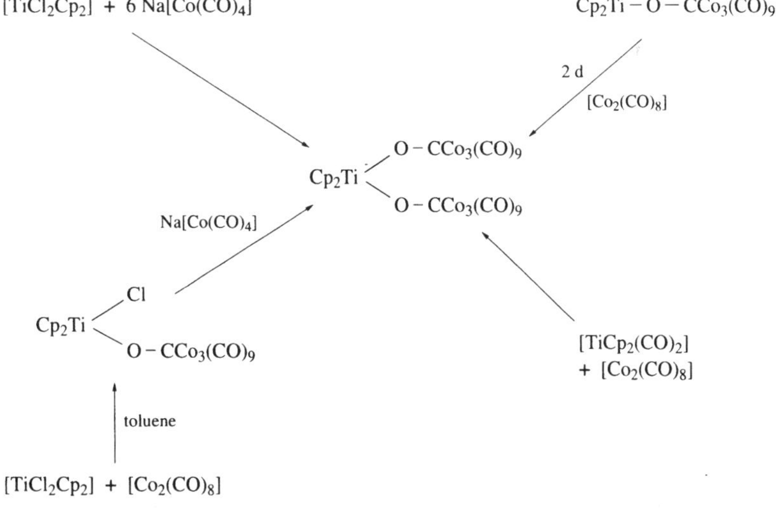

The black cluster compound [TiCp$_2${OCCo$_3$(CO)$_9$}$_2$] is accessible via several routes: the reaction of [TiCl$_2$Cp$_2$] with six equivalents of Na[Co(CO)$_4$], as the product of the reaction of the titanium(III) complex [TiCp$_2${OCCo$_3$(CO)$_9$}] with [Co$_2$(CO)$_8$], from [TiCp$_2$(CO)$_2$] and [Co$_2$(CO)$_8$], and by treating [TiClCp$_2${OCCo$_3$(CO)$_9$}] with Na[Co(CO)$_4$] (Scheme 103).[611] The reaction of [Ti(Cp*)$_2$(CO)$_2$] with [Co$_2$(CO)$_8$] in toluene affords [Ti(Cp*)$_2${OCCo(CO)$_3$}$_2$].[612]

Scheme 103

Treatment of [TiCl$_2$Cp$_2$] with [Cr{=C(Me)OLi}(CO)$_5$] affords (**353**), which reacts with NaN(TMS)$_2$ under HCl abstraction to give the fluxional metallacycle (**354**) (Scheme 104). The more hindered Cp* complex (**355**) is resistant to HCl loss. Compound (**354**) exists as a mixture of two rotamers, which do not interconvert, even on heating to 110 °C. The molecular structure of (**355a**) has been determined; it shows a relatively long Ti–O bond of 0.200(4) nm and a wide Ti–O–C angle of 158.5°, reflecting the steric congestion in the compound.[494] A complex related to (**353**), [TiClCp$_2${OC(R)=Cr(CO)$_5$}], has been prepared from [TiCl$_2$Cp$_2$] and [NMe$_4$][Cr(COR)(CO)$_5$] and crystallographically characterized. It shows a slightly shorter Ti–O distance of 0.1926(2) nm and a C–O bond length of 0.1268(3) nm

(R = C(Me)=CH$_2$).[613] The manganese carbene complex (356) was prepared similarly; it reacts with 3-hexyne under irradiation followed by oxidation with cerium(IV) in dilute nitric acid to give quinones and diones, as shown in Equation (84).[614]

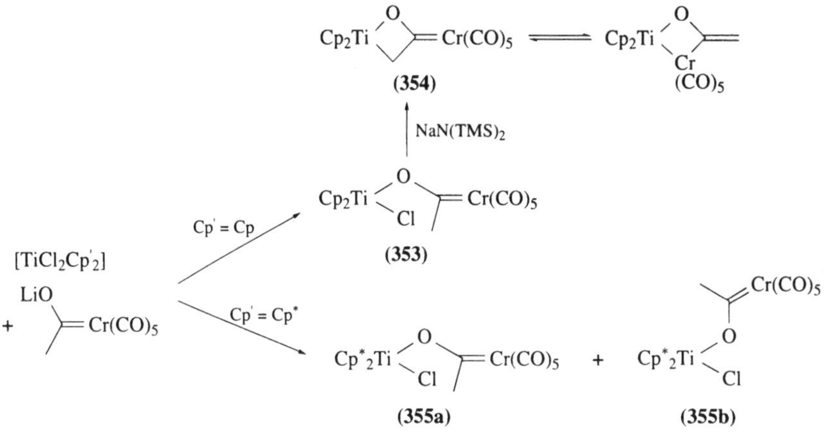

Cp$_2$Ti—O—=Cr(CO)$_5$ ⇌ Cp$_2$Ti—O—Cr(CO)$_5$
(354)

↑ NaN(TMS)$_2$

Cp$_2$Ti(Cl)—O—=Cr(CO)$_5$
(353)

[TiCl$_2$Cp$'_2$] + LiO—=Cr(CO)$_5$

Cp$'$ = Cp →

Cp$'$ = Cp* →

Cp*$_2$Ti(Cl)—O—=Cr(CO)$_5$ +   Cp*$_2$Ti(Cl)—O—Cr(CO)$_5$
(355a)   (355b)

Scheme 104

(356)

$$\xrightarrow{\text{i, Et—≡—Et, } h\nu, \text{ THF} \quad \text{ii, Ce}^{IV}}$$

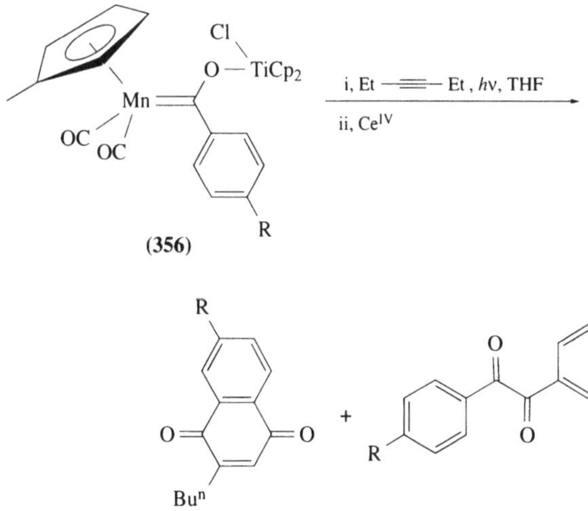

(84)

Titanocene dichloride has been used in a number of studies to modify carbohydrate oligomers and polymers via reaction with the OH groups. Thus, dextran is treated with [TiCl$_2$Cp$_2$] in water/dimethyl sulfoxide, or water/chloroform, to produce TiCp$_2$-containing materials. Phase-transfer conditions have also been used.[615,616] The condensation of sucrose with titanocene dichloride in the presence of triethylamine produces a cross-linked polymeric network with Cp$_2$TiO$_2$ linkages. The temperature stability was poor, however.[617–19] Amylose has been treated similarly to generate materials with improved radiation stability.[620]

5.5.6.2 Titanium carboxylates and related complexes

Numerous titanocene carboxylates have been reported. The benzoato complex [Ti(O$_2$CPh)$_2$Cp$_2$] is prepared from [TiCl$_2$Cp$_2$] and sodium benzoate in water as orange-yellow crystals in 95% yield.[621] The structure of the complex has been determined by x-ray diffraction.[621–3] The carboxylato ligands are η^1-coordinate (Figure 44). Slightly differing geometries have been reported. In two cases[621,623] the Ti–O distances and Ti–O–C angles of the two ligands are reported as significantly different to each other, with the shorter Ti–O bond length being associated with the wider Ti–O–C angle. This bonding mode is indicative of some π-contribution from a lone pair of one of the oxygens, so that the metal centre effectively attains an 18-electron configuration.[622] The complex reacts with bromine to give [TiBr$_2$(O$_2$CPh)$_2$] and pentabromocyclopentane.[623]

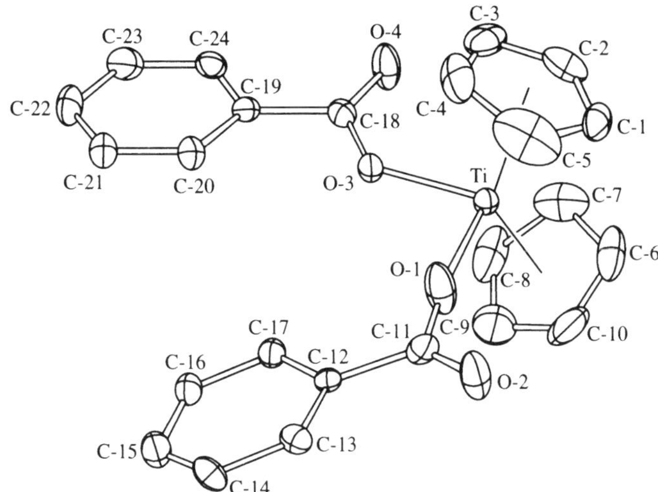

Figure 44 Molecular structure of [Ti(η^1-O$_2$CPh)$_2$Cp$_2$] (reproduced by permission of Elsevier from *J. Organomet. Chem.*, 1986, **303**, 205).

The crystal structures of a number of carboxylato complexes have been determined; they bear out the general η^1-coordination of the carboxylato group. Examples include [Ti(O$_2$CC$_6$H$_4$OMe-3)$_2$Cp$_2$],[624] [Ti(O$_2$CC$_6$H$_4$Me-2)$_2$Cp$_2$],[625] [Ti(O$_2$CCF$_3$)$_2$Cp$_2$][626] and the mixed-metal complex (**357**).[627] The vinyl complex [Ti(C$_2$R$_2$H)(O$_2$CCF$_3$)Cp$_2$] hydrolyses to give the dimer [{Ti(O$_2$CCF$_3$)Cp$_2$}$_2$(μ-O)]; this, too, contains η^1-coordinated carboxylato ligands with relatively long Ti–O distances (0.2037(2) nm), whereas the bonds to the almost linear oxo bridge are short, 0.1836(2) nm.[626]

0.197 2 nm   Cp   Cp   0.192 3 nm

Ti

Fe(CO)$_2$Bz   Fe(CO)$_2$Bz

(**357**)

The mean Ti–O bond enthalpies of a number of titanocene carboxylates [Ti(O$_2$CR)$_2$Cp$_2$] have been determined by reaction solution calorimetry. Values of 455 ± 9 kJ mol^{-1} (R = Ph), 464 ± 9 kJ mol^{-1} (R = CCl$_3$) and 451 ± 9 kJ mol^{-1} (R = CF$_3$) have been found.[544]

Numerous substituted benzoato complexes [Ti(O$_2$CC$_6$H$_4$X)$_2$Cp$_2$] have been reported (e.g., *p*-X = NH$_2$, OMe, OEt, Me, Et, Bu^t, Ph, F, Cl, Br, I, CN or NO$_2$; *m*-X = OMe, Me, F, Cl, Br, I, CN or NO$_2$; *o*-X = Me, Ph, F, Cl, Br, I, MeCO$_2$, NO$_2$; 2,4-Cl$_2$, 3,4-Cl$_2$, 2,5-Cl$_2$, etc.).[625,628–30] A series of compounds [Ti(O$_2$CR)$_2$Cp$_2$] (R = Me, CH$_2$Cl, CHCl$_2$, CCl$_3$, Et or Pr) is also known.[631] With compounds HL = (**358**) the complexes [TiCl(L)Cp$_2$] and [Ti(L)$_2$Cp$_2$] have been isolated.[632] Similar compounds are obtained from heterocyclic carboxylic acids such as thiophene-2-carboxylic acid or nicotinic acid; the heteroatoms are not thought to be involved in coordination to titanium.[633]

(**358**)

The enantiomers of *ansa*-titanocenes with chiral ligands can be resolved by complexation with chiral carboxylates. Thus, the reaction of [TiCl$_2${C$_2$H$_4$(η-C$_9$H$_{10}$)$_2$}] (**151**) with *O*-acetyl-(*R*)-mandelic acid gives two diastereomers, which can be separated by fractional crystallization from toluene/pentane mixtures. The crystal structures of both the C$_2$H$_4$(tetrahydro-(1*S*)-indenyl) mandelate and the (1*R*) isomer have been determined (Figure 45). The diasteromers formed with amino acid derivatives such as *N*-acetyl-L-leucine and *N*-acetyl-L-methionine have been identified by NMR spectra and were oils. The enantiomerically pure chloro complexes are regenerated by alkylation of the mandelato complexes with methylmagnesium chloride to give [TiMe$_2${C$_2$H$_4$(η-C$_9$H$_{10}$)$_2$}], which can be readily separated and hydrolysed with HCl in diethyl ether at 0 °C to give pure (*S*)- or (*R*)-(**151**).[634]

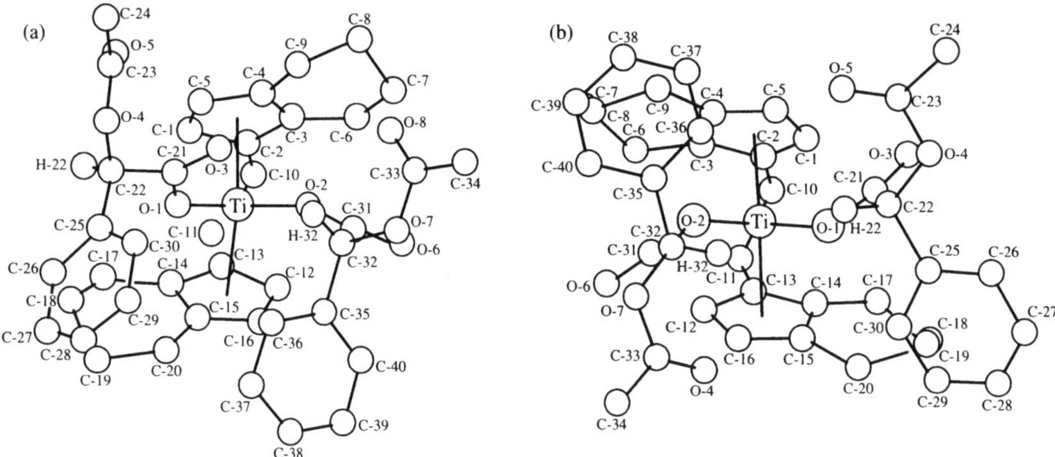

Figure 45 Molecular structures of (a) ethylenebis(4,5,6,7-tetrahydro-(1*S*)-indenyl)titanium-bis(*O*-acetyl-(*R*)-mandelate) and (b) its (1*R*)-indenyl diastereomer (reproduced by permission of Elsevier from *J. Organomet. Chem.*, 1987, **328**, 87).

Titanocene dimethyl reacts with phenylpropiolic acid under irradiation in the presence of hydrogen gas to give (**359**); a molecule of propiolic acid is associated via hydrogen bonds. The thermal reaction between [TiMe$_2$Cp$_2$] and PhC≡CCO$_2$H in toluene produces [Ti(O$_2$CC≡CPh)$_2$Cp$_2$] (Equation (85)).[635] [TiCl$_2$(η-C$_5$H$_4$Me)$_2$] reacts with Na[RC(O)S] to give thiocarboxylato complexes (R = H, Me, Et, Ph, C$_6$H$_4$OMe or C$_6$H$_4$Me).[636] Sulfonato complexes [Ti(O$_3$SR)$_2$Cp$_2$] are similarly accessible (R = *m*- or *p*-C$_6$H$_4$NH$_2$, or C$_6$H$_3$-3-Me-4-NH$_2$).[637] The hydrolysis of [TiCl$_2$Cp$_2$] at pH > 5 in the presence of formic acid proceeds with the loss of one cyclopentadienyl ligand per titanium to give the trinuclear complex (**360**) (Equation (86). The structure was confirmed by x-ray diffraction. The Ti$_2$O moiety is nonplanar. Each metal centre has an approximately octahedral coordination geometry.[638]

$$[\text{TiMe}_2\text{Cp}_2] + \text{Ph}\text{---}\text{CO}_2\text{H} \xrightarrow{\text{H}_2,\ h\nu} \text{Cp}_2\text{Ti} \cdots \bullet\text{Ph}\text{---}\text{CO}_2\text{H} \qquad (85)$$

(**359**)

$$[\text{TiCl}_2\text{Cp}_2] \xrightarrow[\text{ii, HCO}_2\text{H, Et}_2\text{O}]{\text{i, H}_2\text{O, pH} >5} \qquad [\text{HCO}_2]\cdot 2\text{HCO}_2\text{H} \qquad (86)$$

(**360**)

Whereas [TiCl$_2$Cp$_2$] reacts with amino acid derivatives RCO$_2$H to give the expected complexes [Ti(O$_2$CR)$_2$Cp$_2$] (R = CH$_2$NHCOPh or *o*-C$_6$H$_4$NH$_2$), the reaction with cysteine in DMSO in the presence of pyridine leads to the loss of one Cp$^-$ ligand, to give (**361**).[639] Whereas the structurally characterized titanocene carboxylates discussed earlier contain four-coordinate titanium in an approximately tetrahedral environment, the product of the reaction of [TiMe$_2$Cp$_2$] with 2,6-pyridine dicarboxylic acid has the structure (**362**), with the nitrogen atom occupying a fifth coordination site. Due to the nitrogen coordination, π-interactions with the oxygen lone pairs are absent, and the Ti–O distances are unusually long (0.211 nm). All atoms of the pyridinecarboxylato ligand are essentially coplanar. The $Me_4C_2(\eta\text{-}C_5H_4)_2Ti$ derivative has an analogous structure.[640] With 3,5-pyridinedicarboxylic acid, a dimeric complex (363) results, in which the pyridyl moiety is not coordinated. The compound forms an almost perfectly planar 16-membered ring. The nitrogen atoms can act as donors to other metals and form complexes with rhodium and iridium carbonyls.[641] The reaction of $[TiCl_2Cp_2]$ with pyrazinetetracarboxylic acid produces a tetrameric complex (364); the molecular structure is shown in Figure 46. The complex reacts with HCl to give a monomeric complex $[TiCp_2(C_4N_2(CO_2)_2(CO_2H)_2]\cdot 2H_2O\cdot MeNO_2$ with hydrogen-bonded water molecules.[642]

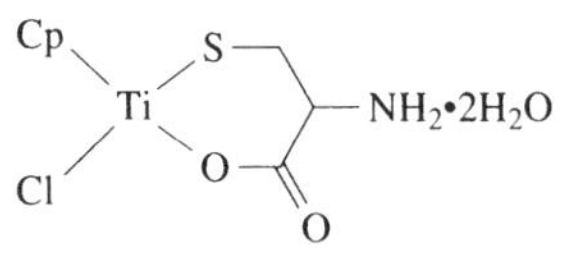

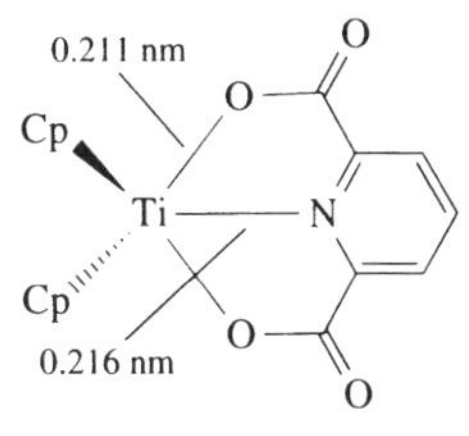

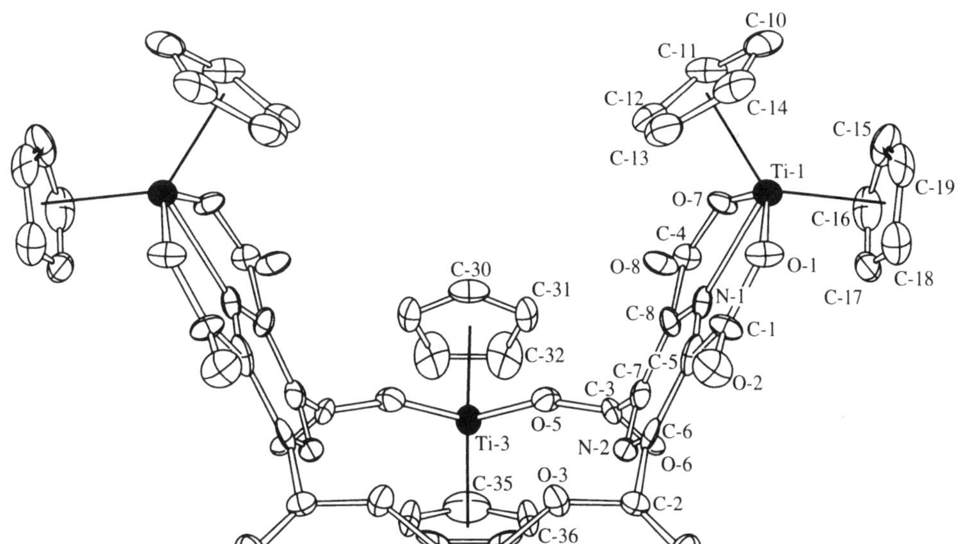

Figure 46 Structure of the pyrazinetetracarboxylatotitanium complex in $[(TiCp_2)_2\{C_4N_2(CO_2)_4\}]_2\cdot 2H_2O\cdot 2CHCl_3\cdot 3MeNO_2$. The Cp groups of Ti-2 have been omitted for clarity (reproduced by permission of Elsevier from *J. Organomet. Chem.*, 1989, **371**, 43).

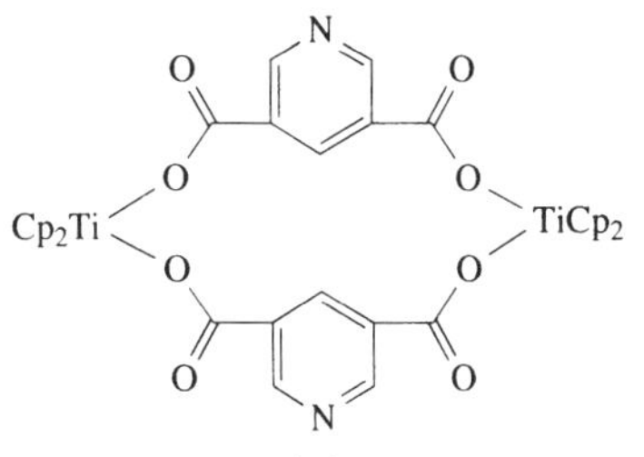

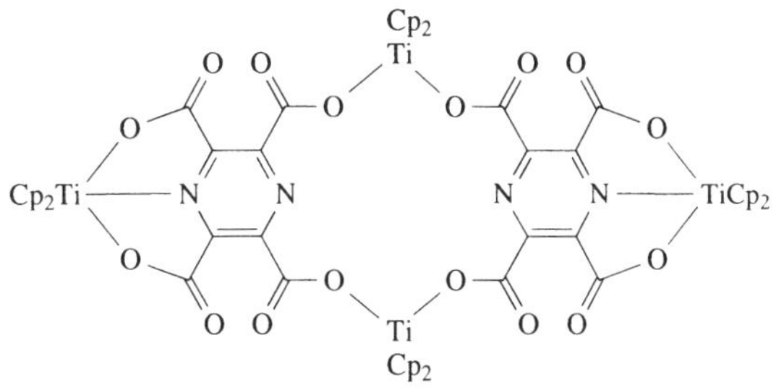

Salicylato complexes (365) (R = H, CHO or NH_2) and analogous phthalato compounds are accessible from titanocene dichloride and the corresponding acid under aqueous conditions.[643] With oxalic acid the deep red oxalate (366) has been isolated and crystallographically characterized, while maleic acid produces $[Ti(\eta^1\text{-}O_2CCH{=}CHCO_2H)_2Cp_2]$.[644] The reaction of $[TiCl_2Cp_2]$ with disodium oxydiacetate in an aqueous biphasic medium leads to a mixture of a yellow five-coordinate monomer (367) with long

Ti–O bonds of 0.2100(9) nm, and an orange dimeric complex (**368**) with shorter Ti–O bonds of 0.1959(7) nm (Equation (87)).[645] Complexes with different degrees of association are also observed with acetylene dicarboxylic acid. The reaction in CH_2Cl_2/H_2O at 4 °C with [NBu$_4$]Br as the phase-transfer agent produces [Ti(O$_2$CC≡CCO$_2$)Cp$_2$]$_4$·5CH$_2$Cl$_2$. At 20 °C the dimer [{Ti(O$_2$CC≡CCO$_2$)Cp$_2$}$_2$] crystallizes. The tetramer is unstable and rearranges slowly to the dimer at room temperature.[646] The reaction of [TiCl$_2$Cp$_2$] with disodium fumarate leads to a number of isolable dimer structures which differ in the degree of solvation, for example, [{Ti(O$_2$CCH=CHCO$_2$)Cp$_2$}$_2$] and [{Ti(O$_2$CCH=CHCO$_2$)Cp$_2$}$_2$]·2CHCl$_3$. The unit cell of the latter contains two independent molecules, one of which contains a nearly planar 16-membered ring, while the other ring is distorted.[647] The analogous terephthalato complex is tetranuclear and forms a folded 36-membered ring which describes a figure of eight (Figure 47).[648]

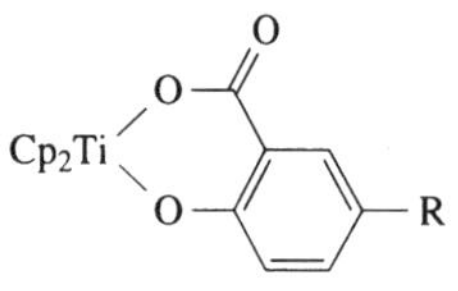

(365) 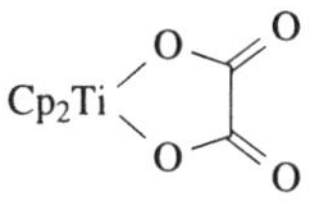   (366)

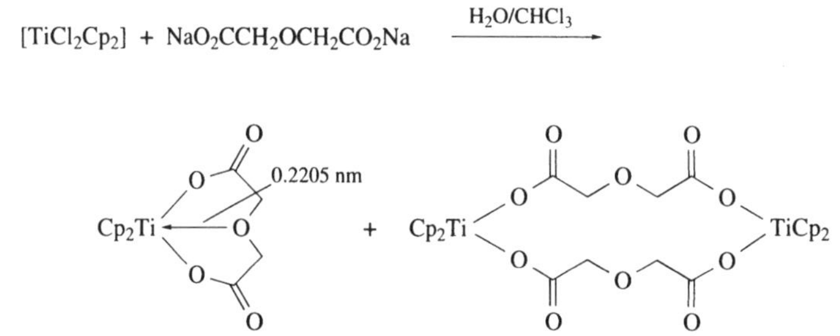

$$[TiCl_2Cp_2] + NaO_2CCH_2OCH_2CO_2Na \xrightarrow{\ H_2O/CHCl_3\ }$$

(367)   +   (368)   (87)

In contrast to the normally η^1-coordinate carboxylato ligands, titanium acetylacetonato complexes contain almost invariably bidentate ligands. Thus the reaction between [TiCl$_2$Cp$_2$] and HL = R^{1}C(O)CH$_2$C(O)R^2 (R^1 = R^2 = Me; R^1 = Me, R^2 = Ph) gives the ionic complexes [Ti(L)Cp$_2$]Cl. Anion exchange affords a number of complexes [Ti(L)Cp$_2$]X (X = Br, I, CuCl$_3$, ZnCl$_3$(H$_2$O), $\frac{1}{2}$ CdCl$_4$, HgCl$_3$, FeCl$_4$ or S$_2$CNR^{3}R^4; R^3, R^4 = alkyl, aryl or benzyl;).[649–52] By contrast, the pyrazolone derivatives (**369**) (R = Me, Et, Ph or C$_6$H$_4$Cl) are formulated as electroneutral; they are soluble in benzene and are monomeric.[653,654]

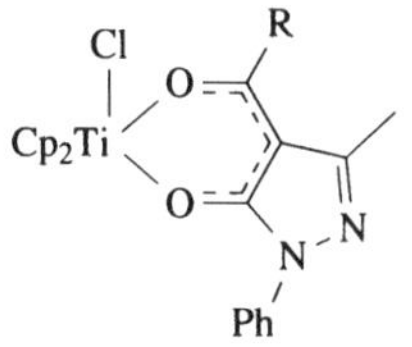

(369)

5.5.7 Complexes with Ti–P and Ti–As Bonds

The bis(phosphido) complex (**370**) is accessible from [TiCl$_2$Cp$_2$] and C$_6$H$_4$(PPhLi)$_2$, or by protolysis of [TiMe$_2$Cp$_2$] with C$_6$H$_4$(PHPh)$_2$ (Scheme 105).[655] The reaction of [W(CO)$_4$(PPh$_2$H)$_2$] with butyllithium, followed by the addition of [TiCl$_2$(η-C$_5$H$_4$Me)$_2$], gives the purple bimetallic complex [Ti(η-C$_5$H$_4$Me)$_2$(μ-PPh$_2$)$_2$W(CO)$_4$]. The crystal structure of the zirconium analogue was determined.[656] Titanocene dichloride reacts with As$_6$Ph$_6$ in the presence of magnesium as a reducing agent according to Equation (88) to give a four-membered metallacycle, [Ti(As$_3$Ph$_3$)Cp$_2$].[657]

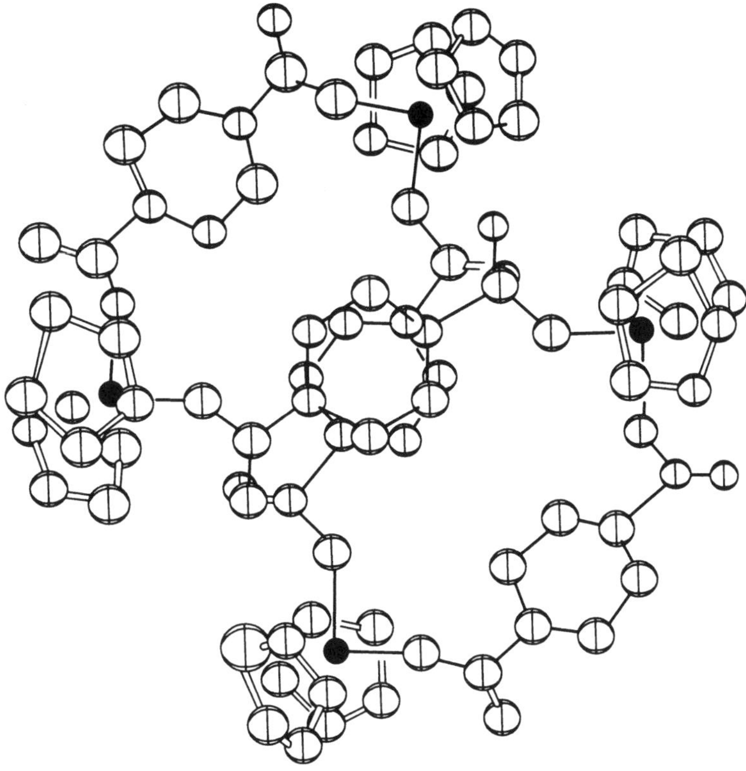

Figure 47 Molecular structure of tetrakis[µ-terephthalato-bis(cyclopentadienyl)titanium] showing the figure-of-eight folding of the 36-membered ring (reproduced by permission of Elsevier from *J. Organomet. Chem.*, 1987, **326**, C37).

Scheme 105

(88)

5.5.8 Complexes with Ti–S, Ti–Se and Ti–Te Bonds

5.5.8.1 *Titanium chalcogenides and hydrochalcogenides*

The chemistry of metal polysulfides, including the complexes of titanium such as [TiS$_5$Cp$_2$] and [TiSe$_5$Cp$_2$], has been summarized in a review.[658] Detailed synthetic procedures have been published for [TiS$_5$Cp$_2$], [TiSe$_5$Cp$_2$], [TiS$_3$(Cp*)$_2$][659] and [TiS$_5$(η-C$_5$H$_4$Me)$_2$].[660] The latter complex has also been synthesized by reacting ZnS$_6$(*N*-methylimidazole)$_2$ with [TiCl$_2$(η-C$_5$H$_4$Me)$_2$].[661] The trisulfido complex [TiS$_3$(Cp*)$_2$] is made from [TiCl$_2$(Cp*)$_2$] and Li$_2$S$_2$ in the presence of elemental sulfur. The compound forms black-purple crystals in 55% yield, whose structure was confirmed by x-ray crystallography.[662,663]

The complex is fluxional. Cooling to −90 °C slows the ring-flipping motion down so that the Cp*⁻ ligands become inequivalent. The barrier to this interchange is ~40 ± 2 kJ mol⁻¹, smaller than in S₅ complexes.[662] The pentachalcogenido complexes [TiE₅Cp₂] have been prepared by the oxidation of [Ti(SH)₂Cp₂] with sulfur, or by the treatment of [TiCl₂Cp₂] with Li₂E$_x$ (E = S or Se; x = 2 or 5). Ring inversion barriers of 80 kJ mol⁻¹ were measured for the TiE₅ complexes, and the value of 40 kJ mol⁻¹ for [TiS₃(Cp*)₂] was confirmed. These inversion barriers decrease in the order Ti > Hf > Zr. The substituted complexes [TiS₅Cp′₂] (Cp′ = C₅H₄Me or C₅H₄TMS) and mixed-ligand compounds such as [TiS₅Cp(Cp*)] form red crystals in high yields. The *ansa*-titanocene derivative [TiS₅{H₂C(C₅H₄)₂}] was only obtained in 9% yield; its ring inversion barrier of 66.5 kJ mol⁻¹ is reduced compared to unbridged analogues.[664]

The electrochemical reduction of grey selenium under ultrasound conditions produces Se₅²⁻, which reacts with [TiCl₂Cp′₂] to give [TiSe₅Cp′₂] as dark purple solids (Cp′ = Cp or C₅H₄Prⁱ).[665] The mixed chalcogenide complexes [Ti(SSe₄)Cp′₂] (Cp′ = C₅H₄Me or C₅H₄TMS) have been obtained by the treatment of [Ti(SH)₂Cp′₂] with selenium in refluxing carbon disulfide.[666] The crystal structures of some of these complexes have been determined, such as [TiS₅(η-C₅H₄TMS)₂][667] and [TiSe₅Cp₂][668,669] and Ti–S bond lengths of 0.2416(5) nm and Ti–Se distances of 0.2539(2)–0.2588(2) nm have been found. In all cases the TiE₅ rings adopt a chair conformation.

The reaction of [TiCl₂(η-C₅H₄Me)₂] with As₂S₃ and S²⁻ in aqueous solution produces a green precipitate, from which after extraction and chromatography the brown crystallographically characterized cage complex (371) is isolated.[670]

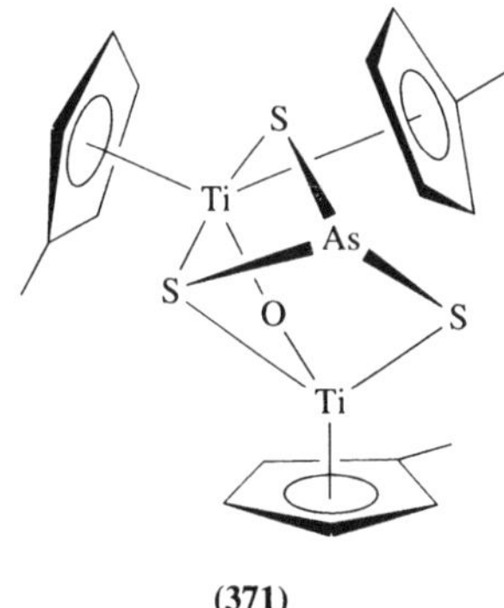

(371)

Several hydrosulfido and hydroselenido complexes have been prepared, and a detailed procedure for the synthesis of [Ti(SH)₂Cp₂] has appeared.[671] Whereas the reaction of [TiCp₂(CO)₂] with H₂S gives the mixed-valence cluster [Ti₅Cp₅S₆], the analogous reaction with [Ti(Cp*)₂(CO)₂] leads to the red complex [Ti(SH)₂(Cp*)₂], whose crystal structure reveals Ti–S bond lengths of 0.2409(2) nm and 0.2418(3) nm, and Ti–S–H angles of 106(3)° and 116(3)°. The orientation of the SH⁻ ligands indicates some π-overlap between the sulfur lone pairs with empty metal orbitals lying in the plane bisecting the Cp–Ti–Cp angle.[672] While [TiCl₂Cp₂] reacts with H₂Se in THF in the presence of NEt₃ to give [Ti₄Cp₄(μ₂-Se)₃(μ₃-Se)₃] (104) (Equation (32)), the reaction in THF/pyridine affords [TiCl(SeH)Cp₂]. With further H₂Se, [Ti(SeH)₂Cp₂] is produced. [TiCp₂(CO)₂] is oxidized by two equivalents of H₂Se under anaerobic conditions to give [Ti(μ-Se)Cp₂]₂, hydrogen and CO, while in the presence of traces of air the well-known pentaselenide [TiSe₅Cp₂] results. [TiCl₂Cp₂] reacts with H₂Se in pyridine in the presence of traces of air to give the crystallographically characterized diselenide [(TiClCp₂)₂(μ-Se₂)].[220] The titanium(III) fulvalene complex [{Ti(μ-H)Cp}₂(η⁵:η⁵-C₁₀H₈)] is oxidized by sulfur to give the dark-brown μ-sulfido complex (372) (Equation (89)).[673]

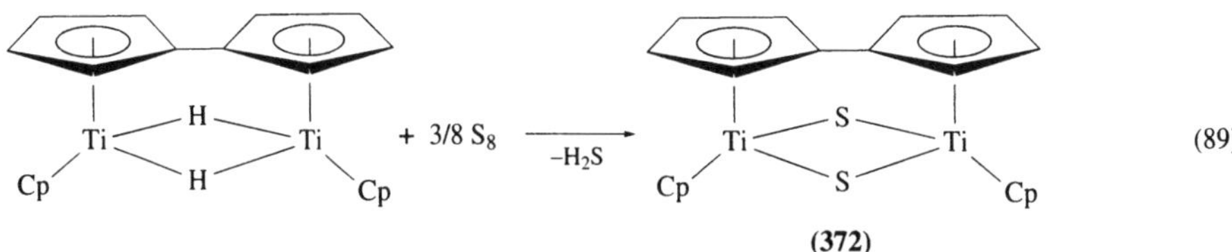

(372)

(89)

The reaction of [Ti(SH)₂Cp₂] with RSX (X = phthalimide or succinimide) in THF at 0 °C leads to the purple SSR and SSSR complexes (373) and (374) (Scheme 106). The compounds are thermally unstable and decompose slowly above 0 °C. The structure of (374) (R = Ph) has been determined by x-ray diffraction (Figure 48).[674] The oxidation of [Ti(SH)$_2$(η-C$_5$H$_4$Me)$_2$] with iodine in the presence of triethylamine gives [{Ti(η-C$_5$H$_4$Me)$_2$(μ-S$_2$)}$_2$]. The Ti$_2$S$_4$ six-membered ring possesses the chair conformation. Similar dinuclear ring complexes are formed from titanocene dichlorides and perthiocarbonate, Li$_2$CS$_4$, under elimination of CS$_2$. The oxidation of [TiCp$_2$(CO)$_2$] with [TiS$_5$(η-C$_5$H$_4$Pr^i)$_2$] gives the analogous unsymmetrical dimer [(TiCp$_2$){Ti(η-C$_5$H$_4$Pr^i)$_2$(μ-S$_2$)$_2$].[675] [Ti(SH)$_2$Cp$_2$] reacts with metal carbonyls such as the norbornadiene complex [Mo(CO)$_4$(C$_7$H$_8$)] to give the blue SH-bridged compound (375). The complex is fluxional. The SH ligands can be deprotonated to give (376), which is methylated by iodomethane to give the μ-SMe complex (377) (Scheme 106).[676]

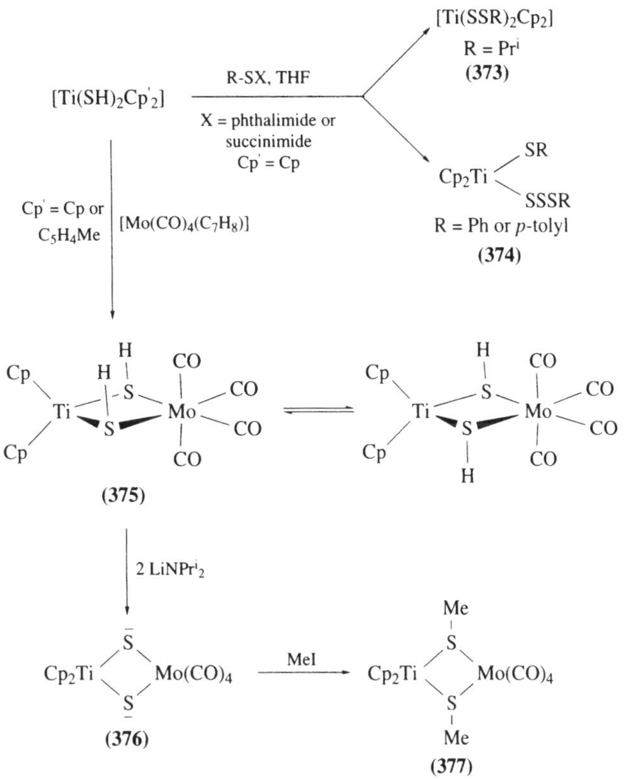

Scheme 106

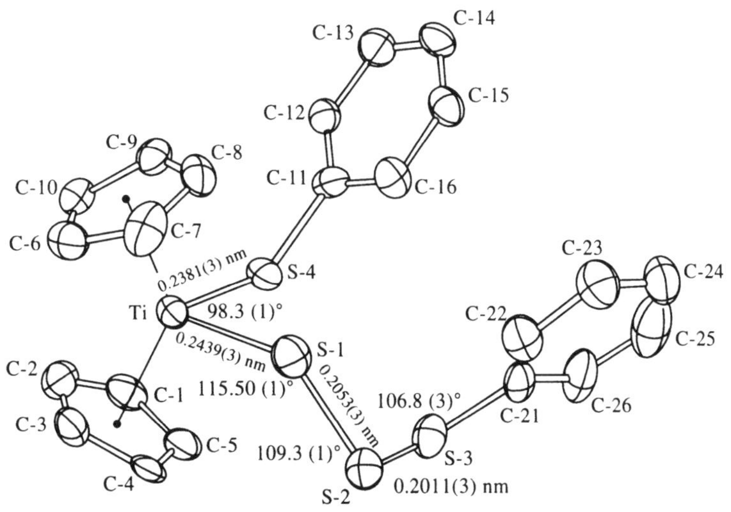

Figure 48 Molecular structure of [Ti(SPh)(SSSPh)Cp$_2$] (374) (reproduced by permission of the American Chemical Society. Copyright (1983) from *Organometallics*, 1983, **2**, 1894).

5.5.8.2 *Titanium–ER complexes (E = S, Se or Te)*

Several titanocene arenethiolato complexes $[Ti(SAr)_2Cp_2]$ have been prepared from $[TiCl_2Cp_2]$ and ArSH in benzene in the presence of NEt_3 as a base (Ar = C_6F_5,[677] o-C_6H_4Me or p-C_6H_4X; X = H,[678] Me,[678,679] Cl or OMe[678]), while $[TiCl_2Cp_2]$ reacts with one equivalent of MSC_6H_4X to give dark-red (X = Me) to black (X = NH_2) crystals of $[TiCl(SC_6H_4X)Cp_2]$. $[TiMeClCp_2]$ and MSC_6H_4X afford the dark red, air-stable alkyl complexes $[TiMe(SC_6H_4X)Cp_2]$ (M = Li or Na; X = o-Me, p-Me[679], o-NH_2 or p-NH_2[680]). With NH_2-substituted thiolates, *N,S*-chelate formation is not observed. Refluxing $[TiCl_2Cp_2]$ with $HSCH_2CH_2NH_2$ in THF affords the zwitterionic complex $[TiCl(SCH_2CH_2NH_3{}^+Cl^-)Cp_2]$. The addition of base leads to the unstable complex $[TiCl(SCH_2CH_2NH_2)Cp_2]$, which reacts to give polycondensation products.[681] The mercaptoacetamides $[TiCl(SR^1)Cp_2]$ and $[Ti(SR^1)_2Cp_2]$ were prepared similarly ($R^1 = CH_2C(O)NHR^2$; R^2 = Ph, α-naphthyl, or o-, m- or p-tolyl).[682,683]

The substitution of bromide by thiophenolate for the chiral complex (**377**) proceeds highly stereoselectively to give 99% (**378**) and only 1% (**379**), whereas the diasteromer (**380**) gives only 10% (**378**) and 90% (**379**) (Scheme 107). The structural assignment was confirmed by the crystal structure of (**378**).[684,685]

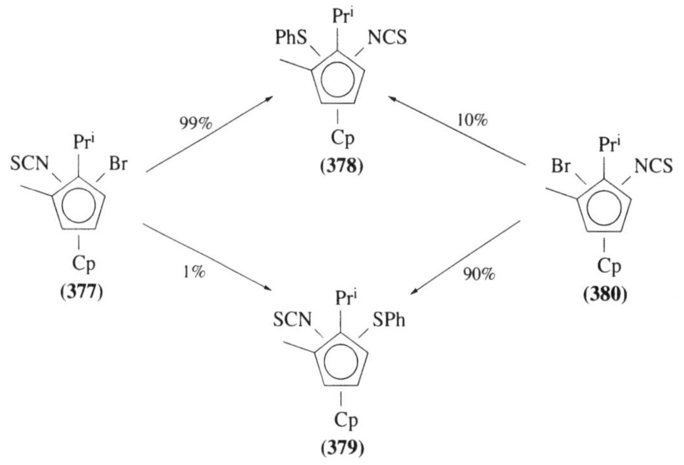

Scheme 107

The tendency of thiolato ligands to act as bridging ligands has been employed for the synthesis of heteronuclear complexes. The dark maroon, air-sensitive arenethiolato compounds $[Ti(SC_6H_4X)_2Cp_2]$ (X = H, Cl, Me or OMe) react with $[Mo(CO)_4(C_7H_8)]$ (C_7H_8 = norbornadiene) to give the fluxional complexes (**381**) (Scheme 108). In solution at $-60\,^\circ C$, two stereoisomers can be distinguished, which interchange with $\Delta G^\ddagger = 50$ kJ mol^{-1}. The compounds can be electrochemically reduced at potentials of -1.028 V (X = Cl) to -1.186 V (X = OMe) (vs. an Ag/AgCl reference electrode).[678] The binuclear complexes $[TiCp_2(\mu\text{-}SEt)_2M(CO)_4]$ (M = Cr, Mo or W) show quasireversible reduction potentials ranging from -0.97 V (M = Cr) to -1.03 V (M = W), and are oxidized at 0.48 V (M = Cr) to 0.63 V (M = W).[686]

The thiolates $[Ti(SR)_2Cp_2]$ react with $[Ni_2Cp_3][BF_4]$ in nitromethane to give nickelocene and $[TiCp_2(\mu\text{-}SR)_2NiCp][BF_4]$. Due to the *cis* and *trans* orientation of the substituents R, two isomers are observed, with an inversion barrier of 67 kJ mol^{-1} (R = Me) and 62 kJ mol^{-1} (R = Ph).[687] Similarly, $[Ti(SEt)_2Cp_2]$ and $[Cu(MeCN)_4][PF_6]$ in the presence of L = PPh_3, $P(C_6H_{11})_3$, PEt_3, $P(CH_2Ph)_3$, pyridine or $\frac{1}{2}$ $Ph_2PCH_2CH_2PPh_2$ give the fluxional orange-red compound $[TiCp_2(\mu\text{-}SEt)_2CuL][PF_6]$. The crystal structures of the PPh_3 and $P(C_6H_{11})_3$ adducts have been determined. The ethyl groups are mutually *cis*. The TiS_2Cu core is folded, with a relatively short Ti–Cu distance of ~0.28 nm, which is thought to indicate a dative Cu $\rightarrow$ Ti interaction.[688] The red-purple methylthiolato complex $[Ti(SMe)_2Cp_2]$ has been prepared from $[TiCl_2Cp_2]$ and NaSMe in 92% yield. Its crystal structure gives a Ti–S bond length of 0.2400(1) nm.[689] A separate structure determination reports a Ti–S bond length of 0.2403(1) nm.[690] The compound reacts with bis(cyclooctadiene)nickel to give black crystals of (**382**), which possess longer Ti–S bonds of 0.2503(2) nm and a Ti–Ni distance of 0.2786(1) nm. This metal–metal distance, together with the small Ti–S–Ni angle of 72.0°, hints at a weak Ni $\rightarrow$ Ti donor interaction. On addition of tricyclohexylphosphine, (**383**) is obtained.[689] The phosphino groups in phosphinothiolato complexes such as $[Ti(SCH_2CH_2PPh_2)_2Cp_2]$ can be quaternized by iodomethane. With $[Cu(MeCN)_4][BF_4]$ the

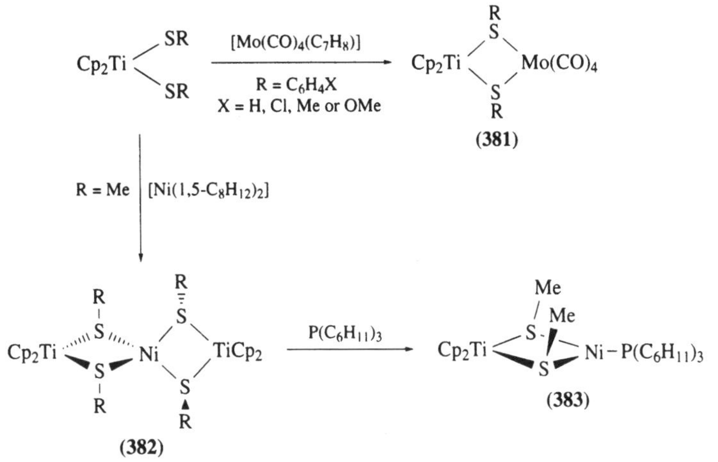

Scheme 108

adduct (**384**) is formed in which the sulfur and phosphorus atoms are tetrahedrally coordinated to copper(I). The TiS_2Cu core is planar, with a Ti–Cu distance of 0.3024(1) nm (Scheme 109).[691] Extending the distance between the sulfur and phosphorus atoms, as in $[Ti\{S(CH_2)_3PPh_2\}_2Cp_2]$, enables a number of metals centres to be encapsulated. With $[Ni(1,5\text{-}C_8H_{12})_2]$ the tetrahedrally coordinated nickel complex (**385**) is formed, which reacts with CO to give (**386**). Both complexes were crystallographically characterized. Compound (**385**) undergoes two one-electron oxidation steps, although the dication could not be isolated. However, with Pd^{2+} the isoelectronic complex $[TiCp_2\{S(CH_2)_3PPh_2\}_2Pd]^{2+}$ is formed.[692] With $[Rh(C_7H_8)_2][BF_4]$ (C_7H_8 = norbornadiene) the square-planar complex (**387**) is formed. The compound is fluxional, and the alkyl bridges may adopt either cisoid or transoid orientation. The compound is reduced reversibly at −0.87 V (vs. SCE), while chemical reduction with cobaltocene gives the very air-sensitive zero-valent complex.[693]

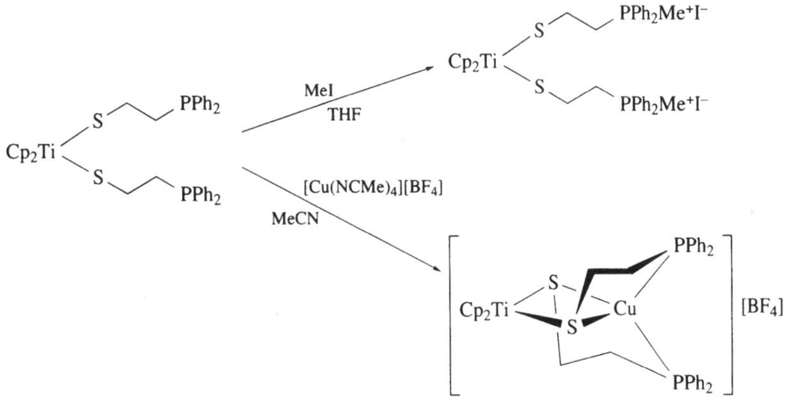

Scheme 109

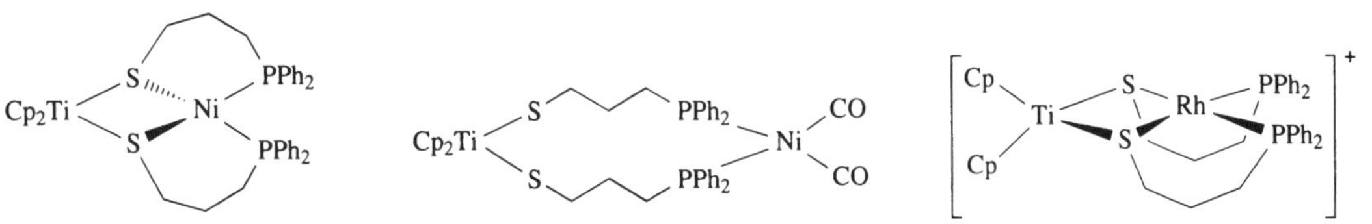

The ethylenedithiolato complex (**388**) is in equilibrium with a number of dimeric structures. The 1,3-propanedithiol analogue (**389**) is accessible from $[TiCl_2Cp_2]$, or in almost quantitative yield by the oxidation of $[TiCp_2(PMe_3)_2]$ with the disulfide (Scheme 110). With $[Cu(MeCN)_4][BF_4]$ the green-black adduct $[Cu(388)_2]^+$ is formed, which is reduced by cobaltocene to $[Cu(388)_2]^0$. The red-black complex

[Ag(**389**)$_2$][BPh$_4$] has been characterized by x-ray diffraction; one-electron reduction gives the paramagnetic Ti$^{III/IV}$ mixed-valence complex [Ag(**389**)$_2$]^0. The reduced species are unstable and could not be isolated.[694]

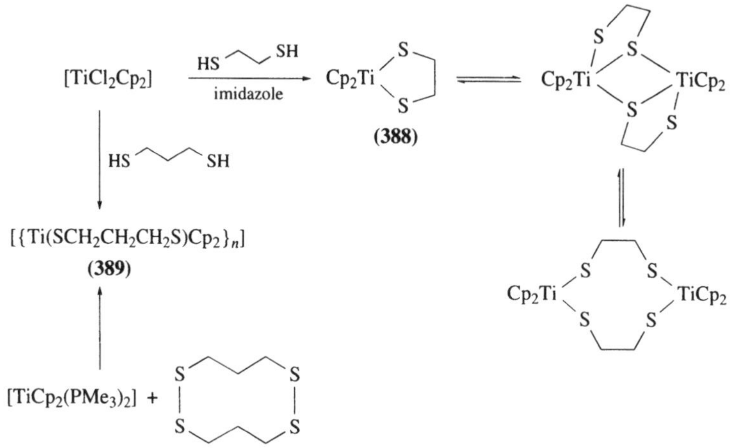

Scheme 110

[TiCl$_2$(η-C$_5$H$_4$Me)$_2$] and Li$_2$S$_2$SiMe$_2$ in THF give the dark green complex [Ti(S$_2$SiMe$_2$)Cp$_2$]. The compound reacts with [PtCl$_2$(PPh$_3$)$_2$] under transfer of the thiolato ligand to platinum.[695] The ring inversion in the fluxional 1,2-dithiolato complexes [Ti(1,2-S$_2$C$_6$H$_3$Me-4)Cp'(η-C$_5$H$_4$Me)] (Cp' = Cp or C$_5$H$_4$Me) proceeds with $\Delta G^{\ddagger}$ ~50 kJ mol^{-1}, irrespective of Cp ring substitution.[552] The selenium analogues, [Ti(1,2-Se$_2$C$_6$H$_4$)Cp'$_2$] (Cp' = C$_5$H$_4$Bu^t or Cp*), have been prepared from [TiCl$_2$Cp'$_2$] and K$_2$Se$_2$C$_6$H$_4$.[696] The tetrathiolato complex (**390**) is similarly accessible,[697] as are the air-sensitive, black-red sulfur, nitrogen-chelate compounds (**391**) (R = H or Me).[698] The aminothiolato complexes (**392**) and analogous bis(aminothiolato) compounds are monomeric and nonionic (R^1 = Pr^n or Pr^i; R^2 = Ph or C$_6$H$_4$X; X = Me, Cl or OMe; R^3 = Ph or Me). According to the IR spectra the ligands are bidentate.[699] The ditellurolato compound [Ti(1,2-Te$_2$C$_6$H$_4$)Cp$_2$] is prepared either from [TiCl$_2$Cp$_2$] and 1,2-(LiTe)$_2$C$_6$H$_4$ at −78 °C, or by treating [TiPh$_2$Cp$_2$] with tellurium. The compound is thermally stable and decomposes at 180 °C. The ring inversion barrier of 53 kJ mol^{-1} is very similar to that of the sulfur and selenium analogues.[700] The reaction of Li$_2$[Fe$_2$(μ-Te)$_2$(CO)$_6$] with [TiCl$_2$Cp$_2$] at −78 °C leads to the telluride-bridged cluster [TiCp$_2${(μ-Te)$_2$Fe$_2$(CO)$_6$}].[701]

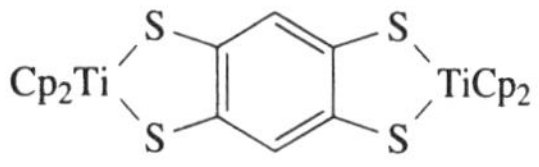

(**390**)   (**391**)   (**392**)

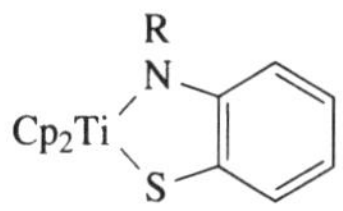

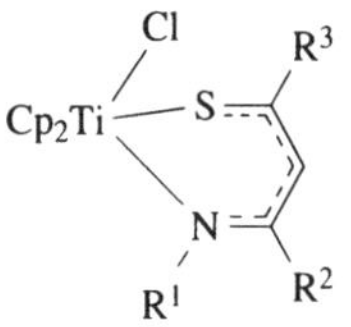

The reaction of R^{1}SCH$_2$SR^2 with butyllithium, followed by the addition of CS$_2$ and [TiCl$_2$Cp$_2$], leads to the purple compound (**393**) (R^1, R^2 = (CH$_2$)$_3$, Ph).[702] The black-green C$_3$S$_5$$^{2-}$ complex (**394**) is obtained according to Equation (90). The compound is a convenient starting material for higher carbon sulfides such as C$_3$S$_8$ and C$_6$S$_{12}$.[703]

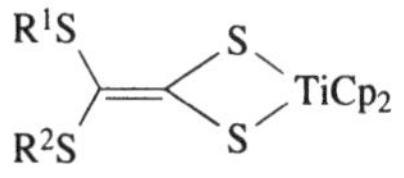

(**393**)

Several dithiocarbamato complexes have been reported. [TiCl$_2$Cp(η-Ind)] reacts with three equivalents of NaS$_2$CNR$_2$ with loss of an indenyl ligand to give compounds of the stoichiometry [Ti(S$_2$CNR$_2$)$_3$Cp],[704] while the reaction with one equivalent of thiocarbamate gives [TiCl(S$_2$CNR$_2$)Cp'$_2$] (R = CH$_2$Ph; Cp' = Cp or C$_5$H$_4$Me).[705]

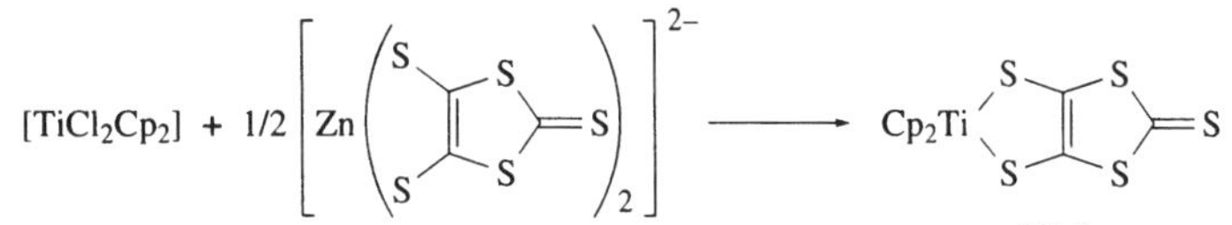

$$[\text{TiCl}_2\text{Cp}_2] + \tfrac{1}{2}\left[\text{Zn}\left(\begin{array}{c}\text{S}\quad\text{S}\\\diagdown\ \ \diagup\\ =\text{S}\\\diagup\ \ \diagdown\\ \text{S}\quad\text{S}\end{array}\right)_2\right]^{2-} \longrightarrow \text{Cp}_2\text{Ti}\begin{array}{c}\text{S}\quad\text{S}\\=\text{S}\\\text{S}\quad\text{S}\end{array} \qquad (90)$$

(394)

The reaction of [TiCl$_2$Cp$_2$] with LiSPR$_2$ affords [Ti(η^1-SPR$_2$)$_2$Cp$_2$] as intense purple crystals (R = cyclohexyl). The structure of the compound was determined by x-ray diffraction. With R = Ph or SePR$_2$ the metal is reduced to titanium(III).[706] The thiophosphoryl complex (**395**) is oxidized with excess oxygen gas to (**396**), while one equivalent of oxygen gas or the addition of DMSO gives the crystallographically characterized complex (**397**) (Scheme 111). The oxidation of [Ti(S$_3$PAr)(η-C$_5$H$_4$Me)$_2$] with [ArPS$_2$]$_2$ affords (**398**); the reaction is reversible (Ar = p-C$_6$H$_4$OMe) (Equation (91)).[707]

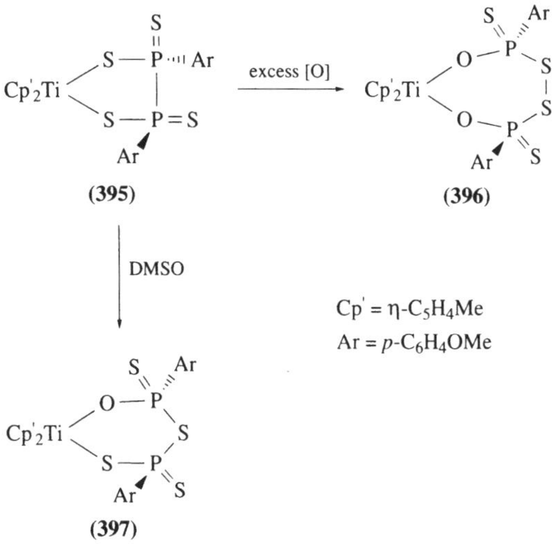

(395)   excess [O]   (396)

DMSO

Cp$'$ = η-C$_5$H$_4$Me
Ar = p-C$_6$H$_4$OMe

(397)

Scheme 111

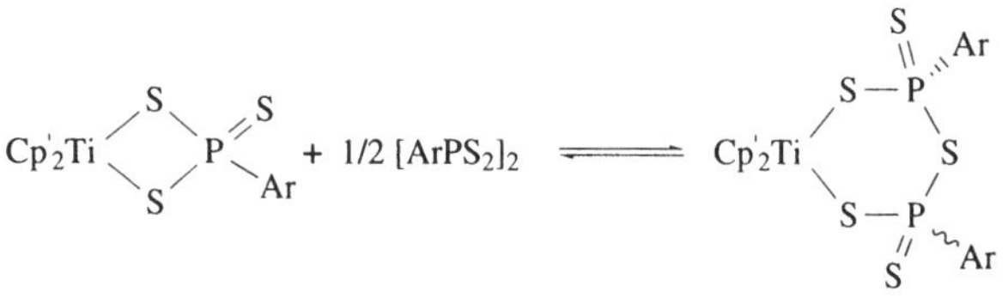

$$\text{Cp}_2'\text{Ti}\begin{array}{c}\text{S}\\ \diagdown\text{P}=\text{S}\\ \text{S}\ \diagup\ \text{Ar}\end{array} + \tfrac{1}{2}\,[\text{ArPS}_2]_2 \;\rightleftharpoons\; \text{Cp}_2'\text{Ti}\begin{array}{c}\text{S}\\\text{S}-\text{P}\\\text{S}-\text{P}\\\text{S}\end{array}\text{Ar} \qquad (91)$$

(398)

The mean Ti–S bond enthalpies for [Ti(SPr^n)$_2$Cp$_2$], [Ti(SPh)$_2$Cp$_2$] and [Ti(1,2-S$_2$C$_6$H$_3$Me)Cp$_2$] are 331 ± 10 kJ mol^{-1}, 358 ± 10 kJ mol^{-1} and 333 ± 10 kJ mol^{-1}, respectively.[708,709] ^{1}H NMR spectral spin–lattice relaxation time measurements allow the barriers of rotation of the cyclopentadienyl ligands in [TiS$_5$Cp$_2$] to be determined. Values of 9.0 ± 1.0 kJ mol^{-1} and 7.7 ± 1.2 kJ mol^{-1}, respectively, were found for the axial and the equatorial Cp$^-$ ligands.[118] The ^{77}Se chemical shifts and Se–Se coupling constants were measured for [TiSe$_5$Cp$_2$] and [Ti(S$_{5-x}$Se$_x$)Cp$_2$].[710,711] In the series [M(SeC$_6$H$_4$R^1-4)$_2$(η-C$_5$H$_4$R^2)$_2$], ^{77}Se NMR spectra has proved to be more sensitive to the influences of the substituents R^1, R^2 and the metal than ^{13}C. The ^{77}Se signals become shifted to low frequency in the sequence M = Ti < Zr < Hf.[712]

Flash photolysis of [TiS$_5$Cp$_2$] leads to the homolysis of a Ti–S bond. Photolysis in CXCl$_3$ (X = Cl or Br) gives [TiX$_2$Cp$_2$] and sulfur.[713] [TiS$_5$Cp$_2$] is a versatile starting material for the synthesis of sulfur and mixed sulfur–selenium rings. The complex reacts with SO$_2$Cl$_2$ in CS$_2$ to give titanocene dichloride and a mixture of separable (S$_5$)$_n$ species;[714] with Se$_2$Cl$_2$[715] or Se$_2$Br$_2$[716] the seven-membered S$_5$Se$_2$ results. Se$_7$ and Se$_6$S$_n$ (n = 1 or 2) are prepared from [TiSe$_5$Cp$_2$].[717] Selenium abstraction from [TiSe$_5$(η-C$_5$H$_4$Me)$_2$] with triphenylphosphine leads to [{Ti(η-C$_5$H$_4$Me)$_2$(μ_2-Se$_2$)}$_2$], with the Ti$_2$Se$_4$ ring in the chair conformation. The compound reacts with S$_2$Cl$_2$ to give isomerically pure (S$_2$Se$_2$)$_2$.[718] With ketones in the presence of [NH$_4$]$_2$S the red-black complex (**399**) is obtained (R^2 = R^3 = Me; R^2 = Me, R^3 = Et; R^{2}R^3 = (CH$_2$)$_5$) (Scheme 112).[719] The dimers [{TiCp$'_2$(μ_2-E$_2$)}$_2$] react with AsMeCl$_2$ to give [Ti(E$_4$AsMe)Cp$'_2$] (Cp$'$ = Cp or η-C$_5$H$_4$Me; E = S or Se) (Equation (92)).[720,721] The pentachalcogenides

[TiE$_5$Cp$_2$] (E = S or Se) act as chalcogen-transfer agents towards rhenium and iridium complexes.[722,723] The reaction of [TiS$_5$Cp'$_2$] (Cp' = Cp or C$_5$H$_4$Me) with activated alkynes ZC≡CZ (Z = CO$_2$Me or CF$_3$) at 140 °C affords the green dithiolene complexes (**400**) (Scheme 113). As the x-ray structure shows, the TiS$_2$C$_2$ ring is folded. The selenium analogue is made similarly but requires milder conditions.[724] In the presence of phosphines as nucleophiles, the formation of (**400**) proceeds more smoothly via nucleophilic opening of the TiS$_5$ ring to give ionic intermediates such as (**401**). Another route to (**400**) is the reaction of [TiCp$_2$(μ$_2$-S$_2$)]$_2$ with alkynes. The primary product is the green, crystallographically characterized vinyl disulfide complex (**402**), which rearranges at 90 °C to (**400**) (Scheme 113).[675] The compounds (**400**) undergo reversible electrochemical reduction at around −1.2 V (vs. Ag/AgNO$_3$). The oxidation at 0.55 V is irreversible. The chalcogen ligand can be transferred to noble metals.[724]

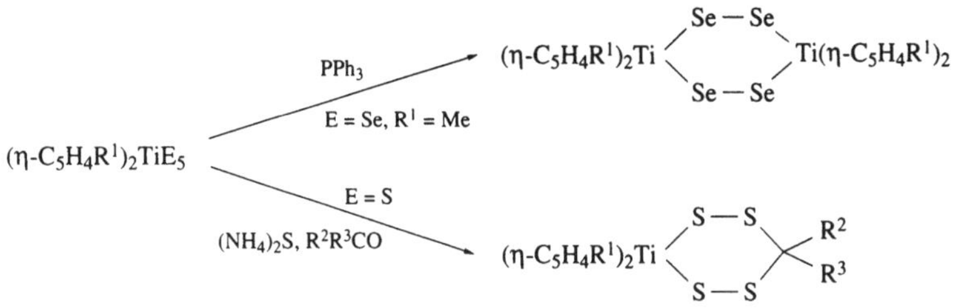

Scheme 112

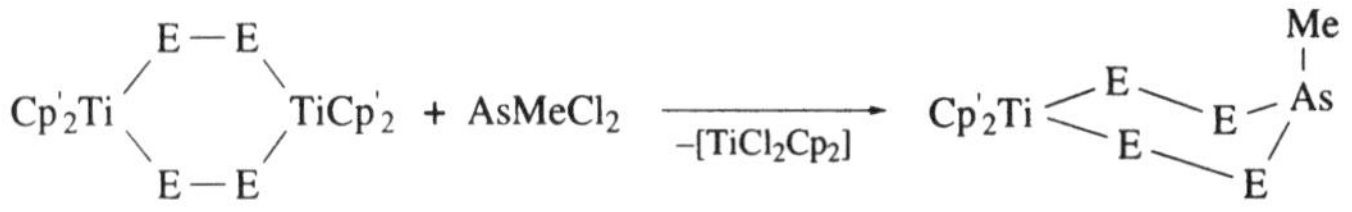

$$(92)$$

E = S or Se

Cp' = Cp or C$_5$H$_4$Me

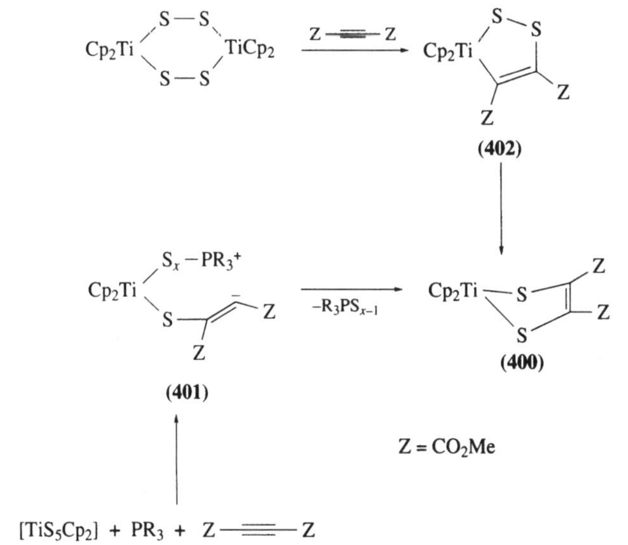

(**402**)

(**401**)

(**400**)

Z = CO$_2$Me

[TiS$_5$Cp$_2$] + PR$_3$ + Z———≡———Z

Scheme 113

   The thermolysis of [TiS$_5$Cp$_2$] in refluxing xylene for prolonged periods of time produces, *inter alia*, red crystals of (**403**) in 40% yield. One S–S bond is broken on treatment with PHPh$_2$ to give (**404**) (Scheme 114).[725]

   The reaction of [Ti(SR)$_2$Cp$_2$] with Br$_2$ gives [TiBr$_2$Cp$_2$] and RSSR (R = CH$_2$Ph, Ph, Pr^i or Bu^n). If R = Ph or Bu^n the reaction is slow enough for an intermediate, [TiBr(SR)Cp$_2$], to be identified.[726] The ethylenetetrathiolato complex (**405**) reacts with S$_x$Cl$_2$ to give (**406**) (x = 1 or 2; Cp' = Cp or C$_5$H$_4$Me). With (CH$_2$SCl)$_2$, titanocene dichloride and bicyclic sulfur compounds are formed (Scheme 115).[727,728]

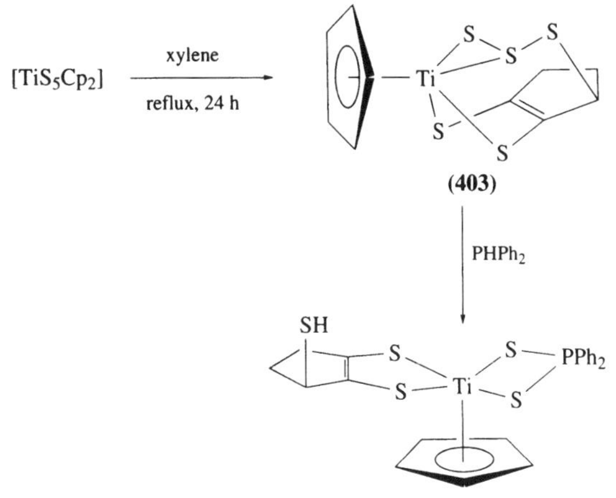

Scheme 114

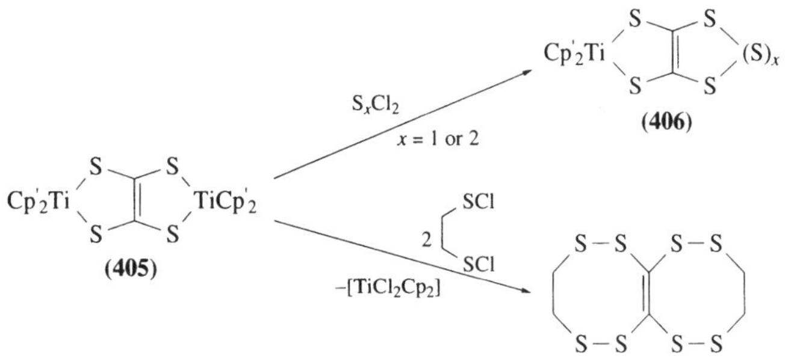

Scheme 115

5.6 TITANIUM(IV) COMPLEXES WITH NONCYCLOPENTADIENYL π LIGANDS

Tetraphenylcyclobutadienyl complexes result when $[TiCl_3(THF)_3]$ is treated with Pr^iMgCl in the presence of an excess of diphenylacetylene. The blue, extremely moisture-sensitive complex (**407**) was characterized by x-ray diffraction (Equation (93)). The average Ti–C distances of 0.224 nm are significantly shorter than in the isoelectronic compound $[TiCl_3Cp]$ (0.231 nm). The complex decomposes in acetone with the formation of tetraphenylbutadiene.[729] The η^6-arene complex (**408**) is formed when a solution of $TiCl_4$ in a noncoordinating solvent such as a chlorocarbon or hexane is treated with 6–10 equivalents of hexamethylbenzene. The yellow crystals are isolated in 70% yield (Equation (94)). The structure was confirmed by x-ray diffraction. The same product is obtained from $[TiCl_4]$ and 2-butyne.[730] The red, diamagnetic and air-stable cyclooctatetraenyl complex (**409**) (X = Y = H) is the product of the reaction between $TiCl_3$, $[C_8H_8]^{2-}$ and $[B_4C_2Et_2H_4]^-$ in THF followed by air oxidation. The crystal structure of the complex was determined. Treatment with $AlCl_3/MeI$ gives the red–brown iodo derivative (**409**) (X = I, Y = H), while iodine in benzene leads to (**409**) (X = Y = I) in 70% yield.[731] Cyclooctatetraene complexes such as $TiCp(\eta\text{-}C_8H_8)$ can be used for the gas-phase deposition of films of titanium metal at 150–500 °C.[732]

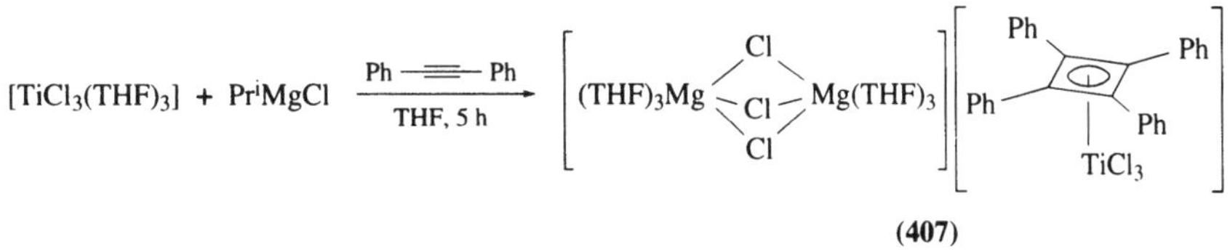

$$[TiCl_3(THF)_3] + Pr^iMgCl \xrightarrow[\text{THF, 5 h}]{Ph \equiv\!\!\!\equiv Ph} \qquad (\mathbf{407}) \qquad (93)$$

$$[TiCl_4] + C_6Me_6 \xrightarrow{CH_2Cl_2} \left[\text{TiCl}_3 \right][Ti_2Cl_9] \tag{94}$$

(408)

(409)

5.7 CYCLOOCTATETRAENE COMPLEXES OF TITANIUM (*by F. G. N. Cloke, University of Sussex, UK*)

The mixed ring sandwich compound $[Ti(\eta\text{-}C_8H_8)(Cp^*)]$ may be obtained from the reaction of $[Ti(Cp^*)Cl_3]$ with $K_2[C_8H_8]$ in THF in 34% yield[733] or by the reduction of $[Ti(Cp^*)Cl_3]$ with magnesium in the presence of cyclooctatetraene in 31–75% yield (Equation (95)).[733,734]

$$[TiCp^*Cl_3] \xrightarrow[\substack{C_8H_8/Mg \text{ in THF}}]{K_2[C_8H_8] \text{ in THF or}} \tag{95}$$

The x-ray crystal structure of $[Ti(\eta\text{-}C_8H_8)(Cp^*)]$ shows the expected parallel ring sandwich structure, very similar to that of the unsubstituted analogue $[Ti(\eta\text{-}C_8H_8)(Cp)]$.[735] The electronic structures of $[Ti(\eta\text{-}C_8H_8)(Cp)]$ and $[Ti(\eta\text{-}C_8H_8)(Cp^*)]$ have been examined by ENDOR[736] and photoelectron spectroscopy,[737] respectively, and are fully consistent with the molecular orbital scheme for a parallel ring sandwich molecule in which the HOMO is the partially occupied d_{z^2} orbital.

Electrochemical studies in THF on a variety of monocyclooctatetraenyltitanium(III) complexes of the general class $[Ti(\eta\text{-}C_8H_8)L]$ (L = Cp*, Cp, $\eta\text{-}C_5MeH_4$, η^5-indenyl or η^5-fluorenyl) have been carried out and show fully reversible one-electron oxidations to the $[Ti(\eta\text{-}C_8H_8)L]^+$ cations;[738,739] in the case of $[Ti(\eta\text{-}C_8H_8)(Cp^*)]$, bulk electrolysis in THF/$[NBu_4][ClO_4]$ affords the orange, stable diamagnetic salt $[Ti(\eta\text{-}C_8H_8)(Cp^*)][ClO_4]$.[739] In no case was one-electron reduction to the 18-electron anion $[Ti(\eta\text{-}C_8H_8)L]^-$ observed.

The reaction of $[TiCl_3(THF)_3]$ with cyclooctatetraene and sodium in the presence of 2,2'-bipyridyl affords the formally Ti^{III} complex $[Ti(\eta\text{-}C_8H_8)(N,N'\text{-}C_{10}H_8N_2\text{-}2,2')Cl]$, which has been structurally authenticated (Equation (96)).[740]

$$[TiCl_3(THF)_3] + 2,2'\text{-bipyridyl} \xrightarrow{C_8H_8/Na \text{ in THF}} \tag{96}$$

5.8 REFERENCES

1. 'Gmelin Handbook of Inorganic Chemistry', System No. 41, New Suppl. Ser. Vol. 40. Mononuclear Compounds 2, 3 and 4, Gmelin Institute, Frankfurt, 1980–1984. Di- and Polynuclear Compounds, Gmelin Institute, Frankfurt, 1990.
2. J. A. Labinger, *J. Organomet. Chem.*, 1982, **227**, 341; J. A. Labinger, *ibid.*, 1988, **357**, 1.
3. D. Cozak and M. Melnik, *Coord. Chem. Rev.*, 1986, **74**, 53.
4. M. T. Reetz, 'Organotitanium Reagents in Organic Synthesis', Springer, Berlin, 1986.
5. Y. Dang and H. J. Geise, *J. Organomet. Chem.*, 1991, **405**, 1.
6. M. Riediker and R. O. Duthaler, *Angew. Chem.*, 1989, **101**, 488; *Angew. Chem., Int. Ed. Engl.*, 1989, **28**, 494.
7. R. O. Duthaler, P. Herold, W. Lottenbach, K. Oertle and M. Riediker, *Angew. Chem.*, 1989, **101**, 490; *Angew. Chem., Int. Ed. Engl.*, 1989, **28**, 495.
8. M. Riediker, A. Hafner, U. Piantini, G. Rihs and A. Togni, *Angew. Chem., Int. Ed. Engl.*, 1989, **28**, 499.
9. (a) R. O. Duthaler and A. Hafner, *Chem. Rev.*, 1992, **92**, 807. (b) K. H. Dötz, *Nachr. Chem. Techn. Lab.*, 1990, **38**, 1244.
10. M. T. Reetz, S. H. Kyung and J. Westermann, *Organometallics*, 1984, **3**, 1716.
11. T. Tanaka, T. Inoue, K. Kamei, K. Murakami and C. Iwata, *J. Chem. Soc., Chem. Commun.*, 1990, 906.
12. A. Mori, J. Fujiwara, K. Maruoka and H. Yamamoto, *J. Organomet. Chem.*, 1985, **285**, 83.
13. Y. Kobayashi, K. Umeyama and F. Sato, *J. Chem. Soc., Chem. Commun.*, 1984, 621.
14. S. Collins, W. P. Dean and D. G. Ward, *Organometallics*, 1988, **7**, 2289.
15. T. K. Hollis, N. P. Robinson and B. Bosnich, *Organometallics*, 1992, **11**, 2745.
16. T. K. Hollis, N. P. Robinson and B. Bosnich, *J. Am. Chem. Soc.*, 1992, **114**, 5464.
17. T. L. Chen, T. H. Cha and A. Shaver, *J. Organomet. Chem.*, 1984, **268**, C1.
18. W. A. Nugent and T. V. Rajan Babu, *J. Am. Chem. Soc.*, 1988, **10**, 8561.
19. T. V. Rajan Babu and W. A. Nugent, *J. Am. Chem. Soc.*, 1989, **111**, 4525.
20. T. V. Rajan Babu, W. A. Nugent and M. S. Beattie, *J. Am. Chem. Soc.*, 1990, **112**, 6408.
21. J. S. Yadav, T. Shekharam and V. R. Gadgil, *J. Chem. Soc., Chem. Commun.*, 1990, 843.
22. Y. Handa and J. Iananaga, *Terahedron Lett.*, 1987, **28**, 5717.
23. C. H. Wong, C. W. Hung and H. N. C. Wong, *J. Organomet. Chem.*, 1988, **342**, 9.
24. S. C. Berk, K. A. Kreutzer and S. L. Buchwald, *J. Am. Chem. Soc.*, 1991, **113**, 5093.
25. L. Clawson, S. L. Buchwald and R. H. Grubbs, *Tetrahedron Lett.*, 1984, **25**, 5733.
26. J. R. Stille and R. H. Grubbs, *J. Am. Chem. Soc.*, 1983, **105**, 1664.
27. P. L. McGrane, M. Jensen and T. Livinghouse, *J. Am. Chem. Soc.*, 1992, **14**, 5459.
28. S. C. Berk, R. B. Grossman and S. L. Buchwald, *J. Am. Chem. Soc.*, 1993, **115**, 4912.
29. K. Tamao, M. Akita, R. Kanatani, N. Ishida and M. Kumada, *J. Organomet. Chem.*, 1982, **226**, C9.
30. F. Sato, H. Watanabe, Y. H. Tanaka, T. Yamaji and M. Sato, *Tetrahedron Lett.*, 1983, **24**, 1041.
31. E. C. Ashby and R. D. Ainslie, *J. Organomet. Chem.*, 1983, **250**, 1.
32. K. Mach, P. Sedmera, L. Petrusova, H. Antropiusova, V. Hanus and F. Turecek, *Tetrahedron Lett.*, 1982, **23**, 105.
33. B. Bogdanovic and M. Maruthamuthu, *J. Organomet. Chem.*, 1984, **272**, 115.
34. K. Mach *et al.*, *J. Organomet. Chem.*, 1985, **289**, 331.
35. K. Mach, F. Turecek, H. Antropiusova and V. Hanus, *Organometallics*, 1986, **5**, 1215.
36. T. J. Katz, S. M. Hacker, R. D. Kendrick and C. S. Yannoni, *J. Am. Chem. Soc.*, 1985, **105**, 2182.
37. G. B. Erskine, N. B. Vanstone and J. D. McCowan, *Inorg. Chim. Acta*, 1983, **75**, 159.
38. M. Schlegel and K. H. Thiele, *Z. Anorg. Allg. Chem.*, 1985, **526**, 43.
39. S. Sobota, J. Utko and Z. Janas, *J. Organomet. Chem.*, 1986, **316**, 19.
40. Z. Dawoodi, M. H. L. Green, V. S. B. Mtetwa and K. Prout, *J. Chem. Soc., Chem. Commun.*, 1982, 1410.
41. Z. Dawoodi, M. H. L. Green, V. S. B. Mtetwa and K. Prout, *J. Chem. Soc., Chem. Commun.*, 1982, 802.
42. A. W. Duff, R. A. Kamarudin, M. F. Lappert and R. J. Norton, *J. Chem. Soc., Dalton Trans.*, 1986, 489.
43. D. C. Bradley *et al.*, *Polyhedron*, 1990, **9**, 1423.
44. C. C. Cummins, R. R. Schrock and W. M. Davies, *Organometallics*, 1992, **11**, 1452.
45. H. tom Dieck, H. J. Rieger and G. Fendesak, *Inorg. Chim. Acta*, 1990, **177**, 191.
46. C. C. Cummins, C. P. Schaller, G. D. van Duyne, P. T. Wolczanski, A. W. E. Chan and R. Hoffmann, *J. Am. Chem. Soc.*, 1991, **113**, 2985.
47. K. A. Hughes, P. G. Dopico, M. Sabat and M. G. Finn, *Angew. Chem.*, 1993, **105**, 603; *Angew. Chem., Int. Ed. Engl.*, 1993, **32** 554.
48. T. V. Lubben, P. T. Wolczanski and G. D. van Duyne, *Organometallics*, 1984, **3**, 977.
49. R. W. Chesmit, L. D. Durfee, P. E. Fanwick, I. P. Rothwell, K. Folting and J. C. Huffman, *Polyhedron*, 1987, **6**, 2019.
50. C. Floriani, F. Corazza, W. Lesueur, A. Chiesi-Villa and C. Guestini, *Angew. Chem.*, 1989, **101**, 93; *Angew. Chem., Int. Ed. Engl.*, 1989, **28**, 66.
51. E. Solari, C. Floriani, A. Chiesi-Villa and C. Rizzoli, *J. Chem. Soc., Dalton Trans.*, 1992, 367.
52. H. J. Gais, J. Vollhardt, H. J. Lindner and H. Paulus, *Angew. Chem.*, 1988, **100**, 1598; *Angew. Chem., Int. Ed. Engl.*, 1988, **276**, 1540.
53. J. E. Hill, P. E. Fanwick and I. P. Rothwell, *Organometallics*, 1990, **9**, 2211.
54. D. Schenke and K. H. Thiele, *J. Organomet. Chem.*, 1990, **382**, 109.
55. R. Beckhaus and K. H. Thiele, *J. Organomet. Chem.*, 1986, **317**, 23.
56. A. Berry *et al.*, *J. Chem. Soc., Chem. Commun.*, 1986, 520.
57. P. Briant, J. C. Green, A. Haaland, H. Møllendal, K. Rypdal and J. Tremmel, *J. Am. Chem. Soc.*, 1989, **111**, 3434.
58. R. L. Williamson and M. B. Hall, *J. Am. Chem. Soc.*, 1988, **110**, 4428.
59. M. L. H. Green and A. K. Hughes, *J. Chem. Soc., Chem. Commun.*, 1991, 1231.
60. S. Akinobu, J. Kojima, T. Sasaki and Y. Kikuzono, *J. Organomet. Chem.*, 1988, **345**, 275.
61. P. Kuappa and N. Rösch, *J. Organomet. Chem.*, 1989, **359**, C5.
62. Z. Dawoodi *et al.*, *J. Chem. Soc., Dalton Trans.*, 1986, 1629.
63. S. Obaru, N. Koga and K. Morokuma, *J. Organomet. Chem.*, 1984, **270**, C33.
64. N. Koga, S. Obaru and K. Morokuma, *J. Am. Chem. Soc.*, 1984, **106**, 4625.
65. N. Koga and K. Morokuma, *J. Am. Chem. Soc.*, 1988, **110**, 108.

66. J. Scholz, M. Schlegel and K. H. Thiele, *Chem. Ber.*, 1987, **120**, 1369.
67. H. B. Abrahamson and M. E. Martin, *J. Organomet. Chem.*, 1982, **238**, C58.
68. H. B. Abrahamson, K. L. Brandenburg, B. Lucero, M. E. Martin and E. Dennis, *Organometallics*, 1984, **3**, 2379.
69. A. Roeder, J. Scholz and K. H. Thiele, *Z. Anorg. Allg. Chem.*, 1983, **505**, 121.
70. S. L. Latesky, A. K. Mullen, I. P. Rothwell and J. C. Huffman, *J. Am. Chem. Soc.*, 1985, **105**, 5981.
71. T. Bartik, B. Happ, A. Sarkau, K. H. Thiele and G. Pályi, *Organometallics*, 1989, **8**, 558.
72. D. Selent, R. Beckhaus and T. Bartik, *J. Organomet. Chem.*, 1991, **405**, C15.
73. R. P. Planalp, R. A. Andersen and A. Zalkin, *Organometallics*, 1983, **2**, 16.
74. G. S. Girolami, J. A. Jensen, D. M. Pollina, W. S. Williams, A. E. Kaloyeros and C. M. Allocca, *J. Am. Chem. Soc.*, 1987, **109**, 1579.
75. T. V. Lubben and P. T. Wolczanski, *J. Am. Chem. Soc.*, 1985, **107**, 701.
76. T. V. Lubben and P. T. Wolczanski, *J. Am. Chem. Soc.*, 1987, **109**, 424.
77. M. T. Reetz, H. Hugel and K. Dresely, *Tetrahedron*, 1987, **43**, 109.
78. M. T. Reetz and S. Maus, *Tetrahedron*, 1987, **43**, 101.
79. K. H. Thiele, C. Krüger, A. Sarkau, I. Ötvös, T. Bartik and G. Pályi, *Organometallics*, 1987, **6**, 2290.
80. S. L. Latesky, A. K. McMullen, G. P. Niccolai and I. P. Rothwell, *Organometallics*, 1985, **4**, 1896.
81. L. R. Chamberlain *et al.*, *J. Am. Chem. Soc.*, 1987, **109**, 390.
82. L. R. Chamberlain *et al.*, *J. Am. Chem. Soc.*, 1987, **109**, 6068.
83. L. D. Durfee, A. K. McMullen and I. P. Rothwell, *J. Am. Chem. Soc.*, 1988, **110**, 1463.
84. L. D. Durfee, J. E. Hill, P. E. Fanwick and I. P. Rothwell, *Organometallics*, 1990, **9**, 75.
85. L. D. Durfee, J. E. Hill, J. L. Kerschner, P. E. Fanwick and I. P. Rothwell, *Inorg. Chem.*, 1989, **28**, 3095.
86. L. D. Durfee, P. E. Fanwick, I. P. Rothwell, K. Folting and J. C. Huffman, *J. Am. Chem. Soc.*, 1987, **109**, 4720.
87. J. E. Hill, P. E. Fanwick and I. P. Rothwell, *Inorg. Chem.*, 1989, **28**, 3602.
88. J. E. Hill, P. E. Fanwick and I. P. Rothwell, *Inorg. Chem.*, 1991, **30**, 1143.
89. J. E. Hill, P. E. Fanwick and I. P. Rothwell, *Organometallics*, 1991, **10**, 15.
90. G. J. Balaich and I. P. Rothwell, *J. Am. Chem. Soc.*, 1993, **115**, 1581.
91. J. E. Hill, G. J. Balaich, P. E. Fanwick and I. P. Rothwell, *Organometallics*, 1991, **10**, 3428.
92. J. E. Hill, P. E. Fanwick and I. P. Rothwell, *Organometallics*, 1992, **11**, 1771.
93. A. Zambelli, L. Oliva and C. Pellecchia, *Macromolecules*, 1989, **22**, 2129.
94. A. Zambelli, P. Ammendola and A. Proto, *Macromolecules*, 1989, **22**, 2126.
95. J. A. Shelburne and G. L. Baker, *Macromolecules*, 1987, **20**, 1212.
96. (a) B. Crociani, M. Nicolini and R. L. Richards, *J. Organomet. Chem.*, 1975, **101**, C1; (b) M. Benham-Dahkordy, B. Crociani and R. L. Richards, *J. Organomet. Chem.*, 1979, **181**, 69; (c) B. Crociani and R. L. Richards, *J. Chem. Soc., Chem. Commun.*, 1973, 127; (d) M. Benham-Dahkordy, B. Crociani and R. L. Richards, *J. Chem. Soc., Dalton Trans.*, 1977, 2015.
97. M. Bottril, P. D. Gavens, J. W. Kelland and J. McMeeking, in 'COMC-I', vol. 3, p. 440.
98. T. Carofiglio, C. Floriani, A. Chiesi-Villa and C. Guastini, *Inorg. Chem.*, 1989, **28**, 4417.
99. I. Jibril, S. Alu-Orabi, S. A. Klaib, W. Imhof and G. Huttner, *J. Organomet. Chem.*, 1992, **433**, 253.
100. J. Okuda, *Chem. Ber.*, 1990, **123**, 87.
101. J. Okuda and E. Herdtweck, *Inorg. Chem.*, 1991, **30**, 1516.
102. H. Qichen, Q. Yanlong, L. Guisheng and T. Youqi, *Transition Met. Chem.*, 1990, **15**, 483.
103. Y. Qian, G. Li, W. Chen, B. Li and X. Jin, *J. Organomet. Chem.*, 1989, **373**, 185.
104. T. Kauffmann, J. Ennen and K. Berghus, *Tetrahedron Lett.*, 1984, **25**, 1971.
105. J. Okuda, *Chem. Ber.*, 1990, **123**, 1649.
106. P. Courtot, R. Pichon, J. Y. Salaun and L. Toupet, *Can. J. Chem.*, 1991, **69**, 661.
107. G. Hidalgo Llinas, M. Mena, F. Palacios, P. Royo and R. Serrano, *J. Organomet. Chem.*, 1988, **340**, 37.
108. D. T. Mallin, M. D. Rausch, E. A. Mintz and A. L. Rheingold, *J. Organomet. Chem.*, 1990, **381**, 35.
109. M. Ogasa, D. T. Mallin, D. W. Macomber, M. D. Rausch, R. D. Rogers and A. N. Rollins, *J. Organomet. Chem.*, 1991, **405**, 41.
110. A. Dormond, A. El Bauadili and C. Moise, *Tetrahedron Lett.*, 1983, **24**, 3087.
111. L. M. Alvaro, T. Cuenca, J. C. Flores, P. Royo, M. A. Pellinghelli and A. Tiripicchio, *Organometallics*, 1992, **11**, 3301.
112. S. Helin and H. Nöth, *Angew. Chem.*, 1988, **100**, 1378; *Angew. Chem., Int. Ed. Engl.*, 1988, **27**, 1331.
113. F. Nief and F. Mathey, *J. Chem. Soc., Chem. Commun.*, 1988, 770.
114. L. M. Engelhardt, R. I. Papasergio, C. L. Raston and A. H. White, *Organometallics*, 1984, **3**, 18.
115. N. W. Alcock, G. E. Toogood and M. G. H. Wallbridge, *Acta Crystallogr., Sect. C*, 1984, **40**, 598.
116. K. Kirschbaum and D. M. Giolando, *Acta Crystallogr., Sect. C, Cryst. Struct. Commun.*, 1991, **C47**, 2216.
117. C. H. Winter, X. X. Zhan, D. A. Dobbs and M. J. Heeg, *Organometallics*, 1991, **10**, 210.
118. D. F. R. Gilson and G. Gomez, *J. Organomet. Chem.*, 1982, **240**, 41.
119. A. Dormond, M. Fauconet, J. C. Leblanc and C. Moise, *Polyhedron*, 1984, **3**, 897.
120. N. Hao, B. G. Sayer, G. Dénes, D. G. Bickley, C. Detellier and M. J. McGlinchey, *J. Magn. Reson.*, 1982, **50**, 50.
121. M. J. McGlinchey and D. G. Bickley, *Polyhedron*, 1985, **4**, 1147.
122. A. Terpstra, J. N. Louwen, A. D. Oskam and J. H. Teuben, *J. Organomet. Chem.*, 1984, **260**, 207.
123. K. Mach, V. Varga, H. Antropiusova and J. Blacek, *J. Organomet. Chem.*, 1987, **333**, 205.
124. D. L. Hughes, M. Jimenez-Tenorio and G. J. Leigh, *Polyhedron*, 1989, **8**, 1780.
125. D. L. Hughes, M. Jimenez-Tenorio and G. J. Leigh, *J. Chem. Soc., Dalton Trans.*, 1989, 2453.
126. D. B. Morse, D. N. Hendrickson, T. B. Rauchfuss and S. R. Wilson, *Organometallics*, 1988, **7**, 496.
127. D. L. Hughes, G. J. Leigh and D. G. Walker, *J. Organomet. Chem.*, 1988, **35**, 113.
128. E. J. M. de Boer, J. de With and A. G. Orpen, *J. Am. Chem. Soc.*, 1986, **108**, 8271.
129. W. A. Herrmann, P. Kiprof and J. K. Felixberger, *Chem. Ber.*, 1990, **123**, 1971.
130. T. E. Patten and B. M. Novak, *Macromolecules*, 1993, **26**, 436.
131. U. Bueschges and J. C. W. Chien, *J. Polym. Sci., Part A, Polym. Chem.*, 1989, **27**, 1525.
132. N. Ishihara, M. Kuramoto and M. Uoi, *Macromolecules*, 1988, **21**, 3356.
133. W. Skupinski and I. Cieslowska-Glinska, *J. Organomet. Chem.*, 1984, **269**, 29.

134. W. Skupinski, in 'Proceedings of the International Symposium on Relationship Homogeneous and Heterogeneous Catalysis', eds. Y. I. Ermakov and V. A. Likholobov, VNU Science Press, Utrecht, 1986, p. 489.
135. G. J. Erskine, G. J. B. Hurst, E. L. Weinberg, B. K. Hunter and J. D. McCowan, *J. Organomet. Chem.*, 1984, **267**, 265.
136. G. J. Erskine, B. K. Hunter and J. D. McCowan, *Tetrahedron Lett.*, 1985, **26**, 1371.
137. P. G. Cozzi, T. Carofiglio, C. Floriani, A. Chiesi-Villa and C. Rizzoli, *Organometallics*, 1993, **12**, 2845.
138. M. Wedler, H. W. Roesky, F. T. Edelmann and U. Behrens, *Z. Naturforsch. Teil B*, 1988, **43**, 1461.
139. M. P. Gomez-Sal, M. Mena, P. Royo and R. Serrano, *J. Organomet. Chem.*, 1988, **358**, 147.
140. M. Mena, M. A. Pellinghelli, P. Royo, R. Serrano and A. Tiripicchio, *J. Chem. Soc., Chem. Commun.*, 1986, 1118.
141. M. Mena, P. Royo, R. Serrano, M. A. Pellinghelli and A. Tiripicchio, *Organometallics*, 1989, **8**, 476.
142. R. Blom, K. Rypdal, M. Mena, P. Royo and R. Serrano, *J. Organomet. Chem.*, 1990, **391**, 47.
143. P. Gomez-Pilar, M. Mena, F. Palacios, P. Royo, R. Serrano and S. Martinez Carreras, *J. Organomet. Chem.*, 1989, **375**, 59.
144. G. M. Johannes, *J. Chem. Soc., Chem. Commun.*, 1983, 550.
145. R. Fandos, A. Meetsma and J. H. Teuben, *Organometallics*, 1991, **10**, 59.
146. S. G. Blanco, P. Royo and R. Serrano, *J. Chem. Soc., Chem. Commun.*, 1986, 1572.
147. H. W. Roesky, Y. Bai and M. Noltemeyer, *Angew. Chem.*, 1989, **101**, 788, *Angew. Chem., Int. Ed. Engl.*, 1989, **28**, 754.
148. M. Bochmann, G. Karger and A. J. Jaggar, *J. Chem. Soc., Chem. Commun.*, 1990, 1038.
149. R. E. Campbell (Dow Chemical Co.), *Eur. Pat. Appl.* EP 421 659 (1991).
150. D. J. Gillis, M. J. Tudoret and M. C. Baird, *J. Am. Chem. Soc.*, 1993, **115**, 2543.
151. D. J. Gillis, in 'Abstracts of the International Symposium on 40 Years of Ziegler Catalysis, Freiburg 1993', p,. 41.
152. C. Pellecchia, P. Longo, A. Proto and A. Zambelli, *Makromol. Chem. Rapid Commun.*, 1992, **13**, 265.
153. J. C. Stevens and D. R. Neithammer (Dow Chemical Co.), *Eur. Pat. Appl.* EP 418 044 (1991).
154. J. C. Flores, M. Mena, P. Royo and R. Serrano, *J. Chem. Soc., Chem. Commun.*, 1989, 617.
155. R. Serrano, J. C. Flores, P. Royo, M. Mena, M. A. Pellinghelli and A. Tiripicchio, *Organometallics*, 1989, **8**, 1404.
156. C. T. Jekel-Vroegop and J. H. Teuben, *J. Organomet. Chem.*, 1985, **286**, 309.
157. D. M. Giolando, K. Kirschbaum, L. J. Graves and U. Bolle, *Inorg. Chem.*, 1992, **31**, 3887.
158. D. J. Brauer, H. Bürger and G. R. Liewald, *J. Organomet. Chem.*, 1986, **307**, 177.
159. Y. Bai, H. W. Roesky and M. Noltemeyer, *Z. Anorg. Allg. Chem.*, 1991, **595**, 21.
160. Y. Bai, H. W. Roesky, H. G. Schmidt, M. Noltemeyer, *Z. Naturforsch., Teil B*, 1992, **47**, 603.
161. J. Scholz, A. Dietrich, H. Schumann and K. H. Thiele, *Chem. Ber.*, 1991, **124**, 1035.
162. K. V. Katti and R. G. Cavell, *Inorg. Chem.*, 1989, **28**, 413.
163. K. V. Katti and R. G. Cavell, *Organometallics*, 1991, **10**, 539.
164. P. Olms, H. W. Roesky, K. Keller and M. Noltemeyer, *Z. Naturforsch., Teil B*, 1992, **47**, 1609.
165. J. R. Dilworth, I. A. Latham, G. J. Leigh, G. Huttner and I. Jibril, *J. Chem. Soc., Chem. Commun.*, 1983, 1368.
166. E. M. R. Kiremire, G. J. Leigh, J. R. Dilworth and R. A. Henderson, *Inorg. Chim. Acta*, 1984, **83**, L83.
167. R. Hemmer, U. Thewalt, D. L. Hughes, G. J. Leigh and D. G. Walker, *J. Organomet. Chem.*, 1987, **323**, C29.
168. I. A. Latham, G. J. Leigh, G. Huttner and I. Jibril, *J. Chem. Soc., Dalton Trans.*, 1986, 385.
169. I. A. Latham, G. J. Leigh, G. Huttner and I. Jibril, *J. Chem. Soc., Dalton Trans.*, 1986, 377.
170. D. L. Hughes, M. Jimenez-Tenorio, G. J. Leigh and D. G. Walker, *J. Chem. Soc., Dalton Trans.*, 1989, 2389.
171. I. A. Latham and G. J. Leigh, *J. Chem. Soc., Dalton Trans.*, 1986, 399.
172. D. L. Hughes, I. A. Latham and G. J. Leigh, *J. Chem. Soc., Dalton Trans.*, 1986, 393.
173. O. P. Pandey and S. K. Sengupta, *J. Sci. Ind. Res.*, 1989, **48**, 43.
174. S. Kher, V. Kumari and R. N. Kapoor, *Acta Chim. Acad. Sci. Hung.*, 1982, **111**, 333.
175. S. Kher, V. Kumari and R. N. Kapoor, *Acta Chim. Hung.*, 1984, **115**, 159.
176. S. Kher, V. Verma and R. N. Kapoor, *Transition Met. Chem. (Weinheim)*, 1983, **8**, 163.
177. O. P. Pandey, S. K. Sengupta and S. C. Tripathi, *Synth. React. Inorg. Metal-Org. Chem.*, 1983, **13**, 867.
178. O. P. Pandey, S. K. Sengupta and S. C. Tripathi, *Transition Met. Chem. (Weinheim)*, 1983, **8**, 362.
179. Y. Singh, R. Sharan and R. N. Kapoor, *Transition Met. Chem. (London)*, 1987, **12**, 335.
180. K. Dey, D. Koner, A. K. Biswas and S. Ray, *J. Chem. Soc., Dalton Trans.*, 1982, 911.
181. O. P. Pandey, S. K. Sengupta and S. C. Tripathi, *Inorg. Chim. Acta*, 1984, **90**, 91.
182. O. P. Pandey, S. K. Sengupta and S. C. Tripathi, *Bull. Chem. Soc. Jpn.*, 1985, **58**, 2395.
183. O. P. Pandey, *Inorg. Chim. Acta*, 1986, **118**, 105.
184. T. Carofiglio, C. Floriani, A. Sgamellotti, M. Rosi, A. Chiesi-Villa and C. Rizzoli, *J. Chem. Soc., Dalton Trans.*, 1992, 1081.
185. P. Gowik, T. Klapötke and J. Pickardt, *J. Organomet. Chem.*, 1990, **393**, 343.
186. J. Okuda, *J. Organomet., Chem.*, 1990, **397**, C37.
187. T. J. Clark, T. A. Nile, D. McPhail and A. T. McPhail, *Polyhedron*, 1989, **8**, 1804.
188. F. Palacios, P. Royo, R. Serrano, J. L. Balcazar, I. Fonseca and F. Florencio, *J. Organomet. Chem.*, 1989, **375**, 51.
189. L. M. Babcock, V. W. Day and W. G. Klemperer, *J. Chem. Soc., Chem. Commun.*, 1987, 859.
190. L. M. Babcock and W. G. Klemperer, *Inorg. Chem.*, 1989, **28**, 2003.
191. S. I. Troyanov, V. Varga and K. Mach, *J. Organomet. Chem.*, 1991, **402**, 201.
192. A. Roth, C. Floriani, A. Chiesi-Villa and C. Guastini, *J. Am. Chem. Soc.*, 1986, **108**, 6823.
193. F. Bottomley and F. Grein, *Inorg. Chem.*, 1982, **21**, 4170.
194. H. P. Klein and U. Thewalt, *J. Organomet. Chem.*, 1982, **232**, 41.
195. P. Gomez-Sal, B. Royo, P. Royo, R. Serrano, I. Saez and S. Martinez-Carreras, *J. Chem. Soc., Dalton Trans.*, 1991, 1575.
196. C. H. Winter, X. X. Zhou and M. J. Heeg, *Organometallics*, 1991, **10**, 3799.
197. R. Martin-Benito, R. Jimenez-Aparicio and M. C. Barral, *J. Organomet. Chem.*, 1984, **269**, 267.
198. H. Aslan, T. Sielisch and R. D. Fischer, *J. Organomet. Chem.*, 1986, **315**, C69.
199. A. Gomez-Carrera, M. Mena, P. Royo and R. Serrano, *J. Organomet. Chem.*, 1986, **315**, 329.
200. T. Liu, H. W. Roesky, H. G. Schmidt and M. Noltemeyer, *Organometallics*, 1992, **11**, 2965.
201. D. M. Choquette, W. E. Buschmann, M. M. Olmstead and R. P. Planalp, *Inorg. Chem.*, 1993, **32**, 1062.
202. R. Taube and P. Knoth, *Z. Anorg. Allg. Chem.*, 1990, **581**, 89.
203. S. R. Wade, M. G. H. Wallbridge and G. R. Willey, *J. Chem. Soc., Dalton Trans.*, 1982, 271.
204. S. Wu, Y. Chen and Y. Zhang, *Synth. React. Inorg. Metal-Org. Chem.*, 1991, **21**, 599.

205. U. Thewalt and K. Doeppert, *J. Organomet. Chem.*, 1987, **320**, 177.
206. W. Skupinski and A. Wasilewski, *J. Organomet. Chem.*, 1981, **220**, 39.
207. J. C. W. Chien, Z. Salajka and S. Dong, *Macromolecules*, 1992, **25**, 3199.
208. R. G. Finke, C. A. Green and B. Rapko, *Inorg. Synth.*, 1990, **27**, 128.
209. R. G. Finke, B. Rapko and P. J. Domaille, *Organometallics*, 1986, **5**, 175.
210. T. M. Che, V. W. Day, L. C. Francesconi, M. F. Fredrich, W. G. Klemperer and W. Skum, *Inorg. Chem.*, 1985, **24**, 4055.
211. T. M. Che *et al.*, *Inorg. Chem.*, 1992, **31**, 2920.
212. J. F. W. Keana and M. D. Ogan, *J. Am. Chem. Soc.*, 1986, **108**, 7951.
213. J. F. W. Keana, M. D. Ogan, Y. Lu and M. V. Beer, *J. Am. Chem. Soc.*, 1985, **107**, 6714.
214. J. F. W. Keana, M. D. Ogan, Y. Lu, M. Beer and J. Varkey, *J. Am. Chem. Soc.*, 1986, **108**, 7957.
215. A. Müller, E. Krickemeyer, A. Sprafke and N. H. Schladerbeck, *Chimia*, 1988, **42**, 68.
216. F. Bottomley, G. O. Egharevba and P. S. White, *J. Am. Chem. Soc.*, 1985, **107**, 4353.
217. C. E. Davies *et al.*, *Inorg. Chem.*, 1992, **31**, 3779.
218. G. A. Zank, C. A. Jones, T. B. Rauchfuss and A. L. Rheingold, *Inorg. Chem.*, 1986, **25**, 1886.
219. F. Bottomley and R. W. Day, *Organometallics*, 1991, **10**, 2560.
220. T. Klapötke, R. Laskowski and H. Köpf, *Z. Naturforsch., Teil B.*, 1987, **42**, 777.
221. T. Klapötke, H. Köpf and P. Gowik, *J. Chem. Soc., Dalton Trans.*, 1988, 1529.
222. H. Köpf and T. Klapötke, *J. Organomet. Chem.*, 1986, **307**, 319.
223. T. T. Nadasdi, Y. J. Huang and D. W. Stephan, *Inorg. Chem.*, 1993, **32**, 347.
224. S. Kumar and N. K. Kaushik, *Acta Chim. Acad. Sci. Hung.*, 1982, **109**, 13.
225. Z. Wang, S. Lu and H. Guo, *Synth. React. Inorg. Metal-Org. Chem.*, 1991, **21**, 1243.
226. O. P. Pandey, S. K. Sengupta and S. C. Tripathi, *Polyhedron*, 1984, **3**, 695.
227. R. C. Fay, J. R. Weir and A. H. Bruder, *Inorg. Chem.*, 1984, **23**, 1079.
228. K. C. Ott, E. J. M. de Boer and R. H. Grubbs, *Organometallics*, 1984, **3**, 223.
229. Q. Huang, Y. Qian, Y. Tang and S. Chen, *J. Organomet. Chem.*, 1988, **340**, 179.
230. M. Kapon and G. M. Reisner, *J. Organomet. Chem.*, 1989, **367**, 77.
231. R. A. Howie, G. P. McQuillan and D. W. Thompson, *Acta Crystallogr., Sect. C, Cryst. Struct. Commun.*, 1985, **C41**, 1045.
232. R. A. Howie, G. P. McQuillan, D. W. Thompson and G. A. Lock, *J. Organomet. Chem.*, 1986, **303**, 213.
233. R. A. Howie, G. P. McQuillan and D. W. Thompson, *J. Organomet. Chem.*, 1984, **268**, 149.
234. Q. Huang, Y. Qian, W. Xu, M. Shao and Y. Tang, *Wuji Huaxue*, 1985, **1**, 166.
235. Q. Huang, Y. Qian and Y. Tang, *J. Organomet. Chem.*, 1989, **368**, 277.
236. H. Qichen, Q. Yanlong and T. Youqi, *Transition Met. Chem. (London)*, 1989, **14**, 315.
237. R. L. Halterman and K. P. C. Vollhardt, *Organometallics*, 1988, **7**, 883.
238. M. D. Rausch, J. F. Lewison and W. P. Hart, *J. Organomet. Chem.*, 1988, **358**, 161.
239. B. L. Booth, G. C. Ofunne, C. Stacey and P. J. T. Tait, *J. Organomet. Chem.*, 1986, **315**, 143.
240. J. C. Leblanc, C. Moise, A. Maisonnat, R. Poilblanc, C. Charrier and F. Mathey, *J. Organomet. Chem.*, 1982, **231**, C43.
241. K. P. Stahl, G. Boche and W. Massa, *J. Organomet. Chem.*, 1984, **277**, 113.
242. E. A. Anslyn and R. H. Grubbs, *Inorg. Synth.*, 1992, **29**, 198.
243. B. G. Conway and M. D. Rausch, *Organometallics*, 1985, **4**, 688.
244. N. Klouras and V. Nastopoulos, *Monatsh. Chem.*, 1991, **122**, 551.
245. J. Okuda, *J. Organomet. Chem.*, 1990, **385**, C39.
246. I. F. Urazowski, I. E. Nifant'ev, A. P. Moravskii, V. I. Ponomarev and L. O. Atovmyan, *Izv. Akad. Nauk. SSR, Ser. Khim.*, 1987, 1438.
247. I. F. Urazowski, V. I. Ponomaryov, O. G. Ellert, I. E. Nifant'ev and D. A. Lemenovskii, *J. Organomet. Chem.*, 1988, **356**, 181.
248. N. Klouras, *Chem. Scri.*, 1984, **24**, 193.
249. J. Okuda, *J. Organomet. Chem.*, 1988, **356**, C43.
250. C. H. Winter, X. X. Zhou and M. J. Heeg, *Inorg. Chem.*, 1992, **31**, 1808.
251. R. K. Sharma and C. P. Sharma, *J. Indian Chem. Soc.*, 1986, **63**, 838.
252. C. Sornay, P. Meunier, B. Gautheron, G. A. O'Doherty and L. A. Paquette, *Organometallics*, 1991, **10**, 2082.
253. J. C. Galluci, B. Gautheron, M. Gugelchuk, P. Meunier and L. A. Paquette, *Organometallics*, 1987, **6**, 15.
254. L. A. Paquette, M. Gugelchuk and M. McLaughlin, *J. Org. Chem.*, 1987, **52**, 4732.
255. L. A. Paquette, K. J. Moriarty and R. D. Rogers, *Organometallics*, 1989, **8**, 1506.
256. L. A. Paquette, J. A. McKinney, M. L. McLaughlin and A. L. Rheingold, *Tetrahedron Lett.*, 1986, **27**, 5599.
257. L. A. Paquette, K. J. Moriarty, J. A. McKinney and R. D. Rogers, *Organometallics*, 1989, **8**, 1707.
258. K. J. Moriarty, R. D. Rogers and L. A. Paquette, *Organometallics*, 1989, **8**, 1512.
259. M. R. Sivik, R. D. Rogers and L. A. Paquette, *J. Organomet. Chem.*, 1990, **397**, 177.
260. V. V. Bhide, P. L. Rinaldi and M. F. Farona, *Organometallics*, 1990, **9**, 123.
261. R. L. Halterman, K. P. C. Vollhardt, M. E. Walker, D. Bläser and R. Boese, *J. Am. Chem. Soc.*, 1987, **109**, 8105.
262. P. Burger, H. U. Hund, K. Evertz and H. H. Brintzinger, *J. Organomet. Chem.*, 1989, **378**, 153.
263. Z. Chen, K. Eriks and R. L. Halterman, *Organometallics*, 1991, **10**, 3449.
264. S. L. Colletti and R. L. Halterman, *Organometallics*, 1991, **10**, 3438.
265. P. Courtot, V. Labed, R. Pichon and J. Y. Salaun, *J. Organomet. Chem.*, 1989, **359**, C9.
266. P. Burger, K. Hortmann and H. H. Brintzinger, *Makromol. Chem., Macromol. Symp.*, 1993, **66**, 127.
267. M. P. Castellani, S. J. Geib, A. T. Rheingold and W. C. Trogler, *Organometallics*, 1987, **6**, 2524.
268. J. Szymoniak, M. M. Kubicki, J. Besançon and C. Moise, *Inorg. Chim. Acta*, 1991, **180**, 153.
269. G. Schmid, S. Amirkhalili, U. Höhner, D. Kampmann and R. Boese, *Chem. Ber.*, 1982, **115**, 3830.
270. G. Schmid, D. Kampmann, W. Meyer, R. Boese, P. Paetzold and K. Delpy, *Chem. Ber.*, 1985, **118**, 2418.
271. F. Nief, L. Ricard and F. Mathey, *Organometallics*, 1989, **8**, 1473.
272. P. G. Gassman and C. H. Winter, *J. Am. Chem. Soc.*, 1986, **108**, 4228.
273. M. D. Rausch, B. H. Edwards, R. D. Rogers and J. L. Atwood, *J. Am. Chem. Soc.*, 1983, **105**, 3882.
274. M. D. Rausch and W. C. Spink, *Synth. React. Inorg. Metal-Org. Chem.*, 1989, **19**, 1093.
275. A. Reissová, Z. Bastl and M. Capka, *Collect. Czech. Chem. Commun.*, 1986, **51**, 1430.

276. A. K. Sharma and N. N. Kaushik, *Acta Chim. Hung.*, 1984, **116**, 361.
277. A. Cano, T. Cuenca, G. Rodriguez, P. Royo, C. Cardin and D. J. Wilcock, *J. Organomet. Chem.*, 1993, **447**, 51.
278. T. Wöhrle and U. Thewalt, *J. Organomet. Chem.*, 1993, **447**, 45.
279. I. E. Nifant'ev, M. V. Borzov, A. V. Churakov, S. G. Mkoyan and L. O. Atovmyan, *Organometallics*, 1992, **11**, 3942.
280. P. Härter, G. Boguth, E. Herdtweck and J. Riede, *Angew. Chem.*, 1989, **101**, 1058; *Angew. Chem., Int. Ed. Engl.*, 1989, **28**, 1008.
281. S. L. Colletti and R. L. Halterman, *Tetrahedron Lett.*, 1989, **30**, 3513.
282. L. A. Paquette, K. J. Moriarty, P. Meunier, B. Gautheron and V. Crocq, *Organometallics*, 1988, **7**, 1873.
283. L. A. Paquette *et al.*, *Organometallics*, 1989, **8**, 2159.
284. L. A. Paquette and M. R. Sivik, *Organometallics*, 1992, **11**, 3503.
285. R. D. Rogers, M. M. Benning, L. K. Kurihara, K. J. Moriarty and M. D. Rausch, *J. Organomet. Chem.*, 1985, **293**, 51.
286. J. Chen, Y. Kai, N. Kasai, H. Yasuda, H. Yamamoto and A. Nakamura, *J. Organomet. Chem.*, 1991, **407**, 191.
287. B. Demerseman, R. Mahe and P. H. Dixneuf, *J. Chem. Soc., Chem. Commun.*, 1984, 1394.
288. C. H. Winter, J. W. Kampf and X. X. Zhou, *Acta Crystallogr., Sect. C, Cryst. Struct. Commun.*, 1990, **C46**, 1231.
289. U. Thewalt and G. Schmid, *J. Organomet. Chem.*, 1991, **412**, 343.
290. J. A. Smith and H. H. Brintzinger, *J. Organomet. Chem.*, 1981, **218**, 159.
291. P. Burger, J. Diebold, S. Gutmann, H. U. Hund and H. H. Brintzinger, *Organometallics*, 1992, **11**, 1319.
292. F. R. W. P. Wild, L. Zsolnai, G. Huttner and H. H. Brintzinger, *J. Organomet. Chem.*, 1982, **232**, 233.
293. S. Collins, B. A. Kuntz, N. J. Taylor and D. G. Ward, *J. Organomet. Chem.*, 1988, **342**, 21.
294. S. Collins, Y. Hong and N. J. Taylor, *Organometallics*, 1990, **9**, 2695.
295. S. Collins, Y. Hong, R. Ramachandran and N. J. Taylor, *Organometallics*, 1991, **10**, 2349.
296. F. Wochner, L. Zsolnai, G. Huttner and H. H. Brintzinger, *J. Organomet. Chem.*, 1985, **288**, 69.
297. H. Schwenlein and H. H. Brintzinger, *J. Organomet. Chem.*, 1983, **254**, 69.
298. S. Gutmann, P. Burger, M. H. Prosenc and H. H. Brintzinger, *J. Organomet. Chem.*, 1990, **397**, 21.
299. A. L. Rheingold, N. P. Robinson, J. Whelan and B. Bosnich, *Organometallics*, 1992, **11**, 1869.
300. T. K. Hollis, A. L. Rheingold, N. P. Robinson, J. Whelan and B. Bosnich, *Organometallics*, 1992, **11**, 2812.
301. S. C. Sutton, M. H. Nantz and S. R. Parkin, *Organometallics*, 1993, **12**, 2248.
302. Z. Chen and R. L. Halterman, *J. Am. Chem. Soc.*, 1992, **114**, 2276.
303. P. Burger, K. Hortmann, J. Diebold and H. H. Brintzinger, *J. Organomet. Chem.*, 1991, **417**, 9.
304. M. S. Erickson, F. R. Fronczek and M. L. McLaughlin, *J. Organomet. Chem.*, 1991, **415**, 75.
305. S. Gutmann, P. Burger, H. U. Hund, J. Hofmann and H. H. Brintzinger, *J. Organomet. Chem.*, 1989, **369**, 343.
306. M. S. Erickson, F. R. Fronczek and M. L. McLaughlin, *Acta Crystallogr., Sect. C*, 1990, **46**, 1802.
307. W. Röll, L. Zsolnai, G. Huttner and H. H. Brintzinger, *J. Organomet. Chem.*, 1987, **322**, 65.
308. P. Burger and H. H. Brintzinger, *J. Organomet. Chem.*, 1991, **407**, 207.
309. M. J. Burk, S. L. Colletti and R. L. Halterman, *Organometallics*, 1991, **10**, 2998.
310. M. E. Huttnloch, J. Diebold, U. Rief, H. H. Brintzinger, A. M. Gilbert and T. J. Katz, *Organometallics*, 1992, **11**, 3600.
311. K. L. Gibis, G. Helmchen, G. Huttner and L. Zsolnai, *J. Organomet. Chem.*, 1993, **445**, 181.
312. H. Köpf and N. Klouras, *Z. Naturforsch., Teil B*, 1983, **38**, 321.
313. G. S. Herrmann, H. G. Alt and M. D. Rausch, *J. Organomet. Chem.*, 1991, **401**, C5.
314. D. T. Mallin, M. D. Rausch, Y. G. Lin, S. Dong and J. C. W. Chien, *J. Am. Chem. Soc.*, 1990, **112**, 2030.
315. C. S. Bajgur, W. Tikkanen and J. L. Petersen, *Inorg. Chem.*, 1985, **24**, 2539.
316. H. Lang and D. Seyferth, *Organometallics*, 1991, **10**, 347.
317. D. M. Curtis, J. J. D'Errico, D. N. Duffy, P. S. Epstein and L. G. Bell, *Organometallics*, 1983, **2**, 1808.
318. P. Jutzi and R. Dickbreder, *Chem. Ber.*, 1986, **119**, 1750.
319. H. Köpf and N. Klouras, *Monatsh. Chem.*, 1983, **114**, 243.
320. N. Klouras, *Z. Naturforsch., Teil B.*, 1991, **46**, 647.
321. P. G. Gassman, W. H. Campbell and D. W. Macomber, *Organometallics*, 1984, **3**, 385.
322. J. A. Smith, J. v. Seyerl, G. Huttner and H. H. Brintzinger, *J. Organomet. Chem.*, 1979, **173**, 175.
323. M. J. Calhorda, A. R. Dias, A. M. Galvao and J. A. M. Simoes, *J. Organomet. Chem.*, 1986, **307**, 167.
324. A. R. Dias, M. S. Salema, J. A. M. Simoes, J. W. Pattiasina and J. H. Teuben, *J. Organomet. Chem.*, 1988, **346**, C4.
325. A. R. Dias, P. B. Dias, H. P. Diogo, A. M. Galvao, M. E. Minas da Piedade and J. A. M. Simoes, *Organometallics*, 1987, **6**, 1427.
326. M. R. M. Bruce, A. Kenter and D. R. Tyler, *J. Am. Chem. Soc.*, 1984, **106**, 639.
327. M. R. M. Bruce and D. R. Tyler, *Organometallics*, 1985, **4**, 528.
328. M. R. M. Bruce, A. Sclafani and D. R. Tyler, *Inorg. Chem.*, 1986, **25**, 2546.
329. T. Vondrak, K. Mach and V. Varga, *J. Organomet. Chem.*, 1989, **367**, 69.
330. P. G. Gassman, D. W. Macomber and J. W. Hershberger, *Organometallics*, 1983, **2**, 1470.
331. J. Müller, F. Lüdemann and H. Köpf, *J. Organomet. Chem.*, 1986, **303**, 167.
332. G. Fu, Y. Qian, Y. Xu and S. Chen, *J. Organomet. Chem.*, 1986, **314**, 113.
333. K. Hortmann and H. H. Brintzinger, *New J. Chem.*, 1992, **16**, 51.
334. G. M. Brown, *Inorg. Chem.*, 1989, **28**, 3028.
335. B. Brindley, A. G. Davies and J. A. A. Hawari, *J. Organomet. Chem.*, 1983, **250**, 247.
336. M. Kopietz, M. D. Lechner, D. G. Steinmeier and H. Franke, *Makromol. Chem.*, 1986, **187**, 2787.
337. B. S. Ault, *Inorg. Chem.*, 1991, **30**, 2483.
338. R. bin Ali, J. Burgess and A. T. Casey, *J. Organomet. Chem.*, 1989, **362**, 305.
339. P. Soni, K. Chandra, R. K. Sharma and B. S. Garg, *J. Indian Chem. Soc.*, 1983, **60**, 419.
340. S. Chen, Q. Liu and J. Wang, *Jiegou Huaxe*, 1983, **2**, 145.
341. T. Klapötke and U. Thewalt, *J. Organomet. Chem.*, 1988, **356**, 173.
342. P. Gowik, T. Klapötke and U. Thewalt, *J. Organomet. Chem.*, 1990, **385**, 345.
343. T. Klapötke, *Polyhedron*, 1988, **7**, 1221.
344. P. Gowik and T. Klapötke, *J. Organomet. Chem.*, 1990, **387**, C27.
345. P. K. Gowik, T. Klapötke and S. Cameron, *J. Chem. Soc., Dalton Trans.*, 1991, 1433.
346. T. Klapötke, *Polyhedron*, 1989, **8**, 311.

347. P. Gowik and T. Klapötke, *J. Organomet. Chem.*, 1991, **402**, 349.
348. P. Köpf-Maier and T. Klapötke, in 'Proceedings of the 1st International Symposium on Metal Ions in Biology and Medicine', ed. P. Collery, Libbey, Paris, 1990, p. 465.
349. P. Köpf-Maier, S. Grabowski and H. Köpf, *Eur. J. Med. Chem.*, 1984, **19**, 347.
350. D. P. Fairlie, M. W. Whitehouse and J. A. Broomhead, *Chem.-Biol. Interact.*, 1987, **61**, 277.
351. J. H. Toney and T. J. Marks, *J. Am. Chem. Soc.*, 1985, **107**, 947.
352. K. Döppert, *J. Organomet. Chem.*, 1987, **319**, 351.
353. K. Döppert, *Naturwissenschaften*, 1990, **77**, 19.
354. C. C. Perry, E. J. Moss and R. J. P. Williams, *Proc. R. Soc. London, Ser. B*, 1990, **241**, 47.
355. H. Yasuda, H. Yamamoto, K. Yokota and A. Nakamura, *Chem. Lett.*, 1989, 1309.
356. P. Sobota and Z. Janas, *J. Organomet. Chem.*, 1983, **243**, 35.
357. E. U. van Raaij, C. D. Schmulbach and H. H. Brintzinger, *J. Organomet. Chem.*, 1987, **328**, 275.
358. N. El Murr and A. Chaloyard, *J. Organomet. Chem.*, 1982, **231**, 1.
359. G. L. Soloveichik, A. B. Gavrilov and V. V. Strelets, *Metallorg. Khim.*, 1989, **2**, 431.
360. S. V. Kucharenko, G. L. Soloveichik and V. V. Strelets, *Metallorg. Khim.*, 1989, **2**, 395.
361. E. Samuel and J. Vedel, *Organometallics*, 1989, **8**, 237.
362. J. J. Singh, G. Singh, N. Kumar, R. K. Sharma and R. K. Multani, *Indian J. Chem., Sect. A*, 1982, **21**, 631.
363. J. E. Anderson and S. M. Sawtelle, *Inorg. Chem.*, 1992, **31**, 5345.
364. S. K. Tyrlik, A. Korda and L. W. Zatorski, *J. Mol. Catal.*, 1988, **45**, 161.
365. G. A. Luinstra and J. H. Teuben, *J. Organomet. Chem.*, 1991, **420**, 337.
366. G. A. Luinstra and J. H. Teuben, *Organometallics*, 1992, **11**, 1793.
367. S. Chen, Y. Liu and J. Wang, *Sci. Sin. (Engl. Ed.)*, 1982, **25**, 341.
368. I. E. Nifant'ev, A. V. Churakov, I. F. Urazowski, S. G. Mkoyan and L. O. Atovmyan, *J. Organomet. Chem.*, 1992, **435**, 37.
369. R. Gómez, T. Cuenca, P. Royo and E. Hovestreydt, *Organometallics*, 1991, **10**, 2516.
370. J. C. W. Chien, G. H. Llinas, M. D. Rausch, G. Y. Lin and H. H. Winter, *J. Am. Chem. Soc.*, 1991, **113**, 8569.
371. J. C. W. Chien *et al.*, *J. Polym. Sci. Part A: Polym. Chem.*, 1992, **30**, 2601.
372. V. Dimitrov, *J. Organomet. Chem.*, 1985, **282**, 321.
373. B. J. J. van de Heisteeg, G. Schat, O. S. Akkerman and F. Bickelhaupt, *Organometallics*, 1985, **4**, 1141.
374. B. J. J. van de Heisteeg, G. Schat, O. S. Akkerman and F. Bickelhaupt, *Organometallics*, 1986, **5**, 1749.
375. G. Erker, R. Schlund and C. Krüger, *J. Organomet. Chem.*, 1988, **338**, C4.
376. H. Schmidbaur, R. Pichl and G. Müller, *Chem. Ber.*, 1987, **120**, 39.
377. T. Cuenca, J. Carlos Flores, P. Royo, A-M. Larsonneur, R. Choukroun and F. Dahan, *Organometallics*, 1992, **11**, 777.
378. F. Senocq, M. Basso-Bert, R. Choukroun and D. Gervais, *J. Organomet. Chem.*, 1985, **297**, 155.
379. D. Steinborn and R. Taube, *J. Organomet. Chem.*, 1985, **284**, 395.
380. Y. A. Ol'dekop and V. A. Knizhnikov, *Zh. Obshch. Khim.*, 1982, **52**, 1571.
381. R. Beckhaus, K. H. Thiele and D. Stroehl, *J. Organomet. Chem.*, 1989, **369**, 43.
382. R. Beckhaus, S. Flatau, S. Trojanov and P. Hofmann, *Chem. Ber.*, 1992, **125**, 291.
383. H. Lang and D. Seyferth, *Z. Naturforsch., Teil B*, 1990, **45**, 212.
384. K. H. Thiele, C. Krüger, T. Bartik and M. Dargatz, *J. Organomet. Chem.*, 1988, **352**, 115.
385. P. H. van Rooyen, M. Schindehutte and S. Lotz, *Organometallics*, 1992, **1**, 1104.
386. P. Binger, P. Müller, R. Wenz and R. Mynott, *Angew. Chem.*, 1990, **102**, 1070; *Angew. Chem., Int. Ed. Engl.*, 1990, **29**, 1037.
387. P. Binger, *et al.*, *Chem. Ber.*, 1992, **125**, 2209.
388. H. G. Alt, K. H. Schwind, M. D. Rausch and U. Thewalt, *J. Organomet. Chem.*, 1988, **349**, C7.
389. H. G. Alt and G. S. Herrmann, *J. Organomet. Chem.*, 1990, **390**, 159.
390. V. B. Shur *et al.*, *J. Organomet. Chem.*, 1983, **243**, 157.
391. J. A. Martinho Simoes and J. L. Beauchamp, *Chem. Rev.*, 1990, **90**, 629.
392. A. R. Dias, M. S. Salema and J. A. Martinho Simoes, *Organometallics*, 1982, **1**, 971.
393. M. J. Calhorda, A. R. Dias, M. S. Salema and J. A. Martinho Simoes, *J. Organomet. Chem.*, 1983, **255**, 81.
394. G. P. McQuillan, D. C. McKean and I. Torto, *J. Organomet. Chem.*, 1986, **312**, 183.
395. J. Scholz, F. Rehbaum, K. H. Thiele, R. Goddard, P. Betz and C. Krüger, *J. Organomet. Chem.*, 1993, **443**, 93.
396. R. Gómez, T. Cuenca, P. Royo, W. A. Herrmann and E. Herdtweck, *J. Organomet. Chem.*, 1990, **382**, 103.
397. C. J. Cardin, D. J. Cardin, D. A. Morton-Blake, H. E. Parge and A. Roy, *J. Chem. Soc., Dalton Trans.*, 1987, 1641.
398. A. Sebald, P. Fritz and B. Wrackmeyer, *Spectrochim. Acta, Part A*, 1985, **41A**, 1405.
399. E. T. Knight, L. K. Myers and M. E. Thompson, *Organometallics*, 1992, **11**, 3691.
400. S. J. Heyes and C. M. Dobson, *J. Am. Chem. Soc.*, 1991, **113**, 463.
401. L. M. Hansen and D. S. Marynick, *J. Am. Chem. Soc.*, 1988, **110**, 2358.
402. L. M. Hansen and D. S. Marynick, *Organometallics*, 1989, **8**, 2173.
403. V. B. Shur, E. G. Berkovich, E. I. Mysov and M. E. Vol'pin, *Izv. Akad. Nauk. SSSR, Ser. Khim.* 1987, 2435.
404. C. McDade, J. C. Green and J. E. Bercaw, *Organometallics*, 1982, **1**, 1629.
405. K. Mach, V. Varga, V. Hanus and P. Sedmera, *J. Organomet. Chem.*, 1991, **415**, 87.
406. M. Pankowski and E. Samuel, *J. Organomet. Chem.*, 1981, **221**, C21.
407. R. F. Johnston and J. C. Cooper, *Organometallics*, 1987, **6**, 2448.
408. P. Courtot, R. Pichon, Y. Raoult and J. Y. Salaun, *J. Organomet. Chem.*, 1987, **327**, C1.
409. B. Klingert, A. Roloff, B. Urwyler and J. Wirtz, *Helv. Chim. Acta*, 1988, **71**, 1858.
410. M. Etienne, R. Choukroun and D. Gervais, *J. Chem. Soc., Dalton Trans.*, 1984, 915.
411. D. J. Cardin, J. M. Kelly, G. A. Lawless and R. J. Trautman, *J. Chem. Soc., Chem. Commun.*, 1982, 228.
412. R. J. Puddephatt and M. A. Stalteri, *Organometallics*, 1983, **2**, 1400.
413. P. Rigollier, J. R. Young, L. A. Fowley and J. R. Stille, *J. Am. Chem. Soc.*, 1990, **112**, 9441.
414. J. R. Young and J. R. Stille, *Organometallics*, 1990, **9**, 3022.
415. M. Bochmann and L. M. Wilson, *J. Chem. Soc., Chem. Commun.*, 1986, 1610.
416. M. Bochmann, L. M. Wilson, M. B. Hursthouse and R. L. Short, *Organometallics*, 1987, **6**, 2556.

417. M. Bochmann, L. M. Wilson, M. B. Hursthouse and M. Motevalli, *Organometallics*, 1988, **7**, 1148.
418. S. L. Borkowsky, N. C. Baenziger and R. F. Jordan, *Organometallics*, 1993, **12**, 486.
419. R. Taube and L. Krukowka, *J. Organomet. Chem.*, 1988, **347**, C9.
420. M. Bochmann and A. J. Jaggar, *J. Organomet. Chem.*, 1992, **424**, C5.
421. M. Bochmann, A. J. Jaggar, L. M. Wilson, M. B. Hursthouse and M. Motevalli, *Polyhedron*, 1989, **8**, 1838.
422. R. F. Jordan, C. S. Bajgur, R. Willet and B. Scott, *J. Am. Chem. Soc.*, 1986, **108**, 7410.
423. D. M. Amrose, R. A. Lee and J. L. Petersen, *Organometallics*, 1991, **10**, 2191.
424. K. Mashima, K. Jyodol, A. Ohyoshi and H. Tanaka, *Organometallics*, 1987, **6**, 885.
425. R. Fandos, A. Meetsma and J. H. Teuben, *Organometallics*, 1991, **10**, 2665.
426. R. Fandos, J. H. Teuben, G. Helgesson and S. Jogner, *Organometallics*, 1991, **10**, 1637.
427. N. A. Petasis and E. I. Bzowej, *J. Am. Chem. Soc.*, 1990, **112**, 6392.
428. S. H. Bertz, G. Dabbagh and C. P. Gibson, *Organometallics*, 1988, **7**, 563.
429. C. P. Gibson, G. Dabbagh and S. H. Bertz, *J. Chem. Soc., Chem. Commun*, 1988, 603.
430. C. P. Gibson and D. S. Bem, *J. Organomet. Chem.*, 1991, **414**, 23.
431. T. T. Nadasdi and D. W. Stephan, *Can. J. Chem.*, 1991, **69**, 167.
432. G. Erker and U. Korek, *Z. Naturforsch., Teil B.*, 1989, **44**, 1593.
433. G. Erker, P. Czisch and R. Mynott, *J. Organomet. Chem.*, 1987, **334**, 91.
434. H. J. R. de Boer *et al.*, *Angew. Chem.*, 1986, **98**, 641; *Angew. Chem., Int. Ed. Engl.*, 1986, **25**, 639.
435. G. Erker, U. Dorf, R. Lecht, M. T. Ashby, M. Aulbach, R. Schlund, C. Krüger and R. Mynott, *Organometallics*, 1989, **8**, 2037.
436. R. Choukroun, A. Iraqi, C. Rifai and D. Gervais, *J. Organomet. Chem.*, 1988, **353**, 45.
437. R. D. Dennehy and R. J. Whitby, *J. Chem. Soc., Chem. Commun.*, 1990, 1060.
438. H. Nöth and U. Wietelmann, *Chem. Ber.*, 1987, **120**, 863.
439. H. Lang, M. Herres, L. Zsolnai and W. Imhof, *J. Organomet. Chem.*, 1991, **409**, C7.
440. H. Lang and L. Zsolnai, *J. Organomet. Chem.*, 1991, **406**, C5.
441. H. Lang and W. Imhof, *Chem. Ber.*, 1992, **125**, 1307.
442. J. R. Berenguer, L. R. Falvello, J. Forniés, E. Lalinde and M. Tomás, *Organometallics*, 1993, **12**, 6.
443. S. Fukada and H. Onuki (Hitachi Ltd.), *Japan Kokai. Tokkyo Koho, Jpn. Pat.*, 88 230 877 (1988) (*Chem. Abstr.*, 1988, **111**, 15 671d).
444. P. J. Krusic and F. N. Tebbe, *Inorg. Chem.*, 1982, **21**, 2900.
445. D. A. Strauss and R. H. Grubbs, *Organometallics*, 1982, **1**, 1658.
446. J. W. Bruin, G. Schat, O. S. Akkerman and F. Bickelhaupt, *Tetrahedron Lett.*, 1983, **24**, 3935.
447. B. J. van de Heisteeg, G. Schat, O. S. Akkerman and F. Bickelhaupt, *Tetrahedron Lett.*, 1987, **28**, 6493.
448. A. M. Piotrowski and J. J. Eisch, in 'Organometallic Syntheses', eds. R. B. King and J. J. Eisch, Elsevier, Amsterdam, 1986, vol. 3, p. 16.
449. B. J. J. van de Heisteeg, G. Schat, O. S. Akkerman and F. Bickelhaupt, *J. Organomet. Chem.*, 1986, **310**, C25.
450. J. D. Meinhart, E. V. Anslyn and R. H. Grubbs, *Organometallics*, 1989, **8**, 583.
451. E. V. Anslyn and R. H. Grubbs, *J. Am. Chem. Soc.*, 1987, **109**, 4880.
452. T. H. Upton and A. K. Rappé, *J. Am. Chem. Soc.*, 1985, **107**, 1206.
453. A. K. Rappé and T. H. Upton, *Organometallics*, 1984, **3**, 1440.
454. W. C. Finch, E. V. Anslyn and R. H. Grubbs, *J. Am. Chem. Soc.*, 1988, **110**, 2406.
455. J. D. Meinhart, B. D. Santarsiero and R. H. Grubbs, *J. Am. Chem. Soc.*, 1986, **108**, 3318.
456. D. A. Strauss and R. H. Grubbs, *J. Am. Chem. Soc.*, 1982, **104**, 5499.
457. S. C. Ho, S. Hentges and R. H. Grubbs, *Organometallics*, 1988, **7**, 780.
458. J. W. Park, L. M. Henling, W. P. Schaefer and R. H. Grubbs, *Organometallics*, 1990, **9**, 1650.
459. A. Kaki-Satpathy, C. S. Bajgur, K. P. Reddy and J. L. Petersen, *J. Organomet. Chem.*, 1989, **364**, 105.
460. T. Butkowskyj-Walkiw and G. Szeimies, *Tetrahedron*, 1986, **42**, 1845.
461. G. S. Bristow, M. F. Lappert, T. R. Martin, J. L. Atwood and W. F. Hunter, *J. Chem. Soc., Dalton Trans.*, 1984, 399.
462. M. F. Lappert, C. L. Raston, B. W. Skelton and A. H. White, *J. Chem. Soc., Dalton Trans.*, 1984, 893.
463. W. P. Leung and C. L. Raston, *J. Organomet. Chem.*, 1982, **240**, C1.
464. L. M. Engelhardt, W. P. Leung, R. I. Papasergio, C. L. Raston, P. Twiss and A. H. White, *J. Chem. Soc., Dalton Trans.*, 1987, 2347.
465. M. J. Earle, A. G. Massey, A. R. Al-Soudani and T. Zaidi, *Polyhedron*, 1989, **8**, 2817.
466. C. R. Lucas, *J. Organomet. Chem.*, 1982, **236**, 281.
467. H. Schreer and H. O. Froehlich, *Z. Chem.*, 1984, **24**, 442.
468. H. Lehmkuhl and R. Schwickardi, *J. Organomet. Chem.*, 1986, **303**, C43.
469. S. M. Yousaf, M. F. Farona, R. J. Shively, Jr. and W. J. Youngs, *J. Organomet. Chem.*, 1989, **363**, 281.
470. A. Famili, M. F. Farona and S. Thanedar, *J. Chem. Soc., Chem. Commun.*, 1983, 435.
471. K. C. Ott, J. B. Lee and R. H. Grubbs, *J. Am. Chem. Soc.*, 1982, **104**, 2942.
472. S. C. H. Ho, D. A. Strauss and R. H. Grubbs, *J. Am. Chem. Soc.*, 1984, **106**, 1533.
473. M. J. Burk, W. Tumas, M. D. Ward and D. R. Wheeler, *J. Am. Chem. Soc.*, 1990, **112**, 6133.
474. W. Tumas, D. R. Wheeler and R. H. Grubbs, *J. Am. Chem. Soc.*, 1987, **109**, 6182.
475. T. Ikariuya, S. C. H. Ho and R. H. Grubbs, *Organometallics*, 1985, **4**, 199.
476. S. L. Buchwald and R. H. Grubbs, *J. Am. Chem. Soc.*, 1983, **105**, 5490.
477. J. M. Hawkins and R. H. Grubbs, *J. Am. Chem. Soc.*, 1988, **110**, 2821.
478. S. L. Buchwald, E. V. Anslyn tand R. H. Grubbs, *J. Am. Chem. Soc.*, 1985, **107**, 1766.
479. M. J. Burk, D. L. Staley and W. Tumas, *J. Chem. Soc., Chem. Commun.*, 1990, 809.
480. L. R. Gilliom and R. H. Grubbs, *Organometallics*, 1986, **5**, 721.
481. P. Paetzold, K. Delpy, R. P. Hughes and W. A. Herrmann, *Chem. Ber.*, 1985, **118**, 1724.
482. J. M. Meinhart and R. H. Grubbs, *Bull. Chem. Soc. Jpn.*, 1988, **61**, 171.
483. K. M. Doxsee and J. K. M. Mouser, *Organometallics*, 1990, **9**, 3012.
484. R. D. Dennehy and R. J. Whitby, *J. Chem. Soc., Chem. Commun.*, 1992, 35.
485. W. Tumas, J. A. Suriano and R. L. Harlow, *Angew. Chem.*, 1990, **102**, 88; *Angew. Chem., Int. Ed. Engl.*, 1990, **29**, 75.

486. K. M. Doxsee and G. S. Shen, *J. Am. Chem. Soc.*, 1989, **111**, 9129.
487. R. D. Barr, M. Green, J. A. K. Howard, T. B. Marder, I. Moore and F. G. A. Stone, *J. Chem. Soc., Chem. Commun.*, 1983, 746.
488. M. R. Awang, R. D. Barr, M. Green, J. A. K. Howard, T. B. Marder and F. G. A. Stone, *J. Chem. Soc., Dalton Trans.*, 1985, 2009.
489. P. B. Mackenzie, R. C. Ott and R. H. Grubbs, *Pure Appl. Chem.*, 1984, **56**, 59.
490. P. B. Mackenzie, R. J. Coots and R. H. Grubbs, *Organometallics*, 1989, **8**, 8.
491. F. Ozawa, J. W. Park, P. B. Mackenzie, W. P. Schaefer, L. M. Henling and R. H. Grubbs, *J. Am. Chem. Soc.*, 1989, **111**, 1319.
492. J. W. Park, P. B. Mackenzie, W. P. Schaefer and R. H. Grubbs, *J. Am. Chem. Soc.*, 1986, **108**, 6402.
493. J. W. Park, L. M. Henling, W. P. Schaefer and R. H. Grubbs, *Organometallics*, 1991, **10**, 171.
494. E. V. Anslyn, B. D. Santarsiero and R. H. Grubbs, *Organometallics*, 1988, **7**, 2137.
495. R. H. Grubbs and W. Tumas, *Science*, 1989, **243**, 907.
496. J. B. Lee, K. C. Ott and R. H. Grubbs, *J. Am. Chem. Soc.*, 1982, **104**, 7491.
497. L. R. Gilliom and R. H. Grubbs, *J. Am. Chem. Soc.*, 1986, **108**, 733.
498. L. F. Cannizzo and R. H. Grubbs, *Macromolecules*, 1988, **21**, 1961.
499. L. R. Gilliom and R. H. Grubbs, *J. Mol. Catal.*, 1988, **46**, 255.
500. W. Risse, D. R. Wheeler, L. F. Cannizzo and R. H. Grubbs, *Macromolecules*, 1989, **22**, 3205.
501. K. J. Ivin, J. J. Rooney, C. D. Stewart, M. L. H. Green and J. R. Mahtab, *J. Chem. Soc., Chem. Commun.*, 1978, 604.
502. L. Clawson, J. Soto, S. L. Buchwald, M. L. Steigerwald and R. H. Grubbs, *J. Am. Chem. Soc.*, 1985, **107**, 3377.
503. G. Fink and D. Schnell, *Angew. Makromol. Chem.*, 1982, **105**, 15.
504. G. Fink and D. Schnell, *Angew. Makromol. Chem.*, 1982, **105**, 31.
505. G. Fink and D. Schnell, *Angew. Makromol. Chem.*, 1982, **105**, 39.
506. R. Mynott, G. Fink and W. Fenzl, *Angew. Makromol. Chem.*, 1987, **154**, 1.
507. G. Fink, W. Fenzl and R. Mynott, *Z. Naturforsch., Teil B*, 1985, **40**, 158.
508. J. Mejzlik, M. Lesna and J. Majer, *Makromol. Chem.*, 1983, **184**, 1975.
509. F. S. Dyachkovskii, in 'Coordination Polymerization', ed. J. C. W. Chien, Academic Press, New York, 1975, p. 199.
510. J. J. Eisch, A. M. Piotrowski, S. K. Brownstein, E. J. Gabe and F. L. Lee, *J. Am. Chem. Soc.*, 1985, **107**, 7219.
511. C. A. Jolly and D. S. Marynick, *J. Am. Chem. Soc.*, 1989, **111**, 7968.
512. J. J. Eisch, K. R. Caldwell, S. Werner and C. Krüger, *Organometallics*, 1991, **10**, 3417.
513. J. J. Eisch, S. I. Pombrik and G. X. Zheng, *Makromol. Chem., Macromol. Symp.*, 1993, **66**, 109.
514. J. J. W. Eshuis, Y. Y. Tan, J. H. Teuben and J. Renkema, *J. Mol. Catal.*, 1990, **62**, 277.
515. M. Bochmann, A. J. Jaggar and J. C. Nicholls, *Angew. Chem.*, 1990, **102**, 830; *Angew. Chem., Int. Ed. Engl.*, 1990, **29**, 780.
516. M. Bochmann and S. J. Lancaster, *J. Organomet. Chem.*, 1992, **434**, C1.
517. J. A. Ewen, *J. Am. Chem. Soc.*, 1984, **106**, 6355.
518. Z. Zambelli, P. Ammendola, A. Grassi, P. Longo and A. Proto, *Macromolecules*, 1986, **19**, 2703.
519. G. Erker and C. Fritze, *Angew. Chem.*, 1992, **104**, 204; *Angew. Chem., Int. Ed. Engl.*, 1992, **31**, 199.
520. P. Longo, A. Grassi, C. Pellecchia and A. Zambelli, *Macromolecules*, 1987, **20**, 1015.
521. G. H. Llinas, R. Day, M. D. Rausch and J. C. W. Chien, *Organometallics*, 1993, **12**, 1283.
522. E. A. Fushman, A. N. Shupik, L. F. Borisova, V. E. L'vovskii and F. S. D'yachkovskii, *Dokl. Akad. Nauk. SSSR*, 1982, **264**, 651.
523. L. I. Chernaya and P. E. Matkovskii, *Metalloorg. Khim.*, 1988, **1**, 849.
524. E. A. Grigoryan, K. R. Gyulumyan and G. N. Menchikova, *Acta Polym.*, 1984, **35**, 33.
525. Y. Sasaki (Showa Denko K. K.), *Japan Kokai Tokkyo Koho, Jpn. Pat.* 88, 168 407 (1988) (*Chem. Abstr.*, 1988, **109**, 211 636u).
526. H. W. Turner (Exxon Chemical Patents, Inc.), *Eur. Pat. Appl.* EP 226 463 (1986) (*Chem. Abstr.*, 1986, **107**, 199 105u).
527. V. Busico, L. Mevo, G. Palumbo, A. Zambelli and T. Tancredi, *Makromol. Chem.*, 1983, **184**, 2193.
528. K. Soga, T. Shiono and H. Yanagihara, *Makromol. Chem. Rapid Commun.*, 1986, **7**, 719.
529. K. Soga and H. Yanagihara, *Makromol. Chem.*, 1988, **189**, 2839.
530. K. Soga, T. Uozumi and T. Shiono, *Makromol. Chem. Rapid Commun.*, 1989, **10**, 293.
531. K. Soga and H. Yanagihara, in 'Proceedings of the International Symposium on Transition Metals in Organometallic Catalysis of Olefin Polymerization', eds. W. Kaminsky and H. Sinn, Springer, Berlin, 1988, p. 21.
532. H. W. Blunt (Hercules Inc.), *US Pat.* 4 408 019 (1983) (*Chem. Abstr.*, 1984, **100**, 7417h).
533. K. Soga and H. Yanagihara, *Macromolecules*, 1989, **22**, 2875.
534. H. Yanagihara and K. Soga (Showa Denko K. K.), *Japan Kokai Tokkyo Koho, Jpn. Pat.* 90 206 602 (1990) (*Chem. Abstr.*, 1990, **113**, 232 268y).
535. H. Yanagihara and K. Soga (Showa Denko K. K.), *Japan Kokai Tokkyo Koho, Jpn. Pat.* 89 254 707 (1989) (*Chem. Abstr.*, 1989, **112**, 99 489g).
536. M. Imayoshi, T. Arai and S. Tsuyama (Asahi Chemical Industry Co.), *Japan Kokai Tokkyo Koko, Jpn. Pat.* 88 268 710 (1988) (*Chem. Abstr.*, 1988, **110**, 135 925r).
537. E. P. Bokaris, M. G. Siskos and A. K. Zarkadis, *Eur. Polym. J.*, 1992, **28**, 1441.
538. B. K. Campion, J. Falk and T. D. Tilley, *J. Am. Chem. Soc.*, 1987, **109**, 2049.
539. V. A. Igomin *et al.*, *J. Organomet. Chem.*, 1989, **371**, 187.
540. J. F. Harrod, A. Malek, F. D. Rochon and R. Melanson, *Organometallics*, 1987, **6**, 2117.
541. N. L. Ermolaev, Z. K. Borisova, A. M. Kosyak and N. V. Kochnev, *Metallorg. Khim.*, 1990, **3**, 553.
542. W. Zheng and D. W. Stephan, *Inorg. Chem.*, 1988, **27**, 2386.
543. M. J. Calhorda, R. Gomes da Costa, A. R. Dias and J. A. Martinho Simoes, *J. Chem. Soc., Dalton Trans.*, 1982, 2327.
544. A. R. Dias and J. A. M. Simoes, *Polyhedron*, 1988, **7**, 1531.
545. P. N. Billinger, P. P. K. Claire, H. Collins and G. R. Willey, *Inorg. Chim. Acta*, 1988, **149**, 63.
546. K. Berhalter and U. Thewalt, *J. Organomet. Chem.*, 1987, **332**, 123.
547. P. A. Seewald, G. S. White and D. W. Stephan, *Can. J. Chem.*, 1988, **66**, 1147.
548. F. Rehbaum, K. H. Thiele and S. I. Trojanov, *J. Organomet. Chem.*, 1991, **410**, 327.

549. U. Thewalt and K. Berhalter, *J. Organomet. Chem.*, 1986, **302**, 193.
550. A. L. Beauchamp, D. Cozak and A. Mardhy, *Inorg. Chim. Acta*, 1984, **92**, 191.
551. J. Scholz, M. Dlokan, D. Ströhl, A. Dietrich, H. Schumann and K. H. Thiele, *Chem. Ber.*, 1990, **123**, 2279.
552. H. Köpf and T. Klapötke, *Z. Naturforsch., Teil B*, 1986, **41**, 667.
553. K. M. Doxsee, J. B. Farahi and H. Hope, *J. Am. Chem. Soc.*, 1991, **113**, 8889.
554. K. M. Doxsee and J. B. Farahi, *J. Chem. Soc., Chem. Commun.*, 1990, 1452.
555. M. Bochmann, A. J. Jaggar, M. B. Hursthouse and M. Mazid, *Polyhedron*, 1990, **9**, 2097.
556. I. P. Parkin and J. D. Woollins, *J. Chem. Soc., Dalton Trans.*, 1990, 519.
557. H. Plenio, H. W. Roesky, F. T. Edelmann and M. Noltemeyer, *J. Chem. Soc., Dalton Trans.*, 1989, 1815.
558. H. Plenio, H. W. Roesky, M. Noltemeyer and G. M. Sheldrick, *J. Chem. Soc., Chem. Commun.*, 1987, 1483.
559. M. Herberhold, F. Neumann, G. Süss-Fink and U. Thewalt, *Inorg. Chem.*, 1987, **26**, 3612.
560. J. Losada and M. Moran, *J. Organomet. Chem.*, 1984, **276**, 13.
561. M. Tacke, C. Klein, D. J. Stufkens and A. Oskam, *J. Organomet. Chem.*, 1993, **444**, 75.
562. H. Plenio and H. W. Roesky, *Z. Naturforsch., Teil B*, 1989, **44**, 94.
563. C. G. Marcellus, R. T. Oakley, A. W. Cordes and W. T. Pennington, *J. Chem. Soc., Chem. Commun.*, 1983, 1451.
564. B. Khera, A. K. Sharma and N. K. Kaushik, *Polyhedron*, 1983, **2**, 1177.
565. A. K. Sharma, B. Khera and N. K. Kaushik, *Synth. React. Inorg. Metal-Org. Chem.*, 1983, **13**, 491.
566. B. Khera and N. K. Kaushik, *Proc. Indian Acad. Sci., Chem. Sci.*, 1984, **93**, 139.
567. B. R. Pirgonde and M. A. Pujar, *Acta Chim. Hung.*, 1989, **126**, 247.
568. B. Khera, A. K. Sharma and N. K. Kaushik, *Bull. Chem. Soc. Jpn.*, 1985, **58**, 793.
569. Y. Singh, R. Sharan and R. N. Kapoor, *Indian J. Chem., Sect. A*, 1985, **24**, 115.
570. Y. Singh, R. Sharan and R. N. Kapoor, *Transition Met. Chem. (Weinheim)*, 1986, **11**, 321.
571. R. V. Singh, *Indian J. Chem., Sect. A*, 1985, **24**, 1064.
572. A. L. Beauchamp, F. Belanger-Gariepy, A. Mardhy and D. Cozak, *Inorg. Chim. Acta*, 1986, **124**, L23.
573. G. S. Sodhi, R. K. Bajaj and N. K. Kaushik, *Inorg. Chim. Acta*, 1984, **92**, L27; S. Kamrah, G. S. Sodhi and N. K. Kaushik, *ibid.*, 1985, **107**, 29.
574. B. Khera, A. K. Sharma and N. K. Kaushik, *Synth. React. Inorg. Metal-Org. Chem.*, 1987, **17**, 25.
575. S. K. Bansal, A. K. Sharma, S. Tikku and R. S. Sindhu, *Synth. React. Inorg. Metal-Org. Chem.*, 1991, **21**, 587.
576. G. S. Sodhi, A. K. Sharma and N. K. Kaushik, *Synth. React. Inorg. Metal-Org. Chem.*, 1982, **12**, 947.
577. A. K. Sharma, G. S. Sodhi and N. K. Kaushik, *Bull. Soc. Chim. Fr., Part I*, 1983, 52.
578. A. K. Sharma and N. K. Kaushik, *J. Indian Chem. Soc.*, 1986, **63**, 242.
579. G. S. Sodhi, A. K. Sharma and N. K. Kaushik, *J. Organomet. Chem.*, 1982, **238**, 177.
580. R. K. Bajaj, G. S. Sodhi, N. K. Kaushik and K. N. Johri, *Polyhedron*, 1984, **3**, 883.
581. K. C. Goyal and B. D. Khosla, *J. Indian Chem. Soc.*, 1983, **60**, 399.
582. V. K. Srivastava, N. K. Kaushik and B. Khera, *Synth. React. Inorg. Metal-Org. Chem.*, 1987, **17**, 15.
583. G. Pneumatikakis, A. Yannopoulos and J. Markopoulos, *Inorg. Chim. Acta*, 1988, **151**, 125.
584. A. R. Dias, M. S. Salema and J. A. M. Simoes, *J. Organomet. Chem.*, 1981, **222**, 69.
585. M. A. V. Ribeiro da Silva, M. D. M. C. Ribeiro da Silva and A. R. Dias, *J. Organomet. Chem.*, 1988, **345**, 105.
586. M. R. Smith, P. T. Matsunaga and R. A. Andersen, *J. Am. Chem. Soc.*, 1993, **115**, 7049.
587. F. Bottomley, G. O. Egharevba, I. J. B. Lin and P. S. White, *Organometallics*, 1985, **4**, 550.
588. H. P. Klein, U. Thewalt, K. Döppert and R. Sanchez-Delgado, *J. Organomet. Chem.*, 1982, **236**, 189.
589. J. Malek, T. Jelinek, J. Holecek and K. Handlir, *Z. Chem.*, 1983, **23**, 189.
590. U. Thewalt and B. Honold, *J. Organomet. Chem.*, 1988, **348**, 291.
591. B. Honold and U. Thewalt, *J. Organomet. Chem.*, 1986, **316**, 291.
592. S. Chen, Y. Liu and J. Wang, *Sci. Sinica (Engl. Ed.)*, 1982, **25**, 227.
593. K. Döppert, *J. Organomet. Chem.*, 1987, **333**, C1.
594. Q. Yang, X. Jin, X. Xu, G. Li, Y. Tang and S. Chen, *Sci. Sinica (Engl. Ed.)*, 1982, **25**, 356.
595. A. K. Sharma and N. K. Kaushik, *Synth. React. Inorg. Metal-Org. Chem.*, 1982, **12**, 827.
596. A. K. Sharma and N. K. Kaushik, *Synth. React. Inorg. Metal-Org. Chem.*, 1984, **14**, 513.
597. S. S. Deshpande, P. A. Awasarkar, S. Gopinathan and C. Gopinathan, *Indian. J. Chem. Sect. A*, 1984, **23**, 95.
598. H. Köpf and T. Klapötke, *Monatsh. Chem.*, 1986, **117**, 1003.
599. C. J. Cardin and A. Roy, *Inorg. Chim. Acta*, 1985, **107**, L37.
600. W. Jabs, W. Gaube and R. Lukowski, *Z. Chem.*, 1988, **28**, 412.
601. R. Schobert, *J. Organomet. Chem.*, 1991, **405**, 201.
602. S. Dürr, U. Höhlein and R. Schobert, *Organometallics*, 1992, **11**, 2950.
603. M. Williams, C. E. Carraher, F. Medina and M. J. Aloi, *Polym. Mater. Sci. Eng.*, 1989, **61**, 227.
604. C. Guimon, G. Pfister-Guillouzo, J. Besançon and P. Meunier, *J. Chem. Soc., Dalton Trans.*, 1987, 107.
605. M. D. Curtis, S. Thanedar and W. M. Butler, *Organometallics*, 1984, **3**, 1855.
606. K. Hortmann, J. Diebold and H. H. Brintzinger, *J. Organomet. Chem.*, 1993, **445**, 107.
607. P. Berno, C. Floriani, A. Chiesi-Villa and C. Guastini, *Organometallics*, 1990, **9**, 1995.
608. G. Beck, P. B. Hitchcock, M. F. Lappert and I. A. MacKinnon, *J. Chem. Soc., Chem. Commun.*, 1989, 1312.
609. U. Thewalt, S. Klima and H. G. Alt, *J. Organomet. Chem.*, 1987, **321**, 209.
610. T. Carofiglio, C. Floriani, M. Rosi, A. Chiesi-Villa and C. Rizzoli, *Inorg. Chem.*, 1991, **30**, 3245.
611. J. Martin, M. Fauconet and C. Moise, *J. Organomet. Chem.*, 1989, **371**, 87.
612. J. S. Merola, K. S. Campo and R. A. Gentile, *Inorg. Chem.*, 1989, **28**, 2950.
613. M. Sabat, M. F. Gross and M. G. Finn, *Organometallics*, 1992, **1**, 745.
614. B. L. Balzer, M. Cazanoue, M. Sabat and M. G. Finn, *Organometallics*, 1992, **11**, 1759.
615. Y. Naoshima and C. E. Carraher, *Polym. Mater. Sci. Eng.*, 1984, **50**, 403.
616. Y. Naoshima, C. E. Carraher, S. Iwamoto and H. Shudo, *Appl. Organomet. Chem.*, 1987, **1**, 245.
617. Y. Naoshima, S. Hirono and C. E. Carraher, *J. Polym. Mater.*, 1985, **2**, 43.
618. Y. Naoshima, S. Hirono and C. E. Carraher, *Polym. Mater. Sci. Eng.*, 1985, **52**, 29.
619. Y. Naoshima, C. E. Carraher, S. Hirono, T. S. Bekele and P. D. Mykytiuk, *Polym. Sci. Technol.*, 1986, **33**, 53.
620. C. E. Carraher, D. Gill, Y. Naoshima and M. Williams, *Polym. Mater. Sci. Eng.*, 1989, **61**, 447.

621. K. Döppert, H. P. Klein and U. Thewalt, *J. Organomet. Chem.*, 1986, **303**, 205.
622. D. M. Hoffman, N. D. Chester and R. C. Fay, *Organometallics*, 1983, **2**, 48.
623. C. R. Lucas, E. J. Gabe and F. L. Lee, *Can. J. Chem.*, 1988, **66**, 429.
624. B. Bracke, Y. Dang, A. T. H. Lenstra and H. J. Geise, *Acta Crystallogr., Sect. C, Cryst. Struct. Commun.*, 1991, **C47**, 2043.
625. Y. Zhou, Z. Wang, X. Wang and Y. Zhu, *Polyhedron*, 1990, **9**, 783.
626. G. S. Herrmann, H. G. Alt and U. Thewalt, *J. Organomet. Chem.*, 1990, **399**, 83.
627. H. M. Gau, C. C. Schei, L. K. Liu and L. H. Luh, *J. Organomet. Chem.*, 1992, **435**, 43.
628. Y. Dang, H. J. Geise, R. Dommisse, E. Esmans and H. O. Desseyn, *J. Organomet. Chem.*, 1990, **381**, 333.
629. Y. Dang, Y. Zhang and S. Shi, *Synth. React. Inorg. Metal-Org. Chem.*, 1987, **17**, 347.
630. Y. Dang, H. J. Geise, R. Dommisse, E. Esmans and H. O. Desseyn, *Polyhedron*, 1989, **8**, 1844.
631. R. K. Tuli, K. Chandra, R. K. Sharma and B. S. Garg, *J. Indian Chem. Soc.*, 1983, **60**, 813.
632. A. K. Saxena, S. Saxena and A. K. Rai, *Indian J. Chem., Sect. A.*, 1990, **29**, 255.
633. S. C. Dixit, R. Sharan and R. N. Kapoor, *Inorg. Chim. Acta*, 1987, **133**, 251.
634. A. Schäfer, E. Karl, L. Zsolnai, G. Huttner and H. H. Brintzinger, *J. Organomet. Chem.*, 1987, **328**, 87.
635. E. Samuel, J. L. Atwood and W. E. Hunter, *J. Organomet. Chem.*, 1986, **311**, 325.
636. R. S. Arora and R. K. Multani, *J. Inst. Chem. (India)*, 1983, **55**, 93.
637. R. S. Arora, S. C. Hari and R. K. Multani, *J. Inst. Chem. (India)*, 1982, **54**, 143.
638. K. Döppert and U. Thewalt, *J. Organomet. Chem.*, 1986, **301**, 41.
639. C. J. Cardin and A. Roy, *Inorg. Chim. Acta*, 1985, **107**, L33.
640. R. Leik, L. Zsolnai, G. Huttner, E. W. Neuse and H. H. Brintzinger, *J. Organomet. Chem.*, 1986, **312**, 177.
641. C. G. Arena, G. Bruno and F. Faraone, *J. Chem. Soc., Dalton Trans.*, 1991, 1223.
642. T. Güthner and U. Thewalt, *J. Organomet. Chem.*, 1989, **371**, 43.
643. M. G. Meirim, E. W. Neuse, M. Rhemtula, S. Schmitt and H. H. Brintzinger, *Transition Met. Chem. (London)*, 1988, **13**, 272.
644. K. Döppert, R. Sanchez-Delgado, H. P. Klein and U. Thewalt, *J. Organomet. Chem.*, 1982, **233**, 205.
645. U. Thewalt and F. Güthner, *J. Organomet. Chem.*, 1989, **379**, 59.
646. T. Güthner and U. Thewalt, *J. Organomet. Chem.*, 1988, **350**, 235.
647. H. P. Klein, K. Döppert and U. Thewalt, *J. Organomet. Chem.*, 1985, **280**, 203.
648. U. Thewalt, K. Döppert, T. Debaerdemaeker, G. Germain and V. Nastopoulos, *J. Organomet. Chem.*, 1987, **326**, C37.
649. B. Khera, A. K. Sharma and N. K. Kaushik, *Polyhedron*, 1983, **2**, 108.
650. A. K. Sharma and N. K. Kaushik, *Synth. React. Inorg. Metal-Org. Chem.*, 1983, **13**, 481.
651. A. K. Sharma and N. K. Kaushik, *Indian. J. Chem., Sect. A*, 1983, **22**, 76.
652. A. K. Sharma and N. K. Kaushik, *Z. Naturforsch., Teil B*, 1984, **39**, 604.
653. S. Saxena, Y. P. Singh and A. K. Rai, *J. Organomet. Chem.*, 1984, **270**, 301.
654. S. Saxena, P. N. Saxena, A. K. Rai and S. C. Saxena, *Toxicology*, 1985, **35**, 241.
655. H. Köpf and V. Richtering, *J. Organomet. Chem.*, 1988, **346**, 355.
656. T. S. Targos, R. P. Rosen, R. R. Whittle and G. L. Geoffroy, *Inorg. Chem.*, 1985, **24**, 1375.
657. P. Mercando, A. J. DiMaio and A. L. Rheingold, *Angew. Chem.*, 1987, **99**, 252; *Angew. Chem., Int. Ed. Engl.*, 1987, **26**, 244.
658. M. Draganjac and T. B. Rauchfuss, *Angew. Chem.*, 1985, **97**, 745; *Angew. Chem., Int. Ed. Engl.*, 1985, **24**, 742.
659. A. Shaver, J. M. McCall and G. Marmolejo, *Inorg. Synth.*, 1990, **27**, 59.
660. J. Darkwa, D. M. Giolando, C. J. Murphy and T. B. Rauchfuss, *Inorg. Synth.*, 1990, **27**, 51.
661. S. Dev, E. Ramli, T. B. Rauchfuss and C. L. Stern, *J. Am. Chem. Soc.*, 1990, **112**, 6385.
662. P. H. Bird, J. M. McCall, A. Shaver and U. Siriwardane, *Angew. Chem.*, 1982, **94**, 375; *Angew. Chem., Int. Ed. Engl.*, 1982, **21**, 384.
663. A. Shaver, J. M. McCall, P. H. Bird and U. Siriwardane, *Acta Crystallogr., Sect. C, Cryst. Struct. Commun.*, 1991, **47**, 659.
664. A. Shaver and J. M. McCall, *Organometallics*, 1984, **3**, 1823.
665. G. Tainturier, B. Gautheron and C. Degrand, *Organometallics*, 1986, **5**, 942.
666. N. Klouras, *Montash. Chem.*, 1991, **122**, 533.
667. N. Klouras, S. Voliotis and G. Germain, *Acta Crystallogr., Sect. C, Cryst. Struct. Commun.*, 1984, **40**, 1791.
668. N. Albrecht and E. Weiss, *J. Organomet. Chem.*, 1988, **355**, 89.
669. D. Fenske, J. Adel and K. Dehnicke, *Z. Naturforsch., Teil B*, 1987, **42**, 931.
670. G. A. Zank, T. B. Rauchfuss, S. R. Wilson and A. L. Rheingold, *J. Am. Chem. Soc.*, 1984, **106**, 7621.
671. A. Shaver, G. Marmolejo and J. M. McCall, *Inorg. Synth.*, 1990, **27**, 65.
672. F. Bottomley, D. F. Drummond, G. O. Egharevba and P. S. White, *Organometallics*, 1986, **5**, 1620.
673. T. Cuenca, W. A. Herrmann and T. V. Ashworth, *Organometallics*, 1986, **5**, 2514.
674. A. Shaver, J. M. McCall, P. H. Bird and N. Ansari, *Organometallics*, 1983, **2**, 1894.
675. D. M. Giolando, T. B. Rauchfuss, A. L. Rheingold and S. R. Wilson, *Organometallics*, 1987, **6**, 667.
676. C. J. Ruffing and T. B. Rauchfuss, *Organometallics*, 1985, **4**, 524.
677. T. Klapötke and H. Köpf, *Inorg. Chim. Acta*, 1987, **133**, 115.
678. M. Y. Darensbourg *et al.*, *Inorg. Chem.*, 1992, **31**, 1487.
679. H. Köpf and S. Grabowski, *Z. Anorg. Allg. Chem.*, 1988, **560**, 163.
680. H. Köpf, S. Grabowski and B. Block, *J. Organomet. Chem.*, 1983, **246**, 243.
681. H. Köpf and S. Grabowski, *Z. Anorg. Allg. Chem.*, 1983, **496**, 167.
682. Y. Singh, R. Sharan and R. N. Kapoor, *Indian J. Chem., Sect. A*, 1986, **25**, 771.
683. Y. Singh, R. Sharan and R. N. Kapoor, *Synth. React. Inorg. Metal-Org. Chem.*, 1986, **16**, 1225.
684. J. Besançon, J. Tirouflet, B. Trimaille and Y. Dusausoy, *J. Organomet. Chem.*, 1986, **314**, C12.
685. J. Besançon, D. Dimitri, B. Trimaille and Y. Dusausoy, *C. R. Acad. Sci., Ser. 2*, 1985, **301**, 83.
686. J. C. Kotz, W. Vining, W. Coco, R. Rosen, A. R. Dias and M. H. Garcia, *Organometallics*, 1983, **2**, 68.
687. H. Werner, H. Ulrich, U. Schubert, P. Hofmann and B. Zimmer-Garser, *J. Organomet. Chem.*, 1985, **297**, 27.
688. T. A. Wark and D. W. Stephan, *Inorg. Chem.*, 1987, **26**, 363.
689. T. A. Wark and D. W. Stephan, *Organometallics*, 1989, **8**, 2836.
690. M. A. A. F. de C. T. Carrondo and G. A. Jeffrey, *Acta Crystallogr., Sect. C, Cryst. Struct. Commun.*, 1983, **39**, 42.

691. G. S. White and D. W. Stephan, *Inorg. Chem.*, 1985, **24**, 1499.
692. G. S. White and D. W. Stephan, *Organometallics*, 1988, **7**, 903.
693. G. S. White and D. W. Stephan, *Organometallics*, 1987, **6**, 2169.
694. T. T. Nadasdi and D. W. Stephan, *Organometallics*, 1992, **11**, 116.
695. D. M. Giolando, T. B. Rauchfuss and G. M. Clark, *Inorg. Chem.*, 1987, **26**, 3082.
696. P. Meunier, B. Gautheron and A. Mazouz, *J. Organomet. Chem.*, 1987, **320**, C39.
697. H. Köpf and H. Balz, *J. Organomet. Chem.*, 1990, **387**, 77.
698. H. Köpf and T. Klapötke, *Chem. Ber.*, 1986, **119**, 1986.
699. D. D. S. Yadav and R. C. Mehrotra, *Inorg. Chim. Acta*, 1985, **96**, 39.
700. H. Köpf and T. Klapötke, *J. Chem. Soc., Chem. Commun.*, 1986, 1192.
701. P. Mathur and V. D. Reddy, *J. Organomet. Chem.*, 1990, **385**, 363.
702. H. G. Raubenheimer, S. Lotz, G. J. Kruger, A. van A. Lombard and J. C. Viljoen, *J. Organomet. Chem.*, 1987, **336**, 349.
703. X. Yang, T. B. Rauchfuss and S. R. Wilson, *J. Am. Chem. Soc.*, 1989, **111**, 3465.
704. P. Soni, K. Chandra, R. K. Sharma and B. S. Garg, *J. Indian. Chem. Soc.*, 1982, **59**, 913.
705. Z. Q. Wang, S. W. Lu, H. F. Guo and N. H. Hu, *Polyhedron*, 1992, **11**, 1131.
706. L. Gelmini and D. W. Stephan, *Organometallics*, 1987, **6**, 1515.
707. G. A. Zank and T. B. Rauchfuss, *Inorg. Chem.*, 1986, **25**, 1431.
708. M. J. Calhorda, A. R. Dias, J. A. M. Simoes and C. Teixei, *J. Chem. Soc., Dalton Trans.*, 1984, 2659.
709. M. J. Calhorda *et al.*, *Inorg. Chem.*, 1988, **27**, 2513.
710. P. Pekonen, Y. Hiltunen and R. S. Laitinen, *Acta Chem. Scand.*, 1989, **43**, 914.
711. P. Pekonen, Y. Hiltunen, R. S. Laitinen and J. Valkonen, *Inorg. Chem.*, 1991, **30**, 1874.
712. P. Granger, B. Gautheron, G. Tainturier and S. Pouly, *Org. Magn. Reson.*, 1984, **22**, 701.
713. A. E. Bruce, M. R. M. Bruce and D. R. Tyler, *J. Am. Chem. Soc.*, 1984, **106**, 6660.
714. R. Steudel and R. Strauss, *J. Chem. Soc., Dalton Trans.*, 1984, 1775.
715. R. Steudel and E. M. Strauss, *Angew. Chem.* 1984, **96**, 356; *Angew. Chem., Int. Ed. Engl.*, 1984, **23**, 362.
716. R. Steudel, D. Jensen and F. Baumgart, *Polyhedron*, 1990, **9**, 1199.
717. R. Steudel, M. Papavassiliou, E. M. Strauss and R. Laitinen, *Angew. Chem.*, 1986, **98**, 81; *Angew. Chem., Int. Ed. Engl.*, 1986, **25**, 99.
718. D. M. Giolando, M. Papavassiliou, J. Pickardt, T. B. Rauchfuss and R. Steudel, *Inorg. Chem.*, 1988, **27**, 2596.
719. D. M. Giolando and T. B. Rauchfuss, *Organometallics*, 1984, **3**, 487.
720. R. Steudel, B. Holz and J. Pickardt, *Angew. Chem.*, 1989, **101**, 1301; *Angew. Chem., Int. Ed. Engl.*, 1989, **28**, 1269.
721. B. Holz and R. Steudel, *J. Organomet. Chem.*, 1991, **406**, 133.
722. J. Kulpe, E. Herdtweck, G. Weichselbaumer and W. A. Herrmann, *J. Organomet. Chem.*, 1988, **348**, 369.
723. J. E. Hoots and T. B. Rauchfuss, *Inorg. Chem.*, 1983, **22**, 2806.
724. C. M. Bolinger and T. B. Rauchfuss, *Inorg. Chem.*, 1982, **21**, 3947.
725. D. M. Giolando and T. B. Rauchfuss, *J. Am. Chem. Soc.*, 1984, **106**, 6455.
726. A. Shaver, S. Morris and A. Desjardins, *Inorg. Chim. Acta*, 1989, **161**, 11.
727. U. Westphal and R. Steudel, *Chem. Ber.*, 1991, **124**, 2141.
728. R. Steudel and U. Westphal, *J. Organomet. Chem.*, 1990, **338**, 89.
729. M. E. E. Meijer-Veldman, J. L. de Boer, H. J. de Liefde Meijer, A. M. M. Schreurs, J. Kroon and A. L. Spek, *J. Organomet. Chem.*, 1984, **269**, 255.
730. E. Solari, C. Floriani, A. Chiesi-Villa and C. Guastini, *J. Chem. Soc., Chem. Commun.*, 1989, 1747.
731. R. G. Swisher, E. Sinn and R. N. Grimes, *Organometallics*, 1984, **3**, 599.
732. N. Awaya and M. Arita (Nippon Telegraph and Telephone Co.), *Japan Kokai Tokkyo Koho, Jpn. Pat.* 89 290 771 (1989) (*Chem. Abstr.*, 1989, **113**, 88 791u).
733. L. B. Kool, M. D. Rausch and R. D. Rogers, *J. Organomet. Chem.*, 1985, **297**, 289.
734. J. Blenkers, P. Bruin and J. H. Teuben, *J. Organomet. Chem.*, 1985, **297**, 61.
735. P. A. Kroon and R. B. Helmholdt, *J. Organomet. Chem.*, 1970, **25**, 451.
736. D. Gourier and E. Samuel, *J. Am. Chem. Soc.*, 1987, **109**, 4571.
737. R. R. Andrea, A. Terpstra, A. Oskam, P. Bruin and J. H. Teuben, *J. Organomet. Chem.*, 1986, **307**, 307.
738. E. Samuel, D. Guery and J. Vedel, *J. Organomet. Chem.*, 1984, **263**, C43.
739. J. E. Anderson, E. T. Maher and L. B. Kool, *Organometallics*, 1991, **10**, 1248.
740. R. Kempe and J. Sieler, *Z. Kristallogr.*, 1992, **201**, 290.

6
Zirconium and Hafnium Complexes in Oxidation States Zero and Below

F. GEOFFREY N. CLOKE
University of Sussex, Brighton, UK

6.1 BIS(η-ARENE) COMPLEXES

The 16-electron bis(η-arene) complexes of zirconium(0) and hafnium(0) may be stabilized by use of an appropriately bulky arene ligand; thus, cocondensation of zirconium or hafnium vapour with 1,3,5-tri-*t*-butylbenzene affords the stable diamagnetic sandwich compounds $[Zr(\eta\text{-}C_6H_3Bu^t_3\text{-}1,3,5)_2]$ and $[Hf(\eta\text{-}C_6H_3Bu^t_3\text{-}1,3,5)_2]$ (Equation (1)).[1]

$$M(at.) + \text{(arene)} \xrightarrow{77\ K} \text{(sandwich complex)} \qquad (1)$$

M = Zr, Hf

Less sterically demanding arenes do not afford isolable products; simultaneous cocondensation of a mixture of trimethylphosphine and toluene with zirconium or hafnium vapour does, however, afford the stable 18-electron adducts $[M(\eta\text{-}C_6H_5Me)_2(PMe_3)]$ (M = Zr or Hf).[2] The compounds $[M(\eta\text{-}C_6H_3Bu^t_3\text{-}1,3,5)_2]$ (M = Zr or Hf) do not react with trimethylphosphine (presumably for steric reasons), but do react with CO to form the 18-electron carbonyls $[M(\eta\text{-}C_6H_3Bu^t_3\text{-}1,3,5)_2CO]$; in the case of M = Zr the reaction is reversible, and $[Zr(\eta\text{-}C_6H_3Bu^t_3\text{-}1,3,5)_2CO]$ may be characterized only in solution under a CO atmosphere by IR and NMR spectroscopy (Equations (2) and (3)).[1]

$$[Hf(\eta\text{-}C_6H_3Bu^t_3\text{-}1,3,5)_2] \xrightarrow{\text{C}_5\text{H}_{12}, \text{ CO, 295 K}} [Hf(\eta\text{-}C_6H_3Bu^t_3\text{-}1,3,5)_2CO] \qquad (2)$$

$$[Zr(\eta\text{-}C_6H_3Bu^t_3\text{-}1,3,5)_2] \underset{-\text{CO}}{\overset{\text{C}_5\text{H}_{12}, \text{ CO, 295 K}}{\rightleftharpoons}} [Zr(\eta\text{-}C_6H_3Bu^t_3\text{-}1,3,5)_2CO] \qquad (3)$$

Similarly, the sterically demanding 2,4,6-tri-*t*-butylpyridine allows access to the 16-electron heteroatom-substituted sandwich complexes $[Zr(\eta\text{-}C_5H_2NBu^t_3\text{-}2,4,6)_2]$ and $[Hf(\eta\text{-}C_5H_2NBu^t_3\text{-}2,4,6)_2]$ in moderate (25–30%) yield, via cocondensation with the metal vapours (Equation (4)).[3] In solution, the compounds $[M(\eta\text{-}C_5H_2NBu^t_3\text{-}2,4,6)_2]$ (M = Zr or Hf) are static on the NMR timescale below 300 K and adopt the conformation shown; above this temperature they are fluxional and barriers to ring rotation of 59 kJ mol^{-1} (M = Zr) and 61 kJ mol^{-1} (M = Hf) may be determined from NMR coalescence experiments.

$$(4)$$

$$M = \text{Zr, Hf}$$

To date, there are no known low (≤0) oxidation state zirconium or hafnium mono(η-arene) complexes, and 'conventional' synthetic procedures for the bis(η-arene) compounds have not been reported.

6.2 BIS(η-CYCLOHEPTATRIENE) COMPLEXES

Sodium amalgam reduction of ZrCl$_4$ in the presence of cycloheptatriene affords the structurally characterized 16-electron bis(triene)complex $[Zr(\eta\text{-}C_7H_8)_2]$ (**1**) (Scheme 1).[4] The latter exhibits a unique nonparallel arrangement of the two triene ligands with a dihedral angle of 25.6° between the two best planes of the coordinated portions of the C$_7$H$_8$ rings.

$[Zr(\eta\text{-}C_7H_8)_2]$ is the precursor for a variety of half-sandwich η-cycloheptatrienyl–zirconium complexes, and of particular synthetic utility is the reaction with $[AlEt_2Cl]_2$ to afford the dimer $[\{Zr(\eta\text{-}C_7H_8)(THF)(\mu\text{-}Cl)\}_2]$ (Scheme 1). If the ligand is classed as C$_7$H$_7^+$, the η-cycloheptatrienyl compounds in Scheme 1 are all formally zirconium(0); however, EHMO calculations and photoelectron spectroscopic studies on $[Zr(\eta\text{-}C_7H_7)(\eta\text{-}C_7H_9)]$ show that the (η-cycloheptatrienyl)–metal interaction requires three essentially metal-based *d* electrons to form a satisfactory bond, and therefore that the compounds may more realistically be classed as zirconium(IV).[4]

The hafnium analogue of (**1**), $[Hf(\eta\text{-}C_7H_8)_2]$, has not yet been isolated, but is detectable by NMR spectroscopy amongst the mixture of products obtained by reduction of HfCl$_4$ in the presence of cycloheptatriene.[4]

6.3 CARBONYL COMPLEXES

6.3.1 Heteroleptic Carbonyl Compounds

Treatment of ZrCpCl$_3$ with sodium naphthalenide yields a thermally unstable product, which reacts as if it were a monocyclopentadienylmetallate. This species reacts readily with CO to afford the tetracarbonylmetalate, isolated as the thermally robust tetraalkylammonium salt $[Et_4N][ZrCp(CO)_4]$ (Equation (5)).[5] Similarly, ZrCp*Cl$_3$ affords $[Et_4N][ZrCp*(CO)_4]$ on reduction and treatment with CO (Equation (6)); the latter exhibits an extremely low field resonance at δ 296.5 for the carbonyl carbon in the ^{13}C NMR spectrum.[6]

Preparation of the analogous hafnium complexes requires the use of potassium naphthalenide as the reducing agent for reasons that are unclear, although it is generally the case that the nature of the counterion (and whether or not it is complexed by a macrocycle) is crucial in the synthesis of stable

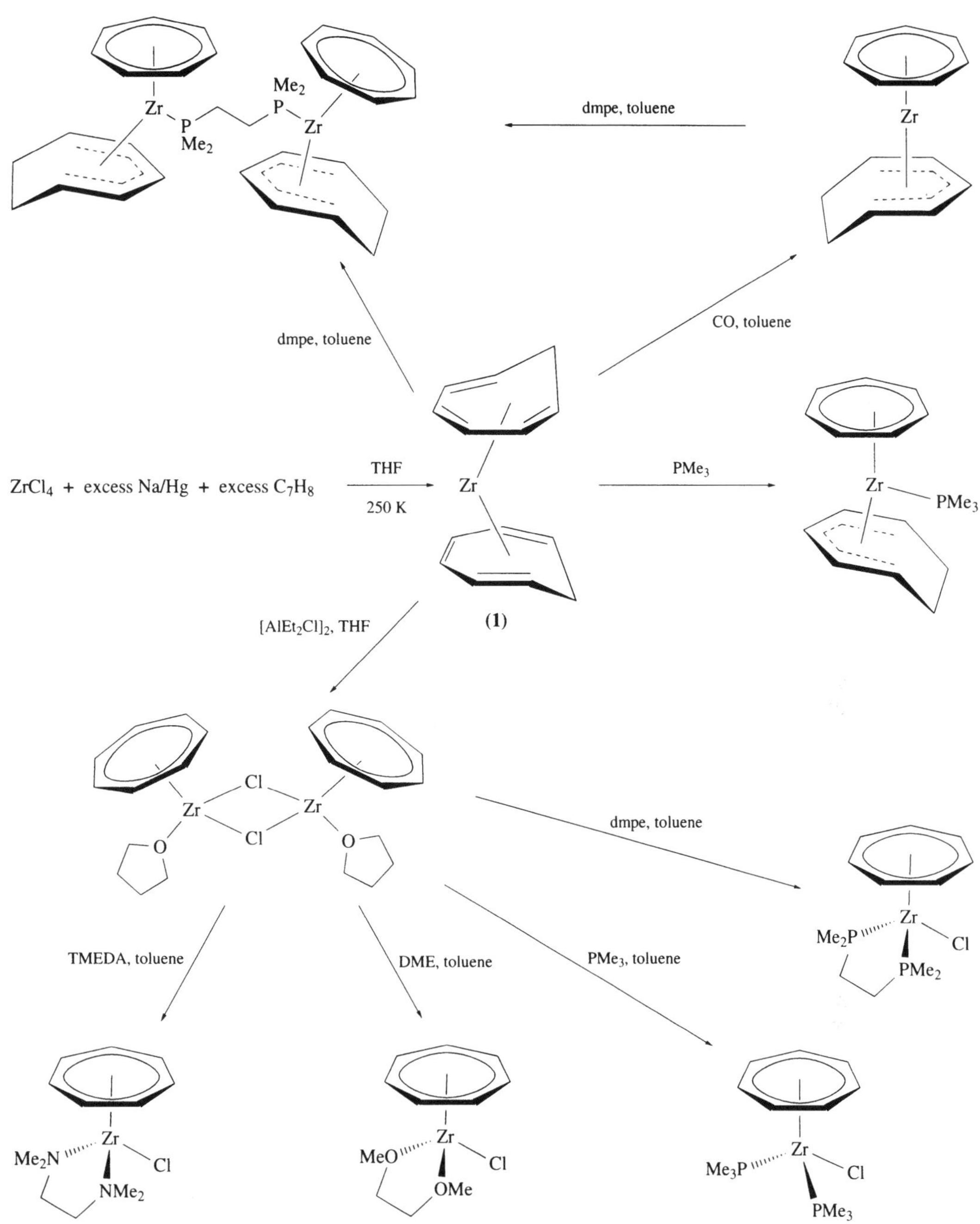

Scheme 1

$$ZrCpCl_3 \;+\; 4\,Na[C_{10}H_8] \xrightarrow[\substack{\text{ii, CO}\\\text{iii, }[Et_4N]Br}]{\text{i, DME, Ar, 203 K}} [Et_4N]^+ \left[\begin{array}{c} OC\cdots Zr \cdots CO \\ OC \quad CO \end{array}\right]^- \qquad (5)$$

$$ZrCp^*Cl_3 \;+\; 4\,Na[C_{10}H_8] \xrightarrow[\substack{\text{ii, CO}\\\text{iii, }[Et_4N]Br}]{\text{i, DME, Ar, 203 K}} [Et_4N][ZrCp^*(CO)_4] \qquad (6)$$

group 4 carbonylmetallates. Thus, $K[C_{10}H_8]$ allows the preparation of the anions $[Hf(\eta\text{-}C_5R_5)(CO)_4]^-$ (R = H or Me), as tetraalkylammonium salts, or cryptand or crown ether complexed potassium salts (Equations (7) and (8)).[7]

$$Hf(\eta\text{-}C_5R_5)Cl_3 + 4\ K[C_{10}H_8] \xrightarrow[\substack{ii, CO \\ iii, [Et_4N]Br}]{i, DME, Ar, 203\ K} [Et_4N][Hf(\eta\text{-}C_5R_5)(CO)_4] \qquad (7)$$

$$R = H, Me$$

$$Hf(\eta\text{-}C_5R_5)Cl_3 + 4\ K[C_{10}H_8] \xrightarrow[\substack{ii, n\ L \\ iii, CO}]{i, DME, Ar, 203\ K} [K(L)_n][Hf(\eta\text{-}C_5R_5)(CO)_4] \qquad (8)$$

$$R = H, Me$$
$$L = 15\text{-crown-}5,\ n = 2$$
$$L = crypt\ 2.2.2,\ n = 1$$

The thermally robust tripodal phosphine M^0 derivatives $[Hf(CO)_4(trmpe)]$ and the structurally characterized $[Zr(CO)_4(trmpe)]$ (**2**) (trmpe = $MeC\{CH_2PMe_2\}_3$) are directly accessible from the metal tetrachlorides via potassium naphthalenide reduction (Equations (9) and (10)).[8] Structure (**2**) may be described as a four-legged piano stool molecule, in which the interatomic distances are consistent with those observed for other zirconium carbonyl and phosphine derivatives.

$$[HfCl_4(THF)_2] + trmpe + 4\ K[C_{10}H_8] \xrightarrow{THF, CO, 210\ K} [Hf(CO)_4(trmpe)] \qquad (9)$$

$$[ZrCl_4(THF)_2] + trmpe + 4\ K[C_{10}H_8] \xrightarrow{THF, CO, 210\ K} \qquad (10)$$

(**2**)

6.3.2 Homoleptic Carbonyl Compounds

Reductive carbonylation of $[M(CO)_4(trmpe)]$ (M = Zr or Hf) yields isolable salts of $[Zr(CO)_6]^{2-}$ and $[Hf(CO)_6]^{2-}$ provided an appropriate crown ether or cryptand ligand is present to effectively complex the alkali metal cations (Equation (11) (M = Zr or Hf)).[9] $[Zr(CO)_6]^{2-}$ is also accessible directly from $[ZrCl_4(THF)_2]$ (Equation (12)), but this reaction fails for the hafnium analogue.[10] Solution IR spectroscopic studies (ν_{CO} 1757 cm^{-1}) for both $[K(cryptand\ 2.2.2)]_2[M(CO)_6]$ (M = Zr or Hf) are consistent with the presence of discrete octahedral $[M(CO)_6]^{2-}$ units,[9] as is the very narrow ^{91}Zr NMR spectral resonance in $[K(15\text{-crown-}5)_2]_2[Zr(CO)_6]$, which displays $J_{Zr\text{-}C}$ of 81 Hz in ^{13}CO enriched samples.[10] Essentially unperturbed octahedral geometry is confirmed by the x-ray structure of $[K(cryptand\ 2.2.2)]_2[Hf(CO)_6]\cdot$pyridine, which shows discrete anions and cations and one pyridine of crystallization.[9]

$$[M(CO)_4(tmpe)] + 2\ K[C_{10}H_8] + 2\ crypt\ 2.2.2 \xrightarrow{THF, CO, 200\ K} [K(crypt\ 2.2.2)]_2[M(CO)_6] \qquad (11)$$

$$M = Zr, Hf$$

$$[ZrCl_4(THF)_2] + 6\ K[C_{10}H_8] + 4\ 15\text{-crown-}5 \xrightarrow{DME, CO, 200\ K} [K(15\text{-crown-}5)]_2[Zr(CO)_6] \qquad (12)$$

6.4 REFERENCES

1. F. G. N. Cloke, K. A. E. Courtney, A. A. Sameh and A. C. Swain, *Polyhedron*, 1989, **8**, 1641; F. G. N. Cloke, M. F. Lappert, G. A. Lawless and A. C. Swain, *J. Chem. Soc., Chem. Commun.*, 1987, 1667.
2. F. G. N. Cloke and M. L. H. Green, *J. Chem. Soc., Dalton Trans.*, 1981, 1938.
3. K. Khan, D. Phil. Thesis, University of Sussex, 1991.
4. J. C. Green, M. L. H. Green and N. M. Walker, *J. Chem. Soc., Dalton Trans.*, 1991, 173.
5. B. A. Kelsey and J. E. Ellis, *J. Am. Chem. Soc.*, 1986, **108**, 1344.

6.　B. A. Kelsey and J. E. Ellis, *J. Chem. Soc., Chem. Commun.*, 1986, 331.
7.　S. R. Frerichs and J. E. Ellis, *J. Organomet. Chem.*, 1989, **359**, C41.
8.　D. W. Blackburn, K. M. Chi, S. R. Frerichs, M. L. Tinkham and J. E. Ellis, *Angew. Chem., Int. Ed. Engl.*, 1988, **27**, 437.
9.　K. M. Chi and J. E. Ellis, *J. Am. Chem. Soc.*, 1990, **112**, 6022.
10.　K. M. Chi, S. R. Frerichs, S. B. Philson and J. E. Ellis, *Angew. Chem., Int. Ed. Engl.*, 1987, **26**, 1190.

7

Metallocene(II) Complexes of Zirconium and Hafnium

PAUL BINGER and STEFAN PODUBRIN
Max-Planck-Institut für Kohlenforschung, Mülheim, Germany

7.1 PREPARATION

7.1.1 General Methods

Ligand-free zirconocene and hafnocene, although frequently referred to as highly reactive intermediates in the literature, have never been isolated as discrete chemical compounds.[1] Depending on the nature of the synthetic approach (e.g., reduction of the dichlorides, photochemical decomposition of diaryl- or dimethylzirconocenes or -hafnocenes, respectively) and applied reaction conditions, the obtained pyrophoric multicoloured products mainly possess the right composition but are clearly best formulated as metallocene(III), rather than metallocene(II), compounds probably containing ($\eta^1{:}\eta^5$-C_5H_4)-bridged metal hydride moieties.[2] Therefore, metallocene(II) complexes of zirconium and hafnium are generally prepared by the reduction of metallocene(IV) compounds in the presence of stabilizing ligands such as carbon monoxide, phosphines, alkenes, alkynes, and so on. In any case the readily available metallocene dichlorides serve as starting materials which may be reduced by either heterogeneous or homogeneous procedures (Scheme 1). In contrast with the heterogeneous reduction which yields the desired metallocene(II) complex directly, using sodium amalgam or activated magnesium, the homogeneous pathways primarily involve a substitution reaction. By reacting alkyl- or aryllithium or Grignard reagents with the metallocene dichlorides, the corresponding metallocene(IV) species are formed in a first step and subsequently decomposed thermally or photochemically via elimination producing the metallocene(II) complex. Although in this context the generation of an intermediate 'free' metallocene has been discussed frequently, in no instance has this been realized; instead this moiety is trapped by coordinated unsaturated ligands which may either be generated in the course of the β-hydrogen elimination or added separately. Whereas numerous zirconocene(II) complexes have been described since the 1980s, only a few examples of the corresponding hafnocene compounds are known (Scheme 1).

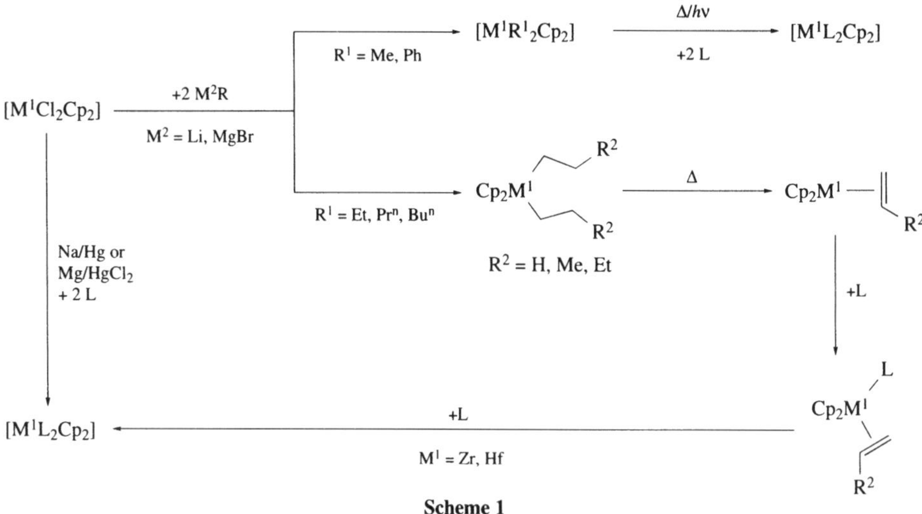

Scheme 1

7.1.2 CO and N_2 Complexes

The synthesis of carbonyl derivatives of zirconocene and hafnocene has been reviewed.[3] Several different procedures for the preparation of the air- and moisture-sensitive metallocene dicarbonyls have been reported,[4] but the most efficient method has been developed by Rausch and co-workers[5] by reducing the metallocene dichlorides with magnesium activated with $HgCl_2$ under an atmosphere of carbon monoxide (Equation (1)).

$$[MCl_2Cp_2] + Mg + 2\,CO \xrightarrow{\text{THF, HgCl}_2} [M(CO)_2Cp_2] + MgCl_2 \qquad (1)$$

M = Zr, 53%
M = Hf, 30%

Similarly, the permethylcyclopentadienyl, indenyl[6] and mixed ligand congeners, such as [M(η-Cp)-(η-Cp*)], are formed under the same conditions, except for the hafnium compounds which require the use of the highly activated Rieke magnesium.[5] The compound [Hf(CO)$_2$(η-C$_9$H$_7$)$_2$] is only obtained in nonreproducible and extremely low yields. [Zr(CO)$_2$(η-Cp*)$_2$] may also be generated by the photochemical decomposition of [ZrPh$_2$(η-Cp*)$_2$] in the presence of CO (1 atm) in toluene at room temperature.[7] The high-pressure carbonylation of [HfH$_2$(η-Cp*)$_2$] in toluene produces [Hf(CO)$_2$-(η-Cp*)$_2$] in 95% yield.[8] The trimethylsilylcyclopentadienyl complex [Zr(CO)$_2$(η-C$_5$H$_4$TMS)$_2$] may either be prepared via the carbonylation of the zirconacyclopentene complex [Zr(o-CH$_2$C$_6$H$_4$CH$_2$)-(η-C$_5$H$_4$TMS)$_2$][9] or by the reductive carbonylation of [ZrCl$_2${η-C$_5$H$_3$(TMS)$_2$}$_2$] using Na/Hg, Mg/HgCl$_2$ or Li$_2$(cot) as reducing agent.[10]

The incorporation of dinitrogen into the metallocene moiety does not, unlike titanocene, yield stable zirconocene or hafnocene complexes, whereas the use of the permethylcyclopentadienyl congeners has led to the isolation of dinuclear complexes. In 1974 Bercaw and co-workers reported that the reduction of [ZrCl$_2$(η-Cp*)$_2$] with excess sodium amalgam in toluene under 1 atm of N_2 over a period of 2 d at room temperature yields the corresponding dinitrogen complex [{Zr(N$_2$)(η-Cp*)$_2$}$_2$(μ-N$_2$)].[4] The hafnium analogue was described by the same author in 1985, starting from [HfI$_2$(η-Cp*)$_2$] which was reduced with Na/K alloy in DME at $-41\,^\circ$C.[8] An exchange of the two terminally bound dinitrogen molecules with CO is achieved by treatment of [{Zr(N$_2$)(η-Cp*)$_2$}$_2$(μ-N$_2$)] with CO in toluene at $-23\,^\circ$C for 1 h, yielding [{Zr(CO)(η-Cp*)$_2$}$_2$(μ-N$_2$)]. The probably isostructural hafnium complex is far more labile than the zirconium compound, as at room temperature it decomposes over the course of several days in C_6D_6 even under 1 atm of N_2, although as a solid it may be stored for extended periods under N_2 at −20 °C.

In contrast with the N_2 complexes bearing end-on coordinated ligands, η^2-coordinated diazo complexes (for a recent review see Ref.11) are formed by the addition of zirconocene dichloride in THF to a solution of dilithio-1,2-diphenylhydrazide at room temperature. This is the most efficient preparation of the THF-stabilized azobenzene complex $[Zr(N_2Ph_2)(THF)(\eta\text{-}Cp)_2]$.[12] Treatment of the latter complex with pyridine, PMe_3, $NCPh$ or $CNBu^t$ leads to the dissociation of THF and the final formation of the corresponding azobenzene complexes (Equation (2)).

$$[ZrCl_2Cp_2] + \quad \underset{Li \quad Li}{\overset{Ph \quad Ph}{N-N}} \quad \xrightarrow[-2\,LiCl]{THF,\ Et_2O} \quad \underset{THF}{\overset{Ph}{\underset{\ }{Cp_2Zr{-}N}}_{Ph}} \quad \xrightarrow[-THF]{+L} \quad \underset{L}{\overset{Ph}{\underset{\ }{Cp_2Zr{-}N}}_{Ph}} \tag{2}$$

$$L = py, PMe_3, NCPh, CNBu^t$$

Despite the crystallographic results for the pyridine complex giving evidence for a diazazirconacyclopropane structure rather than for an η^2-coordinated azobenzene, the observed reactivity towards alkynes (see Section 7.3) corresponds to that of a zirconocene(II) complex involving the reductive coupling of the two organic moieties in order to yield stable diazazirconacyclopentadienes.

7.1.3 Phosphine Complexes

Several bis(phosphine)zirconocene adducts have been prepared using different procedures (for a recent review see Ref. 13), while phosphine complexes of hafnium(II) are still quite rare, reflecting the inherent difficulty in reducing hafnium rather than a lack of research in this area. The bis(phosphine) adducts $[ZrL_2(\eta\text{-}Cp)_2]$ (L = PMe_3 or $P(OMe)_3$; L_2 = dmpe) are accessible by reduction of zirconocene dichloride with either Na/Hg or magnesium in THF; the sodium amalgam reduction is reported to work best for dmpe. As an alternative synthesis for $[ZrL_2(\eta\text{-}Cp)_2]$ (L = PMe_2Ph or $PMePh_2$; L_2 = dmpe or dppe), the phosphine-induced reductive elimination of methylcyclohexane from $[ZrH(CH_2C_6H_{11})(\eta\text{-}Cp)_2]$ has been reported.[14] Zirconocene derivatives of the type $[ZrL_2(\eta\text{-}Cp)_2]$, containing monodentate phosphine ligands, are thermally unstable and readily convert to the dinuclear zirconium(III) complexes $[\{ZrL(\eta\text{-}Cp)\}_2(\mu\text{-}\eta^1{:}\eta^5\text{-}C_5H_4)_2]$ (1).[15] Concerning bis(phosphine)hafnocene compounds, $[Hf(PMe_3)_2(\eta\text{-}Cp)_2]$ is the only representative of this class which has been reported to be formed in traces by reduction of hafnocene dichloride with magnesium metal, the dinuclear hafnocene(III) species $[\{Hf(PMe_3)(\eta\text{-}Cp)_2\}_2(\mu\text{-}\eta^1{:}\eta^5\text{-}C_5H_4)_2]$ being the main product.[16]

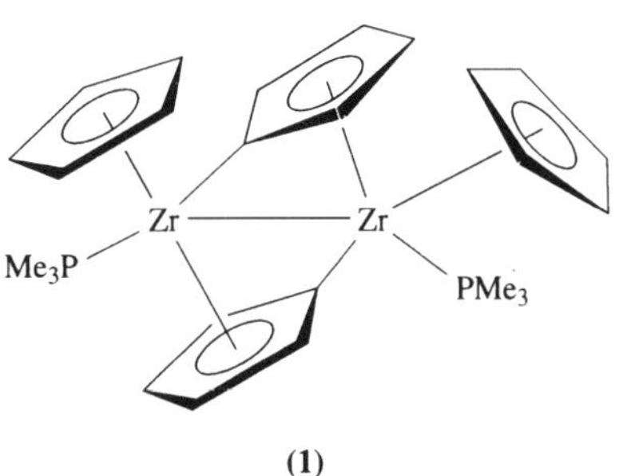

(1)

The closely related phosphine or phosphite carbonyls $[M(CO)(L)(\eta\text{-}Cp)_2]$ (M = Zr and L = PF_3, $PF_2N(Me)PF_2$, PMe_3, PMe_2Ph, $PMePh_2$, PPh_3, PBu_3 or $P(OMe)_3$; or M = Hf and L = PMe_3, PF_3 or dppe) have been prepared by thermally or photochemically induced substitution of one CO of the corresponding $[M(CO)_2(\eta\text{-}Cp)_2]$ compound. The same applies for the preparation of $[Zr(CO)(PMe_3)(\eta\text{-}C_9H_7)_2]$ and $[Zr(CO)(PMe_3)\{\eta\text{-}C_5H_3(TMS)_2\}_2]$. In this context, it is noteworthy that the displacement of both carbonyl ligands by phosphines is not possible and therefore the bis(phosphine)metallocenes may not be obtained using this path.[17]

Addition of PF_3 to $[\{Zr(N_2)(\eta\text{-}Cp^*)_2\}_2(\mu\text{-}N_2)]$ results in the formation of $[\{Zr(PF_3)_2(\eta\text{-}Cp^*)_2\}_2(\mu\text{-}N_2)]$ which is stable only at low temperature.[18] Other phosphines or phosphites, for example, PMe_3 or $P(OMe)_3$, do not react in the same way.

7.1.4 Alkene and Alkyne Complexes

Up to now no zirconocene or hafnocene monoalkene or -alkyne complex has been isolated as such, although their existence as highly reactive 16-electron intermediates is undoubtedly proved (e.g., by NMR spectroscopy[19]). In the presence of ligands such as a phosphine, 4-dimethylaminopyridine (dmap) or THF, this species may effectively be trapped and this has led in recent years to a considerable number of stable, well-characterized 18-electron alkene and alkyne complexes. While this is true for zirconocene, for hafnocene only very few alkene complexes have been reported, due to the greater propensity of hafnium to prefer the +4 oxidation state. Thus, the formation of $[Hf(PMe_3)(C_2H_4)(\eta\text{-}Cp)_2]$ from the metallacyclopentane requires rather severe reaction conditions in contrast with those for the corresponding zirconocene compound (Equation (3)).[20] The intermediacy of a bis(alkene)metallocene is proposed in this context but actually no such complex has been isolated as yet. On the contrary, Hoffmann and co-workers have deduced—although the extended Hückel calculations were carried out for titanocene—that a d^0 metal(IV) metallacycle should be more stable than a d^2 metal(II) bis(alkene)metallocene.[21]

$$
[MCl_2Cp_2] \xrightarrow[\text{BrMg(CH}_2)_4\text{MgBr}]{\text{THF}} Cp_2M \quad \xrightarrow{\text{4 PMe}_3}
\begin{cases}
M = Zr, \ -78\,°C \text{ to RT} \longrightarrow Cp_2Zr(\text{C}_2\text{H}_4)(PMe_3) \\
M = Hf, \ \text{reflux, 6 h} \longrightarrow Cp_2Hf(\text{C}_2\text{H}_4)(PMe_3)
\end{cases}
\tag{3}
$$

Basically the formation of zirconocene alkene or alkyne complexes may be achieved in two different ways: (i) by a reductive elimination process starting from alkyl- or arylzirconocene(IV) complexes, or (ii) by ligand exchange reactions involving the previously discussed carbonyl or phosphine complexes.

Focusing on the first alternative (i), again there are two different possibilities: the starting compound may either be symmetrically or unsymmetrically substituted. In both cases it is necessary that at least one β-hydrogen atom is present in order to generate the corresponding alkene or alkyne. Alkylation of zirconocene dichloride with either a Grignard or an alkyllithium reagent yields thermally unstable dialkylzirconocenes which decompose upon warming from $-78\,°C$ to room temperature, forming stable zirconocene(II) complexes in the presence of a donor, mainly PMe_3 (Equation (4)).[22]

$$
Cp_2Zr(\text{CH}_2\text{CHR})_2 \xrightarrow[\text{$-78\,°C$ to RT}]{\text{4 PMe}_3} Cp_2Zr(\text{CH}_2=\text{CHR})(PMe_3) + Cp_2Zr(\text{CH}=\text{CHR})(PMe_3)
\tag{4}
$$

R = H, Me, Et, Ph

The use of an alkylating reagent other than an ethyl-Grignard or -lithium compound yields two enantiomeric pairs of rotamers. Dynamic NMR spectroscopic experiments indicate that even at room temperature only a slow exchange of the rotamers occurs. The same procedure may be applied for the preparation of the corresponding indenyl complexes $[Zr(alkene)(PMe_3)(\eta\text{-}C_9H_7)_2]$.[23] In contrast with these rather labile alkene complexes, the thermal decomposition of diphenylzirconocene in the presence of a large excess of trimethylphosphine yields the much more stable benzyne complex $[Zr(C_6H_4)(PMe_3)(\eta\text{-}Cp)_2]$ (Equation (5)).[24]

$$
Cp_2Zr(\text{C}_6\text{H}_5)_2 \xrightarrow[\substack{\text{benzene, reflux, 24 h} \\ 90\%}]{\text{PMe}_3} Cp_2Zr(\text{C}_6\text{H}_4)(PMe_3)
\tag{5}
$$

A comparable behaviour has been reported for substituted cyclopropylzirconocene compounds which serve as a source for the corresponding cyclopropene-stabilized zirconocene (Equation (6)).[25] If instead the unsubstituted congener (which may be obtained from 2 equiv. of cyclopropyllithium and zirconocene dichloride) is used, a rather labile dialkylzirconocene complex is generated, which quickly decomposes on warming to room temperature. The already known zirconocene complex (**8**) of cyclopropene may be trapped in the presence of PMe_3.

$$Cp_2Zr \xrightarrow[55\,°C]{PMe_3} Cp_2Zr(PMe_3) \;+\; Cp_2Zr(PMe_3) \qquad (6)$$

$n = 1$ or 2

This pathway has been extended to 1,2-cyclohexa- and heptadiene (**2**)[26] as well as to 1,2,3-cyclohexatriene zirconocenes (**3**).[27]

(**2**) $n = 1, 2$ (**3**)

Trimethylphosphine-stabilized hafnocene complexes may be formed as indicated for the zirconium analogues in Equation (4), but under rather harsh conditions in the presence of PMe_3 thermolysis of $[Hf(Bu^n)_2(\eta\text{-}Cp)_2]$ proceeds at 80 °C causing noticeable decomposition. Therefore a thermally more labile hafnocene complex was used to produce $[Hf(isobutene)(PMe_3)(\eta\text{-}Cp)_2]$, which may be converted to the corresponding alkyne complex when reacted with 1 equiv. of diphenylacetylene (Scheme 2).[28]

$$[HfCp_2Cl_2] \xrightarrow[\text{ii, Bu}^n\text{Li}]{\text{i, Bu}^t\text{Li}} Cp_2Hf(Bu^t)(Bu^n) \xrightarrow[45\,°C,\,21\,h]{10\ \text{equiv. }PMe_3} Cp_2Hf(PMe_3) \xrightarrow{Ph\!\equiv\!Ph} Cp_2Hf(PMe_3)$$

Scheme 2

Similarly, the use of Bu^tLi has been reported to serve as a source of zirconocene alkene complexes. Unlike the hafnocene analogue, zirconocene dichloride reacts in the first step with 1 equiv. of Bu^tLi yielding isobutylzirconocene monochloride and not the expected *t*-butyl derivative (Equation (7)).[29] Alkenes bearing bulky substituents such as *t*-butylethene do not form stable zirconocene complexes.

$$[ZrCp_2Cl_2] \xrightarrow[-78\ \text{to }25\,°C]{Bu^tLi} Cp_2Zr(Cl) \xrightarrow[\text{ii, }PMe_3]{\text{i, Bu}^t\text{Li, }-78\,°C} \qquad (7)$$

$R = H, Et, n\text{-Hex}, Cy, Ph$

A very versatile method has been developed by Buchwald and co-workers to produce unsymmetrically substituted diorganylzirconocenes in order to generate well-defined alkene or alkyne complexes (Scheme 3).[30]

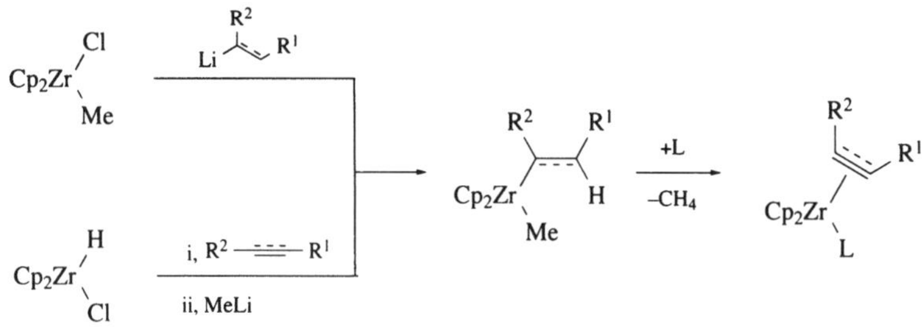

Scheme 3

The use of [ZrCl(Me)(η-Cp)$_2$] or the Schwartz reagent, [ZrCl(H)(η-Cp)$_2$], as a starting material makes this method even more favourable as these reagents are readily accessible even on a multigram scale. Besides the synthesis of the alkyne complexes (**4**) and (**5**)[22a] or (**6**) and (**7**),[31] the scope of the method depicted in Scheme 3 equally includes the generation of the cycloalkene (**8**)[32] and cycloalkyne zirconocenes (**9**)[33] or (**10**).[34] Moreover, even a dinuclear 1,4-benzdiyne-bridged complex is accessible using this method (Equation (8)).[35]

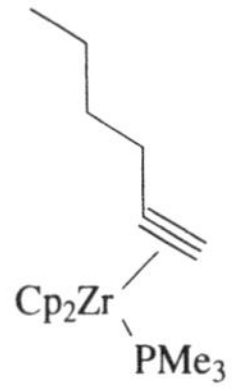

(**4**)

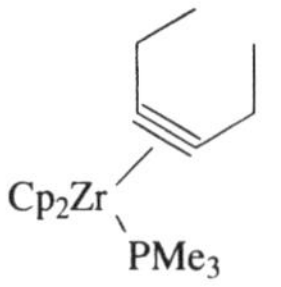

(**5**)

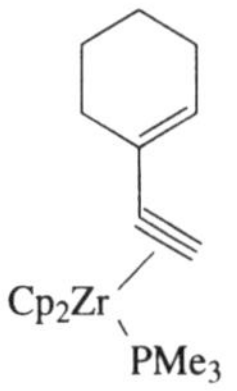

(**6**)

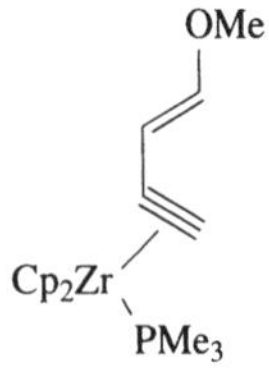

(**7**)

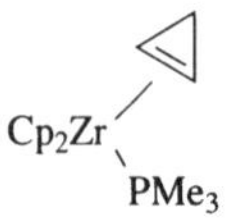

(**8**)

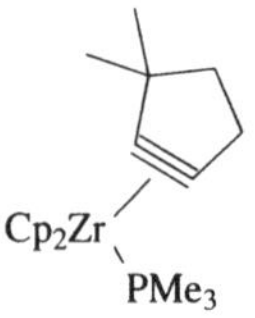

(**9**)

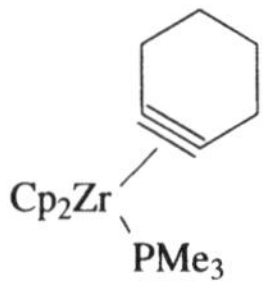

(**10**)

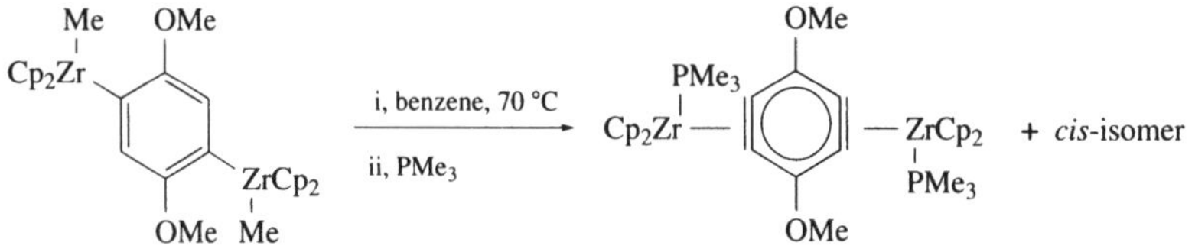

$$\text{(8)}$$

The unsymmetrically substituted dialkylzirconocene precursor may alternatively be generated starting from dimethylzirconocene by addition of 1 equiv. of an alkyl-Grignard or -lithium reagent. It was presumed that prior to β-hydrogen elimination in an initial step, an ate complex of the type M[ZrR$^1_2$R^2(η-Cp)$_2$] (M = Li, MgCl or MgBr; R^1 = Me, R^2 = Et, Pr^n or Bu^n) may be involved.[36]

Concerning method (ii), the readily accessible compounds [Zr(PMe$_3$)$_2$(η-Cp)$_2$] and [Zr(1-butene)(PMe$_3$)(η-Cp)$_2$] have been shown to undergo selective ligand exchange reactions which have led to a variety of alkene and alkyne complexes (Scheme 4).[22c,37]

This pathway is quite intriguing as alkylated alkynes, as well as most of the other tested allenes so far, are so reactive that even if exactly 1 equiv. of alkyne (or allene) is used, the remaining PMe$_3$ ligand is partly displaced, yielding a mixture of the desired alkyne complex and the corresponding five-membered metallacycle which is formed by oxidative coupling of two alkynes (or allenes) at the zirconocene moiety (see Section 7.3). This difficulty may be overcome by using alkynes with bulky or electron-withdrawing substituents. In this context, the recently reported formation of [Zr(THF)(TMS-CC-TMS)(η-Cp)$_2$] by reduction of zirconocene dichloride with magnesium in the presence of bis(trimethylsilyl)acetylene is an illustrative example.[38]

Another straightforward alternative to trap the monoalkyne complex rather than the metallacyclopentadiene complex has recently been described, involving cleavage of the C$_\beta$–C$_{\beta'}$ bond of

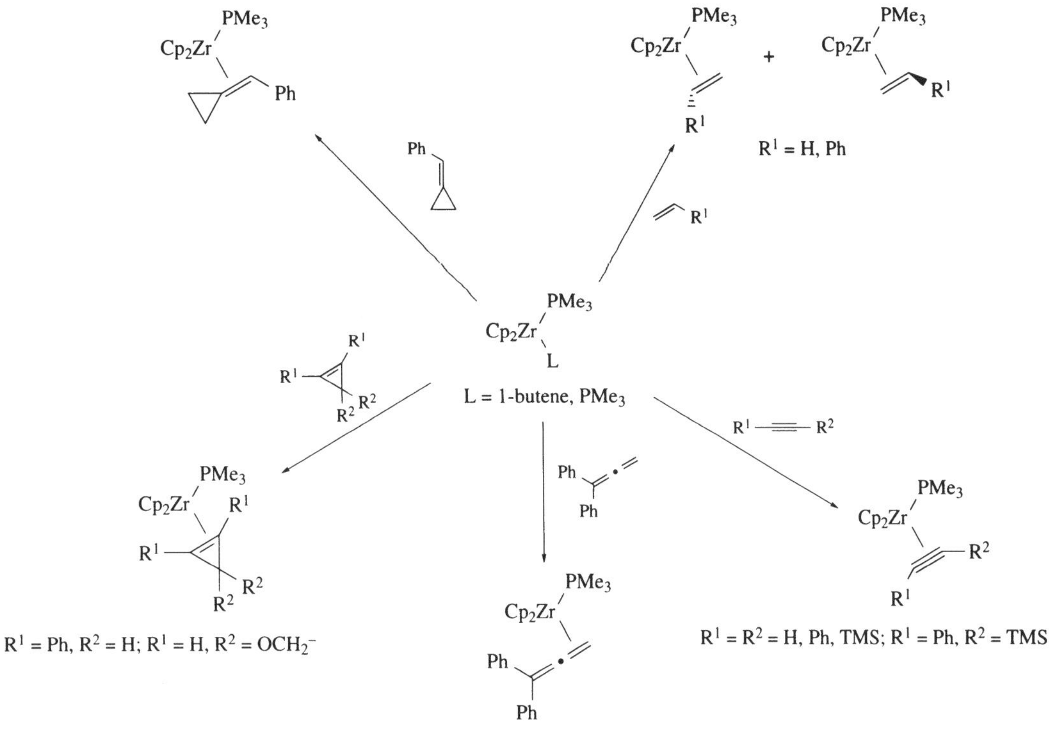

Scheme 4

zirconacyclopentenes which are moderately stable at 0 °C and decompose on being warmed to 20 °C in the presence of PMe_3, liberating ethene (Equation (9)).[39] Although this pathway is quite similar to the decomposition of a metallacyclopentane as shown in Equation (3), its scope is much wider because the cumbersome use of di-Grignard or dilithium reagents is omitted and the substitution pattern is determined by the chosen alkyne.

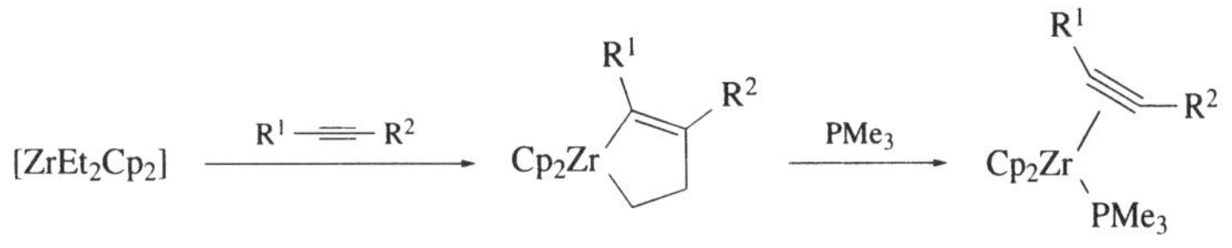

$$[ZrEt_2Cp_2] \xrightarrow{R^1 \equiv R^2} Cp_2Zr \xrightarrow{PMe_3} Cp_2Zr \qquad (9)$$

Generally, zirconocene alkyne complexes are formed more easily than the corresponding alkene complexes. Diphenylacetylene reacts 150 times faster with $[Zr(Bu^n)_2(\eta\text{-}Cp)_2]$ than (E)-stilbene, forming $[Zr(PhCCPh)(PMe_3)(\eta\text{-}Cp)_2]$ by warming from -78 °C to 20 °C in the presence of PMe_3.[40]

A rather exceptional synthesis of a PMe_3-stabilized zirconocene and even the rare hafnocene η^2-cycloheptatrienyl complex has recently been reported by Green and co-workers. Starting from $[M(\eta\text{-}C_7H_8)(PMe_3)_2Cl_2]$ (M = Zr or Hf), the cyclopentadienyl ligands are introduced in the last step (Equation (10)).[41]

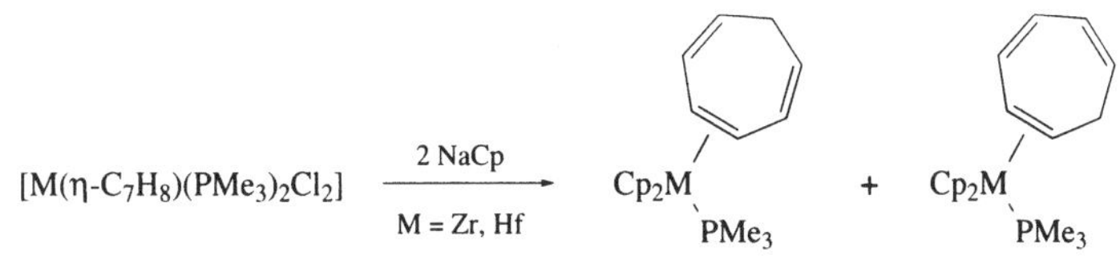

$$[M(\eta\text{-}C_7H_8)(PMe_3)_2Cl_2] \xrightarrow[M = Zr, Hf]{2\, NaCp} Cp_2M \qquad + \qquad Cp_2M \qquad (10)$$

In contrast with the behaviour depicted in Equation (10), the related reaction with lithium indenide does not yield the corresponding bis(indenyl) compound, but instead a mixed $(\eta\text{-}C_7H_7)(\eta\text{-}C_9H_7)$ complex is formed.

Apart from $[Zr(PhCCPh)(\eta\text{-}Cp^*)_2]$,[42] which has been obtained as a stable isolable complex by reacting diphenylacetylene with $[\{Zr(N_2)(\eta\text{-}Cp^*)_2\}_2(\mu\text{-}N_2)]$, the permethylated congeners of zirconocene and hafnocene have not yet been investigated for the formation of alkene or alkyne complexes.

Nevertheless, alkenes such as ethene, 1-butene or 1,2-dimethylallene have been reacted with [Zr-(η-p*)$_2$] derivatives yielding stable zirconacyclopentanes instead of the corresponding alkene complexes, although the CH activation of the η-Cp* ligand is frequently responsible for undesired side reactions.[43] In contrast with the reactivity of alkenes towards the decamethylzirconocene moiety, the synthesis of several bis(indenyl)zirconium complexes of the type [Zr(alkene)(PMe$_3$)(η-C$_9$H$_7$)$_2$] (where the 'alkene' may be ethene, 1-butene or styrene) has been achieved,[23] indicating that to a large extent electronic rather than steric effects are responsible for the observed reactivity.

7.1.5 1,3-Butadiene Complexes

The chemistry of (η^4-1,3-butadiene)zirconocene and -hafnocene has been extensively reviewed, mainly by Erker *et al.*[44] and Nakamura and Yasuda.[45] The most remarkable feature of some of these monomeric compounds is definitely their unique *s-trans*-diene conformation, because the preference of most transition metal fragments for binding 1,3-butadiene is dominated by *s-cis*-diene transition metal complexes. Depending on the method of preparation and on the substitution pattern of the 1,3-butadiene, as well as on the metallocene moiety, for example, [M(η-Cp*)$_2$] or [M(η-C$_5$H$_4$Bu^t)], different amounts of (*s-cis*-diene)- and (*s-trans*-diene)metallocene are formed. Especially at elevated temperatures, isomerization occurs and *s-trans*–*s-cis* equilibria are established. (Diene)zirconocene and -hafnocene complexes are readily available by several routes: (i) reaction of the metallocene dichloride with butadienemagnesium,[46] (ii) generation of a reactive metallocene unit in the presence of a diene as a scavenger,[47] or (iii) coupling of alkenyl ligands at the metallocene moiety[48] (Scheme 5). Moreover, (1,4-diphenyl-1,3-butadiene)zirconocene and -hafnocene have alternatively been prepared by a ligand exchange reaction starting from the (isoprene)metallocene.[45a] In this context it is noteworthy that no such ligand exchange reactions with 1,3-butadienes involving stabilized alkene-, alkyne- or bis(ligand)zirconocenes have yet been reported, although the reactivity of an alkene or alkyne towards (1,3-butadiene)metallocene species has been investigated quite thoroughly (see Section 7.3.2.2). Besides the acyclic 1,3-diene complexes listed in Table 1,[49] some cyclic congeners are also known. The zirconocene complexes (**11**),[50] (**12**)[51] and (**13**)[52] contain exclusively the *s-cis*-diene, as in these cases the conformation is invariably locked by the annelated substituents.

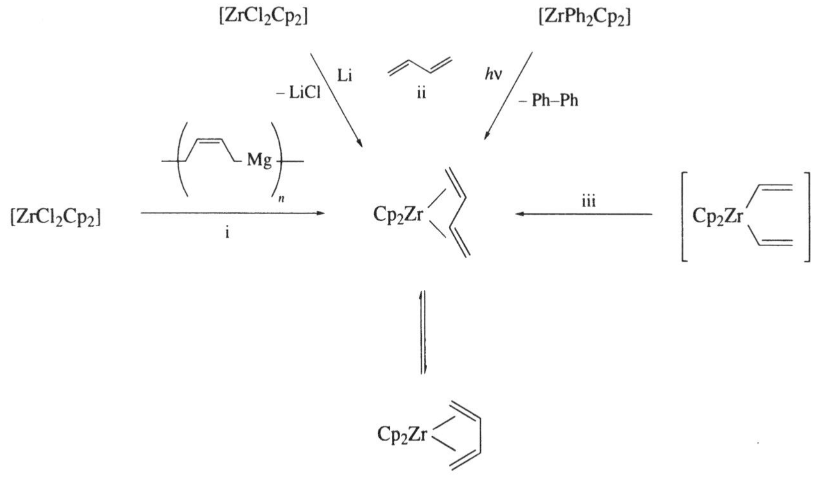

Scheme 5

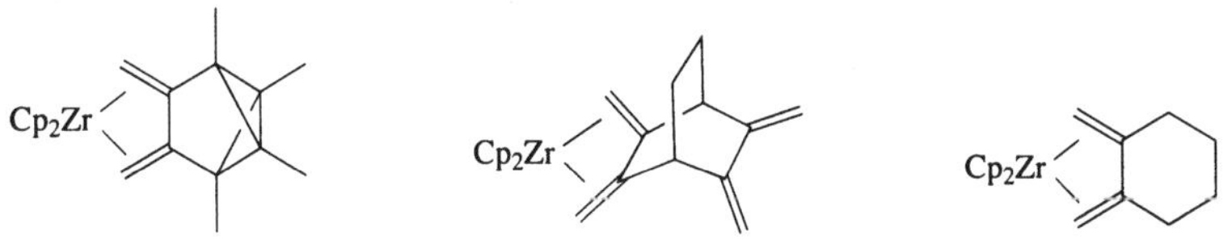

Table 1 Butadiene complexes of the type [ML$_2$(R^{1}CH=CR^2–CR^3=CHR^4] (*s-cis* amount at room temperature).

M	L$_2$	Method	R^1	R^2	R^3	R^4	*s-cis* (%)	Ref.
Zr	(η-Cp)$_2$	i, ii, iii	H	H	H	H	55	44
		i	Me	H	H	H	39	44
		i	H	Me	H	H	99	44
		ii	Me	H	H	Me	40	44
		i	H	Me	Me	H	>99	44
		i, ii, iii	Ph	H	H	Ph	5	44
		ii	H	Ph	Ph	H	>99	44
Hf	(η-Cp)$_2$	i	H	H	H	H	>99	44
		i	H	Me	H	H	>99	44
		i	H	Me	Me	H	>99	45a
		i, ii, iii	Ph	H	H	Ph	>99	49a
Zr	(η-Cp*)$_2$	i	H	H	H	H	60	44
		i	H	Me	H	H	>99	44
Zr	(η-C$_5$H$_4$Bu^t)$_2$	i	H	H	H	H	94	49b
		i	H	Me	H	H	>99	49b
		i	H	Me	Me	H	>99	49b
Hf	(η-C$_5$H$_4$Bu^t)$_2$	i	H	H	H	H	>99	49b
		i	H	Me	H	H	>99	49b
Zr	(η-Cp)(η-C$_5$H$_4$Bu^t)	i	H	H	H	H	85	49b

7.1.6 Heteroalkene and -alkyne Complexes

Several zirconocene aldehyde and ketone complexes have been prepared starting from zirconocene acyl precursors. Thermolysis of η^2-benzoyl(phenyl)zirconocene yields the dimeric (η^2-benzophenone)-zirconocene (Equation (11)).[53] By reaction with [{ZrH$_2$(η-Cp)$_2$}$_x$], η^2-benzoylzirconocene monochloride has been converted to a very similar complex containing formaldehyde and benzaldehyde, both η^2-coordinated (Equation (12)).[54] Carbonylation of [ZrH$_2$(η-Cp)$_2$]$_x$ occurs at room temperature in toluene at 150 bar carbon monoxide, but even after prolonged treatment only a 14% yield of the trimeric formaldehydezirconocene [{Zr(η^2-CH$_2$O)(η-Cp)$_2$}$_3$][55] (**14**) is finally isolated.

$$\text{(11)}$$

$$\text{(12)}$$

$$\textbf{(14)}$$

The preparation of zirconocene ketene complexes has been reported via dehydrohalogenation of zirconium chloroacyl complexes. Again dimeric compounds are formed if zirconocene units are present

(Equation (13)).[56] In contrast, permethylated zirconocene ketene complexes have been isolated as monomers if an additional neutral donor such as pyridine or carbon monoxide is present. The preparation of these compounds is essentially similar to that of the dimeric zirconocene ketene (Equation (13)), but in this case the intermediate anionic complex with sodium or lithium as counterion is isolated first. On treatment with pyridine or carbon monoxide, the latter is decomposed liberating the corresponding sodium or lithium salt (Scheme 6).[57]

$$\text{(13)}$$

Scheme 6

Similarly, such species may be generated by methane elimination from a suitable acyl(methyl)-zirconocene compound. Thus, diphenylketenezirconocene (**15**)[58] was obtained by carbonylation of $[Zr(CHPh_2)Me(\eta\text{-}Cp)_2]$.

(15)

Several monomeric thioaldehydezirconocene complexes were prepared in a high-yield one-pot synthesis by heating *in situ* formed (alkylthio)methylzirconocenes in the presence of trimethylphosphine (Equation (14)).[59] First-order kinetics, independent of trimethylphosphine concentration, were revealed.

$$\text{(14)}$$

$R = Me, 85\%; Ph, 90\%; p\text{-}(CF_3)C_6H_4, 76\%; p\text{-}ClC_6H_4, 79\%; p\text{-}MeOC_6H_4, 76\%; p\text{-}Me_2NC_6H_4, 47\%$

Starting from (methyl)chlorozirconocene and a suitable lithium amide, trimethylphosphine- or THF-stabilized aldiminezirconocene complexes are accessible. This probably depends on the availability of the lone pair on nitrogen; the intermediate methylzirconocene complex liberates methane rapidly at −10 °C or by heating to 110 °C for several hours (Equation (15)). The corresponding aldimine complexes were trapped by donors such as trimethylphosphine or tetrahydrofuran yielding the stable compounds (**16**)–(**19**).[60] This method has been extended to cyclic and α-disubstituted amines yielding the aldimine complex (**20**)[61] and the ketiminezirconocenes (**21**) and (**22**).[62] In both cases, methane elimination requires thermolysis at elevated temperatures (60–80 °C). The insertion of phenyl isocyanide into a zirconacyclopentane yields an iminoacylzirconocene which rearranges on warming to 40 °C in the presence of trimethylphosphine to form a stable ketiminezirconocene complex (Equation (16)).[63] If Bu^{t}NC is used instead of phenyl isocyanide, a thermally stable iminoacyl complex is obtained, which does not rearrange at all to the corresponding ketiminezirconocene.

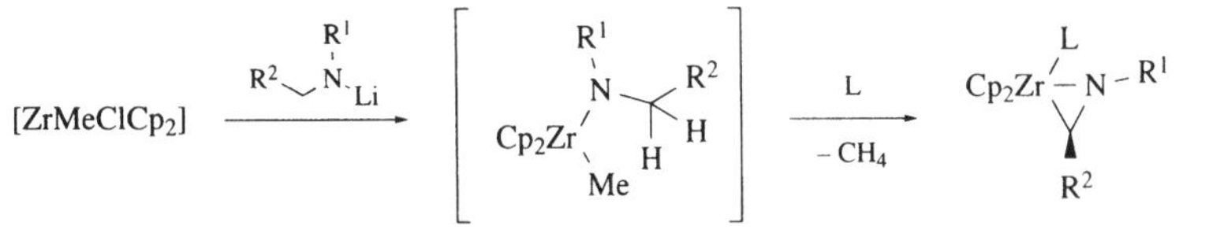

(15)

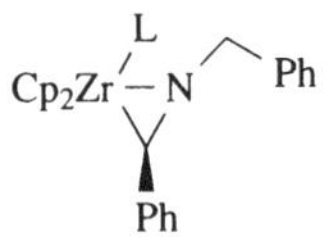

L = THF, PMe$_3$

(**16**)

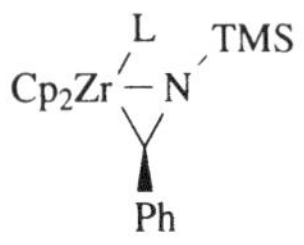

L = THF, PMe$_3$

(**17**)

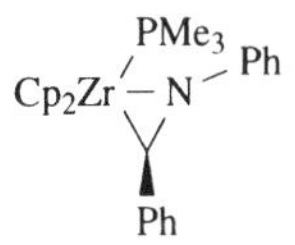

(**18**)

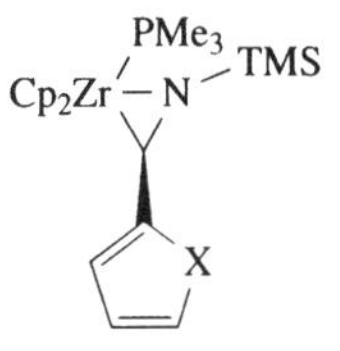

X = O, S

(**19**)

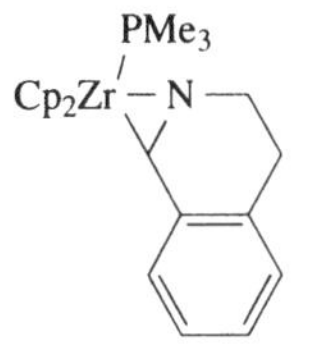

(**20**)

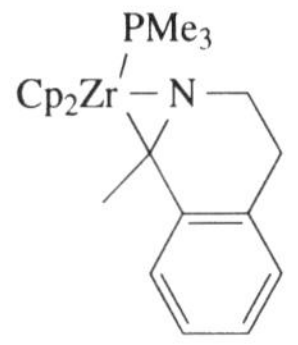

(**21**)

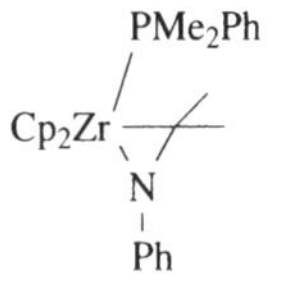

(**22**)

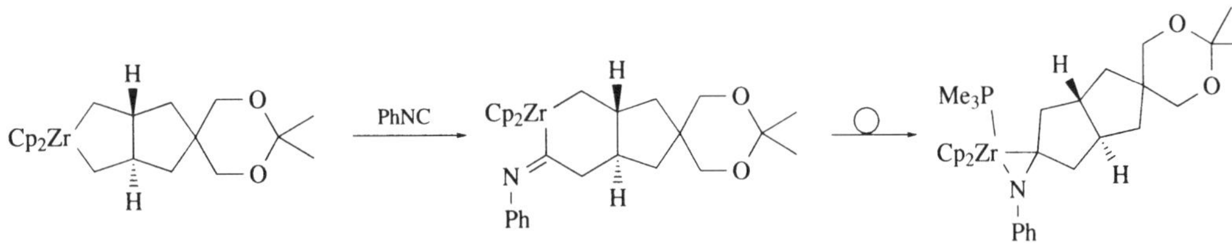

(16)

Selective ligand exchange, as depicted in Scheme 4 (Section 7.1.4), has been shown to yield the ketenimine complex (**23**).[37c]

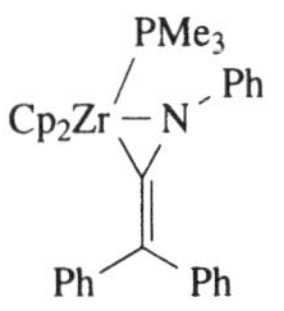

(**23**)

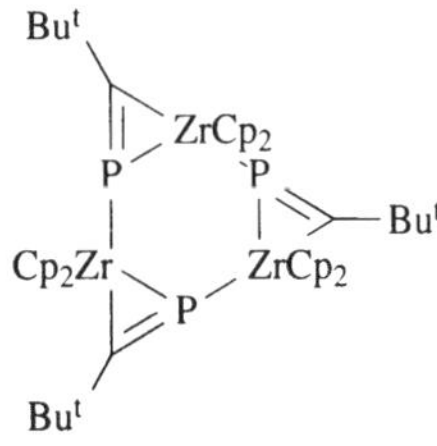

(**24**)

The preparation of phosphaalkynezirconocene complexes proceeds via ligand exchange reactions starting from zirconocene(II) precursors. As shown in Scheme 7, even the strongly coordinating diphenylacetylene is replaced by *t*-butylphosphaalkyne, which indicates that the latter forms an even more stable π-complex.[64] Abstraction of trimethylphosphine by means of triethylborane as a scavenger yields the trimeric phosphaalkynezirconocene (**24**), which is remarkably similar to the formaldehyde complex (**15**). The use of other zirconocene or hafnocene sources led to an intramolecular crossed [2 + 2]-cycloaddition of the phosphaalkyne, yielding a σ-bound 1,3-diphosphabicyclo[1.1.0]butane instead of (**24**) (Scheme 8).[65] Apart from *t*-butylphosphaalkyne, [Zr(1-butene)(PMe₃)(η-Cp)₂] has been reacted with other phosphaalkynes yielding (**25**) and (**26**).[66]

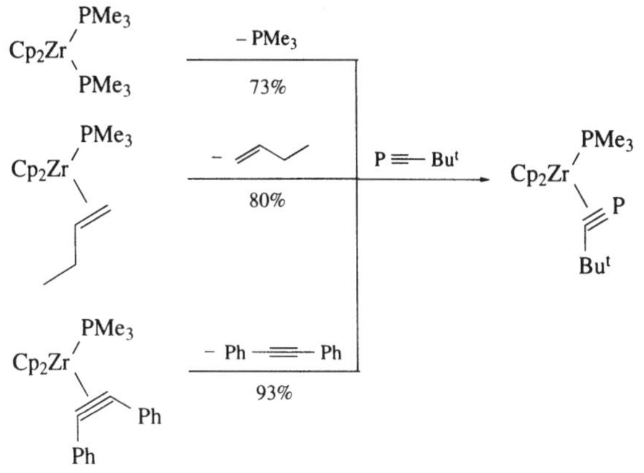

Scheme 7

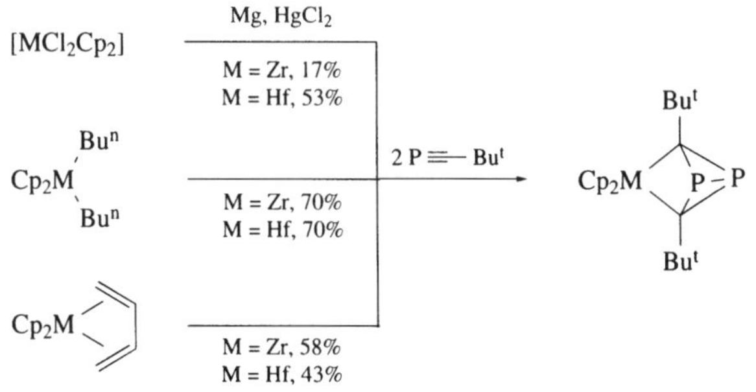

Scheme 8

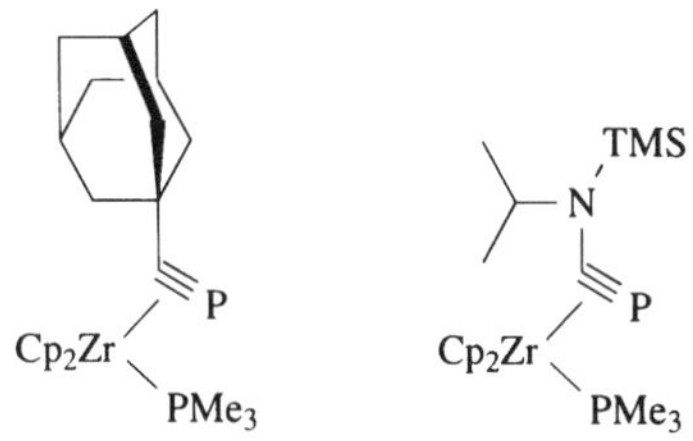

(**25**)   (**26**)

7.2 STRUCTURAL DATA

7.2.1 X-ray Crystallographic Data

Zirconocene and hafnocene complexes with monodentate neutral ligands possess a typical pseudotetrahedral coordination sphere. The angle between the two centroids of the cyclopentadienyl ligands and the metal ranges from 136.6° for [Zr(PMe₃)₂(η-Cp)₂][15] to 148.2° for [Hf(CO)₂(η-Cp*)₂].[67] By contrast, the range for the ligand–metal–ligand angle is smaller, displaying a minimum value of 86.3° for [Zr(CO)₂(η-Cp*)₂][67] and a maximum value of 93.9° for [Zr(PMe₃)₂(η-Cp)₂].[15] Table 2[68–70] summarizes further selected data from crystallographic studies of such complexes.

Numerous crystallographic investigations of zirconocene complexes containing η²-coordinated unsaturated molecules and donors have revealed that, apart from their pseudotetrahedral geometry,

Table 2 Selected bond distances and angles of zirconocene and hafnocene complexes [$ML^1L^2Cp_2$] (L^1, L^2 = monodentate neutral ligands).

Compound	*M–CO* (nm)	*M–PR₃* (nm)	*Cp–M–Cp* (°)	*L¹–M–L²* (°)	*Ref.*
[Zr(CO)₂(η-Cp)₂]	0.218 7(4)		142.2	89.2(2)	68
[Hf(CO)₂(η-Cp)₂]	0.216(2)		141.0	89.3(9)	69
[Zr(CO)₂(η-Cp*)₂]	0.214 5(9)		147.4	86.3(5)	67
[Hf(CO)₂(η-Cp*)₂]	0.214(2)		148.2	87.0	67
[Zr(CO)₂(η-C₉H₇)₂]	0.2204		145.0	86.8(2)	6
[Zr(CO){P(OMe)₃}(η-Cp)₂]	0.216	0.2625	140.6	91.2	70
[Zr(PMe₃)₂(η-Cp)₂]		0.264	136.6	93.9	15

generally the two coordinated atoms of the unsaturated ligand lie in one plane with zirconium and the donor atom. In most cases this plane constitutes the mirror plane of the two cyclopentadienyl rings. The quite strained 1,2,3-cyclohexatriene complex (**3**) (Section 7.1.4) is an exception to this generalization, as the deviation, expressed by the angles between the cyclopentadienyl planes and the previously defined plane, assumes values of 30.5° and 19.4° for each cyclopentadienyl ligand. Moreover, the bond lengths as well as the bond angles of the coordinated unsaturated molecule indicate significant backbonding of the metallocene into the ligand π* orbital. Thus, the carbon–carbon bond lengths for alkenezirconocenes have been found to be in the range of 0.138 nm (for [Zr{(*E*)-PhCH=CHPh}(PMe₃)(η-Cp)₂][71]) to 0.147 nm (for [Zr(1-butene)(PMe₃)(η-Cp)₂]).[72] For alkyne complexes, values between 0.128 nm (for [Zr(1-hexyne)(PMe₃)(η-Cp)₂][22a]) and 0.136 nm (for [Zr(PhCCPh)(PMe₃)(η-Cp)₂][73]) indicate that the situation is best described as intermediate between a carbon–carbon triple and double bond. For this reason, alkene and alkyne complexes (including heteroalkenes or -alkynes) are frequently considered as metallacyclopropanes and -cyclopropenes. Typical examples of an alkene and an alkyne complex are shown in Figures 1 and 2 exhibiting the previously discussed features. Moreover, in [Zr(1-hexyne)(PMe₃)(η-Cp)₂] the C-6–C-7–C-8 bond angle of 135.8(3)° is more typical of a zirconacyclopropene than of an alkyne.

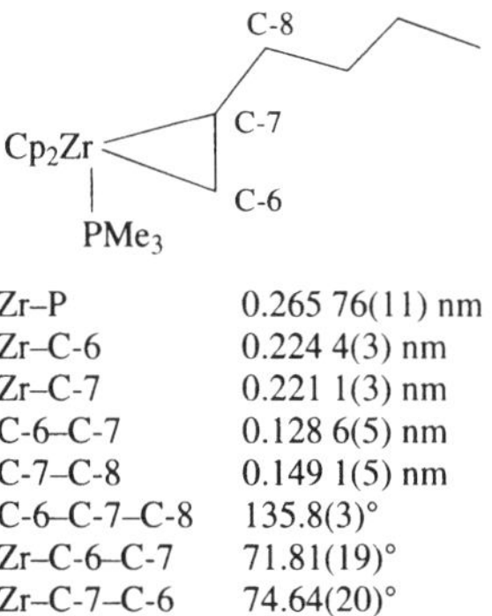

Zr–P	0.265 76(11) nm
Zr–C-6	0.224 4(3) nm
Zr–C-7	0.221 1(3) nm
C-6–C-7	0.128 6(5) nm
C-7–C-8	0.149 1(5) nm
C-6–C-7–C-8	135.8(3)°
Zr–C-6–C-7	71.81(19)°
Zr–C-7–C-6	74.64(20)°

Figure 1 Typical bond angles and lengths for an alkenezirconocene (data from *J. Am. Chem. Soc.*, 1987, **109**, 2544).

The structure of the pyridine-ligated diazobenzenezirconocene complex [Zr(PhNNPh)(py)(η-Cp)₂][74] reveals that both nitrogens are pyramidal, displaying a torsional C_{ipso}–N–N–C_{ipso} angle of 91.32(0.46)°, and that the nitrogen–nitrogen bond of 0.143 4(4) nm is comparable with a single bond (Figure 3).

Butadienehafnocene and -zirconocene complexes also possess a pseudotetrahedral geometry. In *s-cis*-butadiene complexes, all four coordinated carbon atoms lie in one plane; the two adjacent atoms to zirconium are ~0.025–0.035 nm closer to the metal than the internal two carbon atoms. These complexes have also been considered as zirconacyclopentenes with an opened envelope conformation and the double bond coordinated to zirconium. The carbon–carbon bond lengths in the butadiene framework are opposite to those found in the free ligand (Figure 4).[52] The reverse is true for the *s-trans* complex [Zr-(η-1,4-diphenyl-1,3-butadiene)(η-Cp)₂] depicted in Figure 5.[75] Another difference to the *s-cis* isomers appears considering the nonplanarity of the coordinated butadiene ligand, which may be expressed by a torsion angle C-1–C-2–C-3–C-4 of 126.3°.

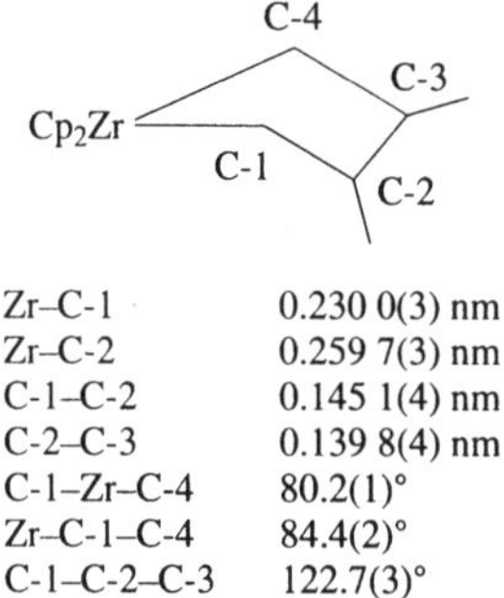

Zr–P	0.267 9(4) nm
Zr–C-1	0.235(1) nm
Zr–C-2	0.235(2) nm
C-1–C-2	0.146(2) nm
C-2–C-16	0.156(2) nm
C-1–C-2–C-16	118.1(1)°
Zr–C-1–C-2	71.8(8)°
Zr–C-2–C-1	72.0(8)°

Figure 2 Typical bond angles and lengths for an alkynezirconocene (data from *Chem. Ber.*, 1989, **122**, 1035).

Zr–N-1	0.216 1(3) nm
Zr–N-2	0.210 5(4) nm
Zr–N-3	0.243 1(4) nm
N-1–N-2	0.143 4(4) nm
N-1–C-11	0.139 6(11) nm
N-2–C-17	0.136 8(6) nm
N-1–N-2–C-17	116.2(4)°
N-2–N-1–C-11	114.6(4)°
Zr–N-1–N-2	68.3(2)°
Zr–N-2–N-1	72.5(2)°

Figure 3 Geometry of the complex [Zr(PhNNPh)(py)(η-Cp)$_2$] (data from *J. Am. Chem. Soc.*, 1990, **112**, 894).

Zr–C-1	0.230 0(3) nm
Zr–C-2	0.259 7(3) nm
C-1–C-2	0.145 1(4) nm
C-2–C-3	0.139 8(4) nm
C-1–Zr–C-4	80.2(1)°
Zr–C-1–C-4	84.4(2)°
C-1–C-2–C-3	122.7(3)°

Figure 4 Pseudotetrahedral geometry of a butadienezirconocene complex (data from *Adv. Organomet. Chem.*, 1985, **24**, 1).

7.2.2 Spectroscopic Properties

Apart from generally laborious crystallographic studies, spectroscopic methods serve as a very convenient tool to characterize the previously discussed metallocene(II) compounds. For the characterization of hafnocene and zirconocene complexes containing at least one carbon monoxide ligand, infrared spectroscopy has frequently been applied. Comparing [M(CO)$_2$(η-Cp)$_2$] and [M(CO)$_2$(η-Cp*)$_2$] (M = Zr or Hf), the carbonyl stretching frequencies for decamethylmetallocene compounds are lower in energy relative to the cyclopentadienyl analogues by about 29–41 cm^{-1}, thus indicating an

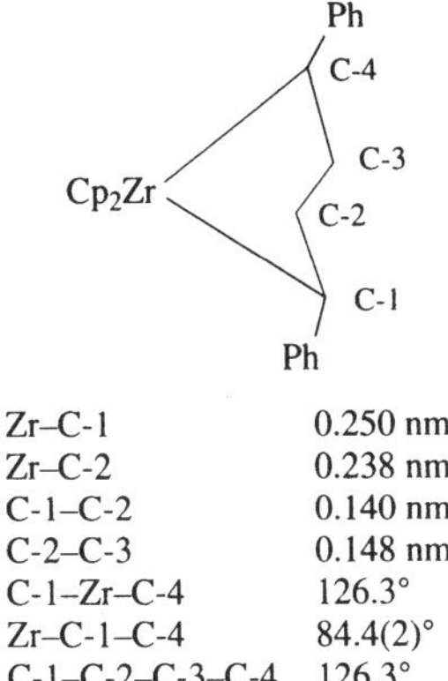

Zr–C-1	0.250 nm
Zr–C-2	0.238 nm
C-1–C-2	0.140 nm
C-2–C-3	0.148 nm
C-1–Zr–C-4	126.3°
Zr–C-1–C-4	84.4(2)°
C-1–C-2–C-3–C-4	126.3°

Figure 5 Geometry of the *s-trans* complex [Zr(η-1,4-diphenyl-1,3-butadiene)(η-Cp)₂] (data from *J. Chem. Soc., Chem. Commun.*, 1982, 191).

enhanced π backbonding of the electron-rich metal centre to the π^* orbitals of CO in the permethylated metallocene. The metal carbonyl bands of such complexes are compiled in Table 3.[10,76,77]

Table 3 Carbonyl stretching frequencies of zirconocene and hafnocene complexes.

Compound	$\nu(CO)$ (cm^{-1})	Ref.
[Zr(CO)₂(η-Cp)₂]	1975, 1885;[a] 1967, 1872[b]	76
[Hf(CO)₂(η-Cp)₂]	1969, 1878;[a] 1960, 1861[b]	76
[Zr(CO)₂(η-Cp*)₂]	1945, 1852[a]	76
[Hf(CO)₂(η-Cp*)₂]	1940, 1844[a]	76
[Zr(CO)₂(η-Cp)(η-Cp*)]	1965, 1875[c]	3
[Hf(CO)₂(η-Cp)(η-Cp*)]	1962, 1868[c]	3
[Zr(CO)₂(η-C₉H₇)₂]	1985, 1899[c]	6
[Hf(CO)₂(η-C₉H₇)₂]	1979, 1892[c]	6
[Zr(CO)₂(η-C₅H₄TMS)₂]	1970, 1880[d]	10
[Zr(CO)₂{η-C₅H₃(TMS)₂}₂]	1962, 1875[d]	10
[Hf(CO)₂{η-C₅H₃(TMS)₂}₂]	1950, 1855[d]	10
[Zr(CO)(PF₃)(η-Cp)₂]	1932,[a] 1921[b]	3
[Zr(CO){PF₂N(Me)PF₂}(η-Cp)₂]	1907,[b] 1925[c]	3
[Zr(CO)(PPh₃)(η-Cp)₂]	1842[b]	77
[Zr(CO)(PMePh₂)(η-Cp)₂]	1840[e]	14
[Zr(CO)(PMe₃)(η-Cp)₂]	1852,[a] 1836[b]	17
[Zr(CO){P(OMe)₃}(η-Cp)₂]	1849[f]	70
[Hf(CO)(PMe₃)(η-Cp)₂]	1824[b]	70
[Hf(CO)(PPh₃)(η-Cp)₂]	1830[b]	70
[Hf(CO)(PF₃)(η-Cp)₂]	1910	70
[Zr(CO)(PMe₃){η-C₅H₃(TMS)₂}₂]	1842[a]	70

[a] Hexane. [b] THF. [c] Pentane. [d] Nujol. [e] Benzene. [f] KBr.

NMR spectroscopy is the method of choice to effectively characterize zirconocene(II) and hafnocene(II) complexes. Nevertheless, the ^{1}H NMR spectra in most cases are rather complex, although the number of signals observed for the cyclopentadienyl ligands (between δ 4.5 and δ 5.5) frequently indicates the number of isomers (e.g., rotamers or *s-cis–s-trans* isomeric 1,3-butadiene complexes) and whether the ligand–metal plane is identical with the mirror plane of the cyclopentadienyl ligands. However, well-resolved ^{1}H NMR spectra have provided H–H coupling constants which are particularly sensitive to changes in hybridization as a result of complexation. Therefore this method has in some instances been valuable to establish the metallacyclopropane character of alkene complexes. In addition, phosphine-stabilized zirconocene alkene complexes display characteristic P–H coupling constants for the coordinated alkene. Thus, the P–H coupling constants (as well as the P–C coupling constants) of the proximal nuclei to the phosphine ligand have larger values than those for the distal nuclei, which has been confirmed by nuclear Overhauser effect spectroscopy (NOESY) and C–H correlation experiments.[22c] These methods, in combination with H–H correlated spectroscopy, are most important for the characterization of static rotamer mixtures. Moreover, significant high-field shifts of these protons relative to the free ligand have been observed for α-alkene complexes even beyond the Me_4Si standard up to δ −3 (for $[Zr(1\text{-butene})(PMe_3)(\eta\text{-}C_9H_7)_2]$[23]).

The ^{13}C NMR spectra provide the most important data for these complexes; shift values for cyclopentadienyl carbons are observed around δ 100. Apart from the already mentioned aspects concerning the cyclopentadienyl substructure, ^{13}C shifts and C–H coupling constants of the coordinated ligands are of particular interest, since they reveal details of the bonding situation when compared with the free ligands. The carbon atoms of alkene complexes coordinated to the metal appear at approximately 80–100 ppm upfield relative to the shifts of the free ligands, indicates the relevance of the metallacyclopropane structure. This fact is also reflected in the reduced $^1J(C\text{–}H)$ values (by ~10–15 Hz) in such complexes. In contrast, carbon atoms of coordinated alkynes display chemical shift values of ~90 ppm downfield from the free ligand, so that again the metallacyclopropene structure has been inferred.

Phosphorus NMR spectroscopy provides an additional tool to characterize the considerable number of phosphine-stabilized (mainly trimethylphosphine) metallocene(II) complexes. Typical values for the ^{31}P NMR spectral chemical shifts (relative to H_3PO_4) of trimethylphosphine-stabilized complexes are at δ ~0 ± 10. The expected downfield shift of trimethylphosphine is thus approximately 60 ppm, although even bigger values are known; for $[Zr(PMe_3)_2(\eta\text{-}Cp)_2]$ a chemical shift of δ ca. +24 is rather exceptional.[15]

Moreover, NMR spectroscopic techniques have proved to be a powerful means to investigate the dynamic features of metallocene(II) complexes. 1,3-Butadienezirconocene and -hafnocene complexes have been extensively investigated regarding *s-cis–s-trans* isomerization, autoisomerization and rotation of the cyclopentadienyl ligands in order to determine the corresponding activation barriers.[49] Stabilized monoalkene complexes also display various other dynamic peculiarities. In some cases the stabilizing donor undergoes exchange reactions, for example, in $[Zr(PhNNPh)(L)(\eta\text{-}Cp)_2]$[12] (Equation (2)). In the latter complex, rotation of the phenyl groups is hindered, as indicated by low-temperature NMR spectroscopy. The same has been found for $[Zr(\eta\text{-}1,2\text{-diphenylcyclopropene})(PMe_3)(\eta\text{-}Cp)_2]$,[78] concomitant with a hindered rotation of the trimethylphosphine ligand around the zirconium–phosphorus bond. Two-dimensional ^{13}C exchange spectra have shown that even at 300 K only a slow exchange of the two rotamers in $[Zr(alkene)(PMe_3)(\eta\text{-}Cp)_2]$ (alkene = styrene or 1-butene) occurs.[22c]

7.3 REACTIONS

7.3.1 Protonolysis and Halogenation

Although protonolysis of zirconocene(II) complexes has no synthetic relevance, it has been used several times either to provide chemical evidence for the existence of unstable alkene or alkyne complexes, or in the context of regio- or stereochemical reactions. The decomposition of dibutylzirconocene in the presence of (*E*)-stilbene yields the unstable zirconocene(II) complex $[Zr\{\eta\text{-}(E)\text{-stilbene}\}(\eta\text{-}Cp)_2]$, which has been identified by protonolysis with 3 N hydrochloric acid to yield a mixture of 1,2-diphenylethane (79%) and (*E*)-stilbene (7%).[19a] Isoprenezirconocene reacts regioselectively with D_2O at the disubstituted double bond yielding the 1,2-dideuterated hydrocarbon.[79]

Alkene and alkyne complexes readily react with alcohols to yield the corresponding alkoxy(alkyl)-[80] or alkoxy(vinyl)zirconocene (Scheme 9).[81] The stoichiometric reaction of a polyfunctional alcohol, phenol or carboxylic acid such as catechol, glycerine, phloroglucinol or citric acid with $[Zr(C_2H_4)(PMe_3)(\eta\text{-}Cp)_2]$ has been investigated, leading to the corresponding di-, tri- and tetranuclear ethylzirconocene complexes.[82] An excess of the polyfunctional protic substrate results in the concomitant formation of five-membered dioxazirconacycles liberating ethane.

Zirconocene(II) complexes have been reacted with CH acidic compounds. Whereas *in situ* generated $[Zr(1\text{-butene})(\eta\text{-}Cp)_2]$ reacts with trimethylsilylacetylene to form the dimeric alkynylzirconocene (**27**)[83] in only 1.1% yield, $[Zr(PhNNPh)(THF)(\eta\text{-}Cp)_2]$ reacts cleanly with phenylacetylene or acetophenone (Equation (17)).[12]

Various element monohalides have been investigated regarding their reactivity towards alkyne complexes. Starting from a trimethylphosphine-stabilized enynezirconocene, the regio- and stereochemical outcome has been found to be governed by the properties of the chosen halide. Boron or tin halides selectively add in a *trans*-fashion to the least-hindered carbon, whereas diphenylchlorophosphine yields exclusively the *cis*-adduct as a mixture of the two possible regioisomers (Scheme 10).[31] The amount of the major product which bears phosphorus at the least-hindered site depends on the bulk of the substituent adjacent to the alkyne carbon, thus leading to a 50:1 mixture of

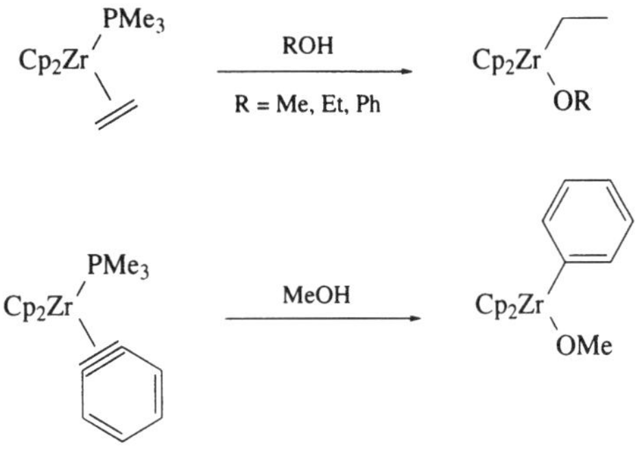

Scheme 9

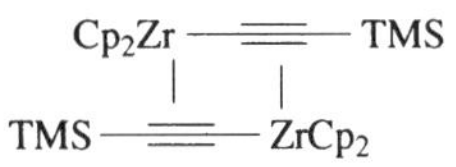

(27)

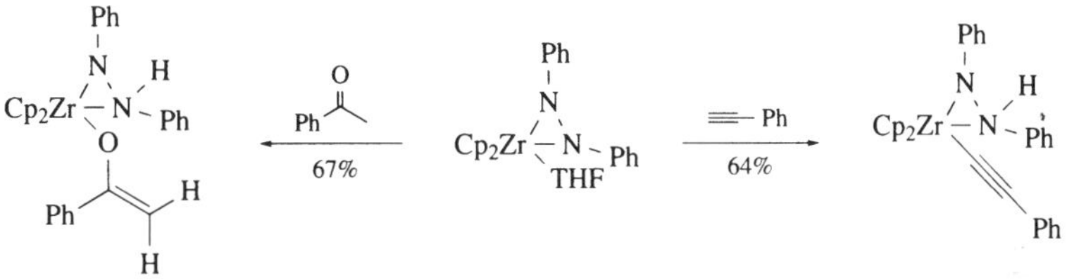

(17)

regioisomers when the *t*-butylacetylenezirconocene complex is used.[84] Further transmetallation with a suitable halide yields 1,2-functionalized alkenes or 1,3-dienes depending on the kind of complex used.

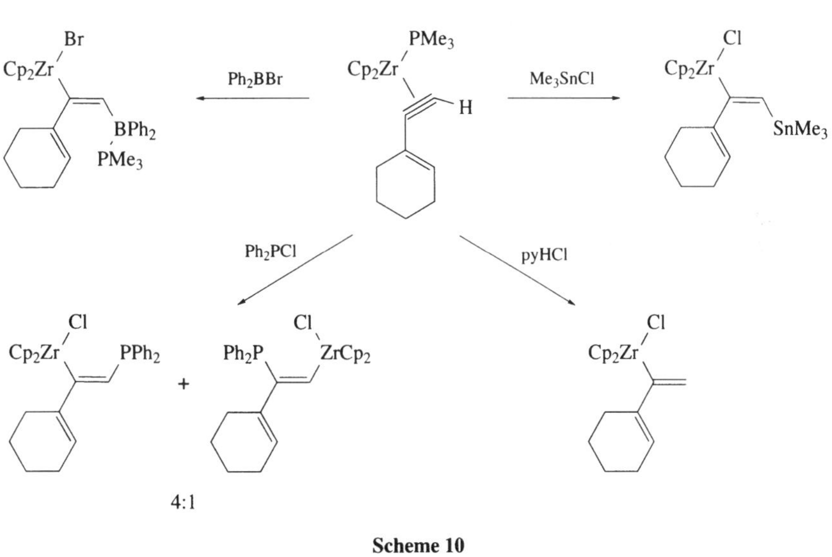

Scheme 10

7.3.2 Oxidative Addition

7.3.2.1 Intramolecular reactions

Oxidative addition reactions between metallocene(II) complexes and unsaturated organic compounds have become very important in organic synthesis. The driving force is the change from the +2 to the most favoured +4 oxidation state, which is especially true for hafnocene compounds and has been discussed in the course of the preceding sections. Intramolecular oxidative addition reactions require additional reactive sites within the metallocene(II) compound. [Zr(1,2-diphenylcyclopropene)(PMe$_3$)-

(η-Cp)$_2$] upon warming to 60 °C for 14 d undergoes a rearrangement reaction concomitant with an intramolecular oxidative addition to yield the corresponding zirconacyclobutene.[78] Similarly, the thermally induced rearrangement of [Zr(benzylidenecyclopropane)(PMe$_3$)(η-Cp)$_2$] leads to a phosphine-free exomethylene-substituted zirconacyclobutane (Scheme 11).[23]

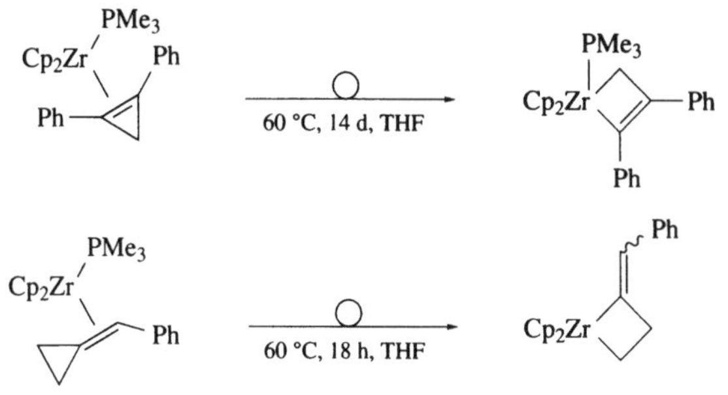

Scheme 11

The zirconocene-mediated cyclization of α,ω-diynes, -dienes, -enynes or heteroatom-substituted congeners has frequently been investigated with a view to its use for organic synthesis (Scheme 12).[22b,85–9] Hence, most of these metallacycles have not been isolated. Instead their characterization was achieved by means of NMR spectroscopy, or the properties of the organic products obtained via chemical reactions (hydrolysis, halogenation or carbonylation). In the case of enyne cyclization, a competitive situation is established since the triple bond should react faster than the double bond. Moreover, the oxidative coupling, which is frequently referred to as a reductive process considering the changes in the organic part, is reversible and the alkene may be displaced by a more reactive species such as a nitrile (Equation (18)).[22b] Due to the remarkable frequently observed regio- and stereoselectivity of such cyclizations, this method has even been used as a key step in the synthesis of natural products, for example, dendrobine (Equation (19)).[90]

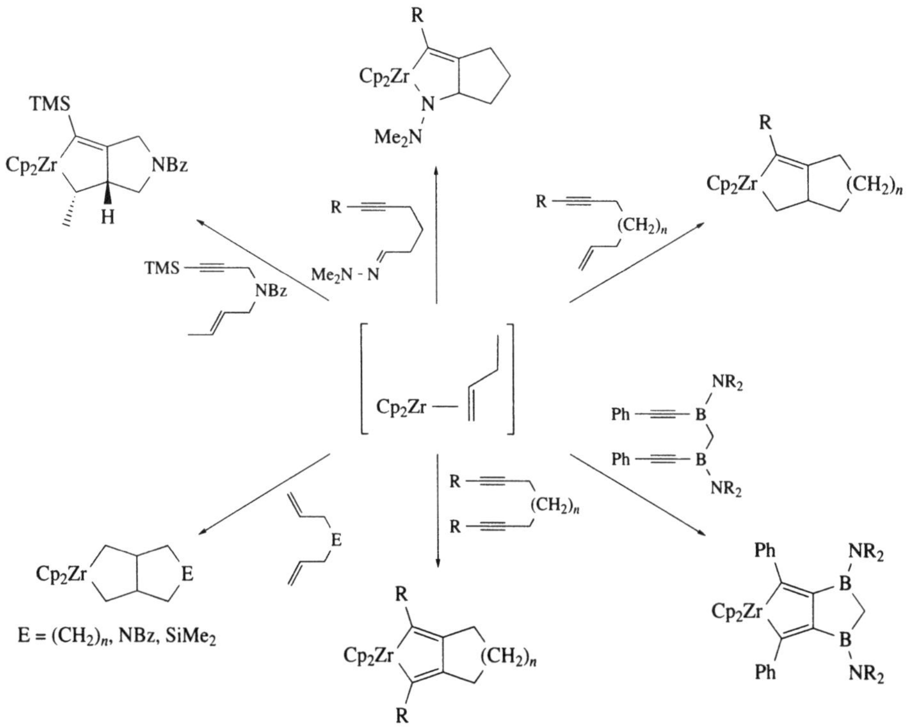

Scheme 12

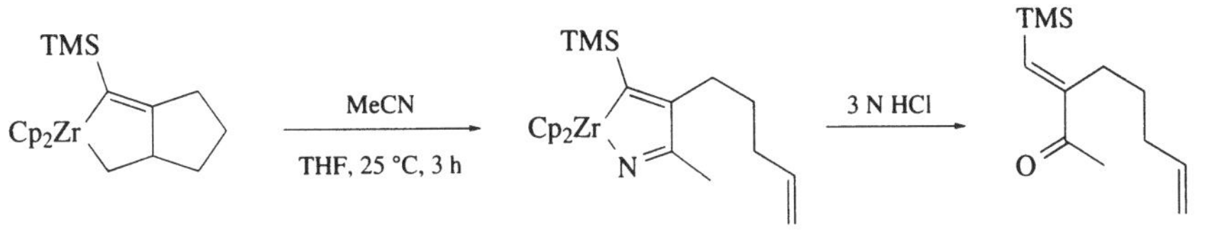

$$(18)$$

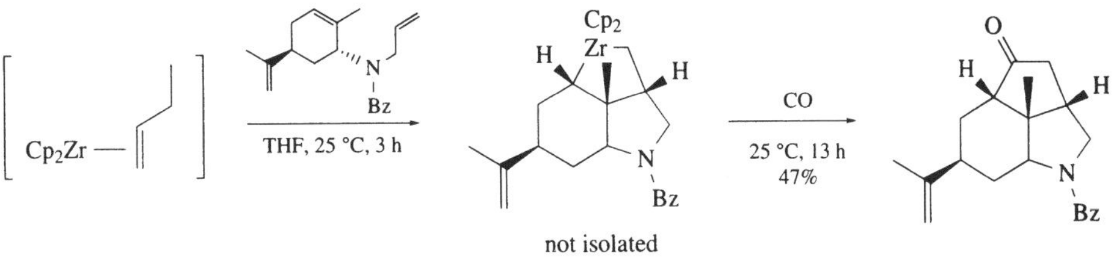

$$(19)$$

not isolated

7.3.2.2 Intermolecular reactions

The vast majority of reactions of zirconocene(II) and hafnocene(II) complexes are dominated by an intermolecular oxidative addition, especially those that are used as a trap for *in situ* generated metallocene(II) complexes. Such intermolecular reactions may be divided into two general classes, one concerning the formation of mononuclear metallocene(IV) compounds and the other leading to di- or even polynuclear complexes containing two or more other metal atoms.

Oxidative coupling reactions of zirconocene heteroalkene and heteroalkyne complexes have been extensively reviewed.[32,40,91] As a representative example, the reactivity of the imine complex (17) is depicted in Scheme 13.[60] The more electronegative atom is invariably bound to the metal, so that phosphaalkyne complexes form metallacycles with the more electronegative carbon adjacent to the metal (Scheme 14).[64] The stereo- and regiochemistry of the zirconocene-mediated coupling of an α-alkene have been found to be very sensitive to the substitution pattern of the alkene. Thus, the coupling reaction of 1-butene and styrene is even pair-selective, yielding thermally labile metallacyclopentanes in a regio- and stereoselective fashion (Scheme 15).[92] In addition, this coupling reaction has been applied to chiral tetrahydroindenylzirconocene complexes of the Brintzinger type for the enantioselective synthesis of allylic amines (Equation (20)).[93]

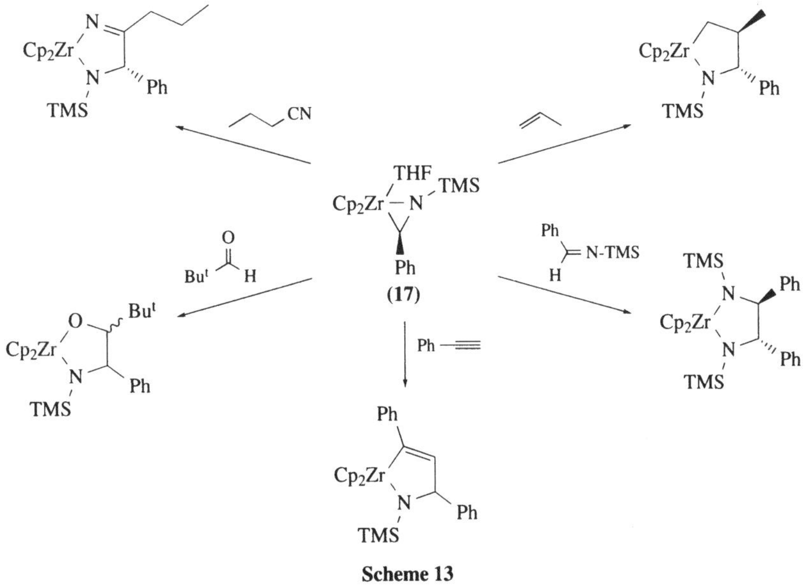

Scheme 13

Apart from these stoichiometric reactions, catalytic coupling reactions of alkenes have also been developed, with an intermediate ethenezirconocene playing a key role in the catalytic cycle (Scheme 16).[88b,94] This method has also been extended to chiral zirconocene complexes for asymmetric carbomagnesiation reactions.[95]

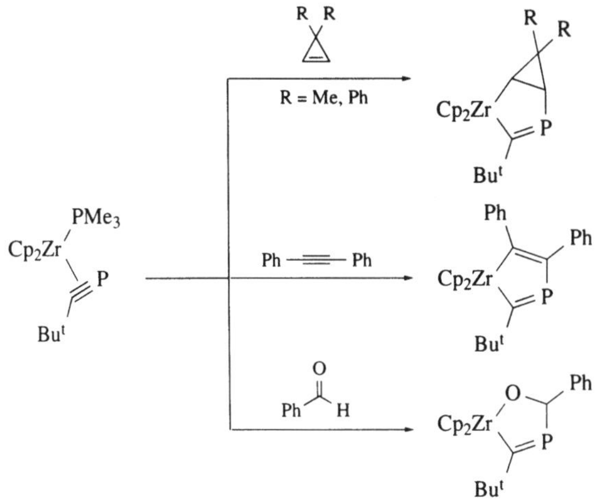

Scheme 14

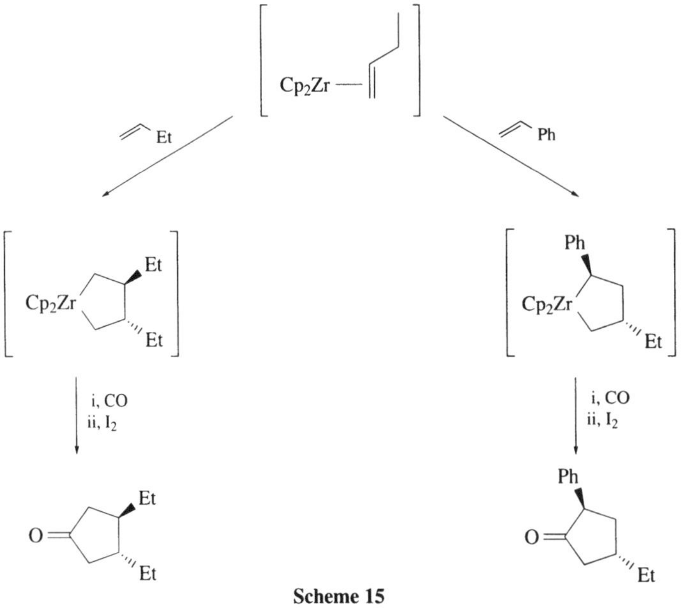

Scheme 15

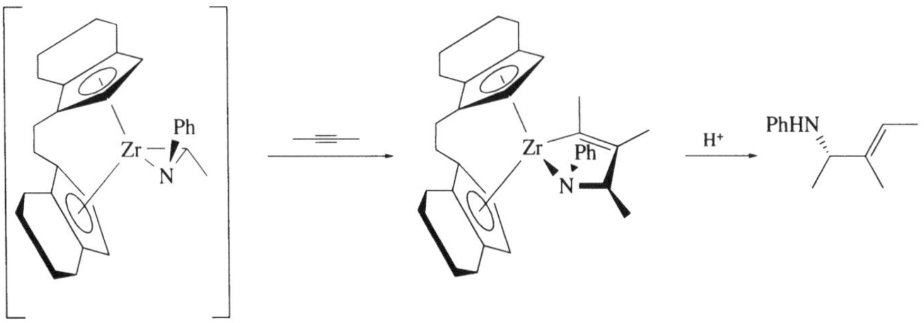

(20)

Ylides have been reacted with metallocene(II) alkene complexes to yield four-membered metallacycles and alkenyl complexes (Scheme 17).[96] Another interesting route to four-membered metallacycles has been found by inserting oxygen into a zirconocene alkyne complex. Depending on the kind of complex used, either dimeric[97] or monomeric[98] metallacycles are formed (Scheme 18). Concerning the monomeric diphenyloxametallacyclobutene complex, it is noteworthy that even by exposing solid [Zr(PhCCPh)(η-Cp*)₂] to an atmosphere of nitrous oxide without any solvent at ambient temperature, the metallacyclic product is formed quantitatively as an analytically pure substance. Insertion of the isoelectronic *p*-tolyl azide and diphenyldiazomethane has yielded the corresponding diphenylazacyclobutenes upon reaction with [Zr(PhCCPh)(η-Cp*)₂] (Equation (21)).[99]

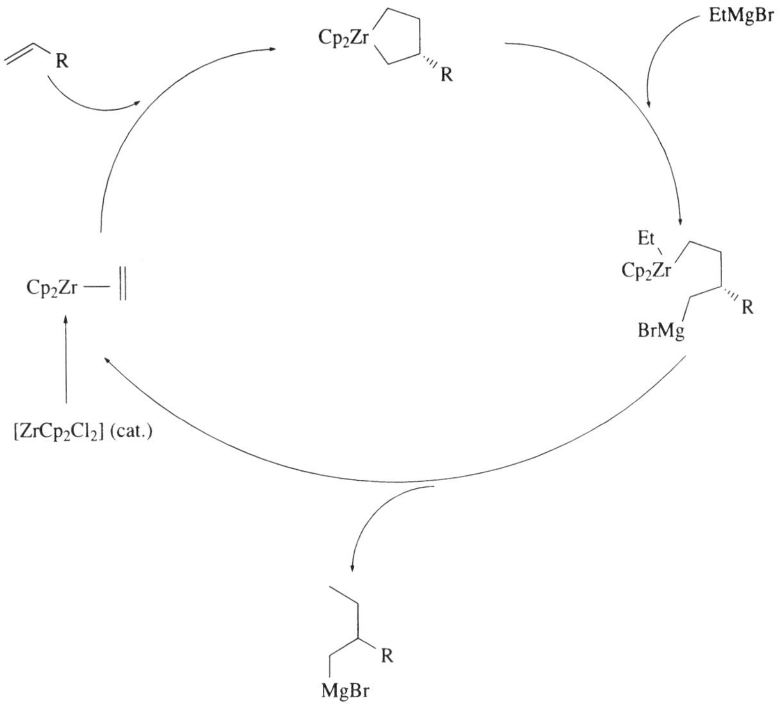

Scheme 16

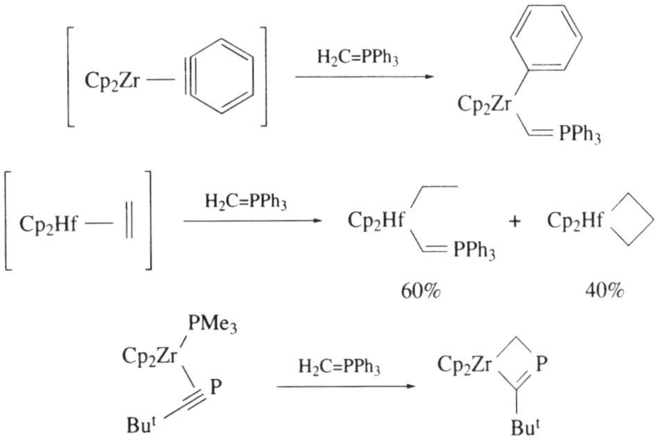

Scheme 17

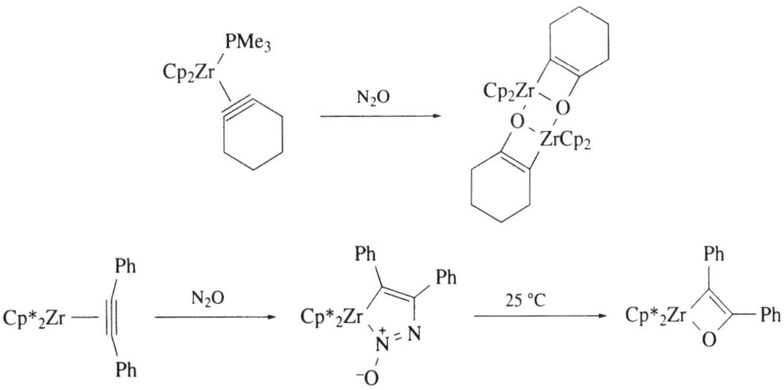

Scheme 18

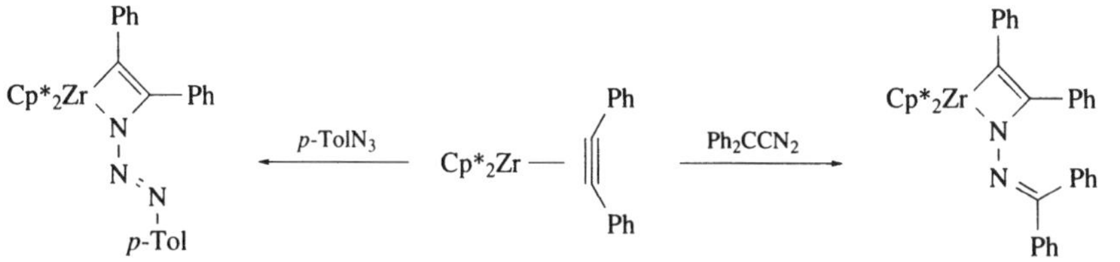

$$(21)$$

Oxidative coupling reactions of (1,3-butadiene)metallocene compounds involving alkenes, alkynes, carbonyl compounds or nitriles have been extensively investigated and reviewed.[44,45] In comparison with the previously discussed reactions of monoalkene and -alkyne complexes, the reactivity of the butadiene system is far more complex, mainly because of the different reactivities of *s-cis-* and *s-trans-*isomers, a situation which is extremely dependent on the kind of metal, the substitution pattern of the butadiene ligand, and of the cyclopentadienyl moiety.

The insertion of an alkyne into a butadienemetallocene often occurs regioselectively, as shown by the reaction of 2-butyne with isoprenezirconocene (Equation (22)).[100] Ketones or aldehydes readily insert into butadienemetallocenes to form seven-membered metallacycles. Mechanistic studies have revealed that the *s-trans-*isomer reacts faster than the corresponding *s-cis-*isomer.[101] Further studies using isoprenezirconocene have led to the conclusion that 3,3-dimethyl-2-butanone predominantly inserts into the more-substituted double bond if reacted at 60 °C, whereas the photochemically induced reaction at −78 °C occurs into the less-substituted part of the coordinated diene (Scheme 19).[102] Similarly, nitriles[103] or esters[104] have been reacted with isoprenezirconocene yielding the corresponding seven-membered metallacycles upon insertion into the more substituted double bond. The reaction of an excess of benzonitrile with butadienezirconocene has recently been used for the synthesis of stable, conjugated primary enamines as a result of double insertion and subsequent hydrolysis (Equation (23)).[105]

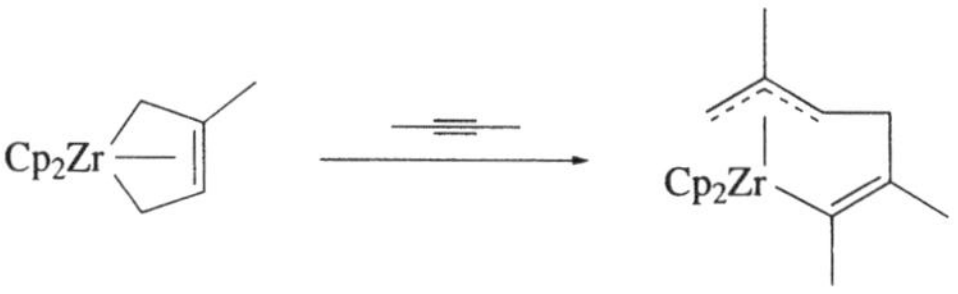

$$(22)$$

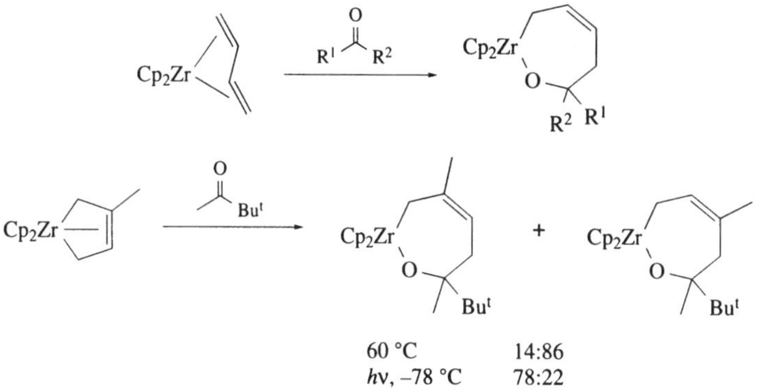

$$
\begin{array}{ccc}
& 60\ °C & 14{:}86 \\
& h\nu,\ -78\ °C & 78{:}22
\end{array}
$$

Scheme 19

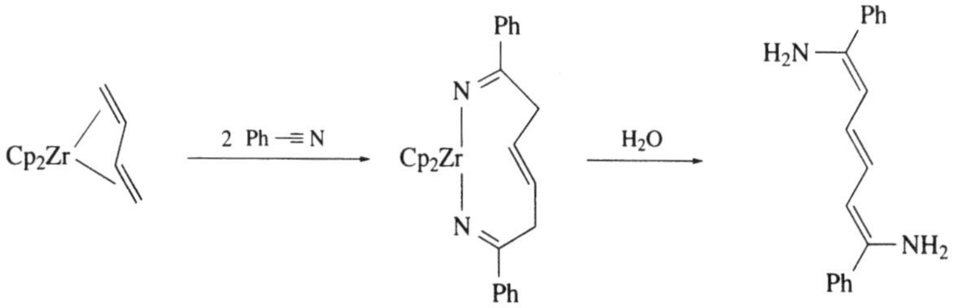

$$(23)$$

Oxidative addition reactions of zirconocene alkyne complexes with main group metal organyls have yielded bimetallic bicyclic zirconocene complexes containing a planar tetracoordinate carbon atom, as shown by means of crystallographic studies (Scheme 20).[106] When butadienezirconocene is treated with

1 equiv. of aluminum trichloride, a bimetallic complex is formed, exhibiting a very unusual coordination of the butadiene ligand (Equation (24)).[107]

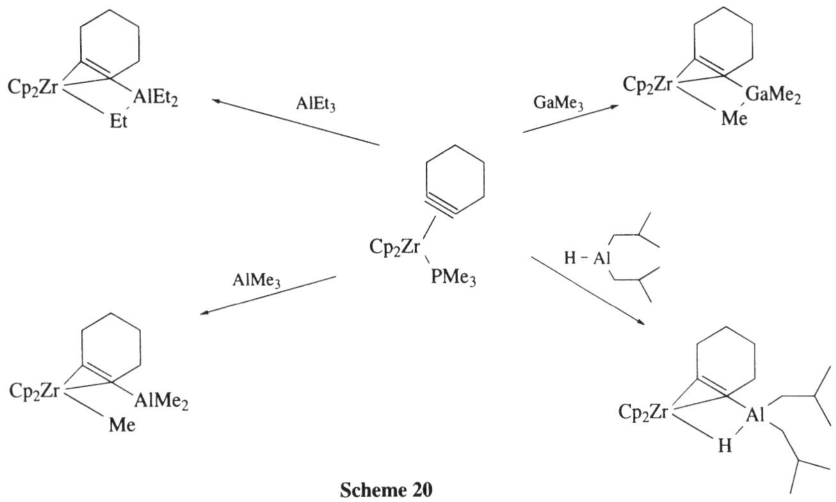

Scheme 20

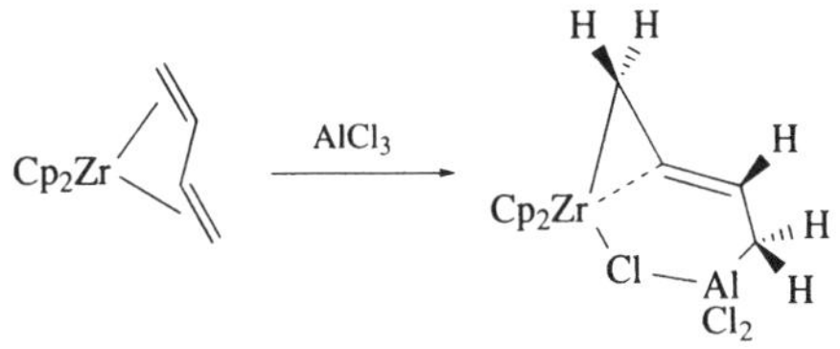

$$(24)$$

Apart from the previously mentioned decomposition of trimethylphosphine-stabilized zirconocene (Section 7.1.3), which has yielded the dinuclear zirconium complex (**1**), a fulvalene-bridged product is obtained when zirconocene dichloride is reacted with $[Zr(PMe_3)_2(\eta\text{-}Cp)_2]$ (Equation (25)).[108]

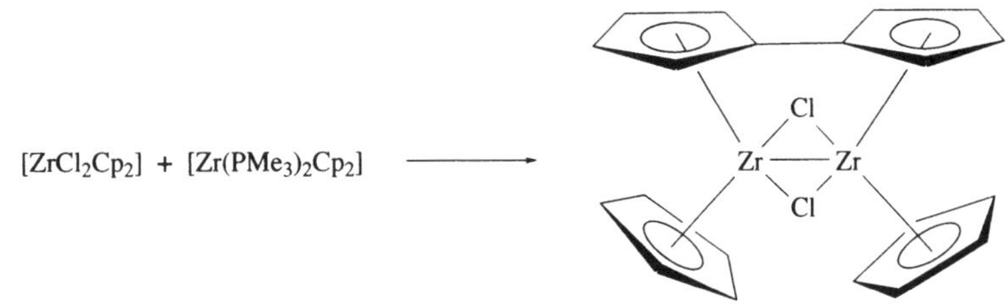

$$[ZrCl_2Cp_2] + [Zr(PMe_3)_2Cp_2] \longrightarrow \qquad (25)$$

The formation of carbene complexes by the reaction of a metal carbonyl with a metallocene butadiene or alkyne complex has been extensively reviewed.[109] Concerning butadiene complexes, it is noted that only the *s-trans*-isomers are reactive (Scheme 21).

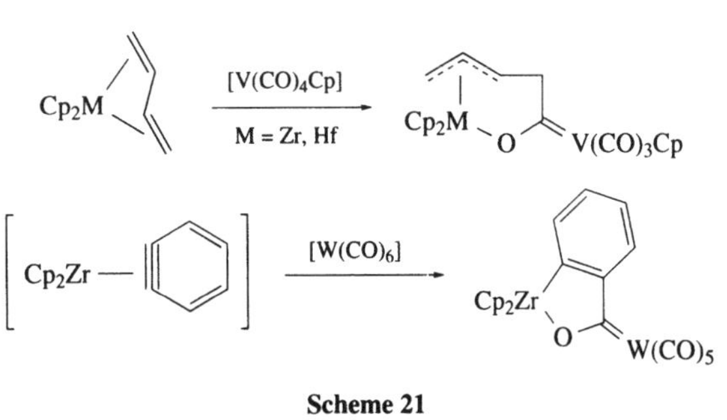

Scheme 21

7.4 REFERENCES

1. G. P. Pez and J. N. Armor, *Adv. Organomet. Chem.*, 1981, **19**, 1.
2. K. I. Gell, T. V. Harris and J. Schwartz, *Inorg. Chem.*, 1981, **20**, 481.
3. D. J. Sikora, D. W. Macomber and M. D. Rausch, *Adv. Organomet. Chem.*, 1986, **25**, 318.
4. D. J. Cardin, M. F. Lappert, C. L. Raston and P. I. Riley, in 'COMC-I', vol. 1, p. 549.
5. D. J. Sikora, K. J. Moriarty and M. D. Rausch, *Inorg. Synth.*, 1990, **28**, 249.
6. M. D. Rausch, K. J. Moriarty, J. L. Atwood, W. E. Hunter and E. Samuel, *J. Organomet. Chem.*, 1987, **327**, 39.
7. H.-S. Tung and C. H. Brubaker, *Inorg. Chim. Acta*, 1981, **52**, 197.
8. D. M. Roddick, M. D. Fryzuk, P. F. Seidler, G. L. Hillhouse and J. E. Bercaw, *Organometallics*, 1985, **4**, 97.
9. G. S. Bristow, M. F. Lappert, T. R. Martin, J. L. Atwood and W. F. Hunter, *J. Chem. Soc., Dalton Trans.*, 1984, 399.
10. A. Antiñolo, M. F. Lappert and D. J. W. Winterborn, *J. Organomet. Chem.*, 1984, **272**, C37.
11. D. Sutton, *Chem. Rev.*, 1993, **93**, 995.
12. P. J. Walsh, F. J. Hollander and R. G. Bergman, *J. Organomet. Chem.*, 1992, **428**, 13.
13. M. D. Fryzuk, T. S. Hadad and D. J. Berg, *Coord. Chem. Rev.*, 1990, **99**, 137.
14. K. I. Gell and J. Schwartz, *J. Am. Chem. Soc.*, 1981, **103**, 2687.
15. L. B. Kool, M. D. Rausch, H. G. Alt, M. Herberhold, B. Honold and U. Thewalt, *J. Organomet. Chem.*, 1987, **320**, 37.
16. L. B. Kool, M. D. Rausch, H. G. Alt, M. Herberhold, U. Thewalt and B. Honold, *J. Organomet. Chem.*, 1986, **310**, 27.
17. L. B. Kool, M. D. Rausch, H. G. Alt, M. Herberhold, B. Wolf and U. Thewalt, *J. Organomet. Chem.*, 1985, **297**, 159.
18. J. M. Manriquez *et al.*, *J. Am. Chem. Soc.*, 1978, **100**, 3078.
19. (a) E. Negishi, F. E. Cederbaum and T. Takahashi, *Tetrahedron Lett.*, 1986, 2829; (b) A. H. Hoveyda and J. P. Morken, *J. Org. Chem.*, 1993, **58**, 4237.
20. T. Takahashi, M. Tamura, M. Saburi, Y. Uchida and E. Negishi, *J. Chem. Soc., Chem. Commun.*, 1989, 852.
21. J. W. Lauher and R. Hoffmann, *J. Am. Chem. Soc.*, 1976, **98**, 1729.
22. (a) S. L. Buchwald, B. T. Watson and J. C. Huffman, *J. Am. Chem. Soc.*, 1987, **109**, 2544; (b) E. Negishi *et al.*, *J. Am. Chem. Soc.*, 1989, **111**, 3336; (c) P. Binger, P. Müller, R. Benn, A. Rufinska, B. Gabor and P. Betz, *Chem. Ber.*, 1989, **122**, 1035.
23. S. Podubrin, Ph.D. Thesis, University of Kaiserslautern, 1993.
24. S. L. Buchwald, B. T. Watson and J. C. Huffman, *J. Am. Chem. Soc.*, 1986, **108**, 7441.
25. J. Yin and W. M. Jones, *Organometallics*, 1993, **12**, 2013.
26. J. Yin, K. A. Abboud and W. M. Jones, *J. Am. Chem. Soc.*, 1993, **115**, 3810.
27. J. Yin, K. A. Abboud and W. M. Jones, *J. Am. Chem. Soc.*, 1993, **115**, 8859.
28. S. L. Buchwald, K. A. Kreutzer and R. A. Fisher, *J. Am. Chem. Soc.*, 1990, **112**, 4600.
29. D. R. Swanson and E. Negishi, *Organometallics*, 1991, **10**, 825.
30. S. L. Buchwald and R. A. Fisher, *Chem. Scr.*, 1989, **29**, 417.
31. M. D. Fryzuk, C. Stone, R. E. Alex and R. K. Chadha, *Tetrahedron Lett.*, 1988, **29**, 3915.
32. S. L. Buchwald and R. B. Nielsen, *Chem. Rev.*, 1988, **88**, 1047.
33. S. L. Buchwald, R. T. Lum, R. A. Fisher and W. M. Davis, *J. Am. Chem. Soc.*, 1989, **111**, 9113.
34. S. L. Buchwald, R. T. Lum and J. C. Dewan, *J. Am. Chem. Soc.*, 1986, **108**, 7441.
35. S. L. Buchwald, E. A. Lucas and J. C. Dewan, *J. Am. Chem. Soc.*, 1987, **109**, 4396.
36. T. Takahashi, Y. Nitto, T. Seki, M. Saburi and E. Negishi, *Chem. Lett.*, 1990, 2259.
37. (a) H. G. Alt, H. E. Engelhardt, M. D. Rausch and L. B. Kool, *J. Organomet. Chem.*, 1987, **329**, 61; (b) H. G. Alt, C. E. Denner, U. Thewalt and M. D. Rausch, *J. Organomet. Chem.*, 1988, **356**, C83; (c) P. Binger, F. Langhauser, P. Wedemann, B. Gabor, R. Mynott and C. Krüger, *Chem. Ber.*, 1994, **127**, 39.
38. U. Rosenthal, A. Ohff, M. Michalik, H. Görls, V. V. Burlakov and B. Shur, *Angew. Chem.*, 1993, **105**, 1228.
39. T. Takahashi, M. Kageyama, V. Denisov, R. Hara and E. Negishi, *Tetrahedron Lett.*, 1993, **34**, 687.
40. E. Negishi, *Chem. Scr.*, 1989, **29**, 457.
41. G. M. Diamond, M. L. H. Green, P. Mountford, N. M. Walker and J. A. K. Howard, *J. Chem. Soc., Dalton Trans.*, 1992, 417.
42. C. McDade and J. E. Bercaw, *J. Organomet. Chem.*, 1985, **279**, 281.
43. (a) J. R. Schmidt and D. M. Duggan, *Inorg. Chem.*, 1981, **20**, 318; (b) L. E. Schock and T. J. Marks, *J. Am. Chem. Soc.*, 1988, **110**, 7701.
44. G. Erker, C. Krüger and G. Müller, *Adv. Organomet. Chem.*, 1985, **24**, 1.
45. H. Yasuda and A. Nakamura, *Angew. Chem.*, 1987, **99**, 745.
46. (a) H. Yasuda, Y. Kajihara, K. Mashima, K. Nagasuna, K. Lee and A. Nakamura, *Organometallics*, 1982, **1**, 388; (b) U. Dorf, K. Engel and G. Erker, *Organometallics*, 1983, **2**, 462.
47. G. Erker, J. Wicher, K. Engel and C. Krüger, *Chem. Ber.*, 1982, **115**, 3300.
48. (a) P. Czisch, G. Erker, H.-G. Korth and R. Sustmann, *Organometallics*, 1984, **3**, 945; (b) R. Beckhaus and K.-H. Thiele, *J. Organomet. Chem.*, 1986, **317**, 23.
49. (a) K. Mashima and A. Nakamura, *J. Organomet. Chem.*, 1992, **428**, 49; (b) G. Erker, T. Mühlenbernd, A. Rufinska and R. Benn, *Chem. Ber.*, 1987, **120**, 507.
50. G. Erker, K. Engel, C. Krüger and A.-P. Chiang, *Chem. Ber.*, 1982, **115**, 3311.
51. G. Erker, K. Engel and P. Vogel, *Angew. Chem.*, 1982, **94**, 791.
52. C. Krüger, G. Müller, G. Erker, U. Dorf and K. Engel, *Organometallics*, 1985, **4**, 215.
53. G. Erker and F. Rosenfeldt, *J. Organomet. Chem.*, 1982, **224**, 29.
54. G. Erker, *Acc. Chem. Res.*, 1984, **17**, 103.
55. K. Kropp, V. Skibbe, G. Erker and C. Krüger, *J. Am. Chem. Soc.*, 1983, **105**, 3353.
56. D. A. Straus and R. H. Grubbs, *J. Am. Chem. Soc.*, 1982, **104**, 5499.
57. E. J. Moore, D. A. Straus, J. Armantrout, B. D. Santarsiero, R. H. Grubbs and J. E. Bercaw, *J. Am. Chem. Soc.*, 1983, **105**, 2068.
58. G. S. Bristow, P. B. Hitchcock and M. F. Lappert, *J. Chem. Soc., Chem. Commun.*, 1982, 462.
59. S. L. Buchwald and R. B. Nielsen, *J. Am. Chem. Soc.*, 1988, **110**, 3171.
60. S. L. Buchwald, B. T. Watson, M. W. Wannamaker and J. C. Dewan, *J. Am. Chem. Soc.*, 1989, **111**, 4486.

61. N. Coles, R. J. Whitby and J. Blagg, *Synlett*, 1990, 271.
62. N. Coles, R. J. Whitby and J. Blagg, *Synlett*, 1992, 143.
63. J. M. Davis, R. J. Whitby and A. Jaxa-Chamiec, *Tetrahedron Lett.*, 1992, **33**, 5655.
64. P. Binger *et al.*, *Chem. Ber.*, 1990, **123**, 1617.
65. (a) P. Binger, B. Biedenbach, C. Krüger and M. Regitz, *Angew. Chem.*, 1987, **99**, 798; (b) S. Barth, Ph.D. Thesis, University of Kaiserslautern, 1993; (c) T. Wettzing, B. Geissler, R. Schneider, S. Barth, P. Binger and M. Regitz, *Angew. Chem.*, 1992, **104**, 761.
66. (a) B. Geißler, Ph.D. Thesis, University of Kaiserslautern, 1993; (b) G. Glaser, Ph.D. Thesis, University of Kaiserslautern, 1994.
67. D. J. Sikora, M. D. Rausch, R. D. Rogers and J. L. Atwood, *J. Am. Chem. Soc.*, 1981, **103**, 1265.
68. J. L. Atwood, R. D. Rogers, W. E. Hunter, C. Floriani, G. Fachinetti and A. Chiesi-Villa, *Inorg. Chem.*, 1980, **19**, 3812.
69. D. J. Sikora, M. D. Rausch, R. D. Rogers and J. L. Atwood, *J. Am. Chem. Soc.*, 1979, **101**, 5079.
70. G. Erker, U. Dorf, C. Krüger and K. Angermund, *J. Organomet. Chem.*, 1986, **301**, 299.
71. T. Takahashi *et al.*, *Chem. Lett.*, 1989, 761.
72. R. Goddard, P. Binger, S. R. Hall and P. Müller, *Acta Crystallogr., Sect. C*, 1990, **46**, 998.
73. T. Takahashi, D. R. Swanson and E. Negishi, *Chem. Lett.*, 1987, 623.
74. P. J. Walsh, F. J. Hollander and R. G. Bergman, *J. Am. Chem. Soc.*, 1990, **112**, 894.
75. Y. Kai *et al.*, *J. Chem. Soc., Chem. Commun.*, 1982, 191.
76. D. J. Sikora, K. J. Moriarty and M. D. Rausch, *Inorg. Synth.*, 1986, **24**, 147.
77. D. J. Sikora and M. D. Rausch, *J. Organomet. Chem.*, 1984, **276**, 21.
78. P. Binger *et al.*, *Chem. Ber.*, 1991, **124**, 2165.
79. H. Yasuda, K. Tatsumi and A. Nakamura, *Acc. Chem. Res.*, 1985, **18**, 120.
80. H. G. Alt and C. E. Denner, *J. Organomet. Chem.*, 1989, **368**, C15.
81. S. L. Buchwald, B. T. Watson and J. C. Huffman, *J. Am. Chem. Soc.*, 1986, **108**, 7411.
82. H. G. Alt, C. E. Denner and R. Zenk, *J. Organomet. Chem.*, 1992, **433**, 107.
83. N. Metzler and H. Nöth, *J. Organomet. Chem.*, 1993, **454**, C5.
84. M. D. Fryzuk, *Chem. Scr.*, 1989, **29**, 427.
85. M. Jensen and T. Livinghouse, *J. Am. Chem. Soc.*, 1989, **111**, 4495.
86. W. A. Nugent, D. L. Thorn and R. L. Harlow, *J. Am. Chem. Soc.*, 1987, **109**, 2788.
87. N. Metzler, H. Nöth and M. Thomann, *Organometallics*, 1993, **12**, 2423.
88. (a) W. A. Nugent and D. F. Taber, *J. Am. Chem. Soc.*, 1989, **111**, 6435; (b) T. Takahashi, T. Seki, Y. Nitto, M. Saburi, C. J. Rousset and E. Negishi, *J. Am. Chem. Soc.*, 1991, **113**, 6266.
89. E. Negishi, D. R. Swanson, F. E. Cederbaum and T. Takahashi, *Tetrahedron Lett.*, 1987, **28**, 917.
90. M. Mori, N. Uesaka and M. Shibasaki, *J. Org. Chem.*, 1992, **57**, 3519.
91. (a) M. E. Maier, *Nachr. Chem. Tech. Lab.*, 1993, **41**, 811; (b) E. Negishi and T. Takahashi, *Rev. Heteroatom Chem.*, 1992, **6**, 177.
92. D. R. Swanson, C. J. Rousset and E. Negishi, *J. Org. Chem.*, 1989, **54**, 3521.
93. R. B. Grossman, W. M. Davis and S. L. Buchwald, *J. Am. Chem. Soc.*, 1991, **113**, 2321.
94. A. H. Hoveyda, J. P. Morken, A.-F. Houri and Z. Xu, *J. Am. Chem. Soc.*, 1992, **114**, 6692.
95. J. P. Morken, M. T. Didiuk and A. H. Hoveyda, *J. Am. Chem. Soc.*, 1993, **115**, 6997.
96. (a) G. Erker, P. Czisch, R. Mynott, Y.-H. Tsay and C. Krüger, *Organometallics*, 1985, **4**, 1310; (b) G. Erker, P. Czisch, C. Krüger and J. M. Wallis, *Organometallics*, 1985, **4**, 2059; (c) A. T. Herrmann, Ph.D. Thesis, University of Kaiserslautern, 1990.
97. G. A. Vaughan, G. L. Hillhouse, R. T. Lum, S. L. Buchwald and A. L. Rheingold, *J. Am. Chem. Soc.*, 1988, **110**, 7215.
98. G. A. Vaughan, C. D. Sofield, G. L. Hillhouse and A. L. Rheingold, *J. Am. Chem. Soc.*, 1989, **111**, 5491.
99. G. A. Vaughan, G. L. Hillhouse and A. L. Rheingold, *J. Am. Chem. Soc.*, 1990, **112**, 7994.
100. Y. Kai *et al.*, *Chem. Lett.*, 1982, 1979.
101. G. Erker, K. Engel, J. L. Atwood and W. E. Hunter, *Angew. Chem.*, 1983, **95**, 506.
102. G. Erker and U. Dorf, *Angew. Chem.*, 1983, **95**, 800.
103. H. Yasuda, Y. Kajihara, K. Mashima, K. Nagasuna and A. Nakamura, *Chem. Lett.*, 1981, 671.
104. M. Akita, H. Yasuda and A. Nakamura, *Chem. Lett.*, 1983, 217.
105. G. Erker, R. Pfaff, D. Kowalski, E.-U. Würthwein, C. Krüger and R. Goddard, *J. Org. Chem.*, 1993, **58**, 6771.
106. (a) G. Erker, *Comments Inorg. Chem.*, 1992, **13**, 111; (b) G. Erker, M. Albrecht, S. Werner, M. Nolte and C. Krüger, *Chem. Ber.*, 1992, **125**, 1953.
107. G. Erker, R. Noe, C. Krüger and S. Werner, *Organometallics*, 1992, **11**, 4174.
108. Y. Wielstra, S. Gambarotta, A. Meetsma and J. L. de Boer, *Organometallics*, 1989, **8**, 250.
109. G. Erker, *Angew. Chem.*, 1989, **101**, 411.

8

Zirconium and Hafnium Compounds in Oxidation State +3

EVELYN J. RYAN
University of Sussex, Brighton, UK

8.1 INTRODUCTION

In 1981, when *COMC-I* was published,[1] the known zirconium(III) and hafnium(III) compounds were few in number and only one, $[Zr(\eta\text{-}Cp)_2(\mu\text{-}C_{10}H_7)(\mu\text{-}H)]$,[2] had been structurally characterized. The only other zirconium(III) compounds to have been isolated in the solid state were the dinitrogen complexes $[Zr(\eta\text{-}C_5H_4R^2)_2(\eta^2\text{-}N_2)R^1]$ ($R^1 = CH(TMS)_2$, $R^2 = H$ or Me).[3,4] With these exceptions, evidence for the formation of Zr^{III} complexes rested upon ESR spectral data alone.

Although the +4 oxidation state still predominates, the number of isolated Zr^{III} complexes has increased considerably, and since that time 19 dinuclear and three neutral, mononuclear metal(III) and two metallate(III) complexes have been structurally characterized. The majority of complexes are intensely coloured and are extremely air and moisture sensitive. The relative abundance of dinuclear complexes illustrates the propensity of these species to dimerize, in order to achieve coordinative saturation, but the use of sufficiently sterically demanding ligands has been shown to reduce this tendency and has resulted in the three known mononuclear complexes. This chapter provides a guide to the literature in this field since 1981, complementing earlier reviews.[5,6]

8.2 DIMERIC ZIRCONIUM(III) COMPOUNDS

8.2.1 Synthesis

Crystalline, dinuclear zirconocene(III) complexes have been synthesized via the reduction of zirconium(IV) precursors. Reducing agents employed have included Na/Hg, potassium metal, and $M[C_{10}H_8]$ (M = Na or K).

Reduction of $[Zr\{\eta\text{-}C_5H_3(TMS)_2\text{-}1,3\}_2X_2]$ (X = Cl, Br or I) by Na/Hg afforded the paramagnetic complexes of Equation (1).[7,8]

$$[Zr\{\eta\text{-}C_5H_3(TMS)_2\text{-}1,3\}_2X_2] \xrightarrow[-NaX]{Na/Hg} [\{Zr(\eta\text{-}C_5H_3(TMS)_2\text{-}1,3)_2(\mu\text{-}X)\}_2] \qquad (1)$$

Reduction of $[Zr(\eta\text{-}Cp)_2Cl_2]$ with 1.5 molar equiv. of Na/Hg afforded the diamagnetic fulvalene complex of Equation (2).[9] Reduction of $[ZrCl_2\{SiMe_2(\eta,\eta\text{-}C_5H_4)_2\}]$ with 1 equiv. of Na/Hg yielded $[\{Zr(\eta\text{-}Cp)(\mu\text{-}Cl)\}_2\{SiMe_2(\mu\text{-}\eta,\eta\text{-}C_5H_4)_2\}]$.[10,11]

$$[Zr(\eta\text{-}Cp)_2Cl_2] \xrightarrow[-NaCl]{Na/Hg} [\{Zr(\eta\text{-}Cp)(\mu\text{-}Cl)\}_2(\mu\text{-}\eta,\eta\text{-}C_5H_4\text{-}C_5H_4)] \qquad (2)$$

Thermolysis of Zr^{II} bis(phospine) compounds resulted in the formation of the diamagnetic complexes of Equation (3).[12] The complex with L = PMe_3 and its hafnocene analogue were prepared by the reaction of $[M(\eta\text{-}Cp)_2Cl_2]$ (M = Zr or Hf) with magnesium in the presence of 2 equiv. of L.[13]

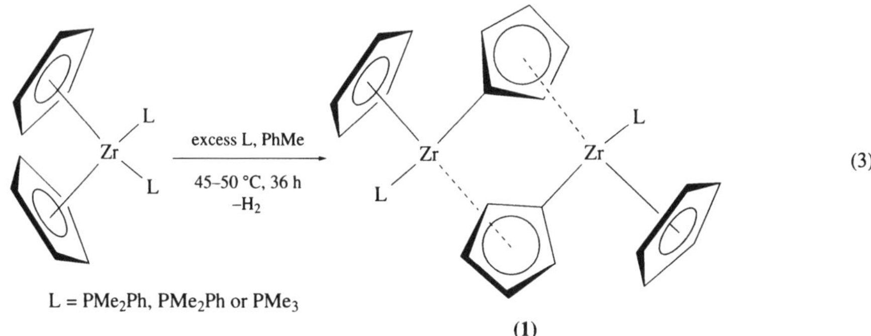

$$ \qquad (3) $$

L = PMe₂Ph, PMe₂Ph or PMe₃

(1)

Thermolysis of $[Zr(\eta\text{-}Cp)_2(PMe_3)_2]$ followed by reaction with stoichiometric amounts of a mild oxidizing agent has provided a convenient synthetic route via **(1)** to several dinuclear Zr^{III} complexes (Scheme 1).[12,14–17]

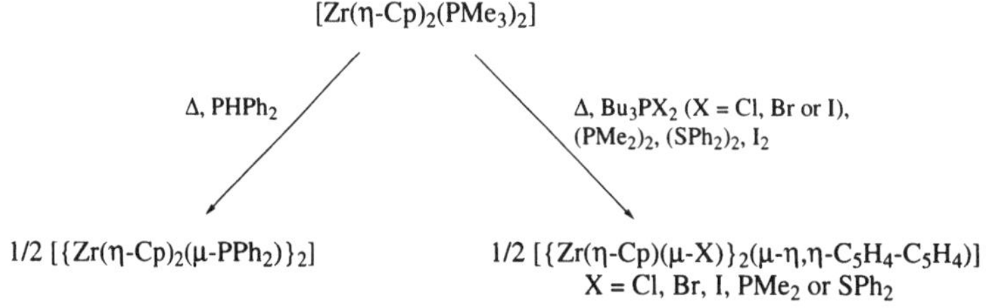

Scheme 1

The phosphido-bridged complex $[\{Zr(\eta\text{-}Cp)_2[\mu\text{-}P(C_6H_{11})H]\}_2]$ was prepared via the reaction of Equation (4).[18,19] The reaction proceeded with oxidative addition of P–H to the transient $[Zr(\eta\text{-}Cp)_2]$ species, followed by intermolecular loss of dihydrogen.

$$[Zr(\eta\text{-}Cp)_2Cl_2] + Mg + P(C_6H_{11})H_2 \longrightarrow [Zr(\eta\text{-}Cp)_2\{P(C_6H_{11})H\}H] \xrightarrow{-H_2}$$

$$1/2\,[\{Zr(\eta\text{-}Cp)_2[\mu\text{-}P(C_6H_{11})H]\}_2] \qquad (4)$$

Treatment of $[Zr(\eta\text{-}Cp)_2]$, generated *in situ* from the reaction of $[Zr(\eta\text{-}Cp)_2Cl_2]$ with 2 equiv. of Bu^nLi,[20] with a primary phosphine PH_2R ($R = Ph$, C_6H_{11} or $C_6H_2Me_3\text{-}2,4,6$), yielded the phosphido-bridged complexes of Scheme 2.[18,19,21] The reaction with PH_2SiPh_3 yielded two products: $[\{Zr(\eta\text{-}Cp)\text{-}(\mu\text{-}PHSiPh_3)\}_2(\mu\text{-}\eta,\eta\text{-}C_5H_4\text{-}C_5H_4)]$ and the unusual phosphido/phosphinidine complex $[\{Li(\mu\text{-}Cl)\}_2\text{-}(THF)_2][\{Zr(\eta\text{-}Cp)\}_2(\mu\text{-}PHSiPh_3)(\mu\text{-}PSiPh_3)(\mu\text{-}\eta,\eta\text{-}C_5H_4\text{-}C_5H_4)]$.[19]

$$[Zr(\eta\text{-}Cp)_2Cl_2] \xrightarrow{Bu^nLi,\ -78\ ^\circ C} [Zr(\eta\text{-}Cp)_2Bu^n_2] \xrightarrow{-C_4H_{10},\ RT} [Zr(\eta\text{-}Cp)_2(\eta^2\text{-}C_4H_8)] + PH_2R \xrightarrow{-78\ \text{to}\ 25\ ^\circ C}$$

$$[Zr(\eta\text{-}Cp)_2(PHR)H] \xrightarrow{-H_2} [Zr(\eta\text{-}Cp)_2(PR)] \longrightarrow [\{Zr(\eta\text{-}Cp)_2(\mu\text{-}PHR)\}_2]$$
$$R = C_6H_{11}\ \text{or}\ SiPh_3$$

$$R = Ph,\ SiPh_3\ \text{or}\ C_6H_2Me_3\text{-}2,4,6$$

$$1/2\ [\{Zr(\eta\text{-}Cp)_2(\mu\text{-}\eta^1,\eta^5\text{-}C_5H_4)(PHR)\}_2] \longrightarrow 1/2\ [\{Zr(\eta\text{-}Cp)_2(\mu\text{-}PHR)\}_2(\mu\text{-}\eta,\eta\text{-}C_5H_4\text{-}C_5H_4)]$$

$$+\ 1/2\ [\{Li(\mu\text{-}Cl)\}_2(THF)_2][\{Zr(\eta\text{-}Cp)\}_2(\mu\text{-}PHSiPh_3)(\mu\text{-}PSiPh_3)(\mu\text{-}\eta,\eta\text{-}C_5H_4\text{-}C_5H_4)]$$
$$R = SiPh_3$$

Scheme 2

The conproportionation reaction between $[Zr(\eta\text{-}Cp)_2X_2]$ ($X = Cl$ or I) and $[Zr(\eta\text{-}Cp)_2(PMe_3)_2]$ resulted in the formation of the fulvalene complex $[\{Zr(\eta\text{-}Cp)(\mu\text{-}X)\}_2(\mu\text{-}\eta,\eta\text{-}C_5H_4\text{-}C_5H_4)]$ ($X = Cl$[14,22] or I[15]) and $[\{Zr(\eta\text{-}Cp)_2(\mu\text{-}I)\}_2]$ (Equation (5)).[15,23] The reaction proceeded with the formation of Zr^{III} intermediates of the type shown in Equation (3).

$$2\ [Zr(\eta\text{-}Cp)_2X_2] + 2\ [Zr(\eta\text{-}Cp)_2(PMe_3)_2] \xrightarrow{-H_2,\ -PMe_3}$$

$$[\{Zr(\eta\text{-}Cp)(\mu\text{-}X)\}_2(\mu\text{-}\eta,\eta\text{-}C_5H_4\text{-}C_5H_4)] + [\{Zr(\eta\text{-}Cp)_2(\mu\text{-}X)\}_2] \qquad (5)$$
$$X = Cl\ \text{or}\ I \qquad\qquad X = I$$

Ambient photolysis of $[\{Zr(\eta\text{-}Cp)Cl(\mu\text{-}H)\}_2\{SiMe_2(\mu\text{-}\eta,\eta\text{-}C_5H_4)_2\}]$ proceeded with the reductive elimination of dihydrogen and the formation of the chloro-bridged complex of Equation (6).[24]

$$[\{Zr(\eta\text{-}Cp)Cl(\mu\text{-}H)\}_2\{SiMe_2(\mu\text{-}\eta,\eta\text{-}C_5H_4)_2\}] \xrightarrow{h\nu} [\{Zr(\eta\text{-}Cp)(\mu\text{-}Cl)\}_2\{SiMe_2(\mu\text{-}\eta,\eta\text{-}C_5H_4)_2\}] + H_2 \quad (6)$$

Thermal decomposition of $[Zr(\eta\text{-}Cp)_2ClH]$ or $[Zr(\eta\text{-}Cp)_2H(C_6H_{11})]$ afforded the fulvalenedizirconium complex of Equation (7).

$$2\ [Zr(\eta\text{-}Cp)_2ClH] \xrightarrow{\Delta,\ -H_2} [\{Zr(\eta\text{-}Cp)Cl(\mu\text{-}\eta^1,\eta^5\text{-}C_5H_4)\}_2] \longrightarrow$$

$$[\{Zr(\eta\text{-}Cp)(\mu\text{-}Cl)\}_2(\mu\text{-}\eta,\eta\text{-}C_5H_4\text{-}C_5H_4)] \qquad (7)$$

Low-temperature photolysis of $[Zr(\eta\text{-}C_5H_4R)_2(Bu^i)X]$[20] ($X = Br$ or I; $R = H$ or Me) yielded the halo-bridged dimers of Equation (8).[15]

$$2\ [Zr(\eta\text{-}C_5H_4R)_2(Bu^i)X] \xrightarrow{h\nu,\ -20\ ^\circ C} [\{Zr(\eta\text{-}C_5H_4R)_2(\mu\text{-}X)\}_2] + Me_2C=CH_2 + Bu^iH \qquad (8)$$
$$R = H\ \text{or}\ Me \qquad\qquad X = Br\ \text{or}\ I$$

The zirconium(III) phosphido-bridged dimers of Equation (9) were obtained from the direct reaction of the appropriate primary phosphine with $[Zr(\eta\text{-}Cp)_2ClH]$.[19]

Several homo- and heterobinuclear phosphinidine-bridged complexes have been prepared.[19,25] The reaction of $[Zr(\eta\text{-}Cp)_2Cl_2]$ with Li_2PPh resulted in the formation of the thermally unstable paramagnetic complex $[\{Li(THF)_3\}_2][\{Zr(\eta\text{-}Cp)_2(\mu\text{-}PPh)\}_2]$.[19]

$$[Zr(\eta\text{-}Cp)_2ClH] \xrightarrow{PRH_2} [Zr(\eta\text{-}Cp)_2Cl(PHR)] \longrightarrow [Zr(\eta\text{-}Cp)_2Cl_2] + [Zr(\eta\text{-}Cp)_2(PHR)_2] \xrightarrow{-Cl_2}$$

$$1/2\,[\{Zr(\eta\text{-}Cp)_2(\mu\text{-}PHR)\}_2] \qquad (9)$$
$$R = Ph \text{ or } SiPh_3$$

Treatment of the primary phosphido-bridged dimer $[\{Zr(\eta\text{-}Cp)_2(\mu\text{-}P(C_6H_{11})H_2)\}_2]$ (**2**) with excess KH afforded the diphosphinidine species $K_2[\{Zr(\eta\text{-}Cp)_2(\mu\text{-}PC_6H_{11})\}_2]$.[19]

Two zirconium(III) phosphinidine dimers have been prepared by the protonation or methylation of $[\{Li(THF)_3\}_2][\{Zr(\eta\text{-}Cp)_2(\mu\text{-}PPh)\}_2]$ (Scheme 3).

Scheme 3

8.2.2 Structural Data

The x-ray molecular structures of various binuclear zirconium(III) complexes have been determined; selected structural data are given in Tables 1 and 2. Various bridging groups have been employed including halides,[7,15,16,22,23] phosphides,[16,18,19,21] phosphinidines[19,25] and sulfides.[24] The coordination geometry around the zirconium atoms is essentially similar, with the typical distorted tetrahedral ligand arrangement found in bent metallocene complexes (**2**).[5,19–21]

Table 1   Selected x-ray diffraction data on neutral, binuclear zirconocene(III) halides.

Compound	Zr–Zr (nm)	Zr–X (nm)	Zr–Cp (nm)	Zr–X–Zr (°)	X–Zr–X (°)	Cp–Zr–Cp (°)	Ref.
$[\{Zr(\eta\text{-}Cp)(\mu\text{-}Cl)\}_2\{SiMe_2(\mu\text{-}\eta,\eta\text{-}C_5H_4)_2\}]$	0.33853(4)	0.25728(7)	0.2203(5)	82.28(2)	97.72(2)	128.0(2)	24
$[\{Zr(\eta\text{-}Cp)(\mu\text{-}Cl)\}_2(\mu\text{-}\eta,\eta\text{-}C_5H_4\text{-}C_5H_4)]$	0.3233(1)	0.2578(2)	0.2183(9)	77.7(1)	100.0(1)	135.2(4)	22
$[\{Zr(\eta\text{-}Cp)_2(\mu\text{-}I)\}_2]$	0.3669(2)	0.2916(1)	d	77.97(3)	102.03(3)	115.68(3)	23
$[\{Zr(\eta\text{-}C_5H_4Me)_2(\mu\text{-}I)\}_2]$	0.3649(1)	0.29161(8)	0.2209(2)	77.47(2)	102.53(2)	103.13(9)	15
$[\{Zr(\eta\text{-}Cp)(\mu\text{-}I)\}_2(\mu\text{-}\eta,\eta\text{-}C_5H_4\text{-}C_5H_4)]$	0.3472(1)	0.29260(9)	0.2197(1)	72.79(2)	103.30(2)	134.27(2)	15
$[\{Zr(\eta\text{-}Cp)_2\}_2(\mu\text{-}Cl)(\mu\text{-}PMe_2)]$	0.3524(2)	0.2591(2)[a] 0.2655(2)[c]	0.224(2)	85.7(1)[a] 83.2(1)[c]	95.6(1)[b]	d	16
$[\{Zr[\eta\text{-}C_5H_3(TMS)_2\text{-}1,3]_2(\mu\text{-}Cl)\}_2]^e$	0.3905(1)	0.2603(2)	0.2182	97.23(6)	82.77(6)	131.0	7
$[Zr\{\eta\text{-}C_5H_3(TMS)_2\text{-}1,3]_2(\mu\text{-}Br)\}_2]$	0.4101(1)	0.2754(1)	0.2199	96.13(3)	83.87(3)	131.0	30
$[Zr\{\eta\text{-}C_5H_3(TMS)_2\text{-}1,3]_2(\mu\text{-}I)\}_2]$	0.4171(1)	0.2900(2)	0.222	91.92(6)	88.08(6)	129.9	30

[a] X = Cl.   [b] This is Cl–Zr–P.   [c] X = PMe$_2$.   [d] Data not available.   [e] Average of two independent molecules in the unit cell.

In the case of the bis(cyclopentadienyl) complexes $[\{Zr(\eta\text{-}C_5H_3R^1R^2)_2(\mu\text{-}X)\}_2]$ ($R^1 = R^2 = H$; X = I, PMe$_2$, PH(C$_6$H$_{11}$), PHSiPh$_3$ or PHPh; or R^1 = H, R^2 = Me, X = I; or $R^1 = R^2 = $ TMS; X = Cl, Br or I), the central cores containing the Zr$_2$X$_2$ four-membered rings are all planar. The ring geometry is governed by a balance between the repulsive forces acting between the two cyclopentadienyl rings, the

Table 2   Selected x-ray diffraction data on zirconocene(III) phosphido, phosphidine and sulfide dimers.

Compound	Zr–Zr (nm)	Zr–X (nm)	Zr–Cp (nm)	Zr–X–Zr (°)	X–Zr–X (°)	Ref.
$[\{Zr(\eta\text{-}Cp)_2(\mu\text{-}PMe_2)\}_2]$	0.3653(2)	0.2672(2)	0.222(2)	86.3(1)	93.8(1)	16
$[\{Zr(\eta\text{-}Cp)_2[\mu\text{-}P(C_6H_{11})H]\}_2]$	0.3606(4)	0.2646(4)	a	86.0(1)	94.0(1)	18
$[\{Zr(\eta\text{-}Cp)_2(\mu\text{-}PHSiPh_3)\}_2]$	0.3674(5)	0.2625(4)	a	88.4(1)	91.6(1)	19
$[\{Zr(\eta\text{-}Cp)_2(\mu\text{-}PHPh)\}_2]$	0.3677(4)	0.2663(7)	a	87.3(2)	92.4(2)	19
$[\{Zr(\eta\text{-}Cp)(\mu\text{-}PPhH)\}_2(\mu\text{-}\eta,\eta\text{-}C_5H_4\text{-}C_5H_4)]$	0.3549(2)	0.2616(4)	a	85.4(1)	90.6(1)	21
$[\{Zr(\eta\text{-}Cp)[\mu\text{-}PH(C_6H_2Me_3\text{-}2,4,6]\}_2\text{-}(\mu\text{-}\eta,\eta\text{-}C_5H_4\text{-}C_5H_4)]$	0.3423(2)	0.2657(5)	a	80.2(1)	97.7(1)	19
$[\{Zr(\eta\text{-}Cp)(\mu\text{-}PHSiPh_3)\}_2(\mu\text{-}\eta,\eta\text{-}C_5H_4\text{-}C_5H_4)]$	0.3585(3)	0.2606(4)	a	86.9(1)	89.7(1)	19
$[Li(TMEDA)(THF)_2][\{Zr(\eta\text{-}Cp)\}_2(\mu\text{-}PHPh)(\mu\text{-}PPh)]$	0.3672(2)	0.2591(4)	a	90.3(1)	89.8(1)	19
$[\{Li(THF)_3\}_2(\mu\text{-}N_2)][\{Zr(\eta\text{-}Cp)_2(\mu\text{-}PPh)\}_2]$	a	0.2572(3)	a	91.10(8)	88.90(8)	25
$[\{Zr(\eta\text{-}Cp)(\mu\text{-}SPh)\}_2(\mu\text{-}\eta,\eta\text{-}C_5H_4\text{-}C_5H_4)]$	0.3420(1)	0.2540(1)	a	84.66(3)	89.78(3)	14

^{a} Data not available.

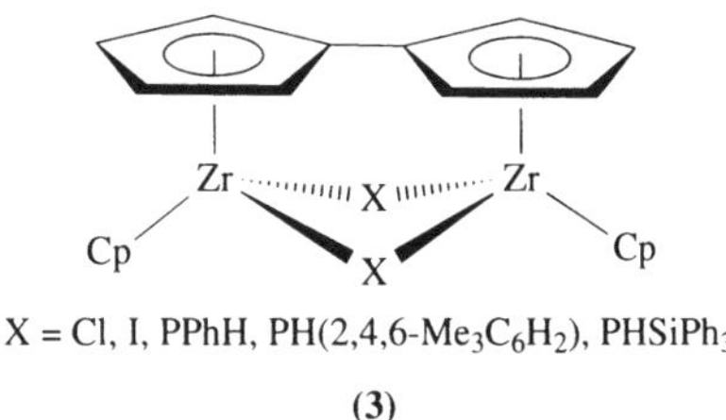

(2)

antibonding interaction of the in-plane halide p orbitals and any restoring force due to a direct Zr–Zr interaction,[26–8] as well as the size of the bridging atom. The exceptionally long Zr–Zr distances (0.3905(1)–0.417 1(1) nm) found in the halide-bridged dimers $[\{Zr[\eta\text{-}C_5H_3(TMS)_2\text{-}1,3]_2(\mu\text{-}X)\}_2]$ (X = Cl, Br or I) illustrate the effect of introducing sterically demanding substituents onto the cyclopentadienyl ring.

For any given bridging X group, the introduction of a bridging fulvalene ligand causes a deviation from planarity in the Zr_2X_2 core with consequent folding along the Zr–Zr axis, resulting in a significant decrease in the intermetallic separation from that seen in a related bis(cyclopentadienyl) dimer (Tables 1 and 2). This is illustrated by a comparison of the Zr–Zr separation in $[\{Zr(\eta\text{-}Cp)_2(\mu\text{-}I)\}_2]$ (0.366 9(2) nm)[23] and $[\{Zr(\eta\text{-}Cp)(\mu\text{-}I)\}_2(\mu\text{-}\eta,\eta\text{-}C_5H_4\text{-}C_5H_4)]$ (3) (0.347 2(1) nm).[15]

X = Cl, I, PPhH, PH(2,4,6-Me₃C₆H₂), PHSiPh₃

(3)

The presence of an $SiMe_2$ group between the cyclopentadienyl rings in $[\{Zr(\eta\text{-}Cp)(\mu\text{-}Cl)\}_2\text{-}\{SiMe_2(\mu\text{-}\eta,\eta\text{-}C_5H_4)_2\}]$ (4) causes an increase in the Zr–Zr separation with a consequent reduction in the folding of the Zr_2X_2 core.[24]

The anionic complexes $[\{Zr(\eta\text{-}Cp)\}_2(\mu\text{-}PHPh)(\mu\text{-}PPh)]^-$ and $[\{Zr(\eta\text{-}Cp)_2(\mu\text{-}PPh)\}_2]^-$ are geometrically similar to the neutral diphosphides (Table 2), the most noticeable difference being the significantly shorter average Zr–P distances of 0.259 1(4) nm and 0.257 2(3) nm, respectively.

An unusual dimeric phosphido–phosphinidine complex $[\{Li(\mu\text{-}Cl)\}_2(THF)_2][\{Zr(\eta\text{-}Cp)\}_2\text{-}(\mu\text{-}PHSiPh_3)(\mu\text{-}PSiPh_3)(\mu\text{-}\eta,\eta\text{-}C_5H_4\text{-}C_5H_4)]$ in which one of the zirconium centres is in the +4 oxidation state and the other in the +3 oxidation state has been reported; the $Zr(\mu\text{-}P)_2Zr$ core is shown schematically as (5).[19] The average Zr–P-2 bond length, of 0.258 0(5) nm, is significantly shorter than the average Zr–P-1 bond length of 0.261 7(5) nm, and is consistent with the presence of a phosphinidine fragment.

Molecular orbital calculations on $[\{Zr(\eta\text{-}Cp)_2(\mu\text{-}X)\}_2]^{26,27}$ (X = I, PH_2 or PMe_2) suggested that a direct Zr–Zr bond could exist for intermetallic separations of up to 0.4 nm, and also that superexchange, postulated to account for the diamagnetism observed in complexes with a Zr–Zr distance of greater than

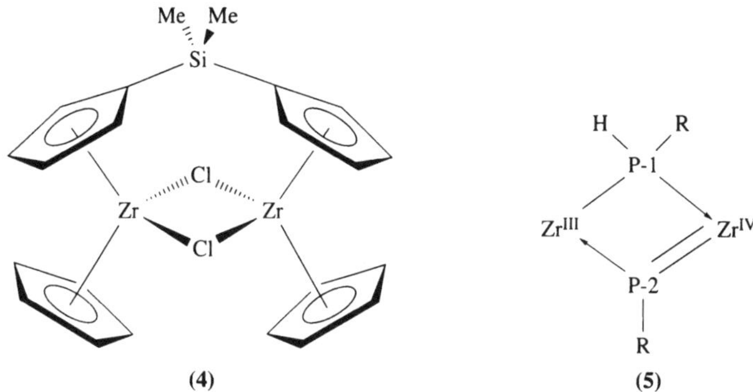

(4)   (5)

0.34 nm, does not occur. A recent theoretical study of $[\{Zr(\eta\text{-}Cp)_2(\mu\text{-}X)\}_2]^{28}$ (X = I or PH_2, Zr–Zr ca. 0.365 nm) again supported the presence of a direct, weak Zr–Zr interaction with a postulated bond order of ca. 0.5. Such orbital overlap, although weak, was considered strong enough to cause spin-pairing of the two electrons, thus supporting the interpretation of the observed diamagnetism in terms of a through-space Zr–Zr coupling.

8.2.3 Spectroscopic Properties

With the exception of $[\{Zr[\eta\text{-}C_5H_3(TMS)_2\text{-}1,3]_2(\mu\text{-}X)\}_2]$ (X = Cl, Br or I), all of the neutral dinuclear complexes were diamagnetic; ^1H, ^{13}C and ^{31}P (as appropriate) NMR spectroscopic data were reported in each case.[10,12,14–16,18,21,23,24,29]

ESR spectral data recorded for the neutral complexes $[\{Zr[\eta\text{-}C_5H_3(TMS)_2\text{-}1,3]_2(\mu\text{-}X)\}_2]$ (X = Cl, Br or I) showed two well-separated isotropic lines in both THF and toluene solutions, each with hyperfine splitting to zirconium, comprising a broad asymmetric signal centred at $g_{iso} = 1.933$–1.944, $(a(^{91}Zr) = 52$–54 G) and a narrow signal at $g_{iso} = 1.971$ $(a(^{91}Zr) = 22$–47 G).[7,30] Similarly, ESR spectral data recorded for the product of the chemical reduction of $[\{Zr[\eta\text{-}C_5H_2(TMS)_3\text{-}1,2,3]_2(\mu\text{-}Cl)\}_2]$ showed two signals ($g_{iso} = 1.965$, $g_{iso} = 1.936$, $a(^{91}Zr) = 56$ G).[31] The low-field resonances were of higher intensity in THF solutions and were attributed to a solvated species; the relative intensities were reversed in toluene solutions where the high-field Lewis base-free species were thought to dominate. Frozen solution spectral data were also reported.[30,31]

The anionic phosphido and phosphinidine dimers $[\{Zr(\eta\text{-}Cp)\}_2(\mu\text{-}PHPh)(\mu\text{-}PPh)]^-$, $[\{Zr(\eta\text{-}Cp)_2(\mu\text{-}PPh)\}_2]^-$, $[\{Zr(\eta\text{-}Cp)_2[\mu\text{-}P(C_6H_{11})]\}_2]^-$ and $[\{Zr(\eta\text{-}Cp)\}_2(\mu\text{-}PMePh)(\mu\text{-}PPh)]^-$ were paramagnetic in THF solutions and ESR spectral data were consistent with the presence of a single dimeric species in each case ($g_{iso} = 2.004$–2.018, $a(^{31}P) = 13.6$–19.5 G, $a(^{91}Zr) = 7$ G).[19,25]

8.2.4 Chemical Properties

All of the dinuclear zirconium(III) complexes were extremely air and moisture sensitive. Various reactions have been carried out, usually resulting in oxidation to zirconium(IV) products although nucleophilic bridge-cleavage and ligand-exchange reactions have also been reported.

The propensity of the less sterically hindered compounds such as $[\{Zr(\eta\text{-}Cp)_2(\mu\text{-}X)\}_2]$ (X = Cl, Br or I) or $[\{Zr(\eta\text{-}C_5H_4Me)_2(\mu\text{-}Cl)\}_2]$ to disproportionate to the appropriate zirconocene(IV) halide and unidentified Zr^{II} coproducts was investigated.[15] Disproportionation was the dominant process in reactions of $[\{Zr(\eta\text{-}Cp)_2(\mu\text{-}X)\}_2]$ with a Lewis base and in reactions aimed at alkyl substitution (Equation (10)).

$$[\{Zr(\eta\text{-}Cp)_2(\mu\text{-}I)\}_2] + LiR \xrightarrow{-LiI} [Zr(\eta\text{-}Cp)_2R_2] + \{Zr(\eta\text{-}Cp)_2\} \qquad (10)$$
R = Me or Bz

The presence of bifunctional bis(cyclopentadienyl) ligands or sterically demanding cyclopentadienyl ligands such as $[\eta\text{-}C_5H_3(TMS)_2\text{-}1,3]^-$ increased the thermal stability of the halide dimers.[7]

Reactions of $[\{Zr[\eta\text{-}C_5H_3(TMS)_2\text{-}1,3]_2(\mu\text{-}Cl)\}_2]$ are summarized in Scheme 4. Nucleophilic bridge-cleavage addition reactions have been demonstrated, including that with a Lewis base (THF, PMe_3 or dmpe) with formation of a paramagnetic zirconium(III) species without any noticeable tendency to accelerate disproportionation,[7,30] and with Cl^- to generate a crystalline paramagnetic zirconocenate(III) complex. Substitution products have been obtained in reactions with MeLi and $[\{Li(THF)_n\}_2][C_6H_4\{P(TMS)\}_2\text{-}1,2\}].$[30,32]

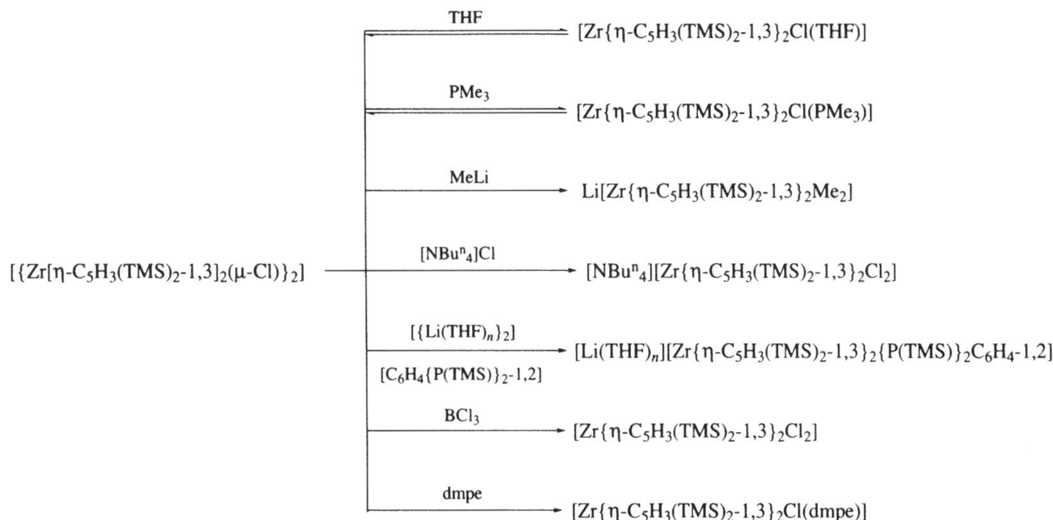

Scheme 4

Addition of the Lewis base THF or PMe_3 (L) to $[\{Zr(\eta\text{-}Cp)(\mu\text{-}Cl)\}_2\{SiMe_2(\mu\text{-}\eta,\eta\text{-}C_5H_4)_2\}\}]$[24] resulted in the formation of the dinuclear paramagnetic species of Equation (11), which was identified by its solution ESR spectrum.

$$[\{Zr(\eta\text{-}Cp)(\mu\text{-}Cl)\}_2\{SiMe_2(\mu\text{-}\eta,\eta\text{-}C_5H_4)_2\}] + 2\,L \xrightarrow{\;L = THF\ or\ PMe_3\;}$$

$$[\{Zr(\eta\text{-}Cp)Cl(L)\}_2\{SiMe_2(\mu\text{-}\eta,\eta\text{-}C_5H_4)_2\}] \qquad (11)$$

Reactions of the fulvalene-bridged complexes $[\{Zr(\eta\text{-}Cp)(\mu\text{-}X)\}_2(\mu\text{-}\eta,\eta\text{-}C_5H_4\text{-}C_5H_4)]$ (X = Cl, I or SPh) are summarized in Scheme 5.[9,14,15,22] The chloride reacted quantitatively with molecular oxygen to yield the structurally characterized Zr^{IV} complex $[\{Zr(\eta\text{-}Cp)Cl\}_2(\mu\text{-}O)(\mu\text{-}\eta,\eta\text{-}C_5H_4\text{-}C_5H_4)].$[17] Addition of X_2 (X = Cl or Br) or CH_2Cl_2 gave the appropriate fulvalenezirconium(IV) halide $[\{Zr(\eta\text{-}Cp)ClX\}_2\text{-}(\mu\text{-}\eta,\eta\text{-}C_5H_4\text{-}C_5H_4)],$[33] in which the zirconium fragments adopted a *trans* geometry about the fulvalene moiety. Treatment of $[\{Zr(\eta\text{-}Cp)(\mu\text{-}X)\}_2(\mu\text{-}\eta,\eta\text{-}C_5H_4\text{-}C_5H_4)]$ with an excess of the oxidizing agent Bu_3PX_2 (X = Cl or I) or PhSSPh (X = SPh) also yielded the appropriate fulvalenezirconium(IV) complex.[14,34]

The *ansa*-zirconocene complex $[\{Zr(\mu\text{-}Cl)[SiMe_2(\eta,\eta\text{-}C_5H_4)_2]\}_2]$ was readily oxidized by molecular oxygen, but in contrast with its fulvalene analogue the μ-oxo product was only identified as a transient intermediate; it rapidly disproportionated to yield the *ansa*-zirconocene(IV) chloride and an unidentified oxo derivative (Equation (12)).[10]

$$[\{Zr(\mu\text{-}Cl)[SiMe_2(\eta,\eta\text{-}C_5H_4)_2]\}_2] \xrightarrow{\;O_2\;} [\{Zr(\mu\text{-}Cl)[SiMe_2(\eta,\eta\text{-}C_5H_4)_2]\}_2(\mu\text{-}O)] \longrightarrow$$

$$[ZrCl_2\{SiMe_2(\eta,\eta\text{-}C_5H_4)_2\}] + \text{an oxo derivative} \qquad (12)$$

Reaction of $[\{Zr(\eta\text{-}Cp)(\mu\text{-}Cl)\}_2(\mu\text{-}\eta,\eta\text{-}C_5H_4\text{-}C_5H_4)]$ with an isocyanide ($CNBu^t$ or $CNC_6H_3Me_2$-2,3)[33,35] or diphenyldiazomethane[36] resulted in the cleavage of both chloride bridges and the oxidative addition of the ligand to give a $\mu\text{-}\eta^1{:}\eta^2$ coordination adduct $[\{Zr(\eta\text{-}Cp)Cl\}_2(\mu\text{-}\eta^1{:}\eta^2NCR)(\mu\text{-}\eta,\eta\text{-}C_5H_4\text{-}C_5H_4)]$. A similar product was obtained from $[\{Zr(\eta\text{-}Cp)(\mu\text{-}Cl)\}_2\{SiMe_2(\mu\text{-}\eta,\eta\text{-}C_5H_4)_2\}]$ and $NCBu^t$.[24] Reaction with $LiC{\equiv}CPh$ gave the zirconium(IV) complex, $[\{Zr(\eta\text{-}Cp)(\mu\text{-}C{\equiv}CPh)\}_2(\mu\text{-}\eta,\eta\text{-}C_5H_4\text{-}C_5H_4)]$, which has not been structurally characterized. Oxidation proceded with retention of the fulvalene bridge in each case.

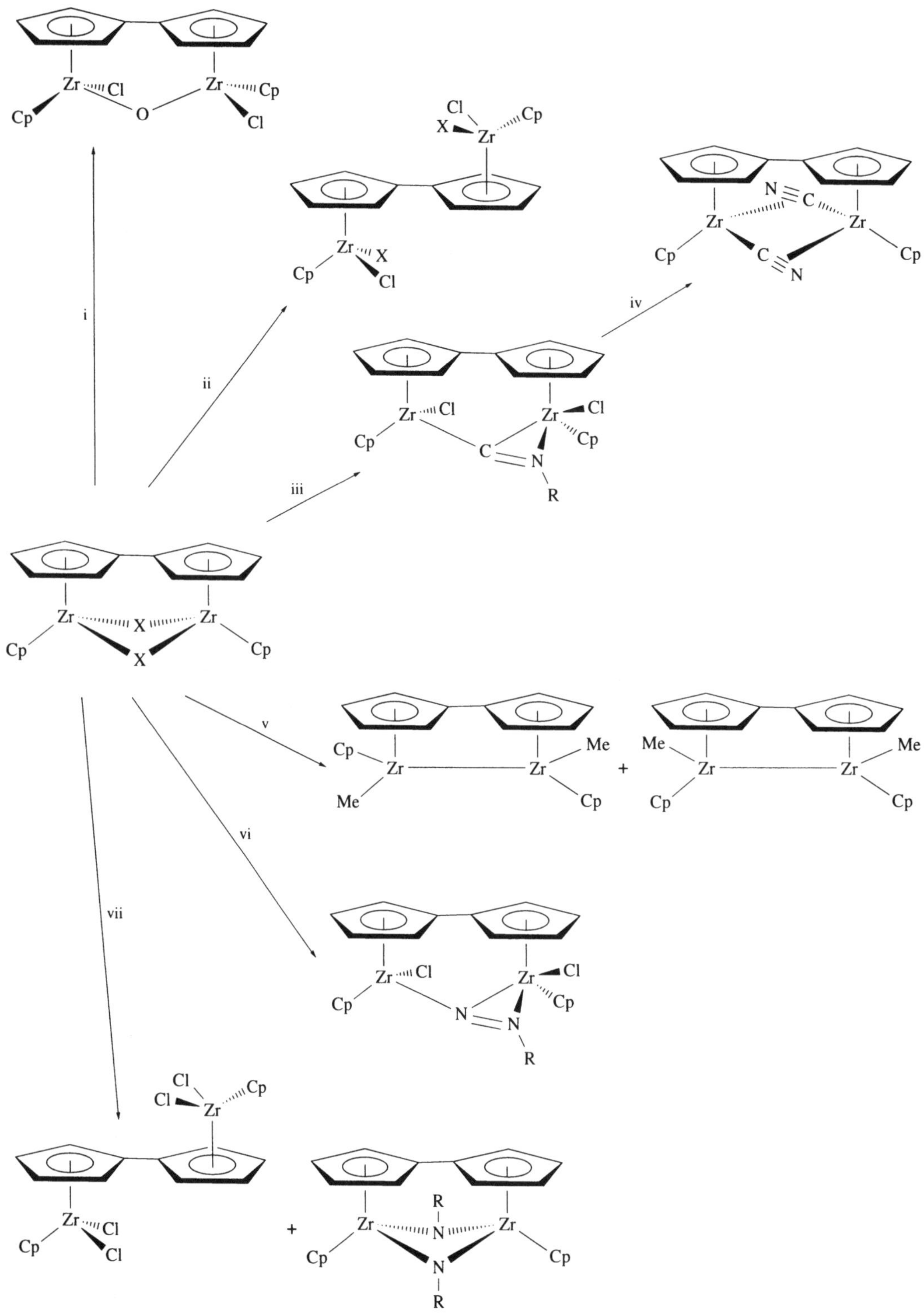

i, (X = Cl), PhMe, 20 °C, 1/2 O$_2$;[17] ii, (X = Cl, Br or I), PhMe, 20 °C, Cl$_2$, Br$_2$, CH$_2$Cl$_2$, Bu$_3$PCl$_2$ or Bu$_3$PI$_2$;
or (X = SPh), pyridine, 20 °C, PhSSPh;[14] iii, (R = Bu^t or C$_6$H$_3$Me$_2$), PhMe, 20 °C, CNBu^t or CNC$_6$H$_3$Me$_2$-2, 5;[33, 35]
iv, PhMe, 110 °C, 5 h, CNBu^t, Na/Hg;[36] v, PhMe, –78 °C to 20 °C, 2 h, LiMe;[28] vi, (R = CPh$_2$), PhMe, 70 °C, 1 h, N$_2$CPh;[36]
vii, (R = TMS), THF, reflux, 7 h, TMS-N$_3$[36]

Scheme 5

Sodium/amalgam reduction of [{Zr(η-Cp)(μ-η^1:η^2-CNBu^t)Cl}$_2$(μ-η,η-C$_5$H$_4$-C$_5$H$_4$)][36] in the presence of CNBu^t resulted in dealkylation of the isocyanide ligand and the formation of a cyano-bridged complex [{Zr(η-Cp)(μ-η^1:η^2-CN)}$_2$(μ-η,η-C$_5$H$_4$-C$_5$H$_4$)].

The reaction of [{Zr(η-Cp)(μ-Cl)}$_2$(μ-η,η-C$_5$H$_4$-C$_5$H$_4$)][36] with TMS-N$_3$ proceeded with the loss of molecular nitrogen and the formation of a mixture of [{Zr(η-Cp)Cl$_2$}$_2$(μ-η,η-C$_5$H$_4$-C$_5$H$_4$)] and the imido-bridged complex [{Zr(η-Cp)(μ-N-TMS)}$_2$(μ-η,η-C$_5$H$_4$-C$_5$H$_4$)].

Methylation of [{Zr(η-Cp)(μ-Cl)}$_2$(μ-η,η-C$_5$H$_4$-C$_5$H$_4$)] with 2 equiv. of MeLi at −78 °C in the dark led to complete substitution of the chlorides and retention of the fulvalene link, a *cis–trans* mixture of the thermally unstable, light sensitive methylzirconium(III) complex [{Zr(η-Cp)Me}$_2$(μ-η,η-C$_5$H$_4$-C$_5$H$_4$)] being obtained.[33]

The ability of [{Zr(η-Cp)(μ-Cl)}$_2${SiMe$_2$(μ-η,η-C$_5$H$_4$)$_2$}] to act as a reducing agent has been demonstrated by its reaction with Ph$_3$PO or CO$_2$, which yielded the dinuclear Zr(μ-O)Zr complex [{Zr(η-Cp)Cl}$_2${SiMe$_2$(μ-η,η-C$_5$H$_4$)$_2$}(μ-O)].[24]

Unlike its titanocene analogue, [{M(η-Cp)$_2$(μ-PMe$_2$)}$_2$] (M = Zr or Hf) was oxidized to M^{IV} products by I$_2$, PPh$_2$Cl or BCl$_3$ (Scheme 6).[37]

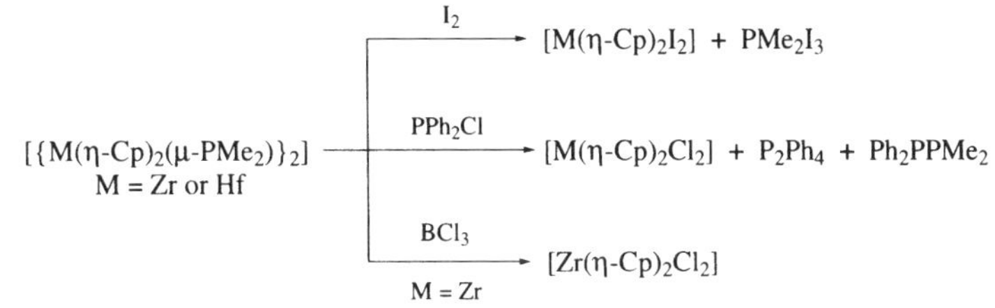

Scheme 6

A diamagnetic, primary phosphido-bridged, heterobimetallic complex was obtained from the reaction of [{Zr(η-Cp)[μ-P(C$_6$H$_{11}$)]H}$_2$] with [{Mo(η-Cp)(CO)$_3$}$_2$] (Scheme 7).[19]

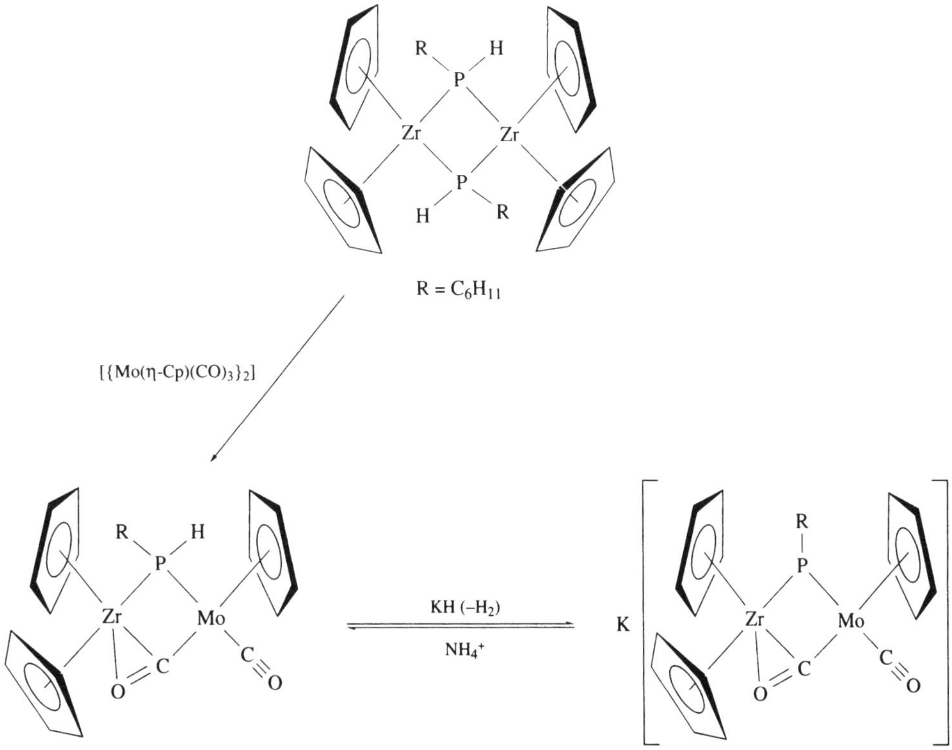

Scheme 7

8.3 NEUTRAL, MONONUCLEAR METAL(III) COMPOUNDS

8.3.1 Synthesis, and Spectroscopic and Chemical Properties

Reduction of $[M(\eta\text{-}C_5H_3Bu^t_2\text{-}1,3)_2Cl_2]$ (M = Zr or Hf) with finely divided potassium metal afforded the zirconocene(III) complex of Equation (13).[38]

$$[M(\eta\text{-}C_5H_3Bu^t_2\text{-}1,3)_2Cl_2] \xrightarrow{\text{K, }-\text{KCl}} [M(\eta\text{-}C_5H_3Bu^t_2\text{-}1,3)_2Cl] \qquad (13)$$
$$M = \text{Zr or Hf}$$

The room-temperature ESR spectrum of a hexane solution of $[Zr(\eta\text{-}C_5H_3Bu^t_2\text{-}1,3)_2Cl]$ showed two well-separated isotropic lines; the line centred at $g_{iso} = 1.971$ was narrower and of lower intensity than that at $g_{iso} = 1.952$ and showed hyperfine splitting to zirconium $(a(^{91}Zr) = 90$ G$)$.[30]

Magnesium reduction of $[Zr(\eta\text{-}Cp^*)Cl_3]$ in the presence of cot yielded the zirconium(III) complex of Equation (14).[39,40] The analogous reduction of $[Hf(\eta\text{-}Cp^*)Cl_3]$ failed to yield the hafnium(III) compound, the reaction proceeding with the formation of the hafnium(IV) hydrido complex $[Hf(\eta\text{-}Cp^*)H(\eta\text{-}C_8H_8)]$.

$$[Zr(\eta\text{-}Cp^*)Cl_3] \xrightarrow{\text{Mg, cot}} [Zr(\eta\text{-}Cp^*)(\eta\text{-}C_8H_8)] \qquad (14)$$

The ESR spectrum of a pentane–toluene solution of $[Zr(\eta\text{-}Cp^*)(\eta\text{-}C_8H_8)]$,[41] recorded at $-140\,°C$, showed a singlet centred at $g_{iso} = 1.976$ with poorly resolved proton superhyperfine coupling $(a(^1H) = 3.8$ G$)$; zirconium hyperfine coupling was not observed. An external nuclear double resonance (ENDOR) spectroscopic study of the $(\eta^8\text{-}C_8H_8)$ electrons was reported.[41]

Sodium/amalgam reduction of $[Zr(\eta\text{-}Cp)\{N(SiMe_2CH_2PPr^i_2)_2\}Cl_2]$ afforded the zirconium(III) chloro complex of Equation (15).[42]

$$[Zr(\eta\text{-}Cp)\{N(SiMe_2CH_2PPr^i_2)_2\}Cl_2] \xrightarrow{\text{Na/Hg, }-\text{NaCl}} [Zr(\eta\text{-}Cp)\{N(SiMe_2CH_2PPr^i_2)\}Cl] \qquad (15)$$

The ESR spectrum of $[Zr(\eta\text{-}Cp)\{N(SiMe_2CH_2PPr^i_2)\}Cl]$ showed a triplet centred at $g_{iso} = 1.955$ with coupling to two equivalent ^{31}P nuclei $(a(^{31}P) = 20.1$ G$)$ flanked by ^{91}Zr satellites $(a(^{91}Zr) = 37.2$ G$)$.

Metathesis reactions of $[Zr(\eta\text{-}Cp)\{N(SiMe_2CH_2PPr^i_2)\}Cl]$ with $LiCH_2TMS$, PhLi, MeMgBr or $Li[BH_4]$ yielded the zirconium(III) alkyl or borohydrido complexes of Scheme 8. Reaction of the alkyl complexes with dihydrogen afforded the hydrido species, which reacted with ethene to form the ethyl complex $[Zr(\eta\text{-}Cp)\{N(SiMe_2CH_2PPr^i_2)\}Et]$, also produced by the metathetical reaction of $[Zr(\eta\text{-}Cp)\text{-}\{N(SiMe_2CH_2PPr^i_2)\}Cl]$ with LiEt.[42]

The ESR spectrum of $[Zr(\eta\text{-}Cp)\{N(SiMe_2CH_2PPr^i_2)\}H]$ comprised a 1:1:2:2:1:1 doublet of triplets centred at $g_{iso} = 1.988$ due to coupling to two equivalent ^{31}P nuclei $(a(^{31}P) = 21.7$ G$)$ and one proton $(a(^1H) = 8.7$ G$)$. The ESR spectrum of $[Zr(\eta\text{-}Cp)\{N(SiMe_2CH_2PPr^i_2)\}CH_2TMS]$ showed coupling to two equivalent ^{31}P nuclei, two inequivalent ^1H nuclei and one ^{14}N nucleus, $(a(^{31}P) = 21.4$ G; $a(^1H) = 9.3$ G; $a(^1H) = 6.2$ G; $a(^{14}N) = 2.0$ G$)$.

Evidence for the identification of mononuclear M^{III} complexes in solution is frequently afforded by ESR spectroscopy.[7,24,30,43–57] Representative ESR spectral data showing typical g_{iso} values and hyperfine coupling constants are summarized in Table 3. The complexes were generated by chemical or electrochemical reduction, photolysis or thermolysis, oxidation of a Zr^{II} precursor or addition of a Lewis base to a Zr^{III} precursor.

$Na[C_{10}H_8]$ reduction of the dialkylmetallocene(IV) complexes $[M(\eta\text{-}Cp)_2R_2]$ (M = Zr or Hf) afforded a series of paramagnetic monocyclopentadienylmetal(III) complexes which were identified by ESR spectroscopy $(g_{iso} = 1.989$ (1.984–1.993); $a(^{91}Zr) = 10.0$–24.0 G$)$.[53,54,58–61] Evolution of PPh_2Me from $[Zr(\eta\text{-}C_5H_4Bu^t)(CH_2PPh_2)_2]$ at room temperature afforded the hydrido complex of Equation (16) $(g_{iso} = 1.989$; $a(Zr) = 7.4$ G; $a(^{31}P) = 25.5$ G; $a(^1H) = 6.0$ G$)$.[58,59]

$$[Zr(\eta\text{-}C_5H_4Bu^t)_2(CH_2PPh_2)] \xrightarrow[-Na[C_5H_4Bu^t]]{[C_{10}H_8]} [Zr(\eta\text{-}C_5H_4Bu^t)(CH_2PPh_2)_2] \xrightarrow{-PPh_2Me}$$

$$[Zr(\eta\text{-}C_5H_4Bu^t)(H)(CH_2PPh_2)] \qquad (16)$$

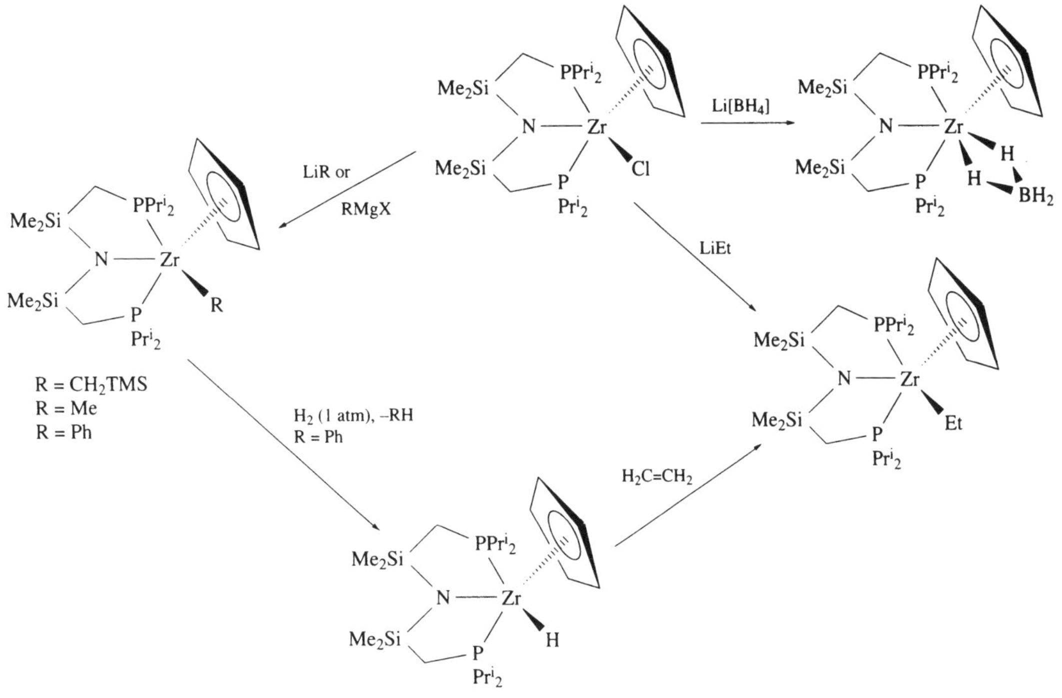

Scheme 8

Table 3   Selected ESR spectral data on some mononuclear zirconocene(III) complexes.

Compound	g_{iso}	$a(^{91}Zr)$ (G)	$a(^{31}P)$ (G)	$a(^{1}H)$ (G)	Ref.
[Zr(η-Cp)$_2$I]	1.992	b	a	a	48
[Zr{η-C$_5$H$_3$(TMS)$_2$-1,3}$_2$(Cl)(PMe$_3$)]	1.978	23.0	14.5	a	7,30
[{Zr(η-Cp)Cl(PMe$_3$)}$_2${SiMe$_2$(μ-η,η-C$_5$H$_4$)$_2$}]	1.982	20.0	14.3	a	24
[Zr(η-Cp)$_2$(CH$_2$PPh$_2$)]	1.987	22.5	16.1	2.8	45,49,50
[Zr(η-Cp)$_2$H]	1.987	b	a	8.0	51
[ZrH{SiMe$_2$(μ-η,η-C$_5$H$_4$)$_2$}]	1.989	b	a	6.8	47
[Zr(η-Cp)$_2$(CH$_2$PMe$_2$)]	c	13.5	19.5	b	52
[Zr(η-Cp)$_2$Cl(PMe$_3$)]	1.979	23.0	19.5	a	48
[Zr(η-Cp)$_2$Br(PMe$_3$)]	1.994	22.7	19.4	a	48
[Zr(η-Cp)$_2$Bu^t(PMePh$_2$)]	1.983	b	28.0	a	46
[Zr(η-Cp)$_2$Bz(PPh$_3$)]	1.984	16.5	20.7	4.0	53
[Zr(η-Cp)$_2$Me(PPh$_3$)]	1.988	a	21.0	b	54
[Zr(η-Cp)$_2$(CH$_2$TMS)]	1.979	23.5	a	ca. 3.0	53
[Zr(η-Cp)$_2$H(PMePh$_2$)]	1.998	24.0	29.0	6.0	45
[Zr(η-Cp)$_2${CH$_2$(C$_6$H$_4$)NMe$_2$-2}]	1.973	<10.0	a	6.1	55
[Zr(η-Cp)$_2${CH$_2$(C$_6$H$_4$)PPh$_2$-2}]	1.984	6.7	19.1	5.5	55
[Zr(η-Cp)$_2${CH(TMS)(C$_6$H$_4$)Me-2}(THF)]	1.997	12.5	a	c	44
[Zr(η-Cp)$_2${CH(TMS)(C$_5$H$_4$N-2)}]	1.992	11.5	a	a	43
[Zr(η-Cp)$_2${CH(TMS)(C$_5$H$_4$N-2)}(PMe$_3$)]	1.992	11.5	a	a	43
[Zr(η-Cp)$_2${CH(TMS)(C$_6$H$_4$PPh$_2$-2)}]	1.979	9.9	16.3	a	43
[Zr(η-Cp)$_2${CH(TMS)(C$_6$H$_4$PPh$_2$-2)}(PMe$_3$)]	1.987	22.5	14.8	a	43
[Zr(η-Cp)$_2${CH(TMS)(C$_6$H$_4$Me-4)}]	1.986	7.2	a	7.6	43
[Zr(η-Cp)$_2${CH(TMS)(C$_6$H$_4$Me-4}(PMe$_3$)]	1.978	25.0	16.0	a	43

[a] Not applicable.   [b] Not resolved.   [c] Data not available.

8.3.2 Structural Data

The x-ray molecular structure of [Zr(η-C$_5$H$_3$Bu$^t_2$-1,3)$_2$Cl] (**6**)[38] revealed a mononuclear structure which was both coordinatively and electronically unsaturated, unique among the structurally characterized neutral zirconocene(III) halide complexes. The Zr–Cl bond length at 0.2423 nm was considerably shorter than that found in the binuclear chlorides (Table 1). In contrast with its titanium analogue, the molecule did not have a C_2 axis coinciding with the Zr–Cl axis.

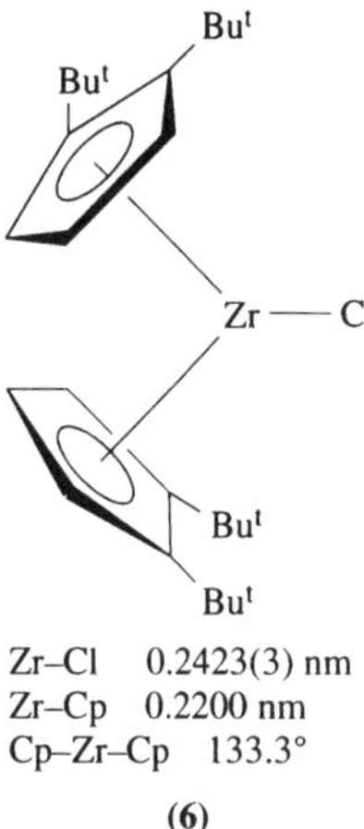

Zr–Cl   0.2423(3) nm
Zr–Cp   0.2200 nm
Cp–Zr–Cp   133.3°

(6)

A representation of the structure of $[Zr(\eta\text{-}Cp^*)(\eta\text{-}C_8H_8)]^{39}$ **(7)** is shown.

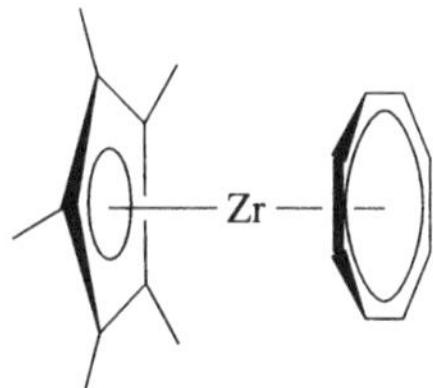

Zr–centroid(C-5)   0.217 nm
Zr–centroid(C-8)   0.159 nm, 0.167 nm
centroid(C-5)–Zr–centroid(C-8)   173.8°

(7)

The structure of $[Zr(\eta\text{-}Cp)\{N(SiMe_2CH_2PPr^i_2)\}(CH_2TMS)]$ is shown schematically as **(8)**.[42] The geometry about zirconium was described as a distorted trigonal bipyramid with the phosphines occupying the axial sites.

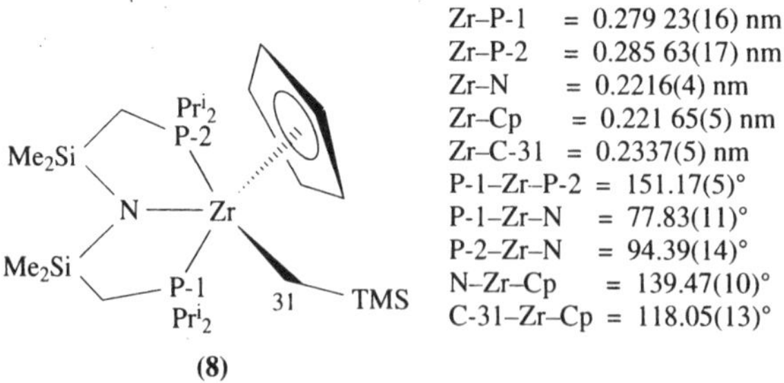

Zr–P-1   = 0.279 23(16) nm
Zr–P-2   = 0.285 63(17) nm
Zr–N   = 0.2216(4) nm
Zr–Cp   = 0.221 65(5) nm
Zr–C-31   = 0.2337(5) nm
P-1–Zr–P-2 = 151.17(5)°
P-1–Zr–N = 77.83(11)°
P-2–Zr–N = 94.39(14)°
N–Zr–Cp   = 139.47(10)°
C-31–Zr–Cp = 118.05(13)°

(8)

8.4 METALLATE(III) COMPOUNDS

8.4.1 Synthesis and Spectroscopic Data

Metallate(III) species have been synthesized by both electrochemical and chemical reduction of the appropriate zirconocene(IV) precursor (Equation (17)).[56,62–73] ESR spectroscopic analysis has again been used as the principal means of characterization. Electrochemical data are given in Table 4.

Reduction of the metallacycles $[\overline{M(C_5H_4R^1)_2\{R^2(C_6H_4)\dot{C}H_2\text{-}2\}}]$ (M = Zr or Hf; R^1 = H, R^2 = CH(TMS); or R^1 = TMS, R^2 = CH$_2$) resulted in cleavage of the M–C σ bond to yield the complexes of Equation (18). Evidence for cleavage of the metallacycle came from ESR spectroscopy, (g_{iso} values in the range 1.983–1.986; $a(^{91}Zr)$ 14.8–15.0 G; $a(^{177}Hf)$ not observed).[62,63]

$$[Zr(\eta\text{-}C_5H_3R^1R^2)_2X_2] \; \underset{-e}{\overset{+e}{\rightleftharpoons}} \; [Zr(\eta\text{-}C_5H_3R^1R^2)_2X_2]^- \tag{17}$$

Table 4 Reduction potential data for metallocene dihalides in THF at a platinum electrode.

Complex	$-E_{1/2}^{red}$ (V)[a]	Ref.
$[Zr(\eta\text{-}Cp)_2Cl_2]$	1.70	61,48
$[Zr(\eta\text{-}C_5H_4Me)_2Cl_2]$	1.73	61
$[Zr(\eta\text{-}C_5H_4Et)_2Cl_2]$	1.74	61
$[Zr\{\eta\text{-}C_5H_3(TMS)_2\text{-}1,3\}_2F_2]$	1.71	67
$[Zr(\eta\text{-}C_5H_4TMS)_2Cl_2]$	1.59	67
$[Zr\{\eta\text{-}C_5H_3(TMS)_2\text{-}1,3\}_2Cl_2]$	1.55	67
$[Zr\{SiMe_2(\eta\text{-}C_5H_4)_2\}Cl_2]$	1.600	72
$[Zr\{SiEt_2(\eta\text{-}C_5H_4)_2\}Cl_2]$	1.602	72
$[Zr\{SiPr_2(\eta\text{-}C_5H_4)_2\}Cl_2]$	1.603	72
$[Zr(\eta\text{-}Cp)_2Br_2]$	1.80	48
$[Zr(\eta\text{-}C_5H_4TMS)_2Br_2]$	1.39	5
$[Zr\{\eta\text{-}C_5H_3(TMS)_2\text{-}1,3\}_2Br_2]$	1.32	67
$[\{Zr(\eta\text{-}C_5H_4Bu^t)_2(\mu\text{-}H)H\}_2]$	1.90	65
$[Hf\{\eta\text{-}C_5H_3(TMS)_2\text{-}1,3\}_2Cl_2]$	1.87	67
$[\overline{Zr(\eta\text{-}Cp)_2\{CH(TMS)C_6H_4}CH(TMS)\text{-}2\}]$	2.02	68
$[\overline{Hf(\eta\text{-}Cp)_2\{CH(TMS)C_6H_4}CH(TMS)\text{-}2\{]$	2.26	68
$[Zr(\eta\text{-}Cp)_2\{1,8(CH\text{-}TMS)_2C_{10}H_6\}]$	1.99	64
$[\overline{Zr(\eta\text{-}Cp)_2\{CH(TMS)C_6H_4}\}_2]$	2.21	64
$[\overline{Zr(\eta\text{-}Cp)_2\{CH_2(C_6H_4)_2}CH_2\text{-}2\}]$	2.09	64

[a] Reduction potentials vs. the standard calomel electrode.

$$[\overline{M(\eta\text{-}C_5H_4R^1)\{R^2(C_6H_4)}CH_2\text{-}2\}] \xrightarrow{Na[C_{10}H_8]} [M(\eta\text{-}C_5H_4R^1)\{R^2(C_6H_4)(CH_2\text{-}2)\}Na] \tag{18}$$

M = Zr or Hf

$R^1 = H$, $R^2 = CH(TMS)$

$R^1 = TMS$, $R^2 = CH_2$

Chemical reduction of $[\overline{Zr(\eta\text{-}Cp)_2\{CH_2(C_6H_4\text{-}C_6H_4)}CH_2\text{-}2\}]$ resulted in a similar cleavage of an M–C σ bond, whereas electrochemical reduction yielded the zirconocenate(III) complex (Scheme 9).[64]

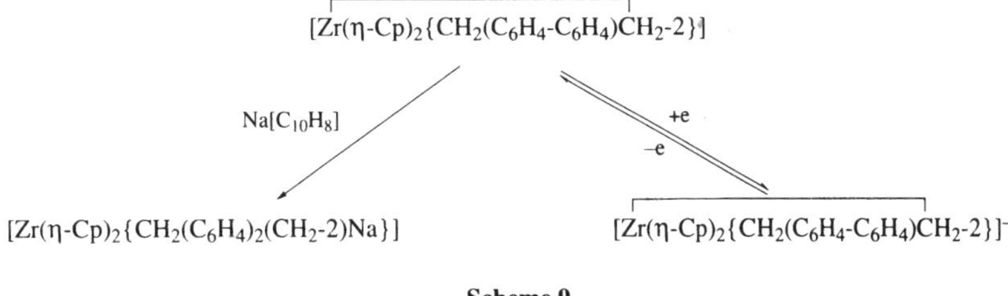

Scheme 9

Electrochemical reduction of $[\{Zr(\eta\text{-}C_5H_4R)_2H(\mu\text{-}H)\}_2]$[65,66] (R = Me or Bu^t) afforded the dihydrido monomeric complex $[Zr(\eta\text{-}C_5H_4R)_2H_2]^-$, characterized by its ESR spectrum which comprised a 1:2:1 triplet due to coupling to two equivalent ^1H nuclei flanked by ^{91}Zr satellites in each case. Reduction of the deuterated compound afforded an ESR spectrum (1:2:3:2:1 quintet) consistent with the formation of the zirconocenate(III) species $[Zr(\eta\text{-}C_5H_4R)_2D_2]^-$. A two-electron reduction process was involved and the mechanism of Scheme 10 (a or b) was postulated to account for these observations. Reduction with $M[C_{10}H_8]$ (M = Li, Na or K) afforded ESR spectra with additional coupling to the alkali metals (^7Li, ^{23}Na or ^{39}K, $I = 3/2$). ESR spectral data are given in Table 5.

$$[\{Zr(\eta\text{-}C_5H_4R)_2H(\mu\text{-}H)\}_2] + e^- \rightleftharpoons [\{Zr(\eta\text{-}C_5H_4R)_2H(\mu\text{-}H)\}_2]^- \rightleftharpoons [Zr(\eta\text{-}C_5H_4R)_2H_2]^- + [Zr(\eta\text{-}C_5H_4R)_2H_2]$$

$$[Zr(\eta\text{-}C_5H_4R)_2H_2] + e^- \rightleftharpoons [Zr(\eta\text{-}C_5H_4R)_2H_2]^-$$

$$2\,[Zr(\eta\text{-}C_5H_4R)_2H_2] \rightleftharpoons [\{Zr(\eta\text{-}C_5H_4R)_2H(\mu\text{-}H)\}_2] \qquad \text{(mechanism a)}$$

$$[\{Zr(\eta\text{-}C_5H_4R)_2H(\mu\text{-}H)\}_2] + e^- \rightleftharpoons [\{Zr(\eta\text{-}C_5H_4R)_2H(\mu\text{-}H)\}_2]^- \xrightarrow{+e^-} [\{Zr(\eta\text{-}C_5H_4R)_2H(\mu\text{-}H)\}_2]^{2-} \rightleftharpoons$$

$$2\,[Zr(\eta\text{-}C_5H_4R)_2H_2]^-$$

(mechanism b)

Scheme 10

Table 5 Selected ESR spectral data on some metallate(III) complexes.[65,66]

Compound	g_{iso}	$a(^{91}Zr)$ (G)	$a(^1H)$ (G)	$a(M)^a$ (G)
$[NBu^n_4][Zr(\eta\text{-}C_5H_4Bu^t)_2H_2]$	1.996	16.2	8.4	b
$Li[Zr(\eta\text{-}C_5H_4Bu^t)_2H_2]$	1.993	c	7.8	3.9
$Na[Zr(\eta\text{-}C_5H_4Bu^t)_2H_2]$	1.992	10.4	8.3	10.1
$K[Zr(\eta\text{-}C_5H_4Bu^t)_2H_2]$	1.994	14.4	8.5	2.0
$[NBu^n_4][Zr(\eta\text{-}C_5H_4TMS)_2H_2]$	1.996	8.0	17.6	b
$Li[Zr(\eta\text{-}C_5H_4TMS)_2H_2]$	1.994	8.1	7.3	3.6
$Na[Zr(\eta\text{-}C_5H_4TMS)_2H_2]$	1.995	12.8	8.9	8.9
$K[Zr(\eta\text{-}C_5H_4TMS)_2H_2]$	1.996	14.7	8.2	1.8

[a] M = Li, Na, or K.   [b] Not applicable.   [c] Not resolved.

Hydrogen–deuterium exchange with dihydrogen was investigated for $[Zr(\eta\text{-}C_5H_4TMS)_2D_2]^-$.[66] Three mechanisms (Schemes 11–13) were proposed to account for hydrogen–deuterium exchange at the metal (Scheme 11), on the ring substituents (Scheme 12) and on the cyclopentadienyl ring itself (Scheme 13). Selective formation of cyclooctene was obtained in the catalytic hydrogenation of 1,3- or 1,5-cyclooctadiene.[66]

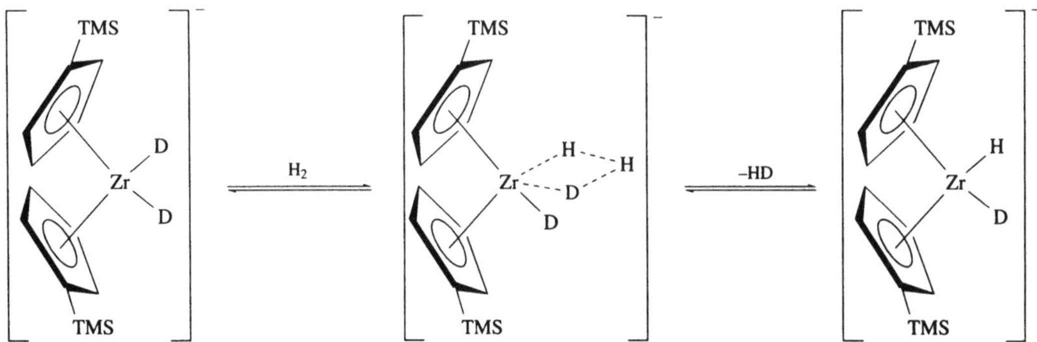

Scheme 11

Scheme 12

Crystalline, paramagnetic zirconocenate(III) complexes $[NBu^n_4][Zr\{\eta\text{-}C_5H_3(TMS)_2\text{-}1,3\}_2X_2]$ (X = Cl or Br) were prepared by the nucleophilic bridge-cleavage addition reaction of Equation (19).[7,30] Electrochemical reduction was reported (Table 4).

$$[\{Zr[\eta\text{-}C_5H_3(TMS)_2\text{-}1,3]_2(\mu\text{-}X)\}_2] + [NBu^n_4]X \longrightarrow [NBu^n_4][Zr\{\eta\text{-}C_5H_3(TMS)_2\text{-}1,3\}_2X_2] \qquad (19)$$

Although extremely air and moisture sensitive, the complexes were indefinitely stable both in toluene solutions and in the solid state at ambient temperature. Their room-temperature ESR spectra comprised a central singlet flanked by ^{91}Zr satellites.

8.4.2 Structures

There are only two structurally characterized metallate(III) complexes, of the form $[NBu^n_4][Zr\{\eta\text{-}C_5H_3(TMS)_2\text{-}1,3\}_2X_2]$ (X = Cl or Br; (9)).[30] They are isostructural and the stereochemistry about the zirconium atoms is unexceptional. The mean Zr–X bond lengths and angles are within the range expected for zirconocene(III) halides (Table 1).

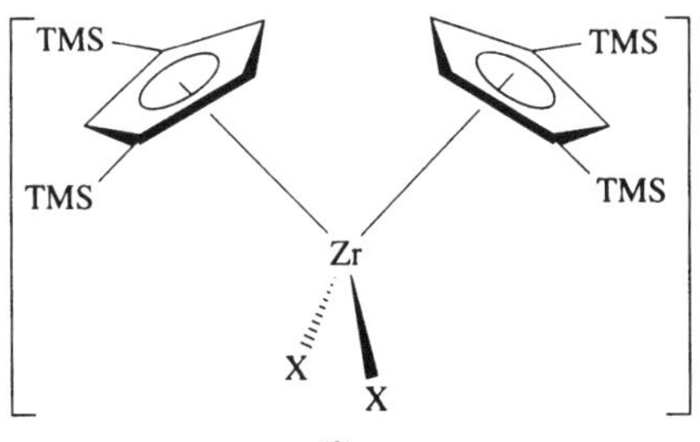

(9)

	X = Cl	X = Br
Zr–X	0.2561(8) nm	0.2740(2) nm
Zr–Cp	0.219 nm	0.2187 nm
X–Zr–X	85.1(3)°	85.91(7)°
Cp–Zr–Cp	131.7°	132.6°

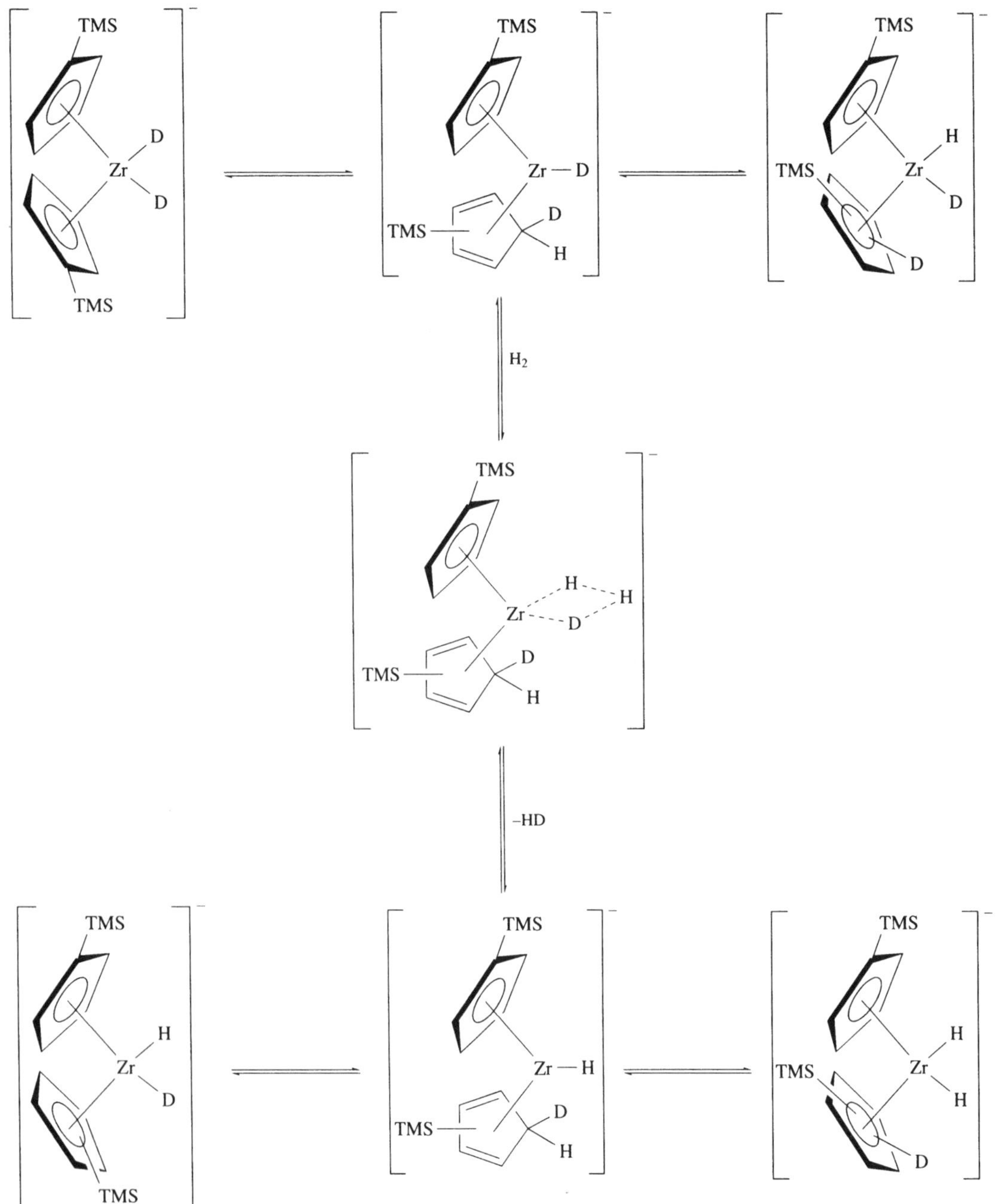

Scheme 13

8.5 REFERENCES

1. D. J. Cardin, M. F. Lappert, C. L. Raston and P. I. Riley, in 'COMC-I', vol. 3, p. 549.
2. G. P. Pez, C. F. Putnik, S. L. Suib and G. D. Stucky, *J. Am. Chem. Soc.*, 1979, **101**, 6933.
3. J. Jeffrey, M. F. Lappert and P. I. Riley, *J. Organomet. Chem.*, 1979, **181**, 25.
4. M. J. S. Gynane, J. Jeffrey and M. F. Lappert, *J. Chem. Soc., Chem. Commun.*, 1978, 34.
5. D. J. Cardin, M. F. Lappert and C. L. Raston, 'Chemistry of Organo-Zirconium and -Hafnium Compounds', Ellis Horwood, Chichester, 1986.
6. Y. Wielstra, S. Gambarotta and M. Y. Chiang, *Recl. Trav. Chim. Pays-Bas*, 1989, **108**, 1.
7. P. B. Hitchcock, M. F. Lappert, G. A. Lawless, H. Olivier and E. J. Ryan, *J. Chem. Soc., Chem. Commun.*, 1992, 474.
8. A. Antiñolo, M. F. Lappert, G. A. Lawless and H. Olivier, *Polyhedron*, 1989, **8**, 395.
9. T. Cuenca, W. A. Herrmann and T. V. Ashworth, *Organometallics*, 1986, **5**, 2514.

10. R. Gomez, T. Cuenca, P. Royo, M. A. Pellingelli and A. Tiripicchio, *Organometallics*, 1991, **10**, 1505.
11. R. Choukroun, Y. Raoult and D. Gervais, *J. Organomet. Chem.*, 1990, **391**, 189.
12. K. I. Gell, T. V. Harris and J. Schwartz, *Inorg. Chem.*, 1981, **20**, 481.
13. K. B. Kool, M. D. Rausch, H. G. Alt, M. Herberhold, V. Thewalt and B. Honold, *J. Organomet. Chem.*, 1986, **310**, 27.
14. Y. Wielstra, S. Gambarotta, A. L. Spek and W. J. J. Smeets, *Organometallics*, 1990, **9**, 2142.
15. Y. Wielstra, S. Gambarotta, A. Meetsma and A. L. Spek, *Organometallics*, 1989, **8**, 2948.
16. M. Y. Chiang, S. Gambarotta and F. van Bolhuis, *Organometallics*, 1988, **7**, 1846.
17. T. V. Ashworth, T. Cuenca, E. Herdtweck and W. A. Herrmann, *Angew. Chem., Int. Ed. Engl.*, 1986, **25**, 289.
18. J. Ho and D. W. Stephan, *Organometallics*, 1991, **10**, 3001.
19. J. Ho, Z. Hou, R. J. Drake and D. W. Stephan, *Organometallics*, 1993, **12**, 3145.
20. D. R. Swanson and E. Negishi, *Organometallics*, 1991, **10**, 825.
21. J. Ho and D. W. Stephan, *Organometallics*, 1992, **11**, 1014.
22. S. Gambarotta and M. Y. Chiang, *Organometallics*, 1988, **6**, 897.
23. Y. Wielstra, S. Gambarotta, A. Meetsma and J. L. de Boer, *Organometallics*, 1989, **8**, 250.
24. J. Cacciola, K. P. Reddy and J. L. Petersen, *Organometallics*, 1992, **11**, 665.
25. J. Ho, R. J. Drake and D. W. Stephan, *J. Am. Chem. Soc.*, 1993, **115**, 3792.
26. M. Bénard and M.-M. Rohmer, *J. Am. Chem. Soc.*, 1992, **114**, 4785.
27. M.-M. Rohmer and M. Bénard, *Organometallics*, 1991, **10**, 157.
28. R. L. DeKock, M. A. Peterson, L. E. L. Reynolds, L.-H. Chen, E. J. Baerends and P. Vernooijs, *Organometallics*, 1993, **12**, 2794.
29. G. Fochi, G. Guidi and C. Floriani, *J. Chem. Soc., Dalton Trans.*, 1984, 1253.
30. E. J. Ryan, D. Phil. Thesis, University of Sussex, 1993.
31. R. Choukroun and F. Dahan, *Organometallics*, 1994, **13**, 2097.
32. Yin Ping, D. Phil. Thesis, University of Sussex, 1991.
33. W. A. Herrmann, T. Cuenca, B. Menjon and E. Herdtweck, *Angew. Chem., Int. Ed. Engl.*, 1987, **26**, 697.
34. Y. Wielstra, D. Phil. Thesis, University of Gröningen, 1990.
35. T. Cuenca, R. Gomez, P. Gomez-Sal, G. M. Rodriguez and P. Royo, *Organometallics*, 1992, **11**, 1229.
36. W. A. Herrmann, B. Menjon and E. Herdtweck, *Organometallics*, 1991, **10**, 2134.
37. S. R. Wade, M. G. H. Wallbridge and G. R. Willey *J. Chem. Soc., Dalton Trans.*, 1983, 2555.
38. I. F. Urazowski, V. I. Ponomaryev, I. F. Nifant'ev and D. A. Lemenovskii, *J. Organomet. Chem.*, 1989, **368**, 287.
39. R. D. Rogers and J. H. Teuben, *J. Organomet. Chem.*, 1989, **359**, 41.
40. D. Gourier, E. Samuel and J. H. Teuben, *Inorg. Chem.*, 1989, **28**, 4663.
41. J. Blenkers, P. Bruin and J. H. Teuben, *J. Organomet. Chem.*, 1985, **297**, 61.
42. M. D. Fryzuk and M. Mylvaganam, *J. Am. Chem. Soc.*, 1993, **115**, 10 360.
43. S. I. Bailey *et al.*, *J. Chem. Soc., Dalton Trans.*, 1986, 603.
44. M. F. Lappert and C. L. Raston, *J. Chem. Soc., Chem. Commun.*, 1981, 173.
45. R. Choukroun and D. Gervais, *J. Chem. Soc., Chem. Commun.*, 1985, 224.
46. G. M. Williams and J. Schwartz, *J. Am. Chem. Soc.*, 1982, **104**, 1122.
47. C. S. Bajgur, S. B. Jones and J. L. Petersen, *Organometallics*, 1985, **4**, 1929.
48. E. Samuel, D. Guery, J. Vedel and F. Basile, *Organometallics*, 1985, **4**, 1073.
49. N. E. Schore, S. J. Young, M. M. Olmstead and P. Hofmann, *Organometallics*, 1983, **2**, 1769.
50. R. Choukroun, Y. Raoult and D. Gervais, *Polyhedron*, 1989, **8**, 1758.
51. R. Choukroun, M. Basso-Bert and D. Gervais, *J. Chem. Soc., Chem. Commun.*, 1986, 1317.
52. S. J. Young, M. M. Knudsen and N. E. Schore, *Organometallics*, 1985, **4**, 1432.
53. A. Hudson, M. F. Lappert and R. Pichon, *J. Chem. Soc., Chem. Commun.*, 1983, 374.
54. J. M. Atkinson, P. B. Brindley, A. G. Davies and J. A.-A. Hawari, *J. Organomet. Chem.*, 1984, **264**, 253.
55. J. J. Koh, P. H. Rieger, I. W. Shim and W. M. Risen, *Inorg. Chem.*, 1985, **24**, 2313.
56. S. B. Jones and J. L. Petersen, *J. Am. Chem. Soc.*, 1983, **105**, 5502.
57. E. Samuel, *Inorg. Chem.*, 1983, **22**, 2967.
58. R. Choukroun, D. Gervais and C. Rifai, *J. Organomet. Chem.*, 1989, **368**, C11.
59. R. Choukroun, F. Dahan, D. Gervais and C. Rifai, *Organometallics*, 1990, **9**, 1892.
60. M. Etienne, R. Choukroun and D. Gervais, *J. Chem. Soc., Dalton Trans.*, 1984, 915.
61. M. F. Lappert, C. J. Pickett, P. I. Riley and P. I. W. Yarrow, *J. Chem. Soc., Dalton Trans.*, 1981, 805.
62. M. F. Lappert, T. R. Martin, C. R. C. Milne, J. L. Atwood, W. E. Hunter and R. E. Pentilla, *J. Organomet. Chem.*, 1980, **192**, C35.
63. G. S. Bristow, M. F. Lappert and T. R. Martin, *J. Chem. Soc., Dalton Trans.*, 1984, 399.
64. L. M. Engelhardt, W. P. Leung, R. I. Papasergio, C. L. Raston, P. Twiss and A. H. White, *J. Chem. Soc., Dalton Trans.*, 1987, 2347.
65. R. Choukroun *et al.*, *Organometallics*, 1991, **10**, 374.
66. A.-M. Larsonneur, R. Choukroun and J. Jaud, *Organometallics*, 1993, **12**, 3216.
67. A. Antiñolo *et al.*, *Polyhedron*, 1989, **8**, 1601.
68. M. F. Lappert and C. L. Raston, *J. Chem. Soc., Chem. Commun.*, 1980, 1284.
69. A. Fakhr, Y. Mugnier, B. Gautheron and E. Laviron, *J. Organomet. Chem.*, 1986, **302**, C7.
70. N. E. Schore, *J. Am. Chem. Soc.*, 1980, **102**, 4251.
71. R. T. Baker, J. F. Whitney and S. S. Wreford, *Organometallics*, 1983, **2**, 1049.
72. C. S. Bajgur, W. Tikkanen and J. L. Petersen, *Inorg. Chem.*, 1985, **24**, 2539.
73. M. F. Lappert, C. L. Raston, B. W. Skelton and A. H. White, *J. Chem. Soc., Dalton Trans.*, 1984, 893.

9

Bis(cyclopentadienyl)zirconium and -hafnium Halide Complexes in Oxidation State +4

EVELYN J. RYAN
University of Sussex, Brighton, UK

9.1 INTRODUCTION

Since the publication of *COMC-I*, interest in organozirconium and organohafnium chemistry has continued to grow, with a broad and diverse range of topics encompassing both theoretical and applied research. The major area of expansion during the 1980s and early 1990s has been the preparation of metallocene complexes containing chiral or prochiral cyclopentadienyl ligands, and their application to stereoselective synthesis and catalysis.[1,2]

This chapter is mainly concerned with the synthesis and characterization of zirconocene and hafnocene dihalides and is intended to be a guide to the literature since 1982. Although they are precursors of a wide range of metal complexes, the chemical reactivity of these compounds will not be discussed in detail since many of these reactions are covered in other chapters of this volume and in comprehensive reviews.[1,3–6]

9.2 BIS(CYCLOPENTADIENYL) COMPLEXES

The work relating to zirconocene and hafnocene dihalides containing simple Cp^- ligands has been widely reported and will only be summarized here. The dichlorides $[MCl_2Cp_2]$ (M = Zr (**1**) or Hf (**2**)) were commonly prepared by the well-established reaction of the lithium, sodium, potassium or thallium cyclopentadienide or cyclopentadienyl-Grignard reagent with the metal tetrahalide (method 1).[3] The metallocene difluorides were prepared via the reaction of the appropriate metallocene bis(amide) complex with $[BF_3(OEt_2)]$, PhCN or $[NH_4]_2[SiF_6]$.[3] The heavier halides were prepared by halogen exchange reactions using the appropriate zirconocene dichloride and boron tribromide or triiodide (method 2).[7] The crystal structures of all eight metallocene dihalides $[MX_2Cp_2]$ (M = Zr or Hf; X = F, Cl, Br or I) have been reported[8] and their spectroscopic properties have been extensively studied.[7,9] Molecular modelling and molecular orbital calculations have shown good agreement with experimental electronic and molecular structures and with vibrational data.[10,11] Research into applications for these complexes has covered a broad spectrum with interest ranging from polymerization catalysis to antitumour effects.[12]

9.2.1 Substituted Cyclopentadienyl Complexes

The introduction of alkyl substituents onto the cyclopentadienyl ring in metallocene complexes results in significant electronic and steric changes as well as conferring increased lipophilicity. Symmetrical bis(alkyl) complexes were most commonly prepared by method 1.[3,13] The reaction between MCl_3Cp^* and the appropriate lithium or sodium cyclopentadienide provided a convenient route to the mixed-ring complexes $[MCl_2CpCp^*]$ (M = Zr (**3**) or Hf (**4**)), $[ZrCl_2\{\eta\text{-}C_5H_2(Me\text{-}1,2,4)_3\}Cp^*]$ (**5**), $[ZrCl_2Cp^*\{\eta\text{-}C_9H_6(Me\text{-}1)\}]$ (**6**), $[HfCl_2Cp\{\eta\text{-}C_5H_2(Me_3\text{-}1,2,4)\}]$ (**7**) $[HfCl_2\{\eta\text{-}C_5H_2(Me_3\text{-}1,2,4)\}Cp^*]$ (**8**).[14–16] Similarly, reaction of $ZrCl_3Cp'''$ ($Cp''' = \eta\text{-}C_5H_2(Bu^t_3\text{-}1,2,4)$) and the appropriate lithium or sodium cyclopentadienide gave $[ZrCl_2CpCp''']$ (**9**), $[ZrCl_2Cp^*Cp''']$ (**10**), $[ZrCl_2\{\eta\text{-}C_5H_3(Bu^t_2\text{-}1,3)\}Cp''']$ (**11**) and $[ZrCl_2\{\eta\text{-}C_5H(Bu^t_4)\}Cp''']$ (**12**).[17]

A wide range of metallocenes has been reported with between one and five alkyl substituents on each cyclopentadienyl ring.[13–15,18–40] In general, structures conformed to the expected distorted tetrahedral geometry found in bent metallocenes, with the precise geometry influenced by the steric demands of the substituents; for example, the centroid(Cp)–Zr–centroid(Cp) angle in unsubstituted (**1**) was 126.0° compared with 138.8° in bulky (**12**); the steroid-substituted complex (**13**) is illustrated.[40] The steric demands of the substituents also had a significant influence on the barrier to ring rotation as evidenced by NMR spectroscopy.[17,41–4] Several chiral metallocene dichlorides containing monosubstituted enantiomerically pure cyclopentadienyl ligands were reported and (**14**) and (**15**) are typical examples.[45–8]

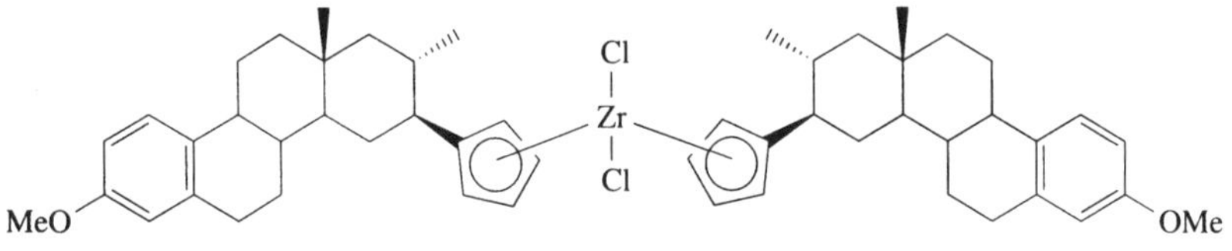

Cl
Zr
Cl
MeO
OMe

(13)

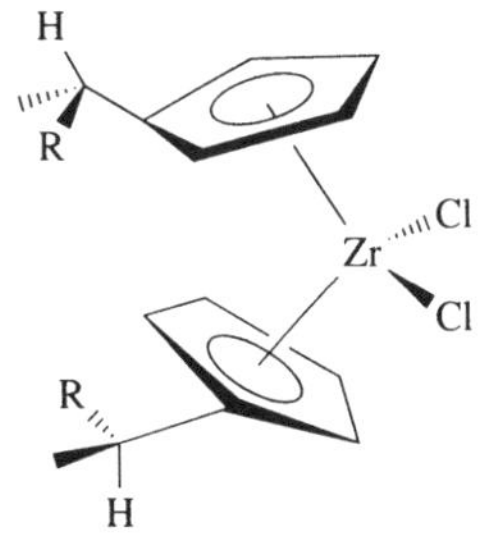

H
R
Zr
Cl
Cl
R
H
rac

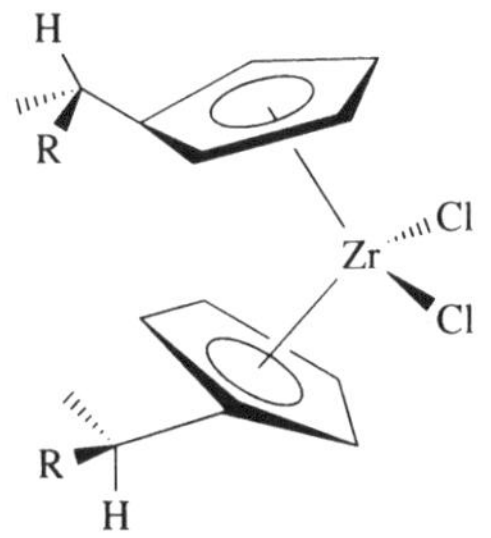

H
R
Zr
Cl
Cl
R
H
meso

(14) R = Ph   (15), R = C_6H_{11}

The electronic influence of methyl substituents was measured by He^I and x-ray photoelectron spectroscopy.[20,49–52] An increase in electron donation leads to a decrease in the binding energy of the core electrons. A linear correlation was observed between the degree of methyl substitution and the binding energy. The electron donation was largely absorbed by the metal and was only slightly transmitted to the halides. Rearrangement processes during electron impact mass spectrometry in bis(alkyl) complexes have been reported.[53,54]

The effect of introducing silyl substituents onto the cyclopentadienyl rings of zirconocene complexes has been studied. Methods 1 and 2 were the most widely used synthetic routes to $[ZrX_2(\eta\text{-}C_5H_4TMS)_2]$ (X = Cl (**16**) or Br (**17**)), $[ZrX_2\{\eta\text{-}C_5H_3(TMS)_2\text{-}1,3\}_2]$ (X = F (**18**), Cl (**19**), Br (**20**) or I (**21**)) and $[MCl_2\{\eta\text{-}C_5H_2(TMS)_3\text{-}1,2,4\}_2]$ (M = Zr (**22**) or Hf (**23**)).[55–7] Complex (**16**) and $[HfCl_2(\eta\text{-}C_5H_4TMS)_2]$

were also conveniently prepared in good yield via the reaction shown in Equation (1). Similarly, reaction of (trimethylsilyl)cyclopentadiene with MCl_4 (M = Zr or Hf) afforded $[MCl_2Cp_2]$.[58]

$$2 \quad \underset{\text{TMS} \quad \text{TMS}}{\bigtriangleup} \quad + \text{ 0.5 equiv. } MCl_4 \quad \xrightarrow{-\,2\ \text{TMS-Cl}} \quad \underset{\text{M = Zr, Hf}}{} \qquad (1)$$

The x-ray crystal structures of (**16**)–(**22**) have been reported.[55,56,59] The steric effect of introducing trimethylsilyl groups onto the cyclopentadienyl ring was reflected in an increase in the centroid–Zr–centroid angle with increased substitution. The Zr–F bond distance was significantly longer in (**18**) than in the unsubstituted complex $[ZrF_2Cp_2]$. The variation in bond length observed in the chlorides was independent of the degree of ring substitution. The monosubstituted complex (**16**) had the longest average Zr–Cl bond (0.249 1(2) nm), while the tris-substituted complex (**22**) had the shortest (0.242 4(3) nm).

The electronic influence of trimethylsilyl substituents was investigated by x-ray photoelectron spectroscopic analysis of the metal centres of (**1**), (**2**), (**16**), (**19**) and (**22**). The trimethylsilyl group was shown to be 1.25 times more electron donating than analogous methyl substituents.[60] These results contrast with reduction potential measurements which were consistent with the trimethylsilyl group being less electron donating than alkyl or hydrogen substituents.[13]

Metallocene complexes containing phosphinocyclopentadienyl ligands, such as $[ZrCl_2\{\eta\text{-}C_5H_4(PR_2)\}_2]$ (R = Ph (**24**) or Me (**25**)) and $[ZrCl_2\{\eta\text{-}C_5H_4(CH_2)_nPR_2\}_2]$ (n = 2 or 3, R = Ph; n = 2, R = Bu^t), are of interest in the preparation of heterobinuclear complexes.[61-5] Various metals including Pt, Fe, Co, Cr, Mo and W have been used as the heteroatom. In a typical reaction, (**24**) and (**25**) combined with a $Mo(CO)_4$ fragment to yield $[ZrCl_2\{\mu\text{-}\eta\text{-}C_5H_4(PR_2)\}_2Mo(CO)_4]$.[62,63,66-71] Examples of a new class of heterobimetallic complexes were prepared via the reaction of the dilithium salts of bis(cyclopentadienyl)- or *rac*-bis(indenyl)ferrocene with $ZrCl_4(THF)_2$, giving (**26**) or *rac*-(**27**); no *meso*-(**27**) was detected.[72] Bridged bis(cyclopentadienyl) ligands have provided a convenient route to homobimetallic complexes such as $[\{ZrCl_2Cp^*\}_2\{\mu\text{-}SiMe_2(C_5H_4)_2\}]$ or (**28**).[73,74]

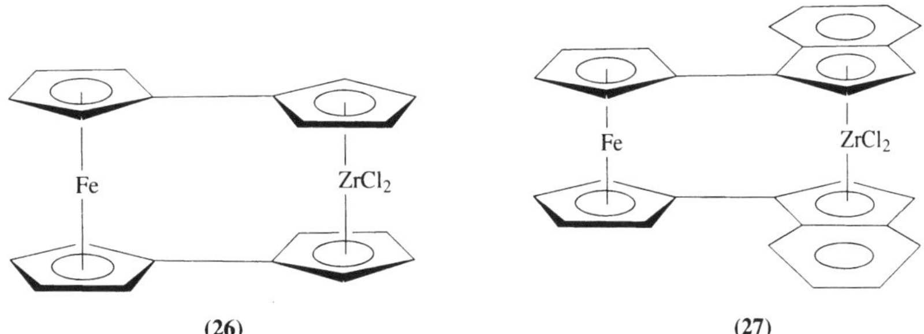

(**26**)   (**27**)

Alkenyl substituents on the cyclopentadienyl ring of metallocene dichlorides (Table 1) have provided a useful reactive site for the preparation of novel functionalized zirconocene dichlorides.[48,75,76] The complexes were prepared in good yield via the reaction of the appropriate alkenyl–cyclopentadienyl lithium reagent with $ZrCl_4(THF)_2$.

Complexes (**37**), (**39**) and (**41**) have been structurally characterized; schematic representations of their structures are given.

Catalytic hydrogenation of (**38**) on a platinum catalyst afforded the cyclohexyl complex *meso/rac*-$[Zr\{\eta\text{-}C_5H_4[CHMe(C_6H_{11})]\}_2Cl_2]$ where both phenyl rings were hydrogenated in addition to the alkenyl groups. A similar reaction on $[\{RhCl(cod)\}_2]/(+)$-diop (diop = 2,3-*O*-isopropylidene-2,3-dihydroxy-1,4-bis(diphenylphosphino)butane) afforded *meso/rac*-$[Zr\{\eta\text{-}C_5H_4[CH(Me)Ph]\}_2Cl_2]$ when the phenyl groups remained intact.[48]

Hydroboration with $H[B(C_8H_{14})]$ (9-BBN) of (**32**) and (**33**) was completely regioselective and gave a single addition product in high yield (Equation (2)).[75]

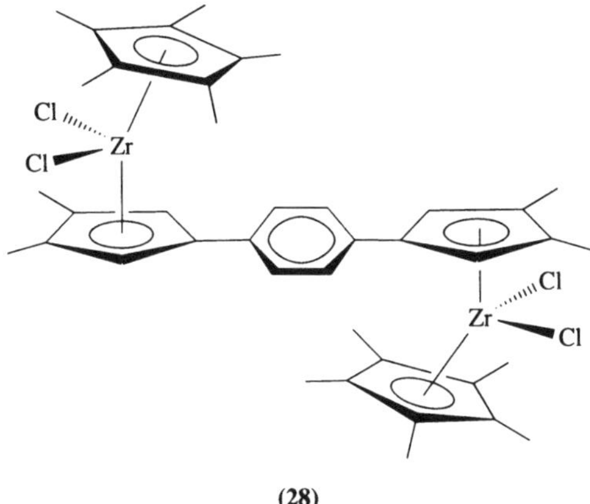

(28)

Table 1 Bis(alkenylcyclopentadienyl)metallocene complexes.

Compound		Characterization	Yield (%)	Ref.
$[Zr\{\eta\text{-}C_5H_4(\overline{CHCH = CH(CH_2)_3})\}_2Cl_2]$	**(29)**	Anal., NMR, IR	73	77
$[Hf\{\eta\text{-}C_5H_4(\overline{CHCH = CH(CH_2)_3})\}_2Cl_2]$	**(30)**	Anal., NMR, IR	55	77
$[Zr\{\eta\text{-}C_5H_4(\overline{C = CH(CH_2)_4})\}_2Cl_2]$	**(31)**	Anal., NMR, IR	74	75
$[Zr\{\eta\text{-}C_5H_4(CH_2CH = CH_2)\}_2Cl_2]$	**(32)**	Anal., NMR, IR	66	75
$[Hf\{\eta\text{-}C_5H_4(CH_2CH = CH_2)\}_2Cl_2]$	**(33)**	Anal., NMR, IR	87	75
$[Zr\{\eta\text{-}C_5H_4(CH = CMe_2)\}_2Cl_2]$	**(34)**	Anal., NMR, IR	66	75
$[Zr\{\eta\text{-}C_5H_4[CH = C(Me)Ph]\}_2Cl_2]$	**(35)**	Anal., NMR, IR	79	75
$[Zr\{\eta\text{-}C_5H_4[CPh = C(Me)H]\}_2Cl_2]$	**(36)**	Anal., NMR, IR	74	75
$[Zr\{\eta\text{-}C_5H_4[C(C_6H_{11}) = CH_2]\}_2Cl_2]$	**(37)**	Anal., NMR, IR, XSD	73	48
$[Zr\{\eta\text{-}C_5H_4[C(Ph) = CH_2]\}_2Cl_2]$	**(38)**	Anal., NMR, IR	78	48
$[Zr\{\eta\text{-}C_5H_4[C(PhMe\text{-}2) = CH_2]\}_2Cl_2]$	**(39)**	Anal., NMR, IR, XSD	90	76
meso/rac-$[Zr\{\eta\text{-}C_5H_3[C(C_6H_{11}) = CH_2\text{-}1](Pr^i\text{-}3)\}_2Cl_2]$	**(40)**	Anal., NMR, IR	90	76
meso/rac-$[Zr\{\eta\text{-}C_5H_3[CPh = CH_2\text{-}1](Pr^i\text{-}3)\}_2Cl_2]$	**(41)**	Anal., NMR, IR, XSD	86	76
meso/rac-$[Zr\{\eta\text{-}C_5H_3[C(C_6H_{11}) = CH_2\text{-}1](Me\text{-}3)\}_2Cl_2]$	**(42)**	Anal., NMR, IR	88	76

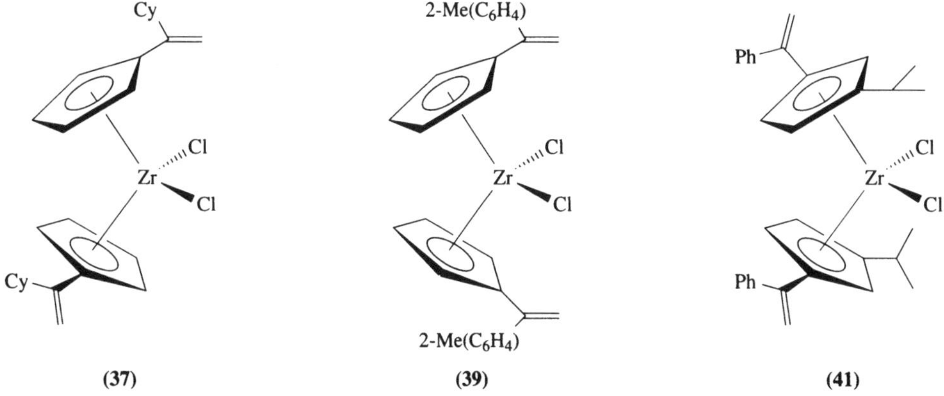

(37)   **(39)**   **(41)**

Hydroboration of the achiral complexes (**36**) and (**37**) (Equation (3)), with 9-BBN afforded a 1:1 *meso:rac* ratio of diastereoisomers of each complex which were resolved. The reaction was regioselective in each case with exclusive addition of the 9-BBN group at the β-position on the alkenyl substituent.[48]

Similarly, hydroboration of (**34**) and (**35**) was regioselective and gave *meso/rac*-$[Zr\{\eta\text{-}C_5H_4[CH(9\text{-}BBN)CHMeR]\}_2Cl_2]$ (R = Me (**43**) or Ph (**44**)) with exclusive addition of the 9-BBN group at the less sterically congested α-position.

A series of cyclobutene-bridged *ansa*-zirconocene complexes was prepared via the intramolecular photochemical [2 + 2] cycloaddition of (**39**), (**40**), (**41**) or (**42**) (Equation (4)).[76] The photostationary equilibrium reached was dependent on the nature of the alkenyl substituents and the wavelength of light used. The highest conversion was achieved with alkyl substituents and irradiation with light at 450 nm.

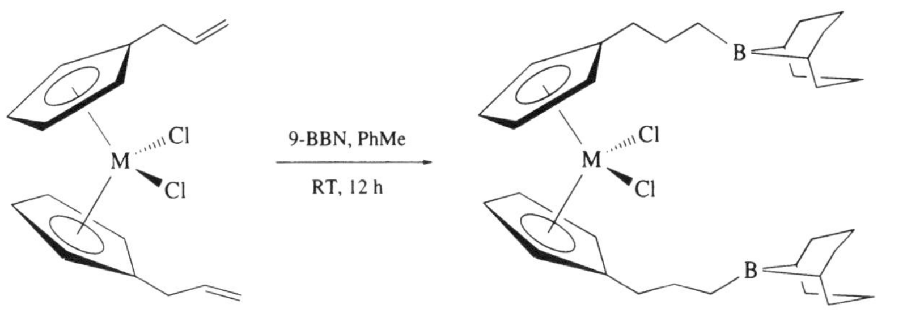

$$(2)$$

(**32**) M = Zr
(**33**) M = Hf

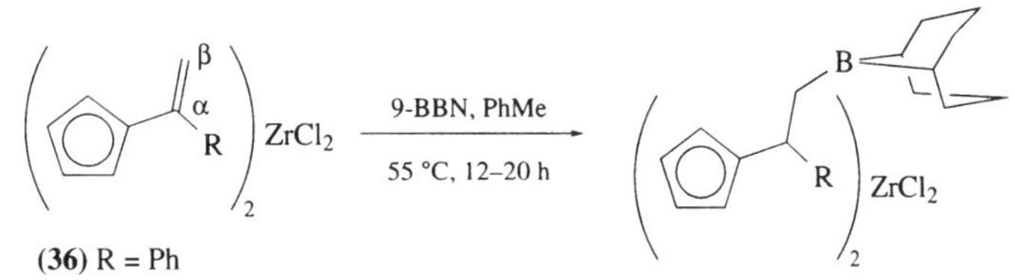

$$(3)$$

(**36**) R = Ph
(**37**) R = C_6H_{11}

meso:rac = 1:1

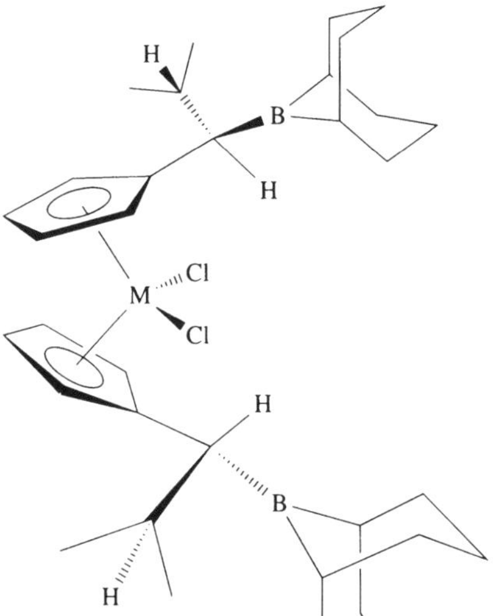

rac-(**43**) 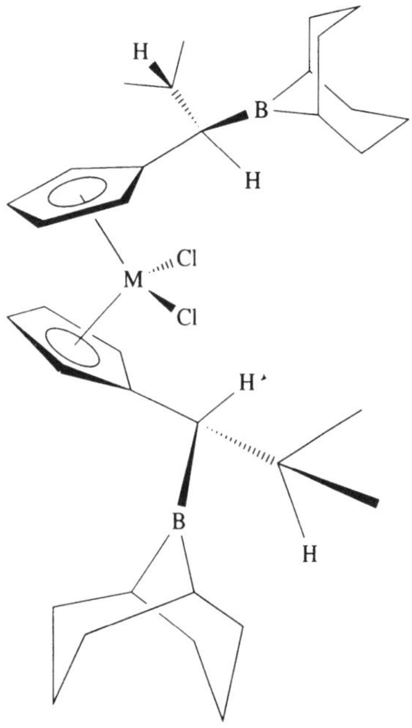   *meso*-(**43**)

The presence of aryl sustituents resulted in much lower rates of conversion to the desired cyclobutene-bridged *ansa*-zirconocene products.

The π-facial stereoselectivity of complexation to zirconium or hafnium of the C_1 symmetrical, annulated, enantiomerically pure, isodicyclopentadienyl ligands (**45**)–(**49**) was investigated.[78–84] Complexes containing bis(π-facial) ligands were reported together with complexes where the steric demands were reduced by the introduction of Cp^- or Cp^{*-} as the second cyclopentadienyl ligand.

The nonequivalence of the two faces of the cyclopentadienyl ligand in each case allows three possible modes of coordination; *exo:exo, exo:endo* or *endo:endo*. A high degree of selectivity was observed in the reaction of Li(**45**) with MCl_4 (M = Zr or Hf) and only the *exo:exo* isomer was formed in 38% (M = Zr) or 27% (M = Hf) yield (Scheme 1).[78] The effect of mixed complexation on *exo/endo* facial preference was examined in the reactions of Li(**45**) with $ZrCl_3Cp\cdot THF$ and $ZrCl_3Cp^*$.[79] Since heating was required to drive these reactions, *exo*-complexation was again favoured, in contrast with the analogous reaction with titanium where the facial preference was temperature dependent (Scheme 1).

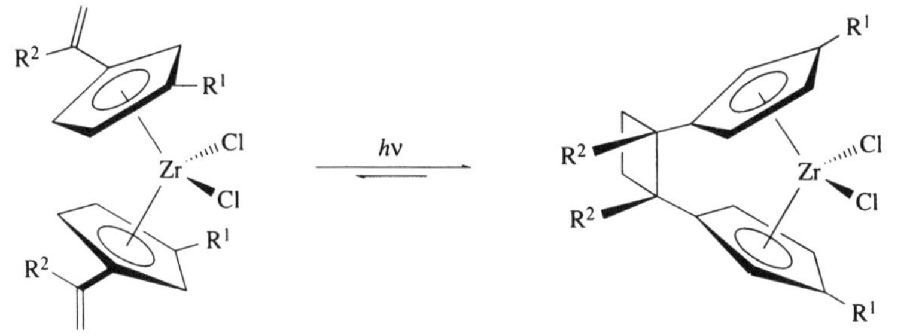

$$(4)$$

R¹ = H; R² = Cy, Ph or *o*-Tol
R¹ = Pr^i; R² = Cy, Ph
R¹ = Me; R² = Cy

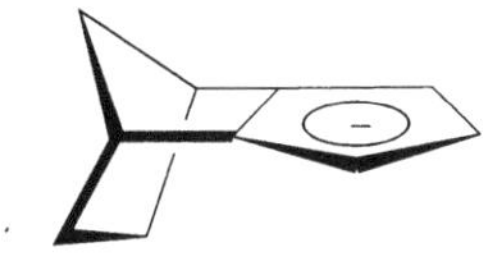

(45)   **(46)**   **(47)**

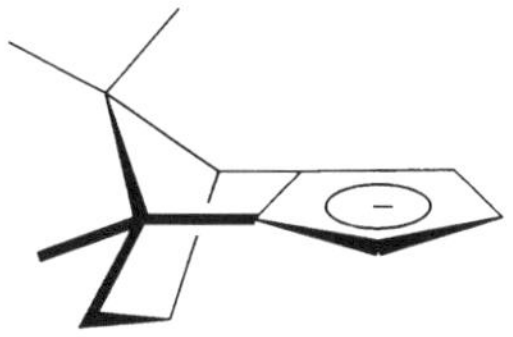

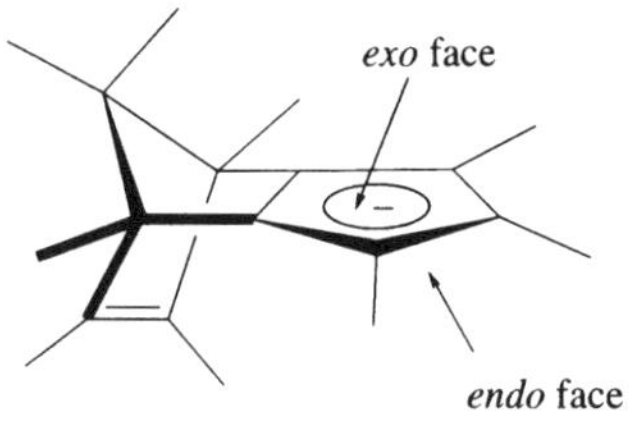

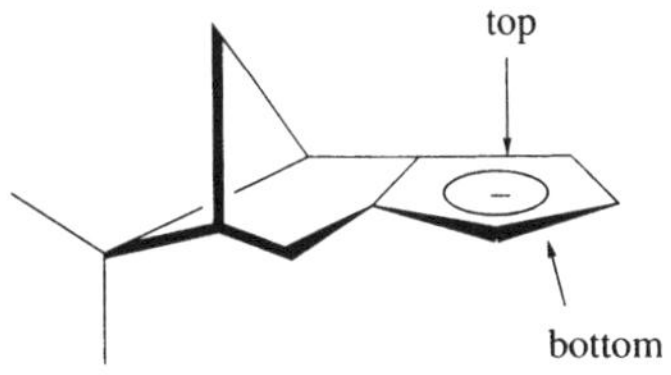

(48)   **(49)**

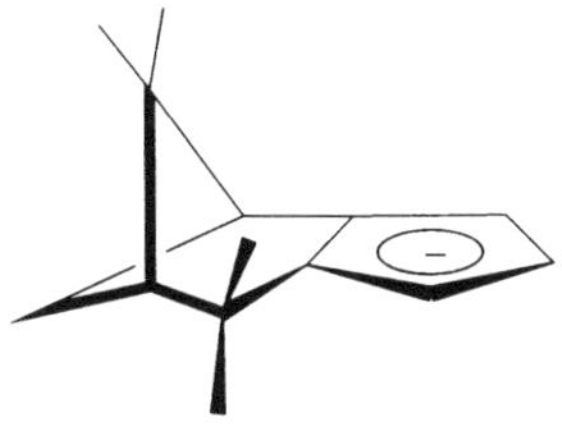

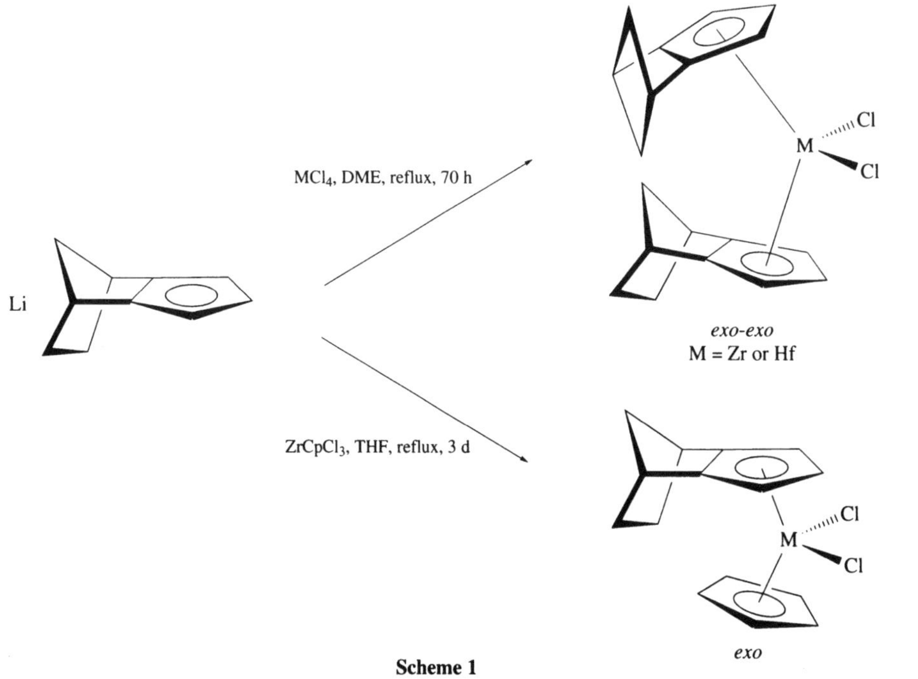

Scheme 1

Metallation of the related camphor-derived ligand (**46**) with $ZrCl_4$, $ZrCl_3Cp\cdot THF$ or $ZrCl_3Cp$ resulted in the formation of a pair of stereoisomeric complexes in each case. In contrast to the equivalent complexes of ligand (**45**), there was a marked preference for *endo*-coordination; the results are summarized in Table 2.[80] The *endo/endo* complex (**50**) has been structurally characterized. Metallation of ligand (**47**) with $ZrCl_4$ or $HfCl_4$ followed a similar pattern; the results are summarized in Table 2.[81]

Table 2   Complexes derived from the metallation of ligands (**46**) and (**47**).

Complex	Yield (%)	Stereoisomers	Ratio
[Zr(**46**)$_2$Cl$_2$]	81	*exo/endo:endo/endo*	1:32
[ZrCp (**46**)Cl$_2$]	50	*exo:endo*	1:2
[ZrCp*(**46**)Cl$_2$]	58	*exo:endo*	1:1.4
[Zr(**47**)$_2$Cl$_2$]	44	*exo/endo:endo/endo*	1:1.6
[Zr(**47**)$_2$Cl$_2$]	40	*exo/exo:exo/endo:endo/endo*	1:35:20

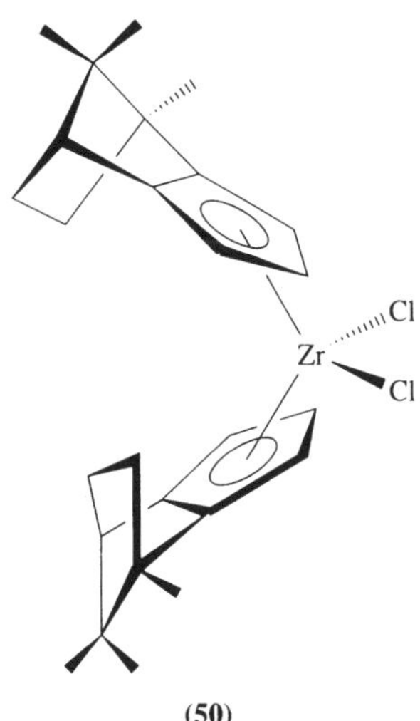

(**50**)

Similarly, the facial stereoselectivity of ligand (**48**) was examined (Scheme 2).[82] The preference for above-plane coordination fell as the steric demands of the second ligand were reduced, and a mixture of above- and below-plane coordination was observed although preference for metallation to the top face persisted. When low-temperature mixing was employed in the reaction with $ZrCl_3Cp\cdot THF$, followed by heating to reflux, the facial preference was reversed with the bottom metallated complex as the exclusive product (Scheme 2).

In contrast, the reaction of the related anion (**49**) with $ZrCl_4$ or $ZrCl_3Cp^*$ demonstrated a strong preference for coordination from its less sterically hindered bottom face to produce a single product in each case, (**50a**) or (**50b**). When (**49**) was metallated with $ZrCl_3Cp\cdot THF$, a 1:1 mixture of the two possible isomers was obtained, (**51a**) and (**51b**).[83]

Metallation of the appropriate cyclopentadiene precursor with $ZrCl_4$ yielded the enantiomerically pure C_2-symmetrical zirconocene dichloride complexes (**52**) and (**53**).[85] Other fused-ring systems reported include the cyclohexyl (**54**)[86] and cholestenyl (**55**)[87] substituted bis(indenyl) complexes and [ZrCl$_2${C$_5$H$_2$[(CH$_2$)$_5$-1,2]Me-3}] (**56**),[88] which contains a fused seven-membered hydrocarbon ring.

9.3 *ANSA*-METALLOCENES

One of the most significant areas of growth has been in the development of cyclopentadienyl ligands linked by an interannular bridge which behave as bis(pentahapto) chelates in metallocene complexes, thus restricting the normally free rotation of the cyclopentadienyl rings. The first examples of *ansa*-metallocenes containing a trimethylene bridge, [MCl$_2${(CH$_2$)$_3$(η-C$_5$H$_4$)$_2$}] (M = Zr[89,90] or Hf[91]), were reported in the early 1970s and since then various alkyl and silyl bridges have been employed in the preparation of such compounds.[92–103] The most common synthetic method has been via the reaction of the dilithium salt of the bridged cyclopentadienyl with the metal tetrahalide.

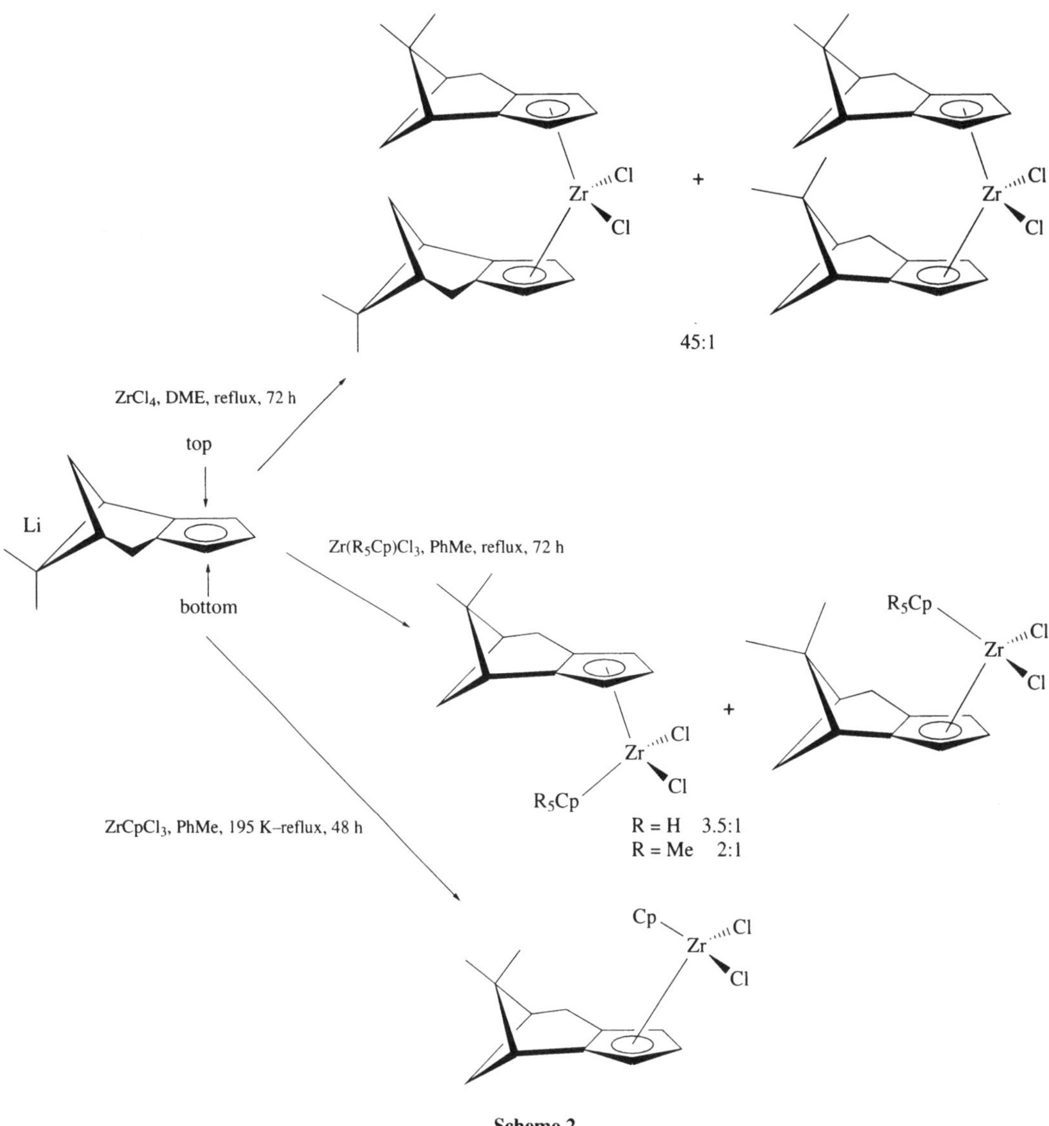

Scheme 2

The introduction of further substituents onto each cyclopentadienyl ring of the bridged ligand can produce chiral metallocene complexes; these are of particular interest as potential catalysts for stereospecific alkene polymerization. They fall into two general classes: those derived from bridged bis(indenes) or bis(fluorenes) and those derived from subtituted bis(cyclopentadienes).

9.3.1 *ansa*-Metallocenes from Substituted Bis(cyclopentadienes)

A series of silyl-bridged *ansa*-metallocene complexes which gave a mixture of *rac*- and *meso*-isomers was reported.[104,105] Representative examples are shown in Figure 1 together with their isomer product ratios. The presence of two alkyl ring substituents in (**57**) and (**58**) resulted in a significant increase in the *rac:meso* ratio. Fractional crystallization was used to separate several of the isomer pairs and *rac*-(**57**) was structurally characterized. Out-of-plane distortions in the bridging group were observed and there was a significant steric interaction between the chloride ligands and the Bu^t substituents.[105] A *rac:meso* isomer ratio of approximately 2:1 was obtained in the tetramethylethanediyl-bridged complexes [ZrCl$_2${C$_2$Me$_4$[C$_5$H$_3$(R)]$_2$}] (R = Bu^t (**59**) or TMS (**60**)).[101] The isomers of (**59**) were separated and *rac*-(**59**) was structurally characterized.[106,107]

The highly stereorigid complexes (**61**)–(**64**) containing two adjacent interannular bridges were reported. X-ray molecular structures were reported for (**61**),[108,109] (**62**)[110] and (**64**).[111] The introduction of a second bridge resulted in a significant reduction in each Cp–Zr–Cp angle compared with those found

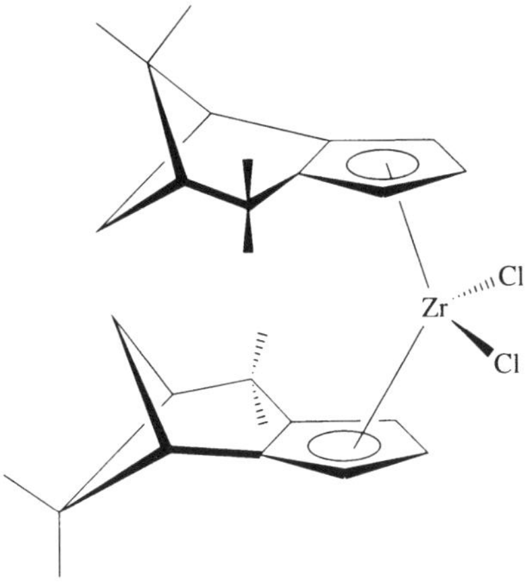

(50a)

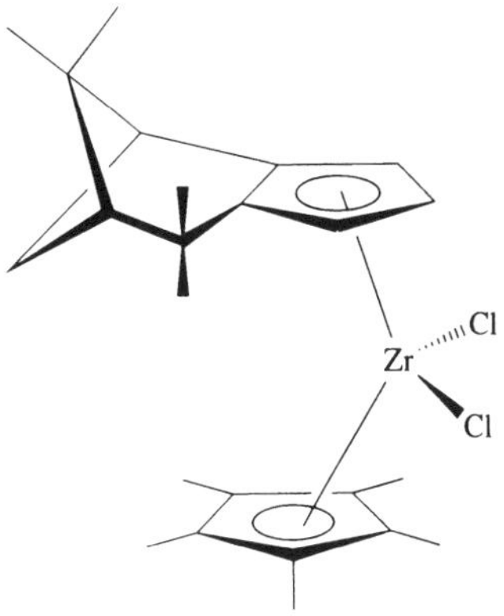

(50b)

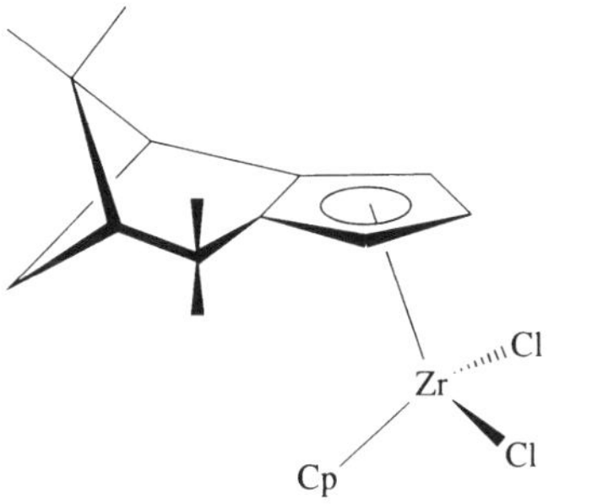

(51a)

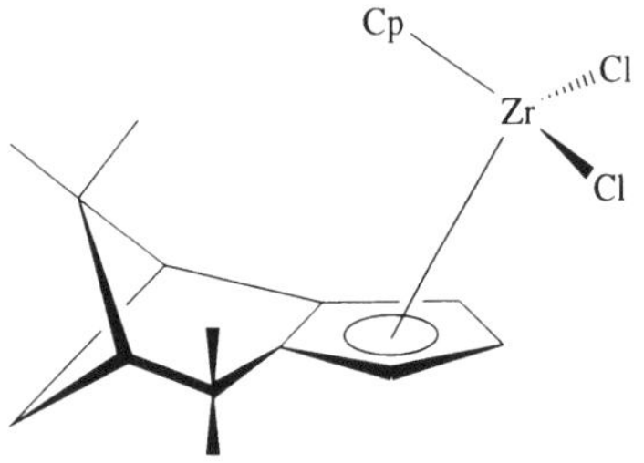

(51b)

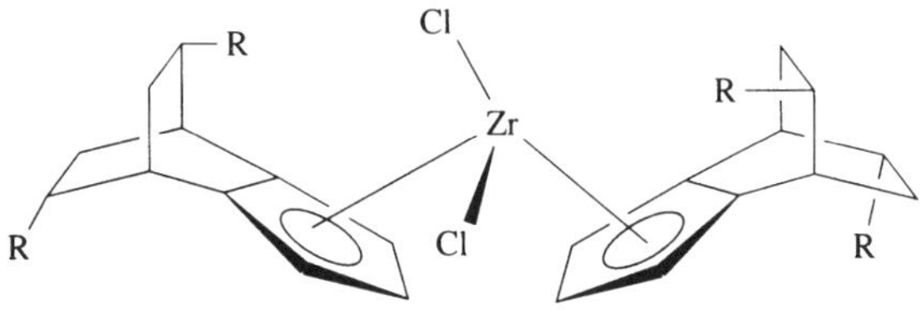

(52) R = Me or Pr^i

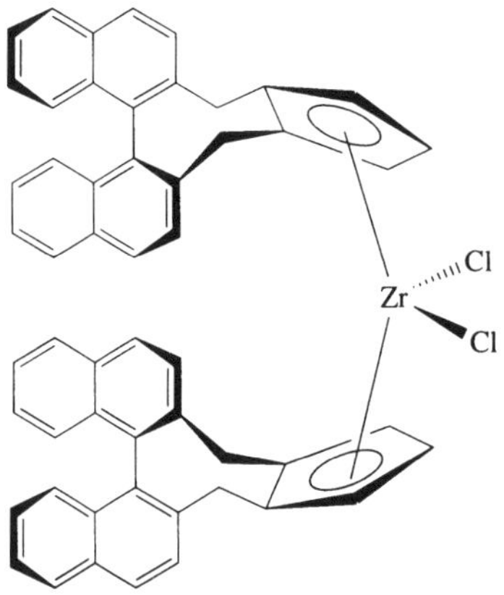

(53)

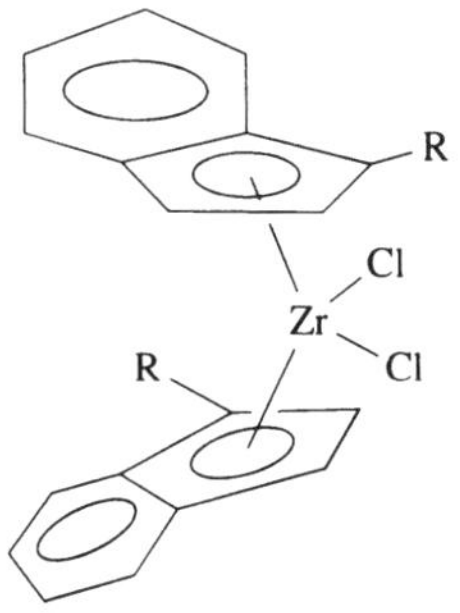

(54) R = C$_6$H$_{11}$

(55) R =

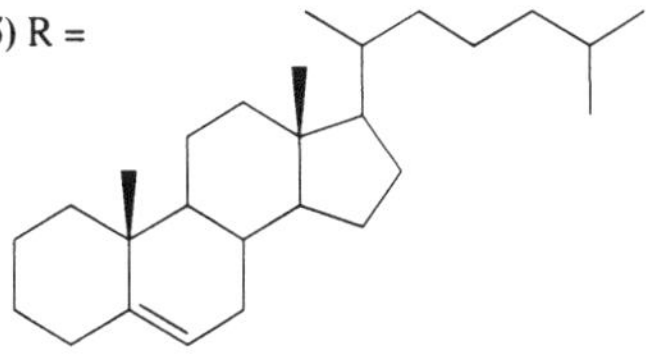

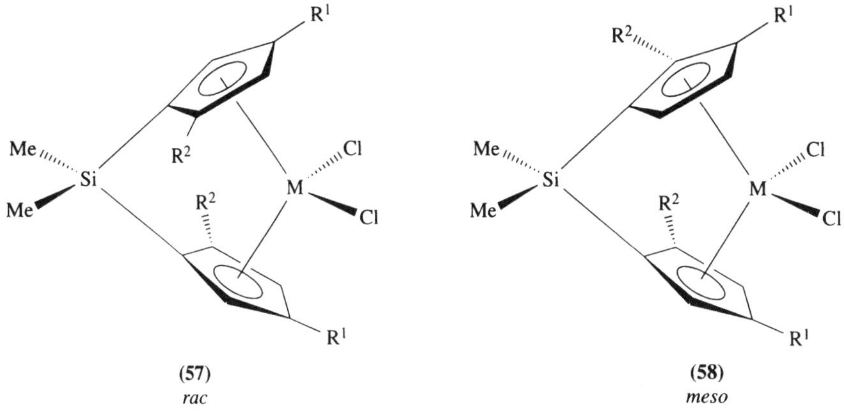

(57)
rac

(58)
meso

M = Zr or Hf, R^1 = H, R^2 = TMS, *rac:meso* = 1:1

M = Zr, R^1 = H, R^2 = Bu^t, CMe$_2$Ph, C(CH$_2$)$_5$Ph, *rac:meso* = 1:1

M = Zr, R^1 = Me, R^2 = Bu^t (44), *rac:meso* = 2:1 or Pr^i (45), *rac:meso* = 6:1

Figure 1   Silyl-bridged *ansa*-zirconocene complexes.

in singly bridged complexes. Complex (64) was chiral with a *rac:meso* isomer ratio of approximately 3:1; the isomers were separated by fractional crystallization.

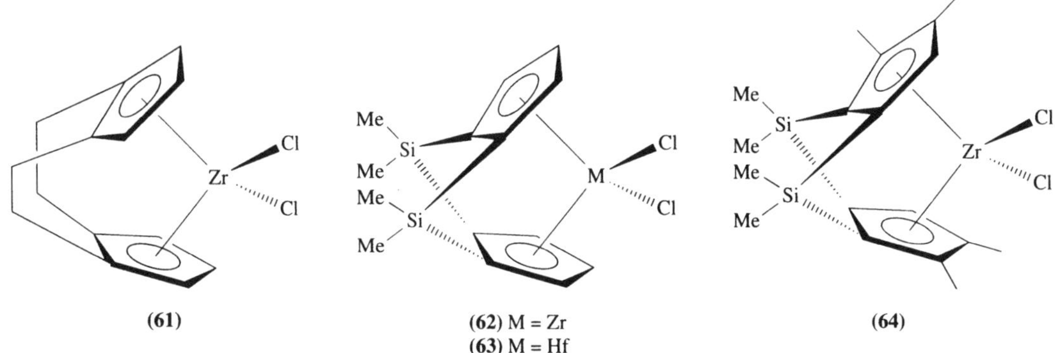

(61)   (62) M = Zr   (64)
                (63) M = Hf

In the highly rigid complex (65) the single hydrocarbyl bridge contains a fused annulated ring. The unusual bridging unit comprises a C-1 connection which is also part of a C-4 *ansa*-chain.[112]

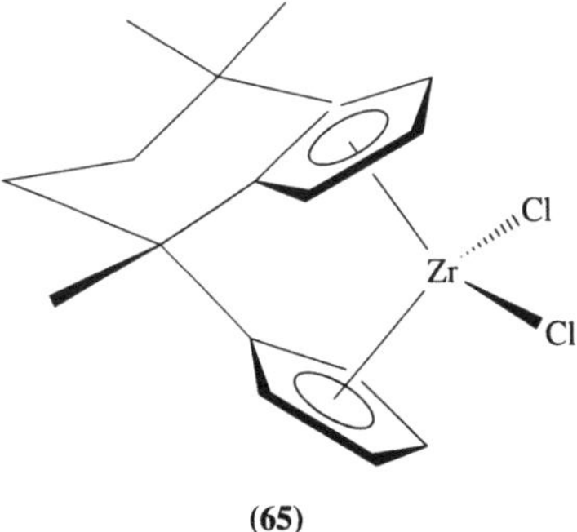

(65)

9.3.2 *ansa*-Metallocenes from Bridged Bis(indenes) and Bis(fluorenes)

Complexes (66) and (67) were prepared via the reaction of either the dilithium or dipotassium salt of the bridged indenyl ligand with MCl$_4$(THF)$_2$ (M = Zr or Hf). Catalytic hydrogenation afforded the tetrahydroindenyl complexes, (68) and (69) (Equation (5)).[113–16] The *rac* and *meso* isomers were separated in each case and the *rac* complexes were structurally characterized; *meso*-(68) was also structurally characterized.[107,114,116] The bond lengths and angles were broadly in accord with other metallocene derivatives.[60] Derivatization of *rac*-(68) with *O*-acetyl-(*R*)-mandelic acid enabled the separation of the two enantiomers to be achieved; conversion of the bis(acetyl mandelate) complexes into their dimethyl derivatives followed by treatment with ethereal HCl regenerated the enantiomerically pure dichlorides.[117] Catalytic deuteration of (66) afforded the tetradeuteroindenyl analogue of (70).[89]

$$(66)\ M = Zr \qquad (68)\ M = Zr \tag{5}$$

(66) M = Zr
(67) M = Hf

(68) M = Zr
(69) M = Hf

A wide range of metallocenes containing bridged indenyl ligands has been reported; representative examples are given in Table 3. Symmetrical bis(indenes) gave a mixture of racemic and *meso* C_2-symmetrical complexes, most commonly in a 1:1 ratio; the isomers were generally separated by fractional crystallization. Selectivity in (79) was influenced by the reaction conditions; mixing of very dilute THF solutions of each component gave a single *rac* product. Complex (81), containing a highly sterically demanding bridging ligand, was found in a *meso:rac* ratio of 7:1. The introduction of a chiral binaphthyl bridge gave an enantiomerically pure complex (80). Where the two bridged ligands were not equivalent, for example, complexes (87)–(98), C_1-symmetrical metallocenes were obtained.[133]

9.4 CATALYTIC ALKENE POLYMERIZATION

A considerable research effort has been invested in the development of homogeneous Ziegler–Natta catalysts based on zirconocenes with methylaluminoxane.[2] Active catalysts derived from chiral metallocenes produced high molecular weight polymers with a narrow molecular weight distribution.[134] Polymer activity and stereoselectivity were sensitive to both electronic and steric effects: electron-donating groups in the metallocene generally giving enhanced activity, *meso*-isomers generally producing atactic polymers and *rac*-isomers tending to produce polymers with high isotacticity. Unsubstituted achiral catalysts produced only atactic polymers.

The nature of ring substituents had a marked influence on polymer activity; for example, both molecular weight and isotacticity decreased in the order (44) > (43) > (15) > (14).[45,48,75] Selectivity was generally increased by rigidity imposed by the bridge in *ansa*-metallocenes. This, together with electronic and steric effects, has conferred a high degree of control on polymerization.[11,120,134–7] The most widely studied system has been the polymerization of propene, the polymer showing molecular weights as high as 920 000 with a 99.1% isotacticity index in (76), compared with 36 000 and 81.7% in (68).[120] In polymerizations involving the very rigid, doubly-bridged complex (64), degradation of the catalyst ligand framework occurred.[111]

Table 3   *ansa*-Metallocenes derived from bridged bis(indenes).

Complex	Ref.

(71) $R^1 = R^2 = H$ — 118
(72) $R^1 = Me, R^2 = H$ — 119
(73) $R^1 = Et, R^2 = H$ — 119
(74) $R^1 = Me, R^2 = Pr^i$ — 119
(75) $R^1 = R^2 = Ph$ — 120
(76) $R^1 = Me, R^2 = C_{10}H_7$ — 120

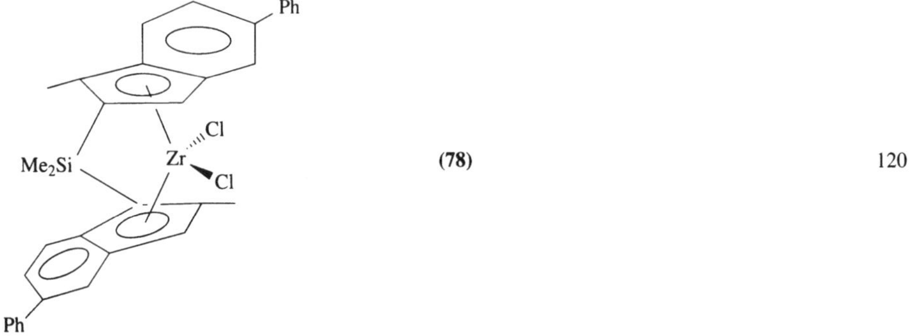

(77) — 118

(78) — 120

(79) — 122

Table 3   (continued)

Complex		Ref.
	(80)	123, 133
	(81)	124
	(82) R = H	120, 125
	(83) R = Me	120, 125
	(84)	121
	(85)	120

Table 3   (continued)

Complex		Ref.
	(86)	126
	(87) M = Zr, R^1 = H, R^2 = Me	127
	(88) M = Zr, R^1 = R^2 = Me	128
	(89) M = Zr, R^1 = R^2 = Ph	128
	(90) M = Hf, R^1 = R^2 = Me	128
	(91) M = Hf, R^1 = R^2 = Ph	128
	(92) M = Zr	128
	(93) M = Hf	128
	(94) M = Zr or **(95)** M = Hf, R = Me	129, 130
	(96) M = Zr or **(97)** M = Hf, R = Ph	131
	(98)	132

9.5 REFERENCES

1. R. L. Halterman, *Chem. Rev.*, 1992, **92**, 965.
2. P. C. Möhring and N. J. Coville, *J. Organomet. Chem.*, 1994, **479**, 1.

3. D. J. Cardin, M. F. Lappert and C. L. Raston, 'Chemistry of Organo-zirconium and -hafnium Compounds', Ellis Horwood, Chichester, 1986.
4. J. J. Byrne and S. M. Draper, *Coord. Chem. Rev.*, 1994, **134**, 171.
5. S. M. Draper and B. Twanley, *Coord. Chem. Rev.*, 1994, **134**, 189.
6. Y. Wielstra, S. Gambarotta and M. Y. Chiang, *Recl. Trav. Chim. Pays-Bas*, 1989, **108**, 1.
7. P. M. Druce, B. M. Kingston, M. F. Lappert, T. R. Spalding and R. C. Srivastava, *J. Chem. Soc. (A)*, 1969, 2106.
8. (a) G. L. Soloveichik, T. M. Arkhireeva, V. K. Bel'skii and B. M. Bulychev, *Metalloorg. Khim.*, 1988, **1**, 226; (b) M. A. Bush and G. A. Sim, *J. Chem. Soc. (A)*, 1971, 2225; (c) K. Prout, T. S. Cameron, R. A. Forder, S. R. Critchley, B. Denton and G. V. Rees, *Acta Crystallogr., Sect. B*, 1974, **30**, 2290.
9. (a) T. Fujii and H. Ishii, *Chem. Phys. Lett.*, 1989, **163**, 69, 203; A. F. Reid and P. C. Wailes, *J. Organomet. Chem.*, 1964, **2**, 329; (b) Z. Zhu, C. Mao, Y. Fang and Y. Qian, *Guangpuxue Yu Guangpu Fenxi*, 1991, **11**, 7; (c) G. Balducci, L. Bencivenni, G. De Rosa, R. Gigli, B. Martini and S. N. Cesaro, *J. Mol. Struct.*, 1980, **64**, 163; (d) M. Spoliti, L. Bencivenni, A. Farina, B. Martini and S. N. Cesaro, *J. Mol. Struct.*, 1980, **65**, 105; (e) X. Yang, X. Kao, Z. Li, S. Liang, S. Chen and Y. Wang, *Gaodeng Xuexiao Huaxue Xuebao*, 1987, **8**, 427; (f) P. Bougard, J. J. McCullough, B. G. Sayer and M. J. McGlinchey, *Inorg. Chim. Acta*, 1984, **89**, 133.
10. (a) C. S. Winter, S. N. Oliver and J. D. Rush, *Spec. Publ., R. Soc. Chem.*, 1989, **69**, 232; (b) L. Zhu and N. M. Kostic, *J. Organomet. Chem.*, 1987, **335**, 395; (c) T. V. Timofeeva, Y. L. Slovokhotov and Y. T. Struchkov, *Dokl. Akad. Nauk SSSR*, 1987, **294**, 1173; (d) J. Tau, G. Xu and L. Li, *Huaxue Xuebao*, 1983, **41**, 680; (e) Y. Huang, S. Lu, J. Hu and B. Wan, *Fundam. Res. Organomet. Chem., Proc. China–Jpn.–US Trilateral Semin. Organomet. Chem.*, 1982, 675; (f) J. Tao, G. Hu and L. Li, *Fenzi Kexue Xuebao*, 1982, **2**, 1; (g) Y. Huang, S. Lu, J. Hu and B. Wan, *Huaxue Xuebao*, 1982, **40**, 311; (h) J. W. Lauher and R. Hoffmann, *J. Am. Chem. Soc.*, 1976, **98**, 1729.
11. U. Höweler, R. Mohr, M. Knickmeier and G. Erker, *Organometallics*, 1994, **13**, 2380.
12. (a) M. K. Mincheva, Z. S. Klemenkova, I. G. Barakovskaya, B. V. Lokshin and E. M. Brainina, *Metalloorg. Khim.*, 1988, **1**, 658; (b) S. G. Ward, R. C. Taylor, P. Koepf-Maier, H. Koepf, J. Balzarini and E. De Clercq, *Appl. Organomet. Chem.*, 1989, **3**, 491; (c) S. Chang, T.-P. Chu, F.-L. Lu, C.-C. Lu, P.-C. Lu and W.-H. Hsu, *Yao Hsueh T'ung Pao*, 1981, **16**, 57; (d) P. Koepf-Maier, W. Wagner and H. Koepf, *Cancer Chemother. Pharmacol.*, 1981, **5**, 237; (e) P. Koepf-Maier, B. Hesse and H. Koepf, *J. Cancer Res. Clin. Oncol.*, 1980, **96**, 43.
13. M. F. Lappert, C. J. Pickett, P. I. Riley and P. I. W. Yarrow, *J. Chem. Soc., Dalton Trans.*, 1981, 805.
14. R. D. Rogers, M. Benning, L. K. Kurihara, K. J. Moriarty and M. D. Rausch, *J. Organomet. Chem.*, 1985, **293**, 51.
15. P. T. Wolczanski and J. E. Bercaw, *Organometallics*, 1982, **1**, 793.
16. P. G. Gassman and C. H. Winter, *Organometallics*, 1991, **10**, 1592.
17. H. Sitzmann, P. Zhou and G. Wolmershäuser, *Chem. Ber.*, 1994, **127**, 3.
18. C. J. Krüger and M. Nolte, *Z. Naturforsch., Teil B*, 1992, **47**, 995.
19. R. A. Newmark, L. D. Boardman and A. R. Siedle, *Inorg. Chem.*, 1991, **30**, 853.
20. E. Cesarotti, H. B. Kagan, R. Goddard and C. J. Krüger, *J. Organomet. Chem.*, 1978, **162**, 297.
21. Y. Wang, X. Zhou, H. Wang and X. Yao, *Huaxue Xuebao*, 1991, **49**, 1107.
22. K. Peters, E. M. Peters, H. G. von Schnering, U. Boehme and K. H. Thiele, *Z. Kristallogr.*, 1992, **198**, 288.
23. K. Peters, E. M. Peters, H. G. von Schnering, U. Boehme and K. H. Thiele, *Z. Kristallogr.*, 1992, **198**, 285.
24. S. Chen and W. Yao, *Wuji Huaxue Xuebao*, 1991, **7**, 267.
25. Q. Huang, Y. Qian and Y. Tang, *J. Organomet. Chem.*, 1989, **368**, 277.
26. P. Courtot, V. Labed, R. Pichon and J. Y. Salaun, *J. Organomet. Chem.*, 1989, **359**, C9.
27. R. F. Jordan, *J. Organomet. Chem.*, 1985, **294**, 321.
28. T. Ji, X. Han, G. Cheng, M. Bei, Y. Wu, W. Zhang and H. Guo, *Kexue Tongbao*, 1984, **29**, 1200.
29. D. M. Roddick, M. D. Fryzuk, P. F. Seidler, G. L. Hillhouse and J. E. Bercaw, *Organometallics*, 1985, **4**, 97.
30. S. Chen, Q. Liu and J. Wang, *Kexue Tongbao*, 1984, **29**, 186.
31. D. J. Mazzo, Z. B. Cheng, P. C. Uden and M. D. Rausch, *J. Chromatogr.*, 1983, **269**, 11.
32. S. Chen, Q. Liu and J. Wang, *Kexue Tongbao*, 1983, **28**, 156.
33. Y. Dong, S. Wu, R. Zhang and S. Chen, *Kexue Tongbao*, 1982, **27**, 1436.
34. H. Koepf and N. Klouras, *Chem. Scr.*, 1982, **19**, 122.
35. J. L. Petersen and J. W. Egan, *Inorg. Chem.*, 1983, **22**, 3571.
36. G. Erker *et al.*, *J. Organomet. Chem.*, 1989, **364**, 119.
37. Y. Dusausoy and J. Protas, *J. Organomet. Chem.*, 1978, **157**, 167.
38. P. R. Schonberg, T. T. Paine and E. N. Duesler, *Organometallics*, 1982, **1**, 799.
39. P. Renaut, G. Tainturier and B. Gautheron, *J. Organomet. Chem.*, 1978, **148**, 35.
40. G. Erker, C. Mollenkopf, M. Grehl and B. Schönecker, *Chem. Ber.*, 1994, **127**, 2341.
41. R. Benn, H. Grondey, G. Erker, R. Aul and R. Nolte, *Organometallics*, 1990, **9**, 2493.
42. R. A. Howie, G. P. McQuillan, D. W. Thompson and G. A. Lock, *J. Organomet. Chem.*, 1986, **303**, 213.
43. D. Braga, F. Grepioni and E. Parasini, *Organometallics*, 1991, **10**, 3735.
44. J. H. Davis, H. N. Sun, D. Redfield and G. D. Stucky, *J. Magn. Reson.*, 1980, **37**, 441.
45. G. Erker, R. Nolte, Y. H. Tsay and C. Krüger, *Angew. Chem.*, 1989, **28**, 628.
46. Q. Huang, Y. Qian and Y. Tang, *Transition Met. Chem. (London)*, 1989, **14**, 315.
47. S. Couturier, G. Tainturier and B. Gautheron, *J. Organomet. Chem.*, 1980, **195**, 291.
48. G. Erker, R. Nolte, R. Aul, S. Wilker, C. Krüger and R. Noe, *J. Am. Chem. Soc.*, 1991, **113**, 7594.
49. P. G. Gassman, D. W. Macomber and J. W. Hershberger, *Organometallics*, 1983, **2**, 1470.
50. E. Ciliberto, G. Condorelli, P. J. Fagin, J. M. Manriquez, I. Fragala and T. J. Marks, *J. Am. Chem. Soc.*, 1981, **103**, 4755.
51. C. Cauletti *et al.*, *J. Electron Spectrosc. Relat. Phenom.*, 1980, **18**, 61.
52. P. J. Fagin, J. M. Manriquez, I. Fragala and T. J. Marks, *J. Electron Spectrosc. Relat. Phenom.*, 1980, **20**, 249.
53. Y. A. Andrianov, V. P. Mar'in and O. N. Druzhkov, *Metalloorg. Khim.*, 1991, **4**, 107.
54. Y. A. Andrianov and V. P. Martin, *J. Organomet. Chem.*, 1992, **441**, 419.
55. A. Antiñolo *et al.*, *J. Chem. Soc., Dalton Trans.*, 1987, 1463.
56. R. Choukroun and F. Dahan, *Organometallics*, 1994, **13**, 2097.
57. C. H. Winter, D. A. Dobbs and X. X. Zhou, *J. Organomet. Chem.*, 1991, **403**, 145.
58. C. H. Winter, X. X. Zhou, D. A. Dobbs and M. J. Heeg, *Organometallics*, 1991, **10**, 210.

59. P. B. Hitchcock, M. F. Lappert, G. A. Lawless, H. Olivier and E. J. Ryan, *J. Chem. Soc., Chem. Commun.*, 1992, 474.
60. P. G. Gassman, P. A. Deck, C. H. Winter, D. A. Dobbs and D. H. Cao, *Organometallics*, 1992, **11**, 959.
61. W. Tikkanen and J. W. Ziller, *Organometallics*, 1991, **10**, 2266.
62. D. Morcos and W. Tikkanen, *J. Organomet. Chem.*, 1989, **371**, 15.
63. D. R. Tueting, S. R. Iyer and N. E. Schore, *J. Organomet. Chem.*, 1987, **320**, 349.
64. R. Kettenbach, W. Bonrath and H. Butenschön, *Chem. Ber.*, 1993, **126**, 1657.
65. I. E. Nifant'ev, L. F. Manzhukova, M. Y. Antipin, Y. T. Struchkov and E. E. Nifant'ev, *Metalloorg. Khim.*, 1991, **4**, 475.
66. W. A. Schenk and C. Labude, *Chem. Ber.*, 1989, **122**, 1489.
67. G. K. Anderson and M. Lin, *Inorg. Chim. Acta*, 1988, **142**, 7.
68. G. K. Anderson and M. Lin, *Organometallics*, 1988, **7**, 2285.
69. W. Tikkanen, Y. Fujita and J. L. Jeffrey, *Organometallics*, 1986, **5**, 888.
70. C. P. Casey and F. Nief, *Organometallics*, 1985, **4**, 1218.
71. A. Seyam, H. Samha and H. Hodali, *Gazz. Chim. Ital.*, 1990, **120**, 527.
72. P. Scott, U. Rief, J. Diebold and H. H. Brintzinger, *Organometallics*, 1993, **12**, 3094.
73. K. P. Reddy and J. L. Petersen, *Organometallics*, 1989, **8**, 2107.
74. S. Jüngling, R. Mülhaupt and H. Plenio, *J. Organomet. Chem.*, 1993, **460**, 191.
75. G. Erker and R. Aul, *Chem. Ber.*, 1991, **124**, 1301.
76. G. Erker, S. Wilker, C. Krüger and M. Nolte, *Organometallics*, 1993, **12**, 2140.
77. S. Chen, R. Wei and J. Wang, *Kexue Tongbao*, 1988, **33**, 648.
78. J. C. Gallucci, B. Gautheron, M. Gugelchuk, P. Meunier and L. A. Paquette, *Organometallics*, 1987, **6**, 15.
79. L. A. Paquette *et al.*, *Organometallics*, 1989, **8**, 2159.
80. L. A. Paquette, K. J. Moriarty, J. A. McKinney and R. D. Rogers, *Organometallics*, 1989, **8**, 1707.
81. V. V. Bhide, P. L. Renaldi and M. F. Farona, *Organometallics*, 1990, **9**, 123.
82. L. A. Paquette, K. J. Moriarty and R. D. Rogers, *Organometallics*, 1989, **8**, 1506.
83. K. J. Moriarty, R. D. Rogers and L. A. Paquette, *Organometallics*, 1989, **8**, 1512.
84. R. K. Sharma and C. P. Sharma, *J. Indian Chem. Soc*, 1986, **63**, 838.
85. Z. Chen, K. Eriks and R. L. Halterman, *Organometallics*, 1991, **10**, 3449.
86. C. Krüger, F. Lutz, M. Nolte, G. Erker and M. Aulbach, *J. Organomet. Chem.*, 1993, **452**, 79.
87. G. Erker, M. Aulbach, D. Wingbermühle, C. Krüger and S. Werner, *Chem. Ber.*, 1993, **126**, 755.
88. W. A. Herrmann, R. Anwander, H. Rieple, W. Scherer and C. R. Whitaker, *Organometallics*, 1993, **12**, 4342.
89. M. Hillman and A. J. Weiss, *J. Organomet. Chem.*, 1972, **42**, 123.
90. C. H. Saldarriaga-Molina, A. Clearfield and I. Bernal, *J. Organomet. Chem.*, 1974, **80**, 79.
91. C. H. Saldarriaga-Molina, A. Clearfield and I. Bernal, *Inorg. Chem.*, 1974, **12**, 2880.
92. W. Röll, H. H. Brintzinger, B. Rieger and R. Zalk, *Angew. Chem., Int. Ed. Engl.*, 1990, **29**, 279.
93. H. Koepf and N. Klouras, *Z. Naturforsch., Teil B*, 1983, **38**, 321.
94. A. Chaloyard, A. Dormond, J. Tirouflet and N. E. Murr, *J. Chem. Soc., Chem. Commun.*, 1980, 214.
95. Y. Wang, X. Zhou, X. Yao and H. Wang, *Gaodeng Xuexiao Huaxue Xuebao*, 1991, **12**, 488.
96. P. Jutzi and R. Dickbreder, *Chem. Ber.*, 1986, **119**, 1750.
97. R. Gomez, T. Cuenca, P. Royo, W. A. Herrmann and E. Herdtweck, *J. Organomet. Chem.*, 1990, **382**, 103.
98. C. S. Bajgur, W. Tikkanen and J. L. Petersen, *Inorg. Chem.*, 1985, **24**, 2539.
99. C. Qian, J. Guo and C. Ye, *Youji Huaxue*, 1991, **11**, 498.
100. F. Wochner, L. Zsolnai, G. Huttner and H. H. Brintzinger, *J. Organomet. Chem.*, 1985, **288**, 69.
101. H. Schwemlein and H. H. Brintzinger, *J. Organomet. Chem.*, 1983, **254**, 69.
102. I. E. Nifant'ev, A. V. Churakov, I. F. Urazowski, Sh. G. Mkoyan and L. O. Atovmyan, *J. Organomet. Chem.*, 1992, **435**, 37.
103. S. Chen, Y. Chen and J. Wang, *Huaxue Xuebao*, 1990, **48**, 298.
104. T. Mise, S. Miya and H. Yamazaki, *Chem. Lett.*, 1989, 1853.
105. H. Wiesenfeldt, A. Reinmuth, E. Barsties, K. Evertz and H. H. Brintzinger, *J. Organomet. Chem.*, 1989, **369**, 359.
106. S. Gutmann, P. Burger, H. U. Hund, J. Hofmann and H. H. Brintzinger, *J. Organomet. Chem.*, 1989, **369**, 343.
107. S. Collins, W. J. Gauthier, D. A. Holden, B. A. Kuntz, N. J. Taylor and D. G. Ward, *Organometallics*, 1991, **10**, 2061.
108. B. Dorer, M.-H. Prosenc, U. Rief and H. H. Brintzinger, *Organometallics*, 1994, **13**, 3868.
109. K. Hafner, C. Mink and H. J. Linder, *Chem. Ber.*, 1994, **127**, 1479.
110. A. Cano, T. Cuenca, P. Gómez-Sal, B. Royo and P. Royo, *Organometallics*, 1994, **13**, 1688.
111. W. Mengele, J. Diebold, C. Troll, W. Röll and H. H. Brintzinger, *Organometallics*, 1993, **12**, 1931.
112. G. Erker, C. Psiorz, C. Krüger and M. Nolte, *Chem. Ber.*, 1994, **127**, 1551.
113. S. Collins, B. A. Kuntz, N. J. Taylor and D. G. Ward, *J. Organomet. Chem.*, 1988, **342**, 21.
114. F. R. W. P. Wild, M. Wasiucionek, G. Huttner and H. H. Brintzinger, *J. Organomet. Chem.*, 1985, **288**, 63.
115. R. B. Grossman, R. A. Doyle and S. O. Buchwald, *Organometallics*, 1991, **10**, 1501.
116. J. A. Ewen, L. Haspeslagh, J. L. Atwood and H. Zhang, *J. Am. Chem. Soc.*, 1987, **109**, 6544.
117. A. Schaefer, E. Karl, L. Zsolnai, G. Huttner and H. H. Brintzinger, *J. Organomet. Chem., Int. Ed. Engl.*, 1987, **328**, 87.
118. W. A. Herrmann, J. Rohrmann, E. Herdtweck, W. Spaleck and A. Winter, *Angew. Chem.*, 1989, **28**, 1511.
119. W. Spaleck *et al.*, *Angew., Chem., Int., Ed., Engl.*, 1992, **31**, 1347.
120. W. Spaleck *et al.*, *Organometallics*, 1994, **13**, 954.
121. A. L. Reingold, N. P. Robinson, J. Whelan and B. Bosnich, *Organometallics*, 1992, **11**, 1869.
122. G. Erker, C. Mollenkopf, M. Grehl and R. Fröhlich, *Organometallics*, 1994, **13**, 1950.
123. M. J. Burk, S. L. Colletti and R. L. Halterman, *Organometallics*, 1991, **10**, 2998.
124. Y.-X. Cheng, M. D. Rausch and J. C. W. Chien, *Organometallics*, 1994, **13**, 748.
125. U. Stehling *et al.*, *Organometallics*, 1994, **13**, 964.
126. H. G. Alt, W. Milius and S. J. Palackal, *J. Organomet. Chem.*, 1994, **472**, 113.
127. G. H. Llinas, R. O. Day, M. D. Rausch and J. C. W. Chen, *Organometallics*, 1993, **12**, 1283.
128. M. L. H. Green and N. Ishihara, *J. Chem. Soc., Chem. Commun.*, 1994, 657.
129. J. A. Ewen, R. L. Jones, A. Razavi and J. D. Ferrara, *J. Am. Chem. Soc.*, 1988, **110**, 6255.
130. A. Razavi and J. Ferrara, *J. Organomet. Chem.*, 1992, **435**, 299.

131. A. Razavi and J. L. Atwood, *J. Organomet. Chem.*, 1993, **459**, 117.
132. B. Rieger, G. Jany, R. Fawzi and M. Steimann, *Organometallics*, 1994, **13**, 647.
133. S. L. Colletti and R. L. Halterman, *Organometallics*, 1991, **10**, 3438.
134. W. Kaminsky, K. Külper, H. H. Brintzinger and R. W. P. Wild, *Angew. Chem., Int. Ed. Engl.*, 1985, **6**, 507.
135. G. W. Coates and R. M. Weymouth, *J. Am. Chem. Soc.*, 1991, **113**, 6270.
136. J. A. Ewen, *J. Am. Chem. Soc.*, 1984, **106**, 6355.
137. G. Guerra, L. Cavallo, G. Moscardi, M. Vacatello and P. Corradini, *J. Am. Chem. Soc.*, 1994, **116**, 2988.

10

Bis(cyclopentadienyl) Metal(IV) Compounds with Si, Ge, Sn, N, P, As, Sb, O, S, Se, Te or Transition Metal-centred Ligands

EVAMARIE HEY-HAWKINS

Universität Leipzig, Germany

10.1 METALLOCENE(IV) COMPLEXES WITH SILYL, GERMYL OR STANNYL LIGANDS

10.1.1 Synthesis

Up to 1980 only six complexes were known that contained a metallocene–group 14 element bond: $[MCl(ER_3)Cp_2]$ (M = Zr, ER_3 = $SiPh_3$, $GePh_3$, $SnMe_3$ or $GeEt_3$; M = Hf, ER_3 = $SiPh_3$ or $GePh_3$), which were obtained by reaction of $[MCl_2Cp_2]$ with $LiER_3$, $NaER_3$ or $[Cd(GeEt_3)_2]$.[1] It was only in 1985 that studies of these complexes were resumed. Some results are summarized in a review.[2] Similarly, the bulky $^-E(TMS)_3$ ligand was introduced by the reaction of $[MCl_2Cp(\eta\text{-}C_5R_5)]$ with $[Li(THF)_3E(TMS)_3]$ (M = Zr or Hf, R = H or Me, E = Si; M = Zr, R = H, E = Ge).[2] Analogously, reaction of $[ZrCl(TMS)Cp_2]$ or $[(ZrClCp_2)_2O]$ with $[Li(THF)_3Si(TMS)_3]$ yielded $[Zr(TMS)\{Si(TMS)_3\}Cp_2]$ and $[Cp_2(Cl)Zr(\mu\text{-}O)Zr\{Si(TMS)_3\}Cp_2]$, respectively.[2] Replacement of the second chloro ligand was not observed for the latter. Both chloro ligands of $[ZrCl_2Cp_2]$ were replaced on reaction with $[Li(SiPh_2)_5Li]$ to give $[Zr\{SiPh_2(SiPh_2)_3SiPh_2\}Cp_2]$ in low yield.[3]

The previously reported $[MCl(EPh_3)Cp_2]$ (M = Zr or Hf, E = Si, Ge or Sn) were described as being difficult to purify due to their instability in solution at room temperature.[1] However, the analogous complexes $[MCl(EPh_3)(\eta\text{-}C_5R_5)Cp^*]$ (M = Zr or Hf, E = Si, Ge or Sn, R = H; M = Zr, E = Si, Ge or Sn, R = Me; M = Hf, E = Si, R = Me) have been obtained as stable, hydrocarbon-soluble materials. Reaction of $[ZrCl_2CpCp^*]$ with $LiSnPh_3$ yielded only the aryl complex $[ZrCl(Ph)CpCp^*]$.[4]

Silylation of $[MCl_2Cp_2]$ with $[Al(TMS)_3(OEt_2)]$ gave $[MCl(TMS)Cp_2]$ as red (M = Zr) or orange (M = Hf) solids; $[ZrCl(X)CpCp^*]$ (X = Cl or $Si(TMS)_3$) failed to react.[2] The reaction of $[ZrCl_2Cp_2]$ with $[Hg(TMS)_2]$ in benzene was earlier reported to give white solids, formulated as $[Zr(TMS)(X)Cp_2]$ (X = Cl or TMS);[5] however, the product with X = Cl was later shown to be $[ZrCl(O\text{-}TMS)Cp_2]$,[2] which was obtained by reaction of $[ZrCl(TMS)Cp_2]$ with dry oxygen.

Application of the above synthetic route to silyl complexes having Si–H bonds is limited by the scarcity of the corresponding lithium reagents. Thus, the only hydridosilyl derivatives obtained by lithium halide elimination are $[ZrMe(SiHMes_2)Cp_2]$, which is unstable above 0 °C, and the thermally stable $[ZrMe(SiHMes_2)Cp^*_2]$.[2] An attempt to replace both chlorine atoms in $[ZrCl_2Cp^*_2]$ by reaction with $[Li(THF)_2SiHMes_2]$ yielded the metallacyclic dehydrohalogenation product (1) and Mes_2SiH_2; (1) was also obtained from $[ZrCl(SiHMes_2)Cp^*_2]$ and $[Li(THF)_2SiHMes_2]$. The compound $[ZrMe(SiHMes_2)Cp^*_2]$ reacted rapidly with gaseous HCl at −78 °C to give methane and the chloro complex.[2]

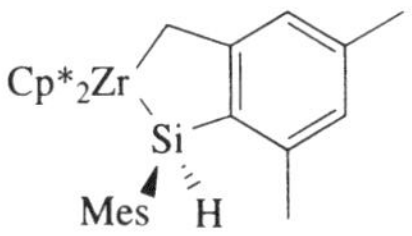

$$Cp^*_2Zr \diagdown \!\!\!\!\!\underset{\underset{Mes}{}}{Si} \!\!\!- H$$

(1)

A more general route to hydridosilyl complexes is the σ-bond metathesis of silyl or germyl complexes with primary or secondary silanes under fluorescent room light.[2,4,6] In contrast, the stannyl derivative $[HfCl(SnPh_3)CpCp^*]$ failed to react with $PhSiH_3$.[4] Unsymmetrically substituted secondary silanes yielded both diastereomers.[2,6] Depending on the nature of the reactants, further dehydrocoupling reactions of the initial hydridosilyl complex occurred with formation of zirconocene hydridochloride and a mixture of polysilanes.[6] Small Lewis bases (e.g., py or PMe_2Ph), but not large ones (e.g., PCy_3), strongly inhibited the photochemical or thermal σ-bond metathesis reactions.[2] Clearly, this process requires an empty coordination site at the metal centre.

The compound $[MCl\{Si(TMS)_3\}CpCp^*]$ (M = Zr or Hf) reacts with Ph_3SnH to give $HSi(TMS)_3$ and the triphenylstannyl derivative.[4] Secondary silanes reacted slower than primary silanes, and tertiary silanes (TMS-H or Et_3SiH) either failed to react or reacted slowly with $[HfCl\{Si(TMS)_3\}CpCp^*]$.[2] No reaction was observed between $[MCl(SiR^1_3)Cp^*(\eta\text{-}C_5R^2_5)]$ (M = Zr or Hf; R^1 = Me or TMS; R^2 = H or Me) and bulky silanes such as TMS-H, Et_3SiH, $Bu^t_2SiH_2$ or $CyMeSiH_2$.[6] Bulky cyclopentadienyl ligands also prevent the formation of isolable silyl derivatives.[4] Generally, hafnocene silyls undergo smooth σ-bond metathesis with primary or secondary silanes that introduce a smaller silyl ligand. An exception is the reaction of $[HfCl\{Si(TMS)_3\}CpCp^*]$ with trichlorosilane, which yielded only hafnocene dichloride together with unidentified silicon-containing products.[6]

An attempt to introduce alkoxysilyl ligands by reaction of $[HfCl\{Si(TMS)_3\}CpCp^*]$ with $HSiMe(OMe)_2$ gave the hydridosilyl complex $[HfCl(SiH_2R)CpCp^*]$ (R = Me); with $HSi(OMe)_3$ the silyl complex (R = H) was formed, together with hafnocene dichloride and hafnocene methoxychloride.[6]

The bimetallic silyl complex [{Cp*Cp(Cl)Hf}$_2${μ-1,4-(SiH$_2$)$_2$C$_6$H$_4$}] was obtained from the reaction of [HfCl{Si(TMS)$_3$}CpCp*] with p-(SiH$_3$)$_2$C$_6$H$_4$. The *trans* arrangement of the Hf–Cl bonds was deduced from NMR spectral data.[6]

A number of zirconium and hafnium silyl complexes of the type [MR^1(SiR$^2_3$)(η-C$_5$R$^3_5$)$_2$] (M = Zr or Hf; R^1 = Cl, alkyl or silyl; R^2 = Me, Ph or TMS; R^3 = H or Me) act as catalyst precursors for the dehydrogenative coupling reaction of silanes R^{1}SiH$_3$ to polysilanes H(SiR^{1}H)$_n$H (n = 10–20).[2] Hydridosilylzirconocene complexes have been proposed as the reactive intermediates in the dehydrocoupling of secondary organosilanes catalysed by metallocene alkyls. In some cases these complexes have been isolated from stoichiometric reactions. Thus, the reaction of [ZrMe$_2$Cp$_2$] with a substituted silane yielded [Cp$_2$(R^1)Zr(μ-H)$_2$Zr(R^2)Cp$_2$] (R^1 = PhH$_2$Si, R^2 = PhMeHSi;[2] R^1 = PhMeHSi, R^2 = PhMe$_2$Si; R^1 = Ph$_2$HSi, R^2 = Ph$_2$MeSi; R^1 = Me, R^2 = Ph$_2$HSi; R^1 = Me, R^2 = Ph$_2$MeSi; R^1 = R^2 = BuMeHSi).[7] The σ-bond metathesis route was also employed in the synthesis of [ZrH(SiHPh$_2$)(PMe$_3$)Cp$_2$], which was obtained as two geometric isomers from the reaction of [ZrH(SiPh$_3$)(PMe$_3$)Cp$_2$] with Ph$_2$SiH$_2$.[8]

The complex [Zr(η^2-1-butene)Cp$_2$] reacts with 2 equiv. of Ph$_2$SiH$_2$ to give [{Cp$_2$(Ph$_2$SiH)Zr(μ-H)}$_2$] and BuPh$_2$SiH.[9] The proposed intermediate in this reaction, [ZrH(SiHPh$_2$)Cp$_2$], resulting from oxidative addition of Ph$_2$SiH$_2$ to zirconocene(II), was trapped as the phosphine adduct [ZrH(SiHPh$_2$)(L)Cp$_2$] (L = PMe$_3$ or PMePh$_2$). However, [{Cp$_2$(Ph$_2$SiH)Zr(μ-H)}$_2$] did not react with PMe$_3$ to give a monomeric hydridosilyl complex.[9] Oxidative addition of Ph$_3$SiH to the zirconocene(II) or hafnocene(II) complexes or of diphenylsilane to the zirconocene(II) complexes [M(η^2-alkene)(PMe$_3$)Cp$_2$] (M = Zr, alkene = 1-butene; M = Hf, alkene = isobutene) also afforded [MH(SiPh$_3$)(PMe$_3$)Cp$_2$] and [ZrH(SiHPh$_2$)(PMe$_3$)Cp$_2$]. The hafnocene derivative is unstable in benzene solution.[8]

[(ZrCp$_2$)$_4$] oxidatively adds trimethylchlorosilane to give [ZrCl(TMS)Cp$_2$] as a pale yellow solid. The reaction with trimethyltin chloride afforded the pale-orange zirconocene(III) complex [{Zr(μ-SnMe$_3$)Cp$_2$}$_2$] as well as [ZrCl$_2$Cp$_2$] and Sn$_2$Me$_6$.[10] Decomposition of [ZrBu$_2$(η-C$_5$H$_4$R)$_2$] (R = H or Me) in the presence of 2 equiv. of [Sn{CH(TMS)$_2$}$_2$] yielded [Zr(Sn{CH(TMS)$_2$})$_2$(η-C$_5$H$_4$R)$_2$], which was characterized by a crystal structure determination (see Chapter 7, this volume).[11]

The η^2-silanimine complex [Zr(η^2-SiMe$_2$NBu^i)(PMe$_3$)Cp$_2$] has been prepared by the reaction of [ZrI(NBu^{i}SiMe$_2$H)Cp$_2$] with LiCH$_2$TMS in the presence of PMe$_3$ (elimination of LiI and SiMe$_4$) (see Sections 10.2.1.1 and 10.2.1.2).[2]

10.1.2 Structural Data

Crystal structures of the four-coordinate complexes [HfCl(SiH$_2$Ph)CpCp*] and [HfCl{Si-(TMS)$_3$}CpCp*],[6] and the five-coordinate complexes [Cp$_2$(SiH$_2$Ph)Zr(μ-H)$_2$Zr(SiHMePh)Cp$_2$],[2,7] [Cp$_2$(SiHMePh)Zr(μ-H)$_2$Zr(SiMe$_2$Ph)Cp$_2$],[7] [{Cp$_2$(Ph$_2$HSi)Zr(μ-H)}$_2$],[9] [ZrH(SiPh$_3$)(PMe$_3$)Cp$_2$],[8] [ZrH(SiHPh$_2$)(PMe$_3$)Cp$_2$][9] and [Zr(η^2-S$_2$CNEt$_2$)(TMS)Cp$_2$][2] have been determined. No structural data for metallocene germyl complexes are available; in [HfCl$_2${Ge(TMS)$_3$}Cp*][2] the Hf–Ge bond length is 0.2740(1) nm (cf. Chapter 11, this volume) and the only crystal structures reported for a metallocene compound with a metal–tin bond are those of [Zr(Sn{CH(TMS)$_2$})$_2$)$_2$Cp$_2$][11] and [TiCl(SnPh$_3$)Cp$_2$].[12]

10.1.3 Spectroscopic Properties

Proton, ^{13}C and ^{29}Si NMR spectral data are available for most complexes. In their ^{1}H NMR spectra the compounds [MCl(EPh$_3$)Cp(η-C$_5$R$_5$)] (M = Zr or Hf; E = Si, Ge or Sn; R = H[1] or Me[4]) exhibit sharp resonances for the phenyl protons, while [ZrCl(EPh$_3$)Cp*$_2$] (E = Si or Ge)[4] show broad signals due to restricted rotation about the E–C bond. Sharp resonances were observed for the stannyl derivative (E = Sn).[4] This observation is consistent with the expected steric demand of the EPh$_3$ group (SiPh$_3$ > GePh$_3$ > SnPh$_3$). Hydrogen-1 NMR spectral studies of [Zr(SiHMes$_2$)(X)Cp*$_2$] (X = Cl or Me) revealed significant steric interaction between the Mes and Cp* ligands, resulting in restricted rotation about both the Zr–Si and Si–C(Mes) bonds. For [ZrMe(SiHMes$_2$)Cp$_2$], unrestricted rotation was observed down to –85 °C.[2]

Silicon-29 NMR spectral data have been reported for several four-[2,4,6] and five-coordinate[6,8] metallocene silyl complexes.

Infrared spectra have been recorded for most compounds, but assignments for the ν(M–E) modes (M = Zr or Hf; E = Si, Ge or Sn) have not been made. The ν(Si–H) stretching frequencies of metallocene hydridosilyl complexes are generally observed in the range 2020–2060 cm^{-1} and are shifted by ca. 100 cm^{-1} to lower wavenumbers, relative to the corresponding hydridosilanes.[2,6]

In general, zirconocene silyl complexes are dark red to orange, and the hafnocene analogues yellow. The colour is associated with the presence of a silyl ligand; related alkyl derivatives are colourless (see Chapter 11, this volume).

10.1.4 Chemical Properties

The metallocene–group 14 element compounds are air and moisture sensitive. The compound [ZrCl(TMS)Cp$_2$] reacted with water to give [(ZrClCp$_2$)$_2$O]; with oxygen the siloxide [ZrCl(O-TMS)Cp$_2$] was formed.[2] The latter may be the product isolated from the reaction of [ZrCl$_2$Cp$_2$] with [Hg(TMS)$_2$].[5]

As was observed earlier,[1] protic species (HCl or HBr) cleave the metal–group 14 element bond. Thus, [ZrCl(EPh$_3$)Cp*$_2$] (E = Si or Ge) reacted with HCl to give zirconocene dichloride and Ph$_3$EH; the stannyl analogue (E = Sn) afforded zirconocene dichloride, but no Ph$_3$SnH was observed due to cleavage of the Sn–C(Ph) bonds.[4] Treatment of [ZrH(SiHPh$_2$)(PMe$_3$)Cp$_2$] with 3 M HCl gave Ph$_2$SiH$_2$, while [{Zr(μ-H)(SiHPh$_2$)Cp$_2$}$_2$] did not react under analogous conditions.[9] Silver triflate oxidatively cleaved the Zr–Si bond in [ZrMe(SiHMes$_2$)Cp*$_2$] with quantitative formation of pale green [ZrMe(OSO$_2$CF$_3$)Cp*$_2$] and silanes.[2]

10.1.4.1 Thermolysis

The compounds are generally stable in the solid state, but slow decomposition in solution was observed for [MCl(EPh$_3$)Cp*(η-C$_5$R$_5$)] (M = Zr or Hf, E = Si, Ge or Sn, R = H; M = Zr or Hf, E = Si, R = Me; M = Zr, E = Ge or Sn, R = Me) with elimination of Ph$_3$EH.[4]

Thermal or photochemical decomposition of [ZrCl{Si(TMS)$_3$}CpCp*] gave HSi(TMS)$_3$ and [{Zr(μ-Cl)Cp*}$_2$(η^5:η^5-C$_{10}$H$_8$)].[6] The methyl derivatives [MMe(SiPh$_3$)Cp*$_2$] (M = Zr or Hf),[4] [ZrMe{Si(TMS)$_3$}CpCp*]^2 and [ZrMe(SiHMes$_2$)Cp$_2$]^2 are thermally unstable in solution at room temperature, while [ZrMe(SiHMes$_2$)Cp*$_2$]^2 decomposed on heating in C$_6$D$_6$ with formation of Mes$_2$SiH$_2$. The hydridosilyl complexes [MCl(SiH$_2$Ph)CpCp*] decomposed at room temperature for zirconium and 75 °C for hafnium, with formation of metallocene hydridochloride and the polysilanes (SiHPh)$_n$.[2]

10.1.4.2 Hydrogenolysis

Hydrogenolysis of [Zr(TMS)(X)Cp$_2$] (X = Cl or Si(TMS)$_3$)^2 and [MCl(SiPh$_3$)Cp*(η-C$_5$R$_5$)] (M = Zr or Hf; R = H or Me)[4] rapidly yielded zirconocene hydridochloride and the corresponding silane. Reaction of [ZrMe(SiHMes$_2$)Cp*$_2$] with hydrogen showed that the Zr–Si bond is cleaved faster than the Zr–Me bond.[2] The reaction of [MMe(SiPh$_3$)Cp*$_2$] with hydrogen gave zirconocene dihydride, while the less labile hafnocene silyl derivative gave the hydridomethyl complex [HfH(Me)Cp*$_2$].[4] The germyl complexes [ZrCl(GePh$_3$)Cp*(η-C$_5$R$_5$)] (R = H or Me) react more slowly with hydrogen than the corresponding silyl complexes, yielding zirconocene hydridochloride and triphenylgermane. The complexes [MCl(SnPh$_3$)Cp*(η-C$_5$R$_5$)] (M = Hf, R = H; M = Zr, R = Me) failed to react.[4] Hydrogenolysis of [Zr(η^2-SiMe$_2$NBu^t)(PMe$_3$)Cp$_2$] yields [ZrH{NBu^t(SiHMe$_2$)}Cp$_2$].[2]

10.1.4.3 Salt elimination

The chloride in [MCl(SiR$^1_3$)Cp$_2$] can be displaced by a salt-elimination route yielding [M(R^2)(SiR$^1_3$)Cp$_2$] (M = Zr or Hf, R^1 = Me, R^2 = η^2-S$_2$CNEt$_2$, (μ-H)$_2$BH$_2$ or OBu^t; M = Zr, R^1 = TMS, R^2 = (μ-H)$_2$BH$_2$).[2,13] The compound [ZrCl(TMS)Cp$_2$] reacted with 2-furyl- or 2-thienyllithium with elimination of LiCl. The substitution products underwent dyotropic rearrangement to give an oxa- or thiazirconacyclohexadiene complex.[13]

Reaction of [MCl(SiR$^1_3$)(η-C$_5$R$^2_5$)(η-C$_5$R$^3_5$)] with MeMgBr gave the corresponding methylsilyl complexes (M = Zr, R^1 = TMS, R^2 = H, R^3 = H or Me;[2] M = Zr or Hf, R^1 = Ph, R^2 = R^3 = Me).[4]

10.1.4.4 Insertion reactions

Pioneering studies by Tilley and by Harrod have demonstrated that metallocene silyl complexes are reactive towards insertion of unsaturated substrates such as CO or isonitriles, and participate in silane polymerization and σ-bond metathesis reactions.[2]

(i) Carbon monoxide

The reactivity of the metal–silicon bond can be influenced by changing the substituents on the metal or silicon atom. While [ZrCl(TMS)Cp$_2$] is readily carbonylated to give [Zr(η^2-CO-TMS)(Cl)Cp$_2$], the Si(TMS)$_3$ analogue is not.[2]

In the reaction of [Zr(TMS){Si(TMS)$_3$}Cp$_2$] or [ZrMe{Si(TMS)$_3$}Cp$_2$] with CO, insertion into the Zr–TMS or Zr–Me bond occurred exclusively, with formation of the corresponding η^2-silaacyl or η^2-acyl complex.[2] The presence of the bulkier cyclopentadienyl ligand, $^-$Cp, in [ZrCl{Si(TMS)$_3$}CpCp*] results in higher reactivity of the metal–silicon bond due to increased steric interaction around the metal centre. Thus, reaction with CO readily gave [Zr{η^2-COSi(TMS)$_3$}(Cl)CpCp*].[2] The same trend in reactivity is observed in the reaction of [ZrCl(SiPh$_3$)Cp*(η-C$_5$R$_5$)] (R = H or Me) with CO. While the complex with R = H failed to give a pure product, the more crowded compound (R = Me) cleanly inserts CO to give [Zr(η^2-COSiPh$_3$)(Cl)Cp*$_2$].[4] On reaction with gaseous HCl these silaacyl complexes formed zirconocene dichloride and the first stable formylsilanes, (TMS)SiCHO[2] and Ph$_3$SiCHO.[4] Like the chloro derivative, [ZrMe{Si(TMS)$_3$}CpCp*] reacted with CO to give [Zr{OC(=CH$_2$)-Si(TMS)$_3$}(H)CpCp*].[2]

The influence of steric effects was also evident in the reaction of [ZrMe(SiHMes$_2$)(η-C$_5$R$_5$)$_2$] (R = H or Me) with CO. While the Cp* complex yielded an η^2-silaacyl derivative (insertion into Zr–Si), the Cp complex formed an η-acyl derivative (insertion into Zr–Me). The former isomerized slowly in benzene solution to give [Zr{OC(=CH$_2$)SiHMes$_2$}(H)Cp*$_2$].[2] Due to the lower reactivity of Zr–Ge and Zr–Sn compared with Zr–Si bonds, [ZrCl(EPh$_3$)Cp*(η-C$_5$R$_5$)] (E = Ge, R = H; E = Ge or Sn, R = Me) failed to react with CO.[4]

(ii) Alkene insertion

The enhanced reactivity of zirconocene silyl complexes with sterically demanding ligands is also demonstrated by the photochemically induced reaction of [ZrCl(SiR$_3$)Cp$_2$] (R = Me or TMS) or [ZrCl{Si(TMS)$_3$}CpCp*] with ethene. Whereas the former did not react,[2] the latter gave the insertion product [Zr{CH$_2$CH$_2$Si(TMS)$_3$}(Cl)CpCp*].[14] Reaction of [ZrH(SiHPh$_2$)(L)Cp$_2$] (L = PMe$_3$ or PMePh$_2$) with 1-octene gave the hydrosilylation product *n*-OctPh$_2$SiH. The complex [{Zr(μ-H)(SiHPh$_2$)Cp$_2$}$_2$] did not react with 1-octene under analogous conditions.[9]

The compound [Zr(η^2-SiMe$_2$NBu^i)(PMe$_3$)Cp$_2$] reacts with ethene or formaldehyde with insertion into the Zr–Si bond to yield the five-membered metallacycles [$\overline{\text{Zr(CH}_2\text{CH}_2\text{SiMe}_2\text{NBu}^i)}Cp_2$] or [$\overline{\text{Zr(OCH}_2\text{SiMe}_2\text{NBu}^i)}Cp_2$].[2]

(iii) Alkynes

To date insertion of alkynes into M–E bonds has not been observed.[2,4] The compound [ZrH(SiPh$_3$)(PMe$_3$)Cp$_2$] reacts with dimethylacetylene to give [$\overline{\text{Zr(CMe=CMeCMe=CMe)}}Cp_2$].[8]

(iv) Isonitriles

The complex [MCl(ER$^1_3$)Cp*(η-C$_5$R$^2_5$)] cleanly inserted CN-2,6-Me$_2$C$_6$H$_3$ to give the η^2-iminosilaacyl or η^2-iminogermaacyl complexes [M{η^2-C(N-2,6-Me$_2$C$_6$H$_3$)ER$^1_3$}(Cl)Cp*(η-C$_5$R$^2_5$)] (M = Hf, E = Si, R^1 = Ph, R^2 = H; M = Zr or Hf, E = Si, R^1 = Ph, R^2 = Me; M = Zr, E = Ge, R^1 = Ph, R^2 = H;[4] M = Zr, E = Si, R^1 = Me or TMS, R^2 = H^2). The stannyl complexes [MCl(SnPh$_3$)Cp*(η-C$_5$R$_5$)] (M = Hf, R = H; M = Zr, R = Me) did not react.[4]

The methyl complex [ZrMe{Si(TMS)$_3$}CpCp*] underwent insertion of the isonitrile exclusively into the Zr–Me bond,[2] while Bu^{i}NC was inserted into the Zr–H bond of [ZrH(SiPh$_3$)(PMe$_3$)Cp$_2$] to give [Zr{η^2-C(H)NBu^i}(SiPh$_3$)Cp$_2$].[8]

(v) Nitriles or pyridine

Insertion of nitriles (RCN; R = Me, CH=CH$_2$ or Ph) into the Zr–Si bond of [ZrMe{Si(TMS)$_3$}CpCp*] gave [ZrMe{N=C(R)Si(TMS)$_3$}CpCp*]. Reaction with pyridine results in 1,2 addition of the Zr–Si bond to pyridine. The silyl complex was much less reactive toward substituted pyridines. The related complexes [Zr{Si(TMS)$_3$}(X)Cp(η-C$_5$R$_5$)] (X = Me, R = H; X = Cl, R = Me) were much less reactive toward a nitrile or pyridine.[2] The complex [ZrH(SiPh$_3$)(PMe$_3$)Cp$_2$] undergoes insertion of Bu^{t}CN into the Zr–H bond yielding [Zr(η^1-N=CHBu^t)(SiPh$_3$)Cp$_2$].[8]

(vi) Others

The compound [ZrH(SiPh$_3$)(PMe$_3$)Cp$_2$] underwent insertion of acetone into the Zr–H bond to give [Zr(OCHMe$_2$)(SiPh$_3$)Cp$_2$].[8] No reaction of [{Zr(μ-H)(SiHPh$_2$)Cp$_2$}$_2$] with acetone was observed.[9]

Activation of the Si–H bond in hydridosilyl complexes (M–SiHR$_2$) to produce terminal silylene (M = SiR$_2$) or bridging silylene (M–SiR$_2$–M) complexes has not been observed to date.

10.2 METALLOCENE(IV) COMPLEXES WITH NITROGEN-CENTRED ANIONIC LIGANDS

10.2.1 Metallocene(IV) Amides

10.2.1.1 Synthesis and spectroscopic properties

A general synthetic approach to metallocene bisamido complexes which was also employed earlier[1] is the reaction of [M^{1}Cl$_2$Cp$_2$] with alkali metal amides M^{2}NR^{1}R^2 (M^2 = Li or Na) (M^1 = Zr, NR^{1}R^2 = pyrrolyl[1] or 2,5-dimethylpyrrolyl;[15] M^1 = Zr, R^1 = R^2 = TMS;[16] M^1 = Zr or Hf, R^1 = H, R^2 = Mes;[17] M^1 = Zr, R^1 = H, R^2 = 2,6-Me$_2$C$_6$H$_3$, 4-Bu^{t}C$_6$H$_4$, o-Tol[18] or Ph[19]). Reaction of zirconocene hydridochloride with LiN(TMS)$_2$ gives [M{CH$_2$SiMe$_2$N(TMS)}Cp$_2$] (M = Zr), formed via a facile γ-hydrogen elimination. Interestingly, [TiCl$_2$Cp$_2$] reacts with the lithium amide with formation of the metallacycle (M = Ti) exclusively, independent of the stoichiometry employed.[16]

The metallocene dichlorides of titanium, zirconium and hafnium react with equivalent amounts of the dilithium salts of 2-aminobenzenethiol or o-phenylenediamine to give [M(1-NH-2-X-C$_6$H$_4$)Cp$_2$] (X = S or NH). Variable-temperature ^{1}H NMR spectral studies suggest a folding of the five-membered MNC$_2$S metallacycles along the S$\cdots$N axis (see Section 10.5.1.2). While the hafnocene derivative (X = S) shows two signals for the Cp ligands below −17 °C, no fluxionality is observed for the titanium and zirconium complexes. All three o-phenylenediamido complexes gave one signal for the Cp protons, suggesting rapid ring inversion.[20]

Alkyl amides [Zr(NHR^2)(R^1)Cp$_2$] (R^1 = Me, R^2 = 2,6-Me$_2$C$_6$H$_3$, 4-Bu^{t}C$_6$H$_4$ or Bu^t;[18,21,22] R^1 = CH$_2$CH$_2$Bu^t, R^2 = o-Tol or SiMe$_2$Bu^t)[18] are accessible from [ZrCl(R^1)Cp$_2$] by the salt elimination route. The methylzirconocene amides (R^1 = Me) react with a further equivalent of amine to give [Zr(NHR^2)$_2$Cp$_2$] (R^2 = 2,6-Me$_2$C$_6$H$_3$, 4-Bu^{t}C$_6$H$_4$ or Bu^t).[18] Alternatively, [ZrMe$_2$Cp$_2$] reacted with 4-t-butylaniline with loss of 1 equiv. of methane and generation of the corresponding amido complex.[21] Methane elimination was also exploited in the synthesis of [Zr(NHR)$_2$Cp$_2$] (R = 2,6-Me$_2$C$_6$H$_3$, 4-Bu^{t}C$_6$H$_4$, Bu^t or SiMe$_2$Bu^t) from dimethylzirconocene and an excess of H$_2$NR.[18]

Mixed zirconocene alkoxo-amides [Zr(NMe$_2$)(OR)Cp$_2$] are obtained by reaction of [Zr(NMe$_2$)$_2$Cp$_2$] with ROH (R = 2,6-Bu$^t_2$C$_6$H$_3$, 2,6-Bu$^t_2$-4-Me-C$_6$H$_2$ or 2,4,6-Bu$^t_3$C$_6$H$_2$). However, the zirconocene alkoxo-chloride [ZrCl(OR)Cp$_2$] did not react with LiNMe$_2$ (R = 2,6-Bu$^t_2$-4-Me-C$_6$H$_2$).[23]

The compound [MCl$_2$Cp$_2$] (M = Zr or Hf) reacts with an amine in the presence of base. Thus, mono- and disubstituted complexes of 3-substituted 2-mercaptoquinazol-4-ones[24] or heterocyclic thiones[25,26] were prepared. The ligands act as N,S-bidentate chelating agents. In the IR spectrum new bands at ca. 460–480 cm^{-1} and 440–480 cm^{-1} were assigned to the ν(Zr–N) and ν(Hf–N) mode, respectively.[25,26]

The complex [ZrCl$_2$Cp$_2$] reacts with base (NEt$_3$) and an aliphatic acid dihydrazide to give [Zr{N(NH$_2$)C(=O)(CH$_2$)$_n$C(=O)N(NH$_2$)}Cp$_2$] (n = 0, 2 or 4).[27] Derivatives of 6-aminopenicillinic acid were obtained in the same way.[28]

Reaction of hafnocene or zirconocene dihydride with excess NH$_3$[29] or of hafnocene dihydride with excess MeNH$_2$[30] gave [MH(NH$_2$)Cp*$_2$] (M = Zr or Hf)[29] and [(HfH(NHMe)Cp*$_2$].[30] The zirconium complex is also obtained from the oxidative addition of ammonia to [{Zr(N$_2$)Cp*$_2$}$_2$(N$_2$)].[29] Hafnocene hydridoamides [HfH(NHR)Cp*$_2$] (R = Ph or p-Tol) are also accessible by the reaction of hafnocene dihydride with a primary amine (elimination of H_2) or thermolysis of the corresponding triazenido complex [HfH(NHNNR)Cp*$_2$] (elimination of N_2).[31] Alternatively, the bisamido complexes [Hf(NHR)$_2$Cp*$_2$] (R = Ph or *p*-Tol) were obtained by reacting hafnocene dihydride, the corresponding monoamido complex, or the hafnocene triazenido complex with an excess of aryl azide RN$_3$ (R = Ph or *p*-Tol). The same product is obtained from the hafnocene monoamido complex and a primary amine at 80 °C.[31]

Zirconocene dihydrido complexes react with isonitriles to give formimidoyl complexes. Subsequent reaction with hydrogen or thermolysis yields methylamidozirconocene complexes.[32] Alternatively, zirconocene hydridochloride or dimethylzirconocene react with benzyl isonitrile to give imidoyl complexes. Subsequent reaction with a zirconocene hydrido derivative yields zirconocene monoamido complexes.[33] The dialky complexes [MR$_2$Cp$_2$] (M = Zr or Hf; R = 6-methylpyridylmethyl) undergo insertion of 2,6-dimethylphenyl isocyanide into the M–C bond to give vinylamido complexes formed by a facile 1,2-hydrogen shift in intermediate η^2-iminoacyl complexes.[34,35] In a comparable reaction, insertion of two isonitriles CNR (R = 2,6-Me$_2$C$_6$H$_3$) into the Zr–C bond of [Cp*(Cl)Zr(η-C$_5$Me$_4$CH$_2$)] is observed; at 140 °C the resulting imidoyl complex undergoes thermal rearrangement to an amido complex. The product is formed via a 1,3-hydrogen shift from the original fulvene methylene group to a carbon atom of the second inserted isonitrile with concurrent formation of a Zr–N σ bond.[36]

Hydrozirconation of phosphaimines $R^1P = NR^2$ (R^1 = N(TMS)$_2$, R^2 = TMS;[37,38] R^1 = N(TMS)Bu^t, R^2 = Bu^t; R^1 = tetramethylpiperidino, R^2 = TMS[37]) or the bis(imino)phosphorane (TMS)$_2$NP(=N-TMS),[37,39] with zirconocene hydridochloride gives [Zr(NR^{2}PHR^1)(Cl)Cp$_2$] or [Zr{N(TMS)P(H){N(TMS)$_2$}N(TMS)}(Cl)Cp$_2$], respectively. Ring opening of [Zr(NR^{2}PR^{1}H)(Cl)Cp$_2$] (R^1 = N(TMS)$_2$, R^2 = TMS; R^1 = N(TMS)Bu^t, R^2 = Bu^t) is observed on reaction with [Fe$_2$(CO)$_9$], S$_8$ or Se. The metallacyclic structure is retained on reaction with RSO$_3$CF$_3$ (R = TMS or Me) or Na[BPh$_4$]/MeCN.[37,38]

The complex [Zr(NBu^t–BPh–NBu^t)Cp$_2$] was prepared from [ZrCl$_2$Cp$_2$] and Li$_2$(NBu^t–BPh–NBu^t).[40] Analogously, [Zr(η^2-SiMe$_2$NBu^t)(PMe$_3$)Cp$_2$] was prepared from [ZrI{NBu^t(SiMe$_2$H)}Cp$_2$] and LiCH$_2$TMS (see Section 10.1.1).[2]

Insertion of an isocyanate RNCO (R = Ph or 1-naphthyl (nap)) or *p*-tolyl carbodiimide into one Zr–C bond of [ZrMe$_2$Cp$_2$] gives [Zr{XC(Me)NR}(Me)Cp$_2$] (X = O, R = Ph or nap; X = NR, R = *p*-Tol).[41] The alkyne complex [Zr(C$_2$Ph$_2$)Cp*$_2$] cleanly inserts Ph$_2$CN$_2$ or TolN$_3$ with formation of [Zr(CPh=CPhNR)Cp*$_2$] (R = N=CPh$_2$ or N=NTol).[42]

Oxidative addition of a diazadiene to zirconocene(II) yields [Zr(NPhCR=CRNPh)Cp$_2$] (R = Ph or Me). The hafnocene analogues were not accessible by this route. The air- and moisture-sensitive products exhibit dynamic NMR spectra indicating a rapid intramolecular migration of the bent Cp$_2$M fragment from one 'face' of the reduced diazadiene to the other.[43] A comparable synthetic approach is the oxidative addition of a 1-azadiene, R^1N=CH–CH=CHR^2, to zirconocene(II) yielding [Zr(NR^{1}CH=CHCHR^2)Cp$_2$] (R^1 = Bz, Ph, CHMePh or NMe$_2$, R^2 = Ph; R^1 = Bz or Bu^t, R^2 = Me). Variable-temperature NMR spectroscopy indicates a ring flipping analogous to that observed in zirconocene enediamido complexes. An alternative route to these complexes starts with an allylic amide LiN(CH$_2$CHCH$_2$)(R) (R = Ph or TMS) and [ZrCl(Me)Cp$_2$]. Loss of methane from [ZrMe{N(CH$_2$CHCH$_2$)R}Cp$_2$] yields an intermediate η^2-imine complex, which rearranges to [Zr(NRCH=CHCH$_2$)Cp$_2$].[44] Similarly, the generation of a transient acyclic zirconocene η^2-imine complex and successive trapping with an alkyne is a useful synthetic approach to 1-zircona-2-azacyclopentenes.[22,45,46]

Isolable metallaaziridenes [Zr(NR^{1}CHR^2)(L)Cp$_2$] (L = THF, R^1 = TMS, R^2 = H, Pr or Ph; L = PMe$_3$, R^1 = TMS, R^2 = Ph, 2-furyl or 2-thienyl; L = PMe$_3$, R^1 = R^2 = Ph) were obtained from methylzirconocene chloride and LiN(CH$_2$R^2)R^1.[46]

10.2.1.2 *Structural data*

Structural data are available for [Zr(η^1-NC$_4$H$_2$Me$_2$)$_2$Cp$_2$],[15] [HfH(NHMe)Cp*$_2$],[30] [Hf{N(2,6-Me$_2$C$_6$H$_3$)(6-methylpyridylvinyl)}$_2$Cp$_2$],[34,35] [Zr(NHPh)(OSO$_2$CF$_3$)Cp$_2$],[19] [Zr{N(TMS)P(H)N(TMS)$_2$}(Cl)Cp$_2$],[38] [Zr(η^2-SiMe$_2$NBu^t)(PMe$_3$)Cp$_2$]^2, [Zr(NPhCPh=CPhNPh)Cp$_2$][43] and [Zr{η^2-N(TMS)CHPh}(THF)Cp$_2$].[46]

The structure of the bis(2,5-dimethylpyrrolyl)zirconocene complex[15] is very similar to that of the previously described pyrrolyl derivative [Zr(η^1-NC$_4$H$_4$)$_2$Cp$_2$].[1] The methylamido group in [HfH(NHMe)Cp*$_2$] is essentially planar, with the methyl group directed toward one $^-$Cp* ligand, and the hydrogen towards the other.[30] The central ZrN$_2$C$_2$ five-membered ring of [Zr(NPhCPh=CPhNPh)Cp$_2$] is folded along the N$\cdots$N axis (dihedral angle 39.2(3)°), thus confirming the assumptions made from variable-temperature NMR spectral data.[43] The crystal structure of [Zr{η^2-N(TMS)CHPh}(THF)Cp$_2$] suggests the presence of a metallaaziridene.[46]

10.2.1.3 Chemical properties

Metal–ligand bond disruption enthalpies have been determined for several metallocene amides.[47] Amides of group 4 metallocenes with a direct M–NR$_2$ linkage exhibit a wide range of chemical behaviour.[48] Loss of electron density at nitrogen through internal p_π–d_π bonding weakens their donor capacity. Facile bond cleavage of the M–N bond with protic reagents[1] and group redistribution or scrambling with metal or metalloid halides are observed. The complex [HfH(NH$_2$)Cp*$_2$] reacts with water to afford [HfH(OH)Cp*$_2$] (see Section 10.4.5) and ammonia; formation of [Hf(OH)(NH$_2$)Cp*$_2$] was not observed.[29]

Group exchange reactions have been observed in the reaction of [Zr(NMe$_2$)$_2$Cp$_2$] with MCl$_4$ (M = Si, Ge, Sn, Ti, Zr or Hf) to give [ZrCl$_2$Cp$_2$] and [MCl$_2$(NMe$_2$)$_2$].[49] [ZrCl(NMe$_2$)Cp$_2$] reacts with [Ru(C≡CH)(PMe$_3$)$_2$Cp] with formation of [Cp$_2$(Cl)Zr–C≡C–Ru(PMe$_3$)$_2$Cp].[50] The reaction between [Zr(NMe$_2$)(R)Cp$_2$] (R = Cl or NMe$_2$) and [ReH$_7$(PPh$_3$)$_2$] results in the formation of the heterobimetallic polyhydrides [Cp$_2$(R)Zr(μ-H)$_n$\{ReH$_{6-n}$(PPh$_3$)$_2$\}] ($n > 1$) (see Section 10.6.1) with elimination of dimethylamine.[51] Compound [Zr(NHPh)$_2$Cp$_2$] reacts with the molybdenum cyclohexadiene complex [Mo(CO)$_2$(η^2-chd)(η^5-indenyl)][OSO$_2$CF$_3$] with ligand exchange and formation of [Zr(NHPh)(OSO$_2$CF$_3$)Cp$_2$]. The latter was also obtained from the zirconocene bisamido complex and CF$_3$SO$_3$H.[19]

The complex [Zr(NMe$_2$)(OR)Cp$_2$] (R = 2,6-Bu$^t_2$C$_6$H$_3$) reacts with PhNCO in hexane with insertion and formation of [Zr{NPhC(=O)NMe$_2$}(OR)Cp$_2$].[23] The zirconocene bisamido complexes [Zr(NHR)$_2$Cp$_2$] (R = 2,6-Me$_2$C$_6$H$_3$ or Bu^t) catalyse the conversion of an alkyne and an amine into an enamine.[18] The η^2-imido complex [Zr{N(TMS)CHR}(THF)Cp$_2$] (R = Ph, H or C$_5$H$_{11}$) undergoes a number of chemo-, regio- and diastereoselective coupling reactions with unsaturated organic compounds forming zirconaazacyclopentane derivatives.[46]

10.2.2 Dinuclear Amido-bridged Metallocene(IV) Complexes

No dinuclear amido-bridged complexes of zirconium and hafnium were known up to 1991. One example of a titanium derivative is [CpTi(μ-NMe$_2$)$_3$M(CO)$_3$] (M = Cr, Mo or W).[52]

10.2.3 Zirconocene Imido Complexes [Zr(=NR)Cp$_2$]: Generation and Trapping

Thermolysis of [Zr(NHR)$_2$Cp$_2$] (R = 2,6-Me$_2$C$_6$H$_3$, 4-Bu^{t}C$_6$H$_4$ or Bu^t) or [Zr(NHR^1)R^{2}Cp$_2$] (R^1 = 2,6-Me$_2$C$_6$H$_3$, 4-Bu^{t}C$_6$H$_4$ or Bu^t; R^2 = Me) in the presence of an alkyne resulted in formation of an azametallacyclobutene.[18,21] The formation of the metallacycle was shown to be reversible. Formation of a transient zirconocene imido complex was suggested. Thermolysis of [ZrMe(NHR)Cp$_2$] yields the intermediate [Zr(=NR)Cp$_2$] (R = 4-Bu^{t}C$_6$H$_4$, Bu^t or 2,6-Me$_2$C$_6$H$_3$).[21] The transient imidozirconocene complex either dimerizes (R = 4-Bu^{t}C$_6$H$_4$), or activation of a C–H bond of the solvent occurs (R = Bu^t). Alternatively, the intermediates can be stabilized in THF or trapped with an alkyne or amine. The molecular structures of [Zr(=NBu^t)(THF)Cp$_2$] and dimeric [{Zr(μ-NR)Cp$_2$}$_2$] (R = 4-Bu^{t}C$_6$H$_4$) were determined.[21]

10.2.4 Dinuclear Imido-bridged Metallocene(IV) Complexes

Only a few dinuclear imido-bridged metallocene(IV) complexes are known. The dimerization product of transient [Zr(=NR)Cp$_2$] (R = 4-Bu^{t}C$_6$H$_4$) is discussed above (see Section 10.2.3). The complex [{Zr(μ-NR)Cp}$_2$(η^5:η^5-C$_{10}$H$_8$)] (R = TMS) is obtained from the reaction of [{Zr(μ-Cl)Cp}$_2$(η^5:η^5-C$_{10}$H$_8$)] with trimethylsilylazide.[53]

Oxidation of [{CpZr(PMe$_3$)(η^1:η^5-C$_5$H$_4$)}$_2$] by TMS-N$_3$ gave the deep blue dinuclear imido-bridged complex [{CpZr(η^1:η^5-C$_5$H$_4$)}$_2$(μ-N-TMS)]. The imido group of the latter is located in the region between the two η^1:η^5 cyclopentadienyl rings.[54]

10.2.5 Metallocene(IV) Alkylideneamido Complexes

10.2.5.1 Synthesis

The most common route to aldimido ($Zr-N = CHR^2$) and ketimido ($Zr-N = CR^2$) derivatives $[Zr(N=CHR^2)X(\eta-C_5R^1_5)_2]$ is the hydrozirconation of a nitrile ($X = H$, $R^1 = Me$, $R^2 = p$-Tol; $X = Cl$, $R^1 = H$, $R^2 = Me$, Ph, Bz or $P(NPr^i_2)_2)^{55,56}$ (for a recent review, see Ref. 55). No reaction was observed between $[ZrMe_2Cp^*_2]$ and p-toluonitrile.[55] The salt elimination route can successfully be employed for the synthesis of the bis(alkylideneamido) complexes $[Zr(N=CR_2)_2Cp_2]$ ($R = Ph$),[55] as shown in 1971 ($R = Ph$, p-Tol or Bu^t).[1] The amine displacement route was employed earlier in the synthesis of zirconocene and hafnocene bis(alkylideneamido) complexes.[1] An alternative synthesis is the reaction of a prereduced metal complex $[ZrLCp_2]$ (L = butadiene or 2CO) with benzophenone azine. Unfortunately, this reaction seems to be limited to benzophenone azine.[55] $[\{Zr(N_2)Cp^*_2\}_2(N_2)]$ reacts with 4 equiv. of p-toluonitrile to form $[Zr\{\overline{NC(R)C(R)N}\}Cp^*_2]$ ($R = p$-Tol). The latter reacts with H_2 to give $[Zr\{N=CH(p\text{-}Tol)\}_2Cp^*_2]$.[55]

10.2.5.2 Structural data

The molecular structures of $[Zr(N=CPh_2)_2Cp_2]$ and (E)-$[ZrCl(N=CHPh)Cp_2]$ have been determined.[55] The short $Zr-N$ bonds (0.2013–0.2063 nm) indicate some multiple bond character. The core atoms of the alkylideneamido ligands are almost coplanar.

10.2.5.3 Spectroscopic properties

The alkylideneamidozirconocene complexes are yellow or brown-yellow. In the IR spectra an absorption of high intensity is observed for the $\nu(C=N)$ mode between 1630 cm^{-1} and 1700 cm^{-1}.[55,56] The solid-state structure of $[Zr(N=CPh_2)_2Cp_2]$ (see Section 10.2.5.2) shows a clear distinction of the (E)- and (Z)-oriented phenyl substituents at the $Ph_2C=N-Zr$ moiety. This is not observed in solution. Even at low temperature (198 K) only one set of phenyl resonances is observed in the ^{13}C NMR spectrum, indicating a rapid equilibrium of (E)- and (Z)-phenyl groups. In $[ZrCl(N=CHPh)Cp_2]$ only the (E) isomer is present in the solid state. The (E) isomer was synthesized stereoselectively and it was shown that the isomerization $(E) \leftrightarrow (Z)$ could not be effected thermally or photochemically.[55]

10.2.5.4 Chemical properties

Complex $[ZrCl(N=CHPh)Cp_2]$ reacts with p-tolyllithium with formation of $[Zr(p\text{-}Tol)(N=CHPh)Cp_2]$. The hydrido complex $[ZrH\{N=CH(p\text{-}Tol)\}Cp^*_2]$ reacts with methyl iodide with elimination of methane and formation of the corresponding iodo complex.[55] The reaction of $[ZrCl\{N=CHP(NPr^i_2)_2\}Cp_2]$ with R^1R^2MCl ($M = P$, $R^1 = R^2 = Ph$; $M = P$, $R^1 = NPr^i_2$, $R^2 = Cl$; $M = P$, $R^1 = R^2 = NPr^i_2$; $M = B$, $R^1 = R^2 = NPr^i_2$; $M = B$, $R^1 = N(TMS)N(TMS)_2$, $R^2 = Cl$; $MR^1R^2 = SnMe_3$) gives $[R^1R^2M-N=C(H)P(NPr^i_2)_2]$ with elimination of $[ZrCl_2Cp_2]$.[56]

10.2.6 Metallocene(IV) Hydrazido and Hydrazonato Complexes

10.2.6.1 Synthesis

Salt elimination reactions have been employed for the synthesis of metallocene hydrazido complexes $[Zr(\overline{NR^1NPhR^2})(R^3)Cp_2]$ ($R^3 = Me$, $R^1 = Ph$, $R^2 = H$, $M = K$;[57,58] $R^3 = CH_2CH_2Bu^i$, $R^1 = H$, $R^2 = Ph$, $M = Li$[59]).

Insertion of a diazoalkane $R^3_2CN_2$ into the $Zr-C$ or $Zr-H$ bond of $[Zr(R^4)(R^1)(\eta-C_5R^2_5)_2]$ yields the hydrazonato(1-) complex $[Zr(\overline{NR^1N=CR^3_2})(R^4)(\eta-C_5R^2_5)_2]$ ($R^4 = R^1 = Me$ or Bz, $R^2 = H$, $R^3 = Ph$; $R^4 = R^1 = Bz$, $R^2 = H$, $R^3 = CO_2Et$;[60] $R^4 = Cl$, $R^1 = H$, $R^2 = H$, $R^3 = Ph$;[61] $R^4 = OH$, $R^1 = Me$, $R^2 = Me$, $R^3 = p$-Tol;[62] $R^4 = R^1 = Me$, $R^2 = H$, $R^3 = Ph$).[61] A mechanism involving coordination of the diazoalkane to zirconium followed by intramolecular migration of R^1 has been proposed. Zirconocene phosphido complexes react with diazoalkanes with insertion into the $Zr-P$ bond and formation of (phosphido)hydrazonido(1-) ligands (see Section 10.3.1.4).[63-5]

The hydrazone derivatives [{Zr(μ-NN=CHR^2)(η-C$_5$H$_4$R^1)$_2$}$_2$] (R^1 = H or Me, R^2 = Ph;[66] R^1 = H, R^2 = 4-MeOC$_6$H$_4$,[66,67] Ph, 4-ClC$_6$H$_4$ or 4-(NC)C$_6$H$_4$)[67] are readily available from zirconocene dichloride and LiN(H)N=CHR^2.[66,67] The crystal structure of [{Zr(μ-N$_2$CHPh)(η-C$_5$H$_4$Me)$_2$}$_2$] revealed an unusual unsymmetrical bridging ligand system (2) (see Section 10.2.6.4), while the ^{13}C NMR spectral data suggest a fluxional behaviour in solution (Scheme 1).[66]

Scheme 1

Bimetallic hydrazido(3-) complexes [Zr{N(H)–N=WMe$_3$Cp*}$_{2-n}$(R)$_n$Cp$_2$] (n = 0, 1) are formed in the reaction of dimethyl- or dibenzylzirconocene with [W(=NNH$_2$)Me$_3$Cp*] (R = Me or Bz). The monosubstituted zirconocene complexes do not readily eliminate methane or toluene to generate hydrazido(4-) complexes.[68]

10.2.6.2 Spectroscopic properties

The η^2 bonding of the hydrazido(1-) ligand in these complexes is apparent from ^{1}H and ^{13}C NMR spectroscopy; coordination of the β nitrogen of the hydrazido ligand freezes the inversion at the nitrogen atom, thus making the cyclopentadienyl ligands chemically inequivalent.

10.2.6.3 Chemical properties

Heating [ZrMe{η^2-NPhN(H)Ph}Cp$_2$] in THF causes elimination of methane and generation of the transient hydrazido complex [Zr(N$_2$Ph$_2$)Cp$_2$], which is trapped as the THF adduct [Zr(N$_2$Ph$_2$)(THF)Cp$_2$].[57,58] This adduct was also prepared by addition of monolithio-1,2-diphenyl-hydrazine to [{ZrCl(H)Cp$_2$}$_n$] or most efficiently by adding [ZrCl$_2$Cp$_2$] to a solution of dilithio-1,2-diphenylhydrazide. The THF ligand can be replaced by a variety of σ-donor ligands to form [Zr(N$_2$Ph$_2$)(L)Cp$_2$] (L = py, PMe$_3$, NCPh or CNBu^t). The pyridine adduct has been structurally characterized (see Section 10.2.6.4). The THF adduct undergoes clean insertion of an alkyne into the Zr–N bond (cf. Zr=O, Section 10.4.7; Zr=S, Section 10.5.3; and Zr=NR, Section 10.2.3).[57,58] Phenylacetylene gives a product with an N–H bond, that is, an η^2-hydrazido(1-) acetylide complex.[57] Acetophenone reacts in a similar fashion to give an enolato complex.[57,58]

Thermolysis of the alkyl hydrazido complex [Zr(CH$_2$CH$_2$Bu^t)(NHNPh$_2$)Cp$_2$] results in loss of 2,2-dimethylbutane and generation of the transient intermediate [Zr(=NNPh$_2$)Cp$_2$] which, in the absence of a trapping reagent, dimerizes.[59] Trapping agents such as 4-dimethylaminopyridine give a monomeric adduct of the hydrazido(2-) complex. Trapping with an alkyne or reaction with CO is also possible.[59]

10.2.6.4 Structural data

In [Zr(η^2-N$_2$Ph$_2$)(py)Cp$_2$],[57,58] the N–N bond length of 0.1434(4) nm lies in the range of known N–N single bond distances. Both nitrogen atoms of the hydrazido(2-) ligand are pyramidal. In [Zr(CCPh)(η^2-NPhNHPh)Cp$_2$] the hydrazido(1-) ligand is bonded in an η^2 fashion with a dative Zr–N(HPh) bond and a Zr–N(Ph) σ-bond.[57]

The molecular structures of the hydrazonato(1-) complexes [Zr(NR^{1}N=CR$^4_2$)(R^2)(η-C$_5$R$^3_5$)$_2$] (R^4 = Ph, R^1 = Me, R^2 = Me, R^3 = H;[60,61] R^4 = CO$_2$Et, R^1 = R^2 = Bz, R^3 = H; R^4 = Ph, R^1 = H, R^2 = Cl, R^3 = H;[60] R^4 = p-Tol, R^1 = Me, R^2 = OH, R^3 = Me)[62] show an η^2-N,N$'$ bonding mode of the ligand. The structural data of the Zr, N-1, N-2, C, R^2 core are very similar in all complexes, with the central Zr–N-1 bond (0.2103(3)–0.221(1) nm) being shorter than the lateral Zr–N-2 bond (0.225(1)–0.2348(4) nm).

The hydrazone derivative (2) (R^1 = H, R^2 = Ph, Cp = C$_5$H$_4$Me) in Scheme 1 shows an unusual unsymmetrical bridging ligand system. The Zr–N bond distances range from 0.2004(7) nm to 0.2232(7) nm.[66,67] The dimerization product of transient $[Zr(=NNPh_2)Cp_2]$ has a structure similar to that of the hydrazone derivative (**2**).[59] The structure of the monomeric hydrazido(2-) complex $[Zr(=NNPh_2)(4-NMe_2-pyridine)Cp_2]$ is similar to that of the monomeric imido complex $[Zr(=NBu^i)(THF)Cp_2]$ (see Section 10.2.3). The hydrazido(2-) ligand is almost linear.[59]

10.3 METALLOCENE(IV) COMPLEXES WITH PHOSPHORUS-, ARSENIC- AND ANTIMONY-CENTRED ANIONIC LIGANDS

10.3.1 Metallocene(IV) Phosphido Complexes

10.3.1.1 Synthesis

In 1966 Issleib *et al.* reported that the reaction of zirconocene dibromide and titanocene dichloride with $LiPR_2$ (R = Et or Bu) occurs with reduction to $[\{M(\mu\text{-}PR_2)Cp_2\}_2]$ with concommitant formation of P_2R_4.[1] The spiro derivative $[\{Cp_2Zr(PPh\overline{CH_2})_2\}_2C]$ was reported in 1967.[1]

Metallocene(IV) phosphido complexes with terminal phosphido ligands have been known since 1983. If metallocene dichlorides are reacted with a lithium phosphide $LiPR_2$ with bulky ligands R, bisphosphido complexes $[M(PR_2)_2Cp_2]$ are formed (M = Zr or Hf, R = Ph; M = Hf, R = Cy;[69] M = Zr or Hf, R = Et, Cy or Ph).[70] However, if the steric demand of the ligand R is low or if the readily reduced titanocene dichloride is employed $[\{M(\mu\text{-}PR_2)Cp_2\}_2]$ is formed (M = Ti, Zr or Hf, R = Me; M = Ti, R = Ph;[69] M = Ti or Zr, R = Et)[1,71] (for properties and reactions of metallocene(III) phosphido complexes see Ref. 69 and Chapter 8, this volume).

Metallocene monophosphido complexes with dialkyl- or diarylphosphido ligands are scarce. The bisphosphido complexes $[M(PPh_2)_2Cp_2]$ (M = Zr or Hf; X = Cl or I) produce equilibrium mixtures with metallocene dihalides. The same mixture is obtained on mixing metallocene dihalide with one equivalent of $LiPPh_2$. However, $[M(PPh_2)(X)Cp_2]$ accounted for only ca. 10% of the total Cp proton intensity (^1H NMR) and could not be isolated. The only isolable mono(dialkylphosphido) halide complex is $[ZrCl(PPh_2)Cp_2]$, which was obtained in 5% yield at 0 °C from the reaction of $[Zr(PPh_2)_2Cp_2]$ with PPh_2Cl.[69]

Using bulky silylphosphido ligands provides access to the complexes $[M\{P(TMS)_2\}_n(X)_{2-n}(\eta\text{-}C_5H_4R)_2]$ (M = Zr, X = Cl, R = H or TMS, n = 1;[63,72] M = Zr[63,72,73] or Hf,[73] X = Me, R = H, n = 1, M = Hf, X = Cl, R = H, n = 1;[73] M = Zr[63,72,73] or Hf,[73] X = Cl, R = H, n = 2; M = Zr, X = Cl, R = TMS, n = 2;[72] M = Zr, X = Cl, R = Me, n = 2).[74] No reaction was observed between $[ZrCl_2\{\eta\text{-}1,3\text{-}(TMS)_2C_5H_3\}_2]$ and the lithium phosphide.[63] $[ZrCl\{P(TMS)(2,4,6\text{-}Bu^t_3C_6H_2)\}Cp_2]$ was obtained from zirconocene dichloride and the corresponding lithium phosphide.[75]

Extrusion of dihydrogen from $[HfH_2Cp^*_2]$ and R^1R^2PH has been employed in the synthesis of hafnocene monophosphido hydrido complexes ($R^1 = R^2 = Ph$; $R^1 = H$, $R^2 = Ph$; $R^1 = H$, $R^2 = Cy$). Even in the presence of excess phosphine only the monophosphido complexes are formed. The bulkier Cy_2PH did not react even at 105 °C.[76]

Chelating *o*-phenylenebisphosphido ligands were employed in the synthesis of $[M\{1,2\text{-}(PR_2)C_6H_4\}Cp_2]$ (M = Zr, R = Ph, $1,3\text{-}(TMS)_2C_5H_3$ instead of Cp;[77] M = Ti, Zr or Hf, R = Ph;[78] M = Zr, R = H).[79] Alternatively, the same metallacycles (M = Ti, Zr or Hf, R = Ph;[78] M = Zr, R = H)[79] are obtained in the reaction of the dimethylmetallocene with *o*-phenylenebisphosphine derivatives.

The product of the reaction of a metallocene dichloride with LiPHR strongly depends on the metal and its reducibility, the steric and electronic nature of the cyclopentadienyl ligand and the bulk of the R group on phosphorus. Thus, complexes of the following types (i) to (vi) were obtained: (i) $[\{Zr(\mu\text{-}PHBu^i)(\eta\text{-}C_5H_4R)_2\}_2]$, (ii) $(R/S,S/R)\text{-}[\overline{M(PR^2\text{-}PR^2\text{-}PR^2)}(\eta\text{-}C_5R^1_5)_2]$, (iii) $[M(PHR^2)(X)(\eta\text{-}C_5R^1_5)_2]$, (iv) $[M(PHR^2)_2(\eta\text{-}C_5R^1_5)_2]$, (v) $[\overline{M(PR\text{-}PR)}L_2]$, (vi) $[\{ZrCl(\eta\text{-}C_5H_4R)_2\}_2(\mu\text{-}PMes)]$ (R = H or Me).

Metallocene(III) complexes of type (i) were first prepared with dialkylphosphido ligands.[69,71] The complex *cis*-$[\{Zr(\mu\text{-}PHBu^i)(\eta\text{-}C_5H_4Me)_2\}_2]$ is obtained from $LiPHBu^i$ and the appropriate zirconocenc dichloride.[80] The structurally characterized *trans*-$[\{Zr(\mu\text{-}PHCy)Cp_2\}_2]$ was obtained from the reaction of zirconocene dichloride with $CyPH_2$ in the presence of Mg in THF.[81]

Complexes of type (ii) were first obtained from the reaction of $[M^1X_2Cp_2]$ with polyphosphido anions (i.e., M^1 = Ti or Zr, X = Cl or Br with $M^2_2(PPh)_4$ (M^2 = Li or Na);[1] M^1 = Ti, Zr or Hf, X = Cl with $K_2(PR)_n$ (R = Ph, Et, Me or Bu^t and n = 3–5)).[82] Alternatively, complexes of type (ii) have been obtained in low yield by reacting a metallocene dichloride with a lithium phosphide, whereby the metal and/or phosphorus atom have sterically less demanding ligands (M = Zr or Hf, R^1 = H, $R^2 = Ph$[83] or Bu^t;[84] M = Zr or Hf, R^1 = Me, $R^2 = Ph$;[84] M = Zr, R^1 = H, $R^2 = Cy$[85] or Ph).[86] Products of type (ii) (M = Zr;

R = Ph) have also been obtained in the reaction of dimethylzirconocene or methylzirconocene chloride with phenylphosphine.[83] When zirconocene dichloride was reacted with 2 equiv. of the bulkier LiPHBu^t, complex (ii) (M = Zr, R^1 = H, R^2 = Bu^t) was formed in 15% yield as well as the reduced type (i) complex (R = H) in 50% yield.[84] Analogously, hafnocene dichloride reacts with LiPHBu^t to give complexes of type (ii) (M = Hf, R^1 = H, R^2 = Bu^t), but the type (iv) bisphosphido complex (M = Hf, R^1 = H, R^2 = Bu^t) was also obtained in 60% yield.[84]

Terminal metallocene phosphido complexes of type (iii) (M = Hf, X = I, R^1 = Me, R^2 = Ph or Cy; M = Zr or Hf, X = Cl, R^1 = Me, R^2 = Cy;[76] M = Zr, X = Cl, R^1 = H, R^2 = 2,4,6-Bu$^t_3$C$_6$H$_2$[87] or Mes)[64] and type (iv) (M = Zr, R^1 = H, R^2 = 2,4,6-Bu$^t_3$C$_6$H$_2$[87] or Mes;[64] M = Zr or Hf, R^1 = Me, R^2 = Bu^t)[84] are obtained when either the metal or the phosphorus atom bears a sterically demanding ligand. [ZrCl$_2${η-1,3-(TMS)$_2$C$_5$H$_3$}$_2$] reacts with 2 equiv. of LiPHBu^t to give the terminal phosphido complex of type (iv) [Zr(PHBu^t)$_2${η-1,3-(TMS)$_2$C$_5$H$_3$}$_2$] (11% yield) and the diphosphene complex of type (v) (M = Zr, R = Bu^t, L = 1,3-(TMS)$_2$C$_5$H$_3$) in 48% yield, while the hafnocene analogue gives only complex type (iv) (M = Hf; R^2 = Bu^t; C$_5$R$^1_5$ = 1,3-(TMS)$_2$C$_5$H$_3$).[84]

Diphosphene complexes of type (v) are obtained from [MCl$_2$L$_2$] and LiPHR (M = Zr or Hf, L = η-1,3-(TMS)$_2$C$_5$H$_3$, R = Ph;[84] M = Zr, L = η-C$_5$Me$_4$Et, R = Mes).[88]

The phosphinidene-bridged complex type (vi) is obtained as dark green crystals on reaction of the corresponding zirconocene dichloride with LiPHMes, together with [ZrCl(PHMes)(η-C$_5$H$_4$R)$_2$] or [Zr(PHMes)$_2$(η-C$_5$H$_4$R)$_2$], depending on the stoichiometry employed.[89] The same product is obtained from the reaction of zirconocene dichloride with magnesium and MesPH$_2$.[81]

Oxidative addition of phosphine or diphosphane to an *in situ* generated zirconocene fragment or zirconocene(II) precursor was employed in the synthesis of zirconocene(III) and (IV) phosphido complexes.[81,90–2] The metallocene(III) complexes [{M(μ-PMe$_2$)Cp$_2$}$_2$] (M = Zr or Hf) are obtained from [MCl$_2$Cp$_2$], Me$_2$PPMe$_2$ and magnesium. In the case of zirconium, the by-product [(MCp$_2$)$_2$(μ-Cl)(μ-PMe$_2$)] was obtained in low yield.[90] The complexity of the ^{1}H NMR spectrum of [{ZrH(PH$_2$)Cp*$_2$}$_2$], which is obtained from [{Zr(N$_2$)Cp*$_2$}$_2$(N$_2$)] and PH$_3$, is best explained by a dimeric structure. However, an unambiguous assignment to bridging or terminal ligands (H/PH$_2$) was not possible.[91]

Activation of C–H and subsequent C–C bond formation in the intermediate [ZrH(PHPh)Cp$_2$] yields the Zr^{III} complex [{Zr(μ-PHPh)Cp}$_2$(η^5:η^5-C$_{10}$H$_8$)] as the final product in the reaction of [ZrCl$_2$Cp$_2$], BuLi, and PhPH$_2$. In the analogous reaction with (2,4,6-Bu$^t_3$C$_6$H$_2$)PH$_2$ and magnesium the isolated product (10% yield) is a trimeric phosphido-capped Zr^{IV} species (Figure 1).[92] However, if MesPH$_2$ is employed instead of the bulkier (2,4,6-Bu$^t_3$C$_6$H$_2$)PH$_2$, the mono(phosphinidene)-bridged dinuclear complex [(ZrClCp$_2$)$_2$(μ-PMes)] was obtained.[81] Zirconocene dichloride reacts with magnesium and CyPH$_2$ with formation of [{Zr(μ-PHCy)Cp$_2$}$_2$]. When the bulky [ZrCl$_2$Cp*$_2$] was used as the starting material, monitoring by ^{31}P NMR revealed the initial formation of [ZrH(PHCy)Cp*$_2$], which is unstable and loses dihydrogen yielding a compound which could not be isolated.[81]

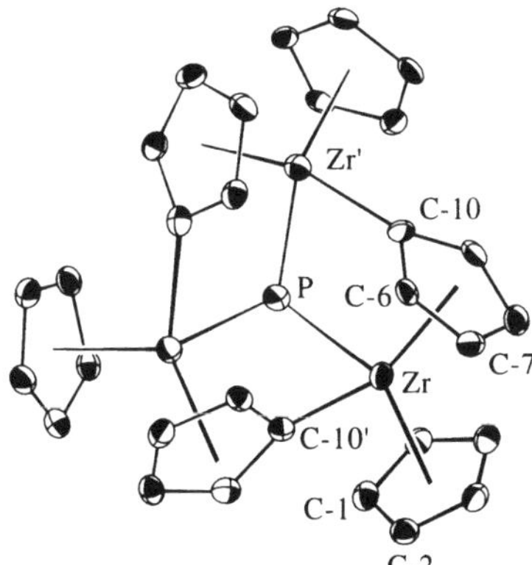

Figure 1 Molecular structure of a trimeric phosphido-capped complex (reproduced by permission of the American Chemical Society from *Organometallics*, 1992, **11**, 1014).

Compounds [M(CO)$_2$(1,3-Bu$^t_2$C$_5$H$_3$)$_2$] (M = Zr or Hf) react with white phosphorus in refluxing xylene to give [M(P–P–P–P)(η-1,3-Bu$^t_2$C$_5$H$_3$)$_2$], which contains the P$_4^{2-}$ anion. The analogous Cp*-substituted complex is only stable in solution and decomposes in the solid state with formation of P$_4$.[93]

10.3.1.2 Spectroscopic properties

A single weak infrared absorption at ca. 350 cm^{-1} for zirconium and ca. 305 cm^{-1} for hafnium in [M(PPh$_2$)$_2$Cp$_2$] was assigned as ν(M–P). Two absorptions were observed for [Hf(PCy$_2$)$_2$Cp$_2$] (299, 270 cm^{-1}).[69] In [Zr{1,2-(PH)$_2$C$_6$H$_4$}Cp$_2$] the ν(Zr–P) mode was observed at 338 cm^{-1}.[79] The monophosphido-chloro complexes [ZrCl(PR$_2$)Cp$_2$] exhibit the ν(Zr–Cl) mode in the range 330–350 cm^{-1} (R = Ph[69] or TMS[63]). The complexes with primary phosphido groups (PHR) show the ν(P–H) mode between 2240 cm^{-1} and 2340 cm^{-1}.[64,76,79]

In complexes with $^-$P(TMS)$_2$ ligands, ^{31}P and ^{1}H NMR spectra show chemical shifts between δ −23 and −154.[63,73,74] The ^{31}P resonance of [ZrCl{P(TMS)(2,4,6-Bu$^t_3$C$_6$H$_2$)}Cp$_2$] (at δ 156) is shifted to low field compared with P(TMS)$_2$ complexes.[75]

The bis(dialkyl- or diarylphosphido) complexes [M(PR$_2$)$_2$Cp$_2$] (M = Zr or Hf; R = Et, Cy or Ph) show a singlet in the ^{31}P NMR spectra between δ 100 and 160.[69] As the crystal structure determination of [Hf(PEt$_2$)$_2$Cp$_2$] showed two different phosphido ligands (see Section 10.3.1.3) it was assumed that the interconversion of the M–P single and double bond is fast on the NMR timescale. Low-temperature freezing of the process is observed only for the bulky $^-$PCy$_2$ ligand.[69,70]

The hafnocene hydrido complexes [HfH(PHR)Cp*$_2$] (R = Ph or Cy) and [HfH(PPh$_2$)Cp*$_2$] show singlets in the ^{31}P{^{1}H} NMR spectrum (δ 30–96).[76] The ^{1}H chemical shift of the hydrido ligand in these complexes reflects the degree of π donation from the phosphido ligand to hafnium. Thus, δ(Hf–H) decreases in the order PPh$_2$ > PHPh > PHCy.[76] The halide derivatives [M(PHR)(X)Cp*$_2$] (M = Hf, X = I, R = Ph or Cy; M = Hf or Zr, X = Cl, R = Cy)[76] and [ZrCl(PHR)Cp$_2$] (R = Mes[64] or 2,4,6-Bu$^t_3$C$_6$H$_2$[87]) show a singlet in the ^{31}P{^{1}H} NMR spectra (δ −6–91).

The bisphosphido complexes [Hf(PHBu^t)$_2$Cp$_2$], [M(PHBu^t)$_2$Cp*$_2$] and [M(PHBu^t)$_2${η-1,3-(TMS)$_2$C$_5$H$_3$}$_2$] (M = Zr or Hf) show a singlet in the ^{31}P{^{1}H} NMR spectrum between δ 101 and 131.[84] The complexes [Zr(PHR)$_2$Cp$_2$] (R = Mes[64] or 2,4,6-Bu$^t_3$C$_6$H$_2$[87]) show a singlet in the ^{31}P {^{1}H} NMR, which splits into a multiplet on coupling to protons.

The diphosphene complexes [M(PR–PR){η-1,3-(TMS)$_2$C$_5$H$_3$}$_2$] (M = Zr or Hf, R = Ph; M = Zr, R = Bu^t, δ 193–271)[84] and [Zr(PMes–PMes)(η-C$_5$Me$_4$Et)$_2$] (δ 139)[88] show a singlet in the ^{31}P NMR spectrum.

Extreme downfield shifts are observed in [{ZrCl(η-C$_5$H$_4$R)$_2$}$_2$(μ-PMes)] (δ ca. 325 for R = H[81,89] or R = Me[89]) and for the μ_3-phosphorus atom in the trimeric complex shown in Figure 1 (δ 783).[92]

The complexes [M(PR–PR–PR)Cp$_2$] (M = Ti, Zr or Hf, R = Ph, Et, Me or Bu^t;[82] M = Zr or Hf, R = Ph;[84] M = Zr, R = Cy)[85] exhibit A$_2$X patterns in the ^{31}P NMR spectra. In the analogous complexes [M(PPh–PPh–PPh)Cp*$_2$] the ^{31}P resonances of the A$_2$ part are shifted to high field.[84] Only one diastereomer (*meso* form) is present in solution. In the ^{1}H NMR spectra two signals are observed for the inequivalent $^-$Cp or $^-$Cp* ligands.[82,84] The presence of a puckered (R = Ph) or planar (R = Bu^t) MP$_3$ metallacycle and a geometry intermediate between that for the Me and Et derivative was concluded on the basis of variable-temperature ^{1}H NMR spectral studies.[82]

The *o*-phenylenebisphosphidometallocene derivatives [M{1,2-(PR)$_2$C$_6$H$_4$}Cp$_2$] (M = Ti, Zr or Hf, R = Ph;[78] M = Zr, R = H)[79] exhibit two resonances in the ^{31}P NMR spectra due to the presence of both the possible isomers, *cis* and *trans*. The single signal observed for [Zr{1,2-(PPh)$_2$C$_6$H$_4$}{η-1,3-(TMS)$_2$C$_5$H$_3$}$_2$] splits into two signals at −70 °C. Thus, the limiting spectrum corresponds to the solid-state structure of the complex (see Section 10.3.1.3).[77]

In [Zr{P(PR$_2$)PP(PR$_2$)P}Cp$_2$] (R = TMS) six multiplets are observed for the six inequivalent phosphorus nuclei.[94] The complexes [Zr(P–P–P–P)(η-1,3-Bu$^t_2$C$_5$H$_3$)$_2$] and the Cp* analogue show A$_2$X$_2$ spectra.[93]

10.3.1.3 Structural data

Crystallographic data are available for [ZrCl{P(TMS)(2,4,6-Bu$^t_3$C$_6$H$_2$)}Cp$_2$],[75] [Zr{1,2-(PPh)$_2$C$_6$H$_4$}{η-1,3-(TMS)$_2$C$_5$H$_3$}$_2$],[77] [ZrCl{P(TMS)$_2$}Cp$_2$],[63,72] [ZrMe{P(TMS)$_2$}Cp$_2$],[63,72] [Hf(PEt$_2$)$_2$Cp$_2$],[70] [Hf{P(TMS)$_2$}$_2$Cp$_2$],[73] [Zr{P(TMS)$_2$}$_2$(η-C$_5$H$_4$Me)$_2$],[74] [HfH(PHPh)Cp*$_2$][76] and [Zr{PH(2,4,6-Bu$^t_3$C$_6$H$_2$)}$_2$Cp$_2$].[87] The hafnium or zirconium atom is coordinated in a distorted tetrahedral fashion in all complexes. The complexes [Hf(PEt$_2$)$_2$Cp$_2$],[70] [Hf{P(TMS)$_2$}$_2$Cp$_2$][73] and [Zr{1,2-(PPh)$_2$C$_6$H$_4$}{η-1,3-(TMS)$_2$C$_5$H$_3$}$_2$][77] show two distinctly different bonding modes of the phosphido ligands. Accordingly, unequal M–P bond lengths are observed (Δ = 0.020–0.009 nm). Molecular structures of [M(PR–PR–PR)Cp$_2$] (M = Zr[86] or Hf;[83] R = Ph; M = Zr, R = Cy)[85] show that the central MP$_3$ core is essentially planar and only one diastereoisomer (*meso* form) is present in the solid state and in solution. In the trimeric phosphido-capped Zr^{IV} complex shown in Figure 1 the phosphorus atom is located on a crystallographic threefold axis.[92] In the phosphinidene-bridged complex [(ZrClCp$_2$)$_2$(μ-PMes)] the central Zr$_2$PC central core is planar.[81] The complex [Zr{P(PR$_2$)PP(PR$_2$)P}Cp$_2$] (R = TMS) (Figure 2) has a planar ZrP$_3$ four-membered ring. In the solid state as well as in solution only one diastereoisomer is present (atoms P-1 to P-4 all (*S*) or all (*R*) configuration).[94]

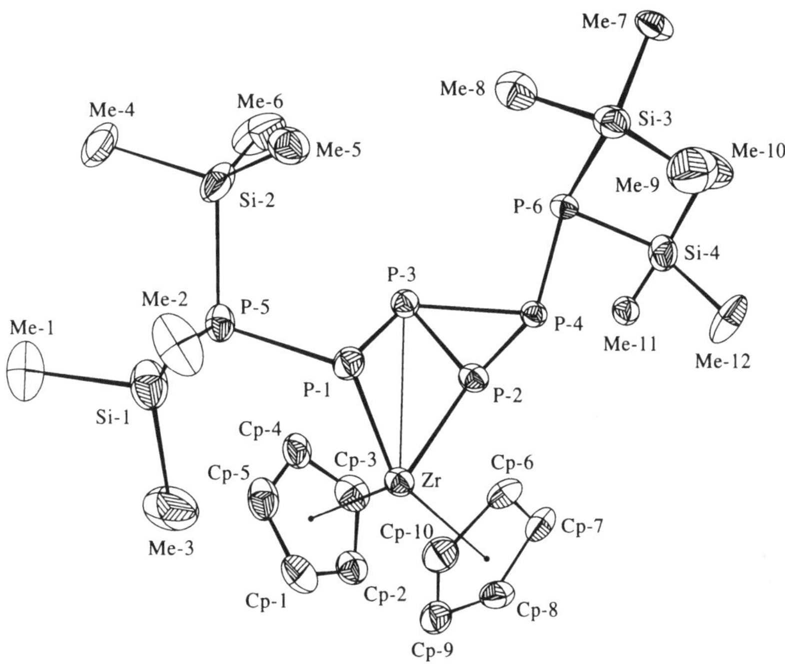

Figure 2 Molecular structure of [Zr{P(PR$_2$)PP(PR$_2$)P}Cp$_2$] (R = TMS) (reproduced by permission of the Royal Society of Chemistry from *J. Chem. Soc., Chem. Commun.*, 1987, 597).

Disorder of the 1,3-Bu$^t_2$C$_5$H$_3$$^-$ ligands in [M(P–P–P–P)(η-1,3-Bu$^t_2$C$_5$H$_3$)$_2$] (M = Zr or Hf) prevented determination of the crystal structure to complete crystallographic satisfaction. However, the presence of a P$_4$-butterfly fragment with a M(1,3-Bu$^t_2$C$_5$H$_3$)$_2$ bridge was shown unambiguously.[93]

10.3.1.4 Chemical properties

The terminal metallocene(IV) phosphido complexes are oxygen and moisture sensitive, especially in solution. They are intensely coloured as solids and in solution (purple, red or blue) and decompose rapidly in halogenated solvents. The diphosphene complexes are green,[84,88] as is the bridging phosphinidene complex.[81]

The complexes [M(PR$_2$)$_2$Cp$_2$] (M = Zr or Hf, R = Ph; M = Hf, R = Cy) react with a variety of protic and halogen-containing species with cleavage of the M–P bond.[69,95] Aliphatic alcohols (MeOH, EtOH or Bu^{t}OH) and Et$_2$NH displace the PR$_2$ groups from [Zr(PPh$_2$)$_2$Cp$_2$] as Ph$_2$PH, but the zirconium products were poorly defined.[69]

Complex [ZrCl{P(TMS)$_2$}Cp$_2$] reacts with polar or protic reagents with displacement of the $^-$P(TMS)$_2$ ligand.[63] Similarly, [ZrMe{P(TMS)$_2$}Cp$_2$] reacts with (2,4,6-Bu$^t_3$C$_6$H$_2$)PCl$_2$ to give [ZrCl(Me)Cp$_2$] and the diphosphenes P$_2$(2,4,6-Bu$^t_3$C$_6$H$_2$)$_2$ and (2,4,6-Bu$^t_3$C$_6$H$_2$)P=P-TMS, the latter being unstable.[73] The reaction of [Zr{P(TMS)$_2$}$_2$Cp$_2$] with MeI led to elimination of MeP(TMS)$_2$ and formation of [ZrI{P(TMS)$_2$}Cp$_2$].[63]

Reactions of [HfH(PPh$_2$)Cp*$_2$] with H$_2$, CO or C$_2$H$_4$ proceed with reductive elimination of Ph$_2$PH and formation of the corresponding hafnocene derivatives [HfL$_2$Cp*$_2$] (L = H or CO or L$_2$ = butadiene).[76] Reaction of CO with [M(PHBu^t)Cp*$_2$] (M = Zr or Hf) led to reductive elimination of the diphosphane (PHBu^t)$_2$ and formation of [M(CO)$_2$Cp*$_2$].[84] However, if a heterocumulene, white phosphorus or phenylacetylene is reacted with [Zr{P(TMS)$_2$}(X)Cp$_2$] (X = Cl or P(TMS)$_2$) insertion into the Zr–P bond(s) occurs with formation of novel phosphorus-functionalized ligands.[63,72,94,96,97] The complex [ZrCl(PHMes)Cp$_2$] inserts diphenyldiazomethane into the Zr–P bond with formation of [ZrCl{η^2-N(=CPh$_2$)N'(PHMes)}Cp$_2$].[64] No reaction was observed between [Zr{P(TMS)$_2$}$_2${η-C$_5$H$_4$(TMS)}$_2$] and Ph$_2$CN$_2$.[72] [ZrCl{P(TMS)$_2$}Cp$_2$] did not react with CO, while CO$_2$ led to decomposition.[63] However,

[HfH(PPh$_2$)Cp*$_2$] undergoes successive insertion of CO$_2$ into the Hf–P and Hf–H bonds, with formation of [HfH(η^2-O$_2$CPPh$_2$)Cp*$_2$] and [Hf{OC(O)H}{OC(O)PPh$_2$}Cp*$_2$], respectively.[76]

Reaction of [Zr{P(TMS)$_2$}$_2$Cp$_2$] or [ZrMe{P(TMS)$_2$}Cp$_2$] with sulfur gives [{Zr(μ-S)Cp$_2$}$_2$] and unidentified phosphorus products.[98] In contrast, [M(PR$_2$)$_2$Cp$_2$] (M = Zr or Hf; R = Cy or Ph) reacts cleanly with sulfur to give [M(SPR$_2$)$_2$Cp$_2$]. The same products are obtained on reaction with SPR$_2$H.[99]

The dimeric complex [{ZrH(PH$_2$)Cp*$_2$}$_2$] selectively forms primary alkylphosphines and zirconocene dihydride when treated with alkene and phosphine at 60–80 °C.[91]

Reduction of [M(PR$_2$)$_2$Cp$_2$] (M = Zr or Hf; R = Et, Cy or Ph) with sodium naphthalenide in THF produced [Cp$_2$M(μ-PR$_2$)$_2$Na(THF)$_n$] as shown by ESR spectroscopy.[70]

Metallocene bisphosphido complexes can be used for the synthesis of heterobi- and heterotrimetallic complexes if the reaction is with a potential precursor for a 12- or 14-electron fragment (see Section 10.3.2.1). However, reaction of [Zr(PR$_2$)$_2$Cp$_2$] (R = Ph or Cy) with [{M(CO)$_3$Cp}$_2$] (M = Mo or W) results in reduction of the M–M bond and oxidation of one of the phosphido ligands with formation of [Cp$_2$Zr(μ-PR$_2$)(μ-OC)Mo(CO)Cp].[100] This complex is also formed in the reaction of [{Zr(μ-PEt$_2$)Cp$_2$}$_2$] with [{Mo(CO)$_3$Cp}$_2$][101] or in the reaction of [MoH(CO)$_3$Cp] with [Zr(PPh$_2$)$_2$Cp$_2$].[100]

The complexes [M($\overline{\text{PR-PR-PR}}$)Cp$_2$] (M = Ti, Zr or Hf; R = Ph, Et, Me or Bu^t) react with CCl$_4$ or CHCl$_3$ with cleavage of the M–P bonds. Reaction with CS$_2$ was also observed, but the products were not characterized.[82] The reaction of [Zr($\overline{\text{PPh-PPh-PPh}}$)Cp$_2$] with ethyl diazoacetate leads to insertion into both Zr–P bonds and formation of (4). The stereochemistry of the P$_3$Ph$_3$ fragment of the starting material is retained in the product.[65]

$$\begin{array}{c}\text{EtO}_2\text{C}\diagdown\\ \| \\ \text{N}\\ | \quad \diagdown \text{N} - \text{P}_\text{A} \diagup \text{Ph}\\ \text{Cp}_2\text{Zr} \diagtimes \qquad \qquad \text{P}_\text{B} - \text{Ph}\\ | \quad \diagup \text{N} - \text{P}_\text{A} \diagdown \\ \text{N} \qquad\qquad\qquad \text{Ph}\\ \| \\ \text{EtO}_2\text{C}\end{array}$$

(4)

The diphosphene complexes [M($\overline{\text{PR-PR}}$){η-1,3-(TMS)$_2$C$_5$H$_3$}$_2$] (M = Zr or Hf, R = Ph; M = Zr, R = Bu^t) are readily hydrolysed to [M(OH)$_2${η-1,3-(TMS)$_2$C$_5$H$_3$}$_2$] and the *meso* and *rac* isomers of (PHR)$_2$.[84]

It was not possible to generate a terminal phosphinidene complex by thermal or photochemical elimination of TMS-Cl from [ZrCl{P(TMS)(2,4,6-Bu$^t_3$C$_6$H$_2$)}Cp$_2$].[75] Similarly, attempts to dehydrohalogenate [M(PHR)(X)Cp*$_2$] (M = Hf, X = I, R = Ph or Cy; M = Zr or Hf, X = Cl, R = Cy) with an anionic base (KH, BuLi or NaN(TMS)$_2$) resulted in product mixtures.[76] However, when [HfI(PHPh)Cp*$_2$] was treated with NaN(TMS)$_2$ at −78 °C, HN(TMS)$_2$ and NaI were liberated and a red product of empirical formula [Hf(PPh)Cp*$_2$]$_x$ was obtained.[76]

10.3.2 Heterobi- and Heterotrimetallic Phosphido-bridged Metallocene(IV) Complexes

10.3.2.1 Synthesis

Although preparation of [Cp$_2$Zr(μ-PR^{1}R^2)$_2$Mo(CO)$_4$] (R^1 = R^2 = Me; R^1 = H, R^2 = Ph; R^1 = Ph, R^2 = TMS) was reported as early as 1977,[1] it was only in 1985 that studies of heterobimetallic phosphido-bridged metallocene complexes were resumed. Some results are summarized in a review article.[102] Two synthetic routes have been described: (i) reaction of a metallocene bisphosphido complex with an appropriate transition metal compound M^{2}LL$_n$ to give [Cp$_2$M^1(μ-PR$_2$)$_2$M^{2}L$_n$] (M^1 = Zr, R = Ph, M^{2}L$_n$ = Rh(η-indenyl), Mo(CO)$_4$, M^2(PPh$_3$) (M^2 = Ni, Pd or Pt), Ni(CO)$_2$, M^2(H)(CO)(PPh$_3$) (M^2 = Rh or Ir) or ReH(CO)$_3$; M^1 = Hf, R = Ph, M^{2}L$_n$ = Rh(η-indenyl), M^2(H)(CO)(PPh$_3$) (M^2 = Rh or Ir); M^1 = Hf, R = Et, M^{2}L$_n$ = Rh(η-indenyl), Ni(CO)$_2$, Fe(CO)$_3$, Mo(CO)$_4$, Ni(cod), M^2(PPh$_3$) (M^2 = Pd or Pt); M^1 = Zr, R = Et, M^{2}L$_n$ = Rh(η-indenyl), Ni(cod), M^2(PPh$_3$) (M^2 = Pd or Pt); M^1 = Zr or Hf, R = Cy, M^{2}L$_n$ = Rh(H)(CO)(PPh$_3$); M^1 = Zr, R = TMS, M^{2}L$_n$ = M^2(CO)$_4$ (M^2 = Cr or Mo))[95,103–5] and (ii) reaction of zirconocene dichloride with a dilithio(bisphosphido) transition metal complex to give [Cp$_2$Zr(μ-PR$_2$)$_2$M^{2}L$_n$] (R = Ph; M^{2}L$_n$ = W(CO)$_4$ or Fe(NO)$_2$; R = Ph or Cy; M^{2}L$_n$ = Fe(CO)$_3$).[95,102] In the reaction between [Hf(PEt$_2$)$_2$Cp$_2$] and [Fe$_2$(CO)$_9$] in hexane at room temperature [Cp$_2$Hf(PEt$_2$)(μ-PEt$_2$)Fe(CO)$_4$] is formed initially. Thermolysis of this product yields $[Cp_2Hf(\mu\text{-}PEt_2)_2Fe(CO)_3]$.[95] Reaction of $[Zr(PPh_2)_2Cp_2]$ with $[Re_2(CO)_{10}]$ gave $[Cp_2Zr(\mu\text{-}PPh_2)_2Re(H)(CO)_3]$ in low yield.[103] The mechanism of formation and the source of the proton are unknown.

The reaction of a metallocene bisphosphido complex with $[Pd(\eta\text{-}2\text{-}Me\text{-}allyl)Cp]$ or $[Pt(cod)_2]$ in the presence of an equimolar amount of a phosphine PR^2_3 yields a variety of heterobimetallic complexes $[Cp_2M^1(\mu\text{-}PR^1_2)_2M^2(PR^2_3)]$ ($M^2 = Pd$, $M^1 = Zr$ or Hf, $R^1 = Cy$, $R^2 = Ph$; $M^1 = Zr$ or Hf, $R^1 = R^2 = Ph$; $M^1 = Zr$, $R^1 = Ph$, $R^2 = Me$, Cy or OMe; $M^1 = Hf$, $R^1 = Ph$, $R^2 = O(o\text{-}Tol)$; $M^2 = Pt$, $M^1 = Zr$ or Hf, $R^1 = Cy$, $R^2 = Ph$; $M^1 = Zr$ or Hf, $R^1 = R^2 = Ph$; $M^1 = Hf$, $R^1 = Ph$, $R^2 = Me$, Cy or OMe; $M^1 = Zr$, $R^1 = Ph$, $R^2 = O(o\text{-}Tol)$). With $P(OMe)_3$, mixtures of $[Cp_2M^1(\mu\text{-}PPh_2)_2M^2\{P(OMe)_3\}_n]$ ($M^1 = Zr$, $M^2 = Pd$, $n = 1$ or 2; $M^1 = Hf$, $M^2 = Pt$, $n = 1$ or 2) were obtained.[104]

Several heterobimetallic phosphido-bridged complexes $[Cp_2M^1(\mu\text{-}PEt_2)_2M^2L_n]$ have been reacted with $[M^1(PEt_2)_2Cp_2]$ ($M^1 = Zr$ or Hf) to give phosphido-bridged hetcrotrimetallic complexes $[\{Cp_2M^1(\mu\text{-}PEt_2)_2\}_2M^2]$ ($M^1 = Zr$ or Hf, $M^2 = Ni$, $L_n = 1,5\text{-}cod$; $M^1 = Zr$ or Hf, $M^2 = Pd$, $L_n = PPh_3$; $M^1 = Zr$ or Hf, $M^2 = Pt$, $L_n = PPh_3$). The complex $[\{Cp_2Hf(\mu\text{-}PPhCH_2CH_2CH_2PPh)\}_2Ni]$ with unsymmetrically substituted phosphido ligands was also prepared.[104]

Heterotrimetallic complexes (Zr/Pd/Hf or Zr/Pt/Hf) have been synthesized by reacting $[Zr(PEt_2)_2Cp_2]$ with $[Pd(PPh_3)_4]$ or $[Pt(PPh_3)_4]$ followed by addition of $[Hf(PEt_2)_2Cp_2]$.[104]

10.3.2.2 Spectroscopic data

The complexes are yellow to red-orange crystalline solids (an exception is $[Cp_2Zr(\mu\text{-}PPh_2)_2Fe(NO)_2]$, which is purple), while the metallocene bisphosphido precursors are deep violet to deep red; this indicates that the charge-transfer system in the latter is disrupted on formation of heterobimetallic complexes.

(i) ^1H NMR spectra

The bimetallic complexes that have a quasi-mirror plane show one signal for the Cp protons,[95,102,104] which is occasionally split into a triplet by coupling to the two bridging phosphido groups.[104] The complexes without a quasi-mirror plane show two signals consistent with their chemical inequivalence.[95,103] In $[Cp_2M(\mu\text{-}PR_2)_2Rh(\eta\text{-}indenyl)]$ ($M = Zr$ or Hf, $R = Et$ or Ph) fluctuation of the indenyl ligand results in the observation of one signal for the Cp protons at 95 °C, while two signals are observed at -60 °C.[95]

(ii) ^{31}P NMR spectra

A singlet, which may be split into a doublet in the presence of coordinated phosphine ligands, is observed for the bridging phosphido ligands of heterobimetallic complexes in the range δ 85–182 (cf. the chemical shift of the corresponding metallocene(IV) bisphosphido complexes at δ 100–160, Section 10.3.1.2).[95,102–4] The phosphido-bridged heterotrimetallic complexes exhibit a singlet (symmetrical substitution) or two triplets for unsymmetrically substituted complexes in the range δ 94–140.[104]

(iii) Infrared spectra

A band at 347 cm^{-1} ($M = Zr$) and 302 cm^{-1} ($M = Hf$) was assigned to the $\nu(M\text{-}P)$ stretching mode in $[Cp_2M^1(\mu\text{-}PPh_2)_2M^2(PPh_3)]$ ($M^1 = Zr$ or Hf; $M^2 = Pd$ or Pt).[104] In $[Cp_2M(\mu\text{-}PR_2)_2Rh(\eta\text{-}indenyl)]$ the $\nu(M\text{-}P)$ mode was assigned to a medium to weak band at 325–338 cm^{-1} ($M = Zr$; $R = Et$ or Ph) or 286–302 cm^{-1} ($M = Hf$; $R = Et$ or Ph).[95] The $\nu(CO)$ modes of the heterobimetallic complexes with $M(CO)_n$ fragments closely resemble those of the depe-substituted analogues.[95]

(iv) Mass spectra

The complexes $[Cp_2Zr(\mu\text{-}PR_2)_2M(CO)_n]$ ($R = Ph$, $M = W$, $n = 4$; $R = Cy$ or Ph, $M = Fe$, $n = 3$)[95] and $[Cp_2Zr(\mu\text{-}PPh_2)_2Fe(NO)_2]$[102] exhibit parent ion peaks in the mass spectra as well as signals corresponding to successive loss of CO and NO, respectively.

10.3.2.3 Structural data

Crystallographic data are available for one heterotrimetallic and for several heterobimetallic phosphido-bridged complexes $[(\eta\text{-}C_5H_4R^1)_2M^1(\mu\text{-}PR^2_2)_2M^2L_n]$ $(M^1 = Zr,\ R^1 = H,\ R^2 = Ph,$ $M^2L_n = Rh(\eta\text{-indenyl}),\ M^3(CO)_4\ (M^3 = Mo\ or\ W),\ Rh(H)(CO)(PPh_3)$ or $Re(H)(CO)_3;\ M^1 = Hf,\ R^1 = H,$ $R^2 = Ph,\ M^2L_n = Rh(H)(CO)(PPh_3),\ Pd(PPh_3)$ or $Pd(dmpe);\ M^1 = Hf,\ R^1 = H,\ R^2 = Et,\ M^2L_n = Mo(CO)_4;$ $M^1 = Zr,\ R^1 = Me,\ R^2 = TMS,\ M^2L_n = Cr(CO)_4).$[95,104,105] In the heterotrimetallic complex $[Cp_2Zr(\mu\text{-}PEt_2)_2Ni(\mu\text{-}PEt_2)_2HfCp_2]$ the zirconium and hafnium atoms are disordered with respect to their crystallographic positions (50:50 occupancy). The MP_2Ni rings are nearly planar, with nickel showing a distorted tetrahedral coordination.[104] In several complexes the $M^1P_2M^2$ core is essentially coplanar, while others exhibit a butterfly arrangement (dihedral angle up to 28°). Extended Hückel molecular orbital calculations performed on the model complexes $[Cp_2Zr(\mu\text{-}PH_2)_2ML_n]$ $(ML_n = Pt(PH_3),\ Pt(dmpe),$ $Rh(\eta\text{-indenyl}),\ Ni(\mu\text{-}PH_2)_2ZrCp_2$ or $Mo(CO)_4)$ indicate the presence of $M \rightarrow Zr$ donor–acceptor metal–metal bonds that become weaker along the series $Pt > Rh > Ni > Mo$.[104]

10.3.2.4 Chemical reactivity

Ligand substitution reactions at the late transition metal centre have been studied.[95,104] The electrophilic substitution reactions of $[Cp_2Zr(\mu\text{-}PPh_2)_2Pt(PPh_3)]$ are typical for Pt^0. However, oxidative addition of MeI and $[Me_3O]^+[BF_4]^-$ was not observed, which is unusual for Pt^0 complexes. The diminished reactivity of platinum in these complexes is possibly due to a metal–metal interaction (see Section 10.3.2.3). However, reaction of $[Cp_2M(\mu\text{-}PEt_2)_2Rh(\eta\text{-indenyl})]$ $(M = Zr\ or\ Hf)$ with MeI proceeds with formation of $[Cp_2M(\mu\text{-}PEt_2)_2Rh(Me)(\eta\text{-indenyl})]^+\ I^-$, while no reaction is observed with H_2 (1 atm), CO (1 atm) or acetyl chloride.[95]

No reaction was observed between $[Cp_2M^1(\mu\text{-}PPh_2)_2M^2(PPh_3)]$ $(M^1 = Hf,\ M^2 = Pd;\ M^1 = Zr,\ M^2 = Pt)$ and CO, H_2, ethene (1 atm), excess Bu^tCN, benzaldehyde, paraformaldehyde or MeI. More reactive oxidative agents such as acetyl bromide react with disruption of the PPh_2 bridge.[104] $[Cp_2Zr(\mu\text{-}PPh_2)_2W(CO)_4]$ did not react with diphenylacetylene, $PMePh_2$ or CS_2, thermally or photochemically. The reaction with LiR (R = Me, Bu, Ph or BEt_3H) led to unidentifiable products.[95]

10.3.3 Metallocene(IV) Arsenido and Antimonido Complexes

10.3.3.1 Synthesis and spectroscopic properties

Only a small number of zirconocene arsenido and antimonido complexes $[Zr(ER^2_2)_2(\eta\text{-}C_5H_4R^1)_2]$ $(E = As\ or\ Sb,\ R^1 = H,\ R^2 = Ph;$[106] $E = As,\ R^1 = H\ or\ Me,\ R^2 = TMS)$[107] is known. The synthesis is based on salt-elimination reactions between zirconocene dichloride and the appropriate lithium reagent. Only one monoarsenido complex $[Zr\{As(TMS)_2\}(Cl)(\eta\text{-}C_5H_4Me)_2]$ has been synthesized to date.[107] The complexes are deeply coloured (red), highly air and moisture sensitive. With protic species and halogenated reagents the Zr–E bond is cleaved. No evidence was found for a mixed halide–EPh_2 complex in the reactions with halogen sources.[106] The crystal structures of $[Zr\{As(TMS)_2\}(Cl)(\eta\text{-}C_5H_4Me)_2]$ and $[Zr\{As(TMS)_2\}_2Cp_2]$ have been determined.[107] The bis(arsenido) complex exhibits two distinctly different Zr–As bond lengths, which indicates the presence of Zr–As single and double bonds.

Although IR spectral data were reported for all complexes, the $\nu(Zr\text{–}E)$ mode (E = As or Sb) was not unambiguously assigned.[106] The ^1H NMR spectra show a single resonance for the Cp protons of $[Zr(EPh_2)_2Cp_2]$ with a downfield shift in the order Sb > As > P, which is the reverse of that expected from electronegativity values and implies that $d_\pi\text{–}p_\pi$ backbonding is more pronounced for Zr–Sb than Zr–P. Mass spectrometric studies on $[Zr(AsPh_2)_2Cp_2]$ failed to detect zirconium-containing fragments; only ions derived from polyarsenic species were observed.[106]

10.4 METALLOCENE(IV) COMPLEXES WITH OXYGEN-CENTRED ANIONIC LIGANDS

Aldehyde and ketone complexes are dealt with in Chapter 7, this volume.

10.4.1 Metallocene Alkoxides and Aryloxides

10.4.1.1 Synthesis

Zirconocene alkoxides were first reported in the patent literature in 1960,[1] but the compounds were not fully characterized. Two major preparative routes have been employed. These are, first the reaction of a metallocene dihalide with excess alcohol in the presence of excess base to give $[M(OR^1)_2(\eta\text{-}C_5H_4R^2)_2]$ (M = Ti, Zr or Hf; R^2 = H or Me; R^1 = alkyl or aryl)[108–10] and second the reaction of a metallocene dichloride with an alkali metal alkoxide or aryloxide M^2OR^1 to give $[M^1Cl(OR^1)(\eta\text{-}C_5H_4R^2)_2]$ (M^1 = Zr, R^2 = H, M^2 = Li, R^1 = 2,6-$Pr^i_2C_6H_3$, 2,6-$Bu^t_2C_6H_3$, 2,6-Bu^t_2-4-MeC_6H_2 or 2,4,6-$Bu^t_3C_6H_2$;[23] M^1 = Hf, R^2 = TMS, M^2 = Na, R^1 = Me).[108]

Excess of alcohol and base is necessary to prevent formation of inseparable mixtures of the mono- and disubstituted products.[108] However, with the bulky α- or β-nap ligand either the mono- or the disubstituted products can be obtained in good yield. The disubstituted complexes have also been obtained from $[ZrCl_2(\eta\text{-}C_5H_4Me)_2]$ and NaOR (R = α- or β-nap).[109,110]

The reaction of zirconocene dichloride with a bulky lithium aryloxide $[Li(OEt_2)(OR)]_2$ (R = 2,6-$Pr^i_2C_6H_3$, 2,6-$Bu^t_2C_6H_3$, 2,6-Bu^t_2-4-MeC_6H_2 or 2,4,6-$Bu^t_3C_6H_2$) is slow in refluxing THF (2–4 d), but addition of TMEDA leads to significant acceleration.[23] No further substitution was observed when an excess of the lithium reagent[23] or the sodium alkoxide[108] was used. No reaction of the lithium reagent was observed with $[ZrCl_2\{\eta\text{-}1,3\text{-}(TMS)_2C_5H_3\}_2]$, $[ZrCl_2\{\eta\text{-}C_5H_4(TMS)\}_2]$ or $[ZrCl_2(\eta\text{-}C_5H_4Bu^t)_2]_2$, and zirconocene dichloride did not react with $[Li(OEt_2)(OR)]_x$ (R = 3,5-$Bu^t_2C_6H_3$ or 2,4-$Bu^t_2C_6H_3$).[23] Zirconocene bisaryloxides with bulky aryloxide ligands $[Zr(OR)_2Cp_2]$ (R = 2,6-$Pr^i_2C_6H_3$ or 3,5-$Bu^t_2C_6H_3$) are obtained from $[Zr(NMe_2)_2Cp_2]$ and the corresponding phenol with facile elimination of dimethylamine. Use of 1 equiv. of phenol also gave the disubstituted product. By contrast, the sterically demanding phenols ROH (R = 2,6-$Bu^t_2C_6H_3$ or 2,6-Bu^t_2-4-MeC_6H_2) gave only the monosubstituted products $[Zr(OR)(NMe_2)Cp_2]$ regardless of the stoichiometry employed. However, reaction of $[Zr(OR)(NMe_2)Cp_2]$ (R = 2,6-$Bu^t_2C_6H_3$) with methanol gave $[Zr(OMe)(OR)Cp_2]$ in 11% yield.[23]

Hydrazone and azine derivatives of zirconium have been prepared by treating $[ZrCl_2Cp_2]$ with the appropriate hydrazone or azine.[111–13]

The reaction of $[ZrMe_2Cp_2]$ with an alcohol, siloxol or phenol proceeds rapidly with elimination of methane and formation of the corresponding alkoxide, siloxide or aryloxide. Thus, R_3SiOH (R = Et or Ph) gives $[Zr(OSiR_3)_2Cp_2]$ (R = Ph or Me),[114] and Bu^tMe_2SiOH yields $[ZrMe(OSiMe_2Bu^t)Cp_2]$.[115] Similarly, (E)-$HOCR^1R^2CH=CHR^3$ gives (E)-$[ZrMe(OCR^1R^2CH=CHR^3)Cp_2]$ (R^1 = R^2 = Me, R^3 = H; R^1R^2 = $(CH_2)_4$, R^3 = H; R^1 = H, R^2 = R^3 = Me).[116] 2,2-Dimethylpropane-1,3-diol and 1,3-benzenedimethanol afford the macrocyclic zirconocene dialkoxides $[Cp_2Zr(\mu\text{-}OCH_2\text{-}X\text{-}CH_2O)_2ZrCp_2]$ (X = CMe_2 or C_6H_4).[117] Elimination of toluene was employed in the synthesis of $[ZrBz(OR)Cp_2]$ from dibenzylzirconocene and ROH (R = Me, 2-Me-6-$Bu^tC_6H_3$ or 2,4-Me_2-6-(α-methylcyclohexyl)C_6H_2).[118,119] Complexes $[Zr\{O(CH_2)_nPhCr(CO)_3\}(X)Cp_2]$ were obtained from dibenzylzirconocene (n = 1, X = Bz) or dimethylzirconocene (n = 1, X = $OCH_2PhCr(CO)_3$) and $[(HOCH_2Ph)Cr(CO)_3]$,[120] or from $[ZrCl_2Cp_2]$ or $[Zr(NMe_2)_2Cp_2]$ and $[NEt_4][OPhCr(CO)_3]$ (n = 0, X = $OPhCr(CO)_3$).[121] The complex $[(\eta\text{-}C_5H_4Bu^t)_2Zr(\eta^1:\eta^5\text{-}C_5H_4)_2Fe]$ reacts with ROH (R = Me, Bz or 3,5-$(MeO)_2Bz$) with successive formation of $[Zr(OR)(Fc)(\eta\text{-}C_5H_4Bu^t)_2]$ and $[Zr(OR)_2(\eta\text{-}C_5H_4Bu^t)_2]$.[122] Oligomeric zirconocene hydridochlorides react with methanol at −75 °C with formation of $[ZrCl(OMe)(\eta\text{-}C_5H_4R)_2]$ (R = H or Me) and elimination of dihydrogen.[123]

Other synthetic routes include the following.

(i) Autoxidation of dialkylzirconocenes

Autoxidation of a dialkylzirconocene gives $[Zr(OR)_2Cp_2]$ (R = Me,[116] Bz or substituted Bz)[124] in good yield. Peroxo complexes $[Zr(O_2R)(R)Cp_2]$ are the proposed intermediates. In the autoxidation of $[ZrR(X)Cp^*_2]$ (R = Me or Bz; X = R or Cl) and $[ZrBz_2CpCp^*]$ formation of alkoxo complexes is not observed. Rates of initiation and oxidation have been compared with those of the corresponding Cp_2Zr complexes.[119] The reaction of Bu^tCO_2H with $[HfH(R)Cp^*_2]$ gives stable t-butyl peroxo complexes $[Hf(O_2Bu^t)(R)Cp^*_2]$ (R = Cl, H, Me, Et, Pr, CH=CHBu^t, Ph, 3,5-$Me_2C_6H_3$[125] or $CH_2CH_2Bu^t$);[126] complexes $[Hf(CH_2CHMeCH_2)Cp^*_2]$ and $[Hf(CH_2CH_2CH_2CH_2)Cp^*_2]$ yield $[Hf(O_2Bu^t)(R)Cp^*_2]$ (R = Bu^i or Bu^n).[125] The compound $[HfH(O_2Bu^t)Cp^*_2]$ decomposes above −35 °C. In the IR spectra the ν(O–O) mode is observed at 835–850 cm^{-1}.[125] The mode of decomposition of these complexes depends on the ligand R^-. Only for R = H, Me, Et, Pr, Bu, Bu^i [125] or $CH_2CH_2Bu^t$ [126] is smooth decomposition to $[Hf(OBu^t)(OR)Cp^*_2]$ observed.

(ii) Insertion of propene oxide

Insertion of propene oxide into the Zr–Cl bond of [ZrCl$_2$Cp$_2$] furnishes both [ZrCl{OCH-(CH$_2$Cl)Me}Cp$_2$] (90%) and [ZrCl(OCH$_2$CHClMe)Cp$_2$] (10%).[127]

(iii) From zirconocene α-alkoxyalkanes

(a) Thermolysis of [ZrCl{CHMe(OEt)}Cp$_2$] gives [ZrCl(OEt)Cp$_2$] and ethene.[128] Similarly, [Zr(CH$_2$OMe)$_2$(η-C$_5$H$_4$R)$_2$] (R = H or Me) eliminates ethene on heating with formation of [Zr(OMe)$_2$(η-C$_5$H$_4$R)$_2$].[123] Rearrangement of [ZrCl{CPh$_2$(OMe)}Cp$_2$] at ambient temperature proceeds with elimination of alkene. At elevated temperature rearrangement to Structures (5)–(7) is observed (Scheme 2, R = Me).[55] Reaction of the ketone complex [{Zr(μ,η^2-OCPh$_2$)Cp$_2$}$_2$] with an alkyl halide RX to give (5)–(7) (R = Me, Et, tetrahydrofurfuryl, allyl, 2-propyl, 2-octyl or Bu^i) is comparable.[55] Ring-opening of [{Zr(μ,η^2-OCPh$_2$)Cp$_2$}$_2$] is also observed upon thermolysis.[55,129] [ZrCl(CH$_2$OCH$_2$Ph)Cp$_2$] underwent a Wittig rearrangement at 110 °C to give [ZrCl(OCH$_2$CH$_2$Ph)Cp$_2$] as the major product.[130] A radical mechanism is discussed for all these reactions.

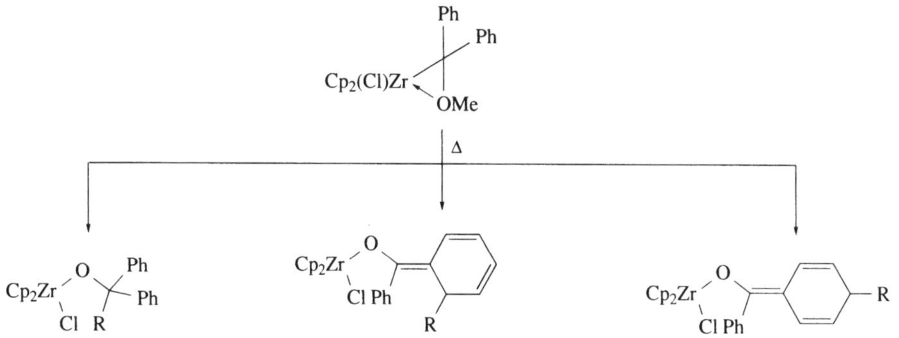

Scheme 2

(b) Reaction of [ZrCl(CH$_2$OMe)Cp$_2$] with zirconocene hydridochloride gives [ZrCl(OMe)Cp$_2$] and methylzirconocene chloride. Similarly, [ZrCl(CPh$_2$OMe)Cp$_2$] gives [ZrCl(OMe)Cp$_2$] and [ZrCl(CHPh$_2$)Cp$_2$], but also [ZrCl$_2$Cp$_2$] and [Zr(CHPh$_2$)(OMe)Cp$_2$].[55]

(iv) Coupling of bis(zirconocenyl)arene complexes

Bis(zirconocenyl)arene complexes undergo a biscoupling reaction with acetone to give bis-oxametallacycles.[131,132] A similar reaction was observed for the alkyne complex [Zr(η^2-TMS–C≡C–Ph)(PMe$_3$)Cp$_2$], which reacts with acetophenone with replacement of the phosphine and CC coupling of the π ligands at the zirconocene template to give the oxametallacyclopentene [Zr{OCMePhCPh=C(TMS)}Cp$_2$].[133]

The diene complexes [Zr(diene)Cp*$_2$] (diene = *s-cis*-isoprene or *s-trans*-butadiene) react with 2,4-dimethyl-3-pentanone regio- and stereoselectively to give [Zr(OCR^{1}R^{2}CH$_2$CR^3=CHCH$_2$)(η-C$_5$R^4$_5$)$_2$] (R^3 = H or Me; R^4 = Me; R^1 = R^2 = Pr^i). The crystal structure analysis revealed the (Z) configuration with respect to the C=C double bond.[134] An isomeric mixture of (*s-trans-* and *s-cis-*η^4-butadiene-)zirconocene reacts with carbonyl compounds R^{1}R^{2}C=O to give [Zr(OCR^{1}R^{2}CH$_2$CR^3=CHCH$_2$)(η-C$_5$R^4$_5$)$_2$] (R^3 = R^4 = H for all complexes, R^1 = R^2 = Ph or Me; R^1 = Me, R^2 = Ph; R^1 = Me, R^2 = Bu^i; R^1 = H, R^2 = Bu^i; CR^{1}R^2 = cyclododecane).[135]

The 1:1 reactions of [Zr(butadiene)(η-C$_5$R$_5$)$_2$] (R = H or Me) with CO$_2$, Bu^{t}NCO, PhNCO or PhMeC=C=O yield Zr–O-bound complexes having (σ, *syn*-η^3-allyl)metal structures (8) (X = NBu^t, NPh, CMePh or O), while [Zr(isoprene)(η-C$_5$R$_5$)$_2$] (R = H or Me) gave seven-membered ring compounds with (Z)-oxametallacyclic structures [Zr{OC(=X)CH$_2$CH=CHCH$_2$}(η-C$_5$R$_5$)$_2$] (X = NBu^t, NPh, CMePh or O) (see Chapter 7, this volume). Complexes [Zr(isoprene)(η-C$_5$R$_5$)$_2$] (R = H or Me) react with Ph$_2$C=C=O to give a six-membered oxametallacycle [Zr{OC(=CPh$_2$)CHMeCMe=CH}(η-C$_5$R$_5$)$_2$].[136]

$$(\eta\text{-}C_5R_5)_2Zr \underset{O}{\overset{}{\diagup}} \diagdown_X$$

(8)

(v) Reaction with ROH

With equimolar amounts of ROH, $[Zr(C_2H_4)(PMe_3)Cp_2]$ undergoes protonation of the ethene ligand to give $[ZrEt(OR)Cp_2]$ (R = Me, Et or Ph);[137,138] excess ROH leads to formation of $[Zr(OR)_2Cp_2]$.[137] Trifunctional alcohols $X(OH)_3$ afford bridged trinuclear complexes $[(Cp_2EtZrO)_3(\mu_3\text{-}X)]$ (X = CH_2CHCH_2, $CH_2CMe(CH_2)CH_2$ or $2,4,6\text{-}C_6H_3$), and tetrahydroxy compounds such as citric acid or pentaerythritol give tetranuclear complexes (see Section 10.4.4.1).[139] However, when the difunctional 1,2-dihydroxybenzene is employed, the reaction products are the dinuclear complex $[\{ZrEtCp_2\}_2(\mu\text{-}1,2\text{-}O_2C_6H_4)]$ and $[Zr(1,2\text{-}O_2C_6H_4)Cp_2]$.[140] The latter was described in 1969 but not fully characterized.

(vi) Oxygen atom transfer

Oxygen atom transfer from N_2O to $[HfH(Ph)Cp^*_2]$ gives $[Hf(OH)(Ph)Cp^*_2]$ and $[HfH(OPh)Cp^*_2]$.[141] An intermediate benzyne complex was implicated in the scrambling process, and it was shown that $[Zr(\eta^2\text{-cyclohexyne})(PMe_3)Cp_2]$ reacts with N_2O to give a dimeric oxametallacyclobutene derivative with elimination of dinitrogen and PMe_3.[142] Similarly, $[Zr(C_2Ph_2)Cp^*_2]$ reacts with N_2O to yield an adduct, which loses dinitrogen at ambient temperature to give $[\overline{Zr(CPh{=}CPh\dot{O})}Cp^*_2]$.[42]

(vii) Elimination and intramolecular hydrozirconation

The reaction of a silylated unsaturated alcohol with zirconocene hydridochloride is dependent on the length and branching of the alkenyl chain. Thus, TMS-H and the yellow complex $[Cp_2(Cl)Zr(CH_2CH_2CY_2O)Zr(Cl)Cp_2]$ (Y = Me) or $[\{\overline{Zr(OCY_2CH_2CH_2)}Cp_2\}_2]$ (Y = H) are obtained.[143] Zirconocene hydridochloride reacts with $KO(CH_2)_nCR{=}CH_2$ ($n = 1$, R = H or Me; $n = 2$, R = H) with elimination of KCl and intramolecular hydrozirconation to give $[\{\overline{Zr(OCH_2CHR^1CR^2R^3)}Cp_2\}_2]$ ($R^1 = R^2 = R^3 = H$; $R^1 = Me$, $R^2 = R^3 = H$; $R^1 = R^2 = H$, $R^3 = Me$; $R^1 = Me$, $R^1 = H$, $R^2 = R^3 = D$).[144,145] However, when $[ZrCl_2Cp^*_2]$ was reacted with $XMg(CH_2)_nOMgX$ ($n = 3$ or 4; X = halide) the monomeric complexes $[\overline{Zr\{CH_2(CH_2)_nO\}}Cp^*_2]$ ($n = 2$ or 3) were obtained due to the presence of the bulky $^-Cp^*$ ligands.[144]

(viii) Reaction with H_2

When $[Hf(CO)_2Cp^*_2]$ is reacted with dihydrogen, clean reduction of one CO ligand is observed with formation of $[HfH(OMe)Cp^*_2]$ and CO.[146]

10.4.1.2 Spectroscopic properties

The $v(C{-}O)$ mode is generally observed in the range 1280–1300 cm^{-1} for aryloxides and 1090–1140 cm^{-1} for alkoxides. The Zr–O stretching modes are observed in the range 420–570 cm^{-1}. The complexes $[Zr(OR)_2Cp_2]$ (R = $2,6\text{-}Bu^t_2C_6H_3$, $2,6\text{-}Bu^t_2\text{-}4\text{-}MeC_6H_2$ or $2,4,6\text{-}Bu^t_3C_6H_2$) and $[Zr(OR)(X)Cp_2]$ (R = $2,6\text{-}Bu^t_2C_6H_3$ or $2,6\text{-}Bu^t_2\text{-}4\text{-}MeC_6H_2$, X = NMe_2; R = $2,6\text{-}Bu^t_2C_6H_3$, X = OMe) show a 1:1 doublet for the o-Bu^t groups in the ^1H NMR spectra, which is attributed to restricted rotation about the carbon–oxygen bond.[23,147]

10.4.1.3 Structural data

Crystal structures have been reported for the following complexes containing either terminal alkoxo or aryloxo ligands: [ZrBz{O-2,4-Me$_2$-6-(α-methylcyclohexyl)C$_6$H$_2$}Cp$_2$],[118] [ZrCl(O-2,6-Bu$^t_2$-4-MeC$_6$H$_2$)Cp$_2$],[147] [Zr(NMe$_2$)(O-2,6-Bu$^t_2$C$_6$H$_3$)Cp$_2$][147] and [Cp$_2$(Cl)Zr(μ-OCMe$_2$CH$_2$CH$_2$)Zr(Cl)Cp$_2$],[143] and the isostructural complexes [M(OR)$_2$(η-C$_5$H$_4$Me)$_2$] (R = 2,6-Cl$_2$C$_6$H$_3$; M = Hf[148] or Zr[149]). The ligands in these complexes occupy positions about the zirconium atom typical of [ZrX(Y)Cp$_2$]-type compounds.

Crystal structures of the macrocyclic complexes [Cp$_2$Zr(μ-OCH$_2$-X-CH$_2$O)$_2$ZrCp$_2$] (X = CMe$_2$ or 1,3-C$_6$H$_4$)[117] and the trinuclear complexes [Zr{OPhCr(CO)$_3$}$_2$Cp$_2$][121] and [Zr{OCH$_2$PhCr(CO)$_3$}$_2$Cp$_2$][120] have been determined. Several complexes which have an oxametallacycloalkane or -alkene fragment have been structurally characterized.[42,134,135,142] Crystallographic data for [HfEt(O$_2$Bu^t)Cp*$_2$] show an η^1-bonding mode for the peroxo ligand.[125]

10.4.1.4 Chemical properties

The metallocene alkoxides or aryloxides are white to yellow, air-sensitive compounds. The M–O bond is readily cleaved by protic reagents. Protonation of [Zr(OR^{1}R^{2}CH$_2$CH=CHCH$_2$)Cp$_2$] (R^1 = R^2 = Ph or Me; R^1 = Me, R^2 = Ph or Bu^t; R^1 = H, R^2 = Bu^t; CR^{1}R^2 = cyclododecane) gives MeCH=CHCH$_2$CR^{1}R^{2}OH and CH$_2$=CH(CH$_2$)$_2$CR^{1}R^{2}OH.[135]

Some reactions that leave the alkoxo ligand unaffected were reported.[23,42] The oxametallacyclobutene complex [Zr(CPh=CPhO)Cp*$_2$] inserts aldehydes, Bu^{t}NC, TolN$_3$, CO or Ph$_2$CN$_2$ to give five- and six-membered metallacycles.[42] [ZrBz{O-2,4-Me$_2$-6-(α-methylcyclohexyl)C$_6$H$_2$}Cp$_2$] reacts with Bu^{t}NC to give the η^2-iminoacyl derivative [Zr(η^2-NBu^{t}CBz){O-2,4-Me$_2$-6-(α-methylcyclohexyl)C$_6$H$_2$}Cp$_2$].[118]

10.4.2 Metallocene Complexes with Transition Metal-substituted Alkoxo Ligands

Two major synthetic approaches are used for the synthesis of metallocene complexes with transition metal-substituted alkoxo ligands: (i) transfer of hydrogen from a transition metal hydride to a metallocene acyl complex and (ii) transfer of hydrogen from a metallocene hydride derivative to a transition metal complex. For a recent review article on metallocene oxycarbene complexes, see Reference;[55] for carbonyl-bridged complexes, see Reference.[95]

(i) Whereas molybdocene dihydride reacts with the acyl complex [Zr(η^2-OCMe)(Me)Cp$_2$] with reduction of the acyl ligand by hydrogen transfer and formation of [ZrMe(OCH$_2$Me)Cp$_2$], rhenocene hydride and tungstocene dihydride give heterodinuclear complexes [Cp$_2$(Me)Zr-O-CHMe-M(X)Cp$_2$] (M(X) = Re, W(H)).[95] The tungsten complex has a limited lifetime of several minutes.

(ii) Zirconoxycarbene complexes [Cp$_2$(R)M=C(H)–O–ZrHCp*$_2$] (M = Nb, R = H; M = Nb, R = H, Me, Bz, CH$_2$C$_6$H$_4$OMe, Ph or CH$_2$OZr(H)Cp*$_2$; M(R) = Cr, Mo or W) are prepared by reacting [NbR(CO)Cp$_2$] (R = H, Me, Bz, CH$_2$C$_6$H$_4$OMe, Ph, CH$_2$OZr(H)Cp*$_2$) or [M(CO)Cp$_2$] (M = Cr, Mo or W)) with [ZrH$_2$Cp*$_2$] at −78 °C in toluene.[95,150,151]

The zirconoxycarbene complex (9) is obtained by thermolysis of diphenylzirconocene in the presence of [W(CO)$_6$]. Compound (9) forms an adduct with acetophenone, and with phenol the Zr–C bond is cleaved with formation of [Zr(OPh)(OC{=W(CO)$_5$}Ph)Cp$_2$]. The latter is also accessible from the reaction of [ZrCl(OPh)Cp$_2$] with [(CO)$_5$W=C(OLi)(Ph)]. Complex (9) was shown to be polymeric in the solid state, while the C$_5$H$_4$Bu^t analogue is monomeric.[152]

Cp$_2$Zr —O— =W(CO)$_5$

(9)

Treatment of [ZrL(CO)Cp$_2$] (L = CO or PMe$_3$) with [ZrH$_2$Cp*$_2$] affords the zirconoxycarbene complexes [Cp$_2$(L)Zr=CHO–Zr(H)Cp*$_2$]. The reaction of the PMe$_3$ adduct with MeI gives methane and [Cp$_2$(L)Zr=CHO–Zr(I)Cp*$_2$]. The latter reacts with HCl to give [ZrCl$_2$Cp$_2$], PMe$_3$ and [ZrI(OMe)Cp*$_2$]; with CO, [Zr{OC{ZrICp*$_2$}C(H)O}Cp*$_2$] is formed. The latter is also obtained from the reaction of [Cp$_2$(CO)Zr=CHO–Zr(H)Cp*$_2$] with MeI (elimination of CH$_4$).[153]

The structurally characterized complex (**10**) was obtained from the reaction of [{Zr(N$_2$)Cp*$_2$}$_2$(N$_2$)] with [{Fe(CO)$_2$Cp}$_2$], or from [ZrI$_2$Cp*$_2$] and Na[Fe(CO)$_2$Cp].[95,154]

(**10**)

(Butadiene)zirconocene reacts with L$_n$M–CO (ML$_n$ = Cr(CO)$_5$, Mo(CO)$_5$, W(CO)$_5$, Fe(CO)$_4$, Fe(CO)$_3$PPh$_3$, Rh(CO)Cp, Co(CO)Cp, Co(CO)(η-C$_5$H$_4$Cl), Ni(CO)$_3$, Zr(CO)Cp$_2$, Hf(CO)Cp$_2$ or V(CO)$_3$Cp) with formation of [Zr{η^3-CH$_2$CHCHCH$_2$C(=ML$_n$)O}Cp$_2$].[55]

Reaction of zirconocene hydridochloride with metal enolates [W(CHR^{2}COX)(CO)$_3$Cp] (X = OEt, Me or Ph) gives [Cp$_2$(Cl)Zr(OCHR^{1}CHR^2)W(CO)$_3$Cp] (X = OEt, R^1 = R^2 = H; X = R^1 = Me or Ph, R^2 = H; X = R^1 = R^2 = Ph). The phenyl derivative (R^1 = R^2 = Ph) exists as two diastereomers which are separable by recrystallization. Photolysis of [Cp$_2$(Cl)Zr(OCHR^{1}CHR^2)W(CO)$_3$Cp] gives [Cp$_2$(Cl)Zr–O–W(CO)$_3$Cp] via elimination of R^{2}CH=CHR^1. The dinuclear complex rapidly decomposes at room temperature.[155]

Comparable heterobimetallic complexes with metal–metal interaction are dealt with in Section 10.6.

10.4.3 Metallocene Enolates and Enediolates

10.4.3.1 Synthesis

Several synthetic routes to metallocene enolates and enediolates are employed.

(i) Salt elimination

The complexes [Zr{OC(=CR^{2}R^3)R^1}XCp$_2$] (X = Me or OCH=CH$_2$ and R^1 = R^2 = R^3 = H; R^1 = H, R^2 = R^3 = Me;[156] R^1 = N(CHMe$_2$)$_2$, R^2 = R^3 = H;[157] R^1 = PPh$_2$, R^2 = R^3 = H;[158] R^1 = Fe(CO)(PPh$_3$)Cp, R^2 = R^3 = H)[159] are obtained from zirconocene chloride derivatives and the appropriate lithium reagent. Where R^2 ≠ R^3 both isomers (*E*) and (*Z*) are obtained in varying ratios.

(ii) Reductive coupling

Reductive coupling of two CO molecules with formation of [M{OC(R)=C(R)O}Cp*$_2$] has been employed (M = Hf, R = CH=CH$_2$;[144] MR$_2$ = $\overline{\text{HfCH}_2\text{(CH}_2\text{)}_2\text{CH}_2}$, RC=CR = $\overline{\text{CH}_2\text{CH}_2\text{C=CCH}_2\text{CH}_2}$[160]). The (η^4-*s-cis*-diene)metallocene complexes [M(CH=CR–CH=CH)Cp*$_2$] (M = Hf, R = H or Me; M = Zr, R = Me) react with CO to give the metal enediolates (**11**).[161] Alternatively, the enediolato complex [Zr{OC(Bu^t)=C(Bu^t)O}Cp*$_2$] is obtained from the oxidative addition of Bu^{t}C(=O)C(=O)Bu^t to [{Zr(N$_2$)Cp*$_2$}$_2$(N$_2$)].[162]

(iii) CO insertion

Carbonylation of [HfH(CH=CBu^{t}H)Cp*$_2$] gives [Hf($\overline{\text{OCH=CHCBu}^t\text{H}}$)Cp*$_2$] (–10 °C) which reacts with further CO at 25 °C to give [Hf($\overline{\text{OCH=CHCBu}^t\text{=CHO}}$)Cp*$_2$].[160] Carbonylation of zirconocene hydridochloride gives [(ZrClCp$_2$)$_2$(μ-CHO)], which reacts slowly with further CO to afford (**12**).[163] Carbonylation of [ZrCl{CH(TMS)R}Cp$_2$] (R = 9-anthryl) stereospecifically yields (*E*)-[ZrCl{OC(TMS)=CHR}Cp$_2$] via 1,2-TMS migration.[164] The insertion of CO into the Zr–C bond(s) of [Zr(CH$_2$SiMe$_2$CH$_2$)(η-C$_5$R$_5$)$_2$] (R = H or Me) has been examined. For the Cp derivative, insertion of one equivalent of CO is followed by an intramolecular 1,2-silyl shift and formation of [Zr{OC(=CH$_2$)SiMe$_2$(CH$_2$)}Cp$_2$]$_n$. Comparable studies of the carbonylation of the Cp* derivative have shown that this reaction can lead to the formation of two different diinsertion products: a cyclic

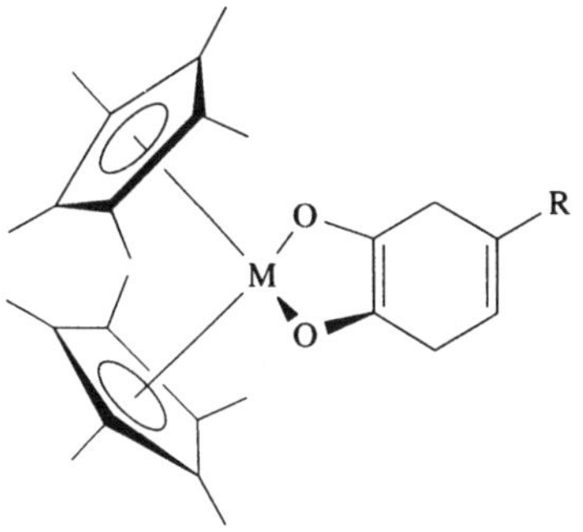

(11)

dienolate, $[Zr\{OC(=CH_2)SiMe_2C(=CH_2)O\}Cp*_2]$ and a bicyclic enediolate $[Zr\{OC(CH_2SiMe_2CH_2)=CO\}Cp*_2].^{165}$

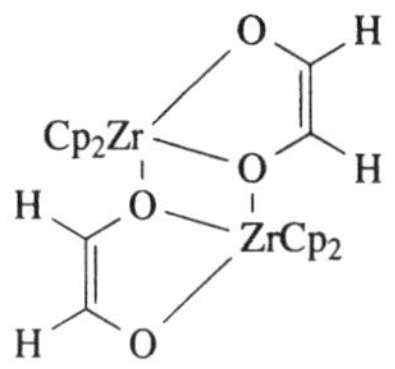

(12)

(iv) Hydrogenation

Hydrogenation of the ketene complexes $[Zr(OC=CHR)(L)Cp*_2]$ gives zirconocene enolates $[Zr\{OCH=CHR\}(H)Cp*_2]$ (R = H, L = py; R = Bu^t, L = CO or CH$_2$PMe$_3$). The enolato geometry for R = Bu^t is >96% *cis*. Treatment of this complex with MeI gives $[ZrI(OCH=CHBu^t)Cp*_2]$, which isomerizes to the *trans*-isomer.[166]

The complex $[Hf(CO)_2Cp*_2]$ reacts with $[MH_2Cp*_2]$ (M = Zr or Hf) under dihydrogen to produce *cis*-$[Cp*_2(H)Hf(\mu\text{-}OCH=CHO)M(H)Cp*_2].^{146}$ Warming of $[HfH_2(CO)Cp*_2]$ above $-10\,^{\circ}C$ under dihydrogen gives *cis*- or *trans*-$[\{HfHCp*_2\}_2(\mu\text{-}OCH=CHO)]$, $[\{HfHCp*_2\}_2(\mu\text{-}OCH_2CH_2O)]$ and $[HfH(OMe)Cp*_2],^{146}$ while the *trans* enediolato dimer is the sole product in the thermal reaction of the zirconium analogue $[ZrH_2(CO)Cp*_2].^{167}$ The oxametallacyclobutene complex $[Zr(CPh=CPhO)Cp*_2]$ reacts with dihydrogen to give the enolate $[ZrH\{OC(Ph)=CHPh\}Cp*_2].^{42}$

(v) Substrates with acidic hydrogen

Substrates R–H with acidic hydrogen (HO–H, PhO–H, MeC(=O)CH$_2$–H, HCC–H, PhCC–H or Bu^{t}CC–H) react with the oxametallacyclobutene complex $[Zr(CPh=CPhO)Cp*_2]$ to give $[Zr(OCPh=CHPh)(R)Cp*_2].^{42}$

(vi) Rearrangement

Complex $[Zr(CO)(\eta^2\text{-}OCHBu^i)Cp*_2]$ slowly loses its coordinated CO and rearranges to $[ZrH(OCH=CHCHMe_2)Cp*_2]$. By contrast, the hafnium derivative rearranges without loss of CO to give $[Hf\{OCH=CBu^tO\}Cp*_2]$. The compound $[Zr(\eta^2\text{-}COBu^i)(H)Cp*_2]$ reacts with ethene or 2-butyne to give $[Zr\{OCHBu^iCH_2CH_2\}Cp*_2]$ or $[Zr\{OCHBu^iCMe=CMe\}Cp*_2].^{160}$

(vii) Insertion of diphenylketene

Insertion of diphenylketene into the Zr–C bonds of $[ZrR_2Cp_2]$ (R = Me, Bz or Ph) gives $[Zr(OCR=CPh_2)_n(R)_{2-n}Cp_2]$ (n = 1 or 2, R = Me; n = 1, R = Bz; n = 2, R = Ph).[41]

10.4.3.2 Spectroscopic properties

The zirconocene enolates are yellow or orange, and the enediolate complexes red, violet or blue. The unusual colour of zirconocene enediolato complexes gives rise to an absorption in the UV–visible spectrum at 433–570 nm,[161,162] which is due to an electronic transition from the HOMO (π) to the LUMO (y^2). This absorption is shifted to higher wavenumber if the ZrO_2C_2 ring is less puckered.[162] All compounds have been characterized by NMR spectroscopy. In the ^{13}C NMR spectrum the resonance of the C=C bond of the enediolato ligand is observed between δ 132 and 170.[157,158,161] In the IR spectrum the ν(C=C) mode is observed between 1600 cm^{-1} and 1700 cm^{-1}, and the ν(C–O) mode between 1200 cm^{-1} and 1270 cm^{-1}.[161,166]

10.4.3.3 Structural data

Structural data are available for $[ZrCl(OCR=CH_2)Cp_2]$ (R = NPh$_2$,[157] PPh$_2$[158] or Fe(CO)(PPh$_3$)Cp[159]), (*E*)-$[ZrCl\{OC(TMS)=CHR\}Cp_2]$ (R = 9-anthryl),[164] $[ZrMe(OCMe=CPh_2)Cp_2]$[41] and $[Zr(O-CMe=CPh_2)_2Cp_2]$.[41] The cyclic dienolate $[Zr\{OC(=CH_2)SiMe_2C(=CH_2)O\}Cp*_2]$ has two exocyclic methylene groups.[165] Crystallographic studies of $[Zr\{OC(R)=C(R)O\}Cp*_2]$ (R = Me or Bu^t) show a planar ZrO_2C_2 ring for R = Bu^t, and a nonplanar ring for R = Me.[162]

10.4.3.4 Chemical reactivity

The nucleophilicity of the methylene carbon in the enolates $[ZrCl(OCR=CH_2)Cp_2]$ (R = NPh$_2$[157] or PPh$_2$[158]) has been exploited to form complexes with Cr(CO)$_5$. The complex $[ZrCl\{OC(=CH_2)PPh_2\}Cp_2]$ reacts rapidly with benzaldehyde to form $[ZrCl\{OCHPhCH_2C(=O)PPh_2\}Cp_2]$.[158]

10.4.3.5 Related complexes

The reaction of zirconocene dichloride with dipotassium dithiooxalate (dto) in CH_2Cl_2 gives $[(ZrClCp_2)_2(dto)]$. In the latter each zirconium is five-coordinate, being bound to an oxygen and a sulfur atom of the dto ligand. The ZrCl(μ-dto)ZrCl fragment is planar.[168]

10.4.4 Metallocene Carboxylates

10.4.4.1 Synthesis

Reaction of metallocene halides with carboxylic acids in the presence of excess base for 1–2 d is a general synthetic route to metallocene carboxylates $[Zr(\eta^2\text{-}O_2CR)_n(Cl)_{2-n}Cp_2]$ (base usually NEt$_3$, n = 1 or 2; R = Pr^i or MeC(=O)NHCHMe–;[169] O_2CR = a substituted pyridinecarboxylic[170,171] or thiophenecarboxylic acid).[171,172] Reaction of $[MCl_2Cp_2]$ (M = Ti or Zr) with 3-indole derivatives in the presence of NEt$_3$ gave the mono- (1:1 reaction) and disubstituted (1:2) carboxylato complexes.[173] Hafnocene dichloride (1:1, 1–2 d) gave the monosubstituted products $[HfCl(O_2CR)Cp_2]$. However, the 1:2 reaction yields $[HfCl(O_2CR)_2Cp]$.[174]

Treatment of $[(ZrClCp_2)_2O]$ with excess RCO_2H gave $[ZrCl(O_2CR)Cp_2]$ (R = H, Me, Bu^t or Ph).[175] The reaction of zirconocene dichloride with TFA yields $[\{Zr(O_2CCF_3)(\mu\text{-}OH)Cp_2\}_2]$, which contains five-coordinate zirconium.[176] The compound $[Zr(\eta^2\text{-}CHPhNMe_2)(Me)Cp_2]$ reacts with TFA with replacement of the Me$^-$ ligand by $CF_3CO_2^-$.[177] In these trifluoroacetato complexes η^1 coordination of the carboxylato ligand is observed (see Section 10.4.4.3).

The reaction of zirconocene dichloride with the sodium salt of a carboxylic acid has been employed for the synthesis of $[ZrCl(\eta^2\text{-}O_2CR^2)(\eta\text{-}C_5H_4R^1)_2]$ (R^1 = H, R^2 = Ph or a substituted phenyl;[178] R^1 = Me, R^2 = nap;[179] R^1 = H or Me, R^2 = nap, $-CH_2C_{10}H_7$, PhCH=CH or p-ClC$_6$H$_4$OCH$_2$[180]), as well as for the synthesis of zirconocene complexes containing ferrocenylcarboxylato ligands $[\{ZrCl(\eta\text{-}C_5H_4R)_2\}_n(Y)]$ (n = 1, R = H or Me, Y = FcCO$_2$, FcCH$_2$CO$_2$, Fc(CH$_2$)$_3$CO$_2$, FcC(O)(CH$_2$)$_2$CO$_2$ or FcC(O)(C$_7$H$_8$)CO$_2$ (C$_7$H$_8$ = 5,6-norbornene);[181] n = 1, R = H, Y = FcC(O)(CH$_2$)$_2$CO$_2$;[182] n = 2, R = H or Me, Y = Fe$\{C_5H_4C(O)(CH_2)_2CO_2\}_2$).[181] The monothiocarbamato complex $[ZrCl(SOCNMe_2)Cp_2]$ was prepared similarly (see Section 10.5.5.1).[183]

The dicarbonyl complex (**13**) (M = Re, L = CO) is obtained by salt elimination from zirconocene dichloride and Na[Re(CO)$_2$(CO$_2$)Cp].[184] Analogously, zirconocene dichloride reacts with Na[M(CO)$_2$(CO$_2$)Cp] (M = Fe or Ru) to give (**13**) (M = Fe or Ru; L = CO).[185] Treatment of methylzirconocene chloride with [Re(CO)(CO$_2$H)(NO)Cp] results in formation of (**13**) (M = Re, L = NO).[186] [(η-C$_5$H$_4$Bu^t)$_2$Zr(η^1:η^5-C$_5$H$_4$)$_2$Fe] reacts with PhCO$_2$H to give [Zr(O$_2$CPh)$_2$(η-C$_5$H$_4$Bu^t)$_2$].[122]

$$Cp_2(Cl)Zr \overset{O}{\underset{O}{\diagdown\,\diagup}} M(CO)(L)Cp$$

(**13**)

Utilization of aromatic dicarboxylato ligands allows the synthesis of monomeric,[170] dimeric[187] or trimeric[188] metallocene complexes, depending on the position of the carboxyl groups. A monomeric product [Zr{OC(=O)CH$_2$OCH$_2$C(=O)O}Cp$_2$] was also obtained from the reaction of zirconocene dichloride with the disodium salt of oxydiacetic acid.[189] The reaction of titanocene dichloride with tetrasodium pyrazinetetracarboxylate gives a tetrametallic complex. Attempts to prepare the zirconium analogue were unsuccessful. However, if the tetrasodium salt is reacted with equimolar amounts of titanocene and zirconocene dichloride a tetrametallic complex (Zr$_2$Ti$_2$) is obtained.[190] The polymeric complexes [Zr(μ-OCOC$\equiv$COCO)Cp$_2$]$_n$ and [Zr(μ-OCOC$\equiv$COCO)Cp$_2$·CHCl$_3$]$_n$ are obtained from zirconocene dichloride and acetylenedicarboxylic acid.[191]

The compound [Zr(C$_2$H$_4$)(PMe$_3$)Cp$_2$] reacts with RCO$_2$H, undergoing protonation of the ethene ligand and formation of [ZrEt(O$_2$CR)Cp$_2$] (R = H, Me or CF$_3$). Excess TFA gives [Zr(O$_2$CCF$_3$)$_2$Cp$_2$].[137] With bifunctional hydrogen-acidic compounds, bridged dinuclear complexes [Cp$_2$(Et)Zr(μ-O$_2$C–X–CO$_2$)Zr(Et)Cp$_2$] are obtained (X = CH$_2$, CH$_2$CH$_2$, or *cis* and *trans* CH=CH or CH=CMe).[192] Reactions with tetrafunctional carboxylic acids yield tetranuclear complexes (see Section 10.4.1.1).[139]

10.4.4.2 Spectroscopic properties

In the IR spectrum, the difference between v_{as}(CO$_2$) and v_s(CO$_2$) is indicative of the bonding mode of the carboxylato ligand. Thus, for η^2-coordination a difference of 60–100 cm^{-1} is expected and observed in most carboxylato complexes, which show the v_{as}(CO$_2$) and v_s(CO$_2$) mode in the range 1495–1640 cm^{-1} and 1430–1525 cm^{-1}, respectively. The v(Zr–O) mode occurs in the range 460–480 cm^{-1}. In (**13**) the v_{as} and v_s(CO$_2$) mode are displaced to lower energies[184,185] with respect to analogous monometallic complexes. In [Zr{OC(=O)CH$_2$OC$_6$H$_4$-p-Cl}(Cl)Cp$_2$] a five-membered ring is observed (by IR) instead of η^2-coordination of the carboxylato ligand.[180] The solution and solid-state IR spectra of [ZrCl(O$_2$CR)Cp$_2$] (R = H, Me, Bu^t or Ph) are consistent with η^2-bonding of the carboxylato ligand, although the formato complex (R = H) is in equilibrium with the unidentate η^1-carboxylate as shown by variable-temperature ^{1}H and ^{13}C NMR spectral studies. In THF solution all these complexes exist as solvated η^1-carboxylato complexes.[175] In the ^{13}C NMR spectra the chemical shift of the carboxylate carbon atom –CO$_2$– is observed between δ 187 and 213.[185,186]

10.4.4.3 Structural data

It was not until 1988 that crystallographic studies on carboxylato complexes were reported. [ZrCl(O$_2$CR)(η-C$_5$H$_4$Me)$_2$] (R = nap) has a bidentate carboxylato ligand.[179] In this complex and in [ZrCl{O$_2$C(CH$_2$)$_2$C(O)Fc}Cp$_2$][182] the Zr–O bond lengths are much greater than the Zr–O distances in oxo-bridged dimeric complexes. A monodentate carboxylato ligand is observed in [Zr(η^2-CHPhNMe$_2$)(O$_2$CCF$_3$)Cp$_2$],[177] [Zr{OC(=O)CH$_2$OCH$_2$C(=O)O}Cp$_2$][189] and [{Zr(O$_2$CCF$_3$)(μ-OH)Cp$_2$}$_2$].[176] The Zr–O bonds are shorter than in complexes with η^2-bonding of the carboxylato group.

The carboxylato ligand acts as a (2 + 1)-dentate ligand in [Zr(μ-OCOC$\equiv$COCO)Cp$_2$]$_n$ and [Zr(μ-OCOC$\equiv$COCO)Cp$_2$·CHCl$_3$]$_n$. The polymeric complexes consist of chains with 2$_1$ symmetry, packed parallel to each other in the crystals.[191]

10.4.4.4 Chemical properties

Reduction of the coordinated carboxylato ligand in [ZrCl(O$_2$CR)Cp$_2$] to the aldehyde (R = H, Me or Ph) or alcohol (R = Me) has been reported.[175] Similarly, the dimeric complexes (**13**) (M = Re, L = NO;[186]

M = Ru, L = CO[184]) react with zirconocene hydridochloride to give [Cp$_2$(Cl)Zr–OCH$_2$–M(CO)(L)Cp]. The reaction of the iminoacyl complex [Zr(η^2-CMe=NBu^t)(O$_2$CCF$_3$)Cp$_2$] with methylzirconocene hydride results in reduction of the trifluoroacetato ligand and formation of [Cp$_2$Zr(η^2-CMe=NBu^t){μ-OC(H)(CF$_3$)O}Zr(Me)Cp$_2$].[177] Thermal degradation of the heterobimetallic carboxylates (13) (M = Fe or Ru; L = CO) gives [{M(CO)$_2$Cp}$_2$] and [(ZrClCp$_2$)$_2$O], indicating partial deoxygenation of CO$_2$.[185]

10.4.4.5 Related complexes

The nitronato complexes [Zr(O$_2$N=CMeR)(X)Cp$_2$] (R = Me or H, X = Cl; R = Me, X = CH$_2$CMe$_2$Ph) were prepared from the reaction of Li(O$_2$N=CMeR) with the appropriate zirconocene derivative (see Section 10.7.2).[193] The ligand 'bite' is similar to that observed in carboxylato and dithiocarbamato complexes.

10.4.5 Metallocene Hydroxides

Monohydroxides of zirconocene derivatives and zirconocene dihydroxide have been known since 1966. However, using bulky ⁻Cp* ligands allows the synthesis and isolation of metallocene hydroxo complexes that are not accessible from the corresponding Cp-substituted complexes. Thus, the reaction of [ZrMe$_2$(η-C$_5$R$_5$)$_2$] or [ZrCl(Me)(η-C$_5$R$_5$)$_2$] with water affords for R = H the well-known oxo-bridged dimers [(ZrXCp$_2$)$_2$O] (X = Cl or Me) and, for R = Me, the complexes [ZrCl(OH)Cp*$_2$] and [Zr(OH)$_2$Cp*$_2$].[194] The former is also obtained on hydrolysis of [ZrCl(R)Cp*$_2$] (R = Bu^t[194] or H[29]); alternatively, the dihydroxo complex is obtained from hydrolysis of [(ZrHCp*$_2$)$_2$O] or by hydrolysis of the reaction mixture of [ZrCl$_2$Cp$_2$] and BuLi.[29] The monohydroxo complex shows the presence of significant intramolecular nonbonding contacts. The mass spectrum of [Zr(OH)$_2$Cp*$_2$] shows the parent ion peak and an intense signal for [Zr(=O)Cp*$_2$].[194] Extended Hückel MO calculations were performed on the model complex [Zr(OH)$_2$Cp$_2$].[117]

The hafnocene analogues [Hf(OH)(X)Cp*$_2$] (X = H or Cl) were prepared from [HfH(X)Cp*$_2$] or [HfH(NH$_2$)Cp*$_2$] (X = H) and water. The monohydroxo complex (X = H) reacts further with H$_2$O to give the dihydroxo complex. The latter is also obtained from hydrolysis of [(HfHCp*$_2$)$_2$O].[29]

The complex [ZrCl$_2$Cp$_2$] reacts with TFA to give the hydroxo-bridged complex [{Zr(O$_2$CCF$_3$)(μ-OH)Cp$_2$}$_2$] (see Section 10.4.4.1). The Zr–O(H) bond lengths are ca. 0.02 nm longer than the Zr–O distances of oxo-bridged complexes (see Section 10.4.6.3).[176]

The alkyne complex [Zr(C$_2$Ph$_2$)Cp*$_2$] reacts with H$_2$O to give the monohydroxide [Zr(CPh=CHPh)(OH)Cp*$_2$][42] Similarly, [(η-C$_5$H$_4$Bu^t)$_2$Zr(η^1:η^5-C$_5$H$_4$)$_2$Fe] reacts with stoichiometric amounts of water to give [Zr(Fc)(OH)(η-C$_5$H$_4$Bu^t)$_2$].[122]

N$_2$O reacts with [HfH$_2$Cp*$_2$] to yield dinitrogen and [HfH(OH)Cp*$_2$] quantitatively. Competitive oxidation of the hydrido and aryl ligands in [HfH(Ph)Cp*$_2$] occurs with N$_2$O at 80 °C, affording [Hf(OH)(Ph)Cp*$_2$] and [HfH(OPh)Cp*$_2$] (3:2). Independently, [Hf(OH)(Ph)Cp*$_2$] has been prepared from [HfH(Ph)Cp*$_2$] and water.[141]

10.4.6 Oxo-bridged Metallocene Dimers

10.4.6.1 Synthesis

The two major synthetic routes employed in the synthesis of oxo-bridged metallocene dimers are hydrolysis of metallocene complexes and oxidation of low-valent metallocene compounds.

(i) Hydrolysis

Hydrolysis of a metallocene dialkyl or dihalide gives [{MX(η-C$_5$H$_4$R)$_2$}$_2$O] (M = Zr, R = H, X = Cl;[195,196] M = Zr, R = Bu^t, X = Me).[197,198] [MCl$_2${(η-C$_5$H$_4$)$_2$Y}] (M = Zr, Y = (SiMe$_2$)$_n$, n = 2 or 3)[195] is hydrolysed to [(MCl{(η-C$_5$H$_4$)$_2$Y})$_2$O]. [ZrEt(O$_2$CR)Cp$_2$] (R = H, Me or CF$_3$) reacts with water to give [{Zr(O$_2$CR)Cp$_2$}$_2$O].[137] Kinetic studies have been reported for the hydrolysis of [ZrCl$_2$Cp$_2$].[199]

Hydrolysis of the dinuclear complexes [(ZrCl$_2$Cp)$_2${(η-C$_5$H$_4$)$_2$Y}] (Y = SiMe$_2$ or CH$_2$)[200] or [(ZrMe$_2$Cp)$_2$(η^5:η^5-C$_{10}$H$_8$)][201] gives [{(ZrXCp)$_2$({η-C$_5$H$_4$}$_2$Y)}$_2$O] (X = Cl, Y = SiMe$_2$ or CH$_2$; X = Me, no Y). The structurally characterized chlorine derivative (X = Cl, no Y) was obtained from oxidation of [{Zr(μ-Cl)Cp}$_2$(η^5:η^5-C$_{10}$H$_8$)] with oxygen.[202]

[MH$_2$Cp*$_2$] (M = Zr or Hf) reacts stepwise with water to afford [MH(OH)Cp*$_2$], [(MHCp*$_2$)$_2$O], and finally [M(OH)$_2$Cp*$_2$]·H$_2$O with evolution of dihydrogen. The dihydrides react with [M(OH)(X)Cp*$_2$] (X = Cl, OH or H) to give [Cp*$_2$(X)M–O–M(H)Cp*$_2$] (M = Zr or Hf).[29]

Hydrolysis of (9) gives [{Cp$_2$ZrOC{=M(CO)$_5$}(Ph)}$_2$O] (M = Mo or W).[203] Similarly, [Zr(C$_2$H$_4$)(PMe$_3$)Cp$_2$] reacts with water with protonation of the ethene ligand and formation of [(ZrEtCp$_2$)$_2$O]; [ZrEt(OH)Cp$_2$] is proposed as an intermediate.[137]

(ii) Oxidation

Oxidation of dimethylzirconocene or [(Zr(μ-Cl){(η-C$_5$H$_4$)$_2$SiMe$_2$})$_2$] with oxygen (air) gave [(ZrMeCp$_2$)$_2$O][204] or [(ZrCl{(η-C$_5$H$_4$)$_2$SiMe$_2$})$_2$O].[205]

Ph$_3$PO or CO$_2$ has been used as oxidizing agent for [{Zr(μ-Cl)Cp}$_2${(η-C$_5$H$_4$)$_2$SiMe$_2$}]. Oxygen-atom abstraction gives [{(ZrXCp)$_2$({η-C$_5$H$_4$}$_2$Y)}$_2$O] (X = Cl; Y = SiMe$_2$).[206] Similarly, [{Zr(μ-O)Cp$_2$}$_3$] was first prepared from [Zr(CO)$_2$Cp$_2$] and CO$_2$ but is also obtained on thermolysis of tris[(η^2-formaldehyde)zirconocene].[55] Reaction of [{Zr(μ-Cl)Cp}$_2$(η^5:η^5-C$_{10}$H$_8$)] and [Zr(μ-Cl)Cp$_2$]$_2$ with [ReO$_3$Cp*] gives [{(ZrXCp)$_2$({η-C$_5$H$_4$}$_2$Y)}$_2$O] (X = Cl, no Y) and [{Zr(μ-O)Cp$_2$}$_3$], respectively.[207]

(iii) Others

Zirconocene dichloride reacts with [WO$_3$Cp*]$^-$ to give the heterobimetallic μ-oxo complex [Cp*WO$_3$Zr(Cl)Cp$_2$].[208] Photolysis of [Cp$_2$(Cl)Zr(OCHR^{1}CHR^2)W(CO)$_3$Cp] (R^1 = R^2 = H; R^1 = Me or Ph, R^2 = H; R^1 = R^2 = Ph) gives [Cp$_2$(Cl)Zr–O–W(CO)$_3$Cp] via elimination of R^{2}CH=CHR^1.[155]

Mixed chalcogenido-bridged metallocene dimers are dealt with in Section 10.5.2.2.

10.4.6.2 Spectroscopic properties

The oxo-bridged metallocene dimers display the characteristic IR spectral band associated with the Zr–O–Zr unit as an intense absorption between 750 cm^{-1} and 790 cm^{-1}. In [Cp$_2$(Cl)Zr–O–W(CO)$_3$Cp] the Zr–O–W mode is observed at 789 cm^{-1}.[155] Oxygen-17 NMR spectra have been reported for [{MHCp*$_2$}O] (M = Zr or Hf).[29]

10.4.6.3 Structural data

Crystal structures of several homodinuclear complexes have been determined: [(ZrRCp$_2$)$_2$O] (R = 4-Cl-phenoxy,[209] OPh,[210] Me,[204] Br[211] or O–C(Ph)=Mo(CO)$_5$)[203] and [{(ZrXCp)$_2$({η-C$_5$H$_4$}$_2$Y)}$_2$O] (X = Cl, no Y;[202] X = Cl, Y = SiMe$_2$; X = Cl, Y = CH$_2$;[200] X = Cl, Y = (SiMe$_2$)$_2$; X = Cl, Y = (SiMe$_2$)$_3$).[195] The relatively short Zr–O bond lengths, typically ca. 0.195 nm, indicate the presence of partial double bond character. Structural data are also available for [Cp$_2$(Cl)Zr–O–W(CO)$_3$Cp] (Zr–O–W 175.7°).[155]

10.4.6.4 Chemical properties

Introduction of a functionalized alkyl ligand in oxo-bridged zirconocene complexes and further reactions with [{Rh(CO)$_2$Cl}$_2$] have been studied.[212] Reactions of [Cp$_2$(Cl)Zr–O–W(CO)$_3$Cp] with acetyl chloride, an alkyne or PMe$_3$ have been carried out. As expected, MeLi reacts with [Cp$_2$(Cl)Zr–O–W(CO)$_2$(PMe$_3$)Cp] with substitution of the chloro ligand.[155]

10.4.7 Zirconocene Oxide [Zr(=O)Cp*$_2$]: Generation and Trapping

Unstable [Zr(=O)Cp*$_2$] was generated from [Zr(OH)(OSO$_2$CF$_3$)Cp*$_2$] and KN(TMS)$_2$ or by thermolysis of [Zr(OH)(Ph)Cp*$_2$] and was trapped with a nitrile or an alkyne.[213-15] The thermolysis of [Zr(OH)(Ph)Cp*$_2$] was also carried out in the presence of donors such as pyridine, substituted pyridines or triphenylphosphine oxide; however, pure products were not isolated from these reactions.[214]

10.5 METALLOCENE(IV) COMPLEXES WITH SULFUR-, SELENIUM- AND TELLURIUM-CENTRED ANIONIC LIGANDS

10.5.1 Metallocene(IV) Thiolates, Selenates and Tellurates

10.5.1.1 Synthesis, spectroscopic and chemical properties

The following synthetic routes have been employed for the synthesis of metallocene(IV) thiolates and selenates.

(i) The triethylamine method

The triethylamine method, which was used earlier for the preparation of $[Zr(EPh)_2Cp_2]$ (E = S or Se)[1] and $[M(SH)_2Cp_2]$ (M = Ti or Zr),[1] has been employed for the synthesis of the thiolatoacetamide complexes $[M\{\overline{SCH_2C(NHR)=O}\}_n(Cl)_{2-n}Cp_2]$ (n = 1 or 2; M = Hf [216] or Zr;[217] R = Ph, o-, m- or p-Tol, nap or $3,5$-$Me_2C_6H_3$) from $[MCl_2Cp_2]$ and $RHNC(=O)CH_2SH$. Spectroscopic data suggest that the thiolatoacetamides behave as bidentate (O and S) ligands. Metalladiselenaferrocenophanes were prepared accordingly from $[Fe(\eta$-$C_5H_4SeH)_2]$ and the appropriate metallocene dichloride.[218]

(ii) Salt elimination

While the triethylamine method affords titanocene and zirconocene thiolates in good yield, this has proved to be unsuitable for the preparation of hafnocene thiolates.[219] An improved synthesis is the reaction of hafnocene dichloride with a lithium thiolate which gave $[Hf(SR)_2Cp_2]$ (R = Ph, o- or p-Tol, or 4-$NH_2C_6H_4$).[219] This method is also applicable to other systems. Thus, the reaction of a metallocene dichloride with a lithium selenate gave the bis(selenato) complexes $[(M(SeR^2)_2(\eta$-$C_5H_4R^1)_2]$ (M = Zr or Hf, R^1 = H or Bu^t, R^2 = Ph;[220] M = Zr, R^1 = Me, R^2 = Bz or Ph;[221] M = Zr or Hf, R^1 = Bu^t, R^2 = p-Tol; M = Hf, R^1 = H, R^2 = p-Tol;[222] M = Ti, Zr or Hf, R^1 = H or Bu^t, R^2 = Ph or p-Tol).[223] The complexes $[Zr(SR^1)_2(\eta$-$C_5H_4R^2)_2]$ (R^1 = Bz, R^2 = Me or Bu^t) are prepared similarly from zirconocene dichloride and a mixture of dibenzyl disulfide with $Li[BEt_3H]$.[221] Reaction of the metallocene dichloride with an equimolar amount of lithium phenylselenate gave $[MCl(SePh)Cp_2]$ (M = Ti, Zr or Hf).[224]

Phospho sulfido complexes were prepared by the reaction of $[MCl_2Cp_2]$ with $LiSPR_2$. Alternatively, the complexes $[M(SPR_2)_2Cp_2]$ were obtained from the reaction of the appropriate metallocene bisphosphido complex and $HSPR_2$ or elemental sulfur (see Section 10.3.1.4).[99]

(iii) Insertion of chalcogen into M–C bonds

Chalcogens readily insert into Zr–Me and Hf–Me bonds to give $[M(EMe)_2(\eta$-$C_5H_4R)_2]$ (M = Zr or Hf, R = H or Bu^t, E = Se;[220,223] M = Zr, R = H, E = S or Te).[221] The rate of insertion decreases in the sequence S > Se > Te, as do the yields. $[HfPh_2(\eta$-$C_5H_4Bu^t)_2]$ cleanly inserts selenium with formation of $[Hf(SePh)_2(\eta$-$C_5H_4Bu^t)_2]$, while $[HfPh_2Cp_2]$ failed to give a clean reaction product.[223] $[ZrPh_2(\eta$-$C_5H_4R)_2]$ (R = H[223] or Me)[221] inserts selenium to give $[Zr(SePh)_2(\eta$-$C_5H_4R)_2]$, while $[ZrPh_2(\eta$-$C_5H_4Bu^t)_2]$ gives $[Zr(1,2$-$Se_2C_6H_4)(\eta$-$C_5H_4Bu^t)_2]$[223] (see Section 10.5.1.2). The reaction mixtures (R = H or Me) also contain the four-membered metallacyclic compounds. This side reaction considerably restricts the use of this procedure for the preparation of metallocene chalcogenato complexes.[221]

(iv) Oxidative addition of a diphenyl dichalcogenide

The reaction of a dimethylmetallocene with a diphenyl dichalcogenide yields $[MMe(EPh)(\eta$-$C_5H_4R)_2]$ (M = Zr or Hf; R = H or Bu^t; E = S or Se).[223,225] The corresponding bis-chalcogenato complexes are formed on UV irradiation of the solutions or by employing two moles of E_2Ph_2.[225] The reaction was much slower if the t-butylcyclopentadienyl derivatives were used or when Se_2Ph_2 was employed in the reaction. Similarly, irradiation of $[MR^2_2(\eta$-$C_5H_4R^1)_2]$ (M = Zr or Hf, R^1 = H or Bu^t, R^2 = Ph;[24,122] M = Hf, R^1 = H or Bu^t, R^2 = p-Tol)[24] in the presence of Se_2Ph_2 gives $[M(SePh)_2(\eta$-$C_5H_4R^1)_2]$ and R^2–R^2.[223]

Oxidation of $[\{Zr(PMe_3)Cp(\eta^1:\eta^5-C_5H_4)\}_2]$ with Ph_2S_2 afforded $[\{Zr(SPh)Cp\}_2(\eta^5:\eta^5-C_{10}H_8)]$.[54,201] Similarly, oxidative addition of a disulfide R_2S_2 (R = Et or Ph) to $[Zr(CO)_2Cp_2]$ gives $[Zr(SR)_2Cp_2]$.[226]

With protic reagents $[Zr(C_2H_4)(PMe_3)Cp_2]$ undergoes protonation of the ethene ligand. Thus, ethanedithiol yields $[Cp_2(Et)Zr(\mu\text{-}SCH_2CH_2S)Zr(Et)Cp_2]$[192] and Bu^tSH gives $[ZrEt(SBu^t)Cp_2]$.[137] Reaction of the latter with HX (X = HC(=O)O or ac) or ROH (R = Me, Et or Ph) gives $[ZrEt(X)Cp_2]$ or $[ZrEt(OR)Cp_2]$, respectively.[137]

Selenium-77 and ^{13}C NMR spectral data are available for $[M(SeR^1)_2(\eta\text{-}C_5H_4R^2)_2]$ (M = Zr or Hf; R^1 = Ph or *p*-Tol, R^2 = H or Bu^t).[222]

(v) Thermal rearrangement of zirconocenyl α-alkylthioalkanes

Zirconocene thiolates $[Zr\{CHR^2(TMS)\}(SR^1)Cp_2]$ were obtained by thermal rearrangement of $[Zr\{CH(SR^1)(TMS)\}(R^2)Cp_2]$ (R^1 = Ph, R^2 = Ph, *p*-Tol, $4\text{-}MeOC_6H_4$ or $4\text{-}ClC_6H_4$;[227] R^1 = Me, R^2 = Ph).[227,228] Similarly, thermolysis of $[Zr\{CH(SPh)Ph\}_2Cp_2]$ afforded $[Zr(SPh)_2Cp_2]$ and stilbene.[229]

(vi) Solvolysis

The reaction of dimethylzirconocene with an alkanethiol in the presence of trimethylphosphine gave $[ZrMe(SCH_2R)Cp_2]$ (R = Me, Ph or a substituted Ph), although the products were not obtained pure.[230,231] Dibenzylzirconocene reacts with benzenethiol with formation of $[Zr(SPh)_2Cp_2]$.[232] When an alkanedithiol $HS(CH_2)_nSH$ (n = 2–4), *o*- or *m*-$(HSCH_2)_2C_6H_4$ was reacted with dimethylzirconocene $[Cp_2Zr(\mu\text{-}S\text{-}R\text{-}S)_2ZrCp_2]$ (R = $(CH_2)_n$, n = 2–4; R = *o*- or *m*-$CH_2C_6H_4CH_2$) was obtained. Alternatively, the same products were obtained by reacting an (*S*)-trimethylsilylalkanedithiol, $HS(CH_2)_nS$-TMS (n = 2 or 3) with dimethylzirconocene or zirconocene hydridochloride. The molecular structures of $[Cp_2Zr(\mu\text{-}S\text{-}R\text{-}S)_2ZrCp_2]$ (R = $(CH_2)_2$ or $(CH_2)_3$) were determined. These compounds act as macrocyclic metalloligands. Thus, the reaction with $Ag[BPh_4]$ led to encapsulation of the silver ion.[233,234]

(vii) Hydrozirconation of thioketones

Aromatic and aliphatic thioketones, $R_2C=S$, undergo hydrozirconation at room temperature with $[\{ZrCl(H)Cp_2\}_n]$ to give yellow $[ZrCl(SCHR_2)Cp_2]$. The products were generated *in situ* for use in organic synthesis.[235]

10.5.1.2 Dichalcogenophenylenemetallocenes

(i) Synthesis

Sulfur and selenium combine with a diarylzirconocene in boiling heptane to give $[Zr(1,2\text{-}E_2\text{-}4\text{-}RC_6H_3)(\eta\text{-}C_5H_4Bu^t)]$ (E = Se, R = H or Me;[222,236] E = S or Se, R = H or OMe; E = Te, R = H;[237] E = S or Se, R = NMe_2 or Br).[238] With tellurium, the reaction proceeded only at lower temperature. It was suggested that upon heating an (aryne)metallocene is generated and undergoes insertion of two chalcogen atoms. Metallacycles were not obtained for zirconocene derivatives with unsubstituted cyclopentadienyl ligands nor for hafnocene derivatives.[237]

A more general route to $[M(1,2\text{-}E_2C_6H_4)(\eta\text{-}C_5H_4R)_2]$ (M = Zr or Hf, R = H, E = Se;[237,239] M = Hf, R = Bu^t, E = Se;[237] M = Hf, R = H, E = Te[240] or S)[219] is the reaction of the metallocene dichloride with dipotassium or dilithium benzene-*o*-dichalcogenate. $[Hf(1,2\text{-}S_2C_6H_3Me\text{-}p)Cp_2]$ was prepared by this procedure. The related *cis*-1,2-enedithiolato complex $[Hf\{S_2C_2(CN)_2\}Cp_2]$ was obtained by the same method.[219] The *o*-thiatelluraphenylenezirconocene complexes $[Zr(1\text{-}S\text{-}2\text{-}TeC_6H_4)(\eta\text{-}C_5H_4R)_2]$ (R = H or Bu^t) were obtained from zirconocene dichloride and dilithium *o*-benzenethiatelluride[241] Thus, the dilithium salt of 2-aminobenzenethiol reacts with $[MCl_2Cp_2]$ (M = Ti, Zr or Hf) to give the corresponding nitrogen–sulfur chelate complexes. The hafnium complex exhibits rapid ring inversion at room temperature (see Section 10.2.1.1).[20] When a metallocene dichloride $[MCl_2(\eta\text{-}C_5H_4R)_2]$ was reacted with 1,2,4,5-tetramercaptobenzene in the presence of triethylamine (R = H) or with 1,2,4,5-tetrasodium tetramercaptobenzene (R = H or TMS) the dimeric chelate complexes $[\{(\eta\text{-}C_5H_4R)_2M\}_2(1,2,4,5\text{-}S_4C_6H_2)]$ (M = Ti, Zr or Hf; R = H[242] or TMS)[243] were obtained.

(ii) Spectroscopic properties

Selenium-77 NMR spectroscopic data are available for several complexes.[222,236] Variable-temperature ^1H NMR spectra indicate that the five-membered ME_2C_2 chelate rings undergo rapid inversion at room temperature. Activation parameters were determined for $[M(1,2-E_2-4-R^2C_6H_3)(\eta-C_5H_4R^1)]$ (M = Ti, Zr or Hf, E = S or Se, $R^1 = R^2 = H$;[219,239,244] M = Ti or Hf, E = S, $R^1 = H$, $R^2 = Me$;[219,245] M = Ti or Hf, E = Te, $R^1 = R^2 = H$;[240,244] M = Zr, E = S or Se, $R^1 = Bu^t$, $R^2 = H$ or OMe; M = Zr, E = Te, $R^1 = Bu^t$, $R^2 = H$).[237]

Complexes $[\{(\eta-C_5H_4R)_2M\}_2(1,2,4,5-S_4C_6H_2)]$ (R = H) are insoluble in common organic solvents, but molecular ions were observed in the mass spectra.[242] For the more soluble derivatives (R = TMS, M = Ti,[242] Zr or Hf[243]) the presence of two conformations (chair and boat) is indicated by variable-temperature ^1H NMR spectra.

(iii) Structural data

A crystal structure determination of $[Zr(1,2-Se_2C_6H_4)(\eta-C_5H_4Bu^t)_2]$ confirmed the 'envelope' conformation deduced from ^1H NMR spectroscopic studies.[236] The molecular structure of $[Zr(1,2-STeC_6H_4)(\eta-C_5H_4Bu^t)_2]$[241] is closely related to that of the *o*-diselenaphenylenezirconocene complex. However, the disorder of the chalcogen atoms in the former prevents an accurate comparison. The molecular structure of $[\{(\eta-C_5H_4R)_2M\}_2(1,2,4,5-S_4C_6H_2)]$ (M = Ti or Hf, R = TMS) shows the presence of the chair conformation in the solid state.[243]

10.5.2 Dimeric Chalcogenido-bridged Metallocene(IV) Complexes

10.5.2.1 Synthesis

Several routes have been employed for the synthesis of chalcogenido-bridged bimetallic complexes.

(i) From metallocene dichlorides and chalcogenide anions

The classical reaction of a metallocene dichloride with a chalcogenide anion provides a facile route to $[\{M(\mu-E)(\eta-C_5H_4R)_2\}_2]$ (M = Zr or Hf; R = H, Me or Bu^t; E = S or Se).[198,246] The rate of reaction is sensitive to steric hindrance, the slowest reaction being observed when R = Bu^t. When a metallocene dichloride was reacted with a lithium phosphinoselenide ($LiSePR_2$, R = Cy or Ph) the complexes $[\{M(\mu-Se)Cp_2\}_2]$ (M = Zr or Hf) were formed immediately, while the reaction with the corresponding lithium phosphinosulfide takes 1–2 d to give the corresponding sulfido-bridged complexes (via phosphino-sulfido complexes, see Section 10.5.1.1).[99]

(ii) From metallocene dihydrides

Dimeric metallocene dihydrides react with grey selenium or tellurium to afford exclusively $[\{M(\mu-E)(\eta-C_5H_4R)_2\}_2]$ (M = Zr or Hf, E = Se or Te, R = Bu^t;[247,248] M = Zr, E = Se, R = H or Me).[247]

When $[\{ZrH(\mu-H)Cp\}_2\{(\eta-C_5H_4)_2SiMe_2\}]$ was reacted with sulfur, only the sulfido-bridged product $[\{Zr(\mu-S)Cp\}_2\{(\eta-C_5H_4)_2SiMe_2\}]$ was obtained. Alternatively, the latter was obtained on refluxing a solution of $[(ZrMe_2Cp)_2\{(\eta-C_5H_4)_2SiMe_2\}]$ and sulfur for 12 h. The sulfido bridges resist cleavage with PMe_3, CNMe or pyridine.[206]

(iii) From metallocene(II) precursors

The complex $[Zr(butadiene)(\eta-C_5H_4R)_2]$ (R = Bu^t or 1,1-dimethylpentyl) reacted with sulfur, grey selenium or tellurium to give $[\{Zr(\mu-E)(\eta-C_5H_4Bu^t)_2\}_2]$ (E = S, Se[249] or Te)[197,250] and $[\{Zr(\mu-Te)(\eta-C_5H_4R)_2\}_2]$ (R = 1,1-dimethylpentyl).[250] Oxidation of $[\{Zr(PMe_3)Cp(\eta^1:\eta^5-C_5H_4)\}_2]$ with elemental sulfur resulted in reductive coupling of the two bridging $\eta^1,\eta^5-C_5H_4$ rings to give $[\{Zr(\mu-S)Cp\}_2(\eta^5:\eta^5-C_{10}H_8)]$. This product was also obtained from the reaction of $[(ZrMe_2Cp)_2(\eta^5:\eta^5-C_{10}H_8)]$ or $[\{Zr(\mu-SPh)Cp\}_2(\eta^5:\eta^5-C_{10}H_8)]$ with sulfur.[201]

(iv) Replacement of chalcogen-containing ligands by chalcogen atoms

The reaction of a zirconocene bisselenate or bisthiolate with elemental sulfur or selenium gave the corresponding compounds [{Zr(μ-E)(η-C$_5$H$_4$Me)$_2$}$_2$] (E = S or Se).[221]

(v) Photochemical or thermal synthesis

Photolysis of a metallocene dialkyl or diaryl in the presence of grey selenium or sulfur gave [{M(μ-E)(η-C$_5$H$_4$Bu^t)$_2$}$_2$] (M = Zr, E = S[246] or Se;[222,223,246,251] M = Hf, E = Se).[223,246,251] The selenido-bridged complexes were also formed by UV irradiation of a solution of diphenylmetallocene and [M(SePh)$_2$(η-C$_5$H$_4$Bu^t)$_2$].[251] Thermally-induced dimerization of a metallocene dithiol [M(SH)$_2$(η-C$_5$H$_4$R)$_2$] with loss of hydrogen sulfide was observed to occur fairly readily when R = H (M = Zr or Hf),[252] while the reaction of the species with R = Bu^t needed more drastic conditions.[248]

Thermolysis of [Zr{η^2-S$_2$CP(TMS)$_2$}(Cl)Cp$_2$] gives [{Zr(μ-S)Cp$_2$}$_2$]. The same product is obtained from [Zr{P(TMS)$_2$}$_2$Cp$_2$] or [ZrMe{P(TMS)$_2$}Cp$_2$] and elemental sulfur in low yield.[98]

10.5.2.2 Mixed species

(i) Mixed chalcogenide ligands

Ultraviolet irradiation of [{ZrMe(η-C$_5$H$_4$Bu^t)$_2$}$_2$O] in the presence of sulfur or grey selenium gave [{Zr(η-C$_5$H$_4$Bu^t)$_2$}$_2$(μ-O)(μ-E)] (E = S[198] or Se).[197,198] The complexes display the characteristic IR spectral band at 790 cm^{-1} associated with the Zr–O–Zr unit. Controlled hydrolysis of [{Zr(μ-Te)(η-C$_5$H$_4$Bu^t)$_2$}$_2$] with CuSO$_4$·5H$_2$O afforded the green complex [{Zr(η-C$_5$H$_4$Bu^t)$_2$}$_2$(μ-O)(μ-Te)].[197]

(ii) Heterobimetallic and unsymmetrical complexes

A facile synthesis of sulfido-bridged bimetallic species or unsymmetrically substituted complexes is the reaction of [M^1(SH)$_2$(η-C$_5$H$_4$Bu^t)$_2$] with [{M^{2}H(μ-H)(η-C$_5$H$_4$R)$_2$}$_2$] to give [(η-C$_5$H$_4$Bu^t)$_2$M^1(μ-S)$_2$M^2(η-C$_5$H$_4$Bu^t)$_2$] (M^1 = M^2 = Zr, R = H; M^1 = Hf, M^2 = Zr, R = H or Bu^t; M^1 = Zr, M^2 = Hf, R = H or Bu^t). Replacement of the dihydride by a dimethyl complex was successful only for the preparation of the derivative with M^1 = M^2 = Zr and R = H.[221] The tungsten–zirconium complex *cis*-[Cp(CO)$_2$(PMe$_3$)W(μ-S)Zr(Cl)Cp$_2$] was prepared by introducing one equivalent of propene sulfide into a mixture of Li[W(CO)$_2$(PMe$_3$)Cp] and [Zr(Cl$_2$Cp$_2$]; one equivalent of propene was evolved.[253]

10.5.2.3 Spectroscopic properties

The dimeric chalcogenido-bridged metallocenes are diamagnetic, high-melting solids. Selenium-77 NMR spectroscopic data are available for [{Zr(μ-Se)(η-C$_5$H$_4$Bu^t)$_2$}$_2$] (δ = 802.6).[222] The mass spectra show the molecular ion peak of the dimeric complexes with the expected isotopic distribution pattern.[223,246,251] Successive loss of $^-$Cp ligands and chalcogen are generally observed.

10.5.2.4 Structural data

Structural data are available for [{Zr(μ-Te)(η-C$_5$H$_4$Bu^t)$_2$}$_2$],[197,250] [{Zr(μ-S)Cp$_2$}$_2$],[98] [{Zr(μ-S)Cp}$_2$(η^5-C$_{10}$H$_8$)],[201] [{Zr(μ-S)Cp}$_2${(η-C$_5$H$_4$)$_2$SiMe$_2$}],[206] [{Zr(μ-C$_5$H$_4$Bu^t)$_2$}$_2$(μ-O)(μ-Te)][197] and *cis*-[Cp(CO)$_2$(PMe$_3$)W(μ-S)Zr(Cl)Cp$_2$].[253] The symmetrically bridged complexes have an almost exactly square-planar M$_2$E$_2$ core. The presence of a SiMe$_2$ group bridging two cyclopentadienyl ligands has only a minor effect on the central Zr$_2$S$_2$ core.[206] The geometry of the Zr$_2$S$_2$ core of the fulvalene complex [{Zr(μ-S)Cp}$_2$(η^5:η^5-C$_{10}$H$_8$)][201] is closely related to that of zirconocene sulfide.[98] However, the former shows considerable deviation from planarity.[201] In contrast with the symmetrical dimeric complexes, [{Zr(η-C$_5$H$_4$Bu^t)$_2$}$_2$(μ-O)(μ-Te)] exhibits a considerably distorted arrangement.[197] The heterobimetallic complex *cis*-[Cp(CO)$_2$(PMe$_3$)W(μ-S)Zr(Cl)Cp$_2$] has completely different structural features due to the presence of only one sulfido bridge. The overall structure is similar to that observed for μ-oxo complexes (see Section 10.4.6.3).[253]

10.5.3 Zirconocene Sulfide [Zr(=S)Cp*$_2$]: Generation and Trapping

The unstable zirconocene sulfide was generated by treating [ZrI(SH)Cp*$_2$] with KN(TMS)$_2$ and trapped with a pyridine, alkyne or PhCN.[213,214] The trapped products [Zr(SCPh=CPh)Cp*$_2$] and [Zr(=S){4-Bu^t(NC$_5$H$_4$)}Cp*$_2$] were structurally characterized.

10.5.4 Metallocene(IV) Complexes with Polychalcogenido Ligands

10.5.4.1 Synthesis

The complexes [MS$_5$Cp$_2$] (M = Ti, Zr or Hf) have been known since 1966 (M = Ti) and 1980 (M = Zr or Hf). While the titanium and zirconium compounds can be prepared from the reaction of [M(SH)$_2$Cp$_2$] with benzimidazole, the highest yield of all three complexes was obtained by the reaction of the metallocene dichloride with a Li$_2$S$_2$–3S mixture.[252] Pentasulfido complexes are formed during the synthesis of sulfido-bridged complexes when the starting materials are treated with excess sulfur (see Section 10.5.2.1). Thus, [MS$_5$(η-C$_5$H$_4$Bu^t)$_2$] (M = Zr or Hf) was formed when (i) [MPh$_2$(η-C$_5$H$_4$Bu^t)$_2$] was photolysed in the presence of excess sulfur;[246] (ii) [{MH(μ-H)(η-C$_5$H$_4$Bu^t)$_2$}$_2$] was treated with excess sulfur;[247,248] or (iii) [Zr(butadiene)(η-C$_5$H$_4$Bu^t)$_2$] was reacted with excess sulfur (see Section 10.5.2.1(iii)).[249] Alternatively, the complexes [MS$_5$(η-C$_5$H$_4$Bu^t)$_2$] (M = Zr or Hf)[246] and [ZrS$_5${(η-C$_5$H$_4$)$_2$SiMe$_2$}][254] are accessible by treatment of the appropriate metallocene dichloride with a mixture of Li[BEt$_3$H] and sulfur (2:5). [ZrS$_5$(η-C$_5$H$_4$Bu^t)$_2$] reacts with the starting material (metallocene dihydride or butadiene complex) to give [{Zr(μ-S)(η-C$_5$H$_4$Bu^t)$_2$}$_2$].[247–9] It was shown that chalcogenido-bridged dimeric complexes react with excess chalcogen with formation of the corresponding pentachalcogenido complexes [ZrE$_5$(η-C$_5$H$_4$R)$_2$] (E = S, R = H, Me or Bu^t; E = Se, R = H or Me).[221]

The general route to pentasulfanes has been extended to the corresponding selenanes [MSe$_5$Cp$_2$] (M = Ti, Zr or Hf).[252] Alternatively, Na$_2$Se$_x$ (x = ca. 5) was used and the same products were obtained. With traces of moisture, the hafnocene derivative underwent hydrolysis to [(HfCp$_2$)$_2$(μ-O)(μ-Se$_4$)], which has been structurally characterized.[255]

While pentasulfido complexes with $^-$Cp ligands are only slightly soluble substituents on the $^-$Cp ligand increase the solubility.[256] However, formation of [MS$_3$Cp*$_2$] was observed when the bis(pentamethylcyclopentadienyl) metal dichloride was reacted with sulfur in the presence of Li[BEt$_3$H] (M = Ti, Zr or Hf)[252] or with Li$_2$S$_2$–S$_8$ (1:3/8) (M = Ti or Zr).[257] The products are high-melting solids, air stable in solution and monomeric (molecular weight determination).

10.5.4.2 Spectroscopic properties

(i) Mass spectra

The pentasulfido, pentaselenido and trisulfido complexes show a molecular ion peak in the mass spectrum. While the first two show metastable ion peaks corresponding to loss of S$_2$ or Se$_2$, the trisulfido complexes do not.[252]

(ii) Variable-temperature NMR spectral studies

It was shown earlier that the complexes [MS$_5$Cp$_2$] undergo chair-to-chair ring flipping.[1] Thus, at low temperature two distinct peaks due to the axial and equatorial Cp groups were observed. The pentaselenido and trisulfido complexes show comparable temperature-dependent ^{1}H NMR spectra, suggesting that a ring inversion takes place in each complex.[252] Activation parameters have been determined for [ME$_5$Cp$_2$] (M = Ti, Zr or Hf; E = S or Se),[252] [MS$_5$(η-C$_5$H$_4$Bu^t)$_2$] (M = Ti, Zr or Hf)[246] and [MS$_3$Cp*$_2$] (M = Ti, Zr or Hf).[252,257]

10.5.4.3 Structural data

While the molecular structure of [TiS$_5$Cp$_2$] has been known since 1971, the structures of the zirconium and hafnium derivatives were reported in 1987.[258] The pentaselenido complexes of zirconium and hafnium have also been structurally characterized.[255] The pentasulfido and pentaselenido complexes exhibit the chair conformation in the solid state. Of the trisulfido complexes, only the titanium derivative was structurally characterized.[257]

10.5.5 Metallocene(IV) Dithiocarbamato Complexes and Related Compounds

10.5.5.1 Synthesis

In general, dithiocarbamato complexes $[MCl(\eta^2\text{-}S_2CNR^2R^3)(\eta\text{-}C_5H_4R^1)_2]$ (M = Zr or Hf; R^1 = H or Me; R^2, R^3 = alkyl or aryl)[259–63] are obtained by treatment of the metallocene dichloride with equimolar amounts of the potassium, sodium or ammonium dithiocarbamate. The metallocene silyl complexes $[MCl(TMS)Cp_2]$ (M = Zr or Hf) reacted with NaS_2CNEt_2 to give a silyl-dithiocarbamato complex (see Section 10.1.4.3).[2] The monothiocarbamato complex $[ZrCl(SOCNMe_2)Cp_2]$ was prepared from zirconocene dichloride and the appropriate sodium salt.[183]

Substituted dithiocarbamato complexes can be obtained by reacting $[ZrCl(X)Cp_2]$ with sodium dithiocarbamate. Thus, complexes $[Zr(\eta^2\text{-}S_2CNMe_2)(X)Cp_2]$ (X = CH_2TMS, CH_2CMe_2Ph, Bz, OPh, 4-hydroxypyridinate or 2,6-dimethylphenoxide) were prepared. For the halide derivatives (X = Br or I) zirconocene dibromide or diiodide were used as starting materials. The methyl derivative (X = Me) was obtained by reaction of the chloro derivative with MeLi.[262]

Insertion of CS_2 into the Zr–P bond of $[Zr\{P(TMS)_2\}(X)Cp_2]$ (X = Cl or Me) yielded $[Zr\{\eta^2\text{-}S_2CP(TMS)_2\}(X)Cp_2]$. The ligand can be regarded as the phosphorus analogue of a dithiocarbamato ligand (see Section 10.3.1.4).[63] The compound $[ZrCl_2Cp_2]$ reacts with $K_2(dto)$ to give $[(ZrClCp_2)_2(dto)]$, in which each zirconium is five-coordinate, bound to one sulfur and one oxygen of the dto ligand (see Section 10.4.3.5).[168]

10.5.5.2 Structural data

Crystal structures have been determined for $[Zr(\eta^2\text{-}S_2CNR_2)(X)Cp_2]$ (X = Cl, R = Et;[263] X = OPh, R = Me;[264] X = Cl, R = Bz;[259] X = TMS, R = Et^2), the related $[ZrCl\{\eta^2\text{-}S_2CP(TMS)_2\}Cp_2]$,[63] and the monothiocarbamato complex $[ZrCl(SOCNMe_2)Cp_2]$.[168] For the dithiocarbamato complexes, the endocyclic paramaters of the ZrS_2X fragment are very similar. The coordination of the nitrogen atom is trigonal planar and the ZrS_2X fragment and the NC_2 fragment are nearly coplanar and lie in a quasi-mirror plane. However, the P–C bond length in the phosphorus analogue indicates a single bond and the PSi_2 plane is roughly orthogonal to the ZrS_2CP plane.

10.5.5.3 Spectroscopic properties

The alkyl- and aryldithiocarbamato complexes of hafnium and zirconium are generally white; they are air stable as solids but moisture sensitive in solution. Infrared and ^1H NMR spectra have been reported for most complexes.

(i) Infrared spectra

The IR spectra unambiguously show the η^2-bonding mode of the dithiocarbamato ligand. One medium to strong band in the range 960–1020 cm^{-1}, ν(C–S), indicates the presence of a four-membered ZrS_2C ring. The strong thioureido band, ν(C–N), in the range 1495–1525 cm^{-1} is characteristic for dithiocarbamato complexes. A medium intensity absorption in the region of 320–370 cm^{-1} was assigned to ν(Zr–S); ν(Zr–Cl) lies in the range 360–390 cm^{-1}.

(ii) Proton NMR spectra

It was shown that the complexes $[Zr(\eta^2\text{-}S_2CNMe_2)(X)Cp_2]$ (X = Cl, Br, I, Me, Bz, OPh, CH_2TMS, CH_2CMe_2Ph, 4-hydroxypyridinate or 2,6-dimethylphenoxide) and the *N,N*-dimethylthiocarbamates $[Zr(SOCNMe_2)(X)Cp_2]$ (X = Cl or 4-pyridyloxy) are fluxional at 30 °C leading to the observation of single resonances. At low temperature these complexes exhibit two signals for the inequivalent methyl groups.[262,263]

10.6 METALLOCENE(IV) COMPLEXES WITH ZIRCONIUM–M OR HAFNIUM–M BONDS (M = TRANSITION METAL)

10.6.1 Synthesis

A widely used route to complexes with metallocene–transition metal bonds is nucleophilic attack of a metal complex anion $M^2[M^1(CO)_2Cp]$ (M^2 = Na or K) on $[ZrCl(X)Cp(\eta\text{-}C_5H_4R)]$ to give $[Cp(\eta\text{-}C_5H_4R)(X)Zr\text{-}M^1(CO)_2Cp]$ (X = Cl, Me or OBu^t, R = H, M^1 = Ru or Fe;[265] X = Me, Cl or octyl, R = H, M^1 = Fe;[266] X = OBu^t, R = PPh$_2$, M^1 = Fe)[267] (for a recent review, see Ref. 95). The derivatives having X = Me or OBu^t, R = H and M^1 = Ru were also obtained from the reaction of the corresponding chloro complex with MeLi or KOBu^t, respectively.[265] Reactions with carbonyl complexes can result in the formation of a Zr–M^1 or Zr–O–C–M^1 linkage, the bonding mode being determined by steric effects. Thus, the low steric requirements of the $M^1(CO)_2Cp$ fragments (M^1 = Fe or Ru) allow Zr–M^1 bond formation,[265-7] whereas with the more crowded $[M^1(CO)_3Cp]^-$ anion, Zr–O bonding to give $[ZrMe\{O=C=M^1(CO)_2Cp\}Cp_2]$ is favoured (M^1 = Mo or W).[95] However, treatment of the latter with CO gives an η^2-acetyl complex, which slowly loses CO to form a metal–metal bonded complex.[95] The complex $[Zr\{C(TMS)=CPh\text{-}CPh=C(TMS)\}Cp_2]$ reacts with $[WH(CO)_3Cp]$ at 70 °C in toluene to give the adduct $[Zr\{C(TMS)=C(H)Ph\}\{O\equiv C\text{-}W(CO)_2Cp\}Cp_2]$.[133]

When $[ZrCl(OBu^t)Cp(\eta\text{-}C_5H_4PPh_2)]$ was reacted with $K[Fe(CO)_2Cp]$, a Zr–Fe-bonded product was obtained. On irradiation, decarbonylation occurred with formation of $[Cp(\eta\text{-}C_5H_4PPh_2)(OBu^t)Zr\text{-}ML_n]$ (ML_n = Fe(CO)Cp). Irradiating a mixture of the zirconocene alkoxo-chloride and $Na[Co(CO)_4]$ gave the analogous complex (ML_n = Co(CO)$_3$) directly.[267]

Reaction of $[ZrI_2Cp_2]$ with two equivalents of $K[Ru(CO)_2Cp]$ gave $[Cp_2Zr\{Ru(CO)_2Cp\}_2]$. The monosubstituted product was observed as an intermediate by ^1H NMR spectroscopy. The corresponding reaction with $[ZrCl_2Cp_2]$ was much slower, and significant amounts of $[\{Ru(CO)_2Cp\}_2]$ were formed.[154] However, the iron analogue $[Cp_2ZrCl\{Fe(CO)_2Cp\}]$ was obtained by reacting $[ZrCl_2Cp_2]$ with $Na[Fe(CO)_2Cp]$. The more soluble alkyl derivatives $[Cp_2ZrR\{Fe(CO)_2Cp\}]$ (R = Me or octyl) have been prepared analogously.[266] The iron complex is not accessible via the reaction of $[ZrI_2Cp_2]$ with two equivalents of $K[Fe(CO)_2Cp]$, which yielded $[\{Fe(CO)_2Cp\}_2]$ and unidentified zirconium products. Variable-temperature ^1H NMR spectroscopic studies indicated that the monosubstituted and the disubstituted compound are formed, but decompose above −20 °C.[154] Reaction of $[ZrI_2Cp^*_2]$ with two equivalents of $Na[Fe(CO)_2Cp]$ or of $[\{Zr(N_2)Cp^*_2\}_2(N_2)]$ with $[\{Fe(CO)_2Cp\}_2]$ gave (**10**) (see Section 10.4.2).[95,154] The isostructural ruthenium compound has also been prepared.[154] The formation of these compounds instead of complexes with a direct Zr–Fe or Zr–Ru bond is probably due to a combination of steric and electronic effects of the $^-$Cp* ligand.

When the dianion $[Fe(CO)_4]^{2-}$ was reacted with $[ZrR(X)Cp_2]$ (R = Me, X = Cl; R = octyl, X = BF$_4$) the complexes $cis\text{-}[Fe(CO)_4(ZrRCp_2)_2]$ were obtained.[268] The reaction of $Na_2[Fe(CO)_4]$ with hafnocene dichloride or diiodide gave the dimeric complex $[Cp_2Hf\{Fe(CO)_4\}]_2$, which contains a four-membered Hf–Fe–Hf–Fe ring.[269]

Zirconocene dihydride or the dinitrogen complex react with a cyclopentadienylmetal dicarbonyl with formation of the metal–metal bonded carbonyl-bridged species (**14**) (M = Co, Rh or RuH). In the reaction of zirconocene dihydride with $[Rh(CO)_2Cp]$, formation of an intermediate zirconoxycarbene complex $[Cp^*_2(X)Zr\{OC(H)=Rh(CO)Cp\}]$ (X = H) was observed by NMR spectroscopy. This complex (X = Cl) was the exclusive product when the zirconocene hydridochloride was used as starting material, because elimination of dihydrogen is then prevented.[95]

$$Cp^*_2Zr \overset{\displaystyle O}{\underset{\displaystyle O}{\diagdown\!\!\diagup\,\,C\,\,\diagdown\!\!\diagup}} MCp$$

(**14**)

The hydrido-bridged complexes $[Cp_2(R^1)Zr(\mu\text{-}H)_n\{ReH_{6-n}(PPh_2R^2)_2\}]$ ($n > 1$)[51,95] with a Zr–M bond were obtained from $[ZrR^1(X)Cp_2]$ and $K[ReH_6(PPh_2R^2)_2]$ or $[ReH_7(PPh_2R^2)_2]$ (R^1 = Cl or NMe$_2$; X = NMe$_2$; R^2 = Ph). All rhenium-bound hydrido ligands appear equivalent at 25 °C and −85 °C in the

^{1}H NMR spectrum, but the low-energy stretching bands (1710, 1565 cm^{-1}) in the IR spectrum suggest the presence of hydrido bridges.[95] [Cp$_2$(R)Zr(μ-H)$_3$Os(PMe$_2$Ph)] (R = Cl or H)[95,270] was obtained from [ZrCl(R)Cp$_2$] and K[OsH$_3$(PMe$_2$Ph)$_3$].

10.6.2 Chemical Properties

The trimetallic complex [Cp$_2$Zr{Ru(CO)$_2$Cp}$_2$] reacts with CO, PMe$_3$ or ethene to give dimetallic complexes.[95,271] Also, the Ru(CO)$_2$Cp groups exchange with the chloro or [Fe(CO)$_2$Cp]$^-$ ligands of [ZrCl$_2$Cp$_2$] or [Cp$_2$(Bu^{t}O)Zr–Fe(CO)$_2$Cp].[154] [Cp$_2$ZrMe{Ru(CO)$_2$Cp}], although thermally stable, is rapidly hydrolysed on exposure to air with cleavage of the metal–metal bonds and formation of [(ZrMeCp$_2$)$_2$O] and [RuH(CO)$_2$Cp].[265] Reaction of the dimetallic carbonyl-bridged species (**14**) (M = Co, Rh or RuH) with CO gives zirconocene dicarbonyl and the corresponding cyclopentadienyl transition metal carbonyl.[95] The dimeric hafnocene–iron(tetracarbonyl) complex reacts rapidly with HCl to give hafnocene dichloride and [Fe$_3$(CO)$_{12}$]. On dissolution in THF, isomerization to an isocarbonyl complex is observed.[269]

10.6.3 Spectroscopic Properties

The two ν(C=O) bands of the Zr–Fe- and Zr–Ru-bonded compounds are ca. 30 cm^{-1} and ca. 120 cm^{-1} higher in energy than the corresponding bands of K[M(CO)$_2$Cp] (M = Fe[154,266] or Ru[154]), as expected for compounds with a Zr–M bond. The ^{13}C NMR spectra of [Cp$_2$ZrX{Ru(CO)$_2$Cp}] (X = Cl, Me or OBu^t)[265] show one CO signal, whereas [Cp$_2$Zr{Ru(CO)$_2$Cp}$_2$] exhibits two CO signals at –60 °C and a single broad peak at 51 °C.[154] The temperature-dependent dynamic process probably involves rotation about the Zr–Ru bonds. The two resonances at low temperature are consistent with the solid-state structure (see Section 10.6.4), which has two sets of equivalent CO ligands.[154]

In [Cp$_2$Hf{Fe(CO)$_4$}]$_2$[269] and *cis*-[Fe(CO)$_4$(ZrRCp$_2$)$_2$] (R = Me or octyl)[268] four terminal CO bands are observed, consistent with the local C_{2v} symmetry of the Fe(CO)$_4$ fragment. In the Raman spectrum the intense peaks at 184 cm^{-1} and 138 cm^{-1} were assigned to ν(Hf–Fe) modes.[269] The complexes [Cp$_2$ZrX{Fe(CO)$_2$Cp}] exhibit the ν(Zr–Fe) band at 187 cm^{-1} (X = Me) and 181 cm^{-1} (X = Cl), respectively.[266] The carbonyl-bridged complexes (**14**) (M = Co, Rh or RuH) exhibit absorptions at 1670–1700 cm^{-1} and 1710–1750 cm^{-1} for the two bridging carbonyl groups.[95]

10.6.4 Structural Data

The zirconocene–ruthenium complexes [Cp$_2$Zr(OBu^t){Ru(CO)$_2$Cp}][265] and [Cp$_2$Zr{Ru(CO)$_2$Cp}$_2$][154] contain direct, unbridged zirconium to late transition metal bonds. In the former, the OBu^t and Ru(CO)$_2$Cp groups reside in a crystallographically imposed mirror plane so that the ruthenium substituents are staggered with respect to those of the zirconium centre. In [Cp$_2$Zr{Ru(CO)$_2$Cp}$_2$], a staggered conformation is associated with each Zr–Ru bond, with the overall molecular symmetry approximating to C_2. Molecular orbital calculations were performed on the model complex [Cp$_2$ZrI{Ru(CO)$_2$Cp}].[272]

In the other complexes, the two metal atoms are also bridged by carbonyl,[95,271] hydrido[51,95] or cyclopentadienyl[95,271] ligands.

10.7 METALLOCENE(IV) COMPLEXES WITH PSEUDOHALIDE OR OXOANION BONDS

10.7.1 Metallocene Complexes with Azido, Triazenido, Sulfur Diimido, Thionylimido and Thiocyanato Ligands

While the reaction of trimethylsilyl azide with zirconocene dichloride was unsuccessful, dimethyl-zirconocene reacts to give tetramethylsilane and white, air-sensitive [ZrMe(N$_3$)Cp$_2$].[273] The IR spectrum shows the ν_{as} and ν_s (N$_3$) modes at 2080 cm^{-1} and 1270 cm^{-1}, respectively. The bisazide, [Zr(N$_3$)$_2$Cp$_2$], is obtained from [Zr(OCHMeCH$_2$Cl)$_2$Cp$_2$] and trimethylsilyl azide.[274]

Interaction of phenyl azide with [ZrR$_2$Cp$_2$] (R = Me or Ph) gives [ZrR{N(R)NNPh}Cp$_2$] with a bidentate triazenido group.[273] Analogously, [HfH$_2$Cp*$_2$] reacts smoothly with RN$_3$ (R = Ph or *p*-Tol) to give moderately stable [HfH{N(H)NNR}Cp*$_2$], which upon thermolysis at 80 °C loses dinitrogen to form an arylamido complex.[31]

The reaction of [MCl$_2$Cp(η-C$_5$R$_5$)] (M = Zr, R = H or Me; M = Hf, R = Me) with Me$_3$SnNSNSnMe$_3$ yields [Cp(η-C$_5$R$_5$)M(NSN)$_2$MCp(η-C$_5$R$_5$)].[275] The unsymmetrically substituted compounds (R = Me) exist as two different isomers with identical cyclopentadienyl groups located on the same or on opposite sides of the M(NSN)$_2$M metallacycle, as deduced from ^{1}H NMR spectra. In the IR spectra, strong absorptions between 1195 cm^{-1} and 1120 cm^{-1} as well as 1100 cm^{-1} and 1200 cm^{-1} were assigned to the ν(NSN) modes.[275]

Thionylimido complexes [M(NSO)$_2$(η-C$_5$R$^1_5$)(η-C$_5$R$^2_5$)] are available by salt elimination (M = Ti, R^1 = R^2 = H; M = Zr, R^1 = H, R^2 = Me; M = Zr or Hf, R^1 = R^2 = Me). In the IR spectra three absorptions (at ca. 1250, 1080 and 520 cm^{-1}) which are characteristic of the NSO$^-$ ligand are observed. The crystal structure of [Zr(NSO)$_2$CpCp*] reveals two nitrogen-bonded NSO$^-$ ligands forming a nearly planar Zr(NSO)$_2$ unit.[276]

The dithiocyanato complexes [M(SCN)$_2$(η-C$_5$H$_4$R^1)$_2$] (R^1 = ER$^2_3$; E = C, Si or Ge; R^2 = alkyl) were prepared by reacting the appropriate metallocene dichloride with potassium thiocyanate.[277]

10.7.2 Metallocene Complexes with Nitrato and Related Ligands

In general, metallocene nitrates are obtained from the reaction of metallocene complexes with nitric acid. Thus, [ZrCl(NO$_3$)Cp$_2$] is obtained from [ZrCl$_2$Cp$_2$] and HNO$_3$.[278] The corresponding bromide [ZrBr(NO$_3$)Cp$_2$] is obtained from the reaction of [ZrBr(X)Cp$_2$] (X = OPh or Br)[279] or [(ZrBrCp$_2$)$_2$O][278,279] with HNO$_3$ or by comproportionation of zirconocene dibromide and zirconocene dinitrate. Infrared and Raman spectra indicate that zirconocene bromo nitrate is a dimer with covalently bound NO$_3^-$ bridging ligands.[279] However, the x-ray structure of [ZrCl(NO$_3$)Cp$_2$] revealed a monomeric molecule with η^2-bonded nitrato ligands.[278] The analogous hafnium complex [HfBr(NO$_3$)Cp$_2$] is available from [(HfBrCp$_2$)O] and HNO$_3$.[280]

If dilute HNO$_3$ is employed, hydrolysis occurs with formation of metallocene hydroxo-nitrates. Thus, [ZrCl$_2$Cp$_2$] reacts with nitric acid in CHCl$_3$ with cleavage of a Zr–Cp bond and formation of [{Zr(NO$_3$)$_2$(μ-OH)Cp}$_2$]·2THF.[281] The same product is obtained on hydrolysis of [Zr(NO$_3$)$_2$Cp$_2$].[282] The crystal structure shows bidentate nitrato ligands and bridging hydroxo groups.[281] [HfCp$_4$] reacts with HNO$_3$ to yield the corresponding hafnocene complex.[280] However, [ZrCp$_4$] reacts with HNO$_3$ to afford [Zr(NO$_3$)(OH)Cp$_2$·H$_2$O], which hydrolyses to yield [Zr(NO$_3$)(OH)$_2$Cp·H$_2$O]$_4$.[282] The IR spectra of solid [Zr(NO$_3$)(OH)Cp$_2$·H$_2$O] and [Zr(NO$_3$)(OH)$_2$Cp·H$_2$O]$_4$ show a strong absorption band at 1385 cm^{-1}, characteristic of the NO$_3^-$ ion. The latter complex also shows bands at 1564 cm^{-1} and 1278 cm^{-1}, which are characteristic of a covalently bonded nitrato group. In solution, only bands due to covalently bonded nitrate are present for both complexes. Infrared and Raman spectra of [Zr(NO$_3$)$_2$Cp$_2$] are different in solution and in the solid state. Molecular weight measurements in THF showed the complex to be trimeric.[282]

The nitronato complexes [Zr(O$_2$N=CMeR)(X)Cp$_2$] (X = Cl, R = Me or H; X = CH$_2$CMe$_2$Ph, R = Me) were prepared from Li(O$_2$N=CMeR) and the appropriate zirconocene chloride derivative. Their IR spectra suggest a symmetrical chelating mode of coordination for the alkanenitronato ligand (ν(CN) mode 1638–1651 cm^{-1}), which was established unambiguously by x-ray analysis of the derivative with X = Cl and R = Me.[193]

The reaction of diphenylzirconocene with NO in benzene gives [Zr(Ph){ON(Ph)NO}Cp$_2$]. The chloro derivative [ZrCl{ON(R)NO}Cp$_2$] (R = Ph or *p*-Tol) is obtained from zirconocene dichloride and Ag[ON(R)NO], whereas with [NH$_4$][ON(*p*-Tol)NO] only [Zr{ON(*p*-Tol)NO}$_4$] was formed. The complexes are white. Infrared spectra indicate the presence of a Zr{ON(R)NO} chelate ring.[283]

Di-*t*-butyl nitroxide readily reacts with [MR$^1_2$Cp$_2$] (M = Zr, R^1 = Cl, Br, Me or CH$_2$R^2 with R^2 = aryl; M = Hf, R^1 = Bz) by displacement of an R^1 group to give high yields of Bu$^t_2$NOR^1 and [M(ONBu$^t_2$)(R^1)Cp$_2$].[232]

10.7.3 Metallocene Complexes with Sulfonato and Sulfone Ligands

The reaction of zirconocene dichloride with Ag[O$_3$SCF$_3$] in THF gives [Zr(O$_3$SCF$_3$)$_2$(THF)Cp$_2$], which has a five-coordinate bent metallocene structure, the THF group being symmetrically surrounded by CF$_3$SO$_3^-$ ligands.[284]

The reaction of dimethylzirconocene with 1 or 2 equiv. of $H_2C(SO_2CF_3)_2$ yields $[ZrMe\{HC(SO_2CF_3)_2\}Cp_2]$ and $[Zr\{HC(SO_2CF_3)_2\}_2Cp_2]$, respectively.[115] The methylhafnocene analogue was prepared accordingly.[204] Analogously, the complexes $[ZrMe_{2-n}(X)_nCp_2]$ ($n = 1$ or 2, $X = PhC(SO_2CF_3)_2$, $N(SO_2CF_3)_2$, CF_3SO_3 or $HC(SO_2)_2(CF_2)_3$; $n = 2$, $X = CH_2{=}CHCH_2C(SO_2CF_3)_2$) were prepared. The x-ray analysis of $[Zr\{HC(SO_2CF_3)_2\}_2Cp_2]$ shows a bidentate and a monodentate ligand. The two types of ligands interconvert rapidly on the NMR timescale. Zirconium-91 NMR spectra of these complexes have been reported. The complex $[ZrMe\{CH(SO_2CF_3)_2\}Cp_2]$ inserts CO into the Zr–Me bond to give the corresponding η^2-acetyl derivative. Fluorine-19 NMR spectra suggest the presence of a monodentate oxygen-bonded $^-CH(SO_2CF_2)_2$ ligand.[115]

10.8 REFERENCES

1. D. J. Cardin, M. F. Lappert, C. L. Raston and P. I. Riley, in *COMC-I*, vol. 3, p. 559.
2. T. D. Tilley, in 'The Silicon-Heteroatom Bond', ed. S. Patai and Z. Rappoport, Wiley, New York, 1991, chaps. 9 and 10.
3. V. V. Dement'ev, S. P. Solodovnikov, B. L. Tumanskii, B. D. Lavrukhin, T. M. Frunze and A. A. Zhdanov, *Metalloorg. Khim.*, 1988, **1**, 1365 (*Chem. Abstr.*, 1990, **112**, 56 257).
4. H. G. Woo, W. P. Freeman and T. D. Tilley, *Organometallics*, 1992, **11**, 2198.
5. A. J. Blakeney and J. A. Gladysz, *J. Organomet. Chem.*, 1980, **202**, 263.
6. H. G. Woo, R. H. Heyn and T. D. Tilley, *J. Am. Chem. Soc.*, 1992, **114**, 5698.
7. Y. Mu, C. Aitken, B. Cote, J. F. Harrod and E. Samuel, *Can. J. Chem.*, 1991, **69**, 264.
8. K. A. Kreutzer, R. A. Fisher, W. M. Davis, E. Spaltenstein and S. L. Buchwald, *Organometallics*, 1991, **10**, 4031.
9. T. Takahashi *et al.*, *J. Am. Chem. Soc.*, 1991, **113**, 8564.
10. H. Köpf and T. Klapötke, *Z. Naturforsch., Teil B*, 1985, **40**, 447.
11. R. M. Whittal, G. Ferguson, J. F. Gallagher and W. E. Piers, *J. Am. Chem. Soc.*, 1991, **113**, 9867.
12. W. Zheng and D. W. Stephan, *Inorg. Chem.* 1988, **27**, 2386.
13. G. Erker, R. Petrenz, C. Krüger, F. Lutz, A. Weiss and S. Werner, *Organometallics*, 1992, **11**, 1646.
14. J. Arnold, M. P. Engeler, F. H. Elsner, R. H. Heyn and T. D. Tilley, *Organometallics*, 1989, **8**, 2284.
15. R. V. Bynum, H. M. Zhang, W. E. Hunter and J. L. Atwood, *Can. J. Chem.*, 1986, **64**, 1304.
16. S. J. Simpson and R. A. Andersen, *Inorg. Chem.*, 1981, **20**, 3627.
17. Y. Bai, H. W. Roesky, M. Noltemeyer and M. Witt, *Chem. Ber.*, 1992, **125**, 825.
18. P. J. Walsh, A. M. Baranger and R. G. Bergman, *J. Am. Chem. Soc.*, 1992, **114**, 1708.
19. J. Feldman and J. C. Calabrese, *J. Chem. Soc., Chem. Commun.*, 1991, 134.
20. H. Köpf and T. Klapötke, *Chem. Ber.*, 1986, **119**, 1986.
21. P. J. Walsh, F. J. Hollander and R. G. Bergman, *J. Am. Chem. Soc.*, 1988, **110**, 8729.
22. N. Coles, R. J. Whitby and J. Blagg, *Synlett*, 1990, 271.
23. A. W. Duff, R. A. Kamarudin, M. F. Lappert and R. J. Norton, *J. Chem. Soc., Dalton Trans.*, 1986, 489.
24. S. K. Sengupta and Nizamuddin, *Ind. J. Chem. Sect. A*, 1982, **21**, 426.
25. V. Srivastava, O. P. Pandey, S. K. Sengupta and S. C. Tripathi, *J. Organomet. Chem.*, 1985, **279**, 395.
26. O. P. Pandey, *Transition Met. Chem. (London)*, 1987, **12**, 521.
27. V. Srivastava, O. P. Pandey, S. K. Sengupta and S. C. Tripathi, *J. Organomet. Chem.*, 1987, **321**, 27.
28. S. Kamrah, G. S. Sodhi and N. K. Kaushik, *Inorg. Chim. Acta*, 1985, **107**, 29.
29. G. L. Hillhouse and J. E. Bercaw, *J. Am. Chem. Soc.*, 1984, **106**, 5472.
30. G. L. Hillhouse, A. R. Bulls, B. D. Santarsiero and J. E. Bercaw, *Organometallics*, 1988, **7**, 1309.
31. G. L. Hillhouse and J. E. Bercaw, *Organometallics*, 1982, **1**, 1025.
32. P. T. Wolczanski and J. E. Bercaw, *J. Am. Chem. Soc.*, 1979, **101**, 6450.
33. W. Frömberg and G. Erker, *J. Organomet. Chem.*, 1985, **280**, 355.
34. S. M. Beshouri, P. E. Fanwick, I. P. Rothwell and J. C. Huffman, *Organometallics*, 1987, **6**, 891.
35. S. M. Beshouri, D. E. Chebi, P. E. Fanwick, I. P. Rothwell and J. C. Huffman, *Organometallics*, 1990, **9**, 2375.
36. R. Fandos, A. Meetsma and J. H. Teuben, *Organometallics*, 1991, **10**, 2665.
37. N. Dufour, A. M. Caminade, M. Basso-Bert, A. Igau and J. P. Majoral, *Organometallics*, 1992, **11**, 1131.
38. N. Dufour, J. P. Majoral, A. M. Caminade, R. Choukroun and Y. Dromzee, *Organometallics*, 1991, **10**, 45.
39. J. P. Majoral, N. Dufour, F. Meyer, A. M. Caminade, R. Choukroun and D. Gervais, *J. Chem. Soc., Chem. Commun.*, 1990, 507.
40. D. Fest, C. D. Habben, A. Meller, G. M. Sheldrick, D. Stahlke and F. Pauer, *Chem. Ber.*, 1990, **123**, 703.
41. S. Gambarotta, S. Strologo, C. Floriani, A. Chiesi-Villa and C. Guastini, *Inorg. Chem.*, 1985, **24**, 654.
42. G. A. Vaughan, G. L. Hillhouse and A. L. Rheingold, *J. Am. Chem. Soc.*, 1990, **112**, 7994.
43. J. Scholz, M. Dlikan, D. Ströhl, A. Dietrich, H. Schumann and K. H. Thiele, *Chem. Ber.*, 1990, **123**, 2279.
44. J. M. Davis, R. J. Whitby and A. Jaxa-Chamiec, *J. Chem. Soc., Chem. Commun.*, 1991, 1743.
45. S. L. Buchwald, M. W. Wannamaker and B. T. Watson, *J. Am. Chem. Soc.*, 1989, **111**, 776.
46. S. L. Buchwald, B. T. Watson, M. W. Wannamaker and J. C. Dewan, *J. Am. Chem. Soc.*, 1989, **111**, 4486.
47. L. E. Schock and T. J. Marks, *J. Am. Chem. Soc.*, 1988, **110**, 7701.
48. M. F. Lappert, P. P. Power, A. R. Sanger and R. C. Srivastava, 'Metal and Metalloid Amides', Ellis Horwood, Chichester, 1980.
49. S. R. Wade and G. R. Willey, *J. Chem. Soc., Dalton Trans.*, 1981, 1264.
50. F. R. Lemke, D. J. Szalda and R. M. Bullock, *J. Am. Chem. Soc.*, 1991, **113**, 8466.
51. W. J. Sartain, J. C. Huffman, E. G. Lundquist, W. G. Streib and K. G. Caulton, *J. Mol. Catal.*, 1989, **56**, 20.
52. D. C. Bradley and A. S. Kasenally, *J. Chem. Soc., Chem. Commun.*, 1968, 1430.
53. W. A. Herrmann, B. Menjon and E. Herdtweck, *Organometallics*, 1991, **10**, 2134.
54. Y. Wielstra, A. Meetsma, S. Gambarotta and S. Khan, *Organometallics*, 1990, **9**, 876.

55. G. Erker, *Angew. Chem., Int. Ed. Engl.*, 1989, **28**, 397.
56. F. Boutonnet, N. Dufour, T. Straw, A. Igau and J. P. Majoral, *Organometallics*, 1991, **10**, 3939.
57. P. J. Walsh, F. J. Hollander and R. G. Bergman, *J. Organomet. Chem.*, 1992, **428**, 13.
58. P. J. Walsh, F. J. Hollander and R. G. Bergman, *J. Am. Chem. Soc.*, 1990, **112**, 894.
59. P. J. Walsh, M. J. Carney and R. G. Bergman, *J. Am. Chem. Soc.*, 1991, **113**, 6343.
60. S. Gambarotta, C. Floriani, A. Chiesi-Villa and C. Guastini, *Inorg. Chem.*, 1983, **22**, 2029.
61. S. Gambarotta, M. Basso-Bert, C. Floriani and C. Guastini, *J. Chem. Soc., Chem. Commun.*, 1982, 374.
62. B. D. Santarsiero and E. J. Moore, *Acta Crystallogr., Sect. C*, 1988, **44**, 433.
63. E. Hey-Hawkins, M. F. Lappert, J. L. Atwood and S. G. Bott, *J. Chem. Soc., Dalton Trans.*, 1991, 939.
64. E. Hey and U. Müller, *Z. Naturforsch., Teil B*, 1989, **44**, 1538.
65. E. Hey and F. Weller, *Chem. Ber.*, 1988, **121**, 1207.
66. G. M. Arvanitis, J. Schwartz and D. Van Engen, *Organometallics*, 1986, **5**, 2157.
67. G. M. Arvanitis, J. Smegal, I. Meier, A. C. C. Wong, J. Schwartz and D. Van Engen, *Organometallics*, 1989, **8**, 2717.
68. T. E. Glassman, A. H. Liu and R. R. Schrock, *Inorg. Chem.*, 1991, **30**, 4723.
69. S. R. Wade, M. G. H. Wallbridge and G. R. Willey, *J. Chem. Soc., Dalton Trans.*, 1983, 2555.
70. R. T. Baker, J. F. Whitney and S. S. Wreford, *Organometallics*, 1983, **2**, 1049.
71. D. G. Dick and D. W. Stephan, *Can. J. Chem.*, 1991, **69**, 1146.
72. E. Hey, M. F. Lappert, J. L. Atwood and S. G. Bott, *Polyhedron*, 1988, **7**, 2083.
73. L. Weber, G. Meine, R. Boese and N. Augart, *Organometallics*, 1987, **6**, 2484.
74. F. Lindenberg and E. Hey-Hawkins, *J. Organomet. Chem.*, 1992, **435**, 291.
75. A. M. Arif, A. H. Cowley, C. M. Nunn and M. Pakulski, *J. Chem. Soc., Chem. Commun.*, 1987, 994.
76. G. A. Vaughan, G. L. Hillhouse and A. L. Rheingold, *Organometallics*, 1989, **8**, 1760.
77. R. Bohra, P. B. Hitchcock, M. F. Lappert and W. P. Leung, *J. Chem. Soc., Chem. Commun.*, 1989, 728.
78. H. Köpf and V. Richtering, *J. Organomet. Chem.*, 1988, **346**, 355.
79. E. Hey, *J. Organomet. Chem.*, 1989, **378**, 375.
80. E. Hey-Hawkins, S. Kurz, J. Sieler and G. Baum, *J. Organomet. Chem.*, in press.
81. J. Ho and D. W. Stephan, *Organometallics*, 1991, **10**, 3001.
82. H. Köpf and R. Voigtlaender, *Chem. Ber.*, 1981, **114**, 2731.
83. E. Hey, S. G. Bott and J. L. Atwood, *Chem. Ber.*, 1988, **121**, 561.
84. B. L. Benac and R. A. Jones, *Polyhedron*, 1989, **8**, 1774.
85. K. Fromm, G. Baum and E. Hey-Hawkins, *Z. Anorg. Allg. Chem.*, 1992, **615**, 35.
86. E. Hey, *Z. Naturforsch., Teil B*, 1988, **43**, 1271.
87. E. Hey-Hawkins and S. Kurz, *J. Organomet. Chem.*, 1994, **479**, 125.
88. E. Hey-Hawkins and S. Kurz, *J. Organomet. Chem.*, 1993, **462**, 203.
89. E. Hey-Hawkins, S. Kurz and H. Schottmüller, *Z. Naturforsch.*, manuscript in preparation.
90. M. Y. Chiang, S. Gambarotta and F. Van Bolhuis, *Organometallics*, 1988, **7**, 1864.
91. S. Nielsen-Marsh, R. J. Crowte and P. G. Edwards, *J. Chem. Soc., Chem. Commun.*, 1992, 699.
92. J. Ho and D. W. Stephan, *Organometallics*, 1992, **11**, 1014.
93. O. J. Scherer, M. Swarowsky, H. Swarowsky and G. Wolmershäuser, *Angew. Chem., Int. Ed. Engl.*, 1988, **27**, 694.
94. E. Hey, M. F. Lappert, J. L. Atwood and S. G. Bott, *J. Chem. Soc., Chem. Commun.*, 1987, 597.
95. D. W. Stephan, *Coord. Chem. Rev.*, 1988, **95**, 41.
96. E. Hey-Hawkins and F. Lindenberg, *Z. Naturforsch., Teil B*, 1993, **48**, 951.
97. E. Hey-Hawkins and F. Lindenberg, *Chem. Ber.* 1992, **125**, 1815.
98. E. Hey, M. F. Lappert, J. L. Atwood and S. G. Bott, *J. Chem. Soc., Chem. Commun.*, 1987, 421; corrigendum ibid., 1987, 1604.
99. L. Gelmini and D. W. Stephan, *Organometallics*, 1987, **6**, 1515.
100. P. Y. Zheng, T. T. Nadasdi and D. W. Stephan, *Organometallics*, 1989, **8**, 1393.
101. D. G. Dick and D. W. Stephan, *Organometallics*, 1990, **9**, 1910.
102. S. A. Yousif-Ross and A. Wojcicki, *Inorg. Chim. Acta*, 1990, **171**, 115.
103. P. Y. Zheng and D. W. Stephan, *Can. J. Chem.*, 1989, **67**, 1584.
104. R. T. Baker, W. C. Fultz, T. B. Marder and I. D. Williams, *Organometallics*, 1990, **9**, 2357.
105. F. Lindenberg, T. Gelbrich and E. Hey-Hawkins, *Z. Anorg. Allg. Chem.*, in press.
106. S. R. Wade, M. G. H. Wallbridge and G. R. Willey, *J. Organomet. Chem.*, 1984, **267**, 271.
107. E. Hey-Hawkins and F. Lindenberg, *Organometallics*, 1994, **13**, 4643.
108. J. B. Hoke and E. W. Stern, *J. Organomet. Chem.*, 1991, **412**, 77.
109. A. K. Sharma and N. K. Kaushik, *Synth. React. Inorg. Metal-Org. Chem.*, 1984, **14**, 513.
110. A. K. Sharma and N. K. Kaushik, *Synth. React. Inorg. Metal-Org. Chem.*, 1982, **12**, 827.
111. G. Gupta, S. K. Sahni, R. Sharan and R. N. Kapoor, *Indian J. Chem., Sect. A*, 1981, **20**, 1033.
112. B. Khera, A. K. Sharma and N. K. Kaushik, *Bull. Soc. Chim. Fr.*, 1984, 172.
113. V. Srivastava, O. P. Pandey, S. K. Sengupta and S. C. Tripathi, *J. Organomet. Chem.*, 1986, **306**, 355.
114. E. A. Babaian, D. C. Hrncir, S. G. Bott and J. L. Atwood, *Inorg. Chem.*, 1986, **25**, 4818.
115. A. R. Siedle, R. A. Newmark, W. B. Gleason and W. M. Lamanna, *Organometallics*, 1990, **9**, 1290.
116. T. V. Lubben and P. T. Wolczanski, *J. Am. Chem. Soc.*, 1987, **109**, 424.
117. D. W. Stephan, *Organometallics*, 1990, **9**, 2718.
118. B. D. Steffey, N. Truong, D. E. Chebi, J. L. Kerschner, P. E. Fanwick and I. P. Rothwell, *Polyhedron*, 1990, **9**, 839.
119. J. M. Atkinson and P. B. Brindley, *J. Organomet. Chem.*, 1991, **411**, 131.
120. H. M. Gau, C. T. Chen and C. C. Schei, *J. Organomet. Chem.*, 1992, **424**, 307.
121. J. A. Heppert, T. J. Boyle and F. Takusagawa, *Organometallics*, 1989, **8**, 461.
122. R. Broussier, A. Da Rold and B. Gautheron, *J. Organomet. Chem.*, 1992, **427**, 231.
123. G. Erker, R. Schlund, M. Albrecht and C. Sarter, *J. Organomet. Chem.*, 1988, **353**, C27.
124. P. B. Brindley and M. J. Scotton, *J. Chem. Soc., Perkin Trans. 2*, 1981, 419.
125. A. Van Asselt, B. D. Santarsiero and J. E. Bercaw, *J. Am. Chem. Soc.*, 1986, **108**, 8291.
126. E. B. Coughlin and J. E. Bercaw, *Organometallics*, 1992, **11**, 465.

127. R. Choukroun, *Inorg. Chim. Acta*, 1982, **58**, 121.
128. S. L. Buchwald, R. B. Nielsen and J. C. Dewan, *Organometallics*, 1988, **7**, 2324.
129. G. Erker, U. Dorf, P. Czisch and J. L. Peterson, *Organometallics*, 1986, **5**, 668.
130. S. L. Buchwald, R. B. Nielsen and J. C. Dewan, *Organometallics*, 1989, **8**, 1593.
131. S. L. Buchwald, E. A. Lucas and J. C. Dewan, *J. Am. Chem. Soc.*, 1987, **109**, 4396.
132. S. L. Buchwald, E. A. Lucas and W. M. Davis, *J. Am. Chem. Soc.*, 1989, **111**, 397.
133. G. Erker and R. Zwettler, *J. Organomet. Chem.*, 1991, **409**, 179.
134. Y. Kai *et al.*, *Bull. Chem. Soc. Jpn.*, 1983, **56**, 3735.
135. G. Erker, K. Engel, J. L. Atwood and W. E. Hunter, *Angew. Chem., Int. Ed. Engl.*, 1983, **22**, 494.
136. H. Yasuda *et al.*, *Organometallics*, 1989, **8**, 1139.
137. H. G. Alt and C. E. Denner, *J. Organomet. Chem.*, 1990, **391**, 53.
138. H. G. Alt and C. E. Denner, *J. Organomet. Chem.*, 1989, **368**, C15.
139. H. G. Alt, C. E. Denner and R. Zenk, *J. Organomet. Chem.*, 1992, **433**, 107.
140. H. G. Alt and C. E. Denner, *J. Organomet. Chem.*, 1990, **390**, 53.
141. G. A. Vaughan, P. B. Rupert and G. L. Hillhouse, *J. Am. Chem. Soc.*, 1987, **109**, 5538.
142. G. A. Vaughan, G. L. Hillhouse, R. T. Lum, S. L. Buchwald and A. L. Rheingold, *J. Am. Chem. Soc.*, 1988, **110**, 7215.
143. E. Uhlig, B. Buerglen, C. Krüger and P. Betz, *J. Organomet. Chem.*, 1990, **382**, 77.
144. K. Mashima, M. Yamakawa and H. Takaya, *J. Chem. Soc., Dalton Trans.*, 1991, 2851.
145. H. Takaya, M. Yamakawa and K. Mashima, *J. Chem. Soc., Chem. Commun.*, 1983, 1283.
146. D. M. Roddick, M. D. Fryzuk, P. F. Seidler, G. L. Hillhouse and J. E. Bercaw, *Organometallics*, 1985, **4**, 97.
147. A. Antinolo *et al.*, *Polyhedron*, 1989, **8**, 1601.
148. S. Dou and S. Chen, *Gaodeng Xuexiao Huaxue Xuebao*, 1984, **5**, 812 (*Chem. Abstr.*, 1985, **102**, 158 434).
149. J. Dai, M. Lou, J. Zhang and S. Chen, *Jiegou Huaxue*, 1982, **1**, 63 (*Chem. Abstr.*, 1983, **98**, 225 629).
150. R. S. Threlkel and J. E. Bercaw, *J. Am. Chem. Soc.*, 1981, **103**, 2650.
151. P. T. Wolczanski, R. S. Threlkel and B. D. Santarsiero, *Acta Crystallogr., Sect. C*, 1983, **39**, 1330.
152. G. Erker *et al.*, *Organometallics*, 1989, **8**, 2037.
153. P. T. Barger, B. D. Santarsiero, J. Armantrout and J. E. Bercaw, *J. Am. Chem. Soc.*, 1984, **106**, 5178.
154. C. P. Casey, R. F. Jordan and A. L. Rheingold, *Organometallics*, 1984, **3**, 504.
155. E. N. Jacobsen, M. K. Trost and R. G. Bergman, *J. Am. Chem. Soc.*, 1986, **108**, 8092.
156. M. D. Curtis, S. Thanedar and W. M. Butler, *Organometallics*, 1984, **3**, 1855.
157. P. Veya, C. Floriani, A. Chiesi-Villa and C. Guastini, *J. Chem. Soc., Chem. Commun.*, 1991, 1166.
158. P. Veya, C. Floriani, A. Chiesi-Villa and C. Guastini, *Organometallics*, 1991, **10**, 2991.
159. I. Weinstock, C. Floriani, A. Chiesi-Villa and C. Guastini, *J. Am. Chem. Soc.*, 1986, **108**, 8298.
160. D. M. Roddick and J. E. Bercaw, *Chem. Ber.*, 1989, **122**, 1579.
161. R. Beckhaus, D. Wilbrandt, S. Flatau and W. H. Boehmer, *J. Organomet. Chem.*, 1992, **423**, 211.
162. P. Hofmann, M. Frede, P. Stauffert, W. Lasser and U. Thewalt, *Angew. Chem., Int. Ed. Engl.*, 1985, **24**, 712.
163. S. Gambarotta, C. Floriani, A. Chiesi-Villa and C. Guastini, *J. Am. Chem. Soc.*, 1983, **105**, 1690.
164. M. F. Lappert, C. L. Raston, L. M. Engelhardt and A. H. White, *J. Chem. Soc., Chem. Commun.*, 1985, 521.
165. J. L. Petersen and J. W. Egan, Jr., *Organometallics*, 1987, **6**, 2007.
166. E. J. Moore, D. A. Straus, J. Armantrout, B. D. Santarsiero, R. H. Grubbs and J. E. Bercaw, *J. Am. Chem. Soc.*, 1983, **105**, 2068.
167. P. T. Wolczanski and J. E. Bercaw, *Acc. Chem. Res.*, 1980, **13**, 212.
168. C. A. Hester, M. Draganjac and A. W. Cordes, *Inorg. Chim. Acta*, 1991, **184**, 137.
169. J. Recht, B. I. Cohen, A. S. Goldman and J. Kohn, *Tetrahedron Lett.*, 1990, **31**, 7281.
170. S. C. Dixit, R. Sharan and R. N. Kapoor, *Inorg. Chim. Acta*, 1989, **158**, 109.
171. S. C. Dixit, R. Sharan and R. N. Kapoor, *Inorg. Chim. Acta*, 1987, **133**, 251.
172. S. C. Dixit, R. Sharan and R. N. Kapoor, *Inorg. Chim. Acta*, 1988, **145**, 39.
173. A. K. Bhagi, B. Singh and R. N. Kapoor, *Indian J. Chem., Sect. A*, 1989, **28**, 1112.
174. P. Lukose, S. C. Bhatia and A. K. Narula, *J. Organomet. Chem.*, 1991, **403**, 153.
175. A. Cutler, M. Raja and A. Todaro, *Inorg. Chem.*, 1987, **26**, 2877.
176. S. Klima and U. Thewalt, *J. Organomet. Chem.*, 1988, **354**, 77.
177. T. V. Lubben, K. Plössl, J. R. Norton, M. M. Miller and O. P. Anderson, *Organometallics*, 1992, **11**, 122.
178. Y. Zhou and H. Chen, *Polyhedron*, 1990, **9**, 2689.
179. Z. Wang, S. Lu, H. Guo, N. Hu and Y. Liu, *Polyhedron*, 1991, **10**, 2341.
180. Z. Wang, S. Lu and H. Guo, *Synth. React. Inorg. Metal-Org. Chem.*, 1992, **22**, 259.
181. Y. Ma, Z. Zeng, G. Huang and C. Ma, *Bull. Soc. Chim. Belg.*, 1990, **99**, 231.
182. Y. Ma, Y. Zhu, X. Wang and C. Ma, *Polyhedron*, 1989, **8**, 929.
183. M. E. Silver and R. C. Fay, *Organometallics*, 1983, **2**, 44.
184. B. D. Steffey, J. C. Vites and A. R. Cutler, *Organometallics*, 1991, **10**, 3432.
185. J. C. Vites, B. D. Steffey, M. E. Giuseppetti-Dery and A. R. Cutler, *Organometallics*, 1991, **10**, 2827.
186. C. T. Tso and A. R. Cutler, *J. Am. Chem. Soc.*, 1986, **108**, 6069.
187. C. G. Arena, G. Bruno and F. Faraone, *J. Chem. Soc., Dalton Trans.*, 1991, 1223.
188. U. Thewalt, S. Klima and K. Berhalter, *J. Organomet. Chem.*, 1988, **342**, 303.
189. U. Thewalt and T. Güthner, *J. Organomet. Chem.*, 1989, **379**, 59.
190. T. Güthner and U. Thewalt, *J. Organomet. Chem.*, 1989, **371**, 43.
191. U. Thewalt and T. Güthner, *J. Organomet. Chem.*, 1989, **361**, 309.
192. H. G. Alt and C. E. Denner, *J. Organomet. Chem.*, 1990, **398**, 91.
193. B. N. Diel and H. Hope, *Inorg. Chem.*, 1986, **25**, 4448.
194. R. Bortolin, V. Patel, I. Munday, N. J. Taylor and A. J. Carty, *J. Chem. Soc., Chem. Commun.*, 1985, 456.
195. X. Zhou, Y. Wang, S. Xu, H. Wang and R. Wang, *Sci. China, Ser. B*, 1991, **34**, 1047 (*Chem. Abstr.*, 1992, **116**, 106 441).
196. P. C. Wailes and H. Weigold, *Inorg. Synth.*, 1990, **28**, 257.
197. G. Erker, R. Nolte, G. Tainturier and A. L. Rheingold, *Organometallics*, 1989, **8**, 454.
198. G. Tainturier, B. Gautheron and M. Fahim, *J. Organomet. Chem.*, 1985, **290**, C4.

199. J. H. Toney and T. J. Marks, *J. Am. Chem. Soc.*, 1985, **107**, 947.
200. K. P. Reddy and J. L. Petersen, *Organometallics*, 1989, **8**, 2107.
201. Y. Wielstra, S. Gambarotta, A. L. Spek and W. J. J. Smeets, *Organometallics*, 1990, **9**, 2142.
202. T. V. Ashworth, T. C. Agreda, E. Herdtweck and W. A. Herrmann, *Angew. Chem., Int. Ed. Engl.*, 1986, **25**, 289.
203. G. Erker, U. Dorf, C. Krüger and Y. H. Tsay, *Organometallics*, 1987, **6**, 680.
204. W. E. Hunter, D. C. Hrncir, R. V. Bynum, R. A. Penttila and J. L. Atwood, *Organometallics*, 1983, **2**, 750.
205. R. Gomez, T. Cuenca, P. Royo, M. A. Pellinghelli and A. Tiripicchio, *Organometallics*, 1991, **10**, 1505.
206. J. Cacciola, K. P. Reddy and J. L. Petersen, *Organometallics*, 1992, **11**, 665.
207. W. A. Herrmann, T. Cuenca and U. Küsthardt, *J. Organomet. Chem.*, 1986, **309**, C15.
208. M. S. Rau, C. M. Kretz, L. A. Mercando, G. L. Geoffroy and A. L. Rheingold, *J. Am. Chem. Soc.*, 1991, **113**, 7420.
209. Q. Yang, X. Jin, X. Xu, G. Li, Y. Tang and S. Chen, *Sci. Sin. (Engl. Ed.)*, 1982, **25**, 356 (*Chem. Abstr.*, 1982, **97**, 64 493).
210. W. Chang, J. Dai and S. Chen, *Jiegou Huaxue*, 1982, **1**, 73 (*Chem. Abstr.*, 1983, **98**, 117 429).
211. L. G. Kuz'mina, Y. T. Struchkov, M. K. Minacheva and E. M. Brainina, *Koord. Khim.*, 1988, **14**, 1257 (*Chem. Abstr.*, 1989, **110**, 31 697).
212. F. Senocq, M. Basso-Bert, R. Choukroun and D. Gervais, *J. Organomet. Chem.*, 1985, **297**, 155.
213. M. J. Carney, P. J. Walsh and R. G. Bergman, *J. Am. Chem. Soc.*, 1990, **112**, 6426.
214. M. J. Carney, P. J. Walsh, F. J. Hollander and R. G. Bergman, *Organometallics*, 1992, **11**, 761.
215. M. J. Carney, P. J. Walsh, F. J. Hollander and R. G. Bergman, *J. Am. Chem. Soc.*, 1989, **111**, 8751.
216. A. K. Narula and P. Lukose, *J. Organomet. Chem.*, 1990, **393**, 365.
217. S. C. Dixit, R. Sharan and R. N. Kapoor, *J. Organomet. Chem.*, 1987, **332**, 135.
218. B. Gautheron and G. Tainturier, *J. Organomet. Chem.*, 1984, **262**, C30.
219. H. Köpf and T. Klapötke, *Z. Naturforsch., Teil B*, 1985, **40**, 1338.
220. B. Gautheron, G. Tainturier and P. Meunier, *J. Organomet. Chem.*, 1981, **209**, C49.
221. G. Tainturier, M. Fahim, G. Trouve-Bellan and B. Gautheron, *J. Organomet. Chem.*, 1989, **376**, 321.
222. P. Granger, B. Gautheron, G. Tainturier and S. Pouly, *Org. Magn. Reson.*, 1984, **22**, 701.
223. G. Tainturier, B. Gautheron and S. Pouly, *Nouv. J. Chim.*, 1986, **10**, 625.
224. H. Köpf and T. Klapötke, *Z. Naturforsch., Teil B*, 1986, **41**, 971.
225. S. Pouly, G. Tainturier and B. Gautheron, *J. Organomet. Chem.*, 1982, **232**, C65.
226. G. Fochi, G. Guidi and C. Floriani, *J. Chem. Soc., Dalton Trans.*, 1984, 1253.
227. A. S. Ward, E. A. Mintz and M. P. Kramer, *Organometallics*, 1988, **7**, 8.
228. E. A. Mintz, A. S. Ward and D. S. Tice, *Organometallics*, 1985, **4**, 1308.
229. E. A. Mintz and A. S. Ward, *J. Organomet. Chem.*, 1986, **307**, C52.
230. S. L. Buchwald and R. B. Nielsen, *J. Am. Chem. Soc.*, 1988, **110**, 3171.
231. S. L. Buchwald, R. B. Nielsen and J. C. Dewan, *J. Am. Chem. Soc.*, 1987, **109**, 1590.
232. P. B. Brindley and M. J. Scotton, *J. Organomet. Chem.*, 1981, **222**, 89.
233. D. W. Stephan, *J. Chem. Soc., Chem. Commun.*, 1991, 129.
234. D. W. Stephan, *Organometallics*, 1991, **10**, 2037.
235. D. E. Laycock and H. Alper, *J. Org. Chem.*, 1981, **46**, 289.
236. B. Gautheron, G. Tainturier, S. Pouly, T. Theobald, H. Vivier and A. Laarif, *Organometallics*, 1984, **3**, 1495.
237. P. Meunier, B. Gautheron and A. Mazouz, *J. Organomet. Chem.*, 1987, **320**, C39.
238. J. Bodiguel, P. Mennier and B. Gautheron, *Appl. Organomet. Chem.*, 1991, **5**, 479.
239. H. Köpf and T. Klapötke, *J. Organomet. Chem.*, 1986, **310**, 303.
240. T. Klapötke, H. Köpf and P. Gowik, *Polyhedron*, 1987, **6**, 1923.
241. P. Tavares *et al.*, *Transition Met. Chem. (London)*, 1992, **17**, 220.
242. H. Köpf and H. Balz, *J. Organomet. Chem.*, 1990, **387**, 77.
243. H. Balz, H. Köpf and J. Pickardt, *J. Organomet. Chem.*, 1991, **417**, 397.
244. H. Köpf and T. Klapötke, *J. Chem. Soc., Chem. Commun.*, 1986, 1192.
245. H. Köpf, *Angew. Chem., Int. Ed. Engl.*, 1971, **10**, 134.
246. G. Tainturier, M. Fahim and B. Gautheron, *J. Organomet. Chem.*, 1989, **362**, 311.
247. G. Tainturier, M. Fahim and B. Gautheron, *J. Organomet. Chem.*, 1989, **373**, 193.
248. M. Fahim and G. Tainturier, *J. Organomet. Chem.*, 1986, **301**, C45.
249. G. Erker, T. Mühlenbernd, R. Benn, A. Rufinska, G. Tainturier and B. Gautheron, *Organometallics*, 1986, **5**, 1023.
250. G. Erker, T. Mühlenbernd, R. Nolte, J. L. Petersen, G. Tainturier and B. Gautheron, *J. Organomet. Chem.*, 1986, **314**, C21.
251. B. Gautheron, G. Tainturier and S. Pouly, *J. Organomet. Chem.*, 1984, **268**, C56.
252. A. Shaver and J. M. McCall, *Organometallics*, 1984, **3**, 1823.
253. J. A. Kovacs and R. G. Bergman, *J. Am. Chem. Soc.*, 1989, **111**, 1131.
254. N. Klouras, *Monatsh. Chem.*, 1991, **122**, 35.
255. N. Albrecht and E. Weiss, *J. Organomet. Chem.*, 1988, **355**, 89.
256. M. Draganjac and T. B. Rauchfuss, *Angew. Chem., Int. Ed. Engl.*, 1985, **24**, 742.
257. P. H. Bird, J. M. McCall, A. Shaver and U. Siriwardane, *Angew. Chem., Int. Ed. Engl.*, 1982, **21**, 384.
258. A. Shaver, J. M. McCall, V. W. Day and S. Vollmer, *Can. J. Chem.*, 1987, **65**, 1676.
259. Z. Wang, S. Lu, H. Guo and N. Hu, *Polyhedron*, 1992, **11**, 1131.
260. N. K. Kaushik, M. S. Yadav, G. S. Sodhi and B. Bhushan, *J. Indian Chem. Soc.*, 1983, **60**, 894.
261. G. S. Sodhi and N. K. Kaushik, *Acta Chim. Acad. Sci. Hung.*, 1982, **111**, 207.
262. D. A. Femec, M. E. Silver and R. C. Fay, *Inorg. Chem.*, 1989, **28**, 2789.
263. M. E. Silver, O. Eisenstein and R. C. Fay, *Inorg. Chem.*, 1983, **22**, 759.
264. D. A. Femec, T. L. Groy and R. C. Fay, *Acta Crystallogr., Sect. C*, 1991, **47**, 1811.
265. C. P. Casey, R. F. Jordan and A. L. Rheingold, *J. Am. Chem. Soc.*, 1983, **105**, 665.
266. J. Ko, *Bull. Korean Chem. Soc.*, 1986, **7**, 334 (*Chem. Abstr.*, 1987, **107**, 7342).
267. C. P. Casey and F. Nief, *Organometallics*, 1985, **4**, 1218.
268. J. Ko, Bull. *Korean Chem. Soc.*, 1986, 7, 413 (*Chem. Abstr.*, 1987, **107**, 236 897).
269. J. Abys and W. M. Risen, Jr., *J. Organomet. Chem.*, 1981, **204**, C5.
270. E. G. Lundquist and K. G. Caulton, *Inorg. Synth.*, 1990, **27**, 26.

271. C. P. Casey, R. E. Palermo, R. F. Jordan and A. L. Rheingold, *J. Am. Chem. Soc.*, 1985, **107**, 4597.
272. B. E. Bursten and K. J. Novo-Gradac, *J. Am. Chem. Soc.*, 1987, **109**, 904.
273. K. W. Chiu, G. Wilkinson, M. Thornton-Pett and M. B. Hursthouse, *Polyhedron*, 1984, **3**, 79.
274. C. Blandy, D. Gervais and M. S. Cardenas, *J. Mol. Catal.*, 1986, **34**, 39.
275. H. Plenio and H. W. Roesky, *Z. Naturforsch., Teil B*, 1988, **43**, 1575.
276. H. Plenio, H. W. Roesky, F. T. Edelmann and M. Noltemeyer, *J. Chem. Soc., Dalton Trans.*, 1989, 1815.
277. H. Köpf and N. Klouras, *Chem. Scr.*, 1982, **19**, 122 (*Chem. Abstr.*, 1982, **97**, 92 469).
278. L. G. Kuz'mina, A. I. Yanovskii, Y. T. Struchkov, M. K. Minacheva and E. M. Brainina, *Koord. Khim.*, 1985, **11**, 116 (*Chem. Abstr.*, 1985, **102**, 220 974).
279. B. V. Lokshin, Z. S. Klemenkova, M. K. Minacheva, G. I. Timofeeva and E. M. Brainina, *Izv. Akad. Nauk SSSR, Ser. Khim.*, 1985, 2599.
280. M. K. Minacheva, O. A. Mikhailova, Z. S. Klemenkova, B. V. Lokshin and G. I. Timofeeva, *Metalloorg. Khim.*, 1991, **4**, 1282 (*Chem. Abstr.*, 1992, **116**, 83 829).
281. W. Lasser and U. Thewalt, *J. Organomet. Chem.*, 1984, **275**, 63.
282. E. M. Brainina, M. K. Minacheva, Z. S. Klemenkova, B. V. Lokshin and O. A. Mikhailova, *Izv. Akad. Nauk SSSR, Ser. Khim.*, 1987, 1848.
283. C. J. Jones, J. A. McCleverty and A. S. Rothin, *J. Chem. Soc., Dalton Trans.*, 1985, 405.
284. U. Thewalt and W. Lasser, *Z. Naturforsch., Teil B*, 1983, **38**, 1501.

11
Zirconium and Hafnium Complexes in Oxidation State +4

J. JUBB, J. SONG, D. RICHESON and S. GAMBAROTTA
University of Ottawa, ON, Canada

11.1 MONOCYCLOPENTADIENYL COMPLEXES

11.1.1 Monodentate Ligands

11.1.1.1 Synthesis, structure and bonding

Mono(Cp) complexes of zirconium possess reactivity profiles which differ considerably from those of their bis(Cp) counterparts and, therefore, it is not surprising that these complexes have assumed an increasingly important role in synthetic[1] and mechanistic[2] endeavours.

Despite the intensifying interest in these compounds, there are only a few practical routes for the preparation of monocyclopentadienylzirconium and -hafnium starting materials. One of the most widely used procedures for the preparation of $ZrCl_3Cp$ is the selective removal of one Cp ligand from $ZrCl_2Cp_2$, via photoinduced free-radical chlorination of zirconocene dichloride (Scheme 1).[3] The reaction is carried out in $CHCl_3$ at room temperature, and is initiated by the photochemical formation of transient $ZrCl_2Cp(\eta^4\text{-}C_5H_5Cl)$. This species is consumed by molecular chlorine to form $ZrCl_3Cp$ and a chain-carrying radical. The other major product of this reaction is 1,2,3,4,5-pentachlorocyclopentadiene. The $ZrCl_3Cp$ obtained by this method is insoluble in common noncoordinating solvents. However, it readily dissolves in benzene or chloroform upon addition of a Lewis base such as ether, THF, NEt_3 or pyridine in excess of 2 molar equiv., which forms the corresponding $ZrCl_3(L)_2Cp$ adducts. Although this procedure is characterized by good yields and can be extended to fairly large-scale preparations of $ZrCl_3Cp$, the method is somewhat exacting and it can on occasions give varying amounts of $ZrCl_4$ as a consequence of overchlorination.[4]

$$ZrCl_2Cp_2 \xrightarrow[h\nu]{[Cl]^\bullet} ZrCl_2(\eta^4\text{-}C_5H_5Cl)Cp \xrightarrow{Cl_2} ZrCl_3Cp + C_5Cl_5H + [Cl]^\bullet$$

Scheme 1

A convenient alternative to the above method is provided by the one-pot synthesis (69–74%) of $ZrCl_3Cp(DME)$.[5] The reaction, performed by treating $ZrCl_4$ with dimethyl sulfide (2 equiv., CH_2Cl_2, 0 °C) and exposing the resulting mixture to 1 equiv. of trimethylsilylcyclopentadiene, forms a new product, presumably $ZrCl_3Cp \cdot 2\ SMe_2$, after only a few minutes. Addition of DME (5 equiv.) affords the white crystalline $ZrCl_3Cp(DME)$ which was characterized by NMR spectroscopic[5] and x-ray crystallographic methods. Mixing equimolar portions of $ZrCl_4$ and $[Zr(Cp')_2Cl_2]$ in toluene, at ambient temperature, rapidly gave a precipitate of $ZrCp'Cl_3$ ($Cp' = Cp$, $C_5H_4Bu^t$ or $C_5H_3Bu^t_2$) in high yield; addition of THF to $Cp' = C_5H_4Bu^t$ or $C_5H_3Bu^t_2$ at ~20 °C gave $[Zr(Cp')_2Cl_2]$ and $[ZrCl_4(THF)_2]$.[7]

Reaction of $HfCl_4(SMe_2)_2$ with $SnBu^n_3Cp$ followed by addition of DME yields $HfCl_3Cp(DME)$ in 55–64% yield.[5] The crystal structure of this compound has shown the same distorted octahedral geometry as the zirconium derivative.[6]

An alternative synthetic route to $ZrCl_3Cp$ consists of the reaction of $ZrCl_4$ with $MgCp_2$ as a ligand-transfer reagent (Equation (1)).[8]

$$ZrCl_4\ +\ 0.5\ Mg(C_5H_3R_2)_2 \quad \xrightarrow[\text{xylene}]{100\ °C,\ 30\ min} \quad ZrCl_3(C_5H_3R_2)\ +\ 0.5\ MgCl_2 \qquad (1)$$

The ability of $ZrCl_3Cp$ to coordinate Lewis bases is well documented. While reaction of $ZrCl_3Cp$ with excess THF gives $ZrCl_3Cp(THF)_2$ in 67% yield, addition of more than 2 equiv. of DMF affords $ZrCl_3Cp(DMF)_2$ in 98% yield.[9] The x-ray crystal structure of both complexes shows the same distorted octahedral coordination geometry around the zirconium atom.

11.1.1.2 Methylated cyclopentadienyl derivatives

The permethylated cyclopentadienyl derivatives MCl_3Cp^* ($M = Zr$ or Hf)[10] have been prepared in 65% yield from the stoichiometric reaction of $LiCp^*$ with MCl_4.[10] The green-yellow complexes are purified by sublimation (140 °C, 0.001 torr) and exhibit a single resonance in the ^1H NMR spectrum ($CDCl_3$) at δ 2.28. However, reaction of $LiCp^*$ and MCl_4 in refluxing toluene[11] yields white solids which were purified by sublimation and which show the Cp^* ^1H NMR resonance (C_6D_6) at δ 1.93.

The interpretation of the photoelectron spectra[12] of these compounds on the basis of the geometry, character of the valence orbitals, shift effects and intensity arguments shows that the mixing between the Cp ligand and d-orbitals of the same symmetry is strong. When compared with the corresponding titanium spectrum, the shifts of the first two halogen bands in the zirconium and hafnium spectra are as expected, given the larger radii of the two heavier elements which force the halide ions further apart and lower the energy of the a_2- and e-orbital contributions.

The monomethylated cyclopentadienyl complex $ZrCl_3(THF)_2(\eta\text{-}MeCp)$ is obtained in 52% yield when equimolar amounts of $ZrCl_4(THF)_2$ and methylcyclopentadienyllithium are reacted at 0 °C in toluene.[9] The x-ray structure[9] of $ZrCl_3(THF)_2(MeCp)$ has revealed the zirconium to be in a distorted octahedral geometry, one coordination site being occupied by the Cp centroid.

Reaction of MCl_3Cp^* ($M = Zr$ or Hf) with 1 equiv. of $LiE(TMS)_3(THF)_3$ ($E = Si$ or Ge) affords the corresponding orange crystalline $[MCl_2\{E(TMS)_3\}Cp^*]$ in 71% and 75% yield for zirconium and hafnium, respectively (Equation (2)).[13] When in the solid state, both compounds are stable indefinitely under nitrogen, while they are pyrophoric in moist air. Exposure of solutions of the silyl derivative to normal room lighting also results in a rapid decomposition. The zirconium silyl complex readily coordinates donor ligands L ($L = PMe_3$, THF or pyridine) in solution to form dark-red adducts $[ZrCl_2\{Si(TMS)_3\}L_xCp^*]$. Similarly, the hafnium analogue reacts readily with pyridine to afford orange crystals (80%) of $[HfCl_2\{Si(TMS)_3\}(pyridine)Cp^*]$. Structural characterization of these complexes shows the same distorted octahedral coordination environment for both metals.[13] The tris(trimethylsilyl)-germyl complex decomposes more slowly in the solid state than the silyl analogue, and it is photolytically and thermally stable at ambient temperatures.

$$ZrCl_3Cp^*\ +\ LiE(TMS)_3(THF)_3 \quad \xrightarrow{Et_2O} \qquad \qquad (2)$$

$$E = Si\ or\ Ge$$

11.1.1.3 Silylated cyclopentadienyl derivatives

Despite the fact that convenient syntheses are available for the TiCl$_3$\{1,3-(TMS)$_2$C$_5$H$_3$\} compounds,[14-17] no facile procedures exist for the preparation of the corresponding zirconium analogue (cf. reference 7). The triboluminescent compound 1,3-bis(trimethylsilyl)cyclopentadienylzirconium trichloride is formed in 73% yield upon reaction of tris(trimethylsilyl)cyclopentadiene with ZrCl$_4$ in CH$_2$Cl$_2$, followed by sublimation (Equation (3)).[18] The complex, obtained as a moderately air-sensitive white powder, is soluble only in relatively polar solvents, for example, acetone or THF, thus implying a polymeric formulation. A similar reaction of HfCl$_4$ with silylated cyclopentadienes has been examined.[18] While with (trimethylsilyl)cyclopentadiene, the mono-Cp complex HfCl$_3$\{1,3-(TMS)$_2$C$_5$H$_3$\} is isolated in 73% yield, in contrast, reaction with either the bis(trimethylsilyl)cyclopentadiene or tris(trimethylsilyl)cyclopentadiene affords the corresponding metallocene dichloride [HfCl$_2$-\{η-C$_5$H$_4$(TMS)\}$_2$] in high yield. This method is a convenient alternative to the MgCp′$_2$ route (Equation (1)),[8] since it is experimentally simpler, entails fewer manipulations or air-sensitive reagents and it gives a similar yield. The only significant drawback of this preparation lies in the time-consuming preparation of (TMS)$_3$CpH.

$$
\text{TMS-C}_5\text{H}_3(\text{TMS})_2 + \text{ZrCl}_4 \xrightarrow{-\text{TMS-Cl}} \text{(TMS)}_2\text{C}_5\text{H}_3\text{ZrCl}_3 \tag{3}
$$

11.1.1.4 Carbon-donor ligands

ZrCl$_3$Cp is an excellent substrate for the preparation of alkyl derivatives via reaction with alkyllithium or Grignard reagents. For example, ZrAr$_3$Cp complexes (Ar = phenyl, *p*-tolyl or *m*-tolyl) are formed by reaction of an ethereal suspension of ZrCl$_3$Cp with the appropriate aryllithium reagent at −30 °C. The resulting pale-yellow ZrAr$_3$Cp·Et$_2$O may be obtained in up to 58% yields upon crystallization from toluene–pentane mixtures.[3]

When HfCl$_3$Cp* is treated with 3 equiv. of MeMgBr, HfMe$_3$Cp* can be isolated (86% yield) as an off-white pentane-soluble solid.[19] Unlike ZrCl$_3$Cp*, which is thermally unstable, HfMe$_3$Cp* shows no decomposition at room temperature over several days in solution. Reaction of HfMe$_3$Cp* with 3 equiv. of C$_6$F$_5$OH gave colourless Hf(OC$_6$F$_5$)$_3$Cp* in 62% yield.[19] Calorimetric measurements of metal–ligand bond disruption enthalpies, D(M–R), determined for ZrMe$_3$Cp*, HfMe$_3$Cp* and Hf(OC$_6$F$_5$)$_3$Cp*[19] show values for HfMe$_3$Cp* (290.5, 296.8 and 305.2 kJ mol^{-1}) slightly greater in magnitude than those of ZrMe$_3$Cp* (270.4, 283.4 and 304.7 kJ mol^{-1}), thus indicating a higher stability of the Hf–C bond with respect to zirconium.

The alkynide–vinyl complex [ZrCl(C≡CBu^t)\{CH=CH(Bu^t)\}(dmpe)Cp] has been obtained by reaction of the zirconium(II) precursor [Zr(dmpe)$_2$ClCp] with the bulky alkyne HC≡CBu^t in ether at room temperature.[20] In this unusual reaction, a hydrogen transfer takes place between two molecules of alkyne rather than the usual oxidative coupling with consequent formation of a metallacycle (Scheme 2). The presence of the –C≡CBu^t and –C(H)=C(H)Bu^t groups is diagnosed by the two intense IR bands at 2070 cm^{-1} and 1550 cm^{-1}. The x-ray structure shows a nearly linear alkynide fragment, the vinyl group bearing the two hydrogens in *trans* positions. Metallacyclopentadiene derivatives have been obtained in the case of less bulky alkynes.

11.1.1.5 Applications

A number of electron-deficient organometallic systems having sufficiently strong metal–carbon bonds are known to be potential candidates as Lewis acidic catalysts in organic synthesis.[21] The d^0 16-electron complex ZrCl$_3$Cp(THF)$_2$ selectively catalyses[9] the Diels–Alder addition of 2,3-dimethylbutadiene and methacrolein at ambient temperature to give 1,3,4-trimethyl-3-cyclohexenyl-1-carbaldehyde in 65% yield. However, the subsequent ZrCl$_3$Cp-catalysed rearrangement to 1,3,4-trimethylbicyclo[2.2.1]heptan-2-one does not take place at a significant rate until 150 °C (Scheme 3).

An efficient mono-Cp zirconium-based catalyst for the epoxidation of cyclohexene (22% in 18 h) is obtained when ZrCl$_3$Cp is incorporated into a polymeric support (polystyrene–divinylbenzene) via

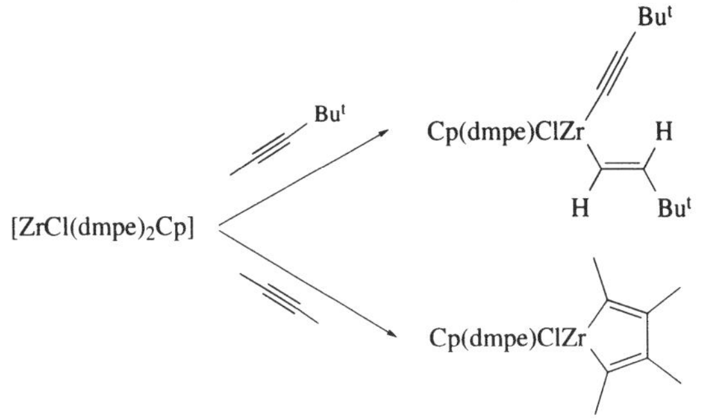

Scheme 2

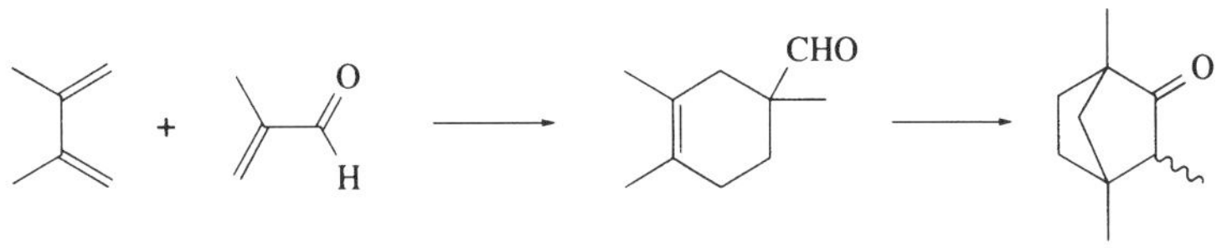

Scheme 3

reaction of $ZrCl_4$ with the LiCp-functionalized polymer (Scheme 4).[22] Hafnium has also been investigated for similar catalytic activities (catalytic isomerization of allylbenzene, *cis*-stilbene and 1,5-cyclooctadiene).[22] The polymer-attached mono-Cp metal centre was found to be a more efficient catalyst than the corresponding polymer-attached analogues of $HfCl_2Cp_2$ and $HfClCp_3$. Polymer-attached $HfCl_3Cp$ catalyses the epoxidation of cyclohexene with *t*-butyl hydroperoxide.

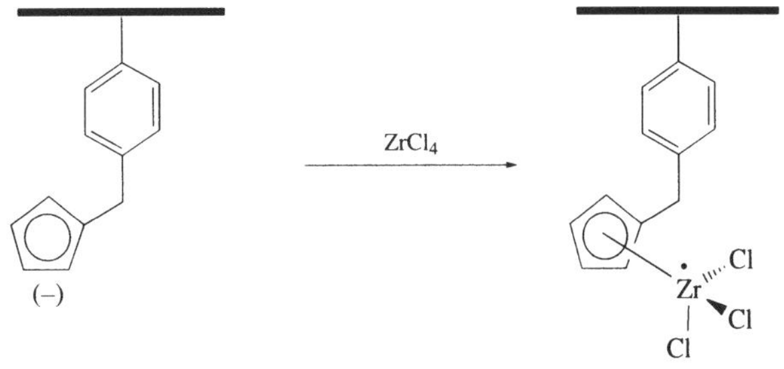

Scheme 4

11.1.2 Bidentate Ligands

11.1.2.1 Nitrogen-donor ligands

The air-sensitive cyclopentadienyl(trimethylhydrazido)dichlorozirconium, $[ZrCl_2(NMeNMe_2)Cp]$, has been prepared in 90% yield by the slow reaction of equimolar quantities of $ZrCl_3Cp$ and trimethylsilyltrimethylhydrazine in diethyl ether at room temperature.[23] The ^1H NMR spectrum exhibits two Cp resonances, and three resonances due to hydrazido methyl groups. The inequivalence of the methyl protons can be accounted for on the basis of a mixture of slowly interconverting side-on and end-on isomers (with respect to the $NMeNMe_2$ moiety).

Mixed tris- and bis(pyrazolyl)boratomonocyclopentadienylzirconium(IV) complexes are readily prepared at $-78\ ^\circ C$ in CH_2Cl_2 as white solids (Equation (4)).[24] These thermally stable but air-sensitive complexes, although isoelectronic with $[ZrCl_2Cp_2]$, are not of sufficient stability to function as precursors for the preparation of hydrido or alkyl derivatives. In order to prepare a similar but more reactive complex with two electrons fewer at the metal centre, a reaction with the bis(pyrazolyl)borate was carried out at high concentrations.[24] The x-ray crystal structure[24] of the resulting colourless crystals obtained from CH_2Cl_2/hexane demonstrated that in the solid state there is a three-centre two-electron B–H–Zr interaction as expected from the presence of bands at 2530 cm^{-1} (v(B–H) terminal) and 2160 cm^{-1} (v(B–H) three-centre) in the IR spectrum. Boron-11 NMR spectra indicate that this interaction is also retained in solution (double doublets with J_{B-H} = 88 and 118 Hz). Owing to the three-centre bond, the $[H_2B(pz)_2]^-$ ligand donates as many electrons to the zirconium as the tris(pyrazolyl)borato ligands. However, the two complexes differ in that the three-centre Zr–H–B bond is a potential site of reactivity. While no appreciable reactivity is observed when $[ZrCl_2Cp\{HB(pz)_3\}]$ is treated with RNC, reaction of $[ZrCl_2Cp\{H_2B(pz)_2\}]$ with Bu^tNC yields a crude beige product $[ZrCl_2Cp\{H_2B(pz)_2\}(NCBu^t)]$ in which the isocyanide has displaced the bridging hydrogen. This result is confirmed by the ^{11}B NMR spectrum showing a triplet ($J_{B-H} = 95$ Hz) for the BH protons.

$$ZrCl_3(DME)Cp + Na[HB(pz)_3] \xrightarrow{CH_2Cl_2} \qquad \xrightarrow{Bu^tNC} [ZrCl_2Cp\{HB(pz)_3\}(NCBu^t)] \qquad (4)$$

11.1.2.2 Sulfur-donor ligands

Stereochemical nonrigidity (fluxional behaviour) is an ubiquitous feature of the chemistry of seven-coordinate complexes because the three common seven-coordinate polytopes (the pentagonal bipyramid, capped octahedron and capped trigonal prism) are easily interconverted by relatively small ligand displacements.[25–7] In this respect, dithiocarbamatomonocyclopentadienylzirconium complexes $[Zr(R_2dtc)_3Cp]$ and $[Zr(R_2dtc)_3(MeCp)]$ (dtc = dithiocarbamate) are of considerable interest since metal-centred rearrangement in these compounds is slow on the NMR timescale at room temperature and therefore enables detailed spectroscopic studies on molecular dynamics to be carried out.[28] The heptacoordination of zirconium in these compounds is supported by the crystal structure of $[Zr(Me_2dtc)_3Cp]$,[28] showing a pentagonal-bipyramidal structure with the Cp occupying an axial position. The complex $[Zr(Me_2dtc)_3Cp]\cdot CH_2Cl_2$ is prepared by reacting $ZrCl_2Cp_2$ with $Na(Me_2dtc)$ in refluxing CH_2Cl_2 for 18 h,[28,29] and is isolated as colourless crystals after repeated recrystallization of the crude brown product from CH_2Cl_2/hexane. This complex is soluble in chlorinated hydrocarbons, less soluble in aromatic hydrocarbons, acetonitrile or CS_2, and insoluble in saturated hydrocarbons. It is stable in the solid state and at room temperature but decomposes slowly in solution at elevated temperatures. The stereochemical rearrangements of pentagonal-bipyramidal $[Hf(Me_2dtc)_3Cp]$ have been investigated by variable-temperature ^1H NMR spectroscopy.[29] The spectra exhibit evidence of the simultaneous presence of three different kinetic processes: (i) exchange of dithiocarbamato methyl groups within the equatorial ligand; (ii) exchange of methyl groups within one ligand; and (iii) exchange of equatorial and unique ligands. The rates of (i) and (ii) are the same within a factor of two. Processes (i) and (ii) are believed to involve rotation about the C–N bond in the dithiocarbamato ligands, perhaps preceded by the rupture of an equatorial metal–sulfur bond.

The Cp–diarylthiobenzamidatozirconium(IV) complexes $[ZrAr_2\{N(R)C(=S)Ar\}Cp]$ (Ar = *p*-tolyl; R = Bz or 1-phenylethyl) are formed by the reaction of $[Zr(p\text{-}tolyl)_3Cp]$ with benzyl or 1-phenylethyl isothiocyanate and can be isolated as yellow products in 30% and 35% yields, respectively (Scheme 5).[30] An NMR spectral analysis has shown that these compounds have achiral molecular geometries. By contrast, the reaction of $[Zr(p\text{-}tolyl)_3Cp]$ with 2 equiv. of benzyl isothiocyanate gives a dark-yellow chiral product in 49% yield. Presumably the different coordination modes of the substituted thiobenzamidato ligands are responsible for the metal chirality observed for this complex. The existence of a rapid intramolecular degenerate rearrangement ($\Delta G^{\ddagger}_{50} = 67$ kJ mol^{-1}), proceeding with loss of metal chirality, has also been shown by NMR spectral studies.[30]

The toluene-3,4-dithiolate chelates $[ZrCl\{4\text{-}MeC_6H_3(1,2\text{-}S)_2\}Cp]$ are synthesized by reaction of $ZrCl_3Cp$ with an equimolar quantity of $(1,2\text{-}LiS)_2C_6H_3Me\text{-}4$ in hexane to give the deep-orange product in 62% yield (Equation (5)).[31]

$$ZrCl_3Cp + \qquad \xrightarrow{-2\,LiCl} \qquad (5)$$

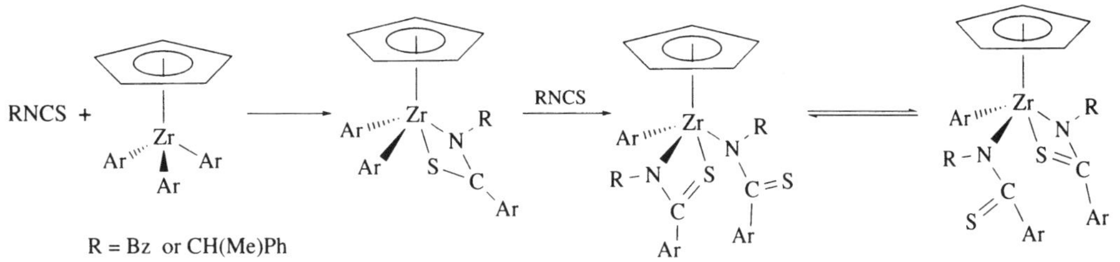

Scheme 5

Room-temperature reaction of $ZrCl_2Cp_2$ in THF with 3 equiv. of thiosemicarbazone TSCH in the presence of 2 equiv. of *n*-butylamine slowly forms (24–30 h) brown-yellow diamagnetic complexes in 60–65% yield (Equation (6)).[32] (TSCH is the monobasic thiosemicarbazone derived from the condensation of thiosemicarbazide with cinnamaldehyde, benzaldehyde, acetophenone or *p*-hydroxyacetophenone.) These compounds, formed through a rather complicated reaction, are insoluble in all common organic solvents, susceptible to hydrolysis and decompose on heating at ca. 182–300 °C.

$$ZrCl_2Cp_2 \ + \ 2 \ TSCH \ \xrightarrow{Bu^nNH_2} \qquad\qquad\qquad (6)$$

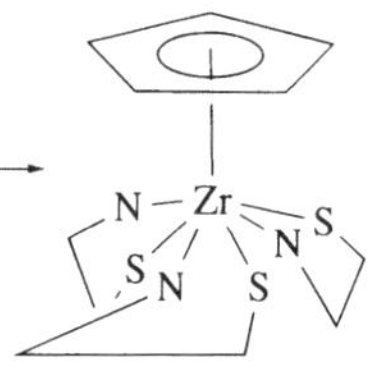

On the basis of elemental analysis, conductance measurements, electronic and IR spectroscopy, the structure shown in Equation (6) was proposed with the thiosemicarbazone coordinating in the enolic form. Electronic spectra in nujol show a single absorption in the region 23 000–23 800 cm^{-1}, which is assigned to a charge-transfer transition.

Similar reactions of $ZrCl_2Cp_2$ with ligands containing a thioamide function, such as heterocyclic thiones, oxadiazolethione and thiohydantoin, form complexes capable of undergoing thione $\rightleftharpoons$ thiol ($NH=C=S \rightleftharpoons N=C-SH$) tautomerism (Scheme 6) so that the bonding to the metal atom may be realized through either nitrogen or sulfur, or through both simultaneously.

$$ZrCl_2Cp_2 \ + \ 3 \ S \ \xrightarrow[2 \ BuNH_2]{THF} \ Zr(S)_3Cp$$

S = 5-substituted 1,3,4-oxadiazole-2-thiones

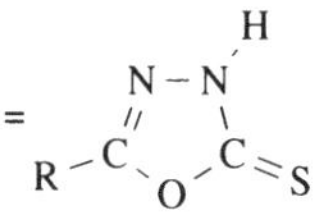

R = Ph	BOxtH (54%)
= *m*-C$_6$H$_4$NO$_2$	MOxtH (58%)
= *p*-C$_6$H$_4$NO$_2$	POxtH (60%)
= *o*-C$_6$H$_4$Cl	COxtH (60%)

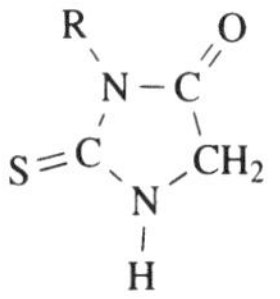

S = 1-aryl-2-thiohydantoins

R = Ph (52%)
= C$_6$H$_4$Me (58%)
= Bz (52%)
= *p*-MeOC$_6$H$_4$ (60%)
= *o*-MeOC$_6$H$_4$ (62%)
= C$_{10}$H$_7$ (60%)

Scheme 6

All these brown complexes are diamagnetic, nonelectrolytes, susceptible to hydrolysis and decompose upon heating in the range 148–237 °C.

11.1.2.3 Oxygen-donor ligands

Monocyclopentadienylzirconium catecholate derivatives $ZrCl(O_2Ar)Cp$ may be prepared by reaction of $ZrCl_3(DME)Cp$ with catechol according to Equation (7).[33]

$$ZrCl_3(DME)Cp + \underset{R = H, Me\ or\ Bu^t}{\text{(catechol)}} \xrightarrow{\ toluene\ } ZrCl(O_2Ar)Cp + 2\,HCl + DME \qquad (7)$$

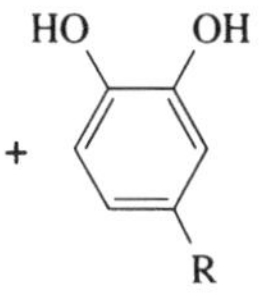

These products are probably polymeric, as suggested by their insolubility in most common solvents, and are obtained in 75–80% yield as white solids. Although the poor solubility has prevented detailed NMR spectroscopic characterization; data obtained in CD_3OD gave evidence for the existence in solution of a monomeric solvated species and confirmed the presence of the various components.

Formation of monocyclopentadienylzirconium derivatives by removal of one cyclopentadienyl ring is achieved from both $ZrCl_2Cp_2$ and $ZrCl_2(MeCp)_2$ upon reaction with 3 equiv. of various straight-chain monocarboxylic acids (palmitic $C_{16}H_{32}O_2$, stearic $C_{18}H_{36}O_2$ or behenic $C_{22}H_{44}O_2$) in CH_2Cl_2. The reaction yields highly soluble, moisture-sensitive, white or yellow diamagnetic complexes of the formula $ZrAc_3Cp'$ (Ac = carboxylate; Cp' = Cp or MeC_4H_4) in near-quantitative yield.[34] Infrared spectral data[34] suggest that the carboxylates are bonded in a bidentate manner to zirconium.

This synthetic pathway is rather versatile and can be employed to anchor the CpZr moiety to other functionalized hydroxy derivatives such as 4-acyl-3-methyl-1-phenyl-2-pyrazoline-5-ones (AcMPPOH) (Ac = acetyl, propionyl, benzoyl or *p*-chlorobenzoyl) (Equation (8)).[35] The yellow complex, obtained in 55% yield, is soluble in common organic solvents and can be purified by recrystallization. Although the absence of IR absorptions in the 1675–1655 cm^{-1} region indicates a bidentate ligand, the spectral evidence is not sufficiently conclusive for either a *cis* or *trans* arrangement of the two ligands to be established.

$$ZrCl_2Cp_2 + \text{(pyrazolone)} \xrightarrow{\ Et_3N\ } ZrCl(OPPMAc)_2Cp_2 + HCl \qquad (8)$$

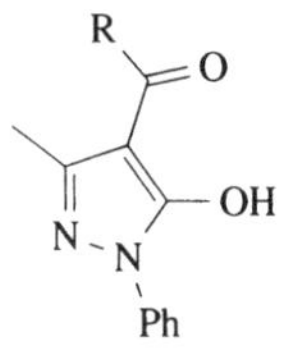

The same types of complexes $ZrCl(L)_2Cp$ (LH = 8-hydroxyquinoline *N*-oxide, 5-nitro-, 5,7-dinitro- or 5-phenylazo-8-quinoline *N*-oxide) have been prepared in good yield (57–62%) in benzene from $ZrCl_2Cp_2$ by using a procedure similar to that for the quinoline derivative.[36] These poorly soluble yellow complexes have been characterized by elemental analysis, IR and ^1H NMR spectra and were found to contain the quinoline ligands coordinated in a bidentate manner.

11.1.3 Polymetallic Systems

11.1.3.1 Oxoclusters

The highly reactive, hydrocarbon-soluble, triangular cluster $[(ZrClCp^*)_3(\mu_3\text{-}O)(\mu_3\text{-}OH)\text{-}(\mu\text{-}OH)_3]\cdot 2\,THF^{37}$ has been prepared in 61% yield upon reaction in THF of $ZrCl_3Cp^*$ with 2 equiv. of H_2O in the presence of 2 equiv. of triethylamine (Scheme 7). An x-ray structure[37] has shown the presence of a cluster containing a nearly equilateral triangular core of three $ZrClCp^*$ units capped on the two sides of the trimetallic plane by μ_3-O and μ_3-OH groups and bridged by three symmetrically placed bridging μ_2-OH units. When the hydrolysis of $ZrCl_3Cp^*$ with a smaller amount of water was carried out in toluene for 12 h at 80 °C, a yellow trinuclear complex $[(ZrClCp^*)_3(\mu\text{-}Cl)_4(\mu_3\text{-}O)]$ was obtained in ca. 70% yield.[38] This compound can also be synthesized by the other routes shown in Scheme 7.[38] The structure[38] consists of three $ZrClCp^*$ fragments with the three zirconium atoms at the vertices of an isosceles triangle. The two longest edges are bridged by Cl^-, a double Cl^- bridge is on the shortest edge, and O^{2-} caps the zirconium triangle. Proton and ^{13}C NMR spectra indicate that a dynamic process may be taking place in solution that interchanges the terminal and bridging Cl^- ligands.

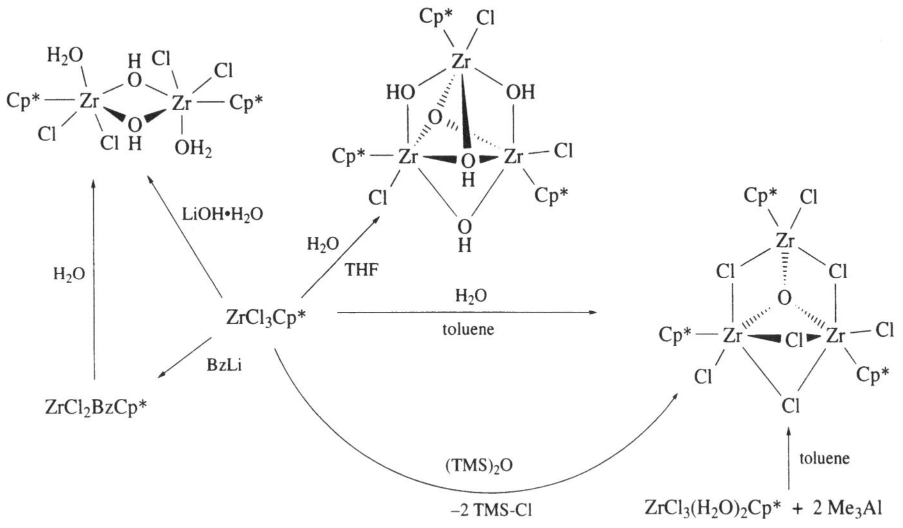

Scheme 7

The dinuclear monocyclopentadienyl (hydroxo-bridged) complex [{ZrCl$_2$Cp*(OH$_2$)(μ-OH)}$_2$], which may be regarded as the first product arising in the hydrolytic sequence from ZrCl$_3$Cp* during the formation of [(ZrClCp*)$_3$(μ_3-O)(μ_3-OH)(μ-OH)$_3$]·2 THF, can be obtained in 45–50% yield as shown in Scheme 7. The dinuclear structure of this compound has been confirmed by its x-ray crystal structure,[38] showing that the molecule is composed of two ZrCl$_2$Cp*(OH)(OH$_2$) units connected via the bridging hydroxy groups.

Alcoholysis of ZrMe$_3$Cp* with *cis*-[PtMe$_2$(PPh$_2$CH$_2$OH)$_2$] has been successfully used for the preparation of the coordinatively unsaturated early/late heterobimetallic complex *cis*-[Zr(Me)Cp*(μ-OCH$_2$Ph$_2$P)$_2$PtMe$_2$].[39] Exchange of methyl groups between the two transition metals was indicated by the thermolysis of the deuterated analogue [Zr(CD$_3$)Cp*(μ-OCH$_2$Ph$_2$P)$_2$PtMe$_2$] which produced [Zr(Me)Cp*(μ-OCH$_2$Ph$_2$P)$_2$PtMe(CD$_3$)]. The methyl-exchange mechanism was elucidated by a series of labelling, kinetic and cross-over studies, some involving *cis*-[ZrCp*(CD$_3$)-(μ-OCH$_2$CH$_2$Ph$_2$P)$_2$PtMe$_2$] isotopomers.

Treatment of ZrMe$_3$Cp* with 2 equiv. of HOCH$_2$Ph$_2$P produces [Zr(OCH$_2$Ph$_2$P)$_2$MeCp*] as a clear oil which can be induced to precipitate as a microcrystalline material (80–90% yield) upon exposure of the solution to trace amounts of air.[40] Phosphorus-31 NMR spectroscopy provides evidence that the phosphines are not chelated to zirconium since the complex shows a sharp resonance at δ −11.24, which compares well with that of the free phosphine (δ −11.84). Furthermore, when [Zr(OCH$_2$Ph$_2$P)$_2$MeCp*] was reacted with 0.5 equiv. of [{Rh(μ-Cl)(CO)$_2$}$_2$] in toluene at −78 °C and the mixture was allowed to warm up to room temperature, the μ_2,η^2-acetyl 'A-frame' complex [{ZrCp*(μ-OCH$_2$Ph$_2$P)$_2$-(μ_2,η^2-O=CMe)}(μ-Cl)Rh(CO)] was obtained as yellow, microcrystalline flakes in 45–65% yield (Equation (9)).

$$[\text{Zr(OCH}_2\text{Ph}_2\text{P)}_2\text{Cp*Me}] + 0.5\,[\{\text{Rh}(\mu\text{-Cl})(\text{CO})_2\}_2] \xrightarrow[\text{toluene}]{25\,^\circ\text{C}} [\{\text{ZrCp*}(\mu\text{-OCH}_2\text{Ph}_2\text{P})_2(\mu_2,\eta^2\text{-O=CMe})\}(\mu\text{-Cl})\text{Rh(CO)}] +$$

$$[\text{ZrCp*}(\mu\text{-OCH}_2\text{Ph}_2\text{P})_2(\mu\text{-Cl})_2\text{Rh(CO)}] \qquad (9)$$

The complex [{ZrCp*(μ-OCH$_2$Ph$_2$P)$_2$(μ_2,η^2-O=CMe}(μ-Cl)Rh(CO)] was sparingly soluble in CH$_2$Cl$_2$, CHCl$_3$ or THF and virtually insoluble in hydrocarbons. Structural elucidation was aided by the preparation of the ^{13}C isotopomers. Infrared and NMR spectral data indicated a trigonal-bipyramidal geometry around the rhodium atom. The reactivity of this bimetallic A-frame compound is summarized in Scheme 8. The crystal structure of [ZrCp*(μ-OCH$_2$Ph$_2$P)$_2$RhMe$_2$] has shown the presence of a short Rh–Zr bond of 0.244 4(1) nm, one of the shortest ever observed. Extended Hückel calculations revealed that the Zr–Rh interaction is 47% σ and 53% π in character, providing a strong indication of multiple bonding.[40]

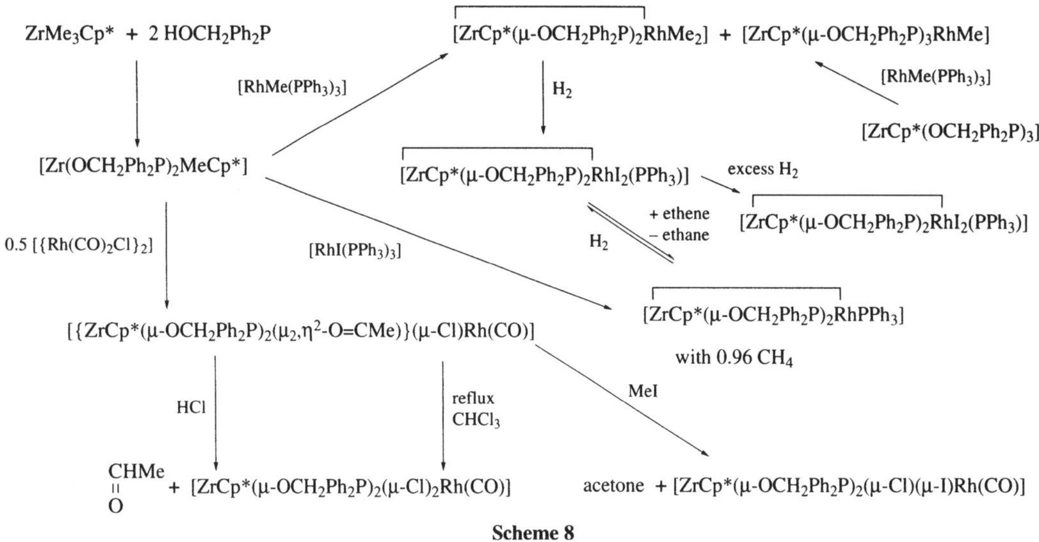

Scheme 8

11.1.3.2 Bridging fulvalene and bridging Cp derivatives

A general synthetic pathway to a series of *trans*-fulvalene complexes of zirconium(IV) has been achieved[41] with the oxidation of the zirconium(II) complex [Zr(PR$_3$)$_2$Cp$_2$] with a variety of oxidizing agents including PR$_3$X$_2$ (X = Cl, Br or I), RN$_3$ and RSSR. The reaction initially forms the zirconium(III) species *cis*-[Zr(PMe$_3$)(η^1:η^5-C$_5$H$_4$)Cp]$_2$ which reacts further to form a series of zirconium(IV) fulvalene derivatives (Scheme 9). In all the reactions the fulvalene ligand is formed by the reductive coupling of the two bridging η^1:η^5-C$_5$H$_4$ rings. The zirconium(III) intermediates may be isolated and characterized by employing only 1 equiv. of oxidant.[41,42] Oxidation carried out with an organic azide yields an intensely blue dinuclear N-TMS^{2-} zirconium(IV) compound, having two ZrCp units bridged by one N-TMS^{2-} moiety and by two η^1:η^5-C$_5$H$_4$ groups.[43] The reaction of the latter with PhSSPh yields a new zirconium(IV) complex in which the dinuclear unit has one bridging imido and one η^1:η^5-C$_5$H$_4^-$ ligand. Reaction of the bridged zirconium(III) halide with Bu^{t}NC forms a curious oxidative addition product where the two zirconium(IV) centres are asymmetrically bonded to the isocyanide group (Scheme 9).[44] A dinuclear monocyclopentadienyl(fulvalenyl)(oxo)zirconium complex [{CpZrCl(μ-O)}$_2$(μ,η^5:η^5-C$_{10}$H$_8$)] is obtained[45] as a brown solid when the oxidation of the dinuclear zirconium(III) complex is carried out with [ReO$_3$Cp*]. During this reaction the [ReO$_3$Cp*] transfers O^{2-} ligands to the organozirconium compound in a fast, clean metathesis reaction in which the zirconium(III) is oxidized to the corresponding μ-oxozirconium(IV) derivative. The same compound may be obtained upon reaction of the halide-bridged fulvalene complex with O$_2$.[46] The complex [(ZrCl$_2$Cp)$_2$(μ,η^5:η^5-C$_{10}$H$_8$)] is an excellent starting material for alkylation reactions.[41,44,47]

11.2 TRIS(CYCLOPENTADIENYL) COMPOUNDS

Monomeric [ZrHCp$_3$] can be prepared[48,49] by the reaction of ZrCp$_4$ with LAH in THF solution (Equation (10)). The deuterated analogue can be prepared in a similar manner by using Li[AlD$_4$]. The IR (4000–100 cm^{-1}) and Raman spectra (Table 1) indicate that the three Cp rings are η^5-coordinated. No crystal structure has been reported for this species.

$$[ZrCp_4] + LAH \longrightarrow [ZrHCp_3] \qquad (10)$$

There is a brief mention[54] of the [ZrClCp$_3$] complex which contains an average Zr–Cp(η-Cp) bond length of 0.258 nm, but no other details are available.

Xα-SW (scattered wave) theoretical calculations[50] have been carried out to compare the bonding of Zr(η-Cp)$_3$ with that of tris(cyclopentadienyl)actinides, concluding that the 4d-orbitals of zirconium are much more effective at bonding the Cp$^-$ ligands than the 6d-orbitals of the actinides.

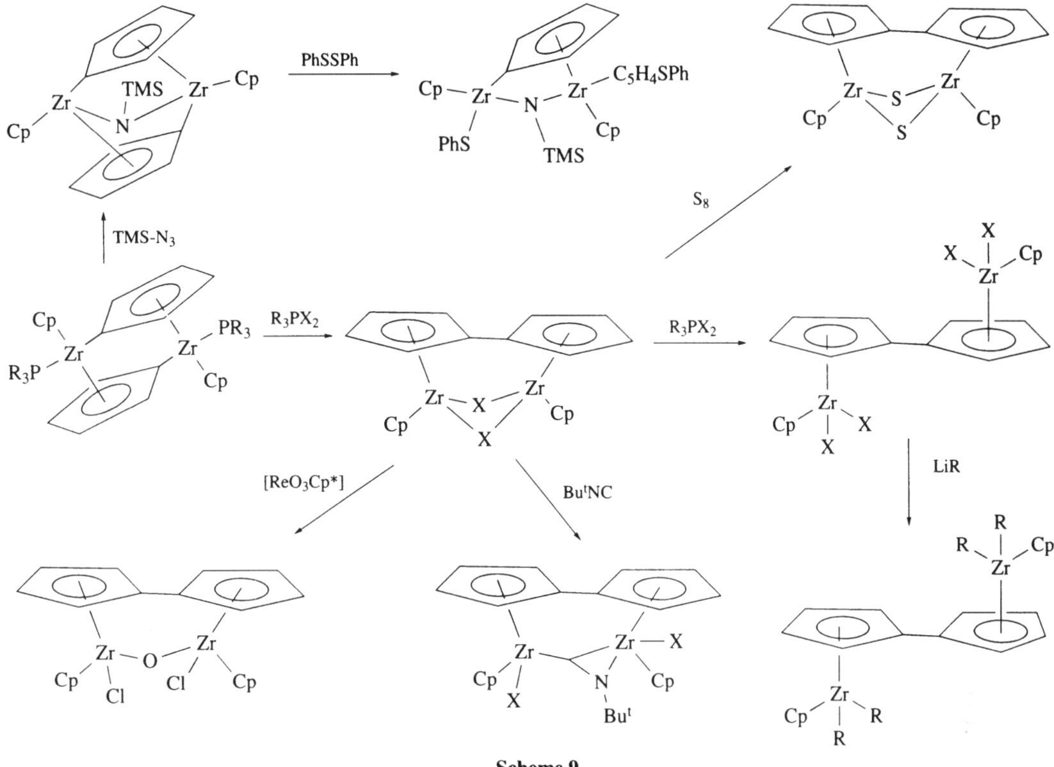

Scheme 9

Table 1 Infrared and Raman data for ZrHCp$_3$ and ZrDCp$_3$.[48,49]

Compound	Raman (cm^{-1})	IR (cm^{-1})
[ZrHCp$_3$]	ν(CH): 3125 (m), 3111 (vw), 3104 (w), 3083 (vw)	ν(CH): 3129 (m), 3118 (w), 3106 (w), 3090 (w)
	ν(ZrH): 1608 (s)	ν(ZrH): 1609 (m)
	ν(CC)E_1: 1457 (m), 1443 (m)	ν(CC)E_1: 1453 (m), 1439 (m)
	ν(CC)E_2: 1372 (s)	ν(CC)E_2: 1372 (m)
		β(CH)A_2: 1278 (w)
	ν(Cp)A_1: 1134 (vs)	ν(Cp)A_1: 1134 (w)
	β(CH)E_2: 1086 (m), 1079 (m)	ν(CH)E_2: 1087 (w), 1069 (m)
	β(CH)E_1: 1028 (w)	β(CH)E_1: 1031 (s), 1019 (s)
	γ(CCC)E_2: 935 (vw), 904 (vvw)	γ(CCC)E_2: 926 (s)
	ρ(CH): 848 (w), 823 (w), 805 (w)	ρ(CH): 851 (s), 838 (vs), 807 (vs)
	β(ZrH): 753 (s), 622 (w)	β(ZrH): 752 (s)
	χ(CCC): 614 (m)	χ(CCC): 612 (w)
	ν(ZrCp)A_1: 285 (vs)	ν(ZrCp)A_1: 290 (s)
	tilt(ZrCp): 267 (s)	tilt(ZrCp): 264 (s)
	ν(ZrCp)E: 252 (vs)	ν(ZrCp)E: 247 (s)
	tilt(ZrCp): 221 (w)	
	δ(CpZrCp): 185 (vs)	δ(CpZrCp): 182 (vw), 157 (m)
[ZrDCp$_3$]	ν(CH): 3126 (m), 3113 (vw), 3105 (w), 3085 (w)	ν(CH): 3127 (m), 3116 (w), 3108 (w), 3088 (w)
	ν(ZrH): 1161 (m), 1144 (m)	ν(ZrH): 1161 (m), 1143 (m)
	ν(CC)E_1: 1457 (w), 1442 (w)	ν(CC)E_1: 1453 (m), 1436 (m)
	ν(CC)E_2: 1370 (m)	ν(CC)E_2: 1370 (m)
		β(CH)A_2: 1278 (w)
	ν(Cp)A_1: 1131 (vs)	
	β(CH)E_2: 1086 (m), 1079 (m)	β(CH)E_2: 1079 (w), 1062 (m)
	β(CH)E_1: 1031 (vw), 1026 (vw)	β(CH)E_1: 1026 (s), 1031 (s)
		γ(CCC)E_2: 920 (m)
	ρ(CH): 846 (w), 823 (w), 795 (w)	ρ(CH): 846 (m), 828 (s), 799 (vs)
	β(ZrH): 538 (s), 620 (vw)	β(ZrH): 538 (m)
	χ(CCC): 615 (w), 611 (m)	χ(CCC): 617 (w)
	ν(ZrCp)A_1: 284 (vs)	ν(ZrCp)A_1: 290 (s)
	tilt(ZrCp): 264 (s)	tilt(ZrCp): 263 (s)
	ν(ZrCp)E: 249 (vs)	ν(ZrCp)E: 246 (s)
	tilt(ZrCp): 222 (w)	
	δ(CpZrCp): 184 (vs)	δ(CpZrCp): 188 (m), 156 (m)

11.3 TETRAHYDROBORATO COMPLEXES

11.3.1 Introduction

Much of the attention concerning the organometallic derivatives of tetrahydroborates, $L_xM(BH_4)_y$, has focused on the bonding of the $[BH_4]^-$ ligand to the metal atom, which can occur in the form of two or three hydrogen bridges per ligand.[51] Determination of the hydrogen-bridge order rests mainly on the results from x-ray diffraction.[52] Electron[53,54] and neutron[55] diffraction have been used to a lesser extent. Evidence has been provided[55] that in some cases the vibrational spectra of bi- and tridentate (two- and three-hydrido bridged) complexes differ sufficiently in the B–H stretching region to allow their identification. It has also been shown[56,57] that the number of bridging hydrogens from each $[BH_4]^-$ ligand can be correlated with the number of vacant orbitals of suitable energy and symmetry on the metal atom.

11.3.2 Synthesis, Structure and Bonding

11.3.2.1 Mono(tetrahydroborates)

The complex $[Zr(BH_4)MeCp_2]$ has been prepared as a white solid either by reaction of $[ZrMeClCp_2]$ with 1.5 equiv. of $Li[BH_4]$ in toluene, or by treatment of $[ZrMe_2Cp_2]$ with $BH_3(THF)$ (Scheme 10).[58] In the latter case, 2 equiv. of $BH_3(THF)$ are necessary to complete the reaction which proceeds via elimination of $MeBH_2(THF)$. The complex $[Zr(BH_4)MeCp_2]$ can be considered as an intermediate in the formation of $[Zr(BH_4)_2Cp_2]$. NMR spectral data for $[Zr(BH_4)MeCp_2]$ are in agreement with the proposed formulation (Table 2).

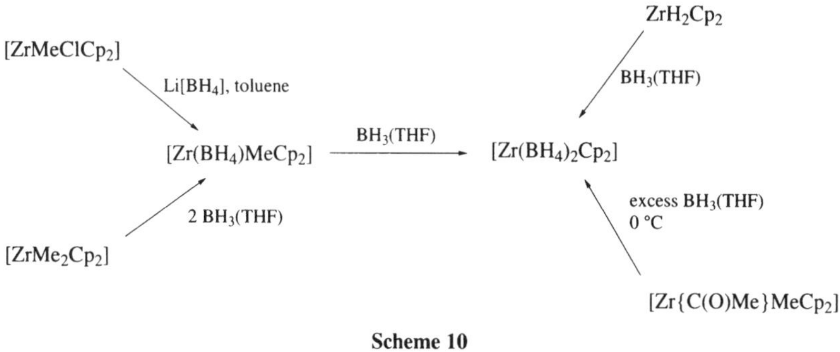

Scheme 10

11.3.2.2 Bis(tetrahydroborates)

Several synthetic routes to $[Zr(BH_4)_2Cp_2]$ are well documented.[59–61] Reaction of ZrH_2Cp_2 with $BH_3(THF)$ has been shown to be an effective synthetic pathway for the preparation of bis(tetrahydroborates).[62] After removal of the solvent, the colourless $[Zr(BH_4)_2Cp_2]$ may be isolated in a 67% yield and purified by sublimation. Infrared and NMR spectral data (Table 2) indicate that the tetrahydroborate adopts the doubly bridged bonding mode.[61] Higher yields, up to 90%, have been obtained upon reaction of excess $BH_3(THF)$ with either $[ZrMe_2Cp_2]$ or $[Zr\{C(O)Me\}MeCp_2]$[58] (Scheme 10). Although the fate of the methyl group was never established, it is believed that $MeBH_2(THF)$ species are formed.[58]

11.3.2.3 Tris(tetrahydroborates)

The compounds $[Zr(BH_4)_3Cp]$ and $[Zr(BD_4)_3Cp]$ have been synthesized as white crystals in excellent yield (95%) from $ZrBr_3Cp$ and $Li[BH_4]$ or $Li[BD_4]$.[63] The IR spectral data (Table 2) indicate that in benzene solution the $[BH_4]^-$ ligands are attached in a tridentate manner.[55] Moreover, gas-phase IR spectra are in accordance with this formulation. The ^1H and ^{11}B NMR spectra are characteristic of fluxional tetrahydroborato systems in terms of peak multiplicities, chemical shifts and coupling constants. The rapid interconversion of terminal and bridging hydrogen environments is reflected in a

1:4:6:4:1 quintet for the ^{11}B resonance and a 1:1:1:1 quartet for the ^1H signal. The pattern remains unchanged even at $-90\,^\circ C$, thus not allowing the distinction between doubly and triply bridged structures. The molecular structure[63] of $[Zr(BH_4)_3Cp]$ in the gas phase was determined by electron diffraction. The salient features of the structure are a Cp group and a trigonal-pyramidal arrangement of the boron atoms around zirconium. The diffraction data fitted equally well for the doubly and triply bridged models. Extended Hückel MO calculations[63] support the notion that a triply bridged $[BH_4]^-$ ligand can be considered to be isolobal with Cp^-.

The permethylated analogue, $[Zr(BH_4)_3Cp^*]$, has been isolated from the reaction of $ZrCl_3Cp^*$ with excess $Li[BH_4]$ in toluene[64] in 82% yield as a white crystalline solid. The IR spectrum (Table 2) indicates that each $[BH_4]^-$ is triply bridging. A typical coupling constant ($^1J_{^{11}B-H} = 86$ Hz)[55] is observed for the quartet (12 H) of broad resonances centred at δ 1.39 in the ^1H NMR spectrum.

Table 2 Infrared and NMR data for tetrahydroborate complexes.

Compound	^1H	^{13}C	^{11}B	IR (cm^{-1})	Ref.
		NMR (δ) in C_6D_6 (ppm)			
$[Zr(BH_4)MeCp_2]$	5.60 (s, 10 H)		(in THF)		58
	0.24 (q, 4 H), $J_{B-H} = 86$ Hz		-3.4 (quintet,		
	0.17 (s, 3 H)*		$^1J_{B-H} = 83$ Hz)		
$[Zr(BH_4)_2Cp_2]$	5.71 (s, 10 H)	114.1	-15.8 (quintet, 83 Hz)	2430 (s), 2380,	58,59,
	0.78 (q, br, 8 H), $J = 85$ Hz			2140	61,62
$[Zr(BH_4)_3Cp]$	5.8 (s, 5 H)		-11.2 (quintet, 1:4:6:4:1)	C_6H_6 solution 2520 (s), 2158 (s),	63
	1.54 (q, 12 H, 1:1:1:1),			2110 (s)	
	$J_{B-H} = 91$ Hz			gas 2536 (s), 2156 (s),	
				2111 (s)	
				nujol 2530 (s), 2155 (s),	
				2100 (s)	
				Raman 2530 (s), 2156 (m),	
				2110 (s)	
$[Zr(BD_4)_3Cp]$				gas 1899 (s), 1580 (w),	63
				1538 (s)	
				Raman 1893 (s), 1578 (w),	
				1549 (s)	
$[Zr(BH_4)_3Cp^*]$	1.76 (s, 15 H)			2509, 2355, 2330,	64
	1.39 (q, 12 H), $^1J_{^{11}B-H} = 86$ Hz			2166, 2113	
$[\{Zr(BH_4)H-$	3.91 (s, Zr–H$_t$)			ν(Zr–BH$_4$) 2448, 2408, 2239,	
$(\mu\text{-}H)Cp^*\}_2]$	3.0–0.0 (br, BH$_4$)			2112, 2052, 1980	64
	2.14 (s, 15 H)			ν(Zr–H$_t$) 1628	
	1.31 (s, Zr–H$_b$)			ν(Zr–H$_b$) 1350–1500 (br)	

11.3.3 Reactivity

Very few reactivity studies have been reported for tetrahydroborates. The tris(tetrahydroborate) $[Zr(BH_4)_3Cp^*]$ reacts with CO over a period of weeks at $80\,^\circ C$ to yield several unidentified products, none of which contained fragments that might be construed as occurring via a reduction of CO.[64]

Since the abstraction of BH_3 from tetrahydroborato complexes by suitable σ donors is a common route to transition metal hydrides, $[Zr(BH_4)_3Cp^*]$ was reacted with Lewis bases such as PMe_3 and NMe_3. Although trimethylphosphine effectively removed BH_3, no zirconium-containing products were isolated. However, when $[Zr(BH_4)_3Cp^*]$ was reacted with a large excess (8–10 equiv.) of NMe_3 (Equation (11)), the hydrido-bridged dimer,[64] $[\{Zr(BH_4)H(\mu\text{-}H)Cp^*\}_2]$, was isolated as white prisms (40% yield) after recrystallization. The IR spectrum (Table 2) clearly indicates the presence of a doubly bridging (1450 cm^{-1}) and a terminal (1628 cm^{-1}) hydride. The tetrahydroborate appears as a broad, almost undetectable, resonance centred near δ 1.3 in the ^1H NMR spectrum (Table 2).

$$2\;[Zr(BH_4)_3Cp^*] + 8\text{–}10\,NMe_3 \longrightarrow \qquad\qquad\qquad (11)$$

The reaction chemistry of this dimer is ill defined; it does not react with CO or ethene, and on heating it decomposes into a multitude of products.[64]

11.4 METALLOCENE HYDRIDES

11.4.1 MHClCp$_2$

Since the initial discovery that the polymeric zirconocene complex ZrHClCp$_2$[65] reacts rapidly with alkenes[66] and alkynes[67] and that it efficiently isomerizes internal double bonds in short-chain alkenes to yield terminal alkylzirconium derivatives,[68] the synthesis and characterization of zirconium hydrides and the applications[69,70] of the hydrozirconation reaction have received much attention.

The most versatile reagents for the preparation of hydrides from zirconocene halides are Na[AlH$_2$(OCH$_2$CH$_2$OMe)$_2$] (Red-Al, Vitride) and LAH. Several dihalides ZrX$_2$Cp$_2$ (X = F or Br), ZrX$_2$CpCp* and ZrX$_2$Cp*$_2$ (X = Cl) have been treated with these agents to yield the monohydrido derivatives analogous to ZrHClCp$_2$.[7] While ZrBr$_2$Cp$_2$ reacts smoothly with Vitride to give ZrHBrCp$_2$, the reaction is not controllable in the case of ZrF$_2$Cp$_2$ and proceeds further to ZrH$_2$Cp$_2$. By contrast, [ZrCl$_2$Cp*$_2$] is unreactive, while [ZrCl$_2$CpCp*] reacts rapidly forming ill-defined hydridic complexes formulated as ZrHClCpCp*.[71] Both ZrHBrCp$_2$ and ZrHClCpCp* rapidly isomerize internal alkenes producing terminal alkylzirconium complexes. ZrHClCpCp* was found to be the most efficient catalyst for the isomerization of (*E*)-5-decene, (*Z*)-5-decene and oleic acid oxazoline.

Hydrogenolysis of ZrRClCp$_2$, although not feasible for large-scale preparations, affords ZrHClCp$_2$.[72] The mechanism is rather intriguing since the zirconium(IV) centre does not allow oxidative addition of H$_2$. The observation that the rate of alkane formation during the hydrogenolysis of ZrX(CH$_2$Bu^t)Cp*$_2$ (X = H, F, Cl or Br) to form ZrHXCp*$_2$ is greatly reduced when the Cp* rings are connected by an ethene bridge, such as in the *ansa*-zirconocene derivative ZrX(CH$_2$Bu^t)(Cp*CH$_2$)$_2$,[72] prompted the proposal that the hydrogenolysis of the Zr–R function proceeds via partial ring hydrogenation followed by hydrogen transfer to the alkyl and consequent alkane elimination.

Reaction of ZrCl$_2$(MeCp)$_2$ with [{Zr(H)$_2$(MeCp)$_2$}$_2$] in THF or toluene at ambient temperature gives a mixture of products where Zr(H)Cl(MeCp)$_2$[73] is present in small amount. Although ill characterized, the latter may be used very efficiently for hydrozirconation reactions of alkenes, alkynes, ketones, nitriles or CO. It also reacts with 1-hexene or acetophenone about six to seven times faster than the commonly used Schwartz reagent. Reaction with (trimethylsilyl)ethyne followed by carbonylation yields the α,β-unsaturated (η^2-acyl)metallocene [Zr(Cl){η^2-OCC(H)=C(H)TMS}(MeCp)$_2$], whose structure was determined by x-ray diffraction.

11.4.2 Reactivity of MHClCp$_2$

11.4.2.1 Alkynes

The hydrozirconation of alkynes is reviewed in Section 11.1.6.4.

11.4.2.2 Alkenes

Treatment of ZrHClCp$_2$ with styrene leads to a mixture of terminal (85%) and internal (15%) insertion products.[74] Attempts to prepare deuteroorganozirconium derivatives resulted in scrambling of deuterium on the β-positions of both isomers, yielding a statistical distribution of isotopomers.

The (1,3-pentadiene)zirconocene complex and its higher homologues of the type [Zr{CR(Me)=CRCR=CRH}Cp$_2$] are prepared by reaction of ZrHClCp$_2$ with alkyl-substituted or nonsubstituted pentadienyl anions through σ–π rearrangements of hydrido-2,4-pentadienylzirconium species.[75] Proton NMR spectral studies carried out on these complexes revealed that the 1,3-pentadiene and 2,4-hexadiene complexes are composed of a ca. 1:1 mixture of *s-cis* and *s-trans* isomers, while the complexes containing 2-methyl-1,3-pentadiene, 3-methyl-1,3-pentadiene, 2,4-dimethyl-1,3-pentadiene, 1-trimethylsilyl-1,3-pentadiene and 1,5-bis(trimethylsilyl)-1,3-pentadiene ligands consist of a single *s-cis* isomer. The process of hydrogen transfer was clarified by deuterium labelling studies.

11.4.2.3 Carbon monoxide and carbon dioxide

Reaction of $ZrHClCp_2$ with CO in THF gives the formaldehyde complex $[\{ZrClCp_2\}_2(\mu\text{-}CH_2O)]$ as light-yellow crystals (Scheme 11).[76] The elongated C–O distance of 0.143 nm is consistent with an IR band at 1015 cm^{-1}. Further slow carbonylation of this complex in THF forms a new product that has weak IR bands at 1970 cm^{-1} and 1605 cm^{-1}. Elimination of $ZrCl_2Cp_2$ from the formaldehyde complex is also observed.

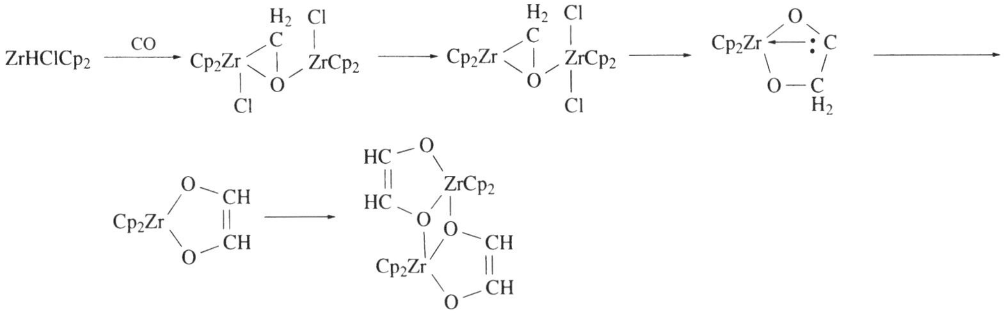

Scheme 11

Stoichiometric reduction of CO_2 by $ZrHClCp_2$ produces formaldehyde and methanol.[77] Two steps have been identified in this reaction: the first leading to the formation of formaldehyde and $[\{ZrClCp_2\}_2O]$, the second involving the insertion of formaldehyde into the Zr–H bond, with consequent formation of $Zr(OMe)ClCp_2$. The oxophilicity of the metal and the high stability of the μ-oxo complex seem to provide the necessary thermodynamic driving force for the reduction of CO_2.

The reduction of the heteroallenes RN=C=O or RN=C=NR has been used to model the CO_2 reduction steps. Insertion of *p*-tolylcarbodiimide into the Zr–H bond forms a complex which is not further reduced by the Zr–H bond. Isocyanates insert into the Zr–H bond, forming successively a formamidino complex which is further reduced by the Zr–H unit to yield the methylamido analogue. The presence of an oxygen on the formamidino ligand would explain its reducibility to imine with simultaneous formation of $[\{ZrClCp_2\}_2O]$.

11.4.2.4 Nitriles

Carbon-substituted (methyleneamido)zirconocene chloride complexes are obtained in high yield as moisture-sensitive yellow solids by hydrozirconation of nitriles (Scheme 12).[78,79] In addition to the typical absorptions of the bent metallocene, an intense –C=N stretching frequency is observed at 1678 cm^{-1} (R = Ph) an 1700 cm^{-1} (R = Me). The x-ray structure of the phenyl-containing complex shows the almost linear Zr–N–C array and a shorter than normal $C=N(sp^2)$ bond distance. Subsequent reaction of $[Zr(N=CHR)ClCp_2]$ with an aryllithium reagent yields $[Zr(N=CHR)ArCp_2]$.[79] $[Zr(N=CHR)ClCp_2]$ also reacts with MH_2Cp_2 (M = Zr or Hf) to form the dinuclear, thermally stable complexes $[ZrCp_2(\mu\text{-}N=CHR)(\mu\text{-}Cl)ZrCp_2]$.

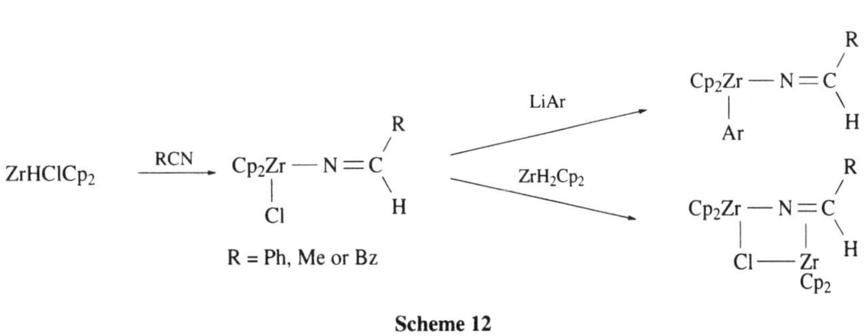

Scheme 12

11.4.2.5 Organophosphorus compounds

Addition of $ZrHClCp_2$ to a THF solution of $(TMS)_2N–P=CH(TMS)$ at low temperature affords the zirconocene phosphirane as a white powder in virtually quantitative yield.[80] The product formally arises from the hydrozirconation of the P=C bond (Equation (12)). The ^{31}P NMR spectrum (doublet at δ 42, $^1J_{P–H} = 346$ Hz) indicates coordination of the phosphorus atom. Similar treatment of the phosphorus-halogenated phosphaalkene with $ZrHClCp_2$ yields the diphosphene as the major product with the diphosphirane and another unidentified by-product, probably $[ZrCl_2Cp_2]$ (Equation (13)).[80] In contrast, the reaction of $ZrHClCp_2$ with $(TMS)_2NP=P(TMS)$ in THF forms a functionalized zirconium-containing phosphoramide species which can be isolated as a white powder in almost quantitative yield (Equation (14)). NMR spectral data have been used to support the cyclic structure.[80] A mixture of unsaturated (^{31}P NMR: δ 84.4, $^1J_{P–H} = 208$ Hz) and saturated (^{31}P NMR: δ –43.5, $^1J_{P–H} = 201$ Hz) P–N compounds was obtained upon reaction of chlorophosphaimine with $ZrHClCp_2$ (Equation (15)).[80]

$$(TMS)_2N–P=CH(TMS) \xrightarrow{\ ZrHClCp_2\ } Cp_2Zr\begin{smallmatrix} N(TMS)_2 \\ | \\ PH \\ | \\ CH \\ | \\ Cl \quad TMS \end{smallmatrix} \qquad (12)$$

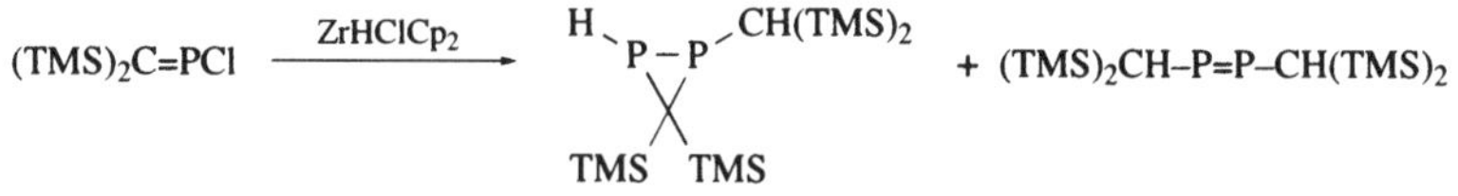

$$(TMS)_2C=PCl \xrightarrow{\ ZrHClCp_2\ } \overset{H}{\underset{TMS}{\diagdown}}P\!-\!P\overset{CH(TMS)_2}{\underset{TMS}{\diagup}} + (TMS)_2CH–P=P–CH(TMS)_2 \qquad (13)$$

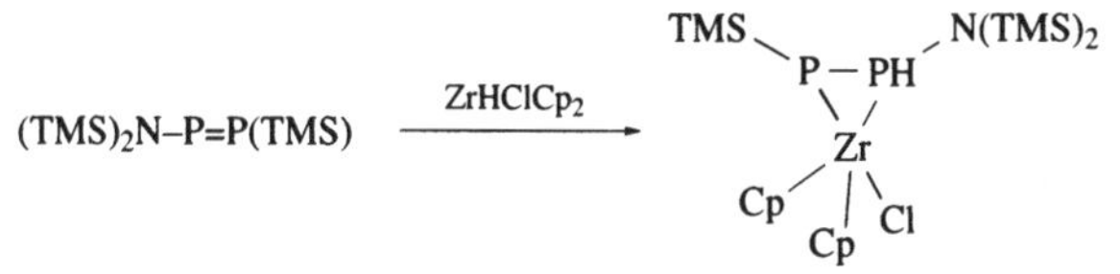

$$(TMS)_2N–P=P(TMS) \xrightarrow{\ ZrHClCp_2\ } \begin{smallmatrix} TMS \diagdown \quad \diagup N(TMS)_2 \\ P\!-\!PH \\ \diagdown \diagup \\ Zr \\ \diagup | \diagdown \\ Cp \; Cp \; Cl \end{smallmatrix} \qquad (14)$$

$$Cl–P=N–Ar \; \xrightarrow[-\,ZrCl_2Cp_2]{+\,ZrHClCp_2} \; H–P=N–Ar + H_2P–NHAr \qquad (15)$$

A synthetically useful precursor for the one-pot formation of *C*-phosphanyl, *N*-phosphanyl or *N*-boranylines[81] has been obtained by reaction of bis(diisopropylamino)cyanophosphine with $ZrHClCp_2$ in THF. The reaction results in the formation of a Zr–N species (Scheme 13), the structure of which was established on the basis of ^1H NMR and ^{13}C NMR spectra.[80]

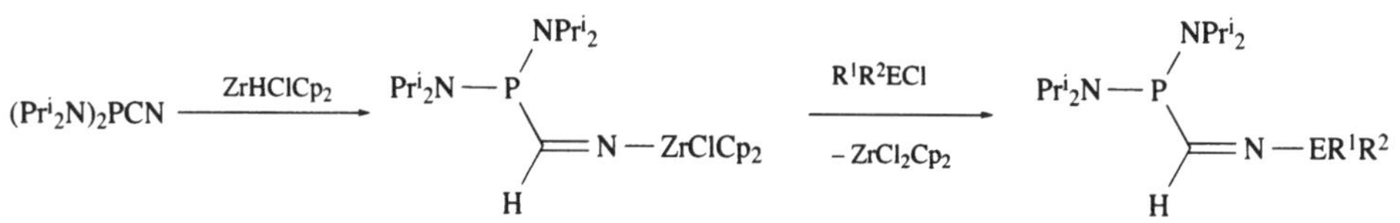

$$(Pr^i_2N)_2PCN \xrightarrow{\ ZrHClCp_2\ } Pr^i_2N\!-\!P\overset{NPr^i_2}{\underset{H}{\diagup\!\!\diagdown}}{=}N\!-\!ZrClCp_2 \xrightarrow[-\,ZrCl_2Cp_2]{R^1R^2ECl} Pr^i_2N\!-\!P\overset{NPr^i_2}{\underset{H}{\diagup\!\!\diagdown}}{=}N\!-\!ER^1R^2$$

$E = P;\ R^1 = R^2 = Ph$
$E = P;\ R^1 = Pr^i_2N,\ R^2 = Cl$
$E = P;\ R^1 = R^2 = NPr^i_2$
$E = B;\ R^1 = R^2 = NPr^i_2$
$E = B;\ R^1 = N(TMS)N(TMS)_2$
$ER^1R^2 = TMS$

Scheme 13

11.4.2.6 Oxygen-containing substrates

Hydrozirconation–deuterolysis of (*E*)-3-methoxy-1-phenyl-1-propene was found to give α- and ω-deuterated propylbenzenes after elimination of the ether function.[82] In an effort to account for the regiochemical outcome of the reaction (a high degree of benzylic substitution in propylbenzene was observed), hydrozirconation of propylbenzenes with mixed reagents has been investigated.[82] Hydrozirconation of (*E*)-3-phenyl-2-propenol results in the loss of the hydroxy group to only a small extent; after hydrozirconation and deuterolysis, moderate yields of benzenepropan-γ-d-ol are formed. Carbonyl insertion into 3-hydroxy-1-phenyl-1-zirconocenepropane gives a low yield of the lactone dihydro-3-phenyl-2(3H)-furanone.

The stereoselectivity of $ZrHClCp_2$ as a reducing agent of cyclic and noncyclic ketones is remarkably good.[83] The reaction forms alcohols at 60 °C, but it does not work fully with ketones bearing an aromatic group in the α- position.

The mixed system $ZrHClCp_2$–Mg serves as an effective catalyst for the regioselective acylation of polyalcohols (including 1,2-diols). However, the reagent has to be used in combination with 3-acyl-2-oxazolones to permit highly preferential protection of primary hydroxy groups.[84]

11.4.3 MHRCp$_2$

The complexes $ZrRHCp_2$ have been prepared by reduction of the corresponding zirconocene chloride with $Li[AlH(OBu^i)_3]$ (Scheme 14).[85] These hydrides undergo hydrogenation under mild conditions to yield the corresponding alkane and ZrH_2Cp_2. Deuterium labelling experiments suggest that these d^0 complexes perform a heterolytic attack on the hydrogen molecule. The hydrogenation rates of a series of alkyl complexes show dependence on the nature of the other groups bonded to zirconium: $ZrRHCp_2 > ZrRClCp_2 \approx [ZrR_2Cp_2] > [(ZrClCp_2)_2(\mu\text{-OCHR})] > [Zr(COR)ClCp_2]$. Since this is the same order as previously determined for carbonylation reactions, a conceptual link between the mechanisms of the two reactions may be inferred.[85]

Scheme 14

Oligomeric $Zr(H)(CH_2PPh_2)Cp_2$ displays a remarkable ability to catalyse selective hydrogenation of conjugated dienes to monoenes.[86] The reaction at 80 °C with cod induces the addition of dihydrogen to each of the cod isomers to give cyclooctene with greater than 98% selectivity.[86] The catalytic hydrogenation is of fractional order with respect to the oligomeric catalyst precursor, thus indicating some rate-limiting dissociative process of the active species.[87] The hydrido complex $[Zr(H)(CH_2PPh_2)Cp_2]$ reacts in THF and at room temperature with $[Zr(\eta^4\text{-butadiene})Cp_2]$ to yield the doubly bridged dinuclear zirconocene complex $[ZrCp_2(\mu\text{-CH=CHEt})(\mu\text{-CH}_2PPh_2)ZrCp_2]$.[87] The latter is also formed when $[Zr(H)(CH_2PPh_2)Cp_2]$ is allowed to react with free butadiene, although in this case methyldiphenylphosphine is obtained as an additional stoichiometric reaction product.

The complex $Zr(H)\{C(Me)=CH(Me)\}Cp^*_2$ rearranges at room temperature in both the solid and solution forming the crotyl hydrido species $[Zr(H)(\eta^3\text{-C}_4H_7)Cp^*_2]$, or in the presence of 2-butyne forming the zirconacyclopentadiene derivative $[\overline{ZrCH_2CH(CH_2Me)C(Me)=C}(Me)Cp^*_2]$.[88] The bis(propenyl) and bis(butenyl) complexes also rearrange, forming zirconacyclopentene species $[\overline{ZrCH_2CH(R)CH=C}RCp^*_2]$ and $[\overline{ZrCH_2CH(R)C(R)=C}HCp^*_2]$ (R = Me or Et). Mechanistic and kinetic studies[88] carried out on these transformations have demonstrated the occurrence of an unusual β-hydrogen elimination from an sp^2-hybridized carbon.

11.4.4 MH(OR)Cp$_2$

Zirconocene hydrides with oxygen-containing functions are extremely rare. An enolato–hydrido species has been obtained via rapid hydrogenation of the monomeric $Zr(L)(OC=CHR)Cp^*_2$ (L = CO or py) under ambient conditions (Equation (16)).[89] In accordance with the proposed stereospecificity of the hydrogenation reaction, the enolate geometry of $Zr(H)\{OC(H)=CHBu^i\}Cp^*_2$ is >96% *cis* as deduced from the vinylic coupling constant ($^3J_{H-H} = 7.9$ Hz).[89] These results helped to support the key postulates dictating a *cis* geometry of the intermediate $[(ZrHCp^*_2)_2(\mu\text{-OCH=CHO})]$ (Scheme 15).

(16)

$$\text{ZrH}_2\text{Cp*}_2 + [\text{Zr(CO)}_2\text{Cp*}_2] \longrightarrow \quad\quad \xrightarrow{\text{H}_2}$$

Scheme 15

11.4.5 MH(SiR$_3$)Cp$_2$

Early transition metal complexes containing silicon ligands are becoming increasingly important in organometallic chemistry.[90–106] In particular, group 4 metal–silicon bonds have been shown to be rather reactive, undergoing insertion reactions[90-2] and participating in silane polymerizations[93–101] and σ-bond metathesis reactions.[100,101]

While metallocene silyl hydrides have been postulated as key intermediates in the polymerization of silanes,[96] very few of these complexes have been characterized. An unusual bimetallic silylzirconocene hydride [ZrCp$_2$(SiPhMeH)(μ-H)$_2$Zr(SiPhMe$_2$)Cp$_2$] has been isolated as orange crystals (44% yield) from the reaction of [ZrMe$_2$Cp$_2$] with phenylsilane.[95] The bimetallic product is produced almost quantitatively (ca. 90%) in the early stages of the reaction as demonstrated by NMR spectral studies. Appreciable quantities of other species, both diamagnetic and paramagnetic, are also produced. The structural assignment was based on ^{1}H and ^{29}Si NMR spectral data and a crystal structure which was unfortunately affected by severe disorder (Equation (17)).[107] A possible intermediate for this reaction, [ZrCp$_2$(Me)-(μ-H)$_2$Zr(SiPh$_2$H)Cp$_2$], has been isolated as orange microcrystals (21% yield) and can also be obtained from the reaction of diphenylsilane with [ZrMe$_2$Cp$_2$]. Its similar structure has been assigned on the basis of ^{1}H NMR spectra. Several other hydrido-bridged species have been identified similarly during the reaction of [ZrMe$_2$Cp$_2$] with phenylmethyl-, diphenylmethyl- or n-butylmethylsilane.

$$2\,[\text{ZrMe}_2\text{Cp}_2] + 2\,\text{PhMeSiH}_2 \longrightarrow \tag{17}$$

Monomeric trimethylphosphine adducts of the zirconocene triphenylsilyl hydride have been prepared[108] by addition of 1 equiv. of triphenylsilane to a THF solution of [Zr(η^2-1-butene)(PMe$_3$)Cp$_2$]. The reaction yields a red-brown solid, in approximately 48% yield. The structure of [Zr(H)(SiPh$_3$)(PMe$_3$)Cp$_2$] was confirmed by an x-ray crystal structure. Proton NMR spectral and deuterium labelling studies[108] suggest that it arises from the addition of the Si–H bond directly across the Zr–C(alkene) bond, followed by β-elimination and loss of alkene. Its reactivity with a variety of unsaturated organic substrates is summarized in Scheme 16. In all cases, the hydrido ligand participates in new bond-forming reactions while the silylated group acts only as a spectator ligand.[108] The Si–Zr bond is easily metathesized by silanes such as Ph$_2$SiH$_2$ to quantitatively produce a mixture of three geometric isomers (3.2:1, based on ^{1}H NMR spectroscopy) along with Ph$_3$SiH (Equation (18)). The structure of one of these isomers was elucidated by an x-ray crystal structure of an identical sample obtained from the reaction of [ZrBu$^n_2$Cp$_2$] with 2 equiv. of H$_2$SiPh$_2$ in the presence of 1.3 equiv. of PMe$_3$ at −78 °C.[109] In a similar manner, the reaction carried out in the presence of 1.3 equiv. of PPh$_2$Me gave Zr(SiHPh$_2$)(H)(PPh$_2$Me)$_2$. The reactivity of these complexes is summarized in Scheme 17. Reaction of [ZrBu$^n_2$Cp$_2$] with 2 equiv. of H$_2$SiPh$_2$ produced not only Bu^{n}SiHPh$_2$ (70%) but also the yellow crystalline [ZrCp$_2$(HSiPh$_2$)(μ-H)$_2$Zr(SiHPh$_2$)Cp$_2$] (85%),[109] which was characterized by x-ray crystallography.

$$\tag{18}$$

High molecular weight, cross-linked poly(disilanylenearylene) can be formed from bis(silyl) monomers via the [(ZrH$_2$CpCp*)$_2$]-catalysed dehydrocoupling of hydrosilanes (Equation (19)).[110] The polymers made by this route have very different properties from the polymers previously prepared by

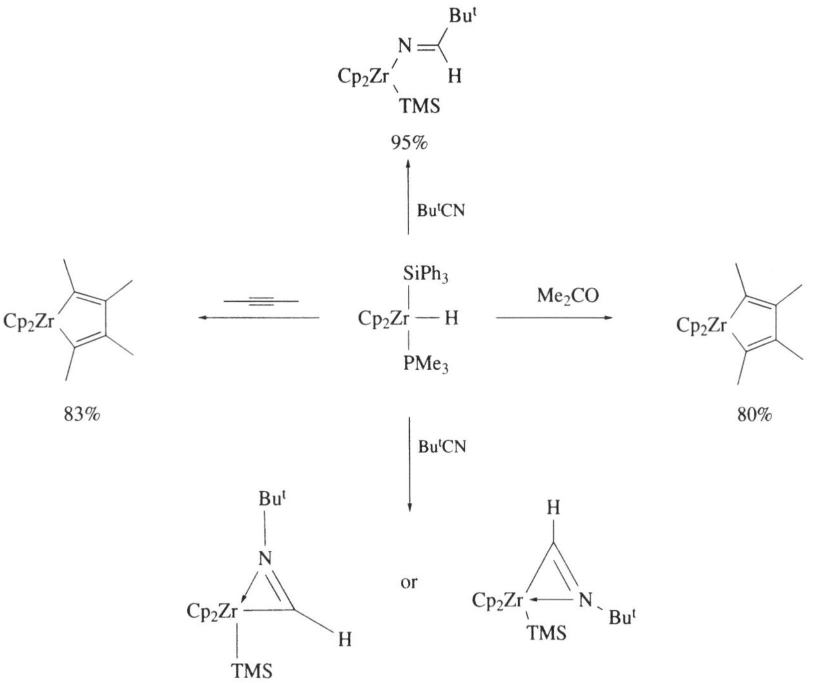

Scheme 16

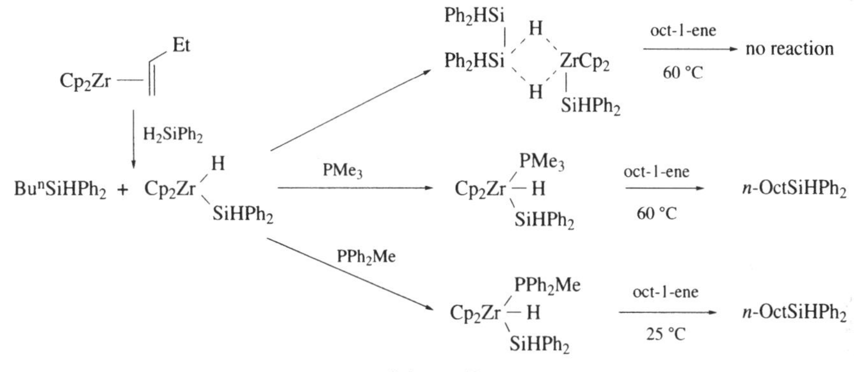

Scheme 17

metal-catalysed dehydropolymerization.[96] Their physical properties are consistent with a substantial degree of cross-linking, while the UV–visible and conductivity data suggest σ–π delocalization in the –(SiArSi)– backbone.[110]

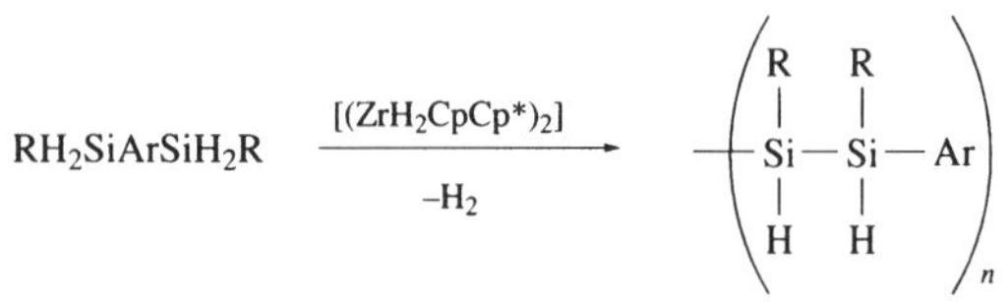

$$RH_2SiArSiH_2R \quad \xrightarrow[-H_2]{[(ZrH_2CpCp^*)_2]} \quad \left(\begin{array}{c} \underset{|}{\overset{R}{}} \quad \underset{|}{\overset{R}{}} \\ -Si-Si-Ar \\ \underset{|}{} \quad \underset{|}{} \\ H \quad\; H \end{array}\right)_n \qquad (19)$$

11.4.6 Polymetallic Hydrides

The reaction of $ZrHClCp_2$ with *fac*-$[OsH_3(PMe_2Ph)_3]K$ or $K[ReH_6(PMePh_2)_2]$ produced the zirconium–osmium or zirconium–rhenium mixed-metal dimeric hydrides.[111] NMR spectroscopic studies on these heterodinuclear species point to these complexes being rather robust. For the zirconium–osmium hydride, the ^1H and ^{31}P NMR spectra show signals arising from an AA'A"XX'X" system. The zirconium–rhenium hydrido signals appear equivalent at 25 °C. In both systems the zirconium-bound hydrides fail to exchange on the NMR timescale with OsH or ReH. Resolution of a 1 Hz coupling between Zr–H and Re–H rules out any dissociative equilibrium involving this dimer.

The reaction of the 14-electron complex [Zr{(CH$_2$)$_3$NMe$_2$}Cl$_2$Cp*] with dihydrogen (1 atm, 20 °C) in aromatic solvents yields a clear yellow solution from which large yellow crystals can be obtained.[112] Characterization by x-ray crystallography was hampered by loss of solvent from the crystal lattice. However, IR spectra (v(Zr–H) = 1575, 1310, 1150 and 901 cm^{-1}; v(Zr–D) = 1136, 976, 802 and 636 cm^{-1} in the D$_2$ reaction product) clearly indicate the presence of the hydridic function.[112] NMR spectral experiments carried out at low temperature indicated a complicated structure 'Cp$_3$Zr$_3$(H)$_4$Cl$_5$' with four nonequivalent hydrides and three nonequivalent [Cp*]$^-$ ligands. Reaction of this polynuclear compound with PMe$_3$ induces disproportionation forming a dinuclear hydrido complex [Zr$_2$(μ-H)$_3$Cl$_3$(PMe$_3$)Cp*$_2$], together with [ZrCl$_3$(PMe$_3$)Cp*]. An x-ray study of the fomer[112] confirmed the structure with the ZrCl(PMe$_3$)Cp* and ZrCl$_2$Cp* moieties in nearly eclipsed conformation. The low-temperature ^{1}H NMR spectrum shows the presence of three nonequivalent hydrido resonances (δ 4.73, 4.44 and 2.97), each with one $^2J_{P-H}$ (12.1 Hz) and two equivalent $^2J_{H-H}$ couplings (8 and 1 Hz). At 75 °C, the hydrido signals coalesce into one broad resonance centred at δ 4.09, thus indicating rapid interchange of the hydrido ligands at higher temperatures.

Reaction of [(ZrH$_2$Cp'$_2$)$_2$] with [{YH(THF)Cp'$_2$}$_2$] (Cp' = C$_5$H$_4$Me) in THF at room temperature for 2 h, followed by recrystallization of the residue from hexanes, yields colourless crystals of the heterotrimetallic tetrahydride [(YHCp'$_2$)$_2$(ZrHCp'$_2$)H] in 32% yield. The formula was inferred from ^{1}H NMR spectral data (δ −1.57 and −2.99 for μ$_2$-H and μ$_3$-H, respectively).[113]

11.4.7 MH$_2$Cp$_2$

ZrH$_2$Cp$_2$ has been extensively characterized by ^{1}H NMR spectral studies and has been shown to adopt a dimeric structure with bridging and terminal hydrido ligands in both benzene-d$_6$ and toluene-d$_8$ solution.[114] The dinuclear complex [{ZrH(μ-H)Cp'$_2$}$_2$] (Cp' = MeC$_5$H$_4$) contains *trans* terminal hydrides,[115] with a Zr–Zr distance of 0.346 nm. On the NMR timescale, the complex is stereochemically rigid in solution at 25 °C. Its degradation in the presence of diphenylethyne, D$_2$ or PPh$_3$ was monitored by EPR spectroscopy, revealing the presence of paramagnetic zirconocene hydrides, which may significantly contribute to its chemical reactivity.[116,117]

The complex [{ZrH(C$_5$H$_4$Bu^t)$_2$(μ-H)}$_2$] was synthesized by the reaction of ZrCl$_2$(Bu^{t}Cp)$_2$ with 2 equiv. of Li[AlH(OBu^t)$_3$] or Na[AlH$_2$(OCH$_2$CH$_2$OMe)$_2$] in either THF or toluene.[40] It can also be formed by hydrogenolysis of [ZrMe$_2$(Bu^{t}Cp)$_2$].[41] The crystal structure[40] shows terminal (0.20 nm) and bridging (0.182 nm) Zr–H bond lengths with a Zr···Zr distance of 0.347 nm. Proton NMR spectral data show hydrido signals at δ 3.84 (terminal) and δ −3.00 (bridging). A cyclic voltammetric study in THF at room temperature showed a pseudoreversible reduction wave at E_p = −1.9 V. Controlled potential electrolysis in the cavity of an EPR spectrometer indicates the formation of a monomeric dihydrido species formulated as the radical anion [ZrH$_2$(Bu^{t}Cp)$_2$]$^{\cdot-}$ (a(H) = 8.4 G, 1:2:1 triplet, a(^{91}Zr) = 16.2 G).

11.4.8 Reactivity of MH$_2$Cp$_2$

The chemistry of ZrH$_2$Cp*$_2$ has been extensively studied and reviewed in *COMC-I*. In particular, it is well known that these species react readily with unsaturated organic functions. The complex ZrH$_2$Cp$_2$ is used as a starting material for the preparation of [Zr(BH$_4$)$_2$Cp$_2$] via reaction in THF with BH$_3$(THF) or with catecholborane.

11.4.8.1 Main-group elements

The reaction of [{MH(μ-H)(Bu^{t}Cp)$_2$}$_2$] (M = Zr or Hf) with selenium or sulfur at room temperature gives the previously known[118,119] four-membered zirconacycles [{M(μ-Se)(Bu^{t}Cp)$_2$}$_2$] and [{M(μ-S)(Bu^{t}Cp)$_2$}$_2$],; respectively.[120] Intermediates have the formula [M(EH)$_2$(Bu^{t}Cp)$_2$]. In the case of E = sulfur, the reaction also gives the zirconahexasulfane species [ZrS$_5$(Bu^{t}Cp)$_2$].[121]

Reduction of ZrH$_2$Cp$_2$ with red or white phosphorus, with grey arsenic or with antimony in toluene results in the formation of the black tetrameric zirconocene [(ZrCp$_2$)$_4$] in 62% yield.[122] This tetramer reacts with TMS-Cl to give [Zr(TMS)ClCp$_2$] as deep-yellow crystals in 79% yield. However, reaction with Me$_3$SnCl gives a mixture of orange-yellow [Zr(SnMe$_3$)$_2$Cp$_2$], ZrCl$_2$Cp$_2$ and Sn$_2$Me$_6$. Reaction of [(ZrCp$_2$)$_4$] with Ph$_2$PCl yields ZrCl$_2$Cp$_2$ and P$_2$Ph$_4$.[122]

11.4.8.2 Alkynes

The reactivity of ZrH_2Cp_2 with alkynes is reviewed in Section 11.1.6.4.

11.4.8.3 Carbon monoxide

Treatment of $ZrH_2Cp^*_2$ with $[Zr(L)(CO)Cp_2]$ (L = PMe$_3$ or CO) affords zirconium oxycarbene complexes, $[Cp_2(L)Zr=CHO–Zr(X)Cp^*_2]$ (L = PMe$_3$, X = H or I; L = CO, X = H) (Scheme 18), which were some of the first examples of group 4 metal–carbon multiple bonds.[123] The x-ray structure[123] was established for $[Cp(PMe_3)Zr=CHO–Zr(I)Cp^*_2]$. Treatment of $[Cp_2(CO)Zr=CHO–Zr(H)Cp^*_2]$ with MeI forms $[Cp_2(CO)Zr=CHO–Zr(I)Cp^*_2]$, which reacts further with CO to afford the zirconium-substituted enediolate zirconacycle $[Cp_2Zr(OCH=\overline{CO})Zr(I)Cp^*_2]$.[123] An isotopic cross-over experiment has demonstrated that the new C–C bond is formed in one intramolecular coupling step. The intermediate ketene species coordinates pyridine and can be isolated as $[Cp_2(py)Zr(O=C=CHO)Zr(H)Cp^*_2]$.

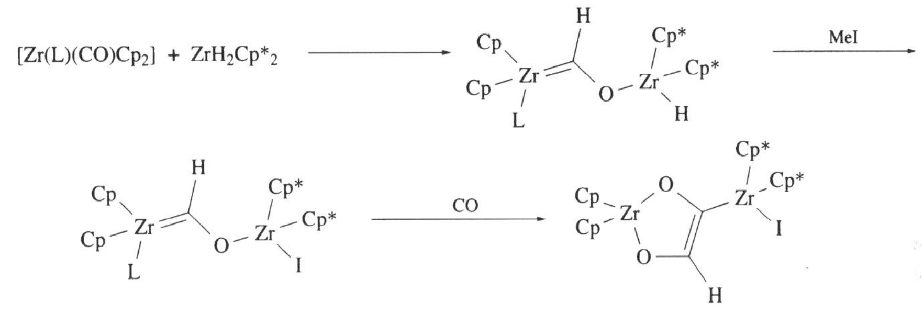

Scheme 18

The reaction of imidoylzirconocene, prepared from $ZrHClCp_2$ and CNR, with zirconocene hydride yields (*N*-alkylamido)zirconocene complexes (Scheme 19).[79]

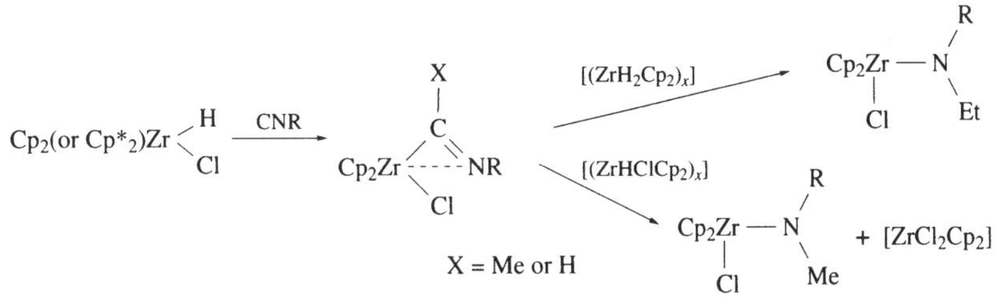

Scheme 19

Treatment of $[ZrMe_2Cp_2]$ with benzyl isonitrile gives (*N*-benzylacetimidoyl)methylzirconocene (ν(C=N) = 1660 cm^{-1} in CDCl$_3$), which when treated with PhICl$_2$ gives (*N*-benzylacetimidoyl)-zirconocene chloride (ν(C=N) = 1614 cm^{-1} in CDCl$_3$ and δ(^{13}C(C=N)) 229.6; Scheme 20).[79] Subsequent reaction with ZrH_2Cp_2 yielded (*N*-benzyl-*N*-ethylamido)zirconocene chloride as the only identified product. A 1:1 mixture of (*N*-benzyl-*N*-ethylamido)zirconocene chloride and methylzirconocene chloride was obtained by treatment of (*N*-benzylacetimidoyl)zirconocene chloride with $ZrHClCp_2$. In contrast, the reaction of $ZrHClCp_2$ with benzyl isonitrile gave an equimolar mixture of $ZrCl_2Cp_2$ and (*N*-benzyl-*N*-methylamido)zirconocene chloride.

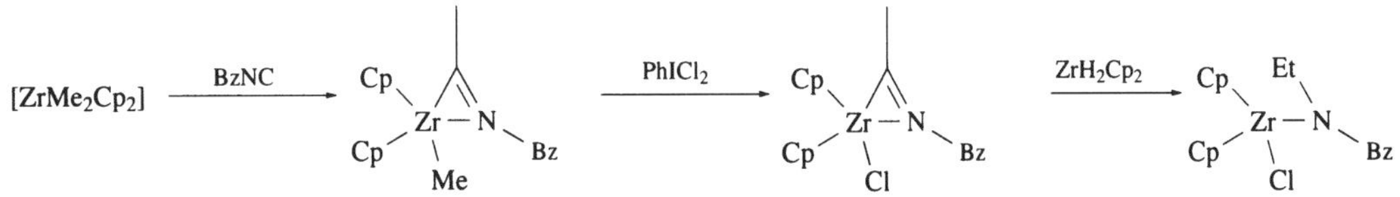

Scheme 20

11.4.8.4 Aldehydes and ketones

The complex MH_2Cp_2 (M = Zr or Hf) catalyses the chemoselective reduction of polycarbonyl compounds to hydroxycarbonyl compounds via a Meerwein–Ponndorf–Verley-type reduction.[124] For instance, with ZrH_2Cp_2, reduction of 3-ketobutanal and 2-phenyl-2-ketoethanal proceeds selectively at the aldehyde group to provide the corresponding hydroxy ketones in 91% and 93% yields. Under similar conditions, however, cyclohexadiones were easily aromatized to benzenediols. On the other hand, this complex also catalyses the selective 1,2-reduction of various types of α,β-unsaturated carbonyl compounds, giving the corresponding allylic alcohols in good to excellent yields. Thus, steroidal dicarbonyl compounds having an enone framework, such as Δ^4-androstene-3,17-dione and Δ^4-progestene-3,20-dione, can be reduced to the essential human hormones 17-hydroxy-Δ^4-androsten-3-one and 20-hydroxy-Δ^4-progest-3-one in 80% and 67% yields, respectively.[124]

In the presence of an appropriate hydrogen acceptor such as benzaldehyde or cyclohexanone, ZrH_2Cp_2 catalyses the Oppenauer-type oxidation of allylic alcohols to α,β-unsaturated carbonyl compounds. For example, primary allylic terpenoid alcohols, such as geraniol and nerol, are oxidized to α- and β-citrals (essential compounds in the perfume industry) in substantial yields. Similarly, secondary allylic alcohols such as 3-hexen-2-ol and 2-cyclohexen-1-ol are also easily oxidized to 3-hexen-2-one and 2-cyclohexen-1-one in 93% and 89% yields, respectively.[125]

11.4.8.5 Water, ammonia and organic azides

Reactions of water or ammonia with several $ML_xCp^*_2$ (M = Zr or Hf) derivatives, including hydrides, have been examined in order to assess some of the fundamental binding properties of hydroxo, oxo and amine ligands.[126] Water reacts in a stepwise manner with $MH_2Cp^*_2$ (M = Zr or Hf) to afford $[M(H)(OH)Cp^*_2]$, $[(MHCp^*_2)_2O]$, $[M(OH)_2Cp^*_2]$ and finally $[M(OH)_2Cp^*_2]\cdot H_2O$, while with $MHClCp^*_2$ the reaction yields $[M(OH)ClCp^*_2]$. The compound $MH_2Cp^*_2$ also reacts with $[M(OH)XCp^*_2]$ (X = Cl, OH or H) to afford $[Cp^*_2(X)M–O–M(H)Cp^*_2]$. In each case, conversion of an M–H to an M–O bond is accompanied by H_2 evolution. Ammonia also reacts rapidly with $MH_2Cp^*_2$ to yield $[M(H)(NH_2)Cp^*_2]$ and H_2. Excess NH_3 or $MH_2Cp^*_2$ does not react further, although exchange with free $^{15}NH_3$ is observed. The reactivity was in general higher for hafnium.

The complex $[(ZrN_2Cp^*_2)_2N_2]$ reacts rapidly and selectively with 1 equiv. of water to produce 3 equiv. of N_2 and $[(ZrHCp^*_2)_2O]$. The stepwise oxidative addition to the two O–H bonds proceeds via formation of a $[Zr(H)(OH)Cp^*_2]$ intermediate.

Hydrozirconation of piperidinoborane using either $ZrHClCp_2$ or ZrH_2Cp_2 gave $[Zr(Cl)Cp_2(\mu\text{-}H)\{N(Bu^t)B(NC_9H_{18})\}]$ (Equation (20)).[127]

$$\text{ZrHXCp}_2 \; + \; \text{(piperidinoborane)} \quad \xrightarrow{42\%} \quad \text{(product)} \tag{20}$$

The compound $HfH_2Cp^*_2$ reacts with organic azides, RN_3 (R = Ph or *p*-tolyl), to produce the moderately stable triazenido complexes $[Hf(H)(NHNNR)Cp^*_2]$.[128] Thermolysis (80 °C) of the latter yields $[Hf(H)(NHR)Cp^*_2]$, which can be prepared independently from $HfH_2Cp^*_2$ and H_2NR. Reaction of $HfH_2Cp^*_2$ with excess RN_3 produces $[Hf(NHR)_2Cp^*_2]$.

11.5 COMPLEXES OF ANIONIC CARBON-CENTRED π-LIGANDS OTHER THAN CYCLOPENTADIENYL

11.5.1 Allyl Complexes

11.5.1.1 Synthesis and bonding

Zirconium allyls are important as active catalysts and catalyst precursors as well as stoichiometric reagents in organic synthesis.[129–32]

[Zr(allyl)$_3$Cp] is prepared by treating a suspension of ZrCl$_3$Cp in ether with an excess (4 equiv.) of allylmagnesium chloride at −40 °C.[10,133] After low-temperature (below −35 °C) workup, large, dark-red crystals are obtained by crystallization. [Zr(allyl)$_3$Cp] is thermally unstable, decomposing above −10 °C even in the solid state. The existence in solution of two different types of ligands (η^1-allyl 1595 cm^{-1}, η^3-allyl 1535 cm^{-1}) rapidly equilibrating on the NMR timescale is suggested by IR spectroscopy. At temperatures below −90 °C, substantial line broadening of the ^{1}H and ^{13}C NMR signals is observed due to decreasing exchange rates. The molecular structure of [Zr(allyl)$_3$Cp] has been determined by a low-temperature (−173 °C) x-ray diffraction analysis,[133] which showed three distinct allyl ligands and one Cp$^-$ group, giving an overall distorted tetrahedral geometry around the zirconium. One allyl group is σ-bonded, a second is slightly η^3-allyl, while a third is symmetrically η^3-allyl.

In the presence of zirconium halides, [Zr(allyl)$_3$Cp] undergoes ligand exchange. A slow comproportionation with 2 equiv. of ZrCl$_3$Cp at −40 °C produces [Zr(η^3-allyl)Cl$_2$Cp] which was not isolated.[134]

Thermally stable pentamethylcyclopentadienyl(allyl) complexes can be synthesized at ambient temperatures from allylmagnesium bromide and ZrCl$_3$Cp*. By contrast, monocyclopentadienylzirconium derivatives of methylated π-allyl ligands have shown an unexplained instability. The compound [Zr(1-methylallyl)$_3$Cp*] was believed to have been prepared by the reaction[10] of ZrCl$_3$Cp* with 2-butenylmagnesium bromide in ether at −78 °C, initially forming a yellow product. Owing to its thermal instability no physical data are available. Upon warming at −40 °C or above, this product decomposed to give the red butadiene–allyl complex [Zr(η^3-MeC$_3$H$_4$)(η^4-C$_4$H$_6$)Cp*] in 83% yield. A similar reaction of HfCl$_3$Cp* with 2-butenylmagnesium bromide resulted in the formation of [Hf(1-methylallyl)$_3$Cp*] as a yellow-orange oil.[10] Infrared, ^{1}H and ^{13}C NMR spectra indicated a series of interconverting isomers. Heating [Hf(1-MeC$_3$H$_4$)$_3$Cp*] in refluxing THF causes the transformation to orange [Hf(η^3-1-MeC$_3$H$_4$(η^4-C$_4$H$_6$)Cp*] (72%) and *trans*-2-butene.[10] A mechanism involving β-elimination of a proton from the methyl group with subsequent reductive elimination of butene was proposed.

Monoallyl(mono-Cp*) complexes (allyl, 1,1,2-trimethylallyl or 1,2,3-trimethylallyl) are thermally stable and may be prepared in higher yield via reaction of allyl Grignard or allyllithium with ZrCl$_3$Cp*.[135] Cyclic voltammetry reveals that the complexes of trimethylated allyl are more difficult to reduce than ZrX$_2$Cp$_2$ (X = Cl or Br). Nevertheless, the former may be reduced by K[M(CO)$_2$Cp] (M = Fe or Ru) yielding [{M(CO)$_2$Cp}$_2$] and intractable oils.

The complexes [ZrBr$_2$Cp*(1,2,3-trimethylallyl)] and [ZrBr$_2$Cp*(1,1,2-trimethylallyl)] have been isolated as orange red and orange crystals in 34% and 24% yields, respectively.[136] X-ray crystallographic investigations[136] show a bent metallocene-type geometry with steric congestion in the asymmetrically methylated compound, causing a significant distortion of the η^3-bound allyl ligand toward an η^1-bonding mode. Similar behaviour has been observed for the bis(cyclopentadienyl) systems. A variable-temperature ^{1}H NMR spectral study indicates that the allyl ligand in [ZrBr$_2$(1,1,2-trimethylallyl)Cp$_2$] undergoes $\eta^3 \rightleftharpoons \eta^1$ isomerization ($\Delta G^{\ddagger}$(−2 °C) = 51.5 ± 1.0 kJ mol^{-1}).[137] The x-ray structures of [ZrBrCp$_2$(η^1-1,1,2-trimethylallyl)] and [ZrBrCp$_2$(η^3-1,2,3-trimethylallyl)] have shown that the employment of more sterically congested η^3-allyls results in greater distortion towards η^1 coordination (Equation (21)).[137] Steric congestion is probably one of the main factors responsible for the thermal instability of [Zr(η^3-1,2,3-trimethylallyl)BrCp$_2$].

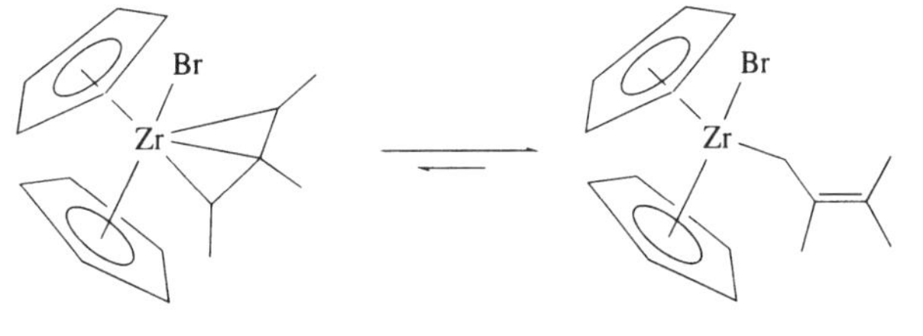

(21)

The butenyl-derived allyls [ZrCl(C$_4$H$_7$)Cp$_2$], [Zr(C$_4$H$_7$)$_2$Cp$_2$] and [Zr(C$_4$H$_7$)$_3$Cp$_2$(MgCl)(THF)] are prepared by analogous procedures from ZrCl$_2$Cp$_2$ and MeC(H)=C(H)CH$_2$MgBr.[138]

11.5.1.2 Reactivity

Oxidation of a series of allylzirconium complexes with H$_2$O$_2$, Bu^{t}O$_2$H, *m*-chloroperoxybenzoic acid or O$_2$ affords allylic alcohols.[139] The butenyl(allyl) complexes[138] have been shown to be effective reagents for the *threo*-selective synthesis of β-methylhomoallyl alcohols.[140]

11.5.2 Fluorenyl Complexes

A series of colourless and diamagnetic compounds of the formula $[Zr(S_2COR)Cl(Flr)_2]$ (Flr = fluorenyl; R = Me, Et or Pr^i) have been synthesized in 70% yield by the reaction of $[ZrCl_2(Flr)_2]$ with $K[S_2COR]$ in dimethoxyethane.[141] The complexes, which are thermally stable, are nonelectrolytes in nitrobenzene, poorly soluble in benzene, nitrobenzene, $CHCl_3$, CH_2Cl_2 or CS_2, and insoluble in water or acetone. The presence of xanthate, $[S_2COR]^-$, was confirmed[141] by the presence of IR bands in the region 1240–1260 cm^{-1} and 1010–1020 cm^{-1}. Electronic spectra show a single charge-transfer band in the 24 800–24 400 cm^{-1} region.

The compound $[ZrCl_2(Flr)_2]$ reacts with the sodium salts of various dialkyl dithiocarbamates in DME to give $[Zr(\eta^2\text{-}S_2CNR_2)Cl(Flr)_2]$ (R = Me, Et or Pr^i)[142] which are light yellow to brown, diamagnetic, nonelectrolytes and soluble in a variety of solvents including $CHCl_3$, CS_2, Et_2O and acetone. Electronic spectra show a single charge-transfer band at 24 980–24 300 cm^{-1}.

Thallous sulfate reacts with sodium fluorenide in THF to give dark-yellow crystalline Tl(Flr).[143] The 1:1 stoichiometric reaction of Tl(Flr) with $ZrCl_3Cp$ yields $[ZrCl_2Cp(Flr)]$ which is slightly soluble in ethanol, benzene, nitrobenzene or THF, and very soluble in acetone, $CHCl_3$ or CS_2.[143]

11.5.3 Indenyl Complexes

11.5.3.1 Monoindenyl derivatives

A number of thermally stable, green complexes of the type $[ZrLCp(Ind)]X$ (Ind = indenyl; $L = C_9H_6NO^-$ (oximate); X = Br, I, $ZnCl_3(H_2O)$, 0.5 $CdCl_4$ or $HgCl_3$) have been isolated.[144] Conductivity measurements show that all the complexes are 1:1 electrolytes,[144] with the exception of the cadmium(II) derivative which is a 1:2 electrolyte. Infrared spectra[144] show medium-intensity absorption bands at 365–345 cm^{-1} assigned to metal–halogen stretching vibrations. The strong band at 1340 cm^{-1} has been attributed to the C–N stretching of the oximate group.

One of the congeners of these compounds, $[ZrL^1L^2(Ind)]Cl$ (L^1 = pyrrolyl; L^2 = acetylacetonate), was prepared from the reaction of $[ZrCl_2(Ind)(pyrrolyl)]$ with acetylacetone.[145] Its further reaction with dithiocarbamate (dtc) salts yields ionic derivatives of the general formula $[Zr(acac)(Ind)(pyrrolyl)][dtc]$ (dtc = Me_2NCS_2, Et_2NCS_2, $Pr^i_2NCS_2$, (methyl)cyclohexylNCS$_2$, (ethyl)cyclohexylNCS$_2$ or (isopropyl)-cyclohexylNCS$_2$).[145] Infrared and NMR spectral studies show that acac adopts the typical chelating bonding mode while the zirconium ion is in a tetrahedral environment.[145]

11.5.3.2 Bis(indenyl) derivatives

Several preparations of bis(η^5-indenyl)dichlorozirconium and -hafnium have been described, each being a slight modification of the original.[146] The zirconium complex is isolated as a dark-brown, crystalline solid, m.p. 210 °C, from the reaction of indenylsodium or Tl(Ind) with $ZrCl_4$ in THF.[147] The most effective method for the preparation of $[MCl_2(Ind)_2]$ (M = Zr or Hf) requires the addition of a solution of 2 equiv. of indenylsodium in THF to a benzene suspension of MCl_4.[148] The resulting yellow solid is isolated and sublimed *in vacuo* affording yellow needles of $[MCl_2(Ind)_2]$ in 50% and 26% yield for zirconium and hafnium, respectively.

The complex $[ZrCl_2(Ind)_2]$ can undergo reductive carbonylation in the presence of magnesium turnings to give a 45% yield of green, diamagnetic $[Zr(CO)_2(Ind)_2]$ ($v(CO)$ = 1985, 1899 cm^{-1}). In contrast, only small and nonreproducible amounts were obtained in the case of the diamagnetic $[Hf(CO)_2(Ind)_2]$ ($v(CO)$ = 1979, 1892 cm^{-1}).[148] The crystal structure of the zirconium derivative shows an average Zr–C(π) distance of 0.2511 nm with the molecule adopting a *gauche* configuration.[148] Kinetic studies[149] performed on the CO substitution reactions of $[Zr(CO)_2(Ind)_2]$, using PMe_2Ph, $PMePh_2$, PPh_3, $P(OEt)_3$, PBu^n_3 and CO as nucleophiles, indicate them to be first order in both substrate and nucleophile. This implies an associative mechanism which probably involves an $\eta^5 \rightarrow \eta^3 \rightarrow \eta^5$ ring slippage in order to maintain the 18-electron count around the zirconium centre. Consistent with the electropositive nature of zirconium(II), equilibrium and rate constants indicate[149] a strong preference for the coordination of the π acid CO over the PR_3 ligand.

The compound $[M(^{13}CH_3)_2(Ind)_2]$ (M = Zr or Hf) undergoes degenerate methyl exchange with the methyl acceptor materials Al_2Me_6, $(MeAlO)_x$ and $AlMe(OC_6H_2Bu^t_2\text{-}2,6\text{-}Me\text{-}4)$.[150,151] The free energies, enthalpies and entropies of activation for these processes have been obtained by ^{13}C NMR spectroscopy, showing that an increased steric hindrance at the metal atom increases the free energy of activation.

Bidentate Schiff base (SB) complexes of bis(indenyl)zirconocene of general formula $[Zr(SB)Cl(Ind)_2]$ have been prepared from $[ZrCl_2(Ind)_2]$ and an SB in 70% yield, and were characterized by IR, NMR and electronic spectra.[147,152,153]

Mono- and bis(phenoxy)bis(indenyl) complexes of zirconium(IV) can be prepared in 60–70% yield by reacting $[ZrCl_2(Ind)_2]$ with a stoichiometric amount of the appropriate ArOH (Ar = Ph, p-ClC$_6$H$_4$, α-C$_{10}$H$_7$ or β-C$_{10}$H$_7$) in refluxing benzene and in the presence of NEt$_3$.[154] The yellow or yellow-brown products are soluble in benzene, nitrobenzene, CH_2Cl_2, $CHCl_3$, CS_2, ethanol and acetone and have been characterized by IR spectroscopy. Their electronic spectra do not show d–d transitions but only a single band in the 24 775–24 225 cm^{-1} region.[154]

The complexes $[Zr(S_2CNR_2)_2(Ind)_2]$, analogous to the monoindenyl dithiocarbamates, have been reported.[155] Similar methods can also be used for the preparation of the green 1:1 electrolytes $[Zr(acac)(Ind)_2][R^1R^2NCS_2]$ ($R^1 = R^2$ = Et or Pr^i; R^1 = Me, R^2 = PhCH$_2$; R^1 = Et, R^2 = C$_6$H$_4$Me; R^1 = Et, R^2 = C$_6$H$_{11}$; R^1R^2 = C$_6$H$_{12}$).

Reaction of $[ZrCl_2(Ind)_2]$ with an equimolar amount of a hydrazone or azine in refluxing THF gives brown complexes of formula $[Zr\{OC_6H_4C(R^1)=NNHR^2\}Cl(Ind)_2]$ and $[Zr\{OC_6H_4C(R^3)=NN=C(R^3)C_6H_4O\}(Ind)_2]$ (R^1 = H or Me; R^2 = H, Ph or C$_6$H$_4$(NO$_2$)$_2$; R^3 = H or Me) in 58% and 70% yield, respectively (Equation (22)).[156] Infrared, NMR and diffuse reflectance spectral studies have established the structures of the hydrazone and azine derivatives.

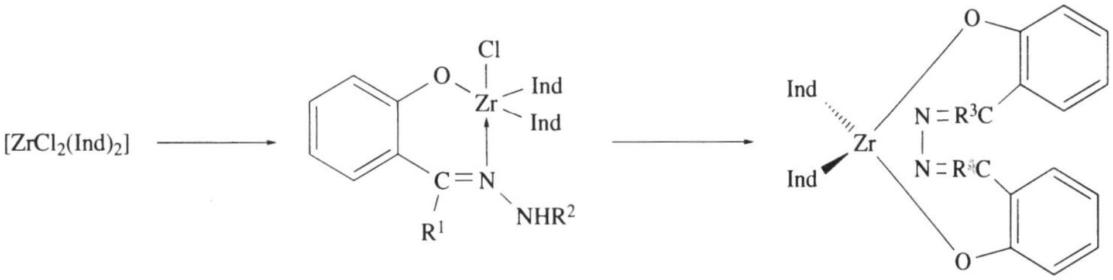

$$[ZrCl_2(Ind)_2] \longrightarrow \qquad \longrightarrow \qquad (22)$$

11.5.4 *ansa*-Derivatives

11.5.4.1 Synthesis and characterization

Zirconocene derivatives with a transannular tetramethylethylene bridge (*ansa*-zirconocenes) can be synthesized in three steps. The reductive coupling of 6,6-dimethylfulvene with sodium amalgam, sodium anthracenide or magnesium metal/CCl$_4$ affords tetramethyldicyclopentadienylethane which can be transformed into the corresponding dilithium salt by treatment with Bu^{n}Li. Reaction of the latter with ZrCl$_4$ forms the zirconocene derivative (Scheme 21).[157] The complex $[ZrCl_2(CH_2CH_2CpCp-1,2)]$ has been characterized by its ^{1}H NMR ((CDCl$_3$) δ 1.45 (s, 12 H), 6.14 (t, 4 H, 2.65 Hz), 6.66 (t, 4 H, 2.65 Hz)) spectrum and by mass spectrometry.

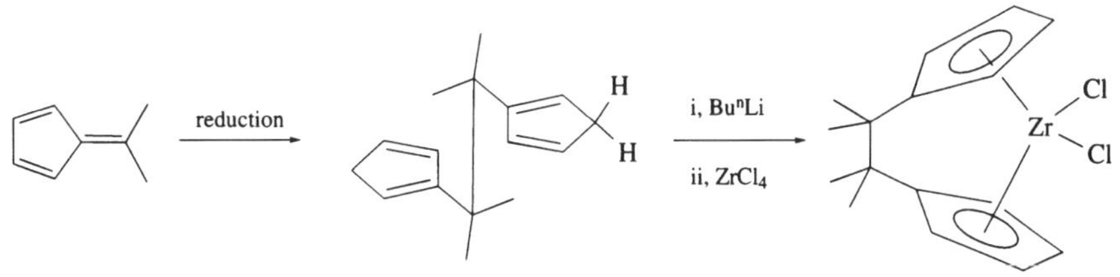

Scheme 21

The rigidity introduced in the bis(cyclopentadienyl) system by the two-carbon of the *ansa*-zirconocenes increases the stability of the ring system compared with ZrX$_2$Cp$_2$ analogues by making rearrangement of the ring ligands (formation of fulvene, fulvalene, etc.) more difficult.[158,159] In order to further decrease the reactivity of the ring system, the *ansa*-permethylcyclopentadienyl, yellow, crystalline, x-ray-characterized complex $[ZrCl_2\{C_2H_4(C_5Me_4)_2\text{-}1,2\}]$ was synthesized in 21% yield by treating dilithium bis(2,3,4,5-tetramethyl-1-cyclopentadienyl)ethane with $ZrCl_4$ first at $-80\ °C$ and then at reflux for 3 d in glyme.[160]

Synthesis of the yellow, air-stable solid $[ZrCl_2\{CH_2CH_2(C_9H_7)_2\text{-}1,2\}]$ (35% yield) was achieved by the room-temperature reaction of the dilithium salt with $[ZrCl_4(THF)_2]$.[161] It was characterized by its mass spectrum.[161] The appearance of only one pair of doublets (δ 6.19, 6.57) in the ^1H NMR spectrum, associated with the two protons of the five-membered ring, indicates only one zirconium ring-linkage isomer.

Catalytic hydrogenation of $[ZrCl_2\{CH_2CH_2(C_9H_7)_2\text{-}1,2\}]$ with PtO_2 in CH_2Cl_2 under a high pressure (100 bar) of H_2 yielded colourless, air-stable crystals (65% yield) of ethylenebis(4,5,6,7-tetrahydroindenyl)zirconium dichloride (Scheme 22).[161] The identity of the product as the *rac* rather than the *meso* isomer was revealed by NMR spectra and by x-ray diffraction.[161] Derivatization of the racemate by crystallization with *O*-acetyl-(*R*)-mandelic acid afforded the diastereoisomeric mixture which was resolved by fractional crystallization (Scheme 22).[162] The crystal structures revealed different chelated ring conformations in the two diastereoisomers. The separated diastereoisomers were converted into the corresponding zirconocene dichloride enantiomers, whose optical purity was demonstrated by their reconversion into the acetyl-(*R*)-mandelate derivatives.

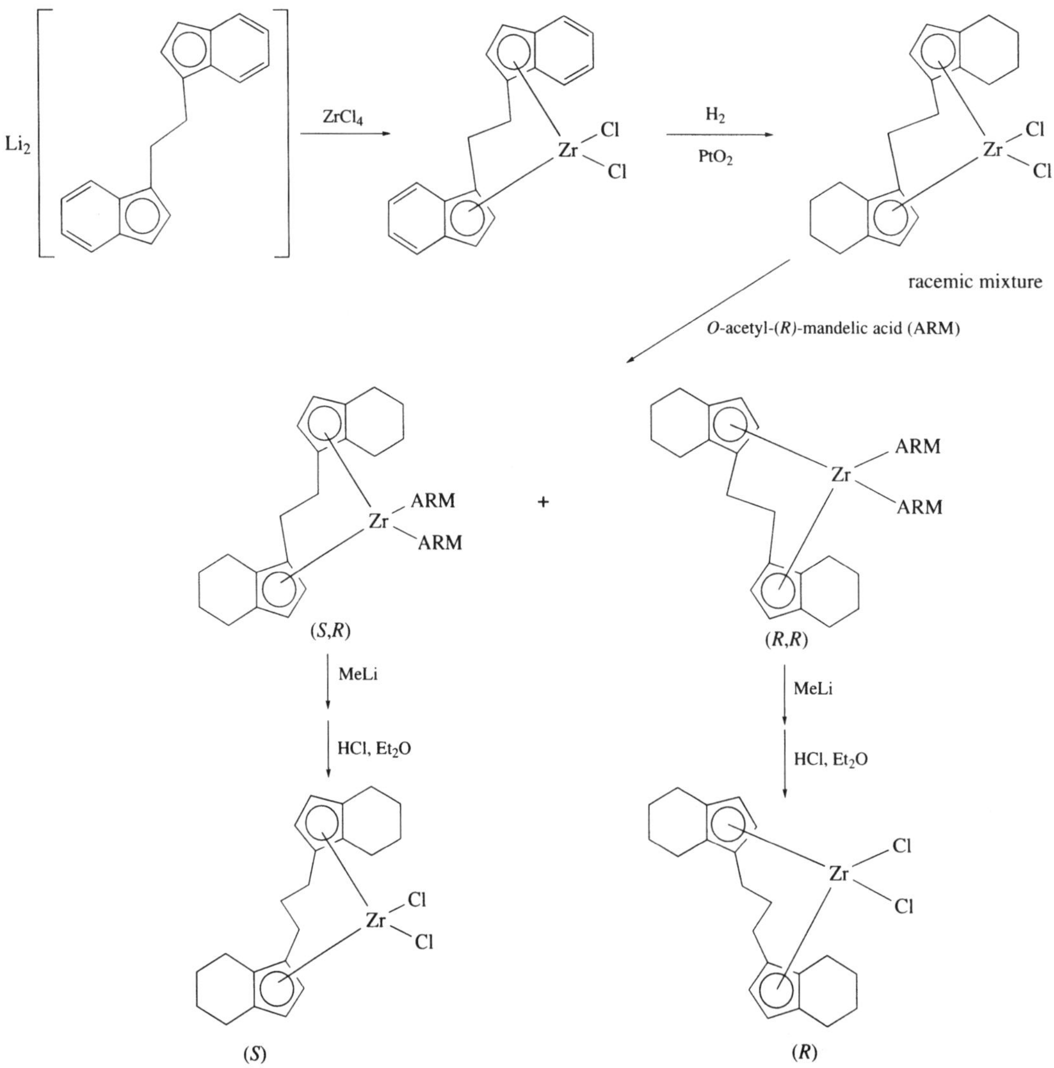

Scheme 22

The air-stable, yellow crystalline (4*S*,5*S*)-*trans*-4,5-bis(1H-inden-1-ylmethyl)-2,2-dimethyl-1,3-dioxolanedichlorozirconium, $ZrLCl_2$, was synthesized from the salt of the chiral indene with $ZrCl_4$ in

THF (Scheme 23).[163] The yields are as low as 10% when the dilithium salt is used, increasing to 16% when the corresponding disodium salt is employed. The molecular structure was confirmed by x-ray diffraction.[163] An analogous procedure was used for the hafnium derivative (8% yield) from $HfCl_4$ and the disodium salt. Hydrogenation under 100 atm of H_2 in the presence of PtO_2 gave the corresponding bis(4,5,6,7-tetrahydroindenyl) derivative (87%) as pale-green crystals. Deuterium labelling studies, carried out to establish the stereochemistry of the hydrogenation reaction,[164] have found that the PtO_2-catalysed hydrogenation occurs with *cis* stereochemistry at the *exo* face of the indenyl ligand.[164]

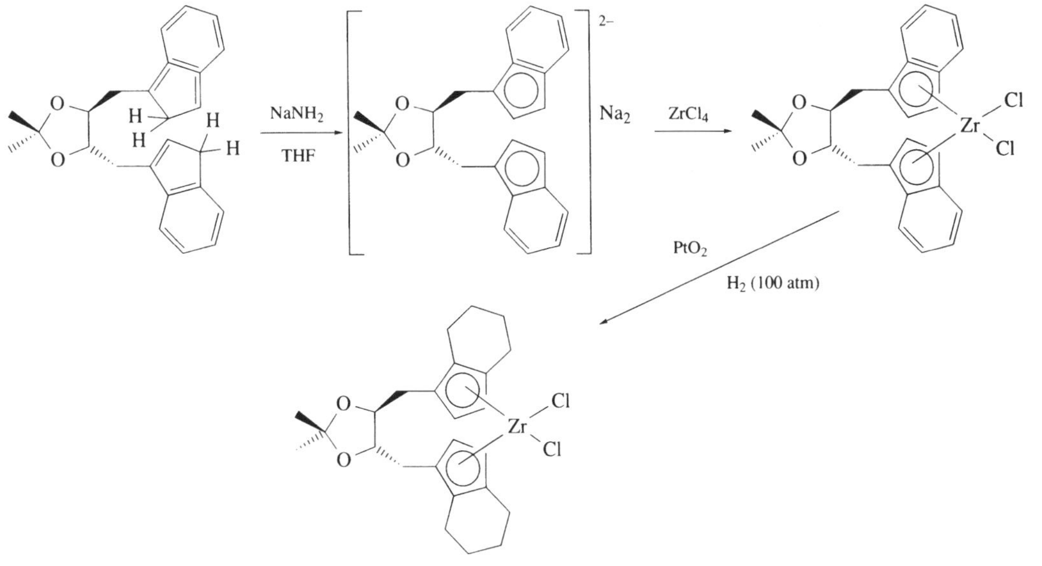

Scheme 23

11.5.4.2 Reactivity

The complex *rac*-[ZrCl$_2${C$_2$H$_4$(4,5,6,7-tetrahydro-1-indenyl)$_2$}] was a highly active and stereoselective catalyst for the polymerization,[165] hydrogenation[166] or hydrooligomerization[167] of α-alkenes in the presence of a methylaluminoxane cocatalyst.[165] Highly isotactic polypropene can be prepared by this procedure.[168]

Styrene, 2-methyl-1-pentene, 2-phenyl-1-butene and *cis*- and *trans*-2-hexene can be hydrogenated in the presence of the catalyst derived from [AlOMe]$_n$ and (−)-ethylenebis(4,5,6,7-tetrahydro-1(*R*)-indenyl)zirconium dichloride.[166] While α-alkenes are readily polymerized by this catalytic system, hydrogenated monomers can be obtained when the reaction is carried out under H_2. Terminal alkenes substituted in the 2- or 3-position and internal alkenes are not polymerized but undergo hydrogenation. Styrene is hydrogenated at 12 turnovers per minute under 20 atm of H_2 at 25 °C. Catalytic deuteration of styrene with [(−)-ethylenebis(tetrahydroindenyl)ZrX$_2$] (X$_2$ = (*R*)-1,1'-bis(bi-2-naphtholate)) yields (−)-(*R*)-1,2-dideuterioethylbenzene in 93% yield and 65% optical purity, indicating that the (*R*) enantioface of styrene is preferentially deuterated.[166] This enantiofacial selectivity is opposite to that observed in the case of the propene oligomers. In the presence of [(−)-ethylenebis(tetrahydro-indenyl)ZrX$_2$], 2-phenyl-1-butene is hydrogenated to give (−)-(*R*)-2-phenylbutane in 95% yield and 36% optical purity. Optically active alkanes can also be synthesized in molar amounts from simple alkenes using this catalyst.[169] The oligomerization can be controlled in order to produce mainly dimers to pentamers.

A wide variety of allylic amines can be prepared with enantiomeric excess up to 99% in moderate to good yields via reaction of alkynes with lithium amides LiNR^{1}R^2 assisted by the enantiomerically pure methyl triflate derivatives [(*S,S*)-ethylenebis(tetrahydroindenyl)ZrMe(OTf)].[170] This method tolerates a wide variety of structures in both the allyl and the amide moieties including substrates with oxygen functionalities.[170]

The oxalane-bridged indenyl derivative (Scheme 23)[163] has also been found to be active in the polymerization of ethene or propene with methylaluminoxane as cocatalyst.

A 'cationic' metallocene derivative $[\{(Ind)CH_2\}_2ZrMe\{CH_2CH_2(C_9H_7)_2\text{-}1,2\}][B(C_6F_5)_4]$ was prepared by reacting $[ZrMe_2\{CH_2CH_2(C_9H_7)_2\text{-}1,2\}]$ with $[Ph_3C][B(C_6F_5)_4]$.[171] Both the polymerization activity and stereospecificity of this complex are very high. The value of the A factor is greater at lower temperatures, which is an unprecedented behaviour in Ziegler–Natta catalysis where A normally decreases sharply with a decrease in T_p.[171] The major drawback in the use of this catalyst is that the average molecular weight of the polymers is too low for many practical applications. This problem has been addressed with the preparation of *ansa*-bridged bis(indenyl) systems, where the two indenyl rings are linked by silyl groups. These complexes have been prepared from indenyllithium and a dichlorosilane affording the corresponding *rac*-zirconium complex.[172] The utilization of an Me_2Si bridging group results in higher activity relative to ethylene-bridged analogues in the polymerization of propene with methylaluminoxane as cocatalyst and yields an average molecular weight twice as high as the ethylene-bridged catalyst, while the narrow molecular weight distribution and high isotacticity are maintained. However, complexes containing longer bridges, such as $-Me_2SiCH_2CH_2SiMe_2-$, are completely inactive towards propene polymerization although they exhibit high activity towards ethene.[172] The crystal structure of $[ZrCl_2\{(C_9H_7)_2SiMe_2\}]$ is similar to that of the ethylene-bridged indenyl complex with a larger angle (62° vs. 55°) between the two halves of the ligand system. This factor probably results in an increased accessibility to the zirconium centre, which is reflected in the higher catalytic activity. In contrast, the complex $[ZrCl_2\{CH_2CH_2(SiMe_2C_9H_7)_2\text{-}1,2\}]$ displays an asymmetrical arrangement of the ligand system, with the interligand angle being 54°.[172] The inactivity of this complex as a propene polymerization catalyst is probably due to shielding of the zirconium by the bulky bridge and by the indenyl ligands which are directed towards the $ZrCl_2$ moiety. These structural features are less detrimental for the sterically less demanding ethene.

11.5.5 Other Cp Derivatives

The optically active pinane-related anionic ligand $[PCp]^-$ has been complexed to $ZrCl_2Cp$ and $ZrCl_2Cp^*$ moieties for the purpose of assessing the level of π-facial stereoselectivity available for this ligand.[173] At or above room temperature the $[Cp]^-$ and $[Cp^*]^-$ ligands exhibit a preference for coordination to the less hindered face (above the plane), although rather high levels of below-plane complexation also appear. By heating $[ZrCl_3(THF)_2Cp]$ in toluene[173] at reflux for 3 d a 56% yield of a 3.5:1 mixture of the above-plane complex and its π-facial stereoisomer $ZrCl_2Cp(PCp)$ was obtained (Scheme 24). However, premixing at $-78\ °C$, followed by refluxing in toluene for 2 d, resulted in the selective formation of the below-plane isomer only, which was isolated as yellow crystals in 66% yield after sublimation. When $ZrCl_3Cp^*$ was used, a 69% yield of a 2:1 ratio of the two isomers was always formed, irrespective of the reaction conditions. The structure of the below-plane complex $ZrCl_2Cp(PCp)$ was confirmed by x-ray crystallography.[173] In the case of the closely related reaction of lithium isodicyclopentadienide (IDCp) with $[ZrCl_3(THF)_2Cp]$ and $ZrCl_3Cp^*$[174] in toluene at reflux for 3 d, only one product (the *exo* form) was obtained (69% and 55% yield for the Cp and Cp* derivatives, respectively) after sublimation (Scheme 25). The identity of the *exo* products was initially ascertained by ^1H and ^{13}C NMR spectroscopy and further confirmed by the crystal structure of the Cp* derivative. Employment of the camphor-derived $[CCp]^-$ ligand yielded a stereoisomeric mixture, the *endo*-complex (below plane) being preferred (Scheme 26).[175] Stirring $[ZrCl_3(THF)_2Cp]$ with LiCCp in THF at 20 °C affords, after recrystallization, a 58% yield of a yellow solid containing a 2:1 *endo:exo* mixture. However, for reaction with $ZrCl_3Cp^*$ to occur, the reagents had to be heated at reflux in toluene in order to give a 1.4:1 mixture of *endo* and *exo* isomers.

The change in π-facial response (i.e., *exo* for IDCp and *endo* for CCp) is attributed to the presence of the apical *syn* methyl group in CCp which serves to make the less hindered space on one of the two sides of the plane more effectively available.

Reactions of the optically active pinane-related cyclopentadienyl anion isodicyclopentadienide, $[IDCp^*]^-$, with $ZrCl_4$ in refluxing DME[176] yield a single *exo* isomer as pale-yellow crystals (**1**). The structure of the dichloride derivative was confirmed by x-ray crystallography. Similarly, the optically active lithium isocyclopentadienide, prepared by chemical modification of the commercially available $(1S,5S)$-$(-)$-verbenone, reacts with $ZrCl_4$ and $ZrCl_3Cp^*$ in DME at $-78\ °C$ followed by reflux for 3 d to form a single above-plane complex (**2**).[177] The two crystalline complexes (**1**) and (**2**) have been characterized by ^1H and ^{13}C NMR spectra, confirming that coordination occurs at the less sterically congested face. When $ZrCl_3Cp$ is the reactant, a 1:1 mixture[177] of both stereoisomers results.

The 'dibornacyclopentadienyl'-containing ligand forms a trichloro complex with zirconium in 40% yield (**3**), which acts as an effective Lewis acid catalyst for the enantioselective C–C coupling of pyruvic esters with a reactive arene.[178]

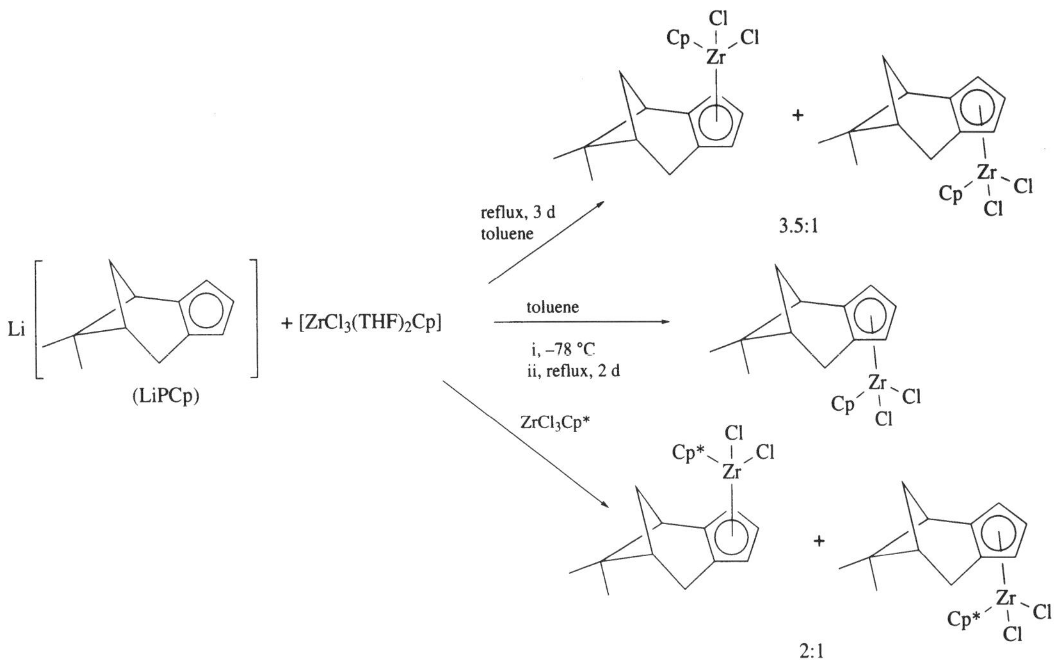

Scheme 24

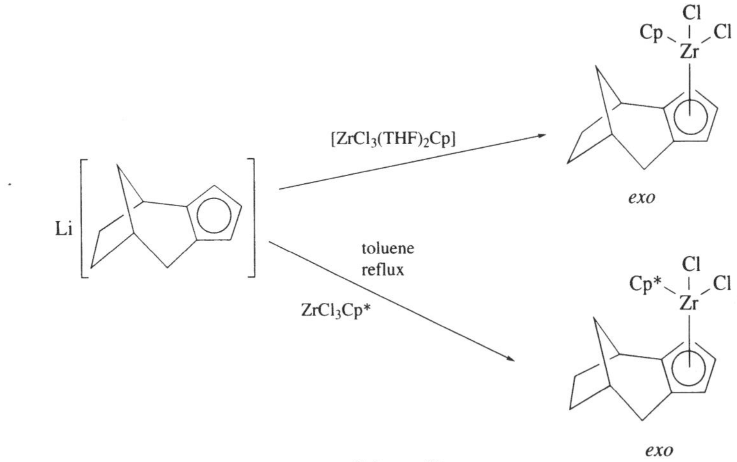

Scheme 25

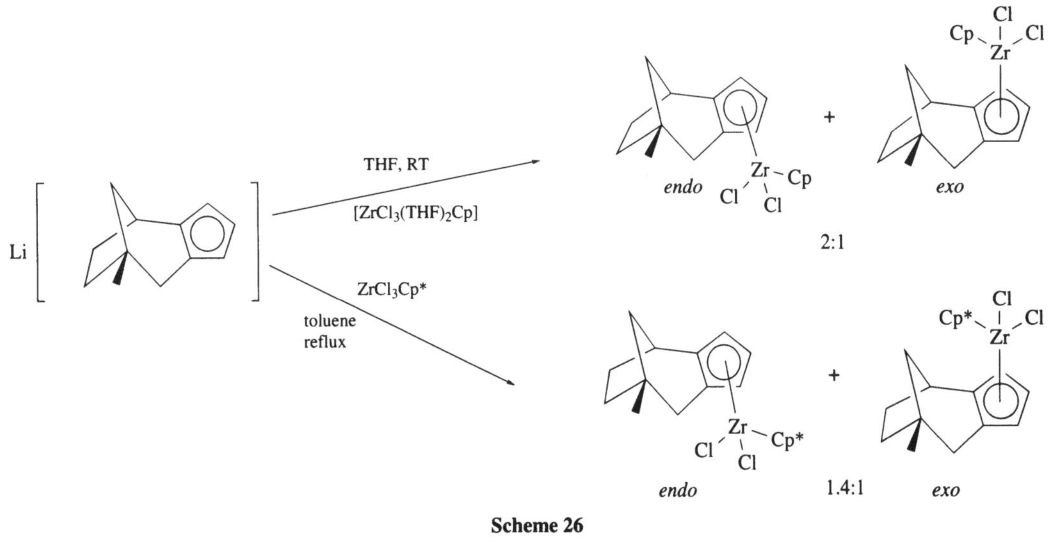

Scheme 26

(1)

(2)

(3)

11.5.6 Miscellaneous

The reaction of equimolar amounts of the carborane $C_2B_9H_{13}$ and [MMe$_3$Cp*] (M = Zr or Hf) affords the neutral 14-electron d^0 bent metallocene [{MMe($C_2B_9H_{11}$)Cp*}$_n$] in 95% and 87% yield, respectively.[179] The possible polymeric nature of this compound is suggested by its very low solubility in common organic solvents. The formal replacement of a uninegative $[C_5R_5]^-$ ligand of $[Zr(R)(C_5R_5)_2]^+$ by the isolobal, dinegative dicarbollide ligand $[C_2B_9H_{11}]^{2-}$ reduces the overall charge by one unit while leaving the gross structural and metal frontier orbital properties unchanged. Both carbollide complexes react with 2-butyne in benzene at 23 °C in less than 5 min to yield the monomeric alkenyl complexes [M{C(Me)=CMe$_2$}Cp*($C_2B_9H_{11}$)] in high yield via a single insertion reaction. An x-ray study of the zirconium compound[179] has established the presence of an $[\eta^5\text{-}C_2B_9H_{11}]^{2-}$ ligand and a bent metallocene geometry at the metal atom (centroid–zirconium–centroid angle 141.3°). The alkenyl group lies in the plane between the two η^5 ligands and is distorted by an agostic interaction involving the β-Me hydrogens, thus resulting in a rather distorted geometry of the alkenyl fragment. This complex does not undergo further insertion chemistry with 2-butyne and it does not coordinate THF. Conversely, [{ZrMe($C_2B_9H_{11}$)Cp*}$_x$] yields the THF adduct [ZrMe($C_2B_9H_{11}$)(THF)Cp*] which does not undergo further exchange with free THF on the NMR timescale at 23 °C. However, the THF complex reacts rapidly with 2-butyne to reform the alkenyl complex.

The complexes [{MMe($C_2B_9H_{11}$)Cp*}$_x$] (M = Zr or Hf) are moderately active catalysts for ethene polymerization, as well as for the oligomerization of propene to 2-methylpentane and 2,4-dimethylheptane. Thermal decomposition of the zirconium complex occurs quantitatively at 45 °C yielding methane and a novel methylidene-bridged complex [{ZrCp*($C_2B_9H_{11}$)}$_2$(μ-CH$_2$)] that has been characterized by x-ray diffraction.[179] It comprises two bent metallocene zirconium centres linked by the μ-CH$_2$ group with centroid–zirconium–centroid angles smaller than in the alkenyl complex (134.9° average).

Anionic heterocyclic ligands, such as pyrrolide anions, may coordinate to zirconium in an η^5 fashion. For example, the yellow, crystalline Tl(pyrrolyl), obtained from the reaction of Tl$_2$SO$_4$ with sodium pyrrolide in THF,[143] reacts in 1:1 stoichiometric ratio with ZrCl$_3$Cp to yield [ZrCl$_2$(π-C_4H_4N)Cp], which is slightly soluble in ethanol, benzene, nitrobenzene or THF, and very soluble in acetone, CHCl$_3$ or CS$_2$. The reaction of the tetralithium salt of octaethylporphyrinogen with [ZrCl$_4$(THF)$_2$] yields [ZrL(THF)] (LH$_4$ = octaethylporphyrinogen) in which two pyrrolyl ligands are bonded σ and two in an η^5 fashion (4).[180]

11.5.7 Cycloheptatrienyl Derivatives

The mixed sandwich compound [ZrCp*(chpt)] (chpt = cycloheptatriene) was prepared as purple crystals in 26% yield (after sublimation) from the reaction of ZrCl$_3$Cp* with magnesium turnings in the presence of cycloheptatriene in THF at room temperature for 15 h.[181]

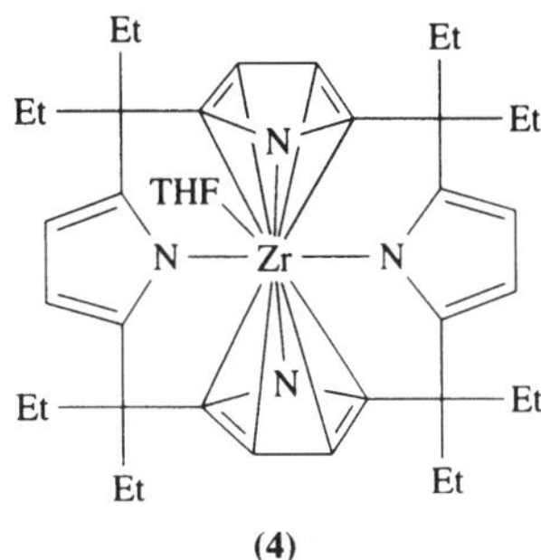

(4)

11.6 METALLOCENE ALKYLS, ARYLS AND ALKENYLS

11.6.1 Monoalkyl Derivatives

Mixed halogen–alkyl metallocenes [MRXCp$_2$] (M = Zr or Hf; X = halogen) can be prepared by reacting ZrX$_2$Cp$_2$ with the appropriate organolithium or organomagnesium reagents through control of reaction stoichiometry and conditions. With this procedure, several monoalkyl complexes [MRClCp$_2$] (M = Zr or Hf; R = CH$_2$C$_6$H$_4$NMe$_2$-o, CH$_2$C$_6$H$_4$PPh$_2$-o or CH$_2$CH=CHMe) have been synthesized.[141,182] The ^{1}H NMR spectra of these complexes exhibit a single resonance for the cyclopentadienyl ligands in the range δ 6.1–6.2, thus indicating that there is no direct coordination of the heteroatom to zirconium.[182]

A series of photolabile [ZrRXCp'$_2$] (Cp' = Cp or MeC$_5$H$_4$; X = Cl, Br or I; R = Bu^n, Bu^i or Et) complexes has been prepared by reacting ZrX$_2$Cp'$_2$ with 1 equiv. of RLi.[183] Upon low-temperature photolysis, the Bu^i complex (which is also obtained by isomerization of the intermediate complex obtained by treatment of ZrX$_2$Cp$_2$ with Bu^{i}Li) eliminates a mixture of isobutene and isobutane to form the corresponding zirconium(III) dimers [{ZrCp'$_2$(μ-X)}$_2$]. Employment of sterically less demanding alkyl groups (Bu^n or Et) gave similar [ZrXRCp$_2$] complexes, but in this case the photolysis produced significant amounts of zirconium(III) fulvalene dimers [{ZrCp'(μ-X)}$_2$(η^5:η^5-C$_5$H$_4$C$_5$H$_4$)] (Scheme 27).

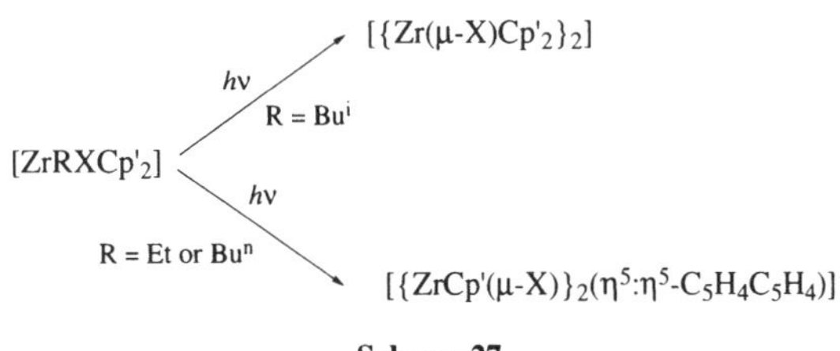

Scheme 27

Sterically hindered, chiral compounds [ZrCl(R)Cp$_2$] have been prepared following the same synthetic procedure with the appropriate organolithium reagent LiR(TMEDA) (R = CH(TMS)(o-PPh$_2$C$_6$H$_4$), 9-[CH(TMS)]C$_{14}$H$_9$, 2-[CH(TMS)]C$_5$H$_4$N or CH(TMS)(p-MeC$_6$H$_4$))[184,185] in THF at −78 °C. Crystal structures have been obtained for all these complexes. Cyclic voltammetry typically shows a one-electron reduction process (−1.82 V vs. SCE) producing zirconium(III) species. In some cases reduction to zirconium(III), by chemical or electrochemical means, leads to Zr–Cl bond rupture, yielding chlorine-free d^1 species Zr(R)Cp$_2$•.

Part of the interest in alkyls which possess a pendant coordinating site (e.g., 2-CH(TMS)py, CH(SPh)Ph or CH(Me)OEt) resides in alternative coordination modes. Variable-temperature ^{1}H NMR spectra of [ZrCl{2-[CH(TMS)]C$_5$H$_4$N}Cp$_2$][184] in toluene show the existence of an equilibrium in solution between four- and five-coordinate complexes. However, in the solid state, the nitrogen of the pyridine ring is bonded to the metal (Zr–N distance 0.234(1) nm), providing a rare example of a five-coordinate metallocene complex with an 18-electron configuration (5).

Reaction of ZrCl$_2$Cp$_2$ with 1 equiv. of LiCH(SR^1)R^2 in THF affords [ZrCl{CH(SR^1)R^2}Cp$_2$] (R^1 = Me or Ph; R^2 = TMS or Ph) in 60–80% yield (characterized by ^{1}H and ^{13}C NMR spectra).[186,187] In sharp contrast with the thermally unstable [ZrCl(CPh$_2$OMe)Cp$_2$] (half-life of 48 h at room temperature), the thio derivatives are indefinitely stable in the dark at room temperature and under an inert atmosphere.[188] The α-zirconocenyl thioethers are readily alkylated in hydrocarbon solution with the appropriate Grignard reagent forming the corresponding alkyl or aryl complexes [ZrR^3{CH(SR^1)R^2}Cp$_2$] (R^3 = Me, Ph or Bz) (60–80% yield).

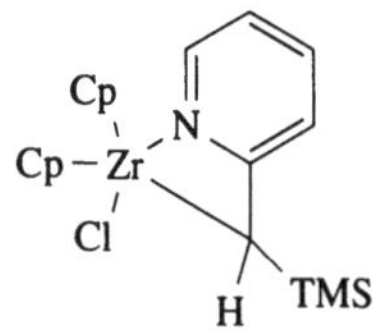

(5)

Thermolysis of [ZrPh{CH(TMS)SMe}Cp$_2$] in toluene gave [Zr(SMe){CH(TMS)Ph}Cp$_2$] (58%) via an α-heteroatom substitution reaction (Equation (23)).[186] Kinetic studies of this transformation by ^{1}H NMR spectroscopy show that it proceeds in a first-order process ($k = 8.14 \times 10^{-6}$ s^{-1}) with a free energy of activation of 122.2 kJ mol^{-1}. Further thermolysis produces BzTMS and an unidentified yellow precipitate. An NMR spectral study, carried out on [Zr{CH(SPh)Ph}$_2$Cp$_2$] and [Zr(CH$_2$SPh)$_2$Cp$_2$], indicates that the (methylthio)methyl and (phenylthio)methyl groups serve as four-electron η^2 ligands,[187] thus suggesting that the thiolato–alkyl rearrangement described above might be mediated by the metal centre.

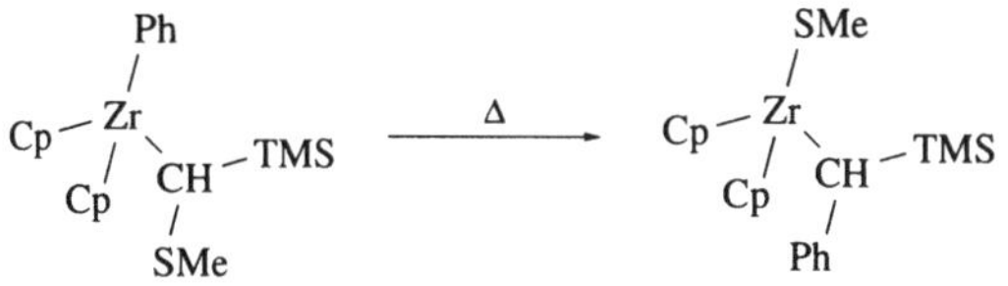

$$\text{(23)}$$

The complex [ZrCl{CH(Me)OEt}Cp$_2$] has been synthesized from zirconocene dichloride with (1-ethoxyethyl)lithium (Scheme 28).[189] The x-ray crystal structure suggests an 18-electron metal configuration, the oxygen atom of the η^2-ethoxyethyl ligand being coordinated to zirconium. Carbon monoxide or an isocyanide inserts into the Zr–C bond to give products without rearrangement of the 1-ethoxyethyl fragment (Scheme 28). Although [ZrCl{CH(Me)OEt}Cp$_2$] is thermally stable it decomposes at high temperatures to give ethene and [ZrCl(OEt)Cp$_2$].

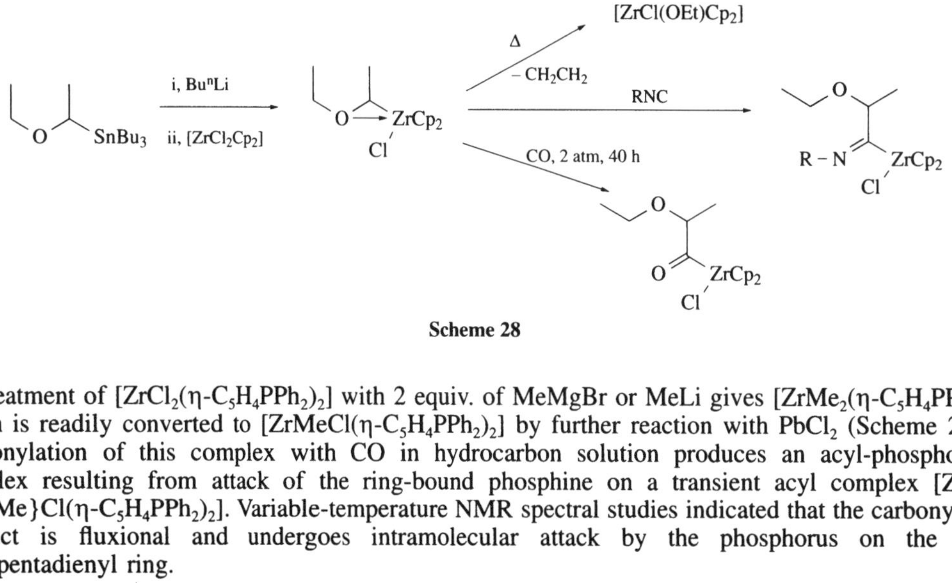

Scheme 28

Treatment of [ZrCl$_2$(η-C$_5$H$_4$PPh$_2$)$_2$] with 2 equiv. of MeMgBr or MeLi gives [ZrMe$_2$(η-C$_5$H$_4$PPh$_2$)$_2$], which is readily converted to [ZrMeCl(η-C$_5$H$_4$PPh$_2$)$_2$] by further reaction with PbCl$_2$ (Scheme 29).[190] Carbonylation of this complex with CO in hydrocarbon solution produces an acyl-phosphonium complex resulting from attack of the ring-bound phosphine on a transient acyl complex [Zr{η^2-(OC)Me}Cl(η-C$_5$H$_4$PPh$_2$)$_2$]. Variable-temperature NMR spectral studies indicated that the carbonylation product is fluxional and undergoes intramolecular attack by the phosphorus on the other cyclopentadienyl ring.

[ZrCl$_2$(η-C$_5$H$_4$PPh$_2$)$_2$] $\xrightarrow{\text{MMe}}$ [ZrMe$_2$(η-C$_5$H$_4$PPh$_2$)$_2$] $\xrightarrow{\text{PbCl}_2}$ [ZrMeCl(η-C$_5$H$_4$PPh$_2$)$_2$] $\xrightarrow{\text{CO}}$

[(η-C$_5$H$_4$PPh$_2$)(η-C$_5$H$_4$)PPh$_2$Zr{η^2-(CO)Me}]Cl

Scheme 29

Ligand redistribution between ZrCl$_2$Cp$_2$ and [Zr(CH$_2$PMe$_2$)$_2$Cp$_2$] yields [ZrCl(CH$_2$PMe$_2$)Cp$_2$] (Scheme 30),[191] which is stable in THF and undergoes carbonylation, similar to that of

[ZrCl(CH$_2$PMePh)Cp$_2$], with consequent loss of PMeR^{1}R^2 and formation of the dinuclear phosphine-functionalized ketene complexes [ZrClCp$_2$\{PR^{1}R^{2}CH=C(O)\}ZrClCp$_2$] (R^1 = Me or Ph; R^2 = Ph).[192] The reaction, which initially was thought to form the normal carbonylation product [ZrCl(COCH$_2$PR^{1}R^2)Cp$_2$],[193] has been demonstrated to proceed via intermolecular proton transfer.[194] Subsequent hydrolysis of the ketene complexes [ZrClCp$_2$\{PR^{1}R^{2}CH=(O)\}ZrClCp$_2$] (R^1 = Me or Ph, R^2 = Ph; R^1 = R^2 = Ph) forms CO and PMePh$_2$ together with [\{ZrClCp$_2$\}$_2$O], indicating that CO may be de-inserted (Scheme 30).[192]

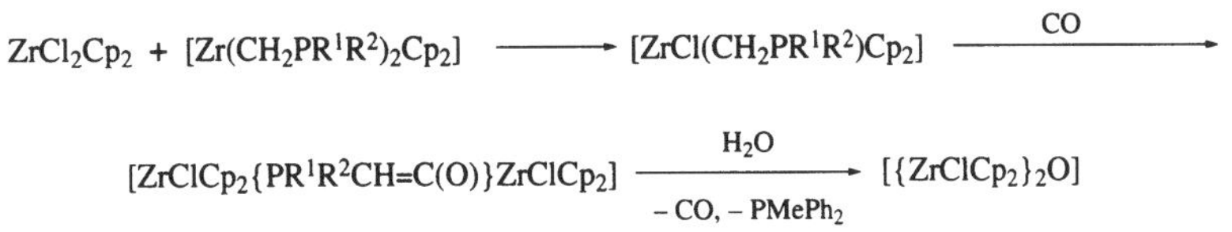

$$\text{ZrCl}_2\text{Cp}_2 + [\text{Zr(CH}_2\text{PR}^1\text{R}^2)_2\text{Cp}_2] \longrightarrow [\text{ZrCl(CH}_2\text{PR}^1\text{R}^2)\text{Cp}_2] \xrightarrow{\text{CO}}$$

$$[\text{ZrClCp}_2\{\text{PR}^1\text{R}^2\text{CH=C(O)}\}\text{ZrClCp}_2] \xrightarrow[-\text{CO, }-\text{PMePh}_2]{\text{H}_2\text{O}} [\{\text{ZrClCp}_2\}_2\text{O}]$$

Scheme 30

Another dinuclear ketene complex has been prepared by reacting [ZrClMeCp$_2$] with Na[ZrMe(OC=CH$_2$)Cp$_2$], obtained by deprotonation of [ZrMe(η^2-OCMe)Cp$_2$] with NaN(TMS)$_2$ (Scheme 31).[195]

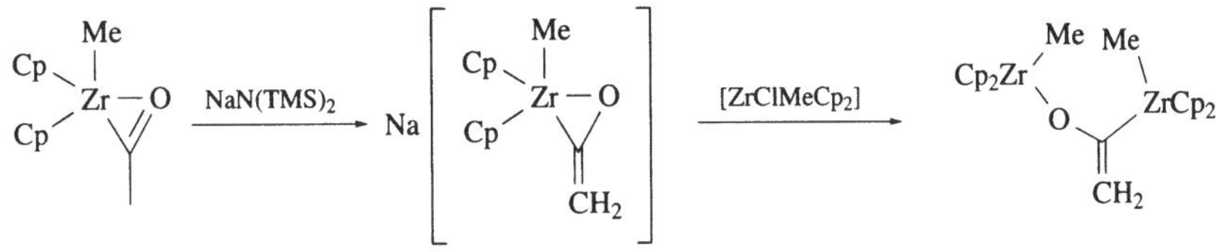

Scheme 31

Reaction of a dimeric zirconocene ketene complex with AlMe$_3$ in hydrocarbon solvents afforded the crystalline adduct given in Equation (24). The x-ray crystal structure of the latter confirmed its trimetallic nature, in which two zirconium centres are bridged by one methyl group, while the two oxygen atoms of the two ketene moieties are bonded to an AlMe$_2$ bridging unit.[196,197] The presence of a sharp singlet at δ −9.24 in the ^{13}C NMR spectrum implies that a single methyl bridges the two zirconium centres, while the C=C stretching mode appears at 1625 cm^{-1}. Although this colourless complex is stable at room temperature, at 60 °C and in aromatic solvents, it slowly decomposes to unidentified products.

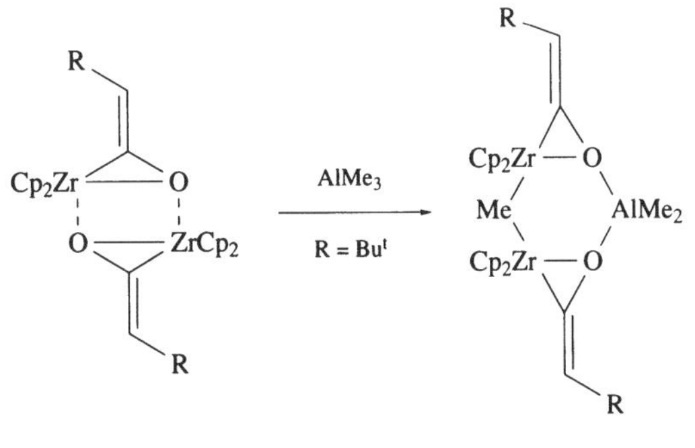

(24)

Attempts to prepare [ZrBr(cyclopropyl)Cp$_2$], via oxidative addition of C$_6$H$_{11}$Br to [Zr(PMePh$_2$)$_2$Cp$_2$] in toluene, were unsuccessful. The reaction afforded [ZrBr(CH$_2$Ph)Cp$_2$] via elimination of cyclopropane.[198]

11.6.2 Dialkyl Derivatives

11.6.2.1 Synthesis and reactivity

Zirconocene dialkyls [ZrR^{1}R^{2}Cp$_2$] (R^1 = R^2 = Bu^n,[199] CH$_2$Bz,[200] Me[201] or Bz[201]; R^1 = Me, R^2 = cyclobutyl[202]) were obtained by treatment of either a dichloro- or chloro(alkyl)zirconocene with either an organolithium or organomagnesium reagent.[199–202] In contrast with the high stability of $[ZrR_2Cp_2]$ (R = Me or Bz), which can be sublimed, a zirconocene dialkyl containing β-hydrogens is often unstable towards β-hydrogen reductive elimination to yield intensely coloured low oxidation state zirconium species which may be synthetically useful.[203]

In some cases, reductive elimination is obtained by treatment simply with PMe_3. For example, reaction of $[Zr(CH_2CH_2R)_2Cp_2]$ with PMe_3 affords $[Zr(CH_2{=}CHR)(PMe_3)Cp_2]$ in good yield.[200] Oxidation of the latter with a phosphorus ylide leads to zirconium(IV) complexes. For example, *trans*-$[Zr(CH_2CH{=}CHMe)(CHPPh_3)Cp_2]$ is formed via intramolecular hydrogen migration when a mixture of (s-cis-η^4-butadiene)- and (s-$trans$-η^4-butadiene)zirconocene complexes is treated with $CH_2{=}PPh_3$ at $-50\,°C$.[204]

The complexes $[ZrR_2Cp_2]$ (R = Bz or CH_2TMS) have been photolysed at low temperature in toluene solution.[205] Loss of alkyl radical (identified by ESR spectra at $-85\,°C$ for the benzyl or trimethylsilylmethyl compounds) leads to $ZrRCp_2{•}$ species which undergo further thermal reactions affording unknown diamagnetic material. In the presence of PPh_3, these radicals may be stabilized to form zirconium(III) species $[ZrR(PPh_3)Cp_2]$. Under these conditions, ready hydrogen transfer from the cyclopentadienyl ring to the metal has been observed with formation of a hydridozirconium(III) species. The latter shows hyperfine coupling in the EPR spectrum ($a(^1H) = 6.6$ G, doublet). Labelling studies show that the metal-bound hydride originates from the Cp^- ligand.

Dialkylzirconocene complexes containing donor atoms in the alkyl function, such as CH_2PMe_2, display higher stability and may be prepared in THF at room temperature by reaction of $ZrCl_2Cp_2$ with $2\,[Li(CH_2PMe_2)(TMEDA)]$.[191] The ^1H NMR spectrum of the product shows a doublet at δ 0.92 attributable to the methylene group; the crystal structure of this complex has confirmed its monomeric nature.[206]

11.6.2.2 Addition reactions

The complex $[Zr(CH_2PPh_2)_2Cp_2]$ can be used as a chelating ligand for the preparation of heterobimetallic complexes. Treatment of $[Zr(CH_2PPh_2)_2Cp_2]$ with $[Cr(norbornadiene)(CO)_4]$ gives cis-$[Cp_2Zr(\mu\text{-}CH_2PPh_2)_2Cr(CO)_4]$[207] which contains a six-membered Zr–C–P–Cr–P–C–Zr ring in a twist–boat conformation. A similar reaction with $[\{Rh(CO)_2(\mu\text{-}Cl)\}_2]$ yields *trans*-$[Cp_2Zr(CH_2PPh_2)_2Rh(CO)Cl]$.[207] The IR data ($v(C{=}O) = 1960$ and $1650\;cm^{-1}$) suggest a strong interaction between the oxophilic zirconium atom and the carbonyl oxygen.

In the absence of UV light $[ZrMe_2Cp_2]$ and $[RuCl_2(PPh_3)_3]$ react quantitatively to produce CH_4, $[ZrMeClCp_2]$ (100%) and $[RuCl(C_6H_4PPh_2)(PPh_3)_2]$.[208,209] The two starting materials react together under UV irradiation to yield the light-yellow complex $[Cp_2Zr^{IV}(\mu\text{-}Cl)(\mu\text{-}CH_2)RuCl(PPh_3)_2]$. The latter is cleaved by an excess of gaseous hydrogen to give $[ZrMeClCp_2]$ and an unidentified ruthenium hydride, $v(Ru{-}H) = 2020\;cm^{-1}$.

11.6.2.3 Insertion reactions

Zirconocene dialkyls are versatile substrates for insertion reactions with a variety of unsaturated reagents. The formation of Zr–O or related bonds resulting from such a reaction is often the driving force.

Nitrogen oxide reacts with $[ZrR_2Cp_2]$ (R = Me or Bz) in hydrocarbon solution by a double NO insertion into the same Zr–C σ-bond. The resulting chelating N-alkyl-N-nitrosohydroxylaminato ligand is bound to zirconium through the two oxygen atoms (Scheme 32).[210] The products have been identified by microanalyses, ^1H NMR spectra and, in the case of the benzyl substituent, by x-ray analysis.

Diphenylketene gives successively $[Zr(R)\{O{-}C(R){=}CPh_2\}Cp_2]$ and $[Zr\{O{-}C(R){=}CPh_2\}_2Cp_2]$ (R = Me, Bz or Ph).[211]

The insertion of phenyl isocyanate into one of the Zr–R bonds yields $[Zr(R)\{O{-}C(R){=}NPh\}Cp_2]$.[211] The chelating bonding mode adopted by the new ligand inhibits the insertion of a second molecule of PhNCO.

A similar insertion reaction is the addition of R_2CN_2 at room temperature to a solution of $[ZrR_2Cp_2]$ (R = Me or Bz).[212] One such monoinsertion product, $[ZrMe\{\eta^2\text{-}N(Me)NCR_2\}Cp_2]$, has been characterized by x-ray diffraction.

Reaction of $[ZrR_2Cp_2]$ (R = Me or Ph) with phenyl azide at room temperature gives $[ZrR(RNNNPh)Cp_2]$ containing the chelating triazenido ligand. However, a similar reaction with azidotrimethylsilane in Et_2O affords tetramethylsilane and $[ZrMe(N_3)Cp_2]$.[213]

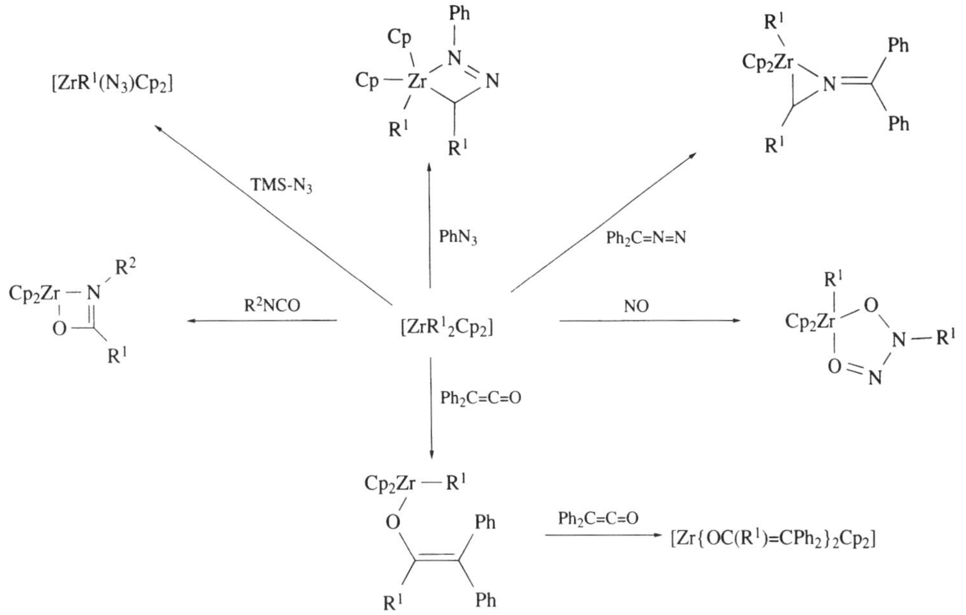

$[ZrR^1(N_3)Cp_2]$

$[Zr\{OC(R^1)=CPh_2\}_2Cp_2]$

Scheme 32

11.6.2.4 Catalysis

Treatment of dehydroxylated or partially dehydroxylated alumina with $[ZrMe_2Cp^*_2]$ or $[ZrMe_2Cp_2]$ provides highly active catalysts for alkene hydrogenation.[214] Cross-polarization–magic angle spinning NMR spectroscopy reveals methyl group transfer to aluminum surface sites and the formation of organozirconium species with cationic character.

Treatment of $ZrCl_2Cp_2$ with 1 equiv. of t-butyllithium at $-78\,°C$, followed by warming to $25\,°C$, quantitatively produces $[ZrClBu^tCp_2]$,[183,215] which upon treatment with an additional equivalent of t-butyllithium in hexane-d^{14} or Et_2O-diol for 1 h cleanly affords $[ZrBu^tBu^tCp_2]$.[215] The latter is only stable below $-60\,°C$ and serves as a source of zirconocene, permitting direct conversion of various alkenes and dienes into the corresponding zirconocene complexes in high yields.[215] The catalyst $[ZrBu^n_2Cp_2]$, generated from $ZrCl_2Cp_2$ and $2\,Bu^nLi$ in THF, also behaves as a versatile source of '$ZrCp_2$', and has been used for a preparation of zirconacycles[216] or zirconium(III) species.[203]

The combination of MCl_2Cp_2 (M = Zr or Hf) with organolithium or organomagnesium reagents is an effective catalyst for the condensation of primary and secondary silanes.[98,199,217] The reaction may be exploited for the synthesis of chains of three or more silicon atoms. Linear oligomers are the initial products, with cyclic polysilanes being increasingly produced as the reaction proceeds. An intermediate of the type $[Zr(H)(SiHR^1R^2)Cp_2]$ is proposed as the chain-growing unit. For secondary silanes, $HfCl_2Cp_2$ produces only disilane and trisilane.

The combination of $ZrCl_2Cp_2$ with $2\,BuLi$ is also an effective catalyst for the hydrosilylation of alkenes such as styrene, 1-hexene or 2-pentene with diphenylsilane.[218]

11.6.3 Diaryl Derivatives

The diarylzirconocenes $[ZrAr_2(\eta\text{-}C_5Me_4H)_2]$ (Ar = Ph or C_6H_4Me), prepared from the corresponding zirconocene dihalide and an aryllithium agent in ether at $-70\,°C$, have been characterized by NMR spectroscopy.[219,220]

The transient benzyne intermediate generated by the thermolysis of $[ZrPh_2Cp_2]$ can be trapped by the introduction of PMe_3 (Scheme 33).[221] The metallocene geometry is relatively regular but the C–C 'triple bond' length was 0.136 4(8) nm. Although this complex is thermally stable, it reacts with unsaturated organic molecules (Scheme 33). Reaction with methanol yields $[Zr(OMe)PhCp_2]$.

As a part of an effort to synthesize bimetallic benzdiynes and trimetallic benztriynes, interesting η^1-bridging arylzirconocene complexes $[(ZrMeCp)_n(C_6H_{6-n})]$ ($n = 2$ or 3) were prepared[222,223] by the reaction between $[ZrMeClCp_2]$ and the appropriate lithiobenzene in ether at $-78\,°C$ (70–90% yield; Scheme 34).

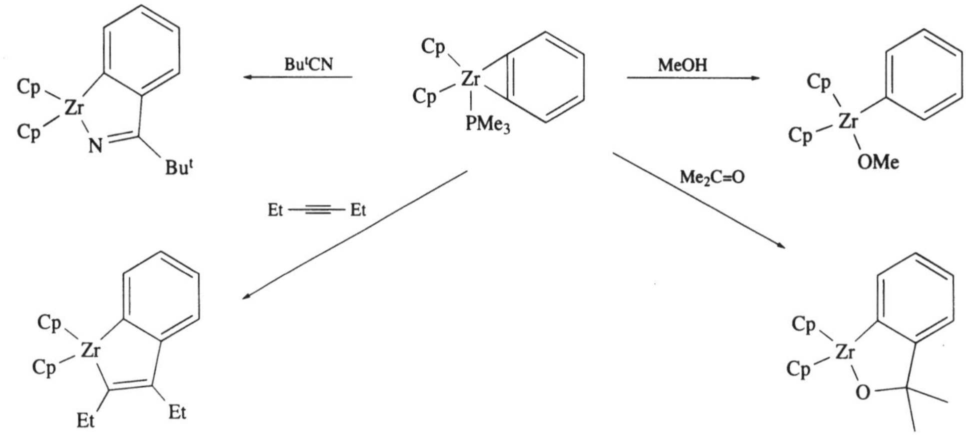

Scheme 33

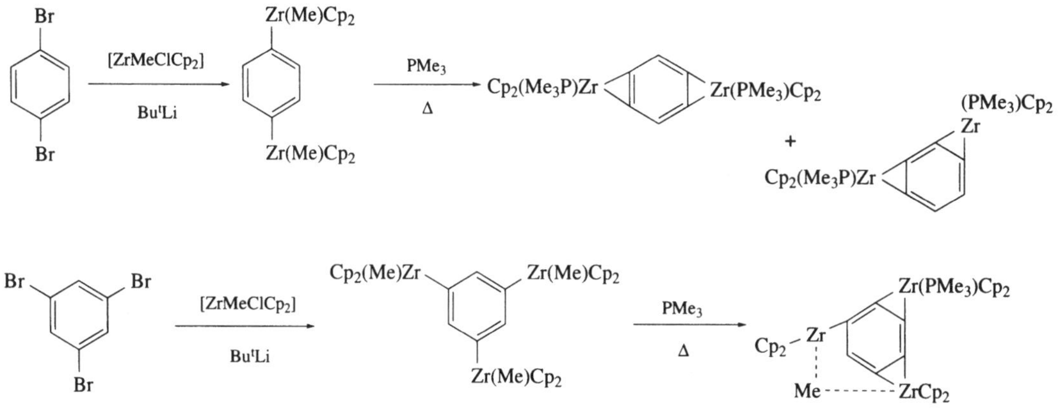

Scheme 34

Thermolysis of [1,4-(ZrMeCp)$_2$(C$_6$H$_4$)] in the presence of PMe$_3$ produces two isomeric bimetallic benzdiynes while, similarly, [1,3,5-(ZrMeCp)$_3$(C$_6$H$_3$)] and an excess of trimethylphosphine gave a benzdiyne species (identified by NMR spectral data).

The complex [ZrPh$_2$Cp*$_2$] undergoes unimolecular thermolysis in toluene at 100 °C ($\Delta H^{\ddagger}$; = 94.2(1) kJ mol^{-1}, $\Delta S^{\ddagger}$ = −11.8(3) eu) to yield an orange ring-metallated fulvene complex (90%; Scheme 35).[224] Its crystal structure has been determined; the ^{1}H NMR spectrum shows two magnetically nonequivalent, coupled ($J_{geminal}$ = 6.6 Hz) methylene protons.

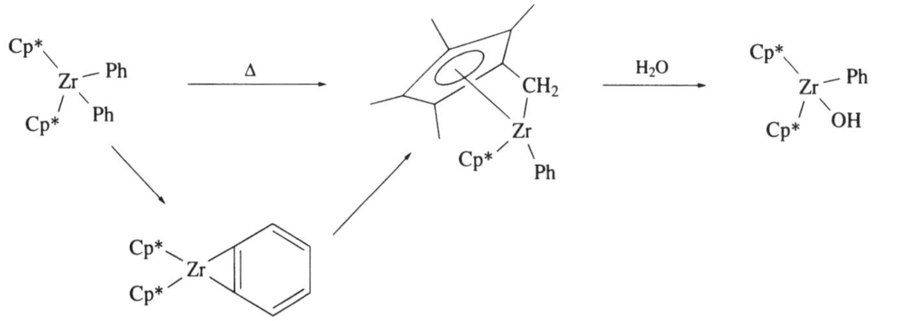

Scheme 35

Labelling and kinetic studies using [Zr(C$_6$D$_5$)$_2$Cp*$_2$] indicate that the thermolysis proceeds by rate-limiting *ortho*-hydrogen atom abstraction to yield a benzyne intermediate. The ring-metallated complex is hydrolysed at room temperature to produce colourless crystals of [Zr(OH)PhCp*$_2$] (Scheme 35).[19]

Like the alkyl counterparts,[225] [ZrPh$_2$Cp$_2$] undergoes NO insertion to form colourless [ZrPh{ON(Ph)NO}Cp$_2$] (44% yield; Scheme 32).[226] Four IR spectral bands characteristic of [ON(R)NO]$^-$ were at 1345, 1295, 1218 and 928 cm^{-1}.

Photolysis of [ZrAr$_2$Cp$_2$] in toluene at -85 °C gave the zirconium(III) species [ZrArCp$_2$] (Ar = Ph or C$_6$H$_4$Me) (g_{av} = 1.979, a(^{91}Zr) = 23.5, a(^{1}H) = 3 G) and an unidentified hydridozirconium(III) species (g_{av} = 1.987, a(^{91}Zr) not observed, a(^{1}H) = 6.6 G).[205] Photolysis in the presence of PPh$_3$ affords [Zr(PPh$_3$)PhCp$_2$] as a major product (g_{av} = 1.988, a(^{91}Zr) = 19–19.5, a(^{31}P) = 23.7–24.2 G) and [Zr(PPh$_3$)Ph$_2$Cp] (g_{av} = 1.996, a(^{91}Zr) = 23.2, a(^{31}P) = 19.5 G) and [Zr(PPh$_3$)Ph$_2$Cp$_2$] as minor products.

11.6.4 Alkenyl Derivatives

Alkenyl(chloro)zirconocenes are generally accessible from the appropriate alkyne, via a highly regio- and stereoselective hydrozirconation reaction. The ν(C=C) values for the alkenyl groups lie in the range 1550–1570 cm^{-1}. Protons on the α-CH exhibit a low-field NMR spectral chemical shift (δ 8.2–8.5). An agostic interaction between the metal centre and a β-CH is often observed in both the IR and ^{13}C NMR spectra, the former showing a broad band around 2600 cm^{-1} and the latter displaying a resonance which has an exceptionally large J_{C-H} value (110 Hz).[227] Reaction of ZrHClCp$_2$ with an internal acyclic alkyne provides a general synthetic route to zirconocene alkenyls (Scheme 36).[221,228,229]

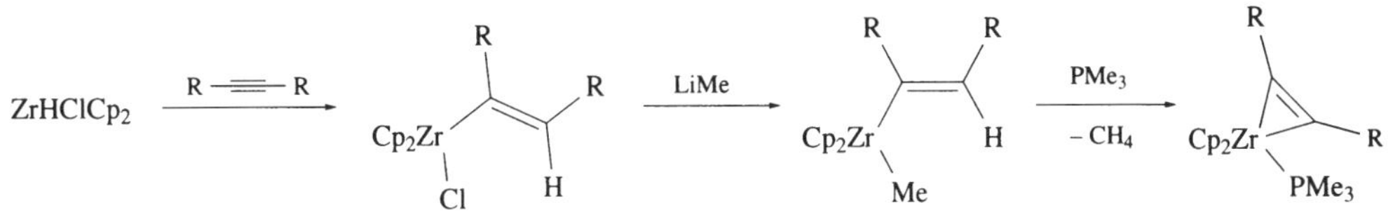

Scheme 36

Hydrozirconation of the alkyne followed by addition of LiMe produces the intermediate methyl–alkenyl complex which, upon coordination of trimethylphosphine, eliminates methane forming an alkyne–phosphine complex. The x-ray structure of the PMe$_3$ adduct of the 1-hexyne complex has been determined.[221] Complexes may be isolated from the reaction mixtures in approximately 70% yield and can be further purified by recrystallization from ether/hexane at low temperature. These alkyne complexes readily couple with nitriles, alkynes, ketones, aldehydes or ethene to give the corresponding insertion products.

Reaction of phenyl(trimethylsilyl)ethyne with ZrHClCp$_2$ yields the β-CH agostic complex [ZrCl{C(TMS)=CHPh}Cp$_2$] (Scheme 37).[230] Halogen exchange with lithium bromide or iodide produces the corresponding alkenylzirconocene bromide or iodide. Hydrozirconation of (trimethylsilyl)ethyne gives [ZrCl{C(H)=CH(TMS)}Cp$_2$] without agostic interaction.[230]

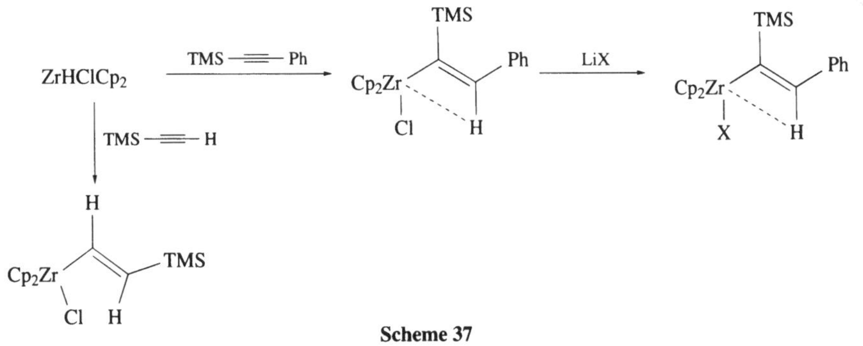

Scheme 37

Other species, including ZrCl$_2$Cp$_2$ and the metallacyclopentadiene [ZrC(TMS)=CPhCPh=C(TMS)Cp$_2$], accompany the formation of [ZrCl{C(TMS)=C(Ph)H}Cp$_2$].[231] The metallacyclic product may be formed via hydride–chloride exchange, producing the intermediate [Zr(H){C(TMS)=CHPh}Cp$_2$] which loses H$_2$ and adds 1 equiv. of (TMS)C≡CPh. Reaction of [ZrCl{C(TMS)=C(Ph)H}Cp$_2$] with LiC≡CR (R = Me or Ph) also produces metallacyclic species. Intramolecular abstraction of the acidic agostic alkenyl proton by the σ-alkynyl ligand leads directly to formation of the corresponding bis(alkyne)metallocene complexes which undergo rapid ring closure to give [ZrC(TMS)=C(Ph)C(H)=C(R)Cp$_2$]. Extended Hückel MO calculations[231] suggested that the 'anomalous' hydrozirconation of diphenylethyne, which cleanly produces [ZrCl$_2$Cp$_2$] and the metallacycle [Zr(Ph)C=C(Ph)C(Ph)=C(Ph)Cp$_2$], also proceeds via similar β-agostic intermediates.

The coupling reaction of a zirconocene alkyne with a second alkyne provides a general method for the preparation of asymmetrically substituted zirconacyclopentadienes.[232] The overall transformation is the chemo- and regioselective intermolecular cross-coupling of two alkynes, a goal that has not been achieved by any other method. Isomerically pure 1,3-dienes or 1,4-diiodo-1,3-dienes are obtained upon treatment of the metallacyclopentadienes with aqueous acid or iodine. Since this synthesis can be carried out in one step starting from the alkynes and $ZrHClCp_2$, it is a convenient alternative to other transition metal based methods for the synthesis of dienes.

A one-pot transmetallation sequence via zirconium intermediates has been developed allowing for the direct low-temperature (−78 °C) conversion of terminal alkynes to vinylcuprates (Scheme 38),[233] providing a short and efficient route on a large scale (1 kg) for the synthesis of prostaglandins, such as the antiulcer misotropol.[234]

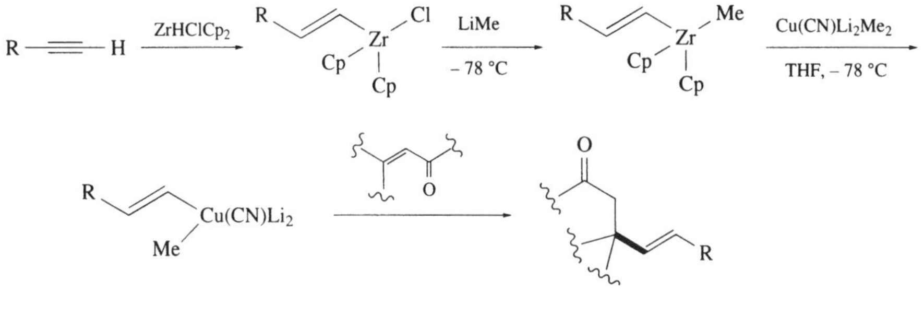

Scheme 38

The reaction of $ZrHClCp_2$ with a 1-ene-3-yne molecule, such as $HC≡C–CH=CHMe$, $HC≡C–CH=CHOMe$ or $HC≡C–C=CH–CH_2–(CH_2)_2Me$ yields within a few hours at room temperature the corresponding dienyl complexes $[ZrCl(CH=CH–CR^1=CR^2R^3)Cp_2]$ ($R^1 = R^3 = H$, $R^2 = OMe$; $R^1R^2 = –(CH_2)_4–$, $R^3 = H$) (Scheme 39).[235] Reaction of the latter with MeLi at −78 °C yields the thermally labile $[ZrMe(CH=CH–CR^1=CR^2R^3)Cp_2]$ ($R^1 = R^3 = H$, $R^2 = OMe$; $R^1R^2 = –(CH_2)_4–$, $R^3 = H$).[236] Addition of 8–10 equiv. of PMe_3 to the chloro alkenyl complex affords a metallacyclopropenezirconocene. These products are susceptible to addition of HCl, from pyHCl, to yield dienyl complexes substituted at the 2-position.

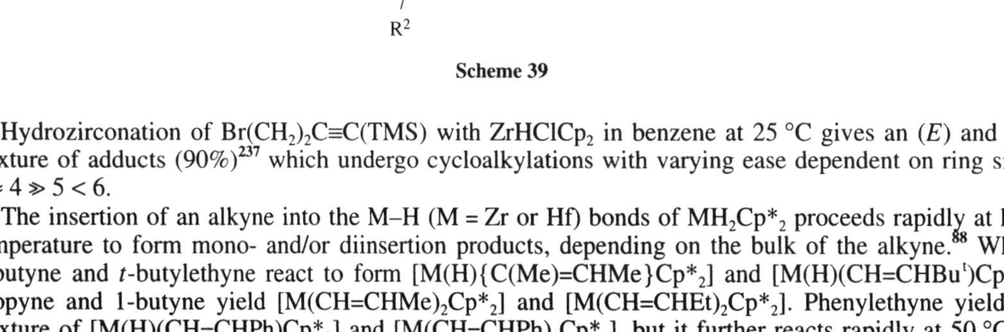

Scheme 39

Hydrozirconation of $Br(CH_2)_2C≡C(TMS)$ with $ZrHClCp_2$ in benzene at 25 °C gives an (*E*) and (*Z*) mixture of adducts (90%)[237] which undergo cycloalkylations with varying ease dependent on ring size: $3 ≈ 4 ≫ 5 < 6$.

The insertion of an alkyne into the M–H (M = Zr or Hf) bonds of $MH_2Cp^*_2$ proceeds rapidly at low temperature to form mono- and/or diinsertion products, depending on the bulk of the alkyne.[88] While 2-butyne and *t*-butylethyne react to form $[M(H)\{C(Me)=CHMe\}Cp^*_2]$ and $[M(H)(CH=CHBu^t)Cp^*_2]$, propyne and 1-butyne yield $[M(CH=CHMe)_2Cp^*_2]$ and $[M(CH=CHEt)_2Cp^*_2]$. Phenylethyne yields a mixture of $[M(H)(CH=CHPh)Cp^*_2]$ and $[M(CH=CHPh)_2Cp^*_2]$, but it further reacts rapidly at 50 °C to give $[Zr\{C(Ph)C(H)C(H)CPh\}Cp^*_2]$ in 50% yield. When the same reaction is carried out in benzene at 60 °C, without removing the H_2, appreciable hydrogenation of the alkyne occurs. For phenylethyne, the principal hydrogenated product is ethylbenzene with a trace amount of styrene. Hence, the chemistry of $[\{ZrH(μ-H)Cp^*_2\}_2]$ involves at least two competing reactions, metallacycle formation vs. alkyne hydrogenation. Ambient-temperature photolysis[238] and prolonged heating of the complex are accompanied by the evolution of over 2 mol of H_2 per mole of complex, resulting in dark-purple solutions. After 400 h of heating, 3 mol of H_2 are eventually obtained.

The reaction of ZrH_2Cp_2 with diphenylethyne under mild conditions has been reinvestigated[238] and found to produce a zirconacyclopentadiene complex, contrary to an earlier suggestion proposing the formation of a bimetallic zirconobenzene.[67] Slow reaction with diphenylethyne produces $[Zr\{C(Ph)C(Ph)C(Ph)CPh\}Cp_2]$ in 80% yield with the evolution of a stoichiometric amount of H_2.[238] Under more drastic conditions (100 °C) and using only the hydrogen present in the system, substantial hydrogenation of the alkyne occurs with formation of *trans*-stilbene (20%)[239] and small amounts of dibenzyl (4%).[239] Reaction carried out under H_2 pressure indicates that a catalytic hydrogenation process occurs[240,266] from which the ZrH_2Cp_2 catalyst may be recovered. A qualitative MO analysis supports a mechanism involving reaction of the dimeric $[\{ZrH(\mu\text{-}H)Cp_2\}_2]$ with diphenylethyne.

Alkenylzirconocenes are known to undergo transmetallation reactions with Al–Cl moieties to yield alkenylalanes. However, the aluminum to zirconium transmetallation has been reported to occur from the lithium alkenylaluminate generated by the addition of LiBu to C_6H_{11}–C(Me)=C(H)AlMe$_2$, ZrX_2Cp_2 (X = Cl or I) giving $[Zr\{C(H)=C(Me)C_6H_{11}\}_2Cp_2]$.[241]

Photolysis of benzyl(β-styryl)zirconocene produces dibenzylzirconocene chloride and (*s-trans*-η^4-*trans,trans*-1,4-diphenylbutadiene)ZrCp$_2$.[242] Based on EPR, cross-over experiments and the kinetic influence of added ligands, a radical reaction mechanism was proposed.

A series of ruthenium–zirconium complexes containing bridging C_2 units having bond orders of three, two and one has been prepared (Scheme 40 and Equation (25)).[243,244] The x-ray-characterized (C≡C)- and (CH=CH)-bridged heterobimetallic complexes $[ZrClCp_2(CC)RuCp(PMe_3)_2]$ and $[ZrClCp_2(CH=CH)RuCp(PMe_3)_2]$, having both an electron-rich and an electron-deficient metal at the two ends of the C_2 moiety, have been prepared (Scheme 40). Both vinyl protons of the latter appear as a doublet of triplets in the ^1H NMR spectrum in CD_2Cl_2 at δ 9.46 (–RuCH) and δ 8.38 (–ZrCH). A particularly intriguing feature of the ^{13}C NMR spectrum was the value of $J_{C-H} = 110$ Hz observed for the carbon bonded to ruthenium. This low J_{C-H} value suggests an agostic interaction between the CH bound to ruthenium and the unsaturated zirconium centre. A broad IR spectral band at 2590 cm^{-1} is assigned to the agostic ν(C–H). The presence of a short Zr$\cdots$H distance (0.208(2) nm) determined in the crystal structure strongly supports the existence of an agostic interaction.

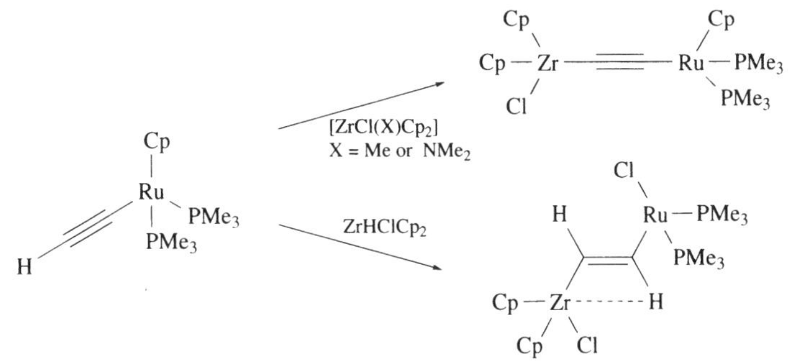

Scheme 40

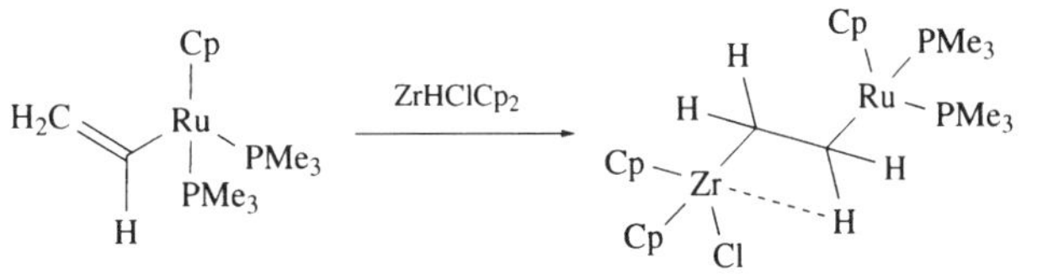

A closely related series of homobimetallic species has been prepared. The compound (μ-ethylene)bis(zirconocene chloride) is rapidly formed (75%) either by hydrozirconation of vinylchlorozirconocene or by reaction of ethyne with ZrHClCp$_2$ (71%; Scheme 41).[245] In contrast, the slow reaction of (β-styryl)zirconocene chloride with ZrHClCp$_2$ gives an equimolar mixture of $ZrCl_2Cp_2$, $[ZrCl(CH_2CH_2Ph)Cp_2]$ and $[ZrClCp_2(CH=CHPh)ZrCp_2]$ (Equation (26)). The last of these was isolated as burgundy-red crystals (34%) and its x-ray crystal structure was determined[245] showing a bridging alkenyl ligand almost coplanar with the two zirconium atoms; its chemistry is dominated by the reactivity of the Zr–C σ-bond. Insertion of CO into the dinuclear structure results in the formation of a dinuclear enolate (Equation (27)).

A homobimetallic complex[227] was obtained (50% yield) by the hydrozirconation of $[ZrCl(C≡CMe)Cp'_2]$ (Equation (28)). Its x-ray crystal structure shows a very short Zr–β-CH distance (0.221(1) nm) indicating a strong agostic interaction between the zirconium and hydrogen.

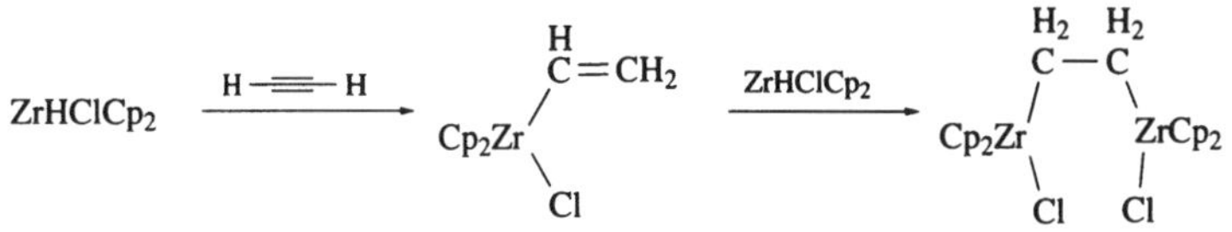

Scheme 41

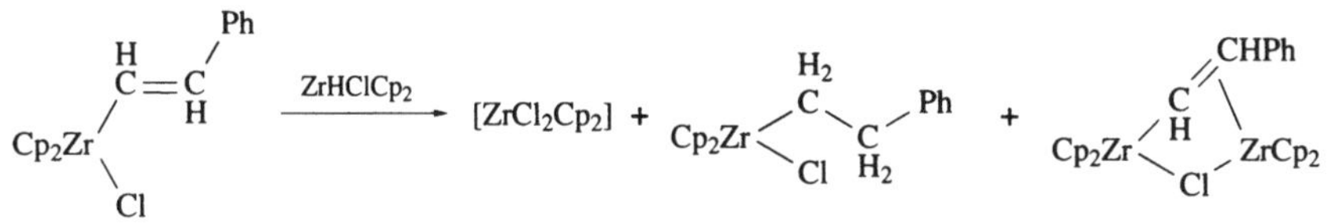

$$\text{(26)}$$

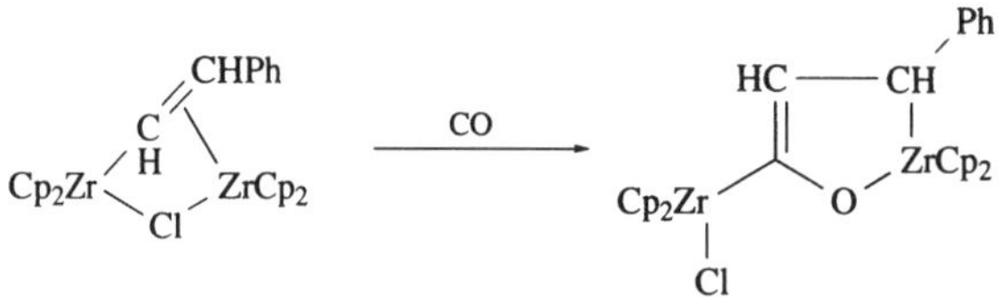

$$\text{(27)}$$

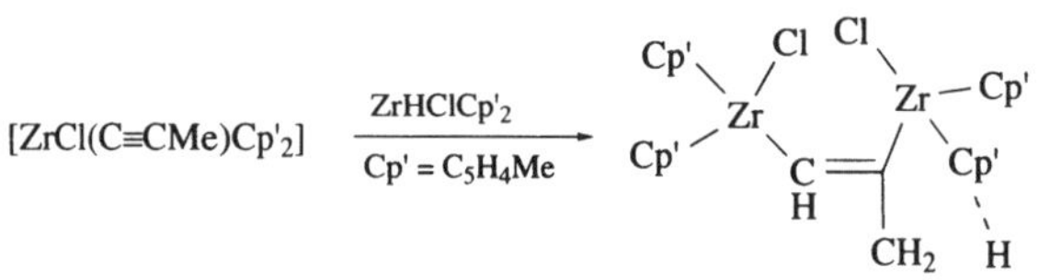

$$\text{(28)}$$

In order to extend the synthetic utility of the hydrozirconation reaction, several methods have been explored for alkyl and alkenyl group transfer with subsequent C–C bond formation. The reaction of 1-hexenylzirconocene chloride with a boron chloride or bromide results in the exchange of halogen and alkenyl groups yielding $ZrCl_2Cp_2$ and the alkenylborane.[246,247] Alkenylzirconocene chloride complexes react with aryl or alkenyl iodides or bromides in the presence of palladium or nickel complexes containing phosphine ligands to give the corresponding cross-coupled (alkenyl–aryl or alkenyl–alkenyl) products.[248] The mechanism of these reactions probably involves a transfer of the alkenyl group from $[Zr(alkenyl)ClCp_2]$ to a palladium or nickel complex as the rate-limiting step.

11.7 METALLOCENE IMINOACYLS

11.7.1 Introduction

In contrast with the extensive information available on the spectroscopic and structural properties of the η^2-acyl function, the chemistry of the isoelectronic η^2-iminoacyl ligand, formed by the migratory insertion of RNC into the metal–alkyl bond, is less understood. The reactivity of η^2-acyl, η^2-iminoacyl and related functional groups has been the subject of a review.[249] Although most of the iminoacyls reported in the literature are bound in an η^2-mode, a few examples of η^1-coordination have been documented.

11.7.2 η^1-Iminoacyl Complexes

Zirconocene iminoacyl complexes are generally prepared from zirconocene alkyl or silyl complexes by insertion of CNR into the Zr–C or Zr–Si bond. For mixed dialkyls, the insertion generally occurs into the more bulky of the two.

The addition of $2,6\text{-}Me_2C_6H_3NC$ to $[Zr(CH_2\text{-}py\text{-}6Me)_2Cp_2]$ ($CH_2\text{-}py\text{-}6Me = 2\text{-}(6\text{-methylpyridyl})\text{-}$methyl) in hydrocarbon solution results in a mono (η^2-iminoacyl) derivative ($\delta[^{13}C] = 246.7$ (CN)), which undergoes a 1,2-hydrogen shift to form a 16-electron species with a *trans*-substituted vinylamide.[250] The latter undergoes a further insertion reaction ultimately to yield a bis(vinylamide) complex (Scheme 42). The reaction pathway has been investigated by ^1H and ^{13}C NMR spectroscopy.

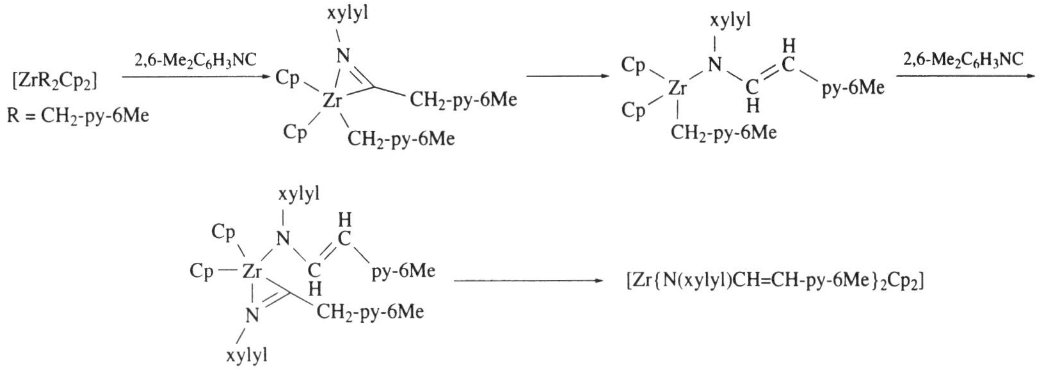

Scheme 42

[Zr(C$_6$H$_4$R)$_2$Cp$_2$] (R = H or Me) reacts cleanly with CH$_2$PPh$_3$ at 80 °C to form the zirconium-substituted ylide [Zr(C$_6$H$_4$R)CHPPh$_3$Cp$_2$] in greater than 90% yield. This complex was characterized by NMR and IR spectroscopy and x-ray diffraction.[251] The reaction proceeds through an (η^2-aryne)zirconocene intermediate as revealed by a deuterium-labelling experiment. Partial multiple-bond character of the Zr–C(ylide) linkage was inferred from the observed conformation and the rather short Zr–C bond distance of 0.215 7(4) nm.

The metallocene ylides [MR(CHPPh$_3$)Cp$_2$] (M = Zr, R = Ph; M = Hf, R = Et) react with benzyl isonitrile by insertion into the M–C(ylide) bond to give the (η^2-iminoacyl)metallocene complexes [MR{η^2-N(CH$_2$Ph)=C(CHPPh$_3$)}Cp$_2$].[252,253] An x-ray diffration study showed for both complexes the (η^2-iminoacyl)metallocene stereoisomer to have an 'RN-inside' arrangement.

The cyclometallated complex [ZrClCp*Fv] (Fv = η^6-C$_5$Me$_4$CH$_2$) reacts instantaneously with 2,6-xylyl isocyanide to give the bright-orange, microcrystalline [ZrCl{C$_5$Me$_4$CH$_2$C(N-2,6-xylyl)}Cp*] in 64% yield.[254] Its ^{1}H NMR spectrum indicates that an insertion of isonitrile took place into the Zr–CH$_2$(fulvene) bond (Equation (29)).[254] The coupling constants of the methylene group protons ($^2J_{H-H}$ = 16.85 Hz) are characteristic for coupling of geminal protons on an sp^3 carbon atom. The IR spectrum shows v(CN) at 1570 cm^{-1}. Conclusive identification of the iminoacyl complex as η^1 or η^2 was not possible solely through IR and NMR spectral data. However, the hapticity was indirectly confirmed by further reactivity studies. The compound decomposes slowly in solution at room temperature to give the [ZrClCp*Fv] starting material together with [ZrClCp*{η^5:η^1-C$_5$Me$_4$CH$_2$C(N-xylyl)C(N-xylyl)}], a product of two consecutive head-to-head insertions of isonitrile. This compound can also be obtained from the reaction of [ZrClCp*{C$_5$Me$_4$CH$_2$C(N-2,6-xylyl)}] with 1 equiv. of (2,6-xylyl)NC. The ^{1}H NMR spectrum again shows a coupling constant (16.50 Hz) characteristic of methylene protons on an sp^3 carbon atom. Although the ^{13}C NMR resonance of the iminoacyl carbon is in the range considered characteristic for an η^2 bond,[255–8] the x-ray structure clearly showed that nitrogen is not bonded to zirconium and that the iminoacyl is therefore η^1-bonded (Scheme 43).[254]

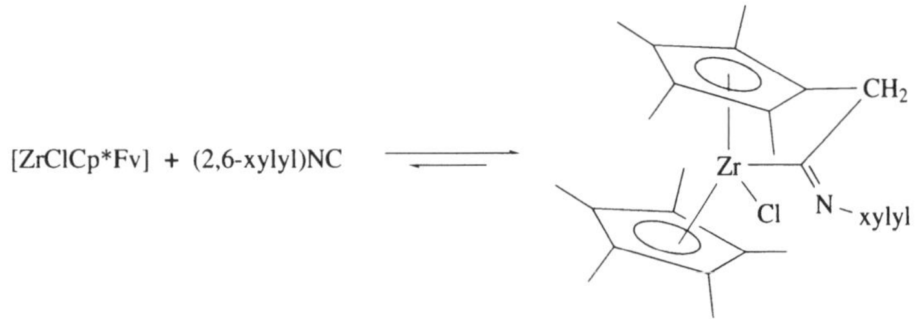

$$[\text{ZrClCp*Fv}] + (2,6\text{-xylyl})\text{NC} \rightleftharpoons \qquad\qquad\qquad (29)$$

The double-insertion product rearranges thermally in toluene at 140 °C via a 1,3-hydrogen shift from the fulvene side chain to form a bidentate permethylcyclopentadienylamido ligand in the complex proposed to be [ZrClCp*{η^5:η^1-C$_5$Me$_4$CH=C(CH=N-xylyl)N-xylyl}] (Scheme 43).

11.7.3 η^2-Iminoacyl Complexes

Reaction of [MR^1$_2$(OAr)$_2$] with R^{2}NC (M = Zr or Hf; Ar = 2,6-Bu^t$_2$C$_6$H$_3$; R^1 = Bz or Me; R^2 = Bu^t) in hexane gives a high yield (81%) of the orange-yellow crystalline mono(η^2-iminoacyl) derivative

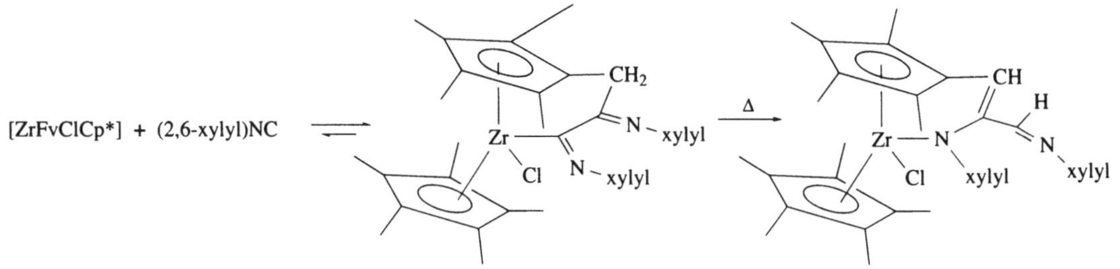

Scheme 43

$[Zr(\eta^2\text{-Bu}^t\text{NCR}^1)(R^1)(OAr)_2]$ (Scheme 44).[256,259,260] When 2 equiv. of isocyanide are used, bright-yellow crystals of the double-insertion product $[M(\eta^2\text{-Bu}^t\text{NCR}^1)_2(OAr)_2]$ separate out of the toluene–hexane reaction mixture in 67% yield. The x-ray molecular structure[256] of one such derivative (M = Zr, R = CH$_2$Ph) showed a pseudotetrahedral geometry about the metal atom with each η^2-iminoacyl unit occupying a single coordination site and lying parallel to each other in a head-to-tail fashion, aligned approximately along the O–Zr–O plane. Correlation of the observed solid-state structures of these complexes with their ^{1}H and ^{13}C NMR spectra indicates that the η^2-iminoacyl groups are fluxional.[256] Whether this fluxionality involves simple rotation while η^2-bound or an η^2–η^1–η^2 process was not distinguished.

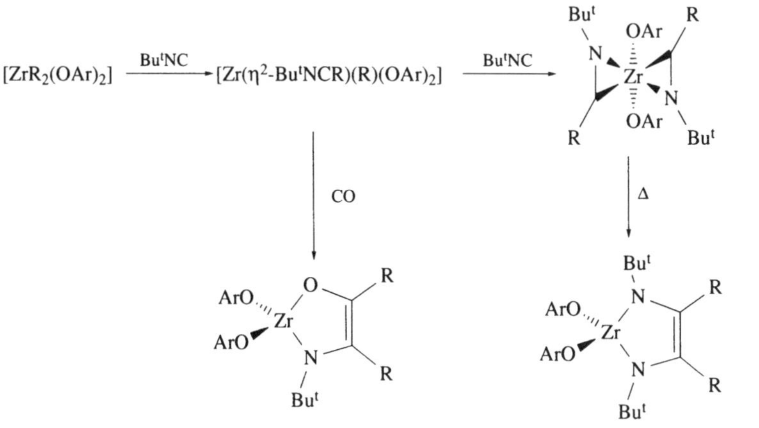

Scheme 44

The tris(alkyl) complex $[ZrBz_3(OAr)]$ reacts with more than 3 equiv. of Bu^{t}NC or (2,6-xylyl)NC in hexane or benzene to give colourless crystals of $[Zr\{\eta^2\text{-BzC=N(R)}\}_3OAr]$ (R = Bu^t or 2,6-xylyl) in 60–80% yield.[259] The intermediate thermally unstable brown-yellow solid $[Zr\{\eta^2\text{-BzC=N(R)}\}_2(Bz)OAr]$ (R = 2,6-xylyl) was isolated when 2 equiv. of (2,6-xylyl)NC were used.[259]

A crystal structure of $[Zr(\eta^2\text{-Bu}^t\text{NCBz})_3OAr]$ has revealed that all three iminoacyl groups are η^2-bound giving a formally seven-coordinate zirconium complex having pseudotetrahedral zirconium, if each of the η^2-iminoacyl units is taken as occupying a single coordination site. As for the single- and double-insertion products, the iminoacyl groups are fluxional on the ^{1}H and ^{13}C NMR spectral timescales.

11.7.4 Reactivity

Reaction of the monoiminoacyl complexes $[Zr\{\eta^2\text{-(xylyl)NC(CH}_2\text{Ph)}\}Bz(OAr)_2]$ with CO at 5–7 MPa yields the colourless $[Zr\{OCBz=CBz(N\text{-xylyl})\}(OAr)_2]$, derived from the formal reductive coupling of one acyl and one iminoacyl group.[256,260] No evidence for the presumed intermediate mixed acyl–iminoacyl was reported. The most notable structural feature of the pseudotetrahedral $[Zr\{OCBz=CBz(N\text{-xylyl})\}(OAr)_2]$, as determined by x-ray crystallography,[260] is the lack of planarity of the five-membered metallacycle. The zirconium lies out of the plane of the OC=CN backbone but, despite a fold angle of 50.0°, solution NMR spectra indicated a facile interconversion of the bent metallacycle.

Thermolysis of the bis(iminoacyl) compounds in toluene at 100 °C gives intramolecular coupling of the two iminoacyl groups forming the corresponding yellow enediamide derivatives.[255,260] An extensive kinetic study,[261] carried out with various substituted phenyl derivatives, has shown that the reaction obeys a first-order law with the rate dependent on the steric and electronic nature of the groups bonded to nitrogen. The use of the bulky 2,6-dimethylphenyl group retards the reaction, while the use of electron-withdrawing substituents (3-F, 3-OMe, 4-OMe, 4-Cl or 4-NMe$_2$) accelerates the reaction.

The solid-state structure of [Zr{OCBz=CBz(N-xylyl)}(OAr)$_2$] shows a pseudotetrahedral coordination about the zirconium with a nonplanar five-membered chelate ring. The ^{1}H NMR resonances are broad at ambient temperature. A low-temperature spectrum in agreement with a nonplanar chelate ring consisting of two equal-intensity sets of resonances was obtained. At higher temperatures, flipping of the chelate rings became rapid on the NMR spectral timescale.

11.8 REFERENCES

1. W. A. Nugent and D. F. Taber, *J. Am. Chem. Soc.*, 1989, **111**, 6435.
2. G. Erker, K. Berg, R. Benn and G. Schroth, *Chem. Ber.*, 1985, **118**, 1383 and references therein.
3. G. Erker, K. Berg. L. Treschanke and K. Engel, *Inorg. Chem*, 1982, **21**, 1277.
4. C. P. Casey and F. Niet, *Organometallics*, 1985, **48**, 1218.
5. E. C. Lund and T. Livinghouse, *Organometallics*, 1990, **9**, 2426.
6. N. J. Wells, J. C. Huffman and K. G. Caulton, *J. Organomet. Chem.*, 1981, **213**, C17; L. M. Engelhardt, R. I. Papasergio, C. L. Raston and A. H. White, *Organometallics*, 1984, **3**, 18.
7. P. B. Hitchcock, M. F. Lappert, D.-S. Lin and E. J. Ryan, *Polyhedron*, 1995, **14**, in press.
8. A. W. Duff, P. B. Hitchcock, M. F. Lappert, R. G. Taylor and J. A. Segal, *J. Organomet. Chem.*, 1985, **293**, 271.
9. G. Erker *et al.*, *J. Organomet. Chem.*, 1990, **382**, 89.
10. J. Blenkers, H. J. de Liefde Meijer and J. H. Teuben, *J. Organomet. Chem.*, 1981, **218**, 383.
11. D. M. Roddick, M. D. Fryzuk, P. F. Seidler, G. L. Hillhouse and J. E. Bercaw, *Organometallics*, 1985, **4**, 97.
12. A. Terpstra, J. N. Louwen, A. Oskam and J. H. Teuben, *J. Organomet. Chem.*, 1984, **260**, 207.
13. J. Arnold, D. M. Roddick, T. D. Tilley, A. L. Rheingold and S. J. Geib, *Inorg. Chem.*, 1988, **27**, 3510.
14. A. M. Cardoso, R. J. H. Clark and S. Moorhouse, *J. Chem. Soc., Dalton Trans.*, 1980, 1156.
15. P. Jutzi and A. Seufert, *J. Organomet. Chem.*, 1979, **169**, 373.
16. P. Jutzi and M. Kuhn, *J. Organomet. Chem.*, 1979, **173**, 221.
17. E. W. Abel and S. Moorhouse, *J. Organomet. Chem.*, 1971, **29**, 227.
18. C. H. Winter, X.-X. Zhou, D. A. Dobbs and M. J. Heeg, *Organometallics*, 1991, **10**, 210.
19. L. E. Schock and T. J. Marks, *J. Am. Chem. Soc.*, 1988, **110**, 7701.
20. Y. Wielstra, S. Gambarotta, A. Meetsma, J. L. de Boer and M. Y. Chiang, *Organometallics*, 1989, **8**, 2696.
21. W. Beck and K. Sünkel, *Chem. Rev.*, 1988, **88**, 1405.
22. B. H. Chang, R. H. Grubbs and C. H. Brubaker, Jr., *J. Organomet. Chem.*, 1985, **280**, 365.
23. I. A. Latham, G. J. Leigh, G. Huttner and I. Jabril, *J. Chem. Soc., Dalton Trans.*, 1986, 385.
24. D. L. Reger, R. Mahtab, J. C. Baxter and L. Lebioda, *Inorg. Chem.*, 1986, **25**, 2046.
25. E. L. Muetterties, *Acc. Chem. Res.*, 1970, **3**, 266; E. L. Muetterties and L. J. Guggenberger, *J. Am. Chem. Soc.*, 1974, **96**, 1748.
26. R. Hofmann, B. F. Beier, E. L. Muetterties and A. R. Rossi, *Inorg. Chem.*, 1977, **16**, 511.
27. M. G. B. Drew, *Prog. Inorg. Chem.*, 1977, **23**, 67.
28. A. H. Bruder, R. C. Fay, D. F. Lewis and A. A. Sayler, *J. Am. Chem. Soc.*, 1976, **98**, 6932.
29. R. C. Fay, J. R. Weir and A. H. Bruder, *Inorg. Chem.*, 1984, **23**, 1079.
30. K. Berg and G. Erker, *J. Organomet. Chem.*, 1984, **263**, 37.
31. H. Köpf and T. Klapötke, *J. Organomet. Chem.*, 1986, **307**, 319.
32. V. Srivastava, S. K. Sengupta and S. C. Tripathi, *Synth. React. Inorg. Metal-Org. Chem.*, 1985, **15**, 163.
33. A. Gómez-Carrera, M. Mena, P. Royo and R. Serrano, *J. Organomet. Chem.*, 1986, **315**, 329.
34. S. K. Sengupta and B. P. Baranwal, *Rev. Roum. Chim.*, 1983, **28**, 487.
35. S. Saxena and A. K. Rai, *Indian J. Chem., Sect. A*, 1985, **24**, 645.
36. S. Sharma, A. Sharma, K. C. Goyal and N. K. Kaushik, *Inorg. Nucl. Chem. Lett.*, 1984, 1327.
37. L. M. Babcock, V. W. Day and W. G. Klemperer, *J. Chem. Soc., Chem. Commun.*, 1988, 519.
38. G. Hidalgo, M. A. Pellinghelli, P. Royo, R. Serrano and A. Tiripicchio, *J. Chem. Soc., Chem. Commun.*, 1990, 1118.
39. S. M. Baxter, G. S. Ferguson and P. T. Wolczanski, *J. Am. Chem. Soc.*, 1988, **110**, 4231.
40. G. S. Ferguson, P. T. Wolczanski, L. Pàrkànyi and M. C. Zonnevylle, *Organometallics*, 1988, **7**, 1967.
41. Y. Wielstra, S. Gambarotta, A. L. Spek and W. J. J. Smeets, *Organometallics*, 1990, **9**, 2142.
42. S. Gambarotta and M. Chiang, *Organometallics*, 1987, **6**, 897.
43. Y. Wielstra, A. Meetsma, S. Gambarotta and S. Khan, *Organometallics*, 1990, **9**, 876.
44. T. V. Ashworth, T. Cuenca, E. Herdtweck and W. A. Herrmann, *Angew. Chem., Int. Ed. Engl.*, 1986, **25**, 289.
45. W. A. Herrmann, T. Cuenca and U. Küsthardt, *J. Organomet. Chem.*, 1986, **309**, C15.
46. W. A. Herrmann, T. Cuenca, B. Menjon and E. Herdtweck, *Angew. Chem., Int. Ed. Engl.*, 1987, **26**, 697.
47. Y. Wielstra, R. Duchateau, S. Gambarotta, C. Bensimon and E. Gabe, *J. Organomet. Chem.*, 1991, **418**, 183.
48. E. M. Brainina, L. I. Strunkina, B. V. Lokshin and M. G. Ezernitskaya, *Izv. Akad. Nauk Khim.*, 1981, 447.
49. B. V. Lokshin, Z. S. Klemenkova, M. G. Ezernitskaya, L. I. Strunkina and E. M. Brainina, *J. Organomet. Chem.*, 1982, **235**, 69.
50. R. J. Strittmatter and B. E. Bursten, *J. Am. Chem. Soc.*, 1991, **113**, 552.
51. N. Edelstein, *Inorg. Chem.*, 1981, **20**, 297.

52. P. H. Bird and M. R. Churchill, *J. Chem. Soc., Chem. Commun.*, 1967, 403.
53. V. P. Spiridonov and G. I. Mamowa, *J. Struct. Chem. (Engl. Transl.)*, 1969, **10**, 120.
54. V. Plato and K. Hedberg, *Inorg. Chem.*, 1971, **10**, 590.
55. T. J. Marks, W. J. Kennelly, J. R. Kolb and L. A. Shrimp, *Inorg. Chem.*, 1972, **11**, 2540.
56. A. P. Hitchcock, N. Hao, N. H. Werstiuk, M. J. McGlinchey and T. Ziegler, *Inorg. Chem.*, 1982, **21**, 793.
57. M. Mancini *et al.*, *Inorg. Chem.*, 1984, **23**, 1072.
58. J. A. Marsella and K. G. Caulton, *J. Am. Chem. Soc.*, 1982, **104**, 2361.
59. B. D. James, R. K. Nanda and M. G. H. Wallbridge, *J. Chem. Soc. (A)*, 1966, 182.
60. R. K. Nanda and M. G. H. Wallbridge, *Inorg. Chem.*, 1964, **3**, 1798.
61. T. J. Marks and J. R. Kolb, *J. Am. Chem. Soc.*, 1975, **97**, 3397.
62. D. Männig and H. Nöth, *J. Organomet. Chem.*, 1984, **275**, 169.
63. A. G. Császár, L. Hedberg, K. Hedberg, R. C. Burns, A. T. Wen and M. J. McGlinchey, *Inorg. Chem.*, 1991, **30**, 1371.
64. P. T. Wolczanski and J. E. Bercaw, *Organometallics*, 1982, **1**, 793.
65. B. Kautzner, P. C. Wailes and H. Weigold, *J. Chem. Soc., Chem. Commun.*, 1969, 1105.
66. P. C. Wailes, H. Weigold and A. P. Bell, *J. Organomet. Chem.*, 1972, **43**, C32.
67. P. C. Wailes, H. Weigold and A. P. Bell, *J. Organomet. Chem.*, 1971, **27**, 373.
68. D. W. Hart and J. Schwartz, *J. Am. Chem. Soc.*, 1974, **96**, 8115.
69. J. Schwartz, *J. Organomet. Chem. Libr.*, 1976, 461.
70. J. Schwartz, *Pure Appl. Chem.*, 1980, **52**, 733.
71. J. Alvhäll, S. Gronowitz and A. Hallberg, *Chem. Scr.*, 1988, **28**, 285.
72. F. Wochner and H. H. Brintzinger, *J. Organomet. Chem.*, 1986, **329**, 65.
73. G. Erker, R. Schlund and C. Krüger, *Organometallics*, 1989, **8**, 2349.
74. J. E. Nelson, J. E. Bercaw and J. A. Labinger, *Organometallics*, 1989, **8**, 2484.
75. H. Yasuda, K. Nagasuna, M. Akita, K. Lee and A. Nakamura, *Organometallics*, 1984, **3**, 1470.
76. S. Gambarotta, C. Floriani, A. Chiesi-Villa and C. Guastini, *J. Am. Chem. Soc.*, 1983, **105**, 1690.
77. S. Gambarotta, S. Strologo, C. Floriani, A. Chiesi-Villa and C. Guastini, *J. Am. Chem. Soc.*, 1985, **107**, 6278.
78. G. Erker, W. Frömberg, J. L. Atwood and W. E. Hunter, *Angew. Chem., Int. Ed. Engl.*, 1984, **23**, 68.
79. W. Frömberg and G. Erker, *J. Organomet. Chem.*, 1985, **280**, 355.
80. J. P. Majoral, N. Dufour, F. Meyer, A.-M. Caminade, R. Choukroun and D. Gervais, *J. Chem. Soc., Chem. Commun.*, 1990, 507.
81. F. Boutonnet, N. Dufour, T. Straw, A. Igau and J. P. Majoral, *Organometallics*, 1991, **10**, 3939.
82. S. Karlsson, A. Hallberg and S. Gronowitz, *J. Organomet. Chem.*, 1991, **403**, 133.
83. E. Cesarotti, A. Chiesa, S. Maffi and R. Ugo, *Inorg. Chim. Acta*, 1982, **64**, L207.
84. T. Kunieda, T. Mori, T. Higuchi and M. Hirobe, *Tetrahedron Lett.*, 1985, **26**, 1977.
85. K. I. Gell, B. Posin, J. Schwartz and G. M. Williams, *J. Am. Chem. Soc.*, 1982, **104**, 1846.
86. R. Choukroun, M. Basso-Bert and D. Gervais, *J. Chem. Soc., Chem. Commun.*, 1986, 1317.
87. Y. Raoult, R. Choukroun, D. Gervais and G. Erker, *J. Organomet. Chem.*, 1990, **399**, C1.
88. C. McDade and J. E. Bercaw, *J. Organomet. Chem.*, 1985, **279**, 281.
89. E. J. Moore, D. A. Straus, J. Armantrout, B. D. Santarsiero, R. H. Grubbs and J. E. Bercaw, *J. Am. Chem. Soc.*, 1983, **105**, 2068.
90. B. K. Campion, J. Falk and T. D. Tilley, *J. Am. Chem. Soc.*, 1987, **109**, 2049.
91. T. D. Tilley, *Organometallics*, 1985, **4**, 1452.
92. F. H. Elsner, T. D. Tilley, A. L. Rheingold and S. J. Geib, *J. Organomet. Chem.*, 1988, **358**, 169.
93. C. T. Aitken, J. F. Harrod and E. Samuel, *J. Am. Chem. Soc.*, 1986, **108**, 4059.
94. J. F. Harrod and S. S. Yun, *Organometallics*, 1987, **6**, 1381.
95. C. A. Aitken, E. Samuel and J. F. Harrod, *Can. J. Chem.*, 1986, **64**, 1677.
96. J. F. Harrod, in 'Inorganic and Organometallic Polymers', eds. M. Zeldin, K. J. Wynne and H. R. Allcock, American Chemical Society, Washington, DC, 1988, pp. 89–100.
97. C. A. Aitken, J.-P. Barry, F. Gauvin, J. F. Harrod, A. Malek and D. Rousseau, *Organometallics*, 1989, **8**, 1732.
98. L. S. Chang and J. Y. Corey, *Organometallics*, 1989, **8**, 1885.
99. W. H. Campbell, T. K. Hilty and L. Yurga, *Organometallics*, 1989, **8**, 2615.
100. H.-G. Woo and T. D. Tilley, *J. Am. Chem. Soc.*, 1989, **111**, 3757.
101. H.-G. Woo and T. D. Tilley, *J. Am. Chem. Soc.*, 1989, **111**, 8043.
102. M. D. Curtis, L. G. Bell and W. M. Butler, *Organometallics*, 1985, **4**, 701.
103. J. Arnold and T. D. Tilley, *J. Am. Chem. Soc.*, 1987, **109**, 3318.
104. D. H. Berry and Q. Jiang, *J. Am. Chem. Soc.*, 1987, **109**, 6210.
105. D. H. Berry and Q. Jiang, *J. Am. Chem. Soc.*, 1989, **111**, 8049.
106. J. Arnold, M. P. Engeler, F. H. Elsner, R. H. Heyn and T. D. Tilley, *Organometallics*, 1989, **8**, 2284.
107. Y. Mu, C. Aitken, B. Cote, J. F. Harrod and E. Samuel, *Can. J. Chem.*, 1991, **69**, 264.
108. K. A. Kreutzer, R. A. Fisher, W. M. Davis, E. Spaltenstein and S. L. Buchwald, *Organometallics*, 1991, **10**, 4031.
109. T. Takahashi *et al.*, *J. Am. Chem. Soc.*, 1991, **113**, 8564.
110. H.-G. Woo, J. F. Walzer and T. D. Tilley, *Macromolecules*, 1991, **24**, 6863.
111. J. W. Bruno, J. C. Huffman, M. A. Green and K. G. Caulton, *J. Am. Chem. Soc.*, 1984, **106**, 8310.
112. J. R. van den Hende, B. Hessen, A. Meetsma and J. H. Teuben, *Organometallics*, 1990, **9**, 537.
113. W. J. Evans, J. H. Meadows and T. P. Hanusa, *J. Am. Chem. Soc.*, 1984, **106**, 4454.
114. D. G. Bickley, N. Hao, P. Bougeard, B. G. Sayer, R. C. Burns and M. J. McGlinchey, *J. Organomet. Chem.*, 1983, **246**, 257.
115. S. B. Jones and J. L. Petersen, *Inorg. Chem.*, 1981, **20**, 2889.
116. R. Choukroun *et al.*, *Organometallics*, 1991, **10**, 374.
117. S. Couturier, G. Tainturier and B. Gautheron, *J. Organomet. Chem.*, 1980, **195**, 291.
118. B. Gautheron, G. Tainturier and S. Pouly, *J. Organomet. Chem.*, 1984, **268**, C56.
119. G. Tainturier, B. Gautheron and M. M. Fahim, *J. Organomet. Chem.*, 1985, **290**, C4.
120. M. Fahim and G. Tainturier, *J. Organomet. Chem.*, 1986, **301**, C45.

121. A. Shaver and J. M. McCall, *Organometallics*, 1984, **3**, 1823.
122. H. Köpf and T. Klapötke, *Z. Naturforsch., Teil B*, 1985, **40**, 447.
123. P. T. Barger, B. D. Santarsiero, J. Armantrout and J. E. Bercaw, *J. Am. Chem. Soc.*, 1984, **106**, 5178.
124. T. Nakano, S. Umano, Y. Kino, Y. Ishii and M. Ogawa, *J. Org. Chem.*, 1988, **53**, 3752.
125. T. Nakano, Y. Ishii and M. Ogawa, *J. Org. Chem.*, 1987, **52**, 4855.
126. G. L. Hillhouse and J. E. Bercaw, *J. Am. Chem. Soc.*, 1984, **106**, 5472.
127. D. Männig, H. Nöth, M. Schwartz, S. Weber and U. Wietelmann, *Angew. Chem., Int. Ed. Engl.*, 1985, **24**, 998.
128. G. L. Hillhouse and J. E. Bercaw, *Organometallics*, 1982, **1**, 1025.
129. G. Wilke, in 'Fundamental Research in Homogenous Catalysis', ed. M. Tsutsui, Plenum, New York, 1979, vol. 3.
130. S. G. Davies, 'Organotransition Metal Chemistry: Applications to Organic Synthesis', Pergamon, Oxford, 1982.
131. B. M. Trost, *J. Organomet. Chem.*, 1986, **300**, 263.
132. J. Tsuji, *J. Organomet. Chem.*, 1986, **300**, 281.
133. G. Erker, K. Berg, K. Angermund and C. Krüger, *Organometallics*, 1987, **6**, 2620.
134. G. Erker *et al.*, *Angew. Chem., Int. Ed. Engl.*, 1984, **23**, 455.
135. P. J. Vance *et al.*, *Organometallics*, 1986, **10**, 917.
136. E. J. Larson *et al.*, *Organometallics*, 1987, **6**, 2141.
137. E. J. Larson *et al.*, *Organometallics*, 1988, **7**, 1183.
138. K. Mashimi, H. Yasuda, K. Asami and A. Nakamura, *Chem. Lett.*, 1983, 219.
139. H. Yasuda, K. Nagasuna, K. Asami and A. Nakamura, *Chem. Lett.*, 1983, 955.
140. Y. Yamamoto and K. Maruyama, *Tetrahedron Lett.*, 1981, **22**, 2895.
141. A. K. Sharma and N. K. Kaushik, *Synth. React. Inorg. Metal-Org. Chem.*, 1983, **13**, 101.
142. A. K. Sharma and N. K. Kaushik, *J. Inorg. Nucl. Chem.*, 1981, **43**, 3024.
143. A. K. Sharma and N. K. Kaushik, *Acta Chim. Hung.*, 1984, **116**, 361.
144. G. S. Sodhi and N. K. Kaushik, *Inorg. Nucl. Chem. Lett.*, 1981, **17**, 211.
145. B. H. Khera, G. S. Sodhi and N. K. Kaushik, *J. Indian Chem. Soc.*, 1983, **30**, 209.
146. E. Samuel and R. Selton, *J. Organomet. Chem.*, 1965, **4**, 156.
147. B. Khera and N. K. Kaushik, *Bull. Soc. Chim. Fr.*, 1983, 278.
148. M. D. Rausch, K. J. Moriarty, J. L. Atwood, W. E. Hunter and E. Samuel, *J. Organomet. Chem.*, 1987, **327**, 39.
149. G. T. Palmer, F. Basolo, L. B. Kool and M. D. Rausch, *J. Am. Chem. Soc.*, 1986, **108**, 4417.
150. A. R. Siedle, R. A. Newmark, W. M. Lamanna and J. N. Schroepfer, *Polyhedron*, 1990, **9**, 301.
151. A. R. Siedle, R. A. Newmark, W. M. Lamanna and J. N. Schroepfer, *Polyhedron*, 1990, **9**, 283.
152. A. K. Sharma, B. Khera and N. K. Kaushik, *Monatsh. Chem.*, 1983, **114**, 907.
153. B. Khera, A. K. Sharma and N. K. Naushik, *Polyhedron*, 1983, **2**, 1177.
154. B. Khera and N. K. Kaushik, *Polyhedron*, 1984, **3**, 611.
155. A. K. Sharma, B. Khera and N. K. Kaushik, *Synth. React. Inorg. Metal-Org. Chem.*, 1983, **113**, 1059.
156. A. K. Sharma, B. Khera and N. K. Kaushik, *Indian J. Chem., Sect. A*, 1986, **25**, 365.
157. H. Schwemlein and H. H. Brintzinger, *J. Organomet. Chem.*, 1983, **254**, 69.
158. J. A. Smith and H. H. Brintzinger, *J. Organomet. Chem.*, 1981, **218**, 159.
159. H. Schwemlein, L. Zsolnai and G. Huttner, *J. Organomet. Chem.*, 1983, **256**, 285.
160. F. Wochner, L. Zsolnai, G. Huttner and H. H. Brintzinger, *J. Organomet. Chem.*, 1985, **288**, 69.
161. F. R. W. P. Wild, M. Wasiucionek, G. Huttner and H. H. Brintzinger, *J. Organomet. Chem.*, 1985, **288**, 63.
162. A. Schäfer, E. Karl, L. Zsolnai, G. Huttner and H. H. Brintzinger, *J. Organomet. Chem.*, 1987, **328**, 87.
163. J. A. Bandy, M. L. H. Green, I. M. Gardiner and K. Prout, *J. Chem. Soc., Dalton Trans.*, 1991, 2207.
164. R. M. Waymouth, F. Bangerter and P. Pino, *Inorg. Chem.*, 1988, **27**, 758.
165. W. Kaminsky, K. Külper, H. H. Brintzinger and F. R. W. P. Wild, *Angew. Chem., Int. Ed. Engl.*, 1985, **24**, 507.
166. R. Waymouth and P. Pino, *J. Am. Chem. Soc.*, 1990, **112**, 4911.
167. P. Pino, P. Cioni and J. Wei, *J. Am. Chem. Soc.*, 1987, **109**, 6189.
168. W. Kaminsky, K. Külper and S. Niedoba, *Makromol. Chem., Macromol. Symp.*, 1986, **3**, 377.
169. W. Kaminsky, A. Ahlers and N. Möller-Lindenhof, *Angew. Chem., Int. Ed. Engl.*, 1989, **28**, 1216.
170. R. B. Grossman, W. M. Davis and S. L. Buchwald, *J. Am. Chem. Soc.*, 1991, **113**, 2321.
171. J. C. W. Chien, W. M. Tsai and M. D. Rausch, *J. Am. Chem. Soc.*, 1991, **113**, 8570.
172. W. A. Herrmann, J. Rohrmann, E. Herdtweck, W. Spaleck and A. Winter, *Angew. Chem., Int. Ed. Engl.*, 1989, **28**, 1511.
173. L. A. Paquette, K. J. Moriarty and R. D. Rogers, *Organometallics*, 1989, **8**, 1506.
174. L. A. Paquette *et al.*, *Organometallics*, 1989, **8**, 2159.
175. L. A. Paquette, K. J. Moriarty, J. A. McKinney and R. D. Rogers, *Organometallics*, 1989, **8**, 1707.
176. J. C. Gallucci, B. Gautheron, M. Gugelchuk, P. Meunier and L. A. Paquette, *Organometallics*, 1989, **8**, 1506.
177. K. J. Moriarty, R. D. Rogers and L. A. Paquette, *Organometallics*, 1989, **8**, 1512.
178. G. Erker and A. A. H. van der Zeijden, *Angew. Chem., Int. Ed. Engl.*, 1990, **29**, 512.
179. D. J. Crowther, N. C. Baenziger and R. F. Jordan, *J. Am. Chem. Soc.*, 1991, **113**, 1455.
180. D. Jacoby, C. Floriani, A. Chiesi-Villa and C. Rizzoli, *J. Chem. Soc., Chem. Commun.*, 1991, 790.
181. J. Blenkers, P. Bruin and J. H. Teuben, *J. Organomet. Chem.*, 1985, **297**, 61.
182. J. J. Koh, P. H. Rieger, I. W. Shim and W. M. Risen, Jr., *Inorg. Chem.*, 1985, **24**, 2312.
183. Y. Wielstra, S. Gambarotta, A. Meetsma and A. L. Spek, *Organometallics*, 1989, **9**, 2948.
184. S. I. Bailey *et al.*, *J. Chem. Soc., Dalton Trans.*, 1986, 603.
185. M. F. Lappert, C. L. Raston, L. M. Engelhardt and A. H. White, *J. Chem. Soc., Chem. Commun.*, 1985, 521.
186. E. A. Mintz, A. S. Ward and D. S. Tice, *Organometallics*, 1985, **4**, 1308.
187. A. S. Ward, E. A. Mintz and M. R. Ayers, *Organometallics*, 1986, **5**, 1585.
188. G. Erker and F. Rosenfeldt, *Tetrahedron*, 1982, **38**, 1285.
189. S. L. Buchwald, R. B. Nielsen and J. C. Dewan, *Organometallics*, 1988, **7**, 2324.
190. W. Tikkanen and J. W. Ziller, *Organometallics*, 1991, **10**, 2266.
191. L. M. Engelhardt, G. E. Jacobsen, C. L. Raston and A. H. White, *J. Chem. Soc., Chem. Commun.*, 1984, 220.
192. R. Choukroun, F. Dahan and D. Gervais, *J. Organomet. Chem.*, 1984, **266**, C33.
193. R. Choukroun and D. Gervais, *J. Chem. Soc., Chem. Commun.*, 1982, 1300.

194. S. J. Young, H. Hope and N. E. Schore, *Organometallics*, 1984, **3**, 1585.
195. S. C. H. Ho, D. A. Straus, J. Armantrout, W. P. Schaefer and R. H. Grubbs, *J. Am. Chem. Soc.*, 1984, **106**, 2210.
196. R. M. Waymouth, B. D. Santarsiero and R. H. Grubbs, *J. Am. Chem. Soc.*, 1984, **106**, 4050.
197. R. M. Waymouth, B. D. Santarsiero, R. J. Coots, M. J. Bronikowski and R. H. Grubbs, *J. Am. Chem. Soc.*, 1986, **108**, 1427.
198. C. R. Randall, M. E. Silver and J. A. Ibers, *Inorg. Chim. Acta*, 1987, **128**, 39.
199. J. Y. Corey and X. H. Zhu, *J. Organomet. Chem.*, 1992, **439**, 1.
200. R. A. Takahashi, Y. Nitto, T. Seki, M. Saburi and E. Negishi, *Chem. Lett.*, 1990, 2259.
201. J. M. Atkinson and P. B. Brindley, *J. Organomet. Chem.*, 1991, **411**, 131.
202. R. A. Fisher and S. L. Buchwald, *Organometallics*, 1990, **9**, 871.
203. J. Ho, Z. Hou, R. J. Drake and D. W. Stephan, *Organometallics*, 1993, **12**, 3145.
204. G. Erker, P. Czisch and R. Mynott, *Z. Naturforsch., Teil B*, 1985, **40**, 1177.
205. A. Hudson, M. F. Lappert and R. Pichon, *J. Chem. Soc., Chem. Commun.*, 1983, 374.
206. N. E. Schore, S. J. Young, M. M. Olmstead and P. Hoffmann, *Organometallics*, 1983, **2**, 1769.
207. R. Choukroun and D. Gervais, *J. Organomet. Chem.*, 1984, **266**, C37.
208. S. Sabo, B. Chaudret and D. Gervais, *Nouv. J. Chim.*, 1983, **7**, 181.
209. S. Sabo, B. Chaudret and R. Poilblanc, *Nouv. J. Chim.*, 1981, **5**, 597.
210. G. Fochi, C. Floriani, A. Chiesi-Villa and C. Guastini, *J. Chem. Soc., Dalton Trans.*, 1986, 445.
211. S. Gambarotta, S. Strologo, C. Floriani, A. Chiesi-Villa and C. Guastini, *Inorg. Chem.*, 1985, **24**, 654.
212. S. Gambarotta, C. Floriani, A. Chiesi-Villa and C. Guastini, *Inorg. Chem.*, 1983, **22**, 2029.
213. K. W. Chiu, G. Wilkinson, M. Thornton-Pett and M. B. Hursthouse, *Polyhedron*, 1984, **3**, 79.
214. K. H. Dahman, D. Hedden, R. L. K. Burwell, Jr. and T. J. Marks, *Langmuir*, 1988, **4**, 1212.
215. D. R. Swanson and E. Negishi, *Organometallics*, 1991, **10**, 825.
216. E. Negishi, F. E. Cederbaum and T. Takahashi, *Tetrahedron Lett.*, 1986, **27**, 2829.
217. J. Y. Corey, X. H. Zhu, T. C. Bedard and L. D. Lange, *Organometallics*, 1991, **10**, 924.
218. M. K. Kesti, M. Absulrahman and R. M. Waymouth, *J. Organomet. Chem.*, 1991, **417**, C12.
219. P. Courtot, V. Labed, R. Pichon and J. Y. Salaün, *J. Organomet. Chem.*, 1989, **359**, C9.
220. P. Courtot, R. Pichon, J. Y. Salaün and L. Toupet, *Can. J. Chem.*, 1991, **69**, 661.
221. S. L. Buchwald, B. T. Watson and J. C. Huffman, *J. Am. Chem. Soc.*, 1987, **109**, 2544.
222. S. L. Buchwald, E. A. Lucas and J. C. Dewan, *J. Am. Chem. Soc.*, 1987, **109**, 4396.
223. S. L. Buchwald, E. A. Lucas and W. M. Davis, *J. Am. Chem. Soc.*, 1989, **111**, 397.
224. L. E. Schock, C. P. Brock and T. J. Marks, *Organometallics*, 1987, **6**, 232.
225. P. C. Wailes, H. Weigold and A. P. Bell, *J. Organomet. Chem.*, 1972, **34**, 155.
226. C. J. Jones, J. A. McCleverty and A. S. Rothin, *J. Chem. Soc., Dalton Trans.*, 1985, 405.
227. G. Erker, W. Frömberg, K. Angermund, R. Schlund and C. Krüger, *J. Chem. Soc., Chem. Commun.*, 1986, 372.
228. S. L. Buchwald, R. T. Lum and J. C. Dewan, *J. Am. Chem. Soc.*, 1986, **108**, 7441.
229. S. L. Buchwald, B. T. Watson and J. C. Huffman, *J. Am. Chem. Soc.*, 1986, **108**, 7411.
230. I. Hyla-Kryspin, R. Gleiter, C. Krüger, R. Zwettler and G. Erker, *Organometallics*, 1990, **9**, 517.
231. G. Erker, R. Zwettler, C. Krüger, I. Hyla-Kryspin and R. Gleiter, *Organometallics*, 1990, **9**, 524.
232. S. L. Buchwald and R. B. Nelson, *J. Am. Chem. Soc.*, 1989, **111**, 2870.
233. B. H. Lipshutz and E. L. Elsworth, *J. Am. Chem. Soc.*, 1990, **112**, 7440.
234. K. A. Babiak *et al.*, *J. Am. Chem. Soc.*, 1990, **112**, 7441.
235. M. D. Fryzuk, G. S. Bates and C. Stone, *Tetrahedron Lett.*, 1986, **27**, 1537.
236. M. D. Fryzuk, C. Stone, R. F. Alex and R. K. Chadha, *Tetrahedron Lett.*, 1988, **29**, 3915.
237. E. Negishi *et al.*, *J. Am. Chem. Soc.*, 1988, **110**, 5383.
238. S. B. Jones and J. L. Petersen, *J. Am. Chem. Soc.*, 1983, **105**, 5502.
239. P. Meunier, B. Gautheron and S. Couturier, *J. Organomet. Chem.*, 1982, **231**, C1.
240. S. Couturier, G. Tainturier and B. Gautheron, *J. Organomet. Chem.*, 1985, **195**, 291.
241. E. Negishi and L. D. Boardman, *Tetrahedron Lett.*, 1982, **23**, 3327.
242. P. Czisch, G. Erker, H. G. Korth and R. Sustmann, *Organometallics*, 1984, **3**, 945.
243. R. M. Bullock, F. R. Lemke and D. J. Szalda, *J. Am. Chem. Soc.*, 1990, **112**, 3244.
244. F. R. Lemke, D. J. Szalda and R. M. Bullock, *J. Am. Chem. Soc.*, 1991, **113**, 8466.
245. G. Erker, K. Kropp, J. L. Atwood and W. E. Hunter, *Organometallics*, 1983, **2**, 1555.
246. T. E. Cole, R. Quintanilla and S. Rodewald, *Organometallics*, 1991, **10**, 3777.
247. E. Negishi, K. Akiyoshi, B. O'Connor, K. Takagi and G. Wu, *J. Am. Chem. Soc.*, 1989, **111**, 3089.
248. E. Negishi, T. Takahashi, S. Baba, D. E. Van Horn and N. Okukado, *J. Am. Chem. Soc.*, 1987, **109**, 2393.
249. L. D. Durfee and I. P. Rothwell, *Chem. Rev.*, 1988, **88**, 1059.
250. S. M. Beshouri, P. E. Fanwick, I. P. Rothwell and J. C. Huffman, *Organometallics*, 1987, **6**, 891.
251. G. Erker, P. Czisch, R. Mynott, Y.-H. Tsay and C. Krüger, *Organometallics*, 1985, **4**, 1310.
252. D. L. Reger, M. E. Tarquini and L. Lebioda, *Organometallics*, 1983, **2**, 1763.
253. G. Erker, U. Korek and J. L. Petersen, *J. Organomet. Chem.*, 1988, **355**, 121.
254. R. Fandos, A. Meetsma and J. H. Teuben, *Organometallics*, 1991, **10**, 2665.
255. S. L. Latesky, A. K. McMullen, G. P. Niccolai and I. P. Rothwell, *Organometallics*, 1985, **4**, 1896.
256. A. K. McMullen, I. P. Rothwell and J. C. Huffman, *J. Am. Chem. Soc.*, 1985, **107**, 1072.
257. L. R. Chamberlain, I. P. Rothwell and J. C. Huffman, *J. Chem. Soc., Chem. Commun.*, 1986, 1203.
258. K. H. den Haan, G. A. Luinstra, A. Meetsma and J. H. Teuben, *Organometallics*, 1987, **6**, 1509.
259. L. R. Chamberlain *et al.*, *J. Am. Chem. Soc.*, 1987, **109**, 390.
260. L. R. Chamberlain *et al.*, *J. Am. Chem. Soc.*, 1987, **109**, 6068.
261. L. D. Durfee, A. K. McMullen and I. P. Rothwell, *J. Am. Chem. Soc.*, 1988, **110**, 1463.

12
Cationic Organozirconium and Organohafnium Complexes

ANIL S. GURAM and RICHARD F. JORDAN
University of Iowa, Iowa City, IA, USA

12.1 INTRODUCTION

The chemistry of cationic group 4 organometallic complexes has developed rapidly since the mid-1980s.[1] The postulated role of cationic metal alkyl species in homogeneous Ziegler–Natta catalysis,[2] which is now well established for cationic biscyclopentadienylmetal alkyl systems, has resulted in an explosive growth of this field. This has been made possible by the development of new synthetic strategies and an increased understanding of the importance of stable, weakly coordinating anions.[3] Most research in this area has focused on the chemistry of $[M(R)(Cp')_2]^+$ complexes of titanium and zirconium (the generic abbreviation Cp' is used to represent cyclopentadienyl ligands in general), which are of interest as alkene polymerization catalysts, models for intermediates in catalytic alkene polymerizations, and reagents or catalysts in organic synthesis.[1] However, interest in the hafnium analogues and in cationic nonmetallocene complexes is rapidly growing. This chapter focuses on the synthesis, structures, and reactivity of cationic zirconium and hafnium organometallics and covers the literature up to early 1993; in general, work reported in the patent literature is not discussed. Selected aspects of the role of cationic group 4 metal complexes in alkene polymerization and in organic synthesis are also discussed.

12.2 CATIONIC BISCYCLOPENTADIENYLMETAL ALKYL, ALLYL, AND BENZYL COMPLEXES

12.2.1 Synthesis

The first cationic zirconocene alkyl complex $[Zr\{CH_2CH(AlEt_2)_2\}Cp_2][Cp]$ (**1**) was isolated from the reaction of $[ZrCp_4]$ and triethylalane (Equation (1)).[4] Subsequently, a variety of general methods for the preparation of cationic metallocene alkyls has been developed.

$$[ZrCp_4] + AlEt_3 \longrightarrow Cp_2Zr \overset{+}{\cdots} \begin{matrix} Et_2 \\ Al \\ \diagdown \diagup \\ \diagup \diagdown \\ Al \\ Et_2 \end{matrix} Cp^- \tag{1}$$

(1)

12.2.1.1 Oxidative M–R bond cleavage reactions of [M(R)₂Cp′₂] complexes

The reaction of $[M(R)_2Cp'_2]$ (M = Zr or Hf) complexes with one-electron oxidants such as $[Fe(\eta\text{-}C_5H_4R)_2][BPh_4]$ in THF proceeds readily via oxidative cleavage of one M–R bond to afford cationic $[M(R)(THF)Cp'_2]^+$ complexes (Equation (2)). This reaction is general for R = alkyl, allyl, or benzyl, and has been employed in the synthesis of many complexes (as $[BPh_4]^-$ salts) including $[Zr(Me)(THF)Cp'_2]^+$ (Cp' = Cp (**2**), C_5H_4Me (**3**), Cp* (**4**)),[5,6] or $[Zr(Me)(THF)(\{\eta\text{-}C_5H_4\}_2SiMe_2)]^+$ (**5**),[7] $[Zr(Bz)(THF)Cp'_2]^+$ (Cp' = Cp (**6**) or C_5H_4Me (**7**)),[8] $[Zr(Bz)(THF)(EBTHI)]^+$ (**8**) (EBTHI = ethylenebis(tetrahydroindenyl)),[9]

$[Zr(Me)(THF)CpCp^*]^+$ **(9)**,[7] $[Hf(Me)(THF)Cp_2]^+$ **(10)**,[7] and $[Zr(\eta^3\text{-allyl})Cp^*_2]^+$ **(11)**.[10] Oxidative cleavage of M–Ph bonds does not occur readily under mild conditions; for example, no reaction is observed in the case of $[Zr(Ph)_2Cp_2]$.[11] The dimethylferrocenium reagent $[Fe(\eta\text{-}C_5H_4Me)_2][BPh_4]$ is preferred over $[FeCp_2][BPh_4]$ due to its greater thermal stability. This reagent reacts with $[Zr(R)_2(Cp')_2]$ complexes in THF, CH_2Cl_2 or toluene.

$$Cp'_2M\!\!\begin{smallmatrix}R\\ \\R\end{smallmatrix} \quad \xrightarrow[-[Fe(\eta\text{-}C_5H_4R)_2],\ -1/2\ \text{R-R}]{[Fe(\eta\text{-}C_5H_4R)_2][BPh_4],\ \text{THF}} \quad Cp'_2\overset{+}{M}\!\!\begin{smallmatrix}R\\ \\O\end{smallmatrix} \ [BPh_4]^- \qquad (2)$$

$$M = Zr,\ Hf$$
$$R = \text{alkyl, allyl, benzyl}$$

Oxidation reactions involving $[FeCp'_2][BPh_4]$ reagents probably proceed via outer-sphere one-electron transfer yielding intermediate radical cation species which undergo M–R bond homolysis (Scheme 1). The resulting $[Zr(R)Cp'_2]^+$ species is rapidly trapped by a Lewis base (L), if present, to afford $[Zr(R)(L)Cp'_2]^+$ complexes. Electrochemical studies of $[Zr(R)_2Cp'_2]$ compounds and reactions with one-electron oxidants (2,3-dichloro-5,6-dicyano-1,4-benzoquinone (DDQ), 7,7,8,8-tetracyanoquinodimethane (TCNQ), etc.) are consistent with an outer-sphere electron transfer mechanism.[12]

$$[M(R)_2Cp'_2] \xrightarrow{[Fe(\eta\text{-}C_5H_4R)_2]^+} [M(R)_2Cp'_2]^+ \xrightarrow{-R\bullet} [M(R)Cp'_2]^+ \xrightarrow{L} [M(R)(L)Cp'_2]^+$$

Scheme 1

Oxidative cleavage reactions of $[Zr(R)_2Cp'_2]$ compounds using $Ag[BPh_4]$ proceed readily in MeCN to afford $[Zr(R)(MeCN)_nCp'_2]^+$ ($n = 1$ or 2) complexes.[13,14] Acetonitrile coordinates strongly to the initially formed $[M(R)Cp'_2]^+$ cation, and may insert into the M–R bond, thereby limiting further chemistry. These reactions may proceed via initial outer-sphere electron transfer, or by R^- abstraction to generate intermediate $[Ag–R]$ species, which undergo subsequent decomposition to Ag^0 and R–R.[15] Other Ag^+ salts have been used to prepare $[M(R)(L)Cp'_2]^+$ complexes. The ion-pairs $[Zr(\eta^2\text{-}Bz)(CB_{11}H_{12})Cp_2]$ **(12)** and $[Zr(Me)(CB_{11}H_{12})(\eta\text{-}C_5H_4Me)_2]$ **(13)** containing the labile $[CB_{11}H_{12}]^-$ ligand were prepared by reaction of $Ag[CB_{11}H_{12}]$ with $[Zr(Bz)_2Cp_2]$ and $[Zr(Me)_2(\eta\text{-}C_5H_4Me)_2]$, respectively (Equation (3)).[16] In **(12)** and **(13)**, the $[CB_{11}H_{12}]^-$ ligand weakly coordinates to zirconium through the B–H bond *para* to the carborane carbon, and is readily displaced by a Lewis base. For example, **(12)** reacts with MeCN to afford $[Zr(\eta^2\text{-}Bz)(MeCN)Cp_2][CB_{11}H_{12}]$ **(14)**, and **(13)** reacts with THF to afford $[Zr(Me)(THF)(\eta\text{-}C_5H_4Me)_2][CB_{11}H_{12}]$ **(15)**. The reaction of $[Zr\{SiH(Mes)_2\}(Me)Cp^*]$ or $[Zr(Me)_2Cp^*_2]$ with $Ag[OSO_2CF_3]$ affords $[Zr(Me)(OSO_2CF_3)Cp^*_2]$ **(16)**.[17]

$$Cp'_2Zr\!\!\begin{smallmatrix}R\\ \\R\end{smallmatrix} \ +\ AgCB_{11}H_{12} \quad \xrightarrow{\text{toluene}} \quad Cp'_2\overset{+}{Zr}\!\!\begin{smallmatrix}R\\ \\HBCB_{10}H_{11}^-\end{smallmatrix} \qquad (3)$$

$$\textbf{(12)}\ R = \eta^2\text{-Bz};\ Cp' = Cp$$
$$\textbf{(13)}\ R = Me;\ Cp' = \eta\text{-}C_5H_4Me$$

Similar oxidative M–R bond cleavage reactions have been exploited in the synthesis of cationic complexes of other metals, including Ti, U, Ir, and Pt.[18]

12.2.1.2 Protolytic M–R bond cleavage reactions of [M(R)₂Cp′₂] complexes

Protolytic M–R bond cleavage reactions of $[M(R)_2Cp'_2]$ complexes provide a general route to $[M(R)Cp'_2]^+$ species. The use of ammonium reagents in such reactions was first described in the preparation of $[Ti(Me)(NH_3)Cp_2][X]$ ($X = [PF_6]$ or $[ClO_4]^-$) via the reaction of $[Ti(Me)_2Cp_2]$ with $[NH_4][X]$ (Equation (4)).[19] Soon thereafter, analogous reactions were used to prepare cationic Zr, Hf, and Th complexes.[20–2] To avoid the formation of amine complexes and decomposition reactions involving the counterion, ammonium reagents based on bulky, weakly coordinating amines and counterions have been developed, of which $[HNMe_2Ph][B(C_6F_5)_4]$ and $[HNMe_2Ph][M(C_2B_9H_{11})_2]$ (M = Co, Fe, or Ni) are the most noteworthy.[23,24] Other $[HNR_3][B(aryl)_4]$ salts (e.g., R = Me, Et, Bu^n; aryl = $p\text{-}C_6H_4Et$, $p\text{-}C_6H_4F$, or $3,5\text{-}(CF_3)_2C_6H_3$) have also been successfully employed. These reagents generally react rapidly with $[M(R)_2Cp'_2]$ complexes in a variety of solvents to yield a hydrocarbon (usually CH_4 or toluene), an amine, and a base-free $[M(R)Cp'_2]^+$ species.

$$\text{Cp}_2\text{Ti}\begin{array}{c}\text{Me}\\\text{Me}\end{array} \xrightarrow[\text{THF}]{[\text{NH}_4][\text{X}]} \overset{+}{\text{Cp}_2\text{Ti}}\begin{array}{c}\text{Me}\\\text{NH}_3\end{array} \text{X}^- \tag{4}$$

$$\text{X} = [\text{PF}_6]^- \text{ or } [\text{ClO}_4]^-$$

(i) Protonolysis reactions in the presence of a Lewis base

Protonolysis reactions in the presence a Lewis base afford base-stabilized cationic complexes. Complexes (**2**), (**7**), and (**10**) were prepared by the reaction of $[\text{HNBu}_3][\text{BPh}_4]$ with the appropriate $[\text{M(R)}_2\text{Cp}_2]$ complex (M = Zr or Hf; R = Me or Bz) in THF.[7,11] Similarly, reaction of $[\text{M(Me)}_2\text{Cp*}_2]$ (M = Zr or Hf) with $[\text{HNEt}_3][\text{BPh}_4]$ in tetrahydrothiophene (THT) yields $[\text{M(Me)(THT)Cp*}_2][\text{BPh}_4]$ (M = Zr (**17**) or Hf (**18**)); Equation (5).[25] The reaction of $[\text{H(OEt}_2)_3][\text{B}\{3,5\text{-}(\text{CF}_3)_2\text{-}\text{C}_6\text{H}_3\}_4]$ with $[\text{Hf(Bu}^n)_2(\text{C}_5\text{H}_4\text{Me})_2]$ in the presence of PMe_3 or THF yields $[\text{Hf(Bu}^n)(\text{L})(\eta\text{-C}_5\text{H}_4\text{Me})_2][\text{B}\{3,5\text{-}(\text{CF}_3)_2\text{C}_6\text{H}_3\}_4]$ (L = THF (**19**) or PMe_3 (**20**)).[26]

$$\text{Cp*}_2\text{M}\begin{array}{c}\text{Me}\\\text{Me}\end{array} \xrightarrow[\text{THT}]{[\text{HNEt}_3][\text{BPh}_4]} \overset{+}{\text{Cp*}_2\text{M}}\begin{array}{c}\text{Me}\\\text{S}\end{array} [\text{BPh}_4]^- \tag{5}$$

(17) M = Zr
(18) M = Hf

Cationic complexes are also formed by protonolysis of metallacycles. The reaction of $[\text{Zr}\{\eta^2\text{-}\text{CH}_2\text{SiMe}_2(\text{CH}_2)\}\text{Cp'}_2]$ (a zirconasilacyclobutane) with $[\text{HNEt}_3][\text{BPh}_4]$ in THF affords $[\text{Zr(CH}_2\text{-TMS)(THF)Cp'}_2][\text{BPh}_4]$ ($\text{Cp'}_2 = \text{Cp}_2$ (**21**), Cp*_2 (**22**) or $\text{Me}_2\text{Si(C}_5\text{H}_4)_2$ (**23**)) in high yield (Equation (6)).[27] Similarly, protonolysis of the hafnacyclobutane $[\text{Hf}\{\eta^2\text{-}\text{CH}_2\text{CH(Me)CH}_2\}\text{Cp*}_2]$ with $[\text{HNBu}_3][\text{BPh}_4]$ in the presence of THF or PMe_3 yields $[\text{Hf(CH}_2\text{CHMe}_2)(\text{L})\text{Cp*}_2]$ (L = THF (**24**) or PMe_3 (**25**)).[26]

$$\text{Cp'}_2\text{Zr}\!\!\overset{\triangle}{\underset{\triangledown}{}}\!\!\text{SiMe}_2 \xrightarrow[\text{THT}]{[\text{HNEt}_3][\text{BPh}_4]} \overset{+}{\text{Cp'}_2\text{Zr}}\begin{array}{c}\text{TMS}\\\text{O}\end{array} [\text{BPh}_4]^- \tag{6}$$

(21) $\text{Cp'}_2 = \text{Cp}_2$
(22) $\text{Cp'}_2 = \text{Cp*}_2$
(23) $\text{Cp'}_2 = \text{Me}_2\text{Si(C}_5\text{H}_4)_2$

(ii) Protonolysis reactions in the absence of a strong Lewis base

Protonolysis of $[\text{M(Me)}_2\text{Cp'}_2]$ with $[\text{HNR}^1_3]^+$ salts of weakly coordinating anions in hydrocarbon or chlorinated hydrocarbon solvents affords highly reactive $[\text{M(Me)}(\eta\text{-Cp'})_2]^+$ species (**26**). NMR spectral studies of (**26**) reveal coordination of anion, solvent, or amine to the $[\text{M(Me)Cp'}_2]^+$ cation, depending on steric factors and experimental conditions. Important examples include $[\text{Zr(Me)Cp}_2][\text{M(C}_2\text{B}_9\text{H}_{11})_2]$ (M = Fe (**27**), Co (**28**) or Ni (**29**)),[24] $[\text{Zr(Me)(NMe}_2\text{Ph)Cp'}_2][\text{B}(p\text{-C}_6\text{H}_4\text{F})_4]$ ($\text{Cp'}_2 = \text{Cp}_2$ (**30**), $(\text{C}_5\text{H}_4\text{-TMS})_2$ (**31**), $(\text{C}_5\text{H}_4\text{Bu}^t)_2$ (**32**), or rac-EBI (EBI = ethylenebis(indenyl)) (**33**),[28] $[\text{Zr(Me)Cp*}_2][\text{B}(p\text{-C}_6\text{H}_4\text{F})_4]$ (**34**),[28] $[\text{Zr(Me)Cp'}_2][\text{BPh}_4]$ ($\text{Cp'}_2 = \text{Cp}_2$ (**35**), rac-EBI (**36**), rac-EBTHI (**37**), $(\text{C}_5\text{H}_4\text{Bu}^t)_2$ (**38**), or $(\text{C}_5\text{H}_4\text{-TMS})_2$ (**39**)),[29,30] $[\text{M(Me)(NMe}_2\text{Ph)(EBTHI)][Co(C}_2\text{B}_9\text{H}_{11})_2]$ (M = Zr (**40**) or Hf (**41**)),[31] and several $[\text{Zr(Me)Cp'}_2][\text{B(C}_6\text{F}_5)_4]$ salts.[23,32,33]

In a related reaction, $[\text{Zr(Me)Cp'}_2(\text{C}_2\text{B}_9\text{H}_{12})]$ ion-pairs ($\text{Cp'} = \text{Cp*}$ (**42**) or $\text{C}_5\text{Me}_4\text{Et}$ (**43**)) were prepared by reaction of $[\text{Zr(Me)}_2\text{Cp'}_2]$ with the acidic carborane $\text{C}_2\text{B}_9\text{H}_{13}$ (Equation (7)).[22] The $[\text{C}_2\text{B}_9\text{H}_{12}]^-$ anion weakly coordinates to the metal cation via a B–H–Zr bridge.

$$\text{Cp'}_2\text{Zr}\begin{array}{c}\text{Me}\\\text{Me}\end{array} + \text{C}_2\text{B}_9\text{H}_{13} \longrightarrow \overset{+}{\text{Cp'}_2\text{Zr}}\begin{array}{c}\text{Me}\\\text{HBC}_2\text{B}_8\text{H}_{11}^-\end{array} \tag{7}$$

(42) $\text{Cp'} = \text{Cp*}$
(43) $\text{Cp'} = \eta\text{-C}_5\text{Me}_4\text{Et}$

The base-free η^3-allyl complex (**11**) was prepared by reaction of the "tuck-in" complex [Zr(η-allyl)(η^6-C$_5$Me$_4$CH$_2$)Cp*] with [HNEt$_3$][BPh$_4$] at low temperature (Equation (8)).[10] Complex (**11**) is also available by oxidation of [Zr(η^3-allyl)(η^1-allyl)Cp*$_2$] with [Fe(η-C$_5$H$_4$Me)$_2$][BPh$_4$] (Section 12.2.1.1).

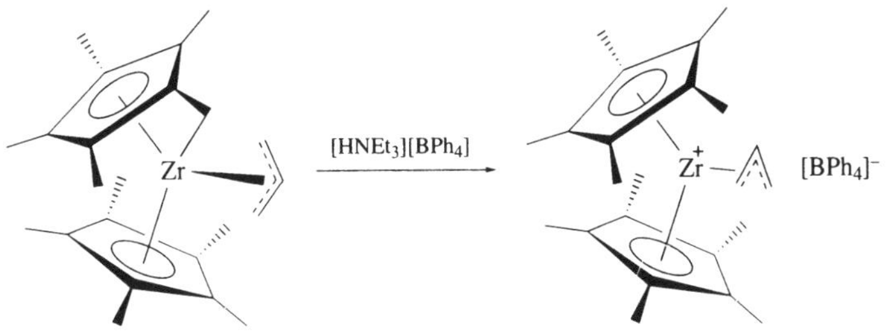

$$\text{(8)}$$

$$\text{(11)}$$

(iii) Protonolysis reactions using a coordinating acid

Protonolysis of [Zr(Me)$_2$Cp$_2$] with H$_2$C(SO$_2$CF$_3$)$_2$, PhCH(SO$_2$CF$_3$)$_2$, or NH(SO$_2$CF$_3$)$_2$ yields [Zr(Me){Y(SO$_2$CF$_3$)$_2$}Cp$_2$] (Y = CH (**44**), CPh (**45**), or N (**46**)); [Hf(Me){HC(SO$_2$CF$_3$)$_2$}Cp$_2$] (**47**) was prepared similarly. Conductivity measurements indicate that (**44**)–(**46**) undergo solvolysis in MeCN to afford [Zr(Me)(MeCN)$_2$Cp$_2$]$^+$.[34] Similarly, the reaction of [Zr(Me)$_2$Cp$_2$] with [Mo(H)(CO)$_3$Cp] yields [Zr(Me)Cp$_2${(μ-OC)Mo(CO)$_2$Cp}] (**48**), which exhibits ionic behavior in MeCN.[35]

12.2.1.3 Alkyl group abstraction from [M(R)$_2$Cp$'_2$] complexes

Alkyl group abstraction from neutral [M(R)$_2$Cp$'_2$] complexes by a strong Lewis acid provides a general route to [M(R)Cp$'_2$]$^+$ species. Often these reactions may be performed under rigorously base-free conditions in low-polarity, weakly coordinating solvents (e.g., benzene or toluene).

The trityl cation [CPh$_3$]$^+$ readily abstracts alkyl groups from early metal complexes, as first demonstrated by the preparation of [Ta(Me)$_2$Cp$_2$][BF$_4$] via reaction of [Ta(Me)$_3$Cp$_2$] with [CPh$_3$][BF$_4$].[36] The use of [CPh$_3$]$^+$ reagents in the synthesis of cationic group 4 metal complexes was established by the preparation of [Zr(Me)(CD$_3$CN)$_2$Cp$_2$][BPh$_4$] (**49**-d$_6$) via reaction of [Zr(Me)$_2$Cp$_2$] with [CPh$_3$][BPh$_4$] in CD$_3$CN.[37] More recently, analogous reactions using [CPh$_3$][B(C$_6$F$_5$)$_4$] have been used for the *in situ* generation of base-free [Zr(R)Cp$'_2$][B(C$_6$F$_5$)$_4$] species (Equation (9)), including [Zr(Me)(EBI)]-[B(C$_6$F$_5$)$_4$] (**50**), [Zr(Me){η-C$_5$H$_4$CMe$_2$(η^5-fluorenyl)}][B(C$_6$F$_5$)$_4$] (**51**) and related species,[33,38,39] [Zr(Bz)Cp$'_2$][B(C$_6$F$_5$)$_4$] (Cp$'_2$ = Cp$_2$ (**52**), (C$_5$H$_4$TMS)$_2$ (**53**), EBI (**54**), or {η-C$_5$H$_4$CMe$_2$(η^5-fluorenyl)} (**55**)).[40] The reaction of [Zr(Me)$_2${η-C$_5$H$_4$CMe$_2$(η^5-fluorenyl)}] with [CPh$_3$][B(C$_6$F$_5$)$_4$] in the presence of PMe$_3$ yields [Zr(Me)(PMe$_3$){η-C$_5$H$_4$CMe$_2$(η^5-fluorenyl)}][B(C$_6$F$_5$)$_4$] (**56**).[41]

$$\text{Cp}'_2\text{Zr}\overset{R}{\underset{R}{\diagdown}} \xrightarrow[-\text{MeCPh}_3]{[\text{CPh}_3][\text{B}(\text{C}_6\text{F}_5)_4]} \left[\text{Cp}'_2\text{Zr}-\text{R}\right]^+ [\text{B}(\text{C}_6\text{F}_5)_4]^- \qquad (9)$$

(**50**) R = Me; Cp$'_2$ = EBI
(**51**) R = Me; Cp$'_2$ = (C$_5$H$_4$)CMe$_2$(fluorenyl)
(**52**) R = Bz; Cp$'_2$ = Cp$_2$
(**53**) R = Bz; Cp$'_2$ = (C$_5$H$_4$TMS)$_2$
(**54**) R = Bz; Cp$'_2$ = EBI
(**55**) R = Bz; Cp$'_2$ = (C$_5$H$_4$)CMe$_2$(fluorenyl)

The reaction of B(C$_6$F$_5$)$_3$ with [Zr(Me)$_2$Cp$'_2$] complexes in benzene yields the "cation-like" species [Zr(Me){MeB(C$_6$F$_5$)$_3$}Cp$'_2$] (Cp$'$ = Cp (**57**), 1,2-Me$_2$C$_5$H$_3$ (**58**), or Cp* (**59**)) which contain a labile [MeB(C$_6$F$_5$)$_3$]$^-$ ligand (Equation (10)).[42] The [MeB(C$_6$F$_5$)$_3$]$^-$ anion is weakly coordinated via a Zr–(μ-Me)–B link.

Halide abstraction from [M(R)(X)Cp$'_2$] complexes also provides a route to cationic species, although halide/alkyl redistribution processes often complicate these reactions. The reaction of [Zr{CH(TMS)$_2$}(Cl)Cp$'_2$] complexes with 2–3 equiv. of AlCl$_3$ yields [Zr{η^2-*C,Cl*-CH(SiMe$_2$Cl)-(TMS)}Cp$'_2$][X] ((**60**), Cp$'$ = Cp, X = [Al$_2$Cl$_7$]$^-$; or (**61**), Cp$'_2$ = *rac*-EBI, X = [Al$_2$Cl$_{6.5}$Me$_{0.5}$]$^-$) via Cl$^-$ abstraction and subsequent Si–Me/Al–Cl redistribution.[43]

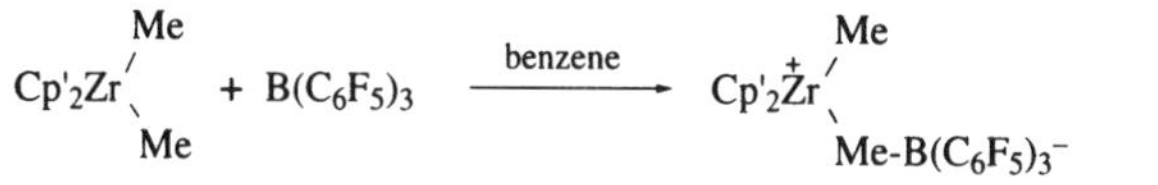

$$\text{Cp'}_2\text{Zr} \Big\langle \begin{array}{c} \text{Me} \\ \text{Me} \end{array} \quad + \quad B(C_6F_5)_3 \quad \xrightarrow{\text{benzene}} \quad \text{Cp'}_2\overset{+}{\text{Zr}} \Big\langle \begin{array}{c} \text{Me} \\ \text{Me-B}(C_6F_5)_3^- \end{array}$$

$$\tag{10}$$

(**57**) Cp' = Cp
(**58**) Cp' = η-1,2-C$_5$H$_3$Me$_2$
(**59**) Cp' = Cp*

12.2.1.4 Alkene and allene insertion reactions of preformed [M(R)(L)Cp'$_2$]$^+$ complexes (R = H or Me)

Cationic alkyl and allyl complexes may be prepared by alkene or allene insertion reactions of preformed [Zr(R)Cp'$_2$]$^+$ species (R = H or Me). The reaction of [Zr(H)(THF)(η-C$_5$H$_4$Me)$_2$][BPh$_4$] (**62**) with alkenes or allene yields [Zr(CH$_2$CH$_2$R)(THF)(η-C$_5$H$_4$Me)$_2$][BPh$_4$] (R = H (**63**), Me (**64**), Et (**65**), Bu^t (**66**), Ph (**67**), or TMS (**68a**)) or [Zr(η-allyl)(THF)(η-C$_5$H$_4$Me)$_2$][BPh$_4$] (**69**) (Scheme 2).[44-6] The soluble hydride (**62**) is available from the reaction of [Zr(R)(THF)(C$_5$H$_4$Me)$_2$][BPh$_4$] ((**3**) or (**7**)) with H$_2$ in THF (Scheme 2). The analogous hydride [Zr(H)(THF)Cp$_2$][BPh$_4$] (**70**) can be prepared similarly, but is less useful due to poor solubility properties.[14]

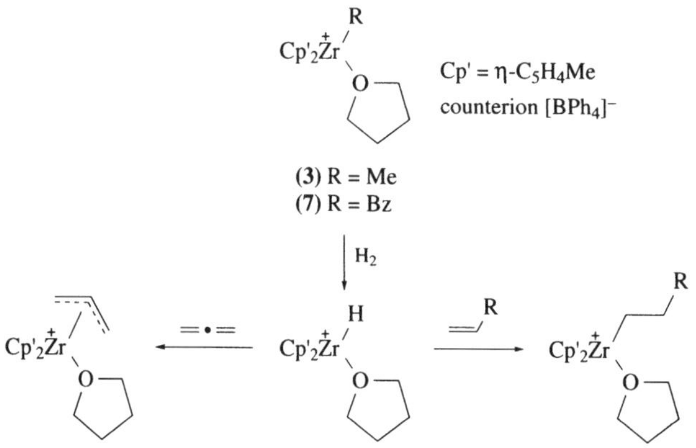

$$\text{Cp'}_2\overset{+}{\text{Zr}} \Big\langle \begin{array}{c} \text{R} \\ \text{O} \end{array} \qquad \text{Cp'} = \text{η-C}_5\text{H}_4\text{Me}$$

counterion [BPh$_4$]$^-$

(**3**) R = Me
(**7**) R = Bz

↓ H$_2$

(**69**) ← == • == (**62**) → (**63**) R = H

(**63**) R = H
(**64**) R = Me
(**65**) R = Et
(**66**) R = Bu^t
(**67**) R = Ph

Scheme 2

The η^5-pentadienyl complexes [Zr{η^5-CH$_2$C(Me)=C(Me)C(Me)=CH(Me)}Cp'$_2$][B(p-C$_6$H$_4$F)$_4$] (Cp'$_2$ = Cp$_2$ (**71**), (C$_5$H$_4$TMS)$_2$ (**72**), (C$_5$H$_4$Bu^t)$_2$ (**73**), or rac-EBI (**74**)) were obtained from the double insertion of 2-butyne into the Zr–Me bond of (**30**)–(**33**) (Equation (11)).[47] Analogous double insertions of 3-hexyne were also reported. In contrast, the reaction of [Zr(Me)Cp*$_2$][B(p-C$_6$H$_4$F)$_4$] with 2-butyne yields the C–H activation product [Zr(CH$_2$CCMe)Cp*$_2$][B(p-C$_6$H$_4$F)$_4$] (**75**).[47] The η^3-allyl complexes [Zr{η-CH$_2$C(Me)CH$_2$}Cp'$_2$][B(p-C$_6$H$_4$F)$_4$] (Cp' = C$_5$H$_4$Bu^t (**76**) or Cp* (**77**)) were prepared by the reaction of (**32**) and (**34**) with allene.[48]

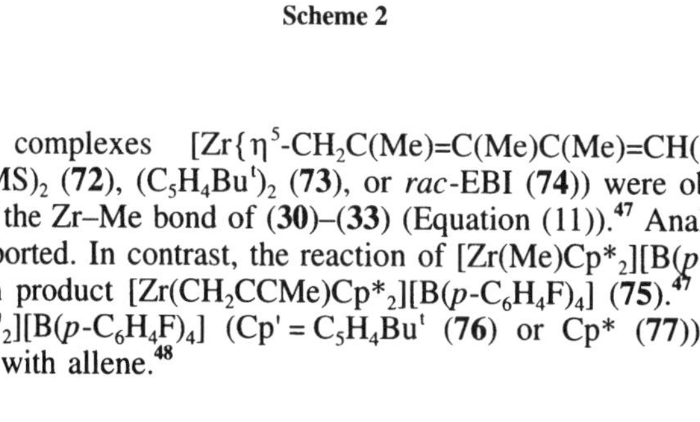

$$\text{Cp'}_2\text{Zr} \Big\langle \begin{array}{c} \text{Me} \\ \text{Me} \end{array} \quad \xrightarrow[\text{PhBr}]{[\text{HNMe}_2\text{Ph}][B(p\text{-C}_6\text{H}_4\text{F})_4]} \quad$$

$$\tag{11}$$

(**71**) Cp' = Cp
(**72**) Cp' = C$_5$H$_4$TMS
(**73**) Cp' = C$_5$H$_4$Bu^t
(**74**) Cp' = rac-EBI

12.2.2 Structures of Lewis Base-stabilized [M(R)(L)Cp'₂]⁺ Complexes

Lewis base-stabilized [M(R)(L)Cp'₂]⁺ complexes adopt normal bent-metallocene structures.[49] The electron deficiency at the formally 16-electron metal center promotes Zr–L π-bonding where possible (e.g., L = THF), agostic M–H–C interactions,[50] and other nonclassical bonding interactions.

12.2.2.1 Structures of [M(Me)(L)Cp'₂]⁺ complexes

The first cation in this class to be structurally characterized was [Zr(Me)(THF)Cp₂][BPh₄] (2) (Figure 1).[5] The cation of (2) adopts a normal bent-metallocene structure with a normal zirconium–methyl distance (0.226 nm), centroid–Zr–centroid angle (129.6°), and Me–Zr–O angle (97.4°). The THF ligand lies perpendicular to the Me–Zr–O plane (C–O–C/C–Zr–O dihedral angle 102.3°) in an orientation in which oxygen–zirconium π-bonding is possible. On the basis of the short zirconium–oxygen distance (0.212 nm) and reactivity trends,[14] it was argued that the THF ligand is a four-electron (σ,π)-donor in this complex. For comparison, in the isoelectronic lanthanide complex [Yb(Me)(THF)Cp₂], in which steric effects are expected to dominate structural preferences, the THF ligand lies nearly in the Me–Yb–O plane.[51] The structure of [Zr(CH₂-TMS)(THF)Cp*₂][BPh₄] (22) has also been determined (Figure 2).[27] In this case, steric interactions between the THF and the bulky Cp*⁻ ligands dictate an in-plane orientation of the THF ligand which precludes zirconium–oxygen π-bonding. Close cation–anion contacts are not observed in (2) or (22).

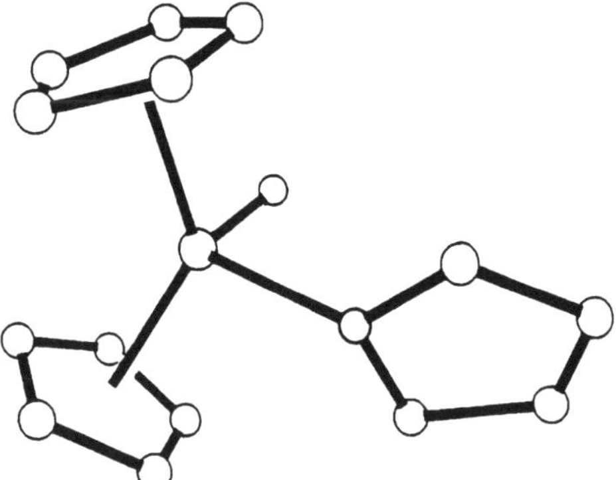

Figure 1   Structure of the [Zr(Me)(THF)Cp₂]⁺ cation of (2). Hydrogen atoms are omitted for clarity.

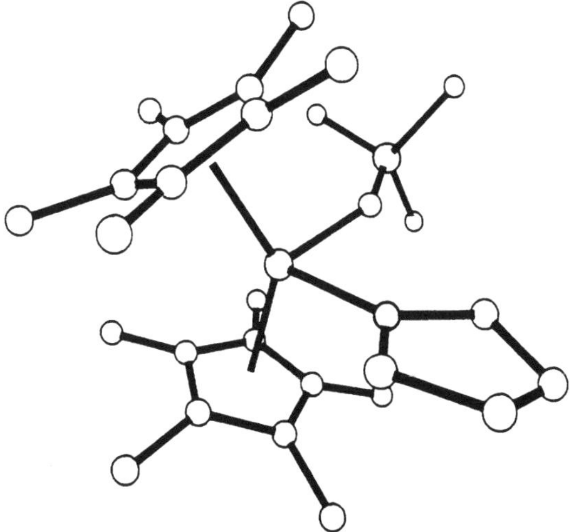

Figure 2   Structure of the [Zr(CH₂TMS)(THF)Cp*₂]⁺ cation of (22). Hydrogen atoms are omitted for clarity.

The structures of [M(Me)(THT)Cp*₂][BPh₄] (M = Zr (17) or Hf (18)) permit direct comparison between cationic zirconium and hafnium analogues.[25a] The cation of (17) adopts a normal bent-metallocene structure with normal Zr–C (0.224 nm) and Zr–S (0.273 nm) bond distances. In contrast, in (18) (Figure 3) the Hf–Cp* bonding is unsymmetrical (Hf–Cp* centroid distances 0.234 and 0.214 nm), and the Hf–S bond is unusually short (0.262 nm).

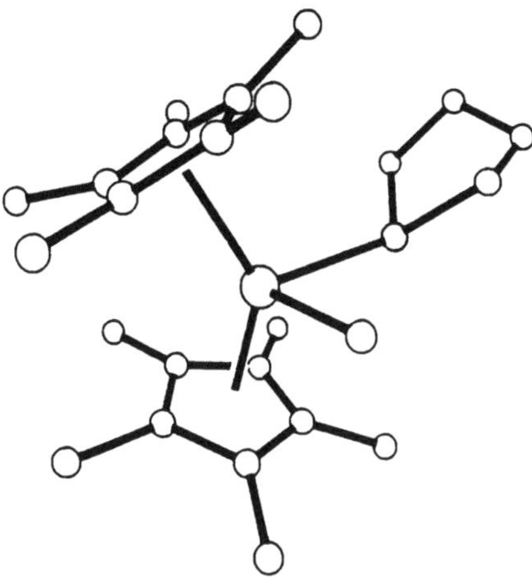

Figure 3   Structure of the [Hf(Me)(THT)Cp*$_2$]$^+$ cation of (**18**). Hydrogen atoms are omitted for clarity.

The structure of [Zr(Me)(PMe$_3$){η-C$_5$H$_4$CMe$_2$(η^5-fluorenyl)}][B(C$_6$F$_5$)$_4$] (**56**) has been determined.[41] As for other early transition metal fluorenyl complexes, there is considerable η^3- character in the zirconium–fluorenyl bonding (Zr–C$_{fluorenyl}$ distances range from 0.238 to 0.264 nm). An interesting feature of this structure is the small Me–Zr–PMe$_3$ angle (88.4°), which is considerably smaller than the Cl–Zr–Cl angle in [Zr(Cl)$_2${η-C$_5$H$_4$CMe$_2$(η^5-fluorenyl)}] (98.2°)[33,52] and X–M–X angles in d^0 [M(X)$_2$Cp'$_2$] complexes in general.[49] The possibility that this results from a zirconium–methyl α-agostic interaction was noted.

12.2.2.2 *Structures of [Zr(Bz)(L)Cp'$_2$]$^+$ complexes*

The η^2-benzyl complex [Zr(η^2-Bz)(MeCN)Cp$_2$][BPh$_4$] (**78**) (Figure 4) has been characterized by x-ray diffraction.[8] The acute Zr–C–Ph angle (84.9°) and short Zr–C$_{ipso}$ distance (0.2648 nm) indicate that the phenyl π-system interacts with zirconium, resulting in the η^2-benzyl structure. The coordinative unsaturation at zirconium and the minimal steric requirements of the MeCN ligand favor this coordination mode. Many related examples are known.[53,54] NMR spectral data establish that the η^2-coordination mode is maintained in solution; in particular, a large $J_{C-H}(CH_2)$ value (145 Hz) characteristic of an acute Zr–C–C angle is observed. In contrast, [Zr(Bz)(THF)Cp$_2$][BPh$_4$] (**6**) adopts an η^1-benzyl structure in solution; evidently π-donation from THF is favored over an η^2-benzyl interaction. At low temperature in CD$_2$Cl$_2$ solvent, the base-free η^2-benzyl species [Zr(Bz)Cp$_2$]$^+$ (or its solvent adduct) formed by THF dissociation from (**6**) was observed by NMR spectroscopy. Analogous chemistry has been noted for [Zr(Bz)(L)(η-C$_5$H$_4$Me)$_2$][BPh$_4$] (L = MeCN (**79**) or THF (**7**)) and [Zr(Bz)(L)(EBTHI)][BPh$_4$] (L = MeCN (**80**) or THF (**8**)) complexes.[6,9,44] NMR spectroscopic data also imply η^2-benzyl structures for the base-free complexes (**52**)–(**54**) (Section 12.2.1.3).[40]

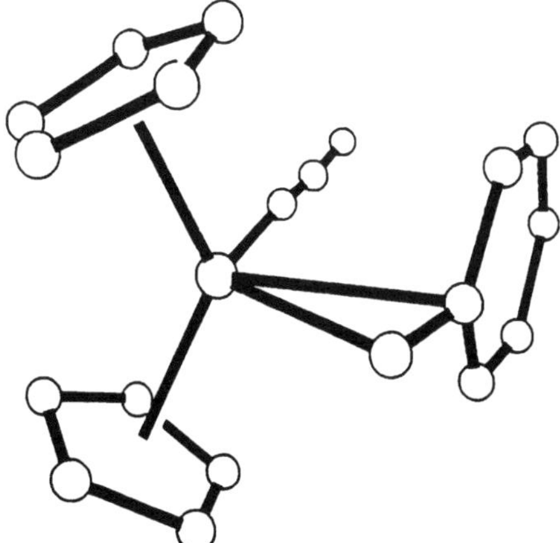

Figure 4   Structure of the [Zr(η^2-Bz)(MeCN)Cp$_2$]$^+$ cation of (**78**). Hydrogen atoms are omitted for clarity.

12.2.2.3 Structures of [Zr(CH₂CH₂R)(L)Cp′₂]⁺ complexes

(i) Agostic Zr–Hβ–C interactions

The solution structures of $[Zr(CH_2CH_2R)(L)(\eta\text{-}C_5H_4Me)_2][BPh_4]$ complexes (L = THF, PMe₃, or MeCN) have been studied in detail.[44–6] The THF complexes $[Zr(CH_2CH_2R)(THF)(\eta\text{-}C_5H_4Me)_2]^+$ (R = H **(63)**, Me **(64)**, Et **(65)**, Bu^i_3 **(66)**, or Ph **(67)**) (Section 12.2.1.4) exhibit NMR and IR spectroscopic properties consistent with undistorted alkyl groups. In contrast, the analogous PMe₃ complexes $[Zr(CH_2CH_2R)(PMe_3)(\eta\text{-}C_5H_4Me)_2]^+$ (**(81)**–**(85)**) obtained by ligand substitution (Scheme 3) adopt agostic Zr–Hβ–C structures. Characteristic spectroscopic features for the distorted alkyl ligands in **(81)**–**(85)** include high-field β-CH_2R ¹H and ¹³C NMR resonances, large $J_{C\alpha-H}$ values (135–145 Hz) characteristic of acute Zr–C–C angles, and in some cases low-frequency ν_{C-H} IR absorbances. In general, exchange of the bridging and terminal β-hydrogens is rapid on the NMR timescale, so averaged β-CH_2R chemical shifts and coupling constants are observed. Isotope perturbation of resonance (IPR) experiments with deuterium-labeled analogues establish a single βH bridge to zirconium. The analogous MeCN complexes $[Zr(CH_2CH_2R)(MeCN)(\eta\text{-}C_5H_4Me)_2]^+$ (**(86)**–**(90)**) also adopt Zr–Hβ–C structures (Scheme 3), although their solution behavior is complicated by the formation of bis(MeCN) adducts with undistorted alkyl ligands.

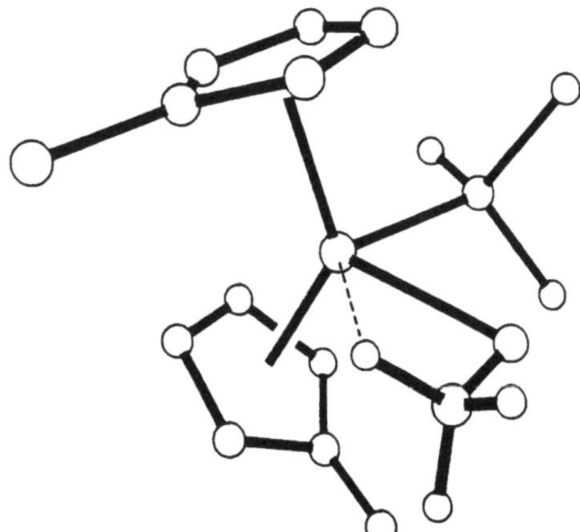

Scheme 3

An x-ray diffraction study of $[Zr(Et)(PMe_3)(\eta\text{-}C_5H_4Me)_2][BPh_4]$ **(81)** (Figure 5) confirmed the presence of a Zr–Hβ–C interaction in the solid state.[45] The Zr–Cα–Cβ angle in **(81)** is acute (84.7(5)°) and the Zr–Cβ (0.2629 nm) and Zr–Hβ (0.216 nm) contacts are short. The agostic β-hydrogen lies in the P–Zr–Cα–Cβ plane where the zirconium LUMO is localized, as required for a three-center, two-electron bonding interaction.[55] The ethyl ligand in **(81)** is structurally similar to that in $[Ti(Cl)_3(Et)(dmpe)]$,[56] but is significantly less distorted than those in $[Co(Et)Cp^*\{P(p\text{-}Tol)_3\}]^+$ and $[Pt\{(Bu^i)_2P(CH_2)_3P(Bu^i)_2\}(Et)]^+$.[57,58]

Figure 5 Structure of the $[Zr(Et)(PMe_3)(\eta\text{-}C_5H_4Me)_2]^+$ cation of **(81)**. Only the β-hydrogens of the ethyl group are shown.

The Zr–Hβ–C agostic interactions in the PMe₃ complexes **(81)**–**(85)** and the MeCN complexes **(86)**–**(90)** reflect the unsaturation of the 16-electron metal centers. In contrast, oxygen–zirconium π-bonding in the THF adducts **(63)**–**(67)** effectively saturates the metal center, and undistorted alkyl ligands are observed.

(ii) Agostic Zr–Cβ–Si interactions

The x-ray structure of [Zr(CH$_2$CH$_2$TMS)(THF)(η-C$_5$H$_4$Me)$_2$][BPh$_4$] (**68a**) (Figure 6) revealed a distorted alkyl ligand (Zr–Cα–Cβ, 84°; Zr–Cβ, 0.257 nm).[59] However, the TMS group lies in the O–Zr–Cα–Cβ plane, indicating that the β-hydrogens lie above and below this plane, suggesting that agostic Zr–Hβ–C interactions are absent. It was proposed that the structure of (**68**) results from overlap of the back lobe of the Cβ–Si bond with the zirconium LUMO. This "agostic Zr–Cβ–Si interaction" is analogous to the stabilization of silyl-substituted carbocations by the "γ-silicon effect,"[60] and is probably fairly strong because it is observed even when a potentially π-donor THF ligand is present. Reaction of (**68a**) with PMe$_3$ or MeCN yields [Zr(CH$_2$CH$_2$TMS)(PMe$_3$)(η-C$_5$H$_4$Me)$_2$]$^+$ (**68b**) and [Zr(CH$_2$CH$_2$-TMS)(MeCN)(η-C$_5$H$_4$Me)$_2$]$^+$ (**68c**), for which spectroscopic data favor analogous agostic Zr–βC–Si structures. Low-temperature studies reveal that (**68c**) is generated as a mixture of "Cβ-inside" and "Cβ-outside" isomers which reverts to the more stable "Cβ-inside" isomer via an intermediate bis(MeCN) adduct.

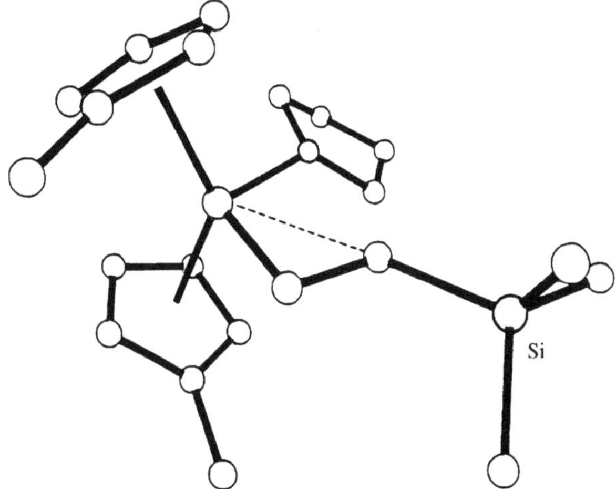

Figure 6   Structure of the [Zr(CH$_2$CH$_2$TMS)(THF)(η-C$_5$H$_4$Me)$_2$]$^+$ cation of (**68a**). Hydrogen atoms are omitted for clarity.

12.2.2.4 Structures of [M(CH$_2$CHMe$_2$)(L)Cp′$_2$]$^+$ complexes

The reaction of hydride (**62**) with isobutene yields [Zr(CH$_2$CHMe$_2$)(THF)(η-C$_5$H$_4$Me)$_2$][BPh$_4$] (**91a**) in which the alkyl group is undistorted. Ligand exchange with PMe$_3$ yields [Zr(CH$_2$CHMe$_2$)(PMe$_3$)(η-C$_5$H$_4$Me)$_2$][BPh$_4$] (**91b**), which adopts an agostic Zr–Hβ–C structure.[26] In contrast, the more crowded hafnium species [Hf(CH$_2$CHMe$_2$)(L)Cp*$_2$][BPh$_4$] (L = THF (**24**) or PMe$_3$ (**25**)) adopt α-agostic structures, as determined by x-ray crystallography for (**24**) (Figure 7) and NMR spectroscopy (including IPR studies) for (**25**).[26] In these cases, steric interactions between the β-Me groups and the Cp* ligands apparently disfavor an agostic Zr–Hβ–C structure.[25]

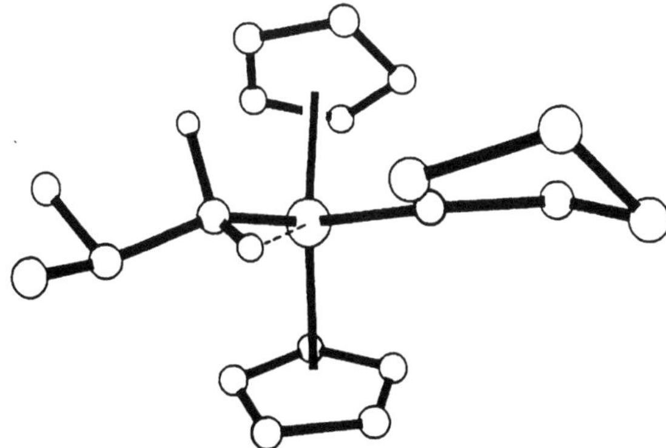

Figure 7   Structure of the [Hf(CH$_2$CHMe$_2$)(THF)Cp*$_2$]$^+$ cation of (**24**). For clarity, the Cp* methyl groups are omitted, and only the α-hydrogen atoms are shown.

12.2.3 Structures of Lewis Base-free [M(R)Cp′₂]⁺ Complexes

The electron deficiency of the formally 14-electron metal centers in base-free [M(R)Cp′₂]⁺ species is usually stabilized by nonclassical bonding interactions involving the hydrocarbyl R⁻ ligand, or by site-specific ion-pairing interactions with the counterion.

12.2.3.1 Complexes with nonclassical bonding interactions

The ionic structure of [Zr{CH₂CH(AlEt₂)₂}Cp₂][Cp] (**1**) was confirmed by x-ray diffraction.[4] The ZrCH₂CH(AlEt₂)₂ group is highly distorted by a Zr–Cβ–Al interaction, which is evident from the acute Zr–Cα–Cβ angle (75.7°), the short Zr–Cβ contact (0.2393 nm), and the placement of the β-substituents, as described earlier for (**68**) (Section 12.2.2.3(ii)).[59,61] The Cp⁻ anion is ion-paired with the cation through Cp–Al interactions.

An unusual Zr–Cl–Si interaction is exhibited in the x-ray structure of [Zr{η²-*C,Cl*-CH(SiMe₂Cl) (TMS)}(*rac*-EBI)][Al₂Cl₆.₅Me₀.₅]⁻ (**61**) (Figure 8).[43] The zirconium–chlorine distance (0.2573 nm), which is only 0.003 nm less than the Zr–μCl distance in [Zr(Cl)(AlCl₄)Cp₂],[62] the placement of Cl in the equatorial plane containing the zirconium LUMO, the acute Zr–Cl–Si angle (98.3°), and the long Si–Cl bond (0.2266 nm) are all consistent with coordination of chlorine to zirconium.

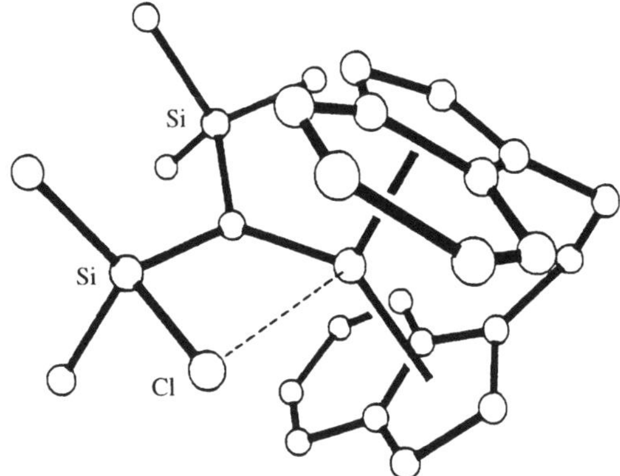

Figure 8 Structure of the [Zr{η²-*C,Cl*-CH(SiMe₂Cl)(TMS)}(*rac*-EBI)]⁺ cation of (**61**). Hydrogen atoms are omitted for clarity.

The x-ray structure of [Zr{η⁵-CH₂C(Me)=C(Me)C(Me)=CH(Me)}(C₅H₄CMe₃)₂][B(*p*-C₆H₄F)₄] (**73**) revealed an η⁵-(σ,π,π) pentadienyl ligand in which the C=C double bonds weakly coordinate to zirconium (Figure 9).[47] There is one normal Zr–C (*sp³*) bond (0.2315 nm) and four long zirconium–carbon contacts (0.26–0.276 nm), and alternating long and short C–C bonds.

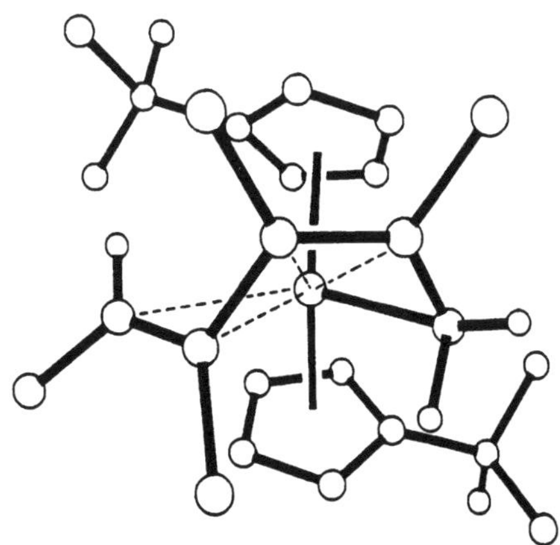

Figure 9 Structure of the [Zr{η⁵-CH₂C(Me)=C(Me)C(Me)=CH(Me)}(C₅H₄Bu′)₂]⁺ cation of (**73**).

12.2.3.2 Complexes with cation–anion interactions

The ion-pair $[Zr(Me)(\eta\text{-}C_5Me_4Et)_2(C_2B_9H_{12})]$ (**43**) was one of the first $[M(R)Cp'_2]^+$ species to be crystallographically characterized.[22] The *nido*-carborane $[C_2B_9H_{12}]^-$ anion weakly coordinates to the $[Zr(Me)(\eta\text{-}C_5Me_4Et)_2]^+$ cation via a B–(μH)–Zr interaction involving a B–H bond on the open carborane face. More recently, x-ray structures of $[Zr(\eta^2\text{-}Bz)(CB_{11}H_{12})Cp_2]$ (**12**) and $[Zr(Me)(CB_{11}H_{12})(\eta\text{-}C_5H_4Me)_2]$ (**13**) (Figure 10), which incorporate the *closo*-carborane $[CB_{11}H_{12}]^-$ have been determined.[16] In (**12**) and (**13**), the $[CB_{11}H_{12}]^-$ anion coordinates via the B–H *para* to the carborane carbon, the most common coordination mode for this anion.[63] An η^2-benzyl ligand is present in (**12**).

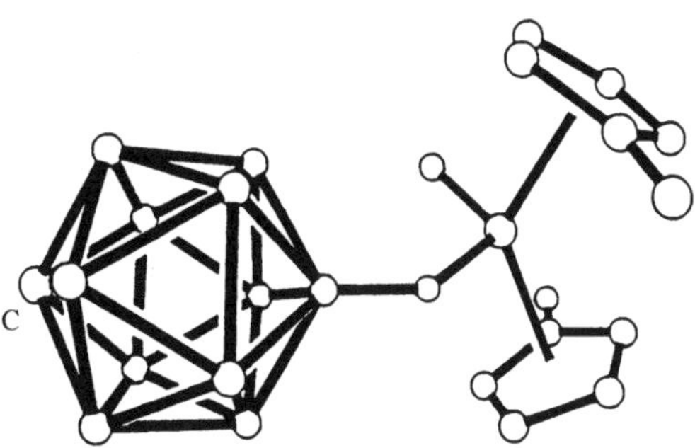

Figure 10   Structure of $[Zr(Me)(CB_{11}H_{12})(\eta\text{-}C_5H_4Me)_2]$ (**13**). For clarity, only the Zr–(μH)–B hydrogen atom is shown.

The ion-pair $[Zr(Me)\{MeB(C_6F_5)_3\}(\eta\text{-}1,2\text{-}Me_2C_5H_3)_2]$ (**58**) has been characterized by x-ray diffraction (Figure 11).[42] The $[MeB(C_6F_5)_3]^-$ counterion weakly coordinates to the $[Zr(Me)(\eta\text{-}1,2\text{-}Me_2C_5H_3)_2]^+$ cation via a nonlinear (161.8°), unsymmetrical Zr–(μ-Me)–B bridge. The Zr–Me(bridge) distance (0.2549 nm) is nearly 0.03 nm longer than the zirconium–methyl(terminal) distance, which is consistent with the ion-pair formulation. The μ-Me hydrogens are bent toward zirconium; however, the closest zirconium–hydrogen contact (0.225(3) nm) is somewhat longer than observed for intramolecular agostic zirconium–hydrogen interactions (ca. 0.215 nm). The μ-Me interaction in (**58**) is reminiscent of that in $[YbCp^*_2(\mu\text{-}Me)BeCp^*]$.[64]

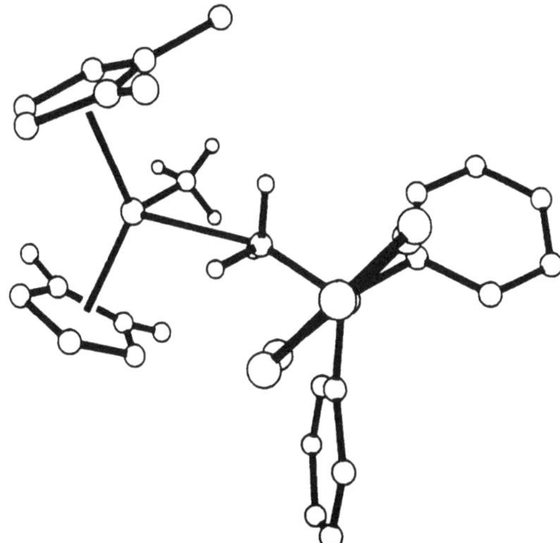

Figure 11   Structure of the ion-pair $[Zr(Me)\{(MeB(C_6F_5)_3\}(\eta\text{-}1,2\text{-}Me_2C_5H_3)_2]$ (**58**). For clarity, only the Zr–Me and Zr–(μ-Me)–B hydrogen atoms are shown. Fluorine atoms are omitted.

12.2.4 Reactivity

Fourteen-electron $[M(R)Cp'_2]^+$ species, and 16-electron $[M(R)(L)Cp'_2]^+$ species which contain labile ligands L, are generally very reactive. The high Lewis acidity of the coordinatively and electronically unsaturated metal center promotes coordination and activation of a variety of substrates. The reactivity is strongly influenced by ion-pairing interactions with the counterion, solvent or Lewis base coordination, and degradation or deactivation processes. In general, $[M(R)Cp'_2]^+$ species with poorly coordinating counterions (especially $[B(C_6F_5)_4]^-$) generated under rigorously base-free conditions in nonpolar solvents are the most reactive; however, base-stabilized $[M(R)(L)Cp'_2]^+$ complexes are also useful for many applications and are easier to handle. This section describes selected aspects of the ligand exchange, σ-bond metathesis, insertion, and elimination chemistry of these systems. Additional examples of these reaction types are discussed in later sections.

12.2.4.1 Ligand exchange and other reactions with nucleophiles

In most cases, the coordinated THF of $[Zr(R)(THF)Cp'_2]^+$ complexes is labile and readily displaced by nitriles, small phosphines, pyridines, and other Lewis bases, as illustrated in Scheme 4. Cationic methyl complexes $[Zr(Me)(THF)Cp'_2]^+$ react with small two-electron donor ligands (L) such as MeCN, PMe_3, and *N*-methylimidazole to form 18-electron $[Zr(Me)(L)_2Cp'_2]^+$ complexes.[6,14,46,65] Dimethylbipyridine coordinates in a bidentate fashion in $[Zr(Me)(4,4'-Me_2bipy)Cp_2]^+$.

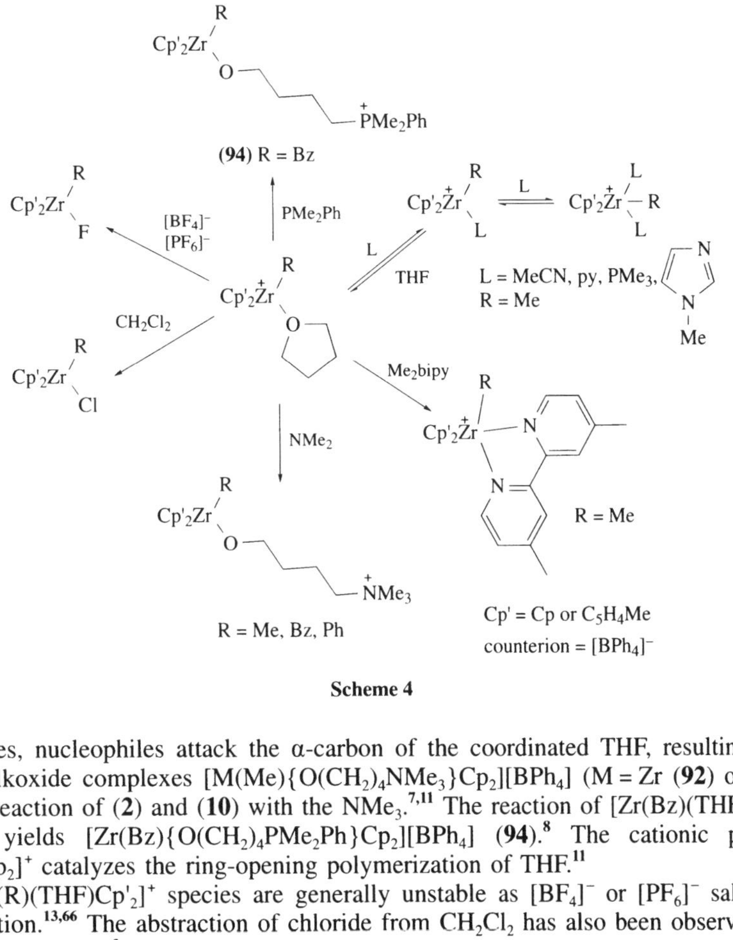

Scheme 4

In some cases, nucleophiles attack the α-carbon of the coordinated THF, resulting in THF ring-opening. The alkoxide complexes $[M(Me)\{O(CH_2)_4NMe_3\}Cp_2][BPh_4]$ (M = Zr (**92**) or Hf (**93**)) were formed by the reaction of (**2**) and (**10**) with the NMe_3.[7,11] The reaction of $[Zr(Bz)(THF)Cp_2][BPh_4]$ (**6**) with PMe_2Ph yields $[Zr(Bz)\{O(CH_2)_4PMe_2Ph\}Cp_2][BPh_4]$ (**94**).[8] The cationic phenyl complex $[Zr(Ph)(THF)Cp_2]^+$ catalyzes the ring-opening polymerization of THF.[11]

Cationic $[M(R)(THF)Cp'_2]^+$ species are generally unstable as $[BF_4]^-$ or $[PF_6]^-$ salts due to facile fluoride abstraction.[13,66] The abstraction of chloride from CH_2Cl_2 has also been observed.[46]

The reactions of the η^3-allyl complex (**11**) with organic nucleophiles have been described.[10] The bulky nucleophiles $KCHPh_2$ and $KCH(Me)Ph$ add regioselectively to the allyl central carbon yielding neutral zirconacyclobutane complexes $[Zr\{\eta^2\text{-}CH_2CH(R)CH_2\}Cp^*_2]$ (R = $CHPh_2$ or CH(Me)Ph), whereas smaller nucleophiles such as MeMgCl, MeLi, and allyl Grignard reagents add to zirconium.

12.2.4.2 Insertion reactions

Cationic $[Zr(R)(L)Cp'_2]^+$ complexes exhibit a rich insertion chemistry, as illustrated for $[Zr(Me)(THF)Cp_2][BPh_4]$ (**2**) in Scheme 5. Complex (**2**) reacts rapidly and irreversibly with carbon monoxide to afford the η^2-acyl complex (**95**) and with Bu^t NC to afford the η^2-iminoacyl complex $[Zr\{C(=NBu^t)Me\}(CNBu^t)][BPh_4]$ (**96**).[13,67] The insertion reactions of benzophenone, acetonitrile, or

2-butyne afford (**97**)–(**99**). Complex (**2**) polymerizes ethene (Section 12.13.1) and oligomerizes propene by an insertion/β-H elimination process.[5] The observation of intermediate MeCN and ketone adducts, and inhibition by added THF, indicate that these reactions proceed via intial displacement of THF by the substrate and subsequent migratory insertion. For comparison, [Zr(R)$_2$Cp$'_2$] and [Zr(R)(X)Cp$'_2$] complexes insert CO reversibly, and do not insert nitriles, ketones, or alkynes under mild conditions.[68]

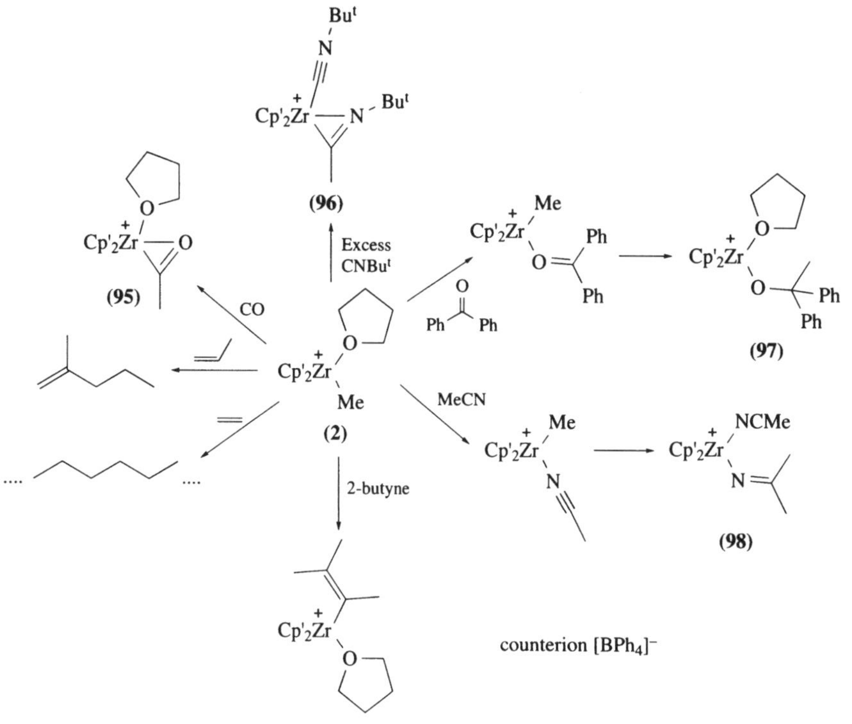

Scheme 5

The MeCN insertion reactions of [Zr(R)(THF)Cp$_2$][BPh$_4$] (R = H (**70**), Ph (**100**), Me (**2**), or η^2-Bz (**6**)) yield azaalkenylidene products [Zr{N=C(R)Me}(MeCN)Cp$_2$]$^+$ (R = H (**101**), Ph (**102**), or Me (**103**)) and have been studied in detail (Scheme 6).[6] These reactions proceed by formation of an equilibrium mixture of (bis)- and (mono)MeCN adducts [Zr(R)(MeCN)$_n$Cp$_2$]$^+$ (n = 1 or 2), followed by migratory insertion of the latter. The order of migratory aptitude is H, Ph (fast at 23 °C) ≫ Me ≫ η^2-CH$_2$Ph (no reaction at 60 °C). The MeCN insertion rate is three times faster for [Zr(Me)(THF)(η-C$_5$H$_4$Me)$_2$]$^+$ than for [Zr(Me)(THF)Cp$_2$]$^+$, which was attributed to greater stabilization of the developing electron deficiency at the metal in the transition state. For [Zr(CH$_2$CH$_2$R)(THF)(η-C$_5$H$_4$Me)$_2$]$^+$ complexes (**63**)–(**67**), β-hydrogen elimination competes with MeCN insertion (Scheme 7) and rates and product distributions are strongly influenced by the β-substituents.[46] The β-agostic mono(MeCN) species (**86**)–(**90**) are key intermediates for both reaction pathways. The hafnium methyl complex [Hf(Me)(MeCN)$_2$Cp$_2$][BPh$_4$] (**104**) does not undergo MeCN insertion, but instead disproportionates to [Hf(Me)$_2$Cp$_2$] and [Hf(MeCN)$_3$Cp$_2$][BPh$_4$]$_2$.[7]

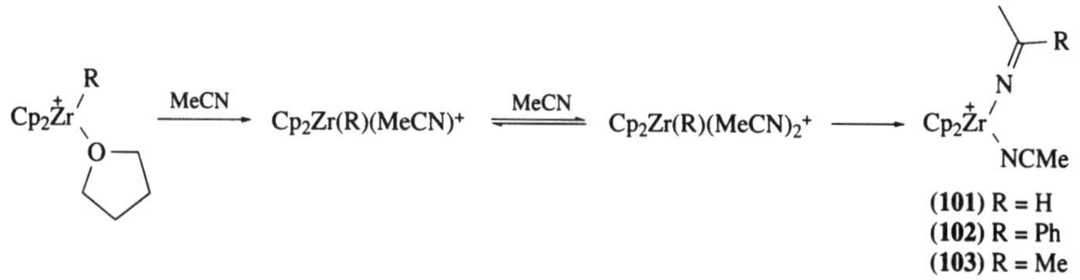

Scheme 6

Alkyne insertion reactions of [M(R)(L)$_n$Cp$'_2$]$^+$ species (n = 0 or 1) were discussed in Section 12.2.1.4.

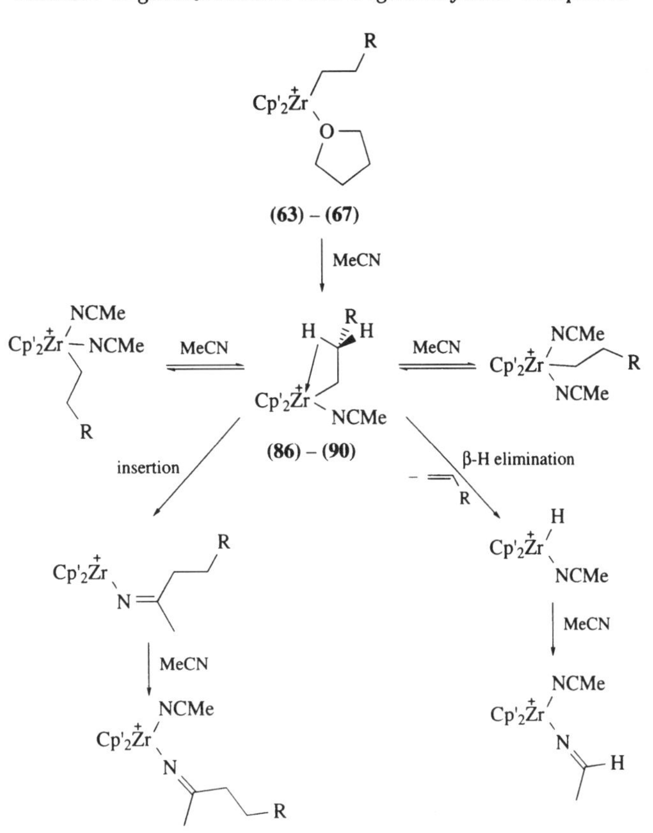

Scheme 7

12.2.4.3 *Sigma bond metathesis reactions*

Cationic $[M(R)(L)_nCp'_2]^+$ species undergo facile σ-bond metathesis reactions with H_2 or ligand C–H bonds, which are believed to proceed via four-center transition states in which the H–H or C–H bond is activated by interaction with the metal LUMO.[69]

(i) *H_2 activation reactions*

The H_2 reactivity of $[Zr(Me)(L)_nCp_2][BPh_4]$ complexes depends strongly on the nature and number of ancillary ligands L.[14] Fourteen-electron species ($n = 0$), and 16-electron species ($n = 1$, L = σ-donor) are highly reactive, while 18-electron species (L = four-electron donor, or $n = 2$) are not. Hydrogenolysis of $[Zr(Me)(THF)Cp_2][BPh_4]$ (**2**) (23 °C, 1 atm) affords $[Zr(H)(THF)Cp_2][BPh_4]$ (**70**) and CH_4. This reaction is eight times faster in CH_2Cl_2 than in THF, which was ascribed to the formation of a reactive 14-electron $[Zr(Me)Cp_2]^+$ species. Hydrogenolysis of (**2**) is considerably faster than that of $[Zr(Me)_2Cp_2]$ under identical conditions, due to the higher metal electrophilicity. In CD_3CN, (**2**) is converted to $[Zr(Me)(CD_3CN)_2Cp_2][BPh_4]$ (**49**-d_6) which undergoes only very slow hydrogenolysis. In this case, CD_3CN insertion leading to $[Zr\{N=C(Me)(CD_3)\}(CD_3CN)Cp_2][BPh_4]$ (**103**-d_6) is faster than reaction with H_2. Hydrogenolysis of (**2**) in THF and CH_2Cl_2 occurs rapidly in the presence of small phosphines (PMe_3 or PMe_2Ph) to afford $[Zr(H)(L)_2Cp_2][BPh_4]$ (L = PMe_3 (**105a**) or PMe_2Ph (**105b**)) and CH_4. This was ascribed to formation of reactive 16-electron $[Zr(Me)(PR_3)Cp_2]^+$ species, in which the zirconium LUMO is not used for π-bonding, as in (**2**). Bulkier phosphines (PPh_3 or $PMePh_2$) do not displace the THF ligand of (**2**) and do not influence the hydrogenolysis rate. The analogous hydride $[Zr(H)(PMe_3)_2(\eta-C_5H_4Me)_2][BPh_4]$ (**106**) has been reported.[45] Similarly, hydrogenolysis of $[M(Me)(NMe_2Ph)(EBTHI)][Co(C_2B_9H_{11})_2]$ (M = Zr (**40**) or Hf (**41**)) proceeds rapidly to yield $[M(H)(NMe_2Ph)(EBTHI)][Co(C_2B_9H_{11})_2]$ (M = Zr (**107a**) or Hf (**107b**)).[31]

The reaction of $[Zr(R)(THF)(\eta-C_5H_4Me)_2]^+$ (R = Me (**3**) or Bz (**7**)) with H_2 yields $[Zr(H)(THF)(\eta-C_5H_4Me)_2]^+$ (**62**), which is a useful reagent for the synthesis of other alkyl complexes (Section 12.2.1.4).

In the corresponding D_2 reactions, H–D exchange into the β-sites of the C_5H_4Me ligands was observed and proposed to result from addition of D_2 across Zr–C_5H_4Me bonds.[70]

The hydrogenolysis of the ion-pair $[Zr(Me)\{MeB(C_6F_5)_3\}Cp*_2]$ (**59**) yields $[Zr(H)\{MeB(C_6F_5)_3\}Cp_2]$ (**108**).[71]

(ii) Ligand C–H activation reactions

Complex (**2**) reacts with α-picoline (2-methylpyridine) via ligand substitution to yield $[Zr(Me)(\alpha\text{-}picoline)Cp_2]^+$ (**109**), which undergoes rapid *ortho*-C–H activation to afford an η^2-picolyl complex (**110**) (two isomers) and CH_4 (Scheme 8).[72,73] Low-temperature NMR spectral studies of (**109**) reveal a Zr–H–C agostic interaction involving the α-picoline *ortho*-C–H bond, which presumably activates it for reaction with the Zr–Me group. The α-Me group accelerates the C–H activation step by forcing the *ortho*-C–H bond closer to zirconium; for comparison, the reaction of (**2**) with pyridine is much slower. This effect[74] has been observed in related reactions of $[Ti(R)(L)Cp_2]$ complexes,[75] and exploited to promote C–H activation in neutral $[Zr(R^1)(R^2)Cp'_2]$ systems.[76]

Scheme 8

Complex (**2**) reacts analogously with other substituted pyridines, as summarized in Scheme 9.[65] A general preference for *ortho*-C–H bond activation leading to three-membered metallacycles is observed. For example, (**2**) reacts with 2-phenylpyridine, 2,5-dimethylpyrazine, quinoline, or 7,8-benzoquinoline to afford (**111**)–(**114**). This selectivity contrasts with the general preference for remote C–H bond activation leading to five-membered metallacycles exhibited by mid- and late transition metal species,[77,78] and was attributed to the shape of the zirconium LUMO (lobes ca. 45° from the Zr–L bond), which favors interaction with proximate C–H bonds. The reaction of (**2**) with (**112**), or double metallation of 2,5-dimethylpyrazine with 2 equiv. of (**2**), affords (**115**).

Complex (**2**) activates more remote C–H bonds of pyridine substrates which lack *ortho*-hydrogens. For example, (**2**) reacts with 2,6-dimethyl- or 2,6-diethylpyridine to yield (**116**) or (**117**), and with acridine or phenazine to afford (**118**) or (**119**) (Scheme 9).[65]

Weakly basic substrates such as 2-methylthiophene, and sterically crowded substrates such as 2-phenylquinoline, do not displace the THF ligand of (**2**) and do not undergo C–H activation. However, it is likely that these substrates would react with base-free $[Zr(R)Cp'_2]^+$ species. *N*-Methylimidazole or 4,4'-dimethylbipyridine reacts with (**2**) to form 18-electron complexes which do not undergo C–H activation (Scheme 4).[65]

Intramolecular reactions involving C–H bonds of R^- and Cp'^- ligands of base-free $[Zr(R)Cp'_2]^+$ species have been observed. The reaction of $[Zr(Me)_2Cp*_2]$ with $[HN(Bu^n)_3][B(p\text{-}C_6H_4R)_4]$ (R = H, Me, or Et) yields zwitterion (**120a–c**) via σ-bond metathesis of the initially formed $[Zr(Me)Cp*_2]^+$ with a *meta*-C–H bond of the $[B(p\text{-}C_6H_4R)_4]^-$ counterion (Equation (12)).[22] The reaction of $[Zr(Me)Cp*_2][B(p\text{-}C_6H_4F)_4]$ (**34**) with excess 1,3-butadiene affords the linked cyclopentadienyl-allyl species $[Zr(\eta^5,\eta^3\text{-}C_5Me_4CH_2CH_2CH=CHCH_2)Cp*][B(4\text{-}C_6H_4F)_4]$ (**121**), via insertion yielding $[Zr(\eta^3\text{-}1,3\text{-}dimethylallyl)Cp*_2]^+$, C–H activation/pentadiene elimination yielding "tuck-in" species $[ZrCp*(\eta^6\text{-}C_5Me_4CH_2)]^+$, and butadiene insertion into the Zr–CH_2 bond.[48]

12.2.4.4 *β-H and β-Me elimination reactions*

In the presence of excess MeCN, $[Zr(CH_2CH_2R)(MeCN)(\eta\text{-}C_5H_4Me)_2]^+$ species (**87**)–(**90**) undergo β-H elimination yielding alkene and $[Zr\{N=CH(Me)\}(MeCN)(\eta\text{-}C_5H_4Me)_2]^+$ (Section 12.2.4.2; Scheme 7). For the ethyl complex (**85**), β-H elimination and MeCN insertion are competitive. Similarly, the PMe_3 analogues (**81**)–(**85**) undergo clean β-H elimination in the presence of excess PMe_3 to yield $[Zr(H)(PMe_3)_2(C_5H_4Me)_2]^+$ (**106**) and alkene.[45] Kinetic studies show that this reaction proceeds by rate-limiting β-H transfer followed by trapping of $[Zr(H)(PMe_3)(\eta\text{-}C_5H_4Me)_2]^+$ by PMe_3.

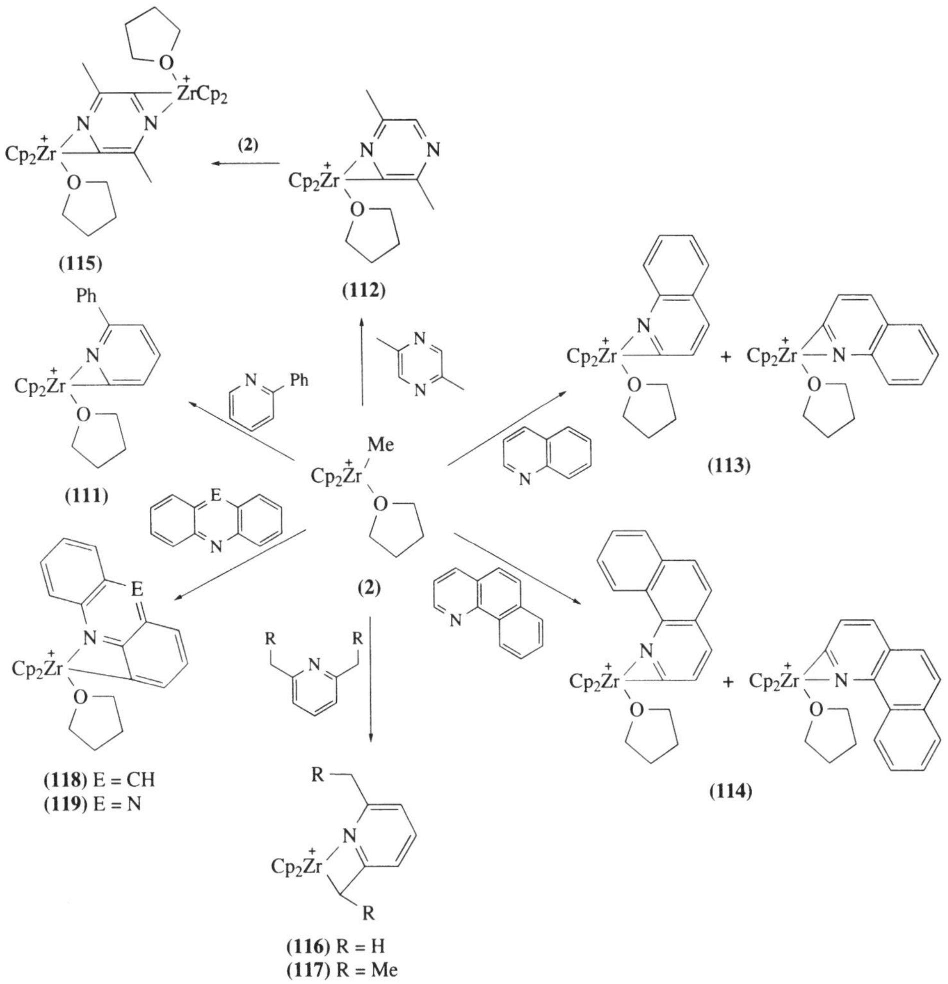

(115)

(112)

(111)

(113)

(118) E = CH
(119) E = N

(114)

(116) R = H
(117) R = Me

Scheme 9

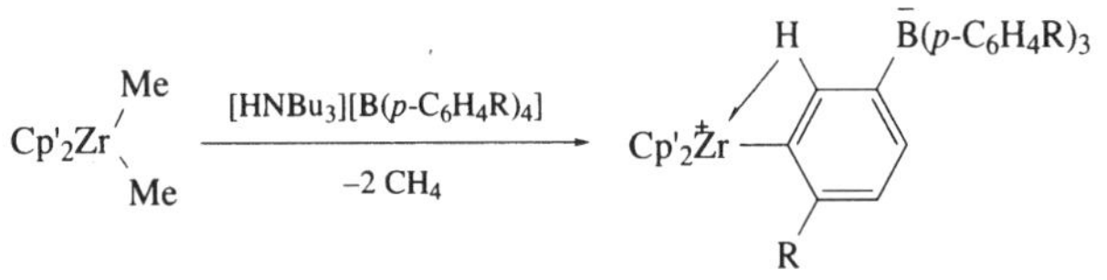

$$Cp'_2Zr\begin{array}{c}Me\\Me\end{array} \xrightarrow[-2\,CH_4]{[HNBu_3][B(p\text{-}C_6H_4R)_4]} Cp'_2Zr^+ \qquad (12)$$

(120a) R = H
(120b) R = Me
(120c) R = Et

β-Me migration is the dominant chain-transfer pathway in propene oligomerization and polymerization catalyzed by $[M(R)Cp^*_2]^+$ species (M = Zr or Hf; Section 12.13.1).[25,79] This was ascribed to destabilization of the conformation leading to β-H elimination by steric interactions between the bulky Cp^{*-} ligands and the β-substituents on the alkyl chain.

12.3 CATIONIC BISCYCLOPENTADIENYLMETAL ALKENYL AND ARYL COMPLEXES

12.3.1 Synthesis and Reactivity

Alkyne insertion reactions of $[Zr(R)(L)_nCp'_2]^+$ ($n = 0$ or 1) species provide general access to zirconium alkenyl complexes. The THF adducts $[Zr(CH_2CH_2R)(THF)(\eta\text{-}C_5H_4Me)_2][BPh_4]$ **(63)**–**(65)**, (Section 12.2.1.4) react with 2-butyne to afford **(122)**–**(124)** (Equation (13)).[44] Similarly, the hydride

(**62**) or methyl complex (**2**) inserts 2-butyne to yield $[Zr\{C(Me)=CHMe\}(THF)(\eta-C_5H_4Me)_2]^+$ (**125**) or $[Zr(CMe=CMe_2)(THF)Cp_2]^+$ (**126**). Complex (**2**) inserts 1-pentyne regioselectively to afford $[Zr\{CH=C(Me)Pr^n\}(THF)Cp_2]^+$ (**127**) in which the alkyl substituent is on the β-carbon; C–H activation leading to an alkynyl species was not observed.[80]

$$ (13) $$

(**63**) – (**65**)

(**122**) R = H
(**123**) R = Me
(**124**) R = Et

Cationic zirconium acyl complexes insert alkynes to yield chelated β-ketoalkenyl complexes (Section 12.5.2); for example, $[Zr\{C(=O)Me\}(THF)Cp_2][BPh_4]$ (**95**) inserts 2-butyne yielding $[Zr\{\eta^2-C(Me)=C(Me)C(=O)Me\}(THF)Cp_2][BPh_4]$ (**128**).[80]

$[Zr(Me)(NMe_2Ph)Cp'_2][B(C_6H_4F)_4]$ species (**30**)–(**33**) (Section 12.2.1.2) generated *in situ* react with methyl(trimethylsilyl)acetylene to afford $[Zr\{C(TMS)=CMe_2\}Cp'_2][B(p-C_6H_4F)_4]$ (**129**)–(**132**) (Equation (14)).[28] The regioselectivity was rationalized on electronic grounds. Analogous reactions of (**30**)–(**34**) with diphenylacetylene yield (**133**)–(**137**). These reactions are much faster than analogous reactions of $[Zr(Me)(THF)Cp'_2]^+$ species, due to the higher lability of NMe_2Ph vs. THF. Complexes (**30**)–(**33**) react with the less bulky 2-butyne via double insertion (Section 12.2.1.4); in contrast, the bulky cation $[Zr(Me)Cp^*_2]^+$ (**34**) undergoes C–H activation with this substrate yielding $[Zr(\eta^3-CH_2CCMe)Cp^*_2]^+$ (Section 12.2.1.4).[47] Cation (**34**) catalyzes the oligomerization of terminal alkynes with very high activity by an insertion/C–H activation process.[81] More complicated chemistry leading to dizirconocene species (**138**) and (**139**)–(**141**) is observed in the reactions of (**30**) with a terminal alkyne (Equation (15)).[82]

$$ (14) $$

$R^1 = TMS, R^2 = Me$
(**129**) $Cp'_2 = Cp_2$
(**130**) $Cp'_2 = (C_5H_4TMS)_2$
(**131**) $Cp'_2 = (C_5H_4Bu^t)_2$
(**132**) $Cp'_2 = rac\text{-}EBI$

$R^1 = R^2 = Ph$
(**133**) $Cp'_2 = Cp_2$
(**134**) $Cp'_2 = (C_5H_4TMS)_2$
(**135**) $Cp'_2 = (C_5H_4Bu^t)_2$
(**136**) $Cp'_2 = rac\text{-}EBI$
(**137**) $Cp' = Cp^*$

$$ (15) $$

(**138**) R = Bu^t

or

(**139**) R = Pr^i
(**140**) R = Pr^n
(**141**) R = p-Tol

The cationic phenyl complex $[Zr(Ph)(THF)Cp_2][BPh_4]$ (**142**) was prepared by reaction of $[Zr(Ph)(Bz)Cp_2]$ with $[Fe(\eta-C_5H_4Me)_2][BPh_4]$ in THF, or more conveniently by reaction of $[Zr(Ph)_2Cp_2]$ with $[HNMe_2Ph][BPh_4]$.[11] The zwitterions (**120a–c**) were isolated from the reactions of $[Zr(Me)_2Cp^*_2]$ with $[HN(Bu^n)_3][B(C_6H_4R)_4]$ in toluene (Section 12.2.4.3(ii), Equation (12)).[22]

Neither the alkenyl complexes (**122**)–(**127**) nor the phenyl complex (**142**) react with 2-butyne.[11,44] This was attributed to the inability of 2-butyne to displace the THF ligand. The more nucleophilic substrate MeCN readily displaces the THF ligand of these species and undergoes subsequent insertion.[6,11] Alkenyl complexes (**99**) and (**127**) insert CO to afford α,β-unsaturated acyl complexes $[Zr\{\eta^2-C(=O)CMe=CMe_2\}(THF)Cp_2][BPh_4]$ (**143**) and $[Zr\{\eta^2-C(=O)CH=CMe(Pr^n)\}(THF)Cp_2][BPh_4]$ (**144**).[80]

12.3.2 Structures

NMR spectral and x-ray diffraction studies of cationic alkenyl and aryl complexes establish that the hydrocarbyl ligands lie in the equatorial plane between the Cp'^- ligands, which precludes strong zirconium–carbon π-bonding. In general, agostic interactions are present. X-ray diffraction revealed a normal, bent metallocene structure for $[Zr\{CMe=C(Me)(Pr^n)\}(THF)(\eta\text{-}C_5H_4Me)_2][BPh_4]$ (**123**) with an agostic interaction involving a C–H bond of the α-Me group (Figure 12).[44] NMR spectral data for (**129**)–(**132**) and an x-ray study of (**132**) revealed agostic interactions involving a Si–Me group (Figure 13).[28] Fluorine-19 NMR spectral data for the diphenylacetylene insertion products (**133**), (**135**), and (**136**) indicate that the anion coordinates weakly via a zirconium–fluorine interaction.[28]

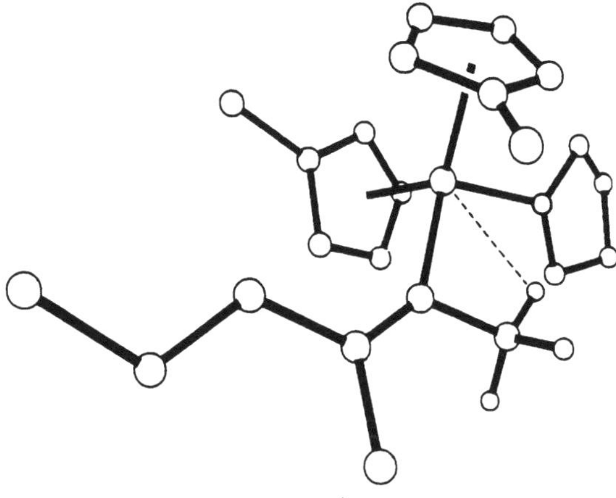

Figure 12 Structure of the $[Zr\{CMe=C(Me)(Pr^n)\}(THF)(\eta\text{-}C_5H_4Me)_2]^+$ cation of (**123**). For clarity only the α-Me hydrogen atoms are shown.

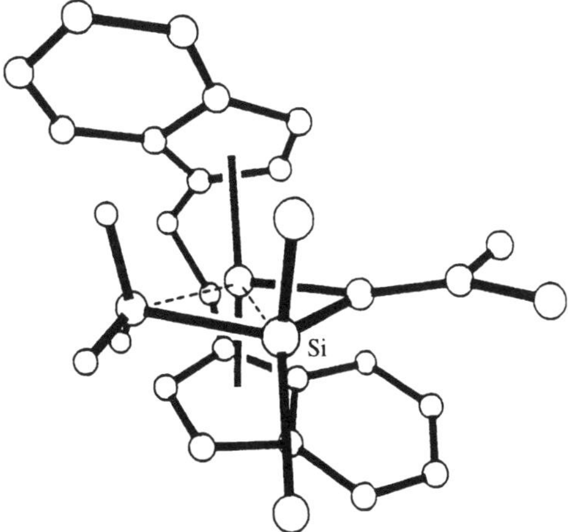

Figure 13 Structure of the $[Zr\{C(TMS)=CMe_2\}(EBI)]^+$ cation of (**132**). For clarity, only the hydrogen atoms on one TMS methyl group are shown.

The zwitterionic structure of (**120c**) was confirmed by x-ray diffraction (Figure 14).[22] The zirconium is σ-bonded to a *meta*-carbon of an anion aryl ring, and a Zr–βH–C agostic interaction is present. The Zr–Hβ distance (0.214 nm) is similar to that in $[Zr(Et)(PMe_3)(\eta\text{-}C_5H_4Me)_2][BPh_4]$ (**81**) (0.216 nm).[45]

12.4 CATIONIC BISCYCLOPENTADIENYLMETAL η²-PYRIDYL AND RELATED COMPLEXES

12.4.1 Synthesis and Structures

Complex (**2**) undergoes *ortho*-C–H activation reactions with pyridines to afford $[Zr(\eta^2\text{-}pyridyl)(L)Cp'_2]^+$ complexes (Schemes 8 and 9, Section 12.2.4.3(ii)). An x-ray diffraction study of $[Zr(\eta^2\text{-}\alpha\text{-picolyl})(PMe_3)Cp'_2][BPh_4]$ (**145**) confirmed the η²-picolyl bonding mode (Figure 15).[73]

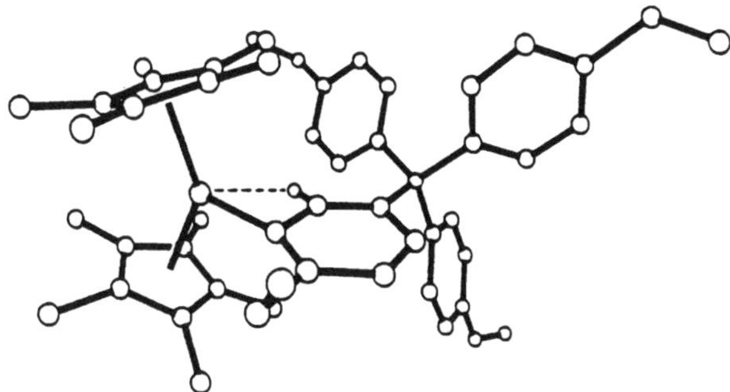

Figure 14 Structure of the zwitterion [ZrCp*$_2$\{3-(B\{C$_6$H$_4$Et-*p*\}$_3$)-6-EtC$_6$H$_3$\}] (**120c**). Only the agostic hydrogen atom is shown.

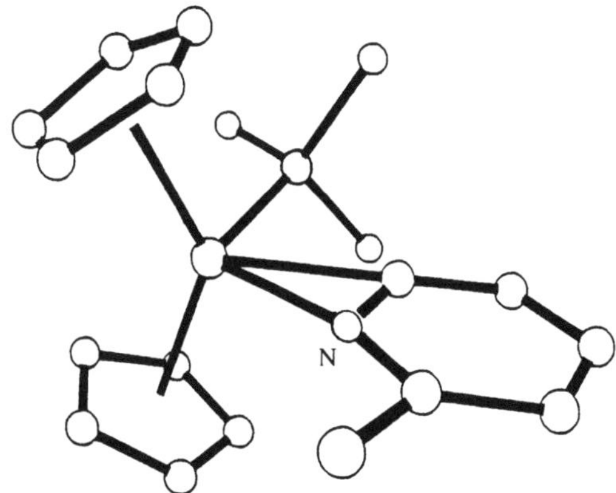

Figure 15 Structure of the [Zr(η^2-α-picolyl)(PMe$_3$)Cp$_2$]$^+$ cation of (**145**). Hydrogen atoms are omitted for clarity.

12.4.2 Reactivity

Cationic [Zr(η^2-pyridyl)(L)Cp$_2$]$^+$ complexes (L = THF or py) undergo regioselective alkene and alkyne insertion to yield five-membered azazirconacycles, as illustrated for (**111**) in Scheme 10.[83] The reaction of (**111**) with propene, allyltrimethylsilane, or allyl ethyl ether affords the β-substituted azazirconacycles (**146**)–(**148**), while reaction with styrene, 2-vinylpyridine, or vinyltrimethylsilanes affords the α-substituted azazirconacycles (**149**)–(**151**). The regioselectivity was rationalized on the basis of electronic effects.[84]

The reaction of (**111**) with terminal alkynes yields α-substituted azazirconacycles (**152**)–(**155**). No regioisomers or C–H bond activation products are observed. The regioselectivity was ascribed to steric effects in the insertion transition states. The reaction of (**111**) with 2-butyne affords (**156**), whereas a mixture of regioisomers (**157a,b**) is obtained from the reaction with 2-hexyne. The reaction with 1-trimethylsilylpropyne affords exclusively the α-TMS substituted azazirconacycle (**158**). Complex (**111**) also reacts with allene to yield a mixture of regioisomers [Zr\{C(=CH$_2$)CH$_2$(6-phenyl-pyridine)\}Cp$_2$]$^+$ (**159a**) and [Zr\{CH$_2$C(=CH$_2$)(6-phenylpyridine)\}Cp$_2$]$^+$ (**159b**).

The related four-membered azazirconacycle [Zr\{η^2-CH$_2$(6-methylpyridine)\}Cp$_2$][BPh$_4$] (**116**), which is derived from the reaction of (**2**) with 2,6-lutidine and is in equilibrium with the corresponding THF adduct [Zr\{η^2-CH$_2$(6-methylpyridine)\}(THF)Cp$_2$][BPh$_4$] (**160**), readily inserts an alkene, alkyne, aldehyde, ketone, or nitrile to afford azazirconacycles (**161**)–(**167**) (Scheme 11).[85] Cation (**116**) is more reactive than related neutral metallacyclobutanes, for example, [$\overline{\text{Zr(CH}_2\text{SiMe}_2\text{CH}_2)}Cp_2$] and [$\overline{\text{Zr(CH}_2\text{CH}_2\text{CH}_2)}Cp_2$] which react only with polar substrates.[86–9] This was ascribed to the high Lewis acidity of the cationic zirconium center in (**116**), and the possibility of Zr–N bond dissociation.

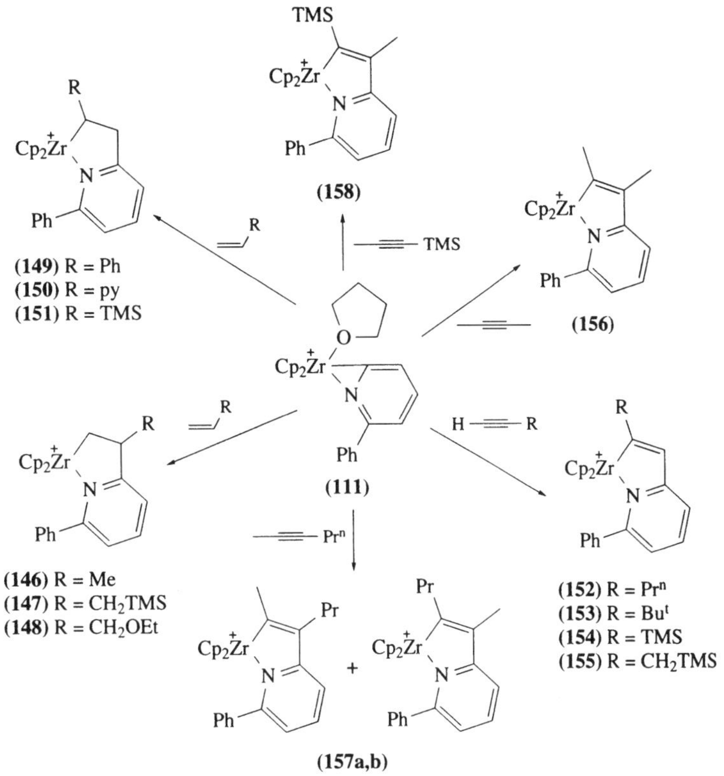

Scheme 10

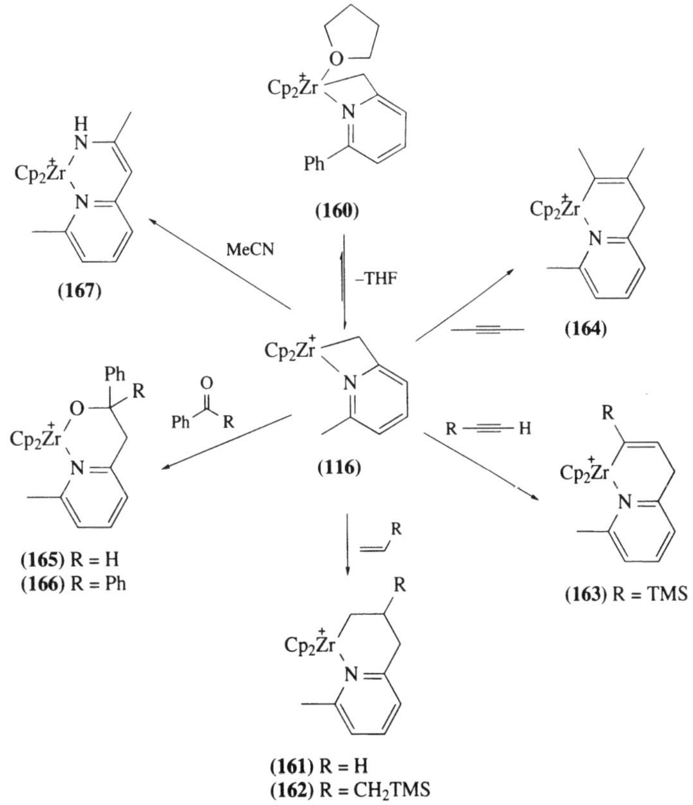

**Scheme 11

12.5 CATIONIC BISCYCLOPENTADIENYLMETAL η^2-ACYL, η^2-IMINOACYL, AND CARBONYL COMPLEXES

12.5.1 Synthesis and Structures

Cationic zirconium η^2-acyl complexes are available via carbonylation of cationic hydrocarbyl complexes. Carbonylation of $[Zr(R)(THF)Cp_2][BPh_4]$ ((2), (126), (127)) yields $[Zr\{\eta^2\text{-}C(=O)R\}(THF)Cp_2][BPh_4]$ (R = Me (95), $(C(Me)=C(Me)_2)$ (143), CH=C(Me)(Pr) (144); Sections 12.2.4.2 and 12.3.1).[13,80] Similarly, reaction of $[Zr(Me)(MeCN)_2Cp_2][BPh_4]$ (49) with CO yields $[Zr\{\eta^2\text{-}C(=O)Me\}(MeCN)Cp_2][BPh_4]$ (168).[13]

Carbonylation of the base-free ion-pairs $[Zr(Me)\{MeB(C_6F_5)_3\}Cp'_2]$ (Cp' = Cp (57) or Cp* (59)) affords $[Zr\{\eta^2\text{-}C(=O)Me\}(CO)Cp'_2][CH_3B(C_6F_5)_3]$ (169) and (170) (Equation (16)) which are unusual examples of Zr^{IV} carbonyl complexes.[67,90,91] Complexes (169) and (170) exist as mixtures of "O-inside" and "O-outside" isomers which differ in the orientation of the η^2-acetyl ligand. The ν_{CO} values for the "O-inside" isomers ((169) 2176 cm^{-1}; (170) 2152 cm^{-1}) are higher than the free CO value (2143 cm^{-1}, gas phase) indicating that the Zr–CO bond is primarily σ-donor in character, as expected for a d^0 metal center. The ν_{CO} values for the "O-outside" isomers ((169) 2123 cm^{-1}; (170) 2105 cm^{-1}) are slightly lower than the free ν_{CO} value as a result of overlap of a filled zirconium–acyl bonding orbital with the CO π*-orbital. An x-ray diffraction study of the isolable "O-outside" isomer of (170) revealed a strong Zr–O$_{acyl}$ interaction and a terminal CO ligand. Consistent with the ν_{CO} value, the carbonyl carbon–oxygen distance (0.113 nm) is not significantly perturbed from that of free CO.

$$Cp_2\overset{+}{Zr}\overset{Me}{\underset{Me-B(C_6F_5)_3^-}{\big\langle}} \quad \xrightarrow{CO} \quad Cp'_2\overset{+}{Zr}\overset{CO}{\underset{}{-}}O \quad + \quad Cp'_2\overset{+}{Zr}\overset{CO}{\underset{O}{-}} \qquad (16)$$

$$[B(C_6F_5)_3Me]^-$$

$$(57)\ Cp' = Cp \qquad\qquad\qquad (169)\ Cp' = Cp$$
$$(59)\ Cp' = Cp* \qquad\qquad\qquad (170)\ Cp' = Cp*$$

The reaction of (2) with t-butyl isocyanide yields the η^2-iminoacyl complex $[Zr\{C(=N\text{-}Bu^t)Me\}(CNBu^t)Cp_2][BPh_4]$ (96) in which an isocyanide ligand is retained in preference to THF (Section 12.2.4.2). The base-free η^2-iminoacyl complex $[Zr\{\eta^2\text{-}C(=NBu^t)Me\}][MeB(C_6F_5)_3]$ (171) was generated by reaction of air-stable $[Zr\{\eta^2\text{-}C(=NBu^t)Me\}(Me)Cp_2]$ (172) with $B(C_6F_5)_3$ (Scheme 12).[92]

$$Cp_2Zr\overset{Me}{\underset{Me}{\big\langle}} \quad \xrightarrow{CNBu^t} \quad Cp_2Zr\overset{Me}{\underset{}{=}}N^{-Bu^t} \quad \xrightarrow{B(C_6F_5)_3} \quad Cp_2\overset{+}{Zr}\overset{}{=}N^{-Bu^t} \quad [B(C_6F_5)_3Me]^-$$

$$(172) \qquad\qquad\qquad\qquad (171)$$

Scheme 12

12.5.2 Reactivity

Unlike neutral zirconocene acyl complexes, cationic $[Zr\{\eta^2\text{-}C(=O)R\}(THF)Cp'_2]^+$ species readily insert alkynes.[80] The acetyl complex $[Zr\{\eta^2\text{-}C(O)Me\}(THF)Cp_2][BPh_4]$ (95) inserts terminal and internal alkynes to afford β-ketoalkenyl complexes (128), (173), and (174) which adopt chelated structures (Scheme 13; see also Section 12.3.1). The regioselectivity of terminal alkyne insertion is analogous to that exhibited by $[Zr(\eta^2\text{-pyridyl})Cp'_2]^+$ species (Section 12.4.2) and was rationalized on steric grounds. Similarly, the α,β-unsaturated acyl complex $[Zr\{\eta^2\text{-}C(=O)C(H)=C(Me)(Pr^n)\}(THF)Cp_2][BPh_4]$ (144) inserts 1-pentyne to afford $[Zr\{C(Pr^n)=CHC(=O)CH=C(Me)(Pr^n)\}(THF)Cp_2][BPh_4]$ (175).[80] Interestingly, added CO strongly inhibits the alkyne insertion reactions of (95). This was ascribed to inhibition of alkyne coordination and insertion via formation of CO adduct (170).[67]

The base-free iminoacyl complex (171) inserts alkenes and alkynes regioselectively to afford β-iminoalkyl and β-iminoalkenyl complexes (176)–(181) (Scheme 14). The regioselectivity is consistent with that observed for cationic zirconium η^2-acyl and η^2-pyridyl complexes.

The insertion chemistry of these cationic acyl and iminoacyl species has been exploited in organic synthesis (Section 12.13.2.2).

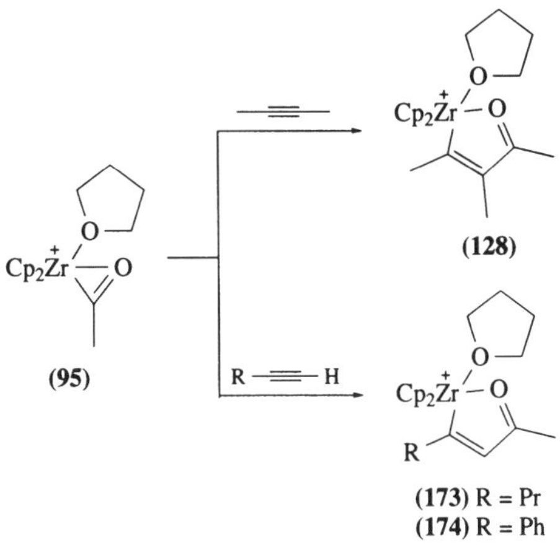

Scheme 13

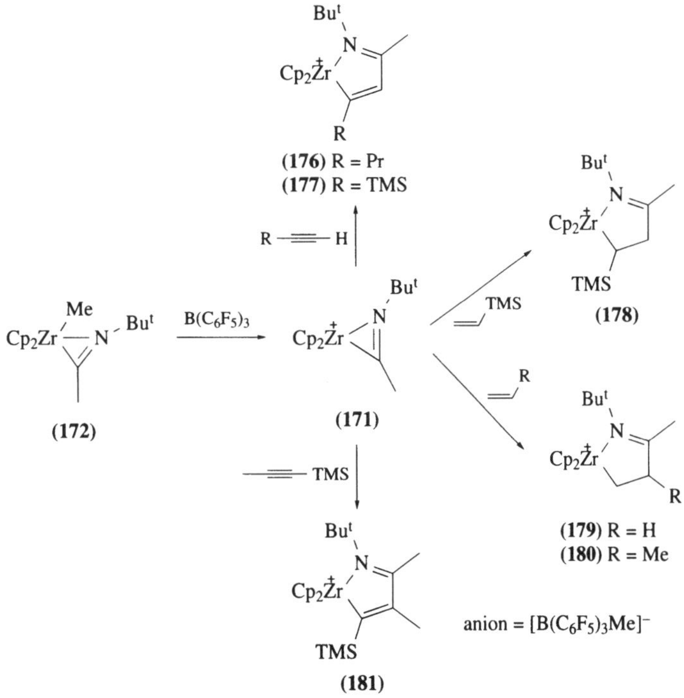

Scheme 14

12.6 CATIONIC BISCYCLOPENTADIENYLMETAL HYDRIDE COMPLEXES

Cationic $[M(H)(L)_n Cp'_2]^+$ species (M = Zr or Hf; $n = 0$ or 1) are of interest due to their role as key intermediates in metallocene-catalyzed alkene polymerizations and hydrogenations, pyridine/alkene coupling, and other reactions (Section 12.13), and as reagents for the synthesis of other $[M(R)(L)_n Cp'_2]^+$ complexes (Section 12.2.1.4).

12.6.1 Synthesis and Structures

The Lewis base-stabilized hydrides $[Zr(H)(THF)Cp'_2][BPh_4]$ (Cp' = C_5H_4Me (**62**) or Cp (**70**)) were prepared by reaction of the corresponding alkyl complexes with H_2 (Sections 12.2.1.4 and 12.2.4.3(i)). Complexes (**62**) and (**70**) were assigned monomeric, terminal hydride structures on the basis of IR $\nu(Zr–H)$ absorbances near 1450 cm^{-1}.[14,44] Hydrogenolysis of $[M(Me)(NMe_2Ph)(EBTHI)][Co(C_2B_9H_{11})_2]$ (M = Zr (**40**) or Hf (**41**)) yields $[M(H)(NMe_2Ph)(EBTHI)][Co(C_2B_9H_{11})_2]$ (M = Zr (**107a**) or (**107b**)).[31]

Complexes (**107a,b**) were also obtained by protonolysis of $[M(H)_2(EBTHI)]_2$ dihydrides (M = Zr (**182**) or Hf (**183**)) with $[PhMe_2NH][Co(C_2B_9H_{11})_2]$ (Scheme 15).

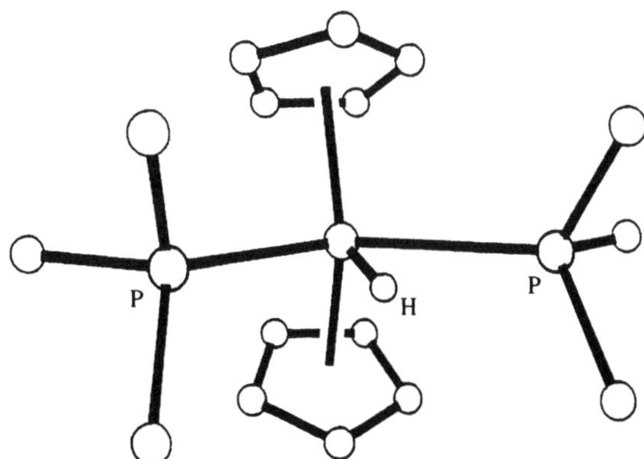

Scheme 15

The base-free hydrides $[Zr(H)Cp^*_2][MeB(C_6F_5)_3]$ (**108**) and $[Zr(H)Cp^*_2][HB(C_6F_5)_3)]$ (**184**) were prepared by reaction of $[Zr(Me)Cp^*_2\{MeB(C_6F_5)_3\}]$ (**59**) with H_2 in pentane and benzene, respectively (Scheme 16).[71] Alternatively, (**184**) can be prepared by reaction of $[Zr(H)_2Cp^*_2]$ with $B(C_6F_5)_3$. An x-ray diffraction study of (**184**) revealed an ion-pair structure in which the $[HB(C_6F_5)_3]^-$ anion coordinates to the $[Zr(H)Cp^*_2]^+$ cation via two $Zr\cdots F-C$ interactions ($Zr\cdots F$ 0.2416 and 0.2534 nm), rather than by a $B-(\mu H)-Zr$ interaction.

Scheme 16

The bis(phosphine) hydride $[Zr(H)(PMe_3)_2Cp_2][BPh_4]$ (**105a**) is prepared by hydrogenolysis of (**2**) in the presence of PMe_3 (Section 12.2.4.3(i)).[14] The C_5H_4Me analogue $[Zr(H)(PMe_3)_2(\eta-C_5H_4Me)_2][BPh_4]$ (**106**) is formed by thermolysis of $[Zr(CH_2CH_2R)(PMe_3)(\eta-C_5H_4Me)_2][BPh_4]$ complexes ((**81**)–(**85**)) in the presence of excess PMe_3 (Section 12.2.4.4).[45] Complexes (**105**) and (**106**) adopt symmetrical structures in which the hydride occupies the central site in the equatorial plane. The x-ray structure of (**105a**) has been determined (Figure 16).[14]

Figure 16 Structure of the $[Zr(H)(PMe_3)_2Cp_2]^+$ cation of (**105a**). For clarity, only the Zr–H hydrogen atom is shown.

12.6.2 Reactivity

As described in Section 12.2.1.4, (62) reacts with an alkene, alkyne, or allene in THF to provide general access to $[Zr(R)(L)(\eta\text{-}C_5H_4Me)_2]^+$ complexes (R = alkyl, alkenyl or η^3-allyl; e.g., (63)–(69)). Hydride (62) also rapidly inserts MeCN to afford $[Zr\{N=C(H)(Me)\}(MeCN)(\eta\text{-}C_5H_4Me)_2]^+$ (101), and in the absence of substrate undergoes slow THF ring opening to yield $[Zr(O\text{-}Bu^n)(THF)(\eta\text{-}C_5H_4Me)_2][BPh_4]$ (185) (Scheme 17).[44,70]

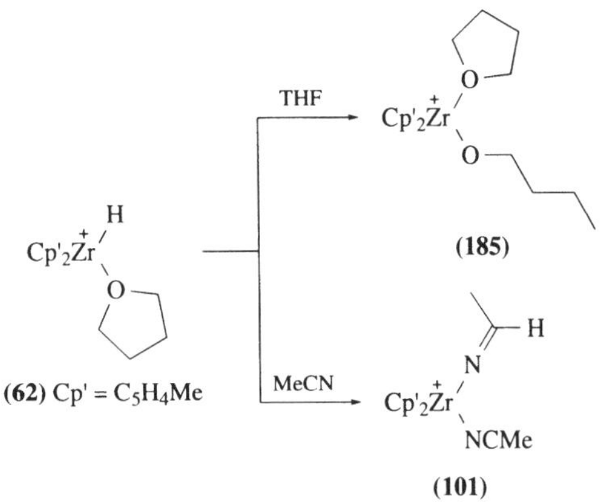

Scheme 17

Cationic hydrides containing labile ligands or anions are more reactive. The chiral hydride (107a) catalyzes alkene hydrogenation, though with low stereoselectivity.[31] Hydrogen–deuterium exchange of (108) and (184) with C_6D_6 was noted.[71]

12.7 CATIONIC BISCYCLOPENTADIENYLMETAL ALKOXIDE AND RELATED COMPLEXES

12.7.1 Synthesis and Structures

Cationic zirconium alkoxide complexes $[Zr(OR)(L)Cp_2]^+$ are available by ketone insertion reactions of $[Zr(R)(L)Cp'_2]^+$ complexes (Section 12.2.4.2), or by THF-ring opening reactions of $[Zr(R)(THF)Cp'_2]^+$ complexes (Section 12.2.4.1 and 12.6.2). Additionally, alcoholysis of $[Zr(R)(L)Cp'_2]^+$ complexes yields alkoxide complexes, cleanly in some cases. The reactions of (2) with t-butanol and t-butanethiol afford $[Zr(OBu^t)(THF)Cp_2][BPh_4]$ (186a) and $[Zr(SBu^t)(THF)Cp_2][BPh_4]$ (186b), respectively;[93] however, similar reactions with less hindered alcohols yield mixtures of unidentified products. The reaction of $[Zr(Me)(OBu^t)(EBTHI)]$ with $[HNEt_3][BPh_4]$ yields $[Zr(OBu^t)(THF)(EBTHI)][BPh_4]$ (187).[94] X-ray structures for (186a) and (187) have been determined. For (186a), the short $Zr\text{-}O(Bu^t)$ bond (0.1899 nm) and the nearly linear $Zr\text{-}O\text{-}Bu^t$ angle (171.0°) indicate significant $Zr\text{-}O$ π-bonding. The $Zr\text{-}O(Bu^t)$ bond distance in (187) is longer (0.1929 nm) and the $Zr\text{-}O\text{-}Bu^t$ bond angle is reduced (161.9°), indicating a weaker $Zr\text{-}O$ π-interaction.

Cationic zirconium quinolato complexes $[ZrCp'(\eta^5\text{-}pyrrolyl)(\eta^2\text{-}quinolato)][X]$ (188) (X = Br⁻, I⁻, $ZnCl_3(H_2O)^-$, $1/2CdCl_4^{2-}$, or $HgCl_3^-$; Cp' = Cp or indenyl) were obtained from the reaction of the dichloride precursors with 8-hydroxyquinoline in aqueous solution, followed by ion exchange.[95] Related triflato, sulfonato, and carboxylato complexes include $[Zr(OSO_2CF_3)(bipy)Cp_2][OSO_2CF_3]$ (189) prepared by reaction of $[Zr(OSO_2CF_3)_2(THF)Cp_2]$ with bipyridine,[96] $[Zr(OSO_2C_7H_7)(H_2O)Cp_2][OSO_2C_7H_7]$ (190) prepared by hydrolysis of $[Zr(OSO_2C_7H_7)_2Cp_2]$,[97] and the unusual trinuclear complex $[Zr_3(\mu_3\text{-}O)(\mu_2\text{-}OH)_3(\mu_2\text{-}Ph)_3(CpCO_2)_3][PhCO_2]$ (191) obtained from the reaction of $[ZrCl_2Cp_2]$ with $Na[O_2CPh]$.[98]

12.7.2 Reactivity

The rate of THF exchange of $[Zr(OBu^t)(THF)Cp_2][BPh_4]$ (186a) is slower than for methyl complex (2). This was attributed to the increased electron density at zirconium in (186a) resulting from alkoxide

π-donation, which impedes associative THF exchange. The THF ligand in [Zr(OBu^i)(THF)-(EBTHI)][BPh$_4$] (187) is displaced by aldehydes, and both (186a) and (187) catalyze Diels–Alder reactions (Section 12.13.2.5).

12.8 CATIONIC BISCYCLOPENTADIENYLMETAL HALIDE COMPLEXES

Oxidation of [{ZrClCp$_2$}$_2$] with Ag[X] or Tl[X] (X = [ClO$_4$]$^-$ or [BF$_4$]$^-$) yields the dinuclear dicationic species [{Zr(μ-Cl)Cp$_2$}]$^{2+}$ (192), which may be trapped by monodentate ligands to yield [Zr(Cl)(L)Cp$_2$][X] (L = OPPh$_3$ (193) or NHPh$_2$ (194)) (Scheme 18).[99] Analogous zirconium complexes with bridging bidentate ligands, and analogous hafnium compounds, were also prepared. The corresponding MeCN complexes appear to be less stable; [Zr(X)(MeCN)$_2$Cp$_2$][BPh$_4$] (X = Br (195) or I (196)) were generated by reaction of [ZrX$_2$Cp$_2$] with Ag[BPh$_4$] in MeCN, and undergo disproportionation to yield equilibrium mixtures of [Zr(X)$_2$Cp$_2$], [Zr(MeCN)$_3$Cp$_2$][BPh$_4$]$_2$ (197), and (195) or (196) (Equation (17)).[100] Analogous chloride complexes [M(Cl)(MeCN)$_2$Cp$_2$][SbCl$_6$] (M = Zr (198) or Hf (199)) have been prepared by chloride abstraction reactions with SbCl$_5$.[101]

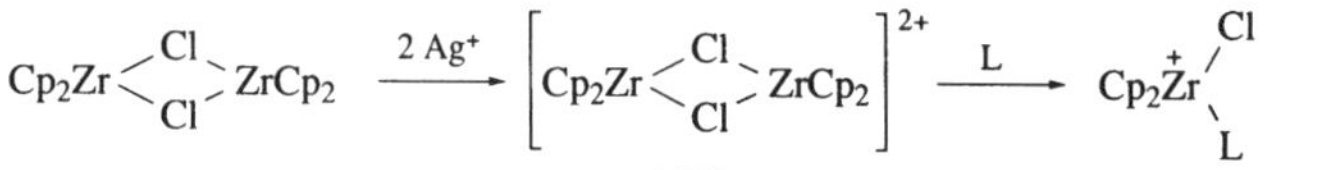

$$\text{Cp}_2\text{Zr}\underset{\text{Cl}}{\overset{\text{Cl}}{<}}{>}\text{ZrCp}_2 \xrightarrow{\;2\,\text{Ag}^+\;} \left[\text{Cp}_2\text{Zr}\underset{\text{Cl}}{\overset{\text{Cl}}{<}}{>}\text{ZrCp}_2\right]^{2+} \xrightarrow{\;L\;} \text{Cp}_2\overset{+}{\text{Zr}}\underset{\text{L}}{\overset{\text{Cl}}{<}}$$

(192)

(193) L = OPPh$_3$
(194) L = NHPh$_2$

Scheme 18

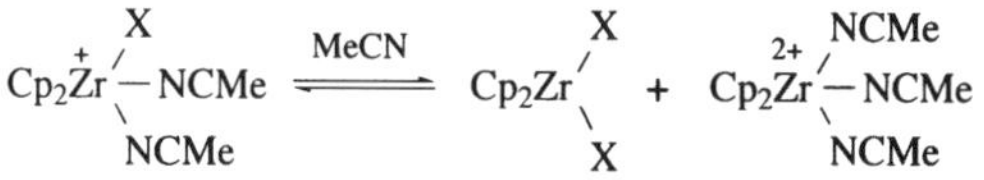

$$\text{Cp}_2\overset{+}{\text{Zr}}\underset{\text{NCMe}}{\overset{\text{X}}{<}}\!\!-\text{NCMe} \;\rightleftharpoons\; \text{Cp}_2\text{Zr}\underset{\text{X}}{\overset{\text{X}}{<}} + \text{Cp}_2\overset{2+}{\text{Zr}}\underset{\text{NCMe}}{\overset{\text{NCMe}}{<}}\!\!-\text{NCMe}$$

(195) X = Br   anion = [BPh$_4$]$^-$ (197)
(196) X = I

(17)

Cationic [M(X)Cp$_2$]$^+$ species generated *in situ* are useful in organic synthesis (Section 12.13.2.4). The hafnium species [Hf(Cl)Cp$_2$][ClO$_4$] (200) was observed spectroscopically in [Hf(Cl)$_2$Cp$_2$]/Ag[ClO$_4$] mixtures.[102] Cationic [Zr(X)(H$_2$O)$_n$Cp$_2$]$^+$ species are possible intermediates in the hydrolysis of [Zr(X)$_2$Cp$_2$] complexes.[103]

12.9 DICATIONIC BISCYCLOPENTADIENYLMETAL COMPLEXES

Several [Zr(L)$_3$Cp′$_2$]$^{2+}$ species are known. The tris(aquo) complex [Zr(H$_2$O)$_3$Cp$_2$][OSO$_2$CF$_3$]$_2$ (201) was prepared by hydrolysis of [Zr(OSO$_2$CF$_3$)$_2$(THF)Cp$_2$] in THF, and undergoes condensation to the dinuclear μ-OH complex [Zr$_2$(μ-OH)$_2$(H$_2$O)$_6$Cp$_2$][OSO$_2$CF$_3$]$_4$ (202).[104,105]

The tris(acetonitrile) complex [Zr(MeCN)$_3$Cp$_2$][BPh$_4$]$_2$ (197) (Section 12.8) was prepared cleanly by reaction of [Zr(I)$_2$Cp$_2$] with 2 equiv. Ag[BPh$_4$] in MeCN (Equation (18)).[100] Complex (197) reacts with 4,4′-dimethylbipyridine to yield [Zr(MeCN)(4,4′-Me$_2$-bipy)Cp$_2$][BPh$_4$]$_2$ (203) and with [Zr(Me)$_2$Cp$_2$] to yield [Zr(Me)(MeCN)$_2$Cp$_2$][BPh$_4$]$_2$ (49). Surprisingly, (197) is extremely air-sensitive even though it lacks an alkyl ligand. The reaction of (197) with dry O$_2$ in MeCN yields biphenyl and [Zr(OPh)$_2$Cp$_2$] (30%) as major products, indicating that anion oxidation occurs.[106] A more stable analogue, [Zr(MeCN)$_3$Cp$_2$][B{3,5-(CF$_3$)$_2$C$_6$H$_3$}$_4$] (204), was prepared by reaction of [Zr(Me)$_2$Cp$_2$] with 2 equiv. of [H(OEt$_2$)][B{3,5-(CF$_3$)$_2$C$_6$H$_3$}$_4$] in MeCN.[106] The analogous [SbCl$_6$]$^-$ salts [M(MeCN)$_3$Cp$_2$][SbCl$_6$]$_2$ (M = Zr (205) or Hf (206)) were prepared by reaction of [MX$_2$Cp$_2$] with SbCl$_5$.[101]

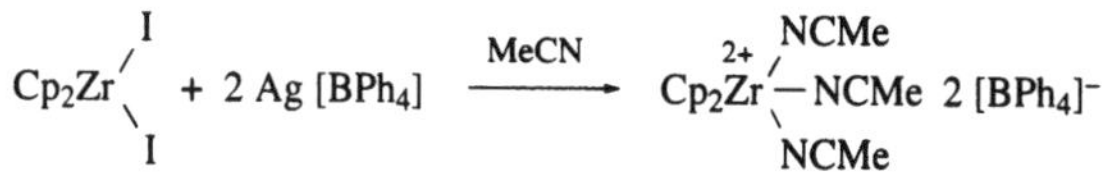

$$\text{Cp}_2\text{Zr}\underset{\text{I}}{\overset{\text{I}}{<}} + 2\,\text{Ag[BPh}_4] \xrightarrow{\;\text{MeCN}\;} \text{Cp}_2\overset{2+}{\text{Zr}}\underset{\text{NCMe}}{\overset{\text{NCMe}}{<}}\!\!-\text{NCMe}\;\; 2\,[\text{BPh}_4]^-$$

(197)

(18)

12.10 CATIONIC MONOCYCLOPENTADIENYLMETAL HYDROCARBYL COMPLEXES

12.10.1 Synthesis and Reactivity

The synthetic methods developed for $[M(R)(L)_n Cp'_2]^+$ systems (Section 12.2.1) have been employed in the preparation of cationic monocyclopentadienyl complexes. The reaction of $[Zr(Me)_3Cp^*]$ with $[Fe(\eta\text{-}C_5H_4Me)_2][BPh_4]$ in THF results in oxidative Zr–Me bond cleavage and formation of $[Zr(Me)_2(THF)_2Cp^*][BPh_4]$ (**207**) (Equation (19)).[107] Similarly, the thermally sensitive complex $[Zr(Bz)_2(THF)Cp^*][BPh_4]$ (**208**) was prepared by oxidation of $[Zr(Bz)_3Cp^*]$ with $[Fe(\eta\text{-}C_5H_4Me)_2][BPh_4]$ at 0 °C (Equation (20)). Complexes (**207**) and (**208**) most probably coordinate additional THF ligands upon dissolution in THF; however, rapid THF exchange precluded determination of precise solution structures. Complex (**207**) reacts with dmpe in THF to afford $[Zr(Me)_2(THF)(dmpe)Cp^*][BPh_4]$ (**209**).

$$[ZrMe_3Cp^*] \xrightarrow[-[Fe(C_5H_4Me)_2],\ -1/2\ Me_2]{[Fe(C_5H_4Me)_2]^+,\ THF} Cp^*\overset{+}{Z}rMe_2(THF)_2 \qquad (19)$$

$$(207)\ \text{anion} = [BPh_4]^-$$

$$[ZrBz_3Cp^*] \xrightarrow[-[Fe(C_5H_4Me)_2],\ -1/2\ Bz_2]{[Fe(C_5H_4Me)_2]^+,\ THF} Cp^*\overset{+}{Z}rBz_2(THF) \qquad (20)$$

$$(208)\ \text{anion} = [BPh_4]^-$$

The ion-pair $[Zr(Me)_2(\eta^3\text{-}CB_{11}H_{12})Cp^*]$ (**210**) was prepared by reaction of $[Zr(Me)_3Cp^*]$ with $Ag[CB_{11}H_{12}]$ in hexane (Equation (21)).[16] The $[\eta^3\text{-}CB_{11}H_{12}]^-$ ligand of (**210**) is displaced by THF yielding $[Zr(Me)_2(THF)_2Cp^*][CB_{11}H_{12}]$; however, (**210**) does not react with propene, 2-butyne, or styrene under mild conditions.

$$[ZrMe_3Cp^*] \xrightarrow[-Ag^0,\ -1/2\ Me_2]{AgCB_{11}H_{12}} Cp^*ZrMe_2(\eta^3\text{-}CB_{11}H_{12}) \qquad (21)$$

$$(210)$$

The cationic η^6-toluene complexes $[MMe_2(\eta^6\text{-}toluene)Cp^*][MeB(C_6F_5)_3]$ (M = Zr (**211**) or Hf (**212**)) were prepared by methyl abstraction from $[M(Me)_3Cp^*]$ with $B(C_6F_5)_3$ in toluene (Scheme 19), and are rare examples of d^0 metal arene complexes.[108] Analogous $[ZrMe_2(\eta^6\text{-}arene)Cp^*][MeB(C_6F_5)_3]$ complexes were isolated (arene = mesitylene (**213**)) or characterized by low-temperature NMR spectroscopy (arene = benzene (**214**) or styrene (**215**)). PMe$_3$ displaces the coordinated toluene of (**211**) and (**212**) to afford $[M(Me)_2(PMe_3)_2Cp^*][MeB(C_6F_5)_3]$ (M = Zr (**216**) or Hf (**217**)).

$$[MMe_3Cp^*] \xrightarrow[arene]{B(C_6F_5)_3} [MMe_2(\eta^6\text{-}toluene)Cp^*][B(C_6F_5)_3Me] \xrightarrow{PMe_3} [MMe_2(PMe_3)_2Cp^*][B(C_6F_5)_3Me]$$

$$\begin{array}{ll} (211)\ M = Zr & (216)\ M = Zr \\ (212)\ M = Hf & (217)\ M = Hf \end{array}$$

Scheme 19

Active alkene polymerization catalysts are generated by reaction of $[Zr(Bz)_3Cp']$ (Cp' = Cp or Cp*) with $[HNMe_2Ph][B(C_6F_5)_4]$ or $B(C_6F_5)_3$. It was suggested that $[Zr(R)_2Cp']^+$ cations analogous to (**211**) are the active species in these systems.[109,110]

12.10.2 Structures

The x-ray structures of $[Zr(Me)_2(THF)_2Cp^*][BPh_4]$ (**207**), $[Zr(Me)_2(THF)(dmpe)Cp^*][BPh_4]$ (**209**), and $[Zr(Me)_2(\eta^3\text{-}CB_{11}H_{12})Cp^*]$ (**210**) have been determined.[16,107] The cation of (**207**) adopts a square pyramidal/four-legged piano-stool structure in which the Cp*$^-$ ligand occupies the apical site and the Me groups are *cis* (Figure 17). The orientation of THF ligands, the short Zr–O bond distances (0.2243, 0.2302 nm vs. 0.231–0.239 nm in related neutral compounds),[111] and orbital overlap considerations suggest that Zr–O π-bonding is important. The cation of (**209**) adopts a distorted octahedral structure in which the Cp*$^-$ ligand occupies an axial position, and the Me groups occupy *cis* equatorial sites. The dmpe ligand coordinates in an axial/equatorial manner as in related neutral complexes.[112] The x-ray structure of (**210**) (Figure 18) reveals that the $[CB_{11}H_{12}]^-$ anion coordinates to zirconium in an unusual tridentate mode via three B–μH–Zr bridges.

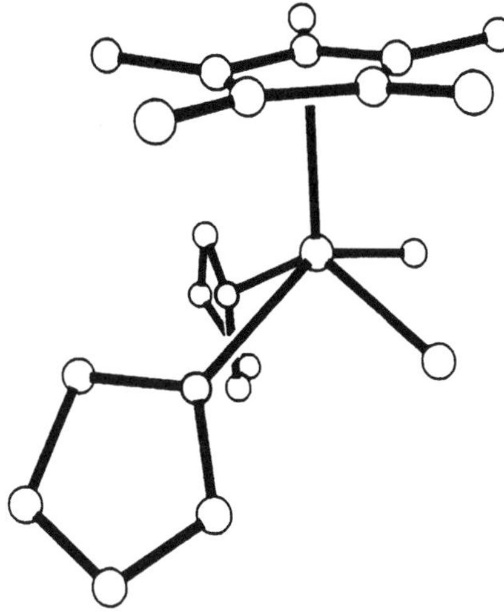

Figure 17	Structure of the $[Zr(Me)_2(THF)_2Cp^*]^+$ cation of (**207**). Hydrogen atoms are omitted for clarity.

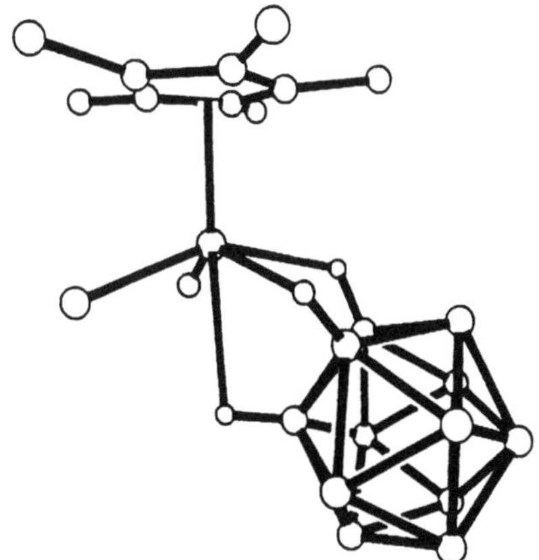

Figure 18	Structure of $[Zr(Me)_2(\eta^3\text{-}CB_{11}H_{12})Cp^*]$ (**210**). Only the B–(μH)–Zr hydrogen atoms are shown.

12.11 CATIONIC TRIS(HYDROCARBYL)METAL COMPLEXES

Protonolysis of $[Zr(Bz)_4]$ with $[HNMe_2Ph][BPh_4]$ in toluene or CH_2Cl_2 affords the zwitterionic complex $[Zr^+(Bz)_3(\eta^6\text{-}Ph)B^-Ph_3]$ (**218**) (Equation (22)).[113] The η^6-coordination of an anionic phenyl ring was inferred from NMR spectral data. The $[(\eta^6\text{-}Ph)BPh_3]^-$ anion is not displaced by toluene or hexamethylbenzene, which suggests that electrostatic factors contribute significantly to the zirconium–arene bond. Complex (**218**) reacts with THF to afford $[Zr(Bz)_3(THF)_n][BPh_4]$ (**219**) (*n* not determined), and with the tridentate amines 1,3,5-trimethylhexahydro-1,3,5-triazine and 1,4,7-trimethyl-1,4,7-triazacyclononane to yield $[Zr(Bz)(L)][BPh_4]$ complexes ((**220**), (**221**)). The closely related zwitterion $[Zr^+(Bz)_2(\eta^2\text{-}Bz)\{\eta^6\text{-}BzB^-(C_6F_5)_3\}]$ (**222**) was prepared by reaction of $[Zr(Bz)_4]$ with $B(C_6F_5)_3$ and characterized by x-ray crystallography.[114] As for (**218**), the coordinated arene of (**222**) is displaced by THF yielding $[Zr(Bz)_3(THF)_3][B(Bz)(C_6F_5)_3]$ (**223**).

$$[ZrBz_4] \xrightarrow[\text{toluene or } CH_2Cl_2]{[HNMe_2Ph][BPh_4]} \quad \text{(218)} \tag{22}$$

Complex (**222**) is a moderately active ethene and propene polymerization catalyst.[115] Interestingly, the reaction with propene at 25 °C and low propene:(**222**) ratios (ca. 5–10) yields the single insertion product [Zr(η^2-Bz)$_2$(η^7-CH$_2$CH(Me)Bz)][B(Bz)(C$_6$F$_5$)$_3$] (**224**), the structure of which was assigned by one- and two-dimensional NMR spectroscopy.[116] Evidently, coordination of the Ph group of the chelated [η^7-CH$_2$CH(Me)Bz]$^-$ ligand inhibits further propene insertion. Warming (**224**) to 50 °C in the presence of propene yields CH$_2$=CH(Me)Bz and polypropene, most probably via β-H elimination followed by propene insertion of the resulting zirconium hydride species.

12.12 CATIONIC (TETRAAZA-MACROCYCLE)METAL HYDROCARBYL COMPLEXES

Cationic zirconium and hafnium hydrocarbyls incorporating tetraaza-macrocyclic ancillary ligands have been reported.[117] Protonolysis of *cis*-[M(R)$_2$(Me$_8$taa)] complexes with [HNMe$_2$Ph][B(C$_6$F$_5$)$_4$] yields [Zr(R)(Me$_8$taa)][B(C$_6$F$_5$)$_4$] (R = CH$_2$TMS (**225**) or Bz (**226**)), and [Hf(Me)(Me$_8$taen)][B(C$_6$F$_5$)$_4$] (**227**) (Equation (23)). In a similar fashion, [Zr(CH$_2$TMS)(Me$_4$taen)][B(C$_6$F$_5$)$_4$] (**228**) was prepared (Equation (24)). Analogous protonolysis reactions using [HNBu$_3$][BPh$_4$] in the presence of a Lewis base (THF, MeCN or PMe$_2$Ph) afford [M(R)(L)(Me$_8$taa)][BPh$_4$] (**229**) and [M(R)(L)(Me$_4$taen)][BPh$_4$] (**230**), in which the R$^-$ and L ligands are *cis*, and the L ligands undergo rapid exchange.

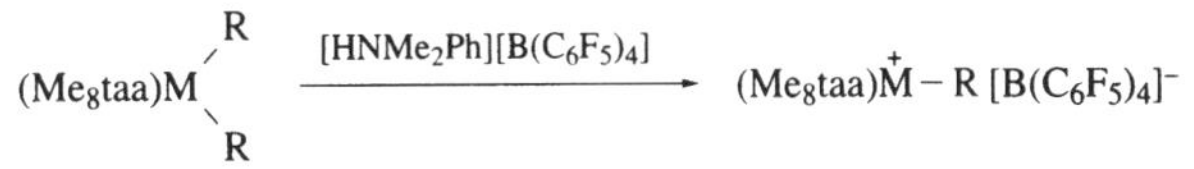

$$(\text{Me}_8\text{taa})\text{M} \overset{R}{\underset{R}{\big\langle}} \quad \xrightarrow{[\text{HNMe}_2\text{Ph}][\text{B}(\text{C}_6\text{F}_5)_4]} \quad (\text{Me}_8\text{taa})\overset{+}{\text{M}} - \text{R} \ [\text{B}(\text{C}_6\text{F}_5)_4]^- \tag{23}$$

$$\text{Me}_8\text{taaH}_2 =$$

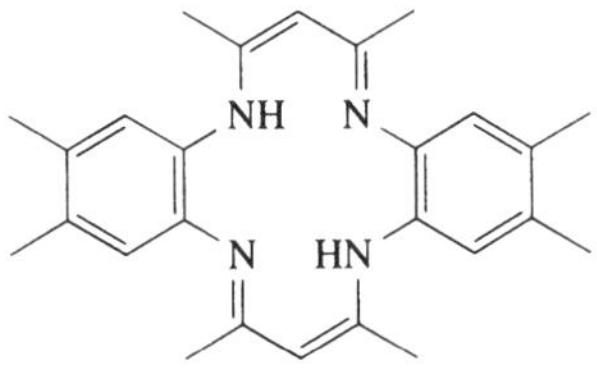

(**225**) M = Zr; R = CH$_2$TMS
(**226**) M = Zr; R = Bz
(**227**) M = Hf; R = Me

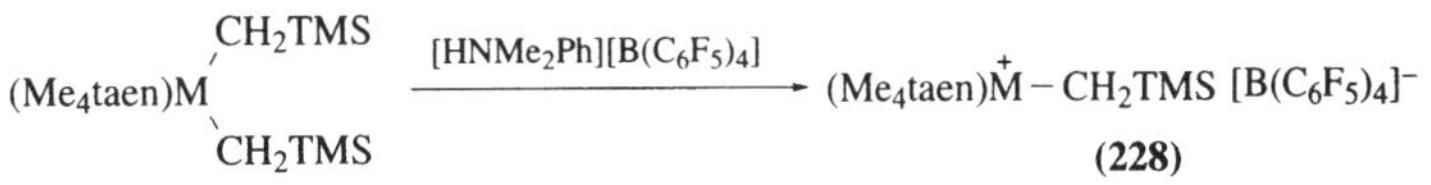

$$(\text{Me}_4\text{taen})\text{M} \overset{\text{CH}_2\text{TMS}}{\underset{\text{CH}_2\text{TMS}}{\big\langle}} \quad \xrightarrow{[\text{HNMe}_2\text{Ph}][\text{B}(\text{C}_6\text{F}_5)_4]} \quad (\text{Me}_4\text{taen})\overset{+}{\text{M}} - \text{CH}_2\text{TMS} \ [\text{B}(\text{C}_6\text{F}_5)_4]^- \tag{24}$$

(**228**)

$$\text{Me}_4\text{taenH}_2 =$$

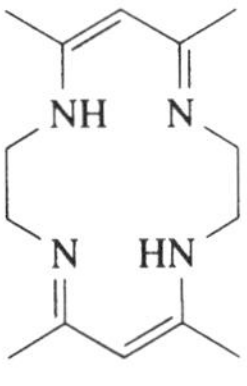

Complexes (**225**)–(**230**) exhibit electrophilic behavior, but are less reactive than [M(R)Cp$'_2$]$^+$ species. Complex (**227**) reacts with 2-butyne to afford the double insertion product [Hf{C(Me)=C(Me)-C(Me)=CMe$_2$}(Me$_8$taa)][B(C$_6$F$_5$)$_4$] (**231**) and with 1-trimethylsilylpropyne yielding the single insertion product [Hf{C(TMS)=CMe$_2$}(Me$_8$taa)][B(C$_6$F$_5$)$_4$] (**232**). In contrast, neither (**225**) and (**226**) nor (**229**) and (**230**) react with these substrates. Interestingly, [Zr(CH$_2$TMS)(Me$_4$taen)][BPh$_4$] (**233**) generated *in situ* reacts with 2-vinylpyridine via remote C–H activation (rather than *ortho*-C–H activation) to afford (**234**) (Equation (25)).

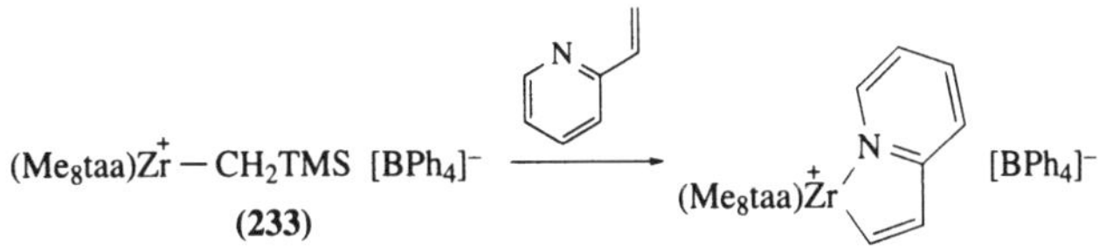

$$(\text{Me}_8\text{taa})\overset{+}{\text{Zr}} - \text{CH}_2\text{TMS} \ [\text{BPh}_4]^- \ \longrightarrow \ (\text{Me}_8\text{taa})\overset{+}{\text{Zr}} \ [\text{BPh}_4]^- \tag{25}$$

(**233**)

(**234**)

12.13 APPLICATIONS OF CATIONIC ORGANOZIRCONIUM AND ORGANOHAFNIUM COMPLEXES

Applications of cationic $[M(R)Cp'_2]^+$ metallocene species (M = Zr or Hf) in catalysis and organic synthesis are summarized in this section. Recent applications of cationic nonmetallocene species in alkene polymerization catalysis were noted in Sections 12.10.1 and 12.11.

12.13.1 Ziegler–Natta Alkene Polymerization Catalysis and Related Reactions

Perhaps the most important application of cationic group 4 organometallic compounds is as catalysts for alkene polymerization.[1,118] Aspects of the historical development of $[M(R)Cp'_2]^+$ catalysts are discussed in papers by Jordan and co-workers.[1] The alkene polymerization chemistry of $[M(R)Cp'_2]^+$ species has been studied extensively and selected aspects are discussed in this section. A detailed discussion is outside the scope of this review.

Cationic $[M(R)Cp'_2]^+$ species (M = Zr or Hf) polymerize ethene, propene, and other α-alkenes, and some cyclic alkenes with high activity[1,5,21–5,38–40,42,79,119] (there is an extensive patent literature on this subject). A concise mechanistic scheme is presented in Scheme 20. Chain growth occurs by repetitive alkene insertion into the M–R bond; the presumed intermediate alkene adducts $[M(R)(alkene)Cp'_2]^+$ have not been detected.[120] There is evidence for the existence of α-agostic interactions in the insertion transition state for some systems ("α-agostic assistance").[121] Both $[M(R)Cp'_2]^+$ and $[M(R)(alkene)Cp'_2]^+$ species probably adopt agostic structures in the ground state (Sections 12.2.2 and 12.2.3).[45,46] Important chain transfer processes, which result in cleavage of the metal–polymer bond and concomitant formation of a catalytically active species, for example, $[M(H)Cp'_2]^+$, include β-H elimination, β-Me elimination in some crowded systems,[25,79] alkyl exchange with Al–R species (in catalyst systems containing aluminum cocatalysts), activation of monomer C–H bonds, and other reactions. Chain transfer can also occur by M–R bond hydrogenolysis with added H_2.[122] Deactivation processes include adventitious hydrolysis and oxidation, irreversible reactions with the counterion leading to unreactive neutral metallocene species, allylic C–H activation leading to cationic allyl complexes,[25] and other reactions. In contrast to titanium systems,[18a] cationic zirconium and hafnium alkyls are resistant to reduction to inactive M^{III} species. Catalyst activities and stereoselectivities are sensitive to Lewis base coordination, ion-pairing interactions with the counterion, solvent effects, and other factors, all of which are omitted in Scheme 20.

Scheme 20

Catalytically active $[M(R)Cp'_2]^+$ species can be generated from $[M(R)_2Cp'_2]$ complexes by protonolysis, oxidative M–R bond cleavage, and R^- abstraction reactions described in Section 12.2.1. Cationic $[M(R)Cp'_2]^+$ species are also believed to form in more complex catalyst systems prepared by a mixture of metallocene complexes with aluminum alkyls,[2,123] or by reaction of $[M(R)_2Cp'_2]$ species with Lewis acidic supports.[124] Of particular interest are catalysts prepared by mixture of $[M(Cl)_2Cp'_2]$ compounds with a large excess of methylaluminoxane (MAO), a complex mixture of oligomeric $[Al(Me)(\mu\text{-}O)]_n$ species formed by controlled partial hydrolysis of $AlMe_3$.[125–8] Early proposals that $[M(R)Cp'_2]^+$ cations are the active species in the $[M(Cl)_2Cp'_2]/AlR_xCl_{3-x}$ and $[M(Cl)_2Cp'_2]/MAO$ systems[1,2,129,130] are supported by extensive model studies with well-characterized $[M(R)Cp'_2]^+$ complexes and isoelectronic group 3 and lanthanide metal complexes,[1,69a,131–4] observations of similar activity and nearly identical stereoselectivity and regioselectivity in propene polymerizations using $[M(R)_2Cp'_2]/[HNMe_2Ph][B(C_6F_5)_4]$ and $[M(Cl)_2Cp'_2]/MAO$ catalysts (Cp'$_2$ = *ansa*-metallocene),[33,39] and direct NMR and x-ray photoelectron spectroscopic studies of MAO catalysts.[134,135]

Metallocene catalysts are "single-site" catalysts, and generally produce polymers with narrow molecular weight distributions ($M_w/M_n = 2$). Polymer properties (molecular weight average and distribution, comonomer incorporation, etc.) and catalyst activity and stereoselectivity can be controlled by modification of the MCp'_2 framework. In particular, polymer tacticity can be controlled, especially with "*ansa*" systems in which the Cp'^- ligands are linked by a bridge.[136] Chiral C_2-symmetric systems, for example, $[M(R)(EBTHI)]^+$, undergo repetitive, stereoselective α-alkene insertions leading to highly isotactic polymer,[33,39,121,137,138] and systems with Cp^- ligands of very different steric bulk, for example, $[M(R)\{\eta\text{-}C_5H_4CMe_2(\eta^5\text{-fluorenyl})\}]^+$, produce syndiotactic polypropylene.[52] Detailed stereocontrol models have been developed.[33,39,120,138] There is a considerable research effort aimed at developing supported metallocene catalysts, and an extensive patent literature.[139]

Cationic $[M(R)Cp'_2]^+$ species, generated as discrete species or in MAO systems, catalyze other reactions in which alkene insertion plays a key role, including alkene oligomerization,[140] alkene hydrogenation,[31,141] alkene hydrooligomerization,[138a,142] and α,ω-diene cyclopolymerization.[143]

Complex (**2**) catalyzes the polymerization of methyl methacrylate in the presence of $[Zr(Me)_2Cp'_2]$. A group-transfer polymerization mechanism involving *in situ* generated cationic zirconium enolate species was proposed.[144]

12.13.2 Organic Synthesis

The ligand substitution, insertion, H_2 and ligand C–H bond activation, and M–C bond cleavage reactions of cationic zirconium and hafnium complexes have been exploited in organic synthesis. In some cases, cationic species are generated *in situ* via addition of catalytic amounts of silver salts to $[M(hydrocarbyl)(Cl)Cp'_2]$ or $[M(Cl)_2Cp'_2]$ species.

12.13.2.1 Zirconium-mediated and catalyzed coupling of alkenes and pyridines

Metallacycles derived from insertion reactions of $[Zr(\eta^2\text{-pyridyl})Cp'_2]^+$ and related species (Schemes 10 and 11; Section 12.4.2) can be hydrolyzed to afford substituted pyridines.[83–5] Presumably, other Zr–C cleavage reactions could be used to prepare more highly functionalized products.

The sequence of pyridine C–H activation, insertion, and hydrolysis was employed in a general synthesis of highly substituted alkylpyrazines.[145] Alkenyl-substituted alkylpyrazines (Scheme 21) were obtained regio- and stereoselectively by sequential one-pot addition of alkylpyrazine, alkyne, and a proton source to a solution of (**2**) in CH_2Cl_2. Conventional synthetic manipulation of the alkenylpyrazine products provided further access to tri- and tetrasubstituted alkylpyrazines, mono- and dibromoalkylpyrazines, and epoxypyrazines.

Scheme 21

In the presence of H_2, $[Zr(R)Cp'_2]^+$ species catalyze the coupling of substituted pyridines and alkenes, as illustrated for α-picoline and propene in Scheme 22.[72] This reaction is catalytic in both Zr and H_2. Hydrogenolysis of the Zr–C bond of (**235**), derived from propene insertion into (**236**), affords the cationic hydride (**237**), which undergoes ligand substitution to release the product disubstituted pyridine and generate (**238**). Subsequent C–H activation regenerates (**236**) and H_2. As many as 50 total turnovers at 23 °C have been observed for this process. Side reactions include alkene hydrogenation, presumably via (**237**), and reversible remote C–H activation which yields (**239**). Pyridine substrates which lack *ortho*-substituents form unreactive 18-electron pyridyl pyridine complexes (**240**); thus β-picoline and γ-picoline are catalyst poisons in Scheme 22.

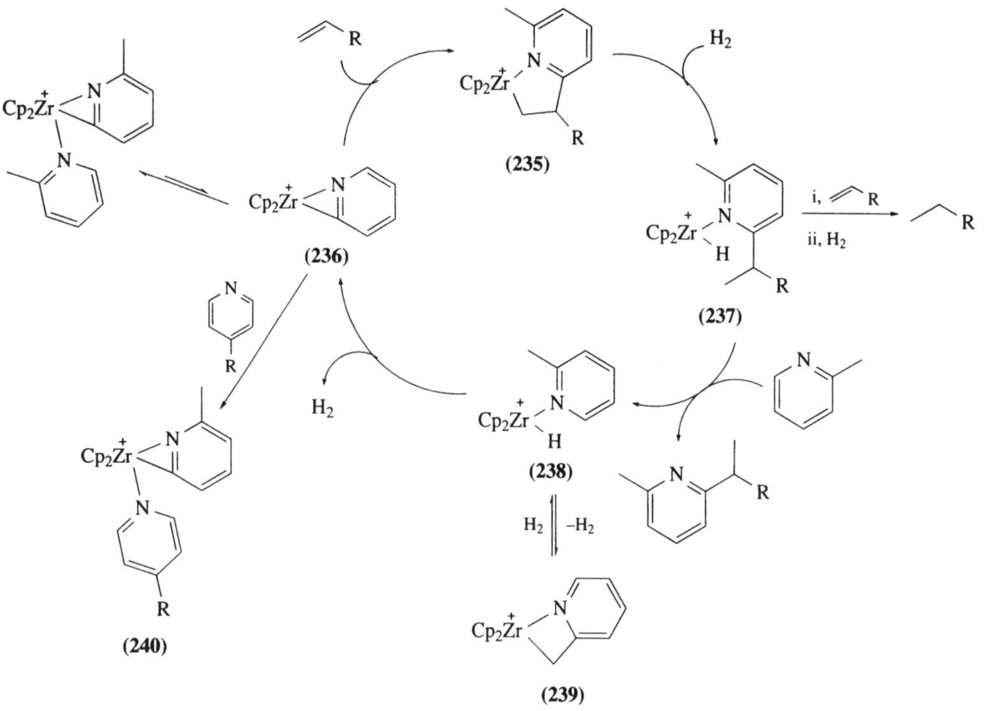

Scheme 22

12.13.2.2 *Zirconium-mediated synthesis of carbonyl compounds*

Multiple, alternating CO and alkyne insertion reactions of [Zr(R)(THF)Cp$'_2$]$^+$ complexes provide access to carbonyl compounds.[80] Syntheses of α,β-unsaturated ketones and 1,4-divinyl ketones are illustrated in Scheme 23. Sequential alternating reaction of (**2**) with 2 equiv. pentyne and CO followed by cyclization yielded the zirconoxyfuran complex (**241**) which was hydrolyzed to the lactone (**242**).

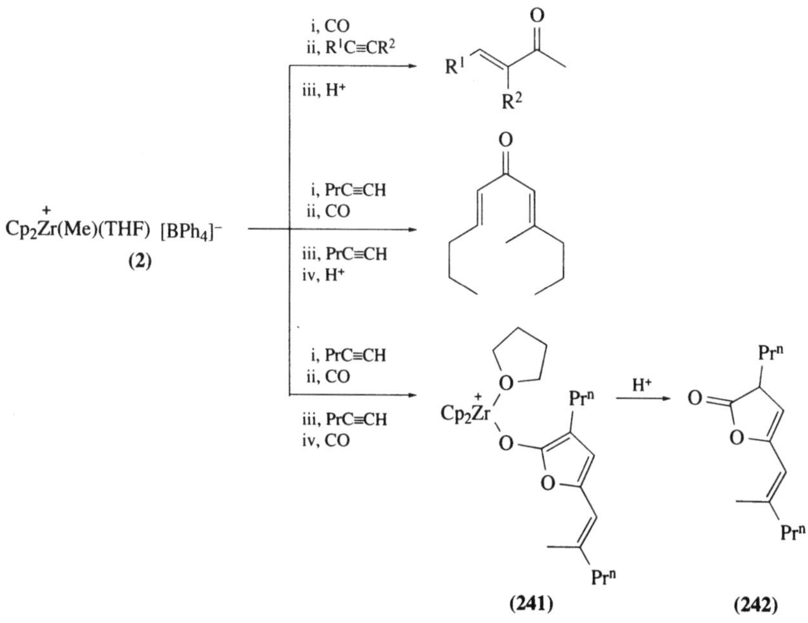

Scheme 23

The synthetic accessibility and high insertion reactivity of the cationic iminoacyl complex $[Zr\{\eta^2\text{-}C(=NBu^i)(Me)\}Cp_2][BMe_3(C_6F_5)_3]$ (**171**) (Schemes 12 and 14, Section 12.5.2) was exploited in a general synthesis of saturated and unsaturated methyl ketones (Equation (26)).[92] The experimental procedure requires sequential one-pot addition of $B(C_6F_5)_3$ and an alkene or alkyne to $[Zr\{\eta^2\text{-}C,N\text{-}C(=NBu^i)Me\}(Me)Cp_2]$, followed by hydrolysis.

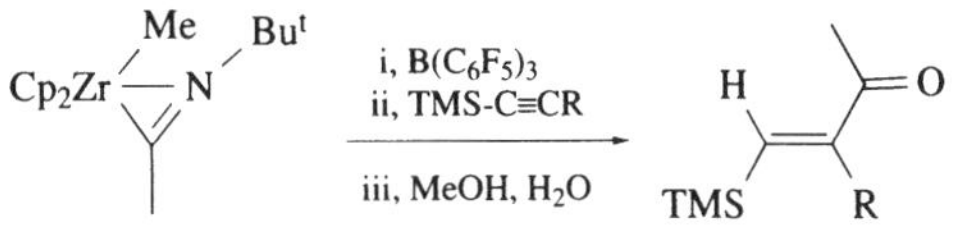

$$(26)$$

12.13.2.3 Zirconium- and hafnium-promoted glycosidations

A reagent system consisting of $[M(Cl)_2Cp_2]$ (M = Zr or Hf) and $Ag[ClO_4]$ (1:1 or 1:2 molar ratio) effects the coupling of glycosyl fluorides and alcohols (Equation (27)).[102,146] NMR spectral studies of the hafnium system reveal the formation of $[Hf(Cl)Cp_2][ClO_4]$ and $[HfCp_2][ClO_4]_2$ from the reaction of $[Hf(Cl)_2Cp_2]$ with 1 and 2 equivalents of $Ag[ClO_4]$. It was proposed that these species abstract fluoride from the glycosyl fluoride, to afford oxonium ions which react with the alcohol. Glycosidation reactions of this type have been exploited in the synthesis of macrolide and glycoside antiobiotics.[147,148] Other silver salts including $Ag[BF_4]$, $Ag[OTf]$, $Ag[PF_6]$, and $Ag[SbF_6]$ can be used in combination with $[Zr(Cl)_2Cp_2]$; yields and stereoselectivities depend strongly on the nature of the counteranion and the solvent.[149]

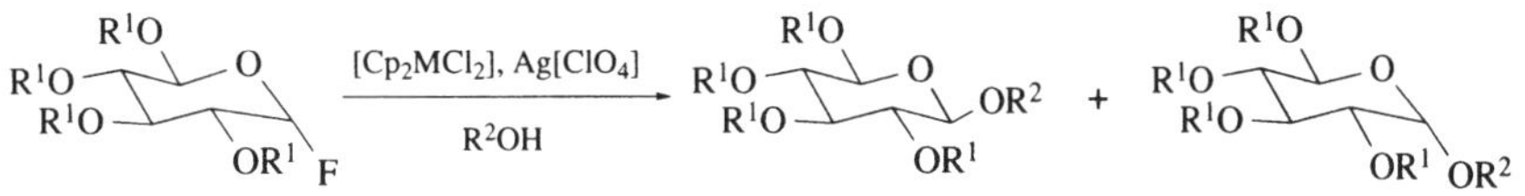

$$(27)$$

12.13.2.4 Zirconium-mediated preparation of alcohols, 1,3-dienes, and polyenes

Neutral $[Zr(alkenyl)(Cl)Cp'_2]$ species, generated *in situ* by alkyne insertion reactions of $[Zr(H)(Cl)Cp]$ (Schwartz's reagent),[150] add to aldehydes in the presence of catalytic amounts of $Ag[ClO_4]$ to afford allylic alcohols after hydrolysis.[151] This reaction probably proceeds via cationic $[Zr(alkenyl)Cp_2]^+$ species which undergo rapid aldehyde insertion and ligand exchange (Scheme 24). An (*E*)-stereoselective synthesis of 1,3-dienes was developed, based on the $Ag[ClO_4]$-catalyzed reaction of $[Zr(Cl)\{(E)\text{-}CH=CHCH_2\text{-}TMS\}Cp_2]$ with aldehydes (Equation (28)).[152] Similarly, $Ag[ClO_4]$-catalyzed addition of $[Zr(Cl)\{(E)\text{-}CH=CHOEt\}Cp_2]$ or $[Zr(Cl)\{(E,Z)\text{-}CH=CHCH=CHOMe\}Cp_2]$ to aldehydes followed by acidic hydrolysis affords two- or four-carbon homologues of the parent aldehyde (Scheme 25).[153] This reaction was exploited in polyene synthesis. The $Ag[ClO_4]$-catalyzed addition of $[Zr(alkenyl)(Cl)Cp_2]$ and $[Zr(alkyl)(Cl)Cp_2]$ species to epoxides provides a general synthesis of secondary alcohols.[154]

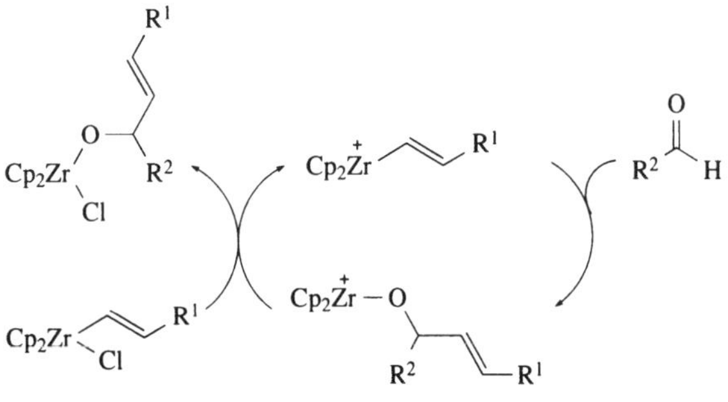

Scheme 24

$$\text{Cp}_2\text{Zr(Cl)} \diagdown \diagup \text{TMS} \xrightarrow[\text{ii, }[\text{H}_3\text{O}]^+]{\text{i, RCHO, cat. Ag[ClO}_4]} R \diagdown \diagup \diagdown \tag{28}$$

$$[\text{Zr(H)(Cl)(Cp)}_2] \xrightarrow{\text{HC}\equiv\text{COEt}} \text{Cp}_2\text{Zr(Cl)} \diagdown \diagup \text{OEt} \xrightarrow[\text{ii, }[\text{H}_3\text{O}]^+]{\text{i, RCHO, cat. Ag[ClO}_4]} R \diagdown \diagup \text{CHO}$$

Scheme 25

12.13.2.5 *Zirconium-catalyzed Diels–Alder and Mukaiyama cross-aldol reactions*

Cationic zirconium complexes have been exploited as Lewis acid catalysts in several reactions. The use of [Zr(OR)(THF)Cp'$_2$]$^+$ complexes (**186a**) and (**187**) as catalysts for Diels–Alder reactions has been demonstrated.[93a,94] Modest enantioselectivity was observed with the chiral catalyst (**187**). Cationic intermediates are also important in [Zr(OSO$_2$CF$_3$)$_2$(THF)Cp$_2$]-catalyzed Diels–Alder and Mukaiyama cross-aldol reactions.[155,156]

12.14 REFERENCES

1. (a) R. F. Jordan, *Adv. Organomet. Chem.*, 1991, **32**, 325; (b) R. F. Jordan, P. K. Bradley, R. E. LaPointe and D. F. Taylor, *New. J. Chem.*, 1990, **14**, 505; (c) R. F. Jordan, *J. Chem. Educ.*, 1988, **65**, 285.
2. (a) A. K. Zefirova and A. E. Shilov, *Dokl. Acad. Nauk. SSSR*, 1961, **136**, 599; (b) F. S. Dyachkovskii, A. K. Shilova and A. E. Shilov, *J. Polym. Sci., Part C: Polym Symp.*, 1967, **16**, 2333.
3. (a) S. H. Strauss, *Chem. Rev.*, 1993, **93**, 927; (b) W. Beck and K. Sünkel, *ibid.*, 1988, **88**, 1405; (c) G. A. Lawrance, *ibid.*, 1986, **86**, 17; (d) M. Bochmann, *Angew. Chem., Int. Ed. Engl.*, 1992, **31**, 1181.
4. (a) W. Kaminsky, J. Kopf, H. Sinn and H.-J. Vollmer, *Angew. Chem., Int. Ed. Engl.*, 1976, **15**, 629; (b) J. Kopf, H.-J. Vollmer and W. Kaminsky, *Cryst. Struct. Commun.*, 1980, **9**, 271.
5. R. F. Jordan, C. S. Bajgur, R. Willett and B. Scott, *J. Am. Chem. Soc.*, 1986, **108**, 7410.
6. Y. W. Alelyunas, R. F. Jordan, S. F. Echols, S. L. Borkowsky and P. K. Bradley, *Organometallics*, 1991, **10**, 1406.
7. S. L. Borkowsky, Ph. D. Thesis, University of Iowa, 1992.
8. R. F. Jordan, R. E. LaPointe, C. S. Bajgur, S. F. Echols and R. Willett, *J. Am. Chem. Soc.*, 1987, **109**, 4111.
9. R. F. Jordan, R. E. LaPointe, N. Baenziger and G. D. Hinch, *Organometallics*, 1990, **9**, 1539.
10. E. B. Tjaden, G. L. Casty and J. M. Stryker, *J. Am. Chem. Soc.*, 1993, **115**, 9814.
11. S. L. Borkowsky, R. F. Jordan and G. D. Hinch, *Organometallics*, 1991, **10**, 1268.
12. M. J. Burk, W. Tumas, M. D. Ward and D. R. Wheeler, *J. Am. Chem. Soc.*, 1990, **112**, 6133.
13. R. F. Jordan, W. E. Dasher and S. F. Echols, *J. Am. Chem. Soc.*, 1986, **108**, 1718.
14. R. F. Jordan, C. S. Bajgur, W. E. Dasher and A. L. Rheingold, *Organometallics*, 1987, **6**, 1041.
15. C. D. M. Beverwijk, G. J. M. van der Kerk, A. J. Leusink and J. G. Noltes, *Organomet. Chem. Rev. A*, 1970, **5**, 215.
16. D. J. Crowther, S. L. Borkowsky, D. Swenson, T. Y. Meyer and R. F. Jordan, *Organometallics*, 1993, **12**, 2897.
17. D. M. Roddick, R. H. Heyn and T. D. Tilley, *Organometallics*, 1989, **8**, 324.
18. (a) S. L. Borkowsky, N. C. Baenziger and R. F. Jordan, *Organometallics*, 1993, **12**, 486; (b) P. B. Hitchcock, M. F. Lappert and R. G. Taylor, *J. Chem. Soc., Chem. Commun.*, 1984, 1082; (c) D. Alvarez Jr. and K. G. Caulton, *Polyhedron*, 1988, **7**, 1285; (d) A. L. Seligson and W. C. Trogler, *J. Am. Chem. Soc.*, 1992, **114**, 7085.
19. M. Bochmann and L. M. Wilson, *J. Chem. Soc., Chem. Commun.*, 1986, 1610.
20. Z. Lin, J. F. Le Marechal, M. Sabat and T. J. Marks, *J. Am. Chem. Soc.*, 1987, **109**, 4127.
21. H. W. Turner and G. G. Hlatky, *Eur. Pat. Appl.* 0 277 003 (1988) (*Chem. Abstr.*, 1989, **110**, 58 290a).
22. G. G. Hlatky, H. W. Turner and R. R. Eckman, *J. Am. Chem. Soc.*, 1989, **111**, 2728.
23. H. W. Turner, *Eur. Pat. Appl.* 0 277 004 (1988) (*Chem. Abstr.*, 1989, **110**, 58 291b).
24. G. G. Hlatky, R. R. Eckman and H. W. Turner, *Organometallics*, 1992, **11**, 1413.
25. (a) J. J. W. Eshuis, Y. Y. Tan, A. Meetsma, J. H. Teuben, J. Renkema and G. G. Evens, *Organometallics*, 1992, **11**, 362; (b) J. J. W. Eshuis, Y. Y. Tan, J. Renkema and J. H. Teuben, *J. Mol. Catal.*, 1990, **62**, 277.
26. Z. Guo, D. C. Swenson and R. F. Jordan, *Organometallics*, 1994, **13**, 1424.
27. D. M. Amorose, R. P. Lee and J. L. Petersen, *Organometallics*, 1991, **10**, 2191.
28. A. D. Horton and A. G. Orpen, *Organometallics*, 1991, **10**, 3910.
29. M. Bochmann, A. J. Jaggar and J. C. Nicholls, *Angew. Chem., Int. Ed. Engl.*, 1990, **29**, 780.
30. A. D. Horton and J. H. G. Frijins, *Angew. Chem., Int. Ed. Engl.*, 1991, **30**, 1152.
31. R. B. Grossman, R. A. Doyle and S. L. Buchwald, *Organometallics*, 1991, **10**, 1501.
32. M. Bochmann and S. J. Lancaster, *J. Organomet. Chem.*, 1992, **434**, C1.
33. J. A. Ewen, M. J. Elder, R. L. Jones, L. Haspeslagh, J. L. Atwood, S. G. Bott and K. Robinson, *Makromol. Chem., Macromol. Symp.*, 1991, **48/49**, 253.
34. A. R. Siedle, R. A. Newmark, W. B. Gleason and W. M. Lamana, *Organometallics*, 1990, **9**, 1290.
35. B. Longato, B. D. Martin, J. R. Norton and O. P. Anderson, *Inorg. Chem.*, 1985, **24**, 1389.

36. R. R. Schrock and P. R. Sharp, *J. Am. Chem. Soc.*, 1978, **100**, 2389.
37. D. A. Straus, C. Zhang and T. D. Tilley, *J. Organomet. Chem.*, 1989, **369**, C13.
38. J. C. W. Chien, W. Tsai and M. D. Rausch, *J. Am. Chem. Soc.*, 1991, **113**, 8570.
39. J. A. Ewen and M. J. Elder, *Makromol. Chem., Macromol. Symp.*, 1993, **66**, 179.
40. M. Bochmann and S. J. Lancaster, *Organometallics*, 1993, **12**, 633.
41. A. Razavi and U. Thewalt, *J. Organomet. Chem.*, 1993, **445**, 111.
42. X. Yang, C. L. Stern and T. J. Marks, *J. Am. Chem. Soc.*, 1991, **113**, 3623.
43. A. D. Horton and A. G. Orpen, *Organometallics*, 1992, **11**, 1193.
44. R. F. Jordan, R. E. LaPointe, P. K. Bradley and N. Baenziger, *Organometallics*, 1989, **8**, 2892.
45. R. F. Jordan, P. K. Bradley, N. C. Baenziger and R. E. LaPointe, *J. Am. Chem. Soc.*, 1990, **112**, 1289.
46. Y. W. Alelyunas, Z. Guo, R. E. LaPointe and R. F. Jordan, *Organometallics*, 1993, **12**, 544.
47. A. D. Horton and A. G. Orpen, *Organometallics*, 1992, **11**, 8.
48. A. D. Horton, *Organometallics*, 1992, **11**, 3271.
49. (a) D. J. Cardin, M. F. Lappert and C. L. Raston, 'Chemistry of Organo-Zirconium and -Hafnium Compounds', Ellis Horwood, Chichester, UK, 1986; (b) K. Prout, T. S. Cameron, R. A. Forder, S. R. Critchley, B. Denton and G. V. Rees, *Acta Crystallogr., Sect. B*, 1974, **30**, 2290.
50. (a) M. Brookhart, M. L. H. Green and L. Wong, *Prog. Inorg. Chem.*, 1988, **36**, 1; (b) R. H. Crabtree and D. G. Hamilton, *Adv. Organomet. Chem.*, 1988, **28**, 299.
51. W. J. Evans, R. Dominguez and T. P. Hanusa, *Organometallics*, 1986, **5**, 263.
52. J. A. Ewen, R. L. Jones, A. Razavi and J. D. Ferrara, *J. Am. Chem. Soc.*, 1988, **110**, 6255.
53. (a) G. R. Davies, J. A. J. Jarvis, B. T. Kilbourn and A. J. P. Pioli, *J. Chem. Soc., Chem. Commun.*, 1971, 677; (b) G. R. Davies, J. A. J. Jarvis, B. T. Kilbourn and A. J. P. Pioli, *ibid.*, 1971, 1511; (c) I. W. Bassi, G. Allegra, R. Scordamaglia and G. J. Chioccola, *J. Am. Chem. Soc.*, 1971, **93**, 3787.
54. (a) E. A. Mintz, K. G. Moloy, T. J. Marks and V. W. Day, *J. Am. Chem. Soc.*, 1982, **104**, 4692; (b) G. S. Girolami, G. Wilkinson, M. Thornton-Pett and M. B. Hursthouse, *J. Chem. Soc., Dalton Trans.*, 1984, 2789; (c) P. G. Edwards, R. A. Andersen and A. Zalkin, *Organometallics*, 1984, **3**, 293; (d) S. L. Latesky, A. K. McMullen, G. P. Niccolai, I. P. Rothwell and J. C. Huffman, *ibid.*, 1985, **4**, 902; (e) J. Scholz, M. Schlegel and K.-H. Thiele, *Chem. Ber.*, 1987, **120**, 1369.
55. J. W. Lauher and R. Hoffmann, *J. Am. Chem. Soc.*, 1976, **98**, 1729.
56. Z. Dawoodi, M. L. H. Green, V. S. B. Mtetwa, K. Prout, A. J. Schultz, J. M. Williams and T. F. Koetzle, *J. Chem. Soc., Dalton Trans.*, 1986, 1629.
57. R. B. Cracknell, A. G. Orpen and J. L. Spencer, *J. Chem. Soc., Chem. Commun.*, 1984, 326.
58. L. Mole, J. L. Spencer, N. Carr and A. G. Orpen, *Organometallics*, 1991, **10**, 49.
59. Y. W. Alelyunas, N. C. Baenziger, P. K. Bradley and R. F. Jordan, *Organometallics*, 1994, **13**, 148.
60. J. B. Lambert, *Tetrahedron*, 1990, **40**, 2677.
61. J. Kopf, W. Kaminsky and H.-J. Vollmer, *Cryst. Struct. Commun.*, 1980, **9**, 197.
62. M. V. Gaudet, M. J. Zaworotko, T. S. Cameron and A. Linden, *J. Organomet. Chem.*, 1989, **367**, 267.
63. (a) D. J. Liston, Y. J. Lee, W. R. Scheidt and C. A. Reed, *J. Am. Chem. Soc.*, 1989, **111**, 6643; (b) K. Shelly, C. A. Reed, Y. J. Lee and W. R. Scheidt, *ibid.*, 1986, **108**, 3117; (c) K. Shelly, D. C. Finster, Y. J. Lee, W. R. Scheidt and C. A. Reed, *ibid.*, 1985, **107**, 5955.
64. C. J. Burns and R. A. Andersen, *J. Am. Chem. Soc.*, 1987, **109**, 5852.
65. R. F. Jordan and A. S. Guram, *Organometallics*, 1990, **9**, 2116.
66. R. F. Jordan, *J. Organomet. Chem.*, 1985, **294**, 321.
67. Z. Gou, D. C. Swenson, A. S. Guram and R. F. Jordan, *Organometallics*, 1994, **13**, 766.
68. (a) G. Fachenetti, G. Fochi and C. Floriani, *J. Chem. Soc., Dalton Trans.*, 1977, 1946; (b) T. Yoshida and E. Negishi, *J. Am. Chem. Soc.*, 1981, **103**, 4985; (c) K. I. Gell, B. Posin, J. Schwartz and G. M. Williams, *ibid.*, 1982, **104**, 1846; (d) J. A. Marsella, K. G. Moloy and K. G. Caulton, *J. Organomet. Chem.*, 1980, **201**, 389; (e) D. B. Carr and J. Schwartz, *J. Am. Chem. Soc.*, 1979, **101**, 3521.
69. (a) P. L. Watson and G. W. Parshall, *Acc. Chem. Res.*, 1985, **18**, 51; (b) M. E. Thompson and J. E. Bercaw, *Pure Appl. Chem.*, 1985, **56**, 1; (b) M. E. Thompson, S. M. Baxter, A. R. Bulls, B. J. Burger, M. C. Nolan, B. D. Santarsiero, W. P. Schaefer and J. E. Bercaw, *J. Am. Chem. Soc.*, 1987, **109**, 203.
70. Z. Guo, P. K. Bradley and R. F. Jordan, *Organometallics*, 1992, **11**, 2690.
71. X. Yang, C. L. Stern and T. J. Marks, *Angew. Chem., Int. Ed. Engl.*, 1992, **31**, 1375.
72. R. F. Jordan and D. F. Taylor, *J. Am. Chem. Soc.*, 1989, **111**, 778.
73. R. F. Jordan, D. F. Taylor and N. C. Baenziger, *Organometallics*, 1990, **9**, 1546.
74. A. J. Cheney, B. E. Mann, B. L. Shaw and R. M. Slade, *J. Chem. Soc. D*, 1970, 1176.
75. (a) B. Klei and J. H. Teuben, *J. Chem. Soc., Chem. Commun.*, 1978, 659; (b) B. Klei and J. H. Teuben, *J. Organomet. Chem.*, 1981, **214**, 53.
76. S. L. Buchwald, R. T. Lum, R. A. Fisher and W. M. Davis, *J. Am. Chem. Soc.*, 1989, **111**, 9113.
77. (a) F. Neve, M. Ghedini, A. Tiripicchio and F. Ugozzoli, *Inorg. Chem.*, 1989, **28**, 3084; (b) M. Lavin, E. M. Holt and R. H. Crabtree, *Organometallics*, 1989, **8**, 99.
78. M. I. Bruce, B. L. Goodall and I. Matsuda, *Aust. J. Chem.*, 1975, **28**, 1259.
79. L. Resconi, F. Piemontesi, G. Franciscono, L. Abis and T. Fiorani, *J. Am. Chem. Soc.*, 1992, **114**, 1025.
80. A. S. Guram, Z. Guo and R. F. Jordan, *J. Am. Chem. Soc.*, 1993, **115**, 4902.
81. A. D. Horton, *J. Chem. Soc., Chem. Commun.*, 1992, 185.
82. A. D. Horton and A. G. Orpen, *Angew. Chem., Int. Ed. Engl.*, 1992, **31**, 876.
83. A. S. Guram and R. F. Jordan, *Organometallics*, 1991, **10**, 3470.
84. A. S. Guram and R. F. Jordan, *Organometallics*, 1990, **9**, 2190.
85. A. S. Guram, R. F. Jordan and D. F. Taylor, *J. Am. Chem. Soc.*, 1991, **113**, 1833.
86. G. Erker, P. Czisch, C. Krüger and J. M. Wallis, *Organometallics*, 1985, **4**, 2059.
87. J. W. F. L. Seetz, G. Schat, O. S. Akkerman and F. Bickelhaupt, *Angew. Chem., Int. Ed. Engl.*, 1983, **22**, 248.
88. K. M. Doxsee and J. B. Farahi, *J. Am. Chem. Soc.*, 1988, **110**, 7239.
89. F. J. Berg and J. L. Petersen, *Organometallics*, 1989, **8**, 2461.
90. A. S. Guram, D. C. Swenson and R. F. Jordan, *J. Am. Chem. Soc.*, 1992, **114**, 8991.

91. J. A. Marsella, C. J. Curtis, J. E. Bercaw and K. G. Caulton, *J. Am. Chem. Soc.*, 1980, **102**, 7244.
92. A. S. Guram and R. F. Jordan, *J. Org. Chem.*, 1993, **58**, 5595.
93. (a) S. Collins, B. E. Koene, R. Ramachandran and N. J. Taylor, *Organometallics*, 1991, **10**, 2092; (b) W. E. Piers, L. Koch, D. S. Ridge, L. R. MacGillivray and M. Zaworotko, *ibid.*, 1992, **11**, 3148.
94. Y. Hong, B. A. Kuntz and S. Collins, *Organometallics*, 1993, **12**, 964.
95. G. S. Sodhi and N. K. Kaushik, *Bull. Soc. Chim. France*, 1982, **1–2**, 45.
96. U. Thewalt and W. Lasser, *J. Organomet. Chem.*, 1989, **363**, C12.
97. W. Lasser and U. Thewalt, *J. Organomet. Chem.*, 1986, **302**, 201.
98. U. Thewalt, K. Doppert and W. Lasser, *J. Organomet. Chem.*, 1986, **308**, 303.
99. T. Cuenca and P. Royo, *J. Organomet. Chem.*, 1985, **293**, 61.
100. R. F. Jordan and S. F. Echols, *Inorg. Chem.*, 1987, **26**, 383.
101. P. N. Billinger, P. P. K. Claire, H. Collins and G. R. Willey, *Inorg. Chim. Acta*, 1988, **149**, 63.
102. K. Suzuki, H. Maeta and T. Matsumoto, *Tetrahedron Lett.*, 1989, **30**, 4853.
103. J. H. Toney and T. J. Marks, *J. Am. Chem. Soc.*, 1985, **107**, 947.
104. U. Thewalt, and W. Lasser, *J. Organomet. Chem.*, 1984, **276**, 341.
105. W. Lasser and U. Thewalt, *J. Organomet. Chem.*, 1986, **311**, 69.
106. D. E. Bowen Ph. D. Thesis, University of Iowa, 1995.
107. D. J. Crowther, R. F. Jordan, N. C. Baenziger and A. Verma, *Organometallics*, 1990, **9**, 2574.
108. D. J. Gillis, M.-J. Tudoret and M. C. Baird, *J. Am. Chem. Soc.*, 1993, **115**, 2543.
109. C. Pellecchia, A. Proto, P. Longo and A. Zambelli, *Makromol. Chem., Rapid Commun.*, 1992, **13**, 277.
110. C. Pellecchia, A. Proto, P. Longo and A. Zambelli, *Makromol. Chem., Rapid Commun.*, 1991, **12**, 663.
111. G. Erker, C. Sarter, M. Albrecht, S. Dehnicke, C. Krüger, E. Raabe, R. Schulund, R. Benn, A. Rufínska and R. Mynott, *J. Organomet. Chem.*, 1990, **382**, 89.
112. (a) B. K. Stein, S. R. Frerichs and J. E. Ellis, *Organometallics*, 1987, **6**, 2017; (b) Y. Wielstra, S. Gambarotta, A. Meetsma, J. L. de Boer and M. Y. Chiang, *Organometallics*, 1989, **8**, 2696.
113. M. Bochmann, G. Karger and A. J. Jaggar, *J. Chem. Soc., Chem. Commun.*, 1990, 1038.
114. C. Pellecchia, A. Grassi and A. Immirzi, *J. Am. Chem. Soc.*, 1993, **115**, 1160.
115. C. Pellecchia, A. Grassi and A. Zambelli, *J. Mol. Catal.*, 1993, **82**, 57.
116. C. Pellecchia, A. Grassi and A. Zambelli, *J. Chem. Soc., Chem. Commun.*, 1993, 947.
117. R. Uhrhammer, D. G. Black, T. G. Gardner, J. D. Olsen and R. F. Jordan, *J. Am. Chem. Soc.*, 1993, **115**, 8493.
118. J. Boor, 'Ziegler–Natta Catalyst and Polymerizations', Academic Press, New York, 1979, p. 251.
119. S. Collins and W. M. Kelly, *Macromolecules*, 1992, **25**, 233.
120. For theoretical calculations see (a) L. A. Castonguay and A. K. Rappé, *J. Am. Chem. Soc.*, 1992, **114**, 5832; (b) H. K. Kawamura-Kuribayashi, N. Koga and K. Morokuma, *ibid.*, 1992, **114**, 8687; (c) C. A. Jolly and D. S. Marynick, *ibid.*, 1989, **111**, 7968.
121. (a) H. Krauledat and H.-H. Brintzinger, *Angew. Chem., Int. Ed. Engl.*, 1990, **29**, 1412; (b) W. E. Piers and J. E. Bercaw, *J. Am. Chem. Soc.*, 1990, **112**, 9406.
122. W. Kaminsky and H. Luker, *Makromol. Chem., Rapid. Commun.*, 1984, **5**, 228.
123. A. Zambelli, P. Longo and A. Grassi, *Macromolecules*, 1989, **22**, 2186.
124. T. J. Marks, *Acc. Chem. Res.*, 1992, **25**, 57.
125. H. Sinn and W. Kaminsky, *Adv. Organomet. Chem.*, 1980, **18**, 99.
126. J. A. Ewen, *J. Am. Chem. Soc.*, 1984, **106**, 6355.
127. M. R. Mason, J. M. Smith, S. G. Bott and A. R. Barron, *J. Am. Chem. Soc.*, 1993, **115**, 4971.
128. L. Resconi, S. Bossi and L. Abis, *Macromolecules*, 1990, **23**, 4489.
129. E. Giannetti, M. Nicoletti and R. Mazzocchi, *J. Polym. Sci., Polym. Chem. Ed.*, 1985, **23**, 2117.
130. J. J. Eisch, A. M. Piotrowski, S. K. Brownstein, E. J. Gabe and F. L. Lee, *J. Am. Chem. Soc.*, 1985, **107**, 7219.
131. B. J. Burger, M. E. Thompson, W. D. Cotter and J. E. Bercaw, *J. Am. Chem. Soc.*, 1990, **112**, 1566.
132. D. G. H. Ballard, A. Courtis, J. Holton, J. McMeeking and R. Pearce, *J. Chem. Soc., Chem. Commun.*, 1978, 994.
133. G. Jeske, H. Lauke, H. Mauermann, P. N. Swepston, H. Schumann and T. J. Marks, *J. Am. Chem. Soc.*, 1985, **107**, 8091.
134. P. G. Gassman and M. R. Callstrom, *J. Am. Chem. Soc.*, 1987, **109**, 7875.
135. C. Sishta, R. M. Hathorn and T. J. Marks, *J. Am. Chem. Soc.*, 1992, **114**, 1112.
136. (a) F. R. W. P. Wild, L. Zsolnai, G. Huttner and H.-H. Brintzinger, *J. Organomet. Chem.*, 1982, **232**, 233; (b) F. R. W. P. Wild, M. Wasiucionek, G. Huttner and H.-H. Brintzinger, *ibid.*, 1985, **288**, 63; (c) T. Mise, S. Miya and H. Yamazaki, *Chem. Lett.*, 1989, **7–12**, 1853; (d) W. Röll, H.-H. Brintzinger, B. Rieger and R. Zolk, *Angew. Chem., Int. Ed. Engl.*, 1990, **29**, 279; (e) J. A. Bandy, M. L. H. Green, I. M. Gardiner and K. Prout, *J. Chem. Soc., Dalton Trans.*, 1991, 2207.
137. (a) W. Kaminsky, K. Külper, H. H. Brintzinger and F. R. W. P. Wild, *Angew. Chem., Int. Ed. Engl.*, 1985, **24**, 507; (b) J. A. Ewen, L. Haspeslagh, J. L. Atwood and H. Zhang, *J. Am. Chem. Soc.*, 1987, **109**, 6544; (c) K. Soga, T. Shiono, S. Takemura and W. Kaminsky, *Makromol. Chem., Rapid Commun.*, 1987, **8**, 305; (d) P. Longo, A. Grassi, C. Pellecchia and A. Zambelli, *Macromolecules*, 1987, **20**, 1015; (e) S. Collins, W. J. Gauthier, D. A. Holden, B. A. Kuntz, N. J. Taylor and D. G. Ward, *Organometallics*, 1991, **10**, 2061.
138. (a) P. Pino, P. Cioni and J. Wei, *J. Am. Chem. Soc.*, 1987, **109**, 6189; (b) L. Cavallo, G. Guerra, L. Oliva, M. Vacatello and P. Corradini, *Polym. Commun.*, 1989, **30**, 16; (c) L. Cavallo, G. Guerra, M. Vacatello and P. Corradini, *Macromolecules*, 1991, **24**, 1784.
139. W. Kaminsky and F. Renner, *Makromol. Chem., Rapid Commun.*, 1993, **14**, 239.
140. W. Kaminsky, A. Ahlers and N. Möller-Lindenhof, *Angew. Chem., Int. Ed. Engl.*, 1989, **28**, 1216.
141. R. Waymouth and P. Pino, *J. Am. Chem. Soc.*, 1990, **112**, 4911.
142. (a) P. Pino and M. Galimberti, *J. Organomet. Chem.*, 1989, **370**, 1; (b) P. Pino, M. Galimberti, P. Prada and G. Consiglio, *Makromol. Chem.*, 1990, **191**, 1677.
143. G. W. Coates and R. M. Waymouth, *J. Am. Chem. Soc.*, 1993, **115**, 91.
144. S. Collins and D. G. Ward, *J. Am. Chem. Soc.*, 1992, **114**, 5460.
145. A. S. Guram and R. F. Jordan, *J. Org. Chem.*, 1992, **57**, 5994.
146. (a) T. Matsumoto, H. Maeta, K. Suzuki and G. Tsuchihashi, *Tetrahedron Lett.*, 1988, **29**, 3567; (b) K. Suzuki, H. Maeta, T. Matsumoto and G. Tsuchihashi, *ibid.*, 1988, **29**, 3571.

147. T. Matsumoto, H. Maeta, K. Suzuki and G. Tsuchihashi, *Tetrahedron Lett.*, 1988, **29**, 3575.
148. T. Matsumoto, T. Hosoya and K. Suzuki, *J. Am. Chem. Soc.*, 1992, **114**, 3568.
149. K. Suzuki, H. Maeta, T. Suzuki and T. Matsumoto, *Tetrahedron Lett.*, 1989, **30**, 6879.
150. J. Schwartz and J. Labinger *Angew. Chem., Int. Ed. Engl.*, 1976, **15**, 333.
151. H. Maeta, T. Hashimoto, T. Hasegawa and K. Suzuki, *Tetrahedron Lett.*, 1992, **33**, 5965.
152. H. Maeta and K. Suzuki, *Tetrahedron Lett.*, 1992, **33**, 5969.
153. H. Maeta and K. Suzuki, *Tetrahedron Lett.*, 1993, **34**, 341.
154. P. Wipf and W. Xu, *J. Org. Chem.*, 1993, **58**, 825.
155. T. K. Hollis, N. P. Robinson and B. Bosnich, *Organometallics*, 1992, **11**, 2745.
156. T. K. Hollis, N. P. Robinson and B. Bosnich, *Tetrahedron Lett.*, 1992, **33**, 6423.

13

Cyclooctatetraene Complexes of Zirconium and Hafnium

F. GEOFFREY N. CLOKE

University of Sussex, Brighton, UK

13.1 BIS(CYCLOOCTATETRAENE) COMPLEXES

Base-free bis(cyclooctatetraene)zirconium may be obtained by heating the adduct $[Zr(\eta\text{-}C_8H_8)(\eta^4\text{-}C_8H_8)(THF)]$[1] followed by recrystallization from toluene or dimethoxyethane;[2] the x-ray structure of bis(cyclooctatetraene)zirconium (**1**) shows the presence of η^8- and η^4-cot rings, as also apparent in the ^{13}C CP MAS NMR spectrum.[3] In solution, $[Zr(\eta\text{-}C_8H_8)(\eta^4\text{-}C_8H_8)]$ is fluxional, and ^1H and ^{13}C NMR spectral studies show both rings to be rendered equivalent on the NMR timescale down to 173 K; similar conclusions are drawn for the hafnium analogue from NMR spectroscopic studies.[3] The ring-substituted compounds $[Zr\{\eta\text{-}C_8H_6(TMS)_2\}\{\eta^4\text{-}C_8H_6(TMS)_2\}]$ and $[Hf\{\eta\text{-}C_8H_6(TMS)_2\}\{\eta^4\text{-}C_8H_6(TMS)_2\}]$ are obtained, starting from 1,4-bis(trimethylsilyl)cyclooctatriene, by the route shown in Scheme 1; the absence of solvation by THF is attributed to steric crowding by the ring substituents.[2] Both display the same fluxionality in solution as the unsubstituted analogues, and the x-ray structure of $[Hf\{\eta\text{-}C_8H_6(TMS)_2\}\{\eta^4\text{-}C_8H_6(TMS)_2\}]$ (**2**) confirms the $\eta^8\!-\!\eta^4$ bonding mode of the rings.

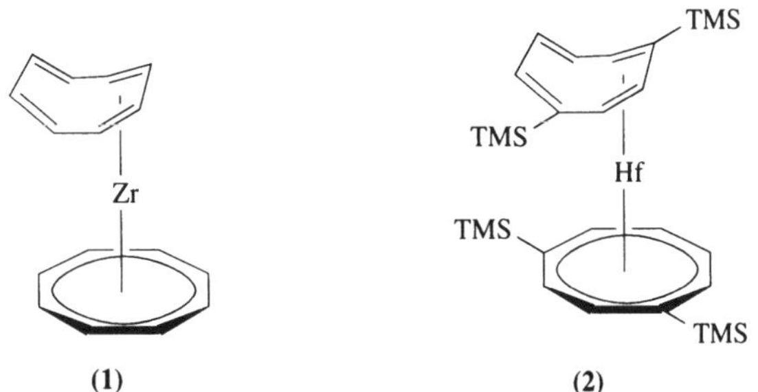

Treatment of $[Zr(\eta\text{-}C_8H_8)(\eta^4\text{-}C_8H_8)]$ with gaseous ammonia yields the stable, sparingly soluble adduct $[Zr(\eta\text{-}C_8H_8)(\eta^4\text{-}C_8H_8)(NH_3)]$ (**3**), characterized by x-ray diffraction; a similar reaction with *t*-butyl isocyanide affords the structurally characterized $[Zr(\eta\text{-}C_8H_8)(\eta^4\text{-}C_8H_8)(CNBu^t)]$ (**4**), and also the hafnium analogue.[2] In solution, a similar fluxional process to that of the $[M(\eta\text{-}C_8H_8)(\eta^4\text{-}C_8H_8)]$ (M = Zr

$$C_8H_8(TMS)_2\text{-}1,4 \xrightarrow[\text{THF}]{2\ Bu^nLi} Li_2[C_8H_6(TMS)_2\text{-}1,4] \xrightarrow[\text{THF}]{1/2\ [MCl_4(THF)_2]} [M(\eta\text{-}C_8H_6\{TMS\}_2)(\eta^4\text{-}C_8H_6\{TMS\}_2)]$$

$$M = Zr\ or\ Hf$$

Scheme 1

or Hf) precursors averages the rings in $[Zr(\eta\text{-}C_8H_8)(\eta^4\text{-}C_8H_8)(CNBu^t)]$ and $[Hf(\eta\text{-}C_8H_8)\text{-}(\eta^4\text{-}C_8H_8)(CNBu^t)]$, but in the latter case the static ^1H NMR spectrum is observed at 203 K.

(3)

(4)

13.2 MONOCYCLOOCTATETRAENE COMPLEXES

13.2.1 Mixed Sandwich Compounds and their Derivatives

The first stable zirconium(III) sandwich compound $[Zr(\eta\text{-}C_8H_8)Cp^*]$ (**5**) was obtained by the reduction of $[ZrCp^*Cl_3]$ in THF with magnesium in the presence of cyclooctatetraene in 75% yield (Equation (1)); the analogous reaction of $[HfCp^*Cl_3]$ affords the hydride $[Hf(\eta\text{-}C_8H_8)Cp^*H]$, attributed to the kinetic lability of the likely intermediate $[Zr(\eta\text{-}C_8H_8)Cp^*Cl]^-$.[4]

$$[ZrCp^*Cl_3] \xrightarrow[\text{THF}]{C_8H_8,\ Mg} \tag{1}$$

(5)

The x-ray crystal structure of $[Zr(\eta\text{-}C_8H_8)Cp^*]$ shows the expected parallel ring-sandwich structure (**5**), very similar to that of the titanium analogue.[5] The electronic structure of $[Zr(\eta\text{-}C_8H_8)Cp^*]$ has been examined by external nuclear double resonance (ENDOR)[6] and photoelectron spectroscopy,[7] and is fully consistent with the molecular orbital scheme for a parallel ring-sandwich molecule in which the HOMO is the partially occupied d_{z^2} orbital.

Sequential addition of $K_2[C_8H_8]$ followed by $Li[C_5Me_4R^1]$ ($R^1 = Me$ or Et) to $ZrCl_4$ gives $[Zr(\eta\text{-}C_8H_8)(\eta\text{-}C_5Me_4R^1)Cl]$; the latter reacts with a range of lithium (LiR^2) or Grignard reagents (R^2MgCl) to afford derivatives of the general type $[Zr(\eta\text{-}C_8H_8)(\eta\text{-}C_5Me_4R^1)R^2]$, and with sodium amalgam to give $[Zr(\eta\text{-}C_8H_8)(\eta\text{-}C_5Me_4R^1)]$, as shown in Scheme 2.[8,9] The compounds $[Zr(\eta\text{-}C_8H_8)\text{-}(\eta\text{-}C_5Me_4R^1)R^2]$ are classed as 18-electron zirconium(IV) species on the basis of the spectroscopic properties of the cot ligands which display singlets in the ^1H NMR spectra in the region $\delta = 5.79\text{--}5.97$, characteristic of a planar η^8-aromatic ring. The thermal stability of the ethyl complexes $[Zr(\eta\text{-}C_8H_8)\text{-}(\eta\text{-}C_5Me_4R^1)Et]$ is noteworthy, and is attributed to the lack of a suitable coordination site to initiate β-elimination.

Thermolysis of compounds $[ZrCp^*R]$ is strongly dependent on the nature of R (see Scheme 3).[10] For $R = Me$ or CH_2TMS, the kinetic product is the structurally characterized $[ZrCp^*(\mu\text{-}\eta:\eta^2\text{-}C_8H_6)\text{-}\{ZrCp^*(\eta^4\text{-}C_8H_8)\}]$ (**6**), in which a cyclooctatrienyne ligand, formed by abstraction of two adjacent hydrogen atoms from a cyclooctatetraene ligand, bridges two zirconium centres. This dinuclear complex is further converted into the more thermodynamically stable fulvene complex $[Zr(\eta\text{-}C_8H_8)\text{-}(\eta\text{-}C_5Me_4CH_2)]$. The coordination mode of the fulvene ligand in the latter is assumed to be η^6, leading to an 18-electron system, and probably via a bent, η^1-methylene-η^5-cyclopentadienyl-type ligation rather than a planar fulvene ligand on the basis of NMR spectral data for other group 4 metal fulvene

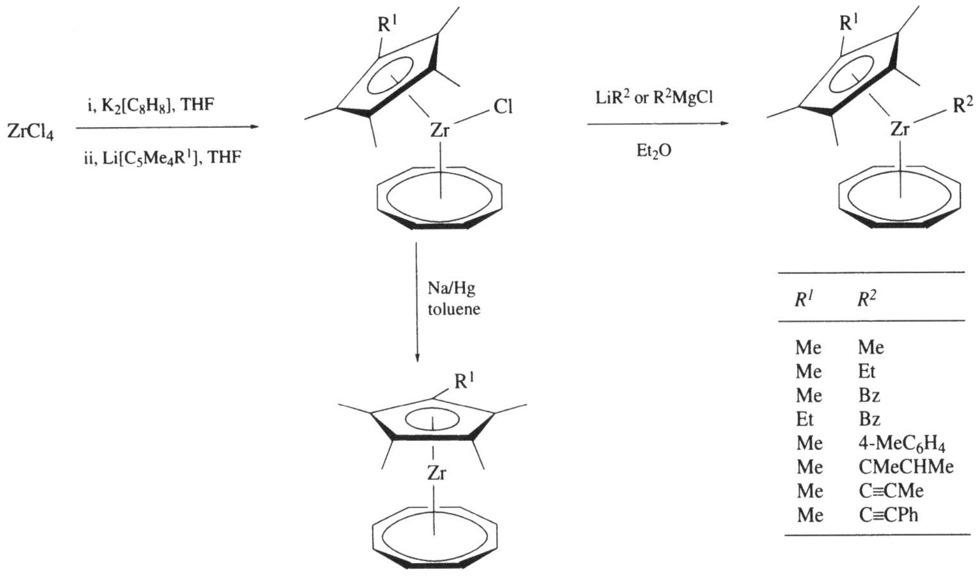

Scheme 2

R^1	R^2
Me	Me
Me	Et
Me	Bz
Et	Bz
Me	4-MeC₆H₄
Me	CMeCHMe
Me	C≡CMe
Me	C≡CPh

complexes. The thermal decomposition of [Zr(η-C₈H₈)Cp*Bz] gives [Zr(η-C₈H₈)(η-C₅Me₄CH₂)] at low temperatures, possibly via a different route. At higher temperatures formation of the dinuclear compound (**6**) predominates. The Zr–Zr distance of 0.419 4(2) nm in (**6**) mitigates against a metal–metal bond but, as drawn, both zirconium centres are zirconium(III) and an electron count gives zirconium(I) and zirconium(II) 17 and 15 electrons, respectively. The observed diamagnetism of the compound is explained by an electron transfer from zirconium(I) to zirconium(II) thus forming a zwitterionic complex with a bridging cyclooctatrienyne ligand connecting the cationic [ZrCp*(η-C₈H₆)]⁺ and anionic [ZrCp*(η⁴-C₈H₈)R₂]⁻ fragments. The resultant differing oxidation states of the metal centres are reflected in two different Cp*–Zr bond lengths.

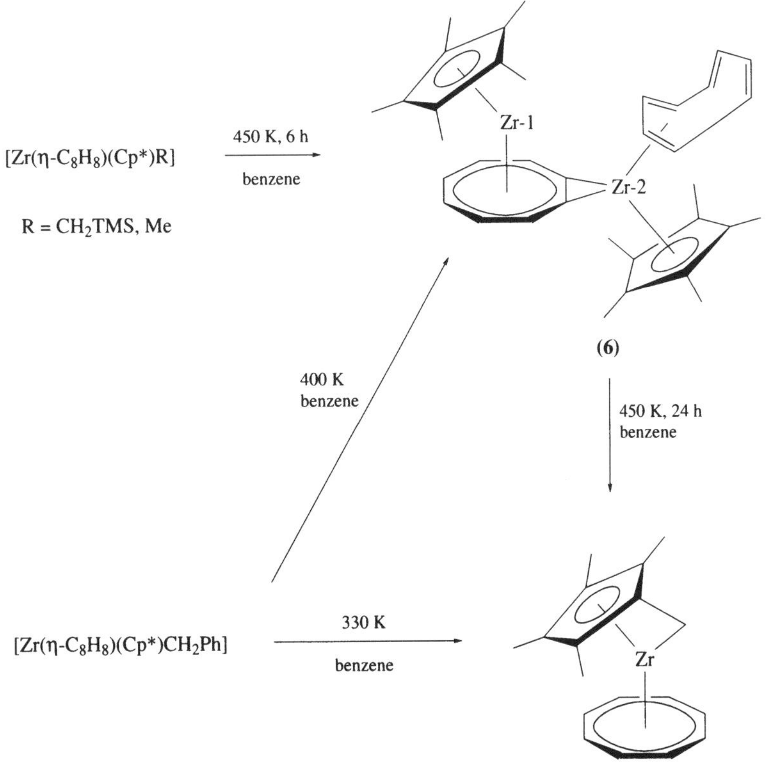

Scheme 3

Treatment of [Zr(η-C$_8$H$_8$)(η-C$_5$Me$_4$R)Cl] with Pr^{i}MgCl results in the formation of the hydride [Zr(η-C$_8$H$_8$)(η-C$_5$Me$_4$R)H], which has been structurally authenticated in the case where R = Me (7) (Equation (2)).[9] The hydrido ligand is characterized by ν_{M-H} at 1500 cm^{-1} in the IR spectrum, and a resonance at δ = 5.18 in the ^{1}H NMR spectrum exhibiting coupling to the cot ring. [Zr(η-C$_8$H$_8$)Cp*H] does not react with ethene or other mono-alkenes, but treatment with buta-1,3-diene affords a mixture of the *E*- and *Z*-isomers of [Zr(η-C$_8$H$_8$)Cp*CH$_2$CH=CHMe]. Disubstituted alkynes are similarly unreactive, but phenylacetylene gives the phenylethynyl complex, which is also prepared using the Grignard reagent as in Scheme 2.

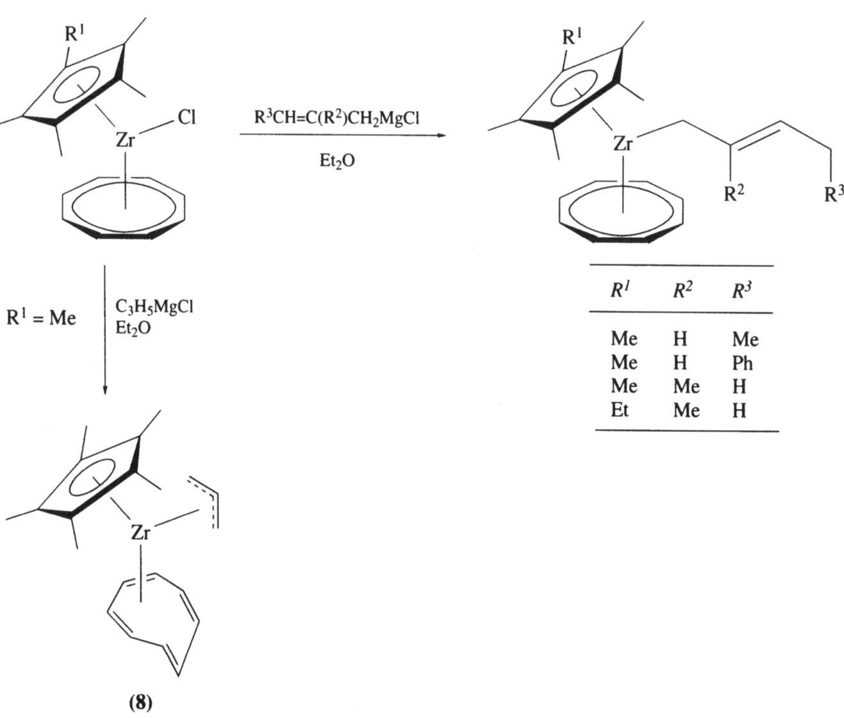

$$(2)$$

(7)

Allylic Grignard reagents react with [Zr(η-C$_8$H$_8$)(η-C$_5$Me$_4$R)Cl] to afford σ-allyl derivatives, except in the case of C$_3$H$_5$MgCl which affords the η^3-allyl complexes [Zr(η^4-C$_8$H$_8$)(η-C$_5$Me$_4$R)(η^3-C$_3$H$_5$)], crystallographically characterized in the case of R = Me (8) (Scheme 4).[8,9] The η^3-allyl–zirconium(II) complexes are intense purple, in contrast to the pale colour of the zirconium(IV) σ-allyl derivatives, and above 233 K the allyl group exhibits an AX$_4$ pattern in the ^{1}H NMR spectrum, consistent with rapid *syn–anti* exchange in these 16-electron molecules. The 18-electron σ-but-2-enyl derivatives [Zr(η-C$_8$H$_8$)(η-C$_5$Me$_4$R^1){CH$_2$C(R^2)=CHR^3}] (R^1 = Me or Et, R^2 = Me, R^3 = H) show facile 1,3-exchange in solution via a 16-electron η^3-but-2-enyl intermediate analogous to (8). [Zr(η^4-C$_8$H$_8$)-Cp*(η^3-C$_3$H$_5$)] reacts rapidly with dihydrogen to afford a mixture of [Zr(η-C$_8$H$_8$)Cp*H] and [Zr(η-C$_8$H$_8$)Cp*Pr].

R^1	R^2	R^3
Me	H	Me
Me	H	Ph
Me	Me	H
Et	Me	H

(8)

Scheme 4

13.2.2 Dihalide, Dialkyl and other Derivatives

An improved synthesis of $[M(\eta\text{-}C_8H_8)Cl_2(THF)]$ (M = Zr or Hf), crystallographically characterized in the case of M = Hf (**9**), results from a ligand redistribution reaction between $[M(\eta\text{-}C_8H_8)(\eta^4\text{-}C_8H_8)]$ and $[MCl_4(THF)_2]$; the unsolvated derivatives $[M(\eta\text{-}C_8H_8)Cl_2]$ may be prepared by heating followed by extraction with an aromatic hydrocarbon (Scheme 5).[11]

$$[M(\eta\text{-}C_8H_8)(\eta^4\text{-}C_8H_8)] + [MCl_4(THF)_2] \xrightarrow{\text{THF}} [M(\eta\text{-}C_8H_8)Cl_2(THF)] \xrightarrow[\text{ii, toluene}]{\text{i, heat 400 K, }10^{-2}\text{ mbar}} [M(\eta\text{-}C_8H_8)Cl_2]$$

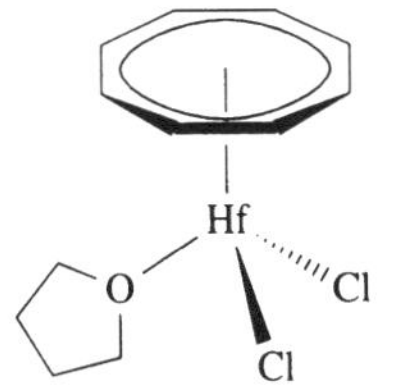

(**9**)

Scheme 5

A similar strategy may also be employed for the synthesis of the crystallographically characterized, dimeric, substituted cyclooctatetraene derivative $[(Zr\{\eta\text{-}C_8H_6(TMS)_2\}Cl)_2(\mu\text{-}Cl)_2]$ (**10**) (Equation (3)).[11]

$$[Zr(\eta^8\text{-}C_8H_6\{TMS\}_2)(\eta^4\text{-}C_8H_6\{TMS\}_2)] + [ZrCl_4(THF)_2] \xrightarrow{\text{toluene}} \qquad (3)$$

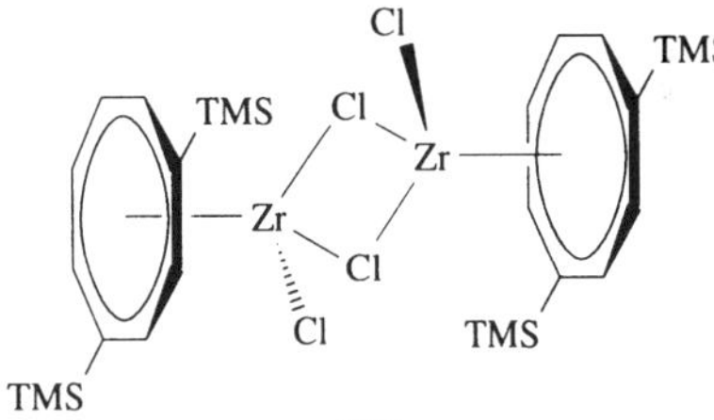

(**10**)

Reaction of $[Zr(\eta\text{-}C_8H_8)Cl_2(THF)]$ with two equivalents of the mesityl Grignard reagent affords the thermally labile $[Zr(\eta\text{-}C_8H_8)(Mes)_2]$ which undergoes rapid carbonylation to form, via double mesityl group migration to carbon monoxide, the structurally characterized diaryl ketone derivative $[Zr(\eta\text{-}C_8H_8)(\eta^2\text{-}OC\{Mes\}_2)]_2$ (**11**) (Scheme 6).[12] The stepwise thermal rearrangement of $[Zr(\eta\text{-}C_8H_8)(Mes)_2]$ leads, via loss of mesitylene, to the dinuclear complex $[\{Zr(\eta\text{-}C_8H_8)\}_2\text{-}(\mu\text{-}C_6H_2CH_2\text{-}6, Me_2\text{-}2,4)_2]$ (**12**), containing two bridging η^3-benzylic groups, which further rearranges to complex (**13**) which contains a bridging mesityl and a bridging alkylidene ligand, as shown by x-ray analysis of each compound (Scheme 6).[13]

Catalytic polymerization of *cis*-piperylene by $[Zr(\eta\text{-}C_8H_8)(\eta\text{-}C_3H_5)_2]$[14] results in the isolation of a crystallographically characterized bisallyl-type intermediate containing a C_{10} chain, $[Zr(\eta\text{-}C_8H_8)(\eta^3,\eta^{3\prime}\text{-}C_{10}H_{16})]$ (**14**).[15] The latter shows significant variations in the Zr–C(cot) bond distances, attributed to steric repulsions between the two hydrocarbyl ligands.

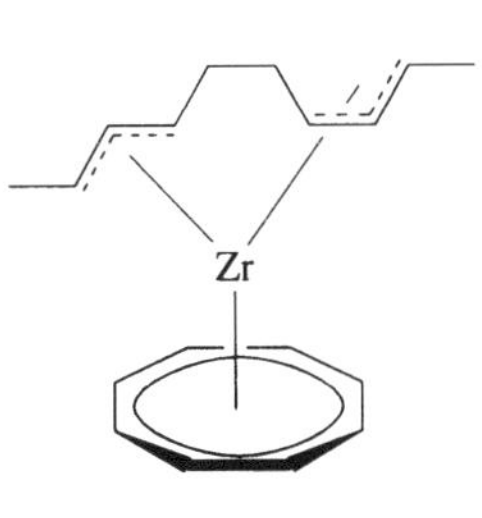

(**14**) 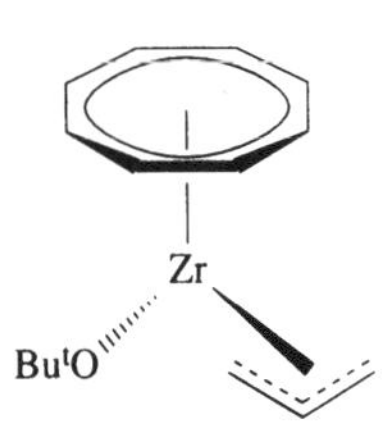   (**15**)

The x-ray structure[16] of the alkoxy derivative $[Zr(\eta\text{-}C_8H_8)(\eta^3\text{-}C_3H_5)(OBu^t)]$ (**15**), in which two distinct conformations of the *t*-butoxy groups are evident, shows short Zr–O bond lengths (average 0.191(1) nm) and large Zr–O–C bond angles [average 169(1)°] indicative of strong Zr–O multiple bonding.

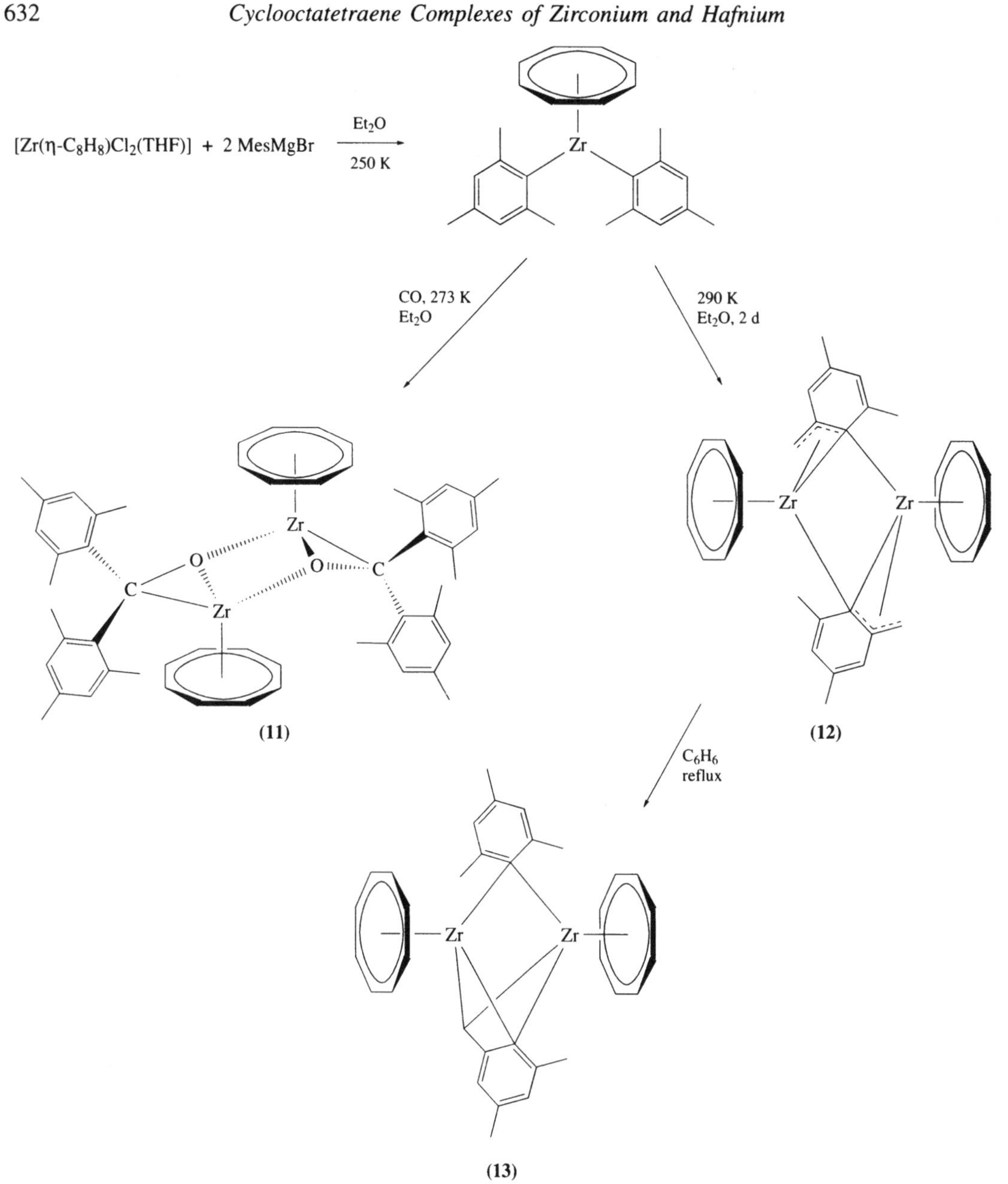

$$[Zr(\eta\text{-}C_8H_8)Cl_2(THF)] \ + \ 2\,MesMgBr \ \xrightarrow[250\ K]{Et_2O}$$

(11)

(12)

CO, 273 K
Et$_2$O

290 K
Et$_2$O, 2 d

C$_6$H$_6$
reflux

(13)

Scheme 6

13.3 REFERENCES

1. H. J. Kablitz and G. Wilke, *J. Organomet. Chem.*, 1973, **51**, 241.
2. P. Berno, C. Floriani, A. Chiesi-Villa and C. Rizzoli, *J. Chem. Soc., Dalton Trans.*, 1991, 3085.
3. D. M. Rogers, S. R. Wilson and G. S. Girolami, *Organometallics*, 1991, **10**, 2419.
4. J. Blenkers, P. Bruin and J. H. Teuben, *J. Organomet. Chem.*, 1985, **297**, 61.
5. L. B. Kool, M. D. Rausch and R. D. Rogers, *J. Organomet. Chem.*, 1985, **297**, 289.
6. D. Gourier, E. Samuel and J. H. Teuben, *Inorg. Chem.*, 1989, **28**, 4663.
7. R. R. Andrea, A. Terpstra, A. Oskam, P. Bruin and J. H. Teuben, *J. Organomet. Chem.*, 1986, **307**, 307.
8. W. J. Highcock, R. M. Mills, J. L. Spencer and P. Woodward, *J. Chem. Soc., Chem. Commun.*, 1982, 128.
9. W. J. Highcock, R. M. Mills, J. L. Spencer and P. Woodward, *J. Chem. Soc., Dalton Trans.*, 1986, 821.
10. P. J. Sinnema, A. Meetsma and J. H. Teuben, *Organometallics*, 1993, **12**, 184.
11. P. Berno, C. Floriani, A. Chiesi-Villa and C. Rizzoli, *J. Chem. Soc., Dalton Trans.*, 1991, 3093.
12. S. Stella and C. Floriani, *J. Chem. Soc., Chem. Commun.*, 1986, 1053.
13. S. Stella, M. Chiang and C. Floriani, *J. Chem. Soc., Chem. Commun.*, 1987, 161.
14. H. J. Kablitz and G. Wilke, *J. Organomet. Chem.*, 1973, **51**, 241.
15. D. J. Brauer and C. Krüger, *Organometallics*, 1982, **1**, 207.
16. D. J. Brauer and C. Krüger, *Organometallics*, 1982, **1**, 204.

Author Index

This Author Index comprises an alphabetical listing of the names of the authors cited in the text and the references listed at the end of each chapter in this volume.

Each entry consists of the author's name, followed by a list of numbers, for example

Templeton, J. L., 366, 385[233] (350, 366), 387[370] (363)

For each name, the page numbers for the citation in the reference list are given, followed by the reference number in superscript and the page number(s) in parentheses of where that reference is cited in the text. Where a name is referred to in text only, the page number of the citation appears with no superscript number. References cited both in the text and in the tables are included.

Although much effort has gone into eliminating inaccuracies resulting from the use of different combinations of initials by the same author, the use by some journals of only one initial, and different spellings of the same name as a result of the transliteration processes, the accuracy of some entries may have been affected by these factors.

Aaliti, A., 207[1018] (134, 192), 207[1020] (134, 192), 207[1022] (134, 192), 207[1027] (134, 187, 192), 208[1084] (143, 192), 208[1103] (146, 167, 190), 208[1104] (146, 167, 190)

Abboud, K. A., 462[26] (443), 462[27] (443)

Abdul Malik, K. M., 207[1017] (133)

Abel, E. W., 585[17] (546)

Abis, L., 195[175] (23, 24, 60), 623[79] (605, 618), 624[128] (618)

Abrahamson, H. B., 422[67] (284), 422[68] (284)

Absulrahman, M., 588[218] (577)

Abys, J., 541[269] (535, 536)

Adachi, G.-Y., 203[744] (97)

Adam, M., 196[262] (32), 197[345] (39), 199[461] (53), 200[500] (56), 200[506] (56, 113), 200[507] (56, 113), 210[1184] (159)

Adam, R., 208[1054] (139), 209[1149] (153)

Adel, J., 430[669] (412)

Adrian, G., 209[1136] (151, 169), 210[1195] (160), 210[1196] (160, 183)

Afanasev, B. N., 194[109] (15)

Afinogenova, L. L., 269[131] (244)

Afzal, D., 209[1162] (155)

Aghabozorg, H., 194[135] (17)

Agreda, T. C., 541[202] (528)

Ahlers, A., 587[169] (569), 624[140] (619)

Ahlrichs, R., 271[265] (265)

Ahmad, S., 206[947] (128)

Ahmed, I., 208[1076] (141, 143)

Ainslie, R. D., 421[31] (277)

Aitken, C., 538[7] (504), 586[107] (560)

Aitken, C. A., 586[95] (560), 586[97] (560)

Aitken, C. T., 212[1331] (191), 271[256] (264), 271[257] (264, 265), 271[258] (265), 271[260] (265), 586[93] (560)

Ajò, D., 207[981] (130, 147)

Akai, S., 206[964] (128)

Akinobu, S., 421[60] (283)

Akita, M., 268[82] (237), 421[29] (277), 463[104] (460), 586[75] (556)

Akiyoshi, K., 588[247] (582)

Akkerman, O. S., 426[373] (358), 426[374] (358), 427[446] (376, 377), 427[447] (376), 427[449] (377), 623[87] (608)

Aknoukh, T., 197[304] (36, 60)

Albrecht, G., 9[12b] (4)

Albrecht, I., 198[370] (40, 78), 201[596] (68, 76), 201[604] (69, 83), 201[605] (69, 76, 83), 201[606] (69), 202[650] (78)

Albrecht, M., 463[106b] (460), 539[123] (519, 520), 624[111] (615)

Albrecht, N., 430[668] (412), 541[255] (533, 534)

Alcock, N. W., 208[1078] (142, 143, 154), 422[115] (291)

Al-Daher, A. G. M., 208[1079] (143, 154)

Aleksandrov, G. G., 270[216] (258)

Alekseev, M. K., 269[163] (248)

Aelyunas, Y. W., 622[6] (590, 596, 601, 602, 606), 623[46] (594, 597, 601, 602, 618), 623[59] (598, 599)

Alex, R. F., 462[31] (444, 454), 588[236] (580)

Alexandrov, Y. A., 210[1201] (161, 189)

Allcock, H. R., 586[96] (560, 561)

Allegra, G., 623[53c] (596)

Allen, F. H., 193[74] (13)

Allocca, C. M., 422[74] (284)

Aloi, M. J., 429[603] (403)

Alper, H., 541[235] (530)

Al-Soudani, A. R., 427[465] (379)

Alt, H. G., 267[19] (224, 225), 267[28] (225, 226), 267[29] (225, 226), 268[54] (230, 231), 268[55] (230, 231), 268[56] (230), 268[58] (231), 268[62] (231, 234), 268[67] (233), 268[68] (233), 268[69] (233), 268[70] (233), 268[71] (233), 268[72] (233, 234), 268[84] (237), 269[142] (244), 270[200] (255), 270[212] (257),

271[266] (265), 425[313], 426[388] (361), 426[389] (361), 429[609] (405), 430[626] (407), 462[15] (441, 450, 451, 454), 462[16] (441), 462[17] (441, 453), 462[37a] (444), 462[37b] (444), 463[80] (454), 463[82] (454), 481[13] (466), 498[126], 540[137] (521, 526–8, 530), 540[138] (521), 540[139] (521, 526), 540[140] (521), 540[192] (526, 530)

Alu-Orabi, S., 422[99] (290, 306)

Alvarez, D., Jr., 201[629] (74, 78, 83, 89, 90, 92, 115), 204[829] (109), 622[18c] (591)

Alvarez, M. M., 204[809] (106)

Alvaro, L. M., 422[111] (291, 294, 295)

Alvhäll, J., 586[71] (556)

Amberger, H.-D., 52, 199[445] (52), 199[447] (52, 55), 199[448] (52, 55), 199[449] (52, 55), 199[450] (52, 55), 199[451] (52, 55), 199[452] (52, 55), 199[453] (52, 55), 199[454] (52, 55), 199[455] (52, 55), 199[456] (52, 55), 199[457] (52, 55), 199[458] (52, 55), 199[478] (55), 199[485] (55, 56), 199[486] (55, 56), 199[487] (55), 200[489] (55), 200[490] (55), 200[513] (57), 209[1113] (147), 210[1183] (159), 210[1188] (159)

Amirkhalili, S., 424[269] (314)

Ammendola, P., 422[94] (287), 428[518] (396)

Amorose, D. M., 427[423] (369), 622[27] (592, 595)

Anand, S. P., 200[514] (58)

Andersen, R. A., 54, 78, 108, 195[188] (25), 198[367] (40, 54, 56, 138, 140, 149), 199[466] (53, 54, 56), 200[503] (56, 139), 201[579] (66, 67), 201[582] (66), 201[583] (66), 201[589] (66), 201[590] (67), 201[591] (67), 201[602] (68), 202[631] (74), 202[648] (78), 202[652] (78), 202[653] (78), 202[655] (78), 203[707] (90),

Bangerter, F., 587[164] (569)
Bansal, S. K., 429[575] (401)
Bao, X., 195[171] (22), 196[264] (32)
Baohu, D., 201[563] (63)
Barakovskaya, I. G., 497[12a] (484)
Baranger, A. M., 538[18] (507, 509)
Baranwal, B. P., 585[34] (550)
Barger, P. T., 540[153] (522), 587[123] (563)
Barker, G. K., 194[139] (17)
Barnhart, D. M., 203[763] (98)
Barr, R. D., 428[487] (384), 428[488] (384)
Barral, M. C., 423[197] (306)
Barron, A. R., 624[127] (618)
Barry, J.-P., 212[1331] (191), 271[258] (265), 586[97] (560)
Barsties, E., 498[105] (490)
Bart, J. C., 268[49] (229)
Barth, S., 463[65b] (450), 463[65c] (450)
Bartik, T., 422[71] (284), 422[72] (284), 422[79] (285), 426[384] (360)
Bartlett, R. A., 194[140] (18, 24)
Basalgina, T. A., 203[742] (97)
Basile, F., 481[48] (474, 475, 477)
Basolo, F., 267[27] (225, 226), 587[149] (566)
Bassi, I. W., 623[53c] (596)
Basso-Bert, M., 426[378] (359), 481[51] (474, 475), 538[37] (508), 539[61] (510, 511), 541[212] (528), 586[86] (559)
Bastl, Z., 424[275]
Bates, G. S., 588[235] (580)
Battaglia, L. P., 270[179] (250)
Bau, R., 198[420] (48), 209[1162] (155)
Baudin, C., 207[999] (130, 136, 137, 144), 207[1033] (136), 208[1090] (143)
Baudry, D., 202[637] (76), 204[799] (106, 110), 208[1038] (138, 143, 144, 146, 177), 208[1052] (139, 141, 146), 208[1087] (143, 144), 208[1088] (143, 144), 208[1089] (143, 144), 208[1092] (144), 208[1105] (146), 209[1134] (151, 157, 189), 211[1273] (177, 180), 211[1274] (177), 211[1275] (177), 211[1277] (178, 179), 211[1280] (180), 211[1302] (184, 186), 211[1305] (185), 211[1306] (185), 212[1319] (189)
Baum, G., 539[80] (512), 539[85] (512, 514)
Baumann, H., 205[848] (111)
Baumgart, F., 431[716] (417)
Baxter, J. C., 585[24] (547)
Baxter, S. M., 585[39] (551), 623[69b] (603)
Bazan, G. C., 201[608] (70)
Beach, D. B., 207[990] (130, 183)
Beattie, M. S., 421[20] (276)
Beauchamp, A. L., 429[550] (399), 429[572] (400)
Beauchamp, J. L., 426[391] (361)
Beck, G., 429[608] (404)
Beck, W., 585[21] (546), 622[3b] (590)
Beckhaus, R., 421[55] (283), 422[72] (284), 426[381] (359), 426[382] (359), 462[48b] (446), 540[161] (523, 525)
Bedard, T. C., 588[217] (577)
Beeckman, W., 210[1209] (163)

Beer, M. V., 424[213] (309), 424[214] (309)
Behrens, U., 197[328] (37), 199[461] (53), 207[1029] (135), 268[44] (228), 268[45] (228), 423[138] (293, 294, 358)
Bei, M., 497[28] (484)
Beier, B. F., 585[26] (548)
Bekele, T. S., 429[619] (406)
Belanger-Gariepy, F., 429[572] (400)
Beletskaya, I. P., 108, 193[47] (13), 194[90] (13, 39, 107, 113), 194[102] (14), 194[103] (14), 194[104] (14), 194[137] (17, 22), 195[201] (26), 196[229] (29), 196[233] (29), 196[234] (29), 196[257] (32), 198[368] (40), 198[375] (41), 198[410] (47, 48), 199[442] (52), 204[816] (107), 204[817] (107), 204[818] (108), 204[824] (108), 204[825] (108), 205[832] (110), 205[834] (110), 205[836] (110), 205[839] (111), 205[841] (111), 205[844] (111), 205[865] (116), 205[896] (122), 206[905] (123), 206[913] (123, 124), 206[914] (123), 206[915] (123), 206[916] (124), 206[917] (124), 206[919] (124), 206[920] (124), 206[930] (125)
Belforte, A., 270[214] (258)
Bell, A. P., 586[66] (556), 586[67] (556, 581), 588[225] (578)
Bell, L. G., 271[277] (225), 425[317], 586[102] (560)
Bel'skii, V. K., 193[69] (13), 196[241] (30), 196[265] (32, 37), 197[305] (36, 37), 197[322] (37, 50, 146), 197[323] (37), 197[324] (37), 197[325] (37), 197[327] (37), 198[414] (47, 49), 198[415] (47, 50), 198[421] (48, 50), 198[425] (49), 198[426] (49), 199[428] (49, 50), 199[429] (49), 199[431] (50), 199[432] (50), 199[433] (50), 199[435] (50), 199[462] (53), 269[171] (249), 270[233] (260, 262), 270[234] (260), 271[235] (260), 271[236] (261), 271[238] (261), 271[240] (261), 497[8a] (484)
Bem, D. S., 427[430] (370)
Benac, B. L., 539[84] (512–16)
Bénard, M., 481[26] (469), 481[27] (469)
Bencivenni, L., 497[9c] (484), 497[9d] (484)
Benecke, J., 268[45] (228)
Benetollo, F., 200[493] (56), 208[1075] (141, 142), 210[1180] (159), 210[1181] (159), 210[1186] (159), 210[1190] (160), 210[1204] (162)
Benham-Dahkordy, M., 422[96b] (288), 422[96d] (288)
Benn, R., 211[1263a] (172), 268[63] (232), 462[22c] (442, 444, 453, 454), 462[49b] (446, 447, 454), 497[41] (484), 541[249] (531, 533), 585[2] (544), 624[111] (615)
Benning, M. M., 425[285], 497[14] (484)
Bensimon, C., 585[47] (552)
Bercaw, J. E., 13, 61, 117, 193[25] (13, 61, 70, 84–6, 94, 114, 116–18), 200[541] (61, 117), 200[542] (61), 200[544] (61), 201[607] (69, 70), 201[608] (70),

201[609] (70, 117), 202[668] (81), 202[689] (85, 114), 202[691] (85), 205[867] (116), 205[870] (117, 118), 205[871] (118), 268[76] (234), 268[77] (234, 235), 426[404] (363), 440, 462[8] (440), 462[42] (445), 462[57] (448), 497[15] (484), 497[29] (484), 538[29] (507, 509, 527, 528), 538[30] (507, 508), 538[31] (508, 537), 538[32] (508), 539[125] (519, 522), 539[126] (519), 540[146] (521, 524), 540[150] (522), 540[153] (522), 540[160] (523, 524), 540[166] (524, 525), 540[167] (524), 585[11] (545), 586[64] (555, 556), 586[74] (556), 586[88] (559, 580), 586[89] (559), 587[123] (563), 587[126] (564), 587[128] (564), 623[69b] (603), 624[91] (610), 624[121b] (618, 619), 624[131] (618)
Berces, A., 202[684] (84)
Berenguer, J. R., 427[442] (374)
Berestenko, V. I., 196[238] (30, 53, 130)
Berg, D. J., 193[42] (13), 201[601] (68), 201[602] (68), 202[652] (78), 202[653] (78), 202[654] (78, 90), 462[13] (441)
Berg, F. J., 623[89] (608)
Berg, K., 585[2] (544), 585[3] (544, 546), 585[30] (548), 587[133] (565)
Bergbreiter, D. E., 201[566] (63)
Berghus, K., 422[104] (290, 292, 356)
Bergman, R. G., 209[1114] (148, 152), 462[12] (441, 454), 463[74] (451), 538[18] (507, 509), 538[21] (507, 509), 539[57] (510, 511), 539[58] (510, 511), 539[59] (510–12), 540[155] (523, 528), 541[213] (528, 533), 541[214] (528, 533), 541[215] (528), 541[253] (532)
Berhalter, K., 428[546] (398), 429[549] (399), 540[188] (526)
Berk, S. C., 421[24] (277), 421[28] (277)
Berkovich, E. G., 268[94] (239), 426[403] (363)
Bernal, I., 498[90] (489), 498[91] (489)
Berno, P., 429[607] (404), 632[2] (627), 632[11] (631)
Berry, A., 219[8] (214), 421[56] (283)
Berry, D. H., 268[61] (231), 586[104] (560), 586[105] (560)
Berryhill, S. R., 211[1287] (182, 183)
Berthet, J.-C., 207[1003] (130), 208[1074] (141), 209[1121] (148), 209[1122] (148), 209[1130] (150), 210[1175] (157), 210[1176] (157), 211[1304] (185, 186), 211[1305] (185), 211[1308] (186)
Berton, A., 208[1085] (143), 208[1101] (145), 209[1106] (146)
Bertoncello, R., 207[981] (130, 147)
Bertz, S. H., 207[967] (129), 427[428] (370), 427[429] (370)
Besançon, J., 267[18] (224, 225), 424[268] (357), 429[604] (404), 430[684] (414), 430[685] (414)
Beshouri, S. M., 207[1032] (136), 208[1044] (138), 208[1045] (138), 208[1046] (138), 538[34] (508), 538[35] (508), 588[250] (582)
Bethune, D. S., 204[803] (106)
Bettinger, B., 267[34] (227)

Subject Index

JOHN NEWTON
David John (Services), Slough, UK

This Subject Index contains individual entries to the text pages of Volume 4. The index covers general types of organometallic compound, specific organometallic compounds, general and specific organic compounds where their synthesis or use involves organometallic compounds, types of reaction (insertion, oxidative addition, etc.), spectroscopic techniques (NMR, IR, etc.), and topics involving organometallic compounds.

Because authors may have approached similar topics from different viewpoints, index entries to those topics may not always appear under the same headings. Both synonyms and alternatives should therefore be considered to obtain all the entries on a particular topic. Commonly used synonyms include alkyne/acetylene, compound/complex, preparation/synthesis, etc. Entries where the oxidative state of a metal has been specified occur after all the entries for the unspecified oxidation state, and the same or similar compounds may occur under both types of heading. Thus $Cr(C_6H_6)_2$ occurs under Chromium, bis(η-benzene) and again under Chromium(0), bis(η-benzene). Similar ligands may also occur in different entries. Thus a carbene–metal complex may occur under Carbene complexes, Carbene ligands, or Carbenes, as well as under the specific metal. Individual organometallic compounds may also be listed in the Cumulative Formula Index in Volume 14.

Edwards Brothers Malloy
Ann Arbor MI. USA
November 20, 2013